# PHYSICAL CONSTANTS

| | |
|---|---|
| Standard gravity | $g = 9.80665 \text{ m/s}^2 = 32.174 \text{ ft/s}^2$ |
| Universal gas constant | $R_u = 0.08205 \text{ L·atm/mol·K}$ |
| | $= 8.314 \text{ kJ/kmol·K}$ |
| | $= 0.08314 \text{ bar·m}^3\text{/kmol/K}$ |
| | $= 8.314 \text{ kPa·m}^3\text{/kmol·K}$ |
| | $= 1545 \text{ ft·lb}_f\text{/lbmol·°R}$ |
| | $= 1.986 \text{ Btu/lbmol·°R}$ |
| | $= 10.73 \text{ psia·ft}^3\text{/lbmol·°R}$ |
| Avogadro's number | $N_A = 6.023 \times 10^{26} \text{ atoms/kmol}$ |
| Boltzmann's constant | $k = 1.380 \times 10^{-23} \text{ J/K·molecule}$ |
| Planck's constant | $h = 6.626 \times 10^{-34} \text{ J·s/molecule}$ |
| Speed of light (vacuum) | $c = 2.988 \times 10^{10} \text{ cm/s}$ |

# THERMODYNAMICS

## SIXTH EDITION

**Kenneth Wark, Jr.**
*Purdue University*

**Donald E. Richards**
*Rose-Hulman Institute of Technology*

Mc Graw Hill WCB McGraw-Hill

Boston   Burr Ridge, IL   Dubuque, IA   Madison, WI   New York   San Francisco   St. Louis
Bangkok   Bogotá   Caracas   Lisbon   London   Madrid
Mexico City   Milan   New Delhi   Seoul   Singapore   Sydney   Taipei   Toronto

# WCB/McGraw-Hill

*A Division of The **McGraw·Hill** Companies*

THERMODYNAMICS

1 2 3 4 5 6 7 8 9 0 DOC/DOC 9 3 2 1 0 9 8

ISBN 0-07-068305-0

Vice president and editorial director:   *Kevin T. Kane*
Publisher:   *Tom Casson*
Senior sponsoring editor:   *Debra Riegert*
Developmental editor:   *Holly Stark*
Marketing manager:   *John T. Wannemacher*
Project manager:   *Jim Labeots*
Production supervisor:   *Michael R. McCormick*
Designer:   *Gino Cieslik*
Supplement coordinator:   *Carol Loreth*
Compositor:   *Publication Services, Inc.*
Typeface:   *10/12 Times Roman*
Printer:   *R. R. Donnelley & Sons Company*

**Library of Congress Cataloging-in-Publication Data**

Wark, Kenneth (date)
    Thermodynamics / Kenneth Warks, Jr., Donald E. Richards. — 6th ed.
      p.    cm.
    ISBN   0-07-068305-0
    Includes bibliographical references and index.
    1. Thermodynamics.   I. Richards, Donald E.   II. Title.
    QC311.W3     1999
    536′.7 21                          98-10720

If you are interested in ordering the text with the EES Problems Disk, please refer to ISBN 0-07-561814-1.
If you are interested in ordering the text with the EES Demo Disk, please refer to ISBN 0-07-561815-X.

http://www.mhhe.com

# McGraw-Hill Series in Mechanical Engineering

CONSULTING EDITORS

**Jack P. Holman,** *Southern Methodist University*
**John R. Lloyd,** *Michigan State University*

| Shigley and Mischke: | *Mechanical Engineering Design* |
| Shigley and Uicker: | *Theory of Machines and Mechanisms* |
| Stiffler: | *Design with Microprocessors for Mechanical Engineers* |
| Stoecker and Jones: | *Refrigeration and Air Conditioning* |
| Turns: | *An Introduction to Combustion: Concepts and Applications* |
| Ullman: | *The Mechanical Design Process* |
| Wark: | *Advanced Thermodynamics for Engineers* |
| Wark and Richards: | *Thermodynamics* |
| White: | *Viscous Fluid Flow* |
| Zeid: | *CAD/CAM Theory and Practice* |

# ABOUT THE AUTHORS

**Kenneth Wark, Jr.** received his B.S. and M.S. degrees in chemical engineering from Purdue University and the University of Illinois, respectively. He joined the faculty of mechanical engineering at Purdue after receiving his Ph.D. degree from that school. In addition to his primary teaching responsibilities in undergraduate and graduate courses in thermodynamics, he has been involved in courses in heat transfer, fluid mechanics, combustion, and design. He was one of the first recipients of the best teacher awards from the Purdue chapter of Tau Beta Pi. His experience outside of Purdue has involved work with the General Electric Company, Boeing Aircraft, Carrier Corporation, U.S. Steel (USX), Allison Division of General Motors, Atlantic-Richfield, Standard Oil of Indiana, NASA-Lewis Field, and the Dow Chemical Company. In addition to this text, Dr. Wark is the author of a graduate text *Advanced Thermodynamics for Engineers,* McGraw-Hill, Inc., New York, 1995, and is a co-author (with Dr. Cecil Warner and Dr. Wayne Davis) of *Air Pollution–Its Origin and Control,* Third Edition, Addison-Wesley, Reading, Massachusetts, 1998. Dr. Wark retired from Purdue in 1996.

**Donald E. Richards** is a Professor of Mechanical Engineering at Rose-Hulman Institute of Technology in Terre Haute, Indiana. He received B.S., M.S., and Ph.D. degrees in mechanical engineering from Kansas State University, Iowa State University, and The Ohio State University, respectively. Before joining Rose-Hulman, he was a member of the mechanical engineering faculty at The Ohio State University. Over the years, he has taught basic and graduate courses in thermodynamics, fluid mechanics, and heat transfer. In addition, he has taught courses in HVAC systems, heat exchanger design, second-law analysis, and turbomachinery. While at Ohio State, he was awarded the Charles F. MacQuigg Outstanding Teaching Award by the students of the College of Engineering. His research has focused on natural convection, augmentation of forced-convection heat transfer, and design of multi-fluid heat exchangers. At Rose-Hulman, he was team leader for a new sophomore engineering science curriculum that uses common concepts—system identification, conservation and accounting of extensive properties, constitutive relations, modeling, and mathematics—as a basis for engineering science education.

# CONTENTS

## 7 EVALUATION OF ENTROPY CHANGE AND THE CONTROL-VOLUME ENTROPY BALANCE 385

## 8 SOME CONSEQUENCES OF THE SECOND LAW 435

## 9 AVAILABILITY (EXERGY) AND IRREVERSIBILITY 487

## A-2  SUPPLEMENTARY TABLES AND FIGURES (USCS UNITS)    1075

## A-3 INTRODUCTION TO EES

## SYMBOLS

## SELECTED PROBLEM ANSWERS

## PHOTO CREDITS

## INDEX

# PREFACE

*This introductory textbook in thermodynamics is designed for undergraduate students in the field of engineering. The primary objectives are to provide an understanding, and to exhibit the wide range of applicability, of the basic laws of thermodynamics and to present a logical development of the relationships among the physical properties of interest in the thermal sciences. The overall aim is to present thermodynamics as a science built upon a group of postulates and concepts which complement one another. An analysis becomes meaningful only through the application and interlocking of these ideas.*

—Preface to *Thermodynamics* (1966) by K. Wark, Jr.

These words are as true today for the sixth edition of *Thermodynamics* as they were for the first edition published in 1966. However, changes in engineering students, engineering education, and the engineering profession over the last thirty years have lead to significant changes in how and what we teach. The sixth edition of *Thermodynamics* has been written to respond to these changes. The authors had two major goals during the preparation of this edition: to clarify further the important concepts and tools of thermodynamics, and to encourage students to develop good problem-solving skills. As a result, major portions of the first nine chapters have undergone considerable changes in writing and format.

In preparing the sixth edition of *Thermodynamics,* Prof. Wark was joined by co-author Prof. Richards from Rose-Hulman Institute of Technology. In addition to his teaching and research experience in the thermal sciences, Prof. Richards brings significant experience in the design and implementation of a new, innovative sophomore engineering science and mathematics curriculum, the Rose-Hulman/Foundation-Coalition Sophomore Engineering Curriculum. This integrated curriculum attempts to improve student learning by stressing an underlying framework—system identification, conservation and accounting of extensive properties, constitutive relations, modeling, and mathematics—as a common basis for engineering science education.

## SIGNIFICANT FEATURES OF THE NEW EDITION

***Problem Solving Approach.*** The authors believe that students need to do more than just "plug-and-chug" solutions using predigested equations from the textbook. As an antidote to this approach, this text provides students with repeated opportunities to develop the necessary equations from general principles by explicitly applying modeling assumptions in the context of a specific problem. In this approach, the emphasis shifts from remembering specialized equations to learning how to pick a system and apply basic modeling assumptions. A specific problem solving methodology is proposed and used for both the example problems and the solutions manual.

***Sign Convention for Work and Heat Transfer.*** One of the constants in this text over the years has been its use of a consistent sign convention for energy transfer by work and heat transfer, i.e., work and heat transfers into a system are positive as written in the energy balance. Once again we have retained this sign convention. The specific sign convention adopted for any term in the energy balance is arbitrary and often dictated by history. At one time work and heat transfer were treated as separate, unrelated concepts that came together in an energy balance. When they were treated as separate concepts, having a different sign convention for each was reasonable. Since steam engines required heat input and did work, it seemed reasonable that heat *input* and work *output* should both be positive.

Today, these two concepts have been unified and are both recognized as *energy transfer mechanisms*. Although we still hear that "work is done" and "heat is added," we now speak about "energy transfer by work" and "heat transfer of energy." Because of this, the authors continue to believe that a consistent sign convention based upon the direction of energy transfer is preferable. Our sign convention, that energy transfer into a system by work and heat transfer are both positive, is consistent with a student's earlier experience in physics, chemistry, and mechanics. This approach is also consistent with current efforts in engineering education to help students integrate material across traditional course boundaries by stressing the similarities between the basic conservation laws of mass, energy, net charge, and momentum plus the accounting principle for entropy.

***Entropy Production and the Second Law of Thermodynamics.*** In recent years there has been an increased emphasis on the concept of entropy production or generation as a tool to help students understand the directional nature of the second law of thermodynamics. The authors applaud this change and have increased the use of the entropy accounting equation or balance as a tool for solving problems.

The development of second law relations (Chap. 6) has undergone extensive changes. The role of internal reversibility and irreversibility are emphasized in the early development, rather than total reversibility. This approach enhances the later introduction to entropy generation (production). To supplement the classical approach based on the Kelvin-Planck statement of the second law in Sec. 6-5, the authors have included in Sec. 6-6 an alternate development of the second law based upon a postulational approach. This approach parallels the development of the first law in Chap. 2 and begins with a statement about entropy transport by heat transfer and entropy production for a closed system. Instructors are encouraged to decide which approach meets their goals and then only assign either Sec. 6-5 or 6-6 to their students. Either approach leads directly to the control-volume entropy balance that then serves as the primary tool for applying the second law throughout the text, including Chaps. 10 and 13, and Chaps. 15, 16, and 17 on cycle analysis. An improved discussion of the loss of work potential associated with heat transfer and the calculation of entropy production in simple, cyclic devices ends Chap. 6.

## CHANGES FROM THE FIFTH EDITION

Specific major changes are listed and discussed below.

1. *Problem solving methodology.* In this edition, the problem solving methodology is introduced in Chap. 1 and used fully beginning with the examples in Chap. 2. The *Solution* of any example problem generally consists of five parts entitled: *Given, Find, Model, Strategy,* and *Analysis.* The *Strategy* section appears in the first eight chapters and asks the student to outline in words a proposed sequence of steps for solving the problem before they begin the *Analysis.* A four- or five-step approach appears in all examples for which it is appropriate throughout the text. A schematic of the system includes all the important input data, and the system boundary is designated by a dashed line. A similar methodology appears in other texts including two thermal science texts written by fellow members of the School of Mechanical Engineering at Purdue University, those by R. W. Fox and A. T. McDonald and by F. P. Incropera and D. P. Dewitt.

2. *Introduction to property data.* In earlier editions, the chapter on ideal gas relations preceded a general discussion of the $PvT$ behavior of substances. In this edition, a general discussion of the properties of a pure, simple-compressible substance appears first (Chap. 3) and now includes a discussion of the $Tv$ diagram. There also is an enhanced discussion of data acquisition and evaluation and of reference states and reference values for tabular data. Chapter 4 then presents two important property models that represent actual behavior under restrictive conditions: the ideal gas and the incompressible substance models. The use of the compressibility chart is also presented here as a method to estimate $PvT$ properties when experimental data are not available and to gauge when the ideal-gas relations are appropriate.

3. *Introduction to the second law.* In Chap. 7 the evaluation and use of the entropy function and entropy generation (production) has now been divided into closed-system analyses followed by steady-flow applications, similar to the energy analyses in Chaps. 3 through 5. The use of the increase in entropy principle has been de-emphasized with preference given to entropy production as a guideline to reversibility and irreversibility. The relation between actual and reversible work is developed early in the chapter, and the chapter concludes with polytropic, steady-state work.

4. *Simple cycles introduced early.* There is now a discussion of simple steam power cycles and refrigeration cycles at the end of Chap. 5 on steady-state systems. This new material serves two purposes. First, it shows the reader applications where several pieces of steady-state equipment operate in series. Second, knowledge of the equipment arrangements for several simple cycles enhances the discussion on cycle performance introduced immediately in Chap. 6 in conjunction with the development of second law theorems. This approach has been class tested over several semesters. Preceding the problems covering these simple cycles in Chap. 5 are other problems that illustrate the use of two pieces of equipment in series.

5. *Choice of units.* This edition continues the use of both SI and English (USCS) units. Early in the text both sets of units are employed in examples, but after Chap. 2 the example problems are predominantly in SI units. Both sets of units are used individually in the problems at the end of each chapter. In the sixth edition, the problems in SI units are 60 to 65 percent of the total in a given chapter. Data in SI units appears in Appendix A-1, where all tables and figures are numbered consecutively from 1 to 32. Data in USCS units appears in Appendix A-2, where tables and figures are numbered from 1E to 26E. In the appendices, data for refrigerant 134a (R-134a) replaces data for refrigerant 12 (R-12), and the tables of specific heat data for liquids and solids now appear as A-4 and A-4E. Former Table A-32M listing the logarithms of the equilibrium constant now appears as Table A-24, immediately following enthalpy of formation data. Also, former Tables A-4M and A-4 for the specific heat data of gases are now Tables A-3 and A-3E, respectively. Other tables are found in their former positions. Finally, data for the specific volume are now reported in $m^3$/kg (or L/kg) in SI tables.

6. *Assigned problems.* Unlike the fifth edition, problems in the sixth edition are not separated into two distinct groups depending on the type of units required. Problems in SI and USCS are intermingled and numbered consecutively, although problems on a given topic will appear together, regardless of the units. Problems involving USCS (English) units are now marked by a capital E after the number, while problems in SI units are unmarked. A new group of problems have replaced a sizable fraction of those in the fifth edition, while some older problems have been upgraded in terms of data and questions. The problems at the end of the chapters now total over 2000. Throughout the text the problems are divided into sections separated by headers that indicate the subject matter, as was done in the fourth edition.

7. *Parametric and design problems.* A group of problems entitled "Parametric and design studies" appears at the end of the application chapters, namely, Chaps. 10 and 13 through 17. Some of the design-oriented problems have been used in the second semester course in thermodynamics at Purdue University, and one

author (Ken Wark) wishes to acknowledge these contributions from a number of his faculty colleagues. These problems are more easily solved by means of a software package that contains property data, such as the EES program described below.

8. *Computer-aided solutions.* Solutions to typical problems in the text have been developed using a software package called EES (Engineering Equation Solver) and are available on an EES Software Problems Disk. Problems included on the problems disk are usually example problems and are denoted in the text with a disk symbol. By using example problems, students can clearly see the differences between hand and computer-aided solutions. Each fully-documented solution is an EES program that is run using the EES engine. These programs illustrate the use of EES and help the student master the important concepts without the computational burden previously required with hand calculations. This type of program is extremely useful in parametric studies as well as for open-ended design problems.

9. *Learning aids.* For emphasis, all basic equations and other important relations are now enclosed in a box, and fundamental terms appear in boldface type when they are defined. In addition to the discussion in the text, important concepts are stressed using notes in the margin and in critical thinking or concept questions (marked by a capital C) at the beginning of the problem sections of the first eight chapters. These questions may be assigned or used for class discussion. Finally, a brief summary has been added at the end of each chapter containing the basic equations and property relations developed for that topic. Equations for specialized applications do not appear in the summaries.

10. *Use of figures.* As a result of the amplification of the problem-solving approach in the sixth edition, the number of figures in the text has increased threefold. There are now about 200 figures which are used in the problem-solving methodology described in item (1) above, and the total number of figures is now around 450.

11. *Availability (exergy) and irreversibility.* The introduction to the concepts of availability and irreversibility, introduced in Chap. 9, has been substantially revised. The major results are now used in the analysis of power and refrigeration cycles (Chaps. 15–17) and of chemically reactive systems (Chap. 13).

12. *Chemical equilibrium.* In Chap. 14 the equations for chemical equilibrium have been changed to include a $K_o$ quantity based on the standard-state Gibbs function change and a $K_p$ term based on ideal-gas partial pressures.

13. *Advanced energy systems.* Chapter 19 in the fifth edition contained discussions on advanced and innovative energy systems such as fuel cells, combined cycles, cogeneration, and geothermal and ocean thermal-energy conversion systems. In the sixth edition, these topics have been moved to Chaps. 14, 16, and 17, where they tie more directly to the basic theory and practice to which they are related.

14. *Arrangement of advanced topics and applications.* An area of difference between thermodynamic text-books is the placement of material on gas mixtures, generalized thermodynamic relationships, combustion, and chemical equilibrium (Chaps. 10–14) versus power and refrigeration cycles (Chaps. 15–17). Some other texts reverse this order. Should the instructor desire a different sequence, there is no conflict in using this text if the chapters on cycle analysis are assigned first.

## SUPPLEMENTS

A **Solution Manual** showing the complete solution to each problem is available; however, solutions to parametric and open-ended design problems are not provided. In addition, adopters of the text may obtain an **Instructor's Resource CD** of selected figures and tables from the text. A **Tables and Figures Supplement** containing information from the two appendices is also available in the format used for the fifth edition.

**EES (Engineering Equations Solver)** is a general program that solves algebraic and initial-value differential equations. EES can also do optimization, parametric analysis, and linear and nonlinear regression and provide publication quality plotting capability. EES has an intuitive interface that is very easy to master. Equations can be entered in any form and in the most efficient manner. The EES engine is available to adopters of the text with the Problems Disk. The book is available with or without the EES Problems Disk. Faculty interested in using the book with the Problems Disk should notify their local WCB/McGraw-Hill representative to obtain information on obtaining the EES engine that drives the Problems Disk.

EES is particularly useful in thermodynamics problems since most property data needed for solving problems in these areas are provided by the program. For example, the steam tables are implemented such that any thermodynamic property can be obtained from a built-in function call in terms of any other properties. Similar capability is provided for all substances. EES also allows the user to enter property data or functional relationships with lookup tables, with internal functions written with EES, or with externally compiled functions written in Pascal, C, C++, or Fortran. Interesting practical problems that may have implicit solutions are often not assigned because of the mathematical complexity involved. EES allows the user to concentrate on concepts by freeing him or her from mundane chores.

## ACKNOWLEDGMENTS

We are grateful to the many reviewers of the sixth edition including:

Kalyan Annamalai, *Texas A&M University*

Charles M. Harman, *Duke University*

E. Ramon Hosler, *University of Central Florida*

David A. Nelson, *Michigan Tech University*

Sudhokar Neti, *Lehigh University*

Edmundo M. Nunes, *Manhattan College*

Eric B. Ratts, *The University of Michigan–Dearborn*

Dave Sree, *Tuskegee University*

Professor Wark welcomes Prof. Richards as co-author to this text. His contributions have been invaluable. He wishes to thank his wife for her patience through six editions. Professor Wark is also indebted to Linda L. Tutin for her outstanding typing of portions of the new manuscript. The input of students and faculty at Purdue University, which influenced a number of the changes in this edition, is also gratefully acknowledged.

Professor Richards would like to thank Prof. Wark for providing the opportunity to collaborate on this text, for sharing his experience, and for his openness to change. Professor Richards would also like to recognize and thank three of his professors—A. O. Flinner, R. C. Fellinger, and M. J. Moran—for sharing their knowledge and love of teaching and thermodynamics. Finally, he wishes to thank his wife and daughters for their continued support, patience, and understanding.

# 1

# BASIC CONCEPTS AND DEFINITIONS

Solar photovoltaic (2.6 MW) and nuclear (900 MW) power stations at
Rancho Seco near Sacramento, CA.

Thermodynamics is a branch of both physics and chemistry. Scientists in these fields have developed the basic principles that govern the physical and chemical behavior of matter with relationship to energy. In addition, basic relationships among the properties of matter which are influenced by energy interactions have evolved. On the other hand, engineers use this basic information as part of an engineering science to study and design energy systems. In practice, these studies also include the use of other engineering sciences such as heat and mass transfer, and fluid mechanics. This text assumes no formal background in the latter engineering sciences. However, the need for them in more complete engineering design studies will become apparent as we progress through the text.

This chapter reviews some of the basic concepts and definitions presented in earlier courses, and presents new ones that are important in our study of engineering thermodynamics. Since many of the topics are review items, the discussions frequently are brief.

## 1-1    THE NATURE OF THERMODYNAMICS

Thermodynamics is a science that includes the study of energy transformations and of the relationships among the physical properties of substances which are affected by these transformations. Engineering thermodynamics traditionally has involved the study of such diverse areas as stationary and mobile power-producing devices, refrigeration and air-conditioning processes, fluid expanders and compressors, jet engines and rockets, chemical processing as in oil refineries, and the combustion of hydrocarbon fuels (coal, oil, and natural gas). More recently, other areas of interest have evolved. For example, the use of passive and active solar energy units, including solar ponds, is undergoing tremendous growth. Commercial power production from fluids heated by geothermal sources beneath the ground is available on a growing scale. In addition, wind power systems continue to be developed and to be added to the electric power grid. Tidal power is under active investigation, as well as the use of the temperature difference between the surface and deeper layers of the oceans as a potential source of power production. This latter energy system is called ocean thermal energy conversion (OTEC). Study continues on a magnetohydrodynamic (MHD) power cycle which generates electricity by passing a high-temperature gas through a magnetic field. Other processes of interest include thermionic and photovoltaic devices, as well as biomedical applications.

Thermodynamic properties and energy relationships can be studied by two methods. **Classical thermodynamics** involve studies which are undertaken without recourse to the nature of the individual particles which make up a substance and to their interactions. This is a *macroscopic viewpoint* toward matter, and it requires no hypothesis about the detailed structure of matter on the atomic scale. Consequently, the general laws of classical thermodynamics are based on macroscopic measurements and are not subject to change as knowledge concerning the nature of matter is discovered.

A second method called **statistical thermodynamics** is based on the statistical behavior of large groups of individual particles. This is a *microscopic viewpoint* of matter. It postulates that the values of macroscopic properties (such as pressure, temperature, and density, among others), which we measure directly or calculate from other measurements, merely reflect some sort of statistical average of the behavior of a tremendous number of particles. This theory has been helpful in the modern development of new, direct energy-conversion methods, such as thermionics and thermoelectrics.

Five laws, or postulates, govern the study of energy transformations and the relationships among properties. Two of these—the first and second laws—deal with energy, directly or indirectly. Consequently, they are of fundamental importance in engineering studies of energy transformations and use. The remaining three statements—the zeroth law, the third law, and the state postulate—relate to thermodynamic properties. The first law of thermodynamics leads to the concept of energy and a conservation of energy

**What is the major difference in the viewpoints of classical and statistical thermodynamics?**

principle. When energy is transferred from one region to another or changes form within a region of space, the total quantity of energy is constant. (In this text we do not consider nuclear transformations of mass to energy.)

The second law of thermodynamics has many ramifications with respect to engineering processes. One of these is that the first law deals with the *quantity* of energy, while the second law deals with the *quality* of energy. The idea of quality arises when one needs to optimize the conversion and transmission of energy. We find that the second law places restrictions on the transformation of some forms of energy to more "useful" types. The second law enables the engineer to measure the "degradation," or change in quality, of energy in quantitative terms. The second law also introduces an important thermodynamic property—entropy.

**What is the difference between the quantity and the quality of energy?**

The use of energy by industrialized countries is an important factor in their continued growth. In addition, the desire of underdeveloped nations to improve their standards of living will lead to continuing studies of improving energy use throughout the world. A move must be made to cut wasteful use of energy in industry, in transportation, and in residential and commercial applications. With the increasing cost and decreasing supply of conventional fossil fuels in the future, it is imperative that engineers look seriously at increasing the efficiency of energy use. As an example several methods are under development for increasing the overall energy conversion efficiency for large electric power generation units. Thus thermodynamics will continue to make a valuable contribution to the study of new energy systems as well as to the revitalization of older energy systems.

## 1-2  DIMENSIONS AND UNITS

*Dimensions* are names that characterize physical quantities. Common examples of dimensions include length $L$, time $t$, force $F$, mass $m$, electric charge $Q_c$, and temperature $T$. In engineering analysis any equation which relates physical quantities must be *dimensionally homogeneous;* that is, the dimensions on one side of an equation equal those of the other side. Such homogeneity also is retained during any subsequent mathematical operation and thus is a powerful tool for checking the internal consistency of an equation.

To make numerical computations involving physical quantities, there is the additional requirement that units, as well as the dimensions, be homogeneous. *Units* are those arbitrary magnitudes and names assigned to dimensions which are adopted as standards for measurements. For example, the primary dimension of length may be measured in units of feet, miles, centimeters, etc. These are all arbitrary lengths which may relate to each other by *unit conversion factors,* or unitary constants. Unit conversion factors include 12 in = 1 ft and 60 s = 1 min. A comparable way of writing

unit conversion factors is

$$\frac{12 \text{ in}}{1 \text{ ft}} = 1 \qquad \frac{60 \text{ s}}{1 \text{ min}} = 1$$

**What is the difference between the dimensions and units of a physical quantity?**

Terms in equations can always be multiplied by unit conversion factors, since it is always permissible to multiply by unity. A given physical quantity may often be measured in several sets of units. Therefore, a tabulation of common unit conversion factors for thermodynamic analyses is presented in Tables A-1 and A-1E, in the Appendix, as well as inside the front cover.

A number of systems of units have been developed over the years. However, in this text we consider only two: the International System of Units (SI) and the United States Customary System (USCS). The latter system is also known as the English engineering system.

### 1-2-1   The Système Internationale (International System)

The fundamental system of units chosen for scientific work is the **Système Internationale,** which is usually abbreviated as SI. The SI employs seven *primary dimensions:* mass, length, time, temperature, electric current, luminous intensity, and amount of substance. Table 1-1 lists the SI *primary units* which are standards for these dimensions. All these units are defined operationally. For example, the SI unit of length is the meter (m), and it is defined as 1,650,763.73 wavelengths of the orange-red line of emission from krypton-86 atoms in vacuum. The unit of time is the second (s), and it is defined as the duration of 9,192,631,770 ± 20 cycles of a specified transition within the cesium atom. There are also precise operational definitions of mass (kg), temperature (K), electric current (A), and luminous intensity (cd).

When very large or small values of a quantity are involved, a set of standard prefixes is used to simplify writing a value in SI units. These prefixes designate certain decimal multiples of the unit. Table 1-2 lists the values of

**Table 1-1**    SI primary dimensions and units

| Physical Quantity | Unit and Symbol |
| --- | --- |
| Mass | kilogram (kg) |
| Length | meter (m) |
| Time | second (s) |
| Temperature | kelvin (K) |
| Electric current | ampere (A) |
| Luminous intensity | candela (cd) |
| Amount of substance | mole (mol) |

**Table 1-2**    Standard SI prefixes

| Factor | Prefix | Symbol |
|--------|--------|--------|
| $10^{12}$ | tera | T |
| $10^9$ | giga | G |
| $10^6$ | mega | M |
| $10^3$ | kilo | k |
| $10^2$ | hecto | h |
| $10^{-2}$ | centi | c |
| $10^{-3}$ | milli | m |
| $10^{-6}$ | micro | $\mu$ |
| $10^{-9}$ | nano | n |
| $10^{-12}$ | pico | p |

the multiplier factor, the name of the prefix, and the symbol for the prefix. For example, kW denotes a kilowatt, that is, $10^3$ watts, while mg denotes a milligram, that is, $10^{-3}$ grams. Although multiples of $10^3$ are used in SI, other multiples are sometimes employed in specific fields.

The seventh SI primary unit, the *mole* (mol), is the amount of substance containing the same number of elementary particles (atoms, molecules, ions, electrons) as there are atoms in 0.012 kg of the pure carbon nuclide $^{12}$C. The number of particles per mole, $6.02205 \times 10^{23}$, is known as an *Avogadro's number* $N_A$. For engineering calculations it is useful to express mass in kilomoles (kmol). A kilomole is 1000 times as large as a mole. For example, a kilomole of diatomic oxygen $O_2$ contains 32 kg of oxygen, and has $1000N_A$ molecules of $O_2$. The number of moles $N$ of a substance is defined as

$$N = \frac{m}{M} \qquad \text{[1-1]}$$

where $M$ is the molar mass. The **molar mass** is the mass of a substance that is numerically equal to its molecular weight, expressed as mass per mole. For example, the molar mass of helium is 4.003 g/mol or 4.003 kg/kmol.

All other SI units are *secondary units* and are derivable in terms of the seven primary units shown in Table 1-1. The secondary SI unit of force is the newton (N), and it is derived from Newton's second law, $F = kma$. The proportionality constant $k$ is chosen to be unity, so that $F = ma$. On the basis of this equation, a derived force of one newton is required to accelerate one kilogram of mass at one meter per second per second. Since $1 \text{ N} = 1 \text{ kg} \times 1 \text{ m/s}^2$,

$$\frac{1 \text{ N·s}^2}{\text{kg·m}} = 1 \qquad \text{[1-2]}$$

**What are the two different sets of SI units commonly used for the molar mass?**

Equation [1-2] is a unit conversion factor which relates the derived force unit to the primary mass, length, and time units in SI. It is useful in converting other secondary units, which are partially defined in terms of force, to a set of primary units containing only the kilogram, the meter, and the second. For example, the secondary SI unit for pressure is the *pascal* (Pa), which is defined in force and length units as $1 \text{ N/m}^2$. Employing Eq. [1-2], we find that

$$1 \text{ Pa} = \frac{1 \text{ N}}{\text{m}^2} \times \frac{\text{kg·m}}{1 \text{ N·s}^2} = 1 \text{ kg/(m·s}^2)$$

As a result, the secondary unit of pressure, the pascal, can be expressed in terms of its *primary equivalent,* $1 \text{ kg/m·s}^2$.

Note that weight $W$ always refers to a force. When it is stated that a body weighs a given amount, this means that this is the force with which the body is attracted toward another body, such as the earth or moon. The acceleration of gravity varies with distance between two bodies (such as the height of a body above the earth's surface). Thus the weight of a body varies with elevation, while the mass of a body is constant with elevation. The value for the **standard gravitational acceleration** on the earth at sea level and 45 degrees latitude is $9.80665 \text{ m/s}^2$. This is shortened to $9.807 \text{ m/s}^2$ in this text.

**Example 1-1**

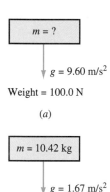

$m = ?$

$g = 9.60 \text{ m/s}^2$

Weight = 100.0 N

(a)

$m = 10.42 \text{ kg}$

$g = 1.67 \text{ m/s}^2$

Weight = ?

(b)

**Figure 1-1**
Mass-weight relationship in SI.

The weight of a piece of metal is 100.0 N at a location where the local acceleration of gravity $g$ is $9.60 \text{ m/s}^2$. Find (*a*) the mass of the metal in kilograms, and (*b*) the weight of the metal on the surface of the moon, where $g = 1.67 \text{ m/s}^2$.

**Solution:**

**Given:**   A piece of metal under two gravity conditions is shown in Fig. 1-1.

**Find:**   (*a*) mass on the earth, in kg, and (*b*) weight on the moon, in N.

**Analysis:**   (*a*) The SI unit system is based on Newton's second law. Since weight $W$ is a force, and $a = g$ in this case, we write the second law as $W = mg$. Hence,

$$m = \frac{W}{g} = \frac{100.0 \text{ N}}{9.60 \text{ m/s}^2} = 10.42 \frac{\text{N·s}^2}{\text{m}} \times \frac{1 \text{ kg·m}}{\text{N·s}^2} = 10.42 \text{ kg}$$

where $1 \text{ N·s}^2/\text{m·kg} = 1$ by definition.

(*b*) The mass of the piece of metal will remain the same regardless of its location. The weight will change, however, with a change in gravitational acceleration. Equating weight to force on the surface of the moon gives

$$\text{Weight} = F_{\text{moon}} = mg = 10.42 \text{ kg} \times 1.67 \frac{\text{m}}{\text{s}^2} \times \frac{1 \text{ N·s}^2}{\text{kg·m}} = 17.4 \text{ N}$$

**Comment:**   Although the mass is the same at the two locations, the weight is quite different.

## 1-2-2 UNITED STATES CUSTOMARY SYSTEM OF UNITS (USCS)

An important system of units employed in the United States is the ***United States Customary System***, or USCS. Unfortunately, it introduces a point of confusion. The unit called a pound is used to denote both a unit of force called a pound-force ($lb_f$) and a unit of mass called a pound-mass ($lb_m$). Table 1-3 lists some basic units for the USCS. If we continue to use the units of mass, length, and time to define a force unit (as we did in the SI), then the unit of force in USCS becomes a secondary unit. As a result we continue to write Newton's second law as $F = ma$, where the proportionality constant $k$ again is taken as unity. However, by definition, a gravitational force of one pound-force ($lb_f$) will accelerate a mass of one pound-mass ($lb_m$) toward the earth at the standard acceleration rate of 32.1740 ft/s$^2$ (9.80665 m/s$^2$).

This definition of a pound-force allows us to develop a unit conversion factor relating the pound-force, a derived unit, to the primary mass, length, and time units in USCS. Since we have defined a pound-force as

$$1 \ lb_f \equiv 32.174 \frac{lb_m \cdot ft}{s^2} \qquad \textbf{[1-3]}$$

it follows that

$$\frac{32.174 \ lb_m \cdot ft}{lb_f \cdot s^2} = 1 \qquad \textbf{[1-4]}$$

Because the term on the left is equal to unity, it can be inserted anywhere into an equation without altering the validity of the equation. This unit conversion factor is used in exactly the same manner as Eq. [1-2], which relates the newton to the kilogram, meter, and second. This unit conversion factor is necessary to ensure that all the terms in an equation have consistent units. For example, a unit conversion factor such as Eq. [1-3] may be necessary when both $lb_f$ and $lb_m$ appear within the same term. Also keep in mind that weight is a force. The weight $W$ of a mass under a local gravitational acceleration $g$ is given by Newton's second law in the form $W = mg$.

In USCS we can also use the mole as a unit for the amount of substance. In this case it is called the pound-mole (lbmol). Since there are approximately 454 g in a pound-mass, a pound-mole is equivalent to 454 mol or

**Note the difference between a pound-mass and a pound-force.**

**Table 1-3** USCS primary dimensions and units

| Physical Quantity | Unit and Symbol |
| --- | --- |
| Mass | pound-mass ($lb_m$) |
| Length | foot (ft) |
| Time | second (s) |
| Temperature | Rankine (°R) |
| Force | pound-force ($lb_f$) |

**Note the molar mass of a given substance has the same numerical value whether expressed as g/mol, kg/kmol, or $lb_m$/lbmol.**

0.454 kmol. Hence a pound-mole of carbon contains 12 $lb_m$, and a pound-mole of diatomic oxygen contains 32.0 $lb_m$. That is, the molar mass $M$ of $O_2$ also can be expressed as 32 $lb_m$/lbmol.

---

**Example 1-2**

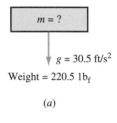

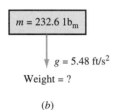

**Figure 1-2**
Mass-weight relationship in USCS.

**The** weight of a piece of metal is 220.5 $lb_f$ at a location where the local acceleration of gravity $g$ is 30.50 ft/s². Find (*a*) the mass of the metal, in pound-mass, and (*b*) the weight of the metal on the surface of the moon, where $g = 5.48$ ft/s².

**Solution:**

**Given:**   A piece of metal under two gravity conditions is shown in Fig. 1-2.

**Find:**   (*a*) mass on earth, in $lb_m$, and (*b*) weight on moon, in $lb_f$.

**Analysis:**   (*a*) Analysis of a system in a gravitational field is based on Newton's second law, $F = ma$. In this case we write the second law as $W = mg$, since $F = W$ and $a = g$. Hence

$$m = \frac{W}{g} = \frac{220.5\ lb_f}{30.50\ ft/s^2} = 7.23\frac{lb_f \cdot s^2}{ft} \times \frac{32.174\ lb_m \cdot ft}{lb_f \cdot s^2} = 232.6\ lb_m$$

where the unit conversion factor between pound-force and pound-mass has been used. (*b*) The mass of the piece of metal will remain the same regardless of its location. Its weight will change, however, with a change in gravitational acceleration. Equating weight to force on the surface of the moon yields

$$\text{Weight} = F_{moon} = mg = \frac{232.6\ lb_m \times 5.48\ ft/s^2}{32.174\ lb_m \cdot ft/(lb_f \cdot s^2)} = 39.6\ lb_f$$

where the unit conversion factor between pound-force and pound-mass again has been used.

**Comment:**   Although the mass is the same at the two locations, the weight is quite different.

---

## 1-3   SYSTEM, PROPERTY, AND STATE

A thermodynamic *system* is a three-dimensional region of space or an amount of matter, bounded by an arbitrary surface. The *boundary* may be real or imaginary, may be at rest or in motion, and may change its size or shape. The region of physical space that lies outside the arbitrarily selected boundaries of the system is called the *surroundings*, or the *environment*. In its usual context the term "surroundings" is restricted to the specific localized region that interacts in some fashion with the system and hence has a detectable influence on the system. These terms are illustrated in Fig. 1-3. *Any thermodynamic analysis begins with a selection of the system, its boundary, and its surroundings.*

Two examples of macroscopic systems and the representative boundaries are shown in Fig. 1-4. Figure 1-4a illustrates the flow of a fluid through a pipe or duct. The dashed line is a two-dimensional representation of one possible choice of a boundary, fixed in space, which outlines the region of space for which a thermodynamic study is to be made. The inside surface of the pipe may be chosen as part of the boundary, and it represents a real obstruction to the flow of matter. However, note that a portion of the boundary is imaginary; i.e., there is no real surface which marks the position of the boundary at the open ends. These latter boundaries are selected for accounting purposes, and they have no real effect or significance in the actual physical process. Thus it is not necessary for any or all of the boundary to be physically distinguishable when thermodynamic analyses are made. Nevertheless, it is extremely important to establish the boundaries of a system clearly before any form of analysis is undertaken.

The analysis of thermodynamic processes includes the study of the transfer of mass and energy across the boundaries of a system. An **open system** is a system for which mass as well as energy may cross the selected boundaries. (The forms of energy transfer are discussed in Chaps. 2 and 5.) The region of space described in Fig. 1-4a, called a **control volume,** would be analyzed as an open system. The boundary of the control volume is called the **control surface.**

In Fig. 1-4b a piston-cylinder assembly is shown. Again the selected boundary is given by the dashed line. In this particular case the boundary is physically well established, for the envelope lies just inside the walls of the cylinder and the piston. For this example note that the shape of the boundary and the volume of the system will change when the position of the piston is altered. The change in shape or volume of the boundary is

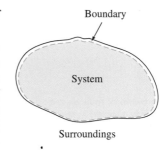

**Figure 1-3**
System and surroundings.

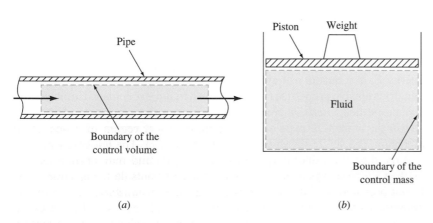

(a)

(b)

**Figure 1-4**    The boundaries of two typical thermodynamic systems. (a) Pipe flow;
(b) piston-cylinder device.

Note that a bottle of milk on the grocery shelf is a closed system, while the same bottle is an open system when in use to fill a glass.

always permissible as long as such changes are recognized in subsequent calculations.

A *closed system* or a *control mass* is a system for which no mass crosses the boundary. Although the quantity of matter is fixed in a closed system, energy is allowed to cross its boundaries, as discussed in Chap. 2. The matter may also change in chemical composition within the boundaries. The control mass within the piston-cylinder assembly in Fig. 1-4*b* is in a closed system. When neither mass nor energy crosses the boundary, the system is known as an *isolated system*.

A system may contain several phases within its boundary. A *phase* is a quantity of matter that is homogeneous throughout in physical structure and chemical composition. Homogeneity in physical structure requires the material to be all gas, all liquid, or all solid. Systems may contain two phases, such as liquid and gas, where the two phases are separated by an internal phase boundary. Homogeneity in chemical composition, on the other hand, does not require a single chemical species. Mixtures of gases, such as atmospheric air, are considered single phases, as well as miscible liquid solutions under certain conditions. A *pure substance* in this text describes a substance with a single chemical (atomic or molecular) structure.

Once a system has been selected for analysis, it can be described further in terms of its properties. A *property* is a characteristic of a system, and its value is independent of the history of the system. Frequently the value of a property is directly measurable. In other instances a property is defined by mathematically combining other properties. A third class of properties includes those defined by laws of thermodynamics. Examples of properties include pressure, temperature, mass, volume, density, electrical conductivity, acoustic velocity, and thermal coefficient of expansion. The value of a property is unique and fixed by the condition of the system at the time of measurement. Note that property values have dimensions associated with them. In addition, the numerical values of properties are dependent on the set of units employed.

What are the three classes of properties?

Properties are classified as either extensive or intensive. Consider a system arbitrarily divided into a group of subsystems. A property is *extensive* if its value for the whole system is the sum of its values for the various subsystems or parts. Examples of extensive properties include volume $V$, energy $E$, and quantity of electric charge $Q_c$. Generally, uppercase letters denote extensive properties (with mass $m$ a major exception). Unlike extensive properties, *intensive* properties have values that are independent of the size or the amount of mass of the system. Intensive properties have a value at a point. If a single-phase system in equilibrium is divided arbitrarily into $n$ parts, then the value of a given intensive property will be the *same* for each of the $n$ subsystems. Thus intensive properties have the same value throughout a system at equilibrium. Examples of intensive properties include temperature, pressure, density, velocity, and chemical concentration.

Figure 1-5 illustrates the concept of extensive and intensive properties. The dashed line represents an arbitrary subdivision of the overall system

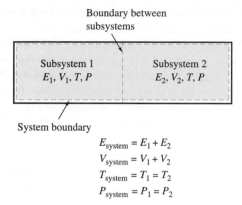

Boundary between subsystems

| Subsystem 1<br>$E_1, V_1, T, P$ | Subsystem 2<br>$E_2, V_2, T, P$ |
| --- | --- |

System boundary

$$E_{system} = E_1 + E_2$$
$$V_{system} = V_1 + V_2$$
$$T_{system} = T_1 = T_2$$
$$P_{system} = P_1 = P_2$$

**Figure 1-5**   Extensive and intensive relationships.

into two subsystems, 1 and 2. When an extensive property $Y$ of the overall system is divided by the mass $m$ of the overall system, the resulting property is called a ***specific property*** and is given the symbol $y$. Thus $y = Y/m$. A specific property is an intensive property. Examples include specific volume $v \equiv V/m$ and specific energy $e$ ($e \equiv E/m$). Lowercase letters denote intensive and specific properties, the most notable exceptions being pressure $P$ and temperature $T$. These two properties are always intensive.

**What is the major difference between extensive and intensive properties?**

The ***state*** of a system is the condition of the system described by the values of its properties. The state of a system frequently can be specified by or identified from the values of only a few of its properties. The values of all remaining properties can be determined from the values of the few used to specify the state.

Properties are defined only when a system is in equilibrium. For this reason, the study of classical thermodynamics places emphasis on equilibrium states and changes from one equilibrium state to another. For an isolated system in equilibrium, the macroscopic state of the system does not change with respect to time. A system is in ***thermodynamic equilibrium*** if it is not capable of a finite, spontaneous change to another state without a finite change in the state of its environment. Hence a system in stable equilibrium cannot change its state without an interaction with its environment. Included among the many classes of equilibrium are thermal, mechanical, phase, and chemical. *Thermal equilibrium* requires that the temperature be uniform throughout the system. In the absence of a gravitational effect, *mechanical equilibrium* implies an equality of forces throughout. *Phase equilibrium* deals with the absence of any net transfer of one or more chemical species from one phase to another in a multiphase system. A mixture of substances is in *chemical equilibrium* if there is no tendency for a net chemical reaction to occur. All of these forms of equilibrium must be met to fulfill the condition of thermodynamic equilibrium.

**What is the key requirement for a system to be in thermodynamic equilibrium?**

A ***process*** is a change of a system from one equilibrium state to another. The ***path*** of a process refers to the specific series of states through which the system passes. A complete description of a process typically involves specification of the initial and final equilibrium states, the path (if identifiable), and the interactions which take place across the boundaries during the process. Processes during which one property remains constant are designated by the prefix *iso-* before the property. An ***isothermal process*** occurs at constant temperature, an ***isobaric process*** is at constant pressure, and an ***isometric* (*isochoric*) *process*** is one of constant volume. A ***cyclic process,*** or ***cycle,*** is a process for which the end states are identical. The change in the value of any property for a cyclic process is zero. A two-process and a four-process cycle are shown in parts *a* and *b* of Fig. 1-6. Hence the change in an intensive property *y* for a cycle is mathematically given by

$$\oint dy = 0 \qquad \text{[1-5]}$$

where the symbol $\oint$ represents integration around a cyclic path. The converse of this equation is also valid. If the integral of a quantity *dy* over an arbitrary cycle is zero, then the quantity *y* is a property.

As noted earlier, any property has a fixed value in a given equilibrium state, regardless of how the system arrives at that state. Therefore,

- *The change in the value of a property that occurs when a system is altered from one equilibrium state to another is always the same, regardless of the process used to bring about the change.*

Conversely,

- *If a quantity evaluated for a process always has the same value between two given equilibrium states, that quantity is a measure of the change in a property.*

Either this latter statement, or the converse to Eq. [1-5], can be used as a test for whether a quantity is a property.

This uniqueness of a property change for a process with given end states 1 and 2 is described mathematically by the ***exact differential*** *dy*, such that

$$\int_1^2 dy = y_2 - y_1 = \Delta y \qquad \text{[1-6a]}$$

or, on an extensive basis,

$$\int_1^2 dY = Y_2 - Y_1 = \Delta Y \qquad \text{[1-6b]}$$

where *dy* again signifies an infinitesimal change in an intensive property *y*, and $Y = my$. For example, the infinitesimal change in the pressure *P* is given by *dP*. The finite change in a property is denoted by the symbol $\Delta$ (capital delta), for example, $\Delta P$. The change in a property value $\Delta Y$ always represents the final value $Y_2$ minus the initial value $Y_1$.

**Note the physical meaning of isothermal, isobaric, and isometric processes.**

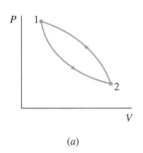

(a)

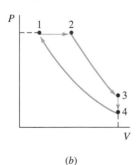

(b)

**Figure 1-6**
Property change for a cyclic process.

Thermodynamic quantities can be divided into two categories: state functions and process functions. All properties are **state functions,** since the change in the value of a property depends solely on the property values at the initial and final states, and not on the path of the process [see Eq. [1-6]]. Quantities whose values depend on the path of the process are called **process** or **path functions.** If $Z$ represents an extensive process function, an infinitesimal (differential) amount of $Z$ is denoted *in thermodynamics* by $\delta Z$, rather than by $dZ$. This use of $\delta$ stresses the fact that $\delta Z$ is not an exact differential in the usual mathematical sense but is an **inexact differential.** Thus its value for a finite process is given by

$$\int_1^2 \delta Z \equiv Z_{12} \qquad \text{[1-7]}$$

where $Z_{12}$ is simply the sum of the $\delta Z$ values along the chosen path between states 1 and 2. Figure 1-7 illustrates this summing of terms. An arbitrary path between states 1 and 2 is shown on a $y$-$x$ diagram, where $y$ and $x$ are two properties, such as $P$ and $V$. A quantity $\delta Z = y\,dx$ is associated with each small increment of the path. The sum of these values of $\delta Z$ for the overall process is $Z_{12}$ for path 1-2. A different path between states 1 and 2 would lead to a different value of $Z_{12}$. *Note that integration of $\delta Z$ does not lead to the notation $\Delta Z$, but simply $Z_{12}$.* The value of a process function only has meaning for the overall process and has no meaning at a state. The two major process functions in thermodynamic studies, work and heat, are discussed in the next chapter.

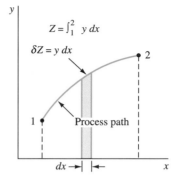

**Figure 1-7**
Property change as a summation process.

**Note the difference between a path function and a state function.**

# 1-4 DENSITY, SPECIFIC VOLUME, AND SPECIFIC GRAVITY

The **density** $\rho$ is defined as the mass per unit volume:

$$\rho \equiv \frac{\text{mass}}{\text{volume}} = \frac{m}{V} \qquad \text{[1-8]}$$

The **specific gravity** (sp. gr.) of a substance is the ratio of its density to that of water at a specified temperature, such as 4 or 20°C or 60°F, and one atmosphere:

$$\text{sp. gr.} \equiv \frac{\rho}{\rho_{H_2O}} \qquad \text{[1-9]}$$

The specific gravities of solids and liquids are frequently listed in handbooks, rather than their densities. The density of water at 4°C is close to 1.00 g/cm$^3$ (kg/L), 1000 kg/m$^3$, or 62.4 lb$_m$/ft$^3$, while that of liquid mercury is close to 13.6 g/cm$^3$, 13,600 kg/m$^3$, or 850 lb$_m$/ft$^3$.

The *specific volume* $v$ is defined in the preceding section as the volume per unit mass of a substance. Thus it is the reciprocal of the density:

$$v \equiv \frac{\text{volume}}{\text{mass}} = \frac{V}{m} = \frac{1}{\rho} \qquad \textbf{[1-10]}$$

The mass units for $v$ or $\rho$ normally are gram or kilogram (in SI) or pound-mass (in USCS). However, sometimes it is useful to use the *mole* (see Sec. 1-2 and Table 1-1) as the unit of mass. Thus typical SI units for specific volume might be $m^3/kg$, $cm^3/g$, or $m^3/kmol$. In USCS units, the choice normally is $ft^3/lb_m$ or $ft^3/lbmol$. The molar density, $N/V$, might be expressed as $kmol/m^3$, for example. Thus any property which is partially defined in terms of the mass of the system may be expressed in conventional mass units or molar units. The *specific weight* $w$ of a substance is the weight per unit volume.

$$w = \frac{\text{weight}}{\text{volume}} = \frac{W}{V} \qquad \textbf{[1-11]}$$

It also may be evaluated from $w = \rho g$, where $g$ is the local acceleration of gravity.

## 1-5  PRESSURE

The *pressure P* is defined as the normal force per unit area acting on the boundary of the system. In static systems the pressure is uniform in all directions around the vicinity of an elemental volume of fluid. However, the pressure may vary throughout the system for a fluid in the presence of a gravitational field. Familiar examples are the variation of pressure with depth of water in a swimming pool or lake, and the variation of atmospheric pressure with elevation. As shown in Fig. 1-8, the variation of pressure with height for liquids is much greater than that for gases of the same height. Thus the pressure of a gas usually may be considered uniform owing to the small height of most systems of interest.

### 1-5-1  PRESSURE UNITS

The basic pressure unit in SI is the *pascal* (Pa). By definition,

$$1 \text{ pascal} = 1 \text{ Pa} = 1 \text{ N/m}^2$$

For engineering studies the pascal often is a relatively small pressure unit. Hence for tabulating or reporting data, the kilopascal (kPa) or megapascal (MPa) is commonly used. In this text the *bar* also is used as a SI pressure unit. Its relationship to other units is

$$1 \text{ bar} \equiv 10^5 \text{ N/m}^2 = 10^2 \text{ kPa} = 0.1 \text{ MPa}$$

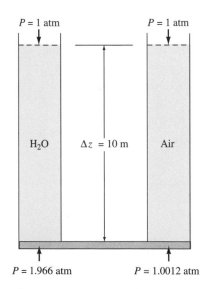

$P = 1$ atm          $P = 1$ atm

H$_2$O     $\Delta z = 10$ m     Air

$P = 1.966$ atm          $P = 1.0012$ atm

**Figure 1-8**
Effect of fluid density on pressure change.

The most commonly used units for pressure in the USCS are pounds-force per square inch ($lb_f/in^2$ or psi), or the atmosphere (atm).

## 1-5-2    THE STANDARD ATMOSPHERE

Although the atmospheric pressure varies with location on earth and with time, a valuable reference value for pressure is the *standard atmosphere*. It is defined as the pressure produced by a column of mercury exactly 760 mm in height at 273.15 K and under standard gravitational acceleration (or 29.92 inHg at 0°C). Some equivalent values in other units are

$$1 \text{ atm } = 1.01325 \text{ bars } = 1.01325 \times 10^5 \text{ N/m}^2 = 14.696 \text{ lb}_f/\text{in}^2 \text{ (psia)}$$

Note that 1 bar is slightly smaller than 1 atm.

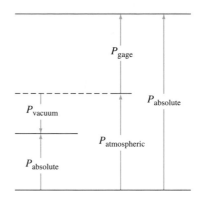

**Figure 1-9**
The relationships among the absolute, atmospheric, gage, and vacuum pressures.

## 1-5-3    ABSOLUTE AND GAGE PRESSURES

The pressure at a given position in a system can be given in reference to zero pressure or in reference to atmospheric pressure. **Absolute pressure** is measured relative to absolute zero pressure. The modifying adjective is necessary since experimentally most pressure-measuring devices indicate what is known as a gage pressure. A **gage pressure** is the pressure difference between the absolute pressure of the fluid and the atmospheric pressure. That is, $P_{gage} = P_{abs} - P_{atm}$. Because a gage reading may be either positive or negative, a general relationship between absolute and gage pressures is

**Note the difference between absolute and positive, or negative, gage pressures.**

$$\boxed{P_{abs} = P_{atm} + P_{gage}} \qquad \textbf{[1-12]}$$

A negative gage pressure, which occurs when the atmospheric pressure is greater than the absolute pressure, is frequently called a **vacuum pressure.** For example, a vacuum pressure of +30 kPa (0.30 bar) is a gage pressure of −30 kPa. If the atmospheric pressure is 100 kPa, then the corresponding absolute pressure is 70 kPa. *If a pressure value in this text is not reported as a gage reading, it is assumed to be an absolute value.* In the technical literature the letters "a" and "g" are frequently added to the abbreviations to signify the difference between absolute and gage values, respectively. For example, the absolute and gage pressures in pounds per square inch are denoted by the terms "psia" and "psig," respectively. In some circumstances the symbols "$lb_f/in^2$ gage" or "mbar gage" may be used. Figure 1-9 illustrates the relations among the various types of pressures.

## 1-5-4    THE MANOMETER AND THE BAROMETER

An instrument for measuring a pressure difference in terms of the height of a liquid column is called a **manometer.** Consider the situation shown in Fig. 1-10. A tube containing a liquid is connected to a tank which is

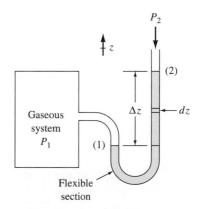

**Figure 1-10**
Measurement of a pressure difference in terms of the height of a liquid column.

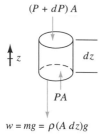

$(P + dP)A$

$dz$

$z$

$PA$

$w = mg = \rho(A\,dz)g$

**Figure 1-11**
Force balance on
a fluid element
within a liquid
column.

Vacuum

Mercury

Height of mercury column

Atmospheric pressure

**Figure 1-12**
Device for mea-
suring the baro-
metric pressure.

filled with a gas at a pressure $P_1$. A pressure $P_2$ is applied to the top of the liquid column. The pressure difference $P_1 - P_2$ can be determined from a knowledge of the height $\Delta z$ of the liquid column. The differential height $dz$ shown in Fig. 1-10 appears as a fluid element in Fig. 1-11. Three forces act on the fluid element in the $z$ direction. Two are normal compressive forces, and the third is the weight of the element in the gravitational field $g$. On the basis of a force balance for the static situation,

$$dP = -\rho g\,dz \qquad \qquad \textbf{[1-13]}$$

This is the general form of the *hydrostatic equation* from basic physics. For liquids the density $\rho$ is essentially independent of pressure. In addition, the variation of $g$ with elevation is negligible for manometer applications. If $g$ and $\rho$ are taken to be constant, then integration of Eq. [1-12] leads to

$$\Delta P = P_2 - P_1 = -\rho g\,\Delta z \qquad \textbf{[1-14]}$$

The negative sign results from the convention that the height $z$ is measured as positive upward, whereas $P$ decreases in this direction.

The actual atmospheric pressure is commonly spoken of as the **barometric pressure.** Its value is not fixed but varies with time and location on the earth. A sketch of a barometer for measuring the barometric or atmospheric pressure is shown in Fig. 1-12. Note that the region in the tube above the mercury column is essentially a vacuum.

**Example 1-3**

**A** manometer is used to measure the pressure in a tank. The fluid is an oil with a specific gravity of 0.87, and the liquid height $\Delta z$ is 45.2 cm. If the barometric pressure is 98.4 kPa, determine the absolute pressure within the tank, in kilopascals and atmospheres, if $g = 9.78$ m/s$^2$.

**Solution:**

**Given:**   A manometer with a 45.2-cm liquid column is shown in Fig. 1-13.

**Find:**   $P_{1,abs}$ in tank, in kPa and atm.

**Analysis:**   The system is the gas within the tank in Fig. 1-13. The absolute pressure within the tank is found from the hydrostatic equation represented by Eq. [1-12], $dP = -\rho g\,dz$. For the short height of the liquid column it is appropriate to *assume* that the density $\rho$ of the liquid and the local acceleration $g$ are constant. Subsequent integration yields Eq. [1-13],

$$P_{1,abs} = P_2 + \rho g\,\Delta z$$

The density of the oil is found from Eq. [1-9], namely, $\rho_{oil} = (sp.gr)\rho_{water}$. The density of water is taken as a room temperature value of 1.00 g/cm$^3$, or 1000 kg/m$^3$ (see Sec. 1-4). Therefore,

$$\rho_{oil} = 0.87(1000 \text{ kg/m}^3) = 870 \text{ kg/m}^3$$

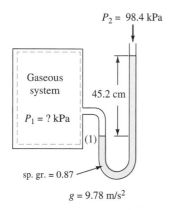

$P_2 = 98.4$ kPa

Gaseous
system

45.2 cm

$P_1 = ?$ kPa

(1)

sp. gr. = 0.87

$g = 9.78$ m/s$^2$

**Figure 1-13**
Schematic and data for Example 1-3.

Use of this value in Eq. [1-13] yields

$$P_{1,abs} = 98.4 \text{ kPa} + \frac{870 \text{ kg}}{m^3} \times \frac{9.78 \text{ m}}{s^2} \times 0.452 \text{ m} \times \frac{1 \text{ N·s}^2}{\text{kg·m}} \times \frac{1 \text{ kPa}}{10^3 \text{ N/m}^2}$$

$$= (98.4 + 3.8) \text{ kPa} = 102.2 \text{ kPa}$$

Since 1 atm $= 101.3$ kPa, $P_1$ also equals 1.009 atm.

---

**A** barometer reads (*a*) 735 mmHg, and (*b*) 28.90 inHg at room temperature. Determine the atmospheric (barometric) pressure of part *a* in millibars and part *b* in $lb_f/in^2$ (psia).

**Example 1-4**

**Solution:**

**Given:** Barometer readings of (*a*) 735 mmHg, (*b*) 28.90 inHg at room temperature.

**Find:** Atmospheric pressure for (*a*) in mbars, and (*b*) in $lb_f/in^2$.

**Analysis:** The conversion of a measured liquid height in a barometer to pressure is found by using Eq. [1-12], $dP = -\rho g \, dz$. Assuming the density and local acceleration of gravity are constant over the small height, this equation becomes Eq. [1-13],

$$\Delta P = P_2 - P_1 = -\rho g \, \Delta z$$

The solution of the above hydrostatic equation requires data for $\rho$ and $g$.

(*a*) In Sec. 1-4 the density of mercury at room temperature is given as 13.6 $g/cm^3$, or 13,600 $kg/m^3$, and we assume this value is relatively independent of temperature. Since the location is not specified, we also assume that $g$ is the standard value of 9.807 $m/s^2$. With reference to Fig. 1-14, for a barometer $P_1$ is the atmospheric pressure and $P_2$ is a complete vacuum, that is, $P_2 = 0$. Then, on the basis of the hydrostatic equation,

$$P_{atm} = \rho g \, \Delta z$$

$$= 13,600 \, \frac{\text{kg}}{m^3} \times 9.807 \frac{m}{s^2}$$

$$\times 735 \text{ mm} \times \frac{m}{10^3 \text{ mm}} \times \frac{1 \text{ N·s}^2}{\text{kg·m}}$$

$$= 0.980 \times 10^5 \text{ N/m}^2$$

$$= 0.980 \text{ bar}$$

$$= 980 \text{ mbars}$$

This value is slightly less than a standard atmosphere, which is 1013 mbars.

(*b*) The density of mercury again is 13.6 $g/cm^3$. Since the location is not specified, we *assume* that $g$ is the standard value of 32.174 $ft/s^2$. First, on the basis of unit conversions in Table A-1 (or data in Sec. 1-4),

$$\rho_{Hg} = 13.6 \text{ g/cm}^3 \times \frac{1 \text{ lb}_m/ft^3}{0.01602 \text{ g/cm}^3} = 850 \text{ lb}_m/ft^3$$

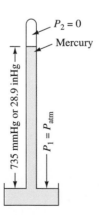

**Figure 1-14**
Schematic and data for Example 1-4.

Application of Eq. [1-14], $P_{atm} = \rho g \Delta z$, then yields

$$P_{atm} = 850\frac{lb_m}{ft^3} \times 32.174\frac{ft}{s^2} \times 28.90 \text{ in}$$

$$\times \frac{1 \text{ ft}^3}{1728 \text{ in}^3} \times \frac{1 \text{ lb}_f \cdot s^2}{32.174 \text{ lb}_m \cdot ft}$$

$$= 14.20 \text{ lb}_f/in^2$$

This value is slightly less than a standard atmosphere, which is 14.696 psia.

---

**Example 1-5**

**If** the barometer reads 734 mmHg, determine (*a*) what absolute pressure, in bars, exists within a system with a vacuum reading of 280 mmHg, and (*b*) what absolute pressure, in psia, is equivalent to a vacuum of 11.0 inHg within a system. Neglect the effect of temperature on the density of mercury.

**Solution:**

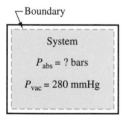

Boundary

System

$P_{abs} = ?$ bars

$P_{vac} = 280$ mmHg

$P_{bar} = 734$ mmHg

**Figure 1-15**
Schematic and data for Example 1-5, part *a*.

**Given:**     Barometer reading of 734 mmHg, vacuum readings of (*a*) 280 mmHg and (*b*) 11.0 inHg within a system. Neglect effect of temperature on the density of mercury.

**Find:**     Absolute pressure in system; (*a*) in bars, (*b*) in psia.

**Analysis:**     (*a*) The system is shown in Fig. 1-15. A vacuum reading is the difference between the barometric or atmospheric pressure and the absolute pressure, on the basis of Eq. [1-12]. In this case the absolute pressure within the system is equal to $734 - 280$, or 454, mmHg. From Table A-1 it is noted that 760 mmHg = 1 atm = 1.013 bars at 0°C. Neglecting the effect of temperature on this unit conversion, we find that

$$P_{abs} = 454 \text{ mmHg} \times \frac{1.013 \text{ bars}}{760 \text{ mmHg}}$$

$$= 0.605 \text{ bar}$$

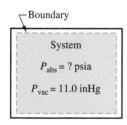

Boundary

System

$P_{abs} = ?$ psia

$P_{vac} = 11.0$ inHg

$P_{bar} = 734$ mmHg

**Figure 1-16**
Schematic and data for Example 1-5, part *b*.

(*b*) The system is shown in Fig. 1-16. A barometric reading of 734 mmHg equals 28.9 inHg. Thus the absolute pressure within the system is equal to $28.9 - 11.0$, or 17.9, inHg. From Table A-1E it is noted that 1 inHg is equal to 0.491 psi at 32°F. Neglecting any temperature effect,

$$P_{abs} = 17.9 \text{ inHg} \times \frac{0.491 \text{ psia}}{1 \text{ inHg}}$$

$$= 8.79 \text{ psia}$$

This value is considerably lower than the barometric value of 14.20 psia.

---

**Example 1-6**

**A** pressure gage reads (*a*) 1.60 bars (gage) when the barometer has a reading of 755 mmHg, and (*b*) 23.0 psig when the barometer reads 28.9 inHg. Find the absolute pressure of the system (*a*) in bars and (*b*) in psia.

**Solution:**

**Given:** (a) Pressure gage reading of 1.60 bars and barometer at 755 mmHg; (b) gage reading of 23.0 psi and barometer at 28.9 inHg.

**Find:** Absolute pressure, (a) in bars, (b) in psia.

**Analysis:** The absolute pressure is the sum of the positive gage pressure and the barometric pressure, as given by Eq. [1-11].

$$P_{abs} = P_{atm} + P_{gage} = P_{bar} + P_{gage}$$

(a) Using a pressure conversion factor from Table A-1 gives

$$P_{bar} = 755 \text{ mmHg} \times \frac{1.013 \text{ bars}}{760 \text{ mmHg}} = 1.01 \text{ bars}$$

and

$$P_{abs} = P_{bar} + P_{gage} = (1.01 + 1.60) \text{ bars} = 2.61 \text{ bars}$$

This latter value is equivalent to 2.58 atm.

(b) On the basis of a pressure conversion factor in Table A-1E,

$$P_{bar} = 28.9 \text{ inHg} \times \frac{0.491 \text{ psi}}{1 \text{ inHg}} = 14.1 \text{ psi}$$

and

$$P_{abs} = P_{gage} + P_{bar} = (23.0 + 14.1) \text{ psia} = 37.1 \text{ psia}$$

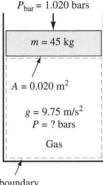

$P_{bar} = 1.020$ bars

$m = 45$ kg

$A = 0.020$ m$^2$

$g = 9.75$ m/s$^2$

$P = ?$ bars

Gas

Gas boundary

**Figure 1-17**
Schematic and data for Example 1-7.

---

**A** piston with a mass of 45 kg and an area of 0.020 m$^2$ encloses a gas in mechanical equilibrium within a vertically positioned piston-cylinder device. A barometric pressure of 1.020 bars acts on the outside of the piston, and the local gravity is 9.75 m/s$^2$. Determine the pressure within the cylinder, in bars.

**Example 1-7**

**Solution:**

**Given:** A piston-cylinder contains a gas as shown in Fig. 1-17.

**Find:** Gas pressure, in bars.

**Analysis:** The gas pressure is found from a vertical force balance on the piston as shown in Fig. 1-18. Thus, for a cross-sectional area $A$,

$$P_{gas}A = P_{bar}A + W_{piston}$$

where the weight $W$ of the piston is given by $mg$. Substitution of numbers yields

$$P_{gas} = P_{bar} + \frac{mg}{A} = 1.020 \text{ bars} + \frac{45 \text{ kg} \times 9.75 \text{ m/s}^2}{0.020 \text{ m}^2} \times \frac{1 \text{ N·s}^2}{1 \text{ kg·m}} \times \frac{1 \text{ bar·m}^2}{10^5 \text{ N}}$$

$$= (1.020 + 0.219) \text{ bars} = 1.239 \text{ bars}$$

**Comment:** Note that the term $mg/A$ has been multiplied by 1 N·s$^2$/kg·m in order to convert kg/m·s$^2$ to a standard pressure unit, N/m$^2$.

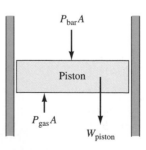

$P_{bar}A$

Piston

$P_{gas}A$

$W_{piston}$

**Figure 1-18**
Force balance on piston in Example 1-7.

## 1-6     THE ZEROTH LAW AND TEMPERATURE

In this section we discuss some important characteristics of temperature. Consider two systems $X$ and $Y$ which initially are individually in equilibrium. If they are brought into contact through a common rigid boundary, there are two possible outcomes with respect to the final states of the systems. One possibility is that the states of $X$ and $Y$ remain unchanged macroscopically. As a second possibility, both systems are observed to undergo a change of state until each reaches a new equilibrium state. These changes of state are due to an interaction between $X$ and $Y$. When two systems that are isolated from the local surroundings undergo no further change of state even when in contact through a common rigid boundary, they are said to be in *thermal equilibrium.* Note that both systems change their state as they proceed toward thermal equilibrium.

It is a matter of experience that the value of a single property is sufficient to determine whether systems will be in thermal equilibrium when they are placed in contact through a common rigid boundary. This property is the *temperature T*. If an interaction occurs, the two systems involved are said to be of unequal temperatures. Such an interaction will continue until the temperatures of the two systems become equal and thermal equilibrium prevails. The concept of thermal equilibrium is shown in Fig. 1-19. Substances $A$ and $B$ are originally isolated from their environment and each other. Removal of the thermal barrier (such as insulation) allows $A$ and $B$ to proceed toward a common temperature when brought into thermal contact.

### 1-6-1     THE ZEROTH LAW

On the basis of experimental observation, it is found that when each of two systems is in thermal equilibrium with a third system, they also will be in thermal equilibrium with each other. This statement is a thermodynamic postulate known as the *zeroth law.* It is a statement of common experience. This law is of importance in thermometry and in the establishment of empirical temperature scales. In practice, the third system in the zeroth law is a thermometer. It is brought into thermal equilibrium with a set of

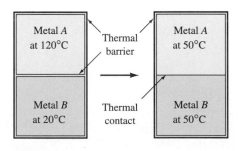

**Figure 1-19**     Thermal equilibrium concept.

temperature standards and is calibrated. At some later time the thermometer is equilibrated with a system at an unknown temperature, and a value is determined. If thermal equilibrium exists during the calibration process and during the test of the system, the temperature of the system must be the same as that set by the calibration standards, on the basis of the zeroth law.

**Why is the zeroth law important in thermometry?**

The application of the zeroth law to temperature measurement is shown in Fig. 1-20. Systems 1 and 2 are at temperatures $T_1$ and $T_2$, respectively. If $T_1 = T_3$ and $T_2 = T_3$ when system 3 is brought independently into thermal contact with the two systems, then $T_1 = T_2$. That is, systems 1 and 2 would be in thermal equilibrium immediately when brought into contact.

## 1-6-2  THERMOMETRIC PROPERTIES AND THERMOMETERS

Temperature is a property of great importance in thermodynamics, and its value can be obtained easily by indirect measurement with calibrated instruments. The temperature of a system is determined by bringing a second body, a thermometer, into contact with the system and allowing thermal equilibrium to be reached. The value of the temperature is found by measuring some temperature-dependent property of the thermometer. Any such property is called a ***thermometric property.*** Commonly used properties of materials employed in temperature-sensing devices include

1. Volume of gases, liquids, and solids
2. Pressure of gases at constant volume
3. Electric resistance of solids
4. Electromotive force of two dissimilar solids
5. Intensity of radiation (at high temperatures)
6. Magnetic effects (at extremely low temperatures)

The first five thermometric properties in the above list are used fairly routinely for temperature measurement. The familiar liquid-in-glass thermometer is an example of the volume change of a liquid used as a thermometric property. A constant-volume gas thermometer, discussed in more detail in Sec. 1-6-4, uses the gas pressure in a bulb as the thermometric property. Electrical resistance sensors involve both normal conductors (such as platinum) as well as semiconductors. This latter type is called a thermistor. Devices based on the use of the electromotive force of dissimilar solids are called thermocouples. Instruments using the intensity of radiation are known as radiation or optical pyrometers.

## 1-6-3  TEMPERATURE SCALES

The absolute temperature scale used by scientists and engineers in the SI is the ***Kelvin scale.*** Since 1954 it has been recommended by an international conference that a reference state value of 273.16 be assigned on the Kelvin temperature scale to that state where solid, liquid, and gaseous water all

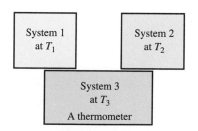

**Figure 1-20**
Zeroth law illustration.

coexist in equilibrium. A state in which three phases of a pure substance co-exist in equilibrium is called a ***triple state*** (or triple point) of the substance. The triple state of water is 0.01 K higher than the normal freezing, or ice, point of water. Thus water freezes at 273.15 K (at 1 atm), which is defined as 0°C on the ***Celsius temperature scale.*** Thus the Celsius scale is related to the Kelvin scale by

**Would the triple state of water have the same value in Washington, D.C., and in Moscow?**

$$T(°C) = T(K) - 273.15 \qquad \textbf{[1-15]}$$

Note that a temperature interval has the same value on either of the two scales. In this text the value of 273.15 frequently will be rounded to 273 when high accuracy is not required.

In USCS there are two additional temperature scales—the ***Rankine scale*** and the ***Fahrenheit scale.*** The temperature in degrees Rankine (°R) is defined arbitrarily as 1.8 times the temperature in kelvins. Consequently

$$T(°R) = 1.8T(K) \qquad \textbf{[1-16]}$$

In terms of temperature intervals, one sees that 1 K = 1.8°R. The triple-state temperature of water is therefore 491.69°R. The Fahrenheit scale $T(°F)$ is defined as

$$T(°F) = T(°R) - 459.67 \qquad \textbf{[1-17}a\textbf{]}$$

In this text the value of 459.67 is frequently rounded to 460. Finally, when Eqs. [1-15] and [1-16] are substituted into Eq. [1-17a], we find that

$$T(°F) = 1.8T(°C) + 32 \qquad \textbf{[1-17}b\textbf{]}$$

This equation shows that water at 1 atm freezes at 32°F. A comparison of these four temperature scales is shown in Fig. 1-21. Note from the figure that recent measurements based on a constant-volume gas thermometer (see subsection below) indicate the normal boiling point of water (at 1 atm) to be 99.97°C, or 211.95°F. In thermodynamic relationships a symbol $T$ for temperature always implies an absolute value (kelvins or degrees Rankine) unless it is specifically stated in terms of the other two scales.

## 1-6-4   CONSTANT-VOLUME IDEAL GAS THERMOMETER

Among the thermometric properties employed in the measurement of temperature is the pressure of a gas maintained at constant volume. The operation of a *constant-volume gas thermometer* is illustrated in Fig. 1-22. The tube on the left is raised or lowered until the mercury level on the right side of the U-tube is at the indicated mark. Then the height $z$ of the mercury column is a measure of the gage pressure of the gas within the thermometer

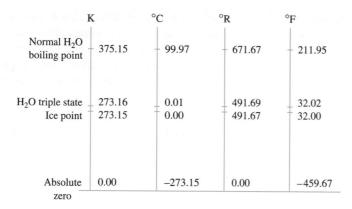

| | K | °C | °R | °F |
|---|---|---|---|---|
| Normal H₂O boiling point | 375.15 | 99.97 | 671.67 | 211.95 |
| H₂O triple state | 273.16 | 0.01 | 491.69 | 32.02 |
| Ice point | 273.15 | 0.00 | 491.67 | 32.00 |
| Absolute zero | 0.00 | −273.15 | 0.00 | −459.67 |

**Figure 1-21**    Comparison of temperature scales.

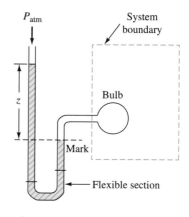

**Figure 1-22**
Constant-volume ideal-gas thermometer.

bulk. The height $z$ will vary as the pressure of the gas within the bulb changes. The analysis below will show that the pressure variation is directly related to the temperature variation.

On the basis of the experimental data shown in Fig. 1-23, we see that all gases exhibit the same value of $Pv$ at a given temperature if the pressure is extremely low. Therefore, the quantity $Pv$ can be used as a thermometric property to measure the temperature, regardless of the nature of the gas. At extremely low pressures, $Pv$ will vary linearly with $T$ to a high degree of accuracy. If a thermometer contains a low-pressure gas, then when the thermometer is placed in contact with two systems at different temperatures, we should find that in the limit as $P \rightarrow 0$,

$$\frac{T}{T^*} = \frac{Pv}{(Pv)^*}$$

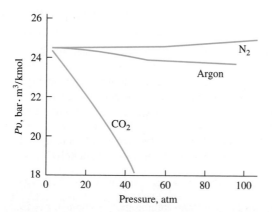

**Figure 1-23**    Experimental data showing the variation of $Pv$ with pressure at a given temperature for several gases.

where the asterisk value represents a reference state. If the gas in the thermometer is maintained at constant volume, the above equation reduces to

$$\frac{T}{T^*} = \frac{P}{P^*}$$                    **[1-18]**

That is, when one brings a constant-volume gas thermometer into contact with a system of interest and then in contact with a system at a reference state, the ratio of the measured gas pressures enables one to evaluate the temperature of the actual system relative to an assigned value $T^*$ at some reference state. If $T^*$ is chosen to be 273.16 K (the triple-state temperature of water), then the Kelvin temperature $T(K)$ at any other state as measured by a gas thermometer is given by

$$T(K) = 273.16\frac{P}{P_{tp}}$$                    **[1-19]**

where $P$ and $P_{tp}$ are absolute pressures. Hence the temperature at a given state of a substance may be determined by measuring the value of $P$ when the gas thermometer is in equilibrium with the substance and also measuring $P_{tp}$ when the thermometer is in equilibrium with the triple point (tp) of water.

The temperature scale based on the constant-volume gas thermometer is commonly referred to as the *gas temperature scale.* We assume that the pressure of the gas within the thermometer remains relatively low so that the pressure reading does not depend on the identity of the gas. However, values on this temperature scale still depend on the behavior of a certain class of materials, namely, gases. However, it is a convenient scale owing to its linearity, and it can be shown that this scale agrees with an "absolute" or thermodynamic temperature scale developed on the basis of the second law of thermodynamics in Secs. 6-5-3 and 6-6-3. Hence the low-pressure gas thermometer is a device that measures a truly thermodynamic temperature.

---

**Example 1-8**

**A** constant-volume gas thermometer containing nitrogen gas is brought into contact with a system of unknown temperature and then into contact with a system maintained at the triple state of water. The mercury column attached to the device has readings of 59.2 and 2.28 cm, respectively, for the two systems. If the barometric pressure is 960 mbars (96.0 kPa), the density of mercury is 13,600 kg/m³, and $g = 9.807$ m/s², find the unknown temperature $T$, in kelvins.

**Solution:**

**Given:** The $\Delta z$ of a constant-volume gas thermometer is 59.2 and 2.28 cm at an unknown temperature $T$ and the triple-state temperature, as shown in Fig. 1-24.

**Find:** $T$, in kelvins.

**Analysis:**    If the nitrogen gas within the thermometer is at a low pressure, the unknown temperature is found by applying Eq. [1-19], namely,

$$T(K) = 273.16 \frac{P}{P_{tp}}$$

The required values of $P$ and $P_{tp}$ must be found in absolute units. The values of these pressures, in turn, are found from Eq. [1-11], namely, $P_{abs} = P_{gage} + P_{atm}$. The positive gage pressure due to $\Delta z$ is determined from Eq. [1-13]. Thus a number of basic principles are involved. First, applying Eq. [1-13] with suitable conversion factors, we find that the gage pressures for the case of the unknown temperature (case 1) and for the triple-state measurement (case 2) are

$$P_{gage,1} = \rho g \, \Delta z$$

$$= 13{,}600 \text{ kg/m}^3 \times 9.807 \text{ m/s}^2 \times 59.2 \text{ cm} \times \frac{m}{100 \text{ cm}} \times \frac{1 \text{ N·s}^2}{\text{kg·m}}$$

$$= 78{,}950 \text{ N/m}^2 = 790 \text{ mbars (79.0 kPa)}$$

and  $P_{gage,2} = 13{,}600 \times 9.807 \times 2.28 \times 10^{-2}$

$$= 3040 \text{ N/m}^2 = 30.4 \text{ mbars (3.04 kPa)}$$

Therefore, the absolute pressures in the two cases are

$$P_{abs} \text{ (unknown temperature)} = (790 + 960) \text{ mbars}$$
$$= 1750 \text{ mbars (175 kPa)}$$

$$P_{abs} \text{ (triple-state temperature)} = (30.4 + 960) \text{ mbars}$$
$$= 990.4 \text{ mbars (99.04 kPa)}$$

Equation [1-19] then yields, for the unknown absolute temperature

$$T = 273.16 \frac{P}{P_{tp}} = 273.16 \text{ K} \times \frac{1750}{990.4} = 483 \text{ K (210°C)}$$

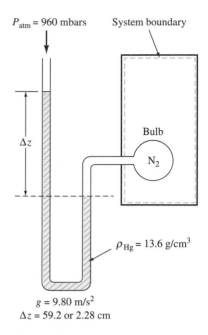

$P_{atm} = 960$ mbars  System boundary

Bulb

$\Delta z$    $N_2$

$\rho_{Hg} = 13.6$ g/cm$^3$

$g = 9.80$ m/s$^2$
$\Delta z = 59.2$ or 2.28 cm

**Figure 1-24**
Schematic and data for Example 1-8.

## 1-7    PROBLEM-SOLVING TECHNIQUES

Engineering design involving thermodynamic principles requires a great deal of problem solving. Hence, it is important to acquire good habits with respect to problem-solving procedures. There are certain steps that are fairly common to most thermodynamic analysis. A problem or example usually begins as a written statement. The information contained in this statement must be translated into sketches, property diagrams, and symbolic equations representing the basic principles. The use of *Given, Find,* and *Analysis* subsections in a problem solution is first suggested in Sec. 1-2, and illustrated throughout this chapter. The breakdown of an analysis into a number of subsections is appropriate for the analysis of all complex systems and processes.

The method suggested by the examples in this chapter will be extended to five subsections throughout the text. In many cases a sixth subsection entitled *Comments* will also appear. Also, some engineers prefer to entitle a preliminary sketch of the system as the *Schematic* subsection of the solution. This is a matter of choice. A more detailed discussion of these subsections in a problem solution is presented below.

1. *Given:* A brief word description of the system is made. A sketch of the system is drawn, with the selected boundary indicated by a dashed line. Input data appear at appropriate places on the sketch. Interactions across the boundaries are indicated, including magnitude and direction if known. This step allows you to visualize the problem without repeated referral to the original problem statement.

2. *Find:* A list of desired answers with units requested.

3. *Model:* Idealizations and assumptions that are thought to be appropriate are listed. For example, is the process one of constant pressure, temperature, or volume? These word statements of how things happen are extremely important. Are conditions at the boundary such that heat transfer can be neglected? What other energy terms might be neglected? Is the system closed or open, steady-state or transient? Some necessary assumptions may not be apparent until the analysis is begun. If they do not appear in the *model* subsection, they must be noted clearly in the *analysis* subsection.

4. *Strategy:* A word description of how the problem might be solved. References are made to basic equations that are applicable. No numerical values are cited. Rather than approach the problem in a random manner, the strategy section requires the problem solver to give some thought to the solution before any actual calculations are carried out. This section should be as brief as possible, but still convey a suggested method of analysis.

5. *Analysis:* The actual symbolic and numerical evaluation is carried out here. This step involves the use of basic principles, property relations and data, and appropriate assumptions and idealizations. The actual order of steps in the analysis may not be clear at the beginning of a complex problem. Care must be taken with respect to units and signs. Steps within the analysis may include the following:
   A. Fundamental principles should be applied. For example, write a suitable *basic* energy balance. Indicate which terms are zero or negligible. Apply a mass balance on the system if necessary. Recognize the correct sign convention on terms.
   B. Make use of property relations. Identify the tables, charts, and equations which contain pertinent property data. The property data employed depend upon the model assumed for the substance within the system.

C. Sketch property diagrams (as discussed in following chapters). Such diagrams are helpful in carrying out the analysis, or in explaining the results. Show the initial and final states clearly and indicate the path of the process if possible. Property diagrams are sometimes useful in determining what type of data may be needed in the numerical solution.

D. Complete the solution. Some engineers prefer to analyze a problem symbolically to the very end, before inserting values. Others evaluate terms as they proceed. It is a matter of preference. In this text data will normally be substituted as the analysis progresses, in order to stress orders of magnitude of various important quantities. Make sure that the units used in various equations are consistent. Is the numerical answer reasonable in the light of your common experience? All problems in this text have been designed to give answers that are consistent with common practice.

6. **Comments:** Frequently problems will end with a *Comment* section. Here the major points of the analysis and/or the results are discussed.

In some textbooks the suggested subsections of a problem solution may include, for example, *Known, Find, Schematic, Assumptions, Analysis,* and *Comments.* The labels are one of choice. The important point is that the method of thermodynamic problem solving is much more effective when the solution is broken down into a series of logical steps.

Note that you may need to check a number of things before ever attempting the numerical solution suggested in item 5-*D* above. Not all these items may be necessary in a simpler problem, and no specific order of the items in the analysis subsection is implied. However, difficulty with a given problem usually arises because one or more of these items have been neglected. Items 1 through 5 on the above checklist are used consistently in the examples which follow, and item 6 appears when appropriate. Examples and problems should be used as major sources for acquiring an understanding of the fundamentals of thermodynamics. In numerical problems, assume that all input data are accurate to three significant figures, even though they frequently are reported to less than this.

## **1-8** SUMMARY

A thermodynamic *system* is a region of space bounded by an arbitrary surface. The region outside of the system boundary is called the *surroundings,* or *environment.* A closed system excludes the passage of mass across its boundary, while the boundary of an isolated system prevents the passage of both mass and energy. When mass transfer may occur, the system is called open, the boundary is called the *control surface,* and the region within the control surface is called the *control volume. Phases* within a system are regions that are homogeneous in structure and chemical composition.

A *property* is a macroscopic characteristic of a system. It is classified as *extensive* or *intensive,* depending on whether its value is dependent on or independent of the mass within the system. A *specific* property is an intensive property. The *state* of a system is described by the values of its properties. A *process* is any transformation of a system from one equilibrium state to another. The *path* of a process specifies the states through which a system passes. Of particular interest is the *cyclic process,* or *cycle,* for which the end states are identical. All properties are *state functions,* since they are independent of the path of a process. Quantities that depend on the path are called *process* or *path functions.*

The major system of units in the world is the *Système Internationale* (SI). The primary dimensions of mass, length, time, and temperature are represented by the kilogram, the meter, the second, and the kelvin. The SI unit of force is the Newton (N), where $1 \text{ N} = 1 \text{ kg·m/s}^2$. In the United States the *United States Customary System* (USCS) of units is frequently employed. In this system the units of the pound-mass, the foot, the second, and the degree Rankine are used. The USCS unit of force is the pound-force ($\text{lb}_f$), where $1 \text{ lb}_f = 32.1740 \text{ lb}_m \text{·ft/s}^2$. The *molar mass* is expressed as g/mol or kg/kmol in SI, and as lb/lbmol in USCS.

The *density* $\rho \equiv m/V$, while the *specific volume* $v \equiv V/m = 1/\rho$. The *specific gravity* sp.gr. $= \rho/\rho_{H_2O}$ at a specified temperature, and the specific weight $w = \rho g$. *Pressure P* is defined as the normal force per unit area. The SI unit for pressure is the pascal ($1 \text{ N/m}^2$), although the kilopascal and the megapascal are used in engineering work. A reference value for pressure is the *standard atmosphere.* The *gage pressure* is related to a local atmospheric pressure by

$$P_{abs} = P_{atm} + P_{gage}$$

The local atmospheric pressure is measured by a barometer and is known as the *barometric* pressure. A negative gage pressure is called a *vacuum* reading.

*Temperature T* is the property that determines whether two systems are in thermal equilibrium. The zeroth law establishes the field of thermometry and empirical temperature scales. Useful temperature relations include

$$T(°C) = T(K) - 273.15 \quad \big| \quad T(°R) = 1.8 \, T(K) \quad \big| \quad T(°F) = T(°R) - 459.67$$

For a constant-volume ideal gas thermometer, the equation

$$T(K) = 273.16 \frac{P}{P_{tp}}$$

yields a *kelvin* (absolute) temperature $T$ relative to the chosen kelvin temperature of 273.16 at the triple state of water.

## PROBLEMS

### GENERAL COMMENT QUESTIONS

1-1C. A 70-kg astronaut weighs 680 N on his bathroom scale before entering a rocket to the moon. Using a spring scale on the moon, where $g = 1.67$ m/s$^2$, does he still weigh 680 N? Has his mass changed?

1-2C. Two different processes $A$ and $B$ connect the same end states 1 and 2. Work $W$ is a process function and volume $V$ is a state function. (a) Is the statement $W_{12,A} = W_{12,B}$ *true* or *false*? (b) Is the statement $\Delta V_{12,A} = \Delta V_{12,B}$ *true* or *false*? Briefly defend your answers.

1-3C. Explain the difference between absolute pressure and gage pressure.

1-4C. At what absolute temperature (a) on the Kelvin scale does the Celsius scale become negative, and (b) on the Rankine scale does the Fahrenheit scale become negative?

1-5C. Can a constant-volume gas thermometer be used to measure degrees Celsius or degrees Fahrenheit?

1-6C. At what temperature, in kelvins and in Rankine, would thermometers calibrated in Celsius and in Fahrenheit read the same value?

1-7C. A hiker at sea level in the morning reaches 2000 m by late afternoon. Discuss the effect of the altitude on the preparation of hot coffee and boiled potatoes that evening.

1-8C. List three devices used daily where temperature contol is extremely important.

### INTENSIVE AND EXTENSIVE PROPERTIES

1-1. Two cubic meters of air at 25°C and 1 bar have a mass of 2.34 kg.
  (a) List the values of three intensive and two extensive properties for this system.
  (b) If local gravity $g$ is 9.65 m/s$^2$, evaluate the specific weight.

1-2. Five cubic meters of water at 25°C and 1 bar have a mass of 4985 kg.
  (a) List the values of two extensive and three intensive properties of the system.
  (b) If the local gravity $g$ for the system is 9.70 m/s$^2$, evaluate the specific weight.

1-3. A metal bucket with a mass of 0.8 kg contains 8 liters of water at 20°C and 1 bar and with a mass of 8.0 kg.
  (a) List the values of two extensive and three intensive properties of the water.

(b) If the local gravity $g$ for the system is 9.60 m/s$^2$, evaluate the specific weight of the combined system of bucket and water.

1-4E. Three cubic feet of water at 60°F and 14.7 psia have a mass of 187 lb$_m$.

    (a) List the values of two extensive and three intensive properties of the system.

    (b) If the local gravity $g$ for the system is 30.8 ft/s$^2$, evaluate the specific weight.

## FORCE, MASS, DENSITY, AND SPECIFIC WEIGHT

1-5. A small experimental rocket which has a mass of 70 kg is accelerated at a rate of 6.0 m/s$^2$. What total force is required, in newtons, if (a) the rocket is moving horizontally and without friction and (b) the rocket is moving vertically upward and without friction at a location where local gravity is 9.45 m/s$^2$?

1-6. The acceleration of gravity as a function of elevation above sea level at 45° latitude is given by $g = 9.807 - 3.32 \times 10^{-6} z$, where $g$ is in m/s$^2$ and $z$ is in meters. Find the height, in kilometers, above sea level where the weight of a person will have decreased by (a) 1 percent, (b) 2 percent, and (c) 4 percent.

1-7E. A rocket with a mass of 200 lb$_m$ is accelerated at a rate of 20.0 ft/s$^2$. What total force is required, in lb$_f$, if (a) the body is moving along a horizontal frictionless plane and (b) the body is moving vertically upward at a location where local gravity is 31.0 ft/s$^2$?

1-8E. The acceleration of gravity as a function of elevation above sea level at 45° latitude is given by $g = 32.17 - 3.32 \times 10^{-6} z$, where $g$ is in ft/s$^2$ and $z$ is in feet. Find the height, in miles, above sea level where the weight of a person will have decreased by (a) 1 percent and (b) 2 percent.

1-9. A mass of 2 kg is subjected to a vertical force of 35 N. The local gravity $g$ is 9.60 m/s$^2$, and frictional effects are neglected. Determine the acceleration of the mass, in m/s$^2$, if the external vertical force is (a) downward and (b) upward.

1-10. The density of a certain organic liquid is 0.80 g/cm$^3$. Determine the specific weight, in N/m$^3$, where local gravity $g$ is (a) 2.50 m/s$^2$ and (b) 9.50 m/s$^2$.

1-11. On the surface of the moon where local gravity $g$ is 1.67 m/s$^2$, 5.4 g of a gas occupies a volume of 1.2 m$^3$. Determine (a) the specific volume of the gas, in m$^3$/kg, (b) the density, in g/cm$^3$, and (c) the specific weight, in N/m$^3$.

1-12. A 7-kg rock is subjected to a vertical force of 133 N. The local gravity $g$ is 9.75 m/s$^2$, and frictional effects are neglected. Determine the acceleration of the rock if the external vertical force is (*a*) downward and (*b*) upward, in m/s$^2$.

1-13E. A 7-lb$_m$ piece of steel is subjected to a vertical force of 8 lb$_f$. The local gravity *g* is 31.1 ft/s$^2$, and frictional effects are neglected. Determine the acceleration of the mass if the external vertical force is (*a*) downward and (*b*) upward, in ft/s$^2$.

1-14E. On the surface of the moon where local gravity *g* is 5.47 ft/s$^2$, 5 lb$_m$ of oxygen within a tank occupies a volume of 40 ft$^3$. Determine (*a*) the specific volume of the gas, in ft$^3$/lb$_m$, (*b*) the density, in lb$_m$/ft$^3$, and (*c*) the specific weight, in lb$_f$/ft$^3$.

1-15. An 11-m$^3$ tank of air is separated by a membrane into side *A* with a volume of 6 m$^3$ and side *B* with an initial specific volume of 0.417 m$^3$/kg. The membrane is broken, and the resulting specific volume is 0.55 m$^3$/kg. Find the initial specific volume of air in side *A*, in m$^3$/kg.

1-16. A 9-m$^3$ tank of nitrogen is separated by a membrane into two sections. Section *A* has an initial density of 1.667 kg/m$^3$, and section *B* has a mass of 6 kg. After the membrane is broken, the density is found to be 1.778 kg/m$^3$. Find the initial density of the gas in section *B*, in kg/m$^3$.

1-17E. A 20-ft$^3$ tank of air is separated by a membrane into section *A* with an initial specific volume of 0.80 ft$^3$/lb$_m$ and section *B* with a mass of 12.0 lb$_m$. The membrane is broken, and the resulting density is 1.350 lb$_m$/ft$^3$. Find the initial specific volume in section *B*, in ft$^3$/lb$_m$.

1-18. A vertical cylinder contains nitrogen at 1.4 bars. A frictionless piston of mass *m* on top of the gas separates the gas from the atmosphere, which is at 98 kPa. If the local gravity is 9.80 m/s$^2$ and the area of the piston is 0.010 m$^2$, determine the mass of the piston in static equilibrium, in kilograms.

1-19. A vertical piston-cylinder device has a piston diameter of 11 cm and a piston mass of 40 kg. The atmospheric pressure is 0.10 MPa and the local gravity is 9.79 m/s$^2$. Determine the absolute pressure of the gas within the device.

1-20. Helium at 0.150 MPa is contained within a vertical cyclinder by a weighted piston of total mass *m* and 400-mm$^2$ cross-sectional area. (See Fig. 1-4*b*.) The atmospheric pressure outside the piston is 1.00 bar. Determine the value of *m*, assuming standard gravity acceleration, in kilograms.

1-21E. A gas at 20.0 psia is contained within a vertical cyclinder by a weighted piston of total mass *m* and 2.40-in$^2$ cross-sectional area. (See Fig. 1-4*b*.) The atmospheric pressure outside the piston is 29.80 inHg. Determine the value of *m*, assuming standard gravity acceleration, in pounds-mass.

1-22. Determine the pressure equivalent to 1 bar in terms of meters of a column of liquid at room temperature, where the liquid is (a) water, (b) ethyl alcohol, and (c) mercury. The specific gravity of ethyl alcohol is 0.789, the specific gravity of mercury is 13.59, and $g = 9.80$ m/s².

1-23. The gage pressure within a system is equivalent to a height of 75 cm of a fluid with a specific gravity of 0.75. If the barometric pressure is 0.980 bar, compute the absolute pressure within the chamber, in millibars.

1-24. If the barometeric pressure is 930 mbars, convert (a) an absolute pressure of 2.30 bars to a gage reading in bars, (b) a vacuum reading of 500 mbars to an absolute value in bars, (c) 0.70 bar absolute to millibars vacuum, and (d) an absolute reading of 1.30 bars to a gage reading in kilopascals.

1-25. If the barometric pressure is 1020 mbars, convert (a) an absolute pressure of 1.70 bars to a gage reading in bars, (b) a vacuum reading of 600 mbars to an absolute value in kilopascals, (c) an absolute pressure of 60 kPa to millibars vacuum, and (d) a gage reading of 2.20 bars to an absolute pressure in kilopascals.

1-26E. Determine the pressure equivalent to 1 atm in terms of height, in feet, of a column of liquid at room temperature, where the liquid is (a) water, (b) ethyl alcohol, and (c) mercury. The specific gravity of ethyl alcohol is 0.789, the specific gravity of mercury is 13.59, and $g = 32.2$ ft/s².

1-27E. The gage pressure within a system is equivalent to a height of 24 in of a fluid with a specific gravity of 0.80. If the barometric pressure is 29.5 inHg, compute the absolute pressure within the chamber, in psia.

1-28E. If the barometeric pressure is 30.15 inHg, convert (a) 35.0 psia to a gage reading in psig, (b) a vacuum reading of 20.0 inHg to psia, (c) 10 psia to inHg vacuum, and (d) 20.0 inHg gage to psia.

1-29E. If the barometric pressure is 29.90 inHg, convert (a) an absolute pressure of 27.0 psia to psig, (b) a vacuum reading of 24.0 inHg to an absolute value, in psia, (c) an absolute pressure of 12.0 psia to inHg vacuum, and (d) a gage reading of 14.0 inHg to an absolute pressure, in psia.

1-30. A vertical storage tank initially contains water ($\rho = 1000$ kg/m³) at a depth of 4 m. Immiscible oil with a specific gravity of 0.88 is added until the total liquid height is 10 m. If the barometric pressure is 97.2 kPa and $g = 9.80$ m/s², determine the absolute pressure at the bottom of the water, in kPa and bars.

1-31. The gage pressure of a gas inside a tank is 25 kPa. Determine the vertical height, in meters, of the liquid within a manometer attached to the system if the fluid at room temperature is (a) water, (b) mercury ($\rho = 13{,}600$ kg/m$^3$), and (c) an oil with a specific gravity of 0.88, and $g = 9.75$ m/s$^2$.

1-32E. The gage pressure of a gas inside a tank is 3.0 lb$_f$/in$^2$. Determine the vertical height, in inches, of the liquid within a manometer attached to the system if the fluid at room temperature is (a) mercury ($\rho = 850$ lb$_m$/ft$^3$), (b) water, and (c) an oil with a specific gravity of 0.90, and $g = 32.0$ ft/s$^2$.

1-33. A manometer has a liquid height difference of 0.87 m, the barometric pressure is 98.0 kPa, and $g$ is 9.80 m/s$^2$. If the system absolute pressure is 0.106 MPa, determine the manometer liquid density, in kg/m$^3$.

1-34. A manometer similar to that shown in Fig. 1-10 contains an immiscible liquid with a density of 700 kg/m$^3$ on top of another liquid with a density of 800 kg/m$^3$. The heights of the top and bottom fluids are 70 and 40 cm, respectively. If the atmospheric pressure is 95 kPa and $g$ is 9.70 m/s$^2$, determine (a) the gage pressure and (b) the absolute pressure of the system, in kPa.

1-35. A pilot notices that the barometric pressure outside his aircraft is 800 mbar. The airport below the plane reports a barometric pressure of 1020 mbar. If the air density averages 1.15 kg/m$^3$ and the local gravity is 9.70 m/s$^2$, determine the height of the aircraft above the ground, in meters.

1-36. Two students are asked to measure the height of a high-rise building. One takes the elevator to the top floor and notes a barometer reading of 993.2 mbar. The student at ground level takes a reading of 1012.4 mbar. The air density is 1.16 kg/m$^3$ and $g$ is 9.68 m/s$^2$. Determine the height, in meters, they reported.

1-37. A submarine is cruising at a depth of 280 m in seawater with a specific gravity of 1.03. If the inside of the submarine is pressurized to standard atmospheric pressure, determine the pressure difference across the hull in (a) kilopascals and (b) bars. The average local gravity is 9.70 m/s$^2$.

1-38. A mountain climber carries a barometer which reads 950 mbars at the base camp for her ascent. During the climb she takes three additional readings which are (a) 904 mbars, (b) 824 mbars, and (c) 785 mbars. Estimate the vertical distance, in meters, she has climbed from the base camp for the three readings if the average air density is 1.20 kg/m$^3$. Neglect the effect of altitude on local gravity.

1-39. Determine the pressure, in kilopascals and bars, exerted on a skin diver who has descended to (a) 10 m and (b) 20 m below the surface of the sea if the barometric pressure is 0.96 bar at sea level and the specific gravity of seawater is 1.03 in this region of the ocean.

1-40E. A submarine is cruising at a depth of 900 ft in seawater with a specific gravity of 1.03. If the inside of the submarine is pressurized to standard atmospheric pressure, determine the pressure difference across the hull in (*a*) psia and (*b*) atmospheres. The average local gravity is 32.10 ft/s².

1-41E. A mountain climber carries a barometer which reads 30.10 inHg at the base camp for his ascent. During the climb he takes three additional readings which are (*a*) 28.95 inHg, (*b*) 27.59 inHg, and (*c*) 26.45 inHg. Estimate the vertical distance, in feet, he has climbed from the base camp for the three readings if the average air density is 0.074 $lb_m$/ft³. Neglect the effect of altitude on local gravity.

1-42E. Determine the pressure, in psia, exerted on a skin diver who has descended to (*a*) 25 ft and (*b*) 65 ft below the surface of the sea if the barometric pressure is 14.5 psia at sea level and the specific gravity of seawater is 1.03 in this region of the ocean.

1-43. If the atmosphere is assumed to be isothermal at 25°C and follows the relationship $Pv = RT$ (an ideal gas), compute the pressure, in bars, and the density, in kg/m³, at (*a*) 2000 m and (*b*) 800 m above sea level. The pressure and density at sea level are taken to be 1 bar and 1.19 kg/m³, respectively.

1-44E. If the atmosphere is assumed to be isothermal at 60°F and follows the relationship $Pv = RT$ (an ideal gas), compute the pressure, in psia, and the density, in $lb_m$/ft³, at (*a*) 5000 ft and (*b*) 2000 ft above sea level. The pressure and density at sea level are taken to be 14.7 psia and 0.077 $lb_m$/ft³, respectively.

## CONSTANT-VOLUME GAS THERMOMETER

1-45. A constant-volume gas thermometer is exposed to an unknown temperature and then to the triple state of water. At the unknown temperature, the fluid column in the manometer is 40.0 cm above the mark. At the triple state the fluid is 3.0 cm below the mark. The fluid in the manometer has a specific gravity of 2.0, the atmospheric pressure is 960 mbars, and the local acceleration of gravity is 9.807 m/s². Determine the unknown temperature in kelvins.

1-46. A constant-volume gas thermometer is brought into contact with a system of unknown temperature *T* and then into contact with the triple state of water. The mercury column attached to the thermometer has readings of +10.7 and −15.5 cm, respectively. Determine the unknown temperature, in kelvins. The barometric pressure is 980 mbars (98.0 kPa), and the specific gravity of mercury is 13.6.

1-47. A constant-volume gas thermometer is brought into contact with a system of unknown temperature $T$ and then into contact with the triple state of water. The mercury column attached to the thermometer has readings of 29.6 and $-12.6$ cm, respectively. The barometric pressure is 975 mbars (97.5 kPa), and the specific gravity of mercury is 13.6. Find the value of the unknown temperature in kelvins.

1-48E. A constant-volume gas thermometer is brought into contact with a system of unknown temperature $T$ and then into contact with the triple state of water. The mercury column attached to the thermometer has readings of $+4.20$ and $-6.10$ in, respectively. Determine the unknown temperature, in degrees Rankine. The barometric pressure is 29.20 inHg, and the specific gravity of mercury is 13.6.

1-49E. A constant-volume gas thermometer is brought into contact with a system of unknown temperature $T$ and then into contact with the triple state of water. The mercury column attached to the thermometer has readings of 14.6 and $-2.6$ in, respectively. The barometric pressure is 29.80 inHg, and the specific gravity of mercury is 13.6. Find the value of the unknown temperature in degrees Rankine.

# CHAPTER

# 2

# THE FIRST LAW OF THERMODYNAMICS

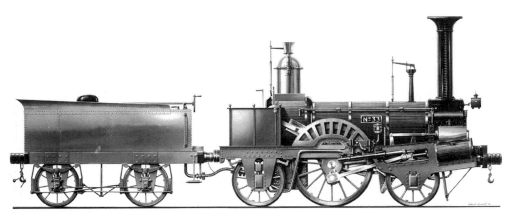

Illustration of a 19th century steam locomotive.

In basic physics, forms of energy such as gravitational potential and kinetic energy are examined, as well as forms of energy associated with electric and magnetic fields. The study of the energy associated with atomic and nuclear binding forces is extremely important to the chemist. A study of thermodynamic principles allows one to relate changes in these and other forms of energy within a system to energy interactions at the boundaries of a system. One of the most important laws of thermodynamics introduces a general conservation of energy principle, which leads to the concept of the internal energy of a substance. The law on which this conservation principle is based is called the *first law of thermodynamics,* which is introduced in this chapter.

## 2-1 CONCEPTS OF WORK AND ENERGY

The concepts of work $W$ and power $\dot{W}$ are usually introduced in the study of mechanics. Through the study of the mechanical work required to move an object, the concepts of kinetic and gravitational potential energy are developed. Because these ideas are so central to the study of thermodynamics, these concepts will be *reviewed* in the following subsections.

### 2-1-1 MECHANICAL WORK AND POWER

*Mechanical work* is defined as the product of a force $F$ and a displacement $\Delta s$ when both are measured in the same direction (collinear). Consider a system of mass $m$, velocity $\vec{V}$, and position vector $\vec{s}$ that is being acted upon by a single force $\vec{F}_{ext}$ as shown in Fig. 2-1. The *general* expression for a differential quantity of mechanical work $\delta W_{mech}$ that results from a differential displacement $d\vec{s}$ is given by the scalar (dot) product of the external force vector $\vec{F}_{ext}$ and the displacement vector $\vec{s}$. Thus

$$\delta W_{mech} = \vec{F}_{ext} \cdot d\vec{s} = F_{ext}(\cos\theta)\,ds \qquad \textbf{[2-1]}$$

where $\theta$ is the angle between the two vectors $\vec{F}_{ext}$ and $\vec{s}$. Work is a scalar quantity. However, there is a sign associated with the work depending upon the magnitude of the angle $\theta$ between the two vectors. If the external force and the displacement have components in the same direction, the calculations produce a positive numerical value, and work is being done on the system. The numerical value of work transfer is negative if the external force and displacement have components in the opposite directions.

The mechanical work for a finite displacement from position 1 to position 2 is obtained by integration of the above equation. That is,

$$W_{mech} = \int_{\vec{s}_1}^{\vec{s}_2} \vec{F}_{ext} \cdot d\vec{s} \qquad \textbf{[2-2]}$$

Note that because the external force may vary with time along the path of the process, mechanical work usually depends upon the path of the process. Because of this, work is a *process function* and *not* a property of the system. Hence the differential of $W$ is inexact and is represented symbolically by $\delta W$, not $dW$. On the basis of Eq. [1-7], the integration of the inexact differential $\delta W$ between states 1 and 2 leads symbolically to $W_{12}$, and not $\Delta W$. That is,

$$W_{12} = \int_1^2 \delta W$$

The symbol $\Delta W$ is inappropriate because work can only be discussed in the context of a process. An examination of the defining equation shows

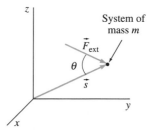

**Figure 2-1**
Schematic of particle of mass $m$ and position vector $\vec{s} = x\hat{i} + y\hat{j} + z\hat{k}$ acted on by external force $\vec{F}_{ext}$.

**Note that a work interaction is a process function—a function of the process path.**

that the primary dimensions for mechanical work in the FLt system would be [force]·[length]. In thermodynamics, the most commonly used SI units for work are *newton-meter* (N·m), *joule* (J), and *kilojoule* (kJ). The USCS (English) units are *foot-pounds force* (ft·lb$_f$) and *British thermal unit* (Btu).

The *time rate at which work is done on or by a system* is defined as the **power** $\dot{W}$. Power, like work, is a scalar quantity. In terms of power, the differential work can be written as

$$\delta W \equiv \dot{W}\, dt \qquad\qquad \text{[2-3]}$$

where work and power have the same signs. Mechanical power *supplied to a system by an external force* is defined as the scalar (dot) product of the external force vector $\vec{F}_{ext}$ and the velocity vector $\vec{V}$. Thus

$$\dot{W}_{mech} = \vec{F}_{ext} \cdot \vec{V} = F_{ext}\mathbf{V}(\cos\theta) \qquad\qquad \text{[2-4]}$$

where **V** represents the scalar value of the velocity. Integration of mechanical power over a specified time period also leads to the work associated with a process. Hence

$$W_{mech} = \int_{t_1}^{t_2} \dot{W}\, dt = \int_{t_1}^{t_2} \vec{F}_{ext} \cdot \vec{V}\, dt \qquad\qquad \text{[2-5]}$$

Use of the definition of velocity, $\vec{V} = d\vec{s}/dt$ in the expression above leads back to Eq. [2-2]. Examination of the defining equation for mechanical power shows that the dimensions for power would be [force]·[length]/[time]. Typical units would be N·m/s or lb$_f$·ft/s. In SI the basic unit of power is a joule per second (J/s), and 1 J/s is called a *watt* (W). In engineering calculations the kilowatt (kW) is used frequently. There are three commonly used USCS or English units for power: horsepower (hp), foot-pound force per second (ft·lb$_f$/s), and British thermal unit per hour (Btu/h). Conversion factors among these latter three quantities are found in Table A-1E.

## 2-1-2 TRANSLATIONAL KINETIC ENERGY AND GRAVITATIONAL POTENTIAL ENERGY

Based on the review of the concepts of mechanical work and power, we shall now calculate the mechanical work and power required to change the velocity and position of a system moving in a gravitational field. This will lead to the important concept of mechanical energy.

Consider a system moving in a gravitational field $\vec{g}$ and acted upon by a surface force $\vec{R}_{net}$ (see Fig. 2-2a). This single force represents the sum of all surface forces acting on the system. Application of Eq. [2-5]

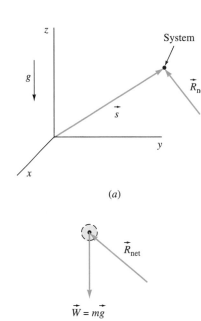

**Figure 2-2**
(a) Schematic diagram of a particle moving in a gravitational field and acted on by a surface force; (b) free-body diagram.

to this situation leads to

$$W_{\text{mech}} = \int_{t_1}^{t_2} \vec{R}_{\text{net}} \cdot \vec{V} \, dt \qquad \qquad \textbf{[2-6]}$$

Further calculations require explicit information about $\vec{R}_{\text{net}}$ and $\vec{V}$ as a function of time. This information can be obtained by applying Newton's second law to the system.

A sketch of the forces on the system is shown in Fig. 2-2$b$. We see that the only external forces acting on the system are the weight $m\vec{g}$ (a body force) due to gravity and the net surface force $\vec{R}_{\text{net}}$. Therefore, $\sum \vec{F}_{\text{ext}} = \vec{R}_{\text{net}} + m\vec{g}$, where $\vec{g}$ is the local acceleration of gravity. Using this expression in Newton's second law, $\sum \vec{F}_{\text{ext}} = m\vec{a}$, leads to

$$\vec{R}_{\text{net}} + m\vec{g} = m\vec{a} = m\frac{d\vec{V}}{dt} \qquad \qquad \textbf{[2-7]}$$

where the term $\vec{a} = d\vec{V}/dt$ is the acceleration of the system. Solving Eq. [2-7] for $\vec{R}_{\text{net}}$ and substituting this result into Eq. [2-6] gives an expression for the mechanical work done on the system by the surface force $\vec{R}_{\text{net}}$.

$$W_{\text{mech},12} = \int_{t_1}^{t_2} \vec{R}_{\text{net}} \cdot \vec{V} \, dt = \int_{t_1}^{t_2} \left( m\frac{d\vec{V}}{dt} - m\vec{g} \right) \cdot \vec{V} \, dt \qquad \qquad \textbf{[2-8]}$$

$$= \int_{t_1}^{t_2} m\frac{d\vec{V}}{dt} \cdot \vec{V} \, dt - \int_{t_1}^{t_2} \left( m\vec{g} \cdot \vec{V} \right) dt \equiv W_{\text{acc}} + W_{\text{grav}}$$

Examination of the terms on the last line shows that the mechanical work done by the external force $\vec{R}_{\text{net}}$ can be separated into two parts—acceleration work $W_{\text{acc}}$ and gravitational work $W_{\text{grav}}$.

The ***acceleration work*** $W_{\text{acc}}$ is the work done on the system to change its velocity. It can be written in a more useful form as follows:

$$W_{\text{acc}} = \int_{t_1}^{t_2} m\frac{d\vec{V}}{dt} \cdot (\vec{V} \, dt) = \int_{\vec{V}_1}^{\vec{V}_2} m\vec{V} \cdot d\vec{V} = \int_{\mathbf{V}_1}^{\mathbf{V}_2} m\mathbf{V} \, d\mathbf{V}$$

Subsequent integration between states 1 and 2 leads to

$$W_{\text{acc},12} = \frac{m\mathbf{V}_2^2}{2} - \frac{m\mathbf{V}_1^2}{2} = \text{KE}_2 - \text{KE}_1 = \Delta\text{KE} \qquad \qquad \textbf{[2-9]}$$

The quantity $m\mathbf{V}^2/2$ is defined as the ***translational kinetic energy*** KE of a system. The value of KE is independent of the identity of the material undergoing acceleration. The specific translational kinetic energy ke is defined by ke $= \text{KE}/m = \mathbf{V}^2/2$.

The ***gravitational work*** $W_{\text{grav}}$ is the work done against gravity to change the elevation of a system. Recall that $\vec{V} \, dt = d\vec{s}$, where in this case $d\vec{s}$ is the change in vertical distance $d\vec{z}$. Consequently, on the basis of Eq. [2-8],

$$W_{\text{grav}} = \int_{t_1}^{t_2} -m\vec{g} \cdot \vec{V}\, dt = \int_{z_1}^{z_2} mg\, dz \qquad \textbf{[2-10a]}$$

If we assume $g$ is constant in the distance between positions $z_1$ and $z_2$, then integration yields

$$W_{\text{grav},12} = mgz_2 - mgz_1 = \text{PE}_2 - \text{PE}_1 = \Delta\text{PE} \qquad \textbf{[2-10b]}$$

The quantity $mgz$ is called the **gravitational potential energy** PE of the system. Like KE, PE is independent of the identity of the material in the system, and the specific gravitational potential energy pe is defined as pe = PE/$m$ = $gz$. When $g$ is not constant, a functional relationship between $g$ and $z$ must be known before Eq. [2-10a] can be integrated.

The energy terms in Eqs. [2-9] and [2-10] have the same primary dimensions as work interactions, and hence the same units as work transfer. The SI units for energy normally are expressed in joules (J) or kilojoules (kJ). In USCS the units typically are foot pound-force (ft·lb$_f$) or British thermal units (Btu).

In summary, we have shown that the mechanical work done by the net surface force to move a system in a gravitational field is given by

$$W_{\text{mech},12} = \int_{\vec{s}_1}^{\vec{s}_2} \vec{R}_{\text{net}} \cdot d\vec{s} = \Delta\text{KE} + \Delta\text{PE} \qquad \textbf{[2-11]}$$

Similarly, the mechanical power supplied by the net surface force to move a system in a gravitational field equals the time rate of change in the translational kinetic and gravitational potential energies of the system. That is,

$$\dot{W}_{\text{mech},12} = \vec{R}_{\text{net}} \cdot \vec{V} = \frac{d}{dt}\left(\frac{m\mathbf{V}^2}{2}\right) + \frac{d}{dt}(mgz) = \frac{d}{dt}(\text{KE} + \text{PE})$$

$$\textbf{[2-12]}$$

Before leaving the discussion of kinetic and gravitational potential energy, it is a useful distinction to separate properties into two classes, called *extrinsic* and *intrinsic* quantities. In general, for any system,

1. An **extrinsic property** is a quantity whose value is *independent* of the nature of the substance within the system boundaries.

2. An **intrinsic property** is a quantity whose value is *dependent* on the nature of the substance composing the system.

Examples of extrinsic properties include the macroscopic translational velocity of a body and the rotational velocity of a body about its center of gravity. Neither of these properties requires information about the substance within the body. Intrinsic properties include pressure, temperature, density, and electric charge. These are properties associated with the substance under study. As was shown above, the linear kinetic energy and the gravitational potential energy are both extrinsic properties.

Collectively, translational kinetic energy and gravitational potential energy are commonly referred to as *mechanical energy*. These two energy terms have several characteristics in common:

**Note the distinction between extrinsic and intrinsic properties, and that both kinetic energy and gravitational potential energy are extrinsic properties.**

- The numerical value of the change in either energy term for any process depends only on the knowledge of the end states. Thus *translational kinetic energy and gravitational potential energy must both be properties*.

- The value of either form of energy is independent of the identity of the material undergoing the change in state, and thus both are *extrinsic* properties.

- The value of both energy terms depends upon the extent of the system, so they are *extensive* properties.

- The numerical values of both KE and PE depend upon establishing a *datum in space* from which the velocity $V$ and the elevation $z$ are measured.

Now we are ready to consider the implications of Eqs. [2-11] and [2-12].

## 2-1-3  WORK, POWER, AND ENERGY—AN INTERPRETATION

Recall what has been done in Sec. 2-1-1 and 2-1-2. First we provided a definition of mechanical work and mechanical power. Then we used these definitions to evaluate the work required to accelerate a system in a gravitational field and derived the two energy quantities called translational kinetic energy and gravitational potential energy. Collectively, these two forms of energy associated with a system are called the *mechanical energy* of the system. Finally, we developed two expressions relating mechanical work, mechanical power, and mechanical energy. Thus Eq. [2-11] for the mechanical work done on a system by a surface force to accelerate the system in a gravitational field can be interpreted as a *mechanical energy balance* for a closed system:[1]

$$\Delta KE_{cm} + \Delta PE_{cm} = W_{mech}$$

where the subscript "cm" stands for "control mass" and clarifies that we are talking about the mechanical energy of the closed system (control mass). This equation states that the change in the mechanical energy of the closed system equals the mechanical work done on the system. Because work can

---

[1] This equation is a very restricted form of the general energy balance discussed later in this chapter. It is valid only under the conditions for which Eqs. [2-11] and [2-12] were developed.

change the amount of energy in a system, and work is an interaction between a system and its surroundings, *work is a mechanism for transporting energy across the boundary of the system.*

Similarly, Eq. [2-12] for the *mechanical power supplied by the external force to accelerate a system in a gravitational field* can be interpreted as the *rate form of the mechanical energy balance* for a closed system:

$$\frac{d}{dt}(\text{KE} + \text{PE})_{\text{cm}} = \dot{W}_{\text{mech}}$$

This equation indicates that the rate of change of the mechanical energy of a closed system equals the mechanical power supplied to the system. Again, to be consistent with our interpretation of work, *power is the rate at which energy is transferred across the boundary of a system by work.*

With this interpretation of work as a transport of energy and power as a transport rate of energy, we must be careful to use a consistent *sign convention* when performing calculations. If we focus on the energy of the system, it seems only natural that interactions that add energy to a system should be positive, and interactions that remove energy from a system should be negative. This is the *sign convention* adopted in this text for work $W$.

- Work transfer of energy *into* a system is given a *positive* value ($W > 0$). Likewise, work transfer of energy *out* of a system is given a *negative* value ($W < 0$).

- A similar convention is adopted for mechanical power; i.e., if $\dot{W} > 0$, then energy is transferred into the system by a work interaction.

This is a natural convention, in that quantities added to a system are positive and quantities removed are negative, and it can be used consistently for transports of all extensive properties, such as mass, energy, momentum, and charge.[2] This sign convention is also consistent with the earlier discussion of mechanical work. As an alternative to the standard sign convention, the sign on a work quantity can be omitted, and its direction explicitly indicated by assigning a directional subscript "in" or "out" to the symbol for work. For example, if work out of a system is 100 kJ, this could be written as $W = -100$ kJ or $W_{\text{out}} = 100$ kJ.

---

**Example 2-1**

**A** car on a roller-coaster track reaches the bottom of the first ramp. There, a chain drive engages and pulls the car up the ramp a distance of 40 m at a constant speed of 1.0 m/s until it reaches the top. At the top of the ramp, the car's elevation has increased 27 m. The mass of the car including its four passengers is 400 kg. Assume all frictional effects are negligible. Determine (*a*) the mechanical work done to lift the car up the first ramp, in N·m, (*b*) the average force applied to the car by the drive

---

[2]The reader should be aware that another sign convention, that work done by a system is positive, is also used by engineers.

chain to pull the car up the ramp, in N, and (c) the average mechanical power supplied by the motor to lift the car, in J/s.

**Solution:**

**Given:** A roller-coaster car is pulled up a ramp by a chain drive, as shown in Fig. 2-3.

**Find:** Determine (a) the work required to pull the car up the ramp, (b) the average force applied by the chain, and (c) the average mechanical power supplied to the car during the lift.

**Model:** Neglect frictional effects; $\Delta KE = 0$; local gravity is constant at $9.81$ m/s$^2$.

**Strategy:** Because we are given no force information, try using the relationship developed between mechanical work and mechanical energy. Then use the definition of mechanical work and power to find the remaining information.

**Analysis:** We must identify a system before we can apply the mechanical energy balance. The car and its contents are taken as the system, as shown by the dashed line in the figure.

(a) The mechanical energy balance for the system requires that

$$\Delta KE_{cm} + \Delta PE_{cm} = W_{mech}$$

Since the speed is constant, only a change in gravitational potential energy need be considered. Solving for $W_{mech}$ and using the definition of PE gives

$$W_{mech} = mg(z_2 - z_1) = 400 \text{ kg} \times \left(\frac{9.81 \text{ m}}{s^2}\right) \times (27 - 0) \text{ m}$$

$$= 105{,}950 \, \frac{\text{kg·m}^2}{s^2} \times \frac{1 \text{ N·s}^2}{1 \text{ kg·m}} = 106 \times 10^3 \text{ N·m}$$

(b) To evaluate the average force applied by the chain to the car, use the definition of mechanical work with a constant average force $F_{av}$:

$$W_{mech} = \int_1^2 F \, ds = F_{av} \, \Delta s$$

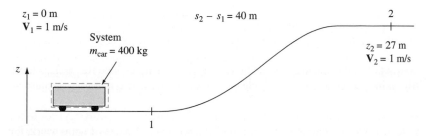

$z_1 = 0$ m
$V_1 = 1$ m/s

System
$m_{car} = 400$ kg

$s_2 - s_1 = 40$ m

2

$z_2 = 27$ m
$V_2 = 1$ m/s

**Figure 2-3**    Schematic and data for Example 2-1.

Solving for the average force gives

$$F_{av} = \frac{W_{mech}}{\Delta s} = \frac{106 \times 10^3 \text{ N·m}}{40 \text{ m}} = 2650 \text{ N}$$

(c) Finally the average power $\dot{W}_{mech,av}$ supplied to the car can be determined from the relation between $W_{mech}$ and $\dot{W}_{mech}$, namely,

$$W_{mech} = \int_1^2 \dot{W}_{mech} \, dt = \dot{W}_{mech,av} \, \Delta t$$

The time interval is given by the distance traveled divided by the constant velocity. Thus

$$\Delta t = \frac{\Delta s}{\mathbf{V}} = \frac{40 \text{ m}}{1.0 \text{ m/s}} = 40 \text{ s}$$

and the average mechanical power becomes

$$\dot{W}_{mech,av} = \frac{W_{mech}}{\Delta t} = \frac{106 \times 10^3 \text{ N·m}}{40 \text{ s}} \times \frac{1 \text{ J}}{\text{N·m}} = 2650 \text{ J/s}$$

**Comments:**     (1) All answers have been rounded off to three significant figures, as will be the practice in this text. In this text, unless provided other information, we will assume that all information provided in a problem is accurate to three significant figures.

(2) The average mechanical power could also have been calculated as $\dot{W}_{mech,av} = F_{av} \cdot \mathbf{V}_{av}$. Check it.

(3) If the average mechanical power is 2650 J/s, the smallest possible electric motor would be 2.65 kW or 3.55 hp.

---

**Example 2-2**

State 1: $t = 0$; $\mathbf{V}_1 = 0$

Process 1–2: $\dot{W}_{mech} = 3$ hp

State 2: $t = ?$ s; $\mathbf{V}_2 = 40$ mph

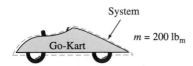

**Figure 2-4**
Schematic and data for Example 2-2.

**A** small race car (a go-kart) has a mass of 200 pound-mass with the rider and is powered by a 3-horsepower engine. *Estimate* how long it would take the go-kart to reach a speed of 40 miles per hour on a level racetrack. Is your estimate high or low? Assume that all of the engine power can be available as mechanical power to accelerate the go-kart.

**Solution:**

**Given:**    A go-kart is powered by a small engine and is accelerating from a dead stop, as shown in Fig. 2-4.

**Find:**    Estimate the time it takes the go-kart to go from a dead stop to 40 mph. Then comment on how your answer relates to actual performance.

**Model:**    Only changes in kinetic energy are important since the track is level. All engine power goes into changing the motion of the car. Engine power is constant.

**Strategy:**    The change in kinetic energy is related to the work transfer to the system. If we can find the work required and we know the rate of doing work, the power, we should be able to find the time.

**Analysis:** The car is a closed system for our model. Applying the mechanical energy balance, and neglecting changes in potential energy, gives

$$\Delta KE_{cm} = W_{mech}$$

But we also know that power and work can be related as $W_{mech} = \int \dot{W}_{mech}\,dt = \dot{W}_{motor}\,\Delta t$ because the motor power is a constant. Thus the mechanical energy balance can be rewritten as

$$\frac{m}{2}(\mathbf{V}_2^2 - \mathbf{V}_1^2) = W_{mech} = \dot{W}_{motor}\,\Delta t$$

Solving for $\Delta t$ and noting that $\mathbf{V}_1 = 0$, we have the result

$$\Delta t = \frac{m}{2}\frac{\mathbf{V}_2^2}{\dot{W}_{motor}}$$

Substituting in the numerical values gives

$$\Delta t = \frac{m\mathbf{V}_2^2}{2\dot{W}_{motor}} = \frac{200\ lb_m}{2(3\ hp)}\left(\frac{40\ miles}{h}\right)^2 \times \left(\frac{5280\ ft}{1\ mile}\right)^2 \left(\frac{1\ h}{3600\ s}\right)^2$$

$$= 114{,}700\ \frac{lb_m\cdot ft^2}{hp\cdot s^2} \times \frac{1\ hp\cdot s}{550\ ft\cdot lb_f} \times \frac{1\ lb_f\cdot s^2}{32.174\ lb_m\cdot ft}$$

$$= 6.48\ s$$

Because we assumed that all of the power went into moving the cart, it would be more realistic to say that $\Delta t_{actual} > 6.48$ s.

**Comments:** (1) Notice the correct use of unit conversion factors. The numerical calculations would have been much easier if we had first converted the cart speed to ft/s and the power to ft·lb$_f$ before substituting them into the symbolic solution. Knowing when to convert units and what to convert them to is a skill learned with experience. The "brute-force" approach will always work, but it will pay you to develop some "finesse" in handling units.

(2) Notice how we carefully distinguished between pound-force and pound-mass. Confusion between these terms always results in *major* errors. Guard against this by never using "pound" without a qualifier.

(3) An alternate approach to this problem would have applied the *rate form* of the mechanical energy balance to find the *rate of change* of the system's kinetic energy in terms of the engine power. This approach would lead to the same required equation, $\Delta t = \Delta KE_{cm}/\dot{W}_{motor}$.

---

Two other well-known forms of work interactions can also be evaluated using the concept of mechanical work as a force moving through a distance. Shaft work occurs when a rotating shaft crosses the boundary of a system, and electrical work occurs when an electric current crosses the boundary of a system. Both of these work transfers of energy will be reviewed in the following subsections.

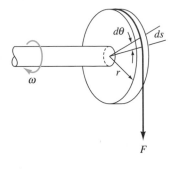

**Figure 2-5**
Schematic showing relationship of torque and angular deflection to rotational mechanical work.

### 2-1-4　SHAFT WORK

Shaft work, sometimes called rotational mechanical work, is evaluated in terms of the external torque transmitted by a rotating shaft. Consider the apparatus shown in Fig. 2-5. An external force $F$ acts through a moment arm $r$ and produces a torque $\tau$. If the angle between the moment arm and the external force is 90°, then the magnitude of the torque $\tau$ is

$$\tau \equiv Fr \qquad \text{or} \qquad F = \frac{\tau}{r}$$

The differential amount of **shaft work** $\delta W_{\text{shaft}}$ done on the system by the external force moving through a differential distance $ds = r\, d\theta$, where the angle $\theta$ is measured in radians, can be written as

$$\delta W_{\text{shaft}} = F\, ds = \frac{\tau}{r} \cdot (r\, d\theta) = \tau\, d\theta$$

Because the rotational motion of the shaft is commonly specified in terms of the number of revolutions per unit time $\dot{n}$, it is often easier to calculate the shaft power before calculating the shaft work. The **shaft power** $\dot{W}_{\text{shaft}}$ transmitted at any instant in time is

$$\dot{W}_{\text{shaft}} = \frac{\delta W_{\text{shaft}}}{dt} = \tau\omega = 2\pi\dot{n}\tau \qquad \textbf{[2-13]}$$

where $\omega$ is the angular or rotational speed $d\theta/dt$ in radians per unit time, $\dot{n}$ is the number of revolutions per unit time, and $\omega = 2\pi\dot{n}$.

　　Shaft work generally depends upon the path of the process. The shaft work can be evaluated from the shaft power as

$$W_{\text{shaft}} = \int_{t_1}^{t_2} \dot{W}_{\text{shaft}}\, dt = \int_{t_1}^{t_2} \tau\omega\, dt$$

if we know how $\omega$ and $\tau$ change with time during the process. If the *torque is constant* during a process, then integration of the above expression leads to

$$W_{\text{shaft}} = \tau\theta = 2\pi n\tau \qquad \textbf{[2-14]}$$

Note that all of the information required to calculate the shaft work (or power) must be evaluated at the system boundary.

　　Shaft work and power are *nonquasiequilibrium* work modes (see Sec. 2-8-6) if the rotation of the shaft and the external torque cannot be directly related to system properties. This is the case, for example, when the shaft is attached to a "paddle wheel" located inside a tank, as shown in Fig. 2-6. For this situation, the shaft work is often referred to as paddle-wheel work.

System boundary

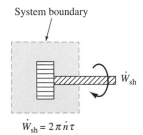

$\dot{W}_{\text{sh}} = 2\pi\dot{n}\tau$

**Figure 2-6**
Rotational shaft work as paddle-wheel work.

## 2-1-5 ELECTRICAL WORK

The evaluation of the electrical work done on a system likewise depends on measurements at the boundary of the system. When a small amount of charge $\delta Q_c$ is moved from point 1 to point 2 in an electrostatic field (for example, an electric circuit), the mechanical work required to move the charge in the field is called the **electrical work** $\delta W_{elec}$. It is given by

$$\delta W_{elec} = V\,\delta Q_c$$

where $V$ (by convention) is the electrical potential difference between the two points in the field. The electric potential difference $V$ has dimensions of [energy]/[charge], and has units of *volts* in both SI and USCS. By definition, 1 volt = 1 joule/coulomb.

The *electric current* $I$ is related to the amount of electric charge $\delta Q_c$ crossing a boundary during a time period $dt$ by the relation $dQ_c = I\,dt$. Electric current $I$ has dimensions of [charge]/[time] and is measured in *amperes* (A). By definition, 1 A = 1 coulomb/s. The differential work required to pass a current $I$ between two points on the system boundary with voltage potential difference $V$ for a time period $dt$ is $\delta W_{elec} = VI\,dt$. The rate of electric work on the system, the **electric power** $\dot{W}_{elec}$, is found from

$$\dot{W}_{elec} = \frac{\delta W}{dt} = VI \qquad \textbf{[2-15]}$$

where the current $I$ and the potential difference $V$ are measured at the boundary of the system, as shown in Fig. 2-7. The electrical power (or work) into the system is positive when the current flows *into* the system at the highest voltage. When a current of one ampere (A) passes through an electrical potential difference of one volt (V), the electrical power for the process is defined as one watt (W), or 1 J/s.

In general, the current $I$ and the voltage difference $V$ can depend upon time. Thus electrical work, like shaft work, depends upon the path of the process, and not just the end states. When the *current and voltage are constant,* the electrical work can be calculated as

$$W_{elec} = \int_{t_1}^{t_2} \dot{W}_{elec}\,dt = \int_{t_1}^{t_2} VI\,dt = VI\,\Delta t \qquad \textbf{[2-16]}$$

Electrical work is another example of a nonquasiequilibrium work interaction if the potential difference and the current cannot be related to the properties of the system. This is the case when electrical power is supplied to an electrical resistor within a system.

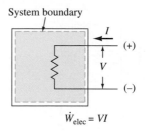

System boundary

$$\dot{W}_{elec} = VI$$

**Figure 2-7**
Electrical work dissipated in a resistor.

**Note that shaft work and electrical work may affect the intrinsic state of a system.**

---

The rigid tank contains a gas, a paddle wheel connected to a shaft, and an electric resistor connected to a battery. The shaft attached to the paddle wheel has an

**Example 2-3**

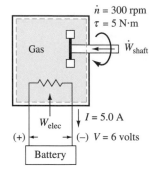

$\dot{n} = 300$ rpm
$\tau = 5$ N·m
$\dot{W}_{shaft}$

Gas

$W_{elec}$     $I = 5.0$ A
(+)          (−) $V = 6$ volts

Battery

**Figure 2-8**
Schematic and data for Example 2-3.

applied torque of 5.0 N·m, and the shaft speed is 300 rpm. At the same time, an electric current of 5.0 A is supplied to the resistor from a battery with a potential of 6.0 V. Determine the net power, in watts, and the net work done on the system, in kilojoules, if the process continues for 1 min.

**Solution:**

**Given:**   A gas within a rigid tank receives energy in the form of paddle-wheel (shaft) work and electrical work. The system and associated data are shown in Fig. 2-8.

**Find:**   (*a*) the net power, in watts, and (*b*) the net work, in joules, if the process continues for 1 h.

**Model:**   Current, voltage, speed, and torque are all assumed to be constant. Closed, rigid system.

**Strategy:**   Evaluate the sum of the electrical and shaft power. Integrate the net power to obtain the net work.

**Analysis:**   The system is the substance within the dashed line shown in Fig. 2-8. The net power for the process is the sum of the shaft and electrical contributions evaluated at the system boundary. That is,

$$\dot{W}_{net} = \dot{W}_{shaft} + \dot{W}_{elec}$$

Energy is being supplied to the system, and the shaft power and electrical power are both positive. If we assume that the applied torque and rotational speed are constant, the shaft power is given by

$$\dot{W}_{shaft} = 2\pi\dot{n}\tau = \left(2\pi \frac{rad}{rev}\right)\left(300 \frac{rev}{min}\right)(5.0 \text{ N·m})$$

$$= 9425 \frac{\text{N·m}}{\text{min}} \times \left(\frac{1 \text{ min}}{60 \text{ s}}\right)\left(\frac{1 \text{ J}}{1 \text{ N·m}}\right)\left(\frac{1 \text{ W·s}}{1 \text{ J}}\right) = 157 \text{ W}$$

Similarly, for the positive electrical power supplied to the system,

$$\dot{W}_{elec} = 6.0 \text{ V } (5.0 \text{ A}) = 30 \text{ VA} \left(\frac{1 \text{ W}}{1 \text{ VA}}\right) = 30 \text{ W}$$

Adding these two results gives the net power input as

$$\dot{W}_{net} = \dot{W}_{shaft} + \dot{W}_{elec} = (157 + 30) \text{ W} = 187 \text{ W}$$

Since the net power does not change with time, the net work transfer is

$$W_{net} = \int_{t_1}^{t_2} \dot{W}_{net} \, dt = \dot{W}_{net} \, \Delta t = 187 \text{ W} \times 1 \text{ h} = 187 \text{ W·h}\left(\frac{1 \text{ J}}{1 \text{ W·s}}\right)\left(\frac{60 \text{ s}}{1 \text{ min}}\right)$$

$$= 11,220 \text{ J} = 11.22 \text{ kJ}$$

**Comments:**   (1) Again note the use of unit conversion factors to obtain the desired units.

(2) The decisions about the direction of the shaft power and electrical power were made after considering what the system was, i.e., the paddle wheel and the resistor. If the system had been the battery, the electrical power would have been $-30$ W, because the external electric current enters the battery at the lowest potential. This could be written as either $\dot{W}_{battery} = -30$ W or $\dot{W}_{battery,out} = 30$ W.

## 2-2   THE FIRST LAW OF THERMODYNAMICS

In the preceding section a few types of work interactions were reviewed. All of these involve specific energy transfer processes that affect the state of a chosen system. When the only energy transfers allowed across a system boundary are work interactions, the boundary is called an ***adiabatic boundary,*** and the system is called an ***adiabatic system.*** An ***adiabatic process*** is one that involves only identifiable work interactions.[3]

**What type of materials might you use to construct an adiabatic boundary?**

In Sec. 2-1 two work interactions, shaft work and electrical work, were introduced. At this point we wish to investigate the separate effect of these two work interactions on a closed system. In each case the system will undergo an identical change of state. To illustrate the point, consider the constant-volume closed system shown in Fig. 2-9. Two different adiabatic processes are employed experimentally to carry out a specified change of state. Process A (see Fig. 2-9a) is carried out by allowing a paddle wheel, driven by a pulley-weight system, to rotate within the constant-volume system. In process B (see Fig. 2-9b) an electric resistor has been placed within the fluid, and it is connected across the boundary of the system to an external battery. Both processes A and B have the same initial state and end at the same final state of the fluid. In addition, one might also consider a process C where both shaft work and electrical work are used to bring about the required change of state. Experiments show that the total work required is the *same* for all three adiabatic processes described above if they begin and end at the same equilibrium states of the constant-volume system.

On the basis of experimental evidence of this type, which began with the work of Joule in the middle of the nineteenth century, it is possible to make a broad assertion. This postulate, based on experimental evidence and called the ***first law of thermodynamics,*** states the following:

*When any closed system (control mass) is altered adiabatically, the net work associated with the change of state is the same for all possible processes between two given equilibrium states.*

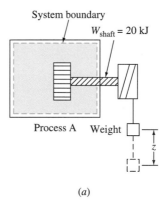

(a)

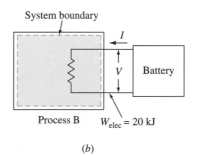

(b)

**Figure 2-9**
Work interactions and the first law.

---

[3] In thermodynamics it is advantageous to define the concept of work in a broad sense that includes within it the traditional definition of mechanical work. The operational definition in thermodynamics is: *Work* is an interaction between a system and its surroundings and is done on its surroundings if the *only* change in the surroundings *could* have been an increase in its gravitational potential energy.

To state it another way, the value of the net work done on or by a closed adiabatic system is dependent solely on the end states of the process. This postulate is true regardless of the type of work interaction involved in the process, the type of process, and the nature of the closed system. The first law of thermodynamics, together with those relationships derivable from it, is so well established that one does not question its validity when it is properly applied to a range of scientific and engineering problems.

It has been postulated that the net work transfer is the same for all adiabatic processes that connect two equilibrium states of a closed system. This first-law statement leads to a general definition of the change in the energy of a closed system between equilibrium states. Recall from Sec. 1-3 and the discussion of properties that *any quantity that is fixed by the end states for all processes between those end states is a measure of the change in a property value.* Because the adiabatic work for a closed-system process is solely a function of the end states, the quantity of adiabatic work defines or measures the change in a property. This property is called the ***energy*** $E$ (or the *total energy*) of the system.

When a closed system undergoes an adiabatic process that changes its state from state 1 to state 2, there may be several different types of work done on or by the system. If we apply the first law of thermodynamics to this adiabatic (ad) process, we may write

$$\Delta E = E_2 - E_1 \equiv W_{\text{net,ad}} \qquad \textbf{[2-17]}$$

where $W_{\text{net,ad}}$ is the net work carried out for any adiabatic process between those two states. The mechanical energy balance developed earlier (Eq. [2-11]) is a special case of this more general result. Equation [2-17] provides an operational definition for the change in energy $\Delta E = E_2 - E_1$ for any closed system. It does not provide any guidance as to the value of the energy at state 1 or state 2. A specific value of $E_1$, for example, can be established only by assigning an arbitrary but fixed value at a specified reference state.

**Note the change in the total energy of a closed system is defined on the basis of experimental evidence.**

## 2-3    A CONSERVATION OF ENERGY PRINCIPLE FOR CLOSED SYSTEMS

The first law of thermodynamics provides an operational definition of energy change. The change in the energy of a closed system between any two states equals the work done on or by the system during an adiabatic process that connects the two states. Experience has shown, however, that it is possible to change the state of a system by nonadiabatic processes.

### 2-3-1    HEAT TRANSFER OF ENERGY FOR A CLOSED SYSTEM

Consider the change in state associated with the stirring of a fluid within an adiabatic, rigid (constant-volume) tank. It has also been shown that the same change of state could be accomplished by bringing the system in contact with another system at a higher temperature. The interaction between the system and its surroundings that has occurred in the latter case is called a **heat interaction** or **heat transfer of energy** $Q$. *Heat transfer and work transfer are the only mechanisms by which energy can be transferred across the boundary of a closed system.*

Consider a closed system undergoing a process between state 1 and state 2 during which both heat and work interactions occur. Because this is a nonadiabatic process, the change in energy of the system will not equal the work for the process. The difference between the energy change and the net work is an operational definition of the heat interaction that took place. Mathematically this relationship is expressed by

$$Q_{net} \equiv (E_2 - E_1) - W_{net} \qquad \textbf{[2-18]}$$

or, on a differential basis,

$$\delta Q \equiv dE - \delta W \qquad \textbf{[2-19]}$$

As used in these equations, the sign convention for heat transfer (and heat transfer rate) is the same as that used for work. Transfers of energy into the system by heat transfer have positive values and transfers of energy out of the system have negative values.

Heat transfer $Q$ is an energy transfer across a system boundary due to a temperature difference between the system and its surroundings. Heat transfer has the dimensions of energy and has units of kilojoules (kJ) in SI and foot-pounds force (ft·lb$_f$) or British thermal units (Btu) in USCS. Experience has shown that *heat transfer of energy occurs spontaneously only in the direction of decreasing temperature.* Heat transfer is zero if no temperature difference exists between the system and its surroundings, if the temperature gradient at the boundary is zero, or if the boundary is *thermally insulated.* A surface or boundary across which there is no heat transfer of energy is called an **adiabatic surface.** Actual adiabatic surfaces are difficult to construct; however, in many situations heat transfer is negligible and the adiabatic surface assumption is realistic.

Based on Eq. [2-18], clearly the sum of $Q$ and $W$ is unique for any process between given end states, because $\Delta E$ is fixed in value by the end states. However, the individual value of $W_{12}$ usually depends on the nature of the process between given end states. Consequently the value of the heat transfer $Q_{12}$ for a process also will usually depend upon the path. Thus *heat transfer is a process or path function,* like work transfer. Recall that the differential of a quantity that depends on the path of a process (such as $Q$ and $W$) is designated by the symbol $\delta$ (rather than the exact differential symbol $d$). Also, a differential amount of heat transfer can be defined in terms of the heat transfer rate $\dot{Q}$ as

$$\delta Q \equiv \dot{Q}\, dt \qquad \textbf{[2-20]}$$

As a result, for a process between states 1 and 2,

$$Q_{12} = \int_1^2 \dot{Q}\, dt = \int_1^2 \delta Q \qquad \textbf{[2-21]}$$

The subscript 12 frequently is omitted when a single-step process takes place. (For a multiple-step process 1-2-3 we would write $Q_{12}$ and $Q_{23}$ for the two steps.) Note again that integration of a process function does not lead to the use of the symbol $\Delta$. For quantities such as heat and work, one speaks not of the change in the quantity (such as one does for energy $\Delta E$), but rather of its overall value for the process. Finally, it cannot be overemphasized that heat transfer is an energy transport that occurs across a boundary. This effect is an interaction that ceases to exist once a process has ended.

### 2-3-2   CONSERVATION OF ENERGY FOR A CLOSED SYSTEM

Equation [2-18], used previously to define heat transfer, contains all of the information required to write a conservation of energy equation for a closed system. Rearrangement of this equation so that the change of energy is on the left-hand side leads to

$$\Delta E_{cm} = E_2 - E_1 = Q + W \qquad \textbf{[2-22]}$$

This equation is the **conservation of energy equation** for a *closed system.* Equation [2-22] also is known as the **general energy balance** for a *closed system.* In words,

$$\begin{pmatrix} \text{Change in} \\ \text{the energy of} \\ \text{the system} \end{pmatrix} = \begin{pmatrix} \text{net transfer of energy} \\ \text{into the system by} \\ \text{a heat interaction} \end{pmatrix} + \begin{pmatrix} \text{net transfer of energy} \\ \text{into the system by} \\ \text{a work interaction} \end{pmatrix}$$

For a differential change of state, Eq. [2-22] is written as

$$dE_{cm} = \delta Q + \delta W \qquad \textbf{[2-23]}$$

Frequently it is convenient to analyze closed systems on a unit-mass basis. If the **heat transfer per unit mass** $q$, the **work per unit mass** $w$, and the **specific energy** $e$ are defined as

$$q \equiv \frac{Q}{m} \qquad w \equiv \frac{W}{m} \qquad e \equiv \frac{E}{m} \qquad \textbf{[2-24]}$$

then the conservation of energy principle for a differential change of state is given by

$$de = \delta q + \delta w \qquad \textbf{[2-25]}$$

For a finite change of state, integration of this equation yields

$$\Delta e = q + w \qquad \textbf{[2-26]}$$

Finally, Eq. [2-22] can be expressed on a *rate basis* by first dividing the equation by a finite time interval $\Delta t$. This yields

$$\frac{\Delta E}{\Delta t} = \frac{Q}{\Delta t} + \frac{W}{\Delta t}$$

Then, in the limit as $\Delta t$ approachs zero, $\Delta E/\Delta t \rightarrow dE/dt$, $Q/\Delta t \rightarrow \delta Q/dt = \dot{Q}$, and $W/\Delta t \rightarrow \delta W/dt = \dot{W}$. As a result the *rate form of the conservation of energy equation* for a *closed system* is

$$\frac{dE_{cm}}{dt} = \dot{Q}_{net} + \dot{W}_{net} \qquad \textbf{[2-27]}$$

In words,

$$\begin{pmatrix} \text{Time rate of change} \\ \text{of the energy} \\ \text{in the system} \end{pmatrix} = \begin{pmatrix} \text{net heat} \\ \text{transfer rate} \\ \text{into the system} \end{pmatrix} + \begin{pmatrix} \text{net work} \\ \text{transfer rate} \\ \text{into the system} \end{pmatrix}$$

The rate form of the energy balance clearly demonstrates how at every instant of time the changes in energy inside the closed system are balanced by the flows of energy across the boundary. Equations [2-22] and [2-27] *are extremely important equations and are the preferred starting points* in the application of the conservation of energy principle to a closed system. In both equations there is an *implicit* sign convention that energy added to a system by heat transfer or work transfer has a *positive* numerical value, and interactions that remove energy from a system have *negative* numerical values. As noted earlier, these implicit sign conventions can be eliminated by using "in" and "out" subscripts on $Q$ and $W$ in the energy equations.

Finally, consider the following two applications of the general conservation of energy equation.

1. Note that since the value of a work interaction and a heat interaction is a function of the path, the separate evaluation of the integral of $\delta W$ or $\delta Q$ around a cycle should not necessarily lead to zero, as it does for the cyclic integration of $dE$. Therefore, for any closed system undergoing a cyclic process, the conservation of energy principle reduces to

$$\oint (\delta Q + \delta W) = \oint \delta Q + \oint \delta W = 0$$

Thus the integrals of $\delta Q$ and $\delta W$ must have the same value, but opposite signs, for the cycle.

2. When a closed system is operating at a *steady-state condition,* time is no longer a variable in the problem. For steady-state operation, any

heat-transfer and work-transfer rates must be independent of time. In addition, the amount of energy in the system is also independent of time, so that the rate of change of the system energy must be zero, i.e., $dE/dt = 0$. This in turn means that the net *rates* of heat transfer and work transfer must be equal in magnitude, but opposite in sign.

**Example 2-4**

Two closed systems, A and B, undergo a process during which work transfer occurs to each system and the total energy of each system increases. (*a*) For system A, work transfer into the system amounts to 100.0 kJ, and the total energy increases by 55.0 kJ. (*b*) For system B, the work transfer out of the system is 77,800 ft·lb$_f$, while the total energy increases by 55 Btu. Find the heat transfer for system A in kilojoules and system B in Btu. Also determine whether it is added to or removed from the system.

**Solution:**

**Given:** A specified work interaction brings about a specified total energy change for two systems. The systems with appropriate input data are shown in Fig. 2-10.

**Find:** Magnitude and direction of $Q$ for (*a*) system A in kJ, and (*b*) system B in Btu.

**Model:** Closed system.

**Strategy:** Apply the basic closed system energy balance to solve for the heat transfer.

**Analysis:** The conservation of energy principle, represented by Eq. [2-22], is $\Delta E_{cm} = Q + W$. Solving for the heat transfer leads to $Q = \Delta E_{cm} - W$.
(*a*) Because work done on a system is positive by convention, $W = 100.0$ kJ and $\Delta E_{cm} = 55.0$ kJ. Substituting into the equation for $Q$ gives

$$Q = \Delta E_{cm} - W = 55.0 \text{ kJ} - 100.0 \text{ kJ} = -45.0 \text{ kJ}$$

Note that the answer is $-45.0$ kJ, and not just $-45.0$. *Answers to engineering problems have units, and they must be stated explicitly.* The negative sign is also important, for it indicates that the heat transfer of energy is *out* of the system. Another way of expressing the answer is to use an explicit directional subscript: $Q_{out} = 45.0$ kJ. When the subscript "out" or "in" is placed on $Q$ (or $W$), a positive numerical value must agree with the indicated direction.
(*b*) The process is illustrated by Fig. 2-10*b*. For system B, $W_{out} = 77,800$ ft·lb$_f$ because the work transfer is from the system, while $\Delta E = 55$ Btu. Although both ft·lb$_f$ and Btu are units of energy, consistent units are required to calculate the heat transfer. To obtain an answer for $Q$ in Btu, the work transfer becomes

$$W_{out} = 77,800 \text{ ft·lb}_f \times \frac{1 \text{ Btu}}{778 \text{ ft·lb}_f} = 100 \text{ Btu}$$

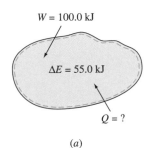

$W = 100.0$ kJ

$\Delta E = 55.0$ kJ

$Q = ?$

(*a*)

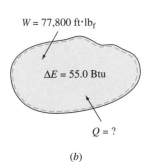

$W = 77,800$ ft·lb$_f$

$\Delta E = 55.0$ Btu

$Q = ?$

(*b*)

**Figure 2-10**
Schematic and data for Example 2-4.

where the conversion factor for the units found in Table A-1E has been rounded to three significant figures. The substitution of values for $W$ and $\Delta E$ into Eq. [2-19] leads to

$$Q = \Delta E_{cm} - W = \Delta E_{cm} - (-W_{out})$$
$$= 55 - (-100) \text{ Btu}$$
$$= 155 \text{ Btu}$$

The positive answer indicates that 155 Btu of energy must be added during the process by heat transfer.

**Comment:** Sign conventions for the energy balance and for heat and work transfers must be followed. If you use an equation that someone else has developed for heat or work transfer, or for an energy balance, you must understand any implicit sign convention in the equation.

---

**Reconsider Example 2-3.** Energy is added to a closed system through an electrical resistor and by a paddle wheel at the rates of 188.5 and 30.0 W, respectively. The equation for the heat-transfer rate to the surroundings is given by $\dot{Q}_{out} = 218.5[1 - \exp(-0.03t)]$, where $t$ is in seconds and $\dot{Q}$ is in watts. (*a*) Determine an equation for the time rate of change of the energy within the system $dE/dt$ as a function of time. (*b*) Plot this functional relationship for the first 3 min of operation.

**Example 2-5**

**Solution:**

**Given:** Electrical and shaft work are supplied to a closed system at constant rates. Heat transfer also occurs at a time-dependent rate. The closed system with appropriate input data is shown in Fig. 2-11*a*.

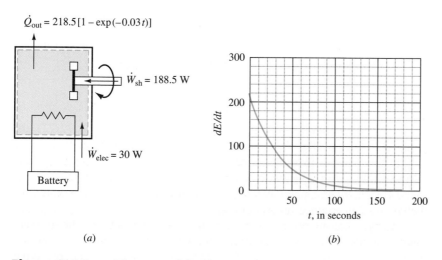

(*a*)

(*b*)

**Figure 2-11** Schematic and data for Example 2-5.

**Find:**     (a) $dE/dt = f(t)$ for the system. (b) Plot equation.

**Model:**     Closed system, constant rates of shaft and electrical power.

**Strategy:**     Apply the closed system energy balance on a rate basis.

**Analysis:**     The appropriate conservation of energy equation for the system is Eq. [2-27], namely,

$$\frac{dE}{dt} = \dot{Q}_{net} + \dot{W}_{net}$$

where the $\dot{W}$ term in this case represents several forms of work interactions.

(a) In terms of known information, and noting that the heat-transfer rate is out of the system,

$$\frac{dE}{dt} = -\dot{Q}_{out} + \dot{W}_{shaft,in} + \dot{W}_{elec,in}$$
$$= -218.5[1 - \exp(-0.03t)] \text{ W} + (188.5 + 30.0) \text{ W}$$
$$= 218.5 \exp(-0.03t) \text{ W}$$

Thus the rate of change of the energy $E$ decays exponentially with time.

(b) A plot of the above equation is shown in Fig. 2-11b. After about 200 s the energy of the system becomes essentially constant with time ($dE/dt$ approaches zero). After this transient time period, the energy balance essentially reduces to $-\dot{Q}_{out} + \dot{W}_{net} = 0$, or

$$\dot{Q}_{out} = (\dot{W}_{shaft} + \dot{W}_{elec}) = 218.5 \text{ W}$$

This value of 218.5 W can also be found by allowing $t$ to approach infinity in the given relation for $\dot{Q}$ as a function of $t$.

**Comment:**     After 3 min the value of $dE/dt$ is only about 0.45 percent of its initial value.

## 2-4     THE NATURE OF THE ENERGY E

In the conservation of energy equation for a closed system, Eq. [2-27], the term $\Delta E$ represents the change in the total energy of the system. It is important at this point to investigate the specific types of energy that contribute to the total energy. Eventually we must decide what types of energy can be neglected when using the conservation of energy equation to model the behavior of an engineering system.

All types of energy E can be classified as either kinetic energy (KE) due to motion of a body or potential energy (PE) due to the position of a body relative to the force-field of other bodies. In addition, energy types can also be classified as being either extrinsic (ext) and intrinsic (int) forms.

Combining these two classifications provides a useful way to look at the total energy of a system. Therefore,

$$E_{\text{tot}} = E_{\text{KE}}^{\text{ext}} + E_{\text{KE}}^{\text{int}} + E_{\text{PE}}^{\text{ext}} + E_{\text{PE}}^{\text{int}} \qquad \textbf{[2-28]}$$

From classical physics it may be shown that the total kinetic energy of a system of particles can be expressed as the sum of three terms.

$$E_{\text{KE,total}} = (E_{\text{KE,trans}}^{\text{ext}} + E_{\text{KE,rot}}^{\text{ext}})_{\text{macro}} + (E_{\text{KE}}^{\text{int}})_{\text{micro}} \qquad \textbf{[2-29]}$$

The first two terms are the familiar translational kinetic energy and rotational kinetic energy of the total system relative to its center of mass. These two contributions are extrinsic and measurable in terms of macroscopic characteristics of the system. The third term is the summation of the kinetic energy of the individual particles within the system, due to the translational, rotational, and vibrational motion of the individual molecules. This form of kinetic energy is intrinsic, but cannot be measured directly. Thus two forms of the kinetic energy of a system are macroscopic (macro) in origin, while a third form is due to microscopic (micro) or molecular motion.

The total potential energy of a system can be expressed as a sum of four separate quantities.

$$E_{\text{PE,total}} = (E_{\text{PE,grav}}^{\text{ext}} + E_{\text{PE,elec}}^{\text{ext}} + E_{\text{PE,mag}}^{\text{ext}})_{\text{macro}} + (E_{\text{PE}}^{\text{int}})_{\text{micro}} \qquad \textbf{[2-30]}$$

One of these is the familiar gravitational potential energy of a body relative to the earth. Two other forms of potential energy are due to the presence of stationary and moving charges, and these are called the *electrostatic* and *magnetostatic* potential energies. These three forms of potential energy are measurable and extrinsic. The fourth form is that due to the forces exerted on a particle by surrounding particles within the system, summed for all particles. This type of potential energy is intrinsic. The last term on the right in Eq. [2-30] requires information on the forces acting between particles within a system; thus it is not directly measurable.

Electrostatic ($E_{\text{PE,elec}}^{\text{ext}}$), magnetostatic ($E_{\text{PE,mag}}^{\text{ext}}$), and macroscopic rotational ($E_{\text{KE,rot}}^{\text{ext}}$) energies are not considered in this text. By neglecting these terms, the substitution of Eqs. [2-29] and [2-30] into Eq. [2-28] yields

$$E = (E_{\text{KE,trans}}^{\text{ext}} + E_{\text{PE,grav}}^{\text{ext}})_{\text{macro}} + (E_{\text{KE}}^{\text{int}} + E_{\text{PE}}^{\text{int}})_{\text{micro}} \qquad \textbf{[2-31]}$$

The last two terms above represent the intrinsic kinetic and potential energies of the particles within the system and are not directly measurable. The sum of these two microscopic contributions to the energy is defined as the **internal energy** $U$ of the substance within the system. That is,

$$U \equiv E_{\text{KE}}^{\text{int}} + E_{\text{PE}}^{\text{int}} \qquad \textbf{[2-32]}$$

The internal-energy function, as defined by Eq. [2-32], is an *extensive, intrinsic property* of a substance in an equilibrium state. In the absence of phase changes, chemical reactions, and nuclear reaction, the internal energy $U$ is sometimes called the *sensible energy* of the system. However, we will always refer to it as the internal energy of the system.

On the basis of the preceding discussion, Eq. [2-28] for the total energy of a system becomes

$$E = U + E_{KE,trans} + E_{PE,grav} = U + \frac{mV^2}{2} + mgz \qquad \textbf{[2-33]}$$

where electrostatic, magnetostatic, and rotational energies are neglected.

Finally, the substitution of Eq. [2-33] into Eq. [2-22] leads to a general conservation of energy principle that considers translational kinetic energy and gravitational potential energy as the only extrinsic forms of interest. The result for such closed systems is

$$\Delta U_{cm} + \Delta\left(\frac{mV^2}{2}\right) + \Delta(mgz) = Q_{net} + W_{net} \qquad \textbf{[2-34]}$$

The rate form then becomes

$$\frac{d}{dt}\left(U + \frac{mV^2}{2} + mgz\right) = \dot{Q}_{net} + \dot{W}_{net} \qquad \textbf{[2-35]}$$

Because $U$ is an extensive property, $U = mu$, where $u$ is the specific internal energy. As shown in the next chapter, the specific internal energy is a function of other measurable intensive, intrinsic properties like pressure and temperature. On an intensive basis, the conservation of energy equation for a closed system becomes

$$\Delta e = \Delta u + \Delta ke + \Delta pe = q + w \qquad \textbf{[2-36]}$$

where $u$ can be on either a mass or molar basis.

## 2-5   HEAT TRANSFER

As defined earlier, heat transfer is a mechanism for transferring energy across the boundary of a system due to a temperature difference. Heat transfer can occur by three different mechanisms: conduction, radiation, and convection. Each of these has associated with it a distinct physical mechanism. No previous knowledge of these mechanisms is assumed in this text. However, it is instructive to introduce the basic concepts underlying heat-transfer calculations.

To compute the heat-transfer rate across any surface without using the energy balance requires information about the heat flux. The **heat flux** $q''$ is the heat-transfer rate per unit area and has dimensions of [energy]/{[length]$^2$·[time]}. In SI the typical units are W/m$^2$ and in USCS the typical units are Btu/(ft$^2$·h). For the situation where the heat flux is uniform across the surface $A$ of interest, the heat-flux rate is $q'' = \dot{Q}/A$. Depending on the situation, the heat flux may be nonuniform across the surface of interest. Thus the heat-transfer rate across any surface can be calculated from the heat flux as

$$\dot{Q} = \int_A q'' \, dA$$

where the $q''$ may vary with position over the surface and the integral is taken over the surface area.

**Conduction heat transfer** is the transfer of energy due to interactions between particles within a material. Conduction heat transfer is directly related to temperature gradients within a body, and is governed by *Fourier's law for heat conduction*. Applying the model to a one-dimensional wall as shown in Fig. 2-12, the equation for the heat-transfer rate at any $x$-location in the wall is

$$\dot{Q}_{\text{cond}} = q''_{\text{cond}}A = -kA \frac{dT}{dx} \qquad \textbf{[2-37]}$$

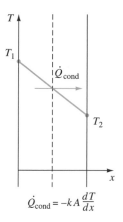

$$\dot{Q}_{\text{cond}} = -kA\frac{dT}{dx}$$

**Figure 2-12**
Conduction heat transfer in a one-dimensional wall.

where $k$ is the *thermal conductivity* of the material and $A$ is the cross-sectional area at location $x$. The negative sign is necessary because by convention the heat flux is assumed to be positive when energy is transferred by conduction in the direction of decreasing temperature. Numerical values for the thermal conductivity can be found in various handbooks. Under steady-state conditions when time is no longer a factor, the temperature distribution in a plane wall with uniform thermal conductivity will be a straight line as shown in Fig. 2-12. Under these conditions the temperature gradient can be written in terms of the two surface temperatures, $T_1$ and $T_2$, and the wall thickness $L$, and the conduction heat-transfer rate becomes

$$\dot{Q}_{\text{cond}} = -kA\left(\frac{T_2 - T_1}{L}\right) \qquad \textbf{[2-38]}$$

This is the heat-transfer rate at surface 1 or surface 2 or at a surface at any arbitrary $x$-location inside the wall.

**Radiation heat transfer** is the transfer of energy by electromagnetic radiation. Radiant energy can be emitted from a surface as well as from within transparent fluids and solids. For our purposes, we will concentrate on radiation heat transfer from surfaces. Unlike conduction (and convection) radiation is the only heat transfer mechanism that can occur across a vacuum. The thermal radiation heat flux emitted from a surface is described by the Stefan-Boltzmann equation:

$$q''_{\text{rad}} = \varepsilon \sigma T_s^4 \qquad \textbf{[2-39]}$$

where $\varepsilon$ is the *emissivity* of the surface, $\sigma$ is the *Stefan-Boltzmann constant*, and $T_s$ is the absolute temperature of the surface. The numerical values of the emissivity can range from 0 to 1 depending upon the condition of the surface. The Stefan-Boltzmann constant is a physical constant equal

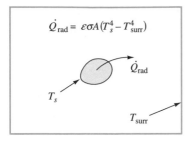

$$\dot{Q}_{rad} = \varepsilon\sigma A(T_s^4 - T_{surr}^4)$$

**Figure 2-13**
Radiation heat transfer from a small object inside an enclosure.

to $5.67 \times 10^{-8}$ W/(m$^2\cdot$K$^4$) [or $0.1714 \times 10^{-8}$ Btu/(h$\cdot$ft$^2\cdot$°R$^4$)]. Radiation heat-transfer calculations can be very complicated because of the nonlinear dependence of properties on the surface temperature and the ability of materials to absorb, transmit, and emit thermal radiation. Consider one special, but widely applicable case—radiation heat transfer between a small convex body and a large enclosure, as shown in Fig. 2-13. Under these conditions, the heat-transfer rate from the body to the enclosure can be written as

$$\dot{Q}_{rad} = \varepsilon\sigma A\left(T_s^4 - T_{surr}^4\right) \qquad \text{[2-40]}$$

where $A$ is the surface area of the body, $T_s$ is the surface temperature of the body, and $T_{surr}$ is the temperature of the enclosure that the small body sees.

**Convection heat transfer** is the transfer of energy between a solid surface and a liquid or gas due to fluid motion. The actual mechanism is a combination of conduction at the solid-fluid interface and fluid motion that carries away the energy. The convection heat-transfer rate from a surface to a fluid is modeled by *Newton's law of cooling*:

$$\dot{Q}_{conv} = hA(T_s - T_{amb}) \qquad \text{[2-41]}$$

where $h$ is the **convection heat transfer coefficient**, $A$ is the surface area, $T_s$ is the surface temperature, and $T_{amb}$ is the temperature of the fluid. The heat-transfer coefficient depends on the fluid and its motion over the surface. It is not a property of the fluid. Numerical values for heat-transfer coefficients can be determined from empirical correlations found in the literature.

In this text, the heat-transfer rate $\dot{Q}$ (and the heat transfer $Q$) will be handled in one of four ways:

- $\dot{Q}$ will be assumed or specified to be zero (adiabatic surface assumption).
- $\dot{Q}$ will be assigned a value based on the problem information.
- $\dot{Q}$ will be found from applying the general energy balance.
- $\dot{Q}$ will be calculated using Newton's law of cooling when sufficient information is provided.

The background for detailed heat-transfer calculations is provided in heat-transfer courses taken by many disciplines. For further information, consult a standard heat-transfer text.

**Example 2-6**

The base of a laundry iron has a surface temperature of 100°C and a surface area of 160 cm$^2$. The convection heat-transfer coefficient for the surface is 6 W/m$^2\cdot$°C, and the ambient air temperature is 25°C. Determine the convection heat-transfer rate from the surface, in watts.

**Solution:**

**Given:** Operating conditions for a laundry iron, as shown in the schematic for the base of the iron in Fig. 2-14.

**Find:** Convective heat-transfer rate from the base, in watts.

**Model:** Heat transfer modeled by Newton's law of cooling.

**Analysis:** Writing down Newton's law of cooling for the convective heat transfer, and substituting in the numerical values, yields

$$\dot{Q}_{conv} = hA(T_s - T_{amb})$$

$$= \left(6\,\frac{W}{m^2 \cdot {}^\circ C}\right)(160\text{ cm}^2)(100 - 25)({}^\circ C)\left(\frac{1\text{ m}}{100\text{ cm}}\right)^2 = 7.2\text{ W}$$

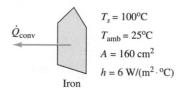

**Figure 2-14**
Schematic and data for Example 2-6.

$T_s = 100^\circ C$
$T_{amb} = 25^\circ C$
$A = 160\text{ cm}^2$
$h = 6\text{ W/(m}^2 \cdot {}^\circ C)$

$\dot{Q}_{conv}$

Iron

**Comments:** (1) Because heat transfer occurs from high temperatures to lower temperatures, we would say that the rate *from* the iron is 7.2 W.

(2) Notice that, as defined above, all of the equations for calculating conduction, radiation, and convection heat transfer give positive values when energy is transferred from a high temperature to a low temperature. You must keep this direction in mind when you apply these equations in an energy balance.

## 2-6   EXPANSION AND COMPRESSION WORK

The values of both shaft and electrical work transfers were determined in Sec. 2-1 from measurements at the system boundary. No knowledge of the properties of the substance within the system was required. There are, however, situations in which intrinsic property values of the substance must be known throughout the process in order to evaluate certain work interactions.

### 2-6-1   QUASIEQUILIBRIUM PROCESSES

Since properties are defined only in equilibrium states, we can conceive of an *idealized* process during which the system internally will be infinitesimally close to a state of equilibrium at all times. Any process carried out in this idealized manner is called a *quasiequilibrium,* or *quasistatic process.* An important consequence of this assumption is that the intensive properties of the system are spatially uniform during a quasiequilibrium process. Although a quasiequilibrium process is an idealization, many actual processes approximate such a condition. This is true since the time required for many substances to reach internal equilibrium is short compared to the time for the overall system change. Thus it is frequently necessary and appropriate in thermodynamic analyses for an actual process to be *modeled* as a quasiequilibrium one.

An additional benefit of modeling a process as quasistatic is that it becomes possible to plot the path of a quasiequilibrium process on a *property diagram*—a two-dimensional diagram where the coordinates represent

**Imagine letting the air out of a balloon very slowly or by popping the balloon. Which of these might qualify as a quasiequilibrium process?**

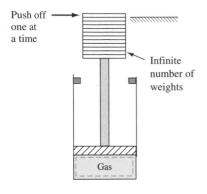

Push off one at a time

Infinite number of weights

Gas

**Figure 2-15**
Illustration of a quasiequilibrium or quasistatic process.

thermodynamic properties. Such diagrams are extremely helpful in the design and analysis of engineering systems. It is common practice on property diagrams to denote quasiequilibrium processes by a *solid* or continuous line, while a nonequilibrium process is represented by a *dashed* line between the given end states. A dashed line emphasizes that the values of the thermodynamic coordinates are not known between the initial and final states. Hence the path of the dashed line is arbitrary. However, the solid or continuous line of a quasistatic process does represent the functional relation between the coordinates during the entire process.

## 2-6-2   EXPANSION AND COMPRESSION WORK

Energy is transferred by a work interaction any time the volume of either a closed or open system is changed. This common type of work is called **compression/expansion work**, or simply $P\,dV$ work, for reasons that will be apparent shortly. As an example, consider the piston-cylinder device shown in Fig. 2-15. A stack of very small weights maintains the gas at an initial pressure. By removing one weight at a time, the pressure on the gas slowly decreases while the volume increases. Such a process is one of quasiequilibrium. If a number of weights were removed simultaneously, the piston would rise rapidly. The pressure of the gas is undefined during a nonequilibrium expansion process like this. However, under quasiequilibrium conditions the pressure and the volume change in a controlled fashion, and the method for determining the work of expansion is well defined.

The evaluation of $P\,dV$ work requires considering the mechanical work done by the motion of a force on the boundary of a system. The prototype closed system for this type of work is the piston-cylinder device shown in Fig. 2-16a. The contents of the piston-cylinder device constitute the closed

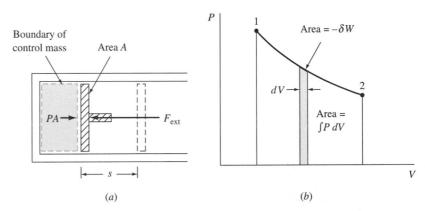

Boundary of control mass

Area $A$

$PA$

$F_{\text{ext}}$

$s$

$(a)$

$P$

1

Area $= -\delta W$

$dV$

2

Area $= \int P\,dV$

$V$

$(b)$

**Figure 2-16**      $(a)$ Mechanical work associated with the moving boundary of a piston-cylinder device; $(b)$ area representation of expansion or compression work for a quasiequilbrium process.

system as indicated by the dashed line. From basic mechanics, the expression for the differential work done on such a system is again given by Eq. [2-1],

$$\delta W = \vec{F}_{ext} \cdot d\vec{s} = F_{ext}\, ds \qquad \textbf{[2-1]}$$

Note that the external force acts at the piston-system interface on the boundary of the system. For the purpose of discussion, assume that the system is being compressed by the external force; thus the displacement vector $d\vec{s}$ is collinear with and in the same direction as $\vec{F}_{ext}$. The differential displacement $ds$ caused by the compression can be written in terms of a differential volume and the piston cross-sectional area, because $dV = A_{piston} \cdot (-ds)$, where the minus sign comes about because of the compression.

The differential work can now be written as

$$\delta W = F_{ext} \cdot ds = F_{ext}\left(-\frac{dV}{A_{piston}}\right) = -\left(\frac{F_{ext}}{A_{piston}}\right)dV$$

If the process is quasistatic, the pressure $P$ is *uniform* everywhere inside the system during the process, and $P = F_{ext}/A_{piston}$. Under this condition, the differential expression for *work done on the system* by the external force in changing the system volume is

$$\boxed{\delta W_{comp/exp} = -P\, dV} \qquad \textbf{[2-42]}$$

where the subscript "comp/exp" indicates compression/expansion work. This is a very significant equation, and it demonstrates the power of the quasiequilibrium assumption. By assuming a quasiequilibrium process, the work done by an external force can be calculated in terms of intrinsic properties of the system.

The total quasiequilibrium work of compression or expansion $W_{comp/exp}$ during a finite change in volume is the sum of the $P\, dV$ terms for each differential change in volume. Mathematically, this is expressed by the relation

$$\boxed{W_{comp/exp} = -\int_{V_1}^{V_2} P\, dV} \quad \text{or} \quad \boxed{w_{comp/exp} = -\int_{v_1}^{v_2} P\, dv} \qquad \textbf{[2-43]}$$

where $w_{comp/exp}$ is the work per unit mass. Note that when the volume decreases, the work is positive in value. This agrees with the adopted sign convention for work as a transfer of energy. The pressure $P$ should be expressed in *absolute* units. Note that the integral of $\delta W$ is simply $W$ and not $\Delta W$. A $P\, dV$ work interaction is associated with a process, and its value is a function of the path of the process.

A plot of a quasistatic process on pressure-volume coordinates is quite useful for describing graphically the expansion or compression work of a process. From integral calculus, the area beneath the curve that represents the quasistatic path of a process is equal to the integral of $P\, dV$ on a

**What would you need to know mathematically to evaluate the integral of $P\, dV$?**

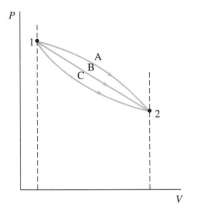

**Figure 2-17**

Illustration that boundary work is a function of the path for different quasiequilbrium processes between the same end states.

pressure-volume diagram. A typical diagram for the evaluation of boundary work is shown in Fig. 2-16b. The differential area represents the work done on or by the gas within the cylinder when the volume changes by the amount $dV$. The entire area beneath the curve from point 1 to point 2 represents the total work done when the gas expands from state 1 to state 2. An infinitely large number of quasistatic paths can be drawn between these same two end states. Three possible paths are shown in Fig. 2-17. The area beneath each of these paths is different. This merely emphasizes the fact that work is a path or process function and, unlike the change in the value of a property, is *not* solely dependent on the end states of the process. (Only in the special case of adiabatic processes does the net value of the work interactions become independent of the path.)

The integration of the equation for compression or expansion work requires a knowledge of the functional relationship between $P$ and $V$. This usually is found either from experimental measurements of $P$ and $V$ during a process, or from knowledge of the specific type of process involved. The general method is illustrated below.

**Example 2-7**

**A** gas contained within a piston-cylinder device is initially at 1.0 MPa and 0.020 m³. It expands to a final volume of 0.040 m³ under the conditions that (*a*) the pressure remains constant and (*b*) $PV$ = constant. Determine the work output, in kilojoules, for the two specified process paths.

**Solution:**

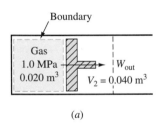

(*a*)

**Given:** A gas is contained within a piston-cylinder device. A schematic of the system and input data are shown in Fig. 2-18a.

**Find:** $W_{out}$ in kJ for (*a*) $P$ = *c*, and (*b*) $PV$ = *C*.

**Model:** Closed system; the process is quasiequilibrium.

**Strategy:** Make use of the definition of $W_{comp/exp}$.

**Analysis:** The system is the gas within the dashed lines in Fig. 2-18a. The expansion work for a closed system during a quasiequilibrium process is given by Eq. [2-43], $W_{comp/exp} = -\int P\,dV$. Information relating $P$ and $V$ is required to evaluate the integral.

(*a*) If the pressure remains constant (isobaric process), then the work $W_{12}$ for path *a* is

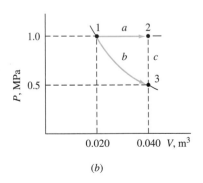

(*b*)

**Figure 2-18**

Schematic and data for Example 2-7.

$$W_{12} = -\int_1^2 P\,dV = -P\int_1^2 dV = -P(V_2 - V_1)$$
$$= -1 \times 10^6 \text{ N/m}^2 \times (0.040 - 0.020) \text{ m}^3 \times 1 \text{ kJ/}(10^3 \text{ N·m})$$
$$= -20.0 \text{ kJ}$$

Since $W_{12}$ is negative, $W_{12,out}$ = 20.0 kJ. The path between states 1 and 2 is shown in Fig. 2-18b.

(b) Under the condition that $PV = C$, then $P = C/V$. Substitution of this latter equation into the equation for $W_{comp/exp}$ and subsequent integration yield, for path $b$ to the final state 3,

$$W_{13} = -\int_1^3 P\,dV = -\int_1^3 \frac{C}{V}\,dV = -C\ln\frac{V_3}{V_1}$$

The constant $C$ can be evaluated at any state along the path. Since both the initial pressure and volume are known, we let $C = P_1 V_1 = 2 \times 10^4$ N·m. The numerical evaluation of the work for path $b$ now becomes

$$W_{13} = -P_1 V_1 \ln\frac{V_3}{V_1} = -2 \times 10^4 \text{ N·m} \times \ln\frac{0.040}{0.020} = -13,860 \text{ N·m}$$

$$= -13.86 \text{ kJ}$$

or $W_{out,13} = 13.86$ kJ. An equation for the type $PV = C$ is a hyperbola on a $PV$ diagram, as shown by path 13 in Fig. 2-18b. The work *output* in part $b$ is numerically only about 70 percent of that for part $a$. This comparison is clearly shown when one compares the areas below the process lines for the two cases. (The pressure at state 3 is exactly one-half that for state 2.)

**Comment:** This example illustrates that the value of quasistatic compression/expansion work (1) depends on the path of the process, and (2) is illustrated by an area on a $PV$ diagram.

---

For many expansion and compression processes, the path of the process may be modeled as a *polytropic process.* For this process the system pressure and specific volume are related by the *polytropic relation*

$$Pv^n = C \qquad\qquad \textbf{[2-44]}$$

where $C$ is a constant and the parameter $n$ is known as the *polytropic constant.* Although $n$ can take on any value, the relation is especially useful when $1 \le n \le 5/3$. The general solution for the quasistatic compression or expansion work for a polytropic process is developed in the following example.

**How does the pressure change in a polytropic process where $n = 0$?**

---

**D**erive a relationship for the compression/expansion work associated with a polytropic process within a closed system.

**Example 2-8**

**Solution:**

**Given:** A polytropic process occurs in a piston-cylinder device. The process path on a $Pv$ diagram is shown in Fig. 2-19.

**Find:** An equation for $w_{comp/exp}$.

**Model:** Closed system, quasiequilibrium process, $Pv^n = c$.

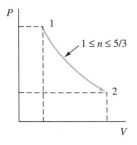

**Figure 2-19**
Polytropic path for Example 2-8.

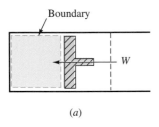

(a)

**Figure 2-20**
Schematic for Example 2-8.

**Strategy:**   Use the polytropic relation in the integration of $w_{\text{comp/exp}} = -\int P\,dv$.

**Analysis:**   The system is the gas within a piston-cylinder device in Fig. 2-20. On an intensive basis, substitution of the relation $P = cv^{-n}$ into Eq. [2-43], $w_{\text{comp/exp}} = -\int P\,dv$, leads to

$$w = -\int_1^2 P\,dv = -\int_1^2 cv^{-n}\,dv = \frac{c(v_2^{-n+1} - v_1^{-n+1})}{n-1}$$

The constant $c = P_1 v_1^n = P_2 v_2^n$. Therefore,

$$w_{\text{comp/exp}} = \frac{[P_2 v_2^n(v_2^{-n+1}) - P_1 v_1^n(v_1^{-n+1})]}{n-1} = \frac{P_2 v_2 - P_1 v_1}{n-1} \quad \text{(polytropic)}$$

The extensive work $W_{\text{comp/exp}}$ is found when the specific volume $v$ in the above equation is replaced by the total volume $V$.

**Comment:**   This equation applies for all values of $n$, except when $n = 1.0$. When $n$ is unity, $P = c/v$, and $w_{\text{comp/exp}} = -\int c\,dv/v = -c\ln(v_2/v_1) = -P_1 v_1 \ln(v_2/v_1)$.

---

**2-6-3**   WORK FOR A CYCLIC, QUASIEQUILIBRIUM PROCESS

Another common model for closed systems with only compression/expansion work pertains to systems that execute a *cyclic, quasiequilibrium process.* Again consider our simple piston-cylinder device shown in Fig. 2-16. One possible *PV* diagram for the working fluid inside the device is shown in Fig. 2-21. The path of the cyclic process consists of four distinct processes 1-2, 2-3, 3-4, and 4-1 that return the system to its initial state. The work for each process can be evaluated as

$$W_{\text{comp/exp},a-b} = -\int_a^b P\,dV$$

For those expansion processes 1-2 and 2-3 in Fig. 2-21, the final volume is greater than the initial volume and work is done by the system on its surroundings. In a similar fashion, the surroundings do work on the system for the compression process along path 4-1, since $\Delta V$ is negative. No $P\,dV$ work occurs during path 3-4 because the volume is constant. The *net work* done by the system for the cycle is the sum of the work interactions for each process in the cycle. In general, for a cycle composed arbitrarily of four different paths,

$$W_{\text{net,cycle}} = W_{12} + W_{23} + W_{34} + W_{41} \qquad \textbf{[2-45]}$$

$$= -\int_1^2 P\,dV - \int_2^3 P\,dV - \int_3^4 P\,dV - \int_4^1 P\,dV = -\oint P\,dV$$

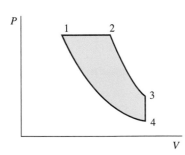

**Figure 2-21**
Area representation of net boundary work during a quasiequilibrium cyclic process.

where the symbol $\oint$ again signifies integration around a cyclic path. For the specific cycle in Fig. 2-21, the value of $W_{34}$ is zero, as noted earlier.

The work for each process in a cycle is represented by the area beneath each process line on the $PV$ diagram. The net work for the cyclic process is therefore represented by the *area enclosed* by the cyclic path. In Fig. 2-21, where the path is clockwise on the $PV$ plot, the net work transfer is out of the system. If the cycle were carried out counterclockwise, the net work would still be shown by the enclosed area. However, in this case the net work would be done by the surroundings on the system (input). Finally, note that in general, *the net work transfer for a cyclic process of a closed system due to volume changes will not be zero.* Only when the enclosed area on a $PV$ diagram is zero will the net work transfer be zero. The change in all properties of the system will be zero for a cycle, but the net boundary work will generally be finite.

**Show that the net work for a clockwise cycle on a Pv diagram that involves two constant-pressure processes at 50 and 100 psia and two constant-volume processes at 1 and 2 ft³, is −7200 ft·lbf.**

---

**A** gas initially at 0.020 m³ and 1.0 MPa is expanded quasistatically in a piston-cylinder device at constant pressure until the volume is 0.040 m³. It is then held at constant volume and cooled until its pressure is one-half of the initial value. Then it is compressed quasistatically to the original state by following the path $PV =$ constant. Determine the net work for the cycle, in kilojoules.

*Example 2-9*

**Solution:**

**Given:** Gas in a piston-cylinder undergoes a three-step, cyclic process involving constant-pressure expansion, constant-volume cooling, and finally compression according to $PV =$ constant to the initial state. The paths of the cyclic process and appropriate data are shown in Fig. 2-22.

**Find:** $W_{\text{net}}$ for cycle, in kJ.

**Model:** Closed system, processes are quasistatic, only $P\,dV$ work is present.

**Strategy:** Treat the cycle as a series of three processes. Evaluate $P\,dV$ work for each process.

**Analysis:** The basic equation for expansion or compression work is

$$W = -\int P\,dV$$

The system is the gas within the piston-cylinder device.

**Process 1-2:** Since pressure is constant,

$$W_{12} = -\int_1^2 P\,dV = -P_1(V_2 - V_1)$$

The value of this work transfer was found in Example 2-7 to be −20.0 kJ, that is, $W_{12} = -20.0$ kJ.

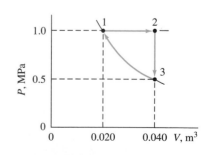

**Figure 2-22**
Pv diagram for cyclic process in Example 2-9.

***Process 2-3:***   Since the volume is constant, $dV = 0$. Hence

$$W_{23} = -\int_2^3 P \, dV = 0$$

***Process 3-1:***   Since $PV = C = P_1 V_1 = P_3 V_3$,

$$W_{31} = -\int_3^1 P \, dV = -C \ln \frac{V_1}{V_3} = -P_1 V_1 \ln \frac{V_1}{V_3}$$

The work-transfer value for path 1-3 was found in Example 2-7 to be $-13.86$ kJ. When a quasistatic path is reversed (in the absence of friction), the work interaction has the same value, but opposite sign. Hence the value of $W_{31}$ is $+13.86$ kJ for the cycle.

Now, to find the net work for the cycle,

$$W_{\text{net,cycle}} = W_{12} + W_{23} + W_{31} = (-20.0 + 0 + 13.86) \text{ kJ} = -6.14 \text{ kJ}$$

Thus the net work transfer into the system is $-6.14$ kJ, *or* the net work transfer *out* of the system during the clockwise, cyclic process shown in Fig. 2-22 is 6.14 kJ.

**Comment:**   The net work transfer for the cycle is not zero, even though the overall changes in pressure and volume for the cycle are zero. Work transfer is a process function and not a state function.

---

### 2-6-4   AN APPLICATION OF $P\,dV$ WORK

Many everyday devices such as air compressors and internal combustion engines involve the expansion and compression of a gas. A realistic model for these devices is the simple piston-cylinder device shown in Fig. 2-23a. The face of the piston is exposed to a gas within the cylinder. The rear of the piston is attached to a connecting rod and also is exposed to the atmosphere. In addition, there is friction between the sliding piston and the cylinder walls. Thus there are three *external* forces acting on the piston, as shown in Fig. 2-23b. One is the moving force $F_{\text{rod}}$ of the connecting rod, and a second is the frictional force $F_{\text{fric}}$ between the cylinder wall and the sliding piston. In addition, a third force $F_{\text{atm}}$ is due to the ambient pressure $P_{\text{atm}}$ on the outside of the piston. In addition, the force $F_{\text{gas}}$ acts on the inside of the piston.

The engineer is interested in determining how the work $W_{\text{rod}}$ transmitted by the connecting rod is related to the $P\,dV$ work for the gas inside the piston-cylinder device. On the basis of a force balance on the piston during a compression process,

$$F_{\text{rod}} + F_{\text{atm}} = F_{\text{gas}} + F_{\text{fric}}$$

where we have assumed that the product of mass times acceleration for the piston itself is small. We now multiply each term of the equation by a differential displacement $ds$ for a compression process, and integrate. As a result, the work transfers associated with the piston as the system are related by

$$W_{rod} + W_{atm} = W_{gas} + W_{fric}$$

Solving for the work done by the force $F_{rod}$, we have

$$W_{rod} = W_{gas} - W_{atm} + W_{fric}$$

The work transfer to the gas is given by $-\int P \, dV_{gas}$. It is usually assumed that the ambient pressure is constant. Hence the work done by the ambient pressure on the piston is

$$W_{amb} = P_{atm} \Delta V_{atm} = -P_{atm} \Delta V_{gas} \qquad \textbf{[2-46]}$$

where $\Delta V_{atm} = -\Delta V_{gas}$. In addition, note that the frictional force always opposes the direction of motion. Thus the value substituted for $W_{fric}$ always is taken as positive. Substitution of these terms into the equation for the connecting-rod work $W_{rod}$ is

$$W_{rod} = W_{gas} - W_{amb} + W_{fric}$$

$$= -\int_1^2 P \, dV_{gas} + P_{atm} \Delta V_{gas} + W_{fric} \qquad \textbf{[2-47]}$$

For the above equation, the change in $V_{gas}$ is negative during compression and positive during expansion. Hence during an expansion process the first term on the right is negative, while the other two terms are positive. In contrast, during a compression process, the first and third terms are positive, while the second term is negative. Overall, the effect of friction is to reduce the useful work output during expansion and to increase the work supplied to the piston during compression.

If the simple piston-cylinder device shown in Fig. 2-23a executes a cycle, the net work transfer by the connecting rod is obtained by integrating Eq. [2-47] over one cycle. Recognizing that the net work transfer to and from the atmosphere is zero, the net work transferred by the connecting rod is found by

$$W_{rod,net,cycle} = -\oint P \, dV + W_{fric,cycle} \qquad \textbf{[2-48]}$$

Again, the effect of friction during a cycle is to reduce the net work output, or to increase the work input, for the same quasiequilibrium changes of the fluid within the system.

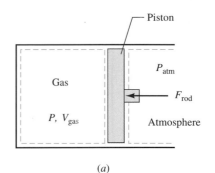

(a)

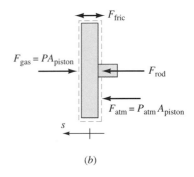

(b)

**Figure 2-23**
(a) Piston-cylinder device; (b) external forces which counterbalance the system pressure on a piston.

**Note that the presence of friction always degrades performance.**

---

**A** gas within a piston-cylinder device is compressed at a constant pressure of 0.50 MPa from 1000 to 400 cm³. The frictional force at the piston-cylinder interface is 200 N, the piston surface area is 100 cm², and the atmospheric pressure is 0.10 MPa. Determine the work transferred by the piston to the gas and the work supplied by the connecting rod, in newton-meters.

**Example 2-10**

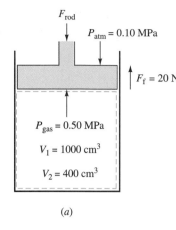

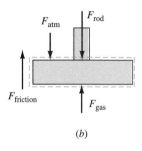

**Figure 2-24**
Schematic and data for Example 2-10.

**Given:**    A gas is compressed in a piston-cylinder device, with friction occurring at the piston-cylinder interface. The systems of interest and known data are shown in Fig. 2-24a.

**Find:**    $W_{comp/exp}$ and $W_{rod}$, in N·m.

**Model:**    Quasiequilibrium compression; frictional force and atmospheric pressure are constant.

**Strategy:**    Identify the system and evaluate the work for the system.

**Analysis:**    To evaluate the work transfer by the piston on the gas, the *gas* is chosen as the system. The only form of work transfer is compression-expansion work. Assuming a quasiequilibrium compression process,

$$W_{comp/exp} = -\int_1^2 P\,dV = -P(V_2 - V_1) = -0.50 \text{ MPa} \times (400 - 1000) \text{ cm}^2$$

$$= 300 \text{ MPa·cm}^3 \times \frac{10^6 \text{ N·m}}{1 \text{ MPa}} \times \frac{1 \text{ m}^3}{10^6 \text{ cm}^3} = 300 \text{ N·m}$$

Since the value is positive, the work transfer is into the system.

To evaluate the work associated with the connecting rod, the system is the *piston* shown by the dashed line in the second figure. From mechanics, Newton's second law requires that $m\,d\vec{V}/dt = \sum \vec{F}$. If the left-hand side, commonly called the inertial force, is assumed to be negligible, then, as shown in Fig. 2-24b,

$$F_{gas} + F_{fric} = F_{atm} + F_{rod}$$

Upon rearrangement,

$$F_{rod} = F_{gas} - F_{atm} + F_{fric}$$
$$= PA - P_{atm}A + F_{fric} = (P - P_{atm})A + F_{fric}$$

The three forces on the right side are all constant, and the piston moves a distance $\Delta s = -\Delta V/A_{pist}$. Therefore, multiplying each term in the force balance by the distance $\Delta s$ and integrating leads to

$$W_{rod} = (P_{atm} - P)\,\Delta V_{gas} + F_{fric}\,\Delta s$$

Substitution of values yields the external work provided by the rod.

$$W_{rod} = (0.10 - 0.50)(400 - 1000) \text{ MPa·cm}^3 \times \frac{10^6 \text{ N}}{\text{MPa·m}^2} \times \frac{1 \text{ m}^3}{10^6 \text{ cm}^3}$$

$$+ \frac{200(1000 - 400) \text{ N·cm}^3}{100 \text{ cm}^2} \times \frac{1 \text{ m}}{100 \text{ cm}}$$

$$= (-60 + 300 + 12) \text{ N·m} = 252 \text{ N·m}$$

The work transfer to the gas is 300 N·m, the work transfer from the atmosphere is $-60$ N·m, and the work against friction is 12 N·m.

**Comments:** (1) The work transfer of 252 N·m is supplied by the piston rod and is a positive number because it is a compression process.

(2) Without any friction, only 240 N·m of piston work would be required to compress the gas. Note how the work of the atmosphere helps reduce the work required from any external source.

## 2-7 ELASTIC SPRING WORK

To change the length of a spring in tension or compression, it is necessary to exert a force $F$ that brings about a displacement $x$. Figure 2-25a shows a material under tension that has undergone a **displacement** $x = L - L_0$ due to a force $F$. The force on an elastic material is related linearly to the displacement of the material by the Hooke's law relation

$$F_{ext} = k_s x = k_s(L - L_0) \qquad \textbf{[2-49]}$$

where $k_s$ is a **spring constant,** $L_0$ is the *unstressed* length of the material, and the displacement $x$ is the change in the length from the unstressed length $L_0$ to the actual length $L$. Figure 2-25b shows the linear force-displacement relationship for two springs, marked $A$ and $B$, which arbitrarily have the same value of $L_0$. Since the slopes of the two lines are different, the two springs have different values for $k_s$. Note that $F$ has a negative value when the actual length is less than the unstressed length. The **spring work** $W_{spring}$ associated with stretching or compressing a spring then becomes, in terms of Eq. [2-1],

$$W_{spring} = \int \vec{F}_{ext} \cdot d\vec{s} = \int F\,dx = \int_1^2 k_s(L - L_0)\,d(L - L_0)$$

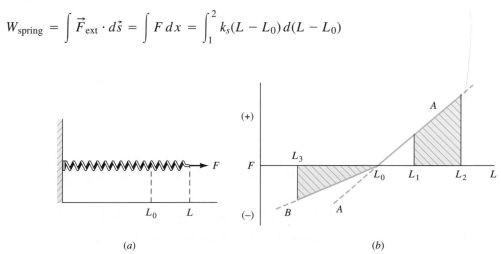

(a)                                                                 (b)

**Figure 2-25**     (a) A spring extended under tension from $L_0$ to $L$; (b) the force-displacement relationship for two elastic materials.

Integrating,

$$W_{\text{spring}} = \frac{k_s}{2}[(L_2 - L_0)^2 - (L_1 - L_0)^2] = \frac{k_s}{2}(x_2^2 - x_1^2) \quad \text{[2-50]}$$

where $x_2 = L_2 - L_0$ and $x_1 = L_1 - L_0$. The work given by Eq. [2-50] is represented by an area in Fig. 2-25b. For material A in the figure, a stretching occurs from length $L_1$ to length $L_2$. The crosshatched area beneath line A is a measure of the work done by some external source on the material. Similarly, material B is compressed from its unstressed length $L_0$ to a final, smaller length $L_3$. The work input to the system in this case is represented by the crosshatched area beneath line B on Fig. 2-25b. For both cases the numerical solution to Eq. [2-50] would yield a positive answer, indicating that work input is required. However, if material A goes from length $L_2$ to $L_1$ or material B increases in length from $L_3$ to $L_0$, work is transferred out of the system and Eq. [2-50] will indicate that the work transfer of energy in both cases is out of the system.

**Note that the sign on *F* depends on whether *L* < *L*₀ or *L* > *L*₀.**

**Example 2-11**

**A**n elastic spring is displaced from a compressed state where $F_1 = -100$ N and $L_1 = 0.40$ m to a final state where $F_2 = -500$ N and $L_2 = 0.20$ m. Determine (*a*) the unstressed length, in meters, (*b*) the spring constant, in N/m, and (*c*) the work for the process, in newton-meters.

**Solution:**

**Given:** The initial and final forces and lengths are known for a compressed elastic spring, as shown in Fig. 2-26.

**Find:** (*a*) $L_0$, in m, (*b*) $k_s$, in N/m, and (*c*) $W$, in N·m.

**Model:** Hooke's law is valid; forces are compressive.

**Strategy:** Use Hooke's law to determine $k_s$ and $L_0$. Then evaluate the spring work.

**Analysis:** (*a*) For an elastic material the force-displacement relationship is given by Eq. [2-50], $F = k_s(L - L_0)$. Using the two sets of data for the system, we find that

$$-100 \text{ N} = k_s(0.40 \text{ m} - L_0) \quad \text{and} \quad -500 \text{ N} = k_s(0.20 \text{ m} - L_0)$$

If the first equation is divided by the second equation, then

$$0.20 = \frac{0.40 \text{ m} - L_0}{0.20 \text{ m} - L_0} \quad \text{or} \quad L_0 = 0.45 \text{ m}$$

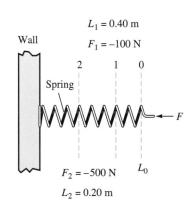

**Figure 2-26**
Schematic and data for Example 2-11.

(b) When the value of $L_0$ is substituted into the first force-displacement equation, then

$$k_s = \frac{-100 \text{ N}}{(0.40 - 0.45) \text{ m}} = 2000 \text{ N/m}$$

(c) The work of compression is given by Eq. [2-50]. Substitution of values yields

$$W_{\text{spring}} = \frac{k_s}{2}[(L_2 - L_0)^2 - (L_1 - L_0)^2]$$

$$= \frac{2000 \text{ N/m}}{2}[(0.20 - 0.45)^2 - (0.40 - 0.45)^2] \text{ m}^2 = 60 \text{ N·m}$$

Thus the work input required is 60 J.

**Comment:** The fact that $F$ is negative indicates that $L_0$ is greater than $L_1$ and $L_2$. Therefore, the spring is in a compressed state throughout. Since the spring length is decreasing, i.e., $L_2 < L_1$, the work transfer of energy is into the system, and hence positive in value.

## 2-8 OTHER QUASIEQUILIBRIUM WORK INTERACTIONS

There are a number of other quasiequilibrium forms of work. A brief summary is presented in this section.

### 2-8-1 EXTENSION OF A SOLID BAR

In addition to springs and wires, other configurations of solid materials can be stretched or compressed. It is often convenient to express the work done on these solids in terms of the stress $\sigma$ and the strain $\varepsilon$ associated with the process. The stress is given by $\sigma = F/A_0$, where $A_0$ is the cross-sectional area of the unstressed material. The strain $\varepsilon$ is given by $d\varepsilon = dx/L_0$, where $L_0$ is the unstressed length. Consider the solid bar under tension shown in Fig. 2-27. For a differential change in length the work $\delta W_{\text{elas}}$ becomes

$$\delta W_{\text{elas}} = F \, dx = (\sigma A_0)L_0 \, d\varepsilon = V_0 \sigma \, d\varepsilon \qquad \textbf{[2-51]}$$

where the volume $V_0 = A_0 L_0$ from geometric considerations. This equation for an elastic bar is the equivalent to $\delta W = -P \, dV$ for a gas.

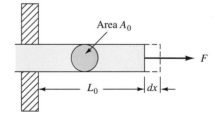

**Figure 2-27**
Illustration of elastic work on a bar.

### 2-8-2 TORSIONAL WORK

The concept of shaft work is evaluated in Sec. 2-1-4 in terms of an external torque $\tau$ transmitted by a rotating shaft. Shaft work is usually considered to be a nonquasistatic work mode because the rotation of the shaft and the external torque cannot be directly related to system properties. This is the case when the shaft is connected to a paddle wheel within a system. However, under other conditions the work done by twisting a shaft could be modeled as a quasiequilibrium process. Consider a system consisting of a cylindrical bar that is fastened at one end and subjected to a twisting torque at the other end. In this new situation the torsion work done by the external torque $\tau$ acting through a differential angular displacement $d\theta$ is given by

$$\delta W_{\text{torsion}} = \tau \, d\theta \qquad \textbf{[2-52]}$$

This equation is also valid for $W_{\text{shaft}}$, but the circumstances are different. A key point in deciding if a work interaction can be modeled as quasiequilibrium or not is whether the applied external "force" can be directly related to an intensive property of the system. In the case of torsion work, the applied torque can be directly related to the torsional shear stress (an intensive property) in the bar, and the angular displacement is an extensive property of the system. *So it is the system and not the work mode that determines if a quasistatic process is possible.*

### 2-8-3   WORK ON AN ELECTRICAL CHARGE

The work required to displace an electric charge across the boundary of a system in the presence of an electrical potential is introduced in Sec. 2-1-5. The energy input to a resistor, for example, is found to be given by $\delta W = V \, \delta Q_c$. Like paddle-wheel work, electrical resistor work is not a quasiequilibrium form. However, in the analysis of chemical cells, batteries, and capacitors, the potential difference is an intensive property of the system. Under these conditions, the equilibrium electric work done on the system is

$$\delta W_{\text{elec}} = \xi \, dQ_c \qquad \textbf{[2-53]}$$

where $dQ_c$ is the electrical charge transported through the electrical potential $\xi$. In electrochemical cells $\xi$ is called the electromotive force (emf), which is the maximum potential of the cell. The quantity $Q_c$ is an extensive property of the system. A schematic of a chemical (electrolytic) cell is shown in Fig. 2-28.

### 2-8-4   WORK IN CHANGING A SURFACE AREA

Work is required in the process of stretching a rubber or plastic sheet of material, of inflating a balloon, and of forming vapor bubbles or liquid droplets. All these processes require a change in the surface area of the material. The stretching of a thin film is shown in Fig. 2-29. An intrinsic property called the *surface tension* $\gamma$ is a measure of the force per unit length required to maintain a surface at a specified area. The change in surface energy for a differential area change is a measure of the *surface* work required, and it is

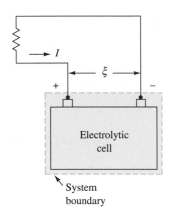

**Figure 2-28**
Illustration of electrochemical work.

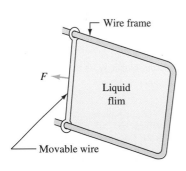

**Figure 2-29**
Illustration of surface work on a film.

given by

$$\delta W_{\text{surf}} = \gamma\, dA_{\text{surface}} \qquad\qquad \textbf{[2-54]}$$

The surface tension, given in dimensions of either force per unit length or energy per length squared, is an intensive property of the system. Typical $\gamma$ values for bubbles and droplets are around $5 \times 10^{-2}$ N/m, or $3 \times 10^{-3}$ lb$_{\text{f}}$/ft.

## 2-8-5   WORK OF POLARIZATION AND MAGNETIZATION

Work is done on a substance contained within an electric or magnetic field when the field is altered. For a dielectric material that lies within a uniform electric field, the work supplied externally to increase the polarization of the dielectric is given by

$$\delta W_{\text{polar}} = V\vec{E} \cdot d\vec{P} \qquad\qquad \textbf{[2-55]}$$

where $V$ is the volume, $\vec{E}$ is the electric field strength, and $\vec{P}$ is the polarization, or electric dipole moment per unit volume, of the dielectric. A similar equation for the work done in increasing the magnetization of a substance due to a change in the uniform magnetic field is expressed by

$$\delta W_{\text{magnet}} = V\mu_0\vec{H} \cdot d\vec{M} \qquad\qquad \textbf{[2-56]}$$

where $\vec{H}$ is the magnetic field strength, $\vec{M}$ is the magnetization per unit volume, $\mu_0$ is the permeability of free space, and $V$ is the volume.

## 2-8-6   SUMMARY OF GENERALIZED QUASIEQUILIBRIUM WORK INTERACTIONS

The equation for mechanical work, as given by Eq. [2-1], has the form

$$\delta W = \vec{F}_{\text{ext}} \cdot d\vec{s} \qquad\qquad \textbf{[2-1]}$$

All the quasiequilibrium work interactions discussed previously in this chapter can be expressed similarly as a product of a *generalized force* $F_{k,\text{eq}}$ and a change in the *generalized displacement* $dX_k$, even though the factors within a given work expression may not bring to mind physical forces and displacements. The subscript $k$ simply refers to the $k$th type of work mode. As a result, the *general formula for any quasiequilibrium work* interaction is given by

$$\delta W_k = \vec{F}_k \cdot d\vec{X}_k \qquad\qquad \textbf{[2-57]}$$

The factor $\vec{F}_{k,\text{eq}}$ is an intensive property, while $\vec{X}_k$ is an extensive property of the system.

**Note that all quasiequilibrium work interactions have the same mathematical form.**

Table 2-1 is a summary of the quasiequilibrium work interactions previously introduced, in terms of the intensive and extensive factors as well as the overall differential expressions. The similarity among these expressions

**Table 2-1**    Examples of quasiequilibrium work interactions

| Type of Work | Generalized Force $F_k$ | Generalized Displacement $X_k$ | Work Equations |
|---|:---:|:---:|:---:|
| Compression/expansion | $P$ | $-V$ | $-P\,dV$ |
| Spring | $k_s x$ | $x(=L-L_0)$ | $k_s x\,dx$ |
| Solid deformation | $\sigma$ | $\varepsilon$ | $V_0\sigma\,d\varepsilon$ |
| Capacitor or chemical cell | $\xi$ | $Q_c$ | $\xi\,dQ_c$ |
| Surface | $\gamma$ | $A$ | $\gamma\,dA$ |
| Torsion | $\tau$ | $\theta$ | $\tau\,d\theta$ |
| Electric polarization | $\vec{E}$ | $\vec{P}$ | $V\vec{E}\cdot d\vec{P}$ |
| Magnetic polarization | $\vec{H}$ | $\vec{M}$ | $V\mu_0\vec{H}\cdot d\vec{M}$ |

is clearly shown. This table emphasizes the fact that most generalized forces do not have the dimension of force, nor do the generalized displacements have the dimension of length. Nevertheless, the product of the dimensions for any corresponding pair of generalized force and displacement has the dimensions of energy.

Quasiequilibrium work interactions have the following characteristics:

1. The value of $F_{k,\text{eq}}$ depends only on the state of the system and is independent of the directions of change of $X_k$. The change in $X_k$ can be either positive or negative.

2. Theoretically a system can be returned to its initial state after a quasistatic work interaction merely by reversing the direction of the original process. As a result, quasistatic work interactions are frequently referred to as *possibly reversible* work modes.

3. The value of $F_{k,\text{eq}}$ remains finite as $dX_k$ approaches zero. In each equilibrium state $F_k$ has a fixed and finite value.

Consider, for example, the compression/expansion work done by a piston-cylinder assembly. For a given initial equilibrium state with a pressure $P$, the magnitude of $P$ in the expression $-P\,dV$ is the same whether $V$ increases or decreases. After a differential increase in $V$, the system can be returned to its initial state by reversing the direction of the piston movement. Whether for a finite or differential change of state, the value of $P$ remains finite and well defined as long as the process is quasistatic.

There are important work interactions that are called nonquasistatic or nonquasiequilibrium work interactions. Nonquasistatic work interactions have a number of characteristics that differentiate them from quasiequilibrium interactions:

1. The force $F$ depends on the rate of change of state.

2. The work interaction is *unidirectional,* and the effect cannot be undone by attempting to reverse the original process. As a result, non-

quasistatic work interactions are frequently referred to as *irreversible* (nonreversible) work modes.

3. The force $F$ approaches zero as the rate of change of state approaches zero. Consequently, the work approaches zero for a differential change of state.

**Note the physical differences between quasiequilibrium (reversible) and nonquasistatic (nonreversible) work modes.**

Typical examples include paddle-wheel and electrical resistance work.

Many engineering systems are designed deliberately to take advantage of unidirectional work interactions. Mechanical mixers in sewage-disposal plants and in the chemical industry, for example, illustrate a practical use of a paddle-wheel type of work. Electric-resistance heaters in homes, business, and industry are in wide use. Both quasiequilibrium and nonquasiequilibrium work interactions can transfer energy across a system boundary. The distinction becomes important when we attempt to evaluate the work in terms of system properties. In the energy analyses that follow, both types of work interactions may play an important role.

## 2-9  SUMMARY

Mechanical work is defined as the scalar product of a force and a displacement, that is, $\delta W = \vec{F} \cdot d\vec{s}$. From this definition and Newton's second law the concepts of linear kinetic energy change and gravitational potential energy change are derived in terms of a work transfer. The power $\dot{W}$ is the time rate at which energy is transferred across the boundary of a system by work, and mechanical power is a scalar product of a force and its velocity at the boundary of a system. During an *adiabatic* process only work interactions occur. The concept of work transfer in conjunction with experimental evidence leads to the first law of thermodynamics:

*When a closed system is altered adiabatically, the total work associated with the change of state is the same for all possible processes between two equilibrium states.*

A major consequence of this law is the operational definition of the total energy change of a closed system, namely,

$$E_2 - E_1 = \Delta E \equiv W_{ad}$$

where $W_{ad}$ is the work done for an adiabatic process.

Energy can be transported across the boundaries of a closed system by two mechanisms—work transfer $W$ and heat transfer $Q$. Heat transfer is defined as the transfer of energy across the boundaries of a system due to a temperature difference. The heat-transfer rate is represented by $\dot{Q}$ and $\delta Q = \dot{Q} \, dt$. More generally, for nonadiabatic processes where heat transfer $Q$ occurs,

$$\Delta E = Q + W$$

For a cyclic process,

$$\oint dE = 0 \qquad \text{and} \qquad \oint \delta Q + \oint \delta W = 0$$

Another important energy relation for closed systems is

$$\frac{dE_{cm}}{dt} = \dot{Q}_{net} + \dot{W}_{net}$$

The total energy $E$ consists of both intrinsic and extrinsic forms. The major intrinsic form is the internal energy $U$. The major extrinsic forms of energy are the gravitational potential energy $mgz$ and the linear kinetic energy $mV^2/2$. Thus if these are the only important forms of energy for a problem, the total energy $E$ becomes

$$E = U + \frac{mV^2}{2} + mgz$$

Then the conservation of energy equation for a closed system becomes

$$\Delta E = \Delta U + m\Delta\left(\frac{V^2}{2}\right) + mg\,\Delta z = Q + W$$

For the instantaneous rate this equation becomes

$$\frac{dE}{dt} = \frac{d}{dt}\left(U + \frac{mV^2}{2} + mgz\right) = \dot{Q} + \dot{W}$$

A *quasiequilibrium* (or quasistatic) process is a series of equilibrium steps. An important form of quasiequilibrium work is compression/expansion work.

## PROBLEMS

2-1C. Under what condition would the work required to accelerate a system become a process or path function?

2-2C. You frequently read or hear the statement concerning "the heat stored in the substance." Why is this a thermodynamically incorrect statement?

2-3C. What are two different physical methods of designing an adiabatic system?

2-4C. Explain why work interactions are generally process or path functions, and why the work of frictionless acceleration is not a path function.

2-5C. Explain in terms of process lines on a $PV$ diagram why boundary work is a path function.

2-6C. Illustrate on a *PV* diagram the net expansion work if a frictional force and the atmospheric pressure are acting on the piston.

2-7C. Explain why the work input to a moving boundary will be less if carried out in quasiequilibrium.

## KINETIC ENERGY, POTENTIAL ENERGY, AND WORK

2-1. A 1-kg piece of lead initially moves horizontally with a velocity of 5 m/s where $g = 9.80$ m/s$^2$. Determine (*a*) the change in velocity for a change in kinetic energy of 10 N·m and (*b*) the change in elevation for a change in potential energy of 10 N·m.

2-2. A bike and its rider with a total mass of 100 kg initially move horizontally with a velocity of 50 m/s at an elevation of 600 m above the ground, where $g = 9.75$ m/s$^2$. Determine (*a*) the final velocity for a kinetic energy change of 500 kJ and (*b*) the final elevation if the potential energy decreases by 500 kJ.

2-3E. A 2-lb piece of iron moves with an initial velocity of 10 ft/s in a location where $g$ is standard gravity. Determine (*a*) the change in velocity for a change in kinetic energy of 10 ft·lb$_f$ and (*b*) the change in potential energy, in ft·lb$_f$, for a 10-ft increase in elevation.

2-4E. A 2000-lb$_m$ sports car initially moves horizontally with a velocity of 100 ft/s at an elevation of 2000 ft above the valley floor, where $g = 32.0$ ft/s$^2$. Determine (*a*) the final velocity for a kinetic energy increase of 180,000 ft·lb$_f$ and (*b*) the final elevation for a potential energy decrease of 180,000 ft·lb$_f$.

2-5. The acceleration of gravity as a function of elevation above sea level is given by $g = 9.807 - 3.32 \times 10^{-6}z$, where $g$ is in m/s$^2$ and $z$ is in meters. A satellite with a mass of 240 kg is boosted 400 km above the earth's surface. Find the work required, in kilojoules.

2-6. Two hundred kilojoules of work are used to accelerate a small rocket at rest to a velocity of 200 m/s.
(*a*) Determine the mass of the body, in kilograms.
(*b*) If an additional 80 kJ of work is done on the body, determine its new velocity, in m/s.

2-7E. Work in the amount of 160,000 ft·lb$_f$ is required to move a small rocket sled at rest to a velocity of 300 ft/s.
(*a*) Determine the mass of the sled, in lb$_m$.
(*b*) If an additional 60,000 ft·lb$_f$ of work is done on the sled, determine its new velocity, in ft/s.

2-8. A 10-kg metal object falls freely from a 100-m elevation with an initial downward velocity of 30 m/s. If $g = 9.75$ m/s$^2$, determine the speed of the object just before it hits the earth, if air resistance is ignored.

2-9.  A bullet with a mass of 30 g leaves a gun aimed vertically upward with a speed of 100 m/s from ground level. If air resistance is neglected and $g = 9.7$ m/s$^2$, find the elevation where the speed reaches zero, in meters.

2-10. A 10-kg piece of steel with an initial velocity of 90 m/s (*a*) is elevated 100 m and accelerated to 120 m/s and (*b*) is decelerated to 60 m/s and elevated 180 m. Determine the net work input or output in kilojoules for the given energy changes if $g = 9.70$ m/s$^2$.

2-11. Find the net work input or output in kilojoules for a 100-kg projectile at an elevation of 40 m and an initial velocity of 60 m/s which (*a*) is elevated to 90 m and decelerated to 20 m/s, and (*b*) is lowered to 10 m and accelerated to 80 m/s. The local $g$ is 9.80 m/s$^2$

2-12E. Find the net work input or output, in ft·lb$_f$, for a mass of 150 lb$_m$ at an elevation of 200 ft and an initial velocity of 150 ft/s which (*a*) is raised to 340 ft and decelerated to 60 ft/s, and (*b*) is lowered to 80 ft and accelerated to 220 ft/s. The local $g$ is 32.0 ft/s$^2$.


## Shaft and Electrical Work

2-13. A torque of 150 N·m is applied to a shaft rotating at 2000 rev/min. (*a*) Find the power transmitted in kilowatts. (*b*) An electrical potential of 115 V is impressed on a resistor such that a current of 9 A passes through the resistor for a period of 5 min. Find the electrical work, in kilojoules, and the instantaneous power, in kilowatts.

2-14. A drive shaft delivers 60 kW when the torque is 120 N·m. (*a*) Determine the shaft rotational speed, in rev/min. (*b*) A 12-V battery is used to pass a current of 4 A through an external resistance for a period of 15 s. Find the electrical work, in kilojoules, and the instantaneous power, in kilowatts.

2-15E. Paddle-wheel and electrical resistor work occur under the following circumstances. (*a*) A torque of 150 lb$_f$·ft is developed by a shaft rotating at 2000 rev/min. Determine the power transmitted, in horsepower. (*b*) A current of 8 A passes through a resistor for 4 min due to an impressed potential of 110 V. Find the electrical work, in Btu, and the instantaneous power, in kilowatts.

2-16E. A drive shaft delivers 40 hp when the torque is 120 lb$_f$·ft. (*a*) Determine the shaft rotational speed, in rev/min. (*b*) A 12-V battery is used to pass a current of 3.5 A through a resistor for 24 s. Find the electrical work, in Btu, and the instantaneous power, in watts.

2-17. A tank of liquid contains both a paddle wheel and an electrical resistor. The paddle wheel is driven by a torque of 9.0 N·m and the shaft speed is 200 rpm. Simultaneously, a current of 6.0 A is supplied from a 12.0-V source to the resistor. Find the total power input to the system, in watts.

2-18. A substance receives energy in the forms of paddle-wheel and electrical resistor work. A torque of 4.0 N·m is applied to the shaft for 300 revolutions. The electrical resistor is supplied for a time period $\Delta t$ with a current of 7.0 A from a 12.0-V source. If the total work added is 22.0 kJ, find the value of $\Delta t$, in minutes.

2-19. Paddle-wheel work is carried out by applying a torque of 7.5 N·m at a shaft speed of 200 rpm for 2 min. Electrical work also occurs due to a current $I$ supplied from a 6.0-V source for 4 min. If the total work is 26 kJ, determine the constant current supplied, in amperes.

2-20. An electric motor draws a current of 8 A from a 110-V source. The output shaft develops a constant torque of 9.4 N·m at a rotational speed of 800 rpm. Determine (a) the net power input to the motor, in kilowatts, and (b) the amount of energy transferred out of the motor by the shaft, in kW·h, during 1.5 h of operation.

2-21E. An electric motor draws a current of 7 A from a 120-V source. The output shaft develops a constant torque of 11 lbf·ft at a rotational speed of 500 rev/min. Determine (a) the net power input to the motor, in horsepower, and (b) the amount of energy transferred out of the motor by the shaft, in Btu, during 30 min of operation.

## GENERAL FIRST-LAW ANALYSIS

2-22. For each of the following cases of a process involving a closed system, fill in the missing data.

| | $Q$ | $W$ | $E_1$ | $E_2$ | $\Delta E$ | | $Q$ | $W$ | $E_1$ | $E_2$ | $\Delta E$ |
|---|---|---|---|---|---|---|---|---|---|---|---|
| (a) | 24 | −15 | | −8 | | (d) | 16 | | 27 | | 12 |
| (b) | −8 | | | 62 | −18 | (e) | −9 | 15 | 29 | | |
| (c) | | 17 | −14 | | 20 | (f) | | −10 | | 6 | −10 |

2-23. A closed system undergoes a cycle composed of processes a, b, and c. Data for the cycle are shown below. Find the missing data for the three processes.

| | $Q$ | $W$ | $\Delta E$ | $E_i$ | $E_f$ |
|---|---|---|---|---|---|
| (a) | 4 | −7 | | 3 | |
| (b) | 1 | | 6 | | |
| (c) | 0 | | | | |

2-24. A closed system undergoes a cycle composed of processes a, b, and c. Data for the cycle appear in the table below. Find the missing data.

|     | $Q$ | $W$ | $E_i$ | $E_f$ | $\Delta E$ |
|-----|-----|-----|-------|-------|------------|
| (a) | 7   | −4  |       | 6     |            |
| (b) |     | 8   |       |       | 3          |
| (c) | 4   |     |       |       |            |

2-25. A closed system undergoes a cycle composed of processes $a$, $b$, and $c$. Data for the processes are shown below. Find the missing data for the three processes.

|     | $Q$ | $W$ | $E_i$ | $E_f$ | $\Delta E$ |
|-----|-----|-----|-------|-------|------------|
| (a) | −3  |     | 4     |       | −2         |
| (b) | 4   |     |       | 5     |            |
| (c) |     | 6   |       |       |            |

2-26. A dc motor draws a current of 50 A at 24 V. The torque applied to the shaft is 6.8 N·m at 1500 rev/min. Determine the rate of heat transfer to or from the steady-state motor, in kJ/h.

2-27. A 12-V storage battery delivers a current of 10 A for 0.20 h. Find the heat transfer, in kilojoules, if the energy of the battery decreases by 94 kJ.

2-28. A 12-V battery is charged by supplying a current of 5 A for 40 min. During the charging period a heat loss of 27 kJ occurs from the battery. Find the change in energy stored in the battery, in kilojoules.

2-29. An experimental energy converter has a heat input of 75,000 kJ/h and a shaft power input of 3.0 kW. The converter produces an electric energy output of 2000 kJ. Calculate the change in energy of the converter, in kilojoules, over 4.0 min.

2-30. A dc motor draws a current of 60 A at 24 V. The rate of heat transfer from the steady-state motor is 390 kJ/h. Determine the torque produced at the output shaft, in N·m, for a shaft speed for 1200 rev/min.

2-31E. A 12-V storage battery delivers a current of 10 A for 0.22 h. Find the heat transfer, in Btu, if the energy of the battery decreases by 98 Btu.

2-32E. A 12-V battery is charged by supplying a current of 5 A for 40 min. During the charging period a heat loss of 26 Btu occurs from the battery. Find the change in energy stored in the battery, in Btu.

2-33E. An experimental energy converter has a heat input of 80,000 Btu/h and a shaft power input of 2.2 hp. The converter produces an electric power output of 18 kW. Calculate the change in energy of the converter, Btu, over a 4.0-min period.

2-34. Heat is removed from nitrogen gas within a rigid tank at a constant rate of 80 W. At the same time paddle-wheel work occurs at a rate given by $\dot{W} = 16t$, where $\dot{W}$ is in watts and $t$ is in minutes. Determine (a) the rate of change of the energy of the gas at $t = 10$ min, in watts, and (b) the net change in energy after 20 min, in kilojoules.

2-35. Paddle-wheel work is carried out at a constant rate of 200 W on a substance within a rigid tank. Simultaneously, heat is removed at a rate given by $\dot{Q} = -6t$, where $\dot{Q}$ is in watts and $t$ is in minutes. Find (a) the rate of change of the energy of the subtance after 12 min, in watts, and (b) the net change in energy after 25 min, in kilojoules.

2-36E. Heat is removed from argon gas in a rigid tank at a constant rate of 5 Btu/min. The only work interaction is electrical resistor work that occurs at a rate given by $\dot{W} = 900t$, where $\dot{W}$ is in ft·lb$_f$/min and $t$ is in minutes. Determine (a) the instantaneous rate of change of the energy of the gas at $t = 8$ min, in Btu/min, and (b) the net change in energy after 15 min, in Btu.

## EXPANSION AND COMPRESSION WORK

2-37. A piston-cylinder assembly contains a gas that undergoes a series of quasiequilibrium processes that make up a cycle. The processes are as follows: 1-2, adiabatic compression; 2-3, constant pressure; 3-4, adiabatic expansion; 4-1, constant volume. The data at the beginning and end of each process appear in Table P2-37. Sketch the cycle on $PV$ coordinates and determine the work and heat interactions, in kilojoules, for each of the four processes.

**Table P2-37**

| State | P, bars | V, cm³ | T, °C | U, kJ |
|-------|---------|--------|-------|-------|
| 1 | 0.95 | 5700 | 20 | 1.47 |
| 2 | 23.9 | 570 | 465 | 3.67 |
| 3 | 23.9 | 1710 | 1940 | 11.02 |
| 4 | 4.45 | 5700 | 1095 | 6.79 |

2-38. A piston-cylinder assembly contains a gas that undergoes a series of quasiequilibrium processes that make up a cycle. The processes are as follows: 1-2, constant pressure expansion; 2-3, adiabatic expansion; 3-4, constant volume; 4-1, adiabatic compression. The data at the beginning of each process appear in Table P2-38. Sketch the cycle on $PV$ coordinates and determine the work and heat interactions, in kilojoules, for each of the four processes.

**Table P2-38**

| State | P, kPa | V, cm³ | T, K | U, kJ |
|-------|--------|--------|------|-------|
| 1 | 950 | 125 | 650 | 0.305 |
| 2 | 950 | 250 | 1300 | 0.659 |
| 3 | 390 | 500 | 1060 | 0.522 |
| 4 | 110 | 500 | 300 | 0.137 |

2-39. A piston-cylinder assembly contains a gas that undergoes a series of quasiequilibrium processes that make up a cycle. The processes are as follows: 1-2, adiabatic compression; 2-3, constant pressure expansion; 3-4, adiabatic expansion; 4-1, constant volume. The data at the beginning of each process appear in Table P2-39. Sketch the cycle on $PV$ coordinates and determine the work and heat interactions, in kilojoules, for each of the four processes.

**Table P2-39**

| State | P, bars | V, liters | T, °C | U, kJ |
|-------|---------|-----------|-------|-------|
| 1 | 1.05 | 3.0 | 27 | 0.78 |
| 2 | 9.83 | 0.6 | 290 | 1.48 |
| 3 | 9.83 | 1.2 | 853 | 3.14 |
| 4 | 2.75 | 3.0 | 515 | 1.35 |

2-40E. A piston-cylinder assembly contains a gas that undergoes a series of quasiequilibrium processes that make up a cycle. The processes are as follows: 1-2, adiabatic compression; 2-3, constant pressure expansion; 3-4, adiabatic expansion; 4-1, constant volume. The data at the beginning of each process appear in Table P2-40. Sketch the cycle on $PV$ coordinates and determine the work and heat interactions, in Btu, for each of the four processes.

**Table P2-40**

| State | P, psia | V, ft³ | T, °R | U, Btu |
|-------|---------|--------|-------|--------|
| 1 | 16 | 0.100 | 540 | 0.736 |
| 2 | 140 | 0.025 | 1180 | 1.635 |
| 3 | 140 | 0.050 | 2360 | 3.540 |
| 4 | 58 | 0.100 | 1950 | 2.860 |

2-41. A piston-cylinder assembly is equipped with a paddle wheel driven by an external motor and is filled with 30 g of a gas. The walls of the cylinder are well insulated, and the friction between the piston and cylinder wall is negligible. Initially the gas is in state 1 (see table.) The paddle wheel is then operated, but the piston is allowed to move to keep the pressure constant. When the paddle wheel is stopped, the system is in state 2. Determine the work transfer, in joules, along the paddle-wheel shaft. See Table P2-41.

**Table P2-41**

| State | $P$, bars | $v$, cm³/g | $u$, J/g |
|-------|-----------|------------|----------|
| 1 | 15 | 7.11 | 22.75 |
| 2 | 15 | 19.16 | 97.63 |

2-42. A piston-cylinder device is maintained at a constant pressure of 7 bars and contains 1.4 kg of air. During a process the heat transfer out is 49 kJ, while the volume changes from 0.15 to 0.09 m³. Find the change in the internal energy of the gas, in kJ/kg.

2-43. A piston-cylinder device contains nitrogen initially at 6 bars, 177°C, and 0.05 m³. The gas undergoes a quasiequilibrium process according to the equation $PV^2$ = constant. The final pressure is 1.5 bars. Determine (a) the work done, in newton-meters, and (b) the change in internal energy, in kilojoules, if the heat input is 5.0 kJ.

2-44E. A piston-cylinder device contains helium that initially is at 100 psia and 350°F and occupies 1.0 ft³. The gas undergoes a quasiequilibrium process according to the equation $PV^2$ = constant. The final pressure is 25 psia. Determine (a) the work done, in ft·lb$_f$, and (b) the change in internal energy, in Btu, if the heat input is 5.0 Btu.

2-45. Oxygen is compressed quasistatically in a piston-cylinder device from an initial state of 0.5 MPa and 25 cm³ to a final state of 2.0 MPa and 5 cm³. The pressure-volume relation is expressed by $P = a+bV$, where $p$ is in megapascals and $V$ is in cubic centimeters.
(a) Determine the values and units of the constants $a$ and $b$.
(b) By integration, determine the magnitude and direction of the work involved, in joules.
(c) Plot the process on a $PV$ diagram, showing clearly the initial and final states.

2-46. Nitrogen expands in a frictionless piston-cylinder device from 0.10 to 0.30 m³. The process is described by $P = 7.4 - 40V + 60V^2$, where $P$ is in bars and $V$ is in cubic meters.

(a) Evaluate $P$ for volumes of 0.1, 0.2, and 0.3 m$^3$, and sketch the process on a $PV$ diagram.

(b) Determine the units on the quantities 40 and 60 in the equation.

(c) Determine the work done, in kilojoules.

2-47. Argon is confined in a frictionless piston-cylinder device surrounded by the atmosphere. Initially, the pressure of the gas is 800 kPa and the volume is 0.010 m$^3$. If the gas expands to a final volume of 0.020 m$^3$, calculate the work done, in newton-meters, along the shaft connected to the piston. The atmospheric pressure is 100 kPa. Assume the processes connecting the end states are of the following types: (a) The pressure is constant, (b) the product $PV$ is constant, and (c) the product $PV^2$ is constant. (d) Compare the processes by plotting the three paths on the same $PV$ diagram.

2-48. One-fifth kilogram of air is contained in a piston-cylinder device at initial conditions of 0.020 m$^3$ and 8 bars. The gas is allowed to expand to a final volume of 0.050 m$^3$. Find the amount of work done, in kJ/kg, for the following quasiequilibrium processes: (a) The pressure is constant, (b) the product $PV$ is constant, and (c) the product $PV^2$ is constant. (d) Compare the processes by plotting the three paths on the same $PV$ diagram.

2-49. One kilogram of a gas with a molar mass of 35 is compressed at a constant temperature of 77°C from a volume of 0.05 m$^3$ to a volume of 0.025 m$^3$. The $PvT$ relationship for the gas is given by $Pv = RT[1+(c/v^2)]$, where $c = 2.0$ m$^6$/kmol$^2$ and $R = 8.314$ kJ/kmol·K. Compute the work done on the gas, in newton-meters.

2-50E. Helium expands in a frictionless piston-cylinder device from 1.0 to 3.0 ft$^3$. The process is described by $P = 740 - 400V + 60V^2$, where $P$ is in psia and $V$ is in cubic feet.

(a) Evaluate $P$ for volumes of 1, 2, and 3 ft$^3$, and sketch the process on a $PV$ diagram.

(b) Determine the units on the quantities 400 and 60 in the equation.

(c) Determine the work done, in ft·lb$_f$.

2-51E. Oxygen is confined in a frictionless piston-cylinder device initially at 160 psia and a volume of 0.10 ft$^3$. If the gas expands to a final volume of 0.20 ft$^3$, calculate the work done, in ft·lb$_f$, along the shaft connected to the piston. The external atmospheric pressure is 1 atm. Assume the processes connecting the end states are of the following types: (a) The pressure is constant, (b) the product $PV$ is a constant, and (c) the product $PV^2$ is a constant. (d) Compare the processes by plotting the three paths on the same $PV$ diagram.

2-52E. One-fifth pound of air is contained in a piston-cylinder device at initial conditions of 0.20 ft$^3$ and 100 psia. The air is allowed to expand to a final volume of 0.50 ft$^3$. Determine the amount of work done by the air, in ft·lb$_f$, for the following quasiequilibrium processes: (a) The pressure is constant, (b) the product $PV$ is constant, and (c) the product $PV^2$ is a constant. (d) Compare the processes by plotting the three paths on the same $PV$ diagram.

2-53. The pressure is related to the volume by $P = a - bV^2$ during a quasiequilibrium piston-cylinder process, where $a = 4.0$ bars and $b = 450$ bars/m$^6$.
   (a) Derive a symbolic equation for $W$ in terms of the quantities $a$, $b$, $V_1$, and $V_2$.
   (b) Evaluate the work required, in N·m, to expand the gas from 0.060 to 0.080 m$^3$.
   (c) Determine the values of $P$, in bars, at 0.06, 0.07, and 0.08 m$^3$, and then sketch the path of the process on $PV$ coordinates.

2-54E. During a process the pressure within a piston-cylinder device varies with volume according to the relation $P = aV^{-3} + b$, where $a = 49.1$ lb$_f$·ft$^7$ and $b = 341$ lb$_f$/ft$^2$.
   (a) Derive a symbolic equation for the work $W$ in terms of the quantities $a$, $b$, $V_1$, and $V_2$.
   (b) Evaluate the work required, in ft·lb$_f$, to compress the gas from 0.30 to 0.20 ft$^3$.
   (c) Determine the values of the pressure $P$ in psia, at 0.20, 0.25, and 0.30 ft$^3$; then sketch the path of the process on $PV$ coordinates.

2-55. A piston-cylinder device contains 0.12 kg of carbon dioxide at 27°C, 1.0 bar, and 0.040 m$^3$. The gas is compressed isothermally to 0.020 m$^3$. The $PVT$ equation of state of the gas is given by $PV = mRT[1 + (a/V)]$, where $R = 0.140$ kJ/kg·K, $V$ is in m$^3$, and $a$ is a constant. Determine
   (a) The value of the constant $a$ in m$^3$.
   (b) By integration, the work done on the gas, in kilojoules.
   (c) Finally, sketch the process on a $PV$ diagram.

2-56E. A gas is compressed in a piston-cylinder device from 15 psia and 0.50 ft$^3$ to a final state of 60 psia. The process equation relating $P$ and $V$ is $P = aV^{-1} + b$, where $a = 25$ psia·ft$^3$, $P$ is in psia, and $V$ is in ft$^3$. Determine
   (a) The value of the constant $b$ in psia.
   (b) By integration, the work done on the gas, in ft·lb$_f$.
   (c) Finally, sketch the process path on $PV$ coordinates.

2-57. The following data are taken during the quasistatic compression of carbon monoxide in a piston-cylinder apparatus:

| $P$, bars | 0.96 | 1.47 | 2.18 | 2.94 | 3.60 |
|---|---|---|---|---|---|
| $v$, m$^3$/kg | 0.928 | 0.675 | 0.503 | 0.403 | 0.346 |

(a) Plot a $Pv$ diagram and graphically estimate the work required, in kJ/kg.

(b) Assume the process equation fulfills the polytropic relation $Pv^n = c$. Use the first and last sets of $Pv$ data to determine the values of the constants $n$ and $c$.

(c) Now use the polytropic relation to determine by numerical integration the work required, in kJ/kg, and compare to part $a$.

2-58. The following data are taken during the quasiequilibrium compression of argon in a piston-cylinder apparatus:

| $P$, bars | 2.0 | 2.5 | 3.0 | 3.5 | 4.0 | 4.5 | 5.0 |
|---|---|---|---|---|---|---|---|
| $V$, m$^3$ | 0.525 | 0.448 | 0.393 | 0.352 | 0.320 | 0.294 | 0.273 |

(a) Assume the process equation fulfills the polytropic relation $PV^n = c$. Use the first and last sets of $PV$ data to determine the values of the constants $n$ and $c$.

(b) Now use the polytropic relation to determine by numerical integration the work required, in kilojoules, and compare to part $a$.

(c) Plot a $PV$ diagram and graphically estimate the work required, in kilojoules.

2-59E. The following data are taken during the quasiequilibrium compression of carbon monoxide in a piston-cylinder apparatus:

| $P$, psia | 15.0 | 26.0 | 37.0 | 50.0 | 62.0 |
|---|---|---|---|---|---|
| $v$, ft$^3$/lb$_m$ | 13.80 | 9.13 | 7.00 | 5.58 | 4.75 |

(a) Plot a $Pv$ diagram and graphically estimate the work required, in ft·lb$_f$/lb$_m$.

(b) Assume the process equation fulfills the polytropic relation $Pv^n = c$. Use the first and last sets of $Pv$ data to determine the values of the constants $n$ and $c$.

(c) Now use the polytropic relation to determine by numerical integration the work required, in ft·lb$_f$/lb$_m$, and compare to part $a$.

2-60E. The following data are taken during the quasistatic compression of helium in a piston-cylinder apparatus:

| $P$, psia | 20.0 | 25.0 | 30.0 | 35.0 | 40.0 | 45.0 | 50.0 |
|---|---|---|---|---|---|---|---|
| $V$, ft$^3$ | 0.540 | 0.460 | 0.404 | 0.362 | 0.329 | 0.303 | 0.281 |

(a) Plot a $PV$ diagram and graphically estimate the work required, in ft·lb$_f$.

(b) Assume the process equation fulfills the polytropic relation $PV^n = c$. Use the first and last sets of $PV$ data to determine the values of the constants $n$ and $c$.

(c) Now use the polytropic relation to determine by numerical integration the work required, in ft·lb$_f$, and compare to part $a$.

2-61. A gas is compressed in a piston-cylinder device from 0.860 to 0.172 m$^3$. The variation of pressure with volume is given by $P = 0.945/V - 8.607 \times 10^{-2}/V^2$, where $P$ is in bars and $V$ is in m$^3$. (a) Calculate the shaft work required. (b) If the atmospheric pressure of 1 bar acts on the other side of the piston, find the shaft work required, in kilojoules.

2-62E. A gas is expanded in a piston-cylinder device from 1.5 to 15 ft$^3$. The process equation relating $P$ and $V$ is $P = 257/V - 33.7/V^2$, where $P$ is in lb$_f$/in$^2$ and $V$ is in ft$^3$. (a) Calculate the work done by the gas, in ft·lb$_f$. (b) If the atmospheric pressure of 14.7 psia acts on the other side of the piston, find the shaft work output, in ft·lb$_f$.

2-63. A vertical piston-cylinder assembly contains air that is compressed by a frictionless piston weighing 3000 N. During an interval of time, a paddle wheel within the cylinder does 6800 N·m of work on the gas. If the heat transfer out of the gas is 8.7 kJ and the change in the internal energy is $-1.0$ kJ, determine the distance the piston moves, in meters. The area of the piston is 50 cm$^2$, and the atmospheric pressure acting on the outside of the piston is 0.95 bar.

2-64. A vertical piston-cylinder assembly contains helium that is constrained by a frictionless piston with a mass of 150 kg. During a 3-min interval, a resistor within the cylinder receives a current of 8 A from an external battery at 6 V. If the heat transfer out of the gas is 5.80 kJ and the change in internal energy of the gas is 2.40 kJ, determine the distance the piston moves, in centimeters. The area of the piston is 30.0 cm$^2$, the atmospheric pressure acting on the outside of the piston is 960 mbar, and local gravity is 9.60 m/s$^2$.

2-65E. A vertical piston-cylinder assembly contains a gas that is compressed by a frictionless piston weighing 684 lb$_f$. During an interval of time, a paddle wheel within the cylinder does 5000 ft·lb$_f$ of work on the gas. If the heat transfer out of the gas is 8.3 kJ and the change in the internal energy is $-1.0$ Btu, determine the distance the piston moves, in feet. The area of the piston is 8.0 in$^2$, and the atmospheric pressure acting on the outside of the piston is 14.5 psia.

2-66E. A vertical piston-cylinder assembly contains argon that is constrained by a frictionless piston with a mass of 330 lb$_m$. During a 2-min interval, a resistor within the cylinder receives a current of

6 A from an external battery at 12 V. If the heat transfer out of the gas is 5.30 Btu and the change in internal energy of the gas is 2.50 Btu, determine the distance the piston moves, in inches. The area of the piston is 5.0 in$^2$, the atmospheric pressure acting on the outside of the piston is 14.4 psia, and local gravity is 31.0 ft/s$^2$.

2-67. A gas expands polytropically from 650 kPa and 0.020 m$^3$ to a final volume of 0.080 m$^3$. Calculate the work done, in kilojoules, for the case where $n = 1.3$.

2-68. A gas at 100 kPa and 0.80 m$^3$ (state 1) is compressed to one-fifth of its original volume (state 2) along a path given by $PV = $ constant. Heat is then added at constant pressure until the original volume is reached (state 3). Finally, the gas is cooled at constant volume back to state 1.
(a) Sketch the process on a $PV$ diagram.
(b) Calculate the net work for the cycle, in kilojoules.

2-69E. A gas at 75 psia and 0.20 ft$^3$ (state 1) is expanded to five times its original volume along the path $PV = $ constant. After reaching state 2, heat is then added at constant volume until the original pressure is attained (state 3). Finally, the gas is then cooled at constant pressure back to state 1.
(a) Sketch the cycle on a $PV$ diagram.
(b) Calculate the net work for the cycle, in ft·lb$_f$.

2-70. The air in a diesel engine is compressed according to the relation $PV^{1.3} = A$, where $A$ is a constant. At the start of compression the state is 100 kPa and 1300 cm$^3$, and at the end state the volume is 80 cm$^3$.
(a) Sketch the path of the process on a $PV$ plane.
(b) Compute the work required to compress the air in kilojoules, under frictionless conditions.
(c) Compute the work required if a frictional force of 160 N is present, the atmospheric pressure outside the device is 100 kPa, and the area of the piston is 80 cm$^2$.

2-71E. The air in a diesel engine is compressed according to the relation $PV^{1.3} = A$, where $A$ is a constant. At the start of compression the state is 14.5 psia and 80 in$^3$, and at the end state the volume is 5.0 in$^3$.
(a) Sketch the path of the process on a $PV$ plane.
(b) Compute the work required to compress the air, in ft·lb$_f$, under frictionless conditions.
(c) Compute the work required if a frictional force of 48 lb$_f$ is present, the atmospheric pressure outside the device is 14.6 psia, and the area of the piston is 15 in$^2$.

2-72. A piston-cylinder apparatus initially contains argon at a volume of 0.8610 m$^3$. During a quasistatic change of state to a volume of 0.04284 m$^3$, the process equation is $P = 0.8610/V - 1.8085 \times 10^{-2}/V^2$, where $P$ is in bars and $V$ is in cubic meters.

(a) Determine the units on the quantity 0.8610 in the equation.
(b) Sketch the process on a $PV$ diagram, roughly to scale.
(c) Calculate the magnitude, in kilojoules, of the work done on the gas.
(d) Calculate the value of the work required if a frictional force of 180 N is present, the atmospheric pressure outside the device is 1 bar, and the piston area is 100 cm$^2$.

2-73E. A piston-cylinder apparatus initially contains air at a volume of 0.15 ft$^3$. During a quasistatic change of state to a volume of 3.0 ft$^3$, the process equation is $P = 43.94/V - 0.0340/V^2$, where $P$ is in psia and $V$ is in cubic feet.
(a) Determine the units on the quantity 43.94 in the equation.
(b) Sketch the process on a $PV$ diagram, roughly to scale.
(c) Calculate the magnitude, in ft·lb$_f$, of the work done on the gas.
(d) Calculate the value of the work required if a frictional force of 40 lb$_f$ is present, the atmospheric pressure outside the device is 14.6 psia, and the piston area is 16 in$^2$.

2-74. One kilogram of a gas with a molar mass of 60 is compressed at a constant temperature of 27°C from a volume of 0.12 to 0.04 m$^3$. The $PvT$ relationship for the gas is given by $Pv = RT[1 + (b/v)]$, where $b$ is 0.012 m$^3$/kg and $R = 8.314$ kJ/kmol·K.
(a) Determine the quasistatic work done on the gas, in newton-meters.
(b) Find the work input needed if the friction force between piston and cylinder is 10,000 N, the piston moves 0.5 m, and the atmospheric pressure is 100 kPa.

2-75E. One-tenth of a pound of a gas with a molar mass of 60 is compressed at a constant temperature of 140°F from a volume of 0.20 to 0.10 ft$^3$. The $PvT$ relationship for the gas is given by $Pv = RT[1 + (b/v)]$, where $b$ is 0.20 ft$^3$/lb$_m$ and $R = 1545$ ft·lb$_f$/lbmol·°R.
(a) Determine the quasistatic work done on the gas, in ft·lb$_f$/lb$_m$.
(b) Find the work input needed, in ft·lb$_f$, if the friction force between piston and cylinder is 180 lb$_f$, the piston moves 0.5 ft, and the atmospheric pressure is 14.6 psia.

## ELASTIC WORK

2-76. An elastic linear spring with a spring constant of 1200 N/m is stretched from its unstressed length to a final length of 12 cm. If the work required is 5.88 J, determine (a) the initial length, in cm, and (b) the final force on the spring in newtons.

2-77. An elastic linear spring with a spring constant of 144 N/cm is compressed from an initial unconstrained length to a final length of 6 cm. If the work required on the spring is (a) 6.48 J and (b) 2.88 J, determine the initial length of the spring in centimeters.

2-78. An elastic linear spring with an unstressed length of 11 cm is compressed by a work input of (a) 20 J and (b) 4.0 J. If the spring constant $k_s$ is 80 N/cm, determine the final length of the spring in centimeters.

2-79. An elastic spring is compressed from its unstressed position to a length $L_1$ of 0.20 m by application of a force of $-100$ N. Later a net work input of 175 N·m is applied so that $L_2$ is greater than $L_0$. Determine (a) the value of $L_0$ and (b) the length $L_2$, both answers in meters, if the spring constant has a value of 1000 N/m.

2-80. A linear spring is extended to a length of 0.60 m by the application of a force of $+800$ N. When the spring is later compressed to a length of 0.20 m, which is less than $L_0$, the force on the system is $-200$ N. Find (a) the unstressed length $L_0$, (b) the spring constant $k_s$ in N/m, and (c) the work required to change the length from 0.60 to 0.20 m, in newton-meters.

2-81. A linear spring is compressed from its unstressed length $L_0$ to 0.40 m by a force of $-100$ N. Later it is maintained at a length of 0.70 m by a force of $+200$ N. Determine (a) the unstressed length $L_0$, in meters, (b) the spring constant $k_s$, in N/m, and (c) the work required to change the length from 0.40 to 0.70 m, in newton-meters.

2-82. Determine the work required, in newton-meters, to increase the length of an unstressed steel bar from 10.00 to 10.01 m if $E_T = 2.07 \times 10^7$ N/cm$^2$ and $A_0 = 0.30$ cm$^2$.

2-83. In Prob. 2-82 use the same values of $E_T$ and $A$, but stretch the 10.00-m wire until the force on the bar is (a) 8000 N and (b) 50,000 N. Find the work required, in newton-meters.

2-84E. An elastic linear spring with a spring constant of 72 lb$_f$/in is compressed from an initial unconstrained length to a final length of 3 in. If the work required on the spring is (a) 54 ft·lb$_f$ and (b) 81 ft·lb$_f$, determine in each case the initial length of the spring in inches.

2-85E. An elastic linear spring with an unstressed length of 8 in is compressed by a work input of (a) 28 ft·lb$_f$ and (b) 14.0 ft·lb$_f$. If the spring constant $k_s$ is 48 lb$_f$/in, determine in each case the final length of the spring in inches.

2-86E. An elastic spring is compressed from its unstressed position to a length $L_1$ of 24 in by application of a force of $-25$ lb$_f$. Later a net work input of 150 ft·lb$_f$ is applied so that $L_2$ is greater than $L_0$. Determine (a) the value of $L_0$ and (b) the length $L_2$, both answers in inches, if the spring constant has a value of 50 lb$_f$/ft.

2-87E. An unstressed linear spring is extended to a length of 20 in by the application of a force of 25 lb$_f$. When the spring is later compressed to a length of 8 in, which is less than $L_0$, the force on the system is 15 lb$_f$. Find (a) the unstressed length $L_0$, (b) the spring constant $k_s$ in lb$_f$/ft, and (c) the work required to change the length from 20 to 8 in, in ft·lb$_f$.

2-88. A linear spring is compressed from its unstressed length $L_0$ to 0.30 m by a force of 50 N. Later it is maintained at a length of 0.70 m, which is greater than $L_0$, by a force of 150 N. Determine (a) the unstressed length $L_0$, in meters, (b) the spring constant $k_s$, in N/m, and (c) the work required to change the length from 0.30 to 0.70 m, in newton-meters.

2-89E. Determine the work required, in ft·lb$_f$, to increase the length of an unstressed steel bar from 20.00 to 20.01 ft if $E_T = 3.0 \times 10^7$ lb$_f$/in$^2$ and $A_0 = 0.10$ in$^2$.

2-90. An insulated piston-cylinder assembly containing a fluid has a stirring device operated externally. The piston is frictionless, and the force holding it against the fluid is due to standard atmospheric pressure and a coil spring. The spring constant is 7200 N/m. The stirring device is turned 100 rev with an average torque of 0.68 N·m. As a result, the 0.10-m-diameter piston moves 0.10 m outward. Find the change in the internal energy of the fluid, in kilojoules, if the initial force on the spring is zero.

2-91E. An insulated piston-cylinder assembly containing a fluid has a stirring device operated externally. The piston is frictionless, and the force holding it against the fluid is due to standard atmospheric pressure and a coil spring. The spring constant is 500 lb$_f$/ft. The stirring device is turned 1000 rev with an average torque of 0.50 lb$_f$·ft. As a result, the 0.20-ft-diameter piston moves outward 2 ft. Find the change in the internal energy of the fluid, in Btu, if the initial force on the spring is zero.

2-92. The pressure of a gas within a piston-cylinder device is counterbalanced on the outside by atmospheric pressure of 100 kPa and an elastic spring, as shown in Fig. P2-92. The initial volume of the gas is 32.0 cm$^3$, the spring is initially uncompressed with a length of 6.0 cm, and the area of the weightless piston is 4.0 cm$^2$. A heat addition of 7.0 J causes the piston to rise 2.0 cm. If the spring constant is 10.0 N/cm, find
(a) The final absolute pressure of the gas, in kPa.
(b) The work done by the gas in the cylinder, in joules.
(c) The change in internal energy of the gas, in joules.

2-93E. The pressure of a gas within a piston-cylinder device is counterbalanced on the outside by atmospheric pressure of 14.7 psia and an elastic spring, as shown in Fig. P2-92. The initial volume of the gas is 8.0 in$^3$, the spring is initially uncompressed with a length of 4.0 in, and the area of the weightless piston is 2.0 in$^2$. A heat addition of 46.1 ft·lb$_f$ causes the piston to rise 1.0 in. If the spring constant is 12.0 lb$_f$/in, find
(a) The final pressure of the gas in psia.
(b) The work done by the gas in the cylinder in ft·lb$_f$.
(c) The change in internal energy of the gas, in ft·lb$_f$.

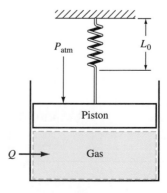

**Figure P2-92**

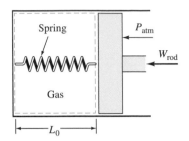

**Figure P2-94**

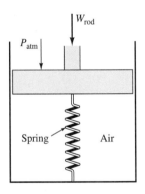

**Figure P2-96**

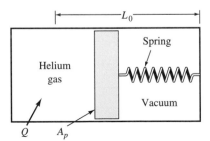

**Figure P2-98**

2-94. A system contains a gas in a piston-cylinder device as well as an elastic spring, as shown in Fig. P2-94. Initially the spring is unstressed and has a spring constant of $1.38 \times 10^7$ N/m, and atmospheric pressure is 0.1 MPa. The process equation for the gas is $PV = $ constant. The gas is compressed to one-half of its initial volume of 0.884 m³. If the initial length of the spring is 0.50 m find, in kilojoules,
(a) The work required to compress the gas alone.
(b) The work done on the spring.
(c) The work done by the atmosphere.
(d) The net connecting-rod work required.

2-95E. A system contains a gas in a piston-cylinder device as well as an elastic spring, as shown in Fig. P2-94. Initially the spring is unstressed and has a spring constant of 11,660 lbf/ft, and atmospheric pressure is 15 psia. The process equation for the gas is $PV = $ constant. The gas is compressed to one-half of its initial volume of 5.0 ft³. If the initial length of the spring is 2.0 ft, find, in ft·lbf,
(a) The work required to compress the gas alone.
(b) The work done on the spring.
(c) The work done by the atmosphere.
(d) The connecting-rod work required.

2-96. Air is contained in a piston-cylinder apparatus initially at 1 bar and 300 K. The cylinder diameter is 0.20 m, and the face of the piston initially is exactly 0.30 m from the base. In the initial position an elastic spring is just touching the piston face as shown in Fig. P2-96. The spring constant is 60 kN/m. The mass of the piston is 20 kg, and assume that the relation $PV = $ constant is valid during the process. Determine how much work, in kJ, is required from an outside source to compress the air within the cylinder to one-half of its initial volume, if the atmosphere outside the cylinder is at 0.1 MPa.

2-97E. Air is contained in a piston-cylinder apparatus initially at 15.0 psia and 80°F. The cylinder diameter is 6.0 in, and the face of the piston initially is exactly 12.0 in from the base. In the initial position an elastic spring is just touching the piston face as shown in Fig. P2-96, and the spring constant is 2400 lbf/ft. Neglect the mass of the piston and assume that the relation $PV = $ constant is valid during the process. Determine how much work, in ft·lbf, is required from an outside source to compress the air within the cylinder to one-half of its initial volume, if the atmosphere outside the cylinder is at 15 psia.

2-98. A piston of area $A_p = 0.02$ m² lies within a closed cylinder, as shown in Fig. P2-98. Helium gas fills one side, while the other side contains a spring within a vacuum. Heat is added slowly as the gas pressure changes from 0.1 to 0.3 MPa. The constant $k_s$ for the spring is $10^4$ N/m. Determine (a) the change in volume, in m³, based on a force balance on the piston, and (b) the work done by the gas, in kilojoules, based on the integral of $P \, dV$.

# CHAPTER
# 3

# PROPERTIES OF A PURE, SIMPLE COMPRESSIBLE SUBSTANCE

Liquified natural gas (LNG) import terminal in Lake Charles, LA.

Knowledge of the values of properties such as $P$, $v$, $T$, and $u$ is indispensable in the application of energy balances to engineering systems. A new property, the enthalpy $h$, is also of significant interest. The values of these properties are generally ascertained in two ways: (1) by experimental measurement coupled with theoretical evaluations, or (2) by use of models under certain limiting conditions. In this chapter, on the basis of method 1 above and the state postulate for simple, compressible substances introduced below, we shall examine the methods of data presentation for gases, liquids, and solids, as well as two-phase systems. We shall rely on tabular and graphical presentation of data in many cases, rather than on an algebraic format. Approximation techniques will also prove useful. Then these methods of data presentation for a pure substance will be used in conjunction with closed-system energy analyses. The use of models for property evaluation (method 2) is discussed in Chap. 4, also in conjunction with closed-system energy analysis.

## 3-1    THE STATE POSTULATE AND SIMPLE SYSTEMS

In Sec. 2-4 it is noted that the value of the internal energy $u$ is not directly measurable. For all properties not directly measurable, a method is needed to evaluate these in terms of directly measurable properties such as $P$, $T$, and $v$. Thus *an important objective of thermodynamics is to develop functional relationships among properties from theoretical and experimental considerations.* To develop these relations, the number of independent properties necessary to fix or specify the state of a system under given conditions must be known.

Functional relationships among properties usually are expressed in terms of *intensive* properties. In addition, because all *intrinsic* properties are characteristics of molecular behavior, it is reasonable to expect intrinsic properties to be functionally related. By functionally related, we mean the general situation where some dependent property $y_0$ is a function of $n$ other intensive, intrinsic properties. Mathematically, this means that $y_0 = f(y_1, y_2, \ldots, y_n)$. Once the values for all $n$ independent properties have been selected, the value of the dependent property $y_0$ is fixed.

On the basis of *experimental* evidence, there is a general rule for determining the number of independent properties $n$ for any system of known mass and composition. Known as the **state postulate,** it asserts

> *The number of independent properties required to completely specify the intensive, intrinsic state of a substance is equal to 1 plus the number of possibly relevant quasistatic work modes.*

"Relevant work modes" are those that have an appreciable or measurable effect on the state of the substance when a process occurs. Objects on earth, for example, are usually under the influence of the natural gravitational, electric, and magnetic force fields of earth. However, their effects on the outcome of most processes are often negligible. So we ignore their effect when determining the number of independent properties. Also, note that only quasiequilibrium work modes are considered.

When one examines past and present-day engineering and scientific studies, one thing becomes apparent. Seldom does one encounter a system where more than one quasiequilibrium work mode is used to alter the state of a substance. On the basis of this observation, it is convenient to classify various systems as simple systems. By definition, a **simple system** is one for which only one quasiequilibrium work mode is relevant to the intrinsic state of the system. On the basis of the state postulate asserted above, the following statement may be made as a state postulate for simple systems.

> *The equilibrium state of a simple substance is fixed by specifying the values of two independent, intrinsic properties.*

Once the two independent properties are fixed in value, the values of all remaining intrinsic properties are also fixed. Thus only two intrinsic properties may be independently varied for any simple substance.

**How does a simple system depend upon the type and number of properties?**

A *simple compressible system* is defined as one for which the only relevant quasistatic work interaction is that associated with volume changes ($P \, dV$ work). For such a substance the effects of capillarity, elasticity, and external force fields are neglected. From a practical viewpoint this means that the system is not influenced by these effects even though they may be present to some small degree. Because of their importance in engineering studies, property relations for simple compressible substances are stressed in this and the following two chapters.

Any two independent, intrinsic properties are sufficient to fix the intensive state of a simple compressible substance. In this case the functional relationship between a set of intrinsic, intensive properties is given by

$$y_0 = f(y_1, y_2)$$

where ($y_1, y_2$) represents a set of two independent variables. This equation expresses some, as yet unknown, relation between two independent, intrinsic, intensive properties and a third dependent property. Various techniques—analytical, graphical, and tabular—for relating and evaluating intrinsic properties of simple compressible substances may be used.

## 3-2 THE $PvT$ SURFACE

In Chap. 2 the state postulate indicated that any intensive, intrinsic property of a simple compressible substance is solely a function of two other independent intrinsic properties. That is, $y_1 = f(y_2, y_3)$, where $y$, in general, is any intensive, intrinsic property. We now wish to examine the relationships among properties for gases, liquids, solids, and two-phase systems of a pure substance. Experimental data reveal a consistent pattern in the behavior of simple compressible substances in the solid, liquid, and gas phases. It is on this consistency that we wish to focus.

The equilibrium states of any simple, compressible substance can be represented as a surface in a rectangular, three-dimensional space. The coordinates are intrinsic properties of interest, for example, $P$, $v$, and $T$. The $PvT$ surface is important because it clearly exhibits the basic structure of matter in a general fashion. More generally, the $PvT$ diagram for a substance which contracts on freezing is shown in Fig. 3-1. The **solid, liquid,** and **gas** (vapor) phases appear as regions on the surface. The state for these single-phase regions is specified by the values of *any two* of the three properties $P$, $v$, and $T$. These single-phase regions on the surface are separated by **two-phase** regions which represent phase changes. Examples of phase changes include melting (or freezing), vaporization (or condensation), and sublimation. (The latter is the transformation of a solid directly to a gas.) The regions for these solid-liquid, liquid-vapor, and solid-vapor mixtures are also shown in Fig. 3-1.

Any state represented by a point lying on a line separating a single-phase region from a two-phase region in Fig. 3-1 is known as a **saturation**

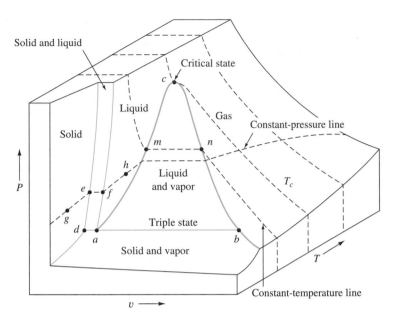

**Figure 3-1**     The *PvT* surface for a substance which contracts on freezing (not to scale).

**Note the significance of saturated-liquid and saturated-vapor states.**

*state.* The curved line that separates the liquid region from the liquid-vapor region, line *a-m-c,* is referred to as the ***liquid-saturation line.*** Any state represented by a point on this line between *a* and *c* is known as a ***saturated-liquid*** state. Similarly, the states represented on the curve *c-n-b* are ***saturated-vapor*** states. A phase change occurs without a change in pressure or temperature. This is shown by the process of vaporization along path *m-n* in Fig. 3-1. Thus, within any two-phase region of a pure substance, the pressure and temperature are not independent properties.

The point on Fig. 3-1 where the saturated-liquid and saturated-vapor lines meet is called the ***critical state.*** Associated with it are certain property values that are commonly signified by the subscript *c*. The three properties of present interest are denoted $P_c$, $v_c$, and $T_c$ at the critical state. A substance which is at a temperature higher than its critical temperature will not be capable of undergoing condensation to the liquid phase, no matter how high a pressure is exerted. Note further in Fig. 3-1 that the liquid and gas (vapor) phases merge in the region above the critical state. All known substances exhibit this behavior. The existence of the critical state demonstrates that the distinction between liquid and gas phases is far from clear-cut, if not impossible, in certain situations. When the pressure is greater than the critical pressure, the state is frequently referred to as a ***supercritical state.*** Many familiar substances have fairly high critical pressures, but critical temperatures which are below normal atmospheric conditions. A limited list of critical data for some common substances is reported in Tables A-2 and A-2E.

**Table 3-1** Triple-state data

| Substance | $T$, K | $P$, atm | $T$, °F |
|---|---|---|---|
| Helium 4 ($\lambda$ point) | 2.17 | 0.050 | −456 |
| Hydrogen, $H_2$ | 13.84 | 0.070 | −435 |
| Oxygen, $O_2$ | 54.36 | 0.0015 | −362 |
| Nitrogen, $N_2$ | 63.18 | 0.124 | −346 |
| Ammonia, $NH_3$ | 195.40 | 0.061 | −108 |
| Carbon dioxide, $CO_2$ | 216.55 | 5.10 | −70 |
| Water, $H_2O$ | 273.16 | 0.006 | 32 |

An additional unique state of matter may be noted from Fig. 3-1. This is represented by the line parallel to the $Pv$ plane marked as the **triple state.** As the term implies, in this state three phases coexist in equilibrium. The triple state in this figure is for equilibrium between a solid (state $d$), a liquid (state $a$), and a gas (state $b$). Recall that the triple state of water is used as the reference point for establishing the Kelvin temperature scale. The triple state of water is assigned a temperature value of 273.16 K. Some triple-state data appear in Table 3-1. Water is an anomalous substance in that it expands upon freezing. Thus the specific volume of the solid phase is greater than that of the liquid phase. The $PvT$ surface modified to take this feature into account is shown in Fig. 3-2.

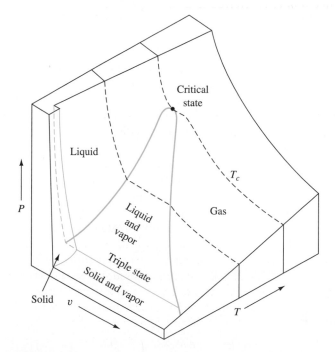

**Figure 3-2** The $PvT$ surface for a substance which expands on freezing.

**Note the difference between the critical state and the triple state.**

Three-dimensional diagrams for the equilibrium states of simple compressible systems are useful in introducing the general relationships between the three phases of matter normally under consideration. The relation of two-phase regions to the single-phase regions is clearly shown, as well as the significance of the critical and triple states of matter. Nevertheless, it is more convenient in the thermodynamic analysis to work with two-dimensional diagrams. All two-dimensional diagrams may be thought of simply as projections of a three-dimensional surface. For example, the surface presented in Fig. 3-1 may be projected on a $PT$, $Pv$, or $Tv$ plane.

## 3-3　THE PRESSURE-TEMPERATURE DIAGRAM

A projection of a $PvT$ surface upon the $PT$ plane is called a **phase diagram**. Since both the temperature and the pressure remain constant during a phase change, the surfaces in Fig. 3-1 (or Fig. 3-2) that represent two phases are parallel to the $v$ axis. Hence these two-phase regions appear as lines when the surface is projected onto the $PT$ plane. A pressure-temperature diagram based on the general characteristics of a substance which contracts on freezing (see Fig. 3-1) is presented in Fig. 3-3.

The line $t$-$d$-$f$-$c$ on the $PT$ diagram which represents the liquid-vapor surface is called the **liquid-vapor saturation line.** It is also known as the **vaporization curve.** The pressure and temperature at any state along this line (such as state $d$) are known as the **saturation pressure** $P_{\text{sat}}$ and the

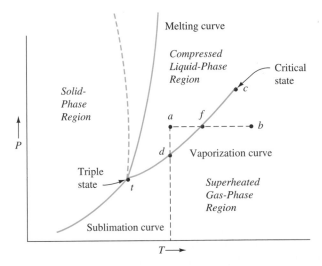

**Figure 3-3**　Phase ($PT$) diagram for a substance that contracts on freezing. (Dashed line is the melting or freezing curve for a substance which expands on freezing.)

*saturation temperature* $T_{sat}$ for that state. The saturation pressure is also known as the *vapor pressure.* The vaporization curve runs from the triple state to the critical state. These two states are designated by the points marked *t* and *c*, respectively, in Fig. 3-3. The liquid-vapor saturation line for water is shown in Fig. 3-4. The rapid rise of vapor pressure with increasing temperature should be noted, because this is a general characteristic of pure substances. The triple state is at 0.006 bar and 0.01°C (32.02°F) and the critical state is at 220.9 bars and 374.1°C (705°F). Recall from Sec. 1-6 that the triple state of water is used as a reference point for establishing the Kelvin temperature scale.

The solid-liquid and solid-vapor surfaces of a *PvT* diagram are shown as the *melting curve* (or freezing curve) and the *sublimation curve*, respectively, in Fig. 3-3. The dashed line in the figure represents the fusion (freezing) curve for a substance such as water which expands on freezing. All other specific properties, except the volume, decrease during the freezing of water. Note on the melting curve for ice an increase in pressure decreases the freezing-point temperature.

Whereas two-phase regions appear as lines on a *PT* diagram, single-phase regions appear as areas. The single-phase liquid and vapor regions in Fig. 3-3 are given special names. For state *a* in the liquid region, for example, the temperature $T_a$ is less than the saturation temperature $T_f$ at state *f* for the same value of pressure. Such a liquid state is called a *subcooled liquid,* since it is achieved by cooling the liquid below its saturation temperature at a given pressure. The pressure $P_a$ of state *a* also is above the saturation pressure $P_d$ at state *d* for the same temperature. Hence state *a* is also called a *compressed liquid,* since it may be achieved by compressing the liquid above its saturation pressure at a given temperature. Thus the terms "subcooled" and "compressed" are synonymous. In a similar fashion, the temperature $T_b$ at state *b* in Fig. 3-3 is greater than the saturation temperature $T_f$ for a given pressure. Hence state *b* is called a *superheated-vapor* state. The process of *superheating* is defined as one for which the temperature of a vapor (gas) is increased at constant pressure.

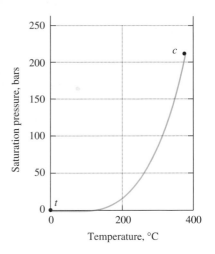

**Figure 3-4**
Liquid-vapor saturation line for water.

**Note the meaning of a subcooled- or compressed-liquid state and of a superheated-vapor state.**

## 3-4 THE PRESSURE–SPECIFIC VOLUME DIAGRAM

The projection of the *PvT* surface onto the *Pv* plane is shown in Fig. 3-5. This figure, like Fig. 3-1, is restricted to substances which contract on freezing. Both single- and two-phase regions appear as areas on this new diagram. The saturated-liquid line represents the states of the substance such that any further infinitesimal addition of energy to the substance at constant pressure will change a small fraction of the liquid to vapor. Similarly, removal of energy from the substance at any state which lies on the saturated-vapor line results in partial condensation of the vapor, whereas addition of energy makes the vapor superheated. The two-phase region

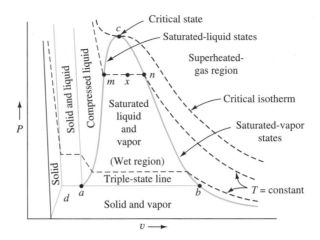

**Figure 3-5**     A *Pv* diagram for a substance which contracts on freezing.

labeled "saturated liquid and vapor," which lies between the saturated-liquid and saturated-vapor lines, is commonly called the ***wet region,*** or the ***wet dome.*** The state at the top of the wet region indicated by point *c* is again the critical state.

A state represented by a point in the liquid-vapor region (wet region), such as *x* in Fig. 3-5, is a mixture of saturated liquid and saturated vapor. The specific volumes of these two phases must be points on the saturation line, such as points *m* and *n* in the figure. Hence the specific volume of state *x* simply represents the *average* property value for the two phases in equilibrium. To find this average value of *v* (or any other specific property of a liquid-vapor mixture, such as *u* and *h*), we need to know the proportions of vapor and liquid in the saturated vapor–liquid mixture. To achieve this, we define the ***quality,*** usually represented by the symbol *x*, as the fraction of the total mixture which is vapor, based on mass (or weight). That is,

$$\text{Quality} = x \equiv \frac{m_{\text{vapor}}}{m_{\text{total}}} = \frac{m_g}{m_g + m_f} \qquad \textbf{[3-1]}$$

In this equation the subscript *g* applies to the saturated-vapor state, while the subscript *f* denotes the saturated-liquid state. These parameters are illustrated in Fig. 3-6, which shows a liquid-vapor mixture at $P_{\text{sat}}$ and $T_{\text{sat}}$. Quality is limited to values between zero and unity. However, it is common to speak of quality also as a percentage. In this case the value of *x* defined above is multiplied by 100. A saturated liquid alone has a quality of 0 percent ($x = 0.00$), and a saturated vapor alone has a quality of 100 percent ($x = 1.00$). The term "quality" applies solely to saturated liquid–vapor mixtures. Figure 3-7 is a *Pv* diagram for water to scale. The *liquid-vapor* saturation line, the critical isotherm of 374.15°C, and quality lines are all represented.

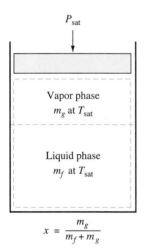

$$x = \frac{m_g}{m_f + m_g}$$

**Figure 3-6**
Illustration of the quality of a liquid-vapor mixture.

**Note the physical meaning of "quality" in terms of a substance in the wet region.**

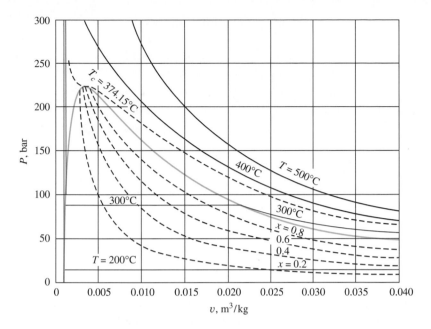

**Figure 3-7** A $Pv$ diagram for water showing temperature and quality lines.

One final comment is appropriate to the $Pv$ diagram shown in Figs. 3-5 and 3-7. It has been noted earlier that other families of curves representing various properties could be shown on any of the two-dimensional or three-dimensional diagrams. For this reason a family of constant-temperature lines is also shown on the $Pv$ diagram. It is important for the student to have some general concept of the position of constant-temperature lines on a $Pv$ diagram since this diagram is useful in problem analysis.

## 3-5 THE TEMPERATURE–SPECIFIC VOLUME DIAGRAM

For the purpose of problem solving it is often useful to describe the path of a process on a temperature–specific volume diagram. Projection of the compressed liquid, liquid-vapor, and superheated regions from either Fig. 3-1 or Fig. 3-2 leads to the $Tv$ diagram shown in Fig. 3-8. This diagram, although specifically for water, is representative of all pure substances. The single- and two-phase regions appear in positions similar to those on a $Pv$ diagram. The liquid-vapor region has been extended to include the triple-state line.

It is important to have pressure lines on a $Tv$ diagram for identifying states and sketching processes. Within the liquid-vapor (wet) region a pressure line follows a temperature line, since pressure and temperature are not

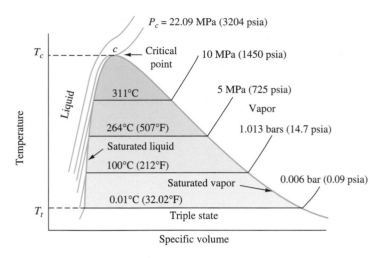

**Figure 3-8**    A *Tv* diagram for water.

independent properties during phase changes. Outside the wet region the temperature increases along a constant-pressure line as the specific volume increases. This is true whether the pressure is less than or greater than the critical pressure.

## 3-6    TABLES OF PROPERTIES OF PURE SUBSTANCES

The preceding sections offered an introduction to the *PvT* behavior of simple compressible, pure substances. In addition, the necessary nomenclature required for future discussions was presented. We now need to investigate the more established methods of assembling or storing data for the properties of primary interest: pressure *P*, specific volume *v*, temperature *T*, specific internal energy *u*, specific enthalpy *h*, and specific entropy *s*. (For the present discussion one is asked to accept the existence of the property entropy. It will not be used to any extent until Chap. 7.) For real gas and saturation states the mathematical relationships among these properties are complex. Hence the property data are presented primarily in the form of tables. As is customary in this text, the tabular data for a given substance are presented in both SI and USCS units. The tables in USCS (English) units use the same numbers as the corresponding tables in SI units, except that the numbers are followed by the letter E.

### 3-6-1    SUPERHEAT TABLES

In a single-phase region, such as the superheat region, two intensive properties are required to fix or identify the equilibrium state. Variables such as

$v$, $u$, $h$, and $s$ are usually tabulated in superheat tables as a function of $P$ and $T$, since the latter are conveniently measurable properties. The format of a superheat table is readily apparent if we again refer to a $PT$ diagram. Figure 3-9 shows the superheated-vapor region of a $PT$ diagram divided into a grid. The lines of the grid represent integer values of the pressure and temperature. A superheat table then reports values of $v$, $h$, and $s$ (and frequently $u$) at the grid points on the figure. The *enthalpy* function $H$ is defined by the relation

$$H \equiv U + PV \qquad \textbf{[3-2]}$$

The specific **enthalpy** $h$ is given by

$$\boxed{h \equiv u + Pv} \qquad \textbf{[3-3]}$$

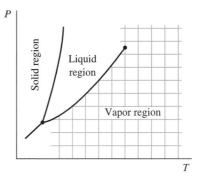

**Figure 3-9**
Sketch of a $PT$ diagram indicating method of tabulating superheat data.

Since $u$, $P$, and $v$ are properties, $h$ also must be a property. In many cases the enthalpy function has no specific interpretation, although it does have the dimension of energy. As a result, it is helpful to tabulate values of the enthalpy along with other properties. The specific entropy $s$ is a property used later in second-law analysis. The values of $u$, $h$, and $s$ are completely arbitrary, each being based on an assigned value at some reference state.

Table 3-2 illustrates one possible format for tabulation of data in the superheat region. A more complete compilation of superheated steam data in SI units is found in Table A-14; Table A-14E presents data in USCS units. The format of both tables is identical. The units for specific properties are usually listed at the top of a table, under the heading. With respect to Table 3-2 and Fig. 3-9, note that the data begin with *saturated-vapor* data and then proceed at a given pressure to higher integral temperatures. The temperature of the saturated-vapor data is indicated by the value in parentheses after the pressure value. For example, from Table 3-2, at 1.0 bar the saturation temperature is 99.63°C, and the specific volume is 1.694 m³/kg.

Many engineering problems involve states of matter that do not fall on the grid of data available for that substance. Interpolation of data then becomes necessary. The intervals for the matrix of data found in unabridged superheat tables are usually chosen so that *linear* interpolation leads to reasonable accuracy. For example, we need to find the specific volume of superheated steam (H$_2$O) at 1.0 bar and 160°C. Although Table 3-2 lists the value of $v$ as 1.984 m³/kg, assume that the data at 160°C are missing from the table. Linear interpolation between 120 and 200°C shows that

$$\frac{v_{160} - v_{120}}{v_{200} - v_{120}} = \frac{T_{160} - T_{120}}{T_{200} - T_{160}}$$

Substitution of tabular data into the equation yields

$$\frac{v_{160} - 1.793}{2.172 - 1.793} = \frac{160 - 120}{200 - 120} = 0.50$$

**Table 3-2**      Properties of superheated steam ($H_2O$)

| Temperature, °C | $v$ | $u$ | $h$ | $s$ |
|---|---|---|---|---|
| | SI: $v$, m³/kg; $u$, kJ/kg; $h$, kJ/kg; $s$, kJ/kg·K | | | |
| | 1.0 bar (0.10 MPa)($T_{\text{sat}}$ = 99.63°C) | | | |
| Sat. | 1.694 | 2506.1 | 2675.5 | 7.3694 |
| 100 | 1.696 | 2506.7 | 2676.2 | 7.3614 |
| 120 | 1.793 | 2537.3 | 2716.6 | 7.4668 |
| 160 | 1.984 | 2597.8 | 2796.2 | 7.6597 |
| 200 | 2.172 | 2658.1 | 2875.3 | 7.8343 |
| | 10.0 bars (1.0 MPa)($T_{\text{sat}}$ = 179.91°C) | | | |
| Sat. | 0.1944 | 2583.6 | 2778.1 | 6.5865 |
| 200 | 0.2060 | 2621.9 | 2827.9 | 6.6940 |
| 240 | 0.2275 | 2692.9 | 2920.4 | 6.8817 |
| 280 | 0.2480 | 2760.2 | 3008.2 | 7.0465 |
| 320 | 0.2678 | 2826.1 | 3093.9 | 7.1962 |

| SOURCE: Abstracted from Table A-14.

| Temperature, °F | $v$ | $u$ | $h$ | $s$ |
|---|---|---|---|---|
| | USCS: $v$, ft³/lb$_m$; $u$, Btu/lb$_m$; $h$, Btu/lb$_m$; $s$, Btu/lb$_m$·°R | | | |
| | 60 psia ($T_{\text{sat}}$ = 292.7°F) | | | |
| Sat. | 7.17 | 1098.3 | 1178.0 | 1.6444 |
| 300 | 7.26 | 1101.3 | 1181.9 | 1.6496 |
| 400 | 8.35 | 1140.8 | 1233.5 | 1.7134 |
| 500 | 9.40 | 1178.6 | 1283.0 | 1.7678 |
| 600 | 10.43 | 1216.3 | 1332.1 | 1.8165 |

| SOURCE: Abstracted from Table A-14E.

The solution to the above equation is

$$v_{160} = 1.793 + 0.50(0.379) = 1.983 \text{ m}^3/\text{kg}$$

Thus the estimated value is off by 0.001 m³/kg, or less than 0.1 percent from the tabulated value. The percentage error in estimating $u$ and $h$ at 160°C by this method is even smaller. Although the superheat tables in the Appendix are condensed versions of the original works, we still assume that linear interpolation is valid. Superheat tables for refrigerant 134a in SI and USCS units appear in Tables A-18 and A-18E. Tables for diatomic nitrogen ($N_2$) appear in Tables A-20 and A-20E and for potassium in Tables A-21 and A-21E.

**Note the use of linear interpolation to determine data not directly listed in a superheat table.**

Example 3-1

**D**etermine the internal energy of superheated water vapor at (*a*) 1.0 bar and 140°C and (*b*) 6 bars and 220°C in kJ/kg.

**Solution:**

**Given:** Superheated water vapor at the two states shown in Fig. 3-10, parts *a* and *b*.

**Find:** The internal energy, in kJ/kg.

**Model:** Superheat Table A-14 for steam.

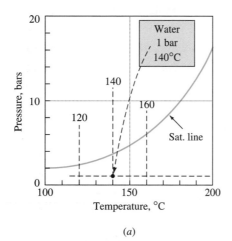

(*a*)

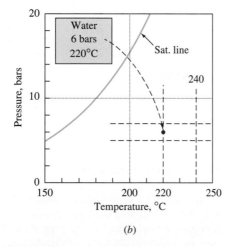

(*b*)

**Figure 3-10** A *PT* diagram showing two states in the superheat region.

**Strategy:**    Use linear interpolation to estimate data in table.

**Analysis:**    (*a*) The position of the state on a *PT* diagram relative to the saturation line is shown in Fig. 3-10*a*. The value of *u* at 1 bar and 140°C can be obtained
from Table 3-2 or the more complete data of Table A-14. With reference to either
of these tables, we find that at 1 bar

$$u = 2537.3 \text{ kJ/kg at } 120°C \qquad \text{and} \qquad u = 2597.8 \text{ kJ/kg at } 160°C$$

Linear interpolation halfway between these two values leads to an internal-energy
value of 2567.6 kJ/kg at 140°C. More complete tables list the value as 2567.7 kJ/kg.

(*b*) The position of the state on a *PT* diagram relative to the saturation line is
shown in Fig. 3-10*b*. To find the value of *u* at 6 bars and 220°C, we must turn to Table
A-14 in the Appendix. In this table we find data only at 5 and 7 bars and only at 200
and 240°C. To find the desired value of *u*, we must carry out double interpolation of
the given data. By linear interpolation first with respect to temperature,

$$\text{At 5 bars and } 220°C: \qquad u = 2675.2 \text{ kJ/kg}$$

$$\text{At 7 bars and } 220°C: \qquad u = 2668.3 \text{ kJ/kg}$$

Consequently, at 6 bars and 220°C the internal energy is the average of these two
latter values, or approximately 2671.8 kJ/kg. A more complete table lists the value
as 2672.1 kJ/kg.

---

**Example 3-2**

**D**etermine the pressure of superheated refrigerant 134a in bars at a state of 40°C
and an enthalpy of (*a*) 285.16 kJ/kg and (*b*) 285.52 kJ/kg.

**Solution:**

**Given:**    Refrigerant 134a at 40°C and (*a*) *h* = 285.16 kJ/kg and (*b*) *h* =
285.52 kJ/kg.

**Find:**    Pressure in bars.

**Model:**    Tabular superheated data.

**Strategy:**    Examine Table A-18 for the variation of the enthalpy value as a
function of pressure.

**Analysis:**    This problem illustrates the point that *any* two independent properties fix the state of a simple compressible substance. In this case we seek *P* as a
function of *T* and *h*.

(*a*) The superheat data for refrigerant 134a are tabulated in Table A-18, with
the temperature running from top to bottom at any pressure. Starting at 40°C and
the lowest pressure (0.6 bar), we note that the enthalpy starts at 288.35 kJ/kg and
decreases as the pressure increases for the same temperature. Continuing to read to
higher pressures at 40°C, we finally find that *h* = 285.16 kJ/kg at 2.4 bars.

(*b*) The value of 285.52 kJ/kg for the enthalpy lies between 2.0 and 2.4 bars at
40°C. Estimating by linear interpolation, we find that

$$P = 2.0 \text{ bars} + (2.4 - 2.0) \text{ bars} \times \frac{285.52 - 285.88}{285.16 - 285.88}$$

$$= (2.0 + 0.20) \text{ bars} = 2.20 \text{ bars}$$

Thus when $h = 285.52$ kJ/kg at 40°C, the pressure is close to 2.20 bars.

---

The preceding examples illustrated the use of the superheat tables in determining property values. In general, superheat tables in the literature are arranged in a similar fashion, although in some cases the positions of the temperature and pressure coordinates are reversed.

## 3-6-2 SATURATION TABLES

A saturation table lists specific property values (such as $v$, $u$, $h$, and $s$) for the saturated-liquid and saturated-vapor states. Recall from Sec. 3-3 that properties for these two phases are denoted by the subscripts $f$ and $g$, respectively. Thus any specific property $y$ in the saturated-liquid or saturated-vapor state is denoted $y_f$ or $y_g$, accordingly. For a unit mass of a mixture of the two phases, an average property value $y_x$ is determined by adding the contributions of the two phases. Recall [see Eq. [3-1]] that the quality $x$ is defined as the ratio of the mass of vapor to the total mass of the liquid plus vapor. Hence for each unit mass of mixture, the contribution by the vapor phase is $xy_g$, while that by the liquid phase is $(1 - x)y_f$. Consequently, the expression for any specific mixture property $y_x$ can be written as

$$y_x = (1 - x)y_f + xy_g \qquad \text{[3-4]}$$

as shown in Fig. 3-11. Alternatively, if we designate the difference between the saturated-vapor and saturated-liquid intensive properties by the symbol $y_{fg}$, that is,

$$y_{fg} \equiv y_g - y_f \qquad \text{[3-5]}$$

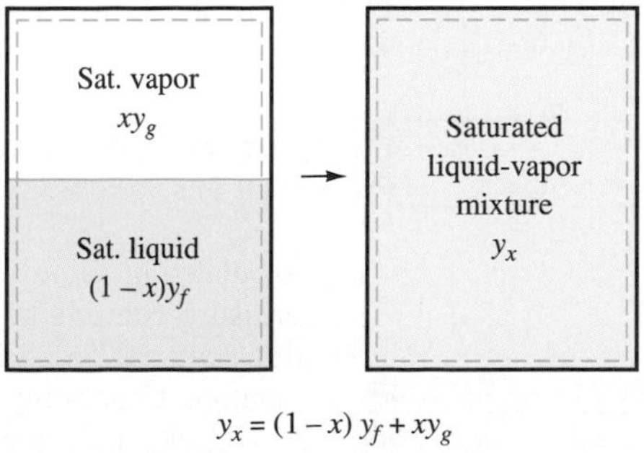

Sat. vapor
$xy_g$

Sat. liquid
$(1 - x)y_f$

Saturated
liquid-vapor
mixture
$y_x$

$$y_x = (1 - x)\, y_f + xy_g$$

**Figure 3-11**  Illustration for evaluating any specific property of a liquid-vapor mixture.

then on the basis of Eq. [3-4] we may also write

$$y_x = y_f + x(y_g - y_f) = y_f + xy_{fg} \qquad \text{[3-6]}$$

Equations [3-4] and [3-6] are equivalent, and either can be used to evaluate an "average" property value for a liquid-vapor mixture in equilibrium. For energy analysis the important formats of Eqs. [3-4] and [3-6] are

$$u_x = (1 - x)u_f + xu_g = u_f + xu_{fg} \qquad \text{[3-7a]}$$

$$h_x = (1 - x)h_f + xh_g = h_f + xh_{fg} \qquad \text{[3-7b]}$$

**Note the significance of the subscript *fg* on tabular saturation data.**

As noted in Eq. [3-5], the difference between saturated-vapor and saturated-liquid properties is symbolized by the subscript $fg$. The quantity $h_{fg}$ is called the **enthalpy of vaporization,** or the *latent heat of vaporization.* It represents the quantity of energy required to vaporize a unit mass of a saturated liquid at a given temperature or pressure. Hence it is an important thermodynamic property.

When Eq. [3-6] is applied to the specific volume, we may write that

$$v_x = (1 - x)\, v_f + xv_g = v_f + xv_{fg} \qquad \text{[3-8a]}$$

or upon rearrangement,

$$x = \frac{v_x - v_f}{v_{fg}} \qquad \text{[3-8b]}$$

This result has a clear interpretation on the $Pv$ diagram shown in Fig. 3-12. The numerator of Eq. [3-8b] is the horizontal distance between points $f$ and

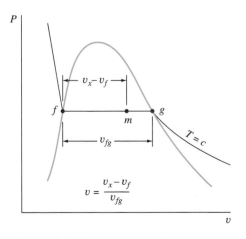

**Figure 3-12**    Interpretation of quality as a function of specific properties on a $Pv$ diagram.

$m$, while the denominator is the distance between points $f$ and $g$. Thus the quality is given by the ratio of horizontal distances within the wet region. A state of 50 percent quality, for example, lies midway between the saturated-liquid and saturated-vapor lines along any constant-pressure line.

The preceding equations which involve the quality of a liquid-vapor mixture also require property data for the saturation states of the two phases. Saturation data for $v$, $u$, $h$, and $s$ are tabulated against either the saturation temperature or the saturation pressure as the independent variable. In these tables the independent variable appears in integer values. Recall that fixing the saturation temperature automatically fixes the saturation pressure, and vice versa, for two-phase systems. In some instances two saturation tables may be available, one with temperature as the independent variable and the other with pressure as the independent variable. Although this may appear as a duplication of data, it is quite convenient when interpolation of either $T$ or $P$ is required.

**Note that saturation data typically are tabulated against integer values of either temperature or pressure.**

Table 3-3 is an abridgment of the properties of saturated water as a function of integral values of temperature, in both SI and USCS engineering units. More complete data for saturated water are given in Tables A-12

**Table 3-3**   Properties of saturated liquid and vapor for water

| SI: $v$, m³/kg; $u$, kJ/kg; $h$, kJ/kg; $s$, kJ/kg·K | | | | | | | |
|---|---|---|---|---|---|---|---|
| | | Specific volume | | Enthalpy | | Entropy | |
| Temperature $T$, °C | Pressure $P$, bars | Sat. liquid* $v_f \times 10^3$ | Sat. vapor $v_g$ | Sat. liquid $h_f$ | Sat. vapor $h_g$ | Sat. liquid $s_f$ | Sat. vapor $s_g$ |
| 20 | 0.02339 | 1.0018 | 57.791 | 83.96 | 2538.1 | 0.2966 | 8.6672 |
| 40 | 0.07384 | 1.0078 | 19.523 | 167.57 | 2574.3 | 0.5725 | 8.2570 |
| 60 | 0.1994 | 1.0172 | 7.671 | 251.13 | 2609.6 | 0.8312 | 7.9096 |
| 80 | 0.4739 | 1.0291 | 3.407 | 334.91 | 2643.7 | 1.0753 | 7.6122 |
| 100 | 1.014 | 1.0435 | 1.673 | 419.04 | 2676.1 | 1.3069 | 7.3549 |

SOURCE: Abstracted from Table A-12.

*At 20°C, for example, $v_f \times 10^3 = 1.0018$, or $v_f = 1.0018 \times 10^{-3}$.

| USCS: $v$, ft³/lb$_m$; $u$, Btu/lb$_m$; $h$, Btu/lb$_m$; $s$, Btu/lb$_m$·°R; $T$, °F; $P$, psia | | | | | | | |
|---|---|---|---|---|---|---|---|
| $T$ | $P$ | $v_f$ | $v_g$ | $h_f$ | $h_g$ | $s_f$ | $s_g$ |
| 60 | 0.2563 | 0.01604 | 1207 | 28.08 | 1087.7 | 0.0555 | 2.0943 |
| 70 | 0.3632 | 0.01605 | 867.7 | 38.09 | 1092.0 | 0.0746 | 2.0642 |
| 80 | 0.5073 | 0.01607 | 632.8 | 48.09 | 1096.4 | 0.0933 | 2.0356 |
| 90 | 0.6988 | 0.01610 | 467.7 | 58.07 | 1100.7 | 0.1117 | 2.0083 |
| 100 | 0.9503 | 0.01613 | 350.0 | 68.05 | 1105.0 | 0.1296 | 1.9822 |

SOURCE: Abstracted from Table A-12E.

and A-12E as a function of temperature and in Tables A-13 and A-13E as a function of pressure. Saturation tables for refrigerant 134a appear as Tables A-16, A-16E, A-17, and A-17E. Saturation data for nitrogen appear in Tables A-19 and A-19E, while data for potassium appear in Tables A-21 and A-21E. The following examples illustrate data retrieval from the saturation tables for water and refrigerant 134a.

**Example 3-3**

Determine the pressure in MPa and the volume change in cubic meters when 2 kg of saturated liquid water is completely vaporized at (a) 100°C and (b) 300°C.

**Solution:**

**Given:**   Two kilograms of saturated liquid water at (a) 100°C and (b) 300°C.

**Find:**   (a) Pressure in megapascals, (b) $\Delta V$ in m$^3$ for both temperatures.

**Model:**   Phase change at constant temperature.

**Strategy:**   Find the change in specific volume $v_{fg}$ from the saturation table, and then multiply by the mass.

**Analysis:**   The pressure is the saturation pressure corresponding to the given temperature. The volume change during vaporization is given by $v_{fg} = v_g - v_f$.
   (a) The saturation pressure at 100°C is found to be 1.014 bars or 0.1014 MPa from Table 3-3 or Table A-12. From the same table we find that

$$\Delta V = m(v_g - v_f) = 2.0 \text{ kg} \times (1673 - 1.04) \times 10^{-3} \text{ m}^3/\text{kg} = 3.344 \text{ m}^3$$

   (b) At 300°C the pressure is 85.81 bars or 8.581 MPa from Table A-12. Therefore,

$$\Delta V = m v_{fg} = 2.0 \text{ kg} \times (21.67 - 1.40) \times 10^{-3} \text{ m}^3/\text{kg} = 0.0406 \text{ m}^3$$

**Comment:**   This problem demonstrates the rapid decrease in $v_{fg}$ as the saturation temperature approaches the critical state at 374°C, where $v_{fg}$ becomes zero. The volume changes at 100°C and 300°C differ by over a factor of 80. This trend, which is typical of substances in general, is also shown in Fig. 3-7 for water.

**Example 3-4**

Two kilograms of water substance at 200°C are contained in a 0.20-m$^3$ vessel. Determine (a) the pressure, in bars, (b) the enthalpy, in kJ/kg, and (c) the mass and volume of the vapor within the vessel.

**Solution:**

**Given:**   Water substance in the state shown in Fig. 3-13a.

**Find:**   (a) P, in bars; (b) h, in kJ/kg; and (c) m and V of the vapor phase.

**Model:** Constant-volume system.

**Strategy:** Use the specific-volume data in the saturation table to determine if the state is within the wet region. If so, this fixes the pressure. Then determine the quality $x$ and other properties.

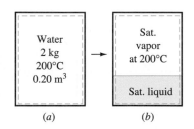

(a)　　　　　　(b)

**Figure 3-13**
Schematic and data for Example 3-4.

**Analysis:** On the basis of the mass and volume data, the specific volume of the overall system is $v = 0.20 \text{ m}/2.0 \text{ kg} = 0.10 \text{ m}^3/\text{kg}$. From Table A-12, the value of $v_f$ is $1.1565 \times 10^{-3} \text{ m}^3/\text{kg}$ and $v_g$ is $0.1274 \text{ m}^3/\text{kg}$ at the given temperature. Because the given value of $0.10 \text{ m}^3/\text{kg}$ lies between $v_f$ and $v_g$, the water is a liquid-vapor mixture, as shown in Fig. 3-13b.

(a) For a two-phase mixture, the pressure is the saturation pressure at the given temperature. In this case this is 15.54 bars, as found in column 2 of Table A-12.

(b) The specific enthalpy of the mixture is found from a form of Eq. [3-5b], namely,

$$h_x = h_f + x(h_g - h_f)$$

However, this equation requires a knowledge of the quality $x$. This may be determined from information on $v_x$, which is $0.1 \text{ m}^3/\text{kg}$. On the basis of Eq. [3-6b],

$$x = \frac{v_x - v_f}{v_g - v_f} = \frac{0.1 - 1.1565 \times 10^{-3}}{0.1274 - 1.1565 \times 10^{-3}} = \frac{0.9884}{0.1262} = 0.783$$

The specific enthalpy, then, is

$$h_x = 852.45 \text{ kJ/kg} + 0.783(1940.7) \text{ kJ/kg} = 2372 \text{ kJ/kg}$$

(c) The mass of the vapor is $m_g = x m_{\text{tot}} = 0.783(2) \text{ kg} = 1.57 \text{ kg}$. Hence the volume occupied by the vapor is

$$V_g = m_g v_g = 1.57 \text{ kg} \times 0.1274 \text{ m}^3/\text{kg} = 0.1995 \text{ m}^3$$

**Comment:** Thus the vapor fills nearly all the volume of the vessel.

---

**T**hree kilograms of saturated liquid water are contained in a constant-pressure system at 5 bars. Energy is added to the fluid until it has a quality of 60 percent. Determine (a) the initial temperature, (b) the final pressure and temperature, and (c) the volume and enthalpy changes.

**Example 3-5**

**Solution:**

**Given:** Energy is added to saturated liquid water. The process is shown in Fig. 3-14.

**Find:** (a) $T_1$, in °C; (b) $P_2$, in bars; and (c) $\Delta V$, in m³, and $\Delta H$, in kJ.

**Model:** Constant-pressure process.

**Strategy:** Use the saturation table and known quality to find initial and final data.

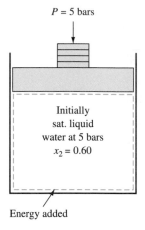

P = 5 bars

Initially
sat. liquid
water at 5 bars
$x_2 = 0.60$

Energy added

**Figure 3-14**
Schematic and data for Example 3-5.

**Analysis:**   (a) There is a unique temperature for a given pressure at a saturation state. From the saturation pressure table for water, Table A-13, we find the saturation temperature corresponding to 5 bars to be 151.9°C.

(b) Since the fluid is not completely vaporized ($x_2 = 0.6$), the pressure and temperature remain equal to their initial values of 5 bars and 151.9°C.

(c) The volume and enthalpy changes are computed from the relations $\Delta V = m(v_2 - v_1)$ and $\Delta H = m(h_2 - h_1)$, where 1 and 2 represent the initial and final states. The initial specific volume and specific enthalpy are read directly from Table A-13, in terms of $v_f$ and $h_f$. These values are

$$v_1 = v_f = 1.0926 \times 10^{-3} \text{ m}^3/\text{kg} \qquad \text{and} \qquad h_1 = h_f = 640.23 \text{ kJ/kg}$$

The values of $v_2$ and $h_2$ must be calculated on the basis of a liquid-vapor mixture of 60 percent quality. Hence

$$
\begin{aligned}
v_2 &= v_f + xv_{fg} \\
&= 1.09 \times 10^{-3} \text{ m}^3/\text{kg} + 0.60(0.3749 - 1.1 \times 10^{-3}) \text{ m}^3/\text{kg} \\
&= 0.225 \text{ m}^3/\text{kg}
\end{aligned}
$$

$$
\begin{aligned}
h_2 &= h_f + xh_{fg} \\
&= 640.2 \text{ kJ/kg} + 0.60(2108.5) \text{ kJ/kg} \\
&= 1905 \text{ kJ/kg}
\end{aligned}
$$

Consequently,

$$\Delta V = 3 \text{ kg} \times (0.255 - 0.001) \text{ m}^3/\text{kg} = 0.672 \text{ m}^3$$

$$\Delta H = 3 \text{ kg} \times (1905 - 640) \text{ kJ/kg} = 3800 \text{ kJ}$$

### 3-6-3   COMPRESSED- OR SUBCOOLED-LIQUID TABLE

There is not a great deal of tabular data for compressed or subcooled liquids in the literature. However, since water is used as the working fluid in electric power plants, considerable data are available for this substance in the liquid region. The data shown in Table 3-4 are taken from more extensive compilations found in Tables A-15 and A-15E. The first row in the first two sets of data (at 80°C and 150°F) is the saturated-liquid data at that temperature. The variation of properties of a compressed liquid with pressure is seen to be *slight* at a given temperature. By using data for water in the saturated-liquid state of 80°C and 0.4739 bar, for example, as an approximation for the compressed-liquid state of 80°C and 100 bars, errors of 0.45, 0.68, 2.3, and 0.61 percent are made in the values of $v$, $u$, $h$, and $s$, respectively.

When compressed-liquid data based on experimental information are available, they should be used. However, in the frequent absence of such data the above comparison indicates a general approximation rule.

*Compressed-liquid data* in most cases can be approximated closely by using the property values of the *saturated-liquid* state at the *given temperature.*

**Table 3-4**  Properties of compressed liquid water

| SI: Data at 80°C | | | | |
|---|---|---|---|---|
| $P$, bars | $v$, m³/kg | $u$, kJ/kg | $h$, kJ/kg | $s$, kJ/kg·K |
| 0.474 (sat.) | 0.001029 | 334.86 | 334.91 | 1.0753 |
| 50 | 0.001027 | 333.72 | 338.85 | 1.0720 |
| 100 | 0.001025 | 332.59 | 342.83 | 1.0688 |

| USCS: Data at 150°F | | | | |
|---|---|---|---|---|
| $P$, psia | $v$, ft³/lb$_m$ | $u$, Btu/lb$_m$ | $h$, Btu/lb$_m$ | $s$, Btu/lb$_m$·°R |
| 3.722 (sat.) | 0.01634 | 117.95 | 117.96 | 0.2150 |
| 500 | 0.01632 | 117.66 | 119.17 | 0.2146 |
| 1000 | 0.01629 | 117.38 | 120.40 | 0.2141 |

| SI: Data at 50 bars | | | | |
|---|---|---|---|---|
| $T$, °C | $v$, m³/kg | $u$, kJ/kg | $h$, kJ/kg | $s$, kJ/kg·K |
| 20 | 0.9995 | 83.65 | 88.65 | 0.2956 |
| 100 | 1.0410 | 417.52 | 422.72 | 1.3030 |
| 200 | 1.1530 | 848.1 | 853.9 | 2.3255 |
| 264 (sat.) | 1.3187 | 1205.4 | 1213.4 | 3.0267 |

SOURCE: Abstracted from Tables A-15 and A-15E.

This statement simply implies that compressed-liquid data are more temperature-dependent than pressure-dependent. The third set of data in Table 3-4 at 50 bars illustrates the *strong* effect of temperature on compressed-liquid property values. For large pressure differences between the saturated-liquid state and the compressed-liquid state, the saturated-liquid enthalpy value should be corrected for its pressure dependence. A simple method for making this correction for $h_f$ data is discussed in Sec. 4-5-2.

**Note the approximation method for determining compressed- or subcooled-liquid data.**

**F**ind the change in the specific internal energy and specific enthalpy of water for a change of state from 40°C, 25 bars, to 80°C, 75 bars, by use of (*a*) the compressed-liquid table, and (*b*) the approximation rule, using saturated data.

**Example 3-6**

**Solution:**

**Given:**   Water substance changes state as shown on the *PT* diagram in Fig. 3-15.

**Find:**   $\Delta u$ and $\Delta h$ by (*a*) compressed-liquid table, and (*b*) saturated-liquid data.

**Model:**   Tabular compressed-liquid data.

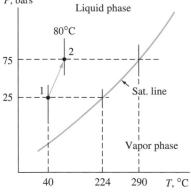

**Figure 3-15**
Illustration for the change of state in Example 3-6.

**Analysis:** From Table A-13 the saturation temperature at 25 bars is 224°C and at 75 bars it is 290°C. At both states the actual temperature is less than the saturation temperature for the given pressure. Hence both states are subcooled (compressed) liquids.

(*a*) The data for compressed-liquid water appear in Table A-15. From this table we determine that

$$u_2 - u_1 = (333.15 - 167.25) \text{ kJ/kg} = 165.90 \text{ kJ/kg}$$

$$h_2 - h_1 = (340.84 - 169.77) \text{ kJ/kg} = 171.07 \text{ kJ/kg}$$

(*b*) As an approximation method, the values of $u_f$ and $h_f$ are used at the specified temperatures, and the pressure data are ignored. Use of Table A-12 leads to

$$u_2 - u_1 = (334.86 - 167.56) \text{ kJ/kg} = 167.50 \text{ kJ/kg}$$

$$h_2 - h_1 = (334.91 - 167.57) \text{ kJ/kg} = 167.34 \text{ kJ/kg}$$

**Comment:** In the absence of compressed-liquid data, the use of saturated-liquid data produces an answer which in this case is 0.84 percent in error for the internal energy, while the error is 2.2 percent for the enthalpy. The error for $v$ would be similar to that for the internal energy.

### 3-6-4   REFERENCE STATE AND REFERENCE VALUES

The properties $P, v,$ and $T$ are directly measurable properties. However, the values of $u, h,$ and $s$ are not found by direct measurement but are determined from property relations based on the first and second laws of thermodynamics. These relations yield only changes in property values. In order to assign a numerical value to a specific state, a *reference value* at a *reference state* must be chosen. In the steam tables, for example, it is common practice to set both $h_f$ and $s_f$ to be zero at the triple state. The common practice for refrigerants is to set both $h_f$ and $s_f$ to be zero at −40°C (−40°F). This results in a negative value for $u_f$ at this temperature. Note that a number of negative values for $u_f$ and $h_f$ appear in Table A-19 in the Appendix for nitrogen due to the choice of a reference state. The selection of the reference value at a reference state is made by the author(s) of the table. The choice is arbitrary and not important, because the first and second laws only require a knowledge of the changes in property values.

### 3-6-5   SELECTION OF APPROPRIATE PROPERTY DATA

Both saturation and superheat tables are available in this text for substances of interest. On the basis of some given property data, it may be difficult to know which table contains the other property data for that state. Often the given data include either the temperature or the pressure and another property value such as $v, u, h,$ or $s$. As a general rule, the best method in our search for data is to examine the saturation tables first.

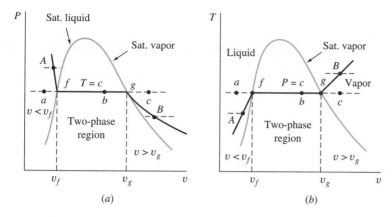

**Figure 3-16**  Illustration of technique for selecting property table for data acquisition.

Sketches of $Pv$ and $Tv$ diagrams in Fig. 3-16 illustrate the technique when $P$ and $v$ or $T$ and $v$ are input data. (The same approach is valid when $u$, $h$, or $s$ is known, instead of $v$.) At the given $P$ (or $T$) the saturation tables are used to determine the values of $v_f$ and $v_g$. If the given value of $v$ lies between $v_f$ and $v_g$, then the system is a two-phase mixture, as shown by point $b$ on the diagrams. The temperature (or pressure) is the corresponding saturation value. The quality and other properties are then calculated from Eq. [3-5]. If the known specific volume is less than $v_f$, then the substance is in the compressed or subcooled state shown as point $a$ in the figures. When $v > v_g$, the state is one of superheated vapor. This is shown as point $c$ in Fig. 3-16. Data for that state are then found in the superheat table. Interpolation of data may be necessary.

Recall that pressure and temperature are not independent properties in a saturation state. If the pressure and temperature are input data, then the condition (state) of the substance is usually either a compressed (subcooled) liquid or a superheated vapor, as shown by states $A$ and $B$ in Fig. 3-16. The following example illustrates data acquisition when the condition of the substance is not identified.

---

Determine the missing properties and the phase condition for various states of water shown in boldface in the following table.

Example 3-7

|     | $P$, bars | $T$, °C | $v$, m³/kg | $u$, kJ/kg | Quality | Condition |
|-----|-----------|---------|------------|------------|---------|-----------|
| (a) | **3.0** | (160) | **0.651** | (2587.1) | (NA) | (superheated) |
| (b) | (4.758) | **150** | (0.3199) | **2200** | (0.814) | (wet) |
| (c) | **25** | **100** | (0.0010423) | (418.24) | (NA) | (compressed) |

**Solution:**

**Given:**   Water substance at (a) 3.0 bars, 0.651 m³/kg; (b) 150°C, $u = $ 2200 kJ/kg; (c) 25 bars, 100°C.

**Find:**   The remaining properties for each state from the list: $P$, $T$, $v$, and $u$; and a phase description (condition).

**Model:**   Tabular saturation, superheat, and compressed-liquid data.

**Strategy:**   Use the data in the saturation table to establish the physical condition of a given state.

**Analysis:**   The answers to the following questions appear in parentheses in the table. An answer marked (NA) means an answer is not appropriate.

(a) From Table A-13, the values of $v_f$ and $v_g$ at 3.0 bars are $1.0732 \times 10^{-3}$ and 0.6058 m³/kg, respectively. The actual specific volume is 0.651 m³/kg. Hence the condition of the fluid is a superheated vapor, since $v > v_g$. At 3.0 bars in Table A-14 for superheated vapor, the given specific-volume value matches a temperature of 160°C. Thus $u = 2587.1$ kJ/kg.

(b) Using Table A-12, at a saturation temperature of 150°C, the values of $u_f$ and $u_g$ are 631.68 and 2559.5 kJ/kg, respectively. Since $u_f < u < u_g$, the condition is a two-phase mixture and the pressure $P_{sat}$ is 4.758 bars. To find the specific volume requires us first to evaluate the quality. From Eq. [3-5a]

$$x = \frac{u - u_x}{u_g - u_f}$$

$$= \frac{2200.0 - 631.68}{2559.5 - 631.68} = 0.814$$

Consequently, from Eq. [3-6a]

$$v = v_f + xv_{fg}$$
$$= [1.09 + 0.814(392.8 - 1.09)] \times 10^{-3} \text{ m}^3/\text{kg}$$
$$= 0.3199 \text{ m}^3/\text{kg}$$

The enthalpy can be found by a similar calculation.

(c) At 25 bars the saturation temperature from Table A-13 is 224°C. Because the actual temperature is 100°C, the fluid is subcooled. Similarly, at 100°C the saturation pressure from Table A-12 is 1.014 bars. Because the actual pressure is 25 bars, the fluid is a compressed liquid. The use of Table A-15 for a compressed or subcooled liquid shows that $v = 1.0423 \times 10^{-3}$ m³/kg and $u = 418.24$ kJ/kg.

**Comment:**   In the absence of a compressed-liquid table, the answers to part c can be closely approximated by saturated-liquid data at the given temperature (and ignoring the pressure value). In this case $v_f = 1.0435 \times 10^{-3}$ m³/kg and $u_f = 418.94$ kJ/kg. These latter values differ by 0.12 and 0.17 percent, respectively, from the compressed-liquid data.

## 3-7 TABULAR DATA AND CLOSED-SYSTEM ENERGY ANALYSIS

In the preceding section we described the format used in the general literature for presenting property data in tabular form for simple compressible
substances. This discussion covers the superheat, compressed liquid, and
saturation states. We are now in a position to employ the data in these tables
for the solution of problems involving the conservation of energy principle for closed systems. The following example problems also illustrate the
usefulness of property diagrams during system analyses.

**Water** substance is maintained in a weighted piston-cylinder assembly at 30 bars
and 240°C. The substance is slowly heated at constant pressure until the temperature
reaches 320°C. Determine (a) the work required to raise the weighted piston and
(b) the heat transfer required, in kJ/kg.

**Solution:**

**Given:** A piston-cylinder device contains water substance. It is slowly heated
to 320°C, as shown in Fig. 3-17.

**Find:** (a) w and (b) q, in kJ/kg.

**Model:** Closed stationary system; quasiequilibrium, constant-pressure process.

**Strategy:** The initial and final states are known, and the pressure is constant.
Calculate the expansion work by the integral of $P\,dv$. Evaluate the internal energy
values from tables, and then use the closed-system energy balance to find the heat
transfer.

**Analysis:** The system is the mass of water within the assembly. The system
boundary is shown by the dashed line in Fig. 3-17. The initial and final states are determined by first checking the saturation table, Table A-13. At 30 bars the saturation
temperature is 233.9°C. Since the system temperature is always above this value,
the fluid is superheated vapor throughout the process. However, the initial state lies
close to the saturation line. A $Pv$ diagram is shown in Fig. 3-18, which illustrates
the path of the slow heat-addition process that is *assumed* to be quasiequilibrium.
    (a) The energy balance for a process on a unit-mass basis is

$$q + w = \Delta u + \Delta ke + \Delta pe$$

The system is stationary; hence $\Delta ke = \Delta pe = 0$. The end states are known; therefore, $\Delta u$ can be found from the superheat table. However, both $q$ and $w$ are unknown.
Because we have no other equation for evaluating $q$, the only solution is to evaluate
$w$ first, by an independent equation. The work during a quasiequilibrium process is
given by the integral of $-P\,dv$. Because the pressure is constant, the work is found
by $-P(v_2 - v_1)$. The specific volume values from Table A-14 are $v_1 = 0.0682$ and

Example 3-8

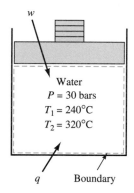

**Figure 3-17**
Schematic and data for
Example 3-8.

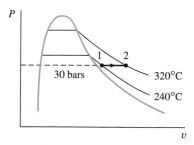

**Figure 3-18**
A $Pv$ plot of the process in Example 3-8.

$v_2 = 0.0850$ m³/kg. Using the relation $w = -P\,\Delta v$, we find that

$$w = -30 \text{ bar} \times (0.0850 - 0.0682) \text{ m}^3/\text{kg} \times 100 \text{ kJ/bar·m}^3$$
$$= -50.4 \text{ kJ/kg}$$

The negative sign indicates that work is done by the system on the surroundings.

(b) On the basis of the energy balance, the heat transfer is given by the relation

$$q = \Delta u - w = \Delta u + P\,\Delta v$$

From Table A-14 the specific internal energy values are $u_1 = 2619.7$ kJ/kg and $u_2 = 2788.4$ kJ/kg. The energy balance now shows that

$$q = 2788.4 - 2619.7 + 50.4 = 219.1 \text{ kJ/kg}$$

The positive value indicates that heat is transferred into the system.

The heat transfer for this quasiequilibrium, constant-pressure process also can be found by using enthalpy data. From the definition $h = u + Pv$, we find in general that

$$dh = du + d(Pv) = du + P\,dv + v\,dP$$

But for a *constant-pressure* process the term $v\,dP$ is zero. Under this condition $dh = du + P\,dv = \delta q$. Hence for a finite process $q = \Delta h$, if the process is quasiequilibrium, at constant pressure, and the only work interaction is due to a volume change. By making use of Table A-14 for enthalpy data, the heat transfer in this case is found by

$$q = h_2 - h_1 = (3043.4 - 2824.3) \text{ kJ/kg} = 219.1 \text{ kJ/kg}$$

This is the same answer as before. Although the heat transfer is found more directly this way, we lose information on the individual contributions of $\Delta u$ and $w$.

**Comment:**  This example introduces the use of the enthalpy function in calculating heat transfer in a quasiequilibrium, constant-pressure process. For closed systems, this is the only special case where the enthalpy function is useful. By always starting from basic principles, one can avoid using restricted equations in the wrong situation.

**Example 3-9**

**R**efrigerant 134a is contained within a rigid tank at an initial state of 2.8 bars and 100°C. Heat is removed until the pressure drops to 2.4 bars. During the process a paddle wheel within the tank is turned with a constant torque of 0.60 N·m for 300 revolutions. If the system contains 0.10 kg, compute the required heat transfer, in kilojoules.

**Solution:**

**Given:**  Refrigerant 134a is within a rigid tank. Heat is removed and shaft work occurs, as shown in Fig. 3-19.

**Find:**  $Q$, in kilojoules.

**Model:**  Stationary, closed system; volume and torque are constant.

**Strategy:** The heat transfer can be found from a closed-system energy balance once the shaft (paddle-wheel) work and internal energy change have been determined independently.

**Analysis:** Considering the refrigerant within the tank as the system, we write an energy balance in the form

$$Q + W_{\text{shaft,in}} + W_{\text{bound}} = \Delta U + \Delta KE + \Delta PE = m(u_2 - u_1)$$

Because $P\,dV$ work is zero when the volume is constant, and $\Delta KE$ and $\Delta PE$ are zero when the system is stationary, the energy balance reduces to

$$Q = m(u_2 - u_1) - W_{\text{paddle}}$$

where 2 and 1 represent final and initial states. The initial state is found by noting that the saturation temperature at 2.8 bars, from Table A-17, is $-5.37°C$. Hence the initial state of 100°C is well into the superheat region. From the superheat table (Table A-18), we find that

$$u_1 = 312.98 \text{ kJ/kg} \quad \text{and} \quad v_1 = 0.10587 \text{ m}^3/\text{kg}$$

The value of $v_1$ is important since, for a rigid system, $v_2 = v_1$. Consequently the final state has a pressure of 2.4 bars and a specific volume of 0.10587 m³/kg. At 2.4 bars, from the saturated data of Table A-17, we find that $v_g$ is only 0.0834 m³/kg. Thus the final state is also in the superheat region. From Table A-18 we find that a specific-volume value of 0.10587 m³/kg corresponds closely to 50.7°C by linear interpolation. By a similar interpolation $u_2 = 269.69$ kJ/kg. The $Pv$ diagram in Fig. 3-20 shows the vertical path above the saturated-vapor line.

To complete the analysis, we need a value for the paddle-wheel work. Based on the discussion in Chap. 2, rotational mechanical work is given by $\tau\theta$ when $\tau$ is a constant; $\theta$ must be expressed in radians. Hence for 300 revolutions,

$$W_{\text{pad}} = 0.60 \text{ N·m} \times 300(2\pi) = 1131 \text{ N·m} = 1.131 \text{ kJ}$$

Substitution of $u$ and $W$ values into the energy equation yields

$$Q = 0.1 \text{ kg} \times (269.69 - 312.98) \text{ kJ/kg} - 1.13 \text{ kJ} = -5.46 \text{ kJ}$$

**Comment:** The heat interaction must account not only for a decrease in the internal energy of the fluid but also for the energy added in the form of paddle-wheel (shaft) work.

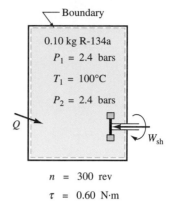

— Boundary

0.10 kg R-134a
$P_1 = 2.4$ bars
$T_1 = 100°C$
$P_2 = 2.4$ bars

$Q$

$W_{\text{sh}}$

$n = 300$ rev
$\tau = 0.60$ N·m

**Figure 3-19**
Schematic and data for Example 3-9.

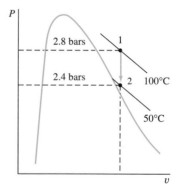

$P$

2.8 bars

1

2.4 bars

2   100°C

50°C

$v$

**Figure 3-20**
A $Pv$ plot of the process in Example 3-9.

**R**efrigerant 134a is cooled within a piston-cylinder device from an initial state of 6 bars and 60°C to a final state of a saturated liquid at the same pressure. Determine (*a*) the work and (*b*) the heat transfer for the process in kJ/kg.

**Solution:**

**Given:** Refrigerant 134a within a piston-cylinder device is cooled until it becomes a saturated liquid. A schematic of the equipment with input data and a $Pv$ diagram are shown in Fig. 3-21.

**Find:** (*a*) $w$ and (*b*) $q$, in kJ/kg.

**Example 3-10**

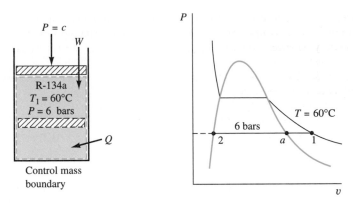

**Figure 3-21**      A schematic and Pv for Example 3-10.

**Model:**      Stationary, closed system; quasistatic, constant-pressure process.

**Strategy:**      The initial and final states are known. Hence the work due to a volume change is found from the integral of $-P\,dv$ at constant pressure, and the internal energy values come directly from the appropriate table. The conservation of energy equation leads to the heat transfer.

**Analysis:**      (a) The only work interaction is due to a volume change and is given by the integral of $-P\,dv$ if we *assume* that the process is quasiequilibrium. In addition, the pressure is held constant. Consequently, a knowledge of the initial and final specific volumes is sufficient to evaluate $w$. At 6 bars the saturation temperature is 21.58°C (Table A-17). Therefore, the initial state of 60°C is a superheated one, as shown on the Pv diagram in Fig. 3-21. From the superheat table (Table A-18), $v_1 = 0.04134$ m³/kg, and from the pressure saturation table (Table A-17), $v_2 = v_f = 0.8196 \times 10^{-3}$ m³/kg at 6 bars. Hence, at constant pressure,

$$w = -\int P\,dv = -P(v_2 - v_1)$$
$$= -6 \text{ bars} \times (41.34 - 0.8196) \times 10^{-3} \text{m}^3/\text{kg} \times 100 \text{ kJ/1 bar·m}^3$$
$$= 24.3 \text{ kJ/kg}$$

(b) For a stationary system ($\Delta ke = \Delta pe = 0$) the conservation of energy principle reduces to

$$q + w_{\text{comp}} = \Delta u$$

The value of $q$ can be found directly from this equation because the internal-energy change is fixed by the end states and these are known. From Table A-18 at 6 bars and 60°C, $u_1 = 273.54$ kJ/kg and from Table A-17 at 6 bars, $u_2 = u_f = 78.99$ kJ/kg. Therefore,

$$q = u_2 - u_1 - w$$
$$= (78.99 - 273.54) \text{ kJ/kg} - 24.3 \text{ kJ/kg} = -218.9 \text{ kJ/kg}$$

The answer is negative, as expected, since heat must be removed to condense the vapor.

**Comment:** The energy equation for the constant-pressure, quasistatic process can be written also as $q = \Delta u + P\,\Delta v = \Delta h$. By recognizing this special form of the energy balance, we do not need to calculate $w$ and $\Delta u$ separately if only the heat-transfer value is required. For this particular problem, $q = h_2 - h_1 = (79.48 - 298.35)$ kJ/kg $= -218.9$ kJ/kg, as before. It might be pointed out that the heat removed between states $a$ and $2$ shown on the $Pv$ diagram is equal to $-h_{fg}$, the negative of the enthalpy of vaporization. Since our energy equation shows that $\Delta h = q$ for a liquid–vapor phase change, it is easy to see why $h_{fg}$ is also called the latent "heat" of vaporization.

---

**⬤**ne-tenth of a kilogram of water at 3 bars and 76.3 percent quality is contained in a rigid tank that is thermally insulated. A paddle wheel inside the tank is turned by an external motor until the substance is a saturated vapor. Determine (*a*) the shaft work necessary, in kilojoules, and (*b*) the final pressure and temperature of the water.

**Example 3-11**

**Solution:**

**Given:** A rigid tank containing 0.1 kg of water is stirred by a paddle wheel. A schematic of the equipment with appropriate data is shown in Fig. 3-22.

**Find:** (*a*) $W_{\text{paddle}}$, in kJ, and (*b*) $P_2$ and $T_2$.

**Model:** Stationary, closed system; thermally insulated, rigid tank.

**Strategy:** Because data on the paddle-wheel operation are not given, the paddle-wheel work must be found from an energy analysis on the system.

**Analysis:** The boundary of the control mass is chosen to lie just inside the tank, as shown in Fig. 3-22. The fluid is assumed to be a simple compressible substance, and during the process the water is in fluid shear and is not necessarily in equilibrium. However, the end states are taken to be equilibrium states. For this stationary system ($\Delta ke = \Delta pe = 0$) the energy equation is

$$Q + W = \Delta U$$

No heat transfer or $P\,dV$ work occurs because of the restrictions on the system (insulated and rigid). Work is done on the system, though, by the action of the paddle wheel, so that the above equation reduces to

$$W_{\text{paddle}} = \Delta U = U_2 - U_1 = m(u_2 - u_1)$$

The shaft work requirement is determined, then, from an evaluation of the initial and final specific internal energies. The initial value of $u$ is found from data in the saturation-pressure table (Table A-13), in conjunction with the equations for intensive properties expressed in terms of the quality $x$. That is, for 3 bars and $x = 0.763$, we find that

$$u_1 = u_f(1 - x) + u_g x = [561.2(0.237) + 2543.6(0.763)]\ \text{kJ/kg} = 2074\ \text{kJ/kg}$$

To find the final state, we recognize that the final specific volume is the same as the initial value, since the tank is rigid and closed. The initial specific volume is

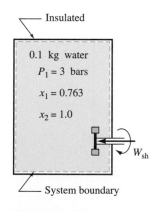

Insulated

0.1 kg water

$P_1 = 3$ bars

$x_1 = 0.763$

$x_2 = 1.0$

$W_{\text{sh}}$

System boundary

**Figure 3-22**
Schematic and data for Example 3-11.

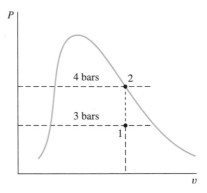

**Figure 3-23**
A $Pv$ plot of the process in Example 3-11.

$$v_1 = v_f(1 - x) + v_g x$$
$$= [1.073(0.237) + 605.8(0.763)] \times 10^{-3} \text{ m}^3/\text{kg}$$
$$= 0.4625 \text{ m}^3/\text{kg}$$

Thus the final state is a saturated vapor with a specific volume of 0.4625 m³/kg.
From the accompanying $Pv$ diagram in Fig. 3-23, we note that to proceed at constant $v$ from the wet region at 3 bars and 76.3 percent quality to the state of a saturated vapor, the final pressure must be greater than 3 bars. From the saturation-pressure table for steam, Table A-13, it is found that a value of $v_g = 0.4625$ m³/kg corresponds to a pressure of 4 bars and a saturation temperature of 143.6°C. At this final state the specific internal energy is 2553.6 kJ/kg. The paddle-wheel work that must be supplied, then, is

$$W_{\text{paddle}} = m(u_2 - u_1)$$
$$= 0.1 \text{ kg} \times (2553.6 - 2074) \text{ kJ/kg} = 48.0 \text{ kJ}$$

The value of the paddle-wheel work is positive, since the work transfer is to the system along the rotating shaft.

**Example 3-12**

**A** piston-cylinder assembly with an initial volume of 0.01 m³ is filled with saturated refrigerant 134a vapor at 16°C. The substance is compressed until a state of 9 bars and 60°C is reached. During the compression process the heat loss amounts to 0.44 kJ. Compute the compression work required, in kilojoules.

**Solution:**

**Given:**    A piston-cylinder assembly contains refrigerant 134a. A schematic of the system with appropriate data is shown in Fig. 3-24.

**Find:**    The work of compression, in kilojoules.

**Model:**    Stationary, closed system; compression process.

**Strategy:**    The compression work cannot be found from the integral of $-P \, dV$ because the relation between pressure and volume is not known. Thus it must be calculated from an energy balance.

**Analysis:**    The system is the region within the dashed line in Fig. 3-24. The conservation of energy principle for a simple compressible substance in a stationary system can be written in the form

$$W_{\text{comp/exp}} = \Delta U - Q = m(u_2 - u_1) - Q$$

The initial and final states of the substance are known. The mass of the system is found from the basic relation $m = V/v$, evaluated at the initial state. For the initial saturation state we find from Table A-16 that $v_1 = v_g = 0.0405$ m³/kg at 16°C. Therefore,

$$m = \frac{V}{v} = \frac{0.01 \text{ m}^3}{0.0405 \text{ m}^3/\text{kg}} = 0.247 \text{ kg}$$

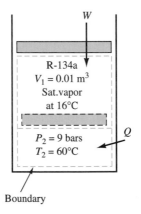

**Figure 3-24**
Schematic and data for Example 3-12.

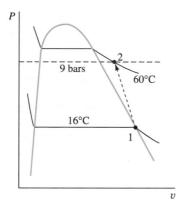

**Figure 3-25**  A $Pv$ plot of the process in Example 3-12.

The value of $u_1$ is $u_g$ at 16°C, or 235.78 kJ/kg. To find state 2 we note that the saturation temperature at 9 bars, from Table A-17, is only 35.53°C. Thus the substance is superheated at 9 bars and 60°C. From Table A-18 we find that $u_2 = 269.72$ kJ/kg. Substitution of the given and acquired data yields

$$
\begin{aligned}
W_{comp/exp} &= m(u_2 - u_1) - Q \\
&= 0.247 \text{ kg} \times (269.73 - 235.78) \text{ kJ/kg} + 0.44 \text{ kJ} \\
&= (8.38 + 0.44) \text{ kJ} = 8.82 \text{ kJ}
\end{aligned}
$$

The heat loss in this case is roughly only 5 percent of the work required. The accompanying $Pv$ sketch of the process in Fig. 3-25 shows the decrease in volume as the vapor is compressed and becomes superheated.

**Comment:**  The path of the process on the $Pv$ diagram is shown as a dashed line, since the actual path is unknown.

---

**Liquid** water in a compressed-liquid state of 75 bars and 40°C is heated quasistatically at constant pressure until it becomes a saturated liquid. Compute the heat input required, in kJ/kg.

**Solution:**

**Given:**  Compressed liquid water is heated at constant pressure. A schematic of the equipment is shown in Fig. 3-26.

**Find:**  $q$, in kJ/kg.

**Model:**  Stationary, closed system; quasiequilibrium, constant-pressure process.

**Strategy:**  The value of a heat interaction must be found from an appropriate energy analysis, which in this case includes a $P\,dV$ work term and an internal energy change.

**Example 3-13**

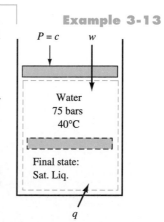

**Figure 3-26**
Schematic and data for Example 3-13.

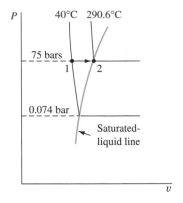

**Figure 3-27**
A $Pv$ plot of the process in Example 3-13.

**Analysis:** The path of the process is shown on the accompanying $Pv$ diagram in Fig. 3-27, where the region close to the saturated-liquid line has been greatly enlarged. At constant pressure the quasiequilibrium expansion work is given by $w = -\int P\,dv = -P\,\Delta v$. Hence the heat input for a stationary system becomes

$$q = \Delta u - w = \Delta u + P\,\Delta v = \Delta h$$

The initial and final states are known. The initial enthalpy for the compressed liquid at 75 bars and 40°C is found from Table A-15 to be 174.18 kJ/kg. In the final saturated-liquid state at 75 bars the temperature is 290.59°C and the enthalpy is 1292.2 kJ/kg from Table A-15 (or interpolated from Table A-13). Consequently,

$$q = h_2 - h_1 = (1292.2 - 174.2) \text{ kJ/kg} = 1118.0 \text{ kJ/kg}$$

This large value of $q$ is due to the fact that the temperature changes by over 250°C during the process. The value of $q$ could also be calculated from the sum of $\Delta u$ and $P\,\Delta v$.

**Comment:** If compressed-liquid data were not available, the initial enthalpy could be approximated by the saturated-liquid enthalpy at 40°C, which is 167.57 kJ/kg. In this case the heat transfer would be $(1292.2 - 167.6)$ kJ/kg, or 1124.6 kJ/kg. This is about 0.6 percent greater than the value using compressed-liquid data.

## 3-8   THE SPECIFIC HEATS

In addition to the properties—pressure $P$, temperature $T$, specific volume $v$, internal energy $u$, and enthalpy $h$—it is necessary to define and relate two other properties of interest—the specific heat at constant pressure $c_p$ and the specific heat at constant volume $c_v$. On the basis of the state postulate for a simple substance, the property relationship among the general property variables $x$, $y$, and $z$ may be expressed as $x = f(y, z)$. The differential change $dx$ is written in the form of a total differential, which contains *partial derivatives*. A brief discussion of a total differential is presented as an appendix to this chapter in Sec. 3-10, if the reader is unfamiliar with partial derivatives and their graphical interpretation.

Since the internal energy is not directly measurable, it is necessary to develop equations for the change in this property in terms of other measurable properties, such as $P$, $v$, and $T$. For simple compressible substances, the sensible internal energy $u$ is a function of two other intensive, intrinsic properties. It is advantageous to select the temperature and the specific volume as the independent variables. If $u = f(T, v)$, then the total differential of $u$ is written as

$$du = \left(\frac{\partial u}{\partial T}\right)_v dT + \left(\frac{\partial u}{\partial v}\right)_T dv \qquad \textbf{[3-9]}$$

The first partial derivative on the right is *defined* as $c_v$, the **specific heat at constant volume**. That is,

$$c_v \equiv \left(\frac{\partial u}{\partial T}\right)_v \qquad \textbf{[3-10]}$$

The value of $c_v$ is experimentally measurable, and tabulated in many references. Consider the following measurement technique. On the basis of Eq. [3-10], $c_v$ is related to the change in the specific internal energy per unit change in temperature for a constant-volume process. At constant volume, the energy balance reduces to $q = \Delta u$. Therefore, values of $c_v$ can be obtained by measuring the heat transfer required to raise the temperature of a unit mass of substance by one degree, while holding the volume constant. The quantity $c_v$ can also be interpreted graphically as the tangent $AB$ to a three-dimensional $uTv$ surface along a constant-volume plane, as shown in Fig. 3-28. Now, when Eq. [3-10] is substituted into Eq. [3-9], we find that

$$du = c_v\, dT + \left(\frac{\partial u}{\partial v}\right)_T dv \qquad \textbf{[3-11]}$$

This is an important relation for the differential change in internal energy for any simple compressible substance in any phase.

Similar to the internal energy, the enthalpy function is not directly measurable but must be related to other measurable quantities. The enthalpy function $h$ is usually expressed mathematically in terms of the temperature and pressure as the independent variables. By letting $h = h(T, P)$, we write

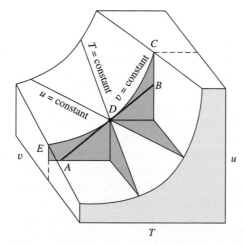

**Figure 3-28**   Graphical interpretation of the specific heat $c_v$ as the slope of a constant-volume line on a $uTv$ surface.

the total differential for $h$ as

$$dh = \left(\frac{\partial h}{\partial T}\right)_P dT + \left(\frac{\partial h}{\partial P}\right)_T dP \qquad \textbf{[3-12]}$$

The first partial derivative on the right is *defined* as the **specific heat at constant pressure** $c_p$. Then we have, analogous to Eq. [3-10],

$$c_p \equiv \left(\frac{\partial h}{\partial T}\right)_p \qquad \textbf{[3-13]}$$

Tables of data for $c_p$ are widely available for many substances over a wide range of temperatures. To measure $c_p$, consider a constant-pressure process for a closed system. An energy balance reduces to $q = \Delta h$, when the only form of work transfer is that due to a volume change. Therefore, values of $c_p$ can be obtained by measuring the heat transfer required to raise the temperature of a unit mass of substance by one degree, while holding the pressure constant. The quantity $c_p$ can also be interpreted as the tangent to a three-dimensional $hTP$ surface along a constant-pressure plane. Now, the substitution of Eq. [3-13] into Eq. [3-12] yields

$$dh = c_p dT + \left(\frac{\partial h}{\partial p}\right)_T dP \qquad \textbf{[3-14]}$$

**If $h = aT + bT^2P$, show that $c_p = a + 2bTP$.**

for the differential change in the enthalpy for any simple compressible substance. Equations (3-11) and (3-14) are quite useful in the development of certain property relations in Chap. 4.

Based on the preceding discussion, we find that *derivatives of properties with respect to other properties are also properties of the system.* The specific heats $c_v$ and $c_p$ are two of the most important derivative-type properties in thermodynamics. The dimensions on the values of the specific heats defined above are in terms of energy/[(mass) (temperature difference)]. It is important to note that the temperature quantity in the denominator involves a change in temperature, and not the value of the temperature itself. Consequently, the specific heat may be expressed in terms of either kelvins or degrees Celsius. The symbols K and °C for these two temperature scales have the same significance in this special case, where only differences in temperature are of interest. The same reasoning obviously applies to degrees Rankine versus degrees Fahrenheit. Common units for specific heats in SI are kJ/kg·K (or kJ/kg·°C) and kg/kmol·K. The USCS units used frequently are Btu/lb$_\text{m}$·°R (or Btu/lb$_\text{m}$·°F) and Btu/lbmol·°R.

**Note that the temperature unit for the specific heats is a temperature difference, and not a temperature value.**

## 3-9  SUMMARY

On the basis of *experimental* evidence, the general rule for determining the number of independent properties $n$ for any system of known mass and composition is based on the *state postulate,* which asserts that: *the number of independent properties required to completely specify the intensive, intrinsic state of a substance is equal to 1 plus the number of possibly relevant quasistatic work modes.* "Relevant work modes" are those that have an appreciable or measurable effect on the state of the substance when a process occurs.

By definition, a *simple system* is one for which only one quasiequilibrium work mode is relevant to the intrinsic state of the system. The state postulate for simple systems is

> *The equilibrium state of a simple substance is fixed by specifying the values of two independent, intrinsic properties.*

In this case the functional relationship between a set of intrinsic, intensive properties is given by $y_0 = f(y_1, y_2)$, where $(y_1, y_2)$ represents a set of two independent variables. A *simple compressible system* is defined as one for which the only relevant quasistatic work interaction is that associated with volume changes ($P\,dV$ work).

A *PvT* surface is a general representation of the equilibrium states of a simple compressible substance. The solid, liquid, and gas phases appear on the surface as regions. These *single-phase regions* are separated by *two-phase regions* which represent phase changes such as melting (or freezing), vaporization (or condensation), and sublimation. Of particular interest are the states which lie on the *liquid-saturation line* and the *vapor-saturation line* which join at the critical state. Vaporization (or condensation) occurs at constant temperature and pressure between the two saturation lines. The liquid-vapor (wet) region ends at the *triple-state line* as the pressure (and temperature) is lowered.

Projections of the *PvT* surface onto two dimensions lead to *Pv*, *PT*, and *Tv* diagrams. The pressure-temperature diagram clearly shows the variation of the vapor pressure as a function of temperature for liquid-vapor mixtures. The nature of a *compressed* or *subcooled liquid* and a *superheated vapor* is also easily explained by means of a *PT* diagram. A *Pv* diagram is useful because it retains the wet region as a primary focal point for analysis. The proportion of saturated vapor and liquid in the wet region is expressed in terms of the *quality x* or the mixture. That is,

$$\text{Quality} = x \equiv \frac{m_{\text{vapor}}}{m_{\text{total}}} = \frac{m_g}{m_g + m_f}$$

The subscripts $g$ and $f$ apply to the saturated-vapor ($x = 1$) and saturated-liquid ($x = 0$) states, respectively. For the purpose of system analysis, knowledge of the position of isothermal lines on a *Pv* diagram is important. The single- and two-phase regions on a *Tv* diagram appear in similar positions to those on a *Pv* diagram.

Property data for the vapor, liquid-vapor, and liquid regions shown on appropriate diagrams are frequently presented in tabular form. The tables for superheated vapor typically list values of $v$, $u$, $h$, and $s$ (the entropy) as a function of temperature for a given pressure. Data for a number of pressures appear in a given table. Since only selected values of temperature and pressure are used, interpolation of data is frequently necessary. A saturation table lists $v$, $u$, and $h$ for both the saturated-liquid and saturated-vapor states, and may be presented for integer values of either temperature or pressure. For a mixture of liquid and vapor, the specific mixture property $y_x$ is calculated by

$$y_x = (1 - x)y_f + xy_g = y_f + xy_g$$

where $y_{fg} = y_g - y_f$. Compressed- or subcooled-liquid data are presented in a manner similar to a superheat table for vapor. Since compressed-liquid data are seldom available, they may be approximated closely by using the values of the saturated-liquid state at the given temperature, while ignoring the pressure. In any of the above-mentioned tables, the values of $u$ and $h$ are arbitrarily set relative to a reference value at a selected reference state.

The specifics heats $c_v$ and $c_p$ are important thermodynamic properties. These are defined as

$$c_v \equiv \left(\frac{\partial u}{\partial T}\right)_v \qquad \text{and} \qquad c_p \equiv \left(\frac{\partial h}{\partial T}\right)_p$$

These definitions are important in determining $du$ and $dh$ in the following relationships:

$$du = c_v\, dT + \left(\frac{\partial u}{\partial v}\right)_T dv \qquad \text{and} \qquad dh = c_p\, dT + \left(\frac{\partial h}{\partial p}\right)_T dP$$

Thermodynamic property data are available on software packages for a number of common substances. This method of data acquisition eliminates the necessity of interpolation. However, an understanding of the nature of tabulating data is extremely important in engineering analysis involving the conservation of energy principles.

## 3-10 APPENDIX: FUNDAMENTALS OF PARTIAL DERIVATIVES

On the basis of the state postulate for simple compressible substances, any dependent variable is expressed as a function of two independent variables. With this in mind, consider three thermodynamic variables (such as $P$, $v$, and $T$) represented in general by $x$, $y$, and $z$. Their functional relationship is expressed in the form $x = x(y, z)$. The total differential when $x$ is the dependent variable is given by the equation

$$dx = \left(\frac{\partial x}{\partial y}\right)_z dy + \left(\frac{\partial x}{\partial z}\right)_y dz \qquad\qquad \text{[3-15]}$$

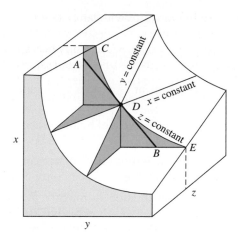

**Figure 3-29**    Graphical interpretation of the partial derivative $(\partial x/\partial y)_z$ at $D$ as the slope of the constant-$z$ line, CDE.

Similar expressions for $dy$ and $dz$ may also be written, depending upon which variable is selected as the dependent variable.

The physical significance of a partial derivative is easily seen by recalling that the equilibrium states of a simple compressible substance can be represented by a three-dimensional surface. Such a surface is shown in Fig. 3-29 for a single-phase region, where a set of three properties is again symbolized by $x$, $y$, and $z$. Sections of the surface in Fig. 3-29 have been cut away around the equilibrium state $D$, so that the curvature of the surface is more clearly seen. Consider a plane of constant $z$ which intersects the surface. The curve of intersection is marked with the points $C$, $D$, and $E$ on the diagram. The quantity $(\partial x/\partial y)_z$ is the slope of the surface at any state along this curve. In particular, the value of this partial derivative at state $D$ is the slope of tangent line $AB$. Similar interpretations of quantities $(\partial x/\partial z)_y$ and $(\partial y/\partial z)_x$ can be made when planes of constant $y$ and $x$, respectively, intersect the surface of equilibrium states. The parameters $x$, $y$, and $z$ represent any combination of three intrinsic, intensive properties, such as $u$, $T$, and $v$, for example.

## PROBLEMS

3-1C. On a $PT$ diagram for water, show approximately the location of constant-volume lines in the three single-phase regions.

3-2C. Describe how one differentiates between a saturated vapor and a superheated vapor at the same pressure.

3-3C. Describe how one differentiates between a saturated liquid and a subcooled liquid at the same pressure.

3-4C. Describe how the pressure varies with temperature for a vaporization process.

3-5C. Describe the difference between the critical state and the triple state of a pure substance.

3-6C. Is it possible for water to appear as a vapor at a temperature below its triple state? Explain.

3-7C. Explain what happens to the process of boiling when the system pressure is supercritical.

3-8C. Does it take more or less energy to boil a kilogram of water as the temperature is increased? Explain.

3-9C. The triple state of carbon dioxide is roughly 5 atm and $-57°C$ ($-70°F$). Explain why dry ice (solid carbon dioxide) sublimes rather than melts when placed at normal room conditions.

## SATURATION, SUPERHEAT, AND COMPRESSED LIQUID DATA

3-1. Complete the following table of properties of water.

|     | $P$, bars | $T$, °C | $v$, m³/kg | $h$, kJ/kg | $x$, % |
|-----|-----------|---------|------------|------------|--------|
| (a) |           | 150     | 392.8      |            |        |
| (b) | 20        | 320     |            |            |        |
| (c) |           | 100     |            | 2100       |        |
| (d) | 50        | 140     |            |            |        |

3-2. Complete the following table of properties of water. Show analysis.

|     | $P$, bars | $T$, °C | $v$, m³/kg | $u$, kJ/kg | $x$, % |
|-----|-----------|---------|------------|------------|--------|
| (a) | 60        |         | 25.0       |            |        |
| (b) | 15        |         |            | 2951.3     |        |
| (c) |           | 290     |            | 2576.0     |        |
| (d) |           | 140     |            | 588.74     |        |

3-3. Complete the following table of properties of water. Show analysis.

|     | $P$, bars | $T$, °C | $v$, m³/kg | $h$, kJ/kg | $x$, % |
|-----|-----------|---------|------------|------------|--------|
| (a) | 4.5       |         |            | 623.25     |        |
| (b) | 10        |         |            |            | 60     |
| (c) | 30        | 400     |            |            |        |
| (d) |           | 140     | 1.0784     |            |        |

3-4. Complete the following table of properties of water. Show analysis.

|     | P, bars | T, °C | v, m³/kg | u, kJ/kg | x, % |
|-----|---------|-------|----------|----------|------|
| (a) |         | 200   | 127.4    |          |      |
| (b) | 40      | 360   |          |          |      |
| (c) | 60      |       |          | 2100     |      |
| (d) | 50      |       |          | 333.72   |      |

3-5E. Complete the following table of properties of water. Show analysis.

|     | P, psia | T, °F | v, ft³/lbm | h, Btu/lbm | x, % |
|-----|---------|-------|------------|------------|------|
| (a) | 250     |       |            |            | 90   |
| (b) | 180     |       |            | 1323.5     |      |
| (c) |         | 250   |            | 218.6      |      |
| (d) |         | 400   | 1.866      |            |      |

3-6E. Complete the following table of properties of water. Show analysis.

|     | P, psia | T, °F | v, ft³/lbm | u, Btu/lbm | x, % |
|-----|---------|-------|------------|------------|------|
| (a) |         | 200   |            | 1000       |      |
| (b) | 80      |       |            | 282.0      |      |
| (c) | 140     |       | 4.86       |            |      |
| (d) |         | 100   | 350.0      |            |      |

3-7. Determine the required data for water for the following specified conditions: (a) the pressure and specific volume of saturated liquid at 20°C, (b) the temperature and enthalpy of saturated vapor at 9 bars, (c) the specific volume and internal energy at 10 bars and 280°C, (d) the temperature and specific volume at 8 bars and a quality of 80 percent, (e) the specific volume and enthalpy at 100°C and 100 bars, (f) the pressure and specific enthalpy at 150°C and 70 percent quality, (g) the temperature and specific internal energy at 15 bars and an enthalpy of 2899.3 kJ/kg, (h) the quality and specific volume at 200°C and an enthalpy of 1822.8 kJ/kg, (i) the internal energy and specific volume at 140°C and an enthalpy of 2733.9 kJ/kg, (j) the pressure and enthalpy at 280°C and an internal energy of 2760.2 kJ/kg, and (k) the temperature and specific volume at 200 bars and an enthalpy of 434.06 kJ/kg.

3-8E. Determine the required data for water for the following specified conditions: (a) the pressure and specific volume of saturated liquid at 150°F, (b) the temperature and enthalpy of saturated vapor at 80 psia, (c) the specific volume and internal energy at 140 psia and 500°F, (d) the temperature and specific volume at 100 psia and a quality of 80 percent, (e) the specific volume and enthalpy at 100°F and 1500 psia, (f) the pressure and specific enthalpy at 300°F and 70 percent quality, (g) the temperature and specific internal energy at 200 psia and an enthalpy of 1268.8 Btu/lb$_m$, (h) the quality and specific volume at 370°F and an enthalpy of 770 Btu/lb$_m$, (i) the internal energy and specific volume at 240°F and an enthalpy of 1160.7 Btu/lb$_m$, (j) the pressure and enthalpy at 500°F and an internal energy of 1172.7 Btu/lb$_m$, and (k) the temperature and specific volume at 2000 psia and an enthalpy of 73.3 Btu/lb$_m$.

3-9. Complete the following table of properties of refrigerant 134a. Show analysis.

|     | P, bars | T, °C | v, m³/kg | u, kJ/kg | x, % |
|-----|---------|-------|----------|----------|------|
| (a) |         | 4     |          | 204.8    |      |
| (b) | 6.0     |       | 0.0341   |          |      |
| (c) | 2.8     | 20    |          |          |      |
| (d) | 4.0     |       |          | 284.75   |      |

3-10. Complete the following table of properties of refrigerant 134a.

|     | T, °C | P, bars | v, m³/kg | u, kJ/kg | x, % |
|-----|-------|---------|----------|----------|------|
| (a) |       | 4.0     | 0.0509   |          |      |
| (b) | 30    |         |          | 248.20   |      |
| (c) |       | 12.0    |          | 182.86   |      |
| (d) | 26    |         | 0.0008309|          |      |

3-11. Complete the following table of properties of refrigerant 134a.

|     | T, °C | P, bars | v, m³/kg | h, kJ/kg | x, % |
|-----|-------|---------|----------|----------|------|
| (a) |       | 0.80    |          | 93.42    |      |
| (b) | 60    | 0.60    |          |          |      |
| (c) | -12   |         |          |          | 80   |
| (d) |       | 0.40    | 0.0509   |          |      |

3-12. Complete the following table of properties of refrigerant 134a.

|      | $T$, °C | $P$, bars | $v$, m³/kg | $h$, kJ/kg | $x$, % |
|------|---------|-----------|------------|------------|--------|
| (a)  |         | 10        |            | 219.17     |        |
| (b)  | 34      |           | 0.0236     |            |        |
| (c)  | 40      | 5         |            |            |        |
| (d)  |         | 3.6       |            | 57.82      |        |

3-13E. Complete the following table of properties of refrigerant 134a.

|      | $T$, °F | $P$, psia | $v$, ft³/lb$_m$ | $u$, Btu/lb$_m$ | $x$, % |
|------|---------|-----------|-----------------|-----------------|--------|
| (a)  |         | 80        |                 | 113.56          |        |
| (b)  | 70      |           | 0.5538          |                 |        |
| (c)  |         | 60        |                 | 75.0            |        |
| (d)  |         | 120       | 0.01360         |                 |        |

3-14E. Complete the following table of properties of refrigerant 134a.

|      | $T$, °F | $P$, psia | $v$, ft³/lb$_m$ | $u$, Btu/lb$_m$ | $x$, % |
|------|---------|-----------|-----------------|-----------------|--------|
| (a)  | 70      |           | 0.01311         |                 |        |
| (b)  |         | 70        | 0.6778          |                 |        |
| (c)  | 140     | 100       |                 |                 |        |
| (d)  | 20      |           |                 |                 | 70     |

3-15. Determine the required data for refrigerant 134a at the following specified conditions: (a) the pressure and specific volume of saturated liquid at 8°C, (b) the temperature and enthalpy of saturated vapor at 6 bars, (c) the specific volume and internal energy at 0.70 MPa and 40°C, (d) the temperature and specific volume at 3.2 bars and a quality of 40 percent, (e) the approximate specific volume and enthalpy at 8°C and 12 bars, (f) the pressure and enthalpy at −16°C and 50 percent quality, (g) the temperature and internal energy at 0.90 MPa and enthalpy of 282.34 kJ/kg, (h) the quality and specific volume at 44°C and an enthalpy of 222.7 kJ/kg, (i) the internal energy and specific volume at 30°C and an enthalpy of 263.50 kJ/kg, (j) the pressure and enthalpy at 40°C and an internal energy of 252.13 kJ/kg, and (k) the approximate enthalpy and specific volume at 10 bars and 20°C.

3-16E. Determine the required data for refrigerant 134a at the following specified conditions: (*a*) the pressure and specific volume of saturated liquid at 60°F, (*b*) the temperature and enthalpy of saturated vapor at 90 psia, (*c*) the specific volume and internal energy at 80 psia and 140°F, (*d*) the temperature and specific volume at 40 psia and a quality of 40 percent, (*e*) the approximate specific volume and enthalpy at 20°F and 60 psia, (*f*) the pressure and enthalpy at 10°F and 35 percent quality, (*g*) the temperature and internal energy at 100 psia and an enthalpy of 132.55 Btu/lb$_m$, (*h*) the quality and specific volume at 80°F and an enthalpy of 90.0 Btu/lb$_m$, (*i*) the internal energy and specific volume at 120°F and an enthalpy of 116.95 Btu/lb$_m$, (*j*) the pressure and enthalpy at 140°F, and an internal energy of 115.58 Btu/lb$_m$, and (*k*) the approximate enthalpy and specific volume at 100 psia and 50°F.

3-17. Determine the internal energy of 0.1 m$^3$ of refrigerant 134a at 0°C if it is known that the specific volume is 0.035 m$^3$/kg in that state, in kilojoules.

3-18. Water vapor at 2.0 MPa and 280°C is cooled at constant volume until the pressure is 0.50 MPa. Determine the internal energy in the final state, and sketch the process on a *Pv* diagram.

3-19E. Determine the internal energy of 0.4 ft$^3$ of refrigerant 134a at 30°F, in Btu, if it is known that the specific volume is 0.80 ft$^3$/lb$_m$ in that state.

3-20E. Water vapor at 300 psia and 450°F is cooled at constant volume until the pressure is 50 psia. Determine the internal energy in the final state, and sketch the process on a *Pv* diagram.

3-21. A wet mixture of water substance is maintained in a rigid tank at 60°C. The system is heated until the final state is the critical state. Determine (*a*) the quality of the initial mixture and (*b*) the initial ratio of volume of vapor to liquid.

3-22. A wet mixture of refrigerant 134a is maintained in a rigid tank at 60°C. The system is heated until the final state is the critical state. Determine (*a*) the quality of the initial mixture and (*b*) the initial ratio of volume of vapor to liquid.

3-23E. A wet mixture of water substance is maintained in a rigid tank at 200°F. The system is heated until the final state is the critical state. Determine (*a*) the quality of the initial mixture and (*b*) the initial ratio of volume of vapor to liquid.

3-24E. A wet mixture of refrigerant 134a is maintained in a rigid tank at 200°F. The system is heated until the final state is the critical state. Determine (*a*) the quality of the initial mixture and (*b*) the initial ratio of volume of vapor to liquid.

3-25. A rigid tank contains water vapor at 15 bars and an unknown temperature. When the vapor is cooled to 180°C, it begins to condense.

Estimate (*a*) the original temperature in degrees Celsius and (*b*) the change in internal energy in kJ/kg. (*c*) Sketch the process on a *Pv* diagram.

3-26. Water substance at 10 bars and 280°C is cooled in a rigid tank until the fluid is a saturated vapor. Determine (*a*) the final pressure and temperature, in bars and degrees Celsius, and (*b*) the change in the internal energy in kJ/kg. (*c*) Sketch the process on a *Pv* diagram.

3-27. A tank with a volume of 0.008 m³ contains a liquid-vapor mixture of refrigerant 134a at 200 kPa with a quality of 20 percent. Determine (*a*) the mass of vapor present, in kilograms, and (*b*) the fraction of the total volume occupied by the liquid phase.

3-28E. A tank with a volume of 0.3 ft³ contains a liquid-vapor mixture of refrigerant 134a at 30 psia with a quality of 15 percent. Determine (*a*) the mass of vapor present, in $lb_m$, and (*b*) the fraction of the total volume occupied by the liquid phase.

3-29. Water substance initially at 0.3 bar and 1.694 m³/kg is heated at constant volume to a final pressure of 1.0 bar. Determine (*a*) the initial quality and (*b*) the change in specific internal energy in kJ/kg. Finally, (*c*) sketch the process on a *Pv* diagram.

3-30. Water substance at 10 bars and 0.02645 m³/kg undergoes a constant-pressure process to a final state of 0.206 m³/kg. (*a*) Determine the specific internal energy change in kJ/kg. (*b*) Sketch the process on a *Pv* diagram.

3-31. Refrigerant 134a undergoes an isothermal process at 40°C. The initial pressure is 4 bars and the final specific volume is 0.010 m³/kg. (*a*) Determine the specific enthalpy change in kJ/kg. (*b*) Sketch the process on a *Pv* diagram.

3-32. Steam initially at 1.5 bars and 200°C is compressed isothermally to two different final states. (*a*) If the final specific volume is 0.30 m³/kg, find the change in internal energy, in kJ/kg. (*b*) If the final internal energy is 2200 kJ/kg, find the change in specific volume in m³/kg. (*c*) Sketch the two processes on the same *Pv* diagram.

3-33. Refrigerant 134a undergoes a change of state at constant pressure from 3.2 bars, 20°C, to a final state of (*a*) 0.030 m³/kg and (*b*) −4°C. For part *a* determine the change in internal energy in kJ/kg, and for part *b* determine the change in enthalpy in kJ/kg. (*c*) Sketch the two processes on the same *Pv* diagram.

3-34. Water substance is contained in a piston-cylinder device initially at 1.0 MPa and 0.2678 m³/kg. It is compressed at constant pressure until it becomes a saturated vapor.
   (*a*)  Find the initial and final temperatures, in degrees Celsius.
   (*b*)  Find the work input required, in kJ/kg.
   (*c*)  If the original volume is 1 liter, determine the internal energy change of the fluid, in kilojoules.

3-35. Initially, a 0.2-m$^3$ rigid tank contains saturated water vapor at 5.0 bars. Heat transfer from the substance results in a drop in pressure to 1 bar. For the final equilibrium state determine (*a*) the temperature, in degrees Celsius, (*b*) the final quality, and (*c*) the ratio of the mass of liquid to the mass of vapor. (*d*) Show the process on a *Pv* diagram.

3-36E. Water substance is contained in a piston-cylinder device initially at 120 psia and 4.36 ft$^3$/lb. It is compressed at constant pressure until it becomes a saturated vapor.
(*a*) Find the initial and final temperatures, in degrees Fahrenheit.
(*b*) Find the work input required, in Btu/lb$_m$.
(*c*) If the original volume is 100 in$^3$, determine the internal energy change of the fluid, in Btu.

3-37E. Initially, a 1.0-ft$^3$ rigid tank contains saturated water vapor at 50 psia. Heat transfer from the substance results in a drop in pressure to 15 psia. For the final equilibrium state determine (*a*) the temperature, in degrees Fahrenheit, (*b*) the final quality, and (*c*) the ratio of the mass of liquid to the mass of vapor. (*d*) Show the process on a *Pv* diagram.

3-38. Refrigerant 134a at a pressure of 0.50 MPa has a specific volume of 0.025 m$^3$/kg (state 1). It expands at constant temperature until the pressure falls to 0.28 MPa (state 2). Finally, the fluid is cooled at constant pressure until it becomes a saturated vapor (state 3).
(*a*) Determine the change in specific volume between states 1 and 2, and states 1 and 3, in m$^3$/kg.
(*b*) Determine the specific internal energy change, in kJ/kg, between states 1 and 2.
(*c*) Determine the specific enthalpy change, in kJ/kg, between states 2 and 3.
(*d*) Show the path of the process on a *Pv* diagram.

3-39. Dry, saturated steam at 30 bars (state 1) is contained within a piston-cylinder device which initially has a volume of 0.03 m$^3$. The steam is cooled at constant volume until the temperature reaches 200°C (state 2). The system is then expanded isothermally until the volume in state 3 is twice the initial value.
(*a*) Determine the pressure at state 2.
(*b*) Determine the pressure at state 3.
(*c*) Determine the change in the internal energy for the two processes 1-2 and 2-3 in kJ.

3-40E. Refrigerant 134a at a pressure of 120 psia has a specific volume of 0.25 ft$^3$/lb$_m$ (state 1). It expands at constant temperature until the pressure falls to 50 psia (state 2). Finally, the fluid is cooled at constant pressure until it becomes a saturated vapor (state 3).
(*a*) Determine the change in specific volume between states 1 and 2, and states 1 and 3.

(b) Determine the specific internal energy change, in Btu/lb$_m$, between states 1 and 2.

(c) Determine the specific enthalpy change, in Btu/lb$_m$, between states 2 and 3.

(d) Show the path of the process on a $Pv$ diagram.

3-41E. Saturated water vapor at 110 psia (state 1) is contained within a piston-cylinder device which initially has a volume of 1.0 ft$^3$. The steam is cooled at constant volume until the temperature reaches 300°F (state 2). The system is then expanded isothermally until the volume in state 3 is twice the initial value.

(a) Determine the pressure at state 2.

(b) Determine the pressure at state 3.

(c) Determine the change in the internal energy for the two processes 1-2 and 2-3 in Btu.

(d) Plot the two processes on a $Pv$ diagram.

3-42. A cylinder having an initial volume of 2 m$^3$ initially contains steam at 10 bars and 200°C (state 1). The vessel is cooled at constant temperature until the volume is 41.95 percent of the initial volume (state 2). The constant-temperature process is followed by a constant-volume process which ends with a pressure in the cylinder of 40 bars (state 3).

(a) Determine the pressure in bars and the enthalpy, in kJ/kg, at state 2.

(b) Determine the temperature, in degrees Celsius, and the enthalpy at state 3.

(c) Sketch the two processes on a $Pv$ diagram, with respect to the vapor dome.

3-43. Steam initially at 3.0 MPa and 400°C (state 1) is cooled at constant volume to a temperature of 200°C (state 2). The fluid is then further cooled at constant temperature to a saturated liquid (state 3). Ascertain (a) the final pressure in bars, (b) the quality after the constant-volume process, (c) the overall specific-volume change, in m$^3$/kg, and (d) the change in specific internal energy, in kJ/kg, between states 2 and 3. Finally, (e) sketch the processes on a $Pv$ diagram.

3-44E. Steam initially at 40 psia and 600°F (state 1) is cooled at constant volume to a pressure of 15 psia (state 2). The fluid is then further cooled at constant temperature to a saturated liquid (state 3). Ascertain (a) the final pressure in psia, (b) the quality after the constant-volume process, (c) the overall specific-volume change, in ft$^3$/lb$_m$, and (d) the change in specific internal energy, in Btu/lb$_m$, between states 2 and 3. Finally, (e) sketch the processes on a $Pv$ diagram.

3-45. A piston-cylinder device initially contains nitrogen at 1.0 MPa, 200 K, and 5.00 L. The fluid is compressed to 10.0 MPa and 0.7706 L. Determine (a) the final temperature in kelvins and the internal-energy change, in kilojoules, based on real-gas data, and (b) the final temperature if the ideal-gas equation, $Pv = R_u T$, is used, where $R_u = 8.314$ kPa·m/kmol·K.

3-46. Saturated liquid water at 40°C is compressed at 80°C and 50 bars.
   (a) Determine the change in specific volume and internal energy on the basis of the compressed-liquid table.
   (b) Determine the same quantities if saturated data were used as an approximation.
   (c) Find the percentage of error involved when the second answers are compared to the first answers.

3-47. Water at 2.5 MPa and 40°C changes states to 10.0 MPa and 100°C.
   (a) Determine the change in specific volume and enthalpy on the basis of the compressed-liquid table.
   (b) Find $\Delta v$ and $\Delta h$ if saturated data were used as an approximation.
   (c) Determine the percentage of error involved in $v_2$ and $\Delta h$ when part $b$ is compared to part $a$.

3-48. Water at 50 bars and 80°C changes states to 200 bars and 100°C.
   (a) Determine the change in internal energy and enthalpy on the basis of the compressed-liquid table.
   (b) Find $\Delta u$ and $\Delta h$ on the basis of saturation data as an approximation.
   (c) Determine the percentage of error involved when the second set of answers is compared to the first set.

3-49E. Water at 500 psia and 50°F changes states to 1500 psia and 100°F.
   (a) Determine the change in specific volume and enthalpy on the basis of the compressed-liquid table.
   (b) Find $\Delta v$ and $\Delta h$ if saturated data were used as an approximation.
   (c) Determine the percentage of error involved in $v_2$ and $\Delta h$ when part $b$ is compared to part $a$.

3-50E. Water at 1000 psia and 100°F changes states to 3000 psia and 150°F.
   (a) Determine the change in internal energy and enthalpy on the basis of the compressed-liquid table.
   (b) Find $\Delta u$ and $\Delta h$ on the basis of saturation data as an approximation.
   (c) Determine the percentage of error involved when the second set of answers is compared to the first set.

## ENERGY ANALYSIS USING SATURATED AND SUPERHEAT DATA

3-51. A piston-cylinder device initially contains water vapor at 200°C. The fluid expands isothermally from 15 to 3 bars.
   (a) Plot the process on a $Pv$ diagram, using tabular data at 3, 5, 7, 10, and 15 bars.

(b) Estimate graphically the work for the process, in kJ/kg, for the real gas.

(c) Determine the work if the fluid is modeled by the equation $Pv = RT$ (ideal gas), for the same initial and final pressures, where $R = 8.314$ kJ/kmol·K.

(d) Estimate the heat transfer, in kJ/kg, for a real gas.

(e) Determine the heat transfer for an ideal-gas model, if $\Delta u = 0$ for the model.

3-52. A piston-cylinder device initially contains refrigerant 134a at 5.0 bars and 40°C. The fluid is compressed isothermally to 9.0 bars.

(a) Plot the process on a $Pv$ diagram, using tabular data at 5, 6, 7, 8, and 9 bars.

(b) Estimate graphically the work for the process, in kJ/kg.

(c) Determine the work if the fluid is modeled by the equation $Pv = RT$ (ideal gas), for the same initial and final pressures, where $R = 8.314$ kJ/kmol·K.

(d) Estimate the heat transfer, in kJ/kg, for the real gas.

(e) Determine the heat transfer for an ideal-gas model for which $\Delta u = 0$.

(f) Find the percentage of error in determining the heat transfer if an ideal-gas model is used.

3-53E. A piston-cylinder device initially contains water vapor at 400°F. The fluid expands isothermally from 200 to 120 psia.

(a) Plot the process on a $Pv$ diagram, using tabular data at 120, 140, 160, 180, and 200 psia.

(b) Estimate graphically the work for the process, in ft·lb$_f$/lb$_m$, for the real gas.

(c) Determine the work if the fluid is modeled by the equation $Pv = RT$ (ideal gas), for the same initial and final pressures, where $R = 1545$ ft·lb$_f$/lbmol·°R.

(d) Estimate the heat transfer, in Btu/lb$_m$, for a real gas.

(e) Determine the heat transfer for an ideal-gas model for which $\Delta u = 0$.

(f) Find the percentage of error in determining the heat transfer if an ideal-gas model is used.

3-54. A piston-cylinder device initially contains 1.2 kg of a liquid-vapor mixture of water at 2 bars and 0.233 m$^3$. The piston rests on stops and will not move until the system pressure reaches 10 bars. Heat is now transferred at a constant rate of 250 kJ/min. Determine (a) the mass of initial liquid, (b) the heat added before the piston moves, in kilojoules, and (c) the time required before the piston moves, in minutes.

3-55. A piston cylinder device contains a liquid-vapor mixture of water initially at 5 bars and 0.356 m$^3$/kg. Expansion occurs along the path

$Pv$ = constant until the pressure reaches 1.5 bars. If the work done by the water is 214 kJ/kg, determine the heat transfer in kJ/kg.

3-56. Nitrogen is cooled in a rigid container from its critical state to a pressure of 4 bars. Find (a) the final temperature, in kelvins, and (b) the heat transferred in kJ/kg. (c) Sketch the process on a $Pv$ diagram.

3-57. A poorly insulated vessel with a volume of 100 L contains liquid nitrogen at 77.24 K. The fluid is 91.5 percent liquid by volume. The filler cap is accidentally sealed off, and the heat-leak rate into the vessel from the surroundings is 5 J/s. If the vessel will rupture when the pressure reaches 400 kPa, find the time to reach this pressure, in hours.

3-58. A piston-cylinder device initially contains water with a quality of 25 percent at 1.5 bars. Heat is added at constant pressure until the volume has increased by 4.09 times the initial value. Determine (a) the work done by the water, in kJ/kg, and (b) the heat transfer required in kJ/kg.

3-59. A rigid, insulated vessel with a volume of 1 m³ contains 2 kg of a liquid-vapor mixture of water at 30°C. Simultaneously, a paddle wheel driven by a motor rotates at 50 rpm with a constant applied torque of 50 N·m and a resistor within the system receives a current of 100 A from a 10-V supply. Determine (a) the time required in minutes to evaporate all of the liquid in the vessel, (b) the pressure in the vessel at that time, in bars, and (c) the cost, in cents, of electricity provided to the motor and resistor, if electricity costs $0.108/kW·h.

3-60. One cubic meter of water substance at 10.0 MPa and 480°C is contained within a rigid tank. The fluid is cooled until its temperature reaches 320°C. Determine the final pressure, in bars, and the heat transfer, in kilojoules.

3-61. One-tenth kilogram of water substance at 3 bars occupies a volume of 0.0303 m³ in a constant-pressure piston-cylinder device. Energy by heat transfer is added in the amount of 122 kJ. Find (a) the final temperature, in degrees Celsius, and (b) the work output, in kilojoules. (c) Sketch the path on a $Pv$ diagram, relative to the saturation line.

3-62. One and one-half kilograms of saturated water vapor at 3 bars are contained in a piston-cylinder device. Energy by heat transfer is added in the amount of 600 kJ, and a paddle wheel within the system is rotated through 2000 rev. If the final temperature is 400°C and the pressure remains constant, determine the constant torque applied to the shaft of the paddle wheel, neglecting energy storage in the paddle wheel, in newton-meters.

3-63. One kilogram of saturated water vapor at 5 bars is contained in a piston-cylinder assembly. Energy by heat transfer is added in the amount of 225 kJ, and some electrical work is done by passing a constant current of 1.5 A through a resistor in the fluid for 0.5 h. If the final temperature of the steam is 400°C and the process is constant-pressure, determine (a) the necessary voltage, in volts, of the power line which supplied the potential for the current, and (b) the cost of the electricity, in cents, if the power company rate is $0.110/kW·h. Neglect energy storage in the resistor.

3-64E. One cubic foot of water substance at 1600 psia and 800°F is contained within a rigid tank. The fluid is cooled until its temperature reaches 600°F. Determine the final pressure, in psia, and the heat transfer, in Btu.

3-65E. A piston-cylinder device with an initial volume of 0.10 ft$^3$ contains water substance initially at 160 psia and 50 percent quality. Energy by heat transfer is added in the amount of 35.6 Btu while the pressure remains constant. Determine (a) the mass within the system, in pounds, and (b) the final temperature, in degrees Fahrenheit. (c) Sketch the process on a $Pv$ diagram.

3-66E. Three pounds of saturated water vapor at 40 psia are contained in a piston-cylinder device. Energy by heat transfer is added in the amount of 600 Btu, and a paddle wheel within the system is rotated through 5000 rev by an electric motor. If the final temperature is 800°F and the pressure remains constant, determine (a) the constant torque applied to the shaft of the paddle wheel, in lb$_f$·ft, and (b) the cost of the electricity, in cents, if the power company rate is $0.104/kW·h. Neglect energy storage in the paddle wheel.

3-67E. Energy by heat transfer is added in the amount of 92 Btu to 1 pound of saturated water vapor at 40 psia contained in a piston-cylinder assembly. In addition, electrical work is done by passing a current of 1.5 A through a resistor in the fluid for 0.5 h. If the final temperature of the steam is 700°F and the process is constant-pressure, determine the necessary voltage, in volts, of the battery which supplied the potential for the current. Neglect energy storage in the resistor.

3-68. Refrigerant 134a is contained in a rigid tank initially at 2 bars, a quality of 50.4 percent, and a volume of 0.10 m$^3$. Energy by heat transfer is added until the pressure reaches 5 bars. Determine (a) the mass within the system, in kilograms, and (b) the quantity of heat added, in kilojoules. (c) Sketch the process path on a $Pv$ diagram.

3-69. A rigid tank contains 6.0 kg of refrigerant 134a at 6 bars and 60°C. A paddle wheel within the tank, which is driven by a motor external to the system, adds energy at a constant torque of 125 N·m for 800 rev. At the same time, the system is cooled to a final temperature of 12°C. Determine (a) the final internal energy, in kilojoules, (b) the direction and magnitude of the heat transfer, in kilojoules, and (c) the cost, in cents, of the electricity to the motor if the power company rate is \$0.094/kW·h. Then, (d) sketch the process on a $Pv$ diagram, relative to the saturation line. Neglect energy storage in the paddle wheel.

3-70. A closed, rigid tank contains 0.5 kg of saturated water vapor at 4 bars. Energy by heat transfer is added in the amount of 70 kJ, and some work is done by means of a paddle wheel until the steam is at 7 bars. Calculate the work required, in kilojoules.

3-71. A rigid vessel with a volume of 0.05 m$^3$ is initially filled with saturated steam at 1 bar. The contents are cooled to 75°C.
(a) Sketch the process on $Pv$ coordinates with respect to the saturation line.
(b) What is the final pressure, in bars?
(c) Find the heat transferred from the steam, in kilojoules.

3-72E. Refrigerant 134a is contained in a rigid tank initially at 30 psia, a quality of 62.9 percent, and a volume of 3.0 ft$^3$. Energy by heat transfer is added until the pressure reaches 80 psia. Determine (a) the mass within the system, in pounds, and (b) the quantity of heat added, in Btu. (c) Sketch the process path on a $Pv$ diagram.

3-73E. Twenty-five pounds of refrigerant 134a at 80 psia and 180°F are contained in a rigid tank. A paddle wheel within the tank adds energy at a constant torque of 120 lb$_f$·ft for 1200 rev. At the same time, the system is cooled to a final temperature of 40°F. Determine (a) the final internal energy, in Btu, and (b) the direction and magnitude of the heat transfer, in Btu. (c) Sketch the process on a $Pv$ diagram, relative to the saturation line. Neglect energy storage in the paddle wheel.

3-74E. Two pounds of saturated water vapor at 60 psia are contained within a rigid tank. Energy by heat transfer is added in the amount of 140 Btu, and some work is done by means of a paddle wheel until the steam is at 100 psia. Calculate the work required, in Btu.

3-75E. A 2.0-ft$^3$ rigid vessel is initially filled with saturated steam at 14.7 psia. The contents are cooled to 150°F. (a) Sketch the process on $Pv$ coordinates with respect to the saturation line. (b) Find the final pressure, in psia. (c) Find the heat transferred from the steam, in Btu.

3-76. A rigid, insulated tank initially is divided into two sections by a partition. One side contains 1.0 kg of saturated-liquid water initially at 6.0 MPa, and the other side is evacuated. The partition is broken, and the fluid expands into the entire container. The total volume is such that the final equilibrium pressure is 3.0 MPa. Determine (a) the initial volume of the saturated liquid, in liters, and (b) the total volume of the entire tank, in liters. (c) Sketch the process on a $Pv$ diagram relative to the saturation line.

3-77. One kilogram of water substance initially at 10 bars and 200°C is altered isothermally until its volume is 50 percent of the initial value.

During the process the compression work is 170 kJ/kg, and paddle-wheel work in the amount of 49 N·m/g also takes place.

(a) Determine the magnitude, in kilojoules, and direction of any heat transfer.

(b) Sketch the process on a $Pv$ diagram, relative to the saturation line.

3-78. One-tenth kilogram of refrigerant 134a initially is a wet mixture with a quality of 50 percent at 40°C. It expands isothermally to a pressure of 5 bars. The measured work output due to the expansion is 19 N·m/g.

(a) Determine the magnitude, in kilojoules, and direction of any heat transfer.

(b) Sketch the process on a $Pv$ diagram, relative to the saturation line.

3-79. Two kilograms of water substance are contained in a piston-cylinder device at 320°C. The substance undergoes a constant-temperature process with the volume changing from 0.02 to 0.17 $m^3$. The measured work output is 889 kJ. Determine (a) the final pressure, in bars, and (b) the heat transfer, in kilojoules. Also sketch the process on a $Pv$ diagram, relative to the saturation line.

3-80. An insulated piston-cylinder assembly contains refrigerant 134a initially as a saturated vapor at 40°C and a volume of 1.194 liters. The pressure is adjusted continuously during a process so that the variation of pressure with volume is linear. The final pressure is 5 bars and the final temperature is 50°C. During the process a resistor within the cylinder is energized by a battery. (a) Sketch the process on a $Pv$ diagram. Then find (b) the mass of R-134a, in kilograms, (c) the change in specific internal energy, in kJ/kg, (d) the boundary work from an area representation on the $Pv$ diagram, in kJ/kg, and (e) the electrical work, in kilojoules.

3-81E. A rigid, insulated tank initially is divided into two sections by a partition. One side contains 1.0 lb of saturated-liquid water initially at 1000 psia, and the other side is evacuated. The partition is broken, and the fluid expands into the entire container. The total volume is such that the final equilibrium pressure is 500 psia. Determine (a) the initial volume of the saturated liquid and (b) the total volume of the entire tank, in cubic feet. (c) Sketch the process on a $Pv$ diagram relative to the saturation line.

3-82E. Water substance initially at 140 psia and 400°F is altered isothermally until its volume is 50 percent of the initial value. During the process the compression work on the 1-pound system is 65,000 ft·lb$_f$, and paddle-wheel work in the amount of 30,000 ft·lb$_f$ also takes place.

(a) Determine the magnitude, in Btu, and direction of any heat transfer.

(b) Sketch the process on a $Pv$ diagram, relative to the saturation line.

3-83E. A piston-cylinder device contains 4 lb of water substance at 500°F. The fluid undergoes a constant-temperature process with the volume changing from 1.40 to 8.60 ft³. The measured work output is 675 Btu. Determine (a) the final pressure, in psia, and (b) the heat transfer, in Btu. (c) Sketch the process on a $Pv$ diagram, relative to the saturation line.

3-84. An insulated piston-cylinder device contains 0.010 kg of saturated-liquid water at 3 bars and $m$ kg of steam at 3 bars and 200°C. Initially the two masses are separated from each other by an adiabatic membrane. The membrane is broken while the pressure is maintained at 3 bars, and the system proceeds toward equilibrium. Determine (a) the mass $m$ of steam, in kilograms, required in order for the final state to be a saturated vapor and (b) the work that occurs, in joules.

3-85. A piston-cylinder device which is maintained at 3 MPa contains 0.025 kg of water substance initially at 280°C. A paddle wheel adds 1800 N·m of energy, while a heat loss occurs. The final volume of the fluid is 60 percent of the initial value. Determine (a) the final temperature, in °C, (b) the final enthalpy, in kJ/kg, (c) the heat transfer out, in kilojoules. (d) Sketch the process path on a $Pv$ diagram.

3-86. Water substance contained in a piston-cylinder device undergoes two processes in series from an initial state of 10 bars and 400°C. In process 1-2 the water is cooled at constant pressure to a saturated-vapor state. In process 2-3 the water is cooled at constant volume to 150°C.
(a) Determine the work for process 1-2, in kJ/kg.
(b) Determine the heat transfer for the overall process, in kJ/kg.
(c) Sketch both processes on a $Pv$ diagram.

3-87. A piston-cylinder assembly initially contains saturated water vapor at 5 bars. The fluid is first heated at constant pressure to 280°C (state 2). It is then cooled at constant volume to 2 bars (state 3).
(a) Determine the work, internal energy change, and heat transfer for process 1-2, in kJ/kg.
(b) Determine the heat transfer for process 2-3 in kJ/kg.
(c) Sketch the two processes on a $Pv$ diagram.

3-88E. Saturated water vapor at 60 psia is contained within a piston-cylinder assembly. The fluid is first heated at constant pressure to 600°F (state 2). It is then cooled at constant volume to 10 psia (state 3). Determine (a) the work, internal energy change, and heat transfer for process 1-2, in Btu/lb$_m$, and (b) the heat transfer for process 2-3. Finally, (c) sketch the two processes on a $Pv$ diagram.

3-89E. A liquid-vapor water mixture initially at 100 psia and a quality of 50 percent is confined to one side of a rigid, well-insulated container by a partition. The other side of the container is initially evacuated.

The partition is removed and the water expands to fill the entire container at 40 psia. Determine the change in volume of the water, in $ft^3/lb_m$.

3-90. A piston-cylinder device contains refrigerant 134a initially at 2.8 bars and 40°C, and the volume is 0.1 m³. The piston is kept stationary, and there is heat transfer to the gas until its pressure rises to 3.2 bars. Then additional heat transfer occurs from the gas during a process in which the volume varies but the pressure is constant. This latter process ends when the temperature reaches 50°C. Assume the processes are quasistatic, and find (a) the mass in the system, in kilograms, (b) the heat transfer, in kilojoules, during the constant-volume process, and (c) the heat transfer for the constant-pressure process, in kilojoules.

3-91. A system having an initial volume of 2.0 m³ is filled with steam at 30 bars and 400°C (state 1). The system is cooled at constant volume to 200°C (state 2). The first process is followed by a constant-temperature process ending with saturated liquid water (state 3). Find the total heat transfer required, in kilojoules, and its direction. Sketch the two processes on a $Pv$ diagram relative to the saturation line.

3-92. Water initially is a saturated vapor at 1.0 bar (state 1). Energy by heat transfer is removed at constant pressure until the volume is 1000 cm³/g (state 2). Heat transfer is then added at constant volume until a pressure of 3.0 bars is reached (state 3). (a) Determine the work, internal-energy change, and heat transfer for process 1-2, in kJ/kg. (b) Determine the same quantities for process 2-3, in kJ/kg. (c) Sketch the two processes on a $Pv$ diagram, relative to the saturation line.

3-93E. Initially, steam at 450 psia and 700°F is contained in a closed system having a volume of 5.0 ft³ (state 1). The system is cooled at constant volume to 400°F (state 2). This process is followed by a constant-temperature process which ends with the fluid as a saturated liquid (state 3). (a) Find the total heat transfer required, in Btu, and its direction. (b) Sketch the two processes on a $Pv$ diagram relative to the saturation line.

3-94E. Water initially is a saturated vapor at 60 psia (state 1). First it is heated at constant pressure to 600°F (state 2). Then it is cooled at constant volume to 10 psia (state 3). (a) Determine the work, internal-energy change, and heat transfer for process 1-2, in $Btu/lb_m$. (b) Determine the same quantities for process 2-3, in $Btu/lb_m$. (c) Sketch the two processes on a $Pv$ diagram, relative to the saturation line.

3-95. An insulated container maintained at 25 bars is subdivided into two sections by an insulated partition. One section contains 0.50 kg of water at 20°C, while the other contains saturated steam. Determine the amount of saturated steam present if, on breaking the partition, the final state is a wet mixture with 30 percent quality.

# CHAPTER
# 4

# THE IDEAL GAS, CORRESPONDING STATES, AND INCOMPRESSIBLE MODELS

Steam-turbine driven generator at a geothermal power plant near Brawley, California.

In the preceding chapter, tables of data representing $P$, $v$, $T$, $u$, $h$, and $s$ were introduced for the saturation, super-heat, and compressed liquid states of pure substances. Values in these tables are established from experimental $PvT$ data measured in the various phase regions, in conjunction with theoretical property relations (as developed in Chap. 13). When such data are available in the literature, they should be used in preference to other information. However, in the absence or scarcity of data, modeling of property relations often is a fruitful approach. As in all modeling techniques, one must be careful to understand the limitations of these models. In this chapter we restrict ourselves to a gas that is modeled as an ideal gas, and to liquids and solids that are modeled as incompressible substances. These models are quite accurate for a number of engineering systems.

## 4-1 IDEAL-GAS EQUATION OF STATE

Of particular interest in thermodynamics are equations that relate the variables $P$, $v$, and $T$. The $PvT$ behavior of many gases at low pressures and moderate temperatures can be *modeled* quite well by the **ideal-gas equation** of state, namely,

$$PV = NR_uT \qquad \textbf{[4-1]}$$

or

$$P\bar{v} = R_uT \qquad \textbf{[4-2]}$$

where $N$ is the number of moles of a gas and $\bar{v}$ is the specific volume on a molar basis. A bar over specific property values is used in Chaps. 4 through 10 to denote properties on a mole basis. The quantity $R_u$ is the **universal gas constant**. The values of $R_u$ in several sets of units are

$$R_u = \begin{cases} 0.08314 \text{ bar·m}^3/\text{kmol·K} \\ 8.314 \text{ kJ/kmol·K} \\ 8.314 \text{ kPa·m}^3/\text{kmol·K} \\ 1545 \text{ ft·lb}_f/\text{lbmol·}^\circ\text{R} \\ 0.730 \text{ atm·ft}^3/\text{lbmol·}^\circ\text{R} \\ 1.986 \text{ Btu/lbmol·}^\circ\text{R} \end{cases}$$

Note that $R_u$ is expressed in both SI and USCS units. The values are also tabulated in Tables A-1 and A-1E.

The ideal-gas equation frequently is used with mass units such as kilograms and pounds instead of kilogram-moles and pound-moles. In such cases one uses a **specific gas constant** $R$ in the ideal-gas equation instead of the universal value $R_u$. Recall that the mass of a mole of a substance is called the *molar mass M*. Hence the universal and specific gas constants are related by

$$R \equiv \frac{R_u}{M} \qquad \textbf{[4-3]}$$

Values of $M$ are given for some elements and common compounds in Tables A-3 and A-3E. Since $R$ depends on the molar mass of a substance, its value is different for each substance, even when it is expressed in the same set of units. Figure 4-1 shows the value of $R$ in kPa·m$^3$/kg·K for six common gases. The equivalent forms of the ideal-gas equation on a mass basis become

| Substance | $R$ |
|-----------|-------|
| Air | 0.287 |
| Ar | 0.208 |
| $N_2$ | 0.297 |
| He | 2.077 |
| $CO_2$ | 0.189 |
| $H_2$ | 4.124 |

$$PV = mRT \qquad \text{or} \qquad Pv = \frac{R_uT}{M} = RT \qquad \text{or} \qquad P = \rho RT$$

$$\textbf{[4-4]}$$

**Figure 4-1**
The specific gas constants in kPa·m$^3$/kg·K for six common substances.

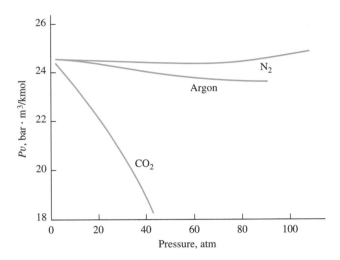

**Figure 4-2**    Experimental data showing the variation of $Pv$ with pressure at a given temperature for several gases.

**Note the difference between $R_u$ and $R$, and how they are related.**

where $v$ is the specific volume on a mass basis, $\rho$ is the density (mass per unit volume), and $m$ is the mass of the system. Recall that the moles $N$ and the mass $m$ are related to the molar mass by the relation $N = m/M$. Carefully note that the temperature in the ideal-gas equation of state is always expressed in either kelvins or degrees Rankine.

On the basis of Eq. [4-2], the quantity $P\bar{v}$ would be constant at a given temperature, independent of pressure, if a gas were ideal at all pressures. A plot of $P\bar{v}$ versus $P$ at a given temperature would be a horizontal line. Typical experimental data on such a plot are shown in Fig. 4-2. Note that nitrogen gas approximates ideal behavior over a wide range of pressures, since the line drawn through experimental data is fairly horizontal out to at least 30 atm. Argon gas begins to deviate after about 10 atm. Carbon dioxide, however, is essentially independent of pressure only at extremely low pressures at a given temperature. This difference in behavior can be explained from a molecular viewpoint in terms of the forces acting between gas particles. As the pressure is reduced at constant temperature, the density of the particles decreases. At low enough pressures the forces between particles become negligible, since the particles are relatively far apart. When the forces are negligible, gases behave like ideal gases. For this reason, all gases behave like ideal gases as the pressure approaches zero. As a *model,* the ideal-gas equation of state is only an approximation at best, strictly valid only at zero pressure. Nevertheless, for monatomic and diatomic gases, the ideal-gas equation is usually a good approximation up to pressures of 10 to 20 atm at room temperature and above, for errors in accuracy not exceeding several percent. The maximum pressure at which a gas can be modeled by the ideal-gas equation of state depends on the degree of accuracy desired.

The data of Fig. 4-2 indicate that the quantity $P\bar{v}$ for gases generally is a function of both $P$ and $T$. However, as the pressure is lowered, the $P\bar{v}$ product approaches the same value, regardless of the nature of the gas. That is, the limiting value of $P\bar{v}$ at zero pressure is the same for *all* gases at the *same* temperature. Since the value of $P\bar{v}$ is a constant, and $P\bar{v} = R_u T$, in the limit as $P$ approaches zero, then

$$\lim_{P \to 0} \frac{P\bar{v}}{T} \equiv R_u$$

Thus this extrapolation of data for $P\bar{v}/T$ to zero pressure provides the experimental means for evaluating the universal gas constant.

---

**Example 4-1**

Determine (*a*) the specific volume of nitrogen gas, in $m^3/kg$, at 27°C and pressures of 1, 10, 50, and 100 bars, and (*b*) the specific volume of water vapor, in $ft^3/lb_m$, at 400°F and pressures of 14.7, 40, 100, and 200 psia based in both cases on (1) an ideal-gas model and (2) a tabular superheat table. Then comment on the significance of the results.

**Solution:**

**Given:** (*a*) Nitrogen gas at 27°C and 1, 10, 50, and 100 bars; (*b*) water vapor at 400°F and 14.7, 40, 100, and 200 psia.

**Find:** $v$ for (*a*) in $m^3/kg$ and for (*b*) in $ft^3/lb_m$ by (1) ideal-gas model and (2) from superheat table.

**Model:** Ideal gas.

**Analysis:** The specific volume of a gas on a mass basis and modeled as an ideal gas is found from Eq. [4-4], $v = R_u T/PM$.

(*a*) In SI units the absolute temperature is $273 + 27 = 300$ K, the molar mass of nitrogen from Table A-2 is 28.01 kg/kmol, and a convenient value of $R_u$ is 0.08314 bar·m³/kmol·K. Use of these values and a pressure of 1 bar in Eq. [4-4] leads to

$$v = \frac{R_u T}{PM} = \frac{0.08314 \text{ bar·m}^3}{\text{kmol·K}} \times \frac{300 \text{ K}}{1.0 \text{ bar}} \times \frac{1 \text{ kmol}}{28.01 \text{ kg}} = 0.8905 \text{ m}^3/\text{kg}$$

The calculated values at the other three pressures, and the tabulated superheat values found in Table A-20, are summarized in Fig. 4-3*a*. Note that the ideal-gas model agrees closely with the tabular data based on experimental measurements. Even at 100 bars the error is only 0.5 percent. Thus at room temperature the ideal-gas model is quite good for nitrogen even at 50 or 100 bars.

(*b*) In USCS units the absolute temperature for water vapor is $400 + 460 = 860$°R. We choose $R_u$ as 10.73 psia·ft³/lbmol·°R, and $M$ is 18.02 lb$_m$/lbmol.

| (*a*) Nitrogen at 300 K | | |
|---|---|---|
| *P*, bars | $v_{ideal}$ | $v_{table}$ |
| 1 | 0.8905 | 0.8902 |
| 10 | 0.0890 | 0.0889 |
| 50 | 0.0178 | 0.0178 |
| 100 | 0.0089 | 0.00895 |
| (*b*) Water vapor at 400°F | | |
| *P*, psia | $v_{ideal}$ | $v_{table}$ |
| 14.7 | 34.84 | 34.67 |
| 40.0 | 12.80 | 12.62 |
| 100.0 | 5.12 | 4.93 |
| 200.0 | 2.56 | 2.36 |

**Figure 4-3**
Summary of data for Example 4-1.

Substitution of these values and the pressure of 14.7 psia into Eq. [4-4] leads to

$$v = \frac{10.73 \text{ psia·ft}^3}{\text{lbmol·°R}} \frac{(400 + 460)°\text{R}}{14.7 \text{ psia}} \frac{1 \text{ lbmol}}{18.02 \text{ lb}_m} = 34.84 \text{ ft}^3/\text{lb}_m$$

The calculated values at the other three pressures, and the tabulated superheat values found in Table A-14E, are summarized in Fig. 4-3b. An 8.4 percent error now occurs at only 200 psia (13.6 bars). To reduce the error made by the ideal-gas model to 0.5 percent for water at 400°F requires that the pressure be 14.7 psia (1 bar) or less.

**Comment:**   The validity of the ideal-gas model depends strongly on the substance as well as the pressure and temperature range. A method for estimating the appropriateness of the ideal-gas model is introduced in Sec. 4-6, when tabular data are not available.

In some calculations involving ideal gases it is not necessary to know the value of the gas constant. A number of thermodynamic relations require a knowledge of ratios of a given property rather than a knowledge of the actual values of the property. For example, information about the value of $v_2/v_1$ might be needed. It is easily seen that for a given ideal gas

$$\frac{v_2}{v_1} = \frac{RT_2/P_2}{RT_1/P_1} = \frac{T_2 P_1}{T_1 P_2}$$

To calculate the individual values of $v_2$ and $v_1$ directly from the ideal-gas equation and then take the ratio of the two values involves considerably more work than using the above relationship directly. In addition, there is the possibility of mathematical error or errors in the units required when separate calculations are made.

One of the interesting consequences of an ideal-gas model is that the equilibrium states of such a gas can be represented by a fairly simple surface in a rectangular coordinate system. Recall from Sec. 3-1 that the functional relationship among properties of a simple compressible substance is given by $y_0 = f(y_1, y_2)$. This relationship can be plotted as a surface with coordinates $y_0$, $y_1$, and $y_2$. If we plot the equation $Pv = RT$ on a $PvT$ coordinate system, then a surface is generated which has the general shape shown in the center of Fig. 4-4. Constant-temperature lines (isotherms) along the surface appear as hyperbolas because $Pv$ is a constant in this case. Conditions of constant pressure or constant volume are represented by straight lines on the surface. The $PvT$ surface shown in Fig. 4-4 represents a portion of the superheat region in Fig. 3-1, where pressure is low and temperature is relatively high. Figure 4-4 also shows the $PT$ and $Pv$ projections of the $PvT$ surface of an ideal gas. The hyperbolic nature of constant-temperature lines on the $Pv$ plane and the straight-line nature of constant-volume lines on the $PT$ plane are clearly shown. Although it is not shown, the surface can also be projected onto the $Tv$ plane. In Chap. 2 we used the $Pv$ plane to exhibit

**Note the position of isothermal lines on a *Pv* diagram and constant-volume lines on a *PT* diagram for an ideal gas.**

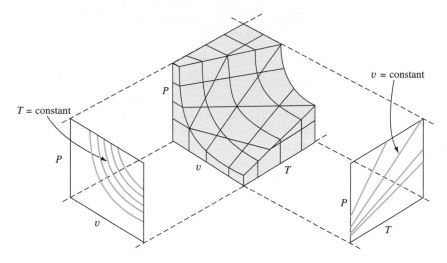

**Figure 4-4**    The *PvT* surface and the *PT* and *Pv* projections for ideal-gas behavior.

and compare the work associated with various quasistatic processes involving compression/expansion work. In that case, the work interaction had an area representation on the *Pv* plane.

## 4-2    INTERNAL ENERGY, ENTHALPY, AND SPECIFIC-HEAT RELATIONS FOR IDEAL GASES

To make suitable energy balances for processes involving ideal gases, it is necessary to evaluate internal-energy and enthalpy changes for these gases. The internal-energy change, $du$, for any simple compressible substance is given by Eq. [3-11], namely,

$$du = c_v\, dT + \left(\frac{\partial u}{\partial v}\right)_T dv \qquad \textbf{[3-11]}$$

The second coefficient, $(\partial u/\partial v)_T$, is a measure of the change in the internal energy of a substance as the volume is altered at constant temperature. From microscopic considerations the internal energy of an ideal gas should not be a function of the volume of the system. As described in the preceding section, there are no forces between particles of an ideal gas. Thus a change in spacing between particles due to a volume change of the system should not affect its energy. This result was confirmed by Joule in 1843. His experiments indirectly indicated that the internal energy of gases at low pressures was essentially a function of temperature only. That is, $(\partial u/\partial v)_T$ is approximately zero at low pressures.

Therefore, when a substance is modeled as an ideal gas, the coefficient $(\partial u/\partial v)_T$ in Eq. [3-11] above may be taken as zero. Thus we may write

$$du = c_v \, dT \qquad \text{for ideal gases} \qquad \textbf{[4-5]}$$

Hence the internal energy of an ideal gas, *unlike that of real gases,* is a function of only one independent variable, the temperature. This equation is misleading, however, in one respect. The presence of the term $c_v$ in the equation often leads one to misconstrue the proper use of the equation. If a gas behaves essentially as an ideal gas, this expression for $du$ is valid for *all processes,* regardless of its path. The use of the equation is *not* restricted to constant-volume processes. Since the specific heat at constant volume is defined for simple compressible substances as $(\partial u/\partial T)_v$, the value of $c_v$ must also be solely a function of temperature for ideal gases.

The extension of these results to the enthalpy function $h$ is straightforward. By definition, $h = u + Pv$, and for an ideal gas $Pv = RT$. Thus we may write

$$dh = du + d(Pv) \qquad \text{and} \qquad d(Pv) = d(RT) = R \, dT$$

The change in the enthalpy for an ideal gas then becomes

$$dh = du + R \, dT \qquad \textbf{[4-6]}$$

The terms on the right-hand side of Eq. [4-6] are solely a function of temperature for an ideal gas. Consequently, the enthalpy of a hypothetical ideal gas is also only a function of temperature. Likewise, since $c_p \equiv (\partial h/\partial T)_p$, the $c_p$ values for ideal gases are a function of temperature only.

To evaluate the enthalpy change of an ideal gas, one starts with Eq. [3-14], which is a general expression for $dh$ for any simple compressible substance:

$$dh = c_p \, dT + \left(\frac{\partial h}{\partial P}\right)_T dP \qquad \textbf{[3-14]}$$

Since the enthalpy of an ideal gas is solely a function of temperature, the above equation reduces to

$$dh = c_p \, dT \qquad \text{for ideal gases} \qquad \textbf{[4-7]}$$

Integration of Eqs. [4-5] and [4-7] for any finite process leads to

$$\Delta u = u_2 - u_1 = \int_1^2 c_v \, dT \qquad \text{for ideal gases} \qquad \textbf{[4-8]}$$

and

$$\Delta h = h_2 - h_1 = \int_1^2 c_p \, dT \qquad \text{for ideal gases} \qquad \textbf{[4-9]}$$

The preceding pair of equations is valid for *all processes* of an ideal gas and is not restricted to either constant-volume or constant-pressure processes. The integration of Eqs. [4-8] and [4-9] for $\Delta u$ and $\Delta h$ requires a knowledge of the specific-heat variation with temperature at low pressures for a given substance. The following section discusses the general behavior of the specific heats of ideal gases in terms of their order of magnitude, as well as their functional dependence on temperature.

**Note that both $u$ and $h$ are solely a function of temperature for a substance modeled as an ideal gas.**

A special relationship between $c_p$ and $c_v$ for ideal gases is obtained by substituting Eqs. [4-5] and [4-7] into Eq. [4-6]. This yields

$$c_p \, dT = c_v \, dT + R \, dT$$

Therefore,

$$\boxed{c_p - c_v = R} \qquad \text{for ideal gases} \qquad \textbf{[4-10]}$$

This simple relationship between $c_p$ and $c_v$ for an ideal gas is an important one, since a knowledge of either $c_p$ or $c_v$ allows the other one to be calculated by the above equation. When the specific heats are given as molar values, the value of $R$ in this equation is $R_u$, the universal gas constant, and $\bar{c}_p - \bar{c}_v = R_u$.

## 4-3 SPECIFIC HEATS OF IDEAL GASES

The specific heats $c_p$ and $c_v$ of gases in general are a function of both temperature and pressure, in accordance with the state postulate. However, as the pressure is lowered and the behavior of a real gas approaches that of an ideal gas, the effect of pressure on the specific heats becomes negligible. As a result, the specific heats of gases at very low pressures are frequently called **ideal-gas,** or **zero-pressure, specific heats.** The symbols $c_{p,0}$ and $c_{v,0}$ signify values at this state of very low pressures. The molar specific heats at constant pressure $\bar{c}_p$ of some common gases are illustrated in Fig. 4-5. In this figure a monatomic gas such as argon has a value of $\bar{c}_{p,0}$ which is essentially constant over the entire range of temperature. Figure 4-5 also shows that molecules with two or more atoms do not have a constant value of $c_p$. These more complex molecules exhibit an increase in $c_p$ with increasing temperature at low pressures. The $\bar{c}_{p,0}$ values of the diatomic gases shown in Fig. 4-5 increase as much as 25 percent over a range of 0 to 1100°C (0 to 2000°F).

### 4-3-1 SPECIFIC HEATS FOR MONATOMIC GASES

Both the kinetic theory of gases and quantum-statistical mechanics predict that $\bar{c}_{p,0}$ of an ideal monatomic gas is $\frac{5}{2} R_u$. This value of 20.8 kJ/kmol·°C or 4.97 Btu/lbmol·°F is characteristic of all *monatomic gases.* On the basis of

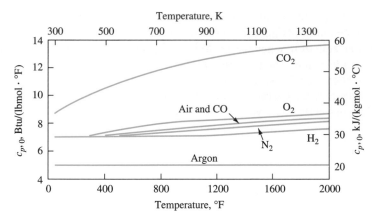

**Figure 4-5**    Values of $c_{p,0}$ for seven common gases. (*Based on data from NBS Circular 564, 1955.*)

Eq. [4-10], it is found that $\bar{c}_{v,0}$ equals $\frac{3}{2}R_u$. Hence $c_v$ for monatomic gases will be approximately 12.5 kJ/kmol·°C or 2.98 Btu/lbmol·°F over a wide range of temperature. These specific-heat values are also shown in Sec. 2 of Tables A-3 and A-3E in the Appendix.

The integration of Eqs. [4-8] and [4-9] for $\Delta u$ and $\Delta h$ is straightforward for monatomic gases, because $c_v$ and $c_p$ are constant and can be brought outside the integral sign. Hence

$$\Delta u = \int_1^2 c_v\, dT = c_v\,\Delta T \qquad \text{(monatomic gas)} \qquad \textbf{[4-11]}$$

and

$$\Delta h = \int_1^2 c_p\, dT = c_p\,\Delta T \qquad \text{(monatomic gas)} \qquad \textbf{[4-12]}$$

for *all monatomic gases*.

**Note that both $c_v$ and $c_p$ of any monatomic ideal gas are independent of temperature.**

**4-3-2**   **INTEGRATION OF ALGEBRAIC EQUATIONS FOR $c_p$ AND $c_v$**

To integrate Eqs. [4-8] and [4-9] for other types of ideal gases, we generally need equations that relate $c_v$ and $c_p$ to temperature. Specific-heat data as a function of temperature are either measured directly or evaluated from theory based on molecular models plus spectroscopic data on molecules. Accurate algebraic equations may then be fitted to the experimental or calculated data by computer techniques. These equations would represent, for example, the lines shown in Fig. 4-5. Equations for $\bar{c}_{p,0}$ are given in Sec. 3 of Tables A-3 and A-3E for a few common gases. The temperature unit is specified by the author of the equations. The units on $\bar{c}_p$ depend on the choice of $R_u$, since the overall equation is written in a dimensionless format.

In many tables of specific-heat equations, the range of temperature over which an equation is appropriate is listed. Also, the maximum percentage error of an equation, when used within the prescribed range of temperature, is frequently noted. The algebraic format of such equations is arbitrary.

Equations for $\overline{c}_{p,0}$ similar to those in Tables A-3 and A-3E are available in the literature for a large number of substances. Use of these equations in the ideal-gas relations given by Eqs. [4-8] and [4-9], namely,

$$\Delta u = \int_{T_1}^{T_2} c_v \, dT \quad \text{and} \quad \Delta h = \int_{T_1}^{T_2} c_p \, dT$$

makes direct integration possible. Such integrations would be carried out for each set of temperature limits of interest. Many repetitive calculations may be required for engineering design purposes, but the use of computers greatly reduces the work required.

### 4-3-3   IDEAL-GAS TABLES

In some situations reasonable accuracy is desired, but repetitive integration of the above equations for $\Delta u$ and $\Delta h$ is not convenient. To meet these needs, the internal-energy and enthalpy data for many gases have been extensively tabulated over relatively small temperature increments. These tables are the result of the integration of accurate specific-heat data, and they greatly simplify the evaluation of $\Delta u$ and $\Delta h$ when repetitive calculations are not essential. To obtain the values of $u$ and $h$ shown in such tables, the integration process is based on the following modification of Eq. [4-9]:

$$h = h_{\text{ref}} + \int_{T_{\text{ref}}}^{T} c_p \, dT \qquad \textbf{[4-13]}$$

That is, to obtain any $h$ value at a given $T$, an arbitrary reference value $h_{\text{ref}}$ is set at a reference temperature $T_{\text{ref}}$. In the ideal-gas tables in the Appendix, the arbitrary reference value of $h = 0$ is chosen to be at 0 K. Therefore, a table of values of $h$ versus $T$ is developed from the expression

$$h = \int_{0}^{T} c_p \, dT \qquad \textbf{[4-14]}$$

where equations for $c_p$ as a function of $T$ must be known throughout the temperature range. A tabulation of the specific internal energy versus temperature is generated from the specific enthalpy data by employing $u = h - Pv = h - RT$ at selected temperature values.

The $h$ and $u$ data for air in Table A-5 are in kJ/kg, and in Table A-5E are in Btu/lb. Tables A-6 to A-11 and Tables A-6E to A-11E list the *molar* specific enthalpy $\overline{h}$ and internal energy $\overline{u}$ for the gases $N_2$, $O_2$, CO, $CO_2$, $H_2O$, and $H_2$ in units of kJ/kmol and Btu/lbmol. Other properties that appear in these tables are introduced in later chapters and should not be used now.

**Note that $u$ and $h$ for air are tabulated on a mass basis, while these data for all other ideal gases are tabulated on a mole basis in this text.**

| $T$, K | Air | Oxygen |
|--------|-------|--------|
| 300 | 1.005 | 0.918 |
| 350 | 1.008 | 0.928 |
| 400 | 1.013 | 0.941 |
| 450 | 1.020 | 0.956 |
| 500 | 1.029 | 0.972 |

**Figure 4-6**
The $c_{p,0}$ values for two common gases versus temperature.

### 4-3-4   AVERAGE SPECIFIC-HEAT APPROXIMATION

In addition to equations for $c_p$ as a function of temperature, Sec. 1 of Tables A-3 and A-3E also tabulates values of $c_{p,0}$ and $c_{v,0}$ as a function of temperature for a few common gases. A condensed list of data from Table A-3 for two of these gases is shown in Fig. 4-6. It is noted from these data that the specific heats are nearly constant over small temperature intervals (several hundred degrees or less). In this situation little error is introduced if the specific heats are assumed to be constant. In addition, the arithmetic average of either $c_v$ or $c_p$ is used for the given temperature interval. When the arithmetic average is used to calculate either $\Delta u$ or $\Delta h$, then Eqs. [4-8] and [4-9] reduce to

$$\Delta u = u(T_2) - u(T_1) = c_{v,\text{av}} \, \Delta T \qquad \textbf{[4-15]}$$

and

$$\Delta h = h(T_2) - h(T_1) = c_{p,\text{av}} \, \Delta T \qquad \textbf{[4-16]}$$

where $c_{v,\text{av}} = (c_{v,1} + c_{v,2})/2$ between states 1 and 2. A similar equation applies to $c_{p,\text{av}}$. Note that the values of $c_v$ and $c_p$ for *monatomic gases,* discussed earlier, are also shown in Sec. 2 of Tables A-3 and A-3E. Equations [4-15] and [4-16] also apply to monatomic gases.

Figure 4-7 shows a typical plot of $c_p$ versus temperature. The *true* average specific heat between states 1 and 2 is designated $c_p^*$ in the figure. It is found by drawing a horizontal line so that the areas shown in color are equal. In this particular case, the value of $c_{p,\text{av}}$ lies below $c_p^*$. However, for small temperature intervals, the use of $c_{p,\text{av}}$ in place of $c_p^*$ will lead to a very small error in $\Delta h$. Alternatively, the specific heat at the average temperature $(T_1 + T_2)/2$ may be used for determining $c_{v,\text{av}}$ and $c_{p,\text{av}}$. There is no reason to prefer one method over the other. As a further rough approxi-

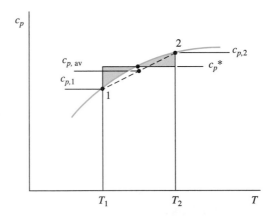

**Figure 4-7**   Illustration showing the difference between the true average specific heat $c_p^*$ and the arithmetic average specific heat for a given temperature interval.

mation method, the specific heat at the *initial* state might be used for the average value. This technique is useful when the value of the final temperature is as yet unknown.

In summary, the evaluation of the integrals of $c_v\, dT$ and $c_p\, dT$ can be carried out by the following four methods. They are in the order of decreasing accuracy.

1. Use an accurate equation for the specific heat versus temperature, based on experimental data.

2. Use tabular data for $u$ and $h$, which are based on accurate specific-heat equations. This may lead to interpolation of data.

3. Use an arithmetically averaged specific heat in the desired temperature range.

4. Use the specific heat at the initial state, and assume it is constant.

Example 4-2 below illustrates and compares various methods of evaluating the enthalpy change of an ideal gas.

---

Find the change in enthalpy of air which is heated at low pressure from 300 to 500 K by use of (*a*) an empirical specific-heat equation, (*b*) the tabular *h* data for air, and (*c*) average tabular specific-heat data, in kJ/kg.

**Example 4-2**

**Solution:**

**Given:** Air is heated from 300 to 500 K at low pressure.

**Find:** $\Delta h$ by (*a*) empirical $c_{p,0}$ equation, (*b*) tabular *h* data, and (*c*) $c_{p,\text{av}}$ data, in kJ/kg.

**Model:** Ideal gas.

**Analysis:** The enthalpy change of an ideal gas in general is given by $\Delta h = \int c_{p,0}\, dT$.

(*a*) The $\bar{c}_p/R_u$ equation for air as an ideal gas is given in Sec. 3 of Table A-3 in the Appendix. Based on an $R_u$ value of 8.314 kJ/kmol·K, the equation for $c_{p,0}$ is

$$\bar{c}_{p,0} = 30.37 - 11.09 \times 10^{-3}\, T$$
$$+ 27.36 \times 10^{-6}\, T^2 - 15.88 \times 10^{-9}\, T^3 + 2.29 \times 10^{-12}\, T^4$$

where $\bar{c}_{p,0}$ is measured in kJ/(kmol·K) and $T$ is in kelvins. If the air is assumed to behave as an ideal gas, then substitution of the above equation for $c_{p,0}$ into Eq. [4-9] and subsequent integration lead to

$$\Delta \bar{h} = \int_1^2 c_{p,0}\, dT = \int_{300}^{500} (30.37 - 11.09 \times 10^{-3}\, T + 27.36 \times 10^{-6}\, T^2$$
$$- 15.88 \times 10^{-9}\, T^3 + 2.29 \times 10^{-12}\, T^4)\, dT$$
$$= (6074 - 887 + 216 + 13)\text{ kJ/kmol} = 5878\text{ kJ/kmol}$$

Since 1 kmol of air contains 28.97 kg,

$$\Delta h = (5878/28.97) \text{ kJ/kg} = 202.9 \text{ kJ/kg}$$

(b) The integral of $c_p \, dT$ can also be found from the $h$ values read directly from Table A-5 in the Appendix, which lists the ideal-gas properties for dry air. Thus,

$$\Delta h = h_2 - h_1 = (503.02 - 300.19) \text{ kJ/kg} = 202.8 \text{ kJ/kg}$$

(c) On the basis of an average specific-heat value, Eq. [4-16] shows that $\Delta h = c_{p,\text{av}} \Delta T$. At 300 and 500 K, from Table A-3M, we note $c_p$ values of 1.005 and 1.029 kJ/kg·K, respectively. Using the arithmetic average of 1.017 kJ/kg·K, we find that

$$\Delta h = c_{p,\text{av}} \Delta T = 1.017(200) \text{ kJ/kg} = 203.4 \text{ kJ/kg}$$

This answer, based on a linear variation of $c_p$ with temperature, differs from the integrated value by roughly 0.2 percent. The close approximation is not unexpected, because the temperature interval is reasonably small. All three answers are substantially the same.

**Comment:**   This problem illustrates the point that use of average specific-heat data leads to reasonably accurate answers if the temperature range is small. In fact, in this particular problem, the use of $c_p$ at 300 K (the initial temperature) of 1.005 kJ/kg·°C leads to a $\Delta h$ equal to 201.0 kJ/kg, which is only 1 percent lower than part $a$. For appreciably larger $\Delta T$ values, the discrepancy in answers will get larger.

---

**Example 4-3**

**W**ater vapor is heated from 600 to 700°F at pressures of 14.7, 100, 250, and 500 psia. Determine the change in internal energy, in Btu/lb$_m$, (a) if the ideal-gas model is used, and (b) if real-gas tabular data are used.

**Solution:**

**Given:**   Water vapor heated from 600 to 700°F at pressures of 14.7, 100, 250, and 500 psia.

**Find:**   $\Delta u$ by (a) ideal-gas data and (b) tabular superheat data.

**Model:**   Ideal- versus real-gas data.

**Analysis:**   (a) The internal energy change of an ideal gas does not depend upon pressure. If ideal-gas Table A-10E is used,

$$\Delta u = \frac{\bar{u}_2 - \bar{u}_1}{M} = \frac{(7163.5 - 6490.0) \text{ Btu/lbmol}}{18.02 \text{ lb/lbmol}} = 37.4 \text{ Btu/lb}_m$$

(b) Data based on experimental measurement are found in Table A-14E. At 14.7 psia,

$$\Delta u = u_2 - u_1 = (1252.8 - 1218.6) \text{ Btu/lb} = 37.5 \text{ Btu/lb}_m$$

The ideal-gas solution is only 0.3 percent lower than the measured value. However, at 100, 200, and 500 psia, the values of $\Delta u$ from Table A-14E are 38.6, 40.6, and 44.9 Btu/lb. In these cases, the ideal-gas model gives an answer that is 3.1, 7.9, and 16.7 percent lower than the measured values.

**Comment:** The higher the pressure, the worse the ideal-gas model becomes, as is to be expected. This example illustrates that one must be careful when using the ideal-gas model for $u$ and $h$ as well as $PvT$ evaluations. Data based on measurement should always be used in preference to data based on other models.

## 4-4 ENERGY ANALYSIS OF CLOSED IDEAL-GAS SYSTEMS

In preceding sections we developed some basic relationships among properties of an ideal gas. The most useful include

$$Pv = RT \qquad \textbf{[4-4]}$$

$$du = c_v\, dT \qquad \textbf{[4-5]}$$

$$dh = c_p\, dT \qquad \textbf{[4-7]}$$

$$c_p - c_v = R \qquad \textbf{[4-10]}$$

In addition, $c_v$ and $c_p$ are solely functions of temperature. For monatomic and diatomic gases, these equations can be used at pressures up to 10 to 20 bars, or several hundred psia and above with reasonable accuracy if the temperature is room temperature and above. If the temperature interval is small (several hundred degrees or less), the internal-energy and enthalpy changes can be estimated accurately by using average specific heats. In this case $\Delta u = c_{v,\mathrm{av}} \Delta T$ and $\Delta h = c_{p,\mathrm{av}} \Delta T$. The equations for $du$ and $dh$ have been integrated for a number of gases by using accurate specific-heat data, and the results have been presented in tabular form at even increments of temperature. For repetitive calculations, specific-heat equations such as found in Tables A-3 and A-3E can be programmed in a computer for accurate evaluations of $\Delta u$ and $\Delta h$ for any desired temperature range.

With the availability of data for $u$, $h$, $c_p$, and $c_v$, we are in a position to employ the conservation of energy principle to closed systems containing gases at relatively low pressures. For simple compressible substances in a *stationary,* closed system,

$$Q + W = \Delta U \qquad \textbf{[2-36]}$$

or, on a unit-mass basis,

$$q + w = \Delta u \qquad \textbf{[2-37]}$$

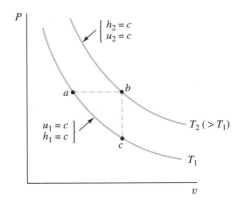

**Figure 4-8**     General *Pv* plot for an ideal
gas, showing position of
isotherms.

Before we proceed, however, it is important to recall the characteristics
of a *Pv* diagram for an ideal gas. This diagram will be a useful problem-
solving device. Figure 4-8 shows a general *Pv* diagram for an ideal gas.
Two isotherms (constant-temperature lines) are shown as hyperbolas, where
$T_2 > T_1$. The change in temperature along the constant-pressure path *ab* is
also the same along the constant-volume path *cb*. Recall that *u* and *h* are
solely functions of temperature. Hence the curve along which *a* and *c* lie is
also a constant-internal-energy and a constant-enthalpy line, $u_1$ and $h_1$. The
curve that contains *b* must also represent $u_2$ and $h_2$, as well as $T_2$. Hence
the five important properties—*P*, *v*, *T*, *u*, and *h*—are all easily represented
on the *Pv* diagram of an ideal gas.

The following two examples illustrate the application of the problem-
solving procedure to ideal-gas systems. The use of the *Pv* diagram is also
demonstrated.

**Example 4-4**

**A** vertical piston-cylinder assembly that initially has a volume of 0.1 m³ is filled
with 0.1 kg of nitrogen gas. The piston is weighted so that the pressure on the diatomic nitrogen is always maintained at 1.15 bars. Heat transfer is allowed to take
place until the volume is 75 percent of its initial value. Determine (*a*) the initial and
final temperature of the nitrogen, in kelvins, and (*b*) the quantity and direction of
the heat transfer, in kilojoules, if quasistatic.

**Solution:**

**Given:**    N₂ gas is maintained at constant pressure in a piston-cylinder device.
Data for the process are shown in the schematic of the system in Fig. 4-9.

**Find:**    (*a*) $T_i$ and $T_f$ in kelvins, (*b*) *Q* in kJ.

**Model:** Ideal gas in a stationary system; quasiequilibrium, constant-pressure process.

**Strategy:** The temperatures are found directly from the ideal-gas equation, since the mass and the initial and final pressures and volumes are known. Then the independent evaluation of boundary work and the internal energy change allow calculation of the heat transfer through the closed-system energy balance.

**Analysis:** The system is the region enclosed by the dashed line in Fig. 4-9. The constant-pressure process is shown in Fig. 4-10 on a $Pv$ plot.

(a) The ideal-gas equation of state is used to obtain the initial temperature $T_i$ of the gas.

$$T_i = \frac{P_i V_i}{N R_u} = 1.15 \text{ bars} \times 0.1 \text{ m}^3 \times \frac{28 \text{ kg/kmol}}{0.1 \text{ kg}} \times \frac{\text{kmol·K}}{0.08314 \text{ bar·m}^3} = 387 \text{ K}$$

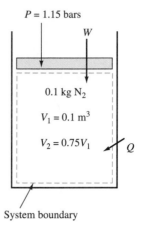

$P = 1.15$ bars

0.1 kg $N_2$

$V_1 = 0.1$ m³

$V_2 = 0.75V_1$

System boundary

The ideal-gas relation for a constant-pressure process also shows that $T_f = T_i(V_f/V_i)$. Thus the final temperature $T_f$ is

$$T_f = T_i \frac{V_f}{V_i} = 387 \text{ K} \left(\frac{0.75}{1.0}\right) = 290 \text{ K}$$

**Figure 4-9**
Schematic and data for Example 4-4.

Thus the range of temperatures from $T_i$ to $T_f$ is only 97°C.

(b) The conservation of energy principle for the stationary closed system is

$$Q + W = \Delta U$$

Since the volume of the control mass is altered, $P\,dV$ work is associated with the process. To evaluate this form of work, it is necessary to *assume* that the process is carried out in quasiequilibrium. At constant pressure, integration of $-P\,dV$ leads to $W_{\text{comp/exp}} = P\,\Delta V$. Hence the energy equation above becomes, upon rearrangement,

$$Q = \Delta U + P\,\Delta V$$

The change in $\Delta U$ can be found either by $c_{v,\text{av}}\Delta T$ or from nitrogen $u$ data in Table A-6. Since $\Delta V$ is known explicitly, $P\,\Delta V$ can be computed directly. Using an average $c_v$ value from Table A-3 for nitrogen, we find that

$$\begin{aligned}
Q &= mc_{v,\text{av}}\,\Delta T + P\,\Delta V \\
&= (0.1) \text{ kg} \times 0.745\frac{\text{kJ}}{\text{kg·°C}} \times (290 - 387)\text{°C} \\
&\quad + 1.15 \text{ bars} \times (0.075 - 0.10) \text{ m}^3 \times \frac{100 \text{ kJ}}{\text{bar·m}^3} \\
&= (7.23 + 2.88) \text{ kJ} = -10.1 \text{ kJ}
\end{aligned}$$

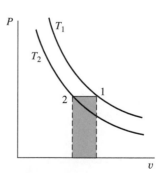

**Figure 4-10**
A $Pv$ plot of the process in Example 4-4.

The contribution of $\Delta U$ found above is 7.23 kJ. In comparison, if Table A-6 were used for tabular $\bar{u}$ data,

$$\Delta U = m\frac{\Delta \bar{u}}{M} = 0.1 \text{ kg}\left(\frac{6021 - 8041}{28.01}\right)\frac{\text{kJ}}{\text{kg}} = -7.21 \text{ kJ}$$

Thus the two methods of evaluating $\Delta U$ are in substantial agreement. The molar mass of 28.01 for nitrogen is needed in the latter calculation, because the $u$ data are on a molar basis. The work required of 2.88 kJ is shown by the crosshatched area in Fig. 4-10 and is transferred into the system.

**Example 4-5**

**An** insulated rigid tank with a volume of 0.80 m³ contains argon gas initially at 300 K and 105 kPa. A paddle wheel is operated within the tank for 25 min with a constant shaft power input of 0.030 kW. Determine (a) the final temperature, in kelvins, and (b) the final pressure in kPa.

**Solution:**

**Given:**    Paddle-wheel work added to argon gas in a tank. A schematic of the system with known data is shown in Fig. 4-11.

**Find:**    (a) $T_2$, in K; (b) $P_2$, in kPa.

**Model:**    Stationary, rigid, insulated tank; monatomic ideal gas; $Q = 0$, $\dot{W} =$ constant.

**Strategy:**    Because the values of heat and work transfer and the initial temperature are known, the closed-system energy balance contains the final temperature as the only unknown. Then the ideal-gas equation can be used to find the final pressure.

**Analysis:**    The closed system shown within the dashed line in Fig. 4-11 has a constant volume, since the tank is rigid. The $Pv$ diagram for such a process is shown in Fig. 4-12. The ideal-gas equation for the final state $(P_2V_2 = NR_uT_2)$ contains two unknowns, namely, $T_2$ and $P_2$. Hence another method must be used to find $T_2$. Another equation that contains $T_2$ (in terms of $u_2$) is the energy balance for the process. For the chosen stationary, closed system, $Q + W = \Delta U$. The tank is insulated; therefore, $Q = 0$. Also, no expansion or compression work occurs, because the volume is constant. However, paddle-wheel work is present. For constant power input, $W = \dot{W}\,\Delta t$. Hence the general energy equation reduces to

$$W = \dot{W}_{\text{pad}}\,\Delta t = \Delta U = m\,\Delta u = mc_v(T_2 - T_1) = N\bar{c}_v(T_2 - T_1)$$

In this equation $\dot{W}_{\text{pad}}$, $\Delta t$, and $T_1$ are known; $T_2$ is the unknown.

(a) Because argon is a monatomic gas, $c_v$ is a constant. This constant, from Sec. 2 of Table A-3 in the Appendix, is 12.8 kJ/kmol·°C. Since $c_v$ is reported on a mole basis, it is convenient to use $N$, the number of moles, rather than $m$ in the energy analysis. On the basis of the ideal-gas equation

$$N = \frac{P_1V_1}{R_uT_1} = \frac{105 \text{ kPa } (0.80 \text{ m}^3)}{300 \text{ K}} \times \frac{\text{kmol·K}}{8.314 \text{ kPa·m}^3} = 0.0337 \text{ kmol}$$

Substitution of data into the energy equation $\dot{W}_{\text{pad}}\,\Delta t = N\bar{c}_v(T_2 - T_1)$ yields

$$(0.030 \text{ kJ/s})(1500 \text{ s}) = 0.0337 \text{ kmol} \times (12.8 \text{ kJ/kmol·°C}) \times (T_2 - 300)°C$$

or

$$T_2 = \left(300 + \frac{45.0}{0.431}\right) \text{K} = 404 \text{ K}$$

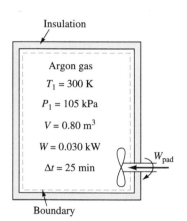

Insulation

Argon gas
$T_1 = 300$ K
$P_1 = 105$ kPa
$V = 0.80$ m³
$W = 0.030$ kW
$\Delta t = 25$ min

$W_{\text{pad}}$

Boundary

**Figure 4-11**
Schematic and data for Example 4-5.

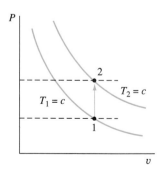

$P$

2

$T_1 = c$

$T_2 = c$

1

$v$

**Figure 4-12**
A $Pv$ plot of the process in Example 4-5.

(b) One method of determining $P_2$ is from the relation $P_2V_2 = NR_uT_2$. Another approach is to note that $P_1/T_1 = P_2/T_2$ for an ideal gas at constant volume. Using this latter relation,

$$P_2 = P_1\left(\frac{T_2}{T_1}\right) = 105 \text{ kPa}\left(\frac{404}{300}\right) = 141 \text{ kPa}$$

**Comment:** No tables of $u$ and $h$ data are presented in the Appendix for monatomic ideal gases. This is unnecessary because both $c_v$ and $c_p$ are independent of temperature.

---

Figure 4-13 shows a physical situation that is quite different from those in the two preceding examples. A gas within one side of a partitioned tank is allowed to expand into an evacuated side with the same volume. The boundary of the closed system, shown by the dashed line in Fig. 4-13, is selected to lie just inside the walls of the entire insulated, rigid tank and includes the section that is initially evacuated. For such a system the energy equation is

$$Q + W = \Delta U + \Delta KE + \Delta PE$$

The $P\,dV$ work associated with a rigid tank is zero, and all other work effects are absent. In addition, $Q$ is zero because of the insulated boundary, and the system is stationary. Because $Q$ and $W$ are zero for this process, we may write for the finite change of state that

$$\Delta U = 0 \qquad \text{or} \qquad U_2 = U_1$$

Consequently, the expansion of the gas in this case is one of constant internal energy; that is, $U_2$ equals $U_1$. This type of expansion is called a "free expansion" for the closed system, because there is no force restraining the flow.

The process is shown in Fig. 4-14 on a $Pv$ diagram. The path between states 1 and 2 in Fig. 4-14 is drawn as a dashed line, since the process is not quasiequilibrium, and constant temperature is not implied by the shape of the dashed line. For an ideal gas the internal energy is solely a function of the temperature for a given equilibrium state. Since the internal energy is constant, the initial and final temperatures of the ideal gas undergoing this "free expansion" must be the same. Hence $T_2$ is 100°C. For an ideal gas at constant temperature, the ideal-gas equation leads to the relation $P_2 = P_1(V_1/V_2) = 2 \text{ bar } (1/2) = 1 \text{ bar}$.

Consider now a situation similar to a free expansion, except the same gas is on both sides of the partition. As shown in Fig. 4-15, the initial state on side $A$ is $(T_1, P_1, V_1)$, whereas on side $B$ the state is $(T_2, P_2, V_2)$. The partition is now broken, and the gases mix until pressure and temperature equilibrium is reached for the entire system at the final state $F$. For this process the energy balance $Q + W = \Delta U$ again reduces to $\Delta U = 0$ if the

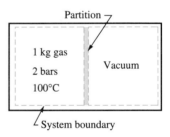

**Figure 4-13**
Schematic for a "free expansion" into a vacuum.

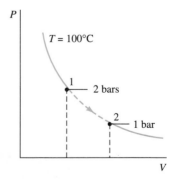

**Figure 4-14**
A Pv plot of a "free expansion" in a closed system.

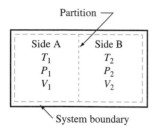

**Figure 4-15**
The mixing of a gas initially at two different states in a closed system.

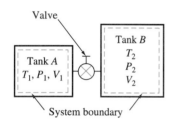

**Figure 4-16**
The mixing of the same gas initially at different states in two tanks.

**Note the physical significance of the compressibility factor Z.**

tank is rigid and insulated. However, the interpretation of $\Delta U = 0$ is different from the free expansion discussed above. Now $\Delta U_A + \Delta U_B = 0$, where $A$ and $B$ represent the gas originally on side $A$ and on side $B$. Note that the final specific internal energy $u_3$ of the mixture ($m_3 = m_1 + m_2$) is measured at the same final equilibrium temperature $T_3$. Use of tabular $u$ data for the substance allows us to calculate $u_3$, and hence to evaluate $T_3$. Another method is to consider $c_v$ to be constant for the overall process. This type of analysis can be extended to include nonadiabatic processes. Also, as shown in Fig. 4-16, these analyses apply equally well to an expansion or mixing that occurs between two tanks connected by a valve.

## 4-5 THE COMPRESSIBILITY FACTOR AND THE CORRESPONDING STATES PRINCIPLE

Unless the pressure is reasonably low and the temperature relatively high, gases cannot be represented accurately by the ideal-gas equation $Pv = RT$. When experimental $PvT$ data also are not available, one method of correcting the ideal-gas equation so that it applies to *nonideal* gases, still maintaining reasonable accuracy, is through the use of a compressibility-factor correction. The departure from ideal-gas behavior may be characterized by a *compressibility factor* $Z$, defined as

$$Z \equiv \frac{Pv}{RT} = \frac{P\bar{v}}{R_u T} \qquad \text{[4-17]}$$

Since $RT/P$ is the ideal-gas specific volume ($v_{ideal}$) of a gas, the compressibility factor may be considered as the ratio of the actual specific volume to the ideal-gas specific volume. That is, $Z = v_{actual}/v_{ideal}$. For an ideal gas, the compressibility factor is unity, but for actual gases it can be either less than or greater than unity. Hence the compressibility factor measures the deviation of an actual gas from ideal-gas behavior.

The specific volume of any gas could be estimated at any desired pressure and temperature if the compressibility factor is known for the gas as a function of $P$ and $T$, since $v = ZRT/P$. A plot of $Z$ versus pressure for nitrogen is shown in Fig. 4-17 for specified values of temperature. Plots for other pure gases would be qualitatively similar. Owing to this similarity, it is found that all such plots can be reduced to a single chart if the coordinates are modified. This modification is one application of what is known as the *principle of corresponding states*. The principle postulates that the $Z$ factor for all gases is approximately the same when the gases have the same reduced pressure and temperature. The *reduced pressure* $P_r$ and *reduced temperature* $T_r$ are defined by

$$P_r \equiv \frac{P}{P_c} \quad \text{and} \quad T_r \equiv \frac{T}{T_c} \qquad \text{[4-18]}$$

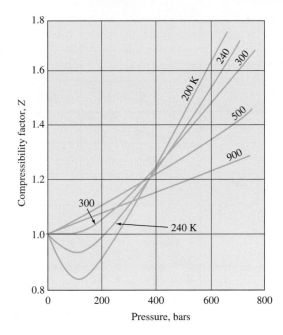

**Figure 4-17**    Isotherms of the compressibility factor versus pressure for nitrogen.

Absolute pressures, and temperatures in kelvins or degrees Rankine, must be used in these equations. Thus the critical pressure and temperature of a substance are used to define a reduced state.

The validity of such a principle must be based on experimental evidence. When reduced isotherms $T_r$ are plotted on a $Z$–$P_r$ chart, the average deviation of experimental data for a number of gases is somewhat less than 5 percent. Figure 4-18 shows a correlation of actual data for 10 gases for a limited number of reduced isotherms. When the best curves are fitted to all the data, a more complete plot results, such as Fig. A-27 in the Appendix. This generalized compressibility chart is for $P_r$ values from 0 to 1.0 and $T_r$ values from 0.60 to 5.0. Once the chart has been drawn from data for a limited number of substances, it is assumed to be generally applicable to all gases. The main virtue of the generalized compressibility chart is that it requires only a knowledge of critical pressures and temperatures to predict the specific volume of a real gas. It must be emphasized that the generalized compressibility chart should not be used as a substitute for accurate experimental $PvT$ data. The major role of a generalized compressibility chart is to provide estimates of $PvT$ behavior in the absence of accurate measurements. In addition to Fig. A-27, Fig. A-28 is a $Z$ chart for $P_r$ values from 0 to 10, and Fig. A-29 is for $P_r$ values from 10 to 40.

**Note the use of the principle of corresponding states to correlate Z versus $P_r$ and $T_r$.**

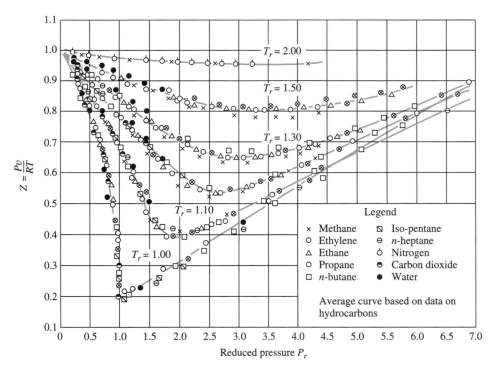

**Figure 4-18**　Experimental data correlation for a generalized Z chart. [*Gour-Jen Su: Modified Law of Corresponding States, Ind. Eng. Chem. (Int. Ed.),* **38**:803 (1946).]

Several additional comments are pertinent at this point. First, a few general characteristics of the compressibility charts should be noted:

1. In the limit as $P_r$ approaches zero, the value of $Z$ approaches unity for all values of the reduced temperature. When $P_r \le 0.05$, the ideal-gas model can be used with an error less than 5 percent.

2. For reduced temperatures greater than 2.5, the value of $Z$ is greater than unity for all pressures. Under these circumstances the actual volume is always greater than the ideal-gas volume for the same pressure and temperature.

3. For reduced temperatures below 2.5, the reduced isotherms go through a minimum at fairly low reduced pressures. In this region the volume is less than the ideal-gas volume, and the deviation from ideal-gas behavior is substantial.

4. When $P_r$ is greater than 10, the deviation from ideal-gas behavior can approach several hundred percent.

Second, it is found experimentally that the gases hydrogen, helium, and neon do not correlate very well on a generalized compressibility chart.

This difficulty is overcome, for temperatures above 50 K, by redefining the reduced pressure and temperature for these three gases in the following manner:

$$P_r = \frac{P}{P_c + C} \quad \text{and} \quad T_r = \frac{T}{T_c + C} \qquad \text{[4-19]}$$

When $P$ is in atmospheres (or bars) and $T$ is in kelvins, then the value of $C$ in both equations is 8.

The compressibility factor can also be found when $vT$ data or $vP$ data are given. For correlation purposes it has been found best to use a pseudocritical volume in the definition of the reduced volume, rather than the actual critical volume. If we define a pseudocritical volume $v'_c$ by the quantity $RT_c/P_c$, then the **pseudoreduced volume** $v'_r$ is

$$v'_r \equiv \frac{vP_c}{RT_c} \qquad \text{[4-20]}$$

Again, only a knowledge of $T_c$ and $P_c$ is required. Lines of constant $v'_r$ are also shown in Figs. A-27, A-28, and A-29.

---

Determine the specific volume of water vapor in m³/kg, at 200 bars and 520°C by (*a*) the ideal-gas equation of state, (*b*) the principle of corresponding states, and (*c*) the experimental value in the superheat table.

Example 4-6

**Solution:**

**Given:**    Water vapor at 200 bars and 520°C.

**Find:**    $v$, in m³/kg, by (*a*) the ideal-gas equation, (*b*) the Z-chart, and (*c*) the superheat table.

**Model:**    Simple compressible substance.

**Analysis:**    (*a*) On the basis of the ideal-gas equation, $v = RT/P$. The specific gas constant $R$, based on Table A-1 and Eq. [4-3], is

$$R = \frac{R_u}{M} = \frac{0.08314 \text{ bar·m}^3}{\text{kmol·K}} \times \frac{\text{kmol}}{18 \text{ kg}} = 4.62 \times 10^{-3} \text{ bar·m}^3/\text{kg·K}$$

Hence

$$v = \frac{RT}{P} = (0.00462 \text{ bar·m}^3/\text{kg·K}) \times \frac{793 \text{ K}}{200 \text{ bars}} = 0.0183 \text{ m}^3/\text{kg}$$

(*b*) The specific volume based on the principle of corresponding states is given by $v_{\text{actual}} = Zv_{\text{ideal}}$. The $Z$ factor is found by computing $T_r$ and $P_r$ and then using Fig. A-27 to find $Z$. The critical data are found in Table A-2. Based on these data

$$T_r = \frac{793 \text{ K}}{647 \text{ K}} = 1.23 \quad \text{and} \quad P_r = \frac{200 \text{ bars}}{220.9 \text{ bars}} = 0.905$$

From Fig. A-27 the $Z$ value is approximately 0.83. Therefore

$$v = 0.83(0.0183) \text{ m}^3/\text{kg} = 0.0152 \text{ m}^3/\text{kg}$$

An alternate method is to use the $v'_r$ value from the chart at the given state. From Fig. A-27 the $v'_r$ value is approximately 1.13. Hence

$$v = \frac{RT_c v'_r}{P_c} = \frac{0.08314(647)(1.13)}{220.9(18.02)} \frac{\text{m}^3}{\text{kg}} = 0.01527 \text{ m}^3/\text{kg}$$

where 18.02 is the molar mass of water. Thus the use of either $Z$ or $v'_r$ leads essentially to the same answer, within the accuracy of reading data from the chart.

(c) The tabulated value, based on experimental data, is found in Table A-14 to be 0.01551 m³/kg.

**Comment:**   In comparison to the tabulated value, the ideal-gas value is in error by nearly 20 percent. However, the corresponding-states principle leads to an error of only about 2 percent. This chosen state is typical of the turbine inlet condition in large steam power plants. Note that $Z$ is considerably different from unity.

The preceding example illustrates that $PvT$ data can be estimated to a fair accuracy by employing the principle of corresponding states. Note, however, that the $Z$ chart should never be used if experimental data are available.

**Example 4-7**

**Ethane** gas ($C_2H_6$) is placed in a rigid tank at a pressure of 34.2 bars and a specific volume of 0.0208 m³/kg. It is heated until the pressure reaches 46.4 bars. Estimate the temperature change for the process, in kelvins, based on the generalized $Z$ chart.

**Solution:**

**Given:**   A rigid tank contains ethane gas. A schematic of the equipment with appropriate data is shown in Fig. 4-19.

**Find:**   The temperature change, in kelvins.

**Model:**   Closed system, constant-volume process.

**Strategy:**   Use a constant $v'_r$ line to determine the value of $T_{r,2}$ on the $Z$ chart.

**Analysis:**   The dashed line in Fig. 4-19 indicates the system of interest. The temperature change can be estimated by finding the values of $T_r$ for the initial and final states. These values are found from a knowledge of $P_r$ and $v'_r$ for each state.

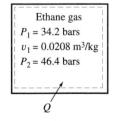

Ethane gas
$P_1 = 34.2$ bars
$v_1 = 0.0208$ m³/kg
$P_2 = 46.4$ bars

$Q$

**Figure 4-19**
Schematic and data for Example 4-7.

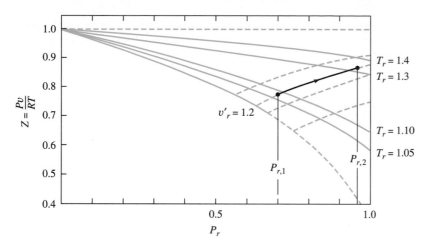

**Figure 4-20**    A Z-chart plot of the process in Example 4-7.

In Table A-2 the critical pressure and temperature of ethane are given as 48.8 bars and 305 K, respectively, and the molar mass is given as 30.07. Hence the reduced properties are

$$v'_r = \frac{vP_c}{RT_c} = \frac{0.0208(30.07)(48.8)}{0.08314(305)} = 1.20$$

$$P_{r,1} = \frac{P_1}{P_c} = \frac{34.2}{48.8} = 0.70 \quad \text{and} \quad P_{r,2} = \frac{P_2}{P_c} = \frac{46.4}{48.8} = 0.95$$

Because the volume is fixed, the process follows a line of constant $v'_r$ of 1.20 from a $P_r$ value of 0.70 to one of 0.95, as shown in Fig. 4-20. From Fig. A-26, we find that $T_{r,1} = 1.07$ and $T_{r,2} = 1.33$, approximately. Hence

$$T_2 = T_c T_{r,2} = 305(1.33) \text{ K} = 406 \text{ K} \quad \text{and} \quad T_1 = T_c T_{r,1} = 305(1.07) \text{ K} = 326 \text{ K}$$

Therefore, the estimated temperature change equals $406 - 326$ K, or 80°C.

**Comment:**    As an alternative solution, at state 2 on the chart we find that $Z_2 = 0.87$. Therefore $T_2 = P_2 v_2 / Z_2 R = 46.4(0.0208)(30.07)/0.87(0.08314) = 401$ K. This value agrees reasonably well with 406 K found above. Both values depend on the accuracy of reading the $Z$ chart.

---

The emphasis in this section has been on the use of the compressibility chart to provide $Z$ factors for $PvT$ calculations. It should be noted that $Z$ values much different from unity also indicate that ideal-gas $u$ and $h$ data are probably not accurate for energy analyses. The corresponding states principle does, however, also provide means for estimating real-gas enthalpy and internal energy data, as shown in Chap. 12.

**Example 4-8**

**W**ater vapor is heated from 600 to 700°F at pressures of 14.7, 100, 250, and 500 psia. Compare the average $Z$ factor for the four processes to the percent error in $\Delta u$ determined in Example 4-3.

**Solution:**

**Given:** Water vapor heated from 600 to 700°F at pressures of 14.7, 100, 250, and 500 psia.

**Find:** $Z_{av}$ for the four processes, and compare to error in $\Delta u$ found in Example 4-3.

**Model:** Ideal-gas model, real-gas $u$ data, and compressibility chart.

**Analysis:** For all four processes $T_{r,1} = 0.91$ and $T_{r,2} = 1.0$. The $P_r$ values for the four given pressures are 0.005, 0.03, 0.078, and 0.156. Based on these reduced temperatures and pressures and Fig. A-27, the average $Z$ values for the processes at 14.7, 100, 250, and 500 psia are roughly 1.0, 0.986, 0.965, and 0.935. In comparison, from Example 4-3 the percentage errors in $\Delta u$ when using ideal-gas data rather than superheat data are 0.3, 3.1, 7.9, and 16.7 percent. As the average compressibility factor decreases, the error in $\Delta u$ using ideal-gas data increases significantly for the states chosen.

**Comment:** Thus the value of $Z$ is also a semiquantitative measure of the error in ideal-gas $\Delta u$ (and $\Delta h$) values when used for real gases.

## 4-6  PROPERTY RELATIONS FOR INCOMPRESSIBLE SUBSTANCES

On the basis of the $PvT$ surface for simple compressible substances (see Fig. 3-1), for many solids and liquids there are wide regions on the $PvT$ surface of equilibrium states where the variation in the specific volume is negligible. Therefore, the assumption that the specific volume (and the density) is constant in the region of interest is often a good approximation to reality, and it leads to no serious error in computations. The equation of state for these two phases often can be represented then by

$$v = \text{constant} \quad \text{or} \quad \rho = \text{constant} \qquad \textbf{[4-21]}$$

By definition, when the specific volume (or density) of a liquid or solid is assumed to be constant, the substance is being *modeled* as an ***incompressible substance***. The boundary or $P\,dV$ work associated with the change of state of an incompressible substance is always zero. So the only ways to change the internal energy of a simple incompressible material are by a heat interaction or by nonquasistatic work interactions. For example, stirring of an incompressible liquid by means of a paddle wheel would be permissible. We noted in Sec. 3-8 that the internal energy of a simple compressible substance can be expressed as $u = u(T, v)$. The total differential of $u$ is

$$du = \left(\frac{\partial u}{\partial T}\right)_v dT + \left(\frac{\partial u}{\partial v}\right)_T dv = c_v dT + \left(\frac{\partial u}{\partial v}\right)_T dv \qquad \textbf{[3-11]}$$

For an incompressible substance, $dv$ is zero. Therefore, we may write for a simple incompressible model that

$$du = c_v dT \qquad \text{(incompressible)} \qquad \textbf{[4-22]}$$

That is, the internal energy of an incompressible substance is solely a function of temperature. As a result, $c_v$ for an incompressible material also is solely a function of the temperature. Hence

$$u_2 - u_1 = \int_1^2 c_v dT \qquad \text{(incompressible)} \qquad \textbf{[4-23]}$$

From the definition of the enthalpy function $h = u + Pv$, we see that

$$h_2 - h_1 = u_2 - u_1 + v(P_2 - P_1) \qquad \text{(incompressible)} \qquad \textbf{[4-24]}$$

**Note that *u* is solely a function of *T*, while *h* is a function of both *T* and *P* for the incompressible model.**

Hence the enthalpy of an incompressible substance is a function of both the temperature and the pressure.

For a differential change of state, $dh = du + d(Pv)$. By combining this relation with Eq. [4-22], we find for an incompressible substance (inc) that

$$dh_{\text{inc}} = du + v\,dP = c_v dT + v\,dP$$

In addition, the total differential of $h$ is given by Eq. [3-14], namely,

$$dh = c_p dT + \left(\frac{\partial h}{\partial P}\right)_T dP \qquad \textbf{[3-14]}$$

A comparison of the $dT$ terms in the above two equations shows that $c_p = c_v$ for incompressible substances. Therefore, both $c_p$ and $c_v$ can be represented by the single specific heat symbol $c$. Hence

$$\boxed{c_p = c_v = c} \qquad \text{(incompressible)} \qquad \textbf{[4-25]}$$

This approximation is now used to evaluate $\Delta u$ and $\Delta h$ for the incompressible model.

## 4-6-1   INTERNAL ENERGY AND ENTHALPY CHANGES

As a consequence of Eq. [4-25], Eqs. [4-23] and [4-24] can be written as

$$\boxed{(u_2 - u_1)_{\text{inc}} = \int_1^2 c\,dT} \qquad \textbf{[4-26]}$$

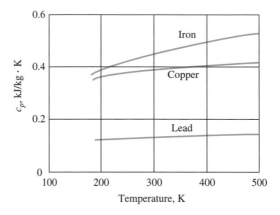

**Figure 4-21**    The constant-pressure specific heat as a function of temperature for three common solids.

and

$$(h_2 - h_1)_{\text{inc}} = \int_1^2 c \, dT + v(P_2 - P_1) \qquad \textbf{[4-27]}$$

Finally, for liquids and many solids, the variation of $c$ with temperature is quite small. This is shown in Fig. 4-21 for three solid elements. The data for liquid water in Table A-4 show a minor variation of $c_p$ with temperature. As a result, Eqs. [4-26] and [4-27] can be written as

$$(u_2 - u_1)_{\text{inc}} \approx c_{\text{av}}(T_2 - T_1) \qquad \textbf{[4-28]}$$

and

$$(h_2 - h_1)_{\text{inc}} \approx c_{\text{av}}(T_2 - T_1) + v(P_2 - P_1) \qquad \textbf{[4-29]}$$

where $c_{\text{av}}$ is the specific heat evaluated at the average temperature. Equations [4-26] through [4-29] require the use of specific-heat data. Tables A-4 and A-4E list the variation of the specific heat of some common liquids and solids with temperature. Engineering and scientific handbooks contain $c_p$ data for a considerable number of such substances.

**Example 4-9**

**A** 2-kg piece of copper initially at 400 K and 0.1 MPa is dropped into 10 L of water initially at 300 K and 0.1 MPa. Estimate the final equilibrium temperature of the water and copper, in kelvins, if the overall system is insulated.

**Solution:**

**Given:**    A piece of copper is dropped into water in a tank, as shown in Fig. 4-22.

**Find:**    The final equilibrium temperature, in kelvins.

**Model:** Copper and water are incompressible; insulated, stationary, closed system.

**Strategy:** The initial states, and heat and work interactions are known. Use the energy analysis on the composite system to find the final temperature.

**Analysis:** The system is chosen to contain both the copper and the water. Determination of the final equilibrium temperature requires an energy balance on the overall process. The basic energy equation for the *stationary* system is

$$Q + W = \Delta U_{\text{system}} = \Delta U_{\text{Cu}} + \Delta U_{\text{water}}$$

If the tank is insulated (or the time for the process is relatively short) any heat loss may be neglected. Because the substances are *assumed* to be incompressible (constant volume), no work effects occur during the process. In addition, the specific internal energy of each substance is given by Eq. [4-28], $\Delta u = c_{\text{av}}\Delta T$. Hence the energy equation reduces to

$$[mc_{\text{av}}(T_2 - T_1)]_{\text{Cu}} + [mc_{\text{av}}(T_2 - T_1)]_{\text{water}} = 0$$

where $T_2$ is the same value for each substance. The unknown data at this point include the mass of water involved as well as the average specific heats to be used.

The mass of the water is found from the relation $m = V/v$. The specific volume of the water is estimated by using $v_f$ data from the saturation table A-12. The final temperature is not known, but $v_f$ is essentially independent of temperature between 300 and 400 K. Hence the initial value of $1.0035 \times 10^{-3}$ m³/kg will be used, and the mass of water is

$$m = \frac{V}{v} = \frac{0.010 \text{ m}^3}{1.0035 \times 10^{-3} \text{ m}^3/\text{kg}} = 9.965 \text{ kg}$$

Although the final temperature is not known, the specific heats of copper and water vary little with temperature. Based on Table A-4, the average $c$ values for copper and water are chosen as 0.391 and 4.2 kJ/kg·K, respectively. On the basis of this information the energy balance may be written as

$$2(0.391)\frac{\text{kJ}}{\text{K}} \times (T_2 - 400 \text{ K}) + 9.965(4.2)\frac{\text{kJ}}{\text{K}} \times (T_2 - 300 \text{ K}) = 0$$

$$T_2 = (12{,}870/42.64) \text{ K} = 301.8 \text{ K}$$

**Comment:** The final temperature in this case is only slightly above the initial temperature of the water. This occurs because both the specific heat and the mass values for water are much larger than those for the copper.

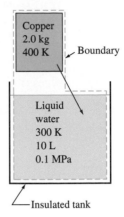

**Figure 4-22**
Schematic and data for Example 4-9.

## 4-6-2 APPROXIMATION EQUATIONS FOR INCOMPRESSIBLE SUBSTANCES

Equation [4-22] supports an approximation introduced in Sec. 3-6-3 regarding compressed- or subcooled-liquid data. Because subcooled liquids are essentially incompressible, the internal energy of a subcooled liquid according to Eq. [4-22] ($du = c_v\, dT$) should be only temperature-dependent. For a selected temperature, the integral of $c\, dT$ must be zero. Therefore, at the same temperature the internal energy for a compressed-liquid state would equal that of a saturated liquid for an *incompressible model*. That is,

$$u_{\text{comp liq at } T} \approx u_{\text{sat liq at } T}(= u_{f,T}) \qquad \textbf{[4-30]}$$

is a good approximation. This type of approximation is not quite as good for the enthalpy of a compressed liquid, since the term $v(P_2 - P_1)$ is present in Eq. [4-24]. However, if state 2 represents the compressed-liquid state and state 1 is the saturated-liquid state, then Eq. [4-24] becomes

$$h(\text{compressed liquid at } T, P) \approx h_{f,T} + v_{f,T}(P - P_{\text{sat}}) \qquad \textbf{[4-31]}$$

where $P_{\text{sat}}$ is the saturation pressure at the given temperature. The value of $v$ is usually quite small, so that the contribution of the last term in Eq. [4-31] is small compared to $h_f$ for the saturated liquid. However, using the last term in Eq. [4-31] does increase the accuracy of estimating $h$ in the subcooled- or compressed-liquid state when experimental data are not available, especially at high pressures. Thus the correction for pressure normally is made for all compressed-liquid enthalpy values based on $h_f$ data.

**Example 4-10**

Determine the value of $h$ for subcooled water at 20°C and 1 bar on the basis of an incompressible fluid, in kJ/kg.

**Solution:**

**Given:**　Subcooled water at 20°C and 1 bar.

**Find:**　$h$, in kJ/kg.

**Model:**　Incompressible fluid.

**Analysis:**　The accompanying $PT$ diagram in Fig. 4-23 shows state 2 lying in the subcooled-liquid region at 20°C and 1 bar. The enthalpy at state 2 on the diagram is approximated by employing Eq. [4-31] for the assumed incompressible fluid. The evaluation will first be carried out along the constant-temperature path 1-2. As a result,

$$h_2 = h_1 + v(P_2 - P_1) = h_{f,1} + v_{f,1}(P_2 - P_1)$$

From Table A-12 the saturated-liquid values are $h_1 = 83.96$ kJ/kg, $v_{f,1} = 1.0018 \times 10^{-3}$ m$^3$/kg, and $P_1 = 0.0234$ bar. Consequently,

$$h_2 = 83.96 \text{ kJ/kg} + 1.0018 \times 10^{-3} \text{ m}^3\text{/kg} \times (1 - 0.0234) \text{ bar} \times \frac{100 \text{ kJ}}{1 \text{ bar·m}^3}$$

$$= (83.96 + 0.10) \text{ kJ/kg} = 84.06 \text{ kJ/kg}$$

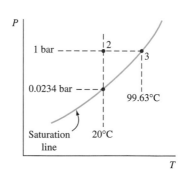

**Figure 4-23**
A $PT$ diagram for states in Example 4-10.

Note that the enthalpy in the subcooled region in this case differs by only 0.1 percent from the saturated-liquid value at the same temperature.

The enthalpy at state 2 could also be found by selecting path 3-2 at a constant pressure of 1 bar. The enthalpy at state 3 is $h_f$ at 1 bar, which is found in Table A-13 to be 417.46 kJ/kg. Use of Eq. [4-29] in this case requires knowledge of $c$. The specific heat $c$ of water between 20 and 99.6°C is relatively constant (see Table A-4) at 4.18 kJ/kg·°C. Hence

$$h_2 - h_3 = c_{av}(T_1 - T_3) = \frac{4.18 \text{ kJ}}{\text{kg·°C}} \times (20 - 99.63)°\text{C} = -332.8 \text{ kJ/kg}$$

$$h_2 = (417.46 - 332.8) \text{ kJ/kg} = 84.66 \text{ kJ/kg}$$

This method gives about the same answer. However, its accuracy is somewhat in doubt since the answer is sensitive to the $c_{av}$ value used.

## 4-7 SUMMARY

The ideal-gas equation of state is given in several formats.

$$PV = NR_uT \qquad Pv = R_uT \qquad PV = mRT$$

where $R_u$ is the universal gas constant and $R = R_u/M$. Another useful format of the ideal-gas equation is $P_1v_1/T_1 = P_2v_2/T_2$. For the ideal-gas model the changes in the specific internal energy and the specific enthalpy are given by

$$\Delta u = u_2 - u_1 = \int_1^2 c_v \, dT \qquad \text{and} \qquad \Delta h = h_2 - h_1 = \int_1^2 c_p \, dT$$

This pair of equations is valid for *all processes* of an ideal gas. The specific heats are solely a function of temperature and are related by

$$c_p - c_v = R \qquad \text{or} \qquad \overline{c}_p - \overline{c}_v = R_u$$

where $R = R_u$ when $c_p$ and $c_v$ are molar values.

Integration of the equations for $\Delta u$ and $\Delta h$ requires information on the variation of the specific heats with temperature. For monatomic gases the ideal-gas specific heats are

$$\overline{c}_{v,0} = 12.5 \text{ kJ/kmol·K} = 2.98 \text{ Btu/lbmol·°R}$$

$$\overline{c}_{p,0} = 20.8 \text{ kJ/kmol·K} = 4.97 \text{ Btu/lbmol·°R}$$

and these values are *independent* of temperature. Thus for a monatomic gas $\Delta u = c_v \Delta T$ and $\Delta h = c_p \Delta T$. For polyatomic gases, algebraic equations are frequently available for $c_{p,0}$ as a function of temperature. Use of such functions leads to *ideal-gas tables* where $u$ and $h$ are tabulated at convenient intervals of temperature and with an arbitrary reference state.

Another approach is to evaluate $\Delta u$ or $\Delta h$ by using an arithmetically averaged specific-heat value in the temperature range of interest. In this case,

$$\Delta u = c_{v,\text{av}}\,\Delta T \qquad \text{and} \qquad \Delta h = c_{p,\text{av}}\,\Delta T$$

where $c_{v,\text{av}} = (c_{v,1} + c_{v,2})/2$ between states 1 and 2. A similar equation applies to $c_{p,\text{av}}$.

Energy analyses of closed systems containing an ideal gas usually require the simultaneous use of several basic ideal-gas relationships. These include

$$Pv = RT \qquad du = c_v\,dT \qquad dh = c_p\,dT \qquad c_p - c_v = R$$

In addition, the plot of a process on a $Pv$ diagram is useful as an aid in problem solving. To this end, the general position of temperature lines on the diagram must be known.

In the absence of superheat data, the $PvT$ relationship of a gas can be approximated by

$$Pv = ZRT$$

where $Z$ is a *compressibility factor*. The values of $Z$ are correlated quite well as a function of the reduced pressure $P_r = P/P_c$ and the reduced temperature $T_r = T/T_c$. Based on the principle of corresponding states and experimental data, generalized charts are available for $Z$ over a wide range of pressure and temperature.

When analyzing solid or liquid systems, the lack of tabular property data is frequently bypassed by assuming the substance to be *incompressible*. When the variation in specific volume is negligible, the following relations are also valid.

$$du_{\text{inc}} = c\,dT \qquad u_2 - u_1 \approx c_{\text{av}}(T_2 - T_1)$$

$$(h_2 - h_1)_{\text{inc}} \approx c_{\text{av}}(T_2 - T_1) + v(P_2 - P_1)$$

In these relations, $c = c_v = c_p$.

## PROBLEMS

4-1C. Under what limiting condition do all gases behave as an ideal gas?

4-2C. Explain whether one can set definite pressures and temperatures as limits for ideal-gas behavior for a given gas.

4-3C. Explain from a molecular viewpoint why the internal energy of an ideal gas is solely a function of temperature, while $u = f(T, v)$ for a real gas.

4-4C. Explain, in terms of the possible molecular forms of energy of a molecule, why $c_v$ of a monatomic gas is independent of temperature.

4-5C. Can the ideal-gas relation $du = c_v \, dT$ be applied to a constant-pressure process, or is it restricted to a constant-volume process? Explain.

4-6C. Can $c_p$ data be used in place of $c_v$ data for an incompressible substance? Explain.

4-7C. In terms of a $PvT$ surface, explain qualitatively why two gases at the same reduced pressure and temperature have approximately the same compressibility factor.

## IDEAL-GAS EQUATION

4-1. A balloon is filled with methane ($CH_4$) at 20°C and 1 bar until the volume is 26.4 m³.
(a) Calculate the mass of gas, in kilograms, in the balloon.
(b) Determine the volume, in cubic meters, if the balloon rises to a height where the state is 0.84 bar and 0°C.

4-2. Carbon monoxide initially is in a 50-L tank at 210 kPa and 127°C.
(a) Determine the mass of the gas, in kilograms. (b) Now 0.02 kg of gas leaks out of the tank and the temperature drops to 27°C. Find the gage pressure of the gas remaining in the tank in kilopascals, if the barometric pressure is 98.8 kPa.

4-3. Propane ($C_3H_8$) in the amount of 1500 kg is to be stored in a gas reservoir at 42°C and 450 kPa. (a) Find the necessary volume of the reservoir, in cubic meters. (b) Later an additional 500 kg of the gas is added, but the temperature remains constant. Find the final pressure, in kilopascals.

4-4. The air inside an automobile tire with a volume of 0.042 m³ has reached 360 kPa at 40°C. Upon cooling to 20°C it is desirable to have the pressure at 300 kPa. Find the volume of air that must be removed, in cubic meters, if measured at 20°C and 100 kPa.

4-5E. A balloon is filled with helium at 70°F and 30.0 inHg until the volume is 900 ft³. (a) Calculate the mass of gas, in pound-mass, if an ideal gas. (b) Determine the volume, in cubic feet, if the balloon rises to a height where the state is 28.0 inHg and 0°F.

4-6E. Methane ($CH_4$) in the amount of 100 lb$_m$ is to be stored in a gas reservoir at 160°F and 250 psia. Find the necessary volume of the reservoir, in cubic feet.

4-7E. One pound of helium is placed in a 15-ft³ rigid tank. If the temperature is 70°F and the barometric pressure is 30.5 inHg, determine the reading, in psi, of a gage attached to the tank.

4-8E. One pound-mass of argon gas is placed in a 7.5-ft³ rigid tank at 26 psia. If the argon behaves as an ideal gas, (a) find its temperature, in degrees Fahrenheit. (*b*) If 0.10 lb$_m$ of gas escapes, find the new pressure at the same temperature, in psia.

4-9. One-half kilogram of helium is placed in a 0.5-m$^3$ rigid tank. If the temperature is 112°C and the barometric pressure is 1.0 bar, determine the reading, in bars, of a gage attached to the tank.

4-10. One kilogram of nitrogen gas is placed in a 0.55-m$^3$ rigid tank at 0.17 MPa. If the nitrogen behaves as an ideal gas, (*a*) find its temperature, in degrees Celsius. (*b*) If 0.20 kg of gas escapes, find the new pressure at the same temperature, in megapascals.

4-11. A rigid tank with a volume of 3.0 m$^3$ contains a gas having a molar mass of 30 at 8 bars and 47°C. Gas leaks out of the tank until the pressure is 3 bars at 27°C. Find the volume, in cubic meters, occupied by the gas which escaped if it is at 1.0 bar and 22°C.

4-12. A tank contains carbon dioxide at 500 kPa and 40°C. A leak occurs in the tank which is not detected until the pressure has dropped to 340 kPa. If the temperature of the gas at the time of detection of the leak is 20°C, determine the mass of carbon dioxide that has leaked out if the original mass was 15 kg.

4-13. A tank contains helium at 600 kPa and 40°C. A kilogram of the gas is removed, which causes the pressure and temperature to change to 340 kPa and 20°C. Find (*a*) the volume of the tank, in liters, and (*b*) the original mass in the tank, in kilograms.

4-14. Two tanks, A and B, are connected by suitable pipes through a valve which initially is closed. Tank A initially contains 0.3 m$^3$ of nitrogen at 6 bars and 60°C, and tank B is evacuated. The valve is then opened, and nitrogen flows into tank B until the pressure in this tank reaches 1.5 bars and the temperature is 27°C. As a result, the pressure in tank A drops to 4 bars, and the temperature changes to 50°C. Determine the volume of tank B, in cubic meters.

4-15. Two insulated tanks, A and B, are connected by suitable pipes through a valve which initially is closed. Tank A initially contains nitrogen at 1.5 bars and 37°C in a volume of 0.030 m$^3$. Tank B contains nitrogen at 2.7 bars and 60°C. The valve is opened and the insulation removed. At equilibrium, the pressure is 2.0 bars and the temperature is the surrounding value of 27°C. Find the volume of tank B in cubic meters.

4-16E. Two insulated tanks, A and B, are connected by suitable pipes through a valve which initially is closed. Tank A initially contains oxygen at 20 psia and 100°F in a volume of 1.0 ft$^3$. Tank B contains oxygen at 40 psia and 140°F. The valve is opened and the insulation is removed. At equilibrium, the pressure is 32 psia and the temperature is the surrounding value of 80°F. Find the volume of tank B in cubic feet.

CHAPTER 4 • THE IDEAL GAS, CORRESPONDING STATES, AND INCOMPRESSIBLE MODELS    181

4-17E. A rigid tank with a volume of 10.0 ft³ contains carbon monoxide gas at 80 psia and 110°F. Gas leaks out of the tank until the pressure is 50 psia at 80°F. Find the volume, in cubic feet, occupied by the gas which escaped if it is at 14.7 psia and 70°F.

4-18E. A tank contains carbon dioxide at 80 psia and 100°F. A leak occurs in the tank which is not detected until the pressure has dropped to 45 psia. If the temperature of the gas at the time of detection of the leak is 70°F, determine the mass of carbon dioxide that has leaked out if the original mass was 50 lb$_m$.

4-19E. A tank contains helium at 8 atm and 100°F. Two pounds-mass of the gas are removed, which causes the pressure and temperature to change to 3.4 atm and 60°F. Find (a) the volume of the tank, in cubic feet, and (b) the original mass in the tank, in pounds-mass.

4-20E. Two tanks, A and B, are connected by suitable pipes through a valve which initially is closed. Tank A initially contains 10 ft³ of nitrogen at 80 psia and 140°F, and tank B is evacuated. The valve is then opened, and nitrogen flows into tank B until the pressure in this tank reaches 20 psia and the temperature is 60°F. As a result, the pressure in tank A drops to 60 psia, and the temperature changes to 110°F. Determine the volume of tank B, in cubic feet.

## DATA ACQUISITION FROM IDEAL-GAS TABLES

4-21. Find the following quantities from appropriate tables:
  (a) The temperature of air in kelvins with an enthalpy of 300 kJ/kg.
  (b) The $c_p$ value of air at 550 K.
  (c) The temperature of oxygen in kelvins with an internal energy of 289.5 kJ/kg.
  (d) The enthalpy of carbon dioxide in kJ/kg with an internal energy of 206.0 kJ/kg.
  (e) The internal energy change based on the arithmetic average specific heat for oxygen between 350 and 500 K, in kJ/kg.

4-22. Find the following quantities from appropriate tables:
  (a) The temperature in kelvins for nitrogen with a specific enthalpy of 405.1 kJ/kg.
  (b) The temperature in kelvins of carbon dioxide with a $c_v$ value of 0.790 kJ/kg·K.
  (c) The internal energy of oxygen in kJ/kg at 500 K.
  (d) The internal energy of air in kJ/kg with an enthalpy of 451.8 kJ/kg.
  (e) The enthalpy change in kJ/kg based on the arithmetic average specific heat for nitrogen between 450 and 600 K.

4-23. Find the following quantities from appropriate tables:
- (a) The temperature in kelvins when nitrogen has a specific enthalpy of about 1076 kJ/kg.
- (b) The temperature in kelvins for hydrogen with a $c_p$ value of 14.501 kJ/kg·K.
- (c) The temperature in kelvins of carbon dioxide with an internal energy of 266.8 kJ/kg.
- (d) The internal energy of carbon monoxide in kJ/kg when the specific enthalpy is 650.5 kJ/kg.
- (e) The internal-energy change based on the arithmetic-average specific heat for air between 350 and 550 K, in kJ/kg.

4-24E. Find the following quantities from appropriate tables:
- (a) The temperature of air in °R when its internal energy is 130 Btu/lb$_m$.
- (b) The temperature of air in °F when $c_v$ is 0.179 Btu/lb$_m$·°R.
- (c) The temperature of nitrogen in °F when its specific enthalpy is 188.7 Btu/lb$_m$.
- (d) The internal energy of carbon dioxide in Btu/lb when its internal energy is 124.2 Btu/lb$_m$.
- (e) The enthalpy change based on the arithmetic average of the specific heat for carbon monoxide between 300 and 600°F, in Btu/lb$_m$.

4-25E. Find the following quantities from appropriate tables:
- (a) The temperature in °R for nitrogen with a specific enthalpy of 133.9 Btu/lb$_m$.
- (b) The temperature in °F for carbon dioxide with a $c_v$ value of 0.202 Btu/lb$_m$·°R
- (c) The internal energy of oxygen in Btu/lb$_m$ at 100°F.
- (d) The internal energy of air in Btu/lb$_m$ when the specific enthalpy is 128.10 Btu/lb$_m$.
- (e) The enthalpy change in Btu/lb$_m$ based on the arithmetic average of the specific heat for nitrogen between 300 and 600°F.

4-26. A quantity of nitrogen initially occupies a volume of 0.890 m$^3$ at a pressure of 2 bar and a temperature of 27°C. It is compressed to a volume of 0.356 m$^3$ and a pressure of 12.5 bar. Find (a) the final temperature, in °C, (b) the mass, in kilograms, and the change in internal energy in kilojoules using (c) Table A-3 and (d) Table A-6.

4-27E. A quantity of nitrogen initially occupies a volume of 18.64 ft$^3$ at a pressure of 150 psia. It is expanded to a volume of 34.47 ft$^3$ at 30 psia and 80°F. Find (a) the initial temperature, in °F, (b) the mass in lb$_m$, and the change in internal energy in Btu using (c) Table A-3E and (d) Table A-6E.

## ENTHALPY AND INTERNAL-ENERGY CHANGES
## FROM SPECIFIC-HEAT DATA

4-28. The temperature of oxygen is increased from (a) 300 to 500 K and (b) 400 to 800 K.
   (1) On the basis of the equation for $c_p$ as a function of temperature at zero pressure (Table A-3), calculate the change in enthalpy, in kJ/kmol, by integration.
   (2) Compare this answer to that obtained by using $h$ data from Table A-7.
   (3) What percentage error will be introduced by using the arithmetic-mean value of $c_p$ from Table A-3 to evaluate $\Delta h$ between the given temperature limits, compared to part 1?

4-29. Do Prob. 4-28, except the gas is carbon dioxide.

4-30. The temperature of oxygen is increased from (a) 300 to 600 K and (b) 300 to 800 K.
   (1) On the basis of the empirical equation for $c_p$ as a function of temperature at zero pressure (Table A-3), calculate the change in internal energy, in kJ/kmol, by integration techniques.
   (2) Compare this answer to that obtained by using $u$ data from Table A-7.
   (3) What percentage error will be introduced by using the arithmetic-mean value of $c_v$ from Table A-3 to evaluate $\Delta u$ between the given temperature limits, compared to part 1?

4-31. Using the specific-heat data cited from monatomic gases in Table A-3, calculate the change in internal energy and enthalpy of (a) helium for the temperature change from 100 to 500°C and (b) argon for the temperature change from 300 to 700 K, in kJ/kg.

4-32E. The temperature of carbon dioxide is increased from (a) 100 to 800°F and (b) 300 to 1200°F. (1) On the basis of the equation for $c_p$ as a function of temperature at zero pressure (Table A-3E), calculate the change in enthalpy, in Btu/lbmol, by integration. (2) Compare this answer to that obtained by using $h$ data from Table A-7E. (3) What percentage error will be introduced by using the arithmetic-mean value of $c_p$ from Table A-3E to evaluate $\Delta h$ between the given temperature limits, compared to part 2?

4-33E. The temperature of oxygen is increased from (a) 100 to 800°F and (b) 200 to 1000°F. (1) On the basis of the empirical equation for $c_p$ as a function of temperature at zero pressure (Table A-3E), calculate the change in internal energy, in Btu/lbmol, by integration techniques. (2) Compare this answer to that obtained by using $u$ data from Table A-7E. (3) What percentage error will be introduced by using the arithmetic-mean value of $c_v$ from Table A-3E to evaluate $\Delta u$ between the given temperature limits, compared to part 1?

4-34E. Using the specific-heat data cited from monatomic gases in Table A-3E, calculate the change in internal energy and enthalpy of (*a*) helium for the temperature change from 100 to 700°F and (*b*) argon for the temperature change from 80 to 800°F, in Btu/lb.

## ENERGY ANALYSIS AT CONSTANT TEMPERATURE

4-35E. A piston-cylinder device initially contains an ideal gas at 50 psia and 3 ft$^3$. The piston travels 2 ft during an isothermal expansion process. Atmospheric pressure is 14.7 psia, a frictional force of 100 lb$_f$ opposes the motion, and the piston area is 36 in$^2$. Find the net work delivered to the output shaft, in ft·lb$_f$.

4-36. One-half kilogram of air is compressed isothermally and quasistatically in a closed system from 1 bar, 30°C to a final pressure of 2.4 bars. Determine (*a*) the internal-energy change and (*b*) the heat transfer, in kilojoules.

4-37. A piston-cylinder device contains 0.12 kg of air at 200 kPa and 123°C. During a quasiequilibrium, isothermal process, heat is removed in the amount of 15 kJ, and electrical work is carried out on the system in the amount of 1.50 W·h. Determine the ratio of the final to the initial volume.

4-38. In a closed system 0.1 kg of argon expands isothermally from 2 bars, 325 K until the volume is tripled and 68.2 kJ/kg of heat is added. If during the process a 12-V battery is operated for 20 s, determine the constant current supplied, in amperes, to a resistor within the system. Neglect energy storage in the resistor.

4-39. Air is contained within a piston-cylinder device at initial conditions of 400 kPa and 0.01 m$^3$. A paddle wheel within the gas is turned under the conditions of an applied torque of 7.50 N·cm for 2400 rev. During the process the temperature remains constant while the volume is doubled. Determine the magnitude, in kilojoules, and the direction of any heat transfer.

4-40. A rigid storage tank for water in a home has a volume of 0.40 m$^3$. Initially the tank contains 0.30 m$^3$ of water at 20°C and 240 kPa. The gas space above the water contains air at the same temperature and pressure. An additional 0.050 m$^3$ of water is slowly pumped into the tank so that the temperature remains constant. Determine (*a*) the final pressure in the tank, in kilopascals, and (*b*) the work done on the air, in newton-meters.

4-41. Carbon dioxide gas initially at 120 kPa and 30°C is compressed isothermally in a piston-cylinder device to 300 kPa.

(a) Compute the work done on the gas, in kJ/kg, and show the area representation on a $Pv$ diagram.

(b) Now consider atmospheric air at 96 kPa acting on the back of the piston, in addition to the connecting rod. Compute the net work required by the outside source in this case, for the same change in state of the $CO_2$ gas, and the $Pv$ area representation.

(c) Finally, in addition to the atmospheric air in part $b$, consider that friction acts between the piston and cylinder. The frictional resistance is equivalent to an effective pressure of 30 kPa which is constant throughout the piston travel. In this case, determine the net work that must be delivered along the connecting rod in kJ/kg. Show the area representation of this final net work on a $Pv$ diagram.

4-42E. One-fourth of a pound of nitrogen is compressed isothermally in a closed system from 15 psia, 80°F to a final pressure of 36 psia. Determine (a) the internal-energy change and (b) the heat transfer, in Btu.

4-43E. A quasiequilibrium isothermal process occurs in a piston-cylinder device which initially contains 0.27 $lb_m$ of air at 35 psia and 240°F. Heat is removed in the amount of 15 Btu, and electrical work is carried out on the system in the amount of 1.76 W·h. Determine the ratio of the final to the initial volume.

4-44E. One-tenth of a pound of argon expands isothermally from 30 psia, 140°F until the volume is doubled and 16.1 Btu/$lb_m$ of heat is added. During the process a 12-V battery is operated for 20 s. Determine the constant current supplied, in amperes, to a resistor within the system. Neglect energy storage in the resistor.

4-45E. Carbon dioxide gas initially at 18 psia and 80°F is compressed isothermally in a piston-cylinder device to 45 psia.

(a) Compute the work done on the gas, in ft·$lb_f$/$lb_m$, and show the area representation on a $Pv$ diagram.

(b) Now consider atmospheric air at 14.4 psia acting on the back of the piston, in addition to the connecting shaft. Compute the net work required by an outside source in this case, for the same change in state of the $CO_2$ gas, and the $Pv$ area representation.

(c) Finally, in addition to the atmospheric air in part $b$, consider that friction acts between the piston and cylinder. The frictional resistance is equivalent to a resistive pressure of 4.0 psi which is constant throughout the piston travel. Determine the net work that must be supplied by the connecting shaft in ft·$lb_f$/$lb_m$. Show the area representation of this final net work on a $Pv$ diagram.

## Energy Analysis at Constant Pressure

4-46. A monatomic gas within a piston-cylinder device undergoes a process where the pressure is fixed at 3 bars while the volume decreases from 0.08 to 0.04 m³. If the initial temperature is 600 K, determine (a) the work done and (b) the heat transfer during the process, in kilojoules.

4-47. A piston-cylinder device which initially has a volume of 0.1 m³ contains 0.014 kg of hydrogen at 210 kPa. Heat is transferred until the final volume is 0.085 m³. If the quasistatic process occurs at constant pressure, determine (a) the final temperature, in degrees Celsius, (b) the heat transfer, in kilojoules, based on data from Table A-11, and (c) the heat transfer based on specific-heat data from Table A-3.

4-48. One-hundredth of a kilogram-mole of helium is contained within an adiabatic piston-cylinder device and is maintained at a constant pressure of 2 bars. The piston is frictionless and the initial temperature of the gas is 20°C. A paddle wheel is rotated for 1000 rev within the gas until the volume of the gas has increased by 25 percent. Compute (a) the constant torque on the paddle wheel, in N·m, and (b) the net work done on the system, in kilojoules.

4-49. One-tenth cubic meter of nitrogen at 0.1 MPa and 25°C is contained in a piston-cylinder device. A paddle wheel within the cylinder is turned until 20,300 N·m of energy has been added. If the process is quasistatic, adiabatic, and at constant pressure, determine the final temperature of the gas in degrees Celsius. Neglect energy storage in the paddle wheel, and use data from Table A-6.

4-50. A piston-cylinder device maintains nitrogen at a constant pressure of 200 kPa. The initial volume and temperature are 0.10 m³ and 20°C, respectively. Electrical work is performed on the system by allowing 6 A at 12 V to pass through a resistor within the system for 9 min. As a result, the volume increases at 0.14 m³. Determine (a) the net work for the process, (b) the change in internal energy of the gas, using specific-heat data, and (c) the magnitude and direction of any heat transfer, all in kilojoules. Neglect energy storage in the resistor.

4-51. An adiabatic piston-cylinder device contains 0.3 kg of air initially at 300 K and 3 bars. Energy is added to the gas by passing a current of 6 A through a resistor within the system for 30 s. The gas expands quasistatically at constant pressure until the volume is doubled. Neglecting energy storage in the resistor, determine the size of the resistor, in ohms. Recall that the power dissipated in a resistor is $I^2R$.

4-52. An ideal gas initially in a piston-cylinder device at 1.5 bars and 0.03 m³ is first heated at constant pressure until the volume is doubled. It is then allowed to expand isothermally until the volume is again doubled. Determine the total work done by the gas, in kJ/kmol, and plot the quasistatic processes on a $PV$ diagram. The initial gas temperature is 300 K.

4-53E. One-hundredth of a pound-mole of helium is maintained at a constant pressure of 30 psia within an adiabatic piston-cylinder device. The piston is frictionless and the initial temperature of the gas is 60°F. A paddle wheel is rotated for 500 rev within the gas until the volume of the gas has increased by 25 percent. Compute (a) the constant torque on the paddle wheel, in $lb_f$·ft, and (b) the net work done on the system, in ft·$lb_f$.

4-54E. A piston-cylinder device initially contains a cubic foot of nitrogen at 15 psia and 80°F. A paddle wheel within the cylinder is turned by an external motor until 6215 ft·$lb_f$ of energy has been added. If the process is quasistatic, adiabatic, and at constant pressure, determine the final temperature of the gas in degrees Fahrenheit. Neglect energy storage in the paddle wheel, and use data from Table A-6E.

4-55E. A piston-cylinder device maintains nitrogen at a constant pressure of 30 psia. The initial volume and temperature are 0.35 ft$^3$ and 60°F, respectively. Electrical work is performed on the system by allowing 2 A at 12 V to pass through a resistor within the system for 4.0 min. As a result, the volume increases at 0.60 ft$^3$. Determine (a) the net work for the process, (b) the change in internal energy of the gas, using specific-heat data, and (c) the magnitude and direction of any heat transfer, all in Btu. Neglect energy storage in the resistor.

4-56E. A monatomic gas within a piston-cylinder device undergoes a quasiequilibrium process where the pressure is fixed at 20 psia while the volume increases from 1.0 to 2.0 ft$^3$. If the initial temperature is 140°F, determine (a) the work done, in ft·$lb_f$, and (b) the heat transfer during the process, in Btu.

4-57E. An adiabatic piston-cylinder device contains 0.66 $lb_m$ of air initially at 60°F and 3 atm. A current of 4 A is then passed through a resistor within the system for 30 s. The gas expands quasistatically at constant pressure until the volume is 75 percent greater. Neglecting energy storage in the resistor, determine the size of the resistor, in ohms. Recall that the power dissipated in a resistor is $I^2R$.

## ENERGY ANALYSIS AT CONSTANT VOLUME

4-58. Nitrogen contained in a rigid tank at 27°C is heated until the pressure increases by a factor of 2. Determine the heat transfer required, in kJ/kg, by using (a) average specific-heat data from Table A-3 and (b) data from Table A-6.

4-59. A rigid tank with a volume of 1.0 m$^3$ contains oxygen at 0.20 MPa and 20°C. A paddle wheel within the gas undergoes 1400 rev with an applied torque of 4.42 N·m. The final temperature is 40°C. Determine (a) the change in the internal energy of the gas, in kilojoules, using specific-heat data, and (b) the magnitude, in kilojoules, and the direction of any heat transfer. Neglect energy storage in the paddle wheel.

4-60. A rigid tank contains nitrogen gas initially at 100 kPa and 17°C. A paddle wheel within the tank is rotated by an external source which provides a torque of 11.0 N·m for 100 rev of the shaft until the final pressure is 130 kPa. During the process a heat loss of 1.0 kJ occurs. Determine (a) the mass within the tank, in kilograms, and (b) the volume of the tank, in cubic meters. Neglect energy storage in the paddle wheel.

4-61. A rigid tank initially contains 0.80 g of air at 295 K and 1.5 bars. An electric resistor within the tank is energized by passing a current of 0.6 A for 30 s from a 12.0-V source. At the same time a heat loss of 156 J occurs.
(a) Determine the final temperature of the gas, in kelvins.
(b) Find the final pressure in bars.

4-62E. Nitrogen, initially at 20 psia and 70°F, occupies a rigid tank with a volume of 4.0 ft³. A paddle wheel within the gas undergoes 180 rev with an applied torque of 3.20 lb$_f$·ft. The final temperature is 120°F. Determine (a) the change in the internal energy of the gas, in Btu, using specific-heat data, and (b) the magnitude, in Btu, and the direction of any heat transfer. Neglect energy storage in the paddle wheel.

4-63E. Oxygen gas, initially at 15 psia and 40°F, fills a rigid tank. An external work source rotates a paddle wheel within the tank with a shaft torque of 8.0 lb$_f$·ft for 100 rev. The final pressure is 17.4 psia, and a heat loss of 2.0 Btu occurs during the process. Determine (a) the mass within the tank, in pounds-mass, and (b) the volume of the tank, in cubic feet. Neglect energy storage in the paddle wheel.

4-64E. A rigid tank contains 0.023 lbmol of nitrogen initially at 120°F and 2 atm. An external 120-V power source is used to pass a current of 1.2 A for 50 s through an electric resistor within the tank. A heat loss of 12.5 Btu occurs during the process. Determine (a) the final temperature of the gas, in degrees Fahrenheit, (b) the final pressure, in atmospheres, and (c) the volume of the tank, in cubic feet.

4-65. A rigid and well-insulated tank contains air in a volume of 0.1 m³. Paddle-wheel work in the amount of 9450 N·m results in a final pressure of 2.5 bars and a temperature of 400 K. Determine (a) the initial temperature in kelvins and (b) the initial pressure in bars.

4-66E. A rigid and well-insulated tank contains air in a volume of 2 ft³ at 20 psia and 500°R. A paddle wheel transfers 5000 ft·lb$_f$ of energy. Determine (a) the final temperature in °R and (b) the final pressure in psia.

## SPECIFIC-HEAT EVALUATION

4-67. Ten grams of an ideal gas having a molar mass of 32 undergo a quasistatic expansion at constant pressure from 1.3 bars, 20°C, to 60°C. During the process 370 J of heat is added to the closed system. Compute (a) the expansion work, in joules, and (b) the average value of $c_v$ for the gas, in kJ/kg·°C.

4-68. One-tenth of a kilogram of an ideal gas ($M = 40$) expands quasistatically at constant pressure in a closed system from 100 kPa, 40°C, to 150°C. If 5200 J of heat is added during the process, find (a) the expansion work, in kilojoules, and (b) the average $c_v$ value, in kJ/kg·°C.

4-69E. An ideal gas with a mass of 0.022 lb and a molar mass of 32 undergoes a quasistatic expansion at constant pressure from 20 psia, 65°F, to 135°F. During the process 0.339 Btu of heat is added to the closed system. Find (a) the expansion work, in Btu, and (b) the average value of $c_v$ for the gas, in Btu/lb$_m$·°F.

## ADDITIONAL ENERGY ANALYSIS

4-70. A reciprocating compressor initially contains air in the piston-cylinder device at 0.10 m³, 0.95 bar, and 67°C. The compression process is quasistatic and represented by the process equation $PV^{1.3} = $ constant. The final volume is 0.02 m³. Determine (a) the mass of the air in kilograms, (b) the final temperature in kelvins, (c) the internal-energy change in kilojoules, (d) the work required in kilojoules, and (e) the magnitude and direction of any heat transfer in kilojoules.

4-71. A rigid, insulated tank with a total volume of 3.0 m³ is divided in half by an insulated, rigid partition. Both sides of the tank contain an ideal monatomic gas. On one side the initial temperature and pressure are 200°C and 0.50 bar, and on the other side the values are 40°C and 1.0 bar. The internal partition is broken, and complete mixing occurs. Determine (a) the final equilibrium temperature, degrees Celsius, and (b) the final pressure, in bars.

4-72. In a tank 0.81 kg of nitrogen is stored at 3 bars and 50°C. Attached to this tank through a suitable valve is a second tank which is 0.50 m³ in volume and completely evacuated. Both tanks are thoroughly insulated. If the valve is opened and equilibrium is allowed to be reached, determine the final pressure, in bars.

4-73. Tank A contains 1 kg of oxygen initially at 100 kPa and 60°C. Attached to this tank through a suitable valve is a second tank B which contains 1 kg of oxygen at 200 kPa and 10°C. Both tanks are insulated. If the valve is opened and equilibrium is allowed to be reached, determine the final pressure, in kilopascals.

4-74. Oxygen is contained within a piston-cylinder device initially at 600 kPa, 200°C, and 0.020 m³. The gas expands according to the process equation $PV^{1.2}$ = constant until the temperature reaches 100°C. Determine (a) the final volume in cubic meters and pressure in kilopascals, (b) the value of the work, in newton-meters, and (c) the heat transfer, in kilojoules.

4-75. A piston-cylinder machine contains nitrogen initially at 2.0 bars, 107°C, and 0.300 m³. The piston moves with negligible friction until the pressure rises to 5.0 bars. The process is described by the equation $V = 0.40 - 0.050\,P$, where $V$ is in cubic meters and $P$ is in bars. Determine (a) the final temperature, in degrees Celsius, (b) the mass present, in kilograms, (c) the work done, in newton-meters, and (d) the heat transfer, in kilojoules.

4-76. Carbon dioxide gas initially at 27°C, 100 kPa, and 0.1 m³ is compressed in a piston-cylinder device. The process equation is given by $Pv^{1.2}$ = constant. If the work input is 103.9 kJ/kg, determine the heat transfer in kJ/kg.

4-77. A piston-cylinder device initially contains 0.5 m³ of nitrogen gas at 4 bars and 27°C. The nitrogen expands in a polytropic process during which $PV^{0.82}$ = constant until the volume doubles. Simultaneously, an electric heater from a 120-V source is turned on for 10 min. A heat loss of 12.8 kJ also occurs during the process. Determine
(a) The final temperature, in °C.
(b) The magnitude and direction of the boundary work, in kilojoules.
(c) The resistance of the heater needed to achieve the final temperature, in ohms.
(d) The cost of electricity, in cents, if the power company rate is $0.10/kW·h.

4-78. A gas with a molar mass of 33 fills a volume of 0.1 m³ within a piston-cylinder device at 3 bars and 360 K. Heat addition causes the gas to expand to 0.2 m³ along the path given by $PV^{1.3}$ = constant. If $c_v = 0.60 + 2.5 \times 10^{-4}T$, where $c_v$ is in kJ/kg·K and $T$ is in kelvins, determine the heat addition in kilojoules.

4-79. Two tanks are connected by a valve. Tank A contains 2 kg of carbon monoxide gas at 77°C and 0.7 bar. Tank B holds 8 kg of the same gas at 27°C and 1.2 bars. The valve is opened and the gases are allowed to mix while receiving energy by heat transfer from the surroundings. The final equilibrium temperature is 42°C. Using the ideal-gas model, determine
(a) The final equilibrium pressure, in bars.
(b) The heat transfer for the process, in kilojoules.

4-80. Two identical tanks, both insulated and having a volume of 1 m³, are connected by a valve. Tank A contains air at 10 bars and 350 K, while tank B contains air at 1 bar and 300 K. The valve is opened and the system allowed to equilibrate. Calculate (a) the final temperature in kelvins and (b) the final pressure in bars.

4-81. One-tenth of a kilogram of air in a piston-cylinder device undergoes a process for which $V = CT^2$, where $C$ is a constant. The initial conditions are 227°C and 0.01 m³. If the volume doubles, find

(a) The final temperature, in kelvins.
(b) The final pressure, in bars.
(c) The work interaction by the gas, in kilojoules.
(d) The heat transferred, in kilojoules.

4-82E. A reciprocating compressor initially contains air in the piston-cylinder device at 60 in³, 14.5 psia, and 100°F. The compression process is quasistatic and represented by the process equation $PV^{1.3}$ = constant. The final volume is 12 in³. Determine (a) the mass of the air, in pounds, (b) the final temperature, in degrees Rankine, (c) the internal-energy change, in Btu, (d) the work required, in ft·lb$_f$, and (e) the magnitude and direction of any heat transfer, in Btu.

4-83E. A rigid, insulated tank is divided in half by an insulated, rigid partition. Both sides of the tank contain 1.5 ft³ of the same ideal monatomic gas. On side A the initial temperature and pressure are 400°F and 1.2 atm, and on side B the values are 120°F and 2.2 atm. The internal partition is broken, and complete mixing occurs. Determine (a) the final equilibrium temperature, in degrees Fahrenheit, and (b) the final pressure, in atmospheres.

4-84E. Tank A contains 0.1 lb$_m$ of nitrogen initially at 20 psia and 100°F. Attached to this tank through a suitable valve is a second tank B which contains 0.20 lb$_m$ of the same gas at 40 psia and 250°F. Both tanks are insulated, and all specific heats are assumed constant. If the valve is opened and equilibrium is allowed to be reached, determine (a) the final mixture temperature, in degrees Fahrenheit, and (b) the final pressure, in psia.

4-85E. Carbon dioxide is contained within a piston-cylinder device initially at 75 psia, 400°F, and 0.50 ft³. The gas expands according to the process equation $PV^{1.2}$ = constant until the temperature reaches 200°F. Determine (a) the final volume, in cubic feet, and pressure, in psia, (b) the work, in ft·lb$_f$, and (c) the heat transfer, in Btu.

4-86E. A piston-cylinder device contains carbon dioxide initially at 340°F, 45.0 psia, and 0.5 ft³. The process equation for the expansion is given by $Pv^{1.2}$ = constant. If the work output is 54.2 Btu/lb$_m$, find the heat transfer in Btu/lb$_m$.

4-87. An insulated, rigid cylindrical chamber contains helium at 10 bars, 27°C, and 0.5 m³ separated by a frictionless piston from air at 1 bar, 327°C, and 0.3 m³, as shown in Fig. P4-87. The piston is released

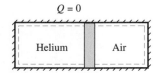

**Figure P4-87**

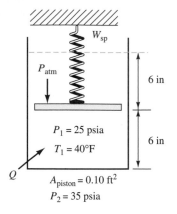

$P_{atm}$

6 in

$P_1 = 25$ psia

$T_1 = 40°F$

6 in

$Q$

$A_{piston} = 0.10$ ft$^2$

$P_2 = 35$ psia

**Figure P4-89**

from the initial position and moves until the pressure equilibrates. The piston conducts heat so that the temperatures also equilibrate. Find (*a*) the final temperature, in °C, (*b*) the final pressure, in bars, and (*c*) the final volume occupied by the air, in m$^3$.

4-88E. An insulated, rigid cylindrical chamber contains helium at 10 psia, 400°F, and 1.0 ft$^3$ separated by a frictionless piston from air at 100 psia, 70°F, and 0.1 ft$^3$, as shown in Fig. P4-87. The piston is released from the initial position and moves until the pressure equilibrates. The piston conducts heat so that the temperatures also equilibrate. Find (*a*) the final temperature, in °F, (*b*) the final pressure, in psia, and (*c*) the final volume occupied by the air, in ft$^3$.

4-89E. Consider the equipment sketch in Fig. P4-89. A piston is initially held in an equilibrium position by an external compressed spring and the atmosphere of 15.0 lb$_f$/in$^2$, so that the total pressure of the *air* within the cylinder initially is 25 lb$_f$/in$^2$. The area of the piston is 0.10 ft$^2$, the vertical height inside the cylinder is 6.0 in, and $T_1 = 40°F$. A heat interaction now occurs such that the spring shortens by 6.0 in and the pressure reaches 35 lb$_f$/in$^2$. From a knowledge of the spring constant, the work done on the spring is known to be 108 ft·lb$_f$ and the piston is massless and frictionless. Find

(*a*) The total work done by the gas, in ft·lb$_f$.
(*b*) The internal energy change of the gas, in Btu.
(*c*) The magnitude, in Btu, and direction of $Q$.

4-90. A cylinder closed by a frictionless piston contains argon initially at 10 bars and 300 K. The piston is retained by a spring, as shown in Fig. P4-90. During a 100-s interval a current of 100 A passes through the resistor in the cylinder with a voltage drop of 20 V. In the same time interval a heat loss occurs at a rate of 1 kW. The initial volume is 0.1 m$^3$ and the final temperature is 450 K. Find the work done on the piston by the gas, in kilojoules.

4-91. Nitrogen gas is contained within a piston-cylinder device initially at 27°C and 0.72 m$^3$/kg (state 1). The gas undergoes an isothermal

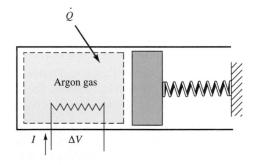

$\dot{Q}$

Argon gas

$I$     $\Delta V$

**Figure P4-90**

process to state 2 where the specific volume is 0.36 m³/kg. It is then expanded at constant pressure to state 3, where $v_3 = v_1$. Finally it is returned to its initial state along a constant-volume path.

(a) Sketch the cyclic process on a $Pv$ diagram.

(b) Determine the value of $P_2$, in bars, and $T_3$, in kelvins.

(c) Determine the work and heat for the three processes, in kJ/kg.

4-92. Air initially at 600 kPa and 660 K (state 1) undergoes a cyclic process composed of a constant-volume process to 250 kPa (state 2), followed by an isothermal process to the initial pressure, and then a constant-pressure process to the initial volume. Find (a) the specific volume at state 1, in m³/kg, (b) the temperature at state 2 in kelvins, (c) the heat and work for the three processes, in kJ/kg, and (d) sketch the process on a $Pv$ diagram.

4-93. A 25-kg piston under standard atmospheric conditions is required to travel a specific distance within the piston-cylinder arrangement shown in Fig. P4-93. The piston, whose area is 0.0050 m², is initially at rest on the lower stops. Air within the cylinder is heated until the piston touches the upper stops. The initial pressure and temperature of the air are 0.101 MPa and 20°C, $c_v = 0.72$ kJ/kg·K, $g = 9.80$ m/s², and $P_{atm} = 101$ kPa. Determine

(a) The mass of air in the cylinder, in kilograms.

(b) The pressure $P_2$ and temperature $T_2$ in the cylinder just as the piston starts to move, in MPa and kelvins.

(c) The heat added before the cylinder will move, in kilojoules.

(d) The temperature $T_3$ when the piston reaches the upper stops, in kelvins.

(e) The work and heat during the movement of the piston, both in kilojoules.

4-94. Carbon monoxide gas is contained within a piston-cylinder device at 100 kPa and 27°C.

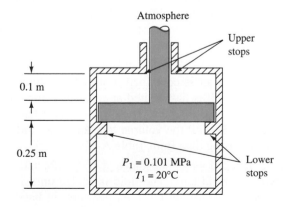

**Figure P4-93**

(*a*) In process A the gas is heated at constant volume until the pressure has doubled. It is then expanded at constant pressure until the volume is three times its initial value.

(*b*) In process B, the same gas in the same initial state is first expanded at constant pressure until the volume has tripled; then the gas is heated at constant volume until it reaches the same final pressure as in process A.

For these two overall processes determine the net heat effect, the net work effect, and the changes in internal energy, in kJ/kg. Compare the results for the two processes.

4-95. A piston-cylinder arrangement contains 1 kg of air initially at 2 bars and 77°C. Two processes are involved: a constant-volume process followed by a constant-pressure process. During the first process, heat is added in the amount of 57,970 J. During the second process, which follows the first, heat is added at constant pressure until the volume is 0.864 m³. If the processes are quasistatic, (*a*) evaluate the total work of expansion, in kilojoules, and (*b*) the total heat transfer, in kilojoules.

4-96E. A piston-cylinder device contains carbon monoxide gas initially at 15 psia and 60°F.

(*a*) In process A the gas is heated at constant volume until the pressure has doubled. It is then expanded at constant pressure until the volume is three times its initial value.

(*b*) In process B, the same gas in the same initial state is first expanded at constant pressure until the volume has tripled; then the gas is heated at constant volume until it reaches the same final pressure as in process A.

For these two overall processes determine the net heat effect, the net work effect, and the changes in internal energy, in Btu/lb. Compare the results for the two processes.

4-97E. Air initially at 30 psia and 0.69 ft³ and with a mass of 0.1 lb expands at constant pressure to a volume of 1.5 ft³. It then changes state at constant volume to a pressure of 15 psia. If the processes are quasistatic, find the total work done, the total heat transferred, and the overall change in internal energy, all in Btu.

4-98E. One cubic foot of air is contained in a piston-cylinder assembly at 20 psia and 40°F. It is first heated at constant volume (process 1-2) until the pressure has doubled. Then it is expanded at constant pressure (process 2-3) until the volume has doubled. Find the total heat added, in Btu, using data from Table A-5E.

4-99. One cubic meter of water substance at 10.0 MPa and 480°C is contained within a rigid tank. The fluid is cooled until its temperature reaches 320°C. Determine the final pressure, in bars, and the heat transfer, in kilojoules, if the fluid is (*a*) a real gas and (*b*) an ideal gas, using tabular data in both cases. (*c*) Find the percentage of error in the heat transfer when the ideal-gas model is used.

## COMPRESSIBILITY FACTOR

4-100. Water vapor exists at 440°C and at 200 bars. Determine the specific volume, in m³/kg, based on (*a*) the ideal-gas equation, (*b*) the corresponding-states principle, and (*c*) the tables of superheat data.

4-101. Refrigerant 134a is maintained at 60°C and 14 bars. Determine the specific volume, in m³/kg, based on (*a*) the ideal-gas equation, (*b*) the principle of corresponding states, and (*c*) data in the superheat table.

4-102. Determine the pressure, in bars, of water vapor at 360°C and 0.03089 m³/kg on the basis of (*a*) the ideal-gas equation, (*b*) the principle of corresponding states, and (*c*) the superheat table.

4-103. A tank with a volume of 12.5 m³ contains propane ($C_3H_8$) initially at 115°C and 37.9 bars. Estimate the mass of the gas, in kilograms, by using (*a*) an ideal-gas model and (*b*) the compressibility chart. Now, gas is withdrawn until the mass remaining is one-half the original amount and the temperature remains the same. On the basis of the result from part *b* and the compressibility chart, (*c*) estimate the final pressure in the tank, in bars. (*d*) Find the final pressure if it is an ideal gas.

4-104E. Water vapor exists at 700°F and at 2500 psia. Determine the specific volume, in ft³/lb$_m$, based on (*a*) the ideal-gas equation, (*b*) the corresponding-states principle, and (*c*) the superheat table.

4-105E. Refrigerant 134a is maintained at 160°F and 200 psia. Determine the specific volume, in ft³/lb$_m$, based on (*a*) the ideal-gas equation, (*b*) the principle of corresponding states, and (*c*) data in the superheat table.

4-106E. Determine the pressure, in psia, of water vapor at 700°F and 0.491 ft³/lb$_m$ on the basis of (*a*) the ideal-gas equation, (*b*) the principle of corresponding states, and (*c*) the superheat table.

4-107E. A tank with a volume of 300 ft³ contains carbon dioxide initially at 140°F and 37.9 atm. Estimate the mass of the gas, in pounds, by using (*a*) an ideal-gas model and (*b*) the compressibility chart. Carbon dioxide is now added until the mass remaining is doubled, but the temperature remains the same. Estimate the final pressure in the tank, in atmospheres, on the basis of (*c*) the compressibility chart, and (*d*) an ideal-gas model.

4-108. Sulfur dioxide ($M = 64.06$) is contained in a 10-m³ tank at 51 bars and 180°C. It is heated to 245°C. Determine (*a*) the mass in the tank in kilograms, (*b*) the final pressure, in bars, by means of the compressibility chart, and (c) the final pressure based on an ideal-gas model.

4-109. A tank truck is filled with methane ($CH_4$) at 7.9 MPa and 200 K. Owing to poor insulation, the gas warms to 286 K after a period of time. If the tank has a volume of 10.0 m$^3$, estimate on the basis of the compressibility chart (a) the mass within the tank, in kilograms, and (b) the final pressure in the tank, in megapascals.

4-110. Steam, initially at 16.0 MPa and 440°C, expands isothermally until its volume is doubled. Determine the final pressure, in megapascals, if (a) the ideal-gas equation applies, (b) the corresponding-states principle is used, and (c) the steam table is used.

4-111. Ethane ($C_2H_6$), initially at 63°C and 34.2 bars, is heated at constant pressure until the temperature reaches 124°C. Estimate the change in volume, in m$^3$/kg, if (a) the ideal-gas equation is used, and (b) the corresponding-states principle is used.

4-112. Steam, initially at 12.0 MPa and 480°C, is compressed isothermally to one-half its initial volume. Determine the final pressure, in megapascals, if (a) the ideal-gas equation applies, (b) the corresponding-states principle is used, and (c) the steam table is used.

4-113E. Propane ($C_3H_8$), initially at 205°F and 29.5 atm, is heated at constant pressure until the temperature reaches 405°F. Estimate the change in volume, in ft$^3$/lb, if (a) the ideal-gas equation is used, and (b) the corresponding-states principle is used.

4-114. Refrigerant 134a is heated in a piston-cylinder device at a constant pressure of 16 bars from 70 to 120°C. On the basis of data from a compressibility chart determine the work required in kJ/kg.

4-115. Steam at 100 bars and 320°C is heated in a rigid container until a pressure of 160 bars is reached. Determine the final temperature in °C using (a) the ideal-gas model, (b) the compressibility factor chart, (c) the steam tables.

4-116. Steam at 180 bars, 520°C, is cooled at constant volume until its temperature reaches 400°C. Determine the initial specific volume, in m$^3$/kg, and the final pressure, in bars, based on (a) the Z chart and (b) the superheat table.

4-117E. Carbon dioxide is heated in a piston-cylinder device at a constant pressure of 965 psia from 90 to 365°F. Determine the work associated with the process, in Btu/lb$_m$, on the basis of (a) the compressibility chart and (b) the ideal-gas model.

## Incompressible Substance

4-118. An electrical resistance heater with a mass of 0.7 kg and a specific heat of 0.74 kJ/kg·K is immersed in an oil with a mass of 2.4 kg and a specific heat of 2.0 kJ/kg·K. Both heater and oil initially are at 20°C. A current of 2.0 A provided from a 220-V source now passes through the heater element for 1 min. Assuming temperature equilibrium occurs fairly rapidly, a thermometer in the oil bath reads 21.74°C. Find (*a*) the heat transfer from the element to the oil and (*b*) the heat transfer from the oil to the surroundings, in kilojoules.

4-119. Ten kilograms of ice at 260 K is dumped into 20 kg of liquid water initially at 300 K in an insulated container. The average specific heats of ice and liquid may be taken to be 2.065 and 4.188 kJ/kg·K, respectively, and 333.5 kJ are required to melt 1 kg of ice at 0°C. Calculate (*a*) the mass of ice melted and (*b*) the final temperature in °C.

4-120. A 25-kg block of copper at 150°C is dropped into an insulated tank containing 0.79 kg of liquid water at 20°C. Steam rises above the liquid as the system equilibrates. If the average specific heat of copper is taken as 0.395 kJ/kg·K, determine the mass of water evaporated at 100°C, in kilograms.

4-121. Three identical copper blocks with a mass of 25 kg, a temperature of 150°C, and a mean specific heat of 0.390 kJ/kg·K, are dropped into three containers of water, each at 20°C. Containers hold (*a*) 5, (*b*) 1, and (*c*) 0.1 kg of water, respectively. If the only heat transfer occurs between the copper and the water, find the final temperature in each case, in degrees Celsius. The *c* value for steam is 2.0 kJ/kg·K.

4-122. A vessel containing 10 L of liquid R-134a at 12 bars and 0°C is used to cool some electronic parts. Each part initially is at 300°C with a mass of 0.1 kg and an average specific heat of 0.14 kJ/kg·K. Find the number of parts which can be cooled to 80°C without causing the temperature of the liquid R-134a to exceed 40°C. Neglect heat losses to the environment and evaporation of liquid R-134a.

4-123. A rigid vessel with an internal volume of 0.03 m$^3$ is filled with nitrogen gas. The vessel is constructed of steel with a specific heat of 0.45 kJ/kg·K and a mass of 6 kg. A high-quality insulation of negligible thermal mass ($mc \approx 0$) is wrapped around the vessel. A paddle wheel of negligible mass rotates at 120 rev/min with an applied torque of 3.15 N·m. The initial temperature of the gas and vessel is 27°C and the initial gas pressure is 6 bars. If the paddle wheel runs for 30 min, find (*a*) the kilograms of nitrogen gas and (*b*) the final temperature in °C.

4-124E. A rigid vessel constructed of steel with a specific heat of 0.107 Btu/lb$_m$·°R, a mass of 15 lb$_m$, and an internal volume of 1 ft$^3$ is filled with nitrogen gas. A high-quality insulation is wrapped around the vessel. A paddle wheel of negligible mass rotates at 120 rev/min with an applied torque of 2.62 ft·lb$_f$. The initial temperature of the gas and vessel is 80°F and the initial gas pressure is 100 psia. If the paddle wheel operates for 20 min, find (*a*) the pounds of nitrogen gas and (*b*) the final temperature of the gas and vessel in degrees Fahrenheit.

4-125. A 16-kg mass of copper is dropped into an insulated tank which contains water at 20.0°C and 1 bar. The initial temperature of the copper is 60°C and the final equilibrium temperature of the copper and water at 1 bar is 21.24°C. Estimate the initial volume of water used, in cubic meters.

4-126. An unknown mass of copper at 77°C is dropped into an insulated tank which contains 0.2 m$^3$ of water at 30°C. If the equilibrium final temperature is 32°C, find the mass of copper, in kilograms.

4-127. Consider three liquids A, B, and C which have specific heats at constant pressure of 2.0, 4.0, and 7.0 kJ/kg·°C, respectively. The three liquids occupy three compartments of an insulated tank, but the walls separating the compartments conduct heat. Initially, one section contains 5 kg of A at 20°C, and the other compartments contain 2 kg of B at 60°C and 1 kg of C at 100°C. Find the temperature of the three fluids at thermal equilibrium, if the overall process is one of constant pressure.

4-128. A 2-kg mass of lead at −100°C is brought into contact with an unknown mass of aluminum initially at 0°C. If the final equilibrium temperature of the two metals is −40°C, determine the mass of aluminum, in kilograms.

4-129. A 4-kW resistance heater is placed in a 0.050-m$^3$ container filled with water at 20°C and 1 bar. The heater is allowed to operate for 20 min. The mass of the heater is 2.0 kg, and its specific heat is 0.75 kJ/kg·°C. If the container is insulated, determine the temperature of the water and heater element in the final state, in degrees Celsius.

4-130. A paddle wheel driven by a 240-W motor is used to agitate liquid water in a closed, insulated tank maintained at 1 bar. The initial temperature is 20°C. Determine the temperature rise of the contents after 0.80 h if the mass of water is 6 kg and the mass and specific heat of the paddle wheel are 0.4 kg and 0.96 kJ/kg·°C, respectively.

4-131. A copper tank with a mass of 18 kg initially is at 27°C. Water at 50°C and 1 bar and with a mass of 4 kg is poured into the tank. The tank is heavily insulated on the outside. Determine (a) the final equilibrium temperature, in °C, and (b) the magnitude and direction of the heat interaction associated with the water, in kilojoules.

4-132. A 30-lb mass of copper with an initial temperature of 150°F is dropped into an insulated tank which contains water initially at 70.0°F and 1 atm. The final equilibrium temperature of the copper and water at 1 atm is 72.80°F. Estimate the initial volume of water used, in cubic feet.

4-133E. A piece of copper initially at 170°F is dropped into an insulated tank which contains 4 ft$^3$ of water at 90°F. If the equilibrium final temperature is 91.60°F, find the mass of copper in pounds.

4-134E. Three liquids A, B, and C occupy separate compartments of a tank which is insulated externally, but the walls separating the compartments conduct heat. The liquids A, B, and C have specific heats at constant pressure of 0.3, 0.5, and 0.8 Btu/lb$_m$·°F, respectively. Initially, one section contains 10 lb$_m$ of A at 80°F, and the other compartments contain 5 lb$_m$ of B at 140°F and 2 lb$_m$ of C at 210°F. Find the temperature of the three fluids at thermal equilibrium in °F, if the overall process is one of constant pressure.

4-135E. A 1.2-ft$^3$ container is filled with water initially at 70°F and 1 atm. A 1-kW electrical-resistance heater in the water is energized from an external source for 20 min. The mass of the heater is 6.0 lb, and its specific heat is 0.32 Btu/lb$_m$·°F. If the container is insulated, determine the temperature of the water and heater element in the final state, in degrees Fahrenheit.

4-136E. Liquid water in a closed, insulated tank maintained at 1 atm is agitated by a paddle wheel driven by a 180-W motor. The initial temperature is 60°F. Determine the temperature rise of the contents after 0.80 h if the mass of water is 14 lb$_m$ and the mass and specific heat of the paddle wheel are 0.9 lb$_m$ and 0.24 Btu/lb$_m$·°F, respectively.

# 5

# CONTROL-VOLUME ENERGY ANALYSIS

First Siemens "3A-Series" gas turbine on a full load test stand has a rating of 170 megawatts at 3600 rpm. (Courtesy of Siemens Power Corporation)

The first law of thermodynamics first presented in Chap. 2 resulted from experimental observations about closed systems. However, a large number of engineering analyses involve *open systems* where matter flows in and out of a region of space. Thus we need a more general formulation of an energy balance, in order to account for energy transported across various parts of the boundary due to mass transfer as well as overall heat and work transports of energy. The analysis of flow processes requires an important modification in one's viewpoint of the system.

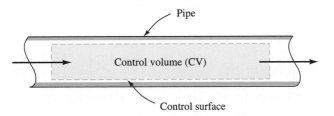

Pipe

Control volume (CV)

Control surface

**Figure 5-1**    Illustration of a control surface for a control volume.

## 5-1    INTRODUCTION

The analysis of flow processes begins by selecting an *open system,* a region of space called a *control volume.* The boundary of the control volume may be in part a well-defined physical barrier (such as a wall), or part or all of it may be an imaginary envelope. The selection of the boundary, or *control surface,* is an important first step in the analysis of any open or closed system. Figure 5-1 shows a control volume within a pipe of constant diameter. Part of the boundary is well defined just inside the wall of the pipe. However, the ends of the control surface where mass enters and leaves the control volume are imaginary surfaces. In our initial development, the control volume will be fixed in size and shape as well as fixed in position relative to the observer.

An energy balance on the control volume requires not only measurements of heat and work transfers but also an accounting for the energy carried into or out of the control volume by mass transferred across the control surface. A general schematic of the process for a control volume with one inlet and one outlet is shown in Fig. 5-2. Such an energy accounting requires a knowledge of the state of the matter as it passes across the control surface. This implies that the condition of equilibrium must be closely met at any open boundary. This condition is fulfilled if the properties of the fluid vary slowly across the control surface. Such a condition is frequently called *local equilibrium.* (This is somewhat analogous to quasiequilibrium changes in closed systems.) This requirement does not prevent the properties at a given open boundary from changing with time, as they do in some unsteady-state processes.

If you were asked to model the process by which you fill an automobile tire, where would you locate the boundaries of your open system?

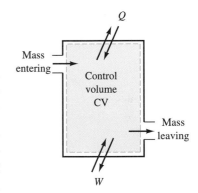

**Figure 5-2**
General schematic of a control volume with one inlet and one outlet.

## 5-2    CONSERVATION OF MASS PRINCIPLE
##       FOR A CONTROL VOLUME

In the absence of nuclear reactions, mass is a conserved property. For any control volume (CV) the conservation of mass principle may be stated in

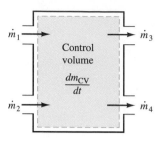

**Figure 5-3**

A control volume illustrating the conservation of mass statement given by Eq. [5-1].

words as

$$
\begin{pmatrix}
\text{Net change in} \\
\text{mass within a CV} \\
\text{during a time period}
\end{pmatrix}
=
\begin{pmatrix}
\text{total mass entering} \\
\text{a CV during} \\
\text{a time period}
\end{pmatrix}
-
\begin{pmatrix}
\text{total mass leaving} \\
\text{a CV during} \\
\text{a time period}
\end{pmatrix}
$$

In engineering analysis the preceding equation is more useful when expressed on a rate basis. In this format it becomes

$$
\begin{pmatrix}
\text{Time rate of change of} \\
\text{mass within a control} \\
\text{volume at time } t
\end{pmatrix}
=
\begin{pmatrix}
\text{total rate of mass} \\
\text{entering a control} \\
\text{volume at time } t
\end{pmatrix}
-
\begin{pmatrix}
\text{total rate of mass} \\
\text{leaving a control} \\
\text{volume at time } t
\end{pmatrix}
$$

Symbolically, the above equation is written in the format

$$
\frac{dm_{\text{cv}}}{dt} = \sum_{\text{in}} \dot{m}_i - \sum_{\text{out}} \dot{m}_e
\qquad\qquad \textbf{[5-1]}
$$

where $\dot{m}$ is defined as the **mass flow rate**. The mass flow rate is the time rate at which mass crosses a boundary. The subscripts $i$ and $e$ represent the *inlet* and *exit* (outlet) states. The summation signs imply that there may be several inlets and outlets, as shown in Fig. 5-3. This equation is called the **conservation of mass equation** or the **mass balance.**

The volume flow rate $\dot{V}$ and the mass flow rate $\dot{m}$ can be expressed in terms of the *local* properties of the fluid at the control surface and the geometry of the control surface. Figure 5-4a shows a fluid element just outside an area $A$ of a control surface. The distance that the fluid element moves during a time interval $\Delta t$ is shown as $\Delta x$. Therefore, the volume of the element is $A \, \Delta x$, and the average volume flow rate is $A \, \Delta x / \Delta t$. In the notation of calculus, the limiting value of $\Delta x / \Delta t$ as $\Delta t$ approaches zero is the instantaneous velocity $\mathbf{V}_n$, where $\mathbf{V}_n$ is the *scalar* value of the velocity component normal to area $A$. (Note that $\mathbf{V}$ in bold type is used as a symbol for velocity,

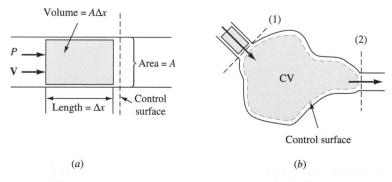

(a)                                          (b)

**Figure 5-4**    Schematic for the development of the conservation of mass.

$V$ in italic type is the volume, and $\dot{V}$ in italic type is the volume flow rate.) Therefore, the instantaneous *volume flow rate* $\dot{V}$ at the control surface is given by

$$\dot{V} = \lim_{\Delta t \to 0} \frac{A \Delta x}{\Delta t} = \mathbf{V}_n A \qquad \text{[5-2a]}$$

when the velocity is uniform over the cross-sectional area. If the velocity is not uniform over the control surface area $A$ open to flow, then we must integrate over the area according to the relation

$$\dot{V} = \int_A \mathbf{V}_n \, dA \qquad \text{[5-2b]}$$

Typical units for the volume flow rate are m³/s in SI and ft³/s, ft³/min, or gal/min in USCS.

The mass flow rate $\dot{m}$ at the control surface in Fig. 5-4a is the product of the volume flow rate and the mass per unit volume, or density, $\rho$. That is,

$$\dot{m} = \rho \dot{V} \qquad \text{[5-3a]}$$

when the density is uniform over the cross section. For the general case where the flow is not uniform over the area $A$, then the quantity $\rho \mathbf{V}$ at area $dA$ must be integrated over the area $A$ of the control surface. In this general case the mass flow rate becomes

$$\dot{m} = \int_A \rho \mathbf{V}_n \, dA \qquad \text{[5-3b]}$$

**What must you assume about the density and velocity in Eq. [5-3b] to recover Eq. [5-3a]?**

If the density is uniform inside the control volume, the mass inside the control volume is

$$m_{\text{cv}} = \rho V_{\text{cv}} \qquad \text{[5-4a]}$$

For nonuniform density, the mass within a differential volume $dV$ of the control volume is the product of the density and the differential volume. Since the density may vary throughout the control volume, the total mass within the control volume at any instant is found by integrating $\rho \, dV$. That is,

$$m_{\text{cv}} = \int_V \rho \, dV \qquad \text{[5-4b]}$$

Substitution of Eqs. [5-3b] and [5-4b] into Eq. [5-1] yields the following conservation of mass equation

$$\frac{d}{dt} \int_V \rho \, dV = \sum_i \left( \int_A \rho \mathbf{V}_n \, dA \right)_i - \sum_e \left( \int_A \rho \mathbf{V}_n \, dA \right)_e \qquad \text{[5-5]}$$

This is the most *general* formulation of Eq. [5-1] for multiple inlets and outlets in terms of property and geometric data.

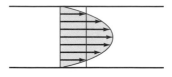

(a) Actual profile

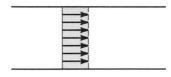

(b) Average profile

**Figure 5-5**
Velocity profiles in a channel for
(a) actual flow and (b) a one-
dimensional flow model.

### 5-2-1 IMPORTANT MODELING ASSUMPTIONS

The solution of any engineering problem requires that a model be developed that simulates the essential feature of the real system. Although each problem is unique, there are certain modeling assumptions used so frequently that they warrant special attention. The first model to consider is a system that operates at steady-state conditions.

- *Any system is defined to be in **steady state** if the system properties are constant with time at every position within and on the boundaries of the system.*

The properties include intrinsic ones, such as temperature, pressure, density, and the specific internal energy, as well as extrinsic ones, such as velocity or elevation. This extremely powerful and widely applied assumption says that the system behavior is independent of time. In terms of the conservation of mass equation, this means that the amount of mass within the system is constant. Hence the rate of change of mass inside the system must be zero, that is, $dm_{cv}/dt = 0$. Note that the amount of mass is constant, but not its identity.

A second model frequently considered is the one-dimensional flow assumption.

- *The flow is called **one-dimensional flow** if the properties at a flow boundary are uniform over the cross-sectional area.*

Thus properties vary only in the direction of flow. Such an idealized condition is an approximation to actual behavior. For example, in pipe flow the velocity usually varies from zero at the wall to a maximum value at the center of the pipe cross section, as illustrated in Fig. 5-5a. (The exact shape of the velocity profile is governed by fluid mechanics.) Thus any single value of the velocity, for example, is assumed to be a suitably averaged value for that particular cross section, as shown for the velocity in Fig. 5-5b.

The first condition above requires that $\rho$ and $\mathbf{V}$ be constant with time at any differential area $dA$. If the flow across any control surface is taken to be one-dimensional and normal to the control surface, Eq. [5-3b] for *one-dimensional flow* reduces to

**Note the difference between steady-state and one-dimensional flow.**

$$\dot{m} = \rho A \mathbf{V} = \frac{A\mathbf{V}}{v} \qquad \text{[5-6]}$$

where $\mathbf{V}_n$ is now represented simply by $\mathbf{V}$.

### 5-2-2 TYPICAL APPLICATIONS

Before looking at specific numerical examples, it is useful to consider several typical applications to demonstrate how the modeling assumptions are applied to simplify the conservation of mass equation. The starting point for most problems will be Eq. [5-1], the rate form of the conservation of mass equation:

$$\frac{dm_{cv}}{dt} = \sum_{in} \dot{m}_i - \sum_{out} \dot{m}_e \qquad \textbf{[5-1]}$$

This equation holds all the information of the general form, [Eq. [5-5]], without the distractions of the various integrals.

To demonstrate how the various assumptions apply, consider a control volume with $i$ inlets and $e$ outlets. The number of inlets and outlets, in practice, would be established by the physical geometry of the control volume. Several cases are presented below, where each case considers a different set of modeling assumptions.

**Case I: Application of Steady-State Assumption.** If a system is operating at steady-state conditions, then the mass within the system is constant, and $dm_{cv}/dt = 0$. Then the conservation of mass equation above becomes

$$0 = \sum_{in} \dot{m}_i - \sum_{out} \dot{m}_e \qquad \textbf{[5-7a]}$$

A rearrangement of the above result leads to

$$\sum_{in} \dot{m}_i = \sum_{out} \dot{m}_e \qquad \textbf{[5-7b]}$$

Thus the sum of the mass flow rates entering a steady-state control volume must equal the sum of the mass flow rates leaving it.

**Case II: Application of One-Dimensional Flow Assumption.** When the flows are one-dimensional, then the density (or specific volume) and velocity are uniform at each flow cross section, and $\dot{m} = \rho A V = AV/v$. The conservation of mass equation then is

$$\frac{dm_{cv}}{dt} = \sum_{in} \dot{m}_i - \sum_{out} \dot{m}_e \qquad \textbf{[5-8]}$$

$$= \sum_{in}(\rho A\mathbf{V})_i - \sum_{out}(\rho A\mathbf{V})_e = \sum_{in}\left(\frac{A\mathbf{V}}{v}\right)_i - \sum_{out}\left(\frac{A\mathbf{V}}{v}\right)_e$$

**Case III: Application of Steady-State and One-Dimensional Flow Assumptions.** Under these two assumptions the conservation of mass equation reduces to

$$\sum_{in}(\rho\mathbf{V}A)_i = \sum_{out}(\rho\mathbf{V}A)_e = \sum_{in}\left(\frac{A\mathbf{V}}{v}\right)_i = \sum_{out}\left(\frac{A\mathbf{V}}{v}\right)_e \qquad \textbf{[5-9]}$$

Note the various formats of the conservation of mass obtained through application of appropriate models.

This form of the conservation of mass equation is often called the *continuity-of-flow equation* or simply the *continuity equation* for steady flow.

For one-dimensional flow the volume flow rate $\dot{V}$ given by Eq. [5-2a] becomes, in conjunction with Eq. [5-7],

$$\dot{V} = \mathbf{V}A = \frac{\dot{m}}{\rho} = \dot{m}v \qquad \text{[5-10]}$$

Thus a direct connection exists between $\dot{m}$ and $\dot{V}$. Volume flow rates are frequently cited in place of mass flow rates in engineering analyses. Note, however, that the volume flow rates in and out of a steady-state control volume generally are different, unless the density of the fluid is constant. Examples of the use of the conservation of mass principle are presented below.

---

**Example 5-1**

**R**efrigerant-134a at 8 bars and 50°C enters a pipe-reducing system in steady state through a pipe of 1.50-cm internal diameter at a velocity of 4.53 m/s. It exits the reducer through a pipe area of 0.35 cm² at 6 bars and 60°C. Determine (*a*) the mass flow rate, in kg/min, and (*b*) the exit velocity, in m/s.

**Solution:**

**Given:**   Refrigerant-134a flows through a pipe system. A schematic of the process with appropriate input data is shown in Fig. 5-6.

**Find:**   (*a*) $\dot{m}$, in kg/min, and (*b*) $\mathbf{V}_2$, in m/s.

**Model:**   Steady-state, one-dimensional flow.

**Strategy:**   Apply the conservation of mass equation.

**Analysis:**   The dashed line in Fig. 5-6 indicates the system boundaries. For this system, the rate form of the conservation of mass equation, Eq. [5-1] can be written as

$$\frac{dm_{cv}}{dt} = \dot{m}_1 - \dot{m}_2$$

Under steady-state conditions, $dm_{cv}/dt = 0$ and thus $\dot{m}_1 = \dot{m}_2$. Note that under steady-state conditions there is no accumulation of mass within the control volume. In addition, steady state also requires that the mass flow rates are independent of time. Assuming one-dimensional flow, we know that $\dot{m} = \rho A V_n = A V_n / v$. If the control surface is normal to the direction of flow across the boundary, as it is in this problem, then the normal velocity $\mathbf{V}_n$ equals the fluid velocity $\mathbf{V}$ and $\dot{m} = A V / v$. Under these constraints the conservation of mass equation reduces to

$$\dot{m}_1 = \frac{A_1 V_1}{v_1} = \frac{A_2 V_2}{v_2} = \dot{m}_2$$

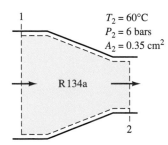

$T_2 = 60°C$
$P_2 = 6$ bars
$A_2 = 0.35$ cm²

R 134a

1

2

$P_1 = 8$ bars
$T_1 = 50°C$
$V_1 = 4.53$ m/s
$D_1 = 1.50$ cm

**Figure 5-6**
Schematic and data for Example 5-1.

(a) The equation for $\dot{m}$ developed above in terms of the specific volume is $\dot{m} = A_1V_1/v_1$. The velocity is given, but the specific volume and area need to be evaluated. At 8 bars the saturation temperature from Table A-17 is found to be 31.33°C. Hence the fluid at 50°C is superheated vapor. From Table A-18 the specific volume at 8 bars and 50°C is given as 0.02846 m³/kg. The inlet cross-sectional area of the pipe is

$$A_1 = \frac{\pi D_1^2}{4} = \frac{\pi(1.5)^2}{4} = 1.77 \text{ cm}^2 = 1.77 \times 10^{-4} \text{ m}^2$$

Substitution of data into the equation for $\dot{m}$ leads to

$$\dot{m} = \frac{A_1V_1}{v_1} = \frac{1.77 \times 10^{-4} \text{ m}^2 \times 4.53 \text{ m/s}}{0.02846 \text{ m}^3/\text{kg}}$$
$$= 0.0282 \text{ kg/s} = 1.69 \text{ kg/min}$$

(b) The final velocity is found from the relation $V_2 = \dot{m}_2v_2/A_2$. At 6 bars the saturation temperature is 21.58°C. Thus the exit state at 60°C is also superheated vapor. The specific volume at this state is found from Table A-18 to be 0.04134 m³/kg. Substitution of data into Eq. [5-6] yields

$$V_2 = \frac{\dot{m}v_2}{A_2} = \frac{0.0282 \text{ kg/s} \times 0.04134 \text{ m}^3/\text{kg}}{0.35 \text{ cm}^2} \times \frac{10^4 \text{ cm}^2}{\text{m}^2} = 33.3 \text{ m/s}$$

An alternative method would be to employ $\dot{m} = V_1A_1/v_1 = V_2A_2/v_2$. This leads to

$$V_2 = \frac{V_1A_1v_2}{A_2v_1} = \frac{4.53(1.77)(0.04134)}{0.35(0.02846)} \frac{\text{m}}{\text{s}} = 33.3 \text{ m/s}$$

**Comment:** Velocities in the range of 4 to 40 m/s (10 to 100 ft/s) are quite common for the flow of gases through commercial processes.

---

**Liquid water at 20°C and 1.5 bars enters a 1.50-cm diameter rubber hose at a rate of 10 L/min. Compute (a) the inlet velocity of the fluid, in m/s, and (b) the mass flow rate, in kg/min.**

**Solution:**

**Given:** Liquid water flows through a constant-area hose, as illustrated in Fig. 5-7.

**Find:** (a) The inlet velocity, in m/s, and (b) $\dot{m}$, in kg/min.

**Model:** One-dimensional flow.

**Strategy:** The velocity is found from the volume flow rate and the area. The mass flow rate is found from the volume rate and the specific volume at the given state.

**Analysis:** The system of interest is indicated by the dashed line in Fig. 5-7.

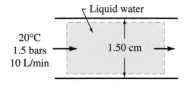

**Figure 5-7**
Schematic and data for Example 5-2.

(*a*) The velocity is found from the equation $\dot{V} = \mathbf{V}A$, which relates velocity to volume flow rate for one-dimensional flow. The area of a 1.50-cm pipe was found in the preceding example to be 1.77 cm². Thus,

$$\mathbf{V} = \frac{\text{volume flow rate}}{\text{area}} = \frac{\dot{V}}{A} = \frac{10 \text{ L/min}}{1.77 \text{ cm}^2} \times \frac{10^3 \text{ cm}^3}{1 \text{ L}} \times \frac{1 \text{ min}}{60 \text{ s}}$$

$$= 94.2 \text{ cm/s} = 0.942 \text{ m/s}$$

(*b*) The mass flow rate can be found by using $\dot{m} = \dot{V}/v$. The only additional value required is the specific volume of the fluid. The saturation pressure of water at 20°C from Table A-12 is 0.023 bar. Because the actual pressure is 1.5 bars, the water is a slightly compressed liquid. We can approximate the specific volume of the fluid by using the saturated-liquid value at the given temperature. From Table A-12 this is given as $1.002 \times 10^{-3}$ m³/kg. As a result the mass flow rate becomes

$$\dot{m} = \frac{\dot{V}}{v} = \frac{10 \text{ L/min}}{1.002 \times 10^{-3} \text{ m}^3/\text{kg}} \times \frac{\text{m}^3}{10^3 \text{ L}} = 9.89 \text{ kg/min}$$

**Comment:**    If the fluid is modeled as incompressible, the velocity and volume flow rate remain constant as long as the area is constant.

**What physical actions could limit the range of velocities in commercial equipment?**

Note that although the velocity of the liquid in Example 5-2 is much less than the gas velocity found in Example 5-1 for the same area of flow, the mass flow rate of liquid is much greater owing to the larger density. Velocities from 0.2 to 3 m/s (or 1 to 10 ft/s) are quite common for liquids in pipe-flow applications.

## 5-3    CONSERVATION OF ENERGY PRINCIPLE FOR A CONTROL VOLUME

There are several methods we could use to develop a conservation of energy principle for a control volume. The general technique chosen here is to extend the closed system (control mass) energy equation presented in Chap. 2 to open systems (control volumes). As a result we will become aware that the energy balances for closed and open systems have a common origin, namely, the first law of thermodynamics presented in Sec. 2-2.

To show this common basis, a control mass will be followed as it passes through a control volume (CV) over a time interval $\Delta t$. The control volume is assumed to be of fixed size and shape. Refer to Fig. 5-8 during the following discussion. At time $t$ the control mass occupies the control volume region marked CV and the small region outside of the inlet $i$ indicated by the symbol $A$. The region $A$ is chosen so that after a period $\Delta t$ all the mass initially in region $A$ has now passed into the control volume. However, in

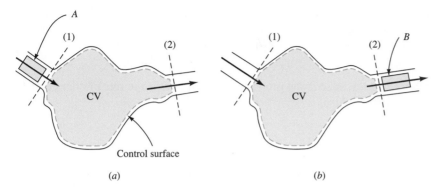

**Figure 5-8**    The development of the conservation of energy principle for a control volume. (a) Control mass at time $t$; (b) control mass at time $t$ and $dt$.

this same time period part of the control mass initially within the control volume has passed into region $B$, which lies just outside the outlet or exit cross section $e$. In the time period $\Delta t$ the conservation of energy principle for the control mass (cm) on a rate basis is given by

$$\frac{dE_{\text{cm}}}{dt} = \dot{Q} + \dot{W} \qquad \textbf{[2-28]}$$

This is the equation we wish to convert to a control volume basis. However, this equation currently contains no terms that account for energy transfer due to mass transfer across any open control surface. To make this transformation we note that the energy associated with the control mass at time $t$ is given by

$$E_{\text{cm},t} = E_{\text{cv},t} + E_A$$

Similarly, at time $t + \Delta t$ we find that

$$E_{\text{cm},t+\Delta t} = E_{\text{cv},t+\Delta t} + E_B$$

Consequently, the change in energy of the *control mass* over time $\Delta t$ is

$$E_{\text{cm},t+\Delta t} - E_{\text{cm},t} = E_{\text{cv},t+\Delta t} - E_{\text{cv},t} + E_B - E_A \qquad \textbf{[a]}$$

We must now express $E_A$ and $E_B$ in terms of local quantities at the control surface.

The energy $E_A$ associated with the mass in region $A$ is equal to the mass $m_A$ of the region times the energy per unit mass $e_A$. Thus $E_A = m_A e_A$. A similar equation holds for region $B$. Substitution of these relations for $E_A$ and $E_B$ into Eq. [a] leads to

$$E_{\text{cm},t+\Delta t} - E_{\text{cm},t} = E_{\text{cv},t+\Delta t} - E_{\text{cv},t} + m_B e_B - m_A e_A \qquad \textbf{[b]}$$

A rate format for this equation is found by first dividing Eq. [b] through by $\Delta t$. Then the limit is taken as $\Delta t \to 0$. Recall from the differential calculus that

$$\lim_{\Delta t \to 0} \frac{Y(t + \Delta t) - Y(t)}{\Delta t} \equiv \frac{dY}{dt} \qquad \text{[c]}$$

where $Y$ in this case is the property $E$, and $dY/dt$ is the rate of change of $E$, that is, $dE/dt$. Also,

$$\lim_{\Delta t \to 0} \frac{Z}{\Delta t} \equiv \dot{Z} \qquad \text{[d]}$$

where $\dot{Z}$ is the instantaneous rate of $Z$ crossing the control surface. Note also that as $\Delta t \to 0$, the regions $A$ and $B$ become differentially small. Thus the properties of regions $A$ and $B$ are the same as those of the control surface at $i$ and $e$, respectively. Hence in the limit as $\Delta t \to 0$ Eq. [b] becomes

$$\frac{dE_{cm}}{dt} = \frac{dE_{cv}}{dt} + \dot{m}_e e_e - \dot{m}_i e_i \qquad \text{[e]}$$

Substitution of Eq. [e] into Eq. [2-17] leads to

$$\dot{Q} + \dot{W} = \frac{dE_{cv}}{dt} + \dot{m}_e e_e - \dot{m}_i e_i \qquad \text{[f]}$$

where again the subscripts $i$ and $e$ represent the inlet and exit conditions, respectively.

Rearrangement of Eq. [f] above leads to

$$\frac{dE_{cv}}{dt} = \dot{Q}_{cv} + \dot{W}_{cv} + \dot{m}_e e_e - \dot{m}_i e_i \qquad \text{[g]}$$

where the subscript cv is added to stress that the heat-transfer rate and the work-transfer rate are summed over the entire surface of the control volume. In a word format,

$$\begin{pmatrix} \text{Time rate of} \\ \text{change of energy} \\ \text{within a control volume} \end{pmatrix}$$

$$= \begin{pmatrix} \text{net rate of energy} \\ \text{crossing boundary} \\ \text{as heat and work} \end{pmatrix} + \begin{pmatrix} \text{total rate of} \\ \text{energy entering} \\ \text{CV with mass} \end{pmatrix} - \begin{pmatrix} \text{total rate of} \\ \text{energy leaving} \\ \text{CV with mass} \end{pmatrix}$$

This clearly shows that the conservation of energy equation is a balance between the rate of energy change within the control volume and the rate at which it is transported into the control volume by various mechanisms.

If the above equation is restricted to simple compressible substances, then the energy $e$ includes the internal energy $u$, the linear kinetic energy $\mathbf{V}^2/2$, and the gravitational potential energy $gz$ of the flow stream. In this case

$$e = u + \frac{V^2}{2} + gz$$

Use of this relation in Eq. [f] leads to

$$\frac{dE_{cv}}{dt} = \dot{Q}_{cv} + \dot{W}_{cv} + \dot{m}_i \left( u + \frac{V^2}{2} + gz \right)_i - \dot{m}_e \left( u + \frac{V^2}{2} + gz \right)_e \quad [5\text{-}11]$$

This is a general rate equation for a control volume with one inlet and one exit and one-dimensional flow. For future applications, we must determine what energy terms to include in the work-transfer term.

### 5-3-1 WORK INTERACTIONS FOR A CONTROL VOLUME

Recall that the work term in a closed-system energy balance, $Q + W = \Delta E$, represents several forms of work. These include compression/expansion work as well as electrical resistor and paddle-wheel (shaft) work. Similarly, in Eq. [5-11] the power term $\dot{W}_{cv}$ represents several possible work interactions. It is useful at this point to make a distinction between work interactions at boundaries where flow occurs and at nonflow boundaries: $\dot{W}_{cv} = \dot{W}_{nonflow} + \dot{W}_{flow}$. The work or power at nonflow boundaries has already been studied extensively in Chap. 2. Any work interaction that can occur for a closed system qualifies as work transfer at a nonflow boundary, as shown in Fig. 5-9. The fundamental relationship for work transfer associated with a flow boundary is developed below.

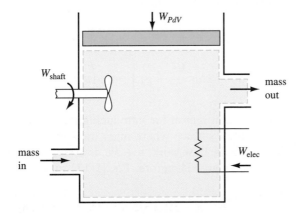

**Figure 5-9** A control volume with expansion/compression work as well as shaft work and electrical work.

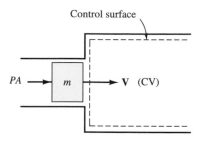

**Figure 5-10**
Illustration of flow work at a control surface.

**Note the existence of flow work at every position on a control surface where mass transfer occurs.**

When mass enters or leaves a control volume, work is required to push the fluid into or out of the system. This term is referred to as *flow work.* In Fig. 5-10 a mass $m$ in region $A$ initially resides just outside a control surface. Recall from Sec. 2-1 that the rate of energy transfer as work can be found from the product of a normal force and velocity. The normal force exerted at the control surface of area $A$ is simply the pressure times the area, or $PA$. Thus the rate of transfer of flow work $\dot{W}_{\text{flow}}$, the flow power, into the control volume at any boundary open to mass transfer is

$$\dot{W}_{\text{flow}} = PA\mathbf{V}$$

It is convenient to transform this expression by recalling from Eq. [5-10] that $A\mathbf{V} = \dot{m}v$. Therefore,

$$\dot{W}_{\text{flow}} = \dot{m}(Pv)$$

We have assumed one-dimensional flow in deriving the above two relations. An expression of this type must be present in the energy balance for every part of the control surface where mass enters or leaves. For a control volume with one inlet and one exit, the $\dot{W}_{\text{cv}}$ term in the energy balance [Eq. [5-11]] can now be written as

$$\dot{W} = \dot{W}_{\text{nonflow}} + \dot{W}_{\text{flow}} \qquad \textbf{[5-12]}$$
$$= \dot{W}_{\text{nonflow}} + \dot{m}_i(Pv)_i - \dot{m}_e(Pv)_e$$

Equation [5-12] summarizes the work-transfer terms associated with any control volume.

### 5-3-2   THE CONTROL-VOLUME ENERGY EQUATION

The substitution of Eq. [5-12] into Eq. [5-11] yields, after rearrangement,

$$\frac{dE_{\text{cv}}}{dt} = \dot{Q} + \dot{W}_{\text{nonflow}} \qquad \textbf{[5-13]}$$
$$+ \dot{m}_i\left(u + Pv + \frac{\mathbf{V}^2}{2} + gz\right)_i - \dot{m}_e\left(u + Pv + \frac{\mathbf{V}^2}{2} + gz\right)_e$$

At this point an important reason for introducing the enthalpy function $h$ is seen. At every control surface where mass transfer occurs, the sum of the internal energy and the flow work, $u + Pv$, is associated with each unit mass crossing the boundary. Therefore, Eq. [5-13] may be written more concisely as

$$\frac{dE_{\text{cv}}}{dt} = \dot{Q} + \dot{W}_{\text{nonflow}} + \dot{m}_i\left(h + \frac{\mathbf{V}^2}{2} + gz\right)_i - \dot{m}_e\left(h + \frac{\mathbf{V}^2}{2} + gz\right)_e$$

$$\textbf{[5-14]}$$

Thus the use of the enthalpy function in control-volume energy analyses is of considerable convenience. In some engineering applications there may be several inlets and exits for mass flow. In these situations Eq. [5-14] takes on the form

$$\frac{dE_{cv}}{dt} = \dot{Q} + \dot{W} + \sum_{in} \dot{m}_i \left(h + \frac{V^2}{2} + gz\right)_i - \sum_{out} \dot{m}_e \left(h + \frac{V^2}{2} + gz\right)_e$$

[5-15]

where the subscript "nonflow" has been removed from the work term, and $\dot{W}$ now represents *all* work mechanisms that can occur at *nonflow* boundaries. One of the most important of these for a control volume is shaft power $\dot{W}_{shaft}$ present in the operation of a turbine, compressor, pump, or fan. Electrical power $\dot{W}_{elec}$ associated with current flow through a resistor within a control volume may also need to be considered, as in the design of electric furnaces. Equation [5-15] represents a ***general conservation of energy principle*** for a control volume with multiple inlets and outlets under one-dimensional flow. Thus the rate of change of energy within the control volume, $dE_{cv}/dt$, is affected by the rates of heat transfer and shaft work, as well as by the energy associated with mass that enters or leaves the control volume.

The last two equations are restricted to one-dimensional flow at the boundaries. A more general formulation which takes into account the variation of local properties at the boundaries and within the control volume is

$$\frac{d}{dt} \int_V \rho e \, dV = \dot{Q} + \dot{W} + \sum_i \left[\int_A \left(h + \frac{V^2}{2} + gz\right)\rho V_n \, dA\right]_i$$

[5-16]

$$- \sum_e \left[\int_A \left(h + \frac{V^2}{2} + gz\right)\rho V_n \, dA\right]_e$$

where $\dot{W}$ again represents all nonflow work forms. This general energy balance is the equivalent of Eq. [5-5] for the conservation of mass for a control volume.

## 5-4 STEADY-STATE CONTROL-VOLUME ENERGY EQUATIONS

As was discussed earlier for the conservation of mass, the general energy balance must be applied to each unique problem by making modeling assumptions that reflect the important features of the physical system. The starting point for our applications of both the energy and mass equations will be Eqs. [5-1] and [5-15]:

$$\frac{dm_{cv}}{dt} = \sum_{in} \dot{m}_i - \sum_{out} \dot{m}_e$$

$$\frac{dE_{cv}}{dt} = \dot{Q} + \dot{W} + \sum_{in} \dot{m}_i \left( h + \frac{\mathbf{V}^2}{2} + gz \right)_i - \sum_{out} \dot{m}_e \left( h + \frac{\mathbf{V}^2}{2} + gz \right)_e$$

These equations are valid only for systems with one-dimensional flow.

A number of practical flow systems operate under steady-state conditions. On the basis of the definition of steady state, *the total energy within a control volume under steady-state conditions remains constant with time.* This requires that $dE_{cv}/dt = 0$. As a result, Eq. [5-15] above reduces to the following expression under *steady-state* conditions:

$$0 = \dot{Q} + \dot{W} + \sum_{in} \dot{m}_i \left( h + \frac{\mathbf{V}^2}{2} + gz \right)_i - \sum_{out} \dot{m}_e \left( h + \frac{\mathbf{V}^2}{2} + gz \right)_e \quad \textbf{[5-17]}$$

In Eq. [5-17] remember that the mass flow rates $\dot{m}$ at each open boundary are *not* usually equal. However, their values are related by noting that since $dm/dt = 0$ in steady state, Eq. [5-1] reduces to

$$0 = \sum_{in} \dot{m}_i - \sum_{out} \dot{m}_e \quad \textbf{[5-7a]}$$

The analysis of steady-state, steady-flow systems frequently requires the use of both the conservation of energy and the conservation of mass principles.

There are numerous applications of the steady-state conservation of energy principle for which there is only one inlet (position 1) and one outlet (position 2), as shown in Fig. 5-11. In this circumstance Eqs. [5-7a] and [5-17] reduce to

$$0 = \dot{m}_1 - \dot{m}_2$$

$$0 = \dot{Q} + \dot{W} + \dot{m}_1 \left( h + \frac{\mathbf{V}^2}{2} + gz \right)_1 - \dot{m}_2 \left( h + \frac{\mathbf{V}^2}{2} + gz \right)_2 \quad \textbf{[5-18]}$$

Under steady-state conditions with one inlet and one outlet the conservation of mass equation shows that $\dot{m}_1 = \dot{m}_2$. Because the mass flow rates are the same, sometimes it is desirable to simplify Eq. [5-18], after rearrangement, to

$$\dot{Q} + \dot{W} = \dot{m} \left[ (h_2 - h_1) + \frac{\mathbf{V}_2^2 - \mathbf{V}_1^2}{2} + g(z_2 - z_1) \right] \quad \textbf{[5-19]}$$

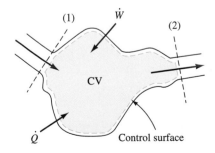

**Figure 5-11**
A control volume with one inlet and one outlet and heat and work transfers.

Keep in mind that all the quantities on the right side of Eqs. [5-17] through [5-19] are evaluated at position 1 or 2 on the control surface where the mass crosses the boundary.

In some applications it is convenient to have a conservation of energy principle on a unit-mass basis for control volumes that have a *single inlet and outlet*. Division of Eq. [5-18] by $\dot{m}$ leads to the desired form, namely,

$$0 = q + w + \left(h + \frac{\mathbf{V}^2}{2} + gz\right)_1 - \left(h + \frac{\mathbf{V}^2}{2} + gz\right)_2 \qquad \textbf{[5-20]}$$

where $\qquad \dot{Q} \equiv \dot{m}q \qquad$ and $\qquad \dot{W} \equiv \dot{m}w \qquad \textbf{[5-21]}$

These latter equations for $\dot{Q}$ and $\dot{W}$ are useful for transforming rate data into unit-mass data, and vice versa. By grouping common terms, the steady-state energy balance on a unit-mass basis is often written as

$$q + w = (h_2 - h_1) + \frac{\mathbf{V}_2^2 - \mathbf{V}_1^2}{2} + g(z_2 - z_1) \qquad \textbf{[5-22]}$$

Note that $\dot{W}$ and $w$ in the steady-state equations represent all forms of nonflow work, such as shaft or electrical work transfer.

where 1 and 2 again represent the inlet and exit states, respectively. Equations [5-17] through [5-22] are important energy statements for a control volume under steady-state conditions.

In addition to the assumption of steady state, there are two other models that may apply to the energy balance. The first of these is the *adiabatic* assumption. This may be applied to the entire system or to just a selected boundary. The other assumption unique to the energy equations is the decision to *neglect the effect of kinetic and gravitational potential energy*. This does not mean that there is no change in elevation or that the velocities are zero, just that their effects are negligible. As discussed in a subsequent section, the relative magnitude of these terms often justifies this assumption.

Since open systems are common in engineering practice, one must be familiar with the use of the two conservation statements presented in this section—those of mass and energy. A number of examples are presented in Sec. 5-6 to illustrate the use of these equations. In addition, these examples give a feeling for the orders of magnitudes of various terms, the handling of units, and the relative importance of various terms in an energy balance when applied to various types of equipment. These problems also will require the use of property relations. Thus we again rely on information from earlier chapters.

## 5-5 COMMENTS ON PROBLEM-SOLVING TECHNIQUES

The major items that need to be considered when solving thermodynamic problems were first presented in Sec. 1-7. They are reviewed below.

1. Sketch the system and indicate the selected control volume.
2. List appropriate idealizations or assumptions.
3. Recognize any specific path of the process.
4. Draw suitable diagram(s) for the process.
5. Write a suitable basic energy balance for the chosen system.
6. Determine what energy interactions are important.
7. Obtain appropriate physical data for the substance under study.
8. Check units of each term in each equation used.

Two items in the above list are extremely important. The first item suggests sketching the system under analysis. On the sketch one should indicate all the relevant data and energy-transfer terms consistent with the boundaries selected for the system. Sketches help the engineer approach a problem in a straightforward and consistent manner. Equally important as an aid in problem solving is the process diagram, which indicates the position of the end states and the path of the process on relevant coordinates.

Before we proceed with the analysis of some specific steady-state devices, it is appropriate to examine the relative magnitude of kinetic-energy and potential-energy terms.

### 5-5-1   TRANSLATIONAL KINETIC ENERGY

The linear kinetic energy ke of a unit mass is $\mathbf{V}^2/2$. Since it is customary in SI units to evaluate energy terms in J/g and kJ/kg, and the velocity is reported usually in m/s; then

$$\text{ke} = \frac{\mathbf{V}^2}{2} \times \frac{1\ \text{N·s}^2}{\text{kg·m}} \times \frac{\text{kJ}}{10^3\ \text{N·m}} = \frac{\mathbf{V}^2}{2000}\frac{\text{kJ/kg}}{(\text{m/s})^2}$$

In USCS units the velocity is expressed in ft/s, and the energy in Btu/lb. Hence

$$\text{ke} = \frac{\mathbf{V}^2}{2} \times \frac{\text{lb}_f \cdot \text{s}^2}{32.2\ \text{lb}_m \cdot \text{ft}} \times \frac{\text{Btu}}{778\ \text{ft·lb}_f} \approx \frac{\mathbf{V}^2}{50{,}000}\frac{\text{Btu/lb}_m}{(\text{ft/s})^2}$$

For many engineering applications the values of various energy terms are at least 2 kJ/kg (or 1 Btu/lb$_m$) in magnitude and more commonly for gas flow in the range of 20 to 200 kJ/kg (or 10 to 100 Btu/lb$_m$).

In the steady-state energy equations, it is the *difference* in the velocity squared that is important. When both $\mathbf{V}_1$ and $\mathbf{V}_2$ are relatively small, then a large $\Delta\mathbf{V}$ leads to a fairly small $\Delta$ke. However, when both $\mathbf{V}_1$ and $\mathbf{V}_2$ are fairly large, the same $\Delta\mathbf{V}$ as before leads to a much larger $\Delta$ke. This is illustrated by the SI data in Table 5-1. Thus when velocities are relatively low (less than 100 m/s), the change in kinetic energy will be fairly small regardless of the actual change in velocity. However, for velocities greater than roughly 100 m/s, reasonably small changes in velocity may lead to fairly large changes in kinetic energy.

**Note from Table 5-1 that the magnitude of a kinetic-energy change for a given change in velocity increases rapidly as the value of the velocities increases.**

**Table 5-1**   A comparison of values of $\Delta$ke for a constant value of $\Delta\mathbf{V}$

| $\Delta V$, m/s | $V_1$, m/s | $V_2$, m/s | $\Delta$ke, kJ/kg |
|---|---|---|---|
| 15 | 75 | 60 | 1.0 |
| 15 | 150 | 135 | 2.1 |
| 15 | 250 | 235 | 3.6 |
| 50 | 75 | 25 | 2.5 |
| 50 | 175 | 125 | 7.5 |
| 50 | 250 | 200 | 11.3 |

## 5-5-2   GRAVITATIONAL POTENTIAL ENERGY

The gravitational potential energy pe of a unit mass relative to the earth's surface is given by $gz$. The value of $g$ on earth's surface is roughly 9.8 m/s$^2$ in SI units. For a potential-energy change of 1 kJ/kg, the elevation change required is

$$\Delta z = \frac{\Delta pe}{g} = 1 \text{ kJ/kg} \times \frac{s^2}{9.8 \text{ m}} \times \frac{1 \text{ kg·m}}{\text{N·s}^2} \times \frac{1000 \text{ N·m}}{\text{kJ}} = 102 \text{ m}$$

For 10 kJ/kg, the height requirement obviously is 1020 m. In USCS units the value of $g$ on the earth's surface is roughly 32.2 ft/s$^2$. For a potential-energy change of 1 Btu/lb$_m$, the elevation change required is

$$\Delta z = \frac{\Delta pe}{g} = 1 \text{ Btu/lb}_m \times \frac{s^2}{32.2 \text{ ft}} \times \frac{32.2 \text{ lb}_m\text{·ft}}{\text{lb}_f\text{·s}^2} \times \frac{778 \text{ ft·lb}_f}{1 \text{ Btu}} = 778 \text{ ft}$$

Since gas flow in many industrial processes seldom has an elevation change of this magnitude, the gravitational potential-energy change frequently is negligible. Some care must be taken, however, about neglecting this term when liquids are pumped through modest elevation changes. Even though $\Delta$pe may be small, for liquid flow the other energy terms may be equally small.

---

**A** fluid enters a piping system with an inlet area of 20 cm$^2$ at 400 kPa and 25°C. The exit area is 5 cm$^2$ and the exit state is 200 kPa and 50°C. The exit lies 25 m above the inlet and $g = 9.70$ m/s$^2$. Determine the changes in kinetic and potential energy and the enthalpy change, in kJ/kg, if the fluid (*a*) is air at 10 kg/min and (*b*) is water at 100 kg/min.

**Example 5-3**

**Solution:**

**Given:**   Either air or water flows through the piping system shown in Fig. 5-12.

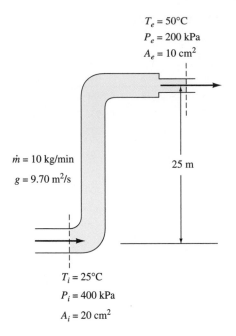

**Figure 5-12**    Schematic and data for Example 5-3.

**Find:**    $\Delta\text{ke}$, $\Delta\text{pe}$, and $\Delta h$ in kJ/kg for (*a*) air and (*b*) water.

**Model:**    Constant-temperature, one-dimensional flow, (*a*) ideal gas, and (*b*) incompressible liquid.

**Strategy:**    States, mass flow rates, and areas are known. Find the velocities from the continuity equation and the enthalpy change from appropriate models for the substances.

**Analysis:**    The changes in kinetic and potential energy are given by $\Delta\text{ke} = (\mathbf{V}_2^2 - \mathbf{V}_1^2)/2$ and $\Delta\text{pe} = g(z_2 - z_1)$. The velocities are found from the mass-flow-rate equation, $\mathbf{V} = \dot{m}v/A$.

(*a*) The specific volume of air is found from the ideal-gas equation, $v = RT/P$. At the inlet and exit conditions,

$$v_1 = \frac{RT_1}{P_1} = \frac{8.314}{29} \times \frac{\text{kPa·m}^3}{\text{kg·K}} \times \frac{298\text{ K}}{400\text{ kPa}} = 0.214\text{ m}^3/\text{kg}$$

$$v_2 = \left(\frac{8.314}{29} \times \frac{298}{200}\right)\frac{\text{m}^3}{\text{kg}} = 0.427\text{ m}^3/\text{kg}$$

Therefore,

$$\mathbf{V}_1 = \frac{\dot{m}v_1}{A_1} = \frac{10\text{ kg/min} \times 0.214\text{ m}^3/\text{kg}}{20\text{ cm}^2} \times \frac{10^4\text{ cm}^2}{\text{m}^2} \times \frac{\text{min}}{60\text{ s}} = 17.8\text{ m/s}$$

$$V_2 = \left[ \frac{10(0.427)}{10} \times \frac{10^4}{60} \right] \frac{m}{s} = 71 \text{ m/s}$$

As a result,

$$\Delta ke = \left[ \frac{(71)^2 - (17.8)^2}{2} \right] \frac{m^2}{s^2} \times \frac{1 \text{ N·s}^2}{kg \cdot m} \times \frac{1 \text{ kJ}}{10^3 \text{ N·m}} = 2.36 \text{ kJ/kg}$$

$$\Delta pe = 9.70(25) \frac{m^2}{s^2} \times \frac{1 \text{ N·s}^2}{kg \cdot m} \times \frac{1 \text{ kJ}}{10^3 \text{ N·m}} = 0.24 \text{ kJ/kg}$$

The enthalpy change of an ideal gas is $\Delta h = c_p \Delta T$. Using $c_p$ data for air from Table A-3,

$$\Delta h = c_p \Delta T = 1.006(25) \text{ kJ/kg} = 25.2 \text{ kJ/kg}$$

(b) For liquid water, $v \approx v_f$ at 25°C $= 1.003 \times 10^{-3}$ m³/kg from Table A-12. Hence

$$V_1 = \frac{\dot{m} v_1}{A_1} = \frac{100 \text{ kg/min} \times 1.003 \times 10^{-3} \text{ m}^3/\text{kg}}{20 \text{ cm}^2} \times \frac{10^4 \text{ cm}^2}{m^2} \times \frac{\text{min}}{60 \text{ s}} = 0.836 \text{ m/s}$$

$$V_2 = \left( \frac{100 \times 1.003 \times 10^{-3}}{10} \times \frac{10^4}{60} \right) \frac{m}{s} = 1.67 \text{ m/s}$$

Therefore,

$$\Delta ke = \left[ \frac{(1.67)^2 - (0.834)^2}{2} \right] \frac{m_2}{s^2} \times \frac{1 \text{ N·s}^2}{kg \cdot m} \times \frac{1 \text{ kJ}}{10^3 \text{ N·m}} = 1.05 \times 10^{-3} \text{ kJ/kg}$$

$$\Delta pe = 9.70(25) \frac{m^2}{s^2} \times \frac{1 \text{ N·s}^2}{kg \cdot m} \times \frac{1 \text{ kJ}}{10^3 \text{ N·m}} = 0.24 \text{ kJ/kg}$$

The enthalpy change based on the saturated-liquid data in Table A-12 is

$$\Delta h = h_{f,2} - h_{f,1} = (209.33 - 104.89) \text{ kJ/kg} = 104.4 \text{ kJ/kg}$$

**Comment:** For reasonable gas and liquid velocities, the kinetic- and potential-energy changes are often quite small, relative to the enthalpy change based on a modest temperature change.

## 5-6 ENGINEERING APPLICATIONS INVOLVING STEADY-STATE CONTROL VOLUMES

In this section a number of specific types of equipment are discussed for which an energy analysis is extremely important. Examples illustrate the use of the various forms of the conservation of energy and conservation of mass equations for steady-state control volumes.

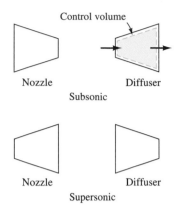

Control volume

Nozzle              Diffuser

Subsonic

Nozzle              Diffuser

Supersonic

**Figure 5-13**
General shapes of nozzles and diffusers for subsonic and supersonic flow.

## 5-6-1  NOZZLES AND DIFFUSERS

In many steady-flow processes there is a need to either increase or decrease the velocity of a flow stream. A device that increases the velocity (and hence the kinetic energy) of a fluid at the expense of a pressure drop in the direction of flow is called a *nozzle*. A *diffuser* is a device for increasing the pressure of a flow stream at the expense of a decrease in velocity. These defining conditions apply for both subsonic and supersonic flow. Figure 5-13 is a set of sketches for the general shapes of a nozzle and a diffuser under the conditions of subsonic and supersonic flow. Note that a nozzle is a converging passage for subsonic flow, whereas the passage is diverging for supersonic flow. The opposite conditions hold for a diffuser. One outgrowth of this that is extremely important in rocket design is that a converging-diverging nozzle must be used if a fluid is to be accelerated from subsonic to supersonic velocities. A common use of a nozzle is as an attachment to a garden or fire hose.

The design and operational conditions for nozzles and diffusers are as follows:

1. No shaft work is involved, since both devices are merely ducts.

2. The change in potential energy is negligible, since the fluid undergoes little or no elevation change.

3. In many cases the heat transfer per unit mass may be quite small compared to the kinetic-energy and enthalpy changes. In some cases the duct may be deliberately insulated. Even without insulation, the fluid velocity may be so large that there is not time enough for significant heat transfer. Also the surface area of the duct is frequently relatively small for effective heat transfer. Thus in many applications the assumption of an adiabatic process is a good approximation for nozzles and diffusers.

Thus in the absence of appreciable heat transfer to the fluid, the kinetic-energy change in either device is due to two effects, namely, a change in the internal energy of the fluid and a change in flow work during the process. The value of the internal-energy change will be larger than the value of the flow (displacement) work. The continuity equation may also be useful because of the variable-area nature of nozzles and diffusers.

**Note the energy forms that change in value owing to the change in kinetic energy in nozzles and diffusers.**

**Example 5-4**

**W**ater vapor enters a subsonic diffuser at a pressure of 0.7 bar, a temperature of 160°C, and a velocity of 180 m/s. The inlet to the diffuser is 100 cm$^2$. During passage through the diffuser the fluid velocity is reduced to 60 m/s, the pressure increases to 1.0 bar, and the heat transfer to the surroundings is 0.6 kJ/kg. Determine (*a*) the final temperature, (*b*) the mass flow rate in kg/s, and (*c*) the outlet area in cm$^2$.

**Solution:**

**Given:** A subsonic diffuser. A schematic with appropriate data is shown in Fig. 5-14.

**Find:** (a) $T_2$, in kelvins, (b) $\dot{m}$, in kg/s, and (c) $A_2$, in cm$^2$.

**Model:** Steady-state, subsonic flow; $w_{shaft}$ and $\Delta pe$ are zero.

**Strategy:** The initial state and both inlet and exit velocities are known. Use an energy balance to find the exit state, and hence the temperature. Then find the mass flow rate and exit area from the conservation of mass principle.

**Analysis:** The system boundary is the dashed line in Fig. 5-14.

(a) The final temperature is a function of the final pressure, which is 1.0 bar, and one other property. The continuity of flow equation is insufficient to help at this point, since both $A_2$ and $v_2$ are unknown. The only other equation available is the general energy balance [Eq. [5-15]]. This is

$$\frac{dE_{cv}}{dt} = \dot{Q} + \dot{W} + \sum_{in} \dot{m}_i\left(h + \frac{V^2}{2} + gz\right)_i - \sum_{out} \dot{m}_e\left(h + \frac{V^2}{2} + gz\right)_e$$

Under the assumption of steady state, $dE/dt = 0$. In addition, $\dot{W}$ and $\Delta pe$ are modeled as zero. Hence, for one inlet and one outlet the energy equation reduces to

$$0 = \dot{Q} + \dot{m}_1\left(h + \frac{V^2}{2}\right)_1 - \dot{m}_2\left(h + \frac{V^2}{2}\right)_2$$

For the conservation of mass equation, $dm/dt = \dot{m}_1 - \dot{m}_2 = 0$ in steady state. Thus $\dot{m}_1 = \dot{m}_2 = \dot{m}$. Since the mass flow rate is yet unknown, converting the rate equation for energy to the unit-mass basis format is appropriate. Dividing the energy equation above by $\dot{m}$, noting that $\dot{Q} = \dot{m}q$, and rearranging leads to

$$q = h_2 - h_1 + \frac{V_2^2 - V_1^2}{2}$$

It is seen that sufficient information is given to evaluate all the quantities except $h_2$. From Table A-14 for steam at 0.7 bar and 160°C, $h_1 = 2798.2$ kJ/kg. Upon substituting the known values, we find that

$$-0.6 \text{ kJ/kg} = (h_2 - 2798.2) \text{ kJ/kg} + \frac{(60)^2 - (180)^2}{2(1000)} \text{ kJ/kg}$$

or $\qquad h_2 = (2798.2 - 0.6 + 14.4) \text{ kJ/kg} = 2812.0 \text{ kJ/kg}$

Knowledge of $P_2$ and $h_2$ establishes the values of all other properties, such as the temperature $T_2$. The enthalpy of saturated vapor at 1.0 bar is given in Table A-13 as 2675.5 kJ/kg. Since $h_2 = 2812.0$ kJ/kg, the final state is in the superheat region. From the 1.0-bar set of data in Table A-14, the calculated value of $h_2$ corresponds to a temperature of 168°C by linear interpolation.

(b) The mass flow rate is found from the continuity relation [Eq. [5-6]]. From Table A-14 the value of $v$ at 0.7 bar and 160°C is 2.841 m$^3$/kg. Therefore, with

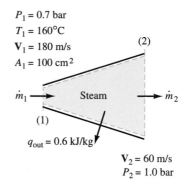

$P_1 = 0.7$ bar
$T_1 = 160°C$
$V_1 = 180$ m/s
$A_1 = 100$ cm$^2$

$\dot{m}_1 \rightarrow$ Steam $\rightarrow \dot{m}_2$

(1)

$q_{out} = 0.6$ kJ/kg

$V_2 = 60$ m/s
$P_2 = 1.0$ bar

**Figure 5-14**
Schematic and data for Example 5-4.

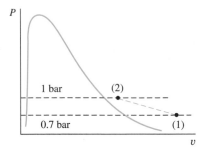

**Figure 5-15**
The process path for Example 5-4 on a $Pv$ diagram.

one-dimensional flow

$$\dot{m} = \frac{\mathbf{V}_1 A_1}{v_1} = \frac{180 \text{ m/s} \times 100 \text{ cm}^2}{2.841 \text{ m}^3/\text{kg}} \times \frac{\text{m}^2}{10^4 \text{ cm}^2} = 0.634 \text{ kg/s}$$

(*c*) The outlet area also is found by application of the continuity equation. One suitable form is Eq. [5-9], namely, $\mathbf{V}_1 A_1/v_1 = \mathbf{V}_2 A_2/v_2$. From the superheated steam Table A-14, the value of $v_2$ at 1.0 bar and 168°C is roughly 2.022 m³/kg by interpolation. Hence the final area from continuity considerations is

$$A_2 = A_1 \frac{\mathbf{V}_1 v_2}{\mathbf{V}_2 v_1} = 100 \text{ cm}^2 \times \frac{180}{60} \times \frac{2.022}{2.841} = 214 \text{ cm}^2$$

The final area can also be found from the continuity equation in the format $A_2 = \dot{m}v_2/\mathbf{V}_2$, using the $\dot{m}$ value found in part *b*.

Figure 5-15 illustrates the process on a $Pv$ diagram, which is not to scale.

## 5-6-2   TURBINES, COMPRESSORS, AND FANS

A ***turbine*** is a device in which the fluid (either a gas or a liquid) does work against some type of blade attached to a rotating shaft. As a result, the device produces shaft work that may be used for some purpose. For example, in steam, gas, and hydroelectric power plants the turbine drives an electrical generator. ***Compressors*** are devices in which shaft work supplied from an external source is done on the fluid, which results in a significant increase in pressure of the fluid, and possibly a significant increase in temperature. A ***fan*** is a device that slightly increases the pressure of a gas; its main purpose is to move the fluid from one location to another. Schematics of an axial-flow turbine and a centrifugal-flow compressor are shown in Fig. 5-16. A picture of a centrifugal fan can be found in Photograph 5-1.

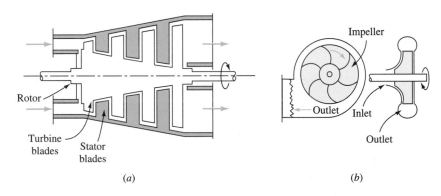

*(a)*                                    *(b)*

**Figure 5-16**   Schematics of (a) an axial-flow turbine and (b) a centrifugal-flow compressor.

The design and operational models for turbines, compressors, and fans are as follows:

1. The potential-energy change is normally negligible.

2. The inclusion of heat transfer depends on the mode of operations. If the device is not insulated, the heat gained or lost by the fluid depends on such factors as whether (1) a large temperature difference exists between the fluid and the surroundings, (2) a small flow velocity exists, and (3) a large surface area is present. In rotating turbomachinery (axial or centrifugal) velocities can be fairly high, and the heat transfer normally is small compared with the shaft work. In reciprocating devices the heat-transfer effects may be reasonably large. Experience and experimental specifications enable the engineer to estimate the relative importance of heat transfer.

3. The change in kinetic energy is usually quite small in these devices, since the velocities at inlet and outlet are frequently less than 100 m/s (330 ft/s). There are exceptions, of course. In a steam turbine the exhaust velocity can be quite high, owing to the large specific volume of the fluid at the low exhaust pressure. On the basis of the continuity equation, velocities may be kept low by selecting large flow areas. However, this may not be a practical choice.

For steady flow through any of these devices the enthalpy change in the direction of flow is a significant factor. In general, the enthalpy decreases for a turbine and increases for a compressor.

**Photograph 5-1**
Centrifugal fans are often used in air-conditioning and ventilation applications. (Courtesy of The Trane Company)

Note whether the volume flow rate increases or decreases from inlet to exit in a turbine.

---

**Air** initially at 1 bar and 290 K is compressed in steady state to 5 bars and 450 K. The power input to the air under steady-flow conditions is 5 kW, and a heat loss of 5 kJ/kg occurs during the process. If the changes in potential and kinetic energies are neglected, determine the mass flow rate, in kg/min.

**Solution:**

**Given:** An air compressor is shown in Fig. 5-17 with appropriate input data.

**Find:** Mass flow rate, in kg/min.

**Model:** Steady state; ideal gas; neglect $\Delta$ke and $\Delta$pe.

**Strategy:** The initial and final states, the power, and the heat transfer are known. Thus the only unknown in the rate form of the energy balance is the mass flow rate.

**Analysis:** The dashed line in Fig. 5-17 indicates the control volume of interest. The mass flow rate, which is the unknown, appears in both the conservation of mass and conservation of energy equations. However, since area information is not

**Example 5-5**

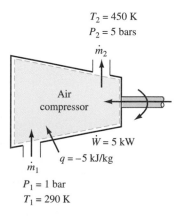

$T_2 = 450$ K
$P_2 = 5$ bars

$\dot{m}_2$

Air compressor

$\dot{W} = 5$ kW

$q = -5$ kJ/kg

$\dot{m}_1$

$P_1 = 1$ bar
$T_1 = 290$ K

**Figure 5-17**
Schematic and data for Example 5-5.

available, the definition of mass flow rate cannot be used. The mass and energy equations are

$$\frac{dm_{cv}}{dt} = \dot{m}_1 - \dot{m}_2$$

$$\frac{dE_{cv}}{dt} = \dot{Q} + \dot{W} + \dot{m}_1 \left( h + \frac{\mathbf{V}^2}{2} + gz \right)_1 - \dot{m}_2 \left( h + \frac{\mathbf{V}^2}{2} + gz \right)_2$$

For steady state $dm/dt = 0$ and $dE/dt = 0$. As a result, $\dot{m}_1 = \dot{m}_2 = \dot{m}$, and the energy equation can be written as

$$\dot{Q} + \dot{W} = \dot{m} \left[ (h_2 - h_1) + \frac{\mathbf{V}_2^2 - \mathbf{V}_1^2}{2} + g(z_2 - z_1) \right]$$

Because we are neglecting changes in kinetic and potential energies, this equation reduces to

$$\dot{Q} + \dot{W} = \dot{m}(h_2 - h_1)$$

Based on Eq. [5-22], $\dot{Q}$ is evaluated from the relation $\dot{Q} = \dot{m}q$, and $q = -5$ kJ/kg. Since the inlet and exit temperatures are known, the ideal-gas enthalpies for air can be read from Table A-5 at 290 and 450 K as 290.2 and 451.8 kJ/kg, respectively. Substitution of known values into the energy above equation leads to

$$\dot{m}(-5 \text{ kJ/kg}) + 5 \text{ kW} = \dot{m} \times (451.8 - 290.2) \text{ kJ/kg}$$

where each rate term is in kilowatts and $\dot{m}$ is in kg/s. Hence the mass flow rate is

$$\dot{m} = \frac{5 \text{ kW}}{(161.6 + 5) \text{ kJ/kg}} = 0.030 \text{ kg/s} = 1.80 \text{ kg/min}$$

**Comment:**   Note that $q$ is only 3 percent of the enthalpy change in this example.

---

**Example 5-6**

**A** steam turbine operates with an inlet condition of 30 bars, 400°C, 160 m/s and an outlet state of a saturated vapor at 0.7 bar with a velocity of 100 m/s. The mass flow rate is 1200 kg/min, and the power output is 10,800 kW. Determine the magnitude and direction of the heat-transfer rate, in kJ/min, if the potential energy change is negligible.

**Solution:**

**Given:**   A steam turbine operates in steady state. A schematic of the equipment with appropriate data are shown in Fig. 5-18. The control volume of interest is shown by the dashed line in the figure.

**Find:**   $\dot{Q}$, in kJ/min.

**Model:**   Steady state; neglect $\Delta$pe.

**Strategy:**   The initial and final states, the mass flow rate, and the power are known. Hence the only unknown in the rate format of the conservation of energy principle is the rate of heat transfer.

**Analysis:** The basic equations are

$$\frac{dm_{cv}}{dt} = \dot{m}_1 - \dot{m}_2$$

$$\frac{dE_{cv}}{dt} = \dot{Q} + \dot{W} + \dot{m}_1\left(h + \frac{\mathbf{V}^2}{2} + gz\right)_1 - \dot{m}_2\left(h + \frac{\mathbf{V}^2}{2} + gz\right)_2$$

If we assume steady-state conditions, then $dm/dt = 0$ and $dE/dt = 0$. Thus for a control volume with one inlet and one exit $\dot{m}_1 = \dot{m}_2 = \dot{m}$, and the energy-rate equation becomes

$$0 = \dot{Q} + \dot{W} = \dot{m}\left[(h_1 - h_2) + \frac{\mathbf{V}_1^2 - \mathbf{V}_2^2}{2}\right]$$

where the potential-energy term was neglected. Enthalpy data are needed to complete the numerical solution. Since the saturation temperature at 30 bars is 233.9°C, the initial state at 400°C is superheated vapor. From the superheat Table A-14, $h_1 = 3230.9$ kJ/kg. The final state is a saturated vapor. Therefore, we find from the saturation-pressure Table A-13 at 0.7 bar that $h_2 = h_g = 2660$ kJ/kg. The mass flow rate of 1200 kg/min is equivalent to 20 kg/s, and the power output is 10,800 kJ/s. Consequently, the energy balance becomes

$$\dot{Q} = -\dot{W} + \dot{m}\left[(h_2 - h_1) + \frac{\mathbf{V}_2^2 - \mathbf{V}_1^2}{2}\right]$$

$$= -\left(\frac{-10,800 \text{ kJ}}{s}\right) + \frac{20 \text{ kg}}{s} \times \left[(2660 - 3230.9) + \frac{100^2 - 160^2}{2(1000)}\right] \text{ kJ/kg}$$

$$= [10,800 + 20(-570.9 - 7.8)] \text{ kJ/s}$$

$$= -460 \text{ kJ/s} = -27,600 \text{ kJ/min}$$

The process is shown on a $Pv$ diagram in Fig. 5-19.

**Comment:** Note that the heat transfer is from the turbine (since the steam is much hotter than the surroundings) and is only about 4 percent of the work interaction. Also, the kinetic-energy change is only 1.4 percent of the enthalpy change.

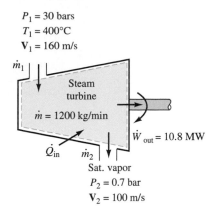

$P_1 = 30$ bars
$T_1 = 400°C$
$\mathbf{V}_1 = 160$ m/s
$\dot{m}_1$
Steam turbine
$\dot{m} = 1200$ kg/min
$\dot{W}_{out} = 10.8$ MW
$\dot{Q}_{in}$  $\dot{m}_2$
Sat. vapor
$P_2 = 0.7$ bar
$\mathbf{V}_2 = 100$ m/s

**Figure 5-18**
Schematic and data for Example 5-6.

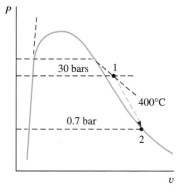

**Figure 5-19**
The process in Example 5-6 illustrated on a $Pv$ diagram.

## 5-6-3 HEAT EXCHANGERS

One of the most important steady-flow devices of engineering interest is the heat exchanger. These devices serve two useful purposes: (1) they are used to remove (or add) energy from (to) some region of space, or (2) they are employed to change deliberately the thermodynamic state of a fluid. The radiator of an automobile is an example of a heat exchanger used for heat removal. Modern gas turbines and electric generators are frequently cooled internally, and their performance is greatly affected by the heat-transfer process. In steam power plants, heat exchangers are used to remove heat

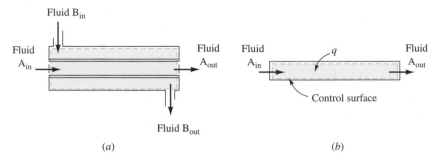

**Figure 5-20**     Two different control surfaces for a concentric-tube heat exchanger.

**Note whether $q_A = -q_B$ or $\dot{Q}_A = -\dot{Q}_B$ for the heat exchanger in Fig. 5-21a.**

**Photograph 5-2**

A hot water coil is a gas-liquid heat exchanger used to heat air moving in a duct. (Courtesy of The Trane Company)

from hot combustion gases and subsequently increase the temperature and enthalpy of the steam in the power cycle. This high-enthalpy fluid is then expanded with a resulting large power output. In the chemical industry, heat exchangers are extremely important in maintaining or attaining certain thermodynamic states as chemical processes are carried out. Modern applications of heat exchangers are numerous.

One of the primary applications of *heat exchangers* is the exchange of energy between two moving fluids. The changes of kinetic and potential energies are usually negligible for each fluid stream, and no work interactions are present. The pressure drop through a heat exchanger is usually small; hence an assumption of constant pressure is often quite good as a first approximation. A heat exchanger composed of two concentric pipes is shown in Fig. 5-20a. Fluid A flows in the inner pipe, and a second fluid B flows in the annular space between the pipes. Two different analyses can be done on this equipment, depending on where the system boundaries are selected. In the first case, the control surface will be placed around the entire piece of equipment, as indicated by the dashed line in Fig. 5-20a. Since the purpose of a heat exchanger is to exchange heat between two fluids, the outside surface is usually heavily insulated. Hence the heat transfer external to the device is frequently negligible for this selection of the system boundary. In the second case, it is sometimes desirable to place a boundary around one of the two fluids. Figure 5-20b shows the system boundary in this latter case for fluid A. In this case a heat-transfer term appears in the energy analysis of the fluid. However, it is easily shown that the heat-transfer rates associated with each fluid are equal in magnitude but opposite in sign when the heat transfer from the overall device is zero. A special application of heat-exchanger design is to equipment called boilers, evaporators, and condensers. As the names suggest, in these types of equipment one of the fluids will change phase. One of the most common liquid-liquid or liquid-vapor heat exchangers is a shell-and-tube heat exchanger. A hot water coil, as shown in Photograph 5-2, is an example of a gas-liquid heat exchanger.

Example 5-7

**A** small nuclear reactor is cooled by passing liquid sodium through it. The liquid sodium leaves the reactor at 2 bars and 400°C. It is cooled to 320°C by passing through a heat exchanger before returning to the reactor. In the heat exchanger, heat is transferred from the liquid sodium to water, which enters the exchanger at 100 bars and 49°C and leaves at the same pressure as a saturated vapor. The mass flow rate of sodium is 10,000 kg/h, its specific heat is constant at 1.25 kJ/kg·°C, and its pressure drop is negligible. Determine ($a$) the mass flow rate of water evaporated in the heat exchanger, in kg/h, and ($b$) the heat-transfer rate between the two fluids, in kJ/h. Neglect kinetic- and potential-energy changes.

**Solution:**

**Given:** A heat exchanger employs liquid sodium and water. A schematic of the equipment with appropriate input data is shown in Fig. 5-21.

**Find:** ($a$) $\dot{m}_{H_2O}$, kg/h, and ($b$) $\dot{Q}$, kJ/h.

**Model:** Steady state; sodium is incompressible; $\dot{W} = 0$; $\Delta ke = 0$; $\Delta pe = 0$; $\Delta P = 0$.

**Strategy:** The initial and final states of the two fluids are known, as well as the mass flow rate of sodium. Thus the only unknown in the energy balance on a rate basis for the overall heat exchanger is the mass flow rate of water.

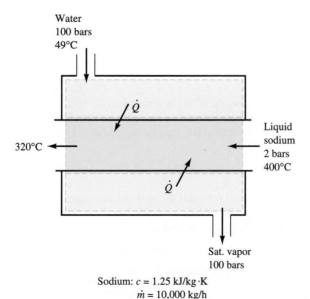

Water
100 bars
49°C

$\dot{Q}$

Liquid
sodium
2 bars
400°C

320°C

$\dot{Q}$

Sat. vapor
100 bars

Sodium: $c = 1.25$ kJ/kg·K
$\dot{m} = 10,000$ kg/h

**Figure 5-21**     Schematic and data for heat exchanger in Example 5-7.

**Analysis:**    (a) The mass flow rate can be determined from an energy balance on the entire heat exchanger. The control volume in this case is shown by the outer dashed line in Fig. 5-21. We begin with the general energy equation [Eq. [5-15]], namely,

$$\frac{dE_{cv}}{dt} = \dot{Q} + \dot{W} + \sum_{in} \dot{m}_i \left( h + \frac{\mathbf{V}^2}{2} + gz \right)_i - \sum_{out} \dot{m}_e \left( h + \frac{\mathbf{V}^2}{2} + gz \right)_e$$

In steady-state $dE/dt = 0$; hence

$$0 = \dot{Q} + \dot{W} + \sum_{in} \dot{m}_i \left( h + \frac{\mathbf{V}^2}{2} + gz \right)_i - \sum_{out} \dot{m}_e \left( h + \frac{\mathbf{V}^2}{2} + gz \right)_e$$

It is assumed that the heat loss from the exterior surface of the exchanger is negligible. In addition, shaft work is zero and changes in kinetic and potential energy are considered negligible. On this basis the above equation reduces to

$$\sum_{out} \dot{m}_e h_e = \sum_{in} \dot{m}_i h_i$$

For the sodium-water heat exchanger under study, the above equation may be written as

$$\dot{m}_{H_2O}(h_{out} - h_{in})_{H_2O} = \dot{m}_{Na}(h_{in} - h_{out})_{Na} = \dot{m}_{Na} c_{Na}(T_{in} - T_{out})_{Na}$$

where sodium has been assumed to be incompressible, so that $\Delta h_{Na} = \Delta h_{inc} = c \Delta T + v \Delta P$, and $\Delta P$ for the sodium stream is assumed negligible. The inlet water stream is a compressed liquid. Its enthalpy is determined from Table A-15 as 176.38 kJ/kg. The outlet enthalpy of saturated water vapor at 100 bars is found in Table A-13 to be 2724.7 kJ/kg. Substitution of these values and the data for liquid sodium into the above equation yields

$$\dot{m}_{H_2O}(2724.7 - 176.4)\ \text{kJ/kg} = 10{,}000(1.25)(400 - 320)\ \text{kJ/h}$$

or

$$\dot{m}_{H_2O} = \frac{10{,}000(1.25)(80)\ \text{kJ/h}}{2548.3\ \text{kJ/kg}} = 392\ \text{kg/h}$$

(b) The heat-transfer rate from the sodium to the water is found by applying an energy balance on either the sodium or water stream. The steady-state energy balance, applicable to either stream, may be written in the form

$$\dot{Q} + \dot{W} = \dot{m} \left[ (h_{out} - h_{in}) + \frac{\mathbf{V}_{out}^2 - \mathbf{V}_{in}^2}{2} + g(z_{out} - z_{in}) \right]$$

For either stream we again note that $\dot{W} = 0$ and changes in kinetic and potential energy are assumed negligible. Note, however, there is heat transfer to or from either fluid. As a result, the above equation becomes

$$\dot{Q} = \dot{m}(h_{out} - h_{in})$$

Using the sodium stream to evaluate the heat-transfer rate, one finds that

$$\dot{Q} = [\dot{m}c(T_{out} - T_{in})]_{Na} = 10{,}000(1.25)(-80)\ \text{kJ/h} = -1.0 \times 10^6\ \text{kJ/h}$$

This is the rate of heat transfer to the sodium. Thus the sodium loses (and the water gains) $1 \times 10^6$ kJ/h by heat transfer.

## 5-6-4 MIXING PROCESSES

Another important type of equipment, loosely classified as a heat exchanger, is used for the **direct mixing** of multiple inlet flows. The resulting mixture stream leaves the device at a single outlet, as shown in Fig. 5-22 for two inlet flows. Air-conditioning ducts frequently form such a system. In power plants these systems are called *open feedwater heaters.* Another common example is the mixing of hot- and cold-water streams in a faucet or shower mixing valve. The idealizations for mixing processes are similar to those for heat exchangers. Heat transfer from the mixing region is frequently negligible owing to insulation on the outside, and $\dot{W}_{shaft} = 0$. In addition, changes in kinetic and potential energy are negligible. Thus the steady-state energy balance reduces to a sum of $\dot{m}h$ terms for the various inlets and outlets. Similarly, the conservation of mass equation is a summation of $\dot{m}$ terms for the inlets and outlets.

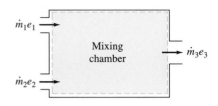

**Figure 5-22**
Schematic and data of a mixing chamber.

Note the necessary conditions for a steady-state mixing process during which the energy of the incoming streams equals the energy of the exit stream.

**Example 5-8**

In a steam power plant an open feedwater heater operates at 5 bars. During the steam cycle a superheated vapor is bled off the turbine and enters the heater at 5 bars and 200°C. A compressed liquid coming from a pump enters the heater at 40°C and 5 bars. For every kilogram of steam bled from the turbine per unit time, determine how many kilograms of compressed liquid enter per unit time, if the leaving stream is a saturated liquid at the heater pressure.

**Solution:**

**Given:** The two inlet water streams to an open heater are compressed liquid (state 1) and superheated vapor (state 2), and the exit stream is a saturated liquid (state 3). A schematic of the equipment with input data is shown in Fig. 5-23, with the control volume indicated by the dashed line.

**Find:** $\dot{m}_1/\dot{m}_2$.

**Model:** $\dot{Q} = \dot{W} = 0$; neglect kinetic- and potential-energy changes.

**Strategy:** The inlet and exit states are known. Thus the only unknowns in the two steady-state rate equations for energy conservation and mass conservation are the three mass flow rates. This permits calculation of the ratio of mass flow rates, but not the actual rates.

**Analysis:** The general energy balance is

$$\frac{dE_{cv}}{dt} = \dot{Q} + \dot{W} + \sum_{in} \dot{m}_i \left( h + \frac{V^2}{2} + gz \right)_i - \sum_{out} \dot{m}_e \left( h + \frac{V^2}{2} + gz \right)_e$$

Under steady-state conditions $dE/dt = 0$. In addition, if the mixing region is essentially adiabatic, $\dot{W} = 0$, and the potential- and kinetic-energy changes of the flow stream are negligible, then the general energy balance on the entire heater becomes

$$0 = \dot{m}_1 h_1 + \dot{m}_2 h_2 - \dot{m}_3 h_3 \qquad \text{or} \qquad \dot{m}_1 h_1 + \dot{m}_2 h_2 = \dot{m}_3 h_3$$

Superheated vapor at 200°C

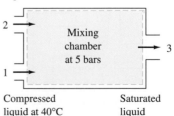

Compressed liquid at 40°C          Saturated liquid

**Figure 5-23**
Schematic and data for the mixing process in Example 5-8.

Also, the general mass balance is

$$\frac{dm_{cv}}{dt} = \sum_{in} \dot{m}_i - \sum_{out} \dot{m}_e$$

In steady state $dm/dt = 0$. On this basis the above equation when applied to the mixing process takes on the form

$$0 = \dot{m}_1 + \dot{m}_2 - \dot{m}_3 \qquad \text{or} \qquad \dot{m}_1 + \dot{m}_2 = \dot{m}_3$$

The energy and mass equations are sufficient to determine the state and flow rates at the different sections, in conjunction with known data.

The first task is to evaluate the enthalpies at the three sections. For section 2 at 5 bars and 200°C the enthalpy from Table A-14 is 2855.4 kJ/kg. At section 1 the compressed liquid enters at 5 bars and 40°C. Since the pressure is not high, we can estimate the enthalpy by using the saturated-liquid value at the given temperature. This is found in Table A-12 at 40°C to be 167.57 kJ/kg. The fluid leaves at section 3 as a saturated liquid at 5 bars. From Table A-13, $h_3$ is 640.23 kJ/kg. Substitution of enthalpy data into the above energy balance leads to

$$\dot{m}_2(2855.4) + \dot{m}_1(167.6) = \dot{m}_3(640.2)$$

From the continuity equation, $\dot{m}_1 + \dot{m}_2 = \dot{m}_3$. When substituted into the preceding equation,

$$\dot{m}_2(2855.4) + \dot{m}_1(167.6) = (\dot{m}_1 + \dot{m}_2)(640.2)$$

Rearrangement of this relation shows that

$$\dot{m}_2(2855.4 - 640.2) + \dot{m}_1(167.6 - 640.2) = 0$$

$$\frac{\dot{m}_2}{\dot{m}_1} = \frac{2855.4 - 640.2}{640.2 - 167.6} = 4.75$$

Thus over 89 percent of the mass entering the feedwater heater is compressed liquid, and the remainder comes from steam bled from the turbine.

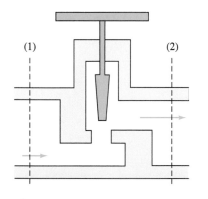

(1)          (2)

**Figure 5-24**
Schematic of a throttling valve.

### 5-6-5   THROTTLING DEVICES

There are circumstances in the design of energy systems where a decrease in pressure is desired but no work effect occurs. This pressure drop is accomplished by inserting a flow restriction into the line. This restriction commonly is a partially opened valve, as shown in Fig. 5-24, or a porous plug or a long capillary tube. A greater flow resistance leads to a greater pressure drop. In some cases the pressure drop may lead to other more desirable effects. These include a temperature change or a phase change.

Regardless of its purpose, a flow restrictor has the following characteristics:

1. Since the control volume is rigid and no rotating shafts are present, no work interactions are involved.

2. The change in potential energy is negligible.

3. Although the velocity may be quite high in the region of the restriction, measurements upstream and downstream from the actual valve area indicate that the change in velocity, and hence the change in kinetic energy, across the restriction can be neglected.

4. In most applications, either the device is deliberately insulated or the heat transfer, by nature of the process, is insignificant. For example, neither the time nor the surface area is sufficient for effective heat transfer.

The steady-state energy equation for the device in Fig. 5-24 is

$$q + w = (h_2 - h_1) + \frac{\mathbf{V}_2^2 - \mathbf{V}_1^2}{2} + g(z_2 - z_1) \qquad \textbf{[5-22]}$$

Based on the flow characteristics listed above, the energy equation reduces to

$$h_1 = h_2 \qquad \textbf{[5-23]}$$

This statement does not say that the enthalpy is constant for the process but merely requires that the initial and final enthalpies be the same. A process that satisfies Eq. [5-23] is called a **throttling process**. The valves in water faucets in the home are examples of **throttling devices**. These devices are also in common use in most home refrigeration units.

---

**Example 5-9**

Refrigerant-134a is throttled through a porous plug from a saturated liquid at 32°C to a final state where the pressure is 2 bars. Determine (*a*) the final temperature and (*b*) the physical state of the fluid at the exit.

**Solution:**

**Given:**  A throttling process through a porous plug is shown in Fig. 5-25.

**Find:**  (*a*) $T_2$ in °C, and (*b*) the physical condition at exit state.

**Model:**  Steady-state, $\dot{Q} = \dot{W} = \Delta\text{ke} = \Delta\text{pe} = 0$.

**Strategy:**  In a throttling process determination of the initial state allows calculation of the final state.

**Analysis:**  The general mass balance is

$$\frac{dm_{cv}}{dt} = \dot{m}_1 - \dot{m}_2$$

In steady state $dm/dt = 0$. Thus for a device with one inlet and one exit, $\dot{m}_1 = \dot{m}_2$. In addition, the general energy balance on a rate basis for a control volume with one

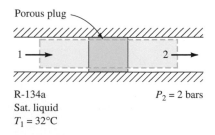

R-134a                                          $P_2 = 2$ bars
Sat. liquid
$T_1 = 32°C$

**Figure 5-25**
A process schematic and data for Example 5-9.

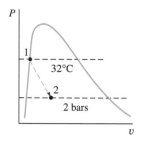

**Figure 5-26**
A $Pv$ diagram showing process path for Example 5-9.

inlet and one exit is Eq. [5-15], namely,

$$\frac{dE_{cv}}{dt} = \dot{Q} + \dot{W}_{shaft} + \dot{m}_i \left(h + \frac{\mathbf{V}^2}{2} + gz\right)_i - \dot{m}_e \left(h + \frac{\mathbf{V}^2}{2} + gz\right)_e$$

In steady state the term $dE/dt = 0$. Also, heat and work effects and changes in kinetic- and potential-energy changes are zero for a throttling process by concept. Thus the above energy balance reduces to $\dot{m}_1 h_1 - \dot{m}_2 h_2 = 0$ or $h_1 = h_2$. The latter equation is used to find the final state.

(*a*) From the saturation-temperature table A-16, the enthalpy of saturated liquid at 32°C is 94.39 kJ/kg. This also must be the enthalpy of a unit mass after throttling. From the saturation-pressure table A-17 at 2 bars, the value of $h_f$ is 36.84 kJ/kg and that of $h_g$ is 241.30 kJ/kg. Because $h_f < h_2 < h_g$, the final state must be a mixture of liquid and vapor, and the exit temperature is the corresponding saturation temperature at 2 bars of $-10.09°C$.

(*b*) The quality in the final state is found by

$$h_2 = h_1 = 94.39 \text{ kJ/kg} = h_f + x_2 h_{fg} = (36.84 + 204.46x_2) \text{ kJ/kg}$$

or

$$x_2 = \frac{57.55}{204.46} = 0.281$$

The final mixture is roughly 28 percent vapor and 72 percent liquid. The approximate path of the process in the wet region is sketched in Fig. 5-26.

### 5-6-6 PIPE FLOW

The transport of fluids, either gaseous or liquid, in **pipes** or **ducts** is of primary importance in many engineering designs. Pipes of different diameters may be used within a given system. In addition, the fluid may undergo a significant elevation change, which is not the case with other pieces of standard equipment. If a piping or duct system includes a pump or fan, then Eq. [5-22] is appropriate for a one-inlet, one-outlet system in steady state:

$$q + w = \Delta h + \Delta\text{ke} + \Delta\text{pe} \qquad \text{[5-22]}$$

In an energy analysis of pipe flow, several terms in the above equation might be negligible.

The following characteristics of pipe flow need to be considered:

1. The pipes or ducts may be uninsulated, with heat transfer occurring between the fluid and the environment. This is especially significant when the flow channel is long or the temperature difference between the fluid and the environment is large. If heat transfer is undesirable, then the pipe or duct is deliberately well insulated.

2. Inclusion of the work term is necessary when a fan or pump is included in the control volume, or an electrical resistance heater is involved.

3. Velocities of liquids usually are low so that kinetic-energy changes are frequently negligible. For gas flow the kinetic-energy change could become relatively large when heat transfer is significant or cross-sectional area changes are large.

4. For liquid flow through pipe the potential-energy change may be significant.

The following example illustrates liquid flow through a pipe.

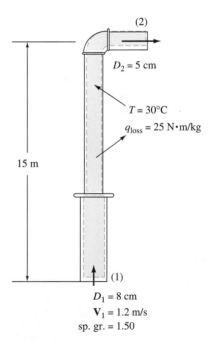

**Figure 5-27**
Schematic and data of piping system in Example 5-10.

**Example 5-10**

**At** a certain position in a piping system a liquid with a specific gravity of 1.50 passes in steady state through an 8-cm pipe with a velocity of 1.2 m/s. At some position downstream, the elevation of the pipe has increased 15.0 m and the pipe size has been reduced to 5 cm. The temperature of the fluid is constant at 30°C, a heat loss of 25 N·m/kg occurs, and local gravity is 9.80 m/s². Determine the change in pressure, in bars and megapascals.

**Solution:**

**Given:** A piping system, as illustrated in Fig. 5-27.

**Find:** $\Delta P$, in bars and MPa.

**Model:** Steady-state, incompressible flow; constant temperature, $g = 9.80$ m/s².

**Strategy:** Using the area data, the conservation of mass and the definition of mass flow rate can be used to find the outlet velocity. Hence the heat transfer, and kinetic- and potential-energy changes are known. In the absence of a work transfer, the only unknown in an energy balance is the enthalpy change, which contains the unknown pressure change.

**Analysis:** The inlet and exit positions on the control volume in Fig. 5-27 are designated as states 1 and 2, respectively. The mass and energy balances written for the control volume containing the fluid within the piping system are

$$\frac{dm_{cv}}{dt} = \dot{m}_1 - \dot{m}_2$$

$$\frac{dE_{cv}}{dt} = \dot{Q} + \dot{W} + \dot{m}_1\left(h + \frac{V^2}{2} + gz\right)_1 - \dot{m}_2\left(h + \frac{V^2}{2} + gz\right)_2$$

In steady state, $dm_{cv}/dt = 0$ and $dE_{cv}/dt = 0$. Thus $\dot{m}_1 = \dot{m}_2 = \dot{m}$. Also $\dot{Q} = \dot{m}q$ and $\dot{W} = \dot{m}w$. On this basis, division of the above energy balance by $\dot{m}$ leads to the unit-mass form of the energy balance, namely,

$$0 = q + w - \Delta h - \Delta ke - \Delta pe$$

If the fluid is assumed *incompressible,* and the temperature is constant, then $\Delta u_{inc} = c\,\Delta T = 0$. Hence the enthalpy change becomes

$$\Delta h_{inc} = \Delta u + v\,\Delta P = v\,\Delta P$$

In the figure: (2), $D_2 = 5$ cm, $T = 30°C$, $q_{loss} = 25$ N·m/kg, 15 m, (1), $D_1 = 8$ cm, $V_1 = 1.2$ m/s, sp. gr. = 1.50

As a result, the energy balance for the control volume reduces to

$$0 = q - v\,\Delta P - \Delta \text{ke} - \Delta \text{pe}$$

Solving for the unknown pressure drop, we find that

$$-v\,\Delta P = \Delta \text{ke} + \Delta \text{pe} - q = \frac{\mathbf{V}_2^2 - \mathbf{V}_1^2}{2} + g(z_2 - z_1) - q$$

The final velocity can be found from the conservation of mass equation. For one-dimensional flow, $\dot{m} = AV/v$. Hence the mass balance becomes

$$0 = \frac{A_1 \mathbf{V}_1}{v_1} - \frac{A_2 \mathbf{V}_2}{v_2}$$

Since the velocity at state 1 is known, and $v_1 = v_2$ for incompressible flow, the above equation leads to $V_2$. Noting that $A = \pi D^2/4$, then

$$\mathbf{V}_2 = \mathbf{V}_1 \frac{A_1}{A_2} = \mathbf{V}_1 \left(\frac{D_1}{D_2}\right)^2 = 1.2\left(\frac{8}{5}\right)^2 \text{ m/s} = 3.1 \text{ m/s}$$

Substitution of the available data into the energy equation, and noting that $g$ is 9.80 m/s², yields

$$-v\,\Delta P = \left[\frac{(3.1)^2 - (1.2)^2}{2} + 15(9.80) - (-25)\right] \text{ N·m/kg} = 176 \text{ N·m/kg}$$

Neglecting the effect of pressure on the specific volume of liquid water at 30°C, $v \approx v_f = 1.004 \times 10^{-3}$ m³/kg from Table A-12. To account for the specific gravity of the fluid, recall that $v_{\text{liq}} = v_{\text{water}}/(\text{sp. gr.})$. As a result,

$$\Delta P = -\frac{176 \text{ N·m}}{\text{kg}} \times \frac{1.50}{1.004 \times 10^{-3}} \frac{\text{kg}}{\text{m}^3} \times \frac{\text{bar·m}^2}{10^5 \text{ N}}$$

$$= -2.64 \text{ bars} = -0.264 \text{ MPa}$$

**Comment:**   The values of $\Delta$ke, $\Delta$pe, and $q$ are 4.12, 147, and 25 J/kg, respectively. Note that $\Delta$pe dominates the calculation for the pressure change, and $\Delta$ke is relatively small.

## 5-7   INTRODUCTION TO THERMODYNAMIC CYCLES

An important application of the devices discussed in the preceding section is their use in power and refrigeration cycles. These cycles involve the passage of a fluid through a sequence of processes which require the use of turbines, compressors, pumps, heat exchangers, and throttling devices. A complete analysis of various cycles requires the use of the second law. However, mass and energy analyses alone provide some important insight into the operation of cycles which affect our lives daily. In addition, a basic knowledge of common cycles is helpful in understanding the role of cycles in the development of second-law concepts in the next chapter. This section examines a simple steam power cycle for electrical power

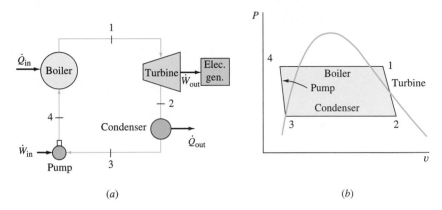

**Figure 5-28**    (a) A schematic and (b) a *Pv* process diagram and data for a simple steam power cycle.

generation and a simple refrigeration cycle for refrigeration and air-conditioning purposes.

## 5-7-1    A SIMPLE STEAM POWER CYCLE

Electrical power available in homes and industry is developed in an electrical generator at a ***steam power plant.*** The generator is driven by the shaft power output of a steam turbine as shown in Fig. 5-28a. The work output of the turbine shaft $w_{turb}$ (or power $\dot{W}_{turb,out}$) is produced by expanding superheated steam at a high pressure and temperature at state 1 adiabatically to a low pressure (possibly less than atmospheric pressure) at state 2. This expansion process is shown on a *Pv* diagram in Fig. 5-28b, where state 2 is shown as a wet mixture of high quality. Next the fluid passes at constant pressure through a heat exchanger called a condenser. The heat transfer out $q_{cond}$ changes the fluid to a saturated liquid at state 3.

A pump with work input $w_{pump}$ then compresses the fluid adiabatically to the turbine inlet pressure at state 4. In this step the fluid is frequently modeled as incompressible and the adiabatic process as isothermal. Finally, a boiler-superheater (heat exchanger) is used to raise the fluid temperature at constant pressure ($P_4 = P_1$) to the desired turbine inlet value. The boiler heat input $q_{boiler}$ is supplied typically from the combustion of coal or natural gas, or from a nuclear source. A more complete discussion of vapor power cycles appears in Chap. 17.

**Note the four processes that comprise a simple steam power cycle, namely, expansion, cooling, compression, and heating.**

**Example 5-11**

**A** steam power cycle operates under the following conditions: (1) 100 bars and 520°C at the turbine inlet, (2) 0.3 bars and 90 percent quality at the turbine exit, (3) saturated liquid at 0.3 bars at the condenser outlet, and (4) 100 bars at the pump outlet. The mass flow rate of water is 40 kg/s. Neglecting Δke and Δpe, determine (*a*) the turbine work, (*b*) the condenser heat loss, (*c*) the pump work, all in kJ/kg, and (*d*) the boiler heat input, in kW.

**Solution:**

**Given:** A simple steam power cycle as illustrated in Fig. 5-29.

**Find:** (a) $w_{turb}$, (b) $q_{cond}$, (c) $w_{pump}$, all in kJ/kg, and (d) $\dot{Q}_{boiler}$, in kW.

**Model:** Steady state, $\Delta ke = \Delta pe = 0$, adiabatic turbine and pump.

**Strategy:** Determine data for the four states in the cycle, and then apply the steady-state energy balance on a unit mass basis to the four processes in the cycle.

**Analysis:** The mass and energy balances written for any control volume with one inlet and one outlet are

$$\frac{dm_{cv}}{dt} = \dot{m}_1 - \dot{m}_2$$

$$\frac{dE_{cv}}{dt} = \dot{Q} + \dot{W} + \dot{m}_1 \left(h + \frac{V^2}{2} + gz\right)_1 - \dot{m}_2 \left(h + \frac{V^2}{2} + gz\right)_2$$

In steady state, $dm_{cv}/dt = 0$ and $dE_{cv}/dt = 0$. Thus the mass balance reduces to $\dot{m}_1 = \dot{m}_2 = \dot{m}$. Dividing the energy balance through by the mass flow rate gives

$$0 = q + w + (h_1 - h_2) + \frac{V_1^2 - V_2^2}{2} + g(z_1 - z_2)$$

where $q = \dot{Q}/\dot{m}$ and $w = \dot{W}/\dot{m}$. In the absence of appreciable kinetic- and potential-energy changes, the above energy equation reduces to $q + w = \Delta h$. This relation is now applied to control volumes placed around the four separate pieces of equipment in the cycle. The turbine and pump are assumed to be adiabatic, and the

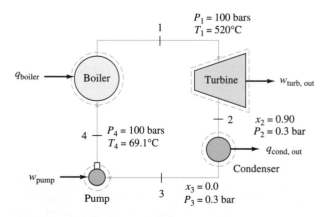

**Figure 5-29** Schematic and data for the simple steam power cycle in Example 5-11.

boiler and condenser are without work transfers. Therefore,

$$w_{turb} = h_2 - h_1 \qquad q_{cond} = h_3 - h_2 \qquad w_{pump} = h_4 - h_3 \qquad q_{boiler} = h_1 - h_4$$

The enthalpy $h_4$ is unknown and must be found by considering the fluid flowing through the pump to be incompressible, and the process to be adiabatic and isothermal. Under these conditions

$$w_{pump} = \Delta h_{inc} = c\,\Delta T + v\,\Delta P = v_{f,3}\,\Delta P$$

Therefore, $h_4 = h_3 + v_{f,3}\,\Delta P$, where $v_f$ again represents a saturated-liquid value.

For the given states specified in Fig. 5-29, the steam tables reveal that

$$h_1 = 3425.1 \text{ kJ/kg (Table A-14)}$$

$$h_2 = 0.1(289.23) + 0.9(2625.3) = 2391.7 \text{ kJ/kg (Table A-13)}$$

$$h_3 = h_f = 289.23 \text{ kJ/kg (Table A-13)}$$

and $\qquad v_{f,3} = 1.0223 \times 10^{-3} \text{ m}^3\text{/kg (Table A-13)}$

Employing these data in the appropriate energy equations cited above, we find the following results.

(a) The turbine work under adiabatic conditions ($q = 0$) is given by

$$w_{turb} = h_2 - h_1 = (2391.7 - 3425.1) \text{ kJ/kg} = -1033.4 \text{ kJ/kg}$$

(b) The heat transfer into the condenser is

$$q_{cond} = h_3 - h_2 = (289.2 - 2391.7) \text{ kJ/kg} = -2102.5 \text{ kJ/kg}$$

(c) The adiabatic pump work is given by $w_{pump} = v_{f,3}\,\Delta P$. Use of this equation yields

$$w_{pump} = 1.0223 \times 10^{-3} \text{m}^3\text{/kg} \times (100 - 0.3) \text{ bars} \times 10^{-2} \text{ kJ/bar·m}^3 = 10.2 \text{ kJ/kg}$$

From this we find that $h_4 = h_3 + w_{pump} = (289.2 + 10.2) \text{ kJ/kg} = 299.4 \text{ kJ/kg}$.

(d) Knowledge of $h_4$ allows us to evaluate the heat transfer into the boiler.

$$q_{boiler} = h_1 - h_4 = (3425.1 - 299.4) \text{ kJ/kg} = 3125.7 \text{ kJ/kg}$$

The rate of heat transfer $\dot{Q}$ into the boiler superheater is simply

$$\dot{Q} = \dot{m}q_{boiler} = 40 \text{ kg/s} \times 3125.7 \text{ kJ/kg} = 125{,}000 \text{ kW}$$

**Comment:** Only 32.7 percent of the heat supplied to the boiler is converted to net shaft work output. This ratio typically is in the range of 30 to 40 percent.

## 5-7-2 A VAPOR COMPRESSION REFRIGERATION CYCLE

The purpose of a *refrigerator* or *air conditioner* is to maintain a region of space at a relatively fixed temperature. Consider the cold region shown in Fig. 5-30a which is to be maintained at a temperature below the environmental value. This region could be the inside of a freezer or the inside of a home or car on a hot summer day. Heat transfer from the environment will begin to raise the temperature of the cold region. To overcome this effect

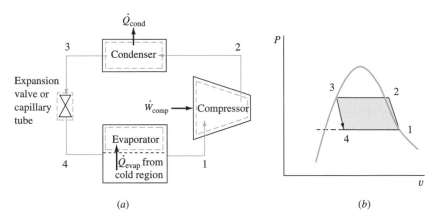

(a)                                                  (b)

**Figure 5-30**     (a) An equipment schematic and (b) a $Pv$ process diagram for a simple refrigeration cycle.

a colder fluid (the refrigerant) circulated through pipes in the cold region removes energy as heat transfer at the rate it enters from the environment. This causes the temperature of the cold region to remain constant.

As a result of the heat transfer $\dot{Q}_{evap}$ from the cold region, the refrigerant undergoes evaporation as it passes through the heat exchanger. Evaporation occurs at constant temperature and pressure from state 4 until the fluid becomes a saturated vapor shown as state 1 on the $Pv$ diagram in Fig. 5-30$b$. This energy added to the refrigerant in the evaporator must be discarded to the environment before the cycle can be repeated. This is accomplished by next compressing the saturated vapor at state 1 to a superheated vapor at state 2 so that $T_2$ is considerably greater than the environmental temperature. This adiabatic compression process requires work input $\dot{W}_{comp}$. Now energy is removed by heat transfer $\dot{Q}_{cond,out}$ to the environment as the refrigerant passes through a condenser. The fluid is assumed to be a saturated liquid at state 3 in Fig. 5-30. Finally the pressure is reduced to the evaporator pressure by passing the refrigerant through a throttling device. The throttling process causes some evaporation of the fluid, so that the refrigerant is a wet mixture of low quality as it enters the evaporator at state 4. The cycle is now repeated. A more detailed presentation of vapor-compression refrigeration cycles appears in Chap. 17.

**Note the four processes that comprise the vapor-compression refrigeration cycle, namely, heating, compression, cooling, and throttling.**

**Example 5-12**

**A**n automotive air-conditioning unit circulates 0.05 kg/s of refrigerant-134a. The fluid enters the compressor as a saturated vapor at 180 kPa and leaves the compressor at 0.70 MPa and 45°C. Determine (a) the power input in kilowatts to the compressor, and (b) the heat-transfer rate in the evaporator in kJ/s.

**Solution:**

**Given:**     A vapor-compression refrigeration cycle is shown in Fig. 5-31.

**Find:**   (*a*) $\dot{W}_{comp}$, kW; and (*b*) $\dot{Q}_{evap}$, kJ/s.

**Model:**   Steady-state, $\Delta ke = \Delta pe = 0$, $P_1 = P_4$, $P_2 = P_3$, $x_3 = 0$, $q_{12} = 0$.

**Strategy:**   Determine the data for the four states in the cycle, and then apply the energy balance on a rate basis to the compressor and the evaporator.

**Analysis:**   The mass and energy balances written for any control volume with one inlet and one outlet are

$$\frac{dm_{cv}}{dt} = \dot{m}_i - \dot{m}_e$$

$$\frac{dE_{cv}}{dt} = \dot{Q} + \dot{W} + \dot{m}_i\left(h + \frac{V^2}{2} + gz\right)_i - \dot{m}_e\left(h + \frac{V^2}{2} + gz\right)_e$$

In steady state, $dm_{cv}/dt = 0$ and $dE_{cv}/dt = 0$. This means the mass balance reduces to $\dot{m}_1 = \dot{m}_2 = \dot{m}$. The energy balance can now be written as

$$0 = \dot{Q} + \dot{W} + \dot{m}\left[(h_i - h_e) + \frac{V_i^2 - V_e^2}{2} + g(z_i - z_e)\right]$$

In the absence of kinetic- and potential-energy changes, this reduces to

$$\dot{Q} + \dot{W}_{shaft} = \dot{m}(h_2 - h_1)$$

This relation is now applied to control volumes placed around each piece of equipment.

   (*a*) The compressor is adiabatic, so that the energy balance on a rate basis is $\dot{W}_{comp} = \dot{m}(h_2 - h_1)$. The values of $h_1$ and $h_2$ are

$$h_1 = h_g \text{ at } 180 \text{ kPa} = 239.71 \text{ kJ/kg (Table A-17)}$$

$$h_2 = \frac{275.93 + 286.35}{2} \text{ kJ/kg} = 281.14 \text{ kJ/kg (Table A-18)}$$

where $h_2$ is found by linear interpolation at 0.70 MPa. Hence the compressor power is

$$\dot{W}_{comp} = \dot{m}(h_2 - h_1) = 0.05 \text{ kJ/s} \times (281.14 - 239.71) \text{ kJ/kg} = 2.07 \text{ kW}$$

The positive answer indicates the power is input.

   (*b*) For the evaporator $\dot{Q}_{evap} = \dot{m}(h_1 - h_4)$, because the shaft power is zero. For the throttling process $h_4 = h_3$, and $h_3$ is modeled as a saturated liquid at the compressor outlet pressure. From Table A-17 $h_{f,3} = 86.78$ kJ/kg at 0.70 MPa. Hence

$$\dot{Q}_{evap} = \dot{m}(h_1 - h_4) = 0.05 \text{ kg/s} \times (239.71 - 86.78) \text{ kJ/kg} = 7.65 \text{ kJ/s}$$

**Comment:**   The ratio of the rate of heat removal in the evaporator to the compressor power input is 3.7. This value is fairly typical for air conditioners. Three or more units of cooling are achieved by the action of one unit of work input.

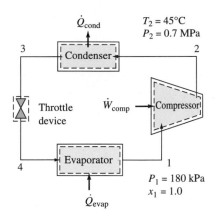

**Figure 5-31**
An equipment schematic and data for the refrigeration cycle in Example 5-12.

## 5-8    TRANSIENT (UNSTEADY) FLOW ANALYSIS

There are engineering processes that are not steady state, owing to variations in mass flow rates, heat-transfer rates, and power crossing the control surface. In addition, properties of the mass crossing a boundary may change with time. Analysis of these situations, called **transient** or **unsteady-state** processes, requires an examination of the conservation of mass and energy principles in their most general form. Several interesting and useful devices operate under such conditions. For example, some wind tunnels operate on the discharge of air that has been stored under high pressure in a large, rigid tank. These are known as blowdown wind tunnels. Compressed air stored in a cylinder might also be used to drive a gas turbine coupled with a generator as an auxiliary electric power source in case of normal power failure. As has been pointed out earlier, in steady-state analyses we usually consider the control volume fixed in space, in size, and in shape. In the following subsections we may wish to allow the size and shape to vary. However, the control volume will still be fixed in location.

### 5-8-1    APPLICATION TO A TRANSIENT CONTROL VOLUME

The typical starting point for an analysis of a transient control volume is the rate form of the conservation of mass and energy equations, Eqs. [5-1] and [5-15],

$$\frac{dm_{\text{cv}}}{dt} = \sum_{\text{in}} \dot{m}_i - \sum_{\text{out}} \dot{m}_e \qquad \text{[5-1]}$$

$$\frac{dE_{\text{cv}}}{dt} = \dot{Q} + \dot{W} + \sum_{\text{in}} \dot{m}_i \left( h + \frac{\mathbf{V}^2}{2} + gz \right)_i - \sum_{\text{out}} \dot{m}_e \left( h + \frac{\mathbf{V}^2}{2} + gz \right)_e$$

$$\text{[5-15]}$$

where one-dimensional flow is assumed at any inlet or exit. For a differential change in state over a time increment $dt$, Eq. [5-1] can be written as

$$dm_{\text{cv}} = \sum_{\text{in}} \delta m_i - \sum_{\text{out}} \delta m_e \qquad \text{[5-24a]}$$

where $\delta m = \dot{m}\, dt$ is the differential amount of mass that flows across the control surface during the time interval $dt$. For the same differential change in state, Eq. [5-15] can be written as

$$dE_{\text{cv}} = \delta Q + \delta W + \sum_{\text{in}} \left( h + \frac{\mathbf{V}^2}{2} + gz \right)_i \delta m_i - \sum_{\text{out}} \left( h + \frac{\mathbf{V}^2}{2} + gz \right)_e \delta m_e$$

$$\text{[5-24b]}$$

where $\delta Q = \dot{Q}\, dt$ and $\delta W = \dot{W}\, dt$.

Given suitable information about the mass and the energy inside the control volume at some initial time $t_1$, it is possible to determine the mass and energy in the control volume at time $t_2$ by integrating Eqs. [5-1] and [5-15] over the time interval $\Delta t = t_2 - t_1$. Integrating the mass equation gives

$$\Delta m_{cv} = \sum_{in} \left[ \int_{t_1}^{t_2} \dot{m}_i \, dt \right] - \sum_{out} \left[ \int_{t_1}^{t_2} \dot{m}_e \, dt \right] \qquad \text{[5-25]}$$

$$= \sum_{in} m_i - \sum_{out} m_e$$

where $\Delta m_{cv} = m_{cv(t_2)} - m_{cv(t_1)}$ is the change in mass within the control volume and $m_i$ (or $m_e$) is the amount of mass that flows into (or out of) the control volume at a given control surface during the time interval $\Delta t$. Integrating the energy equation gives

$$\Delta E_{cv} = \int_{t_1}^{t_2} \dot{Q} \, dt + \int_{t_1}^{t_2} \dot{W} \, dt \qquad \text{[5-26]}$$

$$+ \sum_{in} \left[ \int_{t_1}^{t_2} \left( h + \frac{V^2}{2} + gz \right)_i \dot{m}_i \, dt \right] - \sum_{out} \left[ \int_{t_1}^{t_2} \left( h + \frac{V^2}{2} + gz \right)_e \dot{m}_e \, dt \right]$$

$$= Q + W + \sum_{in} \left[ \int_{t_1}^{t_2} \left( h + \frac{V^2}{2} + gz \right)_i \dot{m}_i \, dt \right] - \sum_{out} \left[ \int_{t_1}^{t_2} \left( h + \frac{V^2}{2} + gz \right)_e \dot{m}_e \, dt \right]$$

where $\Delta E_{cv} = E_{cv(t_2)} - E_{cv(t_1)}$ and $Q$ and $W$ are the net heat transfer and work transfer to the control volume during the time interval. To carry out these integrations, we need to know how the various properties and rates vary with time.

In the analysis of transient control volumes, several assumptions are frequently used to model real situations. Two of the most important assumptions are those of *uniform state* and of *uniform flow*. These are defined as follows:

1. An assumption of **uniform state** for the control volume requires that all intensive properties within the control volume be uniform throughout the control at any instant (or over several different regions of the control volume). However, the state of the control volume may change with time.

2. An assumption of **uniform flow** for an inlet or outlet requires that the state of the mass crossing the control surface be invariant with time. However, the mass flow rate may vary with time.

The condition of uniform state implies rapid or instantaneous approach to equilibrium at all times for the mass within the control volume. This is met by many transient fluid systems in practice. The condition of uniform flow is frequently met when the flow into a transient system is supplied from a

Note the difference between the uniform state and uniform flow assumptions.

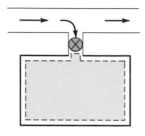

**Figure 5-32**
Schematic for charging of a vessel from a line.

very large reservoir of matter. Because of this, the uniform flow assumption is sometimes referred to as the *reservoir* assumption.

When the control volume is assumed to have a uniform state, then $E_{cv} = m_{cv}e$ and the left-hand side of Eq. [5-15] can be rewritten as

$$\frac{dE_{cv}}{dt} = \frac{d}{dt}(m_{cv}e) = m_{cv}\frac{de}{dt} + e\frac{dm_{cv}}{dt} \qquad \textbf{[5-27]}$$

where $m_{cv}$ is the mass inside the control volume and $e$ is the specific total energy for the mass. This can be combined with the conservation of mass equation, Eq. [5-1], to give

$$\frac{dE_{cv}}{dt} = m_{cv}\frac{de}{dt} + e\left[\sum_{in}\dot{m}_i - \sum_{out}\dot{m}_e\right] \qquad \textbf{[5-28]}$$

It is frequently useful to make this substitution before integrating the energy equation, Eq. [5-15], as was done to obtain Eq. [5-26].

### 5-8-2   The Charging of a Tank

A common and typical example of an unsteady-state flow problem is the charging of a rigid vessel by supplying a fluid from some outside high-pressure source, as shown in Fig. 5-32. The rigid vessel may be initially evacuated or may contain some of the fluid within it before the charging process begins. Generally the mass enters only at one section of the control surface, and no fluid leaves the control volume. A charging process with heat transfer is illustrated in the following example.

**Example 5-13**

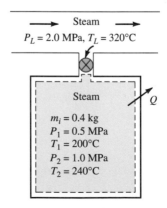

**Figure 5-33**
Schematic and data for Example 5-13.

**A** steam supply line maintained at 2.0 MPa and 320°C is used to charge a rigid tank. The tank initially contains 0.40 kg of steam at 0.5 MPa and 200°C. The connecting valve is opened and steam enters slowly until the steam within the tank reaches 1.0 MPa and 240°C. Determine (*a*) the kilograms of steam that enters the tank and (*b*) the heat transfer across the uninsulated walls of the tank, in kilojoules. Neglect kinetic- and potential-energy effects.

**Solution:**

**Given:**   A steam supply line is attached through a valve to a rigid tank, as shown in Fig. 5-33.

**Find:**   (*a*) $m_L$, in kg, and (*b*) $Q$, in kJ.

**Model:**   Uniform flow, rigid uninsulated tank.

**Strategy:**   Knowledge of the initial and final states and the initial mass permits calculation of the final mass in the rigid tank. Then, a transient energy analysis yields the heat transfer.

**Analysis:** A control volume is selected, as indicated by the dashed line in Fig. 5-33, that includes the volume inside the tank. We assume that the volume in the region of the valve and the entrance pipe is negligible. The rate forms of the conservation of mass and energy equations for this control volume are

$$\frac{dm_{cv}}{dt} = \dot{m}_L$$

$$\frac{dE_{cv}}{dt} = \dot{Q} + \dot{W} + \dot{m}_L \left( h + \frac{V^2}{2} + gz \right)_L$$

where $\dot{m}_L$ is the mass flow rate of steam into the tank from the steam line. These equations are valid at any instant during the charging process. Integrating over the time interval time $t_1$ to the final time $t_2$ gives

$$\Delta m_{cv} = \int_{t_1}^{t_2} \dot{m}_L \, dt = m_L$$

$$\Delta E_{cv} = Q + W + \int_{t_1}^{t_2} \dot{m}_L \left( h + \frac{V^2}{2} + gz \right)_L dt$$

where $m_L$ is the mass that enters the control volume during the time interval.

(a) The mass $m_L$ can be found from the mass balance since $\Delta m_{cv} = m_2 - m_1 = m_L$. The mass of the control volume at state 1 and state 2 can be written as $m_1 = V/v_1$ and $m_2 = V/v_2$ where $V$ is volume of the tank. Since the tank volume is constant, the final and initial volumes can be related by the equation

$$V = m_1 v_1 = m_2 v_2$$

This relation permits an evaluation of the final mass $m_2$. The required steam properties at the initial state 1, the final state 2, and the state of the steam line $L$ are found in Table A-14. These data are:

| | | | |
|---|---|---|---|
| $P_L = 2.0$ MPa | $T_L = 320°C$ | $h_L = 3069.5$ kJ/kg | $v_L = 0.1308$ m³/kg |
| $P_1 = 0.5$ MPa | $T_1 = 200°C$ | $h_1 = 2642.9$ kJ/kg | $v_1 = 0.4249$ m³/kg |
| $P_2 = 1.0$ MPa | $T_2 = 240°C$ | $h_2 = 2692.9$ kJ/kg | $v_2 = 0.2275$ m³/kg |

Using the specific-volume data tabulated above, we find that

$$m_2 = m_1 \frac{v_1}{v_2} = (0.4 \text{ kg}) \frac{0.4249}{0.2275} = 0.747 \text{ kg}$$

And solving for $m_L$ using the conservation of mass we have

$$m_L = m_2 - m_1 = (0.747 - 0.400) \text{ kg} = 0.347 \text{ kg}$$

(b) To solve for the heat transfer, we must use the integrated form of the energy balance. Assuming that kinetic- and potential-energy effects are negligible and that the work is zero because the tank is rigid, the energy equation becomes

$$\Delta U_{cv} = m_2 u_2 - m_1 u_1 = Q + \int_{t_1}^{t_2} \dot{m}_L h_L \, dt$$

If we assume that the quantity of fluid bled from the line is small compared to the quantity flowing through the line, then the properties at the inlet to the valve remain constant, and the charging process is modeled as *uniform flow*. For uniform flow, $h_L$ is a constant and the last term on the right-hand side of the equation above becomes $\int_{t_1}^{t_2} \dot{m}_L h_L \, dt = h_L \int_{t_1}^{t_2} \dot{m}_L \, dt = h_L m_L$. Making this substitution into the energy equation and solving for $Q$ we have

$$Q = (m_2 u_2 - m_1 u_1) - m_L h_L$$

Substitution of data yields

$$
\begin{aligned}
Q &= (m_2 u_2 - m_1 u_1) - m_L h_L \\
&= [0.747(2692.9) - 0.4(2642.9)] \text{ kJ} - 0.347(3069.5) \text{ kJ} \\
&= (2012 - 1057 - 1065) \text{ kJ} = -110 \text{ kJ}
\end{aligned}
$$

The negative sign indicates that the heat transfer occurs *from* the system to the environment.

---

It is important also to examine a rapid charging process. If the tank were allowed to fill rapidly, the amount of heat transferred could be fairly small, even if an appreciable temperature gradient existed between the contents of the tank and the environment. In other situations the tank may be insulated. The example below examines an adiabatic charging process.

**Example 5-14**

**A** steam supply line maintained at 400 psia and 450°F is used to charge an evacuated, rigid tank. The connecting valve is opened and steam enters the insulated tank until the pressure within the tank reaches 200 psia. Determine (*a*) the final temperature in the tank, in degrees Fahrenheit, and (*b*) the pounds-mass of steam that has entered if the volume of the tank is 10.0 ft³.

**Solution:**

**Given:** A steam supply line is connected through a valve to an evacuated tank, as shown in Fig. 5-34.

**Find:** (*a*) $T_f$, in °F, and (*b*) $m_L$, in lb$_m$.

**Model:** Uniform flow, rigid tank, $m_i = 0$, $\Delta ke = \Delta pe = W = Q = 0$.

**Strategy:** The only unknown in a transient energy balance is the final internal energy in the tank at a known pressure. This leads to the final temperature and specific volume from the steam tables. Then the final mass is found from the specific volume and the total volume.

**Analysis:** (*a*) A control volume is selected, as indicated by the dashed line in Fig. 5-34, that includes the volume inside the tank. We assume that the volume in the region of the valve and the entrance pipe is negligible. The rate forms of the conservation of mass and energy equations for this control volume are

$$\frac{dm_{cv}}{dt} = \dot{m}_L$$

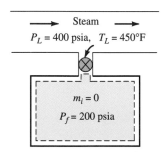

**Figure 5-34**
Schematic and data for Example 5-14.

$$\frac{dE_{cv}}{dt} = \dot{Q} + \dot{W} + \dot{m}_L \left( h + \frac{V^2}{2} + gz \right)_L$$

where $\dot{m}_L$ is the mass flow rate of steam into the tank from the steam line. These equations are valid at any instant during the charging process. Integrating over the time interval time $t_i$ to the final time $t_f$ and neglecting kinetic- and potential-energy effects gives

$$\Delta m_{cv} = m_2 - m_1 = \int_{t_i}^{t_f} \dot{m}_L \, dt = m_L$$

$$\Delta U_{cv} = U_f - U_i = Q + W + \int_{t_i}^{t_f} \dot{m}_L h_L \, dt$$

where $m_L$ is the mass that enters the control volume during the time interval and the subscripts $i$ and $f$ represent the initial and final states, respectively, with the control volume. For *uniform flow* into the control volume the integral in the energy equation becomes $h_L m_L$, where the subscript $L$ designates the uniform line conditions. In the absence of any heat transfer or work transfer for this control volume, the energy balance reduces to

$$\Delta U_{cv} = m_f u_f - m_i u_i = h_L m_L$$

Because the tank is initially evacuated, $m_i = 0$, and $m_f = m_L$. Hence the energy balance above reduces to

$$m_L u_f = m_L h_L$$
$$u_f = h_L$$

This relationship fixes the final state within the control volume. Under the conditions specified, we find that the enthalpy of the fluid in the high-pressure line is numerically equal to the internal energy of the fluid within the control volume at any instant. This latter statement *assumes* that the fluid equilibrates rapidly within the control volume, so that a *uniform state* exists at any instant.

From Table A-14E we find that $h_L = 1209.6$ Btu/lb$_m$. Therefore, the final state in the tank is one for which

$$u_f = 1209.6 \text{ Btu/lb}_m \text{ at } 200 \text{ psia}$$

By linear interpolation in Table A-14E at 200 psia the final temperature is

$$T_f = 600°F + 100°F \frac{1209.6 - 1208.9}{1248.8 - 1208.9} = 602°F$$

This compares to the line temperature of 450°F.

(b) The final mass in the tank is found by $m_f = V/v_f$. The final specific volume is determined by linear interpolation at 200 psia and 602°F to be 3.064 ft³/lb$_m$. Thus

$$m_f = \frac{V}{v_f} = \frac{10 \text{ ft}^3}{3.064 \text{ ft}^3/\text{lb}_m} = 3.26 \text{ lb}_m$$

**Comment:** The final temperatures and masses at other final tank pressures up to 400 psia could be found by a similar analysis.

The result derived in the above example, that $h_L = u_f$ for adiabatic filling of an evacuated tank, is completely general in the sense that it applies to any fluid. Although applied to a real gas in the example, the equation above is of special interest when employed for the flow of an ideal gas into the tank. Since $h = c_p T$ and $u = c_v T$ for an ideal gas (if the values of the two specific heats are assumed constant over the temperature range from $T_L$ to $T_f$), then

$$c_p T_L = c_v T_f$$

or
$$T_f = k T_L$$

where $k$ is the specific-heat ratio. This important result shows that $T_f$ is always greater than $T_L$ for flow of an ideal gas into a rigid tank under the following major idealizations: (1) adiabatic, rigid control volume, (2) equilibrium within the tank at any instant, (3) negligible kinetic energy of the inflowing gas, and (4) fixed properties for the inlet flow. Since $k$ for many gases ranges from 1.25 to 1.40, the absolute temperature rise may be as much as 40 percent.

The problem of charging an evacuated tank from a steady-flow line can also be solved on the basis of a closed-system analysis. It is interesting to pursue this method of analysis, since it provides a striking contrast to the open-system analysis previously used. The closed system selected for analysis is shown in Fig. 5-35. Since we have chosen a closed system, all the mass that eventually enters the tank must initially be included within the system boundaries. Hence the boundaries extend beyond the valve and out into a portion of the line. The conservation of energy equation in this case is

$$Q + W = \Delta E$$

Again we neglect kinetic- and potential-energy terms and assume adiabatic conditions. The energy equation becomes

$$W = \Delta U = m_f u_f - m_i u_i$$

Unlike in the control-volume approach, however, neither the work term nor the initial mass term $m_i$ is zero.

The work done is the compression work related to pushing the mass contained within the system, but outside the valve, into the tank. The pressure in the line in the vicinity of the valve is constant, and the boundary work is simply $-P \Delta V$. Also the initial mass $m_i$ in this closed system is the mass contained outside the valve initially, since the tank is evacuated at the start of the process. Consequently, the energy equation becomes

$$-P(V_f - V_i) = m_f u_f - m_i u_i$$

or
$$-P(0 - m_L v_L) = m_f u_f - m_L u_L$$

Finally, combining the second term on the right with the term on the left, we find that

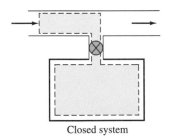

Closed system

**Figure 5-35**

Schematic for closed system analysis of charging an evacuated tank.

$$m_L(P_L v_L + u_L) = m_f u_f$$

$$h_L m_L = m_f u_f$$

and $$h_L = u_f$$

This result is identical to that obtained by a control-volume analysis in Example 5-14, since both were based on comparable idealizations. However, the terms included in the two analyses are quite different. Whether transient flow problems can be solved by both control-volume and closed-system analyses depends largely on the particular conditions. Usually the closed-system approach can be used only in those cases where the position and state of the mass in transit are well defined.

### 5-8-3 TRANSIENT-SYSTEM ANALYSIS WITH A VOLUME CHANGE

Consider the situation shown in Fig. 5-36. A weighted piston-cylinder assembly initially containing a known quantity of a gas at a known equilibrium state is connected through a valve to a high-pressure line. To simplify the problem, the fluid in the line is the same as that in the piston cylinder. We are interested in determining the temperature and the quantity of mass within the cylinder after the volume has changed by a known amount.

The control volume under consideration is shown by the dashed line in Fig. 5-36. Since the upper boundary will move, the control volume changes size in this case, and $P\,dV$ work occurs. It is necessary to place a restriction on the process if the boundary work is to be evaluated. Even though the pressure in the line may be substantially above the pressure within the cylinder (as maintained by the weighted piston), we must assume that the entering gas quickly equilibrates with the gas already present in the control volume, so that a uniform pressure $P$ is maintained on the piston by the gas. Thus a *uniform state* exists within the control volume during the process. Then the work done by the surroundings on the gas is simply $-P\,\Delta V$. In the absence of sufficient information, we assume that the process is adiabatic and requires that the piston be frictionless. Although the center of mass of the control volume changes in elevation, this potential-energy effect can be neglected. The following example illustrates this type of process.

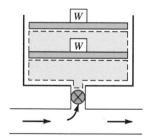

**Figure 5-36**
A control volume with a moving boundary.

**A** piston-cylinder assembly is attached through a valve to a source of air that is flowing in steady state through a pipe. Initially, inside the cylinder the volume is 0.01 m³, the temperature of the air inside is 40°C, and the pressure is 1 bar. The valve is then opened to the air-supply line, which is maintained at 6 bars and 100°C. If the cylinder pressure remains constant, find (*a*) the mass that enters through the valve, in grams, and (*b*) the final equilibrium temperature, in degrees Celsius, when the cylinder volume reaches 0.02 m³.

**Example 5-15**

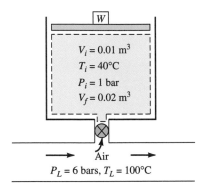

$V_i = 0.01 \text{ m}^3$
$T_i = 40°C$
$P_i = 1 \text{ bar}$
$V_f = 0.02 \text{ m}^3$

Air

$P_L = 6 \text{ bars}, T_L = 100°C$

**Figure 5-37**
Schematic and data for Example 5-15.

**Solution:**

**Given:** A piston-cylinder device is attached through a valve to a line that contains air, as shown in Fig. 5-37.

**Find:** $T_f$, in °C, and $m_L$, in kilograms.

**Model:** Uniform state, uniform flow, adiabatic, ideal gas.

**Strategy:** The combination of the conservation of energy and conservation of mass statements contains the final mass and the final specific internal energy as unknowns. But two other relations are available. First, the final mass is related to the known total volume and the final specific volume. Second, the final specific volume and final specific internal energy are related by the state postulate at the specified pressure. These three relationships contain three unknowns.

**Analysis:** The control volume is selected to lie just inside the solid boundaries of the piston-cylinder assembly (see Fig. 5-37). The integrated form of the conservation of mass and energy equations for this control volume are

$$\Delta m_{cv} = m_f - m_i = \int_{t_i}^{t_f} \dot{m}_L \, dt = m_L$$

$$\Delta E_{cv} = E_f - E_i = Q + W + \int_{t_i}^{t_f} \dot{m}_L \left( h + \frac{V^2}{2} + gz \right)_L dt$$

For this control volume heat transfer is negligible as are kinetic- and potential-energy effects. In addition the only type of work that can occur is compression/expansion work. Under these conditions the energy balance reduces to

$$\Delta U_{cv} = W_{comp/exp} + \int_{t_i}^{t_f} \dot{m}_L h_L \, dt$$

where $W_{comp/exp} = -\int P \, dV = -P_i(V_f - V_i)$ because the pressure in the control volume is constant. Because the flow through the large pipe is steady, it is reasonable to model the filling process as *uniform flow*. Thus $h_L$ is constant and the integral with mass flow rate becomes $h_L m_L$. Combining this information the energy balance becomes

$$m_f u_f - m_i u_i = -P_i(V_f - V_i) + h_L m_L$$

The unknowns in this equation are $m_L$, $m_f$, and $u_f$. However, the mass quantities are related by

$$m_L = m_f - m_i$$

Thus we actually have one equation with two unknowns $m_f$ and $u_f$.

The second required relationship in this case is found from a knowledge of the final volume. Note that $V_f = m_f v_f$. Since $V_f$ is given, the unknowns in this relationship are $m_f$ and $v_f$. Hence another unknown, $v_f$, is introduced. However, $u_f$ and $v_f$ are not independent variables. On the basis of the state principle, for a given pressure within the control volume only $u_f$ or $v_f$ can be varied independently, but not both. Thus a suitable equation of state which uniquely relates $u_f$ and $v_f$ for a given $P$ is the desired third relation which permits the two equations already presented to be satisfied.

(a) If the gas is assumed to behave ideally, then the energy balance developed above for adiabatic conditions takes on the form

$$-P(V_f - V_i) + h_L(m_f - m_i) = m_f c_v T_f - m_i c_v T_i$$

$$= m_f c_v \frac{PV_f}{m_f R} - m_i c_v T_i$$

$$= \frac{c_v PV_f}{R} - m_i c_v T_i$$

$$= \frac{PV_f}{k-1} - m_i c_v T_i$$

Hence by introducing the equation of state for an ideal gas and the relation $c_v = R/(k-1)$, we find that the resulting equation contains only the single unknown $m_f$. Before we substitute numerical values, it is convenient to calculate first the initial mass within the piston-cylinder assembly:

$$m_i = \frac{PV}{RT} = \frac{1 \text{ bar} \times 0.01 \text{ m}^3}{0.08314 \text{ bar·m}^3/(\text{kmol·K}) \times 313 \text{ K}} \times \frac{29 \text{ kg}}{\text{kmol}}$$

$$= 0.0111 \text{ kg}$$

$$= 11.1 \text{ g}$$

Note also that $h_L = c_p T_L$, and $c_p$, $c_v$, and $k$ values are in Table A-3. Substitution of known values into the energy equation then yields (by expressing each term in newton-meters or joules, and $m_f$ in grams)

$$-\left(10^5 \frac{\text{N}}{\text{m}^2}\right)(0.02 - 0.01) \text{ m}^3 + \left(1.01 \frac{\text{J}}{\text{g·K}}\right)(m_f - 11.1 \text{ g})(373 \text{ K})$$

$$= \frac{\left(10^5 \frac{\text{N}}{\text{m}^2}\right)(0.02 \text{ m}^3)}{1.4 - 1} - (11.1 \text{ g})\left(0.72 \frac{\text{J}}{\text{g·K}}\right)(313 \text{ K})$$

$$(-1000 \text{ N·m}) + \left(377 \frac{\text{J}}{\text{g}}\right) m_f - (4180 \text{ J}) = (5000 \text{ N·m}) - (2500 \text{ J})$$

$$m_f = \frac{(5000 - 2500 + 1000 - 4180) \text{ J}}{(377 \text{ J/g})} = 20.4 \text{ g}$$

Thus the mass entering the assembly during the filling process is

$$m_L = m_f - m_i = 20.4 - 11.1 = 9.3 \text{ g}$$

(b) The final equilibrium temperature is found from the ideal-gas relation:

$$T_f = \frac{PV_f}{m_f R} = \frac{1 \text{ bar} \times 0.02 \text{ m}^3}{20.4 \text{ g} \times 0.08314 \text{ bar·m}^3/(\text{kmol·K})} \times \frac{29 \text{ kg}}{\text{kmol}} \times \frac{1000 \text{ g}}{\text{kg}}$$

$$= 342 \text{ K}$$

$$= 69°C$$

Thus the temperature of the resulting mixture in the assembly increases nearly 30°C above the initial temperature.

## 5-9   SUMMARY

After selecting the control volume, the application of the conservation of mass and conservation of energy principles to a control volume (open system) begins with Eq. [5-1] and Eq. [5-15]

$$\frac{dm_{cv}}{dt} = \sum_{in} \dot{m}_i - \sum_{out} \dot{m}_e \qquad \text{[5-1]}$$

$$\frac{dE_{cv}}{dt} = \dot{Q} + \dot{W} + \sum_{in} \dot{m}_i \left( h + \frac{V^2}{2} + gz \right)_i - \sum_{out} \dot{m}_e \left( h + \frac{V^2}{2} + gz \right)_e$$

$$\text{[5-15]}$$

where the subscripts $i$ and $e$ indicate that the properties are evaluated at the inlet or outlet (exit) of the control volume. The only restriction in their use is that Eq. [5-15] assumes that the intensive properties are uniform across any flow boundary. These equations may also be used to study closed systems by recognizing that for a closed system the mass flow rate terms are identically zero.

The mass flow rate $\dot{m}$ and volume flow rate $\dot{V}$ at a control surface (system boundary) are defined in terms of the velocity $V_n$ normal to the surface, the surface area $A$, and the density $\rho$ of the fluid by the relations

$$\dot{m} = \int_{cs} \rho V_n \, dA \qquad \text{and} \qquad \dot{V} = \int_{cs} V_n \, dA$$

where the integral is over the control surface (cs) where the flow occurs. In most applications, the control surface is selected so that flow across the surface is normal to the surface and the normal velocity vector equals the local velocity, i.e. $V_n = V$. For many flows it is reasonable to model the flow as being *one-dimensional* and assume that all properties at the flow boundary are spatially uniform across the flow area. For one-dimensonal flow at a flow boundary, the mass flow rate and volume flow rate can be written as

$$\dot{m} = \rho A V_n = \frac{A V_n}{v} \qquad \text{and} \qquad \dot{V} = A V_n$$

where $\rho = 1/v$. For one-dimensional flow, the mass and volume flow rates can also be related by the algebraic equation $\dot{V} = \dot{m}/\rho = \dot{m}v$.

Open and closed systems often operate under conditions where the system properties and all transport rates of mass and energy are independent of time. Under these conditions, the system is modeled as *steady-state* and the mass and energy in the system remain constant. Thus the rate of change of mass and energy within the system are identically zero, i.e. $dm_{cv}/dt = 0$ and $dE_{cv}/dt = 0$.

Many practical devices, such as turbines, pumps, and compressors, can be modeled as steady-state systems with one inlet (state 1) and one exit (state 2). Under these conditions, the conservation of mass equation, Eq. [5-1], reduces to $\dot{m}_1 = \dot{m}_2$. If the flow is further modeled as one dimensional, the conservation of energy equation, Eq. [5-15], can be written as

$$q + w = (h_2 - h_1) + \left(\frac{V_2^2}{2} - \frac{V_1^2}{2}\right) + g(z_2 - z_1) = \Delta h + \Delta\left(\frac{V^2}{2}\right) + g\Delta z$$

[5-22]

where $q = \dot{Q}/\dot{m}$ and $w = \dot{W}/\dot{m}$.

In the analysis of transient flow systems, it is frequently useful to integrate Eqs. [5-1] and [5-15] over a time interval $\Delta t = t_2 - t_1$. When this is done the equations become

$$\Delta m_{cv} = \sum_{in}\left[\int_{t_1}^{t_2}\dot{m}_i\,dt\right] - \sum_{out}\left[\int_{t_1}^{t_2}\dot{m}_e\,dt\right] = \sum_{in}m_i - \sum_{out}m_e \quad \textbf{[5-25]}$$

$$\Delta E_{cv} = Q + W \qquad\qquad\qquad\qquad\qquad\qquad \textbf{[5-26]}$$

$$+ \sum_{in}\left[\int_{t_1}^{t_2}\left(h + \frac{V^2}{2} + gz\right)_i\dot{m}_i\,dt\right] - \sum_{out}\left[\int_{t_1}^{t_2}\left(h + \frac{V^2}{2} + gz\right)_e\dot{m}_e\,dt\right]$$

To simplify the application of these two equations, the models of *uniform state* for the system and of *uniform flow* for the flow boundaries are frequently employed.

## PROBLEMS

### GENERAL COMMENT QUESTIONS

5-1C. Explain in words the different physical mechanisms for the transfer of energy to a control volume.

5-2C. What are the three possible work transfers associated with the flow of a simple compressible fluid through a fixed-boundary, stationary control volume?

5-3C. Describe in words the concept of flow work, and how it leads to the usefulness of the enthalpy function in control-volume analyses.

5-4C. Describe in words what is required for a steady-state control volume.

5-5C. Describe the conditions necessary for the flow through a control volume to be steady.

5-6C. Explain the conditions necessary for uniform flow.

5-7C. For ideal-gas flow through an adiabatic nozzle, describe whether the temperature will increase or decrease.

5-8C. Explain how heat transfer out of a nozzle will affect the exit velocity for a given inlet state and final pressure.

5-9C. Describe for an adiabatic gas turbine which forms of energy are used to provide a work output.

5-10C. For an adiabatic steam turbine describe the relative values of the inlet and exit volume flow rates. Explain.

5-11C. Describe qualitatively whether you expect the temperature of (*a*) air and (*b*) steam to drop during a throttling process.

## CONTINUITY OF FLOW ANALYSIS

5-1. Water at 80°C and 75 bars enters a boiler tube of constant inside diameter of 2.0 cm at a mass flow rate of 0.760 kg/s. The water leaves the boiler tube at 440°C with a velocity of 90.5 m/s. Determine (*a*) the velocity at the tube inlet, in m/s, and (*b*) the pressure of the water at the tube exit, in bars.

5-2. Oxygen at 180 kPa and 47°C enters a bundle of 200 parallel pipes, each having an internal diameter of 2.00 cm.
   (*a*) Determine the velocity of the gas, in m/s, required as it enters the pipes to ensure a total mass flow rate of 5000 kg/h.
   (*b*) If the exit condition is 160 kPa and 12.5 m/s, determine the exit temperature, in degrees Celsius.

5-3. Air at 300°C and 400 kPa enters a round duct with a mass flow rate of 2.22 kg/s.
   (*a*) Determine the inlet duct diameter, in meters, required for an inlet velocity of 50 m/s.
   (*b*) If the air exits at 240°C and 380 kPa through a duct with a diameter of 20.0 cm, determine the exit velocity, in m/s.

5-4. Steam enters a steady-flow device at 160 bars and 560°C with a velocity of 80 m/s. At the exit the fluid is a saturated vapor at 2 bars, and the area is 1000 cm². If the mass flow rate is 1000 kg/min, determine (*a*) the inlet area, in square centimeters, and (*b*) the exit velocity, in m/s.

5-5E. Water at 150°F and 1000 psia enters a heat-exchanger tube with a constant inside diameter of 1.0 in at a mass flow rate of 1.674 lb$_m$/s. The water leaves the tube at 800°F with a velocity of 268.8 ft/s. Determine (*a*) the velocity at the tube inlet, in ft/s, and (*b*) the pressure of the water at the tube exit, in psia.

5-6E. Carbon monoxide at 20 psia and 140°F enters a bundle of 300 parallel pipes, each having an internal diameter of 1.00 in.
(a) Determine the velocity of the gas, in ft/s, required as it enters the pipes to ensure a total mass flow rate of 15,000 $lb_m$/h.
(b) If the exit condition is 18.5 psia and 23.0 ft/s, determine the exit temperature, in degrees Fahrenheit.

5-7E. Air at 560°F and 60 psia enters a circular duct with a mass flow rate of 4.40 $lb_m$/s.
(a) Determine the inlet duct diameter, in meters, required for an inlet velocity of 150 ft/s.
(b) If the air exits at 420°F and 55 psia through a duct with a diameter of 8.20 in, determine the exit velocity, in ft/s.

5-8. Refrigerant 134a enters a steady-state control volume at 5 bars and 100°C where the diameter of the inlet pipe is 0.10 m and the velocity is 7.0 m/s. At the exit of the control volume the pressure is 0.60 bar, and the quality of the fluid is 70 percent. If the exit diameter is 0.20 m, determine (a) the mass flow rate, in kg/s, and (b) the exit velocity, in m/s.

5-9. Steam enters a turbine at 60 bars and 500°C with a velocity of 100 m/s and leaves as a saturated vapor at 0.60 bar. The turbine inlet pipe has a diameter of 0.60 m, and the exit diameter is 4.5 m. Determine (a) the mass flow rate, in kg/h, and (b) the exit velocity, in m/s.

5-10. In a steady-flow device 0.50 kg/min of saturated refrigerant 134a vapor enters at 5 bars with a velocity of 4.0 m/s. The exit area is 0.90 $cm^2$, and the exit temperature and pressure are 60°C and 4.0 bars, respectively. Determine (a) the inlet area, in square centimeters, and (b) the exit velocity, in m/s.

5-11E. Refrigerant 134a enters a steady-state control volume at 80 psia and 200°F where the diameter of the inlet pipe is 0.30 ft and the velocity is 22 ft/s. At the exit of the control volume the pressure is 10 psia, and the quality of the fluid is 70 percent. If the exit diameter is 0.60 ft, determine (a) the mass flow rate, in $lb_m$/s, and (b) the exit velocity, in ft/s.

5-12E. Steam at 1000 psia and 1000°F enters a turbine with a velocity of 210 ft/s and leaves as a saturated vapor at 2.0 psia. The turbine inlet pipe has a diameter of 1.5 ft, and the exit diameter is 12 ft. Determine (a) the mass flow rate, in $lb_m$/h, and (b) the exit velocity, in ft/s.

5-13. Steam enters a turbine at 40 bars, 440°C, and 100 m/s through an area of 0.050 $m^2$. At the exit the fluid has a quality of 90 percent at 0.30 bar, and the velocity is 200 m/s. Determine (a) the mass flow rate, in kg/s, and (b) the exit area, in square meters.

5-14. Air initially at 0.25 MPa and 80°C flows through an area of 100 $cm^2$ at a rate of 50 kg/min. Downstream at another position the pressure is 0.35 MPa, the temperature is 100°C, and the velocity is 20 m/s.

Determine (*a*) the inlet velocity, in m/s, and (*b*) the outlet area, in square centimeters.

5-15. Air flows through a pipe with a variable cross section. At the pipe inlet the pressure is 6.0 bars, the temperature is 27°C, the area is 35.0 cm$^2$, and the velocity is 60 m/s. At the pipe exit the conditions are 5.0 bars and 50°C, and the cross-sectional area is 20.0 cm$^2$. Find (*a*) the mass flow rate, in kg/s, and (*b*) the exit velocity, in m/s.

5-16. Oxygen enters a control volume in steady state at 18.0 kg/min with a velocity of 20 m/s through an area of 0.0080 m$^2$. The initial temperature is 27°C. The fluid leaves the control volume at 50 m/s and 1.8 bars through an outlet area of 0.0030 m$^2$. Determine (*a*) the inlet pressure, in bars, and (*b*) the outlet temperature, in kelvins.

5-17. Carbon dioxide enters a steady-flow device at 27°C with a velocity of 25 m/s through an area of 4800 cm$^2$. At the exit of the device, the pressure and temperature are 0.14 MPa and 47°C, respectively, and the gas moves with a velocity of 9 m/s through an area of 7500 cm$^2$. Determine (*a*) the mass flow rate, in kg/s, and (*b*) the inlet pressure, in megapascals. Assume ideal-gas behavior.

5-18E. Air at a rate of 100 lb$_m$/min enters a duct through an area of 0.10 ft$^2$ at 30 psia and 140°F. Downstream at another position the state is 50 psia and 140°F, and the velocity is 50 ft/s. Determine (*a*) the inlet velocity, in ft/s, and (*b*) the outlet area, in square feet.

5-19E. At a pipe inlet the state of air is 100 psia, 80°F, 175 ft/s, and the flow area is 6.0 in$^2$. At the pipe exit the conditions are 80 psia and 120°C, and the cross-sectional area is 4.0 in$^2$. Find (*a*) the mass flow rate, in lb$_m$/s, and (*b*) the exit velocity, in ft/s.

5-20. Water at 20°C and 0.20 MPa flows through a garden hose with an inside diameter of 2.50 cm and out of a nozzle with an exit diameter of 0.60 cm. The exit velocity is 6.0 m/s. Determine (*a*) the mass flow rate, in kg/s, and (*b*) the velocity in the hose, in m/s.

5-21. Liquid water at 15°C flows down a long pipe. At position 1 in the pipe the internal diameter is 15 cm, and the velocity is 0.9 m/s. At position 2 downstream from position 1 the velocity is 3.6 m/s. Determine (*a*) the mass flow rate, in kg/min, and (*b*) the internal pipe diameter at position 2, in centimeters.

5-22E. Water enters a garden hose with an inside diameter of 1.0 in at 60°F and 20 psia and leaves at 20 ft/s through a nozzle with an exit diameter of 0.25 in. Determine (*a*) the mass flow rate, in lb$_m$/s, and (*b*) the velocity in the hose, in ft/s.

5-23E. Liquid water at 40°F and 2.5 ft/s enters a long pipe with an initial internal diameter of 12 in (position 1). At position 2 downstream from position 1 the internal diameter is 8 in. Determine (*a*) the mass flow rate, in lb$_m$/h, and (*b*) the fluid velocity at position 2, in ft/s.

5-24. Air enters a fan with a 0.7-m diameter at 22°C and 99 kPa. The air is discharged at 0.6 m³/s at 24°C and 102 kPa. Determine (a) the mass flow rate, in kg/s, (b) the inlet volumetric flow rate, in m³/s, and (c) the inlet and exit velocities, in m/s.

5-25. Refrigerant 134a enters a compressor at 2.4 bars and 0°C at a mass flow rate of 2 kg/min. (a) If the inlet velocity is 10 m/s, determine the inlet tubing diameter, in centimeters. (b) If the exit state is 50°C and 8 bars, and the outlet tubing is the same diameter as the inlet tubing, determine the exit velocity, in m/s.

5-26E. Refrigerant 134a enters a compressor at 40 psia and 40°F at a mass flow rate of 4 lb$_m$/min. (a) If the inlet velocity is 30 ft/s, determine the inlet tubing diameter, in inches. (b) If the exit state is 140°F and 160 psia, and the outlet tubing is the same diameter as the inlet tubing, determine the exit velocity, in ft/s.

5-27. An oil with a specific gravity of 0.90 enters a pump with a volumetric flow rate of 0.1 m³/min through a 10-cm pipe at 20°C. The diameter of the exit pipe is 7 cm, and the flow is isothermal. Determine (a) the mass flow rate in kg/s, (b) the inlet velocity in m/s, and (c) the exit velocity, in m/s.

5-28. Liquid water at 20°C and 1 bar enters a pump with a volumetric flow rate of 0.2 m³/min through a 15-cm pipe. Upon leaving the pump the liquid at 20°C is split into two flow streams which pass through exit pipes with diameters of 5 and 7 cm. The mass flow rate in the 5-cm pipe is 2 kg/s. Determine the velocity in each of the exit pipes, in m/s.

## Nozzles and Diffusers

5-29. Air enters a nozzle at 1.8 bars, 67°C, and 48 m/s. At the outlet the pressure is 1 bar and the velocity is six times its initial value. If the inlet area is 100 cm², determine (a) the outlet temperature, in degrees Celsius, and (b) the exit area of the adiabatic nozzle, in square centimeters.

5-30. Monatomic argon flows through an adiabatic nozzle at 2 kg/s. The inlet state is 500 K and 5 bars, while the exit state is 400 K and 1 bar. If the inlet to exit area ratio is 2, determine the exit velocity, in m/s.

5-31. Air enters a nozzle at 5 bars, 320 K through an area of 0.1 m². It exits at 1 bar, 300 K through a duct area of 0.047 m². Determine (a) the exit velocity, in m/s, and (b) the mass flow rate in kg/s.

5-32. Nitrogen gas enters an insulated nozzle at 200 kPa with a negligible velocity. At the exit of the nozzle the state of the fluid is 120 kPa and 27°C, and the flow area is 10.0 cm³. If the mass flow rate is

0.20 kg/s, determine (a) the exit velocity, in m/s, and (b) the change in temperature, in degrees Celsius.

5-33. Refrigerant 134a enters an adiabatic nozzle at 5 bars and 90 m/s. At the exit the fluid is a saturated vapor at 3.2 bars and has a velocity of 177 m/s. Determine (a) the inlet temperature, in degrees Celsius, and (b) the mass flow rate, in kg/s, if the exit area is 6.0 cm$^2$.

5-34. Steam enters a nozzle at 30 bars and 320°C and leaves at 15 bars with a velocity of 535 m/s. The mass flow rate is 8000 kg/h. Neglecting the inlet velocity and considering adiabatic flow, compute (a) the exit enthalpy, in kJ/kg, (b) the exit temperature, in degrees Celsius, and (c) the nozzle exit area, in square centimeters.

5-35. Air is admitted to an adiabatic nozzle at 3 bars, 200°C, and 50 m/s. The exit conditions are 2 bars and 150°C. Determine the ratio of the exit area to the entrance area $A_2/A_1$.

5-36. An adiabatic diffuser is employed to reduce the velocity of an air stream from 250 to 40 m/s. The inlet conditions are 0.1 MPa and 400°C. Determine the required outlet area, in square meters, if the mass flow rate is 7 kg/s and the final pressure is 0.12 MPa.

5-37. Refrigerant 134a enters an adiabatic diffuser as a saturated vapor at 26°C and with a velocity of 95 m/s. At the exit the pressure and temperature are 7 bars and 30°C, respectively. If the exit area is 50 cm$^2$, determine (a) the exit velocity, in m/s, and (b) the mass flow rate, in kg/s.

5-38E. The inlet state of refrigerant 134a to an adiabatic nozzle is 80 psia and 100 ft/s. At the exit the fluid is a saturated vapor at 60 psia with a velocity of 535 ft/s. Determine (a) the inlet temperature, in degrees Fahrenheit, and (b) the mass flow rate, in lb$_m$/s, if the exit area is 0.016 ft$^2$.

5-39E. The inlet state to a steam nozzle is 400 psia and 600°F. The steam leaves at 250 psia with a velocity of 1475 m/s, and the mass flow rate is 18,000 lb$_m$/h. Neglecting the inlet velocity and considering adiabatic flow, compute (a) the exit enthalpy, in Btu/lb$_m$, (b) the exit temperature, in degrees Fahrenheit, and (c) the nozzle exit area, in square feet.

5-40E. Air expands through a nozzle from 25 psia, 200°F, and 100 ft/s to 15 psia and 80°F. The heat loss is 2.0 Btu/lb$_m$. Determine (a) the exit velocity, in ft/s, and (b) the ratio of the inlet area to the exit area.

5-41. Water in a compressed-liquid state flows through a well-insulated nozzle under steady-state conditions. At the inlet the pressure, temperature, and velocity are 4.0 bars, 15.0°C, and 3 m/s, respectively, and the area is 10.0 cm$^2$. At the exit the area is 2.50 cm$^2$, and the temperature is 15.05°C. Consider the water to be incompressible, with $v = 1.001 \times 10^{-3}$ m$^3$/kg and $c_p = 4.19$ kJ/kg·K. Determine (a) the mass flow rate, in kg/s, (b) the exit velocity, in m/s, and (c) the exit pressure, in bars.

5-42. Water in a compressed-liquid state flows through a well-insulated nozzle under steady-state conditions. At the inlet the pressure, temperature, and velocity are 3.2 bars, 20.0°C, and 4 m/s, respectively, and the area is 16.0 cm². At the exit the area is 4.00 cm², and the pressure is 1.50 bars. Consider the water to be incompressible, with $v = 1.002 \times 10^{-3}$ m³/kg and $c_p = 4.19$ kJ/kg·K. Determine (a) the mass flow rate, in kg/s, (b) the exit velocity, in m/s, and (c) the change in temperature, in degrees Celsius.

5-43. Air enters a diffuser at 0.7 bar, 57°C, with a velocity of 200 m/s. At the outlet, where the area is 20 percent greater than at the inlet, the pressure is 1.0 bar. Determine the outlet temperature, in degrees Celsius, and the outlet velocity, in m/s, if (a) the process is adiabatic and (b) the fluid loses 40 kJ/kg in heat transfer as it passes through.

5-44. Steam enters an adiabatic diffuser as a saturated vapor at 110°C with a velocity of 220 m/s. At the exit the pressure and temperature are 1.5 bars and 120°C, respectively. If the exit area is 50 cm², determine (a) the exit velocity, in m/s, and (b) the mass flow rate, in kg/s.

5-45. Refrigerant 134a enters an adiabatic diffuser at 1.8 bars and 20°C with a velocity of 140 m/s. The inlet area is 10.0 cm². At the outlet the conditions are 2.0 bars and 50 m/s. Determine (a) the mass flow rate, in kg/s, (b) the exit enthalpy, in kJ/kg, (c) the exit temperature, in degrees Celsius, and (d) the exit area, in square centimeters.

5-46E. Water in a compressed-liquid state enters a well-insulated nozzle under steady-state conditions of 60 psia, 50.0°F, and 10 ft/s, and the inlet area is 2.0 in². At the exit the area is 0.50 in² and the temperature is 50.10°F. Consider the water to be incompressible, with $c_p = 1.00$ Btu/lbm·°R. Determine (a) the mass flow rate, in lbm/s, (b) the exit velocity, in ft/s, and (c) the exit pressure, in psia.

5-47E. Water in a compressed-liquid state enters a well-insulated nozzle under steady-state conditions of 50 psia, 70.0°F, and 12 ft/s, respectively, and the inlet area is 3.0 in². At the exit the area is 0.75 in², and the pressure is 20 psia. Consider the water to be incompressible, with $v = 0.0160$ ft³/lbm and $c_p = 1.00$ Btu/lbm·°R. Determine (a) the mass flow rate, in lbm/s, (b) the exit velocity, in ft/s, and (c) the change in temperature, in degrees Fahrenheit.

5-48E. An adiabatic diffuser reduces the velocity of a nitrogen stream from 714 to 120 ft/s. The inlet conditions are 15 psia and 160°F. Determine the required exit area, in square inches, if the mass flow rate is 15 lbm/s and the final pressure is 17.7 psia.

5-49E. Air enters a diffuser of a jet engine at 10 psia, 140°F, with a velocity of 800 m/s. At the outlet, where the area is 28 percent greater than at the inlet, the pressure is 13 psia. Determine the outlet temperature, in degrees Fahrenheit, and the outlet velocity, in ft/s, if the fluid loses 2 Btu/lb$_m$ due to heat transfer.

5-50E. Steam as a saturated vapor at 200°F enters an adiabatic diffuser with a velocity of 1100 ft/s. At the exit the pressure and temperature are 14.7 psia and 250°F, respectively. If the exit area is 8.0 in$^2$, determine (a) the exit velocity, in ft/s, and (b) the mass flow rate, in lb$_m$/s.

5-51E. Refrigerant 134a at 180 psia and 220°F enters an adiabatic diffuser with a velocity of 495 ft/s. The inlet area is 4.0 in$^2$. At the outlet the conditions are 200 psia and 103 ft/s. Determine (a) the mass flow rate, in lb$_m$/s, (b) the exit enthalpy, in Btu/lb$_m$, (c) the exit temperature, in degrees Fahrenheit, and (d) the exit area, in square inches.

5-52. Air enters the diffuser of a high-performance aircraft at 57 mbar, −53°C, and 880 m/s through an area of 0.52 m$^2$. The exit velocity is 18 m/s and the area is 0.93 m$^2$. Find (a) the exit air temperature, in kelvins, (b) the mass flow rate, in kg/s, and (c) the exit pressure, in bars.

## TURBINES

5-53. An adiabatic steam turbine operates with inlet conditions of 120 bars, 480°C, and 100 m/s, and the flow is through an area of 100 cm$^2$. At the exit the quality is 90 percent at 1 bar, and the velocity is 50 m/s. Determine (a) the change in kinetic energy, in kJ/kg, (b) the shaft work, in kJ/kg, (c) the mass flow rate, in kg/s, (d) the power output, in kilowatts, and (e) the exit area, in square meters.

5-54. A small gas turbine operating on hydrogen delivers 28 kW. The gas enters the steady-flow device at 75 m/s through an area of 0.0020 m$^2$. The inlet state is 240 kPa and 500 K. The exit state is 100 kPa and 380 K, and the exit area is 0.00160 m$^2$. Compute (a) the final velocity, in m/s, and (b) the heat-transfer rate, in kJ/min.

5-55. Steam flows steadily through a turbine at 20,000 kg/h, entering at 40 bars, 440°C, and leaving at 0.20 bar with 90 percent quality. A heat loss of 20 kJ/kg occurs. The inlet pipe has an 8-cm diameter, and the exhaust section is rectangular with dimensions of 0.9 by 1.1 m. Calculate (a) the kinetic-energy change and (b) the power output, both answers in kilowatts.

5-56. Air enters a turbine in steady flow at 6 bars, 740 K, and 120 m/s. The exit conditions are 1 bar, 450 K, and 220 m/s. A heat loss of 15 kJ/kg occurs, and the inlet area is 4.91 cm$^2$. Determine (a) the kinetic-energy change, in kJ/kg, (b) the power output, in kilowatts, and (c) the ratio of the inlet- to outlet-pipe diameters.

5-57. An air turbine with an output of 240 kW has inlet conditions of 840 K, 1.0 MPa, and 18 m/s. The outlet state is 420 K and 0.1 MPa.

Both the inlet and outlet ducts have diameters of 0.10 m. Determine (a) the enthalpy change, in kJ/kg, (b) the kinetic-energy change, in kJ/kg, (c) the mass flow rate, in kg/min, and (d) the rate of heat transfer, in kJ/min.

5-58E. A steam turbine operates with inlet conditions of 1000 psia, 800°F, and 300 ft/s, and the flow is through an area of 20 in$^2$. At the exit the conditions are 500°F, 200 psia, and the velocity is 140 ft/s. The rate of heat loss is 17,000 Btu/min. Determine (a) the change in kinetic energy, in Btu/lb$_m$, (b) the mass flow rate, in lb$_m$/min, (c) the shaft work, in Btu/lb$_m$, (d) the power output, in horsepower, and (e) the exit area, in square inches.

5-59E. Helium is used to operate an 18-hp gas turbine. The gas enters the steady-flow device at 220 ft/s through an area of 0.020 ft$^2$. The inlet state is 40 psia and 440°F. The exit state is 15 psia and 220°F, and the exit area is 0.0270 ft$^2$. Compute (a) the final velocity, in ft/s, and (b) the heat-transfer rate, in Btu/min.

5-60E. Steam enters a steady-flow turbine at 700 psia, 700°F, and 300,000 lb$_m$/h. The steam leaves at 1 psia as a saturated vapor, and a heat loss of 4 Btu/lb$_m$ occurs. The inlet pipe has a 12-in diameter, and the exhaust section is rectangular with dimensions of 16 by 7 ft. Calculate (a) the kinetic-energy change, in Btu/lb$_m$, and (b) the power output, in horsepower.

5-61E. Air enters a turbine in steady flow at 90 psia, 940°F, and 480 ft/s. The exit conditions are 15 psia, 440°F, and 240 ft/s. A heat loss of 6 Btu/lb$_m$ occurs, and the inlet area is 31.5 in$^2$. Determine (a) the kinetic-energy change, in Btu/lb$_m$, (b) the power output, in horsepower, and (c) the ratio of the inlet- to outlet-pipe diameters.

5-62. Refrigerant 134a enters a steady-flow turbine at 10 bars, 70°C, and 35 m/s through an inlet area of 32 cm$^2$. At the exit the conditions are 1 bar, 0°C, and 75 m/s. The power developed is 200 kW. Determine (a) the mass flow rate, in kg/s, (b) the rate of any heat transfer, in kJ/min, and (c) the exit area, in square centimeters.

5-63. Steam enters a turbine with inlet conditions of 80 bars, 440°C, and 49 m/s. At the exit the state is 0.2 bar, 90 percent quality, and 80 m/s. The power output is 18,000 kW and the inlet area is 0.0165 m$^2$. Find the rate of heat transfer, in kJ/min, and its direction.

5-64. Steam enters an insulated turbine at 16 MPa, 480°C at a rate of $2.2 \times 10^6$ kg/h. Twenty percent of the steam is extracted from the turbine at 1 MPa and 280°C. The remaining steam expands further until it leaves the turbine as a saturated vapor at 0.008 MPa. The velocity of the extracted steam is 30 m/s, and kinetic- and potential-energy effects are negligible. Determine (a) the diameter of the extraction duct, in m, and (b) the power developed by the turbine, in kilowatts.

COMPRESSORS AND FANS

5-65. Air enters a compressor at 22°C and 1 bar through an area of 0.030 m$^2$ and with a volume flow rate of 4.5 m$^3$/s. At the exit the state is 400 K, 2.4 bars, and 70 m/s. The rate of heat loss is 900 kJ/min. Determine (a) the inlet velocity, in m/s, (b) the mass flow rate, in kg/s, and (c) the power input, in kilowatts.

5-66. Refrigerant 134a enters the compressor of a commercial air conditioner at 0°C and 2 bars, and exits at 60°C and 12 bars. The entering volumetric flow rate is 0.5 m$^3$/min and the heat-transfer loss is 5 kJ/kg. Calculate the dollar cost of operation of the motor driving the compressor for 30 days. Electricity cost $0.12/kW·h, and the motor is on one-fifth of the time.

5-67. A hydrogen compressor has a heat-loss rate of 35 kW. The inlet conditions are 320 K, 0.2 MPa, and 100 m/s. Both the inlet and outlet ducts have diameters of 0.10 m. The exit state is 1.2 MPa and 520 K. Determine (a) the kinetic-energy change, in kJ/kg, (b) the mass flow rate, in kg/min, and (c) the shaft power, in kilowatts, using data from Table A-11.

5-68. Carbon dioxide is compressed from 0.1 MPa, 310 K to 0.5 MPa, 430 K. The required volume flow rate at inlet conditions is 30 m$^3$/min. The kinetic-energy change is negligible, but a heat loss of 4.0 kJ/kg occurs. Determine the required power input, in kilowatts, using data from Table A-9.

5-69. An air compressor handling 300 m$^3$/min at the inlet increases the pressure from 1.0 to 2.3 bars, and heat is removed at a rate of 1700 kJ/min. The inlet temperature and area are 17°C and 280 cm$^2$, respectively, and these quantities for the exit are 137°C and 200 cm$^2$. Find (a) the inlet and exit gas velocities, in m/s, and (b) the required power input, in kilowatts.

5-70. A water-cooled compressor changes the state of refrigerant 134a from a saturated vapor at 1.0 bar to a pressure of 8 bars. The inlet area is 5 cm$^2$, the fluid mass rate is 0.9 kg/min, and the cooling water removes heat at a rate of 140 kJ/min. If the power input is 3.0 kW, determine (a) the exit temperature, in degrees Celsius, and (b) the inlet velocity, in m/s.

5-71. A compressor in a large commercial air conditioner steadily inducts 2000 kg/h of refrigerant 134a at 0.6 bar and 0°C through an inlet pipe with an inside diameter of 7 cm. The gas is discharged at 7 bars and 80°C through a 2-cm-diameter pipe. During the process, a heat loss of 40,000 kJ/h occurs to the surroundings. Determine (a) the inlet and exit velocities, in m/s, and (b) the cost of operating the motor driving the compressor for one day. Electricity costs $0.13/kW·h, and the motor runs one-fourth of the time.

5-72. Air at a rate of 18 kg/min is compressed from 1 bar and 290 K to 5 bars and 450 K. The inlet area is 0.025 m$^2$, the exit area is 0.0025 m$^2$, and a heat loss of 50 kJ/kg occurs. Use specific-heat data at 400 K for property data. Determine (a) the inlet and exit velocities, in m/s, (b) the volume flow rate at the inlet, in m$^3$/min, and (c) the required power input, in kilowatts.

5-73. A compressor is supplied with 50 kg/h of saturated water vapor at 0.04 bar and discharges the fluid at 1.5 bar and 120°C. The power input required is 2.4 kW, and the inlet area is 40 cm$^2$. Determine (a) the rate of heat transfer to or from the fluid, in kJ/min, and (b) the velocity at the inlet, in m/s.

5-74E. The inlet conditions to a hydrogen compressor are 560°R, 2.0 atm, and 250 ft/s. Both the inlet and outlet ducts have diameters of 0.30 ft. The fluid exit state is 12 atm and 940°R, and the heat-loss rate is 2000 Btu/min. Determine (a) the kinetic-energy change, in Btu/lb$_m$, (b) the mass flow rate, in lb$_m$/min, and (c) the shaft power, in horsepower, using data from Table A-11E.

5-75E. An air compressor increases the pressure from 15 to 35 psia, and heat is removed at a rate of 750 Btu/min. The inlet temperature and area are 70°F and 0.96 ft$^2$, respectively, and these quantities for the exit are 280°F and 0.90 ft$^2$. If the inlet volume rate is 10,000 ft$^3$/min, find (a) the inlet and exit gas velocities, in ft/s, and (b) the required power input, in horsepower.

5-76E. Refrigerant 134a changes from a saturated vapor at 20 psia to a final pressure of 120 psia in a water-cooled compressor. The mass rate is 2.0 lb$_m$/min, and the heat-removal rate is 129.4 Btu/min. If the power input is 4.0 hp and the inlet area is 0.5 in$^2$, determine (a) the exit temperature, in degrees Fahrenheit, and (b) the inlet velocity in feet per second.

5-77. A fan receives air at 970 mbar, 20.0°C, and 3 m/s, and discharges it at 1020 mbar, 21.6°C, and 18 m/s. If the flow is adiabatic and 50 m$^3$/min enters, determine the power input in kilowatts.

## HEAT EXCHANGERS

5-78. Liquid water at 5 bars and 140°C flows through the inside of 20 pipes of a shell-and-tube heat exchanger at a total rate of 240 kg/min, and it leaves at 4.8 bars and 60°C. The water is cooled by passing air initially at 110 kPa and 25°C at an inlet volume rate of 1000 m$^3$/min through the shell of the heat exchanger. The exit pressure of the air is 105 kPa. Determine (a) the exit air temperature, in degrees Celsius, (b) the inlet area for the airflow, in square meters, if the air inlet velocity is 25 m/s, and (c) the inlet water velocity, in m/s, if each pipe has a diameter of 2.0 cm.

5-79. Refrigerant 134a at 10 bars, 38°C, flows at 10.0 kg/s through a heat exchanger, leaving at 9.0 bars and 80°C. It exchanges heat with a stream of steam entering at 1 bar and 200°C.
   (a) If the steam leaves the heat exchanger as a saturated vapor at 1 bar, determine its mass flow rate, in kg/s.
   (b) For the same change of state, the refrigerant flow rate is cut to 5.0 kg/s. If the steam flow rate remains the same, determine the exit temperature of the steam, in degrees Celsius.

5-80. Dry air enters an air-conditioning system at 30°C and 0.11 MPa at a volume flow rate of 1.20 m³/s. The air is cooled by exchanging heat with a stream of refrigerant 134a which enters the heat exchanger at −12°C and a quality of 20 percent. Assume the heat transfer takes place at constant pressure for both flow streams. The refrigerant 134a leaves as a saturated vapor, and 22 kJ/s of heat is removed from the air. Find (a) the flow rate of refrigerant 134a required, in kg/s, and (b) the temperature of the air leaving the heat exchanger, in degrees Celsius.

5-81. Steam is condensed on the outside of a heat-exchanger tube by passing air through the inside of the tube. The air enters at 1.20 bars, 20°C, and 10 m/s and exits at 80°C. The steam enters at 3 bars, 200°C, at a mass flow rate of 5 kg/min, and leaves as a saturated liquid. Find (a) the mass flow rate of air required, in kg/min, and (b) the inlet area of the pipe required for the air stream, in square meters.

5-82. Steam enters the tubes of a shell-and-tube heat exchanger at 20 bars and 400°C, and leaves the device at 200°C. The steam is cooled by passing air initially at 1.5 bars and 27°C on the outside of the tubes. The volume flow rate of the entering air is 50 m³/min, and the exit-air temperature is 277°C. The tube cross-sectional area is 4.0 cm², and pressure drops are negligible. Determine (a) the rate of heat transfer between the two fluids, in kJ/min, and (b) the number of tubes required if the velocity of the entering steam is 2.0 m/s.

5-83. Air enters a heat exchanger at 27°C and 130 kPa, and leaves at 227°C and 120 kPa. Steam enters the heat exchanger at 600°C and 1500 kPa. At the exit the steam passes through a throttling device, which changes the exit state to 100 kPa and 80°C. Changes in kinetic and potential energy are negligible. The mass flow rate of air is 165 kg/s. Under these conditions, (a) find the flow rate of steam, in kg/s. (b) Now, all conditions for the air remain the same, as well as the pressures for the water side and the steam mass flow rate found in part a. Find the maximum inlet steam temperature permitted subject to the condition that the water leaving the throttling device is in a liquid state, in degrees Celsius.

5-84E. Liquid water flows through the inside of 28 pipes of a shell-and-tube heat exchanger at a total rate of 250 lb_m/min. The liquid enters at 80 psia and 250°F and leaves at 70 psia and 100°F. The water is cooled by passing air initially at 14.8 psia and 80°F at an inlet volume rate of 17,550 ft³/min through the shell of the heat exchanger. The exit pressure of the air is 14.6 psia. Determine (*a*) the exit-air temperature, in degrees Fahrenheit, (*b*) the inlet area for the airflow, in square feet, if the air inlet velocity is 50 ft/s, and (*c*) the inlet water velocity, in ft/s, if each pipe has a diameter of 0.50 in.

5-85E. Air at a volume flow rate of 30 ft³/s enters an air-conditioning system at 100°F and 14.7 psia. The air is cooled by exchanging heat with a stream of refrigerant 134a which enters the heat exchanger at −20°F and a quality of 10 percent. Assume the heat transfer takes place at constant pressure for both flow streams. The refrigerant 134a leaves as a saturated vapor, and 20 Btu/s of heat is transferred from the air to the refrigerant. Find (*a*) the flow rate of refrigerant 134a required, in lb$_m$/s, and (*b*) the temperature of the air leaving the heat exchanger, in degrees Fahrenheit.

5-86. Air is heated by being passed over tubes containing steam. The steam enters the heat-exchanger tubes at 300 kPa and 200°C at a rate of 8 kg/min, and it leaves at 180 kPa and 100°C. The air enters at 100 kPa and 25°C, and leaves at 47°C. The inlet steam velocity is 5 m/s and the inside tube diameter is 2 cm. Determine (*a*) the rate of heat transfer to the air, in kW, (*b*) the volume flow rate of air at the inlet, in m³/s, and (*c*) the number of tubes required.

5-87. Steam enters a heat exchanger as a saturated vapor at 1.5 bars and leaves at 1 bar with a quality of 10 percent. Air at 1.5 kg/s enters at 1.1 bar, 20°C, and leaves at 1 bar and 80°C. Ten percent of the heat leaving the steam is lost to the surroundings. Determine (*a*) the area ratio of the steam outlet to inlet so that the kinetic-energy change of the steam is zero and (*b*) the mass flow rate of the steam in kg/min.

## MIXING PROCESSES

5-88. Water enters a mixing chamber from two sources. One source delivers steam of 90 percent quality at a rate of 2000 kg/h. The second source delivers steam at 280°C and a rate of 2790 kg/h. The mixing process is adiabatic and at a constant pressure of 10 bars, and the exit velocity is 8.9 m/s. Determine (*a*) the temperature at the exit in degrees Celsius and (*b*) the diameter of the exit pipe, in centimeters.

5-89. Liquid water and steam are mixed in an open feedwater heater. The cold water enters at 50°C and 5 bars at a rate of 100 kg/min. The steam enters at 5 bars and 180°C, and the resulting mixture leaves as a saturated liquid at 5 bars with a velocity of 0.60 m/s. The heater operates with a heat loss to the environment of 151,920 kJ/min. Determine (*a*) the mass flow rate of the entering steam, in kg/min, and (*b*) the diameter of the heater exit pipe, in centimeters.

5-90. An open feedwater heater operates at 7 bars. Compressed liquid water at 35°C enters through one inlet, superheated steam enters through another inlet, and the mixture leaves the device as a saturated liquid. Determine the temperature, in degrees Celsius, of the steam entering if the ratio of the mass flow rate of compressed liquid to superheated steam is 4.52:1.

5-91. Liquid water is heated in an insulated chamber by mixing it with steam. The water enters at a steady rate of 100 kg/min at 20°C and 3 bars. The steam enters at 320°C and 3 bars, and the mixture leaves the mixing chamber at 90°C and 3 bars.
(*a*) Determine the rate of steam required, in kg/min.
(*b*) If the exit area is 25 cm$^2$, determine the exit velocity, in m/s.

5-92E. Liquid water and steam are mixed in an open feedwater heater which operates at 60 psia with a heat loss of 144,000 Btu/h to the environment. The cold water enters at 100°F at a rate of 200 lb$_m$/min, while the steam enters at 300°F. The resulting mixture leaves as a saturated liquid at 60 psia with a velocity of 2.0 ft/s. Determine (*a*) the mass flow rate of the entering steam, in lb$_m$/min, and (*b*) the diameter of the heater exit pipe, in feet.

5-93E. Steam and liquid water flow separately into an insulated mixing chamber. The liquid enters at a steady rate of 200 lb$_m$/min at 60°F and 60 psia. The steam enters at 600°F and 60 psia, and the mixture leaves the mixing chamber at 200°F and 58 psia.
(*a*) Determine the rate of inlet steam required, in lb$_m$/min.
(*b*) If the exit area is 4.0 $in^2$, determine the exit velocity, in ft/s.

## THROTTLING PROCESSES

5-94. Determine the temperature change when water substance is throttled (*a*) from 10 bars, 280°C, to 1 bar and (*b*) from 50 bars, 100°C, to 25 bars. (*c*) Determine the answer for the case where the substance behaves as an ideal gas.

5-95. Refrigerant 134a is throttled from (*a*) 44°C and 12 bars to 2.8 bars, and (*b*) a saturated vapor at 16 bars to 1 bar. In each case find the specific volume in the final state, in m$^3$/kg.

5-96. Nitrogen is throttled from (*a*) a saturated liquid at 20 bars to a pressure of 4.0 bars and (*b*) 100 bars and 200 K to a pressure of 5.0 bars. Determine the temperature, in kelvins, and the specific volume, in L/kg, at the final state.

5-97. Steam flows through a throttling valve under the following conditions: (*a*) enters at 30 bars, 240°C, and exhausts at 7 bars and (*b*) enters as saturated vapor at 8 bars and exits at 3 bars. Determine the final temperature downstream from the valve at the stated exit pressure, in degrees Celsius.

5-98. Steam is throttled from (a) 5 bars to 1 bar and 100°C and (b) 10 bars to 0.7 bar and 100°C. Determine (1) the quality of the steam entering the throttling process and (2) the ratio of exit area to inlet area from the device, if the inlet and exit velocities are essentially the same.

5-99. Refrigerant 134a is throttled from (a) a saturated liquid at 32°C to a temperature of −20°C and (b) a state of 30°C and 9 bars to a temperature of −12°C. Determine (1) the final pressure, in bars, and (2) the final specific volume, in $m^3/kg$.

5-100. Refrigerant 134a is throttled from (a) 48°C and 14 bars to 2.4 bars, and (b) a saturated vapor at 12 bars to a pressure of 2 bars. Find the final specific volume in each case, in $m^3/kg$.

5-101E. Nitrogen is throttled from (a) a saturated liquid at 200°R to a temperature of 140°R. Find the final pressure, in psia, and the final specific volume, in $ft^3/lb_m$. (b) Nitrogen is throttled from 2000 psia and 350°R to 200 psia. Determine the temperature, in degrees Fahrenheit, and the specific volume, in $ft^3/lb_m$, at the final state.

5-102E. Steam is throttled from (a) 80 psia to 10 psia and 200°F and (b) 250 psia to 20 psia and 250°F. Determine (1) the quality of the steam entering the throttling process and (2) the ratio of exit area to inlet area from the device, if the inlet and exit velocities are essentially the same.

5-103E. Refrigerant 134a is throttled from (a) a saturated liquid at 60°F to a temperature of −20°F and (b) a state of 100°F and 120 psia to a temperature of 80°F. Determine (1) the final pressure, in psia, and (2) the final specific volume, in $ft^3/lb_m$.

5-104. Refrigerant 134a at 5 bars and 4°C enters a mixing chamber at a steady rate of 2 kg/s. Another source of refrigerant 134a at 6 bars and 50°C is first throttled before it enters the mixing chamber at 5 bars. The mixture leaves the chamber as a saturated liquid at 5 bars.
   (a) If heat transfer to the chamber from an outside source occurs at a rate of 4 kJ/s, determine the mass flow rate of refrigerant gas entering the chamber, in kg/s.
   (b) If the exit pipe has a diameter of 4.5 cm, determine the exit velocity, in m/s.

5-105. Steam at 30 bars and 280°C is throttled to 20 bars before it enters a mixing chamber. Liquid water at 25 bars and 180°C is throttled to 20 bars before entering the chamber. After mixing within the chamber occurs, a stream of saturated vapor leaves at 20 bars. The chamber is heavily insulated, and kinetic- and potential-energy effects are negligible. For a inlet steam flow of 20,000 kg/h, determine the mass flow rate of liquid water required, in kg/h.

5-106E. Refrigerant 134a enters a mixing chamber at 40 psia, $-10°F$, and a steady rate of 1 lb$_m$/s. Another source of refrigerant 134a at 100 psia and 200°F is first throttled before it enters the mixing chamber at 40 psia. The mixture leaves the chamber as a saturated liquid at 40 psia.

   (*a*) If heat transfer to the chamber from an outside source occurs at a rate of 3 Btu/s, determine the mass flow rate of refrigerant gas entering the chamber, in lb$_m$/s.

   (*b*) If the exit pipe has a diameter of 1.5 in, determine the exit velocity, in ft/s.

## PUMPS AND PIPE FLOW

5-107. An insulated pipe with an 8-cm diameter carries water from a water reservoir to a pump that is 4.0 m above the reservoir level. The water enters the 8-cm pipe at 95 kPa, 15°C, and 3 m/s. The insulated pump discharges into a 5-cm pipe at 480 kPa and 7.5 m/s. The water then flows 20.0 m upward through the 5-cm pipe. A nozzle at the end of the 5-cm pipe reduces the flow diameter, and water discharges from the nozzle at 100 kPa. Local gravity is 9.80 m/s$^2$, and the entire process is isothermal and adiabatic. Determine (*a*) the power required to drive the pump, in kilowatts, and (*b*) the nozzle exit diameter, in centimeters.

5-108. Oil, with a density of 0.85 g/cm$^3$, is stored 700 m below ground level in a cavern at 30°C and 4.0 bars gage. The oil is pumped to the surface through a 6.0-cm-diameter inlet pipe and is discharged through a 15.0-cm-diameter pipe at 30°C and atmospheric pressure. The volume flow rate of oil is 18 m$^3$/h, and local gravity is 9.80 m/s$^2$. For the adiabatic process, determine (*a*) the inlet and exit velocities of the oil, in m/s, (*b*) the required pump power, in kilowatts, and (*c*) the dollar cost of running the motor which drives the pump for 1 day, if electricity costs $0.14/kW·h.

5-109. A water pump operating in steady state has a 7-cm-diameter inlet and exit pipes, the exit lies 4.0 m above the inlet, and local gravity is 9.80 m/s$^2$. For a power input of 1.60 kW, there is a pressure rise of 0.80 bar from inlet to exit. If the water temperature remains constant at 20°C, which is also the ambient temperature, determine the mass flow rate, in kg/s.

5-110E. A 3-in-diameter insulated pipe carries water from a lake to a pump that is 11.0 ft above the lake level. The water enters the 3-in pipe at 14.0 psia, 50°F, and 8 ft/s. The insulated pump discharges into a 2.0-in pipe at 59.0 psia and 18 ft/s. The water then flows 60.0 ft upward through the 2-in pipe. A nozzle at the end of the 2-in pipe reduces the flow diameter, and the fluid discharges from the nozzle at

14.7 psia. Local gravity is 32.2 ft/s$^2$, and the entire process is isothermal and adiabatic. Determine (a) the power required to drive the pump, in horsepower, and (b) the nozzle exit diameter, in inches.

5-111E. A cavern 2000 ft below the ground level is used to store oil at 100°F and 60 psig. The oil is pumped to the surface through a 2.0-in-diameter inlet pipe and is discharged through a 6.0-in-diameter pipe at atmospheric pressure. The volume flow rate of oil, with a density of 53.04 lb$_m$/ft$^3$, is 560 ft$^3$/h. The local gravity is 32.0 ft/s$^2$, and the process is isothermal and adiabatic. Determine (a) the inlet and exit velocities of the oil, in ft/s, and (b) the required pump power, in horsepower.

5-112. Liquid water at 20°C is pumped at 100 L/min through a vertical rise of 100 m in a constant-area pipe. The process is adiabatic and the temperature remains constant. Find the pump power required, in kilowatts, for (a) $P_1 = P_2$ and (b) $\Delta P = -1$ bar. Local gravity is 9.8 m/s$^2$.

5-113. Water at 10°C and 1 bar falls 150 m from the top of a reservoir through a constant-area pipe to the inlet of a hydraulic turbine at the base of a dam. At the turbine inlet the water velocity is 20 m/s, and the water discharges from the turbine to the atmosphere at 1 bar and 5 m/s. The overall process is adiabatic and isothermal, and local gravity is 9.80 m/s$^2$. For a turbine output of 50,000 kW, determine (a) the pressure at the turbine inlet, in bars, (b) the mass flow rate, in kg/s, and (c) the diameter of the circular duct at the turbine inlet, in meters.

5-114. In a building, water flows through a series of pipes from 4.0 m below ground level to 120 m above ground level. At the pipe inlet below ground the conditions are 35°C, 0.070 MPa, and 3 m/s, while at the 120-m level the state is 33°C, 0.64 MPa, and 14 m/s. Determine the heat transfer, in kJ/kg, if no shaft work is done, and local gravity is 9.8 m/s$^2$. Do not neglect any term which can be calculated.

5-115. Water enters a piping system at 25°C and 7 m/s. At a position downstream the conditions are 0.20 MPa, 25°C, and 12 m/s, and the elevation is 10.0 m above the inlet. The local gravity is 9.6 m/s$^2$, and the fluid undergoes a heat loss of 0.010 kJ/kg. If the volume rate at the inlet is 10.0 m$^3$/min, determine (a) the inlet pressure, in megapascals, and (b) the inlet-pipe diameter, in centimeters.

5-116. An oil with a specific gravity of 0.90 enters a piping system at 0.240 MPa, 15°C, and 6.0 m/s. Downstream the conditions are 15°C and 4 m/s at an elevation which is 8.0 m below the inlet. The local gravity is 9.65 m/s$^2$, and the fluid gains 0.0079 kJ/kg due to heat transfer as it passes through the system at a mass flow rate of 2000 kg/min. Determine (a) the outlet pressure, in megapascals, and (b) the outlet-pipe diameter, in centimeters.

5-117. The inlet pipe to a pump extends downward to the free surface of the water in a well at 60 m below ground level. The well water at 15°C is to be raised 5 m above the ground by the pump. The diameter of the pipe is 10 cm at the inlet and 5 cm at the exit. The exit pressure is 250 kPa greater than at the inlet. For a water flow rate of 15 L/s, determine (*a*) the change in kinetic energy, in kJ/kg, and (*b*) the required power input to the pump, in kilowatts. Assume the process is isothermal and adiabatic, and that local gravity is 9.7 m/s$^2$.

5-118E. The fire hose outlet on the top floor of an office complex is 200 ft above ground level. The piping system draws water from an underground tank at an elevation 10 ft below ground level. The water temperature is 60°F, and the pumping process is assumed to be isothermal and adiabatic. The mass flow rate is 100 lb$_m$/s and the hose-nozzle outlet velocity is 40 ft/s. If $g = 32.0$ ft/s$^2$, find (*a*) the pump horsepower required and (*b*) the exit diameter in inches.

5-119E. Water at 14.5 psia falls 450 ft from the top of a water reservoir through a constant-area pipe to the inlet of a hydraulic turbine at the base of a dam. At the turbine inlet the water velocity is 50 ft/s, and the water discharges from the turbine to the atmosphere at 14.5 psia and 10 ft/s. The overall process is adiabatic, the fluid temperature is 50°F throughout, and the local gravity is 32.0 ft/s$^2$. For a turbine output of 50,000 hp, determine (*a*) the pressure at the turbine inlet, in psia, (*b*) the mass flow rate, in lb$_m$/s, and (*c*) the diameter of the circular duct at the turbine inlet, in feet.

5-120E. The volume flow rate of water entering a piping system at 80°F and 20 ft/s is 250 ft$^3$/min. At a position downstream the conditions are 20.0 psia, 80°F, and 30 ft/s, and the elevation is 30.0 ft above the inlet. The local gravity is 31.8 ft/s$^2$, and the fluid undergoes a heat loss of 0.0050 Btu/lb$_m$. Determine (*a*) the inlet pressure, in psia, and (*b*) the inlet-pipe diameter, in inches.

5-121E. An oil with a specific gravity of 0.90 enters a piping system at 18.0 psia, 60°F, and 15.0 ft/s. Downstream at an elevation which is 20.0 ft below the inlet, the conditions are 60°F and 10 ft/s. The local gravity is 32.0 ft/s$^2$, and the fluid gains 0.0040 Btu/lb$_m$ due to heat transfer as it passes through the system at a mass rate of 2400 lb$_m$/min. Determine (*a*) the outlet pressure, in psia, and (*b*) the outlet-pipe diameter, in inches.

## SIMPLE STEAM POWER CYCLE

5-122. Steam in a simple power cycle enters the turbine at 40 bars, 440°C, and exhausts at 0.08 bar and 88 percent quality. Saturated liquid leaves the condenser at 0.08 bars, and the temperature change across the adiabatic pump is negligible. Determine (*a*) the turbine work,

(b) the condenser heat loss, (c) the pump work, in kJ/kg, and (d) the percent of the boiler heat input converted into net work output.

5-123. Steam as a simple power cycle enters the turbine at 6.0 MPa, 540°C, and exhausts at 0.008 MPa and 90 percent quality. The net turbine power is 10 MW. Saturated liquid leaves the condenser at 0.008 MPa, and the temperature change across the adiabatic pump is negligible. Determine (a) the turbine work, in kJ/kg, (b) the pump work, (c) the percent of the boiler heat input converted to net work output, and (d) the rate of heat transfer from the condenser, in kJ/min.

5-124. A simple power cycle with a net power output of 20 MW has steam entering the turbine at 140 bars, 560°C, and leaving at 0.06 bar and 85 percent quality. Saturated liquid leaves the condenser at 0.06 bar, and the temperature change across the adiabatic pump is negligible. Determine (a) the turbine and the pump work, both in kJ/kg, (b) the percent of the heat input which is converted to net work output, and (c) the rate of heat addition, in kJ/s.

5-125. A simple power cycle with a net turbine output of 9 MW has steam entering the turbine at 10 MPa and 560°C, and leaving at 0.010 MPa with a quality of 86 percent. Saturated liquid leaves the condenser at 0.010 MPa, and the temperature change across the adiabatic pump is negligible. Determine (a) the turbine and pump work, both in kJ/kg, (b) the percent of the heat input which is converted into net work output, and (c) the rate of heat removal in the condenser, in kJ/s.

5-126E. A simple power cycle with a net turbine output of 10,000 hp has steam entering the turbine at 1200 psia and 800°F and leaving at 1 psia with a quality of 84 percent. Saturated liquid leaves the condenser at 1 psia, and the temperature change across the adiabatic pump is negligible. Determine (a) the turbine and the pump work, both in Btu/lb$_m$, (b) the percent of the heat input converted to net work output, and (c) the rate of heat removal in the condenser in Btu/s.

## VAPOR COMPRESSION REFRIGERATION CYCLES

5-127. A refrigeration cycle circulates 0.07 kg/s of refrigerant 134a which enters the adiabatic compressor at 1.8 bars as a saturated vapor and leaves at 7 bars and 50°C. The fluid is a saturated liquid at the condenser outlet. Determine (a) the power input to the compressor in kilowatts, (b) the heat-transfer rate in the evaporator, in kJ/s, and (c) the dimensionless ratio of heat transfer in the evaporator to compressor work.

5-128. A refrigeration cycle with a compressor power input of 3 kW has refrigerant 134a entering the adiabatic compressor at 2 bars as a saturated vapor and leaving at 8 bars and 50°C. The fluid is a saturated liquid at the condenser outlet at 8 bars. Determine (a) the mass flow rate of fluid in kg/min, (b) the dimensionless ratio of heat transfer in the evaporator to compressor work, and (c) the heat-transfer rate in the condenser, in kJ/s.

5-129. A refrigeration cycle with an evaporator heat-transfer rate of 10 kJ/s has refrigerant 134a entering the adiabatic compressor at 2.4 bars as a saturated vapor and leaving at 8 bars and 60°C. The fluid is a saturated liquid at the condenser outlet at 8 bars. Determine (a) the mass flow rate in kg/min, (b) the power input in kW, and (c) the dimensionless ratio of evaporator heat transfer to compressor work.

5-130. A refrigeration cycle with a condenser heat-transfer rate of 16 kJ/s has refrigerant 134a entering the adiabatic compressor at 160 kPa as a saturated vapor and leaving at 900 kPa and 50°C. The fluid is a saturated liquid at the condenser outlet. Determine (a) the power input in kilowatts and (b) the dimensionless ratio of the evaporator heat transfer to the compressor work.

5-131E. A refrigeration cycle with a power input of 2.5 hp has refrigerant 134a entering the adiabatic compressor at 20 psia as a saturated vapor and leaving at 140 psia and 140°F. The fluid is a saturated liquid at the condenser outlet. Determine (a) the mass flow rate in $lb_m$/min, (b) the heat-transfer rate in the condenser, in Btu/s, and (c) the dimensionless ratio of evaporator heat transfer to compressor work.

## TWO FLOW PROCESSES IN SERIES

5-132. Air at 0.25 bar and 220 K enters the diffuser of a turbojet aircraft at 300 m/s and 50 kg/s. The air reaches 0.48 bar at the outlet of the diffuser, where the velocity is negligible. It then enters a compressor, where the pressure rise leads to an outlet temperature of 495 K. Determine (a) the temperature of the air at the compressor inlet, in kelvins, and (b) the compressor power required, in kilowatts.

5-133. In a turbojet engine air enters the steady-flow turbine at 4.32 bars and 1320 K and exits the device at 2.22 bars. Turbine velocities are negligible. The air then enters an adiabatic nozzle and discharges at 1040 m/s, 638 K, and 0.25 bar. Determine (a) the nozzle inlet temperature, in kelvins, and (b) the turbine work output, in kJ/kg.

5-134. The inlet conditions to a steady-flow compressor are 0.95 bar and 27°C, and the volume flow at the inlet is 7.0 $m^3$/min. At the exit the pressure and temperature are 2.67 bars and 397 K, respectively. The air now passes through a heat exchanger (an intercooler) until its temperature reaches 27°C. Finally, the air passes through another (second stage) compressor where it undergoes the same pressure and temperature rise as in the first compressor stage. Velocities are negligible. Determine (a) the total power input required to the two stages of the compressor, in kilowatts, and (b) the rate of heat removal in the heat exchanger, in kJ/min.

5-135. Steam enters a steady-flow turbine at 40 bars and 440°C, and exits at 0.08 bar with a quality of 82.7 percent. The wet mixture then enters a heat exchanger where it condenses to a saturated liquid at the same pressure. If the power output of the turbine is 10,000 kW, determine the rate of heat removal in the heat exchanger, in kJ/min.

5-136. Water enters a heat exchanger at 6.0 MPa and 45°C, and is heated at a rate of $1.4 \times 10^6$ kJ/min to a state of 6.0 MPa and 540°C. The fluid then passes through a turbine where it exits at 0.008 MPa and a quality of 83.9 percent. Determine the power output of the turbine, in kilowatts.

5-137. Refrigerant 134a enters a throttling process as a saturated liquid at 8 bars, and exits at 1.8 bars. It then enters a heat exchanger where it is heated to a saturated vapor at 1.8 bars. Determine the rate of heat addition, in kJ/min, if the volume flow rate at the exit of the heat exchanger is 500 L/min.

## TRANSIENT FLOW

5-138. A rigid tank is connected to a pressurized line through which steam flows continuously at 1.0 MPa and 280°C. Initially a valve between the line and tank is closed, and the tank contains 0.20 kg of steam at 300 kPa and 160°C. The valve is opened and steam enters the tank slowly until the steam within the tank is at 500 kPa and 200°C. At that moment determine (a) the mass that has entered the tank, in kilograms, and (b) the heat transfer to or from the tank during the process, in kilojoules.

5-139. A pressurized tank contains 1.5 kg of air at 60°C and 3 bars. Mass is allowed to flow from the tank until the pressure reaches 1 bar. However, during the process, energy as heat transfer is added to the air within the tank to keep it at constant temperature. Assuming constant specific heats, determine how much heat transfer, in kilojoules, was added during the process.

5-140. An insulated tank with a volume of 0.50 m³ contains air at 100 kPa and 25°C. The tank is connected through a valve to a large compressed-air line, which carries air continuously at 700 kPa and 160°C. If the valve is opened and air is allowed to flow into the tank until the pressure reaches 300 kPa, (a) how much mass, in kilograms, has entered and (b) what is the final temperature in the tank, in degrees Celsius?

5-141E. A very large reservoir contains air at 200 psia and an unknown temperature $T_a$. Air flows from this reservoir into a small, insulated tank with a volume of 2 ft³. The small tank initially contains 0.1 lb of air at 80°F. Then a valve is opened to allow air to flow from the reservoir into the tank until the pressure in the tank becomes 50 psia. At this point the temperature in the tank is observed to be 340°F. What is the temperature of the air $T_a$ in the reservoir, in degrees Fahrenheit? Assume room-temperature specific heats.

5-142. A rigid insulated tank is initially evacuated. Atmospheric air at 0.10 MPa and 20°C is allowed to leak into the tank until the pressure reaches 0.10 MPa. (*a*) Find the final temperature of the air within the tank, in degrees Celsius. (*b*) Now consider that the tank initially contains air at 0.05 MPa and 20°C. Determine the final temperature in this case.

5-143E. A pressurized tank contains 1.5 lb$_m$ of air at 140°F and 40 psia. Mass is allowed to flow from the tank until the pressure reaches 15 psia. However, during the process, energy as heat transfer is added to the air within the tank to keep it at constant temperature. Assuming constant specific heats, determine how much heat, in Btu, was added during the process.

5-144E. An insulated tank with a volume of 2.0 ft$^3$ contains air at 15 psia and 80°F. The tank is connected through a valve to a large compressed-air line, which carries air continuously at 100 psia and 300°F. If the valve is opened and air is allowed to flow into the tank until the pressure reaches 80 psia, (*a*) how much mass, in pounds, has entered and (*b*) what is the final temperature in the tank, in degrees Fahrenheit?

5-145E. A pressurized tank contains 1 lb$_m$ of water vapor at 600 psia and 500°F. Mass is allowed to flow from the tank until the pressure reaches 100 psia. However, during the process, energy as heat transfer is added to the steam to kccp it at constant temperature. How much heat transfer, in Btu, was added during the process?

5-146E. A tank with a volume of 90 ft$^3$ contains steam at 300 psia and 500°F. The tank is heated until the temperature reaches 800°F. A relief valve is installed on the tank to keep the pressure constant during the process. Find (*a*) the amount of heat transfer, in Btu, and (*b*) the mass of steam that is bled from the tank, in pounds.

5-147. A pipe contains steam flowing at 30 bars and an unknown temperature $T_L$. To determine $T_L$, an uninsulated tank with a volume of 2.0 m$^3$ is attached to the line. The tank initially contains saturated steam at 1.0 bar. A valve is then opened between the tank and the pipe, and the tank is allowed to fill until its pressure is 20 bars. At this instant a thermometer inside the tank registers 360°C. Determine the value of $T_L$, in degrees Celsius, from this experiment if it is estimated that a heat loss of 397 kJ occurs during the filling process.

5-148. A tank with a volume of 1.0 m$^3$ is half filled with liquid refrigerant 134a and the remaining space is filled with vapor. The pressure is 8.0 bars. Energy as heat transfer is added until one-half of the liquid, by mass, is evaporated, while an automatic valve allows saturated vapor to escape at such a rate that the pressure remains constant within the tank. Determine the amount of heat transfer required, in kilojoules.

5-149. A piston-cylinder device originally contains 0.10 kg of saturated water vapor at 10 bars. A line with steam flowing at 20 bars and 500°C is connected to the cylinder through an initially closed valve. During a constant-pressure process maintained by a weighted piston, steam is admitted to the cylinder until the contents within the cylinder reach 300°C, while a heat loss of 90.0 kJ occurs simultaneously through the cylinder walls. Determine the amount of mass entering the cylinder from the line, in kilograms.

5-150. A very large reservoir contains steam at 30 bars and 360°C. Steam from this reservoir is allowed to enter an insulated piston-cylinder assembly maintained at a constant pressure of 10 bars. The cylinder initially contains saturated vapor in a volume of 0.50 m$^3$. The flow of steam occurs until the temperature in the cylinder reaches 280°C. Determine (a) the final mass in the cylinder, in kilograms, and (b) the final volume of the cylinder, in cubic meters.

5-151. A tank with a volume of 3.0 m$^3$ contains steam at 20 bars and 280°C. The tank is heated until the temperature reaches 440°C. A relief valve is installed on the tank to keep the pressure constant during the process. Find (a) the amount of heat transferred, in kilojoules, and (b) the mass of steam that is bled from the tank, in kilograms.

5-152E. A line with steam flowing at 300 psia and 700°F is connected to a piston-cylinder device through an initially closed valve. The piston cylinder originally contains 0.20 lb$_m$ of saturated water vapor at 160 psia. During a constant-pressure process maintained by a weighted piston, steam is admitted to the cylinder until the contents within the cylinder reach 550°F, while a heat loss of 90.0 Btu occurs simultaneously through the cylinder walls. Determine the amount of mass entering the cylinder from the line, in pounds.

5-153E. Steam from a very large reservoir at 450 psia and 600°F is allowed to enter an insulated piston-cylinder assembly maintained at a constant pressure of 160 psia. The cylinder initially contains saturated vapor in a volume of 2.0 ft$^3$. The flow of steam occurs until the temperature in the cylinder reaches 450°F. Determine (a) the final mass in the cylinder, in pounds, and (b) the final volume of the cylinder, in cubic feet.

5-154E. A rigid insulated tank is initially evacuated. Atmospheric air at 1 atm and 70°F is allowed to leak into the tank until the pressure reaches 1 atm. (a) Find the final temperature of the air within the tank, in degrees Fahrenheit. (b) Now consider that the tank initially contains air at 0.5 atm and 70°F. Determine the final temperature in this case.

5-155. An ideal gas is contained in a rigid tank of volume $V$, initially at $P_1$ and $T_1$. Heat transfer occurs to the contents until the pressure reaches $P_2$. However, a relief valve allows gas to escape so that the temperature remains constant. Derive an expression for the heat transfer in terms of $V$, $P_1$, and $P_2$.

5-156. An ideal gas is contained in a rigid tank of volume $V$, initially at $P_1$ and $T_1$. Heat transfer occurs to the contents until the temperature reaches $T_2$. However, a relief valve allows gas to escape so that the pressure remains constant. Derive an expression for the heat transfer during the process in terms of the quantities $T_1$, $T_2$, $P$, $V$, $c_p$, and $R$.

5-157. A tank with a volume of 10 m$^3$ contains air initially at 5 bars and 40°C. Energy as heat transfer is added at a constant rate of 6 kJ/s while an automatic valve allows air to leave the tank at a constant rate of 0.030 kg/s. Determine (a) the temperature of the air in the tank, in kelvins, and (b) the pressure in the tank, in bars, 7 min after starting from the initial conditions.

5-158. A very large reservoir contains air at 1.2 MPa and an unknown temperature $T_a$. Air flows from this reservoir into a small, insulated tank with a volume of 0.2 m$^3$. The small tank initially contains 0.2 kg of air at 27°C. Then a valve is opened to allow air to flow from the reservoir into the tank until the pressure in the tank becomes 0.3 MPa. At this point the temperature in the tank is observed to be 180°C. What is the temperature of the air $T_a$ in the reservoir, in kelvins? Assume room-temperature specific heats.

5-159E. Air initially at 100 psia and 100°F is contained within a tank with a volume of 100 ft$^3$. Energy as heat transfer is added at a constant rate of 6 Btu/s while an automatic valve allows air to leave the tank at a constant rate of 0.060 lb/s. Determine (a) the temperature of the air in the tank, in degrees Fahrenheit, and (b) the pressure in the tank, in psia, 7 min after starting from the initial conditions.

5-160. Two adiabatic tanks are interconnected through a valve. Tank $A$ contains 0.10 m$^3$ of nitrogen at 3.0 MPa and 100°C. Tank $B$ contains 2.5 m$^3$ of nitrogen at 0.2 MPa and 30°C. The valve is opened until the pressure in $A$ drops isentropically to 2.0 MPa. At this instant, determine (a) the temperature in $A$, in degrees Celsius, (b) the temperature and pressure in tank $B$, (c) the mass remaining in tank $A$, in kilograms, and (d) the entropy production for the process, in kJ/K. Assume constant specific heats. (Requires second-law information.)

5-161E. Two adiabatic tanks are interconnected through a valve. Tank $A$ contains 1.2 ft$^3$ of air at 500 psia and 200°F. Tank $B$ contains 15.0 ft$^3$ of air at 20 psia and 100°F. The valve is opened until the pressure in $A$ drops isentropically to 300 psia. At this instant, determine (a) the temperature in tank $A$, in degrees Fahrenheit, (b) the temperature and pressure in tank $B$, (c) the mass remaining in tank $A$, in pounds, and (d) the entropy production for the process, in Btu/°R. Assume constant specific heats. (Requires second-law information.)

5-162. Two adiabatic tanks are interconnected through a valve. Tank $A$ contains 0.20 m$^3$ of air at 40 bars and 90°C. Tank $B$ contains 2.0 m$^3$ of air at 1 bar and 30°C. The valve is opened until the pressure in $A$ drops isentropically to 15 bars. At this instant, determine (a) the temperature in tank $A$, in degrees Celsius, ($b$) the temperature and pressure in tank $B$, ($c$) the mass remaining in tank $A$, in kilograms, and ($d$) the total entropy production for the process, in kJ/K. Assume constant specific heats. (Requires second-law information.)

5-163E. A tank with a volume of 100 ft$^3$ contains air initially at 100 psia and 100°F. Energy as heat transfer is added at a constant rate of 6 Btu/s while an automatic valve allows air to leave the tank at a constant rate of 0.060 lb$_m$/s. Determine how long it will take, in minutes, for the air in the tank to reach 350°F.

5-164. A piston-cylinder assembly maintained at a constant pressure of 1.2 bars initially contains 0.00070 kmol of air in 0.0130 m$^3$. Air in a line at 6 bars and 167°C is then allowed to flow through a valve and into the cylinder until the volume is doubled. During the process heat enters the cylinder in the amount of 1000 kJ per kmol of gas finally in the tank. For the air let $u = 20.8T$ and $h = 29.1T$, in kJ/kmol, where $T$ is in kelvins. Determine ($a$) the final temperature, in kelvins, and ($b$) the final number of kilogram-moles in the cylinder.

5-165. An insulated tank with a volume of 0.120 m$^3$ contains 0.150 kg of air at 20°C. The tank is connected through a valve to a large compressed-air line, which carries air continuously at 7 bars and 147°C. If the valve is opened and compressed air flows into the tank until the pressure reaches 4.0 bars, determine how much air, in kilograms, has entered. For air, let $c_v = 0.718$ and $c_p = 1.005$ kJ/kg·°C.

5-166. A container of fixed volume $V$ contains air at pressure $P_1$ and temperature $T_a$. It is surrounded by atmospheric air at a pressure $P_a$ and temperature $T_a$. A valve is opened, and atmospheric air is quickly admitted to the container until the pressure reaches the atmospheric value. At that instant the air in the container is at temperature $T_2$. Listing any necessary assumptions, derive an equation for the temperature ratio $T_R = T_2/T_a$ in terms of the specific-heat ratio $k$ and the pressure ratio $P_R = P_1/P_a$.

5-167E. Two adiabatic tanks are interconnected through a valve. Tank $A$ contains 3.0 ft$^3$ of nitrogen at 450 psia and 200°F. Tank $B$ contains 40 ft$^3$ of nitrogen at 50 psia and 100°F. The valve is opened until the pressure in $A$ drops isentropically to 300 psia. At this instant determine ($a$) the temperature in $A$, in degrees Fahrenheit, ($b$) the temperature and pressure in tank $B$, ($c$) the mass remaining in tank $A$, in pounds, and ($d$) the total entropy production for the process, in Btu/°R. Assume constant specific heats. (Requires second-law information.)

5-168E. Air in a line at 100 psia and 340°F is allowed to flow through a valve and into a piston-cylinder assembly until the cylinder volume is doubled. The cylinder, which is maintained at a constant pressure of 14.7 psia, initially contains 0.0010 lbmol of air in 0.365 ft$^3$. During the process heat enters the cylinder in the amount of 1000 Btu/lbmol of gas finally in the tank. For the air let $u = 4.96T$ and $h = 6.96T$, in

Btu/lbmol, where $T$ is in degrees Rankine. Determine (a) the final temperature, in degrees Rankine, and (b) the final number of pound-moles in the cylinder.

5-169E. An insulated tank with a volume of 2.0 ft³ contains 0.150 lb$_m$ of air at 80°F. The tank is connected through a valve to a large compressed-air line, which carries air continuously at 100 psia and 300°F. If the valve is opened and compressed air flows into the tank until the pressure reaches 60 psia, determine how much air, in pounds, has entered. For air, let $c_v = 0.171$ and $c_p = 0.241$ Btu/lb·°F.

5-170. A tank with a volume of 0.5 m³ is half filled with liquid water, and the remainder is filled with vapor. The pressure is 30 bars. Energy as heat transfer is added until one-half of the liquid (by mass) is evaporated while an automatic valve lets saturated water vapor escape at such a rate that the pressure remains constant. Determine the amount of heat transfer, in kilojoules.

5-171. A pressure vessel with a volume of 0.5 m³ contains saturated water at 300°C. The vessel initially contains 50 percent (by volume) of liquid. Liquid is slowly withdrawn from the bottom of the tank, and heat transfer takes place so that the contents are kept at constant temperature. Find how much heat must be added by the time half of the total mass has been removed.

5-172. A tank with a volume of 10.0 m³ contains air initially at 5 bars and 40°C. Energy as heat transfer is added at a constant rate of 6 kJ/s while an automatic valve allows air to leave the tank at a constant rate of 0.030 kg/s. Determine how long it will take, in minutes, for the air in the tank to reach 180°C.

5-173E. A 50-ft³ rigid tank contains saturated water at 600°F. Initially the volume is 40 percent liquid and 60 percent vapor. Liquid is slowly withdrawn from the bottom of the tank, and heat transfer takes place so that the contents are kept at constant temperature. Find how much heat transfer must be added or removed by the time half of the total mass has been removed, in Btu.

5-174. A pressurized tank contains 1 kg of water vapor at 40 bars and 280°C. Mass is allowed to flow from the tank until the pressure reaches 7 bars. However, during the process, energy as heat transfer is added to the steam to keep it at constant temperature. Find the heat transfer added, in kilojoules, during the process.

5-175E. A piston-cylinder assembly is attached through a valve to a source of constant-temperature and constant-pressure air. Initially inside the cylinder the volume is 1 ft³, the temperature of the air is 100°F, and the pressure is 14.7 psia. The valve is slowly opened to the air-supply line which is at 100 psia and 200°F. The piston moves out as air enters in order to maintain the cylinder pressure at the ambient condition of 14.7 psia. When the cylinder volume reaches 2 ft³, find (a) the temperature inside the cylinder, in degrees Fahrenheit, and (b) the mass, in pounds, that has entered through the valve. The process is adiabatic.

5-176E. A 15.0-ft$^3$ tank contains carbon dioxide at 30 psia and 100°F. Nitrogen at a line pressure and temperature of 100 psia and 300°F, respectively, flows from a pipe into the tank until the pressure reaches 75 psia. If the process is adiabatic, determine the final temperature, in degrees Fahrenheit, of the mixture within the tank.

5-177. An insulated piston-cylinder assembly with a volume of 0.1 m$^3$ initially contains air at 30°C and 1 bar. A valve connects the assembly to an air-supply line which is a source of constant-temperature and constant-pressure air. The valve is slowly opened to the air supply which is at 7 bars and 90°C. The piston moves out as air enters in order to maintain the cylinder pressure at the ambient condition of 1 bar. When the cylinder volume reaches 0.2 m$^3$, find (*a*) the mass that has entered the valve, in kilograms, and (*b*) the temperature inside the cylinder, in degrees Celsius.

5-178. Nitrogen at a line pressure and temperature of 8 bars and 150°C, respectively, flows from a pipe into a 0.50-m$^3$ tank until the pressure reaches 5 bars. The tank initially contains carbon dioxide at 2 bars and 30°C. If the entire process is adiabatic, determine the final temperature, in degrees Celsius, of the mixture within the tank.

5-179. An insulated tank of unknown volume initially contains an ideal gas at a known pressure and temperature. The same gas maintained in a large reservoir at $T_R$ and $P_R$ is allowed to enter the tank until the gas within the tank reaches a measured pressure and temperature. Assuming constant specific-heat values, (*a*) derive an equation for the reservoir temperature $T_R$ in terms of the initial and final pressures and temperatures within the tank and the specific-heat ratio $k$ for the gas. (*b*) If air initially is in the tank at 2 bars and 27°C and the final measured state is 4 bars and 87°C, determine the reservoir temperature, in degrees Celsius.

5-180. An insulated 0.060-m$^3$ tank initially contains steam at 80 bars and 400°C. The tank is connected through a closed valve to another tank which is filled with an immiscible liquid at 30 bars. The second tank is fitted with a second valve which is also closed. Both valves are now opened simultaneously. The steam enters the second tank and forces the immiscible liquid out of the second valve at such a rate that the pressure in the second tank remains constant at 30 bars. At pressure equilibrium between the tanks, determine (*a*) the final temperature in the first tank, (*b*) the final temperature of the steam in the second tank (ignoring any heat transfer between the steam and the liquid), in degrees Celsius, and (*c*) the volume of liquid displaced, in cubic meters.

5-181. An ideal gas flows from a line at temperature $T_L$ through a valve into a cylinder fitted with a weightless piston which is constrained solely by a spring. Assume the pressure of the gas in the cylinder is always proportional to the displacement of the piston, and initially no gas is within the cylinder. For the adiabatic cylinder, (*a*) derive an equation for the final temperature of the gas in the cylinder at pressure equilibrium in terms of the line temperature, the specific-heat ratio $k$, and constants. (*b*) If the line contains air at 4 bars and 47°C, determine the final temperature in the cylinder, in degrees Celsius.

# CHAPTER
# 6

# THE SECOND LAW AND ENTROPY

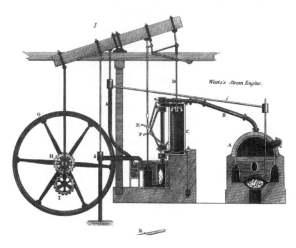

James Watt's steam engine.

In the preceding sections attention has been focused on the conservation of energy, conservation of mass, and thermodynamic property relationships. However, experience has shown that not all processes that satisfy the conservation of energy principle are possible. Real processes must also satisfy the second law of thermodynamics. This chapter serves as an introduction to the second law and its related property called entropy. An entropy balance, which is analogous to the mass and energy balances, is developed as an analytical tool for studying the second-law behavior of systems. Typical applications of the second law and the entropy balance are presented, with special emphasis on thermodynamic cycles that produce either power output or a refrigeration effect.

## 6-1   INTRODUCTION

In nations with a developed or developing technological society, the ability to produce energy in the form of work is of prime importance. Work transformations are necessary to transport people and goods, drive machinery, pump liquids, compress gases, and provide energy input to many other processes that are taken for granted in highly developed societies. Much of the work output in such societies is available as electrical work which is then converted, by means of motors, to rotational mechanical work. Although some of this electrical work is produced by hydroelectric power plants, by far the greatest amount is obtained from fossil or nuclear fuels. These fuels allow the engineer to produce a relatively high-temperature gas or liquid stream that acts as a thermal (heat) source for the production of work. The efficient conversion of thermal energy to work transfer is extremely important, especially in the light of developing shortages of some fossil fuels. It is the second law of thermodynamics that provides guidelines in this area.

There are other important second-law considerations. One is to quantify the concept that energy has *quality* as well as quantity. For example, our experiences indicate that we can use work transfer of energy in many more diverse ways than we can use heat transfer. If work transfer is 100 percent convertible to heat transfer, but the reverse situation is not possible (which we must eventually prove), then work would appear to be a more valuable form of energy transfer than heat transfer. We shall show through second-law arguments that the higher the temperature at which heat transfer occurs, the greater the amount of heat transfer that can be converted into work output. Thus energy available at higher temperatures is generally more useful than that available at lower temperatures. For example, while there is an immense quantity of energy stored in the oceans, its present availability to us for performing useful tasks is quite low because of its low temperature. This implies, in turn, that energy is *degraded* when it is transferred by means of heat transfer from one system to another system of lower temperature. Other examples of energy degradation include energy transformations that involve dissipative effects like friction and electric resistance. Such effects are highly undesirable if the use of energy for practical purposes is to be optimized. Thus every time energy changes form or is transferred from one system to another, it is both conserved and degraded. The first law is concerned with the conservation of energy. The second law provides a means of measuring this energy degradation as well as the quality of energy.

Studies of other well-known phenomena require second-law analysis. Consider the thermal contact between two systems initially at different temperatures. It is our common experience that if the combined system is isolated, the two systems will tend to reach thermal equilibrium, or equality of temperature, after sufficient time. The first law requires that the energy given up by one system must equal the energy acquired by the other. However, conservation of energy places no restriction on the final temperature

**Note the physical significance of the quality and the degradation of energy.**

achieved by each system. The second law is the principle that establishes the nature of the final equilibrium state. As a second example, consider a paddle wheel that stirs a fluid within an insulated container. The paddle wheel might be rotated by a pulley-weight mechanism, for example. The final result, in light of the conservation of energy principle, is a decrease in the potential energy of the weight and an increase in the internal energy of the fluid within the container. However, at some later time we do not expect to observe a spontaneous process that decreases the energy of the fluid and returns the weight to its initial position. The reverse process does not violate the first law, but it will be shown that it does violate the second law. These and many other possible examples illustrate that processes, of their own accord, have a preferred direction of change, irrespective of the first law. Also, after sufficient time, such processes reach an equilibrium state.

In summary, there are a number of phenomena that cannot be explained by conservation principles of any type. Hence we seek another law that, through its generality, will provide guidelines to the understanding and analysis of diverse effects. Among other considerations, the second law is extremely helpful to the engineer in the following ways:

1. It provides the means of measuring the quality (usefulness) of energy.
2. It establishes the criteria for the "ideal" performance of engineering devices.
3. It determines the direction of change for spontaneous processes.
4. It establishes the final equilibrium state for spontaneous processes.
5. It establishes parameters for measuring losses in energy processes.
6. It leads to a "thermodynamic" temperature scale which is independent of the substance used in the measurement.

Because the second law is used to examine the direction of change of a process or to compare its performance against an ideal, it is often presented in terms of a mathematical inequality. Alternatively, the inequality can be eliminated through the introduction of the concept of entropy generation. Regardless of the approach, we will see that entropy, unlike mass and energy, is not a conserved property.

**Note the broad range of applications of the second law.**

The range of applications of the second law, as suggested by the list in the preceding paragraph, is extremely broad. In addition, a number of second-law statements have been presented by scientists and engineers. Fortunately, each of these statements is equivalent. As a result, if we select any one of these statements, then through a series of theorems and corollaries we can develop guidelines for analyzing a diversity of phenomena suggested by the preceding list and not explained by well-known conservations laws (such as mass, momentum, and energy). Since two of the second-law statements discussed in this chapter are associated with devices that exchange heat and work during a cyclic process, the following section describes some of the basic characteristics of cyclic devices known as heat engines, refrigerators, and heat pumps.

## 6-2   HEAT ENGINES, REFRIGERATORS, AND HEAT PUMPS

An important application of the second law is in the analysis of cyclic devices which have general classifications as heat engines, refrigerators, and heat pumps. This section reviews the general operating characteristics of these devices.

### 6-2-1   HEAT ENGINES

A *heat engine* is defined as a *closed* system in which the fluid either undergoes a periodic cycle within a single piece of equipment or passes continuously through a loop of steady-state equipment. Both types of systems produce a net work output due to a net heat transfer into the system. The restriction to a *periodic cycle* or *continuous loop* implies that the substance within either device is returned to its initial state at regular intervals. In many cases, a periodic device is cyclic in a mechanical sense as well.

As an example of a *periodic heat-engine cycle,* consider the frictionless piston-cylinder assembly shown in Fig. 6-1a. The piston and weight rest on supporting pegs at position 1, and a gas is contained in the volume below the piston. At state 1 the pressure of the gas is less than the equivalent pressure exerted by the piston-weight-atmosphere combination above the gas. Heat transfer then occurs from the high-temperature source until the gas pressure just balances this opposing pressure. The constant-volume process from state 1 to state 2 is shown on the *PV* diagram in Fig. 6-1b. If additional heat transfer occurs from the high-temperature source, the gas will expand at constant pressure until the piston hits the upper set of pegs.

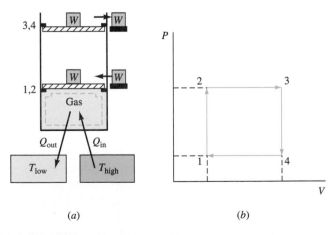

(a)                              (b)

**Figure 6-1**   A simple, cyclic heat engine.

The heat-addition process is halted at this point, and the weight is moved horizontally off the piston. The state of the gas is now designated as state 3 on the *PV* diagram.

Heat transfer now occurs from the gas to a low-temperature sink. Until the gas pressure drops so that it just balances the weight of the piston and the atmospheric pressure, the volume does not change. The condition where the volume just begins to change is state 4. Additional heat transfer out will cause the piston to drop at constant pressure until it reaches its initial position on the lower set of pegs. At this point the cycle has been completed. If another weight were now added on the piston at the lower level, the cycle could be repeated. The net work output $W_{net,out}$ of the cycle is measured either by the enclosed area on the *PV* diagram in Fig. 6-1*b* or by the change in the gravitational potential energy of the weights. In addition, heat transfer takes place between the system and two regions at different temperatures. At any time during the cycle the rate form of the closed-system energy balance requires that $dE_{sys}/dt = \dot{Q} + \dot{W}$. Integration with respect to time over one cycle leads to $\Delta E = Q + W$. Because the process is cyclic, $\Delta E_{cycle} = 0$. Thus an energy balance for the cycle reduces to $W_{net,out} = Q_{net,in}$. Examples of practical "heat-work" converters that are periodic in nature are the automotive and diesel engines.

A practical example of a *continuous-loop* steady-state heat engine is the simple steam power cycle discussed initially in Sec. 5-7-1. The schematic for the cycle is shown again in Fig. 6-2. In this case there are two heat-transfer terms and two work-transfer terms. Some of the energy input $\dot{Q}_{in}$ is converted into turbine power output, shown as $\dot{W}_{T,out}$. Another portion of the energy input appears as heat transfer $\dot{Q}_{out}$ to the environment. The power used to drive the pump is shown as $\dot{W}_{P,in}$. Again applying the rate form of the energy balance to the closed system, $dE/dt = \dot{Q}_{in} - \dot{Q}_{out} + \dot{W}_{P,in} - \dot{W}_{T,out}$. Because $dE/dt = 0$ in steady state, the overall energy balance on the closed

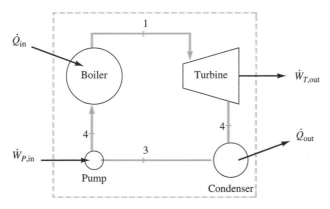

**Figure 6-2**   A continuous heat engine represented by a simple steam power cycle.

**Note the physical difference between the operation of a periodic-cyclic and a continuous-loop heat engine.**

loop on a rate basis reduces to $\dot{W}_{net,out} = \dot{Q}_{net,in}$. Hence the words "periodic cycle" and "continuous loop" describe the nature of equipment used to convert heat-transfer input to work output. In both of these cases the fluid itself undergoes a cyclic process, and there are two heat-transfer processes: one into and one out of the heat engine. This is the common situation for actual heat engines.

The two heat-engine devices outlined above can be represented in a general way by the schematic diagram of Fig. 6-3. The quantity $Q_{in}$ (or $\dot{Q}_{in}$) is the magnitude of the heat transfer (rate) supplied to the heat engine from a high-temperature source (such as hot combustion gases), while $Q_{out}$ (or $\dot{Q}_{out}$) is the magnitude of the heat transfer (rate) rejected by the engine to a low-temperature sink (such as the atmosphere or a lake). The directions of $Q_{in}$ (or $\dot{Q}_{in}$) and $Q_{out}$ (or $\dot{Q}_{out}$) are shown by arrows in the figure. In practical heat engines these heat quantities are transferred to and from a fluid which undergoes a cyclic process within the engine. This fluid is generally referred to as the **working fluid**. As we have shown above, the conservation of energy principle for the closed system denoted by the dashed-line region in Fig. 6-3 is

$$Q_{in} - Q_{out} - W_{net,out} = 0 \qquad \text{or} \qquad \dot{Q}_{in} - \dot{Q}_{out} - \dot{W}_{net,out} = 0 \quad \textbf{[6-1]}$$

where $W_{net,out}\ (= W_{out} - W_{in})$ is the magnitude of the net work output and is shown by an arrow out of the system. The conservation of energy principle requires that $Q_{out}$ (or $\dot{Q}_{out}$) be smaller than $Q_{in}$ (or $\dot{Q}_{in}$). In the case of the simple steam power cycle shown in Fig. 6-2, the net power out is the turbine power output minus the pump power input.

In addition, engineers frequently measure the performance of devices by the ratio of the desired result to the required or costly input. That is,

$$\text{Performance} = \frac{\text{desired result}}{\text{required input}} \qquad \textbf{[6-2]}$$

The parameter used to measure the performance of a heat engine is called the thermal efficiency. For heat engines, the desired result is the net work output, and the required input is the heat transfer into the working fluid. (Note that the required input is *not* the *net* heat transfer.) Thus the **thermal efficiency** $\eta_{th}$ is defined as

$$\eta_{th} \equiv \frac{W_{net,out}}{Q_{in}} = \frac{\dot{W}_{net,out}}{\dot{Q}_{in}} \qquad \textbf{[6-3]}$$

where $\eta_{th}$ is a dimensionless quantity with a value between 0 and 1, or between 0 and 100 percent. If we supply 100 units of energy to a heat engine and find that 70 units are rejected as heat transfer to a low-temperature sink, then the net work out is 30 units and the thermal efficiency is 0.30 (or 30 percent). Note that the thermal efficiency will be unity (100 percent) if $Q_{out}$ in Fig. 6-3 is zero. That is, a heat engine with a thermal efficiency of

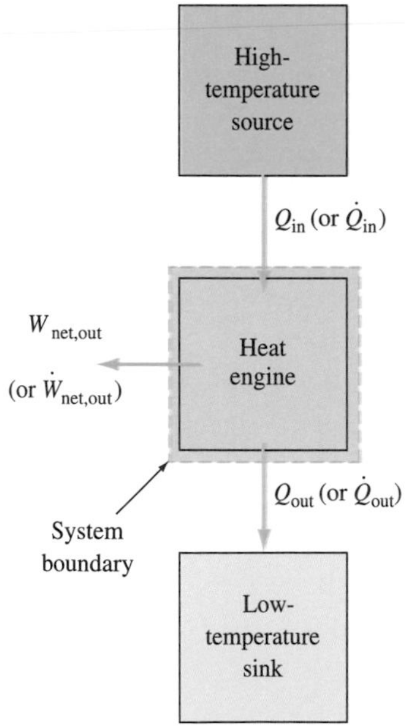

High-temperature source

$Q_{in}$ (or $\dot{Q}_{in}$)

$W_{net,out}$

(or $\dot{W}_{net,out}$)

Heat engine

$Q_{out}$ (or $\dot{Q}_{out}$)

System boundary

Low-temperature sink

**Figure 6-3**
Simple schematic of a heat engine.

100 percent rejects no energy by heat transfer to a low-temperature sink. The conservation of energy principle [see Eq. [6-1]] places no restriction on the conversion of heat transfer to work transfer. However, it will be shown that the second law does place an upper theoretical limit on the heat-work conversion efficiency of heat engines.

**Note the two ways to evaluate the net work output.**

**Example 6-1**

**A** simple steam power cycle receives 100,000 kJ/min as heat transfer from hot combustion gases and rejects 66,000 kJ/min as heat transfer to the environment. If the pump power required is 1400 kJ/min, determine (*a*) the thermal efficiency of the cycle, and (*b*) the turbine power output in kilowatts.

**Solution:**

**Given:** A simple steam power cycle with appropriate data is shown in Fig. 6-4.

**Find:** (*a*) the thermal efficiency and (*b*) the gross turbine power output, in kW.

**Model:** Steady-state closed-loop cycle.

**Strategy:** Apply the conservation of energy principle and the definition of thermal efficiency.

**Analysis:** (*a*) On the basis of Eq. [6-3], the thermal efficiency is $\dot{W}_{net,out}/\dot{Q}_{in}$. In addition, the overall energy balance on the closed system within the dashed line is

$$\frac{dE}{dt} = \sum \dot{Q} + \sum \dot{W} = \dot{Q}_{boil,in} - \dot{Q}_{cond,out} - \dot{W}_{net,out}$$

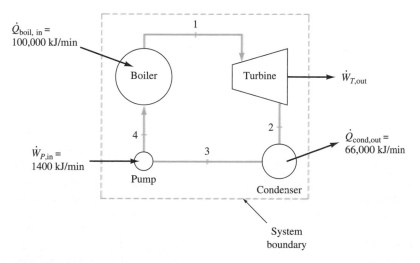

**Figure 6-4** Schematic and data for Example 6-1.

Since $dE/dt = 0$ in steady state, then $\dot{W}_{net,out} = \dot{Q}_{boil,in} - \dot{Q}_{cond,out}$. Therefore,

$$\eta_{th} = \frac{\dot{Q}_{boil,in} - \dot{Q}_{cond,out}}{\dot{Q}_{boil,in}} = \frac{100,000 - 66,000}{100,000} = 0.34 \text{ (or 34 percent)}$$

(*b*) The net power output $\dot{W}_{net,out}$ is the difference between the turbine power output of the cycle and the pump power input. That is, $\dot{W}_{net,out} = \dot{W}_{T,out} - \dot{W}_{P,in}$. Therefore, the above overall energy balance becomes

$$\dot{W}_{T,out} - \dot{W}_{P,in} = \dot{Q}_{boil,in} - \dot{Q}_{cond,out}$$

Therefore, the turbine power output is

$$\dot{W}_{T,out} = \dot{Q}_{boil,in} - \dot{Q}_{cond,out} + \dot{W}_{P,in}$$
$$= 100,000 - 66,000 + 1400 = 35,400 \text{ kJ/min} = 590 \text{ kW}$$

**Comment:** Note that the pump power input is a small fraction of the turbine power output in a simple steam power cycle (0.04 in this case).

## 6-2-2   REFRIGERATORS AND HEAT PUMPS

Two other important classes of cyclic devices of engineering interest are refrigerators and heat pumps. The general operation of a simple refrigeration cycle was discussed initially in Sec. 5-7-2. Both of these cyclic systems are illustrated by the schematic shown in Fig. 6-5. Again the directions of $Q_{in}$, $Q_{out}$, and $W_{net,in}$ are shown by arrows. The schematic is simply the reverse of that shown in Fig. 6-3 for a heat engine. For either a refrigerator or heat pump, a quantity of energy $Q_{in}$ (or $\dot{Q}_{in}$) is transferred from a low-temperature source to the cyclic device, and a quantity of energy $Q_{out}$ (or $\dot{Q}_{out}$) is rejected to a high-temperature sink. Similar to a heat engine, these heat transfers occur to and from a working fluid which undergoes a cyclic process within the device. The net heat transfer is accomplished by providing a net work input $W_{net,in}$ (such as the motor-compressor unit in an actual refrigerator or air conditioner). Owing to the work input to a refrigerator or heat pump, the conservation of energy principle dictates that the magnitude of $Q_{out}$ be larger than $Q_{in}$.

The difference between a refrigerator and a heat pump is merely one of purpose; the operation is the same. The purpose of a *refrigerator* is to maintain a low-temperature region of finite size at a predetermined temperature by removing energy from it by heat transfer. A *heat pump* maintains a region at a selected high temperature by supplying it with energy by heat transfer from a low-temperature source. For example, in the winter a home or business office can be kept warm by means of a heat pump with energy taken from the ground or outside air on a cold winter day. In addition, a heat pump may be used to supply energy to certain industrial processes. It should be recognized that an *air conditioner* is basically a refrigerator, except that the cold region is a room or an entire building and the hot region

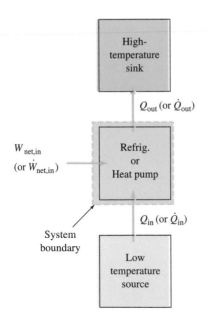

**Figure 6-5**
Simple schematic of a refrigerator or heat pump.

for heat discharge is the outside air. The majority of homes and buildings have separate cooling and heating systems. However, with the proper controls it is possible to design a single unit that operates as an air conditioner in the summer and as a heat pump in the winter. These "combined" units are still called heat pumps in the United States, even though they have a dual purpose.

**Note the difference in purpose of refrigerators, heat pumps, and air conditioners.**

The thermal efficiency is not an appropriate parameter to describe the performance of refrigerators and heat pumps. Instead, the performance of refrigerators and heat pumps is measured by a parameter called the *coefficient of performance,* or COP. Like the thermal efficiency for heat engines, the COP is defined in terms of desired result divided by costly or required input. In refrigeration processes, the desired result is the heat transfer to the cycle from a low-temperature region and the required input is the net work input. Therefore, for any refrigerator (R)

$$\mathrm{COP_R} \equiv \frac{Q_{in}}{W_{net,in}} = \frac{Q_{in}}{Q_{out} - Q_{in}} \qquad \text{or} \qquad \mathrm{COP_R} \equiv \frac{\dot{Q}_{in}}{\dot{W}_{net,in}} = \frac{\dot{Q}_{in}}{\dot{Q}_{out} - \dot{Q}_{in}}$$

[6-4]

The coefficient of performance for refrigerators is a dimensionless number that is always greater than zero. The desired result for a heat pump is the heat transfer from the cycle to a high-temperature sink and the required input is the net work input. Hence, for any heat pump (HP)

$$\mathrm{COP_{HP}} \equiv \frac{Q_{out}}{W_{net,in}} = \frac{Q_{out}}{Q_{out} - Q_{in}} \qquad \text{or} \qquad \mathrm{COP_{HP}} \equiv \frac{\dot{Q}_{out}}{\dot{W}_{net,in}} = \frac{\dot{Q}_{out}}{\dot{Q}_{out} - \dot{Q}_{in}}$$

[6-5]

where again only the magnitudes of $Q_{in}$ and $Q_{out}$ are required in the equation. The value of $\mathrm{COP_{HP}}$ is never less than unity. These definitions of the COP for refrigerators and heat pumps are completely general, in the sense that they apply whether the temperatures of the source and sink are constant or vary during the process.

---

**A** heat pump with a COP of 1.50 is used to supply 270,000 kJ/h of energy to a small industrial process operating at a temperature of a few hundred degrees above the atmospheric air temperature of 2°C. Determine (*a*) the power required in kilowatts to drive the heat pump, (*b*) the rate at which energy is removed from the atmosphere, in kJ/h, and (*c*) the cost of continuous operation for 1 h if electricity costs 12 cents per kilowatt-hour.

**Example 6-2**

**Solution:**

**Given:** A diagram of the heat pump with appropriate data is shown in Fig. 6-6.

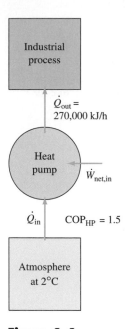

**Figure 6-6**
Heat pump schematic
and data for Example 6-2.

**Find:**    (*a*) $\dot{W}_{net,in}$, in kW, (*b*) $\dot{Q}_{in}$, in kJ/h, and (*c*) cost of 1 h of operation.

**Model:**    Steady-state closed system.

**Strategy:**    Apply the conservation of energy principle and the definition of the heat-pump coefficient of performance.

**Analysis:**    (*a*) Rearrangement of Eq. [6-5] for the heat-pump COP on a rate basis leads to

$$\dot{W}_{net,in} = \frac{\dot{Q}_{out}}{COP_{HP}} = \frac{270,000 \text{ kJ/h}}{1.5}$$

$$= 180,000 \text{ kJ/h} \times \frac{1 \text{ kW}}{3600 \text{ kJ/h}} = 50 \text{ kW}$$

(*b*) The heat-transfer rate $\dot{Q}_{in,HP}$ is found from an energy balance on the heat pump.

$$\frac{dE}{dt} = \dot{Q}_{net,in} + \dot{W}_{net,in} = \dot{Q}_{in} - \dot{Q}_{out} + \dot{W}_{net,in}$$

Since $dE/dt = 0$ for a steady-state process, then

$$\dot{Q}_{in} = \dot{Q}_{out} - \dot{W}_{net,in}$$
$$= (270,000 - 180,000) \text{ kJ/h} = 90,000 \text{ kJ/h}$$

(*c*) The cost of operation for 1 h is

$$\text{Cost} = 50 \text{ kW} \times 1 \text{ h} \times \frac{\$0.12}{\text{kW·h}} = \$6.00$$

**Comment:**    Note that of the 270,000 kJ/h provided to the process, 90,000 kJ/h were obtained without cost from the atmosphere.

## 6-3    SECOND-LAW STATEMENTS

Although there are a number of alternate second-law statements, the Kelvin-Planck and Clausius statements are probably the most frequently used in introductory engineering thermodynamics courses. One reason for this is that the statements are closely connected to the behavior of practical engineering devices such as heat engines, refrigerators, air conditioners, and heat pumps. Second, both statements reflect our common knowledge about certain phenomena observed on a daily basis, although the Clausius statement is much more obvious. In this section we shall introduce the two statements of the second law mentioned above, and show their equivalency. At the end of the section a third statement will be presented which deals with equilibrium states, and is equivalent to the Kelvin-Planck and Clausius statements.

## 6-3-1 THERMAL-ENERGY RESERVOIRS

In the discussion of a heat engine in Sec. 6-2-2, heat-transfer quantities associated with the cyclic device were either transferred from an energy source into the device or transferred from the device into an energy sink. No effort was made to identify any special characteristics of these sources and sinks for heat transfer. Before discussing the Kelvin-Planck statement of the second law, it is necessary to introduce a special type of heat source or sink called a thermal-energy reservoir. By definition, a ***thermal-energy reservoir*** (or simply a ***thermal reservoir***) is a closed system with the following characteristics:

1. The only interaction of interest between a thermal reservoir and its surroundings is heat transfer.
2. Its temperature remains uniform and constant during a process.
3. Heat transfer to or from a thermal-energy reservoir only results in an increase or decrease in the internal energy of the reservoir.

A fourth major characteristic of a thermal-energy reservoir will be introduced in Sec. 6-4. Schematics of thermal-energy reservoirs acting as a source or sink for heat transfer are sketched in Fig. 6-7.

From the definition of a thermal-energy reservoir there is no restriction on the physical makeup of the reservoir. Consequently, the only significant property of a thermal reservoir is its *temperature,* which must remain constant. A thermal-energy reservoir is an *idealization* which in practice can be closely approximated in several ways. If the heat transfer to or from the thermal reservoir is a very small fraction of its total internal energy, its temperature will tend to remain constant during the energy exchange. Large bodies of water, such as oceans and lakes, and the atmosphere around the earth behave essentially as thermal reservoirs. Another example of a practical thermal reservoir is any two-phase system. Although the ratio of the masses of the two phases will change during the heat addition or removal, the temperature is fixed as long as both phases coexist.

## 6-3-2 KELVIN-PLANCK STATEMENT OF THE SECOND LAW

The thermal efficiency of practical heat engines typically ranges from 10 to 40 percent. Thus some portion of the heat supplied from a high-temperature source in practice is always rejected to a low-temperature sink. A basic question with regard to heat engines is whether it is *theoretically* possible to build a cyclic device that is 100 percent efficient. The answer to this

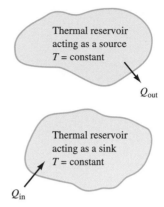

**Figure 6-7**
Illustration of thermal energy reservoirs acting as a source and a sink for heat transfer.

**The only significant thermodynamic property of a thermal-energy reservoir is its temperature.**

question is provided directly by the ***Kelvin-Planck statement*** of the second law, which may be expressed as:

> **I**t is impossible for any system operating in a cyclic manner to receive energy by heat transfer from a single thermal reservoir and deliver an equivalent amount of energy in the form of work to the surroundings.

In terms of heat-engine performance, the Kelvin-Planck statement requires any cyclic device that produces a net work output to exchange energy with a minimum of two thermal reservoirs at different temperatures. It is this restriction that required the simple schematic of a heat engine in Fig. 6-3 to show both a high-temperature source and a low-temperature sink for heat transfer.

The Kelvin-Planck statement requires that heat engines operate between bodies at two (or more) different temperatures. However, the low-temperature body cannot be an energy source like the high-temperature source. If the direction of $Q_{out}$ in Fig. 6-3 were reversed, the overall process would violate the Kelvin-Planck statement, even though the conservation of energy principle would be satisfied. Thus one of the thermal reservoirs must be a source and the other a sink. The Kelvin-Planck limitation on a cyclic heat-to-work converter has a serious ramification on energy usage throughout the world. If any heat engine required only one thermal reservoir, then huge amounts of energy could be taken directly from the environment and converted to work. Thus shortages of energy in the costly form of work would never occur. Unfortunately, the second law prevents this simple solution to energy problems.

In the practical design of power plants the low-temperature sink is usually some portion of the environment such as the atmosphere or cooling water which is taken from a river or lake. Since the thermal efficiency of such plants is 40 percent or less, at least 60 percent of the energy entering the heat engine ultimately appears as heat rejection to the environment. The increase in temperature of portions of the environment due to this heat rejection from the cycle is what constitutes *thermal pollution* from such plants. A heat engine that exchanges heat with a *single* body in an equilibrium state and produces work is called a *perpetual-motion machine of the second kind,* abbreviated PMM2. It is one of the second kind because it violates the second law of thermodynamics. A PMM1 (a perpetual-motion machine of the first kind) is a device that violates the first law; that is, it creates rather than conserves energy. An illustration of a PMM2 is shown in Fig. 6-8. The simple steam power cycle of Fig. 6-2 has been altered by the elimination of the condenser. A portion of the turbine output is still used to drive the pump. As a result the heat transfer input is totally converted to the net work output of the turbine and violates the Kelvin-Planck statement.

**Note that the Kelvin-Planck statement requires a heat engine to have at least one heat source and one heat sink.**

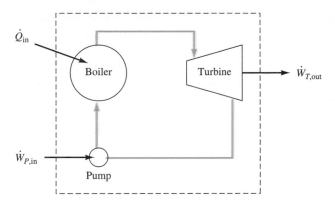

**Figure 6-8**   Elimination of the condenser in a simple steam power plant leads to a PMM2.

An important question arises once the Kelvin-Planck statement is accepted. If theory requires that a portion of the heat transfer to a heat engine must be rejected, is there a maximum theoretical conversion efficiency and how large is the value? If the maximum thermal efficiency is around 95 to 99 percent, and modern heat engines have efficiencies of less than 40 percent, then the current designs are not very good. However, if the theoretical limit is around 60 to 70 percent, for example, then the actual designs look much better. In a following section, a theoretical limit will be developed from second-law considerations, and that limit will be a function solely of the temperatures of the two thermal reservoirs.

### 6-3-3 CLAUSIUS STATEMENT OF THE SECOND LAW

In Sec. 6-2 it was pointed out that actual refrigerators and heat pumps operate with a net work input. Theoretically, it is essential that work be supplied to these devices from an external source. This is due to another form of the second law known as the **_Clausius statement_** of the second law:

> It is impossible to operate any device in such a manner that the sole effect is the transfer of energy by heat transfer from a colder body to another body at a higher temperature.

This statement matches our common experience that heat transfer does not occur spontaneously (by itself) from a cold region to a warmer region.

The Clausius statement excludes the transfer even if some device is employed to aid in the transfer, provided no other effects (changes) are observed in the device itself or in the surroundings. Changes in the device may be avoided by operating the device cyclically, so it always returns to its initial state. However, in this case the Clausius statement still requires

**Note how a refrigerator overcomes the Clausius restriction on heat transfer of energy from a lower to a higher temperature.**

a change in the surroundings if heat transfer is to occur. This requirement simply describes a refrigerator, a device which operates in a cyclic manner, transfers thermal energy from a low- to a high-temperature region, and requires work transfer from the surroundings. Thus the work input observed in the actual operation of refrigerators, air conditioners, and heat pumps is a theoretical requirement. At this point a question similar to that for heat engines arises. Is there a theoretical minimum work input to these devices for specified temperatures of the colder and warmer bodies? Like the heat-engine question posed in the preceding subsection, corollaries to the second law are able to answer both these questions.

### 6-3-4    EQUIVALENCE OF THE KELVIN-PLANCK AND CLAUSIUS STATEMENTS

Since two second-law statements have been proposed, it is essential to prove that they are equivalent. In this way it should not matter which statement is ultimately used to develop theorems and corollaries of the second law. The equivalency of two statements, $X$ and $Y$, occurs if, and only if, the negation of one statement implies the negation of the other statement, and vice versa. A proof that a violation of the Clausius statement leads to a violation of the Kelvin-Planck statement is demonstrated by considering the equipment in Fig. 6-9a. Based on the logic technique outlined above, we assume that a certain cyclic device C can bring about the heat transfer $Q_{low}$ from a low-temperature source at $T_{low}$ to a high-temperature sink at $T_{high}$ without any external effects. A larger transfer of energy $Q_{high}$ then is supplied from the reservoir at $T_{high}$ to a heat engine E which produces net work $W_{net,out}$ and rejects heat transfer $Q_{low}$ at $T_{low}$. Now consider the composite system composed of cyclic device C and heat engine E shown within the dashed-line box in Fig. 6-9a. The heat reservoir $T_{low}$ undergoes no net change, since it

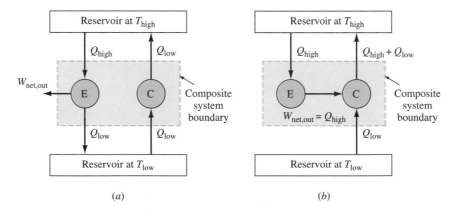

(a)                                              (b)

**Figure 6-9**    Schematics for the proof that the Kelvin-Planck and Clausius statements of the second law are equivalent.

rejects heat $Q_{low}$ to device C and receives heat $Q_{low}$ from engine E. Therefore, the composite system produces net work $W_{net,out}$ while receiving a net heat transfer $Q_{net,in} = Q_{high} - Q_{low}$ from the reservoir at $T_{high}$. This process is a violation of the Kelvin-Planck statement of the second law.

In the proof above, a violation of the Clausius statement led to a violation of the Kelvin-Planck statement. Now consider the arrangement in Fig. 6-9b, where engine E now becomes a Kelvin-Planck violator and cyclic device C becomes an actual refrigerator. On the basis of this arrangement it can be shown that an assumed violation of the Kelvin-Planck statement leads to a violation of the Clausius statement. Thus the two statements are equivalent to each other.

### 6-3-5 HATSOPOULOS-KEENAN STATEMENT OF THE SECOND LAW

It is our experience that systems of all types initially in nonequilibrium will eventually reach equilibrium. If a marble is placed on the side of a bowl and released, we expect the marble eventually to come to rest at the bottom without any change in the environment. If two systems of different temperatures are placed in thermal contact, but they are insulated from the environment, then we observe that the two systems eventually reach a common final temperature. When a gas expands from one tank into another tank, both of which are isolated from the surroundings, soon there will be a common temperature, pressure, and density throughout the two tanks. If two miscible liquids, such as alcohol and water, are poured together, we expect the liquids to mix until a uniform state is reached. Many other examples could be provided. There is no way we can prove that systems left to themselves eventually reach a state of equilibrium. It is simply a matter of common experience. Note, by definition, that a system is in *stable equilibrium* if a finite change in the state cannot occur without a finite change in the state of the environment.[1]

On the basis of this common experience, we can postulate that such behavior is generally to be expected. This postulate is known as the *Hatsopoulos-Keenan statement* of the second law:

---

**A**ny system having certain specified constraints and having an upper bound in volume can reach from any initial state a stable equilibrium state with no effect on the environment.

---

This statement postulates the existence of (stable) equilibrium states. By constraints we mean internal partitions, external conservative force fields,

---

[1] Equilibrium in this text refers to stable equilibrium. Unstable and metastable forms of equilibrium are not considered.

rigid impermeable walls, etc. This statement implies that one particular equilibrium state will be reached. Thus the approach to equilibrium is quite directional in nature. Note that this form of the second law is closely allied to statements 4 and 5 in our original list in Sec. 6-1 of phenomena dealt with by the second law.

An important *corollary* of the Hatsopoulos-Keenan statement is

> *If a system is in stable equilibrium, no change to another stable equilibrium state can have a net work output as its sole effect external to the system.*

To prove the corollary, assume that such a change can have as its sole effect the work transfer of energy to the surroundings. This work transfer then could be taken from the surroundings and used to change the system to a state different from its original state. As a result, the system has changed to a new equilibrium state without a net effect in the surroundings. But this violates the definition of a stable equilibrium state, and hence the corollary is proved.

The three alternative statements of the second law presented above illustrate the wide diversity of observed physical phenomena on which to base a "law of nature." Owing to their equivalency, any statement of the second law will lead to the same set of theorems and corollaries of interest in the engineering design of energy devices.

### 6-3-6    THE EQUIVALENCY OF THE HATSOPOULOS-KEENAN AND CLAUSIUS STATEMENTS

The Hatsopoulos-Keenan (H-K) statement is an alternative form of the second law. Thus the equivalency of this statement to either the Kelvin-Planck or Clausius statements can be demonstrated. To prove the equivalency of the H-K and Clausius statements, consider a system of interest initially in stable equilibrium at temperature $T_{low}$ as shown in Fig. 6-10. Heat transfer $Q_1$ now occurs unaided from the system to another system at a higher temperature $T_{high}$. This process, of course, violates the Clausius statement. Now the same quantity of energy $Q_1$ is transferred to a heat engine. The heat engine during one cycle produces work $W_{net,out}$ and rejects energy equal to $Q_2 = Q_1 - W_{net,out}$ back to the initial system of interest. As a result, the higher-temperature system and the heat engine undergo no net change. Thus the net effect for the entire process is the production of net work output, while the system of interest changes from one equilibrium state to another. But this is a violation of the corollary to the Hatsopoulos-Keenan statement. Thus an assumed violation of the Clausius statement leads to a violation of the H-K statement. The proof that a violation of the H-K statement leads to a violation of the Clausius statement is left to the reader. These two proofs are sufficient to show the equivalency of the Hatsopoulos-Keenan and Clausius statements.

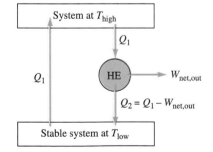

**Figure 6-10**
Schematic for the proof of the equivalence of the Clausius and Hatsopoulos-Keenan statements of the second law.

## 6-4  REVERSIBLE AND IRREVERSIBLE PROCESSES

The Kelvin-Planck statement of the second law requires that the thermal efficiency of all heat engines must be less than 100 percent. However, no limiting maximum value has been established. Thermodynamics also provides an expression for the theoretical upper limit on the thermal efficiency. To derive the general relationship, it is necessary first to describe what is meant by an "ideal" heat engine. This requires an introduction to the concept of reversibility and the reversible process. In general, a ***totally reversible process*** is defined as follows:

> *A process commencing from an initial equilibrium state is called totally reversible if at any time during the process both the system and the environment with which it interacts can be returned to their initial states.*

The concept of total reversibility by its definition requires *restorability*. But the requirement of restorability is quite strong, applying to both the system and its surroundings. Normally, interest is centered solely on the system during a given process. For example, a quasistatic process requires equilibrium conditions within the boundaries of a system, but it places no restriction on the effects which occur in the surroundings. However, a totally reversible process requires something of the environment.

In view of the discussions of work interactions in Chap. 2, any quasistatic form of work, such as boundary work, must be carried out in a series of equilibrium steps. Only in this circumstance will the work output by the system equal the work input for the return path. Consider, as an example, a piston-cylinder assembly which contains a gas and is connected to a flywheel as shown in Fig. 6-11a. The piston is taken to be *frictionless,* and both the cylinder and the piston are perfect heat insulators. If an expansion were carried out by series of small decreases in the external resisting force,

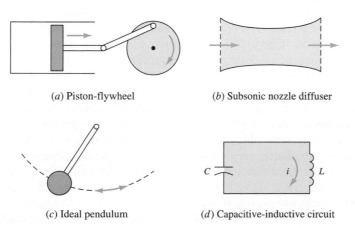

(a) Piston-flywheel        (b) Subsonic nozzle diffuser

(c) Ideal pendulum        (d) Capacitive-inductive circuit

**Figure 6-11**    Some reversible processes.

properties such as the pressure and temperature would change slowly and uniformly throughout the gas. Hence the work output would be equal to the integral of $P\,dV$, and an equivalent amount of energy could be stored in the form of rotational kinetic energy of the flywheel. Now the process could be reversed by removing energy slowly from the flywheel and compressing the gas by a series of small increases in pressure. When the gas reaches its initial pressure the energy removed from the flywheel exactly equals that initially added to it during the expansion process. In addition, the work of compression will equal the work of expansion for the gas. Since the system and surroundings are now back to their initial states, the process described above is a totally reversible one. As a general statement, for the same end states, *the value of a totally reversible work interaction on the reversed path has the same numerical value as that for the original or forward path, but it is of opposite sign.*

This statement regarding work interactions during totally reversible processes can be extended to heat interactions as well. For a closed system, such as the gas inside the piston-cylinder assembly discussed above, the changes in $\Delta U$ for the forward and reverse processes are equal in magnitude, but opposite in sign, because $U$ is a property of the system. The conservation of energy principle for the system states that $\Delta U_{cm} = Q + W$. Hence the following statement must be valid: As a consequence of the definition of a totally reversible process, *it is the nature of a totally reversible process that all heat and work interactions that occurred during the original (forward) process are equal in magnitude but opposite in direction during the reversed process.* Thus no net history is left in the surroundings when the system regains its initial state.

The concept of total reversibility can be applied also to open systems. As a second example, consider the steady flow of a fluid through an adiabatic frictionless nozzle (see Fig. 6-11b). The application of the conservation of energy principle shows that the enthalpy of the fluid decreases as the kinetic energy increases. If the nozzle is now followed by an adiabatic, frictionless diffuser, then the fluid can be returned to a state identical to that at the entrance of the nozzle. The diffuser increases the enthalpy of the fluid at the expense of a decrease in its kinetic energy. Since the surroundings were not involved, the process described is a reversible one. There are other examples of reversible processes, and among these are the ideal pendulum and the pure capacitive-inductive circuit (see Figs. 6-11c and 6-11d) and the perfectly elastic spring. Consequently, many of the basic concepts of introductory physics may be employed to devise totally reversible processes.

The totally reversible process is an *idealization.* It is a concept which can be approximated very closely at times by actual devices, but never matched. There are two fundamental reasons why actual processes never are truly reversible. First, we require the absence of solid friction at the bearing surfaces of pistons, flywheels, and other mechanical devices. A pendulum, in addition, requires the absence of air resistance, which is a type of fluid friction. A capacitive-inductive circuit should not contain a resistive element, and a spring must be made of an ideal elastic material (to fulfill Hooke's law). In actual cases the effects of friction, electric resistance, and inelasticity can be substantially reduced, but their complete elimination is not possible. These physical phenomena are collectively called *dissipative effects,* because in all cases a portion of the energy in the system is converted or dissipated to a less useful form.

A second criterion must be met if a process involving a system and its local environment is to be totally reversible. In the preceding discussion of the piston-cylinder and flywheel combination, it was necessary to carry out the expansion and compression of the gas in a series of equilibrium steps. Otherwise the work output and input would not be the same, and the process would not be reversible. The requirement that a process be quasistatic, in order to be reversible, is a general one. Thus, only infinitesimal unbalanced forces are present during a reversible process. For example, heat transfer is not reversible if its occurrence is due to a finite temperature difference $\Delta T$. Heat transfer is made reversible if the temperature difference between the two interacting systems is of magnitude $dT$.

An overall process is irreversible if either the system or its surroundings cannot be returned to their initial states. Any system that is returned to its initial state after experiencing an irreversible process will leave a history in the surroundings due to the irreversibilities. Irreversibilities arise from two sources:

1. Presence of inherent dissipative effects
2. Presence of a nonquasistatic process

The presence of either class of effects is sufficient to make a process irreversible. Since all actual processes include such effects, the totally reversible process is a *limiting* process toward which all actual processes may approach in performance but never match.

**Note the two possible sources of irreversibilities.**

Whether a given process is totally reversible is probably recognized best by determining whether irreversibilities occur during the process. Most irreversibilities fall into the category of common experience and include

1. Electric resistance
2. Inelastic deformation
3. Viscous flow of a fluid
4. Solid-solid friction
5. Heat transfer across a finite temperature difference
6. Hysteresis effects
7. Shock waves
8. Internal damping of a vibrating system
9. Unrestrained expansion of a fluid
10. Fluid flow through valves and porous plugs (throttling)
11. Spontaneous chemical reactions

12. Mixing of dissimilar gases or liquids

13. Osmosis

14. Dissolution of one phase into another phase

15. Mixing of identical fluids initially at different pressure and temperatures

The foregoing list illustrates two points. First, the processes that are part of one's common experiences are all irreversible. Second, these processes cover a diversity of physical and chemical effects.

The absence of inherent dissipative effects and nonequilibrium effects in a system and its environment leads to the concept of a *totally reversible* process. It frequently is advantageous to consider processes for which *irreversibilities are absent within the system,* but not necessarily absent from the surroundings. Such processes are called **internally reversible.** *Any work interactions that occur during internally reversible processes must be quasistatic.* Thus nonquasistatic work transfers associated with paddle wheels or electric resistors within the system are not permitted, since these effects by nature are irreversible. *An internally reversible process can exactly reverse its path by merely changing the direction of all of its interactions on the boundary.* Finally, an **externally reversible process** is one for which irreversibilities may be present within the boundaries of the system of interest, but the surroundings which interact with the system must undergo only reversible changes. Totally reversible and internally reversible processes both play important roles in the study of thermodynamics. The usefulness of the concept of a reversible process will become apparent by repeated application to different situations.

In Sec. 6-3 the concept of a thermal-energy reservoir was introduced in terms of a constant-temperature source or sink of energy for heat transfer. A more complete definition of a thermal-energy reservoir now becomes:

1. The temperature remains uniform and constant during a process.

2. Changes within the thermal reservoir are internally reversible.

In theory there are no dissipative effects within a thermal reservoir, and the system equilibrates rapidly to maintain a uniform temperature. Therefore, any change in state of a thermal-energy reservoir occurs in an *internally reversible* manner.

**Note the difference between internal and external irreversibilities.**

**Note the two characteristics of a thermal-energy reservoir.**

---

**6-5    THE SECOND LAW AND ENTROPY— CLASSICAL PRESENTATION**

It was noted in Sec. 6-1 that there are numerous phenomena we observe daily that cannot be explained by use of conservation laws such as those for mass and energy. A number of these phenomena were directional in nature.

Then in Sec. 6-3 it was pointed out that a number of second-law statements have arisen historically which can be used to explain our observations. The equivalency of these statements was noted. However, we still need a concept that will unify these second-law statements in an analytical format applicable to all energy processes. The desired mathematical format will be developed in this section by the "classical" or "traditional" method.

This unifying mathematical format of the second law can also be developed from a "postulational" approach. This approach is developed in Sec. 6-6. The authors recommend that in an introductory course in thermodynamics the student be introduced to only one of these approaches. However, they both lead to a common point in the second-law development. The material in Sec. 6-7 and subsequent chapters does not depend on whether the "classical" or "postulational" approach is used initially.

The purpose of Sec. 6-5 is to develop a property called entropy. This property can be used to show why we observe limitations on processes. More importantly, the material we develop below can be used to show how existing energy processes can be improved, and how the performance of new energy systems can be optimized.

## 6-5-1 ANALYTICAL FORMS OF THE KELVIN-PLANCK STATEMENT

Three alternative (and equivalent) statements of the second law were introduced in Sec. 6-3. In theory we can use any one of these statements to develop important second-law deductions. The Kelvin-Planck statement is a convenient one for this purpose because it is easily converted into a useful mathematical form. The concept of reversibility introduced in the preceding section also plays an important role in the development.

Recall that the Kelvin-Planck statement of the second law precludes a heat engine producing work while receiving heat transfer from a *single* thermal reservoir. That is, the *net* work transfer for the above cycle cannot be out of the system. However, there is no restriction on the possibility of the net work being *into* the system or *zero* for any closed cyclic process. Thus an analytical expression for the Kelvin-Planck statement of the second law is

$$W_{\text{net,in}} = \sum_{\text{cycle}} W \geq 0 \qquad \text{(single thermal reservoir)} \quad \textbf{[6-6a]}$$

where the $\sum$ term represent the sum of work transfers during the cycle. Expressing the work transfers on a differential basis, the above equation could also be written as

$$W_{\text{net,in}} = \oint \delta W \geq 0 \qquad \text{(single thermal reservoir)} \quad \textbf{[6-6b]}$$

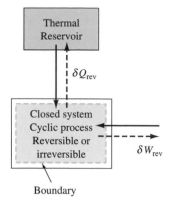

**Figure 6-12**
Schematic of a closed system with reversible work and heat transfer of energy with the local environment.

(Recall that the symbol $\oint$ indicates that the integration is to be carried out over the entire cycle.) In addition, the energy balance for a cyclic process of a closed system reduces to $\oint \delta W = -\oint \delta Q$. As a result, another useful form of the Kelvin-Planck statement in terms of the net heat transfer is

$$Q_{net,in} = \oint \delta Q \leq 0 \qquad \text{(single thermal reservoir)} \qquad \textbf{[6-7]}$$

Hence the net heat transfer during a cycle must be *zero* or *out* of the system. Thus a restriction exists on the direction of the net heat and work transfers associated with a closed system as it undergoes an integral number of cycles while in communication with a single thermal reservoir. The appearance of an inequality sign in the mathematical development of the second law is the first indication that the second law is a *directional* or *nonconservation* law.

The "equality" and "inequality" signs in the preceding equations will now be related to the absence or presence of internal irreversibilities within a cyclic, closed system. In order to focus on the effect of *internal irreversibilities* on the signs in Eqs. [6-6] and [6-7], all effects external to the boundary, including heat and work transfers, will be modeled as reversible, as shown in Fig. 6-12. (The work transfers, for example, could be to or from a spring, a pulley-weight system, or a flywheel.) However, irreversibilities may occur within the closed system. Such an arrangement was described in Sec. 6-4 as an externally reversible one.

Now consider the situation where the closed system in Fig. 6-12 undergoes a cyclic process for which $W_{net,in} = 0$. During a cyclic process several work interactions might occur, either in or out of the closed system. However, the sum of these interactions would be zero for the case under study. For this situation only the equality sign in Eq. [6-6a] [and Eq. [6-6b]] applies. Because the process is cyclic, $\Delta E = 0$ for the closed system. In addition, since $Q_{net} = \Delta E - W_{net}$, the net heat transfer for the cycle is also zero when $W_{net}$ is zero. Hence when the equality sign in Eq. [6-6a] [or Eq. [6-6b]] applies, the equality sign also applies to Eq. [6-7]. In this case there is no net change in the closed system and in the external systems after the cyclic process. But when a system and its total environment are capable of complete restorability, the process is said to be *totally reversible*. Since the heat and work effects were chosen to be externally reversible, the use of the equality sign in Eqs. [6-6] and [6-7] implies that the cyclic system itself is *internally reversible*. It may be shown that, conversely, when a closed system undergoes a cycle in an internally reversible manner, the net work is zero when a single thermal reservoir is present. In summary,

• The equality sign in Eqs. [6-6] and [6-7] is applicable when any closed system, exchanging heat with a single thermal reservoir, undergoes an *internally reversible,* cyclic process.

As the next step, we realize that all processes are either reversible or irreversible. Since all internally reversible, cyclic processes associated with the system in Fig. 6-12 require the use of $W_{\text{net,in}} = 0$ and $Q_{\text{net,out}} = 0$, the *inequality* in these equations must be associated with the presence of *irreversibilities* within the system. Hence

- The inequality sign in Eqs. [6-6] and [6-7] is applicable when any closed system, exchanging heat with a single thermal reservoir, undergoes an *internally irreversible,* cyclic process.

Conversely, for a cyclic process where the net work is in, the process is *internally irreversible,* and heat is rejected by the system to a single thermal reservoir.

The above mathematical rules based on the Kelvin-Planck statement of the second law apply to cyclic closed systems exchanging heat with a *single* thermal reservoir. In addition, the effects *external* to the system boundary are chosen to be reversible. This will allow us to focus attention solely on the cyclic system. The rules represented by Eqs. [6-6] and [6-7] lead to important deductions in subsequent sections regarding the operation of both closed and open systems.

### 6-5-2 GENERAL SECOND-LAW LIMITATIONS FOR HEAT ENGINES

With the introduction of the concepts of reversibility, it is possible now to classify heat engines as internally reversible or internally irreversible. An internally reversible heat engine is free of dissipative and nonequilibrium effects within the engine during its operation.

A schematic of a heat engine with heat transfer $Q_{H,\text{in}}$ into a region of the system at a constant high temperature $T_H$ and heat transfer $Q_{L,\text{out}}$ out of the system from a region at a constant low temperature $T_L$ is shown in Fig. 6-13. The heat engine represented by the circle could be either periodic

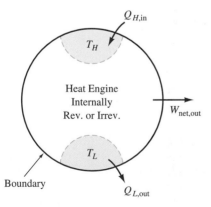

**Figure 6-13**     Schematic of a heat engine operating between thermal reservoirs of $T_H$ and $T_L$.

or closed-loop, steady-state equipment. Whether internally reversible or irreversible, its thermal efficiency on the basis of Eq. [6-3] and the relation $W_{net,out} = Q_{H,in} - Q_{L,out}$ is given by

$$\eta_{th} \equiv \frac{W_{net,out}}{Q_{H,in}} = \frac{Q_{H,in} - Q_{L,out}}{Q_{H,in}} = 1 - \frac{Q_{L,out}}{Q_{H,in}} \qquad \textbf{[6-8]}$$

Now consider the following important corollaries of the second law regarding the thermal efficiency of heat engines.

1. The thermal efficiency of an internally *irreversible* heat engine *is always less* than the thermal efficiency of an internally *reversible* heat engine operating with heat transfer at the same high- and low-temperature regions.

2. The thermal efficiencies of two internally *reversible* heat engines operating with heat transfer at the same high- and low-temperature regions *are equal.*

These two corollaries comprise the *Carnot principle.*

The method of proof of corollary 1 is based on the apparatus shown in Fig. 6-14. An *internally reversible* engine E and an *internally irreversible* engine I are operating between the same two thermal reservoirs. Both engines receive the same quantity of heat $Q_1$. The internally reversible engine produces work $W_E$, while the internally irreversible engine produces work $W_I$. In order to use the analytical form of the Kelvin-Planck statement in our proof, all heat and work transfers for both engines are chosen to be reversible. Since engine E is internally and externally reversible, *its direction of operation may be reversed.* In the reversed direction (as shown by the dashed arrows in Fig. 6-14), the magnitudes of $Q_1$, $Q_2$, and $W_E$ remain the same, but their signs are reversed. Now engine E is operated in the reverse direction as a refrigerator (Ref) simultaneously with heat engine I, and each

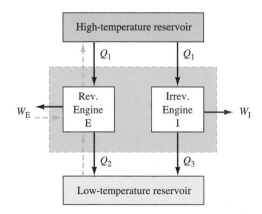

**Figure 6-14**   Sketch of apparatus for proof that $\eta_I < \eta_R$.

goes through an integral number of cycles. In this case the net heat exchange with the high-temperature reservoir is zero. This occurs because this thermal reservoir now receives $Q_1$ from reversed engine $E_{Ref}$ and discharges $Q_1$ to heat engine $I_{HE}$. In addition, a net quantity of heat $Q_2 + Q_3$ is exchanged with the low-temperature thermal reservoir. For the composite system made up of $E_{Ref}$ and $I_{HE}$ (enclosed by the dashed-line box in Fig. 6-14), there is a net exchange of work with the surroundings, while exchanging energy by heat transfer with a single thermal reservoir.

For the system in the dashed box in Fig. 6-14, the Kelvin-Planck analytical statement of the second law requires that $W_{net,in}$ be positive for an internally irreversible cycle with heat exchange with a single thermal reservoir. That is, in this situation Eq. [6-6a] is written as

$$W_{net,in} = W_{E,Ref,in} - W_{I,HE,out} > 0 \qquad \text{[6-9a]}$$

But for the totally reversible heat engine E operated in the reverse direction, $W_{E,Ref,in} = W_{E,HE,out}$. Hence Eq. [6-9a] becomes

$$W_{E,HE,out} - W_{I,HE,out} > 0 \qquad \text{[6-9b]}$$

where the two terms are magnitudes only. Upon rearrangement,

$$W_{E,HE,out} > W_{I,HE,out} \qquad \text{[6-9c]}$$

Thus an internally reversible heat engine E will always produce more work output than an internally irreversible engine I when both receive the same quantity of heat input, and both receive and reject energy by heat transfer at the same two system temperatures. Because each engine receives the same heat input $Q_1$, then on the basis of Eq. [6-8] for the thermal efficiency $\eta_{th} = W_{net,out}/Q_{in}$,

$$\eta_{th,E} > \eta_{th,I} \qquad \text{[6-10]}$$

This confirms Carnot's first corollary. Not unexpectedly, we have shown that internal irreversibilities reduce the thermal efficiency of an actual heat engine in comparison to an internally reversible heat engine operating with the *same* system temperatures during heat supply and heat rejection. Thus internal reversibility is the criterion for "ideal" performance of heat engines.

**Note the restrictions regarding the inequality between the thermal efficiency of an internally reversible and an internally irreversible heat engine.**

The proof of statement 2 of Carnot's principle is carried out in a similar manner. The irreversible engine I in Fig. 6-14 is replaced by an internally reversible engine H. For the composite system consisting of E in reversed operation (as a refrigerator) and H as a heat engine, Eq. [6-6a] requires that

$$W_{net,in} = W_{E,Ref,in} - W_{H,HE,out} = 0$$

The equality sign now applies because both cyclic devices E and H are internally reversible, and the overall process again is taken as externally reversible. Also, as seen earlier, $W_{E,Ref,in} = W_{E,HE,out}$. Thus the above equation can be written as

$$W_{E,HE,out} = W_{H,HE,out}$$

Consequently, for the same heat input $Q_1$ to heat engines E and H, and the same system temperatures during heat supply and heat rejection,

$$\eta_{\text{th,E}} = \eta_{\text{th,H}} \qquad \qquad \textbf{[6-11]}$$

Thus statement 2 of Carnot's principle has been proved. That is, all internally reversible heat engines operating with the same working-fluid temperatures during heat supply and heat rejection have the same thermal efficiency. Cyclic heat engines in general differ from each other only in the series of processes which make up the cycle, and the substance which undergoes the cycle. Corollary 2 implies that the thermal efficiency of an internally reversible heat engine is not a function of the actual processes used for the cycle or of the working fluid used within the heat engine.

The concept of an internally and externally reversible heat engine was used in the development of Carnot's principle. A more detailed schematic of such a device is shown in Fig. 6-15. In order for the heat-transfer processes to be externally reversible, the heat transfer $Q_{H,\text{in}}$ occurs between a thermal reservoir at $T_H + dT$, while the cyclic system temperature is $T_H$. The same differential temperature difference occurs for the heat transfer $Q_{L,\text{out}}$ between $T_L$ and $T_L - dT$. Owing to the differential-temperature differences, the overall process can operate as a heat engine or refrigerator, as demonstrated in the preceding subsection. In either operation the heat- and work-transfer terms are equal in magnitude, and only their direction is reversed. This is a general technique for achieving reversible heat transfer, even though the overall temperature difference $T_H - T_L$ is finite.

**Note that a totally reversible heat engine provides a means for reversible heat transfer across a finite temperature difference.**

### 6-5-3   THE THERMODYNAMIC TEMPERATURE SCALE

On the basis of statement 2 of Carnot's principle, the thermal efficiency of internally reversible heat engines is independent of the construction, cycle design, and working fluid of the engine. As a result the limiting (maximum) value of $\eta_{\text{th,rev}}$ can only be a function of $T_H$ and $T_L$, the temperatures of the working substance during heat addition and removal. They are the only variables left in the design of an internally reversible engine. In the process of seeking the appropriate temperature function, another fundamental question arises. What is the nature of the thermometer to be used for measuring $T_H$ and $T_L$? It is well known that thermometers which rely on different thermometric substances frequently are not in good agreement, even over reasonably small temperature ranges.

Consider now an internally reversible heat engine that operates at steady state and receives energy by heat transfer $Q_{H,\text{in}}$ from a high-temperature source at a boundary temperature $T_H$, as shown in Fig. 6-16. The heat engine also rejects energy by heat transfer $Q_{L,\text{out}}$ at a boundary temperature $T_L$ to its surroundings. The system also produces a net amount of work $W_{\text{net,out}}$. According to the second statement of the Carnot principle, we may write that $\eta_{\text{th}} = f(T_H, T_L)$. In addition Eq. [6-8] shows that $\eta_{\text{th}}$ may also be expressed as

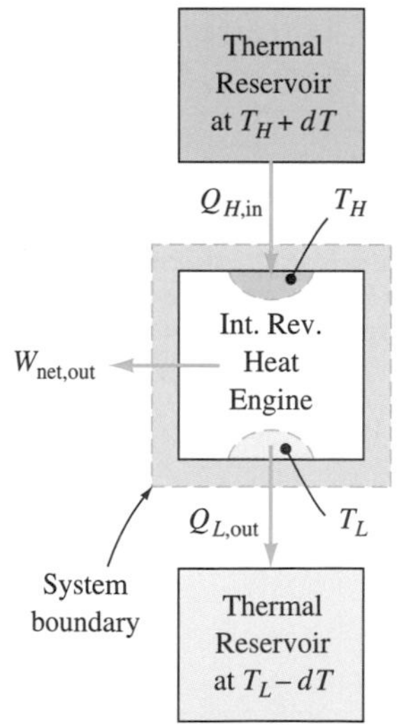

**Figure 6-15**
Schematic of a totally reversible heat engine operating with heat transfer at $T_H$ and $T_l$.

$$\eta_{th} = \frac{W_{net,out}}{Q_{in}} = \frac{Q_{H,in} - Q_{L,out}}{Q_{H,in}} = 1 - \frac{Q_{L,out}}{Q_{H,in}} \qquad \textbf{[6-8]}$$

where again only the magnitudes of $Q_{H,in}$ and $Q_{L,out}$ are required in the equation. Equating the two above expressions for the thermal efficiency, we find that

$$\frac{Q_{L,out}}{Q_{H,in}} = 1 - f(T_H, T_L) \equiv \psi(T_H, T_L)$$

where $f(T)$ and $\psi(T)$ are arbitrary functions. Mathematically, the only acceptable form of the function $\psi(T)$ is $\psi(T_H, T_L) = g(T_L)/g(T_H)$. Therefore, the preceding equation becomes

$$\frac{Q_{L,out}}{Q_{H,in}} = \frac{g(T_L)}{g(T_H)} \qquad \text{(reversible heat-engine cycle)}$$

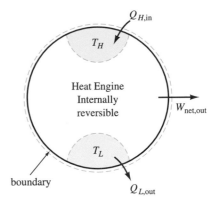

**Figure 6-16**
An internally reversible heat engine with $Q_{H,in}$ at $T_H$ and $Q_{L,out}$ at $T_L$.

where $g(T)$ is another unspecified function. An arbitrary selection of $g(T)$ gives us an opportunity to establish a ***thermodynamic temperature scale*** which is independent of the properties of any substance due to the nature of an internally reversible heat engine. Based on a suggestion by Lord Kelvin, the function $g(T)$ is replaced by $T$ so that

$$\boxed{\frac{Q_{H,in}}{Q_{L,out}} = \frac{T_H}{T_L}} \qquad \text{(reversible heat-engine cycle)} \qquad \textbf{[6-12]}$$

An examination of this result reveals that the ratio of any two temperatures on our thermodynamic temperature scale must have the same ratio as that of the heat transfers for an internally reversible heat engine exchanging energy by heat transfer with its surroundings at these same two boundary temperatures. The temperatures measured on this scale are called ***absolute temperatures***. Equation [6-12] is also valid for internally reversible refrigerators and heat pumps.

The *Kelvin temperature scale* is established by assigning the temperature of a reservoir at the triple state (ts) of water to be 273.16 K. If an internally reversible heat engine is now operated between a reservoir at the triple state of water and another reservoir at an unknown temperature $T$, then this latter temperature is given by

$$T = 273.16 \frac{Q}{Q_{ts}} \qquad \text{(reversible cycle)} \qquad \textbf{[6-13]}$$

where $Q$ and $Q_{ts}$ are the heat transfers that occur at the boundary of the heat engine at temperature $T$ and $T_{ts}$, respectively. Temperatures on the absolute or thermodynamic temperature scale must be expressed in either kelvins or degrees Rankine. The temperature on the Kelvin scale approaches zero as the rejected heat transfer approaches zero.

Since an internally reversible heat engine is an idealization, other techniques must be used to measure numerical values of temperature. Purposely, the temperature scale based on a constant-volume, ideal-gas thermometer (as discussed in Sec. 1-6-4) is identical with the absolute Kelvin scale based on a reversible heat engine. Other empirical measurement techniques enable us to achieve good practical results in those regions of temperature where the ideal-gas thermometer is not applicable.

**6-5-4**    PERFORMANCE STANDARDS
             FOR REVERSIBLE HEAT ENGINES

As noted in Sec. 6-1, the second law establishes the criteria for the "ideal" performance of energy devices. The substitution of Eq. [6-12] into Eq. [6-8] yields the "ideal" thermal efficiency $\eta_{\text{th,rev}}$ for internally reversible heat engines that receive heat transfer at $T_H$ and reject heat transfer at $T_L$.

$$\eta_{\text{th,rev}} = \eta_{\text{th,Carnot}} \equiv 1 - \frac{T_L}{T_H} \qquad \textbf{[6-14]}$$

The thermal efficiency given by Eq. [6-14] is called the ***Carnot efficiency***. According to the Carnot principle, this is the *maximum* efficiency that any heat engine could have when operating between temperatures $T_H$ and $T_L$. It applies to any internally reversible heat-engine device, as well as to a totally reversible heat-engine cycle. Since the first law of thermodynamics prohibits thermal efficiencies greater than 100 percent, it is impossible to have negative temperatures on the absolute or thermodynamic temperature scale.

As noted in Sec. 6-5-2, actual engines would be less efficient than reversible heat engines operating between the same $T_H$ and $T_L$. Common power-producing devices have thermal efficiencies ranging from 10 to 40 percent. These values are low relative to 100 percent However, Eq. [6-14] dictates that actual efficiencies should be compared not to 100 percent but to some lower theoretical maximum value. To achieve high thermal efficiencies in actual heat engines, the second law predicts that $T_H$ should be as high as possible, and irreversibilities should be held to the lowest practical values. Note that a description of the cyclic path of the working fluid in a reversible heat engine is not required. Several ways of operating a theoretical heat engine with a Carnot efficiency are discussed later in the text.

**Example 6-3**

**A** simple steam power cycle receives 100,000 kJ/min as heat transfer to the working fluid at 800 K, and rejects energy as heat transfer from the working fluid at 320 K. If the pump power required is 1400 kJ/min, determine (*a*) the thermal efficiency of an internally-reversible cycle and (*b*) the turbine power output in kilowatts. (*c*) Compare the theoretical thermal efficiency to that of an actual cycle calculated in Example 6-1.

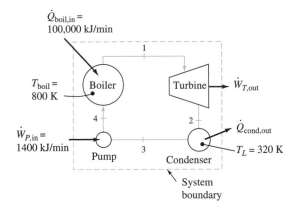

$\dot{Q}_{boil,in} = 100{,}000$ kJ/min

$T_{boil} = 800$ K

$\dot{W}_{P,in} = 1400$ kJ/min

$\dot{W}_{T,out}$

$\dot{Q}_{cond,out}$

$T_L = 320$ K

Boiler

Turbine

Pump

Condenser

System boundary

**Figure 6-17**   Equipment schematic and data for Example 6-3.

**Solution:**

**Given:**   A simple steam power cycle with appropriate data is shown in Fig. 6-17.

**Find:**   (*a*) the theoretical thermal efficiency and (*b*) the gross turbine power output, in kW. (*c*) Compare to Example 6-1.

**Model:**   Steady-state closed-loop cycle.

**Strategy:**   Apply the conservation of energy principle and the Carnot thermal-efficiency equation.

**Analysis:**   (*a*) Employing Eq. [6-14] for the Carnot thermal efficiency, we find that

$$\eta_{th,rev} = \eta_{th,Carnot} = 1 - \frac{T_L}{T_H} = 1 - \frac{320}{800} = 0.625 \text{(or 62.5\%)}$$

(*b*) Now that we know the Carnot thermal efficiency for the closed-loop, steady-state system within the dashed line on Fig. 6-17, we can calculate the net power output from the system using the basic definition of thermal efficiency, $\eta_{th} = \dot{W}_{net,out}/\dot{Q}_{in}$. Thus

$$\dot{W}_{net,out} = \eta_{th}\dot{Q}_{boil,in} = 0.625(100{,}000 \text{ kJ/min}) = 62{,}500 \text{ kJ/min}$$

For this system, the net power output is $\dot{W}_{net,out} = \dot{W}_{T,out} - \dot{W}_{P,in}$. Solving for the turbine power out, we have

$$\dot{W}_{T,out} = \dot{W}_{net,out} + \dot{W}_{P,in} = (62{,}500 + 1{,}400) \text{ kJ/min}$$
$$= (63{,}900 \text{ kJ/min})\left(\frac{1 \text{ kW·min}}{60 \text{ kJ}}\right) = 1065 \text{ kW}$$

Note that the heat transfer rate from the condenser could also be calculated, if required, by writing a steady-state energy balance for the system inside the dashed lines. The results would show that $\dot{W}_{net,out} = \dot{Q}_{net,in} = \dot{Q}_{boil,in} - \dot{Q}_{cond,out}$ and $\dot{Q}_{cond,out} = 37,500$ kJ/min.

(c) The actual thermal efficiency for the same boiler heat-transfer rate and pump work input was found to be 34 percent in Example 6-1. This compares to 62.5 percent for the theoretical (internally reversible) cycle.

**Comment:**    The difference between an actual turbine power output of 590 kW and a theoretical output of 1065 kW reflects the influence of internal irreversibilties on the turbine performance, when pump irreversibilities are not considered.

### 6-5-5    THE CLAUSIUS INEQUALITY

In previous sections we have developed mathematical relations for the Kelvin-Planck statement of the second law, for the measurement of a thermodynamic temperature scale, and for the thermal efficiency of an internally reversible heat engine with constant temperatures at the boundaries during the heat-transfer processes. All these relations solely involve the parameters $\delta Q$ (or $Q$), $\delta W$ (or $W$), and the absolute temperature $T$. We now seek an important relationship among these same variables for a more general situation where a closed system undergoes a cyclic process. The system can experience multiple heat transfers in or out, multiple work transfers in or out, and the temperature at the boundary can change throughout the cycle. The only constraint on the process within the closed system is that it is either internally reversible or internally irreversible.

The system shown by a dotted line on the right side of Fig. 6-18 undergoes an infinitesimal change of state, with heat and work transfers $\delta Q$ and $\delta W$. In order to use the Kelvin-Planck analytical format in our development, the heat-transfer process external to the cyclic, closed system must be reversible. To ensure this, a totally reversible, cyclic device $D$ is placed between a single thermal reservoir at constant temperature $T_R$ and the closed system at a variable temperature $T$ (recall the discussion at the end of Sec. 6-5-2 on totally reversible engines). The solid arrows in the figure representing $\delta Q_R$ into device $D$ and $\delta W_D$ out of device $D$ are for

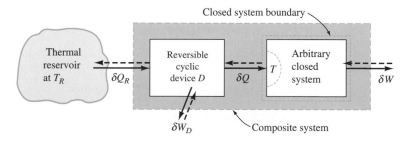

**Figure 6-18**    Schematic for the proof of the Clausius inequality.

cyclic device $D$ operating as a heat engine, and supplying heat transfer $\delta Q$ to the closed system. At the same time work $\delta W$ is transferred out of the closed system. The dashed arrows are for device $D$ operating in the reverse direction (as a refrigerator), and heat transfer $\delta Q$ is out of the closed system.

Recall that for an internally reversible heat engine or refrigerator with heat transfers at constant system temperatures, Eq. [6-12] in general requires that $Q_{in}/T_{in} = Q_{out}/T_{out}$. In terms of Fig. 6-18 and cyclic device $D$, the quantities $\delta Q_R$ *measured relative to cyclic device D* and $\delta Q$ *measured relative to the closed system* always have the same direction. This is true whether device $D$ is operating as a heat engine or in the reverse direction as a refrigerator. Consequently, for each incremental heat transfer $\delta Q_R$ and $\delta Q$ exchanged with the cyclic device, Eq. [6-12] simply becomes

$$\frac{\delta Q_R}{T_R} = \frac{\delta Q}{T} \qquad \text{[6-15]}$$

This equation is applicable regardless of the direction of operation of cyclic device $D$. In addition, consider the *composite system* shown by the *dashed line* in Fig. 6-18. This composite *cyclic* system exchanges heat with the single reservoir $T_R$, and exchanges work transfers $\delta W$ and $\delta W_D$ with the surroundings. Thus the Kelvin-Planck analytical statement, in the format of Eq. [6-7], is applicable to the process within the dashed line. In terms of the symbols in Fig. 6-18, Eq. [6-7] becomes

$$\oint \delta Q_R \le 0 \qquad \text{[6-16]}$$

Noting that $T_R$ is a constant, the substitution of Eq. [6-15] into Eq. [6-16] yields

$$T_R \oint \left(\frac{\delta Q}{T}\right)_b \le 0$$

where the sign on $\delta Q$ is required when its value is inserted into the equation. The subscript b has been added to emphasize that both $\delta Q$ and $T$ must be evaluated at that part of the system *boundary* where the heat transfer occurs. Also, $T$ is an absolute temperature.

Since $T_R$ is a positive number on the Kelvin scale, the above equation can be written as

$$\oint \left(\frac{\delta Q}{T}\right)_b \le 0 \qquad \text{(closed system)} \qquad \text{[6-17]}$$

where $\delta Q$ follows our standard sign convention on heat transfer to the *closed* system. The cyclic integral sign indicates a summation process carried out over the entire cycle and over all parts of the closed-system boundary. This result is known as the **Clausius inequality.** It states that when *any* closed

**Note the physical significance of the Clausius inequality.**

system undergoes a *cyclic process,* the sum of all the $\delta Q/T$ terms at the system boundary for each differential increment of the process will always be equal to or less than zero. It applies equally well to a periodic mechanical device or to a continuous steady-state device. Following previous notation, and recalling that the derivation is based on externally reversible processes, the *equality sign applies to internally reversible cycles* and the *inequality applies to internally irreversible cycles* for any closed system.

### 6-5-6   THE ENTROPY FUNCTION

The internally reversible (int rev) format of Eq. [6-17],

$$\oint \left(\frac{\delta Q}{T}\right)_{\text{int rev}} = 0 \qquad\qquad \textbf{[6-18]}$$

leads to an important result. Recall from Sec. 1-3 that the change in value of any property for a cyclic process is zero. Conversely, if the integral of a quantity is zero over an arbitrary cycle, then the quantity is a property. Therefore, $\delta Q/T$ in the above equation represents the differential change in a thermodynamic property. The property is denoted by $S$ and was named **entropy** by Clausius in 1865. Thus by definition

$$dS \equiv \left(\frac{\delta Q}{T}\right)_{\text{int rev}} \qquad \text{(closed system)} \qquad \textbf{[6-19]}$$

or, for a finite, internally reversible path,

$$\Delta S = S_2 - S_1 = \int_1^2 \left(\frac{\delta Q}{T}\right)_{\text{int rev}} \qquad \text{(closed system)} \quad \textbf{[6-20]}$$

Thus the first and second laws are similar in that both lead to the definition of a *change* in a property ($E$ and $S$) in terms of interactions at the boundary of a closed system. For an internally reversible process the change in entropy ($S_2 - S_1$) is due solely to the heat transfer. The term on the right side of Eqs. [6-19] and [6-20] is interpreted as a measure of the *entropy transfer* associated with heat transfer. To perform the integration of Eq. [6-20], a functional relationship between $\delta Q$ and $T$ must be known and the process must be internally reversible. Since entropy is a property, the change in entropy is the same for all processes between two given end states. Thus, once $\Delta S$ is calculated by Eq. [6-20] for an internally reversible path, the resulting value is the *same* for all other processes between the same end states, whether these processes are internally reversible or irreversible. The dimensions of the entropy function are energy/(absolute temperature). For engineering purposes the SI units are typically kJ/K, while the USCS units are Btu/°R.

**Note that $\delta Q/T$ must be integrated along an internally reversible path to determine the entropy change.**

For any closed system undergoing an internally reversible process, $dS = \delta Q/T$. The relationship between $dS$ and $\delta Q/T$ for internally *irreversible* processes is determined from Eq. [6-17]. Consider a cyclic process composed of an irreversible process (Irr) from state 1 to state 2 and a reversible process (Rev) from 2 to 1 as illustrated in Fig. 6-19. On the basis of Eq. [6-17] we write

$$\oint \frac{\delta Q}{T} = \int_1^2 \left(\frac{\delta Q}{T}\right)_{Irr} + \int_2^1 \left(\frac{\delta Q}{T}\right)_{Rev} < 0$$

Only the inequality applies, because part of the overall cycle is irreversible. Substitution of Eq. [6-20] for the second term on the right in the above relation, and reversing the limits on the last integral, yields

$$\int_1^2 \left(\frac{\delta Q}{T}\right)_{Irr} < S_2 - S_1 = \Delta S \qquad \text{(irreversible process)}$$

Thus for an internally *irreversible* process, the integral of $\delta Q/T$ is *always* less than the value of $\Delta S$. A combination of this equation with Eq. [6-20] leads to the *generalization* that

$$\int_1^2 \frac{\delta Q}{T} \le \Delta S \qquad\qquad\qquad \textbf{[6-21]}$$

or

$$\boxed{\frac{\delta Q}{T} \le dS} \qquad \text{(closed system)} \qquad \textbf{[6-22]}$$

where the equality and inequality signs apply to internally reversible and irreversible processes, respectively. As a consequence, the entropy transfer for a closed system *never* exceeds the value of the entropy change. Note that if the change in entropy for a closed system is found to be less than the integral of $\delta Q/T$, the process is impossible. Hence a comparison of $\Delta S$ with the integral of $\delta Q/T$ is a test as to whether a process involving a closed system is internally reversible, irreversible, or impossible.

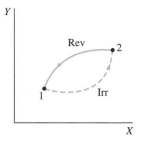

**Figure 6-19**
Cyclic path for a closed system.

**Note how $\delta Q/T$ compares for reversible and irreversible paths with the same end states.**

### 6-5-7 ENTROPY GENERATION AND THE CLOSED-SYSTEM ENTROPY BALANCE

In the preceding section the fundamental relationship $dS \ge \delta Q/T$ was developed for any closed system [see Eq. [6-22]]. This equation indicates that $\delta Q/T$ is a measure of $dS$ when evaluated during an internally reversible process. However, $\delta Q/T$ is not an accurate measure of $dS$ for an internally irreversible process; in fact, its contribution is always too small. This inequality between $dS$ and $\delta Q/T$ during internally irreversible processes can be made into an equality if we consider the concept of entropy generation.

The only difference between internally reversible and irreversible processes is the physical presence of internal irreversibilities in the latter case. In terms of Eq. [6-22] in the preceding section, it is apparent that internal irreversibilties *always make a positive contribution* to the entropy change, independent of any heat-transfer effect. This contribution is called the **entropy generation** or **entropy production** and is noted by the symbol $\sigma$ on an extensive basis. Hence the inequality in Eq. [6-22] can be made an equality if a term $\sigma$ is added to account for the *presence of internal irreversibilities*.

The entropy change of any closed system is due solely to two possible effects, namely, entropy transfer and entropy generation. Thus for a differential change of state a general **entropy balance for a closed system** or control mass (cm) is expressed as

$$dS_{\text{cm}} = \frac{\delta Q}{T} + \delta\sigma$$

[6-23]

For a finite change of state,

$$S_2 - S_1 = \int \frac{\delta Q}{T} + \sigma$$

[6-24]

where $\sigma \geq 0$. These two equations retain the concept that heat transfer always contributes to the entropy change of a closed system, whether the process is reversible or irreversible. This contribution can be positive, negative, or zero, depending on the direction of the actual heat transfer. Note that the symbol $\delta$ is used before $\sigma$ (instead of the differential symbol $d$) in Eq. [6-23]. This symbol is used to emphasize that entropy production, like heat and work, is a function of the path of the process. Entropy generation depends upon the magnitude of the internal irreversibilities present during the process. Since internal irreversibilities represent losses in performance during a process, the entropy-generation term is one measure of these losses.

In some situations heat transfer may occur at several positions on the boundary. If the temperature at these positions is different, then a more general expression for Eq. [6-23] is

$$dS_{\text{cm}} = \sum_{j=1}^{n} \frac{\delta Q_j}{T_j} + \delta\sigma$$

[6-25]

where $T_j$ is the temperature at the point where $\delta Q_j$ enters. Equations [6-23] and [6-25] also may be expressed on a rate basis. For example, Eq. [6-25] becomes

$$\frac{dS_{\text{cm}}}{dt} = \sum_{j=1}^{n} \frac{\dot{Q}_j}{T_j} + \dot{\sigma}$$

[6-26]

where $\dot{\sigma} = \delta\sigma/dt$, the *time rate of entropy production* due to irreversibilities within the system. The term $\sum \dot{Q}_j/T_j$ is the *net rate of entropy transport* by heat transfer. Equation [6-26] may be written in the word format

$$\begin{pmatrix} \text{Rate of entropy} \\ \text{ACCUMULATION} \\ \text{in the system} \\ \text{at time } t \end{pmatrix} = \begin{pmatrix} \text{net rate of entropy} \\ \text{TRANSPORT} \\ \text{into the system with} \\ \text{heat transfer at time } t \end{pmatrix} + \begin{pmatrix} \text{rate of entropy} \\ \text{GENERATION} \\ \text{within the system} \\ \text{at time } t \end{pmatrix}$$

[6-27]

Note the many forms of a closed-system entropy balance. All of them can be obtained from careful application of Eq. [6-26].

Equations [6-23] through [6-26] are all forms of an *entropy balance* for closed-system processes. As a result all of these equations may be written in the format

Entropy change = entropy transfer + entropy production  **[6-28]**

$$\begin{pmatrix} > \\ = 0 \\ < \end{pmatrix} \qquad \begin{pmatrix} > \\ = 0 \\ < \end{pmatrix} \qquad (\geq 0)$$

The terms in the parentheses below Eq. [6-28] indicate the possible values of the quantities in the equation. As a consequence, the only way to decrease the entropy of any closed system is by heat transfer *out* of the system. However, in this case the entropy-transfer contribution must be more negative than the positive contribution of any internal irreversibilities. Note also from the preceding equations that entropy transfer is only associated with heat transfer at a boundary. No entropy transfer is associated with work transfer at a boundary. Some types of work interactions do lead, however, to entropy production within a system.

Note the two mechanisms which can alter the entropy of a closed system.

The introduction of entropy generation into the analysis of energy processes is very important. This term, for example, measures the loss in the work capability or the inefficiencies present during energy processes. Entropy production values generally are used for comparative purposes. For example, the entropy production in various components in a cycle could be compared to determine the relative inefficiencies of operation of the components. On this basis the engineer might first redesign the component with the greatest entropy generation within the cycle. Also, several different cycles with the same design criteria might be compared on the basis of their overall entropy generation.

The evaluation of entropy generation typically requires knowledge of both the entropy change for the process and the entropy transfer which takes place at each portion of the boundary where heat transfer occurs. The entropy-transfer term in some cases may be difficult to evaluate if either the value of the heat transfer or the temperature at the boundary is unknown. In some cases it is necessary to assume a quasiequilibrium process in order to approximate the entropy transfer associated with heat transfer.

## 6-6 THE SECOND LAW AND ENTROPY— ALTERNATE PRESENTATION

It was noted in Sec. 6-1 that there are numerous phenomena we observe daily that cannot be explained by use of conservation laws such as those for mass and energy. A number of these phenomena have a directional nature. Then in Sec. 6-3 it was pointed out that a number of second-law statements have arisen historically which can be used to explain our observations. The equivalency of these statements was noted. However, we still need a concept that will unify these second-law statements in an analytical format applicable to all energy processes. The desired mathematical format will be developed in this section by a "postulational" approach.

This unifying mathematical format of the second law can also be developed from a "classical" or "traditional" approach. This approach is followed in Sec. 6-5. The authors recommend that in an introductory course in thermodynamics the student be introduced to only one of these approaches. They both lead to a common point in the second-law development. The material in subsequent sections and chapters does not depend on whether the "classical" or "postulational" method is used initially.

The goal in this section (Sec. 6-6) is to develop a property called entropy. This property can be used to show why we observe limitations on processes. More importantly, the material we develop below can be used to show how existing energy processes can be improved, and how the performance of new energy systems can be optimized.

### 6-6-1 THE SECOND-LAW POSTULATE

A fourth form of the second law is the second-law postulate. This is just another way of summarizing the second law of thermodynamics. This postulate does three things. First it postulates the existence of a new system property, entropy. Second, it states the conditions under which the property entropy can be transported across the boundaries of a closed system. Third, it places limits upon the allowable changes in entropy for an isolated system. This postulate is stated below:

---

#### THE SECOND-LAW POSTULATE

1. There exists an extensive, intrinsic property called **entropy** $S$.
2. Entropy is transported by heat transfer across the boundaries of a closed system. The **entropy transport rate associated with heat transfer** at a boundary with heat-transfer rate $\dot{Q}$ and uniform temperature $T$ is defined by the quantity $\dot{Q}/T$, and this quantity has the same sign convention as the heat-transfer rate.

3. Entropy can only be produced, and in the limit of an internally reversible process the rate of entropy production reduces to zero. Thus the *rate of entropy production* $\dot{\sigma}$ for any system must satisfy the following relation:

$$\dot{\sigma} > 0 \quad \text{for internally irreversible processes}$$

$$\dot{\sigma} = 0 \quad \text{for internally reversible processes}$$

This statement is equivalent to the other three, as will be demonstrated shortly. The singular advantage of this statement of the second law of thermodynamics is that it immediately provides a means for predicting the behavior of real systems by focusing on the system itself. Using the information in the postulate, the dimensions on the entropy transfer rate must be [energy/time]/[temperature]. This means that entropy must have the physical dimensions of [energy]/[temperature].

Although this postulational approach may seem strange, a similar approach could have been followed for mass and energy. If we had chosen to write postulates for mass and energy, they might have looked like the following:

### CONSERVATION OF MASS POSTULATE

1. There exists an extensive, intrinsic property called mass $m$.
2. The mass of an isolated system remains constant; that is, mass is conserved.

### CONSERVATION OF ENERGY POSTULATE

1. There exists an extensive property called energy $E$.
2. Energy can be transported across the boundaries of a closed system by two mechanisms: work and heat transfer.
3. The energy of an isolated system remains constant; that is, energy is conserved.

Before we can demonstrate the equivalence of the four alternative forms of the second law or can begin to explore the full implications of the second law, we need to turn the postulate into a working equation for the accounting of entropy.

### 6-6-2   AN ENTROPY BALANCE FOR A CLOSED SYSTEM

Using the second-law postulate, it is possible to construct an entropy accounting equation (or balance) for a closed system. (We will extend our results to an open system in the following section.) For a closed system, the time rate of change of any extensive property within the closed system is equal to the net rate at which it is transported across the system boundary into the system plus the rate at which it is produced inside the system. This is a generic accounting principle that is valid for any extensive property. When this accounting principle is applied to a closed system for the extensive property entropy, we obtain the **rate form of the entropy balance for a closed system**:

$$\frac{dS_{cm}}{dt} = \sum_{j=1}^{n} \frac{\dot{Q}_j}{T_j} + \dot{\sigma} \qquad \textbf{[6-29]}$$

where $\dot{\sigma} = \delta\sigma/dt$, the time rate of entropy production due to irreversibilities within the system. The subscript $j$ corresponds to a surface with specified heat-transfer rate $\dot{Q}_j$ and uniform surface temperature $T_j$. Equation [6-29] may be written in the word format

$$\begin{pmatrix} \text{Rate of entropy} \\ \text{ACCUMULATION} \\ \text{in the system} \\ \text{at time } t \end{pmatrix} = \begin{pmatrix} \text{net rate of entropy} \\ \text{TRANSPORT} \\ \text{into the system with} \\ \text{heat transfer at time } t \end{pmatrix} + \begin{pmatrix} \text{rate of entropy} \\ \text{GENERATION} \\ \text{within the system} \\ \text{at time } t \end{pmatrix}$$

$$\begin{pmatrix} > \\ = 0 \\ < \end{pmatrix} \qquad\qquad \begin{pmatrix} > \\ = 0 \\ < \end{pmatrix} \qquad\qquad (\geq 0)$$

$$\textbf{[6-30]}$$

The terms in the parentheses below Eq. [6-30] indicate the possible values of the quantities in the equation. Note from the entropy balance that entropy transfer for a closed system is only associated with heat transfer at a boundary. No entropy transfer is associated with work transfer at a boundary. Some types of work interactions do lead, however, to entropy production within a system. Note also that the only way to decrease the entropy of a closed system is to have heat transfer *out* of the system. However, in this case the entropy-transfer contribution must be more negative than the positive contribution of any internal irreversibilities.

For a differential change of state a general entropy balance for a closed system or control mass (cm) on an *extensive basis* is expressed as

$$dS_{cm} = \frac{\delta Q}{T} + \delta\sigma \qquad \textbf{[6-31]}$$

**Note the two mechanisms which can alter the entropy of a closed system.**

For a finite change of state,

$$(S_2 - S_1)_{cm} = \int \frac{\delta Q}{T} + \sigma \qquad \text{[6-32]}$$

where again $\sigma \geq 0$. Note that the symbol $\delta$ is used before $\sigma$ (instead of the differential symbol $d$) in Eq. [6-31]. This symbol is used to emphasize that entropy production, like heat and work, is a function of the path of the process. Entropy generation depends upon the magnitude of the internal irreversibilities present during the process. Since internal irreversibilities represent losses in performance during a process, the entropy-generation term is one measure of these losses.

In some situations heat transfer may occur at several positions on the boundary. If the temperature at these positions is different, then a more general expression for Eq. [6-31] is

$$dS_{cm} = \sum_{j=1}^{n} \frac{\delta Q_j}{T_j} + \delta\sigma \qquad \text{[6-33]}$$

where $T_j$ is the temperature at the point where $\delta Q_j$ enters, and the summation again is over $n$ different locations on the boundary.

Before we continue there is only one more point that needs to be investigated before we can safely apply the entropy balance. How do we define the temperature in the entropy balance? This will be investigated in the next section.

## 6-6-3 THE THERMODYNAMIC TEMPERATURE SCALE AND THE CARNOT EFFICIENCY

The second-law postulate uses a temperature in its definition of the rate of entropy transport with heat transfer but provides no explicit information about this temperature. As we know, there exist thermometers which rely on different thermometric substances that are frequently not in good agreement, even over reasonably small temperature ranges. For purposes of identification we will call a temperature scale that satisfies the second-law postulate a ***thermodynamic temperature scale***. The temperatures measured on this scale are called ***absolute temperatures.***

To investigate this temperature scale, consider a simple heat-engine cycle that operates at steady-state conditions, receives energy by heat transfer from its surroundings at a rate of $\dot{Q}_{H,in}$ and a boundary temperature of $T_H$, and rejects energy by heat transfer to its surroundings at a rate of $\dot{Q}_{L,out}$ and a boundary temperature of $T_L$. See Fig. 6-20. The system also produces a net amount of work at a rate of $\dot{W}_{net,out}$. Applying the conservation of energy equation and the entropy accounting equation to this closed, steady-state

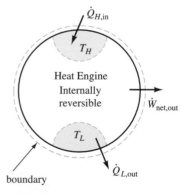

**Figure 6-20**
Schematic of an internally reversible heat engine with heat transfer at $T_L$ and $T_H$.

system gives

$$\frac{dE_{cm}}{dt} = \dot{Q}_{H,in} - \dot{Q}_{L,out} - \dot{W}_{net,out} \rightarrow \dot{W}_{net,out} = \dot{Q}_{H,in} - \dot{Q}_{L,out} \quad \textbf{[6-34]}$$

$$\frac{dS_{cm}}{dt} = \frac{\dot{Q}_{H,in}}{T_H} - \frac{\dot{Q}_{L,out}}{T_L} + \dot{\sigma} \rightarrow \dot{\sigma} = -\left(\frac{\dot{Q}_{H,in}}{T_H} - \frac{\dot{Q}_{L,out}}{T_L}\right) \quad \textbf{[6-35]}$$

where $dE/dt = dS/dt = 0$ in steady state. If the heat engine is operating *internally reversibly* (as shown in Fig. 6-20), then $\dot{\sigma} = 0$, and there exists a unique relationship between the two heat-transfer rates and the related surface temperatures. Under internally reversible conditions, the above entropy accounting equation [Eq. [6-35]] becomes

$$0 = \frac{\dot{Q}_{H,in}}{T_H} - \frac{\dot{Q}_{L,out}}{T_L} \quad \text{or} \quad \left(\frac{\dot{Q}_{H,in}}{\dot{Q}_{L,out}}\right)_{int\ rev} = \frac{T_H}{T_L} \quad \textbf{[6-36]}$$

An examination of Eq. [6-36] reveals that *the ratio of any two temperatures on our thermodynamic temperature scale must have the same ratio as that of the heat-transfer rates* for an internally reversible heat engine exchanging energy by heat transfer at these two temperatures.

The *Kelvin temperature scale* is a thermodynamic temperature scale established by assigning the temperature of the triple state (ts) of water to be exactly 273.16 K. Consider an internally reversible heat engine that operates with one boundary temperature at the triple state of water $T_{ts}$ and another boundary temperature at an unknown temperature $T$. On the basis of Eq. [6-36], this latter temperature $T$ is given by

$$T = 273.16 \frac{\dot{Q}}{\dot{Q}_{ts}} \quad \text{(internally reversible)} \quad \textbf{[6-37]}$$

where $\dot{Q}$ and $\dot{Q}_{ts}$ are the heat-transfer rates that occur at the boundary of an internally reversible heat engine at temperature $T$ and $T_{ts}$, respectively. Temperatures on the absolute or thermodynamic temperature scale must be expressed in either kelvins or degrees Rankine. The temperature on the Kelvin scale approaches zero as the rejected heat transfer approaches zero. All calculations involving entropy transport in the entropy balance *must be done using absolute temperatures.*

Since an internally reversible heat engine is an idealization, other techniques must be used to partially measure temperature. The temperature scale based on a constant-volume ideal-gas thermometer (as discussed in Sec. 1-6-4) is identical with the Kelvin scale based on an internally reversible heat engine. Other empirical measurement techniques enable us to achieve good practical results in those regions of temperature where the ideal-gas thermometer is not applicable.

If the result of Eq. [6-36] for the temperature ratios for an internally reversible heat engine is combined with the definition of thermal efficiency of a heat engine given by Eq. [6-3], we obtain an expression for the thermal efficiency in terms of absolute temperatures.

$$\eta_{th,\text{int rev}} = \frac{\dot{W}_{net,out}}{\dot{Q}_{H,in}} = \frac{\dot{Q}_{H,in} - \dot{Q}_{L,out}}{\dot{Q}_{H,in}} = 1 - \frac{\dot{Q}_{L,out}}{\dot{Q}_{H,in}} = 1 - \frac{T_L}{T_H}$$

Thus the ratio of any two absolute temperatures is directly related to the thermal efficiency of an internally reversible heat engine operating between the same two temperatures. The thermal efficiency given by the above equation is called the **Carnot efficiency.** That is,

$$\eta_{th,\text{int rev}} = \eta_{Carnot} \equiv 1 - \frac{T_L}{T_H} \qquad \textbf{[6-38]}$$

Since the first law of thermodynamics prohibits thermal efficiencies greater than 100 percent, it is impossible to have negative temperatures on the absolute or thermodynamic temperature scale.

Example 6-4

A simple steam power cycle receives 100,000 kJ/min as heat transfer to the working fluid at 800 K, and rejects energy as heat transfer from the working fluid at 320 K. If the pump power required is 1400 kJ/min, determine (*a*) the thermal efficiency of an internally reversible cycle, and (*b*) the turbine power output in kilowatts. (*c*) Compare the theoretical thermal efficiency to that of an actual cycle calculated in Example 6-1.

**Solution:**

**Given:** A simple steam power cycle with appropriate data is shown in Fig. 6-21.

**Find:** (*a*) the theoretical thermal efficiency and (*b*) the gross turbine power output, in kW. (*c*) Compare to Example 6-1.

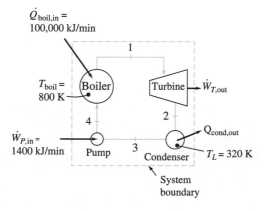

**Figure 6-21** Schematic for Example 6-4.

**Model:**    Steady-state, closed-loop cycle.

**Strategy:**    Apply the conservation of energy principle and the Carnot thermal efficiency equation.

**Analysis:**    (*a*) Employing Eq. [6-38] for the Carnot thermal efficiency, we find that

$$\eta_{\text{th,rev}} = \eta_{\text{th,Carnot}} = 1 - \frac{T_L}{T_H} = 1 - \frac{320}{800} = 0.625 \text{ (or 62.5\%)}$$

(*b*) Now that we know the Carnot thermal efficiency for the closed-loop, steady-state system within the dashed line on Fig. 6-17, we can calculate the net power output from the system using the basic definition of thermal efficiency, $\eta_{\text{th}} = \dot{W}_{\text{net,out}}/\dot{Q}_{\text{th}}$. Thus

$$\dot{W}_{\text{net,out}} = \eta_{\text{th}}\dot{Q}_{\text{boil,in}}$$
$$= 0.625(100{,}000 \text{ kJ/min}) = 62{,}500 \text{ kL/min}$$

For this system, the net power output is $\dot{W}_{\text{net,out}} = \dot{W}_{T,\text{out}} - \dot{W}_{P,\text{in}}$. Solving for the turbine power out, we have

$$\dot{W}_{T,\text{out}} = \dot{W}_{\text{net,out}} + \dot{W}_{P,\text{in}} = (62{,}500 + 1{,}400) \text{ kJ/min}$$
$$= (63{,}900 \text{ kJ/min})\left(\frac{1 \text{ kW·min}}{60 \text{ kJ}}\right) = 1065 \text{ kW}$$

Note that the heat transfer rate from the condenser could also be calculated, if required, by writing a steady-state energy balance for the system inside the dashed lines. The results would show that $\dot{W}_{\text{net,out}} = \dot{Q}_{\text{net,in}} = \dot{Q}_{\text{boil,in}} - \dot{Q}_{\text{cond,out}}$ and $\dot{Q}_{\text{cond,out}} = 37{,}500 \text{ kJ/min}$.

(*c*) The actual thermal efficiency for the same boiler heat-transfer rate and pump work input was found to be 34 percent in Example 6-1. This compares to 62.5 percent for the theoretical (internally reversible) cycle.

**Comment:**    The difference between an actual turbine power output of 590 kW and a theoretical output of 1065 kW reflects the influence of internal irreversibilities on the turbine performance.

---

## 6-6-4    MEASURING THE ENTROPY FUNCTION

The second-law postulate and the entropy balance provide us with a way to define precisely the entropy function. Consider a closed system that is undergoing a differential change in state and is exchanging energy by heat transfer with the surroundings at a boundary temperature $T$. The entropy balance for the closed system is

$$dS_{\text{cm}} = \frac{\delta Q}{T} + \delta\sigma \qquad\qquad \textbf{[6-23]}$$

If we further restrict ourselves to an internally reversible process, the entropy-generation term goes to zero, and the differential change in entropy is defined as

$$dS \equiv \left(\frac{\delta Q}{T}\right)_{\text{int rev}} \qquad \text{(closed system)} \qquad \textbf{[6-39]}$$

The property entropy was named by Clausius in 1865. For a finite change in state this equation can be integrated to give

$$\Delta S = S_2 - S_1 = \int_1^2 \left(\frac{\delta Q}{T}\right)_{\text{int rev}} \qquad \text{(closed system)} \quad \textbf{[6-40]}$$

where the integral must be performed over an internally reversible process that connects the two states of interest. Because entropy is a property, the change in entropy between any two states is independent of the exact path. However, to evaluate this entropy change using the integral of $\delta Q/T$, the process must be internally reversible. Methods of integration are shown in Chap. 7. The dimensions of the entropy function as discussed earlier are [energy]/[absolute temperature]. For engineering purposes the SI units are typically kJ/K, while the USCS units are Btu/°R.

### 6-6-5 Equivalence of the Four Statements of the Second Law

As stated earlier, all four statements of the second law are equivalent. If a system can be devised that violates one of the laws, then all of them are violated.[2]

**A. Kelvin-Planck and the Second-Law Postulate** Imagine a steady-state, closed system that exchanges energy by heat transfer with its surroundings at a single temperature $T_s$ and also transfers energy by work to the surroundings. This would be a Kelvin-Planck violator (see Fig. 6-22). Starting with the rate form of the energy and entropy balance, we have the following equations for the Kelvin-Planck violator:

$$\frac{dE}{dt} = \dot{Q}_{\text{net,in}} - \dot{W}_{\text{net,out}} \qquad \text{and} \qquad \frac{dS}{dt} = \frac{\dot{Q}_{\text{net,in}}}{T_s} + \dot{\sigma}$$

Since $dE/dt = dS/dt = 0$ in steady state, the above equations reduce to

$$\dot{Q}_{\text{net,in}} = \dot{W}_{\text{net,out}} \qquad \text{and} \qquad 0 = \frac{\dot{Q}_{\text{net,in}}}{T_s} + \dot{\sigma}$$

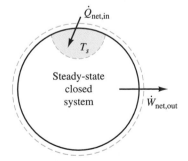

**Figure 6-22**
Schematic for a Kelvin-Planck Violator.

---

[2]This section may be skipped on first reading. The following material shows interesting applications of the second law.

Solving for the heat-transfer rate in the entropy balance and then substituting it into the result of the energy balance gives the following:

$$\dot{W}_{\text{net,out}} = -T_s\dot{\sigma} \leq 0$$

Because the entropy-generation rate must be greater than or equal to zero, the net mechanical power out of the system must be negative or zero in the limit for an internally reversible process. Thus a device that could exchange energy by heat transfer at a single temperature source with the surroundings and produce an equivalent amount of power out of the system would destroy entropy. This violates the second-law postulate and is impossible. Thus a system that violates the Kelvin-Planck statement of the second law also violates the second-law postulate.

**B. Clausius Statement and the Second-Law Postulate**   Imagine that a steady-state system exists that spontaneously transfers energy by heat transfer from a low-temperature region to a high-temperature region. This would be a Clausius violator (see Fig. 6-23). Specifically consider a closed, steady-state system that receives a heat-transfer rate of energy $\dot{Q}_{L,\text{in}}$ from the surroundings at a boundary temperature $T_L$ and rejects energy by heat transfer at a rate of $\dot{Q}_{H,\text{out}}$ to its surroundings at a boundary temperature $T_H$. No other interactions occur between the system and the surroundings, and $T_L < T_H$. Applying both the conservation of energy and the entropy balance rate equations gives the following equations for the Clausius violator:

$$\frac{dE_{\text{cm}}}{dt} = \dot{Q}_{L,\text{in}} - \dot{Q}_{H,\text{out}} \rightarrow \dot{Q}_{L,\text{in}} = \dot{Q}_{H,\text{out}}$$

$$\frac{dS_{\text{cm}}}{dt} = \frac{\dot{Q}_{L,\text{in}}}{T_L} - \frac{\dot{Q}_{H,\text{out}}}{T_H} + \dot{\sigma} \rightarrow \dot{\sigma} = -\left(\frac{\dot{Q}_{L,\text{in}}}{T_L} - \frac{\dot{Q}_{H,\text{out}}}{T_H}\right)$$

since $dE/dt = dS/dt = 0$ in steady state. Now writing the results for the entropy-generation rate in terms of the heat-transfer rate into the system at the low temperature, we obtain the following:

$$\dot{\sigma} = -\left(\frac{\dot{Q}_{L,\text{in}}}{T_L} - \frac{\dot{Q}_{H,\text{out}}}{T_H}\right) = -\frac{\dot{Q}_{L,\text{in}}}{T_L}\left(1 - \frac{T_L}{T_H}\right) \geq 0$$

Examination of these results shows that if the proposed device, the Clausius violator, were possible, the entropy-generation rate for the device would be negative and entropy would be destroyed. This violates the second-law postulate and is impossible. Thus a system that violates the Clausius statement of the second law also violates the second-law postulate.

**C. Hatsopoulos-Keenan Statement and the Second-Law Postulate**   The Hatsopoulos-Keenan statement postulates the existence of stable equilibrium states. In addition, it says that an isolated system will seek a stable equilibrium state. Let's examine what the conservation of

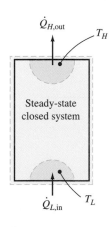

**Figure 6-23**
Schematic for a
Clausius Violator.

mass, conservation of energy, and the entropy balance have to say about the behavior of an isolated system. Recall that an isolated system is one that has no interactions of any kind with the surroundings. Applying this condition to the equations for mass, energy, and entropy gives the following results:

$$\frac{dm_{cm}}{dt} = 0 \qquad \frac{dE_{cm}}{dt} = 0 \qquad \frac{dS_{cm}}{dt} = \dot{\sigma}$$

Now for an isolated system with any initial state, these equations indicate that the mass and energy of the system will remain constant, but the entropy of the system will always increase or remain constant. The state of an isolated system in equilibrium would, by definition, be unchanging with time. The unidirectional nature of the rate of entropy change would seem to support the idea that as time increased the rate of change of entropy would become smaller and smaller until it reaches zero, indicating an equilibrium state.

## 6-7 ENTROPY BALANCE FOR A CONTROL VOLUME

The concepts of entropy generation $\sigma$ and an entropy balance for a closed system have been introduced in sections 6-5 and 6-6. A natural extension of these concepts may be made to a control volume. In Chaps. 2 and 5 the conservation of energy principles for closed and open systems were presented. In order to develop an energy balance for a control volume (open system), energy transport across the system boundary due to mass transfer had to be added to the heat-transfer and work-transfer modes already taken into account for a closed system. In a like manner, an entropy balance for a control volume must include terms which account for the transport of entropy across the control surface due to mass transfer, in addition to those which represent entropy transfer due to heat transfer and entropy generation.

### 6-7-1 THE CONTROL VOLUME ENTROPY BALANCE

In Sec. 5-3, on the basis of Fig. 5-2, the rate of energy change for a control mass is related to the rate of energy change for a control volume by

$$\frac{dE_{cm}}{dt} = \frac{dE_{cv}}{dt} + \sum_{out} \dot{m}_e e_e - \sum_{in} \dot{m}_i e_i$$

where the subscripts $e$ and $i$ represent exit and inlet states. Since the entropy is another extensive property like energy or mass, it too can be transported into and out of a control volume. Thus an equation for the entropy, similar to the one above for energy, takes on the form

$$\frac{dS_{cm}}{dt} = \frac{dS_{cv}}{dt} + \sum_{out} \dot{m}_e s_e - \sum_{in} \dot{m}_i s_i \qquad \textbf{[6-41]}$$

In this equation $s$ represents the specific entropy associated with the mass entering or leaving the control volume. This is analogous to the use of $e$ for the specific energy entering or leaving the control volume. In addition, the entropy balance for a control mass developed earlier states that

$$\frac{dS_{cm}}{dt} = \sum_{j=1}^{n} \frac{\dot{Q}_j}{T_j} + \dot{\sigma} \qquad \textbf{[6-42]}$$

Substitution of Eq. [6-42] into the preceding equation yields an expression for an ***entropy balance for a control volume***, namely,

$$\frac{dS_{cv}}{dt} = \sum_{in} \dot{m}_i s_i - \sum_{out} \dot{m}_e s_e + \sum_{j=1}^{n} \frac{\dot{Q}_j}{T_j} + \dot{\sigma}_{cv} \qquad \textbf{[6-43]}$$

where the terms on the right account for the overall rate of entropy change for the control volume. In the above equation the direction of $Q_j$ is measured relative to the control volume. In words, Eq. [6-43] may be written for a control volume as

$$\begin{pmatrix} \text{Time rate} \\ \text{entropy} \\ \text{change} \end{pmatrix} = \begin{pmatrix} \text{net rate of entropy} \\ \text{transfer due to} \\ \text{mass flow} \end{pmatrix} + \begin{pmatrix} \text{net rate of entropy} \\ \text{transfer due to} \\ \text{heat transfer} \end{pmatrix} + \begin{pmatrix} \text{rate of entropy} \\ \text{production due to} \\ \text{irreversibilities} \end{pmatrix}$$

**Note the three physical effects that can alter the entropy of a control volume.**

Hence the three physical mechanisms for the entropy change of any control volume are mass transfer, heat transfer, and internal irreversibilities. The first two contributions can be positive or negative, when present. The third term is always positive when present. These terms are shown on a control volume schematic with one inlet and one exit in Fig. 6-24.

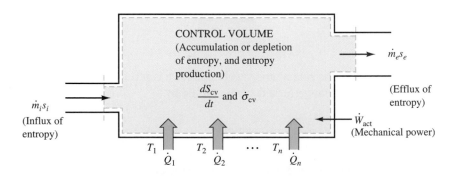

**Figure 6-24**   Schematic for general entropy accounting for a control volume.

## 6-7-2 GENERAL FORM OF THE ENTROPY BALANCE

Generally, Eq. [6-43] and the steady-state forms of this equation will be the formats used for problem solutions in this text. Nevertheless, a more general integral-differential format in terms of local properties is sometimes desirable. In this more general form Eq. [6-43] becomes

$$\frac{d}{dt}\int_V s\rho \, dV = \int_A \left(\frac{q''}{T}\right)_b dA + \sum_i \left(\int_A s\rho V_n \, dA\right)_i - \sum_e \left(\int_A s\rho V_n \, dA\right)_e + \dot{\sigma}_{cv}$$

[6-44]

In this expression $q''$ is the *heat flux* or heat transfer rate per unit of surface area, and $V_n$ is the normal component of velocity relative to the flow area. This equation is analogous to the general format of the energy balance for a control volume given in Sec. 5-3-2.

## 6-8 INCREASE IN ENTROPY PRINCIPLE FOR A CLOSED SYSTEM

One form of the entropy balance developed earlier for a closed system is

$$dS_{cm} = \frac{\delta Q}{T} + \delta\sigma$$

where $\delta\sigma \geq 0$. Processes for which $\delta\sigma < 0$ are *impossible*. This equation leads directly to one form of the *increase in entropy principle*. In the absence of heat transfer, any closed system must fulfill the condition that

$$\boxed{dS_{adia} \geq 0}$$   (closed system)   [6-45]

For a finite change of state the following format is useful:

$$\Delta S_{adia} \geq 0$$   (closed system)   [6-46]

In conclusion, *the entropy of a closed system always increases due to internal irreversibilities during an adiabatic process. In the limiting case of an internally reversible adiabatic process, the entropy will remain constant. Adiabatic closed processes for which $dS_{adia} < 0$ are physically impossible.*

There are a number of engineering processes that are not adiabatic. In such situations Eqs. [6-45] and [6-46] are not directly applicable. For cases where heat transfer occurs, it is possible to include in the second-law analysis every part of the surroundings that is affected by changes in the system of interest. A composite system that includes all the parts that undergo change during a process is an **isolated system.** The entropy change of this composite or isolated (isol) system thus is the sum of the entropy change of the closed system (sys) of interest plus the entropy change of its surroundings (surr)

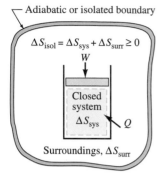

**Figure 6-25**

Illustration of the increase in entropy principle for a closed system interacting with its surroundings.

**Note the effect of a decrease in entropy of one subsystem within an isolated system on the change in entropy of other subsystems.**

which undergo change. That is,

$$dS_{isol} = dS_{sys} + dS_{surr} \qquad \textbf{[6-47]}$$

Since no mass or energy transfer takes place across its boundaries, every isolated system fulfills the condition of being closed and adiabatic. As a result, Eq. [6-45] applies also to an isolated system. Therefore,

$$dS_{isol} = dS_{sys} + dS_{surr} \geq 0 \qquad \textbf{[6-48]}$$

Processes for which $dS_{isol} < 0$ are impossible. For a finite change of state,

$$\Delta S_{isol} = \Delta S_{sys} + \Delta S_{surr} \geq 0 \qquad \textbf{[6-49]}$$

This situation is illustrated in Fig. 6-25. The isolated system is also referred to as the *total* system. For example, consider the isolated system shown in Fig. 6-26. The double arrows indicate heat transfer and work interactions may occur among a number of the subsystems (subsys). On the basis of an adiabatic boundary around all the interacting subsystems (subsys), Eq. [6-49] may be written as

$$\Delta S_{isol} = \sum \Delta S_{subsys} \geq 0 \qquad \textbf{[6-50]}$$

The algebraic sum of the entropy changes of *all* the subsystems participating in a process is zero if all the subsystems undergo reversible changes and all interactions between these parts are reversible. If irreversibilities occur within any subsystem or are associated with any interaction between two or more subsystems, then the entropy of the overall system must increase. When Eq. [6-50] is used, it may be found that $\Delta S$ for one or more subsystems is negative. This is permissible as long as the sum for all subsystems is positive. If the sum of the $\Delta S$ values is negative, the process is physically impossible.

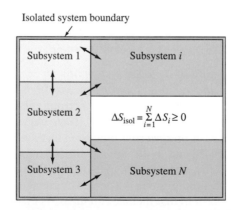

**Figure 6-26** Increase in entropy principle for an isolated system with many interacting subsystems.

These principles are equally valid for a wide range of processes, including the analysis of steady-flow systems. Before we proceed, some important comments must be made with respect to these principles.

1. The increase-in-entropy principles are *directional* statements. They limit the direction in which processes can proceed. A decrease in entropy is not possible for closed adiabatic systems or for a composite of systems which interact among themselves.
2. The entropy function is a *nonconserved* property, and the increase-in-entropy principles are nonconservation laws. Irreversible effects create entropy. The greater the magnitude of the irreversibilities, the greater the entropy generation.
3. Based on the second law, a state of equilibrium is attained when the entropy of a closed system reaches the *maximum* possible value, consistent with the constraints on the system.

In the introductory comments in Sec. 6-1, it was pointed out that some processes are observed to have a preferred direction of change, irrespective of the first law. The increase-in-entropy principles require certain processes to be directional in nature. The various theorems of the second law and the concept of entropy will play an important role throughout the remainder of the text.

Finally, the application of the general entropy balance for a closed system, $dS = \delta Q/T + \sigma$, to an isolated system requires that

$$dS_{\text{isol}} = \delta\sigma_{\text{tot}} \geq 0 \qquad\qquad \textbf{[6-51]}$$

or, for a finite change within an isolated system,

$$\Delta S_{\text{isol}} = \sigma_{\text{total}} \geq 0 \qquad\qquad \textbf{[6-52]}$$

Either of these relations provides a physical reason for the increase in entropy principle for an isolated system. Heat transfer does not occur to or from an isolated system. Hence the only contribution to the overall entropy change of an isolated system is that due to internal entropy generation within subsystems or due to irreversible interactions between subsystems.

## 6-9 SECOND-LAW LIMITATIONS ON THE PERFORMANCE OF HEAT ENGINES, REFRIGERATORS, AND HEAT PUMPS

The entropy balance places restrictions on the heat transfer at the boundary of closed systems. Because all cycles operate as closed systems and typically exchange energy with their surroundings by heat transfer, this should

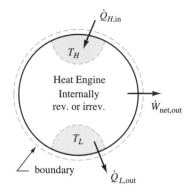

**Figure 6-27**

A heat engine with $\dot{Q}_{H,\text{in}}$ at $T_H$ and $\dot{Q}_{L,\text{out}}$ at $T_L$.

be an area where the second law and the entropy accounting equation can provide us additional information about the performance of these devices.

### 6-9-1 HEAT-ENGINE CYCLE

Consider a heat-engine cycle that operates at steady-state conditions, receives energy by heat transfer from its surroundings at a rate of $\dot{Q}_{H,\text{in}}$ and a temperature of $T_H$, and rejects energy by heat transfer to its surroundings at a rate of $\dot{Q}_{L,\text{out}}$ and a temperature of $T_L$. See Fig. 6-27. The system also produces a net amount of work at a rate of $\dot{W}_{\text{net,out}}$. Applying the conservation of energy equation and the entropy balance to this *closed steady-state* system leads to

$$\frac{dE_{\text{cm}}}{dt} = \dot{Q}_{H,\text{in}} - \dot{Q}_{L,\text{out}} - \dot{W}_{\text{net,out}} \rightarrow \dot{W}_{\text{net,out}} = \dot{Q}_{H,\text{in}} - \dot{Q}_{L,\text{out}} \quad \textbf{[6-53]}$$

$$\frac{dS_{\text{cm}}}{dt} = \frac{\dot{Q}_{H,\text{in}}}{T_H} - \frac{\dot{Q}_{L,\text{out}}}{T_L} + \dot{\sigma} \rightarrow \dot{\sigma} = -\left(\frac{\dot{Q}_{H,\text{in}}}{T_H} - \frac{\dot{Q}_{L,\text{out}}}{T_L}\right) \quad \textbf{[6-54]}$$

where $dE/dt = dS/dt = 0$ in steady state. Using the results of the steady-state entropy balance format in Eq. [6-54] to eliminate the heat-transfer rate $\dot{Q}_{L,\text{out}}$ from the steady-state energy balance in Eq. [6-53] yields

$$\dot{W}_{\text{net,out}} = \dot{Q}_{H,\text{in}} - \dot{Q}_{L,\text{out}} = \dot{Q}_{H,\text{in}} - \left(\dot{Q}_{H,\text{in}}\frac{T_L}{T_H} + T_L\dot{\sigma}\right)$$

Upon rearrangement,

$$\dot{W}_{\text{net,out}} = \dot{Q}_{H,\text{in}}\left(1 - \frac{T_L}{T_H}\right) - T_H\dot{\sigma} \quad \textbf{[6-55]}$$

Notice that the last term on the right-hand side of the last equation contains the product of a temperature and the entropy production rate. This term must of necessity always be greater than or equal to zero. This means that if there is any entropy production (any internal irreversibilities), the power out of the system is reduced below that of an internally reversible cycle.

Another way to consider the cycle performance is to calculate its thermal efficiency. A rearrangement of Eq. [6-55] leads to the thermal efficiency of the heat-engine cycle described above.

$$\eta_{\text{th}} = \frac{\dot{W}_{\text{net,out}}}{\dot{Q}_{H,\text{in}}} = \left(1 - \frac{T_L}{T_H}\right) - \frac{T_H\dot{\sigma}}{\dot{Q}_{H,\text{in}}} \quad \textbf{[6-56]}$$

Since $\dot{\sigma} \geq 0$, we must conclude from Eq. [6-56] that

$$\eta_{\text{th}} \leq 1 - \frac{T_L}{T_H} \quad \textbf{[6-57]}$$

The above equation clearly indicates that the *maximum thermal efficiency* for a heat-engine device that exchanges energy by heat transfer at absolute temperatures $T_H$ and $T_L$ occurs when the heat engine operates internally reversibly. That is, $\eta_{\text{int rev}} > \eta_{\text{int irrev}}$. This maximum efficiency is referred to as the **Carnot efficiency.** Restating the above equation, we find that

$$\eta_{\text{Carnot}} = 1 - \frac{T_L}{T_H} \qquad \textbf{[6-58]}$$

Since the first law of thermodynamics prohibits thermal efficiencies greater than 100 percent, *it is impossible to have negative temperatures on the thermodynamic temperature scale.* Common power-producing devices have thermal efficiencies ranging from 10 to 40 percent, as seen in Table 6-1. To achieve high thermal efficiencies in actual heat engines, $T_H$ should be as high as possible, and irreversibilities should be held to the lowest practical values.

A closer examination of the equations above reveals three significant points. First, the maximum efficiency always occurs for an internally reversible engine. Second, the efficiency of every internally reversible engine operating between boundary temperatures $T_H$ and $T_L$ is identical. And third, the maximum efficiency of a heat engine is solely a function of the temperatures at the boundaries where it receives and rejects energy by heat transfer. These results, derived from an energy balance and an entropy balance, were first proposed by Sadi Carnot and together are known as the *Carnot principles.*[3]

**Table 6-1** Thermal efficiencies of some power-producing devices

| Type | Efficiency |
|---|---|
| Automobile | |
|    Spark ignition, gasoline | 12–25 |
| Truck | |
|    Compression ignition, diesel | 30–35 |
| Locomotive | |
|    Diesel | 30 |
| Gas turbine (100 hp) | |
|    Without regeneration | 12 |
|    With regeneration | 16 |
| Gas turbine (>7500 kW) | |
|    Without regeneration | 25 |
|    With regeneration | 34 |
| Steam power plant | 38–41 |

**Heat** transfer to the working fluid of an internally reversible heat engine occurs at $T_H = 750$ K, and heat rejection occurs at $T_L = 300$ K. Determine (*a*) the thermal efficiency, (*b*) the work produced, in kilojoules, and (*c*) the energy rejected as heat transfer, in kilojoules, if the heat transferred to the engine is 1200 kJ. Then, (*d*) determine the thermal efficiency and net work output, in kilojoules, for temperatures of 350, 400, 500, 600, 800, and 1000 K during heat addition, and plot the thermal efficiency versus $T_H$ for a fixed value of $T_L = 300$ K.

**Example 6-5**

**Solution:**

**Given:** A schematic with appropriate data for an internally reversible heat engine operating between fixed working-fluid temperatures is shown in Fig. 6-28*a*.

**Find:** (*a*) $\eta_{\text{th}}$, (*b*) $W_{\text{net,out}}$, in kJ, (*c*) $Q_{L,\text{out}}$, in kJ. Then (*d*) plot $\eta_{\text{th}}$ versus $T_H$ for $T_H = 350$, 400, 500, 600, 800, and 1000 K, if $T_L = 300$ K.

**Model:** Internally reversible heat engine.

**Strategy:** Apply the first- and second-law relations for internally reversible heat engines.

---

| [3]The Carnot principles were derived from the Kelvin-Planck statement of the second law in Sec. 6-5-2.

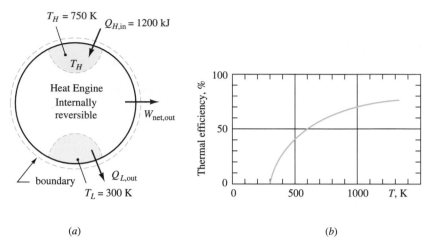

$(a)$                                                      $(b)$

**Figure 6-28**    $(a)$ Equipment schematic and data for Example 6-3; $(b)$ thermal efficiency of a Carnot heat engine as a function of $T_H$ for a fixed $T_L$
value of 300 K.

**Analysis:**    $(a)$ Employing Eq. [6-58] for the thermal efficiency of a Carnot heat
engine, we find that

$$\eta_{th} = 1 - \frac{T_L}{T_H} = 1 - \frac{300}{750} = 0.600$$

Therefore, the thermal efficiency is 60.0 percent.
     $(b)$ The thermal efficiency by definition is $\eta_{th} = W_{net,out}/Q_{H,in}$. Hence,

$$W_{net,out} = \eta_{th}Q_{H,in} = 0.600(1200) = 720 \text{ kJ}$$

     $(c)$ From the conservation of energy principle, $Q_{H,in} - Q_{L,out} - W_{net,out}$ and
$\Delta E = 0$ for a cycle. Hence the heat rejected is

$$Q_{L,out} = Q_{H,in} - W_{net,out} = 1200 - 720 = 480 \text{ kJ}$$

     $(d)$ Repetition of parts $(a)$ and $(b)$ for other source temperatures yields

| $T_H$, K | 350 | 400 | 500 | 600 | 800 | 1000 |
|---|---|---|---|---|---|---|
| $\eta_{th}$, % | 14.3 | 25.0 | 40.0 | 50.0 | 62.5 | 70.0 |
| $W$, kJ | 172 | 300 | 480 | 600 | 750 | 840 |

Figure 6-28$b$ is a plot of thermal efficiency versus $T_H$ for a fixed rejection temperature $T_L$ of 300 K.

**Comment:**    The thermal efficiency and work out are the maximum values possible for the given boundary temperatures. Note that the thermal efficiency rises
rapidly with an early increase in $T_H$ above $T_L$, but this effect diminishes at higher
values of $T_H$.

## 6-9-2 REFRIGERATION CYCLE

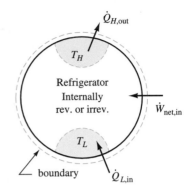

**Figure 6-29**
A refrigerator or heat pump with heat transfer at $T_H$ and $T_L$.

The basic concepts behind Carnot's principle and Carnot's efficiency for heat engines can be extended to refrigerators and heat pumps. As might be anticipated, the maximum COP of either a refrigerator or a heat pump is obtained when these cyclic devices are internally reversible. Thus internally irreversible refrigerators and heat pumps have a COP which is less than that for an internally reversible device. Again, these results apply for given values of $T_H$ and $T_L$. The proof can be obtained by applying energy and entropy balances to either of these two devices.

Consider a simple refrigerator cycle that operates at steady-state conditions and receives energy by heat transfer from its surroundings at a rate $\dot{Q}_{L,\text{in}}$ at a system temperature $T_L$. In addition, the system rejects energy by heat transfer $\dot{Q}_{H,\text{out}}$ at a system temperature $T_H$, as shown in Fig. 6-29. The system also receives a net amount of work at a rate of $\dot{W}_{\text{net,in}}$. Application of the conservation of energy to this closed steady-state system gives

$$\frac{dE}{dt} = \dot{Q}_{L,\text{in}} - \dot{Q}_{H,\text{out}} + \dot{W}_{\text{net,in}} \rightarrow \dot{W}_{\text{net,in}} = \dot{Q}_{H,\text{out}} - \dot{Q}_{L,\text{in}}$$

and the accounting equation for entropy yields

$$\frac{dS}{dt} = \frac{\dot{Q}_{L,\text{in}}}{T_H} - \frac{\dot{Q}_{H,\text{out}}}{T_L} + \dot{\sigma} \rightarrow \dot{\sigma} = \frac{\dot{Q}_{H,\text{out}}}{T_H} - \frac{\dot{Q}_{L,\text{in}}}{T_L}$$

where $dE/dt = dS/dt = 0$ in steady state. Use of the result of the steady-state entropy balance to eliminate the heat-transfer rate $\dot{Q}_{H,\text{out}}$ from the steady-state energy balance leads to

$$\dot{W}_{\text{net,in}} = \dot{Q}_{H,\text{out}} - \dot{Q}_{L,\text{in}} = T_H \left( \dot{\sigma} + \frac{\dot{Q}_{L,\text{in}}}{T_L} \right) - \dot{Q}_{L,\text{in}}$$

Upon rearrangement,

$$\dot{W}_{\text{net,in}} = T_H \dot{\sigma} + \dot{Q}_{L,\text{in}} \left( \frac{T_H}{T_L} - 1 \right) \qquad \text{(refrigeration)} \qquad \textbf{[6-59]}$$

Since $\dot{\sigma} \geq 0$, the above result indicates that the power input required for an internally irreversible refrigeration cycle is always greater than that for an internally reversible cycle, for fixed values of $\dot{Q}_{L,\text{in}}$, $T_H$, and $T_L$.

Equation [6-59] can be rearranged into a general expression for the COP of a refrigeration cycle. The result is

$$\text{COP}_R \equiv \frac{\dot{Q}_{L,\text{in}}}{\dot{W}_{\text{net,in}}} = \left( \frac{T_L}{T_H - T_L} \right) - \frac{T_H T_L}{T_H - T_L} \frac{\dot{\sigma}}{\dot{W}_{\text{net,in}}}$$

Because the last term after the minus sign in this equation is always positive, we arrive at the general conclusion that

$$\text{COP}_{R,\text{int rev}} > \text{COP}_{R,\text{int irrev}} \qquad \textbf{[6-60]}$$

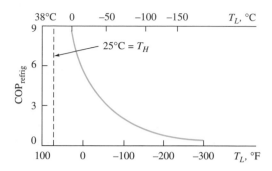

**Figure 6-30**     Variation of the Carnot coefficient of performance for a reversible refrigerator with $T_L$ for a fixed value of $T_H$ of 25°C (77°F).

Thus the COP of an internally irreversible refrigerator is always less than the COP of an internally reversible refrigerator, when operating with the same system temperature during heat supply and the same temperature during heat rejection.

When two internally reversible refrigerators operate between the same temperatures $T_H$ and $T_L$, their COPs are equal. This maximum value is found by setting $\dot{\sigma}$ equal to zero in the general $\text{COP}_R$ expression above. Consequently, for an *internally reversible refrigerator,*

$$\text{COP}_{R,\text{rev}} = \text{COP}_{R,\text{Carnot}} = \frac{T_L}{T_H - T_L} \qquad \textbf{[6-61]}$$

The smaller the temperature difference $T_H - T_L$, the greater the coefficient of performance for a given value of $T_L$. Note that, although the limiting value of the thermal efficiency is less than unity, the coefficient of performance theoretically can be much larger than unity.

Figure 6-30 illustrates the variation of $\text{COP}_{\text{refrig,Carnot}}$ as $T_L$ is changed for a fixed value of the rejection (sink) temperature $T_H$ of 25°C or 77°F. Note that the coefficient of performance increases rapidly as the temperature difference between source and sink is decreased. The colder the low-temperature source of a refrigerator is maintained, the larger the work input per unit of heat transfer $Q_L$ removed from the low-temperature source.

**Example 6-6**

An internally reversible refrigerator is used to maintain foodstuffs in a refrigerated area. Heat transfer into the refrigerator fluid occurs at 2°C, and heat transfer out of the device occurs at 27°C. It is also desired to maintain some frozen foods in a freezer. In this case heat transfer into the cyclic device occurs at −17°C, and heat transfer out occurs again at 27°C. What percentage increase in work input will be required for the frozen-food unit over the refrigerated unit for the same quantity of heat $Q_L$ removed from the cold regions?

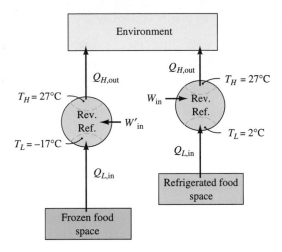

**Figure 6-31** Illustration for Example 6-6 showing two refrigerators operating between different cold regions.

**Solution:**

**Given:** A schematic of two refrigeration units appears in Fig. 6-31.

**Find:** Percent increase in work required to maintain frozen food when compared to nonfrozen refrigerated food.

**Model:** Internally reversible refrigeration cycles.

**Strategy:** Apply energy and second-law relations to internally reversible refrigeration cycles.

**Analysis:** The entropy and energy balances for each refrigerator are $Q_{H,\text{out}}/Q_{L,\text{in}} = T_H/T_L$ and $Q_{H,\text{out}} - Q_{L,\text{in}} = W_{\text{net,in}}$. The combination of these equations gives

$$W_{\text{net,in}} = Q_{L,\text{in}}\left(\frac{T_H}{T_L} - 1\right)$$

Denoting the net work input for the refrigerated area by $W$ and the net work input for the freezer operating between $-17$ and $27°C$ by $W'$, we find that

$$W = Q_{L,\text{in}}\left(\frac{300}{275} - 1\right) = 0.0909Q_L$$

and

$$W' = Q_{L,\text{in}}\left(\frac{300}{256} - 1\right) = 0.1719Q_L$$

The percentage increase of $W'$ over $W$ is given by

$$\frac{W' - W}{W} = \frac{0.1719Q_L - 0.0909Q_L}{0.0909Q_L}$$

$$= 0.891 \text{ (or 89.1\%)}$$

**Comment:** By changing the temperature difference between source and sink from 25 to 44°C, the required theoretical work input is nearly doubled.

As another application, consider the use of an internally reversible heat engine to provide the work input to an internally reversible refrigerator or heat pump. Figure 6-32 illustrates the general scheme of energy transfers. Two additional features are shown. First, if the heat engine E produces more net work than is necessary to drive the refrigerator R (or heat pump), then the excess work $W_{surr}$ delivered to the surroundings can be used to drive other devices. Second, to provide the proper shaft speeds for the engine and refrigerator, the two shafts operate through a transmission. The work transferred through an *ideal* transmission is 100 percent. However, $W_{R,net,in}$ would be less than $W_{E,net,out} - W_{surr}$ if irreversibilities occurred within the transmission.

The analysis of the combination of devices is based on the fundamental first- and second-law equations, and on the definitions of thermal efficiency and coefficient of performance. Some reality can also be introduced into the calculations if corrections to the ideal thermal efficiency, coefficient of performance, and transmission efficiency are specified. The following example illustrates a situation in which all devices are idealized.

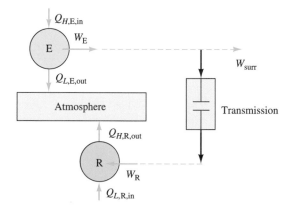

**Figure 6-32**     Illustrations of a Carnot heat engine used to drive a Carnot refrigerator or heat pump.

**Example 6-7**

In order to maintain the temperature in a refrigerated region, an internally reversible refrigerator operates with a temperature of 0°C where heat transfer of 1000 kJ/h occurs into the device. The heat transfer from the refrigerator R to the atmosphere takes place at 22°C. The work to drive the refrigerator is provided by an internally reversible heat engine E which operates with heat transfer into the device at 282°C and heat rejection at 22°C to the atmosphere. Determine the heat-transfer rate which must be supplied to the heat engine, in kJ/h, if all the work output of the heat engine is used to drive the refrigerator.

**Solution:**

**Given:** A heat engine drives a refrigerator, as shown in Fig. 6-33.

**Find:** $\dot{Q}_{H,E,in}$, in kJ/h.

**Model:** Internally reversible refrigerator and heat engine, ideal transmission.

**Strategy:** Apply the energy and entropy balances to the heat engine and to the refrigerator separately.

**Analysis:** Since we are looking for the heat transfer that must be supplied to the engine, begin by writing the energy balance and entropy balance for the steady-state, internally reversible heat engine

$$0 = \dot{Q}_{H,E,in} - \dot{Q}_{L,E,out} - \dot{W}_{E,out} \quad \text{and} \quad 0 = \frac{\dot{Q}_{H,E,in}}{T_{H,E}} - \frac{\dot{Q}_{L,E,out}}{T_{L,E}}$$

where we recognize that $dE_{cv}/dt = dS_{cv}/dt = 0$ and $\dot{\sigma}_E = 0$. These two equations have five variables; thus at least three variables must be known to obtain numerical answers for the device. Unfortunately, we only know two temperatures for this device and that the power out from the heat engine must equal the power input to the refrigerator ($\dot{W}_{E,out} = \dot{W}_{R,in}$).

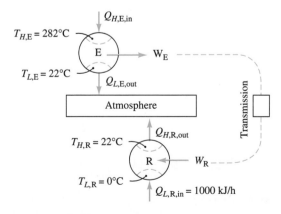

**Figure 6-33** Schematic for Example 6-7.

Similarly for the steady-state, internally reversible refrigerator we have

$$0 = \dot{Q}_{L,R,in} - \dot{Q}_{H,R,out} + \dot{W}_{R,in} \quad \text{and} \quad 0 = \frac{\dot{Q}_{L,R,in}}{T_{L,R}} - \frac{\dot{Q}_{H,R,out}}{T_{H,R}}$$

Again there are five variables, but we know three values for the refrigerator and can solve for the remaining two. Using the entropy balance, we can solve for the heat transfer from the refrigerator as

$$\dot{Q}_{H,R,out} = \dot{Q}_{L,R,in}\left(\frac{T_{H,R}}{T_{L,R}}\right) = (1000 \text{ kJ/h})\left(\frac{22 + 273 \text{ K}}{0 + 273 \text{ K}}\right) = 1080 \text{ kJ/h}$$

and using the energy balance, we can now solve for the power required by the refrigerator as

$$\dot{W}_{R,in} = \dot{Q}_{H,R,out} - \dot{Q}_{L,R,in} = 1080 \text{ kJ/h} - 1000 \text{ kJ/h} = 80 \text{ kJ/h}$$

We now know three values for the heat engine and can solve for the other two. Starting with the entropy balance we have

$$\frac{\dot{Q}_{H,E,in}}{\dot{Q}_{L,E,out}} = \frac{T_{H,E}}{T_{L,E}} = \frac{282 + 273 \text{ K}}{22 + 273 \text{ K}} = 1.88 \quad \text{or} \quad \dot{Q}_{L,E,out} = 0.532\dot{Q}_{H,E,in}$$

Combining this with the energy balance and noting that $\dot{W}_{E,out} = \dot{W}_{R,in}$, we have

$$\dot{Q}_{H,E,in} = \dot{W}_{E,out} + \dot{Q}_{L,E,out} = 80 \text{ kJ/h} + 0.532\dot{Q}_{H,E,in}$$

$$\dot{Q}_{H,E,in} = \frac{80 \text{ kJ/h}}{1 - 0.532} = 171 \text{ kJ/h}$$

**Comment:**  Note in this idealized case (no internal irreversibilities) that a refrigeration effect of 1000 kJ/h is accomplished by a small quantity (171 kJ/h) of heat supplied to the engine. It is easily shown that the thermal efficiency of the heat engine is 47 percent, and the COP of the refrigerator is 12.5.

### 6-9-3   HEAT-PUMP CYCLE

Heat pumps can be analyzed in a manner similar to refrigerators. The steady-state energy and entropy balances are the same as before, namely,

$$\dot{W}_{net,in} = \dot{Q}_{H,out} - \dot{Q}_{L,in} \quad \text{and} \quad \dot{\sigma} = \frac{\dot{Q}_{H,out}}{T_H} - \frac{\dot{Q}_{L,in}}{T_L}$$

Use of the entropy balance to eliminate the heat transfer rate $\dot{Q}_{L,in}$ from the energy balance leads to

$$\dot{W}_{net,in} = \dot{Q}_{H,out}\left(1 - \frac{T_L}{T_H}\right) + T_L\dot{\sigma} \qquad \textbf{[6-62]}$$

Therefore, since $\dot{\sigma} \geq 0$ and $T_H > T_L$,

$$\dot{W}_{HP,int\ irrev} > \dot{W}_{HP,int\ rev}$$

The equation above for $\dot{W}_{\text{net,in}}$ also leads to a general expression for the COP of a heat pump. This is

$$\text{COP}_{\text{HP}} = \frac{\dot{Q}_{H,\text{out}}}{\dot{W}_{\text{net,in}}} = \left(\frac{T_H}{T_H - T_L}\right) - \frac{T_H T_L}{T_H - T_L}\frac{\dot{\sigma}}{\dot{W}_{\text{net,in}}} \qquad \textbf{[6-63]}$$

Because the last term after the minus sign is always positive, then

$$\text{COP}_{\text{HP,int rev}} > \text{COP}_{\text{HP,int irrev}} \qquad \textbf{[6-64]}$$

Finally, the maximum value for $\text{COP}_{\text{HP}}$ is found by setting $\dot{\sigma}$ equal to zero in the general equation above. Hence, for an *internally reversible heat pump*

$$\text{COP}_{\text{HP,rev}} = \text{COP}_{\text{HP,Carnot}} = \frac{T_H}{T_H - T_L} \qquad \textbf{[6-65]}$$

Regardless of the method of construction or the working fluid used, all internally reversible heat pumps operating with heat addition at $T_H$ and heat rejection at $T_L$ will have the COP given by Eq. [6-65]. Note that although the limiting value of the thermal efficiency is less than unity, the coefficient of performance theoretically can be much larger than unity. Hence, in using a heat pump to heat a building, the costly (electrical) work input may be only a fraction of the thermal energy delivered.

An internally reversible or Carnot-type COP value is a *standard of performance* against which the values for actual refrigerators and heat pumps may be compared, for specified values of $T_L$ and $T_H$. All actual devices have a COP which is less than that given by Eqs. [6-61] and [6-65] for the given temperature limits of $T_L$ and $T_H$. Also, the COP of reversible refrigerators, for example, decreases as $T_L$ decreases for a given $T_H$. This implies that more work input will be required to accomplish a desired quantity of heat transfer. This is not surprising since the temperature difference $T_H - T_L$ has increased when $T_L$ is lowered. Intuitively, it should require more effort (work) to transfer a given quantity of heat as the temperature difference increases. The same result occurs when one raises $T_H$ in heat-pump operation for a given $T_L$.

Note the two temperature effects that reduce the overall performance of a refrigerator or heat pump.

### 6-9-4 THE CLAUSIUS INEQUALITY

One other limitation on *cyclic devices,* called the Clausius inequality, can also be developed from the closed-system entropy balance,

$$\frac{dS}{dt} = \sum_{j=1}^{n} \frac{\dot{Q}_j}{T_j} + \dot{\sigma} \qquad \textbf{[6-42]}$$

In steady state $dS/dt = 0$, and in general $\dot{\sigma} \geq 0$. In this case we can write the entropy balance for the cyclic device in the format

$$\sum_{j=1}^{n} \frac{\dot{Q}_j}{T_j} \le 0 \quad \text{or} \quad \oint \left(\frac{\delta Q}{T}\right)_b \le 0 \qquad \textbf{[6-66]}$$

Both formats above indicate a summation process carried out over the entire cycle and over all parts of the closed-system boundary. This result is known as the ***Clausius inequality,*** and it applies equally well to a periodic, mechanical device or to a continuous, steady-state device. The *equality sign applies to internally reversible cycles* and the *inequality sign applies to internally irreversible cycles* of any closed system.

**Example 6-8**

**A** steam power cycle operates under the following conditions. Saturated water vapor enters the turbine at 10 bars (state 1) and leaves at 0.50 bar with a quality of 90.0 percent (state 2). The fluid leaves the condenser as a wet mixture with 10.0 percent quality at 0.50 bar (state 3). State 4 is a saturated liquid at 10 bars. Determine whether the cycle satisfies the Clausius inequality.

**Solution:**

**Given:**   A steam power cycle is shown in Fig. 6-34.

**Find:**   Whether the Clausius inequality is fulfilled.

**Model:**   Continuous cycle, turbine and pump are adiabatic.

**Strategy:**   Apply the Clausius inequality to the steam power cycle.

**Analysis:**   Evaluation of the Clausius inequality requires the determination of the quantity $q/T$ at all system boundaries. For the steam power cycle, heat transfer is into the boiler and out of the condenser. Condensation occurs at 81.33°C and boiling at 179.9°C, based on saturation pressures of 0.50 and 10.0 bars. For the four pieces of equipment in the cycle the steady-state energy

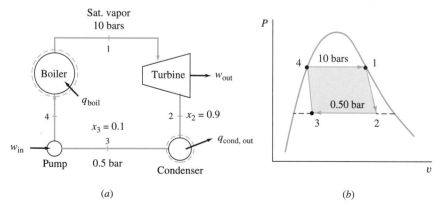

**Figure 6-34**    Schematic for Example 6-8.

balance is $q + w_{sh} = \Delta h + \Delta ke + \Delta pe$. If changes in kinetic and potential energies are neglected, $q = \Delta h$ for the steady-state boiling and condensation processes. The enthalpies at the four states are:

$$h_1 = h_g \text{ at 10 bars} = 2778.1 \text{ kJ/kg} \qquad \text{and} \qquad h_4 = h_f \text{ at 10 bars} = 762.8 \text{ kJ/kg}$$

$$h_2 = (h_f + xh_{fg}) \text{ at 0.50 bar} = 340.49 + 0.90(2305.4) = 2415.4 \text{ kJ/kg}$$

$$h_3 = (h_f + xh_{fg}) \text{ at 0.50 bar} = 340.49 + 0.10(2305.4) = 571.0 \text{ kJ/kg}$$

As a result the heat transfer in the boiler and condenser becomes

$$q_{boil} = h_1 - h_4 = 2778.1 - 762.8 = 2015.3 \text{ kJ/kg}$$

$$q_{cond} = h_3 - h_2 = 571.0 - 2415.4 = -1844.4 \text{ kJ/kg}$$

Therefore, checking the inequality of Clausius on a unit mass basis yields

$$\oint \frac{\delta q}{T} = \left(\frac{q}{T}\right)_{boil} + \left(\frac{q}{T}\right)_{cond} = \frac{2015.3}{453.05} + \frac{-1844.4}{354.48} = -0.755 \text{ kJ/kg·K}$$

**Comment:** The cycle satisfies the inequality of Clausius, and hence the data are valid.

## 6-10 HEAT TRANSFER AND THE $TS$ DIAGRAM

In this section we shall determine the entropy change for a thermal energy reservoir, and then apply this result to determine the entropy generation or production associated with heat transfer between two thermal reservoirs.

### 6-10-1 ENTROPY CHANGE FOR A THERMAL-ENERGY RESERVOIR

An immediate application of the defining equation, $dS \equiv (\delta Q/T)_{int \ rev}$, for the entropy change of any closed system is to the concept of a thermal-energy reservoir. A thermal reservoir is defined as a closed system which undergoes only internally reversible changes at constant temperature $T_R$, while heat transfer $Q_R$ is exchanged with another system. In this case, the integration of the equation for $dS$ leads to

$$\Delta S_R = \int \left(\frac{\delta Q}{T}\right)_{int \ rev} = \frac{Q_R}{T_R} \qquad \text{(thermal reservoir)} \qquad \textbf{[6-67]}$$

where the heat transfer $Q_R$ is measured *relative* to the reservoir and $T_R$ is an absolute temperature. When heat transfer is into a thermal reservoir, its entropy increases, while heat transfer from the thermal reservoir always decreases the entropy, as illustrated in Fig. 6-35. Equation [6-67] may also be written on a rate basis as $dS_R/dt = \dot{Q}_R/T_R$.

**Figure 6-35**    Illustration of entropy changes for thermal energy reservoirs.

An important coordinate system for plotting processes as a part of second-law analyses is the $TS$ diagram. For internally reversible processes, the equation $dS = \delta Q/T$ lends itself to a significant geometric interpretation. This equation can be rearranged to show that

$$\delta Q_{\text{int rev}} = T\, dS \qquad\qquad \textbf{[6-68]}$$

For a finite change of state this becomes

$$Q_{\text{int rev}} = \int T\, dS \qquad\qquad \textbf{[6-69]}$$

Thus for a process which is carried out in an *internally reversible* manner, the heat transfer $Q$ *is represented by the area under the process path on a TS diagram* when the temperature is on an *absolute* scale. Figure 6-36a illustrates this fact for some arbitrary internally reversible process between states 1 and 2. For a system at constant temperature, such as a thermal reservoir, the rearrangement of Eq. [6-67] shows that

$$Q_R = T_R\, \Delta S_R \qquad\qquad \textbf{[6-70]}$$

Figure 6-36b illustrates heat transfer to a thermal reservoir, where the path on the $TS$ diagram is a horizontal line. When the reservoir changes from state 1 to state 2, $\Delta S$ is positive and heat transfer is positive. For process 2-1, $\Delta S$ is negative and heat transfer is out of the thermal reservoir.

**Note the representation of heat transfer as an area on a *TS* diagram for a thermal reservoir.**

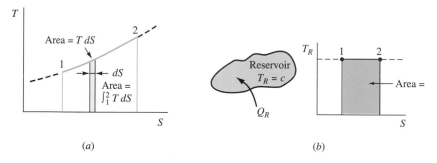

(a)                                              (b)

**Figure 6-36**    (a) Heat-transfer representation on a *TS* diagram for an internally reversible process with a variation of system temperature; (b) area representation on a *TS* diagram of heat transfer for a thermal energy reservoir.

## 6-10-2 ENTROPY GENERATION ASSOCIATED WITH HEAT TRANSFER

As discussed in Sec. 6-4, heat-transfer processes are generally not reversible owing to the presence of a finite temperature difference. Thus entropy generation is associated with the process of irreversible heat transfer. Figure 6-37 shows the passage of heat $\delta Q$ between two systems of fixed temperatures $T_A$ and $T_B$, where $T_A > T_B$. Entropy generation must be associated with the space where the temperature gradient exists. This is the region shown by a dashed line in Fig. 6-37. From a physical viewpoint this region might consist of a solid, heat-conducting wall. When heat transfer occurs between systems of constant temperature, the entropy balance for a closed system given by Eq. [6-42] may be integrated with respect to time to yield

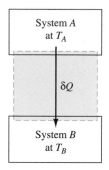

**Figure 6-37**
Schematic showing the heat transfer $\delta Q$ between two reservoirs with $T_A > T_B$.

$$\Delta S_{cm} = \sum_j \frac{Q_j}{T_j} + \sigma \qquad \textbf{[6-71]}$$

This equation will now be applied to the heat-transfer process. In this case there are only two entropy-transfer terms in the summation on the right.

Once the two thermal reservoirs are brought into contact with the solid material within the dashed region, heat transfer begins and a temperature gradient is set up within the material. The heat transfer, the temperature profile, and the entropy transfer are shown at the top, middle, and bottom portions of Fig. 6-38. As long as the rate of heat transfer remains constant, the wall is in steady state and its properties are constant at any given position. Hence the entropy of the overall region does not change with time, and $\Delta S_{cm}$ in Eq. [6-71] above is zero. However, there is entropy transfer $Q/T_A$ entering and entropy transfer $Q/T_B$ leaving the region. Applied specifically to the dashed-line region in Figs. 6-37 and 6-38, Eq. [6-71] shows that the entropy production $\sigma_Q$ due to *heat transfer between two thermal reservoirs* is given by

$$\sigma_Q = -\sum_j \frac{Q_j}{T_j} = -\left(\frac{Q_{in}}{T_A} + \frac{-Q_{in}}{T_B}\right) = Q_{in}\left(\frac{1}{T_B} - \frac{1}{T_A}\right) \geq 0 \quad \textbf{[6-72]}$$

where $T_A > T_B$, and $\sigma_Q$ is a special symbol for ***entropy production due to heat transfer*** between a finite temperature difference.

A graphical representation of Eq. [6-72] is shown at the bottom of Fig. 6-38. This type of representation is known as a *band* diagram. The magnitudes of quantities in a given balance are represented by vertical distances on the band diagram. In this case the entropy transfer $Q/T_A$ entering the wall plus the entropy generation $\sigma_Q$ within the wall equals the entropy transfer $Q/T_B$ leaving the wall.

There are several important conclusions to be drawn from the Eq. [6-72]:

1. The inequality sign applies for irreversible processes. Heat transfer cannot occur unaided from low temperature to high temperature.

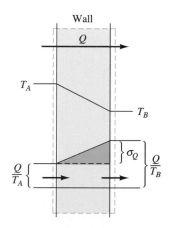

**Figure 6-38**
Graphical representation of Eq. [6-72] in terms of a band diagram.

This requires $\sigma_Q$ to be negative, in violation of the concept of entropy generation.

2. If $T_B$ is held fixed and $T_A$ is made larger, then $\sigma_Q$ becomes larger. As $\sigma_Q$ increases, the heat-transfer process becomes more irreversible.

3. As $T_A$ approaches a fixed value of $T_B$, $\sigma_Q$ decreases. In the limit as $T_A = T_B$, $\sigma_Q$ becomes zero and the process is reversible. However, in practice, a temperature difference must exist for a reasonable rate of heat transfer to occur. Therefore, heat transfer is considered to be reversible when the temperature difference is made infinitesimally small. That is, $T_A = T_B + dT$ for reversible heat transfer from $A$ to $B$.

---

**Example 6-9**

Heat transfer occurs between two thermal energy reservoirs at temperatures of 300 and 1000 K. Treating the two thermal energy reservoirs plus the region through which the heat transfer occurs as a single closed system, determine (*a*) the rate of entropy production and (*b*) the rate of entropy change, in kJ/(K·min), for the system if the heat transfer rate between the two reservoirs is 1000 kJ/min.

**Solution:**

**Given:**   Heat transfer occurs between two thermal energy reservoirs, as shown in Fig. 6-39.

**Find:**   (*a*) $\dot{\sigma}$ and (*b*) $dS_{cv}/dt$, in kJ/(K·min) for the combined system.

**Model:**   Two thermal energy reservoirs (closed systems) with steady-state heat transfer between them.

**Strategy:**   Apply the energy and entropy balance to the three subsystems and the combined system.

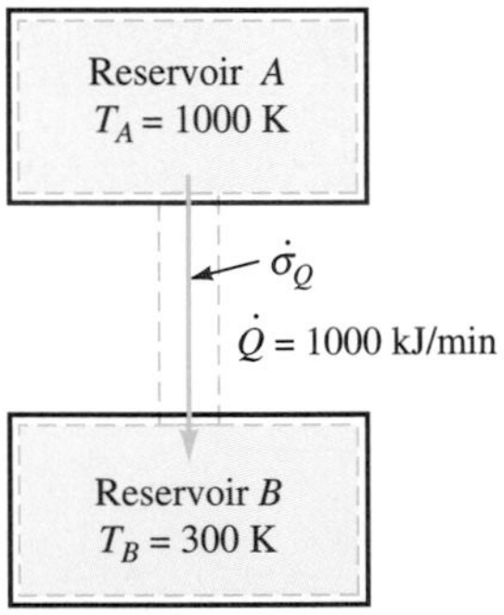

**Analysis:**   The three subsystems of interest are the two thermal energy reservoirs and the heat transfer region between them, as shown in Fig. 6-39.

(*a*) The entropy production rate for the combined system is the sum of the entropy production rates in the three subsystems: $\dot{\sigma} = \dot{\sigma}_A + \dot{\sigma}_Q + \dot{\sigma}_B$. By definition, the entropy production within the thermal energy reservoirs is zero. To solve for the entropy production in the heat transfer region, we write the steady-state energy and entropy balance for this closed system

$$0 = \dot{Q}_{in} - \dot{Q}_{out} \quad \text{and} \quad 0 = \frac{\dot{Q}_{in}}{T_A} - \frac{\dot{Q}_{out}}{T_B} + \dot{\sigma}_Q$$

Combining these two equations and noting that $\dot{Q}_{in} = \dot{Q}_{out} = 1000$ KJ/min gives the entropy production rate for the heat transfer region as

$$\dot{\sigma}_Q = \dot{Q}_{in}\left(\frac{1}{T_A} - \frac{1}{T_B}\right) = (1000 \text{ kJ/min})\left(\frac{1}{300 \text{ K}} - \frac{1}{1000 \text{ K}}\right) = 2.333 \text{ kJ/(K·min)}$$

Reservoir A
$T_A = 1000$ K

$\dot{\sigma}_Q$

$\dot{Q} = 1000$ kJ/min

Reservoir B
$T_B = 300$ K

**Figure 6-39**
Schematic and data for Example 6-9.

Thus the overall entropy production rate for the combined system is 2.333 kJ/(K·min).

(b) The rate of entropy change for the entire system can be determined by applying the entropy balance to the combined system which is closed and adiabatic; thus the rate of change of entropy for the combined system is

$$\frac{dS_{cv}}{dt} = \dot{\sigma} = 2.333 \text{ kJ/(K·min)}$$

It is instructive to examine the behavior of the three subsystems. For thermal energy reservoir A, the rate of change of entropy can be calculated from the entropy balance as

$$\frac{dS_A}{dt} = -\frac{\dot{Q}_{A,out}}{T_A} = -\frac{1000 \text{ kJ/min}}{1000 \text{ K}} = -1.000 \text{ kJ/(K·min)}$$

Similarly for thermal energy reservoir B, the rate of change of the entropy can be calculated as

$$\frac{dS_B}{dt} = \frac{\dot{Q}_{B,in}}{T_B} = \frac{1000 \text{ kJ/min}}{300 \text{ K}} = 3.333 \text{ kJ/(K·min)}$$

Finally we know that the entropy of the heat transfer region is constant because it is at steady-state conditions, $dS/dt = 0$. Thus finally we have

$$\frac{dS_{cv}}{dt} = \frac{dS_A}{dt} + \frac{dS}{dt} + \frac{dS_B}{dt}$$

$$= (-1.000 + 0 + 3.333) \text{ kJ/(K·min)} = 2.333 \text{ kJ/(K·min)}$$

Note that the rate of entropy increase in reservoir B is greater than the rate of entropy decrease in reservoir A by an amount equal to the rate of entropy generation inside the heat transfer region. Figure 6-40 shows the area representations for the two thermal reservoirs on a $TS$ diagram.

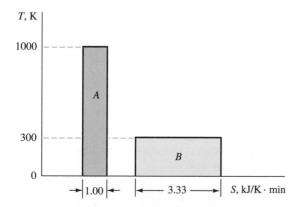

**Figure 6-40**   Heat transfer representations on a $TS$ diagram for the thermal energy reservoirs in Example 6-9.

**Comment:** Note that the total rate of entropy production and the total rate of entropy change are equal, as expected. It is easily seen that as $T_A$ is lowered in value toward 300 K, both the entropy production and entropy change approach zero for the overall process. This indicates that the heat-transfer process is approaching reversibility as the temperature difference approaches zero.

### 6-10-3 LOSS IN WORK POTENTIAL ASSOCIATED WITH HEAT TRANSFER

Heat transfer across a finite temperature difference is one of many possible examples of an irreversible process leading to a degradation of energy. If a heat engine is inserted between two systems which have a finite temperature difference between them, work could be produced as a result of the heat exchange. For the same $\Delta T$ between the systems, direct irreversible heat transfer leads to no work production at all. Hence, any time direct heat transfer occurs between two systems, the capability for producing some work from the interaction is lost. This loss of work production is equivalent to a degrading of the energy involved, or a loss in the *work potential* of the energy.

The *work potential* of a quantity of energy $Q$ taken from a thermal reservoir at $T_A$ is defined as the work that could be produced from $Q$ by a reversible heat engine operating between $T_A$ and the environmental temperature $T_0$. Rearrangement of Eq. [6-14] for the Carnot heat engine efficiency shows the work potential ($W_{\text{pot}}$) associated with $Q$ taken from a region at $T_A$ to be

$$W_{\text{pot}} = Q\eta_{\text{Carnot}} = Q\left(1 - \frac{T_0}{T_A}\right) \qquad \textbf{[6-73]}$$

where the sink temperature is $T_0$ of the environment, and $Q$ and $W_{\text{pot}}$ are positive values.

Now consider a quantity of energy $Q$ transferred irreversibly from a region of temperature $T_A$ to a region at $T_B$, as shown in Fig. 6-41. The work potentials $W_{\text{pot},A}$ of $Q$ at $T_A$ and $W_{\text{pot},B}$ of $Q$ at $T_B$, based on Eq. [6-73], are

$$W_{\text{pot},A} = Q\left(1 - \frac{T_0}{T_A}\right) \qquad \text{and} \qquad W_{\text{pot},B} = Q\left(1 - \frac{T_0}{T_B}\right)$$

The *difference* between $W_{\text{pot},H}$ and $W_{\text{pot},A}$ is the *loss in work potential* $W_{\text{loss},Q}$ associated with $Q$ due to the irreversible heat-transfer process. After algebraic manipulation this difference is

$$W_{\text{loss},Q} = T_0 Q\left(\frac{1}{T_B} - \frac{1}{T_A}\right) \qquad \textbf{[6-74]}$$

The greater the difference between $T_A$ and $T_B$, the larger the loss in work potential. In general, a given quantity of thermal energy $Q$ has a higher

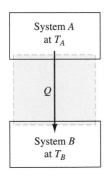

**Figure 6-41**
Loss in work potential with heat transfer between two systems at different temperatures.

potential for conversion to work as the temperature of its source $T_A$ increases for a fixed value of $T_B$. The higher the temperature of the heat source available to the user, the higher the work potential of the energy. There is a tremendous quantity of energy in the ocean, for example, but its low temperature gives that energy a low work potential, or low potential usefulness.

Equations [6-72] and (6-74) provide a direct link between the loss in work potential and the increase in entropy and the entropy production during direct heat transfer from reservoirs at $T_A$ and $T_B$. If Eq. [6-72] is rewritten in terms of $T_A$ and $T_B$, and then substituted into Eq. [6-74], we find that

$$W_{\text{loss},Q} = T_0 \sigma_Q \qquad \textbf{[6-75]}$$

Thus the loss in work potential due to irreversible heat transfer is directly proportional to the entropy production within the heat transfer region.

---

Reconsider Example 6-9, where 1000 kJ/min of heat transfer occurs between constant temperatures of 1000 and 300 K. Determine the loss in work potential associated with the process, in kJ/min, if the environmental temperature is 300 K.

**Example 6-10**

**Solution:**

**Given:** Irreversible heat transfer between constant-temperature regions, as shown in Fig. 6-42.

**Find:** Loss in work potential, in kJ.

**Model:** Irreversible heat transfer $T_0 = T_B = 300$ K.

**Strategy:** Apply equation for the loss in work potential.

**Analysis:** The equation for the loss in work potential for irreversible heat transfer, based on the concept of the work potential of a single source of heat transfer at temperature $T_A$ to the environment at temperature $T_0$ (which equals $T_B$), is

$$W_{\text{loss},Q} = T_0 Q \left( \frac{1}{T_B} - \frac{1}{T_A} \right)$$

This equation is also valid on a rate basis. Substitution of data leads to

$$\dot{W}_{\text{loss},Q} = 300 \text{ K} \times 1000 \text{ kJ/min} \times \left( \frac{1}{300} - \frac{1}{1000} \right) \text{K}^{-1}$$

$$= 700 \text{ kJ/min}$$

**Comment:** Alternatively, the entropy generation for the process was found to be 2.333 kJ/K·min in Example 6-9. Therefore, $\dot{W}_{\text{loss},Q} = T_0 \dot{\sigma}_Q = 300(2.333)$ kJ/min $= 700$ kJ/min.

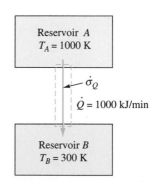

Reservoir $A$
$T_A = 1000$ K

$\dot{\sigma}_Q$

$\dot{Q} = 1000$ kJ/min

Reservoir $B$
$T_B = 300$ K

**Figure 6-42**
Schematic and data for Example 6-10.

The concepts of entropy generation and the loss in work potential with irreversible heat transfer will be applied to heat engines, refrigerators, and heat pumps in Secs. 6-11-2 and 6-11-3.

## 6-11  APPLICATIONS

### 6-11-1  CLOSED STEADY-STATE SYSTEMS

In this section we examine several closed steady-state systems. Under these conditions, we will show that knowledge about the entropy of the system is unnecessary. The entropy production rate is directly related to the rate of entropy transport by heat transfer at the system boundary. Many practical devices satisfy this simple model.

**Example 6-11**

**An** inventor claims to have built a steady-state device that provides refrigeration and is powered by a flame. Her published performance data include a heat-transfer input rate of 400 Btu/min at an evaporator temperature of 510°R, a heat-transfer input rate of 200 Btu/min at a boundary temperature of 960°R from a fire source, and heat transfer from the device at a condenser temperature of 550°R. Modeling the device as a closed system, determine the heat-transfer rate from the condenser in Btu/min and the rate of entropy production in Btu/min·°R.

**Solution:**

**Given:**  A heat-powered refrigerator with performance shown in Fig. 6-43.

**Find:**  $\dot{Q}_{cond}$ in Btu/min and $\dot{\sigma}$ in Btu/min·°R.

**Model:**  Device is a closed steady-state system.

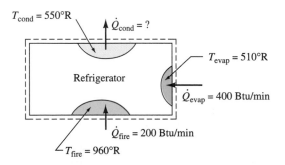

**Figure 6-43**     Heat-powered refrigerator for Example 6-11.

**Strategy:** Apply energy and entropy balances to the system.

**Analysis:** Starting with the energy balance for a closed system,

$$\frac{dE_{cm}}{dt} = \dot{Q}_{fire} + \dot{Q}_{evap} - \dot{Q}_{cond}$$

But for a steady-state system, $dE_{sys}/dt = 0$. Solving for $\dot{Q}_{cond}$,

$$\dot{Q}_{cond} = \dot{Q}_{fire} + \dot{Q}_{evap} = (200 + 400) \text{ Btu/min} = 600 \text{ Btu/min}$$

Thus the device rejects energy by heat transfer at the rate of 600 Btu/min. Now writing an entropy balance

$$\frac{dS_{cm}}{dt} = \frac{\dot{Q}_{fire}}{T_{fire}} + \frac{\dot{Q}_{evap}}{T_{evap}} - \frac{\dot{Q}_{cond}}{T_{cond}} + \dot{\sigma}$$

Since $dS_{cm}/dt = 0$, we can solve for $\dot{\sigma}$.

$$\dot{\sigma} = \frac{\dot{Q}_{cond}}{T_{cond}} - \frac{\dot{Q}_{fire}}{T_{fire}} - \frac{\dot{Q}_{evap}}{T_{evap}}$$

$$= \left(\frac{600}{550} - \frac{200}{960} - \frac{400}{510}\right)\frac{\text{Btu/min}}{°R} = 0.0983 \text{ Btu/min·°R}$$

**Comment:** The device appears to satisfy both the first and second laws of thermodynamics. If the calculated $\dot{\sigma}$ was negative, the reported performance would be very suspect.

---

**A** mechanical transmission operates at steady-state conditions. The power supplied to the input shaft is 35 kW and the measured power at the output shaft is 32 kW. Heat transfer occurs from the surface of the transmission at 50°C. Calculate the rate of entropy production inside the transmission, in kW/K.

Example 6-12

**Solution:**

**Given:** A mechanical transmission operates as shown in Fig. 6-44.

**Find:** $\dot{\sigma}$, in kW/K.

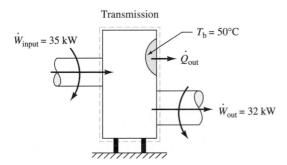

**Figure 6-44** Mechanical transmission for Example 6-12.

**Model:**    Transmission is a closed steady-state system.

**Strategy:**    Apply entropy balance.

**Analysis:**    Starting with the entropy balance,

$$\frac{dS_{cm}}{dt} = -\frac{\dot{Q}_{out}}{T_b} + \dot{\sigma}$$

For steady-state conditions, $dS_{cm}/dt = 0$. Now solving for $\dot{\sigma}$,

$$\dot{\sigma} = \frac{\dot{Q}_{out}}{T_b}$$

To find $\dot{\sigma}$ we must find $\dot{Q}$. Applying an energy balance and recalling that $dE/dt = 0$ gives

$$0 = \dot{W}_{input} - \dot{W}_{output} - \dot{Q}_{out}$$

Solving for $\dot{Q}_{out}$ and substituting back into the equation for $\dot{\sigma}$,

$$\dot{\sigma} = \frac{\dot{W}_{input} - \dot{W}_{output}}{T_b} = \frac{(35-32)\ kW}{(50+273)\ K} = 0.00929\ kW/K$$

**Comment:**    Rearranging the $\dot{\sigma}$ equation we find that $\dot{W}_{output} = \dot{W}_{input} - T_b\dot{\sigma}$. This clearly shows how the entropy production degrades performance. For an ideal transmission, there are no losses and $\dot{\sigma} = 0$.

---

**Example 6-13**

**M**easurements on an electric resistor indicate a direct current of 2 A and a voltage drop across the resistor of 100 V. Determine the heat-transfer rate in watts and the entropy production rate in kW/K for the resistor. The surface temperature of the resistor is 30°C.

**Solution:**

**Given:**    Electrical resistor as shown in Fig. 6-45.

**Find:**    $\dot{Q}$ in W and $\dot{\sigma}$, in kW/K.

**Model:**    Resistor is a closed steady-state system.

**Strategy:**    Apply energy and entropy balances.

**Analysis:**    Starting with the energy balance,

$$\frac{dE_{cm}}{dt} = \dot{Q} + \dot{W}$$

where $\dot{W} = VI$ ($V$ is the potential difference), and $dE_{cm}/dt = 0$. Solving for $\dot{Q}$,

$$\dot{Q} = -\dot{W} = -VI = -(100\ V)(2\ A) \times \frac{1\ W}{1\ A\cdot V} = -200\ W$$

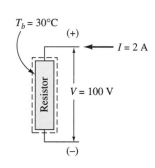

$T_b = 30°C$

(+)

$I = 2\ A$

Resistor

$V = 100\ V$

(−)

**Figure 6-45**
Electric resistor for
Example 6-13.

Note that the negative sign indicates the heat transfer is out of the system. Now, writing an entropy balance,

$$\frac{dS_{cm}}{dt} = \frac{\dot{Q}}{T_b} + \dot{\sigma}$$

where $dS_{cm}/dt = 0$. Solving for $\dot{\sigma}$,

$$\dot{\sigma} = -\frac{\dot{Q}}{T_b} = -\frac{-200 \text{ W}}{(30 + 273) \text{ K}} = 0.660 \text{ W/K}$$

**Comment:** As you might expect, an electric resistor is an irreversible device.

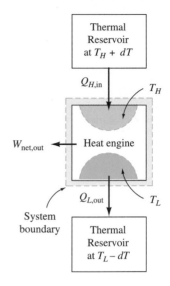

**Figure 6-46**
Schematic of a totally reversible heat engine operating with heat transfer at $T_H$ and $T_L$.

## 6-11-2 EFFECT OF IRREVERSIBLE HEAT TRANSFER ON CYCLE PERFORMANCE

The equations for the thermal efficiency and coefficient of performance of Carnot-type cyclic devices, namely,

$$\eta_{th,Carnot} \equiv 1 - \frac{T_L}{T_H} \quad COP_{R,Carnot} = \frac{T_L}{T_H - T_L} \quad COP_{HP,Carnot} = \frac{T_H}{T_H - T_L}$$

$$[6\text{-}76]$$

were developed earlier for *internally reversible*, cyclic devices. Now consider these same devices operating in a totally reversible manner. This requires that the overall system be internally and externally reversible. This concept was used in the development of Carnot's principle in Sec. 6-5-2. A detailed schematic of such a device is shown in Fig. 6-46. In order for the heat-transfer processes to be externally reversible, the heat transfer $Q_H$ occurs between a thermal reservoir at $T_H + dT$, while the cyclic system temperature is $T_H$. The same type of differential temperature difference occurs for the heat transfer $Q_L$ between $T_L$ and $T_L - dT$. Because of the differential temperature difference $dT$ involved during both heat-transfer processes, the relationships given by Eq. [6-76] *are also valid for totally reversible heat engines, refrigerators, and heat pumps*. Also, Fig. 6-46 is valid for totally reversible refrigerators and heat pumps, if the directions of the arrows and the signs on $dT$ are reversed.

We now wish to examine the performance of internally reversible heat engines which operate with *irreversible heat transfer* between the heat engine and the two thermal reservoirs. Figure 6-47 shows a situation where the working-fluid temperature within the device is $T_H$ when $Q_H$ enters from a thermal reservoir at $T_A$, and the temperature is $T_L$ when $Q_L$ is discharged to a reservoir at $T_B$. In this case the heat engine is operating internally reversibly between $T_H$ and $T_L$. As a result its thermal efficiency is given by $(1 - T_L/T_H)$. Because the temperature difference $(T_H - T_L)$ for the heat engine is less than $(T_A - T_B)$ for the thermal reservoirs, the maximum thermal

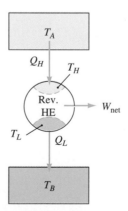

**Figure 6-47**
Reversible heat engine with irreversible heat transfer.

**Note the negative effect of irreversible heat transfer on the performance of a heat-engine cycle.**

efficiency will be less than that when the heat-transfer processes were also reversible between thermal reservoirs at temperatures $T_A$ and $T_B$. Hence the presence of these finite temperature differences external to the heat engine reduces the overall performance of the cycle. A similar argument can be made for refrigerators and heat pumps which operate with irreversible heat transfers between the energy sources and sinks and the working fluid of the cyclic device. When irreversible heat transfer is present, the COP of refrigerators and heat pumps is less than that for a totally reversible cycle.

**Example 6-14**

**A** heat engine operates between thermal reservoir temperatures $T_A$ and $T_B$ of 1000 and 300 K, respectively. For a heat addition $Q_{H,in}$ of 1000 kJ to the heat engine, find the heat rejection, work output, thermal efficiency, and loss in work potential values when (*a*) the device is totally reversible and (*b*) the device is internally reversible but receives and rejects heat at 800 and 400 K rather than 1000 and 300 K. Let $T_0 = 300$ K.

**Solution:**

**Given:**    A heat engine operates under two sets of conditions: (*a*) a totally reversible cycle and (*b*) the engine is internally reversible, but the heat-transfer processes are irreversible, as shown in Fig. 6-48.

**Find:**    Heat rejection, work output, thermal efficiency, and loss in work potential for the conditions of (*a*) total, and (*b*) internal reversibility.

**Model:**    Heat-engine cycle with external irreversibilities.

**Strategy:**    Apply first- and second-law relations for cyclic engines.

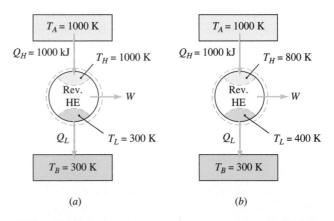

(*a*)                                                                (*b*)

**Figure 6-48**    Schematic for Example 6-14. (*a*) Totally reversible heat engine; (*b*) internally reversible heat engine with irreversible heat transfer.

**Analysis:** (a) For a totally reversible heat engine $Q_{L,out} = Q_{H,in}(T_B/T_A)$, and $Q_{H,in} - Q_{L,out} - W_{net,out} = \Delta E = 0$. Hence

$$Q_{L,out} = 1000\frac{300}{1000} \text{ kJ} = 300 \text{ kJ}$$

and

$$W_{net,out} = Q_{H,in} - Q_{L,out} = (1000 - 300) \text{ kJ} = 700 \text{ kJ}$$

The thermal efficiency is

$$\eta_{th} = W_{net,out}/Q_{H,in} = 700/1000 = 0.70 \qquad \text{(or 70 percent)}$$

Since the heat-transfer processes are reversible, there is no loss in work potential associated with them.

(b) Because the device is internally reversible, the heat rejected is found by

$$Q_{L,out} = Q_{H,in}\frac{T_L}{T_H} = 1000\frac{400}{800} \text{ kJ} = 500 \text{ kJ}$$

and

$$W_{net,out} = Q_{H,in} - Q_{L,out} = (1000 - 500) \text{ kJ} = 500 \text{ kJ}$$

The thermal efficiency is

$$\eta_{th} = W_{net,out}/Q_{H,in} = 500/1000 = 0.50 \qquad \text{(or 50 percent)}$$

In case (b) the work output is diminished by 200 kJ when compared to the totally reversible heat engine in part (a). This loss in work output is due to external irreversibilities. The actual origin of this 200 kJ loss can be found by applying the concept of loss in work potential to the two heat-transfer processes. A loss in work potential occurs during both irreversible heat-transfer processes. For the high-temperature and low-temperature processes Eq. [6-74] shows that

$$W_{loss,high} = T_0 Q_{H,in}\left(\frac{1}{T_H} - \frac{1}{T_A}\right) = 300(1000)\left(\frac{1}{800} - \frac{1}{1000}\right) = 75 \text{ kJ}$$

$$W_{loss,low} = T_0 Q_{L,out}\left(\frac{1}{T_0} - \frac{1}{T_L}\right) = 300(500)\left(\frac{1}{300} - \frac{1}{400}\right) = 125 \text{ kJ}$$

**Comment:** The total loss found from first- and second-law considerations is 200 kJ. However, the last calculations enable us to find the individual contributions to the overall loss in work output. Note that the region of smallest temperature difference (between $T_L$ and $T_0$) produces the largest loss in work potential, even though the heat-transfer quantity is one-half of that on the high-temperature side of the heat engine. Generally, losses on the low-temperature side of a heat engine are sensitive to the temperature difference during heat rejection.

In the next subsection we shall compare irreversibilities associated with simple cyclic devices in terms of entropy generation.

## 6-11-3 ENTROPY PRODUCTION IN SIMPLE CYCLIC DEVICES

In the preceding subsection heat engines, refrigerators, and heat pumps which had external irreversibilities were compared to totally reversible

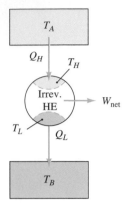

**Figure 6-49**
Irreversible heat engine
with irreversible heat
transfer.

devices. We now wish to study the magnitude of irreversibilities in the heat-transfer processes and within the cyclic devices themselves during actual operation on the basis of entropy generation. Figure 6-49 illustrates the situation where the heat transfers are irreversible as well as the heat engine. Heat transfer $Q_H$ occurs between $T_A$ and $T_H$, and heat transfer $Q_L$ occurs between $T_L$ and $T_B$. Hence the heat engine effectively operates only between $T_H$ and $T_L$, and not $T_A$ and $T_B$. Based on Sec. 6-9-1, the actual efficiency must be less than that of a totally reversible heat engine operating between the reservoir temperatures. For the heat engine (HE) the entropy balance is

$$\Delta S_{HE} = \sum \frac{Q_j}{T_j} + \sigma_{HE}$$

where $Q_j$ and $T_j$ are measured at the *boundary* of the engine. If we consider that the heat engine undergoes an integral number of cycles, then $\Delta S_{HE} = 0$. Then the above equation becomes

$$\sigma_{engine} = -\sum \frac{Q_j}{T_j} = -\left(\frac{Q_{H,in}}{T_H} - \frac{Q_{L,out}}{T_L}\right) \qquad \textbf{[6-77]}$$

In addition,

$$Q_{H,in} - Q_{L,out} - W_{net,out} = 0 \qquad \text{and} \qquad \eta_{th} = W_{net,out}/Q_{H,in} \qquad \textbf{[6-78]}$$

The terms $Q_{H,in}$, $Q_{L,out}$, and $W_{net,out}$ represent magnitudes only.

Consequently the sum of the entropy generation terms for the two heat-transfer processes and the engine process is

$$\sigma_{total} = \sigma_{Q,high} + \sigma_{HE} + \sigma_{Q,low} \qquad \textbf{[6-79]}$$

$$= -\left(\frac{Q_{H,in}}{T_A} - \frac{Q_{H,out}}{T_H}\right) - \left(\frac{Q_{H,in}}{T_H} - \frac{Q_{L,out}}{T_L}\right) - \left(\frac{Q_{L,in}}{T_L} - \frac{Q_{L,out}}{T_B}\right)$$

Several $Q/T$ terms in the above equation will cancel out simply by the process of addition. In its present format, however, Eq. [6-79] will yield the values of the separate entropy-production quantities due to heat transfer and to heat-engine performance. A comparison of these values will show which process contains the largest (or smallest) losses. In practice, this would help the engineer to decide which portion of the overall process needs improvement.

It is obvious that if a boundary is drawn around the entire heat-engine process, then

$$\sigma_{total} = -\left(\frac{Q_{H,in}}{T_A} - \frac{Q_{L,out}}{T_B}\right) \qquad \textbf{[6-80]}$$

This format would be useful in comparing several different heat engines overall, but it gives no information on the contributions of processes associated with the cycle itself. A similar analysis can be made for refrigerators and heat pumps operating between two thermal reservoirs.

Example 6-15

**A** heat engine operates between two thermal reservoirs, the temperatures of which are 1000 and 300 K. However, the heat-engine temperature during heat addition is 900 K and during heat rejection is 350 K. The measured thermal efficiency of the engine is 40 percent, and the heat supplied per cycle is 100 kJ. Determine the entropy production for all the irreversibilities present in the overall process, in kJ/K.

**Solution:**

**Given:** An irreversible heat engine with irreversible heat transfers from two thermal reservoirs, as shown in Fig. 6-50.

**Find:** $\sigma$ for the engine and the heat-transfer processes.

**Model:** System consists of an irreversible heat engine with irreversible heat transfer.

**Strategy:** Apply an entropy balance to the heat-transfer processes and to the heat engine.

**Analysis:** The measured thermal efficiency is 40 percent. For a heat input of 100 kJ, the net work output is $W_{net,out} = \eta_{th}Q_{H,in} = 0.40(100) = 40$ kJ, and the heat rejection $Q_{L,out}$ is 60 kJ. In this situation the entropy production for the engine based on Eq. [6-77] is

$$\sigma_{HE} = -\left(\frac{Q_{H,in}}{T_H} - \frac{Q_{L,out}}{T_L}\right) = -\left[\frac{100}{900} - \frac{60}{350}\right] = 0.0603 \text{ kJ/K}$$

The heat-transfer processes themselves are irreversible, because a finite temperature difference exists in each case. The entropy generation for the heat-transfer process to the engine is given by Eq. [6-72]. In terms of $Q_H$ this becomes

$$\sigma_{Q,H} = -\left(\frac{Q_{H,in}}{T_A} - \frac{Q_{H,out}}{T_H}\right) = -\left(\frac{100}{1000} - \frac{100}{900}\right) = 0.0111 \text{ kJ/K}$$

In addition, for the engine-heat rejection process the entropy generation becomes

$$\sigma_{Q,L} = -\left(\frac{Q_{L,in}}{T_L} - \frac{Q_{L,out}}{T_B}\right) = -\left(\frac{60}{350} - \frac{60}{300}\right) = 0.0286 \text{ kJ/K}$$

Hence for the overall process, $\sigma_{total} = 0.0603 + 0.0111 + 0.0286 = 0.1000$ kJ/K. [This overall value is also given directly by Eq. [6-80].]

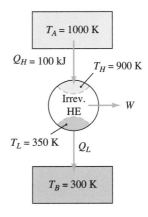

**Figure 6-50**
Schematic and data for Example 6-15.

**Comment:** The engine operation is the major contributor to the entropy production (60 percent). Note also that $\sigma_{Q,L}$ is nearly three times the size of $\sigma_{Q,H}$, even though the temperature difference during heat rejection is only one-half of that for heat supply. Thus entropy production during heat rejection is much more sensitive to the temperature difference than during heat addition.

The above example illustrated the relative losses due to irreversibilities solely in terms of entropy generation. These losses could also be examined in terms of the loss in work potential associated with each process in a cycle.

## 6-12     ENTROPY IN TERMS OF RANDOMNESS AND PROBABILITY

The last three sections have shown some of the usefulness of entropy. First, heat transfer is related to entropy change by the relation $\delta Q = T\,dS$ for internally reversible processes. Thus the value of $Q$ is measured by an area on a $TS$ diagram. Second, we found that entropy generation $\sigma$ is a measure of irreversibilities, such as that associated with irreversible heat transfer. Third, the increase in entropy principle helps explain why every day we observe many unidirectional processes. Last, from the entropy balance, relations can be developed for the performance limitations of heat engines, refrigerators, and heat pumps. In the following chapters we shall develop the entropy concept as a powerful tool in the analysis of engineering processes. However, since entropy is not a familiar property, compared to energy, we need to stop for a moment and consider entropy from a microscopic or molecular viewpoint.

Entropy is a property which measures the randomness (or disorder) of matter on a molecular level. Before examining a thermodynamic system, consider the following two examples. First, two students occupy adjacent rooms in a dormitory. One room is neatly kept, while the other room is messy, without cleaning or organizing for several months. There should be no problem locating things in the neatly kept (orderly) room, unlike the messy (nonorderly) room. One could say there is a great deal of randomness in the messy room. As a second example, consider two textbooks containing the same subject matter except one has no table of contents, index, or even chapter headings. One is written in an orderly fashion, and it is easy to locate a particular subject. The other is almost worthless, because of the randomness of the subject matter.

In a thermodynamic sense this randomness occurs with respect to the position and the energy of the particles of matter. Consider a tank initially separated into two regions by a membrane. A gas fills one side, while the other side contains a vacuum. When the membrane is broken, we expect the gas to quickly fill the entire tank. The location of a given particle at a given time is now more *random,* because it has a larger volume in which to move. This increased randomness of all the gas particles is marked by an increase in the entropy of the gas phase. In a similar fashion, consider adding energy to a solid so that it first melts to a liquid, and then vaporizes into the gas phases. From the strict location of particles in the lattice structure of a solid, the substance proceeds to a state of increased movement of the particles in a liquid, and finally to a great disorder in the gas phase. This is accompanied by a relatively small increase in entropy for the melting process, and a large increase for the vaporization process.

Another measure of disorder or randomness in the molecular level is with respect to the energies of individual particles. Since the early 1800s the concept of discrete energy levels for molecules has been developed through quantum theory. The results of quantum theory indicate a whole series of discrete energy values designated by $\varepsilon_i$. The series begins at some minimum value called the *ground level* $\varepsilon_0$. The series then progresses through distinct steps to larger and larger energy values, or energy levels. A typical diagram for energy levels $\varepsilon_i$ is shown in Fig. 6-51. At a given time a certain number of particles $n_i$ have an energy $\varepsilon_i$. The total internal energy of $n$ particles is given by

$$U = \sum_i n_i \varepsilon_i$$

Consider now a system with fixed values of $\varepsilon_i$. When energy as heat transfer is added, for example, the values of $n_i$ change, with more particles having higher energy values. Thus there are more ways the total energy can be distributed among the $\varepsilon_i$ energy levels. Thus the randomness of the system increases as energy is added by heat transfer, and the entropy increases.

Carry this situation one step further by considering heat transfer from a hotter region to a colder region. The randomness or disorder and the entropy of the hotter region will decrease, while randomness and entropy of the colder region will increase. The increase in entropy principle for the overall process requires that total entropy change be positive. This in turn requires that the increase in the randomness of the colder region be more than the decrease in the hotter region. Also, recall the example above of the expansion of a gas into a vacuum. Quantum theory dictates that the values of the permitted energy levels $\varepsilon_i$ decrease with the increase in volume. Thus there are many more ways to distribute the same total energy among the gas particles, and the disorder increases (as does its entropy).

As a final example, consider the addition of electrical resistor work from the discharge of a battery to a gas within a rigid tank, as shown in Fig. 6-52. The energy in the battery initially is in a highly ordered state. The result of the process is a decrease in the randomness within the battery. But a much larger increase in randomness occurs with the rise in temperature of the gas in the container. Thus molecular disorder is created for the overall process, and this is accompanied by an increase in overall molecular disorder.

Thus macroscopic entropy changes are a reflection of microscopic changes in the molecular randomness of the system. The entropy function and molecular randomness can be related quantitatively through the concepts of macrostates and microstates. We define a *macrostate* as any possible microscopic state of an assembly of particles described in terms of the number of particles in each energy level at a given instant. The description

| Number of a given level | Levels of given energy | Energy of the level |
|---|---|---|
| $i$ | ——— | $\varepsilon_i$ |
| . | - - - | . |
| 3 | ——— | $\varepsilon_3$ |
| 2 | ——— | $\varepsilon_2$ |
| 1 | ——— | $\varepsilon_1$ |
| 0 | ——— | $\varepsilon_0$ |

**Figure 6-51**
A set of discrete energy levels.

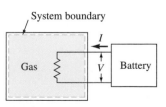

**Figure 6-52**
Electrical work added to a gas.

**Table 6-2**    The nine possible macrostates of four particles having a total energy of six units, for equally spaced energy levels

| Energy Level | Distribution | | | | | | | | |
|---|---|---|---|---|---|---|---|---|---|
| | $A$ | $B$ | $C$ | $D$ | $E$ | $F$ | $G$ | $H$ | $I$ |
| $\varepsilon_6$ | 1 | | | | | | | | |
| $\varepsilon_5$ | | 1 | | | | | | | |
| $\varepsilon_4$ | | | 1 | 1 | | | | | |
| $\varepsilon_3$ | | | | | 1 | 2 | 1 | | |
| $\varepsilon_2$ | | | 1 | | 1 | | | 2 | 3 |
| $\varepsilon_1$ | | 1 | | 2 | 1 | | 3 | 2 | |
| $\varepsilon_0$ | 3 | 2 | 2 | 1 | 1 | 2 | | | 1 |
| $W$ | 4 | 12 | 12 | 12 | 24 | 6 | 4 | 6 | 4 |

of a macrostate does not require enumeration of which particles have a given energy, but only an accounting of how many particles have a given energy. In Table 6-2 the nine distributions labeled $A$ to $I$ will now be identified as the nine possible macrostates for a system composed of four particles and having a total energy of six units. Each of these macrostates satisfies the macroscopic description of the state of the system.

Further analysis of Table 6-2 leads to the concept of a microstate if we assume that the particles are distinguishable. Consider any one of the nine macrostates, e.g., column $A$. Here three particles have a ground level of energy $\varepsilon_0$ and the remaining one is in the sixth quantum level of energy $\varepsilon_6$ above the ground level. Because of the random exchange of energy, the particle having six units of energy could be any one of the four particles in the system. Over a sufficient time macrostate $A$ will occur a number of times. However, each of the four distinguishable particles will share in occupying the sixth level of energy for this particular macrostate. Thus macrostate $A$ really consists of four microstates, all of which are equally likely to occur over a sufficiently long time period. A *microstate* then is a description of which particles have certain energies. Of course, we cannot specify which microstate exists at a given time, or even which macrostate. Microstates do exist, however, and they are continually changing with time. The other distributions, $B$ to $I$, in Table 6-2 likewise may be formed in a number of ways, depending upon the placement of certain particles in various energy levels. The last horizontal row in the table, marked $W$, lists the number of microstates for each macrostate.

The number of microstates $W$ for a given macrostate is called the *thermodynamic probability* of that macrostate. Unlike the usual mathematical probability which varies from zero to unity, the thermodynamic probability is never less than unity and is usually an extremely large number for thermodynamic systems. The value of $W$ in the case of Table 6-2 is obtained from a permutation formula

$$W = \frac{n!}{n_1!n_2!\cdots n_k!} \qquad \textbf{[6-81]}$$

This equation gives the number of microstates possible for $n$ particles, where the identity of each particle as residing in a particular energy level is recognized. However, the order of particles within an energy level is immaterial. The symbols $n_1$, $n_2$, ... represent, respectively, the number of particles in energy levels 1, 2, etc. As an example, four particles are placed in three energy levels containing 2, 1, and 1 particle, respectively. For this situation Eq. [6-81] gives for the possible number of microstates that

$$W = \frac{4!}{2!1!1!} = 12$$

Hence the distributions $B$, $C$, and $D$ in Table 6-2 all have 12 possible microstates. The other values of $W$ in the table are easily verified. It is seen from Table 6-1 that the total number of microstates is 84. Since all the microstates are assumed to occur with equal likelihood, over a long time each microstate would be formed $\frac{1}{84}$ of the time.

For a system of $n$ particles and total energy $U$, there are a fixed number of macrostates for each set of allowed energy levels. As a result, the number of microstates is also fixed. The total number of microstates for a given thermodynamic state is represented by $W_{tot}$. This simply is the sum of the $W$ values for each possible macrostate. In our example $W_{tot}$ would be 84. The number of microstates thus is a unique and identifiable number (at least in theory) for every equilibrium state. If $W_{tot}$ has a fixed value for each equilibrium state, it partially fulfills the concept of a property in at least one respect. There is one and only one value for each equilibrium state.

Although the value of $W_{tot}$ is fixed, it is not a useful measure of a thermodynamic property. Properties are extensive or additive in nature. However, from the theory of combinations and permutations, the thermodynamic probability $W$ is multiplicative. For example, if system A has 3 microstates and system B has 2 microstates, then 6 possible microstates occur when both systems are considered simultaneously. Thus,

$$W_{AB} = W_A W_B$$

Although $W$ is multiplicative, the logarithm of $W$ is an additive function. Taking the logarithm of each side of the above equation, we find that

$$\log W_{AB} = \log W_A + \log W_B$$

Therefore, we can define a new thermodynamic property in terms of the thermodynamic probability of the system. This property is the *statistical entropy S* of the system and is related to the total thermodynamic probability $W_{tot}$ by the definition

$$S \equiv k \ln W_{\text{tot}} \qquad \textbf{[6-82]}$$

where $k$ is a proportionality constant. It is necessary to choose the logarithm to the base $e$ in order to make the numerical values of the entropy function calculated from Eq. [6-82] agree with values calculated from *macroscopic* measurements. This equation is known as the *Boltzmann-Planck equation,* and its general validity is well established. The constant $k$, which is called the Boltzmann constant, determines both the magnitude and the units for the statistical entropy function, since the ln $W_{\text{tot}}$ term is a pure number. Quantitatively, $k = R_u/N_A$, where $R_u$ is the universal gas constant and $N_A$ is Avogadro's number. The value of $k$ in SI units is listed in Table A-1.

Thus the physical nature of entropy can be interpreted in terms of its molecular origin. The thermodynamic probability is minimized as a substance becomes a pure crystalline solid at 0 K. Thus, by convention, tables of *absolute* entropy values list the entropy to be *zero* at this particular state. In addition, tables for the entropy of an ideal gas are typically based on statistical calculations. As entropy changes of substances are evaluated in the next chapter, one might try to visualize the overall change in the entropy in terms of the randomness or disorder of the system on a molecular basis.

## 6-13   SUMMARY

The second law of thermodynamics can be stated in many different ways. Among these are the Kelvin-Planck statement, the Clausius statement, and the Hatsopoulos-Keenan statement. All of these statements are equivalent and provide the basis for a new extensive property *entropy S*. Unlike mass and energy, entropy is not conserved and provides important information about the directional nature of physical processes.

To help in our understanding of the directional nature of physical processes it is useful to classify them as either reversible or irreversible. A *totally reversible process* is one for which both the system and its surroundings can be returned to their original states by reversing the direction of the process. If either cannot be restored, the process is called *irreversible.* Irreversibilities are due to dissipative effects, such as friction, or nonquasistatic effects such as the free expansion of a gas. Focusing only on the system, we can say that a system is *internally reversible* if the system can be returned to its initial state by merely changing the direction of all the interactions between the system and its surroundings.

The starting point for applying the second law of thermodynamics to analyze the behavior of any system is the *control volume entropy balance* Eq. [6-43]

$$\frac{dS_{\text{cv}}}{dt} = \sum_{j=1}^{n} \frac{\dot{Q}_j}{T_j} + \left[ \sum_{\text{in}} \dot{m}_i s_i - \sum_{\text{out}} \dot{m}_e s_e \right] + \dot{\sigma} \qquad \textbf{[6-43]}$$

where $T_j$ is an *absolute temperature* and the right-hand side is the sum of the net rate of transport of entropy by heat transfer, the net rate of transport of entropy with mass flow, and the rate of entropy generation (production). All terms in the entropy balance can take on positive and negative values except for the entropy production rate. For any system the entropy production rate must be greater than or equal to zero:

$$\dot{\sigma} = \begin{cases} > 0 & \text{for an internally irreversible process} \\ = 0 & \text{for an internally reversible process} \\ < 0 & \text{then the process is impossible} \end{cases}$$

When applied to a closed system, the mass flow rate terms in Eq. [6-43] are zero. Experience shows that to improve the thermodynamic performance of an energy system, the rate of entropy generation should be minimized.

When applied to a closed system exchanging energy by heat transfer at a single temperature, the entropy balance reduces to

$$\frac{dS_{cm}}{dt} = \frac{\dot{Q}}{T} + \dot{\sigma} \qquad \text{or} \qquad dS = \frac{\delta Q}{T} + \delta\sigma$$

If we further require that the process be internally reversible, then a differential entropy change can be defined for a closed system by the expression

$$dS = \frac{\delta Q_{\text{int rev}}}{T} \qquad \text{(internally reversible, closed system)}$$

Under these conditions, the heat transfer for a closed system can be evaluated from the integral

$$Q_{\text{int rev}} = \int \delta Q_{\text{int rev}} = \int T \, dS \qquad \text{(internally reversible, closed system)}$$

where the heat transfer is represented by the area under the process path on a TS diagram. If we only restrict ourselves to adiabatic, closed systems we find that

$$\frac{dS_{cm}}{dt} = \dot{\sigma} \geq 0 \qquad \text{(adiabatic, closed system)}$$

and the entropy of an adiabatic, closed system can never decrease in any real process.

One important class of closed-system devices that are technologically important are cycles. Structurally cycles can be classified as either *periodic cycles* or *closed-loop, steady-state* cycles. Operationally, they can be classified as *heat engines* (HE), *refrigerators* (R), or *heat pumps* (HP) depending on whether they have a net power output (HE) or a net power input (R and HP). The performance parameters for these three types of cycles are

$$\eta_{\text{th, HE}} = \frac{W_{\text{net, out}}}{Q_{\text{in}}} \qquad \text{COP}_R = \frac{Q_{\text{in}}}{W_{\text{net, in}}} \qquad \text{COP}_{\text{HP}} = \frac{Q_{\text{out}}}{W_{\text{net, in}}}$$

If the cycles operate internally reversibly and the only heat transfers occur at either a high temperature $T_H$ or a low temperature $T_L$, then it can be shown that the performance parameters can be written in terms of the two temperatures as

$$\eta_{\text{th,int rev}} = 1 - \frac{T_L}{T_H} \qquad \text{COP}_{\text{R,int rev}} = \frac{T_L}{T_H - T_L} \qquad \text{COP}_{\text{HP,int rev}} = \frac{T_H}{T_H - T_L}$$

These values are often referred to as the Carnot efficiency and the Carnot coefficients of performance. Carnot demonstrated that for any cycle exchanging energy by heat transfer at only two temperatures two things are true: (1) no cycle has better performance than an internally reversible cycle operating between the same two temperatures, and (2) all internally reversible cycles have the same performance if they operate between the same two temperatures. Physically, an internally reversible cycle can only be approached in the limit as the entropy generation for the cycle approaches zero.

## PROBLEMS

### GENERAL COMMENT QUESTIONS

6-1C. Describe four practical examples of unidirectional processes one observes daily.

6-2C. Describe two examples of practical "working fluids" used in conventional cyclic devices.

6-3C. Describe the differences among the operation of a refrigerator, a heat pump, and an air conditioner.

6-4C. Explain why the given energy input to a heat pump is so much less than that to an electrical resistance element for the same heating requirement.

6-5C. Derive a general relation between $\text{COP}_R$ and $\text{COP}_{HP}$ that does not contain $Q_H$, $Q_L$, or $W$.

6-6C. Consider the cooking of a potato by means of (1) boiling in water on the stove and (2) baking in an oven. Describe any constant-temperature sources of energy involved in these two processes.

6-7C. Prove the equivalence of the Kelvin-Planck and the Hatsopoulos-Keenan statements of the second law.

6-8C. Prove the equivalence of the Clausius and the Hatsopoulos-Keenan statements of the second law.

6-9C. Describe in your own words the differences between total, internal, and external reversibility.

6-10C. Consider internally reversible heat engines $A$ and $B$ operating in series with heat addition to $A$ at $T_H$ and heat rejection from $B$ at $T_L$. An internally reversible heat engine $C$ also operates between $T_H$ and $T_L$. The heat-rejection temperature from $A$ and the heat-supply temperature to $B$ are the same, and denoted by $T_x$. Show mathematically for this situation that the relation $Q_L/Q_H = \Psi(T_H, T_L)$ used in Sec. 6-5-3 must reduce to $Q_L/Q_H = g(T_L)/g(T_H)$.

6-11C. Explain whether the integral of $\delta Q/T$ is the same for all processes of a closed system between the same end states.

6-12C. Explain how the values of the integral of $\delta Q/T$ compare for an irreversible and a reversible process of a closed system between the same end states.

## GENERAL ANALYSIS OF HEAT ENGINES, REFRIGERATORS, AND HEAT PUMPS

6-1. Geothermal steam is under consideration as an energy source for power generation at a remote site. The steam enters at 2 bars and 65 percent quality, and condenses to a saturated liquid before exiting the heat engine. Groundwater at 15°C is to be used as the cold reservoir, and the heat engine is to generate 3.70 kW. If the mass flow rate of steam is 50 kg/h, determine
(a) The rate at which heat is removed from the steam, in kJ/h
(b) The rate of heat rejection to the cold reservoir, in kJ/h
(c) The thermal efficiency

6-2. A steam power plant, with 456 MW of net power output and a thermal efficiency of 38 percent, is cooled by water from a nearby river. The mass flow rate of required river water is 6500 kg/min per megawatt of power produced. Determine (a) the rate of heat rejection to the river water, in kJ/s, and (b) the temperature rise of the river water in passing through the power plant, in °C. The initial temperature of the water is 18°C, and the change in pressure of the water is negligible.

6-3. A gas turbine auxiliary power cycle develops a power output of 4400 kW and has a thermal efficiency of 22 percent. Determine (a) the rate of fuel consumption, in L/min, if the energy released by the fuel is 43,500 kJ/kg and the fuel density is 0.82 g/cm³, and (b) the dollar cost of operating for 8 h if the liquid-fuel cost is $0.30 per liter.

6-4. A heat engine operates by using solar collectors for heat input and lake water as the low-temperature sink. The collectors have a total area of 900 m² and absorb sunlight at a rate of 650 J/m²·s. If the engine produces 180 kW of power, determine (a) the thermal efficiency of the engine, and (b) the rate of heat rejection to the lake, in kJ/min.

6-5. A car engine has a thermal efficiency of 24.0 percent and a power output of 48 kW. The energy released by combustion of the fuel is

44,000 kJ/kg and the fuel density is 0.75 g/cm$^3$. Determine (a) the volume rate of fuel consumed, in L/h, and (b) the dollar cost of operating for 2 h if the fuel cost is $1.40/gal.

6-6. A small household refrigerator is used to cool 20 L of water initially at 22°C. The COP of the unit is 2.40, and the power input is 360 W. Find the time required, in minutes, (a) if the final water temperature is 10°C, and (b) if the final state is ice at 0°C. (Note that 333.4 kJ/kg are required to freeze water at 0°C.)

6-7. An air-conditioning unit maintains a home at 22°C. The heat generated in the home from lighting, appliances, and people is 6000 kJ/h, and heat that leaks through the structure from the environment amounts to 18,000 kJ/h. If the air conditioner has a COP of 3.2, (a) find the required power input in kilowatts. (b) If electricity costs 9.8 cents/kW·h and the unit runs one-half of the time, find the daily cost of operation.

6-8. A household refrigerator with a COP of 2.9 cools water placed in the food compartment. If the device rejects 2520 kJ/h of heat to the environment, determine (a) the power input required, in kilowatts, and (b) the time required to cool 5 L of water in the compartment from 22 to 15°C, assuming heat leakage into the compartment is negligible, in minutes.

6-9. A refrigerator cycle with a COP of 2.7 is used to maintain the food compartment at 3°C. The compartment continuously receives 1260 kJ/h of heat from the environment. The cost of electricity is $0.094/kW·h, and the refrigerator motor runs one-third of the time. (a) Determine the shaft power, in kW, that the cycle requires when running, and (b) the cost of operating the unit in cents per day.

6-10. A house is to be maintained at a temperature of 20°C by means of a heat pump which removes energy from the outside atmosphere at −10°C. Heat transfer rate $\dot{Q}$ through the walls of the house is estimated to be (0.65 kW/K)$\Delta T$, where $\Delta T$ is the temperature difference between the inside of the house and the atmosphere. If the actual COP of the heat pump is 3.10 and electricity costs 10 cents/kW·h, determine
(a) The dollar cost of 1 h of continuous operation
(b) The heat rate supplied from the atmosphere, in kJ/min

6-11. The cooking tanks in a food-processing plant must be heated to 120°C. Your company proposes to use a groundwater-coupled heat pump that is driven by a steam turbine. Steam enters the turbine at 320°C and 10 bars and exits as a saturated vapor at 1 bar. The turbine supplies 45 hp to the heat pump which uses energy from groundwater at 20°C. The COP under these conditions is 3.0 for the heat pump. Determine (a) the mass flow rate of steam, in kg/s, required to run the turbine, and (b) the rate of heat transfer to the cooking tanks.

6-12. A heat pump maintains a building at 22°C. The maximum rate of heat loss through the walls and windows is 45,000 kJ/h, while energy generated within the home by lighting, appliances, and people is 7000 kJ/h. Determine (a) the required power input, in kW, if the average COP is 2.5 and the unit runs two-thirds of the time, (b) the cost of operation for 6 h if the price of electricity is 10.8 cents/kW·h, and (c) the money saved by the homeowner in 6 hours by using the heat pump rather than electrical resistance heating.

6-13. A home has an average heat loss of 1820 kW·h during the month of January. (a) Determine how much money the homeowner could save if a heat pump were installed instead of electrical resistance heating, if the COP of the heat pump averages 2.6, and the cost of electricity is 11.6 cents/kW·h. (b) Find the percent of the energy required for heating that comes from the cold atmosphere in the case of heat-pump operation.

6-14E. A small industrial power plant heats 25,000 $lb_m$/h of steam in the boiler and has a power output of 3800 kW. The plant uses 3200 $lb_m$/h of coal, which has an energy content of 12,500 Btu/$lb_m$. (a) Determine the thermal efficiency. (b) If the energy added to the steam in the boiler from the combustion of coal is 1150 Btu/$lb_m$ of steam, find the fraction of the energy released from the coal that is added to the steam.

6-15. A steam power plant generates 310,000 kg/h of steam and has an output of 95,000 kW of power. The plant consumes 34,000 kg/h of coal, which has an energy content of 30,000 kJ/kg.
  (a) Determine the overall plant thermal efficiency, in percentage.
  (b) If the energy added to the steam from the combustion of coal in the steam-generating unit is 2730 kJ/kg, what fraction of the energy released from the coal is added to the steam?

6-16E. An automobile running steadily at 55 mi/h for 0.95 h consumes 2.1 gal of gasoline. The energy content of the gasoline is 20,500 Btu/lb, and its density is 47 $lb_m$/ft$^3$. The power delivered to the drive shaft of the car is 33 hp. If the entire energy content of the fuel is released by combustion, determine (a) the thermal efficiency of the engine, (b) the fuel economy in mi/gal, and (c) the dollar cost of 2 h of continuous operation on the interstate, if gasoline costs $1.40/gal.

6-17E. A heat pump with a COP of 3 receives energy from the outdoors at 30°F. If the device receives 72,000 Btu/h of heat from outside air, determine (a) the power of the motor required, in hp, and (b) the rate of thermal energy delivered to a home, in Btu/h.

6-18E. Two refrigerators operate in series. Refrigerator A receives 1500 Btu/h from a region at 0°F while operating with a COP of 2.5. The heat rejected by refrigerator A is supplied to refrigerator B, which operates with a power input of 0.35 hp and rejects to the environment at 70°F. Determine (a) the horsepower input required for refrigerator $A$, ($b$) the rate of heat transfer to the environment, in Btu/h, and ($c$) the COP of refrigerator $B$.

## INTERNALLY REVERSIBLE AND IRREVERSIBLE HEAT ENGINES

6-19. The thermal efficiency of an internally reversible heat engine is 62 percent. The working fluid at 23°C rejects 100 kJ/min as heat transfer from the engine. Determine ($a$) the power output of the engine, in kilowatts, and ($b$) the temperature of the fluid during heat addition, in degrees Celsius.

6-20. An internally reversible heat engine receives 4000 kJ as heat addition at 337°C and rejects energy as heat transfer at the triple state of water. Compute ($a$) the thermal efficiency, and ($b$) the net work output in kilojoules. If the efficiency of an irreversible engine is 70 percent of the reversible heat engine, ($c$) find the percent change in heat rejected for the same heat-transfer input and fluid temperatures.

6-21. The heat input to a proposed power cycle is provided by condensing saturated water vapor to saturated liquid at pressure $P$. The heat engine discharges energy as heat transfer at an average fluid temperature of 21.5°C. The expected thermal efficiency is 35 percent, and the power output is to be 1.0 MW. Determine ($a$) the minimum steam pressure required, in kilopascals, and ($b$) the mass flow rate of steam required, in kg/h.

6-22. A geothermal power plant, utilizing an underground source of hot water, receives energy as heat transfer at 160°C. Find ($a$) the maximum possible thermal efficiency of a heat engine which uses this source and has heat transfer out at 15°C and ($b$) the maximum net work output per kilojoule of heat rejection from the cycle. Finally, if the thermal efficiency of an actual heat engine is 60 percent of an internally reversible one, ($c$) find the percent change in $Q_H$ for the same $Q_L$ as the reversible case and the same working-fluid temperatures.

6-23. An internally reversible heat engine operates between 537 and 27°C. Compute ($a$) the ratio of the heat transfer added to the work output and ($b$) the thermal efficiency. Finally, if an actual heat engine has a work output which is 60 percent of the reversible engine, ($c$) determine the thermal efficiency of the actual engine for the same heat supplied and same temperatures of the working fluid during heat supply and rejection.

6-24. An internally reversible heat engine operates with working-fluid temperatures of 627 and 17°C during heat supply and heat rejection, respectively. For every kilowatt of net power output, calculate ($a$) the rate of heat transfer supplied and rejected, in kJ/h, and ($b$) the thermal efficiency. ($c$) An actual heat engine operating between the same temperatures has a heat rejection which is 3.5 times that of the reversible engine. For the same work output as the reversible case, find the thermal efficiency.

6-25. The thermal efficiency of an internally reversible heat engine is 60 percent. A cooling pond receives $10^3$ kJ/min of heat transfer from the working fluid at 17°C. Determine (a) the power output of the engine, in kilowatts, and (b) the temperature of the fluid during heat addition, in degrees Celsius. (c) An actual heat engine operating between the same temperatures has a work output which is one-half that of the internally reversible engine. Find the percent change in the heat rejection, relative to the reversible case, for the same heat input.

6-26. An internally reversible heat engine has a thermal efficiency of 40 percent and the working-fluid temperature is 15°C during heat rejection. Find (a) the net power output, in kilowatts, and (b) the temperature of the working fluid during heat addition, in degrees Celsius, if the heat supplied is 6000 kJ/h. (c) An actual heat engine operating between the same temperatures has a thermal efficiency of 25 percent. For the same heat input, find the percent change in the rate of heat rejection relative to the reversible case.

6-27E. An internally reversible heat engine operates between heat-supply and heat-rejection temperatures of 1240 and 60°F. For every horsepower of net output, calculate (a) the heat transfer supplied and heat transfer rejected, in Btu/h, and (b) the thermal efficiency. If an actual engine operates between the same thermal reservoirs and has a thermal efficiency of 36 percent, (c) find the percent change in $Q_H$ compared to the reversible case, for the same power output.

6-28. An internally reversible heat engine has a thermal efficiency of 60 percent, with 600 kJ/cycle of heat transfer added at a temperature of 447°C. Calculate (a) the heat-rejection temperature, in degrees Celsius, and (b) the heat transfer rejected, in kJ/cycle. For the same values of $Q_H$ and $T_H$, if an irreversible heat engine rejects 420 kJ/cycle, (c) find its thermal efficiency.

6-29E. An internally reversible heat engine has a thermal efficiency of 40 percent and heat removal at 50°F. Find (a) the net power output, in horsepower, and (b) the temperature of the working fluid during heat addition, in degrees Fahrenheit, if the rate of heat addition is 6000 Btu/h.

6-30E. Two internally reversible heat engines are arranged in series. The first heat engine A has heat addition at 1200°F and heat rejection at temperature T. The second heat engine B receives at temperature T the heat transfer rejected by the first engine, and in turn has heat rejection at 100°F. Calculate the temperature T, in degrees Fahrenheit, for the situation where (a) the work outputs of the two engines are equal and (b) the thermal efficiencies of the two engines are equal.

6-31. Two internally reversible heat engines are arranged in series. The first engine A has heat addition at 927°C and heat rejection at temperature

*T*. The second heat engine *B* receives at temperature *T* the heat rejected by the first engine, and in turn has heat rejection at 27°C. Calculate the temperature *T*, in degrees Celsius, for the situation where (*a*) the work outputs of the two engines are equal and (*b*) the thermal efficiencies of the two engines are equal.

6-32. Two internally reversible heat engines of equal thermal efficiency operate in series. Engine *A* receives 2500 kJ of heat transfer at 1000 K, and all the heat transfer from heat engine *A* is used as energy input to heat engine *B*, which rejects energy to a 300 K sink. Determine (*a*) the intermediate temperature of rejection for engine *A* (and supply temperature for engine *B*), in kelvins, (*b*) the work output of engine *A*, and (*c*) the work output of engine *B*, in kilojoules.

6-33. An internally reversible heat engine operates between 300 and 1200 K, and the heat transfer out is at a rate of 400 kJ/s. Determine whether the following three situations reflect a reversible, irreversible, or impossible operation by second-law analysis. (*a*) $\dot{Q}_H = 1500$ kJ/s and $\dot{W} = 1100$ kJ/s, (*b*) $\dot{Q}_H = 1800$ kJ/s and $\dot{W} = 1400$ kJ/s, and (*c*) $\dot{Q}_H = 1600$ kJ/s and $\dot{W} = 1200$ kJ/s.

## INTERNALLY REVERSIBLE AND IRREVERSIBLE REFRIGERATORS AND HEAT PUMPS

6-34. An internally reversible refrigerator absorbs 450 kJ/min from a cold space and requires 3.0 kW to drive it. If the machine is reversed and the heat-transfer input is 1600 kJ/min, how much power does it produce, in kilowatts?

6-35. An internally reversible refrigerator is used to produce ice at 0°C. The heat-rejection temperature is 30°C, and the enthalpy of freezing for water is 335 kJ/kg. Find the kilograms of ice formed per hour per kilowatt of power input, if the water enters the unit at 20°C.

6-36. A ton of refrigeration is defined as a heat-absorption rate of 211 kJ/min from a cold source. It is desired to operate an internally reversible refrigerator between temperature limits of −23 and +27°C so that 8 tons of refrigeration are produced. Calculate (*a*) the kilowatts of power required to operate the cycle and (*b*) the coefficient of performance.

6-37. A computer room is to be kept at 22°C. The computer transfers 1500 kJ/h to the room air. The heat-rejection temperature of the cooling unit is 28°C, and the temperature of the working fluid during the heat supply is 19°C.
  (*a*) Find the minimum power input, in watts, required for the cooling process.
  (*b*) If the actual power required is 42.0 W, determine the percentage increase in heat rejected to the environment.

6-38. A computer room is to be kept at 23°C. The computer and lighting transfers 2200 kJ/h to the room air, and the heat-transfer rate from the outside environment amounts to 12,000 kJ/h into the room. The heat rejection from the cooling unit is at 33°C, and heat transfer into the working fluid occurs at 20°C.
(a) Find the minimum power input required, in kilowatts.
(b) If the actual power required is 0.80 kW, determine the percentage increase in heat rejected to the environment, and the actual COP.

6-39. The working fluid passing through the food compartment of a home refrigerator is maintained at 4°C during heat transfer to an internally reversible cycle. Heat rejection from the fluid occurs at 24°C. The power input is 600 W. Determine (a) the COP of the reversible device, and the rate of heat input to the 4°C fluid, in kJ/min. Then, for the same power input, if the actual COP is 2.5, (b) determine the percent decrease in cooling rate, compared to the reversible cycle.

6-40. A refrigerator freezes 0.80 kg of liquid water at 0°C. The power input is 400 W, and the rejection temperature from the working fluid is 24°C. If the actual refrigerator has a COP which is one-quarter that of a reversible refrigerator, determine the time required, in minutes. The enthalpy of freezing is 333.4 kJ/kg.

6-41. A bucket of ice cubes is in a freezer unit of a refrigerator that has a COP which is one-third that of a reversible refrigeration cycle. (a) Determine the maximum possible temperature for heat transfer from the refrigeration cycle, in °C and °F, so that the cubes remain frozen. The heat-leakage rate into the freezer is 78 kJ/min and the power input is 0.50 kW. (b) For the same given data, let the temperature for heat rejection be 40°C. If the enthalpy of melting of ice is 333.4 kJ/kg, determine the time required to melt 1 kg of ice inside the freezer, in minutes.

6-42. An internally reversible refrigeration device receives 180 kJ/min of heat transfer from a cold region and requires 1.20 kW to drive it. The cyclic machine is now reversed, and as a heat engine receives 1400 kJ/min as heat transfer at the same boundary temperature that applied to the refrigerator, find the power produced by the heat-engine operation, in kilowatts.

6-43. A refrigeration cycle operates between boundary temperatures of 250 and 312.5 K. For the three cases below, determine whether the cycle is internally reversible, irreversible, or impossible. (a) $Q_L = 900$ kJ and $W = 300$ kJ, (b) $Q_H = 1600$ kJ and $W = 300$ kJ, (c) $Q_H = 1500$ kJ and $Q_L = 1200$ kJ.

6-44. A refrigeration cycle receives 135 kJ/min as heat transfer at a boundary temperature of $-30$°C and has heat transfer out at a boundary temperature of 27°C. If the actual COP is 60 percent of the maximum possible value, (*a*) determine the power required, in kilowatts. (*b*) If the low temperature is raised to $-10°C$, find the percent change in the work required.

6-45E. An internally reversible refrigerator absorbs 250 Btu/min as heat transfer from a cold space, and requires 2.0 hp to drive it. If the machine is reversed and receives 800 Btu/min at the hot boundary, find the power, in horsepower, that it produces.

6-46E. An internally reversible refrigerator is used to produce ice at 32°F. The heat-rejection temperature is 78°F, and the enthalpy of freezing for water is 144 Btu/lb. Find the pounds of ice formed per hour per horsepower of input, if the water enters the unit at 60°F.

6-47E. A ton of refrigeration is defined as a heat-absorption rate of 200 Btu/min from a cold source. It is desired to operate an internally reversible refrigerator between temperature limits of $-10$ and $+85°F$ so that 8 tons of refrigeration are produced. Calculate (*a*) the horsepower input required to operate the cycle and (*b*) the coefficient of performance. (*c*) If the required temperature at the cold boundary is raised to 25°C, find the horsepower required.

6-48E. A computer room is to be kept at 70°F. The computer transfers 7500 Btu/h to the room air, and the boundary temperature at heat rejection is 90°F.

(*a*) Find the minimum power input, in horsepower, required for the cooling process.

(*b*) If the actual power required is 0.32 hp, determine the percentage increase in heat rejected at the boundary for the actual case compared to the theoretical case.

6-49. An internally reversible heat pump is used to supply 120,000 kJ/h to a building. On a cold winter day the lower boundary temperature of the heat pump is $-6°C$, and heat transfer is supplied at 26°C to the building. Find (*a*) the heat transfer from the cold outside air, in kJ/h, (*b*) the power input required, in kilowatts, and (*c*) the cost of operation, in dollars, for 30 min if electricity costs 9.5 cents per kilowatt hour.

6-50. A building is to be supplied energy at 22°C by means of an internally reversible heat pump which receives energy at $-10°C$ from the atmosphere. The heat-loss rate through the walls of the building is estimated to be 0.64 kW for each kelvin of temperature difference between the inside and outside of the building. Find (*a*) the power required to drive the heat pump, in kilowatts, and (*b*) the COP of the device. Now the same heat pump is used to cool the building in the summer. For the same building temperature, heat-gain rate of 0.64 kW per degree temperature difference, and power input, find (*c*) the minimum temperature permissible for the heat rejection, in degrees Celsius, and (*d*) the COP of the cooling device when the temperature of heat rejection is 10°C above the reversible temperature of rejection.

6-51. An internally reversible heat pump operating between boundary temperatures of −7 and 29°C has a power input of 3.5 kW. Determine (a) the coefficient of performance and (b) the heat rate supplied at 29°C, in kJ/s. (c) If an actual heat pump requires 10.0 kW of power input for the same temperatures and same heat transfer supplied at −7°C, determine the actual COP and the actual rejection heat rate at 29°C, in kJ/s.

6-52E. An internally reversible heat pump is used to supply 200,000 Btu/h to a building from the outside atmosphere. The low-temperature boundary is at 22°F, and heat rejection occurs at 76°F. Find (a) the rate of heat transfer from the outside, in Btu/h, (b) the power input required, in horsepower, and (c) the cost of operation, in dollars, for 1.5 h if electricity costs 8.5 cents per kilowatt-hour. Now, if the work input for an actual heat pump is 3 times that for an internally reversible heat pump, (d) find the ratio of the actual heat to the theoretical heat supplied from the environment, for the same heat supplied to the building.

6-53. An actual heat pump operates between boundary temperatures of 600 and 1100 K. The heat pump has a heat-rejection rate of 15,000 kJ/min and a net power input of 120 kW.
   (a) Determine numerically whether the actual heat pump violates the second law in terms of its coefficient of performance.
   (b) Determine the heat-rate input, in kilowatts, for the actual heat pump and for an internally reversible heat pump having the same heat output and operating between the same boundary temperatures.
   (c) Now operate the actual cyclic device as a heat engine. The heat input rate at 1100 K is 250 kW, and the net power output is 120 kW. Determine the thermal efficiency of the actual heat engine and of an internally reversible engine operating between the same boundary temperatures.
   (d) Discuss whether this actual engine operation is feasible by consideration of the thermal efficiencies found in part c.

6-54. An actual heat engine operates between boundary temperatures of 1200 and 500 K. A heat input of 36,000 kJ/min produces a net power output of 270 kW.
   (a) Determine numerically whether the heat engine violates the second law in terms of its thermal efficiency.
   (b) Determine the heat rejection, in kilowatts, for the actual engine and for an internally reversible engine having the same heat input and operating between the same boundary temperatures.
   (c) Now reverse the actual cyclic device, so that it operates as a heat pump. The work input is 270 kW, and the heat rejected

at 1200 K is 36,000 kJ/min. Determine the coefficient of performance for the actual heat pump and for an internally reversible heat pump operating between the same boundary temperatures.

(d) On the basis of part c, is the actual heat-pump operation feasible considering the COP values?

6-55. A heat engine operates between boundary temperatures of 500 and 1400°C. The heat engine requires a heat input of 1300 kW to produce a net work output of 600 kW.

(a) Show from thermal efficiency data whether this engine violates the second law.

(b) Determine the heat-rejection rate, in kW, for the actual heat engine and for an internally reversible engine having the same heat-input rate and operating between the same boundary temperatures during heat addition and heat removal.

(c) What is the COP for an internally reversible heat pump operating between the same boundary temperatures?

(d) If the data for the actual heat engine are valid when the device is operated as a heat pump, what would be its COP? Determine whether or not such operation is feasible.

6-56E. An internally reversible heat pump operates between boundary temperatures of 20 and 80°F and has a power input of 4.2 hp. Determine (a) the coefficient of performance and (b) the heat transfer supplied at 80°F, in Btu/s. (c) If an actual heat pump requires 12.0 hp of power input for the same temperatures and same heat supplied from 20°F, determine the actual COP and the actual heat rejected at 80°F, in Btu/s.

6-57. A heat pump with a COP of 2.5 is used to meet the heating requirements for a house. It supplies energy by heat transfer at a boundary temperature of 25°C. On a winter day, when the lower boundary temperature for the cycle is required to be $-2$°C, the house is estimated to lose heat at a rate of 80,000 kJ/h.

(a) Determine the power consumed by the heat pump, in kW.

(b) Calculate the rate at which heat transfer occurs from the cold outside air, in kJ/h.

(c) What is the minimum power required to drive this heat pump, in kW?

(d) How much power would be consumed if resistance heating were used rather than the heat pump, in kW?

(e) Why is the power consumed when using the heat pump so much less than that when using resistance heating?

6-58. Heat enters a heat pump at a boundary temperature of $-6$°C, and 960 kJ/min as heat transfer is supplied at 25°C to a building.

(a) Determine the COP if the actual power input is 6.6 kW.

(b) Determine the COP and the minimum power, in kW, for an internally reversible heat pump operating between the same two boundary temperatures and supplying the same rate of heat transfer to the building.

(c) Determine the percent change in $Q_L$ for the actual case compared to the reversible case.

6-59. An internally reversible heat pump is used to supply 120,000 kJ/h to a building. The outside atmosphere is the cold source, and the lower boundary temperature for the cycle is $-8°C$. Heat transfer at a boundary temperature of 26°C is used to supply energy to the building. Find (a) the heat rate from the outside air, in kJ/h, (b) the power input required, in kilowatts. (c) If an actual heat pump has a COP which is 40 percent of the theoretical value, determine the cost of operation for 2 h, in dollars, if electricity costs 9.4 cents per kilowatt-hour.

6-60. The rate of heat loss from a building is 2800 kJ/h per °C temperature difference between the inside- and outdoor-air temperatures. A power input of 7 kW to a heat pump is used to supply heat transfer at 27°C to the building. Determine the lowest outdoor temperature, in °C, for which the heat pump can satisfy the heating requirement, if the COP of the actual heat pump is one-third that of a reversible heat pump.

6-61. An internally reversible heat engine operates between boundary temperatures of 375 and 300 K. An internally reversible refrigerator receives heat transfer at temperature $T$ (in °C), and discharges at a boundary temperature of 300 K. If the COP for the refrigerator is 0.20 of the percentage efficiency for the engine, find (a) the COP of the refrigerator and (b) the temperature $T$, in °C.

6-62. Two internally reversible refrigerators are arranged in series. The first receives 300 kJ/cycle as heat transfer at 300 K from an energy source. The heat transfer rejected from this refrigerator serves as the heat-transfer input to a second refrigerator, which delivers its heat-transfer output at a boundary temperature of 1000 K. If the two refrigerators have the same coefficients of performance, determine (a) the heat rejection at 1000 K, in kJ/cycle, (b) the rejection temperature from the first refrigerator (and supply temperature for the second one), in kelvins, and (c) the COP of the two refrigerators.

6-63. Two internally reversible heat pumps are arranged in series. The first receives energy as heat transfer at 280 K. The energy rejection from this heat pump serves as the heat input to a second heat pump, which delivers 2000 kJ/min at a boundary temperature of 1000 K. If the two heat pumps have the same coefficients of performance, determine (a) the heat transfer received by the low-temperature heat pump from the source at 280 K, in kJ/min, (b) the rejection temperature, in kelvins, from the first heat pump (and supply temperature for the second one), (c) the COP of the two heat pumps, and (d) the total power required, in kilowatts.

6-64E. Two internally reversible refrigerators are arranged in series. The first receives 300 Btu/cycle at 80°F. The rejection heat transfer from this refrigerator serves as the heat input to a second refrigerator, which delivers its heat output at a boundary temperature of 1340°F. If the two refrigerators have the same coefficients of performance, determine (*a*) the heat-transfer rejection at 1340°F, in Btu, (*b*) the temperature during heat rejection by the first refrigerator (and during heat supply to the second one), in degrees Fahrenheit, and (*c*) the COP of the two refrigerators.

6-65E. Two internally reversible heat pumps are arranged in series. The first receives heat transfer at 40°F. The heat rejection from this heat pump serves as the heat input to a second heat pump, which delivers 2000 Btu/min at 1540°F. If the two heat pumps have the same coefficients of performance, determine (*a*) the heat-transfer input at 40°F by the first heat pump, in Btu/min, (*b*) the rejection temperature, in degrees Fahrenheit, for the first heat pump (and supply temperature for the second one), (*c*) the COP of the two heat pumps, and (*d*) the total power required, in horsepower.

## COMBINED HEAT ENGINES, REFRIGERATORS, AND HEAT PUMPS

6-66. A heat engine operates with boundary temperatures at 1000 and 300 K. It is used to drive a refrigerator which operates between 250 and 300 K. The heat-transfer input at 1000 K is 1000 kJ, and the heat transfer rejected from the refrigerator at 300 K is 2700 kJ. These values remain fixed.
   (*a*) If the heat engine and refrigerator are both internally reversible devices, find the percentage of the heat-engine work output used to drive the refrigerator.
   (*b*) If the COP of the refrigerator is 80 percent of the value for an internally reversible device operating between the given temperatures, determine the actual work, in kilojoules, required to operate the refrigerator.
   (*c*) Relative to part *b*, find the minimum fraction of the reversible heat-engine efficiency which an actual heat engine must have to just drive the actual refrigerator.

6-67. An internally reversible heat engine operates between supply and rejection temperatures of 600 and 40°C. Part of the work output of the heat engine is used to drive an internally reversible refrigerator which operates between boundary temperatures of 40 and −20°C. The heat input to the engine is 2100 kJ and the net work out (engine work minus refrigerator work) is 370 kJ. Determine (*a*) the heat transfer to the refrigerator at −20°C and (*b*) the net heat transfer at 40°C, in kilojoules.

6-68. An internally reversible heat engine receives 900 kJ of heat transfer at 627°C. Heat rejection occurs at 27°C. One-fifth of its work output is used to drive an internally reversible refrigerator. The refrigerator rejects 600 kJ at a boundary temperature of 27°C. Find (a) the work output of the heat engine, (b) the efficiency of the heat engine, (c) the boundary temperature for heat supply to the refrigerator, in degrees Celsius, and (d) the COP of the refrigerator.

6-69. An internally reversible heat engine operates between boundary temperatures of 397 and 7°C and the heat transfer rejection rate is 120 kJ/min. The total net work output of the engine is used to drive an internally reversible heat pump, which is supplied by heat transfer at 7°C and rejects heat at a temperature of 40°C. Determine (a) the net power delivered by the engine, in kilowatts, (b) the heat supplied to the heat pump, in kJ/min, and (c) an overall COP for the combined devices, which is defined as the energy delivered to the home divided by the initial energy supplied to the engine.

6-70E. An internally reversible heat engine receives 90 Btu as heat transfer at 1100°F. It rejects energy as heat transfer at 60°F. One-third of its work output is used to drive an internally reversible refrigerator. The refrigerator rejects 120 Btu as heat transfer at 60°F. Find (a) the work output of the heat engine, in Btu, (b) the efficiency of the heat engine, (c) the temperature of the low-temperature boundary for the refrigerator, in degrees Fahrenheit, and (d) the COP of the refrigerator.

6-71E. An internally reversible heat engine operates between 740 and 20°F and rejects 40 Btu/min to the environment. The total net work output of the engine is used to drive a heat pump. Heat transfer enters the heat pump at 20°F and heat transfer is rejected at 140°F. Determine (a) the net work delivered by the engine, in Btu/min, (b) the heat supplied to the heat pump, in Btu/min, and (c) an overall COP for the combined devices, which is defined as the energy rejected to the home divided by the initial energy supplied to the engine.

6-72. An internally reversible heat engine operates between 827 and 17°C and rejects 100 kJ/min to the environment. One-half of the net work output of the engine is used to drive an internally reversible heat pump which is supplied with heat transfer at 17°C and has heat transfer out at 37°C to a home. Determine (a) the net power delivered by the engine to the environment, in kW, (b) the heat rejected by the heat pump in kJ/min, and (c) the *net* heat-transfer rate with the 17°C reservoir, in kJ/min.

6-73. An irreversible heat engine receives energy as heat transfer at 800 K and rejects some energy as heat transfer at a boundary temperature of 400 K. The energy delivered as work from the actual heat engine is delivered to an irreversible transmission, and the work output of the transmission is used to drive an irreversible refrigerator. This refrigerator receives heat transfer at a boundary temperature of 240 K, and heat transfer leaves at 400 K. In each cycle the refrigerator removes 240 kJ as heat transfer and the *net* heat transfer occurring at 400 K for the two cycles is 800 kJ/cycle. The actual heat engine has a thermal efficiency which is 80 percent of that of an internally reversible heat engine operating between the same temperatures, and the COP of the actual refrigerator is 80 percent of an internally reversible refrigerator operating between the same temperatures as the actual refrigerator. Determine (*a*) the work input required for the actual refrigerator, in kJ/cycle, (*b*) the heat input from the 800 K reservoir into the actual engine, in kilojoules, and (*c*) the percentage of the energy (work) into the transmission which is lost (does not appear as work output).

6-74. An irreversible heat pump is designed to remove heat at 7°C that is received from the atmosphere and to supply 43,200 kJ/h as heat transfer at 420 K to an industrial process. The COP of the actual heat pump is 60 percent of that of an internally reversible heat pump operating between the same two temperatures. The heat pump is driven, through a transmission, by the output of a heat engine. It receives heat transfer at 1050 K from a high-temperature source and rejects energy by heat transfer at 420 K to the same industrial process. The thermal efficiency of the actual heat engine is 75 percent of that of an internally reversible heat engine operating between the same two temperatures. In addition, the transmission which delivers the actual work output of the heat engine to the heat pump is only 80 percent efficient. Determine (*a*) the power input required for the actual heat pump, in kilowatts, (*b*) the rate of heat input at 1050 K to the actual engine, in kJ/h, and (*c*) the percentage of the total energy supplied to the industrial process which is delivered by the heat pump.

## CLAUSIUS INEQUALITY AND CARNOT PRINCIPLE

6-75. The working fluid of a heat engine receives 2975 kJ/kg as heat transfer at an average temperature of 450°C and rejects 1850 kJ/kg as heat transfer at 47°C to cooling water at 17°C. Determine whether this engine violates the second law on the basis of (*a*) the Clausius inequality and (*b*) the Carnot principle.

6-76. In a heat-engine cycle the working fluid receives 3000 kJ/min as heat transfer at an average temperature of 487°C and rejects 1120 kJ/min as heat transfer at an average temperature of 22°C. Determine whether this engine violates the second law on the basis of (*a*) the Clausius inequality, and (*b*) the Carnot principle.

6-77. Saturated water vapor enters the turbine of a steam power cycle at 40 bars (state 1) and leaves at 1 bar with a quality of 90 percent (state 2). See Fig. 6-18. The fluid leaves the condenser at 1 bar with a quality of 29 percent (state 3), and the fluid is a saturated liquid at 40 bars at state 4. Determine whether the cycle satisfies the second law on the basis of (a) the Clausius inequality and (b) the Carnot principle.

6-78. Saturated R-134a vapor enters the turbine of a power cycle at 30 bars (state 1) and leaves at 1 bar with a quality of 90 percent. See Fig. 6-18. The fluid leaves the condenser at 1 bar with a quality of 60 percent (state 3) and the fluid is a saturated liquid at 30 bars at state 4. Determine whether the cycle satisfies the second law on the basis of (a) the Clausius inequality, and (b) the Carnot principle.

6-79. In a given refrigeration cycle the working fluid receives 114.0 kJ/kg as heat transfer at −5°C and rejects 140.3 kJ/kg as heat transfer at an average temperature of 35°C. If there are no other heat interactions, does the cycle violate the second law on the basis of (a) the Clausius inequality, and (b) the Carnot principle extended to refrigerators?

6-80. The working fluid in a heat-pump cycle is reported to receive 136.0 kJ/kg as heat transfer at 4°C and reject 145.2 kJ/kg as heat transfer at an average temperature of 34°C. If there are no other heat interactions, does the cycle violate the second law on the basis of (a) the Clausius inequality, and (b) the Carnot principle extended to heat pumps?

6-81E. The working fluid in a heat engine cycle receives 2440 Btu/lb$_m$ as heat transfer at an average temperature of 900°F and rejects 870 Btu/lb$_m$ as heat transfer at an average temperature of 60°F. Determine whether this engine violates the second law on the basis of (a) the Clausius inequality, and (b) the Carnot principle.

6-82E. Saturated water vapor enters the turbine of a steam power cycle at 550 psia (state 1) and leaves at 1 psia with a quality of 95 percent (state 2). See Fig. 6-18. The fluid leaves the condenser at 1 psia with a quality of 35 percent (state 3), and the fluid is a saturated liquid at 550 psia at state 4. Determine whether the cycle satisfies the second law on the basis of (a) the Clausius inequality and (b) the Carnot principle.

6-83. A heat engine operates between working-fluid temperatures of 1000 and 300 K, and the heat transfer in is 500 kJ. For each case below, determine on the basis of both the Clausius inequality and the Carnot principle whether the cycle is reversible, irreversible, or impossible: (a) $Q_L = 150$ kJ, (b) $W = 400$ kJ, and (c) $W = 300$ kJ.

6-84. A heat engine operates between temperatures of 840 and 280 K for the working fluid, and the heat transfer in is 900 kJ. For each case below, determine on the basis of both the Clausius inequality and the Carnot principle whether the cycle is reversible, irreversible, or impossible: (*a*) $Q_L$ = 250 kJ, (*b*) $W$ = 400 kJ, and (*c*) $W$ = 600 kJ.

6-85E. A refrigeration device has heat transfer in at 470°R and heat-transfer rejection at 560°R, and the heat-rejection rate is 40 Btu/min. For each case below, determine from both the Clausius inequality and the Carnot principle whether the cycle is reversible, irreversible, or impossible: (*a*) $\dot{Q}_L$ = 35.0 Btu/min, (*b*) $\dot{W}$ = 6.43 Btu/min, and (*c*) $\dot{W}$ = 7.0 Btu/min.

6-86. A refrigeration device receives heat transfer at 260 K and heat-transfer rejection at 310 K, and the heat-input rate is 20 kJ/min. For each case below, determine from both the Clausius inequality and the Carnot principle whether the cycle is reversible, irreversible, or impossible: (*a*) $\dot{Q}_H$ = 25.0 kJ/min, (*b*) $\dot{W}$ = 3.85 kJ/min, and (*c*) $\dot{W}$ = 3.5 kJ/min.

6-87. A heat pump has heat-transfer addition at 280 K and heat transfer out at 420 K, and the heat-transfer rate out at 420 K is 900 kJ/min. For each case below, determine from both the Clausius inequality and the Carnot principle for a heat pump whether the cycle is reversible, irreversible, or impossible: (*a*) $\dot{Q}_L$ = 580 kJ/min, (*b*) $\dot{W}$ = 280 kJ/min, and (*c*) $\dot{Q}_L$ = 600 kJ/min.

## HEAT-TRANSFER ANALYSIS

6-88. A heat transfer of 100 kJ occurs between thermal reservoirs at 680 and 490 K. The environment is at 300 K. Determine
   (*a*) The change in entropy of each thermal reservoir, in kJ/K. Is the sum of these changes in agreement with the increase in entropy principle?
   (*b*) The entropy generation within the heat-transfer region, in kJ/K.
   (*c*) The percent reduction in work potential of the 100 kJ due to its transfer to the 490 K reservoir.
   (*d*) The percent increase in work potential of the 100 kJ at 680 K if $T_0$ is lowered to 275 K.

6-89. One thousand kilojoules of energy are transferred from a thermal reservoir at 880 K by heat transfer to a second thermal reservoir at 360 K. The environmental temperature is 300 K.
   (*a*) Calculate the change in entropy of each thermal reservoir, in kJ/K.
   (*b*) Calculate the entropy production for the heat-transfer region, in kJ/K.
   (*c*) Determine the loss in work potential of the 1000 kJ due to the irreversible heat-transfer process, in kJ.
   (*d*) Find the percent increase in work potential of 1000 kJ at 880 K when $T_0$ is lowered to 280 K.

6-90. Energy in the amount of 3000 kJ is available from a thermal reservoir at 800 K, and the environmental temperature is 300 K. The energy is now transferred as heat transfer from the 800 K reservoir to another thermal reservoir at 600 K. Determine
(a) The entropy change for each thermal reservoir, in kJ/K
(b) The entropy production within the heat-transfer region, in kJ/K
(c) The percent reduction in work potential of the given heat quantity due to its transfer to the 600 K reservoir

6-91. Energy in the amount of 5000 kJ is available from a thermal reservoir at 900 K, and the environmental temperature is 280 K. The energy is now transferred by heat transfer from the 900 K thermal reservoir to another thermal reservoir at 500 K. Determine
(a) The entropy change for each thermal reservoir, in kJ/K
(b) The entropy production within the heat-transfer region, in kJ/K
(c) The percent reduction in work potential of the given heat quantity due to its transfer to the 600 K reservoir

6-92E. Energy in the amount of 3000 Btu is available from a thermal reservoir at 1500°R, and the environmental temperature is 500°R. The heat is now transferred from the 1500°R thermal reservoir to another thermal reservoir at 1100°R. Determine
(a) The entropy change for each thermal reservoir, in Btu/°R
(b) The entropy production within the heat-transfer region, in Btu/°R
(c) The percent reduction in work potential of the given heat quantity due to its transfer to the 1100°R reservoir

6-93E. Energy in the amount of 5000 Btu is available from a thermal reservoir at 1600°R, and the environmental temperature is 510°R. The heat is now transferred from the 1600°R thermal reservoir to another thermal reservoir at 900°R. Determine
(a) The entropy change for each thermal reservoir, in Btu/°R
(b) The entropy production within the heat transfer region, in Btu/°R
(c) The percent reduction in work potential of the given heat quantity due to its transfer to the 900°R reservoir

6-94. A heat transfer of 5000 kJ occurs between a thermal-energy reservoir at 1000 K and four other thermal reservoirs at (1) 800, (2) 600, (3) 400, and (4) 300 K. The atmospheric temperature $T_0$ is 300 K. (a) Determine the entropy generation, in kJ/K, for the separate heat-transfer processes from the 1000 K reservoir to the other four thermal reservoirs. Plot the results versus the low-temperature value. (b) Find the work potential, in kJ, of the 5000 kJ of energy when it resides at the five different temperature levels. Plot the loss in work potential for each of the four heat exchanges versus the low-temperature value on the same diagram as part a.

6-95. A heat transfer of 2000 kJ occurs between a thermal-energy reservoir at 700 K and four other thermal reservoirs at (1) 600, (2) 500, (3) 400, and (4) 300 K. The atmospheric temperature $T_0$ is 300 K. (*a*) Determine the entropy generation, in kJ/K, for the separate heat-transfer processes from the 700 K thermal reservoir to the other four thermal reservoirs. Plot the results versus the low-temperature value. (*b*) Find the work potential, in kJ, of the 2000 kJ of energy when it resides at the five different temperature levels. Plot the loss in work potential for each of the four heat exchanges versus the low-temperature value on the same diagram as part *a*.

6-96. A heat transfer of 3000 kJ occurs between a thermal-energy reservoir at 900 K and four other thermal reservoirs at (1) 750, (2) 600, (3) 450, (4) 300 K. The atmospheric temperature $T_0$ is 300 K. (*a*) Determine the entropy generation, in kJ/K, for the separate heat transfer processes from the 900 K thermal reservoir to the other four thermal reservoirs. Plot the results versus the low-temperature value. (*b*) Find the work potential, in kJ, of the 3000 kJ of energy when it resides at the five different temperature levels. Plot the loss in work potential for each of the four heat exchanges versus the low-temperature value on the same diagram as part *a*.

## CLOSED STEADY-STATE SYSTEMS

6-97. An electric transformer is used to step down the voltage from 220 to 110 V. The current on the high-voltage side is 23 A and on the low-voltage side is 43 A. The transformer operates under steady-state conditions with a surface temperature of 40°C. Determine (*a*) the heat-transfer rate for the device, in watts, and (*b*) the entropy production rate, in W/K.

6-98. An electric motor operates under steady-state conditions and draws 3 kW of electrical power. Ten percent of the electrical power supplied is lost by heat transfer to the surroundings. The surface temperature of the motor is 45°C. Determine (*a*) the shaft power delivered by the motor in horsepower and (*b*) the rate of entropy production in kW/K.

6-99E. A soldering iron under steady-state conditions draws 0.10 A from a 110-V circuit. The operating temperature of the soldering iron is 220°F. Determine the entropy production rate for the soldering iron in kW/K.

6-100E. A transmission consists of two gearboxes connected by an intermediate shaft. A torque of 220 ft·lb$_f$ is applied to the input shaft which rotates at 200 rpm (revolutions per minute). The intermediate shaft and the output shaft rotate at 160 and 128 rpm respectively. *Each* gearbox transmits only 95 percent of the shaft power supplied to it. The rest is lost by heat transfer at a surface temperature of 120°F.

Determine (a) the torque for the intermediate and output shafts, in ft·lb$_f$, and (b) the entropy production rate for each gearbox and for the transmission, in Btu/h·°R.

6-101. The outlet valve for a centrifugal fan has been shut so that no flow is going through the fan. The $\frac{1}{4}$-hp fan motor runs continuously. If the fan is modeled as a closed steady-state system, determine (a) the heat-transfer rate from the fan, in watts, and (b) the entropy production rate in the fan, in W/K, if the fan reaches a temperature of 40°C at steady-state conditions.

## IRREVERSIBLE HEAT TRANSFER IN SIMPLE CYCLES AND WORK POTENTIAL

6-102. A heat engine operates between thermal-reservoir temperatures $T_A$ and $T_B$ of 1500 and 300 K, respectively. For a heat-transfer input rate of 20,000 kJ/s, find the rejected heat-transfer rate, the net work output, the thermal efficiency, and the loss in work potential when (a) the device is totally reversible, and (b) the device is internally reversible but receives and rejects energy as heat transfer at 700 and 350 K rather than 1500 and 300 K. $T_0 = 300$ K.

6-103. A refrigerator operates between thermal-reservoir temperatures $T_A$ and $T_B$ of 320 and 260 K, respectively. For a heat-transfer input of 100 kJ, find the rejected heat transfer, the net work input, the coefficient of performance, and the loss in work potential when (a) the device is totally reversible, and (b) the device is internally reversible but receives and rejects energy as heat transfer at 330 and 250 K rather than 320 and 260 K. $T_0 = 320$ K.

6-104. A heat pump operates between thermal-reservoir temperatures $T_A$ and $T_B$ of 600 and 300 K, respectively. For a heat-transfer output of 1000 kJ/min, compare the heat transfer supplied, the net work input, the coefficient of performance, and the loss in work potential when (a) the device is totally reversible, and (b) the device is internally reversible but receives and rejects energy as heat transfer at 620 and 280 K rather than 600 and 300 K. $T_0 = 300$ K.

6-105E. A refrigerator operates between thermal-reservoir temperatures $T_A$ and $T_B$ of 560 and 470°R, respectively. For a heat-transfer input of 1000 Btu, find the rejected heat transfer, the net work input, the coefficient of performance, and the loss in work potential when (a) the device is totally reversible, and (b) the device is internally reversible but receives and rejects energy as heat transfer at 450 and 630°R rather than 470 and 600°R. $T_0 = 560$°R.

6-106E. A heat pump operates between thermal-reservoir temperatures $T_A$ and $T_B$ of 1200 and 540°R, respectively. For a heat-transfer output of

1000 Btu/min, find the heat transfer supplied and the net work input, the coefficient of performance, and the loss in work potential when (*a*) the device is totally reversible, and (*b*) the device is internally reversible but receives and rejects energy as heat transfer at 520 and 1250°R rather than 540 and 1200 K. $T_0 = 540°R$.

## ENTROPY PRODUCTION IN SIMPLE HEAT ENGINE, REFRIGERATION, AND HEAT PUMP CYCLES

6-107. A heat-engine cycle operates between two thermal-energy reservoirs with temperatures of 700 and 290 K.

(*a*) The working fluid of the heat engine itself has a temperature of 670 K during heat addition from the 700 K thermal reservoir, and a temperature of 310 K during heat rejection to the 290 K reservoir. The thermal efficiency of the actual unit is 52 percent, and the heat-transfer rate from the 700 K reservoir is 2000 kJ/min. Determine the entropy production rate, in kJ/min·K, for (1) the heat engine itself, (2) the low-temperature heat-transfer process, and (3) the high-temperature heat-transfer process.

(*b*) Now for the same thermal-reservoir temperatures, heat-transfer rate, and working-fluid temperatures, the thermal efficiency of the heat engine is 50 percent. As before, determine the entropy production rate for (1) the heat engine itself, (2) the low-temperature heat-transfer process, and (3) the high-temperature heat-transfer process. (4) Which process now has the largest irreversibility?

(*c*) Finally, for the same thermal-reservoir temperatures, heat-transfer supply rate, and thermal efficiency cited in part *b*, a lower-cost heat exchanger will require a decrease in the working-fluid temperature from 670 to 645 K during heat addition. For this new condition, determine the entropy production rate for (1) the heat engine itself, (2) the low-temperature heat-transfer process, and (3) the high-temperature heat-transfer process. (4) Which process now has the largest irreversibility?

6-108. A heat-engine cycle operates between two thermal-energy reservoirs with temperatures of 560 and 280 K.

(*a*) The heat engine itself has a temperature of 560 K during heat addition from the 560 K reservoir, and a temperature of 280 K during heat rejection to the 280 K reservoir. The thermal efficiency of the actual unit is 40 percent, and the heat-transfer rate from the 560 K reservoir is 1000 kJ/min. Determine the entropy-production rate, in kJ/min·K, for (1) the heat engine itself, (2) the low-temperature heat-transfer process, and (3) the high-temperature heat-transfer process.

(b) Now for the same reservoir temperatures, heat-transfer rate, and engine thermal efficiency, the engine itself receives heat at 540 K and rejects heat at 300 K. As before, determine the entropy production rate for (1) the heat engine itself, (2) the low-temperature heat-transfer process, and (3) the high-temperature heat-transfer process. (4) Which process has the largest irreversibility?

(c) Finally, for the same reservoir temperatures, heat-transfer rate, and engine temperatures cited in part b, the thermal efficiency is increased to 42 percent. For these conditions, determine the entropy production rate for (1) the heat engine itself, (2) the low-temperature heat-transfer process, and (3) the high-temperature heat-transfer process. (4) Which process now has the largest irreversibility?

6-109. A refrigeration unit operates between two thermal reservoirs with temperatures of 250 and 330 K. The working fluid within the refrigeration unit itself has a temperature of 240 K during heat addition from the 250 K reservoir, and a temperature of 350 K during heat rejection to the 330 K reservoir. The COP of the actual unit is 2.00, and the heat-transfer rate from the 250 K reservoir is 200 kJ/min. (a) Determine the entropy production rate, in kJ/min·K, for (1) the refrigeration unit itself, (2) the low-temperature heat-transfer process, and (3) the high-temperature heat-transfer process. (4) Which process has the largest irreversibility? (b) Now, for the same temperatures and heat-transfer rate of 200 kJ/min, consider a less costly refrigeration unit with an actual COP of only 1.60. Again, determine the entropy production rate in kJ/min·K for (1) the refrigeration unit itself, (2) the low-temperature heat-transfer process, and (3) the high-temperature heat-transfer process. (4) Which process now has the largest irreversibility?

6-110. A refrigeration unit operates between two thermal reservoirs with temperatures of 240 and 320 K. The refrigeration unit itself has a temperature of 230 K during heat addition from the 240 K reservoir, and a temperature of 340 K during heat rejection to the 320 K reservoir. The COP of the actual unit is 2.00, and the heat-transfer rate from the 240 K reservoir is 1000 kJ/min. (a) Determine the entropy production rate, in kJ/min·K, for (1) the refrigeration unit itself, (2) the low-temperature heat-transfer process, and (3) the high-temperature heat-transfer process. (4) Which process has the largest irreversibility? (b) Now, for the same temperatures and a heat-transfer rate of 1000 kJ/min, install a less costly refrigeration unit with an actual COP of only 1.50. Again determine the entropy production rate in kJ/min·K for (1) the refrigeration unit itself, (2) the low-temperature heat-transfer process, and (3) the high-temperature heat-transfer process. (4) Which process now has the largest irreversibility?

6-111. A heat-pump unit operates between two thermal reservoirs with temperatures of 270 and 330 K. The heat pump itself has a temperature of 255 K during heat addition from the 270 K reservoir, and a temperature of 350 K during heat rejection to the 330 K reservoir. The COP of the actual unit is 3.40, and the heat-transfer rate from the 270 K reservoir is 2000 kJ/min. (*a*) Determine the entropy production rate, in kJ/min·K, for (1) the heat pump unit itself, (2) the low-temperature heat-transfer process, and (3) the high-temperature heat-transfer process. (4) Which process has the largest irreversibility? (*b*) Now, for the same temperatures and a heat-transfer rate of 2000 kJ/min, reduce the actual COP to 3.10. Again determine the entropy production rate for (1) the heat-pump unit itself, (2) the low-temperature heat-transfer process, and (3) the high-temperature heat-transfer process. (4) Which process now has the largest irreversibility?

6-112. An industrial process requires the use of a heat pump operating between two thermal reservoirs with temperatures of 300 and 420 K. The working fluid of the heat pump itself has a temperature of 285 K during heat addition from the 300 K reservoir, and a temperature of 445 K during heat rejection to the 420 K reservoir. The COP of the actual unit is 2.60, and the heat-transfer rate to the 420 K reservoir is 3000 kJ/min. (*a*) Determine the entropy production rate, in kJ/min·K, for (1) the heat-pump unit itself, (2) the low-temperature heat-transfer process, and (3) the high-temperature heat-transfer process. (4) Which process has the largest irreversibility? (*b*) Now, for the same temperatures and a heat-transfer rate of 300 kJ/min, reduce the actual COP to 2.40. Again determine the entropy-production rate for (1) the heat-pump unit itself, (2) the low-temperature heat-transfer process, and (3) the high-temperature heat-transfer process. (4) Which process now has the largest irreversibility?

6-113E. A heat engine operates between two thermal energy reservoirs with temperatures of 1020 and 520°R.

(*a*) The working fluid of the heat engine itself has a temperature of 960°R during heat addition from the 1020°R reservoir, and a temperature of 600°R during heat rejection to the 520°R reservoir. The thermal efficiency of the actual unit is 34 percent, and the heat-transfer rate from the 1020°R reservoir is 1000 Btu/min. Determine the entropy-production rate in Btu/min·°R, for (1) the heat engine itself, (2) the low-temperature heat-transfer process, and (3) the high-temperature heat-transfer process. (4) Which process has the highest irreversibility?

(*b*) Now for the same reservoir temperatures, heat-transfer rate, and working-fluid temperatures, the thermal efficiency of the heat engine is 25 percent. As before, determine the entropy production rate for (1) the heat engine itself, (2) the low-temperature heat-transfer process, and (3) the high-temperature heat-transfer process. (4) Which process now has the largest irreversibility?

6-114E. A refrigeration unit operates between two thermal reservoirs with temperatures of 440 and 570°R. The refrigeration unit itself has a temperature of 420°R during heat addition from the 440°R reservoir, and a temperature of 600°R during heat rejection to the 570°R reservoir. The COP of the actual unit is 2.00, and the heat-transfer rate from the 440°R reservoir is 1000 Btu/min. (*a*) Determine the entropy production rate, in Btu/min·°R, for (1) the refrigeration unit itself, (2) the low-temperature heat-transfer process, and (3) the high-temperature heat-transfer process. (4) Which process has the largest irreversibility? (*b*) Now, for the same temperatures and a heat-transfer rate of 1000 Btu/min, install a less costly refrigeration unit with an actual COP of only 1.50. Again determine the entropy production rate in kJ/min·K for (1) the refrigeration unit itself, (2) the low temperature heat-transfer process, and (3) the high-temperature heat-transfer process. (4) Which process now has the largest irreversibility?

## MOLECULAR PARTICLE DISTRIBUTIONS

6-115. Determine the number of ways that seven distinguishable particles can be arranged into three groups so that the groups contain (*a*) 3, 2, and 2 particles and (*b*) 4, 2, and 1 particles.

6-116. Determine the number of ways that eight distinguishable particles can be arranged into three groups so that the groups contain (*a*) 3, 3, and 2 particles and (*b*) 4, 2, and 2 particles.

6-117. Four distinguishable balls are dropped at random into two boxes. After repeated tests, find the fraction of the time we would expect to find that two balls are in each box if (*a*) the boxes are indistinguishable and (*b*) the boxes are distinguishable.

6-118. Five distinguishable balls are dropped at random into three boxes. After repeated tests, find the fraction of the time we would expect to find (*a*) that the distribution of balls is 3:1:1 if the boxes are indistinguishable and (*b*) that 3 balls are in the first box, 1 ball is in the second box, and 1 ball is in the third box.

6-119. An isolated system contains three particles with a total energy of 3 units. The energy levels for the energy mode under consideration are equally spaced 1 unit apart, and the ground level of energy is taken as zero.

(a) Find the number of macrostates possible for this system. Then find the number of microstates, if the particles are distinguishable. (Use a diagram similar to those used in the text to help develop your answers.)

(b) Determine whether there is a most probable macrostate in part a.

6-120. A system of three indistinguishable particles has a total energy of 3 units. The ground level is taken to be zero, and the spacing between levels is 1 unit. The degeneracy for the bottom four levels, starting from the ground level, is 1, 2, 2, and 2. Determine the number of macrostates and microstates.

6-121. Consider a system of three indistinguishable particles. The energies of each are restricted to values of 0, 1, 2, 3, and 4. Determine the number of macrostates and microstates if each of the energy levels has a degeneracy of unity. The total energy is 6.

6-122. Consider a system of three distinguishable particles with a total energy of 9 units. The degeneracy of each level is unity, and the particles are restricted to energy values of 0, 1, 2, 3, and 4. Determine the number of macrostates and microstates.

6-123. An isolated system contains three particles with a total energy of 5 units. The energy levels are equally spaced 1 unit apart, and the ground-level energy is taken as zero. Find the number of macrostates and microstates that are possible if the particles are distinguishable.

6-124. Consider a system of four distinguishable particles with a total energy of 6 units. The ground-level energy is taken to be zero, the spacing between levels is $i$ energy unit, and the levels are nondegenerate. Determine the number of macrostates and microstates.

6-125. Consider a system of 10 distinguishable particles for which the energy levels are quantized according to the relation $\varepsilon_i = i(i+1)$, where $i = 0, 1, 2, \ldots$. The total energy is 18 units.

(a) Find the number of possible macrostates.

(b) Find the number of possible microstates.

(c) Find the number of microstates for the most probable macrostate.

(d) Find how many particles are in the various energy levels, beginning at the ground level, for the most probable macrostate.

6-126. Consider a system of six distinguishable particles with a total energy of 12 units. The energy levels are quantized according to the relation $\varepsilon_i = i + \frac{1}{2}$, where $i = 0, 1, 2, \ldots$.

(a) Find the number of macrostates and microstates.

(b) Find whether there is a most probable macrostate.

# CHAPTER
# 7

# EVALUATION OF ENTROPY CHANGE AND THE CONTROL-VOLUME ENTROPY BALANCE

Solar One thermal power plant near Barstow, California. Heliostats track the sun focussing its rays on a 20-story tower to create steam. The steam then drives a turbine-generator set.

In Chapter 6 attention is focused on developing the concepts of entropy and entropy generation. Applications are directed toward cyclic devices, such as heat engines, refrigerators, and heat pumps, and toward heat-transfer processes. In this chapter emphasis will be placed on methods of evaluating entropy changes for different classes of substances. These methods, in conjunction with the concept of entropy production, will be applied to many of the engineering devices discussed previously in terms of energy analysis.

## 7-1    GRAPHICAL AND TABULAR PRESENTATION OF ENTROPY DATA

With the introduction of the entropy function in the preceding chapter, we now need to investigate the established methods for evaluating entropy change for various classes of substances and processes. Similar to the development of property data in Chaps. 3 and 4, the tables and equations for entropy are presented on an intensive basis. By definition, $\Delta S \equiv m\,\Delta s$. Using this result, the equation $\Delta S = \int (\delta Q/T)_{\text{int rev}}$ presented in Chap. 6 can also be written on an intensive basis as

$$\Delta s = \int \left(\frac{\delta q}{T}\right)_{\text{int rev}} \tag{7-1}$$

where $q = Q/m$. The value of $s$ relative to a reference state $s_0$ is given by

$$s = \int \left(\frac{\delta q}{T}\right)_{\text{int rev}} + s_0 \tag{7-2}$$

This format shows that the integration of $\delta q/T$ gives the specific entropy $s$ only within an arbitrary constant $s_0$. Therefore, tables and graphs may be constructed for which the reference value $s_0$ is arbitrary. Dimensions for the specific entropy are [energy]/([mass][temperature]). Units in SI for the specific entropy are typically kJ/kg·K or kJ/kmol·K, while USCS units commonly are Btu/lb$_m$·°R or Btu/lbmol·°R.

The $Pv$ diagram has been stressed in problem solutions in earlier chapters as an aid in visualizing energy processes. In conjunction with second-law analyses, visualization is aided by use of diagrams that use entropy as one of the coordinates. The *temperature-entropy* and the *enthalpy-entropy* diagrams are extremely useful for two reasons. First, they can provide a qualitative feeling for a process without requiring a great deal of information. Second, *Ts* and *hs* diagrams may also be employed for the quantitative information they might provide. Diagrams of this type, especially when they are enlarged, can be used to predict property changes for a specified process with reasonable accuracy.

### 7-1-1    THE TEMPERATURE-ENTROPY DIAGRAM

Recall that when a closed system undergoes an *internally reversible process,* the heat transfer is given by

$$Q_{\text{int rev}} = \int T\,dS \tag{6-51}$$

On a unit-mass basis this equation becomes

$$q_{\text{int rev}} = \int T\,ds \tag{7-3}$$

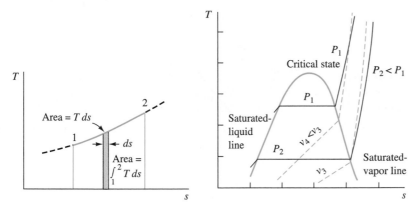

**Figure 7-1**
Heat transfer as an area on a *Ts* diagram for an internally reversible process with variable system temperature.

**Figure 7-2**
A general *Ts* diagram.

The integral may be represented by an area under a process line on a *Ts* diagram, as shown in Fig. 7-1. To plot a *Ts* diagram simply for the qualitative information it provides, the general characteristics of process lines on the diagram relative to the saturated liquid–vapor line are important. Figure 7-2 illustrates a *Ts* diagram in a simplified format. Only a few characteristic property lines in the gas and liquid regions are shown. As an aid in problem solving, one should become familiar with the placement of constant-volume and constant-pressure lines on a *Ts* diagram. Note that in the gas-phase region of Fig. 7-2 constant-volume lines have a slightly steeper slope than constant-pressure lines through any given state. A more detailed *Ts* diagram for $CO_2$ which includes all three phase regions is presented as Fig. A-27E, and a *Ts* diagram for refrigerant 134a in the superheat region appears as Fig. A-32 in the Appendix.

**Note the position of constant-volume lines relative to constant-pressure lines on a *Ts* diagram.**

### 7-1-2    THE ENTHALPY-ENTROPY DIAGRAM

The coordinates of an *hs* diagram represent the two major properties of interest in the first- and second-law analysis of control volumes. For adiabatic, steady-state processes, the *vertical* distance $\Delta h$ between two states on this diagram is related to the work and/or kinetic-energy changes for turbines, compressors, nozzles, for example, and the *horizontal* distance $\Delta s$ between two states is a measure of the degree of irreversibility. This comparison is shown is Fig. 7-3. Thus an *enthalpy-entropy* diagram, as well as a *Ts* diagram, is quite helpful in visualizing processes when analyzing steady-flow processes. This will be explored further later in this chapter.

In addition to its use in process visualization, the *hs* diagram is also a means of presenting data quantitatively. Enthalpy-entropy diagrams for steam are included in the Appendix as Figs. A-26 and A-26E. A schematic

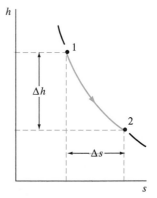

**Figure 7-3**
Energy effects versus degree of irreversibility for adiabatic processes.

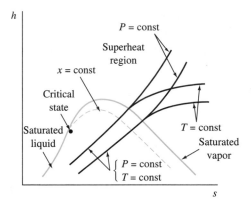

**Figure 7-4**　　Schematic of a Mollier or *hs* diagram.

of an *hs* diagram, commonly called a *Mollier diagram,* is shown in Fig. 7-4. On an *hs* plot, constant-pressure and constant-temperature lines are straight in the liquid-vapor region. Lines of constant quality within the wet region lie roughly parallel to the saturated-vapor line. Another feature of this diagram is that constant-temperature lines tend to become horizontal the farther they extend into the superheat region. This occurs because the fluid is becoming more like an ideal gas as the pressure decreases at constant temperature. For an ideal gas the enthalpy and temperature are directly proportional. Constant-volume lines are not normally shown on an *hs* diagram, simply because the diagram becomes too crowded to distinguish among all the various lines.

**Note the position of constant-pressure and constant-temperature lines on an *hs* diagram.**

### 7-1-3　TABULAR ENTROPY DATA

Tables of data representing superheated-vapor, saturation, and compressed-liquid states were introduced in Chap. 3. We must now reconsider these tables in light of the entropy function. The numerical values of *s* (relative to an arbitrary reference state) are determined from fairly complex numerical techniques which use experimental *PvT* data. For water, the entropy of saturated liquid is commonly assigned a zero value at the triple state (0.01°C or 32.02°F). For refrigerants, it is quite common to assign a zero entropy value to the saturated liquid at −40°C (−40°F). For the superheat region, the entropy is tabulated as a function of temperature and pressure. For the saturation states, the values of $s_f$ and $s_g$ are given as a function of either temperature or pressure. For wet mixtures with a quality *x*,

$$s = (1 - x)s_f + xs_g = s_f + x(s_g - s_f) \qquad \textbf{[7-4]}$$

In summary, the specific entropy *s* of a pure, simple compressible substance is an intensive, intrinsic property like *v*, *u*, and *h*. Numerical values can be found in the appropriate property tables once the state of the substance is known.

The specific entropy of a *compressed liquid* may be tabulated as a function of temperature and pressure. Unfortunately, compressed-liquid data are not available for many substances. In the absence of compressed-liquid data, the value of $s$ in that region can be estimated fairly accurately by using $s_f$ at the given temperature. That is,

$$s_{\text{liq},P,T} \approx s_{f,T} \qquad \textbf{[7-5]}$$

In general, when compressed-liquid data are unavailable, the numerical values of $u$, $v$, and $s$ can be approximated by using saturated-liquid values at the given temperature.

A phase change from liquid to vapor occurs at constant temperature and constant pressure. Equation [7-1] states that $\Delta s \equiv \int (\delta q/T)_{\text{int rev}}$ for a closed system. For an internally reversible phase change, this equation reduces to $\Delta s = q/T$ on a unit-mass basis. In addition, for a closed system undergoing a constant-pressure phase change, $q = \Delta u + P\Delta v = \Delta h$. As a result, the entropy change during a phase change becomes $\Delta s = \Delta h/T$. Therefore, in terms of saturation data,

$$s_g - s_f = \frac{h_g - h_f}{T} = \frac{h_{fg}}{T}$$

where $s_g - s_f \equiv s_{fg}$. Thus a direct connection exists between $s_g - s_f$ and $h_g - h_f$ in a saturation table at a given temperature or pressure.

Several examples below illustrate the use of saturation and superheat tables and property diagrams in conjunction with the entropy balance to analyze closed-system processes.

---

**A**n uninsulated, rigid tank contains 5 kg of steam initially at 40 bars and 280°C. The air surrounding the tank is at a uniform temperature of 15°C. After a period of time, the pressure within the tank has been reduced to 9 bars due to heat transfer from the water substance to the air. Determine (*a*) the heat transfer from the steam, in kJ, (*b*) the entropy change for the steam, in kJ/K, (*c*) the entropy production for an enlarged system composed of the steam and the heat-transfer region, which includes the tank and the layer of air adjacent to the tank, in kJ/K, and (*d*) whether the overall process is reversible, irreversible, or impossible.

**Example 7-1**

**Solution:**

**Given:** Steam is cooled at constant volume and exchanges energy with a constant-temperature environment, as illustrated in Fig. 7-5.

**Find:** (*a*) $Q$, in kJ, (*b*) $\Delta s$ for the steam, in kJ, (*c*) $\sigma_{\text{tot}}$, for the composite steam and heat-transfer region, in kJ/K, and (*d*) whether the process is reversible, irreversible, or impossible.

**Model:** Constant-volume process for a closed system.

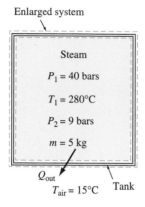

Enlarged system

Steam

$P_1 = 40$ bars

$T_1 = 280°C$

$P_2 = 9$ bars

$m = 5$ kg

$Q_{out}$

$T_{air} = 15°C$   Tank

**Figure 7-5**
Schematic and data for
Example 7-1.

**Strategy:**  Apply the conservation of energy equation, the entropy balance, and required property data.

**Analysis:**  The basic closed-system equations are

$$\Delta E = Q + W \qquad \text{and} \qquad \Delta S = \int \frac{\delta Q}{T} + \sigma$$

where it is assumed that the boundary temperature is uniform.

(a) For the steam as the chosen system, there is no work transfer because the volume is constant and no other work mechanisms are present. If we assume that kinetic- and potential-energy changes are negligible, then the energy balance reduces to $Q = \Delta U = m(u_2 - u_1)$. Information about $u$ is obtained once the states of the system are determined.

**State 1:**  The given values of $T_1$ and $P_1$ fix the initial state. The saturation temperature at 40 bars from Table A-13 is 250.5°C. Therefore, the initial state of 280°C is superheated vapor, as shown on the $Ts$ diagram in Fig. 7-6. From Table A-14 the specific internal energy, specific volume, and specific entropy are read directly as

$$u_1 = 2680.0 \text{ kJ/kg} \qquad v_1 = 55.46 \times 10^{-3} \text{ m}^3\text{/kg} \qquad s_1 = 6.2568 \text{ kJ/kg·K}$$

**State 2:**  The only information given for state 2 is $P_2$. However, since the process is one of constant volume and constant mass, $v_2 = v_1$. Starting with the saturation table based on pressure (Table A-13), it is found that $v_2$ lies between $v_f$ and $v_g$ at 9 bars. Hence state 2 is a two-phase (wet) mixture at 175.4°C (see Fig. 7-6). Because specific properties in the wet region are primarily a function of the quality, we first calculate $x_2$ by

$$x_2 = \frac{v_2 - v_f}{v_g - v_f} = \frac{55.46 - 1.12}{215.0 - 1.12} = 0.254 \text{ (or } 25.4\%)$$

A knowledge of the quality enables us to evaluate $u_2$ as

$$u_2 = (1 - x)u_f + xu_g = [0.746(741.83) + 0.254(2580.5)] \text{ kJ/kg} = 1208.9 \text{ kJ/kg}$$

Therefore, the heat transfer is found by

$$Q = m(u_2 - u_1) = (5 \text{ kg})(1208.9 - 2680.0) \text{ kJ/kg} = -7355.5 \text{ kJ}$$

For the steam, the heat transfer is *out* of the system, i.e., $Q_{out} = 7355.5$ kJ.

(b) The entropy change of the steam is found from $\Delta S = m(s_2 - s_1)$. The specific entropy at state 2 is found by

$$s_2 = s_f + xs_{fg} = [2.0946 + 0.254(6.6226 - 2.0946)] \text{ kJ/kg·K} = 3.245 \text{ kJ/kg·K}$$

Thus the entropy change of the steam is

$$\Delta S = m(s_2 - s_1) = 5 \text{ kg}(3.245 - 6.257) \text{ kJ/kg·K} = -15.060 \text{ kJ/K}$$

The entropy of the steam decreases during the process. This decrease is shown on the $Ts$ plot of Fig. 7-6.

(c) The system is now enlarged to include the steam, the tank wall, and a thin layer of air next to the tank where the air temperature decreases to a uniform value of 15°C. (See the enlarged boundary in Fig. 7-5.) For this composite (comp) system,

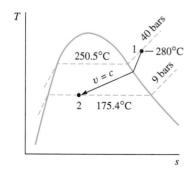

$T$

40 bars

250.5°C

$v = c$

1 — 280°C

9 bars

2   175.4°C

$s$

**Figure 7-6**
A $Ts$ plot for process in Example 7-1.

the boundary temperature is constant. Therefore, integration and rearrangement of the entropy balance leads to

$$\sigma_{\text{comp}} = \Delta S_{\text{comp}} - \frac{Q_{\text{comp}}}{T}$$

where $T = 15 + 273 = 298$ K at the system boundary. It is a common assumption to neglect the mass of the tank wall and air layer in the heat-transfer region. As a result, changes in internal energy and entropy are neglected in this region. Thus the heat transfer leaving the steam is also the heat transfer leaving the heat-transfer region at 15°C; that is, $Q_{\text{comp}} = -7355.5$ kJ. Employing the appropriate data, the entropy production for the composite region is

$$\sigma_{\text{comp}} = -15.060 \text{ kJ/K} - \frac{-7355.5 \text{ kJ}}{(15 + 273) \text{ K}}$$

$$= (-15.060 + 25.540) \text{ kJ/K} = 10.480 \text{ kJ/K}$$

Thus the entropy production for the enlarged system is positive, because the entropy transported out of the system with heat transfer is greater than the decrease in the entropy of the system.

(d) Because the entropy production for the composite system is greater than zero, the overall cooling process is possible, but internally irreversible.

**Comment:** Note that the entropy production within the steam has not been calculated. To do this requires an evaluation of the entropy transfer associated with heat transfer at the boundary. That is, we must integrate $\int \delta Q/T$ for the steam as the system. This integration requires that we know how the heat transfer $\delta Q$ and the boundary temperature vary with time. Unfortunately, we do not have this information. We also have not made any assumption about whether the steam system is internally reversible or irreversible.

---

**A** vertical piston-cylinder device with an internal electrical resistor contains 0.80 kg of refrigerant 134a as a saturated liquid. The pressure of the R-134a is maintained at 10 bars by a weighted piston. It is proposed to completely vaporize the liquid by using two different methods. (a) Process A involves only heat transfer from an external source. (b) Process B involves the use of electrical work with no heat transfer. Assume the mass of the resistor is negligible. For each process, determine the energy transfer, in kJ, and the entropy production and entropy change, in kJ/K.

**Example 7-2**

**Solution:**

**Given:** Refrigerant 134a is vaporized within a piston-cylinder device shown in Fig. 7-7a by two processes. Process A is isobaric heating and Process B is adiabatic and isobaric with electrical work.

**Find:** $Q$ or $W_{\text{elec}}$, in kJ, and $\Delta S$ and $\sigma$, in kJ/K.

**Model:** Closed, quasiequilibrium, constant-pressure processes; $\Delta ke = \Delta pe = 0$; frictionless piston; mass of resistor is neglected.

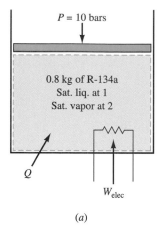

$P = 10$ bars

0.8 kg of R-134a
Sat. liq. at 1
Sat. vapor at 2

$Q$

$W_{elec}$

(a)

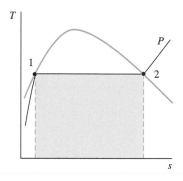

(b)

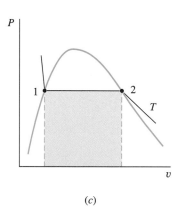

(c)

**Figure 7-7**

Schematic, data, and process path for Example 7-2.

**Strategy:**     Apply energy and entropy balances for each process.

**Analysis:**     The closed-system energy balance and entropy balance are

$$\Delta E = Q + W \qquad \text{and} \qquad \Delta S = \int \frac{\delta Q}{T} + \sigma$$

In general, there are two types of work interactions to consider for the system: $W_{exp,out}$ and $W_{elec,in}$. Using this information, and assuming that $\Delta ke$ and $\Delta pe$ are neglected, the energy balance for the vaporization process becomes

$$Q + W_{elec,in} - W_{exp,out} = \Delta U$$

If the process is assumed to be a *quasiequilibrium* one, then at constant pressure, $W_{exp,out} = P \Delta V$. Therefore,

$$Q + W_{elec,in} = \Delta U + P \Delta V = \Delta H = mh_{fg}$$

where $\Delta U$ refers only to the fluid because the internal energy change of the electrical resistor is zero, because its mass is negligible. In either process the energy transfer $Q$ or $W_{elec}$ to be added is directly related to $h_{fg}$, the enthalpy of vaporization. Thus the above energy equation becomes

$$Q + W_{elec,in} = 0.8 \text{ kg}(162.68 \text{ kJ/kg}) = 130.1 \text{ kJ}$$

The entropy changes for process A and process B are the same, since the end states for both processes are the same. As a result,

$$\Delta S = m \Delta s = m(s_g - s_f)$$
$$= 0.8 \text{ kg}(0.9043 - 0.3838) \text{ kJ/kg·K} = 0.4164 \text{ kJ/K}$$

(a) Because $W_{elec,in} = 0$ for process A, the heat transfer is $Q = 130.1$ kJ. Since the heat transfer process occurs in quasiequilibrium and without friction, process A is internally reversible. Therefore, the entropy production $\sigma$ is zero for the fluid. From the entropy balance in this case, the entropy transfer simply equals the change in entropy of the fluid.

(b) For process B, $Q = 0$. Hence $W_{elec,in} = 130.1$ kJ. The adiabatic work transfer process is internally *irreversible*. Also, there is no entropy transfer at the boundary, since there is no heat transfer. Hence the entropy balance for the fluid in this case becomes

$$\sigma = \Delta S = 0.4162 \text{ kJ/K}$$

The positive $\sigma$ value reflects the irreversible nature of electrical-resistance work.

**Comments:**     (1) Both processes are shown on *Ts* and *Pv* coordinates in Figs. 7-7*b* and 7-7*c*, respectively. On the *Ts* diagram, the area under the process line equals *q* for process A but has no meaning for process B. On the *Pv* diagram, the area under the process line represents $w_{exp,out}$ for both processes.

(2) Note that the entropy production is quite different for the two processes. This result clearly demonstrates that entropy production, along with heat and work transfers, is a *process function*, and not a state function. In addition, note that the replacement of heat transfer by electrical work transfer greatly increases the entropy production of the process. Thus using work transfer in place of heat transfer to accomplish the same change of state of the fluid is an inefficient use of energy.

## 7-2 THE $T\,dS$ EQUATIONS FOR PURE, SIMPLE COMPRESSIBLE SUBSTANCES

In this section it will be shown that the maximum work output or minimum work input occurs when a process is internally reversible. Then equations for evaluating the entropy change of a simple, compressible substance will be developed in terms of properties which are experimentally measurable.

### 7-2-1 COMPARISON OF REVERSIBLE AND IRREVERSIBLE WORK INTERACTIONS FOR A CLOSED SYSTEM

In the discussion of quasistatic work interactions for closed systems in Chap. 2, it is pointed out that such interactions lead to the maximum work output or minimum work input. With the second law as background, we are now in a position to prove this claim. For *any* closed system, the energy balance and the entropy balance can be written as

$$dU = \delta Q + \delta W_{\text{act}} \qquad \text{and} \qquad dS = \frac{\delta Q}{T} + \delta \sigma$$

if changes in kinetic and potential energy are negligible. The subscript "act" emphasizes that $\delta W_{\text{act}}$ is the actual work and applies to both internally irreversible and irreversible processes. To compare the actual and internally reversible work for a closed system, we use the entropy balance to eliminate $\delta Q$ from the energy balance. This leads to

$$dU = T(dS - \delta \sigma) + \delta W_{\text{act}} \qquad \textbf{[7-6]}$$

Solving this equation for the net actual work supplied to a closed system yields

$$\delta W_{\text{act}} = dU - T\,dS + T\,\delta \sigma \qquad \textbf{[7-7]}$$

If a process is internally reversible, the entropy production is zero. Hence the *internally reversible* work $\delta W_{\text{rev}}$ is found by setting $\delta \sigma = 0$ in the above equation. As a result,

$$\delta W_{\text{rev}} = dU - T\,dS \qquad \textbf{[7-8]}$$

When Eq. [7-8] is subtracted from Eq. [7-7], we find that

$$\delta W_{\text{act}} - \delta W_{\text{rev}} = T\,\delta \sigma \qquad \text{or} \qquad \delta W_{\text{act}} = \delta W_{\text{rev}} + T\,\delta \sigma$$

Since $T$ is always a positive value and $\delta \sigma \geq 0$, then

$$\delta W_{\text{act}} \geq \delta W_{\text{rev}} \qquad \textbf{[7-9]}$$

This important relation is valid regardless of the direction of change of the closed system. Hence integrating the above relation and taking into account the adopted sign convention for work interactions, we may write

$$W_{\text{in,act}} \geq W_{\text{in,rev}} \qquad \text{and} \qquad W_{\text{out,act}} \leq W_{\text{out,rev}} \qquad \textbf{[7-10]}$$

where the explicit directional subscripts are used to supplement the adopted sign convention for work. Noting that the inequality sign on Eq. [7-9] denotes the presence of internal irreversibilities, we have shown that irreversibilities within a closed system reduce the work output and increase the work input for a specified change of state of a substance.

**Note the effect of internal irreversibilities on work-producing and work-absorbing devices.**

Another useful form of Eq. [7-10] is on a rate basis, given by

$$\dot{W}_{in,act} \geq \dot{W}_{in,rev} \quad \text{and} \quad \dot{W}_{out,act} \leq \dot{W}_{out,rev} \qquad \textbf{[7-11]}$$

Thus the power supplied to a closed system in an actual process must always be greater than or equal to the power supplied in an internally reversible process.

### 7-2-2   DEVELOPMENT OF THE $T\,dS$ EQUATIONS

As noted in Chap. 4, when tabular data are not available for a substance, certain models can be used to estimate the required information. To derive the basic property relations for the evaluation of the specific entropy change $\Delta s$ of any simple compressible substance, we start with Eq. [7-8] from the preceding subsection.

$$\delta W_{rev} = dU - T\,dS \qquad \textbf{[7-8]}$$

For a simple compressible substance, the only reversible (quasistatic) work mode is compression/expansion work; that is, $\delta W_{rev} = -P\,dV$. Substituting this into Eq. [7-8] gives $-P\,dV = dU - T\,dS$. Dividing through by the mass of the system and rearranging gives the desired form of the equation:

$$T\,ds = du + P\,dv \qquad \textbf{[7-12]}$$

This equation is referred to as the *first T ds*, or *Gibbsian, equation* for simple compressible substances. Since $dh = du + P\,dv + v\,dP$, by definition, Eq. [7-12] also can be written as

$$T\,ds = dh - v\,dP \qquad \textbf{[7-13]}$$

**Note that the *T ds* equations apply whether the actual process is internally reversible or irreversible.**

This is the *second T ds equation* for simple compressible substances. These two equations generally are applied to homogeneous systems in the absence of chemical reactions. They are derived on the basis of a closed system undergoing an internally reversible process. However, the $T\,ds$ equations will lead to values for the specific entropy change that are valid even if the actual process is internally irreversible. Because entropy is a property, its change between two equilibrium states does not depend on the process that actually occurs.

Upon rearrangement, the $T\,ds$ equations can be written in the format

$$ds = \frac{du}{T} + \frac{P\,dv}{T} \qquad \textbf{[7-14]}$$

$$ds = \frac{dh}{T} - \frac{v\,dP}{T} \qquad \textbf{[7-15]}$$

Integration of the above two equations requires a knowledge of the functional relationship among properties. Specifically, the relationship between either $u$ or $h$ and $T$ must be known, as well as the $PvT$ behavior of the substance. Such relationships have been developed in Chap. 4 for ideal gases and incompressible substances. These same relationships will be used in the following sections to derive equations for $\Delta s$ based on these same models.

## 7-3   ENTROPY CHANGE OF AN IDEAL GAS

The $T\,ds$ equations introduced in Sec. 7-2 allow one to evaluate the entropy change of a substance once the fundamental relationships among the properties $u$, $h$, $P$, $v$, and $T$ are known. For the *ideal-gas model, $du = c_v\,dT$, $dh = c_p\,dT$, and $Pv = RT$*. Substitution of these relations into Eqs. [7-14] and [7-15] leads to

$$ds = \frac{du + P\,dv}{T} = c_v\frac{dT}{T} + R\frac{dv}{v} \qquad \text{[7-16]}$$

and

$$ds = \frac{dh - v\,dP}{T} = c_p\frac{dT}{T} - R\frac{dP}{P} \qquad \text{[7-17]}$$

For a finite change of state these two equations become

$$\Delta s = \int_1^2 \frac{c_v\,dT}{T} + R\ln\frac{v_2}{v_1} \qquad \text{(ideal gas)} \qquad \text{[7-18]}$$

and

$$\Delta s = \int_1^2 \frac{c_p\,dT}{T} - R\ln\frac{P_2}{P_1} \qquad \text{(ideal gas)} \qquad \text{[7-19]}$$

We noted in Chap. 4 that both $c_v$ and $c_p$ are functions solely of temperature for an ideal gas. Thus the integrals in Eqs. [7-18] and [7-19] can be evaluated once the temperature dependence of $c_v$ and $c_p$ is known.

In the application of the second law to an ideal-gas process, it is helpful in any analysis to sketch the process on a $Ts$ diagram. Figure 7-8 illustrates the general position of pressure and specific-volume lines on the diagram for the ideal-gas region. Again note that constant-volume lines have a slightly steeper slope than constant-pressure lines in the gas region. With this background information, two approaches to the evaluation of the integrals in Eqs. [7-18] and [7-19] are discussed below.

### 7-3-1   USE OF CONSTANT OR ARITHMETICALLY AVERAGED SPECIFIC-HEAT DATA

The values of $c_v$ and $c_p$ of ideal monatomic gases are independent of temperature. In addition, the variation of $c_v$ and $c_p$ with temperature over a

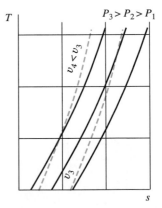

**Figure 7-8**
Pressure and specific volume lines for an ideal gas illustrated on a $Ts$ diagram.

several-hundred-degree range is quite small, and the variation is almost linear for many common polyatomic gases. In the latter situation, it is reasonable to use an arithmetic average of the specific-heat value for the temperature range between $T_1$ and $T_2$. When this is done, $c_v$ and $c_p$ may be considered as constants for the integration steps in Eqs. [7-18] and [7-19], whether the substance is monatomic or polyatomic. Integration of Eqs. [7-18] and [7-19] in this case show that

$$\Delta s = c_{v,av} \ln \frac{T_2}{T_1} + R \ln \frac{v_2}{v_1} \qquad \text{(ideal gas)} \qquad \textbf{[7-20]}$$

and

$$\Delta s = c_{p,av} \ln \frac{T_2}{T_1} - R \ln \frac{P_2}{P_1} \qquad \text{(ideal gas)} \qquad \textbf{[7-21]}$$

where $c_{v,av}$ and $c_{p,av}$ are *average* values over the given temperature range. These equations provide a fairly accurate method for evaluating the entropy change of an ideal gas over a fairly small temperature range. If the variation in the specific heat is large, the method introduced in the following section is preferred.

Example 7-3

**A** rigid, insulated tank contains 1.2 kg of nitrogen gas at 350 K and 1 bar. A paddle wheel inside the tank is driven by a pulley-weight mechanism. During the experiment 25 kJ of work is done on the gas through the pulley-weight mechanism. Assume ideal-gas behavior. Find (*a*) the entropy generation for the nitrogen, in kJ/K, and (*b*) whether the process is reversible, irreversible, or impossible. Then, (*c*) sketch a $Ts$ diagram for the process.

**Solution:**

**Given:**   Nitrogen is stirred by a paddle wheel in a rigid, insulated tank, as shown in Fig. 7-9.

**Find:**   (*a*) $\sigma$ for N$_2$, in kJ/K, (*b*) second-law analysis. (*c*) Sketch process on a $Ts$ diagram.

**Model:**   Stationary, rigid tank; ideal gas.

**Strategy:**   Use an entropy balance to find the entropy production.

**Analysis:**   (*a*) The entropy production $\sigma$ due to the irreversible stirring process is found from the basic entropy balance for a closed system,

$$S_2 - S_1 = \int \frac{\delta Q}{T} + \sigma$$

Since $Q = 0$, then $\sigma = S_2 - S_1 = m\,\Delta s$. The entropy change for an ideal gas is most easily calculated for this process from Eq. [7-20], because the volume is constant. That is,

$$\Delta s = c_{v,av} \ln \frac{T_2}{T_1} + R \ln \frac{v_2}{v_1} = c_{v,av} \ln \frac{T_2}{T_1}$$

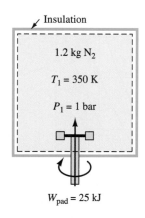

Insulation

1.2 kg N$_2$

$T_1 = 350$ K

$P_1 = 1$ bar

$W_{pad} = 25$ kJ

**Figure 7-9**
Schematic and data for Example 7-3.

To use this equation, however, the final state must be determined. Since energy information is given, the application of the general energy balance on the stationary closed system, $\Delta U = Q + W$, will provide additional information about the final state. The tank is rigid and insulated, so $P\,dV$ work and heat transfer are zero. Thus the energy equation reduces to $W_{paddle} = \Delta U$, and $\Delta U = mc_{v,av}\,\Delta T$ for an ideal gas. Combining these two equations and solving for $T_2$ yields

$$T_2 = \frac{W_{paddle}}{mc_{v,av}} + T_1$$

A $c_v$ value from Table A-4 at the initial temperature will be used, since the final temperature is not yet known. Substitution of known data into the above equation yields

$$T_2 = \frac{25\ \text{kJ}}{1.2\text{kg} \times 0.744\ \text{kJ/kg·K}} + 350\ \text{K}$$

$$= (28.0 + 350)\ \text{K} = 378\ \text{K}$$

Because $\Delta T$ is only 28°C for the process, the use of the $c_v$ value at the initial state for $c_{v,av}$ is appropriate in this case. Therefore, the calculation of the entropy generation becomes

$$\sigma = \Delta S = m\,\Delta s = mc_v \ln\frac{T_2}{T_1} = 1.2(0.744)\ \text{kJ/K} \times \ln\frac{378}{350} = 0.0687\ \text{kJ/K}$$

(b) The entropy change and the entropy production are positive. In accordance with the second law, the adiabatic process is internally irreversible.

(c) A $Ts$ plot of the process is shown in Fig. 7-10. Since the process is internally irreversible, it is shown as a dashed line, and the area under the curve is *not* equal to the heat transfer. From the ideal-gas equation, $P_2/P_1 = T_2/T_1$. Since $T_2 > T_1$, the pressure increased during the process. We also noted in Fig. 7-8 that constant-volume lines have a steeper positive slope than constant-pressure lines in the gas region of a $Ts$ diagram. Hence the diagram also shows that the entropy change is positive for the specified process.

**Comment:** If $\Delta T$ had been larger, it would have been necessary to find a new $c_{v,av}$ and recalculate $T_2$. Typically only one or two iterations are required before $T_2$ and $c_{v,av}$ stop changing.

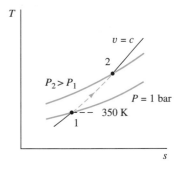

**Figure 7-10**
A $Ts$ plot of process in Example 7-3.

## 7-3-2 USE OF INTEGRATED SPECIFIC-HEAT DATA

A more accurate evaluation of the integrals in Eqs. [7-18] and [7-19] is attained in the following manner. Recall that Tables A-3 and A-3E list $c_p$ data for some common gases as a function of temperature. In addition, $c_v = c_p - R$ for an ideal gas, so that $c_v$ is known also as a function of temperature. Use of these specific-heat equations in Eqs. [7-18] and [7-19] leads to a highly accurate evaluation of the integrals of either $c_v\,dT/T$ or $c_p\,dT/T$. To avoid repetitive integration whenever the limits of integration are changed, the integral of $c_p\,dT/T$ has been evaluated relative to a reference temperature $T_0$ for a number of gases. A function $s_T^0$ is defined such that its value at temperature $T$ relative to a reference temperature $T_0$ is

$$s_T^o \equiv \int_{T_0}^{T} c_p \frac{dT}{T}$$ [7-22]

The results of this integration are then tabulated versus temperature in ideal-gas property tables. On the basis of Eq. [7-22],

$$\int_{T_1}^{T_2} c_p \frac{dT}{T} = \int_{T_0}^{T_2} c_p \frac{dT}{T} - \int_{T_0}^{T_1} c_p \frac{dT}{T} = s_2^o - s_1^o$$

As a result, Eq. [7-19] for the entropy change of an ideal gas between states 1 and 2 becomes

$$s_2 - s_1 = s_2^o - s_1^o - R \ln \frac{P_2}{P_1}$$ [7-23a]

where the units on $s^o$ and $R$ must be the same. On a unit-mole basis, the entropy change is

$$\bar{s}_2 - \bar{s}_1 = \bar{s}_2^o - \bar{s}_1^o - R_u \ln \frac{P_2}{P_1}$$ [7-23b]

where the units on $\bar{s}^o$ and $R_u$ are the same. For air, the value of $s^o$ in Table A-5 has units of kJ/kg·K, while in Table A-5E the units are Btu/lbm·°R. For the remaining ideal gases in Tables A-6 through A-11, the units for $\bar{s}^o$ are kJ/kmol·K; the USCS data in Tables A-6E through A-11E have units of Btu/lbmol·°R.

**Recognize the importance of $s^o$ data in evaluating ideal-gas entropy changes.**

When tables of $s^o$ or $\bar{s}^o$ values are available, they provide a fast and accurate means of evaluating the effect of temperature on the ideal-gas entropy. The use of average specific-heat data leads to an increasing error in evaluating the integral of $c_p \, dT/T$ as the temperature interval increases. Note that evaluating $\Delta s$ for any process only requires information about the end states. The following example illustrates the use of the ideal-gas tables.

---

**Example 7-4**

**O**xygen gas is heated at constant pressure from 300 K to (a) 500 K and (b) 800 K. Evaluate the molar entropy change, in kJ/kmol·K, for both sets of temperature increments by means of (1) a specific-heat value at the average temperature and (2) ideal-gas $\bar{s}^o$ data in Table A-7 for oxygen.

**Solution:**

**Given:**   Oxygen is heated at constant pressure from (a) 300 to 500 K and (b) 300 to 800 K.

**Find:**    $\Delta \bar{s}$, in kJ/kmol·K, by (1) average $c_{p,\text{av}}$ data, and (2) $\bar{s}^o$ data.

**Model:**   Ideal gas.

**Analysis:**   Since the pressure is constant, Eq. [7-21] reduces to $\Delta \bar{s} = \bar{c}_{p,\text{av}} \ln(T_2/T_1)$ and Eq. [7-23] becomes $\Delta \bar{s} = \bar{s}_2^o - \bar{s}_1^o$. Oxygen data for $c_p$ and

$\bar{s}°$ are found in Tables A-4 and A-7, respectively. To find $\bar{c}_p$, use $\bar{c}_p = c_p M$, where $M$ is the molar mass. The molar mass of oxygen is 32.

(a) For the temperature increment from 300 to 500 K:

(1) $\Delta\bar{s} = \bar{c}_{p,av} \ln \dfrac{T_2}{T_1} = 0.941 \text{ kJ/kg·K} \times (32 \text{ kg/kmol}) \times \ln \dfrac{500}{300}$

$\qquad = 15.38 \text{ kJ/kmol·K}$

(2) $\Delta\bar{s} = \bar{s}_2° - \bar{s}_1° = (220.589 - 205.213) \text{ kJ/kmol·K} = 15.38 \text{ kJ/kmol·K}$

(b) For the temperature increment from 300 to 800 K:

(1) $\Delta\bar{s} = 0.988 \text{ kJ/kg·K} \times 32 \text{ kg/kmol} \times \ln \dfrac{800}{300} = 31.01 \text{ kJ/kmol·K}$

(2) $\Delta\bar{s} = (235.810 - 205.213) \text{ kJ/kmol·K} = 30.60 \text{ kJ/kmol·K}$

**Comment:** For the 300 to 500 K interval, use of an average $c_p$ value leads to essentially no percentage error over the integrated value. For the 300 to 800 K interval, this error increases to 1.3 percent.

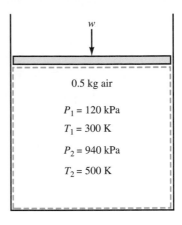

**Figure 7-11**
Schematic and data for Example 7-5.

---

One-half kilogram of air is compressed in a piston-cylinder device from 300 K and 120 kPa to 500 K and 940 kPa. (a) Determine the entropy change, in kJ/K. (b) Determine the direction of any heat transfer by using the entropy balance.

**Example 7-5**

**Solution:**

**Given:** Air is compressed in a piston cylinder, as shown in Fig. 7-11.

**Find:** (a) $\Delta S$, in kJ/K, and (b) direction of $Q$ by an entropy balance.

**Model:** Ideal gas, closed system.

**Strategy:** Use ideal-gas relation for $\Delta s$.

**Analysis:** The change in entropy $\Delta S = m\,\Delta s$.
(a) The specific entropy change for an ideal gas is given by

$$\Delta s = s_2° - s_1° - R \ln \frac{P_2}{P_1}$$

Therefore, the total entropy change for the process is

$$\Delta S = m\,\Delta s = 0.5 \text{ kg} \left(2.21952 - 1.70203 - \frac{8.314}{29} \ln \frac{940}{120}\right) \frac{\text{kJ}}{\text{kg·K}} = -0.0363 \text{ kJ/K}$$

This process is shown on the $Ts$ diagram in Fig. 7-12.
(b) The entropy balance for the closed system is

$$\Delta S = \int \frac{\delta Q}{T} + \sigma$$

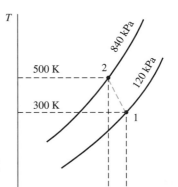

**Figure 7-12**
Process path for Example 7-5 plotted on a $Ts$ diagram.

Because $\Delta S$ is negative and $\sigma \geq 0$, the above relation requires that $Q$ be *negative,* that is, out of the system during compression. If the compression process was internally reversible, then the entropy change would equal the entropy transfer. However, $\sigma > 0$ if the process was irreversible. The larger the value of $\sigma$ (the greater the irreversibility), the greater the heat transfer from the system.

## 7-4  ENTROPY CHANGE OF AN INCOMPRESSIBLE SUBSTANCE

On the basis of the first $T\,ds$ equation, the general equation for the entropy change of any simple compressible substance is

$$ds = \frac{du}{T} + \frac{P\,dv}{T} \qquad \text{[7-14]}$$

An incompressible substance is defined as a substance of constant specific volume. Since $dv$ is zero in this case, Eq. [7-5] above reduces to

$$ds = \frac{du}{T}$$

In addition, recall from Chap. 4 that the internal energy of an incompressible substance is given by

$$du = c_v\,dT \qquad \text{[4-8]}$$

and the specific heats at constant volume and constant pressure are equal; that is,

$$c_p = c_v = c \qquad \text{[4-11]}$$

Employing the relations above for an incompressible substance, we find that the entropy change of an incompressible substance is given by

$$\Delta s = \int_1^2 \frac{c\,dT}{T} \qquad \text{(incompressible)} \qquad \text{[7-24]}$$

In many cases it is appropriate to use an average value of the specific heat over the temperature range of interest, so that

$$s_2 - s_1 = c_{av} \ln \frac{T_2}{T_1} \qquad \text{(incompressible)} \qquad \text{[7-25]}$$

In those cases where the specific heat varies significantly with temperature, it would be necessary to integrate Eq. [7-24] by first inserting a functional relationship for $c$ as a function of $T$.

**The entropy change of an incompressible substance is a function solely of what intensive property?**

**Example 7-6**

Two pieces of copper, $A$ and $B$, have masses of 1 and 3 kg and initial temperatures of 0 and 200°C, respectively. They are brought into thermal contact and allowed to reach an equilibrium temperature while insulated from the surroundings. Determine (a) the entropy change of each piece of copper, and (b) the entropy generation for the process, in kJ/K.

**Solution:**

**Given:** Two pieces of copper are brought into thermal contact, as illustrated in Fig. 7-13.

**Find:** (a) $\Delta S$ for each piece, in kJ/K, and (b) $\sigma$, in kJ/K.

**Model:** Incompressible substance; stationary, adiabatic, closed system.

**Strategy:** Use an energy analysis to find the final temperature. Calculate $\Delta S$ for each piece, and apply the entropy balance to find $\sigma$.

**Analysis:** Although the process is internally *irreversible,* the entropy change of an incompressible solid is still given by Eq. [7-25],

$$\Delta S = mc_{av} \ln \frac{T_2}{T_1}$$

The unknown in this equation for each piece of copper is the final equilibrium temperature $T_2$, which is found from an energy analysis. The basic energy equation for a stationary closed system composed of the two pieces of copper is

$$\Delta U = Q + W$$

The overall system is insulated, so $Q = 0$. If the copper pieces are taken to be incompressible, then $W = 0$. As a result, for the composite system, $\Delta U = \Delta U_A + \Delta U_B = 0$. Recalling that $\Delta u = c_{av} \Delta T$ for an incompressible substance, we have

$$[mc_{av}(T_2 - T_1)]_A + [mc_{av}(T_2 - T_1)]_B = 0$$

Data in Table A-4 indicate that the specific heat of copper varies only from 0.381 to 0.403 kJ/kg·°C over the range from 0 to 200°C. To simplify the analysis initially, we assume that $c_{av}$ is the same for both solids. As a result, $c_{av}$ cancels out in the above energy equation. Substitution of known values leads to

$$1(T_2 - 0) + (T_2 - 200) = 0$$
$$T_2 = 150°C$$

If the average specific heats for $A$ between 0 and 150°C and for $B$ between 150 and 200°C of 0.390 and 0.400 kJ/kg·°C, respectively, were used in the energy balance, the value of $T_2$ would change only to 151°C.

(a) On the basis of $T_2 = 150°C$, the entropy changes, using average $c$ data, are

$$\Delta S_A = 1(0.390) \text{ kJ/K} \times \ln \frac{423}{273} = 0.171 \text{ kJ/K}$$

$$\Delta S_B = 3(0.400) \text{ kJ/K} \times \ln \frac{423}{473} = -0.134 \text{ kJ/K}$$

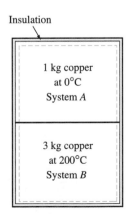

**Figure 7-13**
Schematic and data for Example 7-6.

(b) The reversibility or irreversibility of the overall process is determined by a general entropy balance on the composite closed system.

$$\Delta S = \int \frac{\delta Q}{T} + \sigma$$

For the composite system the heat transfer is zero; hence the entropy transfer is zero. As a result, the entropy generation $\sigma = \Delta S$ for the overall process. The entropy change is just the sum for subsystems $A$ and $B$. Thus

$$\sigma = \Delta S = [0.171 + (-0.134)] \text{ kJ/K} = 0.037 \text{ kJ/K}$$

On the basis of Eq. [6-52], the process is possible, but irreversible.

**Comment:**   The values of the entropy generation for subsystems $A$ and $B$ cannot be evaluated separately, because the value of the entropy transfer at the point of thermal contact is not known.

---

**Example 7-7**

**S**aturated-liquid water at 40°C is heated and compressed to 80°C and 75 bars. Estimate the entropy change by (a) assuming an incompressible fluid, (b) using saturation data for the liquid, and (c) using the compressed-liquid table, in kJ/kg·K.

**Solution:**

**Given:**   Saturated-liquid water at 40°C is heated and compressed to 80°C and 75 bars. The process is shown on the $Ts$ diagram in Fig. 7-14.

**Find:**   $\Delta s$, in kJ/kg·K, by (a) an incompressible model, (b) using saturation data, and (c) using compressed-liquid data.

**Model:**   Those specified in Find statement.

**Analysis:**   (a) Assuming an incompressible-substance model, $\Delta s = c \ln(T_2/T_1)$. From Table A-4, the specific heat of the liquid in this range of temperature is roughly 4.187 kJ/kg·K. Therefore,

$$\Delta s = 4.187 \text{ kJ/kg·K} \times \ln \frac{353}{313} = 0.504 \text{ kJ/kg·K}$$

(b) In the absence of compressed-liquid data for state 2, saturated-liquid data at the given temperature may be used. Employing the data in Table A-12, we find that

$$\Delta s = s_{f,80} - s_{f,40} = (1.0753 - 0.5725) \text{ kJ/kg·K} = 0.5028 \text{ kJ/kg·K}$$

This answer is fairly accurate, since property data in the compressed-liquid region are temperature-dependent and not very pressure-dependent.

(c) The entropy of state 2 in part $b$ is found more accurately by use of the compressed-liquid data in Table A-15. Hence

$$\Delta s = (1.0704 - 0.5725) \text{ kJ/kg·K} = 0.4979 \text{ kJ/kg·K}$$

This last answer is the most accurate, since it is based on experimental data.

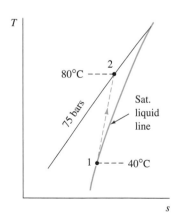

**Figure 7-14**
A $Ts$ process diagram for Example 7-7.

**Comment:** Use of saturated data in part $b$ leads to a 1 percent error, while the assumption of an incompressible fluid in part $a$ leads to a 1.2 percent error. Thus the incompressible-substance relation for $\Delta s$ leads to a reasonable answer when tabular data are unavailable for the liquid state.

## 7-5 APPLICATIONS OF THE STEADY-STATE ENTROPY BALANCE FOR A CONTROL VOLUME

The general entropy balance developed for a control volume in Sec. 6-7-1 is now discussed in terms of modeling assumptions, and then applied to various steady-state processes.

### 7-5-1 ENTROPY BALANCES FOR A CONTROL VOLUME

A general expression for an entropy balance for a control volume with uniform flow is developed in Sec. 6-7-1. The result [Eq. [6-43]] is repeated below.

$$\frac{dS_{cv}}{dt} = \sum_{in} \dot{m}_i s_i - \sum_{out} \dot{m}_e s_e + \sum_{j=1}^{n} \frac{\dot{Q}_j}{T_j} + \dot{\sigma}_{cv} \qquad \textbf{[7-26]}$$

where the terms on the right account for the transport and generation of entropy for the control volume. In the above equation the direction of $\dot{Q}_j$ is measured relative to the control volume. In words, Eq. [7-26] may be written for a control volume as

$$
\begin{pmatrix} \text{Time rate} \\ \text{of change} \\ \text{of entropy} \end{pmatrix} = \begin{pmatrix} \text{rate of entropy} \\ \text{transfer due to} \\ \text{mass flow} \end{pmatrix} + \begin{pmatrix} \text{rate of entropy} \\ \text{transfer due to} \\ \text{heat transfer} \end{pmatrix} + \begin{pmatrix} \text{rate of entropy} \\ \text{production due to} \\ \text{irreversibilities} \end{pmatrix}
$$

Hence the three physical contributors to the entropy change of any control volume are mass transfer, heat transfer, and internal irreversibilities. The first two contributions can be positive or negative, when present. The third term is always positive when present. These terms are shown on a control-volume schematic with one inlet and one exit in Fig. 7-15.

**Note the three physical phenomena that contribute to the entropy change of a control volume.**

The summation term $\sum \dot{Q}_j/T_j$ in Eq. [7-26] may be difficult to evaluate numerically, because the values of $\dot{Q}_j$ and $T_j$ are not known at every position on the control surface. This calculation difficulty can be overcome by two approaches. First, when the variation of $T_j$ along the control-volume boundary is fairly small, it is reasonably accurate to replace $T_j$ by a *constant boundary temperature* $T_b$. That is,

$$
\begin{pmatrix} \text{Rate of entropy change} \\ \text{due to heat transfer} \end{pmatrix} = \sum_{j=1}^{n} \frac{\dot{Q}_j}{T_j} \approx \frac{\dot{Q}}{T_b}
$$

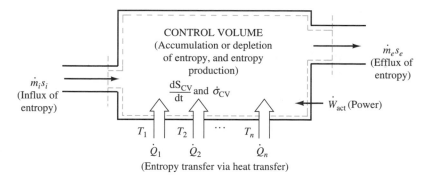

**Figure 7-15**    Schematic for general entropy accounting for a control volume.

The value of $T_b$ frequently is taken to be the arithmetic average in the inlet- and exit-fluid temperatures $T_{b,av}$.

The second approach is to locate the control surface where the boundary temperatures are known. If this enlarges the control volume to include additional mass, this mass must be included in the analysis. This additional region of the control volume may contain a wall of the equipment as well as a thin layer of air (or some other fluid) adjacent to the wall. It is often reasonable to neglect the additional mass in this region in the energy and entropy balances. Figure 7-16 shows an enlarged control volume with heat transfer across the wall of the equipment and a thin layer of air outside the wall (frequently referred to as the boundary layer of the fluid). The dashed line shows the boundary of an enlarged control volume that contains both the volume where flow occurs and the region where heat transfer occurs. In many situations, this heat-transfer region may surround the entire control surface where mass transfer is not occurring.

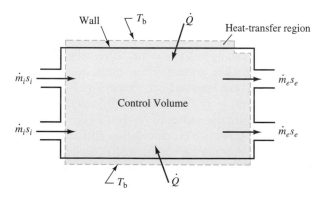

**Figure 7-16**    Enlarged control volume including heat transfer region.

## 7-5-2 MODELING ASSUMPTIONS AND THE STEADY-STATE, STEADY-FLOW ENTROPY BALANCE

Equation [7-26] is the basic rate form of the entropy balance for a control volume and is repeated here for easy reference in the following discussion.

$$\frac{dS_{cv}}{dt} = \sum_{in} \dot{m}_i s_i - \sum_{out} \dot{m}_e s_e + \sum_{j=1}^{n} \frac{\dot{Q}_j}{T_j} + \dot{\sigma}_{cv} \qquad \textbf{[7-26]}$$

This equation is the usual starting place for most control-volume analyses. As with the rate forms of the conservation of mass and conservation of energy, there are numerous assumptions that can be made to simplify this general equation for any specific applications.

Some of the more common modeling assumptions are described below, and their effect on the entropy balance is demonstrated.

***A. Closed System:*** This is the same as assuming no mass flow across the boundary of the system ($\dot{m} = 0$). Under these conditions Eq. [7-26] reduces to the familiar closed-system form of the entropy balance

$$\frac{dS_{cm}}{dt} = \sum_{j=1}^{n} \frac{\dot{Q}_j}{T_j} + \dot{\sigma}_{cm}$$

***B. Steady-State System:*** The steady-state assumption requires that the intensive properties at any point inside the system do not change with time, and hence time is no longer a variable in the analysis. Under this condition the amount of entropy in the control volume $S_{cv}$ is independent of time and $dS_{cv}/dt = 0$. Thus Eq. [7-26] becomes

$$0 = \sum_{in} \dot{m}_i s_i - \sum_{out} \dot{m}_e s_e + \sum_{j=1}^{n} \frac{\dot{Q}_j}{T_j} + \dot{\sigma}_{cv} \qquad \textbf{[7-27]}$$

This equation can be rearranged in any number of ways depending upon what information is available for the analysis.

***C. Steady-State System with One Inlet and One Outlet:*** Under these conditions the summation signs on the mass-flow terms in the entropy balance are unnecessary. Labeling the inlet as 1 and the outlet as 2, and recalling that under these same conditions $\dot{m}_1 = \dot{m}_2$, Eq. [7-26] can be simplified to give

$$0 = \dot{m}_1 s_1 - \dot{m}_2 s_2 + \sum_{j=1}^{n} \frac{\dot{Q}_j}{T_j} + \dot{\sigma}_{cv} \qquad \text{or} \qquad 0 = s_1 - s_2 + \sum_{j=1}^{n} \frac{q_j}{T_j} + \sigma_{m,cv}$$

$$\textbf{[7-28]}$$

Again, this equation can be rearranged depending upon the objective of the analysis and the known information.

Each of these special cases is the result of correctly applying one or more modeling assumptions to the basic rate form of the entropy balance (Eq. [7-26]). The applicable form of the equation should always be developed for the specific context of an analysis. This approach will be demonstrated in the following examples.

### 7-5-3  EXAMPLES OF THE SECOND-LAW ANALYSIS OF STEADY-FLOW PROCESSES

The following examples illustrate the use of the entropy balance in the study of some steady-flow processes.

**Example 7-8**

**Air** is compressed from 1 bar, 27°C, to 3.5 bars, 127°C, in a steady-flow device. Any heat transfer occurs between the air and the environment at 27°C, and the actual shaft work input is 170.0 kJ/kg. (*a*) Determine the heat transfer, in kJ/kg, for the air flowing through the compressor. Then determine the entropy production, in kJ/kg·K, for (*b*) the control volume for the air flowing through the compressor, (*c*) the control volume which includes the air flow and the immediate surroundings where heat transfer occurs, and (*d*) the control volume containing only the heat-transfer region.

**Solution:**

**Given:**   Steady-state compression of air, as shown with input data in Fig. 7-17.

**Find:**   (*a*) $q$ for the air, in kJ/kg, and $\sigma_m$ for (*b*) the air flow, (*c*) the composite air and heat-transfer region, and (*d*) the heat-transfer region alone, in kJ/kg·K.

**Model:**   Steady-state control volumes with negligible Δke and Δpe, ideal gas with constant specific heats.

**Strategy:**   Apply the mass, energy, and entropy balances, along with appropriate equations for ideal-gas properties.

**Analysis:**   (*a*) The heat transfer can be found from an energy balance on the compressor. If the compressor is modeled as a steady-state control volume with one inlet (1) and one outlet (2), the mass balance becomes $\dot{m}_1 = \dot{m}_2$, and the energy balance becomes

$$0 = \dot{Q} + \dot{W} + \dot{m}_1(h + ke + pe)_1 - \dot{m}_2(h + ke + pe)_2$$

Dividing the energy balance by $\dot{m}$ and noting that Δke and Δpe are negligible, we obtain the relation $0 = q + w - \Delta h$. For an ideal gas with constant specific heats, $\Delta h = c_{p,av}\,\Delta T$. Solving for the heat-transfer term gives

$$q = \Delta h - w = c_{p,av}\Delta T - w$$

From Table A-3, the average value of $c_p$ at $T_{av} = 350$ K is 1.008 kJ/kg·K. Substitution of known data yields

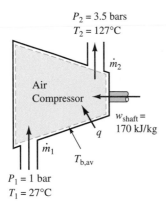

$P_2 = 3.5$ bars
$T_2 = 127°C$

$\dot{m}_2$

Air
Compressor

$w_{shaft} = 170$ kJ/kg
$q$
$\dot{m}_1$
$T_{b,av}$

$P_1 = 1$ bar
$T_1 = 27°C$

**Figure 7-17**
Schematic and data for Example 7-8.

$$q = 1.008 \text{ kJ/kg·K} \times (400 - 300)\text{K} - 170 \text{ kJ/kg} = -69.2 \text{ kJ/kg}$$

Thus the compressor air loses energy by heat transfer

(*b*) The entropy production for the air can be determined by using the entropy balance for the same control volume shown in Fig. 7-17, namely,

$$0 = s_1 - s_2 + \sum_{j=1}^{n} \frac{q_j}{T_j} + \sigma_{m,cv}$$

The change in specific entropy for an ideal gas with constant specific heats can be calculated as

$$\Delta s = s_2 - s_1 = c_{p,av} \ln \frac{T_2}{T_1} - R \ln \frac{P_2}{P_1}$$

$$= 1.008 \frac{\text{kJ}}{\text{kg·K}} \times \ln \frac{400}{300} - 0.287 \frac{\text{kJ}}{\text{kg·K}} \times \ln \frac{3.5}{1} = -0.0696 \text{ kJ/kg·K}$$

where $R = R_u/M = 8.314/28.97 = 0.287$ kJ/kg·K. Figure 7-18 shows the compression process on a $Ts$ diagram. As described in Sec. 7-1-1, constant-pressure lines run from lower left to upper right in the gas region on a $Ts$ plot.

Detailed surface-temperature data are required for evaluating the entropy transport term. As an approximation, assume that all the heat transfer occurs at the arithmetic average of the inlet and outlet temperatures, $T_{b,av} = 350$ K. Then the entropy balance can be rearranged to solve for the entropy generation within the compressor.

$$\sigma_{m,air} = \Delta s - \frac{q_{cv}}{T_{b,av}} = -0.0696 \frac{\text{kJ}}{\text{kg·K}} - \frac{-69.2 \text{ kJ/kg}}{350 \text{ K}}$$

$$= -0.0696 + 0.1977 = 0.1281 \text{ kJ/kg·K}$$

Note that because this answer is strongly dependent upon the *estimate* of the boundary temperature, this answer is only an approximation.

(*c*) The entropy generation found within a control volume containing both the compressor and the immediate surroundings where heat transfer occurs is found from an entropy balance on this enlarged system, as shown in Fig. 7-19. In this case the enlarged system has a uniform and constant boundary temperature equal to the environment temperature of 27°C. In addition, in steady state the heat transfer from the air is also the heat transfer to the environment. As a result, the entropy balance for this composite control volume (comp) is

$$\sigma_{m,comp} = s_2 - s_1 - \frac{q}{T_{surr}} = \left(-0.0696 \frac{\text{kJ}}{\text{kg·K}} - \frac{-69.2 \text{ kJ/kg}}{300 \text{ K}}\right)$$

$$= (-0.0696 + 0.2307) \text{ kJ/kg·K} = 0.161 \text{ kJ/kg·K}$$

This value is exact for the specified control volume because the boundary temperature was constant and known

(*d*) The control volume in this case is the heat-transfer region between the air stream and the surroundings. The same heat transfer enters the region at an average temperature of 350 K, and leaves at the surrounding temperature of 27°C. Hence the steady-state entropy balance reduces to

$$0 = \left(\frac{\dot{Q}}{T_{surr}} - \frac{\dot{Q}}{T_{b,av}}\right) + \dot{\sigma}_Q$$

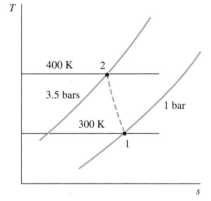

**Figure 7-18**

Process diagram for Example 7-8.

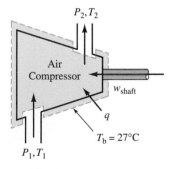

**Figure 7-19**

Enlarged system for Example 7-8.

Dividing through by $\dot{m}$ and solving for the entropy production gives

$$\sigma_{m,Q} = q\left(\frac{1}{T_{\text{surr}}} - \frac{1}{T_{\text{b,av}}}\right)$$

Substitution of data leads to

$$\sigma_{m,Q} = 69.2 \text{ kJ/kg} \times \left(\frac{1}{300} - \frac{1}{350}\right)\frac{1}{\text{K}}$$

$$= 0.0330 \text{ kJ/kg·K}$$

where $\sigma_{m,Q}$ is the symbol for entropy generation in a heat-transfer region developed in Sec. 6-10-2.

**Comment:**   The second control volume selected is simply the composite of the first and third control volumes analyzed. Thus the entropy-generation terms should be related by the equation $\sigma_{\text{comp}} = \sigma_{\text{air}} + \sigma_Q$. The $\sigma$ values calculated above support this result. Although the calculations for the composite system are the most exact, they give no information on the relative contributions of the various sources of irreversibilities. By using the two smaller subsystems and appropriate assumptions to model the compressor and the immediate surroundings, we find that approximately 80 percent of the entropy production occurs during the compression process, and the remaining 20 percent occurs with the heat transfer through a finite temperature difference. As with any calculation, the actual values depend upon the assumptions made to model the system.

**Example 7-9**

**R**efrigerant 134a enters an air-cooled condenser as a superheated vapor at 12 bars and 60°C, and leaves as a saturated liquid at 12 bars. Atmospheric air at 35°C is blown over the condenser tubes and leaves at 45°C. The heat transfer between the two fluid streams equals 25 MJ/h. Changes in kinetic and potential energy are negligible. Determine (*a*) the mass flow rates for the R-134a and the air, in kg/h, (*b*) the entropy production rate in the condenser, in kJ/h·K, and (*c*) the change in kinetic energy for R-134a if the pipe diameter is 2.0 cm, in kJ/h. (*d*) Is neglecting the change in kinetic energy a reasonable modeling assumption?

**Solution:**

**Given:**   R-134a is condensed in the air-cooled condenser shown in Fig. 7-20.

**Find:**   (*a*) $\dot{m}$ for R-134a and air, in kg/h, (*b*) $\dot{\sigma}$, in kJ/h·K, and (*c*) $\Delta$ke for R-134a, in kJ/h.

**Model:**   Air is an ideal gas; steady-state open system; changes in ke and pe are negligible.

**Strategy:**   Apply mass, energy, and entropy balances. Use property data for R-134a, and ideal-gas data for air.

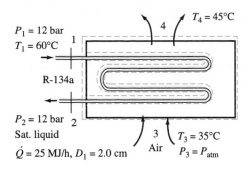

**Figure 7-20** Schematic and data for Example 7-9.

**Analysis:** (a) For the mass flow rates, consider two separate open systems—the refrigerant side and the air side. For the R-134a stream at $P_1 = P_2 = 12$ bars, $T_1 = 60°C$:

*Mass balance:* $\quad\quad\quad\quad\quad\quad 0 = \dot{m}_1 - \dot{m}_2 \quad$ or $\quad \dot{m}_1 = \dot{m}_2 = \dot{m}_{ref}$

*Energy balance:* $\quad\quad\quad\quad\quad\quad 0 = -\dot{Q}_{ref,out} + \dot{m}_1 h_1 - \dot{m}_2 h_2$

$\quad\quad$ or $\quad\quad\quad\quad\quad\quad\quad \dot{Q}_{ref,out} = \dot{m}_{ref}(h_1 - h_2)$

*Data:* $\quad\quad\quad\quad\quad\quad\quad\quad h_1 = 287.44$ kJ/kg (Table A-18)

$\quad\quad$ and $\quad\quad\quad\quad\quad\quad h_2 = h_f = 115.76$ kJ/kg (Table A-17)

As a result,

$$\dot{m}_{ref} = \frac{\dot{Q}_{out}}{-\Delta h} = \frac{25{,}000 \text{ kJ/h}}{(287.44 - 115.76) \text{ kJ/kg}} = 145.6 \text{ kg/h}$$

For the air stream at $T_1 = 35°C$, $T_2 = 45°C$:

*Mass balance:* $\quad\quad\quad\quad\quad\quad 0 = \dot{m}_3 - \dot{m}_4 \quad$ or $\quad \dot{m}_3 = \dot{m}_4 = \dot{m}_{air}$

*Energy balance:* $\quad\quad\quad\quad\quad\quad 0 = -\dot{Q}_{air,in} + \dot{m}_3 h_3 - \dot{m}_4 h_4$

$\quad\quad$ or $\quad\quad\quad\quad\quad\quad\quad \dot{Q}_{air,in} = \dot{m}_{air}(h_4 - h_3)$

*Data:* $\quad\quad\quad\quad\quad\quad\quad\quad h_4 - h_3 = c_{p,av}(T_4 - T_3)$

$\quad\quad$ and $\quad\quad\quad\quad\quad\quad c_{p,av} = 1.005$ kJ/kg·K (Table A-3)

As a result,

$$\dot{m}_{air} = \frac{\dot{Q}_{in}}{c_{p,av}\Delta T} = \frac{25{,}000 \text{ kJ/h}}{1.005 \text{ kJ/kg·°C} \times (45 - 35)°C} = 2488 \text{ kg/h}$$

(b) To find the entropy production, select the complete heat exchanger as the open system. The steady-state entropy balance is

$$0 = \sum_{j=1}^{n} \frac{\dot{Q}_j}{T_j} + \dot{m}_1 s_1 + \dot{m}_3 s_3 - \dot{m}_2 s_2 - \dot{m}_4 s_4 + \dot{\sigma}$$

Assuming no heat transfer at the boundaries, rearrangement yields

$$\dot{\sigma} = \dot{m}_{ref}(s_2 - s_1) + \dot{m}_{air}(s_4 - s_3)$$

For the R-134a stream,

$s_1 = 0.9526$ kJ/kg·K (Table A-18)     and     $s_2 = 0.4164$ kJ/kg·K (Table A-17)

For air as an ideal gas, and assuming negligible pressure drop on the air side,

$$s_4 - s_3 = c_{p,av} \ln \frac{T_4}{T_3} - R \ln \frac{P_4}{P_3} = 1.005 \text{ kJ/kg·K} \ln \frac{45 + 273}{35 + 273} = 0.03211 \text{ kJ/kg·K}$$

As a result, the entropy production rate is

$$\dot{\sigma} = 145.6 \, \frac{\text{kg}}{\text{h}} \times (0.4164 - 0.9527) \, \frac{\text{kJ}}{\text{kg·K}} + 2488 \, \frac{\text{kg}}{\text{h}} \times 0.03211 \, \frac{\text{kJ}}{\text{kg·K}}$$

$$= (-78.09 + 79.89) \text{ kJ/kg·K} = 1.80 \text{ kJ/(K·h)}$$

(c) The change in kinetic energy on a rate basis is given by $\Delta\dot{\text{KE}} = \dot{m}(V_2^2 - V_1^2)/2$. Also, from the definition of $\dot{m}$ from the continuity equation, $V = \dot{m}v/A$. The area at inlet and exit for the R-134a flow is

$$A_1 = A_2 = \frac{\pi D^2}{4} = \frac{\pi(0.020)^2}{4} \text{ m}^2 = 3.142 \times 10^{-4} \text{ m}^2$$

At the inlet state, $v_1 = 0.01835$ m³/kg. Hence the inlet velocity for R-134a is

$$V_1 = \frac{\dot{m}v_1}{A_1} = \frac{145.6 \text{ kg/h}}{3.142 \times 10^{-4} \text{ m}^2} \times 0.01835 \, \frac{\text{m}^3}{\text{kg}} = 8504 \text{ m/h} = 2.36 \text{ m/s}$$

Similarly, for the exit state where $v_2 = v_f = 0.0008928$ m³/kg, the exit velocity is

$$V_2 = \frac{\dot{m}v_2}{A_2} = \frac{145.6 \text{ kg/h}}{3.142 \times 10^{-4} \text{ m}^2} \times 0.0008928 \, \frac{\text{m}^3}{\text{kg}} = 414 \text{ m/h} = 0.0115 \text{ m/s}$$

Solving for the kinetic-energy change on a rate basis,

$$\Delta\dot{\text{KE}} = 145.6 \, \frac{\text{kg}}{\text{h}} \left[ \frac{(0.115)^2}{2} - \frac{(2.36)^2}{2} \right] \frac{\text{m}^2}{\text{s}^2}$$

$$= -404.5 \, \frac{\text{kg·m}^2}{\text{h·s}^2} \times \frac{\text{kJ·s}^2}{1000 \text{ kg·m}^2} = -0.40 \text{ kJ/h}$$

(d) Compared to the heat-transfer rate of 25,000 kJ/h, the change in kinetic energy is negligible.

**Comment:**     The entropy production rate inside the heat exchanger due to heat transfer can be estimated by assuming the heat transfer from the R-134a occurs at the saturation temperature at 12 bars, namely, 46.32°C or 319.47°K, and the air receives the energy at the average air temperature of 40°C or 313.15 K. The steady-state entropy production is estimated as

$$\dot{\sigma} = \frac{\dot{Q}_{out}}{T_{air}} - \frac{\dot{Q}_{in}}{T_{ref}} = 25,000 \text{ kJ/h} \left( \frac{1}{313.15} - \frac{1}{319.47} \right) \frac{1}{\text{K}} = 1.58 \text{ kJ/h·K}$$

This estimate predicts almost 90 percent of the actual amount.

Consider the open feedwater heater discussed in Example 5-8. Energy and mass balances on the steady-state device show superheated steam (flow stream 1) at 5 bars; 200°C enters at 1 kg/s. Subcooled water (flow stream 2) enters at 4.75 kg/s at 5 bars, 40°C, and saturated-liquid water (flow stream 3) leaves at 5.75 kg/s and 5 bars. Determine the rate of entropy generation for the mixing process, in kJ/K·s.

**Solution:**

**Given:** Mixing of two flow streams as shown in Fig. 7-21.

**Find:** $\dot{\sigma}_{cv}$, in kJ/K·s.

**Model:** Steady state, adiabatic.

**Strategy:** Use a control-volume entropy balance.

**Analysis:** The steady-state entropy balance for the control volume shown in Fig. 7-21 is

$$0 = \sum_{in} \dot{m}_i s_i - \sum_{out} \dot{m}_e s_e + \sum_{j=1}^{n} \frac{\dot{Q}_j}{T_j} + \dot{\sigma}_{cv}$$

Solving for the entropy-generation rate within the adiabatic open heater with two inlets and one outlet,

$$\dot{\sigma}_{cv} = s_3 \dot{m}_3 - s_1 \dot{m}_1 - s_2 \dot{m}_2$$

The specific entropies at states 1, 2, and 3 are found from Tables A-14, A-12, and A-13 to be 7.0592, 0.5725, and 1.8607 kJ/kg·K, respectively. Substitution of these values yields

$$\dot{\sigma}_{cv} = [1.8607(5.75) - 7.0592(1) - 0.5725(4.75)] \text{ kJ/K·s}$$
$$= (10.70 - 7.06 - 2.72) \text{ kJ/K·s} = 0.92 \text{ kJ/K·s}$$

**Comment:** The rate of entropy generation is positive, in accordance with the second law, and is due to the irreversible mixing of flow streams at different temperatures.

---

Liquid water at 5 bars and 20°C is throttled to 1 bar as it flows steadily through an insulated valve. Assume changes in kinetic and potential energy are negligible. Determine (a) the temperature change and (b) the entropy production, in kJ/kg·K.

**Solution:**

**Given:** Water is throttled, as shown in Fig. 7-22.

**Find:** (a) $\Delta T$, and (b) $\sigma$, in kJ/kg·K.

**Model:** Steady-state, adiabatic process.

**Example 7-10**

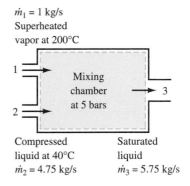

$\dot{m}_1 = 1$ kg/s
Superheated
vapor at 200°C

Compressed
liquid at 40°C
$\dot{m}_2 = 4.75$ kg/s

Saturated
liquid
$\dot{m}_3 = 5.75$ kg/s

**Figure 7-21**
Schematic of mixing process in Example 7-10.

**Example 7-11**

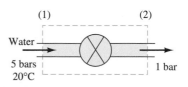

**Figure 7-22**
Schematic and data of throttling process in Example 7-11.

**Strategy:**   Apply energy and entropy balances to the flow process.

**Analysis:**   The steady-state mass, energy, and entropy balances for the control volume shown in Fig. 7-22 are

$$0 = \dot{m}_1 - \dot{m}_2$$

$$0 = \dot{Q} + \dot{W} + \dot{m}_1(h + \text{ke} + \text{pe})_1 - \dot{m}_2(h + \text{ke} + \text{pe})_2$$

$$0 = \dot{m}_1 s_1 - \dot{m}_2 s_2 + \sum_{j=1}^{n} \frac{\dot{Q}_j}{T_j} + \dot{\sigma}_{\text{cv}}$$

For this process, $\dot{Q} = \dot{W} = 0$, and $\Delta\text{ke} = \Delta\text{pe} = 0$. Thus the steady-state energy equation becomes

$$0 = h_1 - h_2 \quad \text{or} \quad h_2 = h_1$$

Solving for the entropy production term in the entropy balance, we have on a unit-mass basis

$$0 = s_1 - s_2 + \sigma_{m,\text{cv}} \quad \text{or} \quad \sigma_{m,\text{cv}} = s_2 - s_1$$

(*a*) To evaluate the temperature change, it is necessary to relate the enthalpy change to the temperature. At 5 bars and 20°C water is a compressed liquid. We shall assume that the water remains a liquid and thus the incompressible substance model can be applied. Applying this model to the energy equation,

$$\Delta h_{\text{inc}} = c\,\Delta T + v\,\Delta P = 0 \quad \text{or} \quad \Delta T = \frac{v(P_1 - P_2)}{c}$$

The specific heat of water from Table A-4 is around 4.186 kJ/kg·K, and the specific volume $v \approx v_f$ at 20°C $= 1.002 \times 10^{-3}$ m³/kg from Table A-12. Hence

$$\Delta T = \frac{1.002 \times 10^{-3}\ \text{m}^3/\text{kg}}{4.186\ \text{kJ/kg·K}} \times (5-1)\ \text{bars} \times \frac{100\ \text{kJ}}{\text{bar·m}^3} = 0.096\ \text{K}$$

Thus the assumption of liquid water at the outlet is correct and the incompressible substance model is acceptable.

(*b*) The rate of entropy generation is $\sigma_{m,\text{cv}} = s_2 - s_1$. For an incompressible fluid, $\Delta s = c \ln T_2/T_1$. Therefore,

$$\sigma_{m,\text{cv}} = c \ln \frac{T_2}{T_1} = 4.186\ \text{kJ/kg·K} \times \ln \frac{293.25}{293.15} = 0.00143\ \text{kJ/kg·K}$$

The answer is positive as expected for the irreversible process.

**Comment:**   (1) This process is equivalent to the release of water through a faucet. Note that the temperature rise would be undetectable to the human skin. The numerical answer for the entropy generation is sensitive to the significant figures used for the two temperatures.

(2) Under some conditions, the exit state would be two-phase.

(3) If the fluid had been an ideal gas, the analysis would lead to $T_2 = T_1$, and $\sigma = -R \ln (P_2/P_1)$.

## 7-6   STEADY-FLOW WORK RELATIONSHIPS

In this section a general equation for shaft work during an internally reversible, steady-flow process is developed. This relationship is then extended to polytropic processes.

### 7-6-1   A RELATIONSHIP FOR STEADY-FLOW MECHANICAL WORK

As was done in Sec. 7-2-1 for a closed system, it is possible to examine the relationship between the internally reversible work and the actual work for a steady-state process. As a starting point, the energy and entropy balances for a steady-state control volume with one inlet and one outlet are

$$0 = q + w - \Delta h - \Delta \text{ke} - \Delta \text{pe}$$

$$0 = s_1 - s_2 + \sum_{j=1}^{n} \frac{q_j}{T_j} + \sigma_m$$

These equations are now written for a differential change in distance $dx$ along the direction of flow, and rearranged. The energy equation becomes, upon rearrangement,

$$\delta w = dh + d\text{ke} + d\text{pe} - \delta q$$

and the entropy balance is written as

$$\delta q = T \, ds - T \, \delta \sigma_m$$

Substituting $\delta q$ from the entropy balance back into the energy balance yields

$$\delta w_{\text{act}} = dh + d\text{ke} + d\text{pe} - (T \, ds - T \, \delta \sigma_m)$$
$$= (dh - T \, ds) + d\text{ke} + d\text{pe} + T \, \delta \sigma_m$$

This equation can be simplified further by using Eq. [7-13], the second $T \, ds$ equation: $T \, ds = dh - v \, dP$, to replace $dh - T \, ds$. The result is

$$\delta w_{\text{act}} = v \, dP + d\text{ke} + d\text{pe} + T \, \delta \sigma_m \qquad \textbf{[7-29]}$$

If we integrate along the length of the control volume, we obtain the actual work for a unit mass passing through a steady-state device.

$$w_{\text{act}} = \int v \, dP + \Delta \text{ke} + \Delta \text{pe} + \int T \, \delta \sigma_m \qquad \textbf{[7-30]}$$

For an *internally reversible* process, $\sigma_m = 0$,

$$\delta w_{\text{sf,rev}} = v \, dP + d\text{ke} + d\text{pe} \qquad \textbf{[7-31]}$$

and integration leads to

$$w_{sf,rev} = \int_1^2 v\,dP + \frac{\mathbf{V}_2^2 - \mathbf{V}_1^2}{2} + g(z_2 - z_1) \qquad \textbf{[7-32]}$$

If the changes in the kinetic and potential energies are negligible, Eq. [7-32] reduces to

$$w_{sf,rev} = \int v\,dP \qquad \textbf{[7-33]}$$

Finally, if Eq. [7-31] is substituted into Eq. [7-29], then

$$\delta w_{sf,act} = \delta w_{sf,rev} + T\,\delta\sigma_m$$

As a result, similar to the closed-system relation found earlier,

$$\delta w_{sf,act} \geq \delta w_{sf,rev} \qquad \textbf{[7-34]}$$

Thus in the analysis of compressors, pumps, and fans, with *work input*,

$$w_{sf,in,act} \geq w_{sf,in,rev} \qquad \text{(compressor, pump, fan)} \qquad \textbf{[7-35]}$$

and for turbines with *work output*,

$$w_{sf,out,act} \leq w_{sf,out,rev} \qquad \text{(turbine)} \qquad \textbf{[7-36]}$$

Although Eq. [7-32] was developed from an energy balance on a control volume, it can also be developed solely from the principles of mechanics. These equations for reversible steady-flow shaft work may be used independently or in conjunction with an energy analysis for a control volume. Note that Eqs. [7-32] and [7-33] are useful in a quantitative sense only if a functional relationship between $v$ and $P$ is known, and the process is internally reversible.

Two other comments must be made with regard to Eqs. [7-32] and [7-33]:

1. For a given pressure change in a work-producing or work-absorbing steady-flow device, the amount of reversible work is greatly affected by the specific volume of the fluid. For example, the larger the specific volume, the greater the reversible work output in a turbine expansion process. Hence gas flows generally lead to much greater work output than liquid flows. In addition, the smaller the specific volume, the less the work input required. Hence pumps handling liquids require much less reversible work input than gas compressors for the same $\Delta P$. These general conclusions are equally valid for actual (irreversible) flows.

2. The equation $w = \int v\,dP$ is only valid for *reversible, steady-flow processes*. It should not be confused with $w = -\int P\,dv$, which is the work associated with reversible comp/exp work on a closed system, as shown in Fig. 7-23.

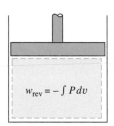

$$w_{rev} = -\int P\,dv$$

Closed system

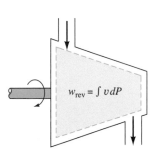

$$w_{rev} = \int v\,dP$$

Steady flow

**Figure 7-23**
Equipment schematics for steady-flow work and closed-system compression/expansion work.

The first item is illustrated by the simple steam power cycle. The expansion and compression processes occur through the same $\Delta P$. The turbine expands a vapor, while a pump compresses a liquid in the cycle. Hence the turbine work output (for driving an electrical generator) greatly exceeds the pump work input for the cycle.

## 7-6-2   REVERSIBLE, STEADY-FLOW MECHANICAL WORK FOR POLYTROPIC PROCESSES

In Sec. 2-6 a polytropic process for gases is defined as one in which the relationship between $v$ and $P$ can be modeled by the relation $Pv^n = c$, where $n$ is a constant known as the polytropic exponent. In Example 2-7 it is shown that the work transfer $w_{comp/exp}$ for a polytropic process in a closed system is given by

$$w_{comp/exp} = -\int_1^2 P\,dv = \frac{P_2 v_2 - P_1 v_1}{n-1} \qquad \textbf{[7-37]}$$

The steady-flow compression of gases frequently is modeled as polytropic. In this case, Eq. [7-33] for steady-state shaft work must be integrated in conjunction with the polytropic relation. The result is

$$w_{sf,rev} = \int_1^2 v\,dP = \int_1^2 c^{1/n} P^{-(1/n)}\,dP = \frac{nc^{1/n}(P_2^{(n-1)/n} - P_1^{(n-1)/n})}{n-1}$$

The constant $c = P_1 v_1^n = P_2 v_2^n$. Therefore

$$w_{sf,rev} = \frac{n[P_2^{1/n}v_2(P_2^{(n-1)/n}) - P_1^{1/n}v_1(P_1^{(n-1)/n})]}{n-1} = \frac{n(P_2 v_2 - P_1 v_1)}{n-1}$$
$$\textbf{[7-38]}$$

This relation applies to any process modeled as polytropic, except when $n = 1$. Note the similarity between Eqs. [7-37] and [7-38], which are valid for closed and steady-flow polytropic processes, respectively. The graphical interpretations of these two equations, however, are quite different. As shown in Fig. 7-24, steady-flow mechanical work is represented on a $Pv$ diagram by the area to the *left* of the process curve. Recall that comp/exp work for a closed system is represented by the area *beneath* the curve.

In the case of an ideal gas, the relation $Pv = RT$ is also valid, and Eq. [7-38] takes the form

$$w_{sf,rev} = \frac{nR(T_2 - T_1)}{n-1} = \frac{nRT_1}{n-1}\left(\frac{T_2}{T_1} - 1\right) \qquad \text{(ideal gas)} \quad \textbf{[7-39]}$$

In addition, a combination of the polytropic and ideal-gas relations leads to

$$\frac{T_2}{T_1} = \left(\frac{P_2}{P_1}\right)^{(n-1)/n} \qquad \textbf{[7-40]}$$

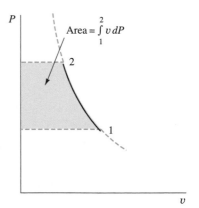

**Figure 7-24**
Area representation of steady-flow mechanical work on a $Pv$ diagram.

Use of this pressure-temperature relationship in the preceding equation leads to a useful expression for *steady-flow shaft work* with negligible changes in ke and pe.

$$w_{sf,rev} = \frac{nRT_1}{n-1}\left[\left(\frac{P_2}{P_1}\right)^{(n-1)/n} - 1\right] \qquad \text{(ideal gas)} \qquad \textbf{[7-41]}$$

The above equation is restricted to *polytropic processes* involving *ideal gases*. The following example illustrates polytropic calculations for an ideal gas.

**Example 7-12**

**A** steady-state compressor changes air polytropically from 25°C and 1 bar to a final state of 4 bars. The constant $n = 1.3$ for the process. Changes in kinetic and potential energy are negligible. If air is modeled as an ideal gas, determine (*a*) the exit temperature, in kelvins, and (*b*) the work and the heat transfer, both in kJ/kg.

**Solution:**

**Given:** Air is compressed polytropically, as shown in Fig. 7-25.

**Find:** (*a*) $T_2$, in K, and (*b*) $w$ and $q$, in kJ/kg.

**Model:** Steady-state, polytropic, ideal gas, $\Delta ke = \Delta pe = 0$.

**Strategy:** Use polytropic property and work-transfer relations, then an energy analysis.

**Analysis:** (*a*) The final temperature is found from the polytropic-property relation given by Eq. [7-40].

$$T_2 = T_1\left(\frac{P_2}{P_1}\right)^{(n-1)/n} = 298\text{ K}\left(\frac{4}{1}\right)^{(1.3-1)/1.3} = 410\text{ K}$$

(*b*) The value of $w_{sf,rev}$ can be found directly from either Eq. [7-39] or Eq. [7-41]. The first of these equations requires only temperature data. Hence

$$w_{sf,rev} = \frac{nR(T_2 - T_1)}{n-1} = \frac{1.3(8.314)(410 - 298)}{28.97(1.3 - 1)}\frac{\text{kJ}}{\text{kg}} = 139.3\text{ kJ/kg}$$

The use of Eq. [7-41] yields the same answer; however, its use does not require the calculation of $T_2$. The path of the process is shown in Fig. 7-26. The area to the left of the process line represents the steady-state work of 139.3 kJ/kg.

The heat transfer is found from the steady-state energy balance on the control volume with $\Delta ke = \Delta pe = 0$,

$$0 = q + w + h_1 - h_2$$

Solving for $q$, the energy equation reduces to

$$q = h_2 - h_1 - w$$

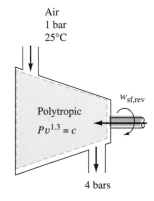

Air
1 bar
25°C

$w_{sf,rev}$

Polytropic

$Pv^{1.3} = c$

4 bars

**Figure 7-25**
Schematic and data for Example 7-12.

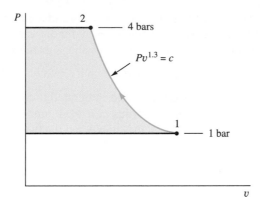

**Figure 7-26**     Steady-flow work representation on a $Pv$ diagram for Example 7-12.

The enthalpy values at 298 and 410 K are found from Table A-5. As a result

$$q = [(411.12 - 298.18) - 139.3]\ \text{kJ/kg} = -26.4\ \text{kJ/kg}$$

**Comment:**     The values of both $w$ and $q$ depend on the value of $n$. When $n = 1.4$, for example, $T_2 = 433$ K, $w = 145.6$ kJ/kg, and $q = 0.9$ kJ/kg. Thus when $n = 1.4$, the process is essentially adiabatic. The adiabatic, polytropic process is discussed further in Sec. 8-1-1.

## 7-7   SUMMARY

Entropy $S$ is an intrinsic property of a substance and like other extensive properties can be written in an intensive form, $s = S/m$. Using the state principle for a simple, compressible substance, the specific entropy $s$ can be related to any two other independent, intrinsic, intensive properties, e.g., $P, v, T, u,$ and $h$. Thus values of the specific entropy are tabulated in property tables.

In second-law analyses of a process or device, familiarity with the general details of $Ts$ and $hs$ diagrams is quite useful. For example, the value of $q$ for an *internally reversible process* is given by

$$q_{\text{int rev}} = \int T\,ds$$

On a $Ts$ diagram, this integral is represented by an area under a process line. The location and shape of constant-pressure and constant-volume process lines relative to the saturation line are important. The $hs$ diagram is particularly helpful in the analysis of steady-flow processes.

The $T\,ds$ equations are valid for any process and are basic relations for the evaluation of the change in specific entropy for any substance. The two important $T\,ds$ equations for simple compressible substances are

$$T\,ds = du + P\,dv \qquad \text{and} \qquad T\,ds = dh - v\,dP$$

These equations can be solved for $ds$ and integrated to find $\Delta s$ if the functional relationships among the properties $u$, $h$, $P$, $V$, and $T$ are known. For an ideal-gas model, these relations are known and the two equations can be integrated to give

$$\Delta s = \int_1^2 c_v \frac{dT}{T} + R \ln \frac{v_2}{v_1} \qquad \text{and} \qquad \Delta s = \int_1^2 c_p \frac{dT}{T} - R \ln \frac{P_2}{P_1}$$

In practice, the change in specific entropy for an ideal gas is usually calculated using information from the ideal-gas tables, where

$$\Delta s = s_2^0 - s_1^0 - R \ln \frac{P_2}{P_1}$$

or by assuming an average specific-heat value, so that

$$\Delta s = c_{v,\text{av}} \ln \frac{T_2}{T_1} + R \ln \frac{v_2}{v_1} \qquad \text{and} \qquad \Delta s = c_{p,\text{av}} \ln \frac{T_2}{T_1} - R \ln \frac{P_2}{P_1}$$

For an incompressible substance model, the equation for $\Delta s$ becomes

$$\Delta s_{\text{inc}} = \int \frac{c\,dT}{T} \approx c_{\text{av}} \ln \frac{T_2}{T_1}$$

where $c = c_v = c_p$.

The entropy balance for a control volume is

$$\frac{dS_{\text{cv}}}{dt} = \sum_{\text{in}} \dot{m}_i s_i - \sum_{\text{out}} \dot{m}_e s_e + \sum_{j=1}^n \frac{\dot{Q}_j}{T_j} + \dot{\sigma}_{\text{cv}}$$

where $\dot{Q}_j$ is measured at the control surface of temperature $T_j$. This equation can be used along with mass and energy balances to study the behavior of various devices. This is the most general form of the entropy balance and should be used as the starting point for most applications.

Using the mass, energy, and entropy balances in conjunction with the $T\,ds$ equations, it can be shown that the work per unit mass for an internally reversible, steady-state device with one inlet and one outlet is

$$w_{\text{sf,rev}} = \int_1^2 v\,dP + \Delta\text{ke} + \Delta\text{pe}$$

This equation can be used independently or in conjunction with an energy analysis of a control volume. Evaluation of the integral requires a functional relationship between $v$ and $P$ for the process. This functional relationship is often modeled as a *polytropic* process using an equation of the form $Pv^n = c$, where $c$ and $n$ are constants.

# PROBLEMS

## GENERAL COMMENT QUESTIONS

7-1C. What is the major restriction when interpreting the area under a process curve on a $Ts$ diagram as a measure of heat transfer?

7-2C. If compressed-liquid data are not available, how can one evaluate fairly accurately the specific entropy of the liquid?

7-3C. The $T\,ds$ equations are derived for an internally reversible process. Explain whether they can be applied as well to a system undergoing an internally irreversible process.

7-4C. The $T\,ds$ equations are derived on the basis of a closed-system energy balance. Explain whether they can be applied as well to a mass passing through a control volume.

7-5C. Why are the specific entropy values of steam and refrigerant 134a, for example, tabulated rather than determined from simple algebraic relations?

7-6C. Explain the significance of the function $s^o$ found in ideal-gas tables.

7-7C. Consider an ideal gas that undergoes two processes between two specified temperatures. In the first process the specific volume is constant, while in the second one the pressure is constant. Explain which process leads to the smaller specific entropy change.

7-8C. What are the three physical effects that affect the rate of change of entropy within a control volume?

7-9C. What is meant by a polytropic process?

## ENTROPY CHANGE AND GENERATION USING TABULAR DATA

7-1. A rigid tank with a volume of 1.0 $m^3$ initially contains water at 30 bars and 600°C. The substance is cooled until its temperature reaches 200°C. The environment, which receives the rejected heat, is at 1.1 bars and 27°C. Determine (a) the final pressure, in bars, (b) the entropy change of the water substance, in kJ/K, and (c) the total entropy production for the overall process, including the heat-transfer region. (d) Is the process reversible, irreversible, or impossible? (e) Sketch the path of the process on a $Ts$ diagram, relative to the saturation line.

7-2. Saturated water vapor in the amount of 0.045 kg at 7 bars is contained in a piston-cylinder assembly. During a constant-pressure process, a paddle wheel within the fluid undergoes 200 rev at a torque of 1.68 N·m. At the same time 5.48 kJ of heat is added from a constant-temperature region at 327°C. The temperature of the surface of the cylinder where heat transfer occurs may be taken to be the average fluid temperature during the process. Determine (*a*) the final temperature of the fluid, in degrees Celsius, (*b*) the entropy change of the fluid, in kJ/K, (*c*) the entropy generation within the fluid, and (*d*) the entropy generation for the composite water vapor plus heat-transfer regions. (*e*) Is the process reversible, irreversible, or impossible? (*f*) Sketch the path of the process on a $Ts$ diagram, relative to the saturation line.

7-3. Refrigerant 134a at 2.8 bars and 60°C is compressed in a closed system to 14 bars. The process is isothermal and internally reversible, and the environmental temperature is 25°C. Determine (*a*) the heat transfer and the work done, in kJ/kg, (*b*) the entropy transfer to or from the fluid, (*c*) the entropy generation in the heat-transfer process, and (*d*) the entropy generation for the overall refrigerant plus heat-transfer regions, in kJ/kg·K. (*e*) Sketch the process on a $Ts$ diagram.

7-4. A piston-cylinder device initially contains refrigerant 134a at 6.0 bars and 80°C. It is compressed quasistatically and at constant pressure, with a boundary-work input of 13.63 kJ/kg. Determine (*a*) the final specific volume, in cm³/g, (*b*) the final specific entropy, in kJ/kg·K, and (*c*) the heat transfer, in kJ/kg. (*d*) If the surrounding temperature is 20°C, determine the total entropy generation within the composite refrigerant and heat-transfer regions, in kJ/kg·K. (*e*) Sketch the process on a $Ts$ diagram. (*f*) Is the process reversible, irreversible, or impossible?

7-5. Two tanks of the same volume are connected by a pipe containing a valve. Initially the valve is closed, one tank is evacuated, and the other tank contains 40 g of steam at 15 bars and 280°C. The valve is opened, and the steam flows into the evacuated tank until pressure equilibrium is reached. During the process heat transfer occurs from a reservoir at 500°C until the steam temperature at the end of the process is 440°C in both tanks. Determine (*a*) the final equilibrium pressure, in bars, (*b*) the heat transfer to the steam, in kilojoules, (*c*) the entropy change of the steam, in kJ/K, and (*d*) the entropy production within the composite steam and heat-transfer regions, in kJ/K.

7-6. Two tanks of the same volume are connected by a pipe containing a valve. Initially the valve is closed, one tank is evacuated, and the other tank contains 200 g of refrigerant 134a at 7 bars and 100°C. The valve is opened, and the refrigerant 134a flows into the evacuated tank until pressure and temperature equilibrium prevails. During the process heat is removed from the refrigerant 134a to the atmosphere at 20°C until the pressure at the end of the process is 3.2 bars in both tanks. Determine (*a*) the final equilibrium temperature, in degrees Celsius, (*b*) the heat transfer from the refrigerant 134a, in kilojoules, (*c*) the entropy change of the refrigerant 134a, in kJ/K, and (*d*) the total entropy generation for the process, in kJ/K.

7-7E. A rigid tank with a volume of 10.0 ft³ initially contains water at 500 psia and 1000°F. The substance is cooled until its temperature reaches 600°F. The environment, which receives the rejected heat, is at 15 psia and 40°F. Determine (a) the final pressure, in psia, (b) the entropy change of the water, in Btu/°R, and (c) the total entropy generation within the composite water and heat-transfer regions. (d) Is the process reversible, irreversible, or impossible? (e) Sketch the path of the process on a $Ts$ diagram, relative to the saturation line.

7-8E. One-tenth pound of saturated water vapor at 100 psia is contained in a piston-cylinder assembly. During a constant-pressure process, a paddle wheel within the fluid undergoes 200 rev at a torque of 1.26 lb$_f$·ft. At the same time 4.54 Btu of heat is added from a thermal reservoir at 500°F. The temperature of the surface of the cylinder where heat transfer occurs may be taken to be the average fluid temperature during the process. Determine (a) the final temperature of the fluid, in degrees Fahrenheit, (b) the entropy change of the fluid, in Btu/°R, (c) an estimate for the entropy generation within the fluid, and (d) the entropy generation of the overall process. (e) Is the process reversible, irreversible, or impossible? (f) Sketch the path of the process on a $Ts$ diagram, relative to the saturation line.

7-9. One-tenth kilogram of water substance, initially at 3 bars and 200°C, is contained within a closed system. A heat transfer of 7.7 kJ to the environment at 20°C and a boundary work input of 17,500 N·m results in a final pressure of 15 bars. Determine (a) the final temperature, in degrees Celsius, (b) the entropy change of the water, in kJ/K, and (c) the total entropy generation within the composite water and heat-transfer regions, in kJ/K.

7-10E. A piston-cylinder device initially contains refrigerant 134a at 50 psia and 120°F. It is compressed quasistatically and at constant pressure, with a boundary-work input of 3520 ft·lb$_f$/lb$_m$. Determine (a) the final specific volume, in ft$^3$/lb$_m$, (b) the final specific entropy, in Btu/lb$_m$·°R, and (c) the heat transfer, in Btu/lb$_m$. (d) If the surrounding temperature is 70°F, determine the total entropy generation for the overall process, in Btu/lb$_m$·°R. (e) Sketch the process on a $Ts$ diagram. (f) Is the process reversible, irreversible, or impossible?

7-11E. Two tanks of the same volume are connected by a pipe containing a valve. Initially the valve is closed, one tank is evacuated, and the other tank contains 0.10 lb$_m$ of steam at 250 psia and 450°F. The valve is opened, and the steam flows into the evacuated tank until pressure equilibrium is reached. During the process heat transfer occurs from a thermal reservoir at 940°F until the steam temperature at the end of the process is 900°F in both tanks. Determine (a) the final equilibrium pressure, in psia, (b) the heat transfer to the steam, in Btu, (c) the entropy change of the steam, in Btu/°R, and (d) the total entropy generation for the overall process, in Btu/°R.

7-12E. Two tanks of the same volume are connected by a pipe containing a valve. Initially the valve is closed, one tank is evacuated, and the other tank contains 0.50 lb of refrigerant 134a at 100 psia and 200°F.

The valve is opened, and the refrigerant 134a flows into the evacuated tank until pressure and temperature equilibrium prevails. During the process heat is removed from the refrigerant 134a to the atmosphere at 60°F until the pressure at the end of the process is 50 psia in both tanks. Determine (a) the final equilibrium temperature, in degrees Fahrenheit, (b) the heat transfer from the refrigerant 134a, in Btu, (c) the entropy change of the refrigerant 134a, in Btu/°R, and (d) the total entropy generation for the overall process, in Btu/°R.

7-13. A rigid vessel contains 0.20 kg of steam at 7 bars and 280°C. Heat is transferred out of the system until the temperature drops to 200°C. The surrounding temperature is 17°C. Find (a) the entropy change of the steam, (b) the total entropy generation for the overall process, in kJ/K, and (c) the entropy generation due to the heat-transfer process. (d) Is the process reversible, irreversible, or impossible? (e) Sketch the process path on a $Ts$ diagram.

7-14E. One-tenth pound of steam at 40 psia and 400°F is compressed isothermally to 200 psia in a piston-cylinder device. During this process 6450 ft·lb$_f$ of work is done on the gas, and heat is lost to the environment, which is at 80°F. Determine (a) the heat transfer, in Btu, (b) the entropy change of the steam, in Btu/°R, (c) the total entropy production for the overall process, in Btu/°R, and (d) the entropy generation for the steam. (e) Is the process reversible, irreversible, or impossible?

7-15E. Two pounds of refrigerant 134a, initially at 40 psia and 80°F, is contained in a piston-cylinder device. The fluid is compressed to 120 psia and 140°F, which requires 22.5 Btu of work input. There is heat transfer also during the process to the surroundings at 70°F, and it leaves the fluid at an average temperature of 110°F. Determine (a) the heat transfer in Btu, (b) the entropy change of the R-134a, in Btu/°R, (c) the entropy generation within the fluid, and (d) the total entropy generation from the overall process. (e) Is the process irreversible or impossible?

7-16. Water substance contained in a rigid tank at 50°C is heated from a thermal reservoir at 400 K until it reaches a pressure of 0.7 bar at 100°C. Find (a) the specific entropy change for the water, (b) the entropy generation associated with heat transfer, in kJ/kg·K, if the average control-surface temperature for heat transfer is the average water temperature, and (c) the entropy production for the overall process. (d) Is the process irreversible or impossible? (e) Sketch the process on a $Pv$ and a $Ts$ diagram.

## ENTROPY CHANGE AND GENERATION FOR IDEAL GASES

7-17. A rigid tank contains 0.10 kg of air initially at 1.0 bar and 300 K. Electrical work is carried out on the air in the amount of 4.80 kJ.

At the same time a heat loss of 0.5 kJ occurs to the environment, which is at 7°C. The average boundary temperature at which heat transfer occurs may be taken as the average air temperature during the process. Use the air table to determine (*a*) the final air temperature, in kelvins, and the final pressure, in bars, (*b*) the entropy change of the air in the tank, in kJ/K, and (*c*) the entropy production within the tank, in kJ/K, and (*d*) the entropy production for the entire process, in kJ/K. (*e*) Is the process reversible, irreversible, or impossible? (*f*) Sketch a $Ts$ diagram for the air change of state.

7-18. It is claimed that an input of 28.0 kJ/kg of compression work to air within a piston-cylinder device will lead to a change of state from 100 kPa, 300 K, to 250 kPa, 330 K. The local environment is at 300 K. Through the use of representative specific-heat data, determine (*a*) the entropy change of the air, in kJ/kg·K, (*b*) the heat transfer, in kJ/kg, and (*c*) the total entropy generation of the overall process, in kJ/kg·K. (*d*) Is the claim valid on the basis of the second law?

7-19. A 214-L rigid tank contains 0.024 kmol of oxygen initially at 37°C. The tank is placed in an environment at a constant temperature of 197°C. The tank remains in this environment until the oxygen reaches 117°C. For this change of state, use the oxygen table to find (*a*) the entropy change of the oxygen, (*b*) the entropy change of the environment, and (*c*) the total entropy production for the overall process, all answers in kJ/K. (*d*) Does the answer to part *c* satisfy the second law? (*e*) Sketch a $Ts$ diagram for the oxygen process.

7-20. Nitrogen, contained in a closed system initially at 6.0 bars and 117°C, undergoes a reversible, constant-temperature process during which 1500 kJ/kmol of heat is transferred from a thermal reservoir at 423°C. Find (*a*) the work occurring during the process, in kJ/kmol, (*b*) the entropy change of the nitrogen, in kJ/kmol·K, (*c*) the total entropy generation for the overall process, and (*d*) the final pressure, in bars. (*e*) Sketch a $Ts$ diagram for the nitrogen part of the process.

7-21. A piston-cylinder device maintained at a constant pressure of 0.1 MPa contains nitrogen gas, which is cooled from 150 to 40°C. Find (*a*) the heat transfer out, in kJ/kg, (*b*) the work done, in kJ/kg, (*c*) the change in entropy of the nitrogen, in kJ/kg·K, and (*d*) the entropy generation for an enlarged system that includes the heat-transfer region, if the temperature of the environment is 22°C, in kJ/kg·K.

7-22. Contained in a constant-pressure closed system is 0.5 kg of hydrogen gas at 6 bars and 17°C. Heat transfer in the amount of 798 kJ is added to the gas from a thermal reservoir at 450 K. Calculate (*a*) the final temperature, in kelvins, (*b*) the entropy change of the hydrogen, in kJ/K, and (*c*) the entropy production within an enlarged system that includes the heat transfer region, in kJ/K. (*d*) Is the process reversible, irreversible, or impossible?

7-23. A rigid tank with a volume of 0.04 m$^3$ contains oxygen at an initial state of 87°C and 1.5 bars. During a process paddle-wheel work is carried out by applying a torque of 13 N·m for 25 revolutions, and a heat loss of 3.74 kJ occurs to the surroundings at 18°C. The temperature of the enclosure where heat transfer occurs is taken as the average of the oxygen temperatures during the process. Determine (a) the final temperature, in kelvins, and (b) the entropy change of the oxygen, (c) the entropy generation within the tank, and (d) the total entropy generation for the overall process, all answers in kJ/K. (e) Is the overall process reversible, irreversible, or impossible?

7-24. Carbon monoxide at 1 bar and 37°C is compressed in a closed system to 3 bars and 147°C. The required work input is 88.0 kJ/kg, and the surroundings temperature is 25°C. The boundary temperature where heat transfer occurs is taken as the average temperature of the CO during the process. Determine (a) the entropy change of the CO, in kJ/kg·K, (b) the heat loss to the surroundings, in kJ/kg, (c) the entropy production within the closed system, and (d) the entropy generation for the overall process, in kJ/kg·K.

7-25E. A piston-cylinder assembly contains 0.50 lb$_m$ of air initially at 14.5 psia and 100°F. Paddle-wheel work is carried out on the air in the amount of 6225 ft·lb$_f$. At the same time a heat loss of 2.0 Btu occurs to the surroundings at 70°F. If the process is one of constant temperature, determine (a) the comp/exp work, in Btu, (b) the ratio of the final to the initial volume, (c) the entropy change of the air in the assembly, in Btu/°R, (d) the entropy production within the air, and (e) the total entropy production for the process, in Btu/°R. (f) Determine whether the process is irreversible or impossible. (g) Sketch a $Ts$ diagram for the air change of state.

7-26E. A 7.5-ft$^3$ rigid tank contains 0.072 lbmol of oxygen initially at 320°F. The tank is placed in an environment at a constant temperature of 60°F. The tank remains in this environment until the oxygen reaches 120°F. For this change of state, use the oxygen table to find (a) the entropy change of the oxygen, in Btu/°R, (b) the heat transfer, in Btu, and (c) the total entropy production for the overall process, including the heat-transfer region, in Btu/°R. (d) Determine whether the overall process is reversible, irreversible, or impossible. (e) Sketch a $Ts$ diagram for the process.

7-27E. Nitrogen, contained in a closed system initially at 90 psia and 240°F, undergoes a reversible, constant-temperature process during which 3000 Btu/lbmol of heat is transferred from a thermal reservoir at 540°F. Find (a) the expansion work occurring during the process, in Btu/lbmol, (b) the entropy change of the nitrogen, in Btu/lbmol·°R, (c) the total entropy production for the process, and (d) the final pressure, in psia. (e) Sketch a $Ts$ diagram for the nitrogen part of the process.

7-28E. A piston-cylinder device maintained at a constant pressure of 1 atm contains nitrogen gas, which is cooled from 300 to 100°F. Find (a) the heat removed, in Btu/lb$_m$, (b) the work done, in Btu/lb$_m$, (c) the change in entropy of the nitrogen, in Btu/lb$_m$·°R, and (d) the total entropy generation for the overall process, if the temperature of the environment is 70°F, in Btu/lb$_m$·°R.

7-29E. Contained in a constant-pressure closed system is 0.5 lb of hydrogen gas at 100 psia and 40°F. Heat transfer in the amount of 520 Btu is added to the gas from a thermal reservoir at 550°F. Calculate (a) the final temperature, in °R, (b) the entropy change of the hydrogen, in Btu/°R, and (c) the entropy production for the overall process, in Btu/°R. (d) Is the process reversible, irreversible, or impossible?

7-30E. A rigid tank with a volume of 1.0 ft$^3$ contains oxygen at an initial state of 200°F and 1.5 atm. During a process paddle-wheel work is carried out by applying a torque of 9.55 lb$_f$·ft for 25 revolutions, and a heat loss of 3.52 Btu occurs to the surroundings at 65°F. The temperature of the enclosure where heat transfer occurs is taken as the average of the oxygen temperatures during the process. Determine (a) the final temperature, in °R, and (b) the entropy change of the oxygen, (c) the entropy generation within the tank, and (d) the total entropy generation for the overall process, all answers in Btu/°R. (e) Is the overall process reversible, irreversible, or impossible?

7-31. A 2-m$^3$ tank contains argon initially at 2 bars and 400 K. Heat transfer occurs between the argon and the environment at 0°C until the argon pressure falls to 1.4 bars. Calculate (a) the final argon temperature, in °C, (b) the entropy change of the argon in kJ/K, (c) the heat transfer, in kilojoules, (d) the total entropy production of the overall process, in kJ/K, and (e) the entropy production due to heat transfer.

7-32. Two kilograms of nitrogen gas is heated in a piston-cylinder apparatus from 0 to 250°C, the pressure remaining constant at 1.013 bar. During the process a heat transfer of 228 kJ occurs to the nitrogen from a constant-temperature heat source held at 300°C. Determine (a) the entropy change of the nitrogen, (b) the entropy change of the heat source, and (c) the total entropy production for the overall process, in kJ/K.

7-33. Air, in the amount of 0.5 kg, in a constant-volume system has an initial state of 0.10 m$^3$ and 27°C. Measurements during the process indicate that 9.50 kJ of energy as heat transfer were removed and paddle-wheel work was carried out by applying a torque of 15.7 N·m to the shaft for 400 revolutions. Find (a) the final temperature of the gas in the tank, in degrees Celsius, (b) the entropy change of the air, in kJ/K, (c) the entropy production within the tank in kJ/K, and (d) the total entropy production for the overall process, if the temperature of the environment to which heat is transferred is 22°C and the boundary temperature of the air is taken as the arithmetic average of its initial and final temperatures.

ENTROPY CHANGE AND GENERATION
FOR INCOMPRESSIBLE SUBSTANCES

7-34. A 50-kg piece of iron at 500 K is thrown into a large lake that is at 285 K. If the average specific heat for iron is 0.45 kJ/kg·K, determine (a) the heat transferred in kilojoules and (b) the total entropy production in kJ/K.

7-35. A kilogram of ice at 0°C is added to 7 kg of water at 30°C and 1 bar. The enthalpy of melting of ice is 335 kJ/kg. If the process is adiabatic, calculate (a) the equilibrium temperature in °C and (b) the entropy production for the process in kJ/K.

7-36. Heat transfer occurs between a 50-kg piece of iron at 500 K and a 100-kg piece of copper initially at 300 K. Eventually the overall insulated system reaches a common temperature. If the iron has an average specific heat of 0.45 kJ/kg·K and that of copper is 0.40 kJ/kg·K, determine (a) the equilibrium temperature in kelvins and (b) the overall entropy production in kJ/K.

7-37E. An ice cube at 32°F and 1 in. on a side is dropped into a 3-in.-diameter cylindrical and insulated glass filled with water at 50°F to a 3-in. depth. Calculate (a) the equilibrium temperature in °F and (b) the total entropy production for the ice cube and water system, in Btu/°R, if the enthalpy of melting is 143 Btu/lb.

7-38. A constant-temperature region at 80°C receives heat transfer from a 3-kg piece of copper that cools from 200 to 100°C. Determine (a) the entropy change of the copper, in kJ/K, (b) the total entropy generation associated with the composite copper and heat-transfer regions, and (c) whether the process is reversible, irreversible, or impossible.

7-39. Fifty kilograms of water at 1 bar and 20°C are mixed with 20 kg of water at 1 bar and 90°C. If the mixing process is adiabatic, determine (a) the equilibrium final temperature, in °C, and (b) the entropy generation for the 70 kg of water mixture, in kJ/K.

7-40. Ten kilograms of water at 20°C are mixed adiabatically at 1 bar with 6 kg of water at 100°C. Determine (a) the equilibrium final temperature, in °C, (b) the entropy generation for the 16 kg of water mixture, in kJ/K, and (c) whether the process is reversible, irreversible, or impossible.

7-41. An 18-kg lead casting at 200°C is quenched in a tank containing 0.03 m³ of water initially at 25°C. Assuming no heat loss to the surroundings, determine (a) the final equilibrium temperature, in kelvins, and (b) the entropy generation for the lead and water system, in kJ/K. (c) Is the process reversible, irreversible, or impossible?

7-42. A 1-kg block of copper at 150°C and a 1-kg block of aluminum at 50°C initially are isolated from each other and from the local surroundings. They are then placed in thermal contact with each other but are still isolated from the surroundings. Determine (*a*) the final equilibrium temperature, in °C, and (*b*) the entropy production for the composite copper and aluminum system, in kJ/K, when they attain thermal equilibrium. (*c*) Is the process reversible, irreversible, or impossible? The specific heat of copper averages 0.385 kJ/kg·K, and for aluminum the value is 0.90 kJ/kg·K.

7-43. A resistor of 40 $\Omega$, initially at 25°C, is maintained at a constant temperature while a current of 5 A is allowed to flow for 5 s. (*a*) Determine the entropy change of the resistor and the entropy production of the overall process, in J/K. (*b*) Now insulate the resistor and carry out the same experiment. In this case, determine the entropy change of the resistor and the entropy production for the overall process, in J/K. The mass of the resistor is 50 g, and the specific heat of the resistor is 1.05 kJ/kg·K. The environmental temperature is 20°C.

7-44. An electric resistor of 30 $\Omega$ is maintained at a constant temperature of 17°C while a current of 6 A is maintained for 3 s. (*a*) Determine the entropy change of the resistor and the entropy generation for the overall process, in J/K. (*b*) Now insulate the resistor and carry out the same experiment. From the latter case, determine the entropy change of the resistor and the entropy generation for the overall process. The mass of the resistor is 19.0 g, and the specific heat of the resistor is 1.10 kJ/kg·K. The environmental temperature is 17°C.

7-45E. A thermal reservoir at 150°F receives heat from a 6-lb$_m$ piece of copper that cools from 390 to 200°F. Determine (*a*) the entropy change of the copper, in Btu/°R, (*b*) the heat transfer, in Btu, (*c*) the total entropy generation within the composite copper and heat-transfer regions, in Btu/°R, and (*d*) whether the process is reversible, irreversible, or impossible.

7-46E. One hundred pounds of water at 1 atm and 60°F are mixed with 40 lb$_m$ of water at 1 atm and 200°F. If the mixing process is adiabatic, determine (*a*) the final equilibrium temperature, in °F, and (*b*) the entropy generation for the 140 lb of water mixture, in Btu/°R.

7-47E. Twenty pounds of water at 180°F are mixed adiabatically at 1 atm with 12 lb$_m$ of water at 60°F. Determine (*a*) the final equilibrium temperature, in °F, (*b*) the entropy generation for the 140 lb$_m$ of mixture, in Btu/°R, and (*c*) whether the process fulfills the increase in entropy principle.

7-48E. A 25-lb$_m$ lead casting at 570°F is quenched in a tank containing 1 ft$^3$ of water initially at 75°F. Assuming no heat loss to the surroundings, determine (*a*) the final equilibrium temperatuere, in °F, and (*b*) the entropy production for the overall process, in Btu/°R. (*c*) Is the process reversible, irreversible, or impossible?

7-49E. A 1-lb$_m$ block of copper at 250°F and a 1-lb$_m$ block of aluminum at 120°F initially are isolated from each other and from the local surroundings. They are then placed in thermal contact with each other but are still isolated from the surroundings. Determine (*a*) the final equilibrium temperature, in °F, and (*b*) the entropy production for the overall process when thermal equilibrium is attained, in Btu/°R. (*c*) Is the increase in entropy principle satisfied? The specific heat of copper averages 0.093 Btu/lb$_m$·°R, and for aluminum the value is 0.215 Btu/lb$_m$·°R.

7-50E. A resistor of 40 $\Omega$, initially at 80°F, is maintained at a constant temperature while a current of 5 A is allowed to flow for 2 s. (*a*) Determine the entropy change of the resistor and the entropy generation for the overall process, in Btu/°R. (*b*) Now insulate the resistor and carry out the same experiment. For the latter case, determine the entropy change of the resistor and the entropy generation for the overall process, in Btu/°R. The mass of the resistor is 0.040 lb and the specific heat of the resistor is 0.55 Btu/lb$_m$·°R. The environmental temperature is 80°F.

7-51E. An electric resistor of 30 $\Omega$ is maintained at a constant temperature of 70°F while a current of 6 A is maintained for 3 s. (*a*) Determine the entropy change for the resistor and the entropy generation for the overall process, in Btu/°R. (*b*) Now insulate the resistor and carry out the same experiment. For the latter case, determine the entropy change of the resistor and the entropy generation for the overall process, in Btu/°R. The mass of the resistor is 0.032 lb$_m$, and the specific heat of the resistor is 0.62 Btu/lb$_m$·°R. The environmental temperature is 70°F.

## ENTROPY CHANGE AND GENERATION IN STEADY-FLOW PROCESSES

7-52. Nitrogen at 2.910 kg/s flows steadily through a horizontal pipe that is uninsulated. At the inlet the conditions are 2.0 MPa, 220 K, and 10 m/s through a diameter of 11 cm. At the outlet the conditions become 0.2 MPa, 280 K, and 100 m/s through a 14-cm-diameter pipe. The pipe is surrounded by the atmosphere at 300 K. Determine (*a*) the heat-transfer rate, in kJ/s, (*b*) the rate of entropy change of the nitrogen, in kJ/K·s, (*c*) the rate of entropy generation within the composite flow and heat-transfer regions, and (*d*) whether the process is reversible, irreversible, or impossible.

7-53. Air enters a turbine at 6 bars and 277°C and leaves at 1 bar. The flow rate is 50 kg/min, and the power output is 180 kW. If the heat removed, which appears in the environment at 22°C, is 28.5 kJ/kg, find (*a*) the final temperature, in degrees Celsius, (*b*) the entropy change of the air, in kJ/kg·K, and (*c*) the entropy generation within the composite flow and heat-transfer regions, in kJ/kg·K.

7-54. A steam turbine operates with an inlet condition of 100 bars and 520°C and an outlet state of 1 bar and 100°C. The device produces 2010 kW of power with a mass flow rate of 10,000 kg/h. The atmospheric temperature is 27°C. Determine the entropy generation within the composite flow and heat-transfer regions, in kJ/kg·K.

7-55. Steam enters a steady-flow compression process at 1.0 bar, 100°C, and exits at 10.0 bars, 200°C. The work input is measured to be 400 kJ/kg, and the kinetic- and potential-energy changes are negligible. The environment has a temperature of 27°C. Determine (a) the magnitude and direction of any heat transfer, in kJ/kg, (b) the entropy change of the fluid passing through the compressor, in kJ/kg·K, and (c) the total entropy generation for the overall process, in kJ/kg·K.

7-56. Refrigerant 134a initially at 6 bars and a quality of 50 percent flows through one side of a heat exchanger. As it passes through the equipment at constant pressure, it becomes a saturated vapor. Air initially at 1.10 bars and 42°C passes through the other side at 10 kg/min, and it emerges at 1.05 bars and 22°C. Determine (a) the mass flow rate of refrigerant, in kg/min, (b) the specific entropy change of the refrigerant, in kJ/kg·K, (c) the specific entropy change of the air stream, in kJ/kg·K, and (d) the entropy production for the overall process, in kJ/K·min.

7-57. Superheated steam enters one section of an open feedwater heater at 5 bars and 240°C. Compressed liquid water at the same pressure and 35°C enters another inlet. The mixture of these two streams leaves as a saturated liquid at the same pressure. If the heater is adiabatic, calculate the entropy production for the process per kilogram of mixture.

7-58. A well-insulated diffuser is supplied with carbon dioxide at 110 kPa, 300 K, and 300 m/s. It is claimed that this diffuser will deliver the gas at 240 kPa and 52 m/s. On the basis of Table A-9 determine (a) the exit temperature, in kelvins, (b) the entropy generation for the gas, in kJ/kmol·K, and (c) whether the process is reversible, irreversible, or impossible.

7-59. Air enters a compressor at 1 bar and 27°C at a rate of 2 kg/min and leaves at 5.8 bars and 227°C. The power required to operate the compressor is 7.12 kW. The average temperature of the turbine surface where heat transfer occurs may be taken to be the average air temperature at the inlet and exit, and the surrounding temperature is 22°C. Determine (a) the rate of heat transfer, in kJ/h, (b) the entropy change of the air, in kJ/K·min, (c) the entropy production within the compressor, and (d) the total entropy production for the composite flow and heat-transfer region, in kJ/K·min. (e) Is the process irreversible or impossible?

7-60. Oxygen is throttled from 2 bars and 600 K to 1.2 bars. Find the entropy generation for the gas, in kJ/kg·K.

7-61. Saturated-liquid refrigerant 134a at 0°C enters a constant-diameter pipe with a velocity of 10 m/s. Heat is transferred so that the fluid leaves the pipe with a specific volume of 0.01481 m³/kg at 20°C. If the pipe diameter is 0.675 cm and the surroundings are at 50°C, determine (*a*) the pressure or quality of the fluid leaving the pipe, (*b*) the velocity at the exit, in m/s, (*c*) the rate of heat transfer, in kJ/s, (*d*) the entropy change of the fluid, in kJ/kg·K, and (*e*) the total entropy generation for the overall process. (*f*) Is the process irreversible or impossible?

7-62. Air enters a nozzle at 3 bars, 127°C, and 30 m/s. During its passage a heat loss of 4.40 kJ/kg occurs, and the fluid leaves at 2 bars and 270 m/s. The heat loss appears in the environment at 17°C, and the average nozzle surface temperature is 380 K. Determine (*a*) the final temperature, in degrees Celsius, (*b*) the entropy change of the air, in kJ/kg·K, (*c*) the entropy production within the nozzle, and (*d*) the entropy production for the overall process. (*e*) Sketch the process on a *Ts* diagram.

7-63. Refrigerant 134a enters a compressor at 2.8 bars and 20°C at a rate of 0.10 kg/s. A power input of 4 kW leads to a final state of 14 bars and 70°C. The surrounding temperature is 30°C and the average compressor surface temperature is 45°C. Determine (*a*) the entropy change of R-134a, in kJ/kg·K, (*b*) the rate of heat transfer to or from the device, in kJ/s, (*c*) the entropy generation for the refrigerant, in kJ/kg·K, and (*d*) the total rate of entropy generation for the overall process, in kJ/K·s.

7-64. Steam enters a turbine at 60 bars and 400°C with a velocity of 180 m/s. Saturated vapor at 100°C leaves the device with a velocity of 100 m/s. The measured work output is 490 kJ/kg. Heat transfer between the turbine and the environment at 300 K occurs at an average outer surface temperature of 550 K for the turbine. Determine (*a*) the heat transfer in kJ/kg, (*b*) the entropy generation within the fluid, in kJ/kg·K, and (*c*) the entropy generation associated with the heat transfer. (*d*) Sketch the path of the process on a *Ts* diagram.

7-65. Monatomic argon flows through an adiabatic nozzle at 2 kg/s. The inlet state is 500 K and 4 bars, and the outlet state is 405 K and 325 m/s. If the inlet to outlet area ratio is 2, determine (*a*) the exit pressure, in bars, and (*b*) the entropy generation within the nozzle, in kJ/kg·K.

7-66. Refrigerant 134a enters a heavily insulated heat exchanger as a saturated liquid at 0°C and a rate of 5 kg/min. Air, which enters at 5 bars, 127°C, and 1 kg/min is cooled to 27°C at constant pressure. Determine (*a*) the exit temperature of the refrigerant stream, in °C, and (*b*) the entropy generation for the overall process in kJ/K·min.

7-67. At one inlet to an adiabatic mixing chamber maintained at 800 kPa steam enters with a quality of 90 percent. At the second inlet water enters at 30°C. At the exit the temperature is 150°C and mass flow rate is 2 kg/s. Determine the rate of entropy production for the mixing process, in kJ/K·s.

7-68E. A refrigerant-134a compressor operates with an inlet condition of saturated vapor at 10°F and an outlet state of 100 psia and 120°F. The device requires 2.42 hp of power input with a mass flow rate of 5 lb/min. The atmospheric temperature is 60°F. Determine the total entropy production for the overall process, in Btu/lb$_m$·°R.

7-69E. Steam enters a steady-flow compression process at 14.7 psia, 250°F, and exits at 160 psia, 400°F. The work input is measured to be 200 Btu/lb$_m$, and the kinetic- and potential-energy changes are negligible. The environment has a temperature of 70°F. Determine (a) the magnitude and direction of any heat transfer, in Btu/lb$_m$, (b) the entropy change of the fluid passing through the compressor, in Btu/lb$_m$·°R, and (c) the total entropy production for the overall process, in Btu/lb$_m$·°R.

7-70E. A well-insulated diffuser is supplied with carbon dioxide at 16 psia, 60°F, and 920 ft/s. It is claimed that this diffuser will deliver the gas at 35 psia and 165 ft/s. On the basis of Table A-9, determine (a) the exit temperature, in degrees Fahrenheit, (b) the entropy generation for the gas, in Btu/lbmol·°R, and (c) whether the process is reversible, irreversible, or impossible.

7-71E. Air enters a nozzle at 40 psia, 240°F, and 100 ft/s. During its passage a heat loss of 1.90 Btu/lb$_m$ occurs, and the fluid leaves at 30 psia and 800 ft/s. The heat loss appears in the environment at 80°F. Determine (a) the final temperature, in degrees Fahrenheit, (b) the entropy change of the air, in Btu/lb$_m$·°R, (c) the total entropy generation of the overall process, and (d) whether the process is reversible, irreversible, or impossible. (e) Sketch the process on a $Ts$ diagram.

7-72E. Oxygen at 1.952 lb$_m$/s flows steadily through a horizontal pipe that is uninsulated. At the inlet the conditions are 100 psia, 540°F, and 300 ft/s through a diameter of 2 in. At the outlet the conditions become 10 psia, 40°F, and 10 ft/s through a 24.5-in.-diameter pipe. The pipe is surrounded by the atmosphere at 40°F. Determine (a) the heat-transfer rate, in Btu/s, (b) the rate of entropy change of the nitrogen, in Btu/°R·s, (c) the rate of entropy production for the overall process, and (d) whether the process is reversible, irreversible, or impossible.

7-73E. Air enters a turbine at 90 psia and 540°F and leaves at 15 psia. The flow rate is 100 lb$_m$/min, and the power output is 240 hp. If the heat removed, which appears in the environment at 70°F, is 14.6 Btu/lb$_m$, find (a) the final temperature, in degrees Fahrenheit, (b) the entropy change of the air, in Btu/lb$_m$·°R, and (c) the total entropy generation for the overall process, in Btu/lb$_m$·°R.

7-74E. Air enters a steady-state compressor at 15 psia, 40°F, and 200 ft/s at a rate of 2 lb$_m$/s, and leaves at 30 psia, 160°F, and 400 ft/s. The power required to operate the compressor is 111 hp. Determine (a) the rate of heat transfer, in Btu/h, (b) the entropy change of the air, in Btu/min·°R, and (c) the total entropy generation of the overall process if heat transfer occurs to the environment at 50°F, in Btu/min·°R. (d) Is the process reversible, irreversible, or impossible?

7-75E. Refrigerant 134a enters a steady-flow compressor at 40 psia and 140°F at a rate of 0.10 lb$_m$/s. It is claimed that a power input of 934 ft·lb$_f$ will lead to a final state of 200 psia and 140°F. Determine (a) the change in entropy of refrigerant 134a as it passes through the compressor, in Btu/lb$_m$·°R, (b) the rate of heat transfer to or from the device, in Btu/s, (c) the total entropy generation for the process, and (d) whether the process is reversible, irreversible, or impossible, if the surrounding temperature is 100°F.

7-76E. Refrigerant 134a as a saturated vapor at 0°F enters a constant-diameter pipe with a velocity of 60 ft/s. Heat transfer occurs so that the fluid leaves the pipe with a specific volume of 1.352 ft$^3$/lb$_m$ at −10°F. If the pipe diameter is 2.22 in and the surroundings are at −35°F, determine (a) the pressure or quality of the refrigerant 134a leaving the pipe, (b) the velocity at the exit, in ft/s, (c) the rate of heat transfer, in Btu/s, (d) the entropy change of the fluid, in Btu/lb$_m$·°R, and (e) the overall entropy generation, in Btu/lb$_m$·°R. (f) Is the process reversible, irreversible, or impossible?

7-77E. Air enters an insulated heat exchanger at 50 psia, 0°F, at 0.10 lb$_m$/s, and it is heated to 160°F. Refrigerant 134a enters the other side as saturated vapor at 300 psia and a rate of 0.50 lb$_m$/s. Neglecting pressure drops, determine (a) the condition of the refrigerant at the exit and its temperature, (b) the entropy change of the air, (c) the entropy change of the refrigerant, and (d) the overall entropy generation, all answers in Btu/°R·s. (e) Does the process violate the second law?

7-78E. Superheated steam enters one section of an open feedwater heater at 60 psia and 300°F. Compressed liquid water at the same pressure and 100°F enters another inlet. The mixture of these two streams leaves as a saturated liquid at the same pressure. If the heater is adiabatic, calculate the entropy generation for the process, in Btu/°R, per pound of superheated steam entering the heater.

7-79E. Refrigerant 134a enters a compressor at 40 psia and 100°F at a rate of 0.10 lb$_m$/s. It is claimed that a power input of 934 ft·lb$_f$ will lead to a final state of 200 psia and 160°F. The surrounding temperature is 80°F and the average compressor surface temperature is 130°F. Determine (a) the entropy change of R-134a, in Btu/lb$_m$·°R, (b) the rate of heat transfer to or from the device, in Btu/s, (c) the entropy generation for the refrigerant, in Btu/lb$_m$·°R, and (d) the total rate of entropy generation for the overall process, in Btu/°R·min. (e) Is the overall process reversible, irreversible, or impossible?

7-80. Air, liquid water, and glycerin separately pass through a complex steady-flow process at rates of 8 kg/s, 8 L/min, and 2.8 kg/s, respectively. The inlet and exit temperatures for air are 22 and 30°C, those for water are 22 and 50°C, and those for glycerin are 22 and 10°C.

The rate of heat loss is 30 kW, shaft power from the process is 80 kW, and electrical power input is also required. The container surrounding the equipment is at 25°C. Determine (*a*) the electrical power, in kilowatts, and (*b*) the rate of entropy generation for the overall process, in kJ/K·min.

## POLYTROPIC PROCESSES AND REVERSIBLE, STEADY-FLOW WORK

7-81. Air is compressed polytropically and reversibly in steady flow from 100 to 500 kPa. The initial temperature is 300 K and *n* is 1.28. Determine (*a*) the work required, and (*b*) the heat transfer, in kJ/kg. Finally, (*c*) find the total entropy generation if the environmental temperature is 300 K.

7-82. Nitrogen gas is expanded polytropically through a turbine from 0.4 to 0.1 MPa. The initial temperature is 400 K and $n = 1.30$. Determine (*a*) the steady-flow work output and (*b*) the heat transfer, in kJ/kg.

7-83. Carbon dioxide is expanded in steady flow through a turbine from 0.9 to 0.1 MPa. The initial temperature is 587°C, and the expansion process follows the relation $Pv^2 = 0.0294$ MPa·m$^6$/kg$^2$.
(*a*) If the process is frictionless, determine the shaft work, in kJ/kg.
(*b*) Determine the final temperature for the ideal gas, in degrees Celsius.
(*c*) Determine the magnitude and direction of any heat transfer, in kJ/kg, using Table A-9.
(*d*) Sketch the path of the process on a $Pv$ diagram and indicate the area representation of work.

7-84. Argon gas is compressed in steady flow from an initial state of 1 bar and 63°C to 5 bars. The process follows the relation $P = av + b$, where $a = -10$ bar·kg/m$^3$ and $b = 8$ bars.
(*a*) If the process is frictionless, determine the work required, in kJ/kg.
(*b*) Determine the final temperature for the ideal gas, in degrees Celsius.
(*c*) Determine the magnitude and direction of any heat transfer, in kJ/kg.
(*d*) Sketch the path of the process on a $Pv$ diagram, and indicate the area representation of the work.

7-85. Carbon monoxide flows frictionlessly and isothermally through a 75-cm$^2$ constant-area duct. Upstream the pressure is 1.6 bars, and the temperature is 100°C. After flowing against an impeller that produces shaft work, the gas reaches a downstream position where the pressure is 1 bar. If the mass flow rate is 0.9 kg/s, determine (*a*) the shaft work and (*b*) the heat transfer, both in kJ/kg. Finally, (*c*) find the total entropy generation if the heat is supplied from a thermal reservoir at 200°C.

7-86. Air is compressed isothermally from 96 kPa and 7°C to 480 kPa. Flow through the compressor is steady at 0.95 kg/s. Kinetic and potential energies are negligible. Calculate (a) the power input, in kilowatts, and (b) the rate of heat removal, in kJ/s, if the process is frictionless. Finally, (c) find the total entropy generation if the environment is at 7°C, in kJ/kg·K.

7-87E. Carbon dioxide is expanded in steady flow through a turbine from 162 to 18 psia. The initial temperature is 1040°F, and the expansion process follows the relation $Pv^2 = 826$ psia·ft$^6$/lb$_m^2$.
  (a) For a frictionless process, determine the shaft work, in ft·lb$_f$/lb$_m$.
  (b) Determine the final temperature for the ideal gas in degrees Fahrenheit.
  (c) Determine the magnitude and direction of any heat transfer, in Btu/lb$_m$, using Table A-9E.
  (d) Sketch the path of the process on a $Pv$ diagram and indicate the area representation of work.

7-88E. Nitrogen gas is compressed in steady flow from an initial state of 20 psia and 520°R to 80 psia. The process follows the relation $P = av + b$, where $a = -10$ psia·lb$_m$/ft$^3$ and $b = 120$ psia.
  (a) If the process is frictionless, determine the work required, in ft·lb$_f$/lb$_m$.
  (b) Determine the final temperature of the ideal gas in degrees Fahrenheit.
  (c) Determine the magnitude and direction of any heat transfer, in Btu/lb$_m$, using data in Table A-6E.
  (d) Sketch the path of the process on a $Pv$ diagram, and indicate the area representation of the work.

7-89E. Carbon monoxide flows frictionlessly and isothermally through a constant-area duct of 12 in.$^2$. Upstream the pressure is 20 psia, and the temperature is 200°F. After flowing against an impeller which produces shaft work, the gas reaches a downstream position where the pressure is 15 psia. If the mass flow rate is 1.95 lb$_m$/s, determine (a) the shaft work and (b) the heat transfer, both in ft·lb$_f$/lb$_m$. Finally, (c) find the total entropy generation, in Btu/lb$_m$·°R, if the heat transfer is supplied from a thermal reservoir at 400°F.

7-90E. Air is compressed isothermally from 14 psia and 40°F to 70 psia. Flow through the compressor is steady at 2.0 lb$_m$/s. Kinetic and potential energies are negligible. Calculate (a) the power input and (b) the rate of heat removal, both in Btu/s, if the process is frictionless. The inlet area is 40 in$^2$ and the exit area is 30 in$^2$. (c) Find the total entropy generation, in Btu/lb$_m$·°R, if the temperature of the environment is 40°F.

# CHAPTER

# 8

# SOME CONSEQUENCES
# OF THE SECOND LAW

Generators connected to hydraulic turbines in the powerhouse of Ocoee Dam No. 2 can produce 18,000 kW of electricity. (Courtesy of TVA)

The preceding chapter presented the methods for evaluating the entropy change for various classes of substances. These methods were then used to determine the entropy generation for both closed and open systems. In this chapter, special attention is directed at some consequences of the second law for adiabatic processes. In addition, a special internally reversible heat-engine cycle called a Carnot cycle is described.

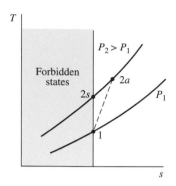

**Figure 8-1**

Isentropic modeling of an actual process involving an increase in pressure.

**Note that isentropic processes are standards for real processes.**

## 8-1    ISENTROPIC PROCESSES

An internally reversible process has a special significance to engineers. Because it is free of dissipative and nonequilibrium effects, the internally reversible process can be used as a standard against which all real (irreversible) processes may be compared. The entropy change of any fixed mass undergoing an internally reversible process is given by the expression $dS = (\delta Q/T)_{\text{int rev}}$. Therefore, the entropy change of a fixed mass in an adiabatic, internally reversible process must be zero. When the entropy of a substance does not change during a process, the process is called **isentropic** (constant-entropy). Many devices of engineering interest are essentially adiabatic. Hence an isentropic process is used as an idealized model to which a real adiabatic process can be compared.

The $Ts$ diagram in Fig. 8-1 illustrates the modeling of a real adiabatic process that involves an increase in pressure. If an internally reversible process occurs adiabatically, the final state $2s$ lies directly above the initial state 1, and the process is isentropic. However, if irreversibilities are present, the actual final state $2a$ must lie to the right of state 1 on the $P_2$ line; that is, $s_2 > s_1$. This increase in entropy is dictated by the second-law statement $dS_{\text{adia}} = \delta\sigma \geq 0$. (In this discussion and others later in the text, the subscript $s$ stands for an isentropic final state, while the subscript $a$ represents an actual final state that occurs after an irreversible process.) The position of state $2a$ on line $P_2$ depends on the extent of the irreversibilities, that is, on the magnitude of $\sigma$. State $2a$ cannot lie to the left of the vertical line through state 1, since all states to the left of this line are forbidden states for an adiabatic process. Consequently, the isentropic process is the *limiting* process as irreversibilities are reduced to zero under adiabatic conditions. As a limiting condition, the isentropic process is a standard of performance against which real adiabatic processes may be compared. Consequently, it is useful to study how $P$, $T$, and $v$ change during an isentropic process. Special isentropic relationships for various classes of substances are developed below.

### 8-1-1    ISENTROPIC RELATIONS FOR AN IDEAL GAS

In Sec. 7-3 the entropy change of an ideal gas was evaluated either by accurately accounting for the variation of the specific heats with temperature or by using constant or average specific-heat data. The same approach may be applied to the development of *isentropic* relations for an ideal gas.

**Use of Variable Specific-Heat Data.**    To attain fairly high accuracy in an isentropic calculation involving an ideal gas, it is necessary to account for the variation of the specific heats with temperature. The most direct way to accomplish this is through the use of Eq. [7-21], namely,

$$\Delta s = s_2^o - s_1^o - R \ln \frac{P_2}{P_1} \qquad \text{[7-21]}$$

For an *isentropic process for an ideal gas,* this equation reduces to

$$s_2^o - s_1^o - R \ln \frac{P_2}{P_1} = 0 \qquad \text{isentropic, ideal gas} \qquad \text{[8-1]}$$

where $s^o$ values are solely a function of temperature. Thus Eq. [8-1] is a relationship among the variables $T_1, T_2, P_1$, and $P_2$. When three of the variables are known for an isentropic process, the fourth can be found.

The use of $s^o$ data, in conjunction with Eq. [8-1], provides a general method of evaluating isentropic changes for ideal gases. In the case of the air table in the Appendix, the method has been extended one step further. A rearrangement of Eq. [8-1] leads to the format

$$\left(\frac{P_2}{P_1}\right)_s = \exp \frac{s_2^o - s_1^o}{R} = \frac{\exp (s_2^o/R)}{\exp (s_1^o/R)} \qquad \text{[8-2]}$$

where the subscript $s$ on the pressure ratio emphasizes the isentropic restriction. A new function called the ***relative pressure*** $p_r$ is defined for a given state as

$$p_r \equiv \exp (s^o/R) \qquad \text{[8-3]}$$

Note that the function $p_r$ is *a function of temperature only,* since $s^o$ is solely a function of temperature. As a result of the above definition of $p_r$, Eq. [8-2] becomes

$$\left(\frac{P_2}{P_1}\right)_s = \frac{p_{r2}}{p_{r1}} \qquad \text{isentropic, ideal gas} \qquad \text{[8-4]}$$

The values of the relative pressure $p_r$ are tabulated as a function of temperature in the air tables A-5 and A-5E. [Because values of $p_r$ determined from Eq. [8-3] are very large, they are all multiplied by a scaling factor to give a more convenient range of values in a table.] Because $p_r$ data are derived from $s^o$ data, use of either Eq. [8-1] or Eq. [8-4] should lead to the same answer within the accuracy of the air table.

In some cases it is necessary to work with volume data instead of pressure data as input for isentropic processes. Such a problem can be solved through the use of Eq. [8-1], but it involves iteration. A simpler method is formulated by noting that

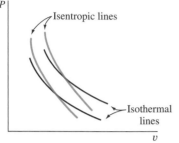

**Figure 8-2**
Isentropic and isothermal lines for an ideal gas on a pressure-volume diagram.

**Note the relative position of isothermal and isentropic lines on a Pv diagram.**

$$\left(\frac{v_2}{v_1}\right)_s = \frac{P_1 T_2}{P_2 T_1} = \frac{p_{r1} T_2}{p_{r2} T_1} = \frac{T_2}{p_{r2}} \frac{p_{r1}}{T_1}$$

Note that the quantity $T/p_r$ is solely a function of temperature. Then a new property called the **relative volume** $v_r$ is defined by the equation $v_r \equiv CT/p_r$, where $C$ is an arbitrary scaling factor. Consequently,

$$\boxed{\left(\frac{v_2}{v_1}\right)_s = \frac{v_{r2}}{v_{r1}}} \qquad \text{isentropic, ideal gas} \qquad \textbf{[8-5]}$$

for an *isentropic process involving an ideal gas*. The air tables A-5 and A-5E also list $v_r$ data for air. (Relative pressure and relative volume data also appear in the literature for a limited number of other ideal gases.)

Note that the above set of relations is valid for any fixed mass of an ideal gas that undergoes a process at constant entropy. This mass may be in a closed system or flowing through a steady-state control volume. Isothermal and isentropic lines are plotted on a $Pv$ diagram for comparative purposes in Fig. 8-2. Isentropic lines have a larger negative slope than isothermal lines through the same state point on a $Pv$ diagram for ideal gases.

---

**Example 8-1**

**A**ir enters a steady-flow adiabatic compressor at 17°C and is compressed through a pressure ratio of 8.6 : 1. If the compressor is assumed to be internally reversible, and air is an ideal gas, determine the work input required, in kJ/kg, using both $s^o$ and $p_r$ data.

**Solution:**

**Given:**   Adiabatic, internally reversible compression of air, as shown in Fig. 8-3.

**Find:**   $w$, in kJ/kg, using $s^o$ and $p_r$ data.

**Model:**   Internally reversible, steady-state process. Ideal gas, neglect kinetic- and potential-energy changes.

**Strategy:**   Apply mass, energy, and entropy balances along with appropriate property relations.

**Analysis:**   The control volume under study is shown by the dashed line in Fig. 8-3. The energy and entropy balances for a steady-state, one-inlet, one-outlet control volume are

$$0 = q + w + (h_1 - h_2) - \Delta\text{ke} - \Delta\text{pe}$$

$$0 = \sum_{j=1}^{n} \frac{q_j}{T_j} + s_1 - s_2 + \sigma_m$$

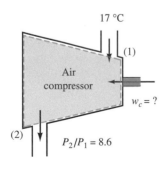

**Figure 8-3**
Equipment schematic and data for Example 8-1.

For an adiabatic compressor with negligible kinetic- and potential-energy changes, these equations simplify further to

$$w = h_2 - h_1 \quad \text{and} \quad 0 = s_2 - s_1$$

If air is modeled as an ideal gas, the enthalpy values are a function only of temperature. $T_1$ is known, but $T_2$ needs to be found. The final temperature is found by using the equation for the entropy change of an ideal gas, $\Delta s = s_2 - s_1 = s_2^o - s_1^o - R \ln P_2/P_1$. Since $\Delta s = 0$, and $s_1^o$ depends on $T_1$, then

$$s_2^o = s_1^o + R \ln \frac{P_2}{P_1}$$

In Table A-5 the value of $s_1^o$ at 290 K is found to be 1.66802 kJ/kg·K. Substitution of the known data into the above equation yields

$$s_2^o = \left(1.66802 + \frac{8.314}{28.97} \ln 8.6\right) \text{kJ/kg·K} = 2.286 \text{ kJ/kg·K}$$

In Table A-5, this value of $s_2^o$ corresponds to a temperature $T_{2s}$ of 533 K by linear interpolation. Further interpolation at 533 K shows that $h_{2s}$ is 536.8 kJ/kg; $h_1$ at 290 K is 290.16 kJ/kg. Hence the compressor work $w_c$ is

$$w_c = h_{2s} - h_1 = (536.8 - 290.2) \text{ kJ/kg} = 246.6 \text{ kJ/kg}$$

A second method of evaluation involves Eq. [8-4], $(P_2/P_1)_s = p_{r2}/p_{r1}$, and $p_r$ data. In Table A-5, $p_{r1}$ is found to be 1.2311. Then the final state is found by

$$p_{r2} = p_{r1} \frac{P_2}{P_1} = 1.2311(8.6) = 10.59$$

In Table A-5, $T_{2s}$ is found by linear interpolation to be 533 K, which agrees with the previous calculation using $s^o$ data. As a result, the work input again is 246.6 kJ/kg. The path of the isentropic process is shown on $Ts$ and $Pv$ diagrams in Fig. 8-4.

**Comment:** Although $s$ is a constant for an isentropic process, the mere fact that $s_2 = s_1$ is not sufficient to ensure that a process is isentropic. The assumptions of internally reversible and adiabatic *are sufficient* to ensure that a process is isentropic.

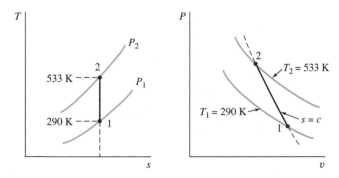

**Figure 8-4**     Ts and Pv process diagrams for Example 8-1.

**Use of Constant or Average Specific-Heat Data.** For some processes involving ideal gases, it is appropriate to assume that the specific heats are constant or that an average value may be used. In the development of isentropic relations under this condition, it is useful to introduce another intrinsic property, the **specific-heat ratio k.** This ratio is defined as

$$k \equiv \frac{c_p}{c_v} \qquad \textbf{[8-6]}$$

The value of $k$ lies in the narrow range between 1 and 1.67 for most substances. Many common diatomic gases have a specific-heat ratio of, roughly, 1.4 at room temperature (see Tables A-3 and A-3E). The specific-heat ratio is related to $c_v$ and $c_p$ by the following relations:

$$c_v = \frac{R}{k-1} \qquad \text{ideal gas} \qquad \textbf{[8-7]}$$

$$c_p = \frac{Rk}{k-1} \qquad \text{ideal gas} \qquad \textbf{[8-8]}$$

As noted, these equations are restricted to ideal gases.

In Sec. 7-3 the $T\,dS$ equations are applied to ideal gases. In the case where the specific heats are assumed constant or an average value over the given temperature interval is used, the following equations are valid:

$$\Delta s = c_{v,\text{av}} \ln \frac{T_2}{T_1} + R \ln \frac{v_2}{v_1} \qquad \textbf{[7-11]}$$

$$\Delta s = c_{p,\text{av}} \ln \frac{T_2}{T_1} - R \ln \frac{P_2}{P_1} \qquad \textbf{[7-12]}$$

For an isentropic process, $\Delta s = 0$. If Eqs. [7-11] and [7-12] are set equal to zero, the following relations result:

$$\boxed{\frac{T_2}{T_1} = \left(\frac{v_1}{v_2}\right)^{k-1}} \qquad \text{isentropic, ideal gas} \qquad \textbf{[8-9a]}$$

$$\boxed{\frac{T_2}{T_1} = \left(\frac{P_2}{P_1}\right)^{(k-1)/k}} \qquad \text{isentropic, ideal gas} \qquad \textbf{[8-10a]}$$

The specific-heat ratio $k$ appears in the above equation through the use of Eqs. [8-7] and [8-8]. If Eq. [8-9a] is substituted into Eq. [8-10a] so that $T$ is eliminated as a variable, then

$$\boxed{\frac{P_2}{P_1} = \left(\frac{v_1}{v_2}\right)^{k}} \qquad \text{isentropic, ideal gas} \qquad \textbf{[8-11a]}$$

Another way of expressing these three isentropic relationships among $P$, $v$, and $T$ is

$$T(v)^{k-1} = \text{constant} \qquad \textbf{[8-9b]}$$

$$T^k P^{1-k} = \text{constant} \qquad \textbf{[8-10b]}$$

$$Pv^k = \text{constant} \qquad \textbf{[8-11b]}$$

Equations [8-9] to [8-11] represent *process* equations for the *isentropic* change of state for *ideal gases with constant or average specific heats*. The equations are fairly accurate when the temperature change during a process does not exceed a few hundred degrees. Again, the above sets of equations are valid for ideal gases in a closed system or flowing through a steady-state control volume.

Note that isentropic property relations are valid for both closed and open systems.

For turbine and compressors with isentropic flow and negligible $\Delta ke$ and $\Delta pe$ values, the steady-state energy balance $q + w = \Delta h + \Delta ke + \Delta pe$ reduces to $w_{sf} = \Delta h$. An alternative equation for $w_{sf}$ may be developed from the relations for polytropic processes presented in Sec. 7-6-2. An isentropic process is simply a polytropic process for which $n = k$. Therefore, on the basis of Eq. [7-37],

$$w_{sf,\text{isen}} = \frac{k(P_2 v_2 - P_1 v_1)}{k-1} = \frac{kRT_1}{k-1}\left[\left(\frac{P_2}{P_1}\right)^{(k-1)/k} - 1\right] \quad \textbf{[8-12]}$$

where $\Delta ke = \Delta pe = 0$. For an isentropic process of an ideal gas within a piston-cylinder device, Eq. [7-36] for $P\,dv$ work becomes

$$w_{\text{comp/exp,isen}} = \frac{P_2 v_2 - P_1 v_1}{k-1} = \frac{R(T_2 - T_1)}{k-1} \quad \textbf{[8-13]}$$

In either case, an appropriate average $k$ value must be chosen.

---

**A**ir enters a steady-flow adiabatic compressor at 17°C and is compressed through a pressure ratio of 8.6:1. If the process is assumed to be internally reversible, determine the work input required, in kJ/kg, using specific-heat data. Compare the answer to that of Example 8-1, where $s°$ and $p_r$ data are used.

**Example 8-2**

**Solution:**

**Given:** Steady-flow compression of air, as shown in Fig. 8-5.

**Find:** $w$, in kJ/kg, using specific-heat data.

**Model:** Adiabatic, internally reversible, steady-state. Ideal gas, neglect $\Delta ke$ and $\Delta pe$, $c_v$ and $c_p$ are constant.

**Strategy:** Apply control-volume energy and entropy balances along with appropriate property relations.

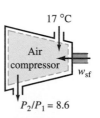

**Figure 8-5**
Equipment schematic and data for Example 8-2.

**Analysis:**   This problem is identical to Example 8-1, except for how the ideal-gas model is implemented. From Example 8-1, the simplified forms of the energy and entropy balances are

$$w = h_2 - h_1 \quad \text{and} \quad 0 = s_2 - s_1$$

For an ideal gas with constant specific heats, these two equations become

$$w_{sf} = h_2 - h_1 = c_{p,av}(T_2 - T_1)$$

$$0 = s_2 - s_1 = c_{p,av} \ln \frac{T_2}{T_1} - R \ln \frac{P_2}{P_1}$$

The final temperature is found by rearranging the above entropy equation to obtain

$$\frac{T_2}{T_1} = \left(\frac{P_2}{P_1}\right)^{R/c_{p,av}} = \left(\frac{P_2}{P_1}\right)^{(k-1)/k}$$

which is based on constant or average specific heats. Since $T_2$ is not known, the inlet temperature of $17°C$ is used to evaluate $k$. From Table A-3, $k$ is found to be 1.40; hence $(k-1)/k = 0.286$. Based on Eq. [8-10a] above, then,

$$T_2 = T_1(P_2/P_1)^{(k-1)/k} = 290 \text{ K} \times (8.6)^{0.286} = 537 \text{ K}$$

Therefore, using the inlet $c_p$ value of 1.005 kJ/kg·K, we have

$$w_{sf,\text{isen}} = c_p(T_2 - T_1) = 1.005(537 - 290) \text{ kJ/kg} = 248 \text{ kJ/kg}$$

The average value of $k$ between 290 and 537 K is around 1.39, so only a small error is introduced when the value of 1.40 at $T_1$ is used.

On the basis of Eq. [8-12], a second method for finding $w_{sf,\text{isen}}$ is

$$w_{sf,\text{isen}} = \frac{kRT_1}{k-1}\left[\left(\frac{P_2}{P_1}\right)^{(k-1)/k} - 1\right]$$

$$= \frac{1.4\,(8.314)(290) \text{ kJ/kmol}}{(1.4 - 1)(28.97) \text{ kg/kmol}}\left[(8.6)^{(1.4-1)/1.4} - 1\right]$$

$$= 247.6 \text{ kJ/kg}$$

where 28.97 is the molar mass of air. These are equivalent methods for evaluating shaft work. The general shape of the process on $Pv$ and $Ts$ diagrams is again shown in Fig. 8-4.

**Comment:**   In Example 8-1, $T_2$ was found to be 533 K, and $w$ was 246.6 kJ/kg. Owing to the small temperature range involved, the use of constant specific-heat data leads to answers that are quite close to answers obtained by accounting for variable specific heats.

## 8-1-2   ISENTROPIC RELATIONS FOR INCOMPRESSIBLE SUBSTANCES

If the specific heat of an incompressible substance is essentially constant, its specific entropy change is given by Eq. [7-23]:

$$\Delta s = c_{av} \ln \frac{T_2}{T_1} \qquad \textbf{[7-23]}$$

Consequently, on the basis of the above equation, an isentropic process involving an incompressible substance is one where $T_2 = T_1$. That is, the *temperature does not change* if the entropy does not change.

As a result, since $du = c\, dT$ for an incompressible substance, $\Delta u = 0$ for an isentropic process. In summary, when a fluid is modeled as incompressible and a process is modeled as isentropic, then the specific volume, specific entropy, temperature, and specific internal energy are all constant. However, properties such as pressure, enthalpy, velocity, and elevation could vary significantly during flow processes. It may appear that having $v$, $s$, $T$, and $u$ all remain constant during a process violates the state postulate for simple compressible substances. Keep in mind, however, that this situation is a *model* which represents a limiting condition. It turns out that the model is quite useful in predicting actual performance.

The transport of liquids in pipes is of importance in numerous engineering designs. A pump may or may not be part of the control volume, and the pipes may have different diameters at various sections of the flow system. In addition, the fluid may undergo a considerable elevation change. In steady state, $dE/dt = 0$. Hence an appropriate conservation of energy principle for pipe flow, based on Eq. [5-22], has the form

$$q + w_{shaft} = \Delta h + \Delta ke + \Delta pe \qquad \textbf{[5-22]}$$

For incompressible flow, $\Delta h = \Delta u + v\,\Delta P$. Therefore, the above equation can be modified to the form

$$w_{shaft} = \Delta u + v\,\Delta P + \Delta ke + \Delta pe - q \qquad \textbf{[8-14]}$$

This equation reduces to various formats depending on whether the process involves heat and/or work interactions, as well as whether the process is internally reversible or not.

Note the properties that remain constant during an isentropic process for an incompressible substance.

---

**Example 8-3**

**A** pump draws a solution with a specific gravity of 1.50 from a storage tank through an 8-cm pipe. The velocity in the inlet pipe is 1.2 m/s. The height of the liquid in the storage tank is maintained constant by the addition of makeup liquid. The open end of the 5-cm discharge pipe is 15 m above the top of the liquid in the storage tank. If the process is assumed to be internally reversible and adiabatic, and the fluid to be incompressible, determine the power supplied to the pump in the piping system, in kilowatts.

**Solution:**

**Given:** A fluid is pumped within a piping system with elevation and velocity changes, as shown in Fig. 8-6.

**Find:** Power supplied to the pump, in kilowatts.

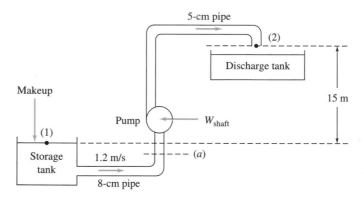

**Figure 8-6**     Schematic and data for flow process for Example 8-3.

**Model:**    Steady-state, internally reversible, adiabatic flow; incompressible fluid.

**Strategy:**    Use a steady-state energy balance and property relations for an incompressible fluid.

**Analysis:**    The top level of the fluid in the storage tank and the discharge of the 5-cm pipe are designated as states 1 and 2 in Fig. 8-6. State $a$ is an intermediate state in the 8-cm pipe, where the velocity is 1.2 m/s. The control volume includes the fluid at the top of the storage tank and within the piping system. Since $dE/dt = 0$ for a steady-state control volume, the energy equation reduces to

$$q + w_{\text{shaft}} = \Delta h + \Delta \text{ke} + \Delta \text{pe}$$

For incompressible flow, $\Delta h = \Delta u + v\,\Delta P$. Thus, the energy balance becomes

$$w_{\text{shaft}} = \Delta u + v\,\Delta P + \Delta \text{ke} + \Delta \text{pe} - q$$

For the same control volume, the entropy balance is

$$0 = \sum_{j=1}^{n} \frac{q_j}{T_j} + s_1 - s_2 + \sigma_m$$

Under the restriction of adiabatic internally reversible flow ($\sigma_m = q_j = 0$), the entropy balance reduces to $s_2 - s_1 = 0$. But for an incompressible substance, $\Delta s = c \ln T_2/T_1$, and $\Delta u = c\,\Delta T$. Because $\Delta s = 0$, $T_2 = T_1$, and $\Delta u = 0$. Thus the energy equation reduces to

$$w_{sf,\text{rev}} = v\,\Delta P + \Delta \text{ke} + \Delta \text{pe}$$

For the selected control volume, $P_1 = P_2$, and therefore $v\,\Delta P$ is zero. As a result,

$$w_{sf,\text{rev}} = \frac{\mathbf{V}_2^2 - \mathbf{V}_1^2}{2} + g(z_2 - z_1)$$

The initial velocity $\mathbf{V}_1$ is essentially zero since the diameter $D_1$ of the storage tank is much larger than the diameter $D_a$ of the inlet pipe. The velocity at state 2 can be found from the continuity equation because the velocity at state $a$ is known, and

$\dot{m}_a = \dot{m}_2$. Hence

$$V_2 = V_a \frac{A_a}{A_2} = V_a \left( \frac{D_a}{D_2} \right)^2 = 1.2 \left( \frac{8}{5} \right)^2 = 3.1 \text{ m/s}$$

If $g$ is now taken to be 9.8 m/s$^2$, then substitution of the available data into the energy equation yields

$$w_{sf,\text{isen}} = \left[ \frac{(3.1)^2}{2} + 15(9.8) \right] \frac{\text{m}^2}{\text{s}^2} \times \frac{1 \text{ N} \cdot \text{s}^2}{\text{kg} \cdot \text{m}} = (4.8 + 147) \text{ N} \cdot \text{m/kg} = 152 \text{ N} \cdot \text{m/kg}$$

Note that the change in potential energy of the water is responsible for nearly all the work required.

The power requirement is found from the relation $\dot{W} = \dot{m}w$. If the density of water is taken as 1000 kg/m$^3$ (the room-temperature value), then the density of the given liquid is 1500 kg/m$^3$. Thus the mass flow rate is

$$\dot{m} = \rho A V = (1.5 \times 10^3) \frac{\text{kg}}{\text{m}^3} \times \frac{\pi(8)^2}{4} \text{ cm}^2 \times 1.2 \frac{\text{m}}{\text{s}} \times 10^{-4} \frac{\text{m}^2}{\text{cm}^2}$$

$$= 9.05 \text{ kg/s}$$

Finally, the power input requirement is

$$\dot{W} = \dot{m}w = 9.05 \text{ kg/s} \times 152 \text{ N} \cdot \text{m/kg} = 1376 \text{ W} = 1.38 \text{ kW}$$

## 8-1-3 ISENTROPIC EVALUATION USING SUPERHEAT AND SATURATION DATA

There are no special relations for the evaluation of isentropic changes for fluids in the superheat or saturation region, other than $s_2 = s_1$. However, this process information, in conjunction with data on the initial and final states, is usually sufficient. An appropriate use of an energy balance or the continuity equation may also be necessary to complete the analysis. Several examples of this type of evaluation are provided below.

**Refrigerant 134a** passes adiabatically through a nozzle in reversible steady flow until its pressure reaches 1.4 bars. At the inlet to the nozzle the pressure and temperature are 5 bars and 70°C, respectively. Determine the exit velocity, in m/s, if the inlet velocity is small.

**Solution:**

**Given:** Refrigerant-134a passes through an adiabatic nozzle as shown in Fig. 8-7.

**Find:** $V_2$, in m/s.

**Model:** Adiabatic, internally reversible flow; $V_1 \approx 0$, $\Delta pe = 0$.

**Example 8-4**

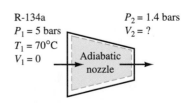

R-134a
$P_1 = 5$ bars
$T_1 = 70°C$
$V_1 = 0$
Adiabatic nozzle
$P_2 = 1.4$ bars
$V_2 = ?$

**Figure 8-7**
Nozzle schematic and data for Example 8-4.

**Strategy:**   Use data from the tables in the steady-state energy and entropy equations for a control volume.

**Analysis:**   The energy and entropy balances for this steady-state control volume with one inlet and one exit are

$$0 = q + w + (h_1 - h_2) + \frac{V_1^2}{2} - \frac{V_2^2}{2} + g(z_1 - z_2)$$

$$0 = \sum_{j=1}^{n} \frac{q_j}{T_j} + s_1 - s_2 + \sigma_m$$

The flow is internally reversible ($\sigma_m = 0$) and adiabatic ($q_j = 0$). Hence the entropy balance reduces to $s_2 = s_1$, and the flow is isentropic. In addition, if the change in potential energy is neglected, and $w = q = 0$, then the energy balance becomes

$$0 = h_1 - h_2 + \frac{V_1^2 - V_2^2}{2}$$

If $V_1 \ll V_2$, then the energy equation can be written as

$$V_2^2 = 2(h_1 - h_2)$$

$V_2$ can be found from the above equation if $h_2$ can be determined. From Table A-18 at 5 bars and 70°C, the value of $h_1$ is 309.92 kJ/kg, and that of $s_1$ is 1.0825 kJ/kg·K. The final state is determined from the values of $P_2$ and $s_2$, which must equal $s_1$. At 1.4 bars the value of $s_g$ is 0.9322 kJ/kg·K, so the final state also is superheated vapor. From Table A-18, it is found that $s = 1.0828$ kJ/kg·K at 1.4 bars and 30°C. This is close to the value of $s_1$ found above. Hence at this state of 1.4 bars and 30°C, $h_{2s} = 277.97$ kJ/kg. Substitution of these $h$ values into the energy balance leads to

$$\frac{V_2^2}{2} = (309.92 - 277.97) \text{ kJ/kg} \times \frac{1000 \text{ N·m}}{1 \text{ kJ}} \times \frac{1 \text{ kg·m}}{\text{N·s}^2}$$

Thus,

$$V_2 = (63,900 \text{ m}^2/\text{s}^2)^{1/2} = 253 \text{ m/s}$$

This is the maximum exit velocity under the specified conditions. The isentropic process is shown on the $Ts$ diagram in Fig. 8-8.

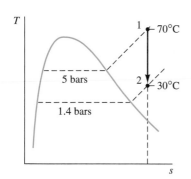

**Figure 8-8**
$Ts$ process diagram for Example 8-4.

---

**Example 8-5**

**S**team initially at 600 psia and 900°F is expanded adiabatically in a piston-cylinder device through a pressure ratio $P_1/P_2$ of 10:1. If the process is internally reversible, calculate the work output, in Btu/lb$_m$.

**Solution:**

**Given:**   Steam is expanded adiabatically in a piston-cylinder device, as shown in Fig. 8-9.

**Find:**   $w_{\text{out}}$, in Btu/lb$_m$.

**Model:** Closed system; adiabatic, internally reversible expansion.

**Strategy:** Use closed-system energy and entropy balances in conjunction with tabular data.

**Analysis:** The work is found from the closed-system energy balance $q + w = \Delta u$. Because the process is adiabatic, this equation reduces to

$$w = u_2 - u_1$$

The entropy balance for this system is

$$\Delta s = \int \frac{\delta q}{T} + \sigma_m$$

Because the process is adiabatic and internally reversible, $\Delta s = 0$, and the process is isentropic. We must now evaluate $u_1$ and $u_2$ from tabular data. The value of $u_1$ is found from a knowledge of $P_1$ and $T_1$. From steam table A-14E, for the initial state,

$$P_1 = 600 \text{ psia} \qquad T_1 = 900°\text{F} \qquad u_1 = 1318.4 \text{ Btu/lb}_m \qquad s_1 = 1.6766 \text{ Btu/lb}_m\cdot°\text{R}$$

Since the specific entropy remains constant during the process, the final state is at 60 psia and a specific entropy of 1.6766 Btu/lb$_m$·°R. The saturated-vapor entropy at 60 psia is 1.6433 Btu/lb$_m$·°R, and $s_2 > s_g$. Therefore, the fluid is superheated vapor at the final state. Entering Table A-14E again, we find that at 60 psia the final temperature lies between 300 and 350°F. Linear interpolation yields a final temperature of

$$T_2 = 300°\text{F} + 50°\text{F} \times \frac{1.6766 - 1.6496}{1.6830 - 1.6496} = 340°\text{F}$$

On the same basis, the internal energy is 1117.4 Btu/lb$_m$. Consequently, the work transferred is

$$w_{\text{isen}} = u_2 - u_1 = (1117.4 - 1318.4) \text{ Btu/lb}_m = -201.0 \text{ Btu/lb}_m$$

This is the maximum work output possible for the gas within the piston-cylinder device under the specified operating conditions. The process is illustrated on the $Pv$ diagram in Fig. 8-10.

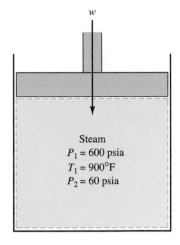

**Figure 8-9**

Equipment schematic and data for Example 8-5.

*Steam*
$P_1 = 600$ psia
$T_1 = 900°$F
$P_2 = 60$ psia

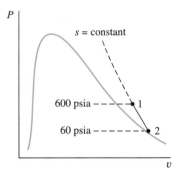

**Figure 8-10**

Isentropic process for Example 8-5 illustrated on a $Pv$ diagram.

---

The discussion in this section has centered on isentropic behavior of substances in various closed and open systems. Although no process in practice is isentropic, it is an extremely important concept with respect to devices that are essentially adiabatic. Since internally reversible processes by concept are free of nonequilibrium and dissipative effects, the isentropic performance of a device may be used as a standard against which the real behavior of adiabatic devices is compared. Isentropic conditions lead to a *maximization* or *minimization* of important system variables, such as work effects or final state properties such as velocity, temperature, or pressure.

## 8-2    ADIABATIC EFFICIENCIES OF STEADY-FLOW DEVICES

Irreversibilities inherently accompany the flow of fluids through actual steady-flow devices and degrade the performance of such devices. It is useful to have parameters that compare the actual performance to that which would be achieved under *ideal* conditions. In the development of these parameters, it should be recognized that the actual flow through many engineering devices is approximately adiabatic. The ideal performance of adiabatic equipment occurs when the flow is also internally reversible and thus isentropic. Hence one appropriate measure of achievement is a comparison of the actual performance to the performance that would occur under isentropic conditions. This comparison is expressed in a parameter known as the **adiabatic,** or **isentropic, efficiency** of a device. (An adiabatic efficiency is sometimes called a *first-law* efficiency. Second-law efficiencies are introduced in Chap. 9.) The definition and use of adiabatic efficiencies for turbines, nozzles, compressors, and pumps are developed below.

Before beginning our discussion of these devices, it is helpful to summarize the basic equations that will govern their performance. We will restrict the discussion to devices that can be modeled as a *steady-state* control volume with *one inlet* and *one outlet,* such as turbines, compressors, pumps, and nozzles. For these conditions, the energy and entropy balances can be written on a unit-mass basis as

$$0 = w + (h_1 - h_2) + \left(\frac{\mathbf{V}_1^2}{2} - \frac{\mathbf{V}_2^2}{2}\right) + g(z_1 - z_2) \qquad \textbf{[8-15]}$$

$$0 = s_1 - s_2 + \sigma_m \qquad \textbf{[8-16]}$$

These equations will now be used in the development of the adiabatic efficiencies of important flow devices.

### 8-2-1    ADIABATIC TURBINE EFFICIENCY

The purpose of a turbine is to produce work output. Therefore, the *adiabatic* (or *isentropic*) *turbine efficiency* $\eta_T$ is defined as the ratio of the actual work output $w_{a,\text{out}}$ (or power output $\dot{W}_{a,\text{out}}$) to the isentropic work output $w_{s,\text{out}}$ (or power output $\dot{W}_{s,\text{out}}$) that could be achieved if the fluid was expanded from the *same inlet state* to the *same exit pressure*. That is,

$$\eta_T \equiv \frac{w_{a,\text{out}}}{w_{s,\text{out}}} = \frac{\dot{W}_{a,\text{out}}}{\dot{W}_{s,\text{out}}} \qquad \text{(turbine)} \qquad \textbf{[8-17]}$$

where the subscripts $a$ and $s$ again represent the actual and the adiabatic internally reversible (isentropic) processes, respectively. If the actual inlet state is identified as state 1, then the actual outlet conditions are identified as state 2a, and the *hypothetical* outlet state for the ideal (isentropic) process is identified as state 2s. When the kinetic- and potential-energy changes across a turbine are negligible, the energy balance given by Eq. [8-15] reduces to $w_{out} = h_1 - h_2$. In this case the adiabatic turbine efficiency becomes

$$\eta_T = \frac{h_1 - h_{2a}}{h_1 - h_{2s}} \qquad \text{(turbine with } \Delta\text{ke} = \Delta\text{pe} = 0) \qquad \textbf{[8-18]}$$

The relation between the actual work term $h_1 - h_{2a}$ and the isentropic work term $h_1 - h_{2s}$ that appear in Eq. [8-18] is illustrated in the *hs* diagram in Fig. 8-11. The entropy balance for adiabatic processes requires that the change in entropy be equal to or greater than zero. For the same initial state and the same final pressure, state 2a for the irreversible expansion must lie to the right of state 2s but on the same $P_2$ line. This requires that $h_1 - h_{2a}$ always be less than $h_1 - h_{2s}$ for an irreversible expansion. Thus the adiabatic efficiency of a real turbine is always less than unity. The value of $\eta_T$ for actual turbines ranges from 80 to 90 percent.

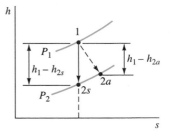

**Figure 8-11**
Comparison of the turbine enthalpy change for actual and isentropic processes on a *hs* diagram.

---

**A** gas turbine operates on hydrogen gas initially at 480 K. The value of $P_1/P_2$ is 2.27 : 1. The process is adiabatic, but irreversibilities reduce the work output to 80.0 percent of the isentropic value. Neglecting kinetic- and potential-energy changes, determine (*a*) the maximum possible work output, in kJ/kg, (*b*) the actual work output, in kJ/kg, (*c*) the actual final temperature, in kelvins, and (*d*) the entropy generation, in kJ/kg·K.

**Example 8-6**

**Solution:**

**Given:** A gas turbine operates with hydrogen gas, as shown in Fig. 8-12.

**Find:** (*a*) $w_{max}$, in kJ/kmol, (*b*) $w_{act}$, and (*c*) $T_{2a}$, in kelvins.

**Model:** Steady state, adiabatic, $\Delta$ke $= \Delta$pe $= 0$, ideal gas.

**Strategy:** Apply the steady-state energy balance, isentropic relations, and the entropy balance.

**Analysis:** In steady state, $dE/dt = 0$. Therefore, the energy balance on a unit-mass basis for a control volume in steady state is

$$q + w = \Delta h + \Delta\text{ke} + \Delta\text{pe}$$

(*a*) For a turbine under adiabatic conditions and negligible kinetic- and potential-energy changes, the energy balance reduces to $w = \Delta h$. The *maximum* work is obtained under isentropic flow. Hence $h_2 = h_{2s}$, and

$$w_{s,out} = h_1 - h_{2,s}$$

Modeling the fluid as an ideal gas, the inlet molar enthalpy found from Table A-11 at 480 K is 13,764 kJ/kmol, and $\bar{s}_1^\circ$ is 144.432 kJ/kmol·K. The final isentropic state

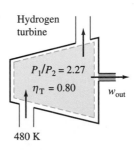

Hydrogen turbine

$P_1/P_2 = 2.27$

$\eta_T = 0.80$

$w_{out}$

480 K

**Figure 8-12**
Turbine schematic and data for Example 8-6.

is determined by applying the isentropic relation for an ideal gas [Eq. [8-1]].

$$\bar{s}_2^o = \bar{s}_1^o + R_u \ln \frac{P_2}{P_1}$$

$$= \left(144.432 + 8.314\ln\frac{1}{2.27}\right) \text{kJ/kmol·K} = 137.616 \text{ kJ/kmol·K}$$

This value of $\bar{s}_2^o$ corresponds closely to 380 K in Table A-11. At this temperature $\bar{h}_{2s} = 10{,}843$ kJ/kmol. Therefore,

$$w_{s,\text{out}} = h_1 - h_{2s} = \frac{\bar{h}_1 - \bar{h}_{2s}}{M}$$

$$= \frac{(13{,}764 - 10{,}843) \text{ kJ/kmol}}{2.016 \text{ kg/kmol}} = 1449 \text{ kJ/kg}$$

where the molar mass $M = 2.016$ for hydrogen.

(b) The actual work $w_a$ is determined from the adiabatic efficiency of the turbine. Since $\eta_T = w_{a,\text{out}}/w_{s,\text{out}}$,

$$w_{a,\text{out}} = \eta_T w_{s,\text{out}} = 0.80(1449) \text{ kJ/kg} = 1159 \text{ kJ/kg}$$

(c) The final temperature is found from the final enthalpy. On the basis of an energy balance on the actual turbine,

$$\bar{h}_{2a} = \bar{h}_1 - w_{a,\text{out}}M$$

$$= 13{,}764 \text{ kJ/kmol} - (1159 \text{ kJ/kg})(2.016 \text{ kg/kmol}) = 11{,}427 \text{ kJ/kmol}$$

This value of $\bar{h}_{2a}$ corresponds to a temperature of 400 K in Table A-11. This is 20°C higher than an isentropic expansion.

(d) The entropy production for the actual expansion process is found from the entropy balance per unit mass,

$$0 = \sum_{j=1}^{n} \frac{q_j}{T_j} + s_1 - s_2 + \sigma_m$$

For an adiabatic expansion, $\sigma_m = \Delta s$. Therefore, for an ideal-gas model,

$$\sigma_m = s_2 - s_1 = \left(\bar{s}_2^o - \bar{s}_1^o - R_u \ln\frac{P_2}{P_1}\right)\frac{1}{M}$$

$$= \left(139.106 - 144.432 - 8.314\ln\frac{1}{2.27}\right)\text{kJ/kmol·K} \times \frac{1}{2.016 \text{ kg/kmol}}$$

$$= 0.739 \text{ kJ/kg·K}$$

This value of $\sigma_m$ would decrease as the adiabatic efficiency increases, causing $T_{2a}$ to approach $T_{2s}$.

The path of the process is shown on the $Ts$ diagram in Fig. 8-13.

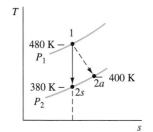

**Figure 8-13**
Process path on $Ts$ and $hs$ diagrams for Example 8-6.

When a liquid flows through a turbine, the turbine is commonly called a *hydraulic* turbine and the fluid can be modeled as an incompressible substance, for which $\Delta h = c\,\Delta T + v\,\Delta P$. Assuming that changes in kinetic and potential energy are negligible, Eq. [8-15] applied to an adiabatic hydraulic turbine can be written as

$$w_{\text{out}} = h_1 - h_2 = c(T_1 - T_2) + v(P_1 - P_2) \quad \text{(incompressible)}$$

[8-19]

where $P_1 > P_2$ for a turbine. Recalling that $\Delta s = c \ln(T_2/T_1)$ for an incompressible substance, the entropy balance [Eq. [8-16]] can be solved for the entropy production to give

$$\sigma_m = s_2 - s_1 = c \ln \frac{T_2}{T_1} \geq 0 \quad \text{(incompressible)} \quad \text{[8-20]}$$

From the entropy balance, it is apparent that $T_2 > T_1$ as a result of entropy production due to irreversibilities. The larger the irreversibility within the turbine, the greater the temperature rise. As a result, on the basis of Eq. [8-19], the shaft-work output must decrease as irreversibilities increase, for a fixed value of $\Delta P$.

When the flow is internally reversible, the process is isentropic and $T_2 = T_1$. Using Eq. [8-19], the expression for the *isentropic* work out of a hydraulic turbine is

$$w_{s,\text{out}} = v(P_1 - P_2) \quad \text{(incompressible)}$$

Note that the isentropic work out of a hydraulic turbine depends only on the pressure change across the turbine. Using this result for the isentropic work, the adiabatic efficiency for a hydraulic turbine with negligible kinetic- and potential-energy changes can be written as

$$\eta_{T,\text{hydraulic}} = \frac{w_{a,\text{out}}}{w_{s,\text{out}}} \quad \quad \text{[8-21]}$$

$$= \frac{c\,\Delta T + v\,\Delta P}{v\,\Delta P} \quad \text{(incompressible, } \Delta\text{ke} = \Delta\text{pe} = 0)$$

In this equation $c\,\Delta T$ and $v\,\Delta P$ are of opposite signs; hence $0 \leq \eta_T \leq 1$. The greater the value of $\Delta T$ due to irreversibilities, the smaller the turbine efficiency. Even a small change in the temperature can result in a sizable decrease in efficiency.

## 8-2-2 ADIABATIC NOZZLE EFFICIENCY

A nozzle is a flow channel constructed to accelerate a fluid to a higher velocity, that is, to increase its kinetic energy. Therefore, the *adiabatic,* or *isentropic, nozzle efficiency* $\eta_N$ is defined as

$$\eta_N \equiv \frac{\Delta\text{ke}_a}{\Delta\text{ke}_s} = \frac{V_{2a}^2/2 - V_1^2/2}{V_{2s}^2/2 - V_1^2/2} \quad \text{nozzle} \quad \text{[8-22]}$$

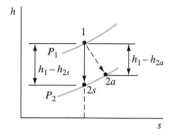

**Figure 8-14**
Comparison of the nozzle enthalpy change for actual and isentropic processes on an $hs$ diagram.

where $\mathbf{V}_{2a}^2/2$ is the actual specific kinetic energy at the nozzle exit and $\mathbf{V}_{2s}^2/2$ is the specific kinetic energy at the nozzle exit for isentropic flow from the *same initial state* to the *same exit pressure*. Nozzle efficiencies generally range upward from 90 percent. For converging nozzles used in subsonic flow, efficiencies of 0.95 or higher are common.

The steady-state energy balance for an adiabatic nozzle with a negligible potential-energy change is

$$h_2 - h_1 + \frac{\mathbf{V}_2^2 - \mathbf{V}_1^2}{2} = 0$$

Use of this equation in Eq. [8-22] leads to

$$\eta_N = \frac{h_1 - h_{2a}}{h_1 - h_{2s}} \qquad\qquad \textbf{[8-23]}$$

This expression is analogous to Eq. [8-18] for a turbine. The enthalpy-entropy diagram in Fig. 8-14 applies to an adiabatic nozzle with negligible inlet velocity. [It should be noted that sometimes the nozzle efficiency is defined as the ratio of the actual to the isentropic exit kinetic energy. That is, $\eta_N = (\mathbf{V}_{2a}^2/2)/(\mathbf{V}_{2s}^2/2)$.]

In an approach paralleling that used for the hydraulic turbine, it is instructive to consider *incompressible* flow through a nozzle. Because a nozzle has no work transfer, the basic energy balance given by Eq. [8-15] for flow through an adiabatic nozzle with negligible potential-energy change can be written as

$$\frac{\mathbf{V}_2^2 - \mathbf{V}_1^2}{2} = h_1 - h_2 \qquad\qquad \textbf{[8-24]}$$

$$= v(P_1 - P_2) + c(T_1 - T_2) \qquad \text{(incompressible nozzle)}$$

where $P_1 > P_2$ for a nozzle. Similar to the result from the entropy balance for a hydraulic turbine, $\sigma_m = s_2 - s_1 = c \ln(T_2/T_1)$. For an internally reversible, adiabatic nozzle, the flow is isentropic, $T_2 = T_1$, and the change in kinetic energy becomes

$$\frac{\mathbf{V}_{2s}^2 - \mathbf{V}_1^2}{2} = v(P_1 - P_2) \qquad \text{(isentropic, incompressible nozzle)}$$

$$\textbf{[8-25]}$$

Using this result, the adiabatic efficiency for an incompressible nozzle with fixed $\Delta P$ can be written as

$$\eta_N = \frac{(\mathbf{V}_{2a}^2 - \mathbf{V}_1^2)/2}{v(P_1 - P_2)} \qquad \text{(incompressible)} \qquad \textbf{[8-26]}$$

where $\mathbf{V}_{2a}$ is the actual outlet velocity, which is less than the isentropic outlet velocity. In order for $\mathbf{V}_{2a} < \mathbf{V}_{2s}$ when $\Delta P$ is constant, the continuity equation requires that the exit area must be larger than that required for isentropic flow. On the other hand, if the nozzle is of fixed geometry, then on the basis of continuity of flow the kinetic-energy change is the same for both reversible and irreversible flow. In this case the pressure drop through the nozzle is always greater than would occur for isentropic flow.

**Example 8-7**

Liquid water enters a nozzle at 17.67 psia, 100°F, and with a negligible velocity. The fluid expands to 15.0 psia. (a) Find the maximum exit velocity under an adiabatic condition, in ft/s. (b) If the nozzle adiabatic efficiency is 90.0 percent, determine the temperature rise of the fluid, in degrees Fahrenheit, and the final velocity, in ft/s.

**Solution:**

**Given:** Liquid water flows through a nozzle as shown in Fig. 8-15.

**Find:** (a) $V_{2s}$, in ft/s, and (b) $\Delta T$ in °F, and $V_{2a}$ in ft/s.

**Model:** Steady state, adiabatic, incompressible, $w = \Delta pe = 0$, $\eta_N = 0.90$, $V_1 \approx 0$.

**Strategy:** Apply the steady-state energy and entropy balances and the nozzle-efficiency definition.

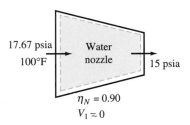

**Figure 8-15**
Nozzle schematic and data for Example 8-7.

**Analysis:** (a) The steady-state energy equation for the control volume on a unit-mass basis is

$$0 = q + w - \Delta h - \Delta ke - \Delta pe$$

In addition, $\Delta h = c\,\Delta T + v\,\Delta P$ for an incompressible fluid. If $\Delta pe$ is negligible, and $w = 0$, then the energy balance for adiabatic incompressible flow through the nozzle is given by

$$0 = c\,\Delta T + v\,\Delta P + \frac{V_2^2 - V_1^2}{2}$$

Finally, with negligible inlet velocity, the energy balance can be rearranged into the format

$$\frac{V_2^2}{2} = c(T_1 - T_2) + v(P_1 - P_2)$$

The entropy balance on a unit-mass basis for steady-state flow is

$$0 = \sum_{j=1}^{n} \frac{q_j}{T_j} + s_1 - s_2 + \sigma_m$$

The entropy change of an incompressible substance is given by $\Delta s = c\ln(T_2/T_1)$. Hence the entropy balance for the adiabatic steady-state process can be written as

$$\sigma_m = s_2 - s_1 = c\ln\,(T_2/T_1)$$

For an internally reversible nozzle, $\sigma_m = 0$, the flow is isentropic, $T_2 = T_1$, and the energy equation becomes $V_{2s}^2/2 = v(P_1 - P_2)$. Solving for $V_{2s}$,

$$\mathbf{V}_{2s}^2 = [2v(P_1 - P_2)]^{1/2}$$

$$= \left(2(0.01613)\ \frac{\text{ft}^3}{\text{lb}_m} \times (17.67 - 15.0)\ \frac{\text{lb}_f}{\text{in}^2} \times \frac{144\ \text{in}^2}{\text{ft}^2}\right)^{1/2}$$

$$= \left(\frac{12.40\ \text{ft·lb}_f}{\text{lb}_m} \times \frac{32.174\ \text{lb}_m\text{·ft}}{\text{lb}_f\text{·s}^2}\right)^{1/2}$$

$$= 20.0\ \text{ft/s}$$

Thus the maximum exit velocity is 20.0 ft/s.

(b) The equation for the nozzle efficiency is

$$\eta_N = \frac{(\mathbf{V}_2^2 - \mathbf{V}_1^2)/2}{v(P_1 - P_2)}$$

In addition, the steady-state energy balance developed above for adiabatic incompressible flow is

$$\frac{\mathbf{V}_2^2}{2} = c(T_1 - T_2) + v(P_1 - P_2)$$

where $\mathbf{V}_1$ is negligible. Substitution of this equation into the numerator of the nozzle efficiency equation yields

$$\eta_N = \frac{c(T_1 - T_2) + v(P_1 - P_2)}{v(P_1 - P_2)}$$

Solving this equation for the temperature rise, and assuming that the specific heat of water is 1.00 Btu/lb$_m$·°R, we find that

$$T_2 - T_1 = \frac{v(P_1 - P_2)(1 - \eta_N)}{c}$$

$$= \frac{0.01613\ \dfrac{\text{ft}^3}{\text{lb}_m} \times (17.67 - 15.0)\ \dfrac{\text{lb}_f}{\text{in}^2} \times \dfrac{144\ \text{in}^2}{\text{ft}^2} \times (1 - 0.90)}{1.00\ \dfrac{\text{Btu}}{\text{lb}_m\text{·°R}}}$$

$$= 0.6202\ \frac{\text{ft·lb}_f\text{·°R}}{\text{Btu}} \times \frac{1\ \text{Btu}}{778\ \text{ft·lb}_f}$$

$$= 8.0 \times 10^{-4}\text{°R}$$

The actual exit velocity is found from $\eta_N = \mathbf{V}_{2a}^2/\mathbf{V}_{2s}^2$ when $\mathbf{V}_1$ is negligible. Therefore,

$$\mathbf{V}_{2a} = [\eta_N \mathbf{V}_{2s}^2]^{1/2} = [0.90(20)^2]^{1/2}\ \text{ft/s} = 19.0\ \text{ft/s}$$

**Comment:**    Although the temperature rise is immeasurable, there is a 5 percent change in the exit velocity due to irreversibilities. This change in velocity will require a change in the exit area for the irreversibile nozzle, as compared to the area of the reversible nozzle. In addition, since $\Delta T$ is extremely small, the entropy generation is very small in value.

## 8-2-3 ADIABATIC COMPRESSOR EFFICIENCY

The *adiabatic*, or *isentropic*, *compressor efficiency* $\eta_C$ is defined as the ratio of the isentropic work $w_s$ (or power $\dot{W}_s$) required to the actual work $w_a$ (or power $\dot{W}_s$) required for the *same inlet state and the same exit pressure*. That is,

$$\eta_C \equiv \frac{w_{s,in}}{w_{a,in}} = \frac{\dot{W}_{s,in}}{\dot{W}_{a,in}} \qquad \text{(compressor)} \qquad \textbf{[8-27]}$$

When the kinetic- and potential-energy changes across the compressor are negligible, Eq. [8-15] permits us to write Eq. [8-27] as

$$\eta_C \approx \frac{h_{2s} - h_1}{h_{2a} - h_1} \qquad \text{(compressor)} \qquad \textbf{[8-28]}$$

where 1 and 2 again represent the inlet and exit states, respectively. The value of $\eta_C$ ranges roughly from 75 to 85 percent for actual gas compressors.

Figure 8-16 is an enthalpy-entropy diagram for a gas compressor. The ideal model is represented by the isentropic line 1-2s between pressures $P_1$ and $P_2$. For an actual irreversible compressor the entropy increases if the process is adiabatic. Thus line 1-2a for the actual process lies to the right of the isentropic line, and the enthalpy $h_{2a}$ must be greater than $h_{2s}$. As in the turbine analysis, vertical lengths on Fig. 8-16 represent the enthalpy changes $h_{2s} - h_1$ and $h_{2a} - h_1$ in Eq. [8-34]. The figure shows that the adiabatic compressor efficiency is always less than unity for a real compressor. Also, for $\sigma_m > 0$, s increases owing to irreversibilities, and the compressor efficiency decreases.

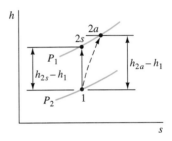

**Figure 8-16**
A *hs* diagram for a gas compressor.

---

**A** refrigeration system contains an adiabatic compressor which handles refrigerant 134a. Inlet conditions are 140 kPa and $-10°C$ and the exit state is 1.4 MPa and 80°. The changes in potential and kinetic energies are negligible. Find (*a*) the adiabatic efficiency of the compressor and (*b*) the entropy generation, in kJ/kg·K.

**Example 8-8**

**Solution:**

**Given:** An adiabatic compressor operates with R-134a, as shown in the schematic of Fig. 8-17.

**Find:** (*a*) Compressor adiabatic efficiency $\eta_C$ and (*b*) $\sigma_m$, in kJ/kg·K.

**Model:** Steady-state, adiabatic, $\Delta pe = \Delta ke = 0$.

**Strategy:** Apply the definition of compressor efficiency and the entropy balance.

$T_1 = -10°C$
$P_1 = 140$ kPa

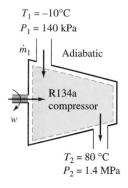

$T_2 = 80\ °C$
$P_2 = 1.4$ MPa

### Figure 8-17
Compressor schematic and data for Example 8-8.

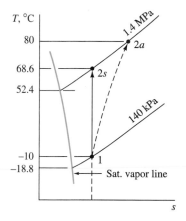

### Figure 8-18
Actual compressor path for Example 8-8 on a *Ts* diagram.

**Analysis:**   (*a*) The adiabatic efficiency is given by

$$\eta_C = \frac{w_{s,in}}{w_{a,in}} = \frac{h_{2s} - h_1}{h_{2a} - h_1}$$

The values of $h_1$ and $h_{2a}$ are found directly from the given data for the inlet and exit states, which are both superheated vapor states. From Table A-18,

$P_1 = 1.4$ bars    $T_1 = -10°C$    $h_1 = 243.40$ kJ/kg    $s_1 = 0.9606$ kJ/kg·K
$P_2 = 14.0$ bars   $T_{2a} = 80°C$   $h_{2a} = 307.10$ kJ/kg   $s_{2a} = 0.9997$ kJ/kg·K

Note that the entropy has increased, in general agreement with the second law for adiabatic processes. The value of $h_{2s}$ is found at $s_{2s} = s_1 = 0.9606$ kJ/kg·K and $P_2 = 1.4$ MPa (14.0 bars). Linear interpolation in Table A-18 between 60 and 70°C shows that

$$h_{2s} = \left[283.10 + (295.31 - 283.10)\frac{0.9606 - 0.9297}{0.9658 - 0.9297}\right] \text{kJ/kg} = 293.55 \text{ kJ/kg}$$

Therefore, the adiabatic efficiency is

$$\eta_C = \frac{293.55 - 243.40}{307.10 - 243.40} = \frac{50.15}{63.70} = 0.787 \text{ (or 78.7 percent)}$$

(*b*) The steady-state entropy balance per unit mass for the control volume is

$$0 = \sum_{j=1}^{n} \frac{q_j}{T_j} + s_1 - s_2 + \sigma_m$$

For an adiabatic process this reduces to $\sigma_m = s_2 - s_1$. Thus for the real gas within the compressor,

$$\sigma_m = s_{2a} - s_1 = (0.9997 - 0.9606) \text{ kJ/kg·K} = 0.0391 \text{ kJ/kg·K}$$

**Comment:**   Further linear interpolation in part *b* would show that $T_{2s} = 68.6°C$. Thus the fluid irreversibility raises the exit temperature about 11.4°C above that for isentropic compression, as shown in Fig. 8-18.

### 8-2-4   ADIABATIC PUMP EFFICIENCY

The *adiabatic* (or *isentropic*) *pump efficiency* $\eta_P$ is defined similarly to that of a compressor. For the *same inlet state and the same exit pressure,*

$$\eta_P \equiv \frac{w_{s,in}}{w_{a,in}} = \frac{\dot{W}_{s,in}}{\dot{W}_{a,in}} \qquad \text{(pump)} \qquad \textbf{[8-29]}$$

If Δke and Δpe are negligible, then on the basis of Eq. [8-15] the above equation becomes

$$\eta_P = \frac{h_{2s} - h_1}{h_{2a} - h_1} \qquad \text{(pump)} \qquad \textbf{[8-30]}$$

Pump efficiencies may range from 50 to 90 percent.

The basic analysis of incompressible flow through a pump is analogous to that for a hydraulic turbine. For isentropic flow the temperature remains constant, and the shaft work again is related solely to the flow work, namely, $w_{P,s} = v(P_2 - P_1)$. The shaft-work input results in an increase in the pressure of the fluid, with no other effect. If the process is adiabatic and internally *irreversible,* the increase in entropy leads to a rise in temperature of the fluid. The effect of this on the pressure rise is seen from Eq. [8-19] written for a pump:

$$w_{P,\text{in}} = h_2 - h_1 = c(T_2 - T_1) + v(P_2 - P_1) \qquad \textbf{[8-31]}$$

For a given work input, the energy added is split between an increase in internal energy and an increase in flow work. For the same work input, the pressure rise is greatest for an isentropic process. If the same pressure rise is desired, the work input will be greater for the irreversible process than for the isentropic process.

The adiabatic pump efficiency is defined in Eq. [8-29] as $\eta_P = w_s/w_a$, when $\Delta P$ is the same for the actual and isentropic cases. In a manner similar to the development of Eq. [8-21] for a hydraulic turbine, the efficiency expression for a pump can be written as

$$\eta_P = \frac{v(P_2 - P_1)}{w_{\text{in}}} \qquad \textbf{[8-32]}$$

When $P_2 - P_1$ is fixed, then $w_{\text{in}}$ represents work transfer for the irreversible case, and $w_{a,\text{in}} \geq w_{s,\text{in}}$. Alternatively, when the work input to the pump is fixed, then $\Delta P$ represents the actual pressure rise, and $\Delta P_a < \Delta P_s$, as shown in the following example.

**For turbines, nozzles, compressors, and pumps, what outlet property is the same for both the actual and the isentropic processes used in the definition of an adiabatic efficiency?**

Water enters a pump at 1 bar and 30°C. Shaft work is done on the fluid in the amount of 4.5 kJ/kg. Neglect kinetic- and potential-energy changes. (*a*) Determine the pressure rise if the process is isentropic, in bars. (*b*) If the fluid temperature rises by 0.20°C during the process, determine the pressure rise and the adiabatic efficiency.

**Solution:**

**Given:** A water pump, as shown in Fig. 8-19.

**Find:** (*a*) $\Delta P_{\text{isen}}$ and (*b*) $\Delta P_{\text{act}}$ for $\Delta T = 0.20°C$.

**Model:** Steady state, incompressible, $\Delta \text{ke} = \Delta \text{pe} = 0$.

**Strategy:** Apply the steady-state energy balance and the equation for the adiabatic pump efficiency.

**Analysis:** (*a*) The steady-state energy equation for flow in a pump is

$$0 = q + w + (h_1 - h_2) - \Delta \text{ke} - \Delta \text{pe}$$

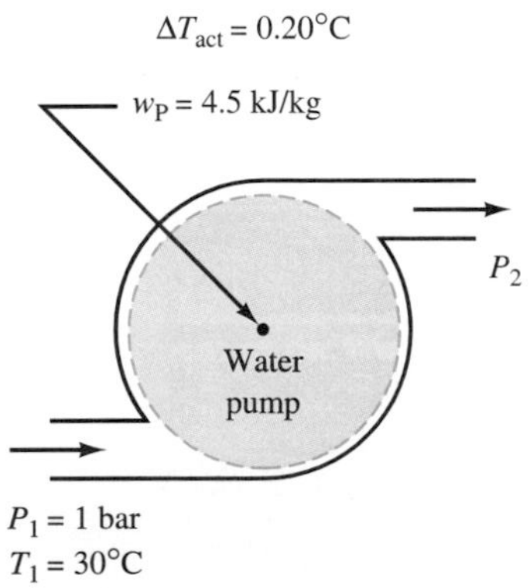

**Example 8-9**

$\Delta T_{\text{act}} = 0.20°C$

$w_P = 4.5 \text{ kJ/kg}$

$P_2$

Water
pump

$P_1 = 1 \text{ bar}$
$T_1 = 30°C$

**Figure 8-19**
Pump schematic and data for Example 8-9.

For incompressible flow, $\Delta h = c\,\Delta T + v\,\Delta P$. The entropy balance for adiabatic internally reversible flow reduces to $\sigma_m = \Delta s = 0$. But $\Delta s = c\ln T_2/T_1$ for an incompressible substance; therefore, $\Delta T = 0$ for isentropic flow. As a result the energy equation reduces to $w_{s,P} = v\,\Delta P$, when $\Delta ke$ and $\Delta pe$ are neglected. Solving for the isentropic pressure rise, we find that

$$\Delta P_s = \frac{w_P}{v} = \frac{4.50 \text{ kJ/kg}}{10^{-3} \text{ m}^3/\text{kg}} \times \frac{10^{-2} \text{ bar·m}^3}{\text{kJ}} = 45.0 \text{ bars}$$

(b) For an irreversible process the energy balance is

$$w = c\,\Delta T + v\,\Delta P$$

For a temperature rise of 0.20°C, substitution of values into the energy equation yields

$$4.5 \text{ kJ/kg} = 4.18 \text{ kJ/kg·C} \times (0.2)\,°\text{C} + 10^{-3} \text{ m}^3/\text{kg} \times \Delta P_a \times (100 \text{ kJ/bar·m}^3)$$

$$\Delta P_a = [10(4.50 - 0.84)] \text{ bars} = 36.6 \text{ bars}$$

Thus $\Delta P_a$ is considerably less than $\Delta P_s$. Insertion of the work terms for internally reversible and irreversible flow into the definition of pump efficiency for a fixed work input gives

$$\eta_P = \frac{v\,\Delta P_a}{w_{a,\text{in}}}$$

Substitution of values into this equations leads to

$$\eta_P = \frac{10^{-3}\,\dfrac{\text{m}^3}{\text{kg}} \times (36.6) \text{ bars} \times \dfrac{100 \text{ kJ/kg}}{\text{bar·m}^3}}{4.50 \text{ kJ/kg}} = \frac{3.66}{4.50} = 0.81 \quad \text{(or 81\%)}$$

**Comment:**  When water is compressed from atmospheric conditions to around 40 or 45 bars, every 0.2°C rise in temperature due to irreversibilities results in roughly a 15 percent reduction in the pressure rise under adiabatic flow.

---

Both the entropy production $\sigma_m$ and the adiabatic efficiency $\eta$ are a measure of the internal irreversibilities within a device. Hence the adiabatic efficiency can be related numerically to the entropy generation. For a pump this relationship is developed by starting with the pump adiabatic efficiency $\eta_P$ given by Eq. [8-32], and writing it in the format

$$\eta_P = \frac{v\,\Delta P}{w_{\text{in},a}} = \frac{v\,\Delta P}{c\,\Delta T + v\,\Delta P} = \frac{v\,\Delta P}{c(T_{2a} - T_1) + v\,\Delta P} \qquad \textbf{[8-33]}$$

For incompressible flow, $\sigma_m = c\ln(T_{2a}/T_1)$ on the basis of Eq. [7-16]. This equation may be written as $T_{2a} = T_1\,e^{\sigma_m/c}$. Substitution of $T_{2a}$ into the above equation for $\eta_P$ yields

$$\eta_P = \frac{v\,\Delta P}{cT_1(e^{\sigma/c} - 1) + v\,\Delta P} \qquad \text{(incompressible)} \qquad \textbf{[8-34]}$$

where $c$, $v$, $T_1$, and $\Delta P$ are input data. As $\sigma$ approaches 0, $\eta_P$ approaches unity. Figure 8-20 shows $\sigma$ versus $\eta_P$ for liquid water which enters a pump at 27°C and 100 kPa and leaves at either 600 or 1100 kPa. The relationship between $\sigma_m$ and $\eta$ is developed for other steady-flow devices in the following subsection.

### 8-2-5 RELATIONSHIP OF ADIABATIC EFFICIENCIES TO ENTROPY PRODUCTION

The effect of internal irreversibilities during adiabatic processes is to increase the entropy of the fluid due solely to entropy generation within the fluid. This increase in entropy is also measured quantitatively by the decrease in the adiabatic efficiency of the process. Hence the adiabatic efficiency is related numerically to the entropy generation. In the special cases of ideal gases and incompressible fluids we can derive specific relationships between $\eta$ and $\sigma$.

The derivation below for the case of an ideal gas is based on a constant specific-heat ratio $k$. The result is reasonably accurate in general, because the percent variation of $k$ with temperature is fairly small in most applications. For all of the adiabatic steady-state devices considered here, the entropy balance reduces to $\sigma_m = \Delta s$, as shown in Eq. [8-16]. Combining this with the relation for the specific-entropy change of an ideal gas with constant specific heats developed in Chap. 7 gives

$$\frac{\sigma_m}{R} = \frac{c_p}{R} \ln \frac{T_{2a}}{T_1} - \ln \frac{P_2}{P_1} = \ln \frac{(T_{2a}/T_1)^{k/(k-1)}}{P_2/P_1} \qquad \text{(ideal gas)} \quad \textbf{[8-35]}$$

where use of the ideal-gas relation $c_p/R = k/(k-1)$ has been made. Solving for $T_{2a}/T_1$, we find that

$$\frac{T_{2a}}{T_1} = \left[\left(\frac{P_2}{P_1}\right) e^{\sigma_m/R}\right]^{(k-1)/k} \qquad \text{(ideal gas)} \qquad \textbf{[8-36]}$$

When the process is internally reversible, $\sigma_m = 0$, and the term $e^{\sigma_m/R} = 1$. In this case Eq. [8-36] reduces to Eq. [8-10a] for an isentropic process, where $T_{2a} = T_{2s}$. Because $\sigma_m \geq 0$, the term $e^{\sigma_m/R}$ in Eq. [8-36] is always equal to or greater than unity. Thus $T_{2a} \geq T_{2s}$, regardless of whether $P_2/P_1$ is less than or greater than unity.

Using the above information, it is now possible to relate the specific-entropy production for an ideal gas with constant specific heats to the adiabatic efficiency for several devices. Recall that $\Delta h = c_{p,av} \Delta T$ for an ideal gas. As a result, both Eq. [8-18] for the adiabatic efficiency of a turbine with negligible changes in kinetic and potential energy, and Eq. [8-23] for a nozzle with negligible changes in potential energy, can be written as $\eta = (T_{2a} - T_1)/(T_{2s} - T_1)$. Using Eq. [8-36] for the actual and isentropic temperature ratios, the adiabatic efficiencies of these two devices can be

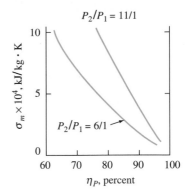

**Figure 8-20**
Entropy generation versus pump efficiency for water initially at 27°C and 100 kPa and final pressures of 600 and 1100 kPa.

written as

$$\eta_T = \eta_N = \frac{1 - [(P_2/P_1)\, e^{\sigma_m/R}]^{(k-1)/k}}{1 - (P_2/P_1)^{(k-1)/k}} \qquad \text{(ideal gas)} \quad \textbf{[8-37]}$$

where $P_2/P_1 < 1$ for a practical turbine or nozzle. When the process is internally reversible, the specific-entropy production $\sigma_m = 0$, and the adiabatic efficiency reaches its maximum value of unity. When the adiabatic efficiency reaches its minimum value of zero, the specific-entropy production for the turbine or nozzle reaches a maximum value $\sigma_{m,\text{max}} = -R \ln (P_2/P_1)$. Physically, this maximum value corresponds to the turbine or nozzle acting as a simple throttling valve with no work output and negligible changes in kinetic energy.

When Eq. [8-36] for $T_{2a}/T_1$ is substituted into the Eq. [8-28] for the adiabatic efficiency of an ideal-gas compressor, the result is

$$\eta_C = \frac{(P_2/P_1)^{(k-1)/k} - 1}{[(P_2/P_1)\, e^{\sigma_m/R}]^{(k-1)/k} - 1} \qquad \text{(ideal gas)} \quad \textbf{[8-38]}$$

where $P_2/P_1 > 1$ for a practical compressor. Unlike the turbine and the nozzle, there is no maximum value of the entropy production. For a fixed pressure ratio, the efficiency decreases as the specific-entropy production increases, and the efficiency approaches zero in the limit as $\sigma_m$ gets very large. Figure 8-21a shows how the adiabatic compressor efficiency for an ideal gas with $k = 1.4$ depends upon the normalized specific-entropy production, $\sigma_m/[R \ln (P_2/P_1)]$, and the pressure ratio. The trends for $\eta_C$ and $\sigma_m$ versus the exit temperature $T_{2a}$ for refrigerant 134a, a real gas, are shown in Fig. 8-21b. The data represent compression from a saturated vapor at 2.4 to 9 bars. A temperature $T_2$ of 40°C represents the isentropic case.

The relationship between $\sigma_m$ and $\eta$ for a pump is developed in Sec. 8-2-4. For hydraulic turbines with negligible changes in kinetic and potential energy and for hydraulic nozzles with negligible changes in potential energy, the adiabatic efficiencies (Eq. [8-21] and [8-29] respectively) can be written as

$$\eta_{T,\text{hydraulic}} = \eta_{N,\text{hydraulic}} \qquad\qquad\qquad\qquad\qquad \textbf{[8-39]}$$

$$= \frac{v(P_1 - P_2) - cT_1(e^{\sigma_m/c} - 1)}{v(P_1 - P_2)} \qquad \text{(incompressible)}$$

where $P_1 > P_2$.

The relationship between the entropy production in an adiabatic steady-flow device and the adiabatic efficiency, a defined performance parameter, has been demonstrated in this section. Entropy production is a universal measure of the internal irreversibilities within a device and provides information on how efficiently energy conversion is occurring. Because it can be calculated for any system or device, it can be used to compare the performance of dissimilar devices. By comparison, the adiabatic efficiency is

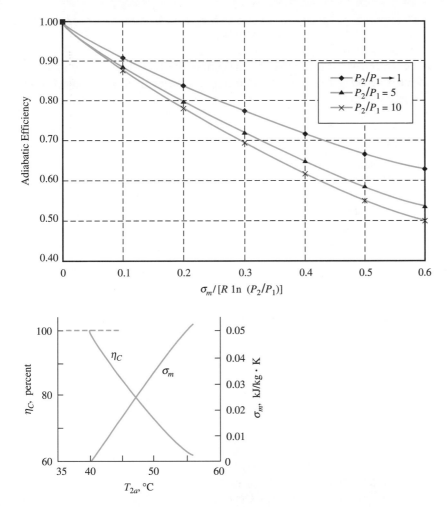

**Figure 8-21**  (a) A plot of adiabatic efficiency versus a normalized specific entropy production for an air compressor with $k = 1.4$ and pressure ratios of 1, 5, and 10. (b) Entropy generation and compressor efficiency versus exit temperature for R-134a operating between 2.4 and 9 bars.

a defined performance parameter that focuses on the useful energy transfers of a specific steady-flow device. As such, it can be easily calculated from measured performance data, correlated with other performance data, and used to compare and predict the behavior of a specific steady-flow device under realistic operating conditions. The adiabatic efficiency will always be a useful parameter; however, as engineers begin to optimize complete energy systems, minimizing the entropy production with the system will be of primary importance.

## 8-3   THE CARNOT CYCLE

In Sec. 6-9-1 we showed that the maximum thermal efficiency of any heat engine operating between two thermal reservoirs is given by the Carnot efficiency, namely,

$$\eta_{\text{th,Carnot}} = 1 - \frac{T_L}{T_H} \qquad \text{[6-58]}$$

There are several theoretical cycles, composed of a series of internally reversible processes, which have efficiencies equal to the Carnot efficiency. One of the best known is called the *Carnot cycle*. The Carnot cycle, which can operate as either a nonflow or a steady-flow system, is composed of two *internally reversible, isothermal* processes and two *internally reversible, adiabatic* processes. The sequence of the four processes is as follows:

1-2. An internally reversible, isothermal expansion during which heat transfer $Q_{H,\text{in}}$ is supplied to the working fluid which is at a boundary temperature $T_H$

2-3. An internally reversible, adiabatic expansion of the working fluid until it reaches a lower temperature $T_L$

3-4. An internally reversible, isothermal compression during which heat transfer $Q_{L,\text{out}}$ occurs at a boundary temperature $T_L$

4-1. An internally reversible, adiabatic compression of the working fluid until it reaches the original high temperature $T_H$

Consider the following example of a Carnot heat-engine cycle. Figure 8-22 shows a gas contained within a piston-cylinder device. Heat transfer $Q_H$ is supplied to the engine working medium, which is at a temperature $T_H$. (See the schematic of the engine in Fig. 8-22b.) During heat addition, the working medium expands isothermally, shown as process 1-2. Work output occurs during the expansion. From state 2 the working medium expands adiabatically and internally reversibly to state 3. During the isentropic expansion additional work is produced. State 3 corresponds to a temperature of $T_L$. The system is now compressed isothermally to state 4. During the compression process 3-4, energy is removed by heat transfer $Q_L$ at a boundary temperature $T_L$. State 4 is selected so that, by the final internally reversible, adiabatic compression, the working medium is returned to the initial state. This is process 4-1 in Fig. 8-22a. During processes 3-4 and 4-1 work is performed *on* the system. Overall, heat transfer occurs between the working medium and the system surroundings, and net work transfer out occurs during each cycle.

The heat transfer supplied to and rejected from a Carnot heat engine is easily shown on a $Ts$ diagram. Figure 8-22c illustrates this point, where the Carnot cycle appears as a rectangular area, regardless of the working fluid. For a closed system, $Q = T\,\Delta S$ during an isothermal, internally re-

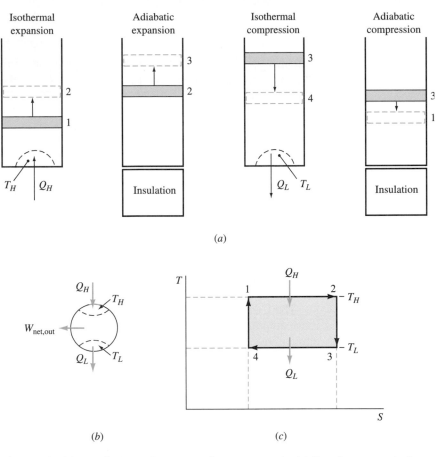

**Figure 8-22**    Illustration for a Carnot heat-engine cycle. (a) Use of a piston cylinder device; (b) schematic of the heat engine; (c) TS diagram for a Carnot heat engine.

versible process. Therefore, the area beneath the horizontal line connecting states 1 and 2 represents the heat transfer $Q_H$ supplied. In a similar fashion, the heat transfer $Q_L$ rejected is given by the area beneath line 3-4. From an energy balance on the working fluid, the difference between $Q_H$ and $Q_L$ is the net work produced by the engine during a cycle. Hence the area enclosed by the two isothermal lines and the two isentropic lines on the TS plot in Fig. 8-22c is a measure of the net work output of the Carnot cycle. For a piston-cylinder device the same net work output also can be shown as an area on a PV diagram. Figure 8-2 shows the general position of isothermal and isentropic lines for an ideal gas on a Pv diagram. On this basis the general appearance of a Carnot heat-engine cycle involving an ideal gas in a piston-cylinder device is shown in Fig. 8-23. The shape of the cycle on a PV diagram for other working fluids could be quite different. The

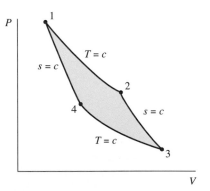

**Figure 8-23**
A typical PV diagram for an ideal gas undergoing a Carnot cycle.

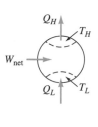

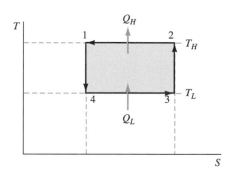

**Figure 8-24**    Illustration for a Carnot refrigerator cycle.
(a) Schematic of the refrigerator; (b) TS diagram
for a Carnot refrigerator.

numbering system in Fig. 8-23 is the same as that used in Fig. 8-22. Two
other internally reversible cycles that fulfill the Carnot efficiency, the Ericsson and Stirling cycles, are discussed in Sec. 16-15.

The Carnot thermal efficiency given by Eq. [6-58] is theoretical, because attainment of internally reversible conditions is impossible. Inherent
dissipative effects within the heat engine lead to internal irreversibilities.
However, the message given by Eq. [6-58] is clear. To increase the thermal
efficiency for both ideal and real cycles:

1. The *average* temperature at which energy is added to the system by
   heat transfer should be as high as possible.

2. The *average* temperature at which energy is rejected from the system
   by heat transfer should be as low as possible.

Usually the highest temperature in the cycle is limited by metallurgical considerations, and the lowest temperature is limited by the temperature of the
cooling medium in the environment.

Since a Carnot cycle is an internally reversible cycle, the direction of
its operation can be reversed. Such a reversed Carnot cycle is known as a
*Carnot refrigerator* cycle, or a *Carnot heat-pump* cycle. A schematic and a
$TS$ diagram for a Carnot refrigerator are shown in Fig. 8-24. As discussed
in Chap. 6, internally reversible cycles are used frequently as a standard to
which real cycles may be compared.

**Example 8-10**

**A** piston-cylinder device that contains water substance is used to execute a
Carnot engine cycle. From an initial state of 240°C and 20 percent quality, the
fluid is expanded isothermally until the pressure reaches 30 bars. This process
is followed by an isentropic expansion to 150°C. For the cycle determine (a) the
thermal efficiency, (b) the heat transfer into and the heat transfer out of the working
fluid, in kJ/kg, and (c) the net work output, in kJ/kg.

**Solution:**

**Given:** A Carnot heat-engine cycle is executed by water substance in a piston-cylinder device. A schematic with data is shown in Fig. 8-25.

**Find:** (a) $\eta_{th}$, (b) $q_H$ and $q_L$, in kJ/kg, and (c) $w_{net}$, in kJ/kg.

**Model:** Closed-system, internally reversible Carnot cycle.

**Strategy:** Apply the closed-system energy and entropy balances to isothermal and isentropic processes.

**Analysis:** (a) The thermal efficiency is found from Eq. [6-58], where $T_H$ and $T_L$ represent the fluid temperatures during the isothermal processes, and only internal reversibility is assumed. The result is

$$\eta_{th,Carnot} = 1 - \frac{T_L}{T_H} = 1 - \frac{423}{513} = 0.175$$

(b) The evaluation of heat-transfer and work-transfer quantities requires knowledge of $s$ and $u$ data for the cycle. For 20 percent quality and 240°C, the properties at the initial state are found from data in Table A-12:

$$u_1 = u_f + xu_{fg}$$
$$= 1033.2 + 0.20(2604.0 - 1033.2) = 1347 \text{ kJ/kg}$$

$$s_1 = s_f + xs_{fg}$$
$$= 2.7015 + 0.20(6.1437 - 2.7015) = 3.3899 \text{ kJ/kg·K}$$

State 2 is at 240°C and 30 bars. Since the saturation pressure at 240°C is 33.33 bars, state 2 is in the superheat region. From Table A-14,

$$u_2 = 2619.9 \text{ kJ/kg} \qquad \text{and} \qquad s_2 = 6.2265 \text{ kJ/kg·K}$$

State 3 is at 150°C and the entropy of state 2. Since $s_g$ at 150°C is 6.8379 kJ/kg·K, state 3 is a wet mixture of high quality. First the quality is found by

$$x_3 = \frac{s_3 - s_f}{s_g - s_f} = \frac{6.2265 - 1.8418}{6.8379 - 1.8418} = 0.878$$

Then the specific internal energy at state 3 is

$$u_3 = u_f + xu_{fg}$$
$$= 631.68 + 0.878(2559.5 - 631.68) = 2324.3 \text{ kJ/kg}$$

Finally, state 4 is at 150°C and the same entropy as state 1. This is a wet mixture of low quality. So, similar to the calculations for state 3,

$$x_4 = \frac{3.3899 - 1.8418}{6.8379 - 1.8418} = 0.310$$

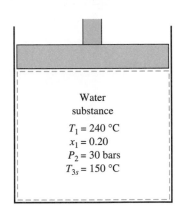

**Figure 8-25**
Equipment schematic and data for Example 8-10.

Water substance
$T_1 = 240 \text{ °C}$
$x_1 = 0.20$
$P_2 = 30 \text{ bars}$
$T_{3s} = 150 \text{ °C}$

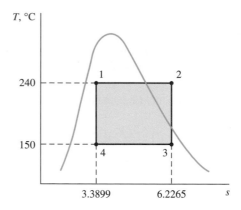

**Figure 8-26**    $Ts$ diagram for the Carnot cycle in Example 8-10.

Then the specific internal energy at state 4 is

$$u_4 = 631.68 + 0.310(2559.5 - 631.68) = 1229.3 \text{ kJ/kg}$$

The $Ts$ diagram for the cycle is shown in Fig. 8-26.

With these data the heat-transfer quantities can be evaluated. Because $q_{\text{rev}} = T \Delta s$ for the fluid during the isothermal heat-addition and heat-rejection processes,

$$q_{12} = T_H(s_2 - s_1)$$
$$= 513(6.2265 - 3.3899) = 1455.2 \text{ kJ/kg}$$

$$q_{34} = T_L(s_4 - s_3)$$
$$= 423(3.3899 - 6.2265) = -1199.9 \text{ kJ/kg}$$

(c) The net work output can be found from an energy balance for a cyclic, closed system, namely, $\sum q + \sum w = \Delta u = 0$. Therefore,

$$w_{\text{net}} = -q_{12} - q_{34} = -1455.2 + 1199.9 = -255.3 \text{ kJ/kg}$$

The negative sign denotes net work *output*.

**Comment:**    As an alternative method, the thermal efficiency based on energy quantities is

$$\eta_{\text{th}} = 1 - \frac{q_{\text{out}}}{q_{\text{in}}} = 1 - \frac{1199.9}{1455.2} = 0.175$$

This agrees with the value based on the Carnot formula. Also, the work-transfer terms could be calculated from an energy balance for each separate process and then added. In this case

$$w_{12} = u_2 - u_1 - q_{12} = -182.9 \text{ kJ/kg}$$
$$w_{23} = u_3 - u_2 = -295.4 \text{ kJ/kg}$$
$$w_{34} = u_4 - u_3 - q_{34} = 104.9 \text{ kJ/kg}$$
$$w_{41} = u_1 - u_4 = 118.1 \text{ kJ/kg}$$

The sum of these four terms again is $-255.3$ kJ/kg.

## 8-4 THE TRANSIENT ADIABATIC-DISCHARGE PROCESS

The general analysis of the charging and discharging of a fluid to or from a vessel is discussed in Chap. 5. With the discussion of isentropic processes presented earlier in this chapter, we are now in a position to analyze a special discharging process. We again shall neglect the kinetic and potential energies associated with either the control volume or the mass leaving the vessel. Thus $e = u$. The absence of shaft work across the boundary of the control volume again is noted. Although heat transfer could occur, we wish to examine at this point only those discharging processes for which heat transfer is negligible. These processes might include a rapid discharge from insulated vessels, for example, as shown in Fig. 8-27$a$. In addition, we shall assume that the process within the control volume is essentially in quasiequilibrium at all times. That is, the process is modeled as one of *uniform state,* so that property values such as $v$, $h$, $u$, and $s$ for the mass within the control volume are well defined at any instant. On the basis of relationships developed from the second law, such an adiabatic internally reversible process is one of constant specific entropy. That is,

$$s = \text{constant} \qquad \text{(adiabatic discharge)}$$

Therefore, for any fluid which flows from a pressurized vessel under the stated assumptions or idealizations, the specific entropy within the control volume remains constant. The total entropy of the control volume decreases, since the amount of mass decreases. The overall process is highly irreversible, so the net entropy change of the control volume plus its environment must be positive.

Figure 8-27$b$ illustrates the process path for the discharging of a fluid from a pressurized tank. The path is for the fluid that remains inside the tank. Note that such a process must fulfill the following idealizations:

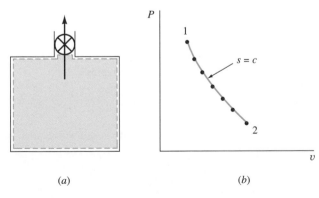

(a)  (b)

**Figure 8-27** (a) Schematic diagram; (b) $Pv$ process diagram for flow from a pressurized vessel.

(1) adiabatic, rigid control volume, (2) equilibrium within the tank at any instant, and (3) negligible kinetic- and potential-energy changes. The fact that the process is isentropic is not surprising, since it is adiabatic and quasistatic and all internal dissipative effects and finite gradients have been ruled out. That is, the process within the tank is adiabatic and internally reversible. As a specific example of such a process, consider below the flow of an ideal gas from a pressurized tank.

**Example 8-11**

**A** tank with a volume of 1.5 m³ is initially filled with air at 7.0 bars and 220°C. Determine (*a*) the final temperature, (*b*) the percentage of mass left in the tank, and (*c*) the quantity of mass, in kilograms, that left the tank if the air is permitted to leave the tank under adiabatic conditions until the pressure drops to 1.0 bar.

**Solution:**

**Given:**   Air is bled from a pressurized tank, as shown in Fig. 8-28.

**Find:**   (*a*) $T_2$, (*b*) $m_2/m_1$, and (*c*) $\Delta m$.

**Model:**   Transient flow, adiabatic, uniform state, ideal gas.

**Strategy:**   In addition to the relevant isentropic relation between $T$ and $v$ for an ideal gas, we must make use of the fact that the volume is constant in order to relate mass to specific volume within the tank.

**Analysis:**   It is realistic to assume ideal-gas behavior because of the low pressures involved, and the gas equilibrates rapidly so the process is one of uniform state within the tank. Under these conditions the mass within the tank expands isentropically. Since the volume of the control volume remains constant, from the basic relationship $V = mv$ we find that

$$dV = m\,dv + v\,dm = 0$$

Therefore

$$\frac{dm}{m} = -\frac{dv}{v}$$

For an isentropic process of an ideal gas, $T(v)^{k-1} = $ constant if one assumes constant specific heats. Differentiation of this relation yields

$$\frac{dT}{T} = (1 - k)\frac{dv}{v}$$

Elimination of $dv/v$ between the two above equations, and subsequent integration, leads to

$$\frac{m_2}{m_1} = \left(\frac{T_2}{T_1}\right)^{1/(k-1)}$$

A comparable equation for the mass ratio as a function of the pressure ratio is found from the isentropic relation $T_2/T_1 = (P_2/P_1)^{(k-1)/k}$.

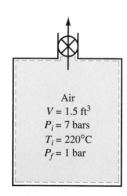

**Figure 8-28**
Schematic and data for Example 8-11.

Air
$V = 1.5 \text{ ft}^3$
$P_i = 7 \text{ bars}$
$T_i = 220°C$
$P_f = 1 \text{ bar}$

(a) Since $k$ is roughly 1.4 for the range of temperature involved, the final temperature is found from the isentropic relation

$$T_2 = T_1 \left( \frac{P_2}{P_1} \right)^{(k-1)/k} = 493 \left( \frac{1}{7} \right)^{0.285} = 283 \text{ K} = 10°\text{C}$$

(b) The percentage of mass left in the tank is found from the relation developed above,

$$\frac{m_2}{m_1} = \left( \frac{T_2}{T_1} \right)^{1/(k-1)} = \left( \frac{283}{493} \right)^{2.5} = 0.250$$

Therefore, only 25 percent of the mass is left when the pressure of 1 bar is reached, and the temperature at that instant is 10°C.

(c) The mass initially in the tank is given by the ideal-gas relation

$$m = \frac{PV}{RT} = \frac{7.0(1.5)(28.97)}{0.08314(493)} = 7.42 \text{ kg}$$

Hence the mass bled from the tank is $0.75(7.42) = 5.57$ kg.

**Comment:** Note that an energy balance was not used in this solution. If an energy balance were used, it would only lead to an analytical proof that the process within the tank is isentropic.

---

Another general process involving discharge from a tank is the situation where energy transfer in the form of heat or work transfer occurs at a rate sufficient to maintain either the temperature or pressure constant within the tank. There are many other types of transient flow problems of theoretical as well as practical interest. The solution of each type depends on the restrictions and idealizations placed on the process. In every case it is best to begin with basic equations and to proceed logically from that point.

## 8-5  SUMMARY

During an adiabatic internally reversible process the entropy remains constant, that is, $\Delta s = 0$. These *isentropic* processes are standards of performance and special property relations exist. For an *ideal gas model with variable specific heats,* these relationships are

$$s_2^0 - s_1^0 - R \ln \frac{P_2}{P_1} = 0 \qquad \frac{P_2}{P_1} = \frac{P_{r2}}{P_{r1}} \qquad \frac{v_2}{v_1} = \frac{v_{r2}}{v_{r1}}$$

When *constant* specific heats are used in an ideal-gas analysis, the isentropic relations become

$$\frac{T_2}{T_1} = \left( \frac{v_1}{v_2} \right)^{k-1} = \left( \frac{P_2}{P_1} \right)^{(k-1)/k} \qquad \text{and} \qquad \frac{P_2}{P_1} = \left( \frac{v_1}{v_2} \right)^{k}$$

where $k = c_p/c_v$. In the case of an *incompressible* substance model, the temperature is constant during an isentropic process.

The performance of adiabatic steady-flow devices is predicted through the use of *adiabatic,* or *isentropic, efficiencies.* For turbines, nozzles, compressors, and pumps, the efficiencies are

$$\eta_T \equiv \frac{w_{a,\text{out}}}{w_{s,\text{out}}} \qquad \eta_N \equiv \frac{\Delta \text{ke}_a}{\Delta \text{ke}_s} \qquad \eta_C \equiv \frac{w_{s,\text{in}}}{w_{a,\text{in}}} \qquad \eta_P \equiv \frac{w_{s,\text{in}}}{w_{a,\text{in}}}$$

where the subscripts $a$ and $s$ represent the actual and the isentropic flows for the same initial state and the same final pressure.

A Carnot cycle is composed of two internally reversible, isothermal processes and two internally reversible, adiabatic processes. Any such device, operating as either a closed or a steady-flow system, has a thermal efficiency given by the Carnot efficiency

$$\eta_{\text{th,Carnot}} = 1 - \frac{T_L}{T_H}$$

This equation applies to an internally reversible (or a totally reversible) heat-engine device. When the direction of operation of a Carnot heat-engine cycle is reversed, the reversed cycle is known as a Carnot refrigeration cycle or a Carnot heat-pump cycle.

## PROBLEMS

### GENERAL COMMENT QUESTIONS

8-1C. Explain the two limitations on the use of the functions $p_r$ and $v_r$ found in tables.

8-2C. Explain how cooling of a gas as it passes through a compressor should reduce the power input required.

8-3C. Explain whether cooling a liquid as it passes through a pump would signficantly lower the power input required.

8-4C. Explain how the isentropic process is used as a standard of performance in the development of adiabatic efficiencies for various flow processes.

### IDEAL GASES

8-1. Oxygen is compressed adiabatically in a piston-cylinder device from an initial state of 27°C and 100 kPa. The work input is 2142 kJ/kmol and the process is internally reversible. Using Table A-7 for property data, find (*a*) the final temperature, in kelvins, (*b*) the final pressure, in kilopascals.

8-2. A piston-cylinder device with a diameter of 5.0 cm and an initial length of 12.0 cm initially contains air with a gage pressure reading of 10 MPa. The piston moves outward until the volume doubles. The atmospheric pressure is 0.1 MPa, and $k = 1.4$ for air. Determine (a) the final pressure, in bars, and (b) the maximum work delivered, in joules, if the process is adiabatic.

8-3. Nitrogen at 3 bars, 400 K, and 120 cm$^3$ is allowed to expand adiabatically and reversibly to 1.70 bars in a closed system. Determine (a) the final temperature, in kelvins, (b) the work output, in kilojoules, and (c) the final volume, in cubic centimeters.

8-4. Air, in the amount of 0.40 kg, is compressed isentropically from an initial state of 1.20 bars and 0.30 m$^3$ to a final volume of 0.050 m$^3$ in a piston-cylinder device. Determine (a) the final pressure, in bars, (b) the work required, in kilojoules, and (c) the final temperature, in degrees Celsius.

8-5. Air is contained in a piston-cylinder device initially at 157°C, 0.307 MPa, and 0.760 L. The gas is expanded isentropically to 0.097 MPa. Determine (a) the final temperature, in degrees Celsius, (b) the mass of air, in kilograms, (c) the final volume, in liters, and (d) the work output, in kilojoules. Use the air table for data.

8-6. A piston-cylinder device contains carbon dioxide at 1.05 bars, 310 K, and 1000 cm$^3$. The gas is compressed isentropically to 2.11 bars. Determine (a) the final temperature, in kelvins, and (b) the work required, in kilojoules. Use tabular data.

8-7E. A piston-cylinder device with a diameter of 2.0 in and an initial length of 5.0 in initially contains air with a gage pressure reading of 1000 psig. The piston moves outward until the volume doubles. The atmospheric pressure is 14.7 psia, and $k = 1.4$ for air. Determine (a) the final pressure, in psia, and (b) the maximum work delivered, in ft·lb$_f$, if the process is adiabatic.

8-8E. Nitrogen at 45 psia, 240°F, and 20 in$^3$ is allowed to expand adiabatically and reversibly to 25.0 psia in a closed system. Determine (a) the final temperature, in degrees Rankine, (b) the work output, in Btu, and (c) the final volume, in cubic inches.

8-9E. Air, in the amount of 0.50 lb$_m$, is compressed isentropically from an initial state of 15.0 psia and 6.0 ft$^3$ to a final volume of 1.0 ft$^3$ in a piston-cylinder device. Determine (a) the final pressure, in psia, (b) the work required, in Btu, and (c) the final temperature, in degrees Fahrenheit.

8-10E. Air is contained in a piston-cylinder device initially at 300°F, 42.5 psia, and 46.3 in$^3$. The gas is expanded isentropically to 14.5 psia. Determine (a) the final temperature, in degrees Fahrenheit, (b) the mass of air, in pounds, (c) the final volume, in cubic inches, and (d) the work output, in Btu. Use the air table for data.

8-11E. A piston-cylinder device contains nitrogen at 20 psia, 100°F, and 200 in$^3$. The gas is compressed isentropically to 58.5 psia. Determine (a) the final temperature, in degrees Fahrenheit, and (b) the work required, in Btu. Use tabular data.

8-12. Air with a volume of 0.36 m$^3$ at 1.03 bars and 15°C (state 1) is compressed reversibly and adiabatically to 10 bars (state 2). It is then cooled at constant pressure to its original temperature (state 3). Finally, it expands isothermally to the original pressure (state 1). Determine the heat and work for (a) process 1-2 and (b) process 2-3, and (c) the net work for the cycle, all values in kilojoules.

8-13. A horizontal, insulated, and rigid cylinder is separated into two sections (A and B) by a frictionless, nonconducting piston. Each section initially contains a monatomic gas at 100 kPa and 300 K in a volume of 2.70 L. An electric resistor on side A is energized from an external battery until the pressure on both sides reaches 232 kPa. Determine (a) the final temperature on side B, in kelvins, (b) the work done on the gas on side B, in kilojoules, (c) the final temperature on side A, in kelvins, and (d) the amount of electrical work added, in kilojoules.

8-14E. Air with a volume of 2.36 ft$^3$ at 15.0 psia and 75°F (state 1) is compressed reversibly and adiabatically to 117 psia (state 2). It is then cooled at constant pressure to its original temperature (state 3). Finally, it expands isothermally to the original pressure (state 1). Determine the heat and work for (a) process 1-2 and (b) process 2-3, and (c) the net work for the cycle, all values in Btu.

8-15. A piston-cylinder device contains nitrogen initially at 20 bars and 250 K in a volume of 1.0 L. The fluid is compressed isentropically to 400 K. If the gas is considered a real gas, determine (a) the final pressure, in bars, (b) the work required, in kilojoules, and (c) the final volume, in liters, using data from Table A-20. (d) If the substance is assumed to be an ideal gas, determine the work required, in kilojoules.

8-16. Argon, initially at 640 kPa and 280°C, flows at a rate of 5 kg/s through an insulated, frictionless nozzle. The initial velocity is negligible, and the outlet pressure is 140 kPa. Determine (a) the final temperature, in degrees Celsius, (b) the final velocity, in m/s, and (c) the exit area, in square centimeters.

8-17. Air enters a diffuser at 0.70 bar and 7°C with a velocity of 300 m/s. The outlet temperature is 320 K, and the process is adiabatic and frictionless. Determine (a) the final velocity, in m/s, (b) the final pressure, in bars, and (c) the ratio of inlet area to outlet area.

8-18. Air enters a diffuser at 0.60 bar, −3°C, and 260 m/s. The air stream leaves the diffuser at a velocity of 130 m/s. For isentropic flow, find (a) the temperature at the outlet, in degrees Celsius, (b) the outlet pressure, in bars, and (c) the ratio of the outlet to the inlet area.

8-19. Carbon dioxide at 800 K and 2.0 MPa enters a turbine with a velocity of 100 m/s through an area of 10.0 cm$^2$. The gas expands isentropically to 500 K and leaves through an area of 60 cm$^2$. Find (a) the work output, in kJ/kmol, and (b) the mass flow rate in kmol/s.

8-20. Air expands through a turbine from 1.5 MPa to 0.10 MPa and 27°C. Changes in kinetic and potential energy are negligible, and the process is adiabatic and internally reversible. The mass flow rate is 40 kg/min, and the exit velocity is 50 m/s. Determine (a) the initial temperature, in degrees Celsius, using Table A-5 for data, (b) the power output, in kilowatts, and (c) the outlet area, in square centimeters.

8-21. Air at 20 m/s, 1.0 bar, and 27°C enters the compressor of a gas-turbine power cycle. The isentropic compressor delivers air at 8.0 bars and 60 m/s at a mass flow rate of 1.4 kg/s. Determine (a) the air exit temperature, in degrees Celsius, (b) the diameter of the circular inlet, in centimeters, and (c) the power required, in kilowatts. Use air table data.

8-22. Nitrogen gas flows steadily through a compressor at a rate of 5 kg/s and undergoes an isentropic change from 1 bar and 17°C to 2.7 bars. The measured kinetic-energy change is 5 kJ/kg. Compute (a) the power input to the gas, in kilowatts, and (b) the inlet diameter, in centimeters, if the inlet velocity is 120 m/s.

8-23E. Argon, initially at 50 psia and 540°F, flows at a rate of 10 lb/s through an insulated, frictionless nozzle. The initial velocity is negligible, and the outlet pressure is 20 psia. Determine (a) the final temperature, in degrees Fahrenheit, (b) the final velocity, in ft/s, and (c) the exit area, in square inches.

8-24E. Air enters a diffuser at 10 psia and 40°F with a velocity of 900 ft/s. The outlet temperature is 104°F, and the process is adiabatic and frictionless. Determine (a) the final velocity, in ft/s, (b) the final pressure, in psia, and (c) the ratio of inlet area to outlet area.

8-25E. Air enters a diffuser at 8 psia, 20°F, and 800 ft/s. The air stream leaves the diffuser at a velocity of 400 ft/s. For isentropic flow, find (a) the temperature at the outlet, in degrees Fahrenheit, (b) the outlet pressure, in psia, and (c) the ratio of the outlet to the inlet area.

8-26E. Carbon dioxide at 1440°R and 20 atm enters a turbine with a velocity of 300 ft/s through an area of 65.0 in$^2$. The gas expands isentropically to 900°R and leaves through an area of 395 in$^2$. Find (a) the work output, in Btu/lbmol, and (b) the mass flow rate in lbmol/s.

8-27E. Air expands through a turbine from 13.42 atm to 1 atm and 80°F. Changes in kinetic and potential energy are negligible, and the process is adiabatic and internally reversible. The mass flow rate is 80 lb/min, and the exit velocity is 150 ft/s. Determine (a) the initial temperature, in degrees Fahrenheit, using Table A-5E for data, (b) the horsepower output, and (c) the outlet area, in square inches.

8-28E. Air at 40 ft/s, 14.7 psia, and 80°F enters the compressor of a gas-turbine power cycle. The isentropic compressor delivers air at 112.5 psia and 120 ft/s at a mass flow rate of 1.8 $lb_m$/s. Determine (a) the air exit temperature, in degrees Fahrenheit, (b) the diameter of the circular inlet, in inches, and (c) the horsepower required. Use air table data.

8-29E. Nitrogen gas flows steadily through a compressor at a rate of 10 $lb_m$/s and undergoes an isentropic change from 15 psia and 60°F to 40 psia. The measured kinetic-energy change is 2.5 Btu/lb. Compute (a) the horsepower input to the gas and (b) the inlet diameter, in inches, if the inlet velocity is 140 ft/s.

8-30. Air at a flow rate of 0.7 $m^3$/s enters a shop air compressor at 32°C and 0.95 bars and leaves at 15 bars. Determine the minimum power requirement, in kilowatts, to drive the adiabatic compressor.

8-31. Monatomic argon initially at 300 kPa and 277°C flows at 4 kg/s through an insulated, frictionless nozzle. The initial velocity is negligible, and the outlet pressure is 220 kPa. Determine (a) the final temperature, in kelvins, (b) the final velocity, in m/s, and (c) the exit area, in square centimeters.

8-32. Nitrogen at 1 bar and 300 K is compressed isentropically in steady flow to 15 percent of its initial volume. Using data from Table A-6, determine (a) the final temperature, in kelvins, by an iterative procedure, and (b) the required work input, in kJ/kg.

## LIQUID SUBSTANCES

8-33. Saturated liquid water at 1 bar enters an ideal water pump at 1 m/s and is discharged at 10 bars and 5 m/s. If the power input to the isentropic device is 2.6 kW, determine the volume flow rate in L/min.

8-34. Water enters a piping system at 20°C and 10 m/s through a pipe of 4.0-cm diameter. At a position downstream the conditions are 0.150 MPa and 20 m/s, and the elevation is 22.0 m above the inlet. The local gravity is 9.70 $m/s^2$. For adiabatic and frictionless flow, determine (a) the inlet pressure, in megapascals, (b) the mass flow rate, in kg/min, and (c) the exit pipe diameter, in centimeters.

8-35. Water at 20°C is pumped at 9.2 kg/s from the surface of an open tank into a constant-diameter piping system. At the discharge from the pipe, the velocity is 10 m/s and the gage pressure is 2.0 bars. The discharge is 15.0 m above the water surface in the tank, and local gravity is 9.60 $m/s^2$. For isentropic flow, determine (a) the power required for the pump, in kilowatts, and (b) the pipe diameter, in centimeters.

8-36. Oil with a specific gravity of 0.85 is pumped from a pressure of 0.70 bar to a pressure of 1.20 bars, and the outlet lies 4.0 m above the inlet. The fluid at 15°C flows at a rate of 0.10 $m^3$/s through an inlet cross-sectional area of 0.050 m$^2$, and the outlet area is 0.020 m$^2$. The local gravity is 9.8 m/s$^2$, and the flow is assumed to be adiabatic and frictionless. Determine the power input to the pump, in kilowatts.

8-37. Water is pumped from an initial state of 2 bars, 15°C, and 2 m/s to a final state of 6.0 bars and 8 m/s. The pipe diameter at the exit is 2.0 cm, and the exit position lies 20.0 m vertically above the inlet position. If the pipe is insulated and the flow frictionless, determine (a) the mass flow rate, in kg/s, (b) the shaft work required, in kJ/kg, and (c) the power input, in kilowatts. Do not neglect any energy term if sufficient information is available. Local gravity is 9.80 m/s$^2$.

8-38. A hydraulic turbine is located 120 m below the surface of water behind a dam. Water is taken to the turbine through a 2.0-m-diameter pipe. At the outlet of a pipe of the same diameter which follows the turbine, the conditions are 1.0 bar, 15°C, and 12 m/s. If the flow is assumed to be isentropic and local gravity is 9.70 m/s$^2$, determine the power output of the turbine, in kilowatts.

8-39. Water enters a piping system at 3.20 bars, 15°C, and 4 m/s at a rate of 10.0 kg/s. Downstream at a given position the pressure is 1.80 bars, and the elevation is 12.0 m above the inlet. The local gravity is 9.75 m/s$^2$. Determine (a) the velocity at the downstream position, in m/s, and (b) the pipe diameter downstream, in centimeters, if the flow is assumed to be adiabatic and frictionless.

8-40E. Water enters a piping system at 70°F and 10 ft/s through a pipe of 2.0-in diameter. At a position downstream the conditions are 20.0 psia and 20 ft/s, and the elevation is 40.0 ft above the inlet. The local gravity is 32.0 ft/s$^2$. For adiabatic and frictionless flow, determine (a) the inlet pressure, in psia, (b) the mass flow rate, in lb$_m$/min, and (c) the exit pipe diameter, in inches.

8-41E. Water at 70°F is pumped at 9.5 lb/s from the surface of an open tank into a constant-diameter piping system. At the discharge from the pipe, the velocity is 30 ft/s and the gage pressure is 25 psig. The discharge is 48.0 ft above the water surface in the tank, and local gravity is 31.8 ft/s$^2$. For isentropic flow, determine (a) the power required for the pump, in horsepower, and (b) the pipe diameter, in inches.

8-42E. Oil with a specific gravity of 0.85 is pumped from a pressure of 8 inHg vacuum to a pressure of 18.0 psig. The outlet lies 18.0 ft above the inlet. The fluid at 60°F flows at a rate of 3.0 ft$^3$/s through an inlet cross-sectional area of 0.50 ft$^2$, and the outlet area is 0.30 ft$^2$. The local gravity is 32.0 ft/s$^2$, and the flow is assumed to be adiabatic and frictionless. Determine the required horsepower input to the pump.

8-43E. Water is pumped from an initial state of 20 psia, 60°F, and 4 ft/s to a final state of 80 psia and 20 ft/s. The pipe diameter at the exit is 3.0 in, and the exit position lies 90.0 ft vertically above the inlet position. If the pipe is insulated and the flow frictionless, determine (a) the mass flow rate, in lb$_m$/s, (b) the shaft work required, in ft·lb$_f$/lb$_m$, and (c) the power input, in horsepower. Do not neglect any energy term if sufficient information is available. Local gravity is 32.0 ft/s².

8-44E. Water enters a piping system at 45 psia, 60°F, and 6 ft/s at a rate of 20.0 $lb_m$/s. Downstream at a given position the pressure is 27.0 psia, and the elevation is 35.0 ft above the inlet. The local gravity is 31.8 ft/s². Determine (a) the velocity at the downstream position, in ft/s, and (b) the pipe diameter downstream, in inches, if the flow is assumed to be adiabatic and frictionless.

## USE OF TABULAR DATA

8-45. Saturated water vapor at 8 bars, in the amount of 9.0 g, expands in a reversible, adiabatic process to a pressure of 1.5 bars in a closed system. Calculate (a) the work done, in kilojoules, and (b) the change in volume, in liters. (c) Sketch the process on a $Ts$ diagram.

8-46. Refrigerant 134a is contained initially in a piston-cylinder device at 1.0 bar and a quality of 96.6 percent. It is compressed isentropically to a final pressure of 6.0 bars. Determine (a) the final temperature, in degrees Celsius, and (b) the required work input, in kJ/kg. (c) Sketch the process on a $Ts$ diagram.

8-47. Steam at 1.5 bars and 120°C expands isentropically through a nozzle to 1.0 bar.
(a) If the inlet velocity is negligible, find the discharge velocity in m/s.
(b) For a flow rate of 20 kg/min, find the exit area of the nozzle, in square centimeters.

8-48. Refrigerant 134a at 1.0 bar and 20°C enters a diffuser with a velocity of 200 m/s. The process is isentropic, and the exit temperature is 30°C. Determine (a) the exit pressure, in bars, (b) the exit velocity, in m/s, and (c) the inlet area, in square centimeters, if the mass flow rate is 40 kg/min.

8-49. An adiabatic steam turbine operates with inlet conditions of 140 bars, 480°C, and 150 m/s. The exit conditions are 10 bars and 250 m/s, and the mass flow rate is 10,000 kg/h. If the process is internally reversible, determine (a) the outlet temperature, in degrees Celsius, (b) the work output, in kJ/kg, (c) the power output, in kilowatts, and (d) the inlet area, in square centimeters. (e) Sketch the process on $Pv$ and $Ts$ diagrams, relative to the saturation line.

8-50. Refrigerant 134a enters a steady-flow turbine at 10 bars and 120°C. It passes through isentropically and leaves at 1 bar. Neglecting kinetic- and potential-energy changes, determine the work output, in kJ/kg.

8-51. Refrigerant 134a is compressed from a saturated vapor at 5 bars to a final pressure of 9 bars in an adiabatic steady-flow process.
(a) Determine the lowest final temperature possible, in degrees Celsius.

(b) Determine the minimum shaft work required, in kJ/kg.

(c) If the actual final temperature is 12.7°C higher than the minimum temperature due to irreversibilties, find the percentage increase in shaft work compared to the minimum value.

8-52. Saturated water vapor at 1.5 bars is compressed in an isentropic steady-flow process to 7 bars.

(a) Determine the final temperature, in kelvins.

(b) Calculate the work required, in kJ/kg.

(c) For a mass flow rate of 10,000 kg/h, find the power input required, in kilowatts.

(d) Find the inlet velocity if the inlet area is 0.20 m², in m/s.

8-53. Steam at 3 bars and 200°C enters an adiabatic nozzle with negligible velocity. At the exit the pressure is 1 bar and the area is 10 cm². Determine (a) the maximum exit velocity, in m/s, and (b) the corresponding mass flow rate, in kg/s.

8-54E. Saturated water vapor at 100 psia, in the amount of 0.1 lb$_m$, expands in a reversible adiabatic process to a pressure of 20 psia in a closed system. Calculate (a) the work done, in Btu, and (b) the change in volume, in cubic feet. (c) Sketch the process on a $Ts$ diagram.

8-55E. Steam at 60 psia and 350°F expands isentropically through a nozzle to 35 psia.

(a) If the inlet velocity is negligible, find the discharge velocity in ft/s.

(b) For a flow rate of 100 lb$_m$/min, find the exit area of the nozzle, in square inches.

8-56E. Steam at 40 psia and 300°F enters a diffuser with a velocity of 1200 ft/s. The process is isentropic, and the exit temperature is 350°F. Determine (a) the exit pressure, in psia, (b) the exit velocity, in ft/s, and (c) the inlet area, in square inches, if the mass flow rate is 80 lb$_m$/min.

8-57E. An adiabatic steam turbine operates with inlet conditions of 600 psia, 800°F, and 150 ft/s. The exit conditions are 15 psia and 250 ft/s, and the mass flow rate is 20,000 lb$_m$/h. If the process is internally reversible, determine (a) the quality of the steam at the exhaust, (b) the power output, in horsepower, and (c) the inlet area, in square inches. (d) Sketch the process on $Pv$ and $Ts$ diagrams, relative to the saturation line.

8-58E. Refrigerant 134a enters a steady-flow turbine at 80 psia and 220°F. It passes through isentropically and leaves at 20 psia and 100 ft³/min. Determine the horsepower output, if the inlet and exit areas are 0.02 and 0.01 ft².

8-59E. Refrigerant 134a is compressed from a saturated vapor at −20°F to a final pressure of 120 psia in an adiabatic steady-flow process.

(a) Determine the lowest final temperature possible, in degrees Fahrenheit.

(b) Determine the minimum shaft work required, in Btu/lb$_m$.

(c) If the actual final temperature is 21.4°F higher than the minimum temperature due to irreversibilities, find the percentage increase in shaft work compared to the minimum value.

8-60E. Water vapor at 20 psia and 300°F is compressed in an isentropic steady-flow process to 120 psia.

(a) Determine the final temperature, in degrees Fahrenheit.

(b) Calculate the work required, in Btu/lb$_m$.

(c) For a mass flow rate of 20,000 lb$_m$/h, find the power input required, in horsepower.

(d) Find the inlet velocity, in ft/s, if the inlet area is 2.0 ft$^2$.

## ACTUAL TURBINE PERFORMANCE

8-61. Steam at 20,000 kg/h passes through an adiabatic turbine from an initial state of 100 bars, 520°C to a final state of 15 bars and 280°C. Determine (a) the isentropic efficiency of the turbine, (b) the actual power output, in kilowatts, and (c) the entropy production within the turbine, in kJ/K·min.

8-62. Air enters a turbine at 1000 K and 475 kPa. The turbine adiabatic efficiency is 88 percent, and the work output is 235.7 kJ/kg. Determine (a) the actual outlet temperature, (b) the isentropic outlet temperature, in kelvins, (c) the outlet pressure in kilopascals, and (d) the entropy production, in kJ/kg·K, for the actual process. Use Table A-5.

8-63. A small hydraulic turbine uses 25 kg/s of water at an inlet state of 20°C and 1.4 MPa and discharges the fluid at 0.10 MPa. The turbine adiabatic efficiency is 76 percent. Determine (a) the power output, in kilowatts, and (b) the entropy production for the actual process, in kJ/kg·K.

8-64. An air turbine with an adiabatic efficiency of 80 percent is required to produce 100 kJ/kg of actual work. The fluid inlet temperature is 460 K, and the exhaust pressure is 0.10 MPa. Find (a) the actual exhaust temperature, in kelvins, (b) the isentropic outlet temperature, (c) the required inlet pressure, in megapascals, and (d) the entropy production for the actual process, in kJ/kg·K.

8-65. Air expands in an adiabatic turbine from 3 bars, 117°C, and 70 m/s to a final pressure of 1 bar. The mass flow rate is 2.0 kg/s.

(a) Determine the maximum work output, in kJ/kg.

(b) If the actual outlet temperature is 30°C, find the turbine isentropic efficiency.

(c) Determine the turbine inlet area, in square centimeters.

(d) Determine the entropy production for the actual process, in kJ/kg·K.

8-66. Steam at 20 bars, 440°C, and 80 m/s enters an adiabatic turbine and expands to 0.7 bar at a rate of 10,000 kg/h.
(a) Determine the turbine inlet area, in square centimeters.
(b) If the isentropic efficiency of the turbine is 80 percent, determine the steam outlet temperature, in degrees Celsius.
(c) Find the power output, in kilowatts.
(d) Determine the entropy generation, in kJ/kg·K.

8-67. Liquid water enters a turbine at 5.0 bars and 15°C with a velocity of 1.4 m/s through an opening of 0.60 m². The exit conditions of the water are 1.0 bar and 4.8 m/s. If the adiabatic efficiency of the turbine is 88 percent, determine (a) the actual power output, in kilowatts, and (b) the temperature change of the fluid, in degrees Celsius, for the adiabatic process.

8-68. Liquid water enters a turbine at 0.720 MPa, 20°C, and 4.2 m/s. The exit conditions are 0.098 MPa and 1.2 m/s. If the temperature of the fluid increases by 0.0120°C, determine (a) the actual work output, in kJ/kg, and (b) the adiabatic efficiency of the turbine.

8-69E. A small hydraulic turbine uses 50 lb$_m$/s of water at an inlet state of 60°F and 200 psia and discharges the fluid at 14.7 psia. The turbine adiabatic efficiency is 76 percent. Determine the power output, in horsepower.

8-70E. An air turbine with an adiabatic efficiency of 80 percent is required to produce 50 Btu/lb$_m$ of actual work. The fluid inlet temperature is 340°F, and the exhaust pressure is 15 psia. Find (a) the isentropic outlet temperature, (b) the required inlet pressure, in psia, (c) the actual exit temperature, in degrees Fahrenheit, and (d) the entropy generation, in Btu/lb$_m$·°R.

8-71E. Air expands in an adiabatic turbine from 45 psia, 240°F, and 200 ft/s to a final pressure of 15 psia. The mass flow rate is 4.0 lb$_m$/s.
(a) Determine the maximum work output, in Btu/lb$_m$.
(b) If the actual outlet temperature is 90°F, find the turbine isentropic efficiency.
(c) Determine the turbine inlet area, in square inches.

8-72E. Steam at 2000 psia and 1000°F enters an adiabatic turbine and expands to 10 psia and 94 percent quality. Determine (a) the adiabatic efficiency of the turbine and (b) the entropy production, in Btu/lb$_m$·°R.

8-73E. Liquid water enters a turbine at 150 psia and 50°F with a velocity of 4.4 ft/s through an opening of 6.2 ft². The exit conditions of the water are 15 psia and 4.8 ft/s. If the adiabatic efficiency of the turbine is 90 percent, determine (a) the actual power output, in horsepower, and (b) the temperature change of the fluid, in degrees Fahrenheit, for the adiabatic process.

8-74E. Liquid water enters a turbine at 105 psia, 60°F, and 3.6 ft/s through a cross-sectional area of 5.4 ft². The exit conditions are 16 psia and

9.8 ft/s. If the actual work output is 415 hp, determine (*a*) the adiabatic efficiency of the turbine and (*b*) the temperature rise of the fluid, in degrees Fahrenheit, for the adiabatic process.

## ACTUAL NOZZLE PERFORMANCE

8-75. Air enters a nozzle at 1.6 bars and 67°C. The final pressure is 1.0 bar, the initial velocity is negligible, and the mass flow rate is 2.0 kg/s. If the actual exhaust velocity is 283 m/s, determine (*a*) the nozzle adiabatic efficiency, (*b*) the actual outlet temperature, in kelvins, (c) the exit area, in square centimeters, and (*d*) the entropy generation in kJ/kg·K.

8-76. The nozzle in a turbojet engine receives air at 180 kPa and 707°C, with a velocity of 70 m/s. The air expands adiabatically to 70 kPa. If the mass flow rate is 3.0 kg/s and the nozzle adiabatic efficiency is 93 percent, determine (*a*) the discharge velocity, in m/s, (*b*) the inlet area, in square centimeters, (*c*) the actual discharge temperature, in degrees Celsius, and (*d*) the entropy generation, in kJ/kg·K.

8-77E. Air enters a nozzle at 24.2 psia and 140°F. The final pressure is 14.7 psia, the initial velocity is negligible, and the mass flow rate is 4.0 lb$_m$/s. If the actual exhaust velocity is 925 ft/s, determine (*a*) the nozzle adiabatic efficiency, (*b*) the actual outlet temperature, in degrees Fahrenheit, (c) the exit area, in square inches, and (*d*) the entropy generation in Btu/lb$_m$·°R.

8-78E. The nozzle in a turbojet engine receives air at 27 psia and 1300°F, with a velocity of 200 ft/s. The air expands adiabatically to 10.5 psia. If the mass flow rate is 6.0 lb$_m$/s and the nozzle adiabatic efficiency is 90 percent, determine (*a*) the discharge velocity, in ft/s, (*b*) the inlet area, in square inches, and (*c*) the actual discharge temperature, in degrees Fahrenheit.

8-79. An oil with a specific gravity of 0.83 enters a nozzle at 3.2 bars, 20°C, and 0.60 m/s. The final velocity is 16.9 m/s. If the increase in internal energy is 0.020 kJ/kg for the process, determine (*a*) the exit pressure, in bars, and (*b*) the nozzle efficiency for the calculated exit pressure.

8-80. A fluid with a specific gravity of 0.86 enters a nozzle at 3.9 bars, 25°C, and 0.75 m/s. The exit conditions are 16.3 m/s and 2.66 bars. For the specified pressure drop, determine (*a*) the change in internal energy for the actual process, in kJ/kg, and (*b*) the nozzle efficiency for the adiabatic process.

8-81E. An oil with a specific gravity of 0.83 enters a nozzle at 48 psia, 65°F, and 3 ft/s. The final velocity is 55 ft/s. If the increase in internal energy is 0.0080 Btu/lb$_m$ for the process, determine (*a*) the exit pressure, in psia, and (*b*) the nozzle efficiency for the calculated exit pressure.

8-82E. Water enters a nozzle at 60 psia, 70°F, and 2.0 ft/s. The exit conditions are 45 ft/s and 45 psia. For the specified pressure drop, determine (a) the change in internal energy for the actual process, in Btu/lb$_m$, and (b) the nozzle efficiency for the adiabatic process.

8-83. Liquid water enters a nozzle at 4 bars and 30°C with a velocity of 5 m/s. The exit pressure is 1 bar. Irreversibilities lead to a temperature rise of 0.015°C. Determine (a) the adiabatic efficiency and (b) the exit velocity, in m/s.

## ACTUAL COMPRESSOR PERFORMANCE

8-84. Refrigerant 134a at 140 kPa and −10°C is compressed to an exit state of 700 kPa and 60°C. The power input is 0.5 kW. Neglecting the changes in kinetic and potential energies, determine (a) the adiabatic efficiency of the compressor, (b) the volume flow rate of the refrigerant at the compressor inlet, in L/min, and (c) the maximum volume flow rate (L/min) at the inlet conditions that this 0.5-kW compressor can handle without violating the second law. In addition, (d) sketch the actual process on a $Ts$ diagram.

8-85. An adiabatic compressor draws 5 kg/s of air at 1 bar, 27°C, and delivers air at 5 bars, 247°C. Find (a) the required power input, in kW, (b) the adiabatic efficiency, and (c) the rate of entropy production in kJ/K·s.

8-86. Air is compressed adiabatically in steady flow from 1 bar and 17°C to 6 bars. If the compressor adiabatic efficiency is 82 percent, determine (a) the outlet temperature, in degrees Celsius, (b) the temperature rise, in degrees Celsius, due to irreversibilities, (c) the actual work input, in kJ/kg, and (d) the entropy generation for the actual process, in kJ/kg·K.

8-87. Refrigerant 134a is compressed from a saturated vapor at −4°C to a final pressure of 9 bars. If the process is adiabatic and the compressor efficiency is 70 percent, determine (a) the outlet temperature, in degrees Celsius, and (b) the entropy production for the process, in kJ/kg·K, for the steady-flow process.

8-88. An ideal gas with constant specific heats enters a steady-flow compressor at 0.10 MPa and 300 K, is compressed adiabatically, and is discharged at 1.0 MPa. For the gas $c_p$ = 2.05 kJ/kg·K and $c_v$ = 1.46 kJ/kg·K. Determine the discharge temperature, in kelvins, and the work, in kJ/kg, if (a) the process is internally reversible and (b) the compressor adiabatic efficiency is 80 percent. (c) Sketch both processes on a $Pv$ and a $Ts$ diagram.

8-89E. Air is compressed adiabatically in steady flow from 14.5 psia and 40°F to 98 psia. If the compressor adiabatic efficiency is 82 percent, determine (a) the actual work input, in Btu/lb$_m$, (b) the outlet temperature, in degrees Fahrenheit, and (c) the entropy generation for the actual process, in Btu/lb$_m$·°R.

8-90E. Refrigerant 134a is compressed from a saturated vapor at 0°F to a final pressure of 200 psia. If the process is adiabatic and the compressor efficiency is 77.3 percent, determine (a) the outlet temperature, in degrees Fahrenheit, and (b) the entropy production for the process, in Btu/lb$_m$·°R, for the steady-flow process.

8-91E. An ideal gas with constant specific heats enters a steady-flow compressor at 15 psia and 500°R, is compressed adiabatically, and is discharged at 150 psia. For the gas $c_p = 0.49$ Btu/lb$_m$·°R and $c_v = 0.35$ Btu/lb$_m$·°R. Determine the discharge temperature, in degrees Rankine, and the work, in Btu/lb$_m$, if (a) the process is internally reversible and (b) the compressor adiabatic efficiency is 80 percent. (c) Sketch both processes on a $Pv$ and a $Ts$ diagram.

## ACTUAL PUMP PERFORMANCE

8-92. Water enters a pump at 1.0 bar and 20°C with a velocity of 2.6 m/s through an opening of 22.0 cm$^2$. The exit conditions of the water are 6.0 bars and 7.8 m/s. If the required power input is 4.0 kW, determine (a) the adiabatic efficiency of the pump and (b) the rise in temperature of the fluid for the adiabatic process.

8-93. A liquid hydrocarbon with a specific gravity of 0.82 enters a pump at 0.10 MPa and 25°C. Shaft work is done on the fluid in the amount of 2.40 kJ/kg. The specific heat of the fluid is 2.20 kJ/kg·K. Find the pressure rise, in megapascals, if (a) the process is isentropic and (b) the fluid temperature rises 0.070°C during the process.

8-94E. Water enters a pump at 15 psia and 70°F with a velocity of 7.8 ft/s through an opening of 3.40 in$^2$. The exit conditions of the water are 90 psia and 24 ft/s. If the required power input is 5.0 hp, determine (a) the adiabatic efficiency of the pump and (b) the rise in temperature of the fluid, in degrees Fahrenheit, for the adiabatic process.

8-95E. A liquid hydrocarbon with a specific gravity of 0.82 enters a pump at 16 psia and 80°F. Shaft work is done on the fluid in the amount of 1.20 Btu/lb$_m$. The specific heat of the fluid is 0.53 Btu/lb$_m$·°R. Find the pressure rise, in psi, if (a) the process is isentropic and (b) the fluid temperature rises 0.125°F during the process.

8-96. Liquid water enters a pump at 2 bars and 27°C, and leaves at 8 bars. The temperature rise of the fluid during the adiabatic process is 0.050°C, and the potential- and kinetic-energy changes are negligible. Find (a) the actual work input, in kJ/kg, (b) the adiabatic efficiency, in percent, and (c) the entropy production, in kJ/kg·K.

8-97. A liquid-water pump with an efficiency of 70 percent is driven by a steam turbine with an adiabatic efficiency of 80 percent. Saturated liquid water enters the pump at 1.5 bars and 1 m/s, and leaves at

10 bars and 5 m/s. Steam enters the turbine at 0.1 kg/s, 5 bars, and 200°C, and leaves at 1.5 bars. Neglect the kinetic-energy change in the turbine. Find (*a*) the power output of the turbine, in kilowatts, and (*b*) the volume flow rate of liquid water at the inlet in L/min.

## LIQUID PIPING SYSTEMS AND THROTTLING

8-98. Water enters a 5.0-cm-diameter pipe at 15°C and a velocity of 3 m/s. The liquid leaves through a 3.0-cm-diameter pipe at an elevation 10.0 m above the inlet. The flow is adiabatic, and the local gravity is 9.70 m/s². Determine the pressure change in kilopascals if (*a*) the flow is internally reversible and (*b*) friction causes a temperature rise of 0.004°C.

8-99. Water enters a piping system at 10 m/s and 20°C. At a position downstream the velocity is 20 m/s, and the elevation is 22 m above the inlet. The local gravity is 9.70 m/s², and the flow is adiabatic. Determine the percentage increase in the pressure drop when the fluid experiences a 0.010°C temperature rise compared to frictionless flow.

8-100. Liquid water at 50 bars and 100°C is throttled to 25 bars. Estimate the temperature change and the entropy production, in kJ/kg·K, (*a*) if the flow is incompressible and (*b*) if linear interpolation of compressed-liquid data is used.

8-101. Liquid water at 75 bars and 80°C is throttled to 25 bars. Estimate the temperature change and the entropy production, in kJ/kg·K, (*a*) if the flow is incompressible and (*b*) if linear interpolation of compressed-liquid data is used.

8-102. Liquid water at 75 bars and 100°C is throttled to 50 bars. Estimate the temperature change and the entropy production, in kJ/kg·K, (*a*) if the flow is incompressible and (*b*) if linear interpolation of compressed-liquid data is used.

8-103E. Water enters a 3.0-in-diameter pipe at 50°F and a velocity of 6 ft/s. The liquid leaves through a 2.0-in-diameter pipe at an elevation 30.0 ft above the inlet. The flow is adiabatic, and the local gravity is 31.9 ft/s². Determine the pressure change, in psi, if (*a*) the flow is internally reversible and (*b*) friction causes a temperature rise of 0.008°F.

8-104E. Water enters a piping system at 10 ft/s and 70°F. At a position downstream the velocity is 20 ft/s, and the elevation is 40 ft above the inlet. The local gravity is 32.0 ft/s², and the flow is adiabatic. Determine the percentage increase in the pressure drop when the fluid experiences a 0.020°F temperature rise compared to frictionless flow.

8-105E. Liquid water at 1000 psia and 100°F is throttled to 500 psia. Estimate the temperature change, in degrees Fahrenheit, and the entropy production, in Btu/lb$_m$·°R, (*a*) if the flow is incompressible and (*b*) if linear interpolation of compressed-liquid data is used.

## ADIABATIC EFFICIENCY AND ENTROPY PRODUCTION

8-106. An axial-flow pump was tested under steady-state conditions to determine its performance characteristics as a function of mass flow rate $\dot{m}$ through the pump. The following equations were developed from fitting test data with $0 \le \dot{m} \le 300$ kg/s:

$$\eta_{pump} = (7.33 \times 10^{-3} \text{ s/kg})\dot{m} - (16.67 \times 10^{-6} \text{ s}^2/\text{kg}^2)\dot{m}^2$$

$$P_2 - P_1 = (49.0 \text{ kPa}) - (0.115 \text{ kPa·s/kg})\dot{m}$$

where $P_2 - P_1$ is the pressure rise across the pump. Assume the data were obtained for water at 25°C. Plot the following information for the pump as a function of mass flow rate: adiabatic pump efficiency $\eta_{pump}$, specific entropy production $\sigma_m$, and the entropy production rate $\dot{\sigma}$. Determine (a) if the pump efficiency has a local maximum and (b) if the specific entropy production and the entropy production rates have local minimums over the indicated range of mass flow rates. (c) If they exist, do the maximum and minimum values occur at or near the same flow rates? (d) What is the significance of this?

8-107. Two adiabatic steady-state gas compressors are connected in series and operate as a single unit. The pressure ratio across the first compressor is $P_2/P_1$ and the pressure ratio across the second compressor is $P_3/P_2$. The pressure ratio across the combined compressors is $P_3/P_1 = 10$. The adiabatic efficiency of each compressor is 85 percent. Determine the optimum value for the pressure ratio across each compressor to maximize the specific entropy production $\sigma_m = \dot{\sigma}/\dot{m}$ for the combined compressors.

8-108. Repeat Prob. 8-107, only now consider whether the actual value of the adiabatic compressor efficiency influences the optimum pressure ratio.

8-109. For very small pressure drops, Eq. [8-37] developed for an ideal-gas turbine should give the same results as Eq. [8-39] for a hydraulic turbine. Show analytically or numerically that this is true.

8-110. A three-stage gas turbine has an inlet pressure of 10 bars and an outlet pressure of 1 bar. It can be modeled as three separate gas turbines connected in series. Each turbine has an adiabatic efficiency of 90 percent. Assume that the working fluid is air and it can be modeled as an ideal gas. Find the optimum value of the two intermediate pressures that will maximize the specific entropy production $\sigma_m$ for the combined three-stage turbine.

## CARNOT CYCLES

8-111. Water substance undergoes a Carnot heat-engine cycle while contained within a piston-cylinder device. From an initial state of saturated liquid at 280°C, the fluid is expanded isothermally until the pressure is 40 bars. This process is followed by an isentropic expansion to 100°C. For the cycle determine (a) the thermal efficiency, and then (b) the heat supplied, (c) the heat rejected, and (d) the work during isentropic expansion, all in kJ/kg.

8-112. Steady-flow equipment is used with water substance to execute a Carnot heat-engine cycle. From an initial state of 320°C and 10 percent quality, the fluid is expanded isothermally to 80 bars. This process is followed by an isentropic expansion to 10 bars. For the cycle determine (a) the thermal efficiency, and then (b) the heat supplied, (c) the heat rejected, and (d) the work during isentropic expansion, all in kJ/kg.

8-113. Refrigerant 134a undergoes a Carnot refrigeration cycle while contained within a piston-cylinder device. From an initial state of 4 bars and 40°C, the fluid is condensed isothermally to a quality of 50 percent. This process is followed by an isentropic expansion to −20°C. For this reverse Carnot cycle determine (a) the coefficient of performance, and then (b) the heat rejected, (c) the heat supplied from the cold region, and (d) the net work input, all in kJ/kg.

8-114. Steady-flow equipment is used with refrigerant 134a as the working fluid to execute a Carnot heat-pump cycle. From an initial state of 14 bars and 60°C, the fluid is condensed isothermally to a saturated liquid. This process is followed by isentropic expansion to 5 bars. For the reverse Carnot cycle, determine (a) the coefficient of performance, and then (b) the heat rejected at 60°C, (c) the heat supplied from the cold region, and (d) the isentropic work of compression, all in kJ/kg.

8-115E. A piston-cylinder device that contains water substance is used to execute a Carnot heat-engine cycle. From an initial state of saturated liquid at 500°F, the fluid is expanded isothermally until the pressure is 600 psia. This process is followed by an isentropic expansion to 200°F. For the cycle determine (a) the thermal efficiency, and then (b) the heat supplied, (c) the heat rejected, and (d) the work during isentropic expansion, all in Btu/lb$_m$.

8-116E. Steady-flow equipment is used with water substance as the working fluid to execute a Carnot heat-engine cycle. From an initial state of 600°F and 10 percent quality, the fluid is expanded isothermally to 1200 psia. This process is followed by an isentropic expansion to 150 psia. For the cycle determine (a) the thermal efficiency, and then (b) the heat supplied, (c) the heat rejected, and (d) the work during isentropic expansion, all in Btu/lb$_m$.

8-117E. Refrigerant 134a undergoes a Carnot refrigeration cycle while contained within a piston-cylinder device. From an initial state of 100 psia and 80°F, the fluid is condensed isothermally to a quality of 50 percent. This process is followed by an isentropic expansion to 0°F. For this reverse Carnot cycle determine (a) the coefficient of performance, and then (b) the heat rejected, (c) the heat supplied from the cold region, and (d) the isentropic work input, all in Btu/lb$_m$.

## TRANSIENT ADIABATIC DISCHARGE

8-118. Steady-flow equipment is used with refrigerant 134a as the working fluid to execute a Carnot refrigeration cycle. From an initial state of 10 bars and 40°C, the fluid is condensed isothermally to a quality of 20 percent. This process is followed by isentropic expansion to 2 bars. For the reverse Carnot cycle, determine (a) the coefficient of performance, and then (b) the heat rejected at 40°C, (c) the heat supplied from the cold region, and (d) the isentropic work of expansion, all in kJ/kg.

8-119. An insulated tank with a volume of 0.12 m$^3$ initially contains argon at 3.0 MPa and 240°C. A valve is opened until one-half of the mass within the tank has escaped. Determine (a) the final temperature in the tank, in °C, and (b) the final pressure in kPa.

8-120. An insulated tank with a volume of 0.10 m$^3$ initially contains helium at 2 MPa and 227°C. It is connected through a valve to a second tank which is insulated and initially evacuated. The valve is opened until the pressure in the tank drops to 1.0 MPa, and the pressure in the second tank reaches 0.50 MPa at the same time. Find (a) the kilomoles of gas that leaves the first tank, (b) the final temperature in the second tank, in kelvins, (c) the volume of the second tank, in cubic meters, and (d) the entropy production for the overall process, in kJ/K.

8-121. An insulated tank with a volume of 0.14 m$^3$ initially contains argon at 3 MPa and 127°C. It is connected through a valve to a second tank which is insulated and initially evacuated. The valve is opened until the pressure in the tank drops to 2 MPa, and the pressure in the second tank reaches 0.7 MPa at the same time. Find (a) the kilomoles of gas that leaves the first tank, (b) the final pressure in the second tank, in bars, (c) the volume of the second tank, in cubic meters, and (d) the entropy production for the overall process, in kJ/K.

8-122E. An insulated tank with a volume of 3.0 ft$^3$ initially contains air at 300 psia and 140°F. It is connected through a valve to a second tank which is insulated and initially evacuated. The valve is opened until the pressure in the tank drops to 200 psia, and the pressure in the second tank reaches 100 psia at the same time. Find (a) the pound-moles of gas that leaves the first tank, (b) the final temperature in the second tank, in °R, (c) the volume of the second tank, in cubic feet, and (d) the entropy production for the overall process, in Btu/°R. The average values of $c_v$ and $c_p$ may be taken as 0.173 and 0.242 Btu/lb$_m$·°R.

# CHAPTER
# 9

# AVAILABILITY (EXERGY) AND IRREVERSIBILITY

Biomass power plant in Whitefield, NH burns wood chips to generate
steam that powers a steam turbine producing electricity.

Experience has taught us that energy has both *quality* and *quantity*. In this chapter the important concepts of availability and irreversibility are used to explain more fully what is meant by the quality of energy. Both concepts are formulated from the first- and second-law constraints on thermodynamic processes. Availability (or exergy) is a thermodynamic property that quantifies the *quality of energy,* but unlike energy it is not conserved. Irreversibility, like entropy production, is a measure of thermodynamic losses in a system and helps us locate and quantify wasteful uses of energy in engineering processes.

## 9-1  INTRODUCTION

One of the major goals of engineering design is the optimization of a process within given constraints. In the energy field this implies the optimal use of energy during transfer or transformation. As in other engineering areas, we need some way of measuring optimum performance. In Chap. 8 this was done by introducing the concept of equipment efficiencies, such as those for turbines, compressors, and nozzles. For devices that operate essentially adiabatically, the "best performance" is considered to be an isentropic process. Equipment efficiencies based on isentropic conditions are frequently called *first-law efficiencies,* and they are very important for comparing performance. However, such efficiencies have a serious drawback. The end state of the actual process and the ideal process to which it is compared are different. A more fundamental approach would be to compare the actual process against an optimum process between the same end states. Such a comparison leads to the concept of *second-law efficiencies.*

Optimization of energy usage is based on the concept that energy has both quantity and quality. Historically, the quality of a given amount of energy is a measure of its potential to produce useful work. If this potential for doing useful work is reduced during a process, then we say that the energy has been degraded. Experience shows that this always occurs during irreversible processes. Thus the second law of thermodynamics is a law concerning the degradation of energy. Whenever energy is transformed or transferred in real processes, its potential for producing useful work is reduced forever. This chapter is devoted to establishing criteria for optimizing energy processes and for measuring the degradation of energy that occurs during real processes.

## 9-2  WORK AND ENTROPY PRODUCTION

The work transfer or power for a system is usually calculated from an energy balance and the entropy production is calculated from an entropy balance. As we demonstrated in Chaps. 6, 7, and 8, production of entropy within a system is usually directly related to a degradation in the performance of the system. The greater the entropy production, the farther the system performance deviates from the ideal performance. To clearly demonstrate the impact of entropy production on work transfer, it is useful to develop an equation that directly relates these two concepts. Because a closed system can be treated as a special case (i.e., $\dot{m} = 0$) of an open system (control volume), we shall begin our discussion with an open system.

A general expression for the reversible work for either a closed or open system is found by combining the unsteady-state equations for an energy balance and an entropy balance. Figure 9-1 shows a control volume which

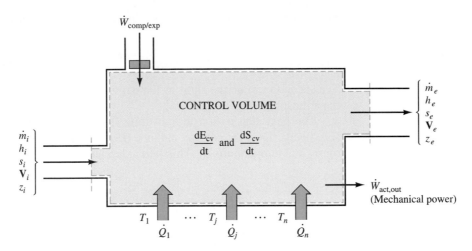

**Figure 9-1** A control volume which exchanges energy by heat transfer $\dot{Q}_j$ at surface temperature $T_j$ along the control surface.

exchanges energy by heat transfer $\dot{Q}_j$ at a surface temperature $T_j$ along the control surface. On the basis of Eq. [7-26], the entropy balance for the control volume is

$$\frac{dS_{cv}}{dt} = \sum_{in} s_i \dot{m}_i - \sum_{out} s_e \dot{m}_e + \sum_{j=1}^{n} \frac{\dot{Q}_j}{T_j} + \dot{\sigma}_{cv} \qquad \textbf{[9-1]}$$

where $\dot{Q}_j/T_j$ is the entropy transfer rate by heat transfer. The appropriate general energy balance, based on Eq. [5-15], is

$$\frac{dE_{cv}}{dt} = \dot{Q} + \dot{W} + \sum_{in} \left( h + \frac{\mathbf{V}^2}{2} + gz \right)_i \dot{m}_i - \sum_{out} \left( h + \frac{\mathbf{V}^2}{2} + gz \right)_e \dot{m}_e$$

$$\textbf{[9-2]}$$

where $\dot{W}$ represents the rates for *all* possible forms of work transfer, including compression or expansion work.

However, it is important to determine the rate of the net *useful* work transfer $\dot{W}_u$ associated with a process. The **useful work transfer** is defined as *the total work transfer minus the work done by the atmosphere.* If we designate the atmospheric or environmental pressure by $P_0$, which is taken to be constant, then the work done by the atmosphere on the system is $-P_0 \, dV$, and

$$\delta W_u = \delta W - (-P_0 \, dV) \qquad \textbf{[9-3]}$$

For a finite change of state,

$$W_u = W + P_0 \Delta V \qquad \text{or} \qquad \dot{W}_u = \dot{W} + P_0 \frac{dV_{cv}}{dt} \qquad \textbf{[9-4]}$$

This leads to the following relationship we need for the energy balance:

$$\dot{W} = \dot{W}_u - P_0 \frac{dV_{cv}}{dt}$$ [9-5]

The second term on the right of the expression above is the work done by the atmosphere.

The next step is first to substitute Eq. [9-5] into the energy balance (Eq. [9-2]). Then the entropy balance (Eq. [9-1]) is multiplied by the environmental temperature $T_0$. This latter step leads to

$$\frac{T_0 dS_{cv}}{dt} = T_0 \sum_{in} s_i \dot{m}_i - T_0 \sum_{out} s_e \dot{m}_e + T_0 \sum_{j=1}^{n} \frac{\dot{Q}_j}{T_j} + T_0 \dot{\sigma}_{cv}$$

and makes the entropy balance dimensionally consistent with the energy equation. Then the new format of the entropy balance above is subtracted from the new format of the energy balance, and rearranged. The final result for the rate of net useful work $\dot{W}_u$ is

$$\dot{W}_u = \sum_{out} \left( h + \frac{V^2}{2} + gz - T_0 s \right)_e \dot{m}_e - \sum_{in} \left( h + \frac{V^2}{2} + gz - T_0 s \right)_i \dot{m}_i$$
$$- \sum_{j=1}^{n} \dot{Q}_j \left( 1 - \frac{T_0}{T_j} \right) + \frac{d(E + P_0 V - T_0 S)_{cv}}{dt} + T_0 \dot{\sigma}_{cv}$$

[9-6]

Equation [9-6] is the desired relationship that directly relates the net useful power into a system with the rate of entropy production within the system. This general expression is in terms of mass-transfer effects, heat-transfer effects, changes within the control volume, and irreversibilities. It is this equation we wish to convert into a form where each term is expressed in terms of an *availability* function.

The application of Eq. [9-6] to an internally reversible process, for which the entropy production is zero, leads to a general expression for the rate of *reversible useful work* $\dot{W}_{rev,u}$. When $\dot{\sigma}_{cv} = 0$, Eq. [9-6] reduces to

$$\dot{W}_{rev,u} = \sum_{out} \left( h + \frac{V^2}{2} + gz - T_0 s \right)_e \dot{m}_e - \sum_{in} \left( h + \frac{V^2}{2} + gz - T_0 s \right)_i \dot{m}_i$$
$$- \sum_{j=1}^{n} \dot{Q}_j \left( 1 - \frac{T_0}{T_j} \right) + \frac{d(E + P_0 V - T_0 S)_{cv}}{dt}$$

[9-7]

This equation will be used in the following sections to develop various forms of the availability function.

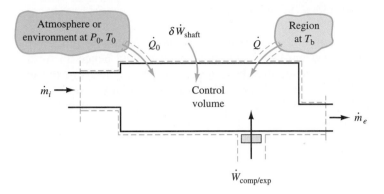

**Figure 9-2** Schematic for the development of reversible useful work associated with a control volume which exchanges heat solely with the environment and/or another thermal reservoir.

As discussed in Sec. 7-5-1, the summation term $\sum \dot{Q}_j/T_j$ in the entropy balance may be difficult to evaluate numerically, because the values of $\dot{Q}_j$ and $T_j$ are not known at every position on the boundary. The same difficulty exists with the term $\sum \dot{Q}_j[1 - (T_0/T_j)]$ in Eqs. [9-6] and [9-7]. To overcome this problem, it may be reasonably accurate to replace $T_i$ by a *constant* or *uniform* temperature $T_b$ at the boundary of the flow region. In this case Eq. [9-6] may be written as

$$\dot{W}_u = \sum_{\text{out}} \left(h + \frac{V^2}{2} + gz - T_0 s\right)_e \dot{m}_e - \sum_{\text{in}} \left(h + \frac{V^2}{2} + gz - T_0 s\right)_i \dot{m}_i$$

$$- \dot{Q}\left(1 - \frac{T_0}{T_b}\right) + \frac{d(E + P_0 V - T_0 S)_{\text{cv}}}{dt} + T_0 \dot{\sigma}_{\text{cv}}$$

$$[9\text{-}8]$$

This situation with one region of temperature $T_b$ and another region consisting of the atmosphere at temperature $T_0$ is shown in Fig. 9-2. The case where the environment acts as the *sole* sink or source for heat transfer is of major importance in the development of the concept of availability.

## 9-3 AVAILABILITY

The equations for actual and reversible work are frequently formulated in terms of the *availability functions* for closed and open systems. To this end it is necessary first to determine the work potential of a system at a given state as it proceeds toward a state of equilibrium with the environment while any heat transfer is solely with the environment.

### 9-3-1   THE DEAD STATE

Once a system and environment are in equilibrium, no further change of state of the system can occur spontaneously, and hence no further work is performed. Hence the above-described process leads to the *maximum reversible work* or *work potential* associated with the *state* of a system. When a system and its environment are in equilibrium with each other, the system is said to be in its **dead state.** Specifically, a system in a dead state is in thermal and mechanical equilibrium with the environment at $T_0$ and $P_0$. The numerical values of $(T_0, P_0)$ recommended for the dead state are those of the *standard atmosphere,* namely, 298.15 K and 1.01325 bars (1 atm).

Additional requirements for the dead state are that the velocity of a closed system or fluid stream be zero and that the gravitational potential energy be zero. This latter requirement is fulfilled by arbitrarily setting some elevation on earth, such as sea level or ground level, as zero. These restrictions of temperature, pressure, velocity, and elevation characterize a *restricted dead state* associated with thermomechanical equilibrium with the atmosphere. It is restricted in the sense that chemical equilibrium with the environment, for example, is not considered. That is, the control mass is not allowed to pass into or react chemically with the environment. The work potential of a system relative to its dead state, which exchanges heat solely with the environment, is called the thermomechanical *availability* of that state.

The methods for evaluating the availability and availability transfer for closed and open steady-state systems, as well as for heat-transfer processes, are presented in several of the following sections. The availability transfer associated with a work interaction is immediately obvious. Work interactions by concept are reversible at the point where they occur at the boundary. Consequently, the availability transfer associated with work transfer (excluding work against the environment) *equals* the value of the useful work itself.

### 9-3-2   AVAILABILITY FOR A CLOSED SYSTEM

The general situation for any closed system is shown in Fig. 9-3. Heat transfer $\delta Q_j$ crosses the system boundary at temperature $T_j$. Because there are no flow streams associated with the closed system [control mass (cm)], Eq. [9-7] developed in Sec. 9-2-2 for the net useful work reduces to

$$\dot{W}_u = \frac{d(E + P_0 V - T_0 S)_{\text{cm}}}{dt} - \sum_{j=1}^{n} \dot{Q}_j \left( 1 - \frac{T_0}{T_j} \right) + T_0 \dot{\sigma}_{\text{cm}} \quad \textbf{[9-9]}$$

Note that $E$ may be replaced by $U$ for a *stationary* system, and the symbol cv has been replaced by cm.

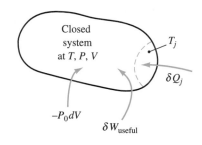

**Figure 9-3**
A closed system that exchanges energy by heat transfer $\delta \overline{Q}_j$ at surface temperature $T_j$ and by work.

For a differential change of state,

$$\delta W_u = d(E + P_0 V - T_0 S) - \sum_{j=1}^{n}\left(1 - \frac{T_0}{T_j}\right)\delta Q_j + T_0\delta\sigma_{cm} \quad \textbf{[9-10]}$$

Integration of this expression between states 1 and 2 of a closed system leads to

$$W_u = \Delta E + P_0\Delta V - T_0\Delta S - \int_1^2\left(1 - \frac{T_0}{T_b}\right)\delta Q + T_0\sigma_{cm}$$

$$= E_2 - E_1 + P_0(V_2 - V_1) - T_0(S_2 - S_1) - \int_1^2\left(1 - \frac{T_0}{T_b}\right)\delta Q + T_0\sigma$$

**[9-11]**

where $T_i$ has been replaced by $T_b$, a uniform boundary temperature where heat transfer occurs. This is the only restriction in the development of the above expression. On a unit-mass basis this may be written as

$$w_u = \Delta e + P_0\Delta v - T_0\Delta s - \int_1^2\left(1 - \frac{T_0}{T_b}\right)\delta q + T_0\sigma_m \quad \textbf{[9-12]}$$

$$= e_2 - e_1 + P_0(v_2 - v_1) - T_0(s_2 - s_1) - \int_1^2\left(1 - \frac{T_0}{T_b}\right)\delta q + T_0\sigma_m$$

The latter two equations allow evaluation of the *reversible useful work* by setting $\sigma = 0$. Under this condition, the equations predict the maximum useful work output or the minimum useful work input associated with a given change of state.

An application of Eqs. [9-11] and [9-12] of fundamental interest is to determine the reversible useful work transfer which occurs when a closed system *exchanges energy as heat transfer solely* with the environment at $T_0$. This situation is shown in Fig. 9-4, where the boundary is drawn around the closed system and the heat-transfer region. Note that the boundary

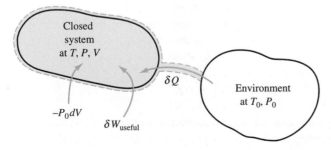

**Figure 9-4**   Schematic for the development of the reversible work associated with a closed system that exchanges energy by heat transfer only with the environment.

temperature $T_b$ where heat transfer occurs is now uniform and constant with a value of $T_0$. Therefore, the integral term in Eqs. [9-11] and [9-12] is zero. If the overall process is reversible, so that $\sigma$ is also zero, then Eq. [9-11] reduces to

$$W_{\text{rev},u} = E_2 - E_1 + P_0(V_2 - V_1) - T_0(S_2 - S_1) \qquad \textbf{[9-13]}$$

This expression leads to a relation for the availability of a closed system.

On the basis of the general discussion in Sec. 9-2-3, we have the following definition.

> The *availability* (or *exergy*) of a closed system in a given state is defined as the maximum useful work output that may be obtained from a system-atmosphere combination as the system proceeds from a given equilibrium state to the *dead state* by a process where any heat transfer occurs only with the atmosphere.

For a closed system that proceeds from a given state to the dead state in a process where any heat transfer occurs only with the environment, the reversible useful work is obtained directly from Eq. [9-13]. The result is

$$W_{\text{rev},u} = E_0 - U + P_0(V_0 - V) - T_0(S_0 - S) \qquad \textbf{[9-14]}$$

where $E_0$, $V_0$, and $S_0$ are properties of the closed system at the dead state. According to our standard sign convention, $W_u = W_{u,\text{in}} = -W_{u,\text{out}}$. Therefore, the *reversible useful work output* is given by the negative of Eq. [9-14], namely,

$$W_{\text{rev},u,\text{out}} = E - U_0 + P_0(V - V_0) - T_0(S - S_0) \qquad \textbf{[9-15]}$$

where "reversible output" implies "maximum output." Note that this result is strictly valid only for the restricted dead state. This equation then also measures the availability of a closed system. The availability of a closed system is given the symbol $\Phi$ and is calculated from the relation

$$
\begin{aligned}
\Phi &= E - U_0 + P_0(V - V_0) - T_0(S - S_0) \\
&= (E + P_0V - T_0S) - (U_0 + P_0V_0 - T_0S_0)
\end{aligned}
\qquad \textbf{[9-16]}
$$

where $E = U + \text{KE} + \text{PE}$ is the total energy of the closed system. The specific availability $\phi$ can be written as

$$\phi = \frac{\Phi}{m} = e - u_0 + P_0(v - v_0) - T_0(s - s_0) \qquad \textbf{[9-17]}$$

The dimensions and units of availability and specific availability are the same as for energy and specific energy, respectively. By applying Eq. [9-16] to the initial and final states 1 and 2 of a closed-system process, we find that

$$\Delta\Phi = \Delta U + P_0\Delta V - T_0\Delta S = m(\Delta u + P_0\Delta v - T_0\Delta s) \qquad \textbf{[9-18]}$$

Equation [9-18] will be used in the development of an availability balance.

Although we speak of the availability of a system in a given state, note that the availability is a function of both the state of the system and the local environment. Because its value depends on $T_0$ and $P_0$ as well as the properties of the system, availability is actually a property of the system-environment combination. Once the environmental conditions are *standardized,* however, availability is usually treated as a property of the system alone. We also observe that the specific availability in the dead state is zero, i.e., $\phi_{\text{dead state}} = 0$. This is consistent with our definition of the dead state. Finally, because of the manner in which $\phi$ is defined, we find that $\phi \geq 0$ for all possible equilibrium states. This result means that *any closed system at an equilibrium state not the dead state has the potential to do useful work on the surroundings.*

Although the concept of availability or exergy is introduced in terms of evaluating the maximum useful work output, it is useful in the analysis of any general process. Even for processes where no useful work is present during the actual change of state, the value of $\Delta\phi$ is still a measure of the change in the useful work potential of the closed system. Hence the availability function is of considerable importance in the analysis of all types of processes of a closed system. The use of the equations developed in this section is illustrated in the following example.

---

**Example 9-1**

**A** frictionless piston-cylinder device contains 0.10 kg of saturated water vapor maintained at 6.0 bars. Heat transfer occurs from the fluid to the atmosphere at 295 K and 1.0 bar until the water is a saturated liquid. Determine (a) the work, (b) the actual useful work, (c) the heat transfer, and (d) the reversible useful work, all in kilojoules.

**Solution:**

**Given:** Heat is transferred from saturated water vapor within a piston-cylinder device to the atmosphere, as shown in Fig. 9-5.

**Find:** (a) $W$, (b) $W_u$, (c) $Q$, and (d) $W_{\text{rev},u}$, all in kJ.

**Model:** Frictionless, constant-pressure process; closed system.

**Analysis:** The system is shown by the dashed line in Fig. 9-5.
(a) To find the actual work for a frictionless process at constant pressure,

$$W = -\int_1^2 P\,dV = -P(V_2 - V_1) = -mP(v_2 - v_1)$$

From Table A-13 we find that $v_1 = v_g = 0.3157$ m³/kg. Also, $v_2 = v_f = 0.0011$ m³/kg. Therefore

$$W = -0.10 \text{ kg} \times (6 \times 10^5)\frac{\text{N}}{\text{m}^2} \times (0.0011 - 0.3157)\frac{\text{m}^3}{\text{kg}}$$

$$= 18{,}880 \text{ N·m} = 18.88 \text{ kJ}$$

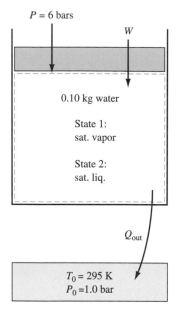

P = 6 bars

W

0.10 kg water

State 1:
sat. vapor

State 2:
sat. liq.

$Q_{out}$

$T_0 = 295$ K
$P_0 = 1.0$ bar

**Figure 9-5**
Schematic and data for Example 9-1.

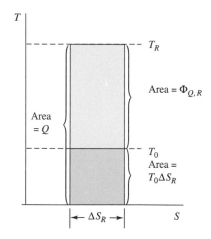

**Figure 9-6**
A $TS$ plot showing area representations
of availability transfer associated with
heat transfer from a closed system at a
uniform constant temperature $T_R$.

(*b*) The useful work input to the system is

$$W_u = W + P_0 m\, \Delta v$$
$$= 18,880 \text{ N·m} + (1 \times 10^5) \text{ N/m}^2 \times (0.1) \text{ kg} \times (-0.3146) \text{ m}^3/\text{kg}$$
$$= 18,880 - 3150 = 15,730 \text{ N·m} = 15.73 \text{ kJ}$$

(*c*) The heat transfer is found from an energy balance on the closed system:

$$Q = m(u_2 - u_1) - W = m(u_f - u_g) - W$$
$$= 0.10 \text{ kg} \times (699.90 - 2567.4) \text{ kJ/kg} - 18.88 \text{ kJ} = -205.63 \text{ kJ}$$

Thus the heat transfer is out of the system. This value is also given by $mh_{fg}$ for the process.

(*d*) Because no heat transfer occurs other than to the atmosphere, the reversible useful work is calculated from

$$W_{rev,u} = \Delta U + P_0\, \Delta V - T_0\, \Delta S$$

The first two terms on the right have already been evaluated. Since $\Delta S = -ms_{fg}$,

$$W_{rev,\,u} = -186.75 \text{ kJ} - 3.15 \text{ kJ} - 295 \text{ K} \times 0.1 \text{ kg} \times (1.9312 - 6.7600) \text{ kJ/kg·K}$$
$$= -47.45 \text{ kJ}$$

The reversible useful work is negative, indicating that useful work output is possible in a reversible process between the same end states, if the heat transfer only occurs with the environment.

**Comment:**    Owing to the irreversible heat transfer between the water and the atmosphere, the potential of the system to do useful work is reduced by 63.2 kJ [15.73 − (−47.45) kJ] for the overall process.

### 9-3-3    AVAILABILITY TRANSFER ASSOCIATED WITH HEAT TRANSFER

Entropy transfer associated with heat transfer $Q_j$ across a system boundary at $T_j$ is represented by the quantity $Q_j/T_j$. A transfer of *availability* is also associated with heat transfer. We have already found in Sec. 6-10-3 the work potential associated with $Q$ taken from a region of constant temperature $T_R$ to be

$$W_{pot} = Q\eta_{Carnot} = Q\left(1 - \frac{T_0}{T_R}\right) \qquad \textbf{[6-73]}$$

where the sink temperature is $T_0$ of the environment and $W_{pot}$ is a positive value. But the work potential of energy relative to its dead state is its availability. Hence

$$\Phi_{Q,R} \equiv Q\left(1 - \frac{T_0}{T_R}\right) \qquad \textbf{[9-19]}$$

where $\Phi_{Q,R}$ is the symbol for *availability transfer associated with heat transfer* $Q$ into or out of any closed system at a uniform constant temperature $T_R$.

The equation for $\Phi_{Q,R}$ has the following interpretation on a $TS$ diagram. First, Eq. [9-19] can be rewritten as

$$\Phi_{Q,R} = Q - T_0\left(\frac{Q}{T_R}\right) = Q - T_0\Delta S_R$$

where $\Delta S_R$ is the entropy change during an internally reversible process of a closed system at constant temperature $T_R$. Each of the terms $\Phi_{Q,R}$, $Q$, and $T_0\Delta S_R$ are represented by rectangular areas in Fig. 9-6. The heat transfer $Q = T_R\Delta S_R$ is the total shaded area.

Now consider the more general situation where the temperature of a closed system varies during a process, such as the constant-pressure change shown on the $TS$ diagram in Fig. 9-7a. In this case we must consider a differential increment of heat transfer $\delta Q_j$ removed from the system at a uniform temperature $T_j$. The reversible work for this increment of heat transfer is found by rewriting Eq. [6-73] above in a differential format, that is,

$$\delta W_{\text{rev}} = \left(1 - \frac{T_0}{T_j}\right)\delta Q_j = \delta Q_j - \frac{T_0\,\delta Q_j}{T_j}$$

For the finite process between end states 1 and 2 we must integrate this equation. The general result is

$$\boxed{\Phi_Q \equiv \int_1^2\left(1 - \frac{T_0}{T_j}\right)\delta Q_j} \qquad \textbf{[9-20]}$$

where $\Phi_Q$ is defined as the *availability transfer associated with heat transfer* $Q$ to or from a closed system of uniform but variable temperature $T_j$. To emphasize that the boundary temperature is uniform where heat transfer occurs, the boundary temperature may be represented by $T_b$. In this case the above equation is written as

$$\Phi_Q = \int_1^2\left(1 - \frac{T_0}{T_b}\right)\delta Q \qquad \textbf{[9-21]}$$

For the availability transfer $\phi_Q$ on a *unit mass basis,* we rewrite Eq. [9-20] in the format

$$\phi_Q \equiv \int_1^2\left(1 - \frac{T_0}{T_j}\right)\delta q_j \qquad \textbf{[9-22]}$$

An important point must be noted with respect to Eqs. [9-20] and [9-22]: If the system temperature $T_A$ is *greater* than $T_0$, then the system gains availability when heat transfer is into the system, and vice versa. However, if the system temperature $T_A$ is less than $T_0$, then the system *loses* availability when heat transfer is into the system and *gains* availability when heat transfer is out. That is, the energy flow and availability flow are in *opposite* directions in this case, as shown in Fig. 9-8.

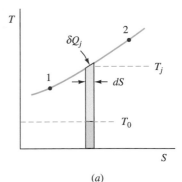

(a)

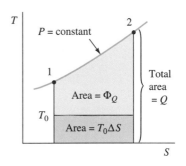

(b)

**Figure 9-7**
A $TS$ plot showing area representation of availability transfer associated with heat transfer from a system with a varying temperature.

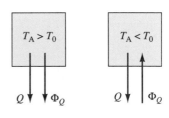

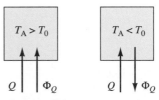

**Figure 9-8**
Schematics showing directions of heat transfer $Q$ and availability transfer $\Phi_Q$ in relation to $T_A$ and $T_0$.

**Example 9-2**

**A**ir in a closed system at 15 psia and 80°F is heated at constant pressure until its temperature reaches 400°F. The heat transfer is supplied from a constant-temperature closed system at 1040°F. The lowest surrounding temperature is 40°F. Assume the constant-pressure process is internally reversible. Determine the availability transfer associated with heat transfer (*a*) into the heat-transfer region, and (*b*) out of the heat-transfer region, and (*c*) the loss in work potential of the heat-transfer process, all answers in Btu/lb.

**Solution:**

**Given:**   Air in a constant-pressure system is heated from a thermal reservoir, as shown in Fig. 9-9.

**Find:**   (*a*) $\phi_{Q,R}$, (*b*) $\phi_{Q,air}$, and (*c*) $i_Q$, in Btu/lb.

**Model:**   Closed constant-pressure system; ideal gas.

**Analysis:**   The heat transfer from the constant-temperature system and to the air is found from an energy balance on the constant-pressure process. The system is the air within the dashed line in Fig. 9-9. If air is an ideal gas and air table A-5E is used for data, then

$$q = \Delta u - w = \Delta u + P\,\Delta v = h_2 - h_1$$
$$= (206.46 - 129.06)\ \text{Btu/lb} = 77.40\ \text{Btu/lb}_m$$

(*a*) The availability transfer into the heat-transfer region from the region of constant temperature is given by

$$\phi_{Q,R} = \int\left(1 - \frac{T_0}{T_j}\right)\delta q_j = q_R\left(1 - \frac{T_0}{T_R}\right)$$

Since $q_R$ represents the heat input into the heat-transfer region, substitution of data for the problem yields

$$\phi_{Q,R} = 77.40\ \text{Btu/lb} \times \left(1 - \frac{500}{1500}\right) = 51.6\ \text{Btu/lb}_m \ (\text{in})$$

(*b*) The availabilty of the heat transfer *out* of the heat-transfer region equals the availability of the heat transfer into the air undergoing the constant-pressure process. The air receives 77.40 Btu/lb of energy as heat transfer. The availability transfer to the air is

$$\phi_Q = \int\left(1 - \frac{T_0}{T_b}\right)\delta q = q - T_0\int\frac{\delta q}{T_b}$$

For a closed internally reversible system we know that $ds_{cm} = \delta q/T$. Making this substitution in the above equation, we can evaluate $\phi_Q$ by

$$\phi_{Q,air} = q - T_0\int ds_{cm} = q - T_0\Delta s_{cm}$$

$$= 77.40\ \text{Btu/lb}_m - (500°\text{R})(0.11245\ \text{Btu/lb}_m\cdot°\text{R}) = 21.2\ \text{Btu/lb}_m$$

Therefore, the availability transfer from the heat-transfer region is 21.2 Btu/lb$_m$. This part illustrates the evaluation of $\phi_Q$ when the temperature of the system varies

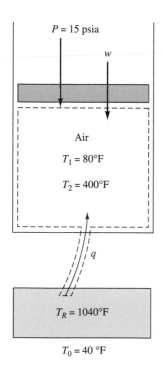

**Figure 9-9**
Schematic and data for Example 9-2.

during the process. The heat-transfer process is shown within the second set of dashed lines representing the heat-transfer region in Fig. 9-9.

(c) The loss in work potential due to irreversible heat-transfer is given by the difference in the availability transfer into and out of the heat transfer region. Thus,

$$w_{loss,pot} = \phi_{Q,in} - \phi_{Q,out} = 51.6 - 21.1 = 30.4 \text{ Btu/lb}_m$$

**Comment:** A theoretical loss of 30.4 Btu/lb of availability occurs owing to the finite temperature difference. The relative change in availability due to irreversible heat transfer is shown on the $TS$ diagram in Fig. 9-10. The total area under the upper line equals 77.4 Btu/lb.

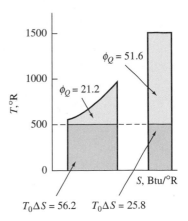

$$T_0 \Delta S = 56.2 \qquad T_0 \Delta S = 25.8$$

**Figure 9-10**
A $Ts$ plot of the process in Example 9-2.

## 9-3-4 AVAILABILITY BALANCE FOR A CONTROL MASS

Equation [9-9] is developed in Sec. 9-3-1 for the rate of useful work $\dot{W}_u$ associated with a closed system where heat transfer $Q_j$ crosses the control surface at a uniform temperature $T_b$. The result is

$$\dot{W}_u = -\sum_{j=1}^{n} \dot{Q}_j \left(1 - \frac{T_0}{T_b}\right) + \frac{d(E + P_0 V - T_0 S)_{cm}}{dt} + T_0 \dot{\sigma}_{cm}$$

For a *stationary* closed system, the kinetic and potential energy do not change, and $E$ can be replaced by $U$. For a finite change of state in this case

$$W_u = -\int_1^2 \left(1 - \frac{T_0}{T_b}\right) \delta Q + (\Delta U + P_0 \Delta V - T_0 \Delta S) + T_0 \sigma_{cm} \quad \textbf{[9-23]}$$

But the first term on the right is defined by Eq. [9-20] as $\Phi_Q$, and the second term on the right is defined by Eq. [9-18] as $\Delta \Phi$. In addition, the term $T_0 \delta \sigma_{cm}$ is defined as the **irreversibility** $I_{cm}$ *within* the closed system. As a result, after rearrangement,

$$\boxed{\Delta \Phi_{cm} = \Phi_Q + W_u - I_{cm}} \qquad \textbf{[9-24]}$$

where $I_{cm}$ measures *availability destruction* within the closed system. In words, the above equation states that

$$\begin{pmatrix} \text{Availability} \\ \text{change of a} \\ \text{control mass} \end{pmatrix} = \begin{pmatrix} \text{availability} \\ \text{transfer with} \\ \text{heat transfer} \\ \text{into system} \end{pmatrix} + \begin{pmatrix} \text{availability} \\ \text{transfer with} \\ \text{useful work} \\ \text{into system} \end{pmatrix} - \begin{pmatrix} \text{availability} \\ \text{destruction} \\ \text{within CM} \end{pmatrix}$$

Thus Eqs. [9-23] and [9-24] represent **availability balances for a control mass** undergoing heat transfer and work interactions.

Any irreversibility always destroys availability and can be evaluated directly from the availability balance by rewriting Eq. [9-24] in the format

$$I_{cm} = W_u - (\Delta \Phi_{cm} - \Phi_Q) \qquad \textbf{[9-25]}$$

In addition, the above equation can be written solely in terms of work inter-actions. For a reversible process, where $I_{cm} = 0$,

$$W_{rev,u} = (\Delta\Phi_{cm} - \Phi_Q)$$

Substitution of the above equation for $W_{rev,u}$ back into Eq. [9-25] for $I_{cm}$ yields

$$I_{cm} = W_u - W_{rev,u} \qquad \textbf{[9-26]}$$

The above relation for $I$ is equivalent to Eq. [9-24]. It is also useful to point out a specific equation for the *irreversibility of a heat-transfer process* which follows directly from Eq. [9-24]. For a heat-transfer region, both $\Delta\Phi_{cm}$ and $W_u$ are zero. Hence for heat transfer between two regions of known temperature, the availability balance reduces to

$$I_Q = \Phi_{Q,in} - \Phi_{Q,out} \qquad \textbf{[9-27]}$$

Thus availability transfers associated with heat transfer at the boundaries of the heat-transfer region solely determine the irreversibility within the region.

The increase in entropy principle for an isolated system states that

$$\Delta S_{isol} = \sigma_{isol} \geq 0$$

Similarly, since $Q$ and $W$ are each zero for an isolated system, the availability balance given by Eq. [9-24] reduces to

$$\Delta\Phi_{isol} = -I_{isol} \qquad \textbf{[9-28]}$$

Since $I_{isol}$ must always be positive for an actual process, *the availability change for an isolated system must always be negative.* That is

$$\Delta\Phi_{isol} \leq 0 \qquad \textbf{[9-29]}$$

This is a counterpart of the increase in entropy principle for an isolated system.

---

**Example 9-3**

**A** tank of air at 50 psia and 660°F has a volume of 20.0 ft³. Heat transfer occurs from the air to the atmosphere until the temperature of the air in the tank is 100°F. The surrounding atmosphere is at 60°F and 14.7 psia. Determine (*a*) the availability change of the air, and (*b*) the irreversibility of the process including the heat-transfer region, both answers in Btu.

**Solution:**

**Given:**   Air within a tank is cooled by heat transfer to the atmosphere, as shown in Fig. 9-11.

**Find:**   (a) $\Delta\Phi_{air}$, and (b) $I$, in Btu.

**Model:**   $V$ = constant ideal gas, closed system.

**Analysis:** (a) The availability change for a closed system is given by

$$\Delta\Phi = \Delta U + P_0\,\Delta V - T_0\,\Delta S = m(\Delta u + P_0\,\Delta v - T_0\,\Delta s)$$

This equation requires information on $m$, $u$, $v$, and $s$. Using an ideal-gas model and data for the initial state,

$$m = \frac{PV}{RT} = \frac{25 \text{ psia } (20.0 \text{ ft}^3)(28.97 \text{ lb}_m/\text{lbmol})}{10.73 \text{ psia·ft}^3/\text{lbmol·°R } (560°R)} = 2.41 \text{ lb}_m$$

and

$$v_1 = \frac{V}{m} = \frac{20 \text{ ft}^3}{2.41 \text{ lb}_m} = 8.30 \text{ ft}^3/\text{lb}_m$$

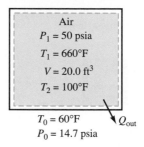

For the final state, $v_2 = v_1$, since the tank is assumed rigid. Also, $P_2 = P_1(T_2/T_1) = 50(560/1120) = 25$ psia. In addition, using average specific-heat data from Table A-3,

$$s_2 - s_1 = c_{p,av}\ln\frac{T_2}{T_1} - R\ln\frac{P_2}{P_1} = \left(0.246\ln\frac{560}{1120} - \frac{1.986}{28.97}\ln\frac{25}{50}\right)\frac{\text{Btu}}{\text{lb}_m\cdot°R}$$

$$= -0.1075 + 0.0475 = -0.1230 \text{ Btu/lb}_m\cdot°R$$

Finally, the internal energy change for the air in the tank is found from the ideal-gas relation

$$\Delta u = c_{v,av}\Delta T = 0.177 \text{ Btu/lb}_m\cdot°R \, (100 - 660)°R = -99.1 \text{ Btu/lb}_m$$

Substituting the above values into the equation for the availability change, we find on a unit-mass basis that

$$\Delta\phi = \Delta u + P_0\Delta v - T_0\,\Delta s$$
$$= -99.1 \text{ Btu/lb}_m + 0 - 520 \,°R \times (-0.1230 \text{ Btu/lb}_m\cdot°R) = -35.14 \text{ Btu/lb}_m$$

The availability change for the total mass within the tank is

$$\Delta\Phi = m\,\Delta\phi = 2.41 \text{ lb}_m(-35.14) \text{ Btu/lb}_m = -84.7 \text{ Btu}$$

The negative sign indicates that the *maximum* useful work output would be 84.7 Btu/lb. The path of the process is shown on the $Ts$ diagram in Fig. 9-12.

(b) The irreversibility for the process can be found by using the availability balance,

$$\Delta\Phi_{cm} = \Phi_Q + W_u - I$$

Because the tank is rigid, $W_u = 0$. By including the heat-transfer region inside our system we know that $\Phi_Q = 0$, because the boundary temperature equals the atmospheric temperature $T_0$. Solving for the irreversibility we find that

$$I = -\Delta\Phi = -(-84.7 \text{ Btu}) = 84.7 \text{ Btu}$$

The irreversibility also can be found from $I = T_0\sigma_{cm}$, where $\sigma_{cm}$ is the entropy production for the enlarged system.

**Comment:** The irreversibility in this case is interpreted as measuring a *loss* in the opportunity to produce useful work during the heat transfer process. Also, the value of $I$ for the tank of air alone (excluding the heat-transfer region) cannot be calculated accurately, because the evaluation of the term $Q[1 - (T_0/T_b)]$ involves a widely varying value of $T_b$.

**Figure 9-11**
Schematic and data for Example 9-3.

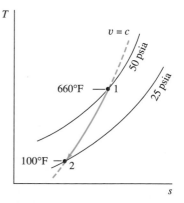

**Figure 9-12**
Path of the process in Example 9-3.

**Example 9-4**

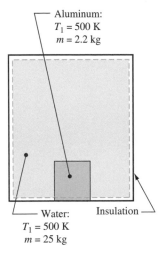

Aluminum:
$T_1 = 500$ K
$m = 2.2$ kg

Water:
$T_1 = 500$ K
$m = 25$ kg

Insulation

**Figure 9-13**
Schematic and data for Example 9-4.

**A** 2.2-kg piece of aluminum at 500 K is dropped into an open insulated tank containing 25 kg of water at 300 K and 1 bar. The two masses reach an equilibrium temperature while insulated from the environment, which is at 300 K and 1 bar. Determine (a) the final equilibrium temperature, in kelvins, and (b) the change in the availability of each substance and (c) the irreversibility of the process, both answers in kilojoules.

**Solution:**

**Given:** A piece of aluminum and liquid water are brought into thermal contact while insulated from the environment, as illustrated in Fig. 9-13.

**Find:** (a) $T_2$, in K; (b) $\Delta\Phi$ for each substance, in kJ, and (c) $I$, in kJ.

**Model:** Incompressible substances; adiabatic closed system.

**Analysis:** (a) The final equilibrium temperature is found from the basic energy equation $Q + W = \Delta U$. For the incompressible substances $W = 0$; for an insulated system $Q = 0$. As a result, for the composite system shown by the dashed line in Fig. 9-13, $\Delta U = \Delta U_{Al} + \Delta U_{water} = 0$. Since $\Delta u = c_{av}\Delta T$ for an incompressible substance, the energy balance becomes

$$[mc_{av}(T_2 - T_1)]_{Al} + [mc_{av}(T_2 - T_1)]_{water} = 0$$

Because the quantity $mc_{av}$ for water will be so much larger than that for aluminum, the final temperature will be close to 300 K. Thus data in Table A-4 indicate that the average specific heat of aluminum is around 0.950 kJ/kg·K, and that for water is 4.18 kJ/kg·K. Therefore, the above energy equation shows that

$$2.2 \text{ kg} \times \left(\frac{0.950 \text{ kJ}}{\text{kg·K}}\right) \times (T_2 - 500 \text{ K}) + 25 \text{ kg} \times \left(\frac{4.18 \text{ kJ}}{\text{kg·K}}\right) \times (T_2 - 300 \text{ K}) = 0$$

$$T_2 = 303.9 \text{ K}$$

Thus basing our estimate for the average specific heats on a final temperature close to 300 K is appropriate.

(b) The availability change for each substance is found from the relation

$$\Delta\Phi = m(\Delta u + P_0 \Delta v - T_0 \Delta s)$$

The values of $m \Delta u$ from the above energy analysis are

$$\Delta U_{Al} = 2.2(0.95)(303.9 - 500) = -409.8 \text{ kJ}$$

$$\Delta U_{water} = 25(4.18)(303.9 - 300) = 407.6 \text{ kJ}$$

These values of $\Delta U$ should be equal; the slight difference is due to a lack of enough significant figures in the calculation. The term $P_0 \Delta v = 0$ since each substance is incompressible. Finally, the entropy changes are found from Eq. [7-25], $\Delta S = mc_{av} \ln(T_2/T_1)$. Using the average $c$ data from above, we find that

$$\Delta S_{Al} = 2.2 \text{ kg} \times (0.950 \text{ kJ/kg·K}) \times \ln\frac{303.9}{500} = -1.041 \text{ kJ/K}$$

$$\Delta S_{water} = 25 \text{ kg} \times (4.18 \text{ kJ/kg·K}) \times \ln\frac{303.9}{300} = 1.351 \text{ kJ/K}$$

The availability changes then are

$$\Delta\Phi_{Al} = -409.8 \text{ kJ} + 0 - 300 \text{ K}(-1.041 \text{ kJ/K}) = -97.5 \text{ kJ}$$

$$\Delta\Phi_{water} = 407.6 \text{ kJ} + 0 - 300 \text{ K}(1.351 \text{ kJ/K}) = 2.3 \text{ kJ}$$

Owing to the modest increase in water temperature, its availability increases slightly. The availability of the aluminum decreases dramatically as it cools nearly 200 K. It should be apparent from the equation for $\Delta\Phi$ in *this particular process* that

$$\Delta\Phi_{tot} = -T_0(\Delta S_{Al} + \Delta S_{water})$$

where four terms in $\Delta\Phi_{tot}$ drop out owing to the modeling of the process.

(c) The irreversibility of the process is found by using the availability balance

$$\Delta\Phi_{cm} = \Phi_Q + W_u - I$$

The value of $\Phi_Q$ is zero because the system is adiabatic, and $W_u = 0$ because there is no work transfer. Thus

$$I = -(\Delta\Phi_{Al} + \Delta\Phi_{water}) = -(-97.5 + 2.3) \text{ kJ} = 95.2 \text{ kJ}$$

**Comment:** Although the energy is conserved during the process, the availability of the overall process decreases, owing to the irreversible nature of the process (heat transfer between the blocks through a finite temperature difference).

---

**Example 9-5**

Carbon dioxide gas is contained in a 1.0-m³ tank initially at 1.2 bars and 300 K. The temperature is increased to 400 K by two different processes: (a) stirring with a paddle wheel and (b) heat transfer from an energy source at 500 K. The environmental conditions are 1.0 bar and 300 K. Determine which method is better, solely from an irreversibility viewpoint.

**Solution:**

**Given:** Carbon dioxide in a tank undergoes a change of state by (a) work transfer and (b) heat transfer, as shown in Fig. 9-14.

**Find:** The better method, based on $I$ values.

**Model:** Rigid, closed system; ideal gas.

**Analysis:** From a thermodynamic viewpoint, the better method is the one with the smaller irreversibility.

(a) When the paddle wheel is used, as shown in Fig. 9-14a, the work transfer is found from the conservation of energy principle $Q + W = \Delta U$. Because $Q = 0$, $W_{pad} = N(\bar{u}_2 - \bar{u}_1)$. Assuming the gas is an ideal gas, the moles within the system is found by

$$N = \frac{PV}{RT} = \frac{1.2 \text{ bars} \times 1.0 \text{ m}^3}{0.08314 \text{ bar·m}^3/\text{kmol·K} \times 300 \text{ K}} = 0.0481 \text{ kmol}$$

Using data from Table A-9, the actual paddle-wheel (pad) work then is

$$W_{pad} = 0.0481 \text{ kmol} \times (10,046 - 6939) \text{ kJ/kmol} = 149.1 \text{ kJ}$$

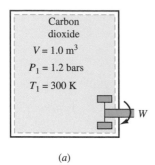

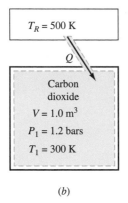

(a)

(b)

**Figure 9-14**
Schematic and data for Example 9-5.

This is also the useful work, because $P_0 \, \Delta V = 0$. The irreversibility is found from the availability balance

$$\Delta\Phi_{cm} = \Phi_Q + W_u - I$$

Since the process is adiabatic,

$$I = W_u - \Delta\Phi_{cm}$$

and $\Delta\Phi_{cm}$ is given by

$$\Delta\bar{\Phi}_{cm} = N(\bar{\phi}_2 - \bar{\phi}_1) = N(\Delta\bar{u} + P_0 \, \Delta\bar{v} - T_0\Delta\bar{s})$$

The entropy change needed for this equation is found for the ideal gas from

$$\Delta\bar{s} = \bar{s}_2^0 - \bar{s}_1^0 - R_u \ln \frac{P_2}{P_1}$$

$$= \left(225.225 - 213.915 - 8.314 \ln \frac{400}{300}\right) \text{kJ/kmol·K} = 8.92 \text{ kJ/kmol·K}$$

where $P_2/P_1 = T_2/T_1$. As a result,

$$\Delta\Phi_{cm} = \Delta U + P_0 \, \Delta V - NT_0 \, \Delta\bar{s}$$
$$= 149.1 \text{ kJ} + 0 - 0.0481 \text{ kmol} \times 300 \text{ K} \times 8.92 \text{ kJ/kmol·K}$$
$$= 20.4 \text{ kJ}$$

where $\Delta V = 0$. Therefore, the irreversibility of the process is

$$I = W_u - \Delta\Phi_{cm} = (149.1 - 20.4) \text{ kJ} = 128.7 \text{ kJ}$$

The irreversibility could also be evaluated by

$$I = T_0\sigma = T_0(m \, \Delta s_{sys}) = 300 \text{ K} \times 0.0481 \text{ kmol} \times 8.92 \text{ kJ/(kmol·K)} = 128.7 \text{ kJ}$$

(b) When the gas is heated from a constant-temperature source at 500 K, the actual useful work is zero. For the enlarged control volume shown in Fig. 9-14b, the irreversibility is obtained by employing the availability balance

$$\Delta\Phi_{cm} = \Phi_Q + W_u - I$$

We know that $W_u = 0$ and

$$I = \Phi_Q - \Delta\Phi_{cm} = Q\left(1 - \frac{T_0}{T_b}\right) - \Delta\Phi_{cm}$$

because $T_b$ is constant. We know that $Q = \Delta U = 149.1$ kJ from part $a$ and similarly $\Delta\Phi_{cm} = 20.4$ kJ. Therefore,

$$I = 149.1 \text{ kJ}\left(1 - \frac{300}{500}\right) - 20.4 \text{ kJ} = (59.6 - 20.4) \text{ kJ} = 39.2 \text{ kJ}$$

**Comment:**    This irreversibility of 39.3 kJ for the heating process compares to 128.7 kJ for the paddle-wheel process. Hence the heat-transfer process is much better thermodynamically than the work-transfer process. In general, solely from a thermodynamic viewpoint, work should not be used if a heat-transfer process could accomplish the same goal.

## 9-4    CONTROL-VOLUME AVAILABILITY ANALYSIS

Expressions for reversible work and stream availability lead to an availability balance for a control volume in steady state in this section.

### 9-4-1    REVERSIBLE WORK FOR A STEADY-STATE CONTROL VOLUME

The general development of the reversible work for a steady-state control volume follows directly from the general equation developed in Sec. 9-2. Figure 9-1 shows a control volume which has heat transfer $Q_j$ at a boundary temperature $T_j$. The general equation for the useful work in this situation is given by Eq. [9-6],

$$\dot{W}_u = \sum_{\text{out}} \left( h + \frac{\mathbf{V}^2}{2} + gz - T_0 s \right)_e \dot{m}_e - \sum_{\text{in}} \left( h + \frac{\mathbf{V}^2}{2} + gz - T_0 s \right)_i \dot{m}_i \qquad \textbf{[9-6]}$$

$$- \sum_{j=1}^{n} Q_j \left( 1 - \frac{T_0}{T_j} \right) + \frac{d(E + P_0 V - T_0 S)_{\text{cv}}}{dt} + T_0 \dot{\sigma}_{\text{cv}}$$

If we now restrict ourselves to *steady-state, steady-flow* (sf) processes, the next-to-last term in Eq. [9-6] is zero, by definition of steady state. Consequently,

$$\dot{W}_{\text{sf}} = \sum_{\text{out}} \left( h + \frac{\mathbf{V}^2}{2} + gz - T_0 s \right)_e \dot{m}_e \qquad \textbf{[9-30]}$$

$$- \sum_{\text{in}} \left( h + \frac{\mathbf{V}^2}{2} + gz - T_0 s \right)_i \dot{m}_i - \sum_{j=1}^{n} \dot{Q}_j \left( 1 - \frac{T_0}{T_j} \right) + T_0 \dot{\sigma}_{\text{cv}}$$

where the subscript sf indicates a "steady-state, steady-flow" restriction. Note also that useful work and shaft work are the same in this case, since $P_0 \Delta V = 0$. The *reversible work* is associated with an internally reversible process for which $\dot{\sigma}_{\text{cv}}$ is zero. Hence for a control volume under steady state,

$$\dot{W}_{\text{sf,rev}} = \sum_{\text{out}} \left( h + \frac{\mathbf{V}^2}{2} + gz - T_0 s \right)_e \dot{m}_e \qquad \textbf{[9-31]}$$

$$- \sum_{\text{in}} \left( h + \frac{\mathbf{V}^2}{2} + gz - T_0 s \right)_i \dot{m}_i - \sum_{j=1}^{n} \dot{Q}_j \left( 1 - \frac{T_0}{T_j} \right)$$

Finally, for a system where mass enters at state 1 and leaves the control volume at state 2, the above expression can be written on a unit-mass basis as

$$w_{sf,rev} = \left(h_2 + \frac{V_2^2}{2} + gz_2 - T_0s_2\right) - \left(h_1 + \frac{V_1^2}{2} + gz_1 - T_0s_1\right) - \sum_{j=1}^{n} q_j\left(1 - \frac{T_0}{T_j}\right)$$

**[9-32]**

Equations [9-31] through [9-32] enable one to evaluate the reversible work for a steady-state control volume. These equations give the maximum work output or minimum work input as a simple compressible fluid passes through a control volume.

### 9-4-2   AVAILABILITY FUNCTION FOR A CONTROL VOLUME

A *dead state for flow* through a control volume implies not only thermal and mechanical equilibrium of the fluid with the atmosphere at $T_0$ and $P_0$ but also that the kinetic energy at the dead state is zero relative to the environment (the fluid is at rest). In addition, its potential energy must be a minimum. That is, its elevation is the same as that of the ground level of the environment. On this basis:

> The *stream availability* of a fluid in steady flow is defined as the maximum work output that can be obtained as the fluid is changed reversibly from the given state to a dead state in a process where any heat transfer occurs solely with the atmosphere.

On the basis of Eq. [9-32] the stream availability is measured by the quantity $(h + ke + pe - T_0s)$ at the given state relative to the dead state. The stream availability is given the symbol $\psi$ for a unit mass and $\Psi$ for the total mass, and $\Psi = m\psi$. That is,

$$\psi = (h + ke + pe - T_0s) - (h_0 + pe_0 - T_0s_0) \qquad \textbf{[9-33]}$$
$$= h - h_0 - T_0(s - s_0) + \frac{V^2}{2} + gz$$

where $z$ is measured relative to $z_0$ and $pe_0 = z_0 = 0$. Although we speak of the stream availability of a fluid in a certain state, the stream availability is a function of the state of the local atmosphere as well as of the state of the fluid. The value of $\psi$ can be greater than or less than zero for any state other than the dead state.

In the case of a control volume with more than one inlet or outlet, Eq. [9-31] can be written on a rate basis in the following form:

$$\dot{W}_{sf,rev} = \sum_{out} \dot{m}_e\psi_e - \sum_{in} \dot{m}_i\psi_i - \sum_{j=1}^{n} \dot{Q}_j\left(1 - \frac{T_0}{T_j}\right) \qquad \textbf{[9-34]}$$

where the subscripts $e$ and $i$ in the summation terms again stand for the exit and inlet streams, respectively.

### 9-4-3 AVAILABILITY BALANCE FOR A STEADY-STATE CONTROL VOLUME

Equation [9-6] is developed in Sec. 9-2 for the rate of net useful work $\dot{W}_{\text{net},u}$ associated with an unsteady-state control volume where heat transfer $Q_j$ crosses the control surface at temperature $T_j$ at a number of locations. When this equation is applied to a steady-state situation, the properties within the control volume are invariant with time. As a result, the general equation reduces to

$$\dot{W}_{\text{act},u} = \sum_{\text{out}} \left( h + \frac{\mathbf{V}^2}{2} + gz - T_0 s \right)_e \dot{m}_e - \sum_{\text{in}} \left( h + \frac{\mathbf{V}^2}{2} + gz - T_0 s \right)_i \dot{m}_i$$

$$- \sum_{j=1}^{n} \dot{Q}_j \left( 1 - \frac{T_0}{T_j} \right) + T_0 \dot{\sigma}_{\text{cv}} \qquad \text{(steady state)}$$

$$\text{[9-30]}$$

The summation terms for the inlets and exits can be replaced by the stream availability function. The next-to-last term is the availability transfer associated with heat transfer $\dot{\Phi}_Q$, and the last term $T_0 \dot{\sigma}_{\text{cv}}$ measures the irreversibility within the control volume. As a result, after rearrangement,

$$\boxed{\sum_{\text{out}} \dot{m}_e \psi_e - \sum_{\text{in}} \dot{m}_i \psi_i = \dot{\Phi}_Q + \dot{W}_{\text{act}} - \dot{I}_{\text{cv}}} \qquad \text{(steady state)} \quad \text{[9-35]}$$

In words, the above equation states that

$$\begin{pmatrix} \text{Net rate of} \\ \text{transfer of} \\ \text{availability out of} \\ \text{a control volume} \\ \text{with mass flow} \end{pmatrix} = \begin{pmatrix} \text{rate of} \\ \text{availability} \\ \text{transfer with} \\ \text{heat transfer} \\ \text{into the CV} \end{pmatrix} + \begin{pmatrix} \text{rate of} \\ \text{availability} \\ \text{transfer with} \\ \text{work transfer} \\ \text{into the CV} \end{pmatrix} - \begin{pmatrix} \text{rate of} \\ \text{availability} \\ \text{destruction within} \\ \text{a control volume} \end{pmatrix}$$

Both Eqs. [9-30] and [9-35] represent an *availability balance for a steady-state control volume.* Shaft work into a control volume always increases the availability of the mass passing through, while internal irreversibilities always decrease the stream availability. On a unit-mass basis for a control volume with one inlet and one exit the preceding equation becomes

$$\psi_2 - \psi_1 = \phi_Q + w_{\text{act}} - i_{\text{cv}} \qquad \text{[9-36]}$$

where $\phi_Q = \sum q_j(1 - T_0/T_j)$. Similar to the development for a closed system, Eq. [9-36] can be used to show that

$$i_{\text{sf}} = w_{\text{sf}} - w_{\text{rev,sf}} \qquad \text{or} \qquad \dot{I}_{\text{sf}} = \dot{W}_{\text{sf}} - \dot{W}_{\text{rev,sf}} \qquad \text{[9-37]}$$

Thus one method of determining the irreversibility within a steady-state control volume is by evaluating the difference between the actual shaft work and the reversible work.

The summation term $\sum \dot{Q}_j(1 - T_0/T_j)$ in Eq. [9-30] (or the equivalent term $\phi_Q$ in Eq. [9-36]) may be difficult to evaluate because the values of $\dot{Q}_j$ and $T_j$ are not known at every position on the boundary of the control volume. To circumvent this, we assume it is reasonably accurate to replace the variable surface temperature $T_j$ by a *constant* boundary temperature $T_b$ (see Sec. 7-5-1 on entropy transfer). That is,

$$\dot{\Phi}_Q = \sum_j \dot{Q}_j\left(1 - \frac{T_0}{T_j}\right) \approx \dot{Q}\left(1 - \frac{T_0}{T_b}\right) \qquad \textbf{[9-38]}$$

The value of $T_b$ typically is the average of the inlet and outlet fluid temperatures. The following examples illustrate the use of stream availability and the irreversibility for control-volumes analysis.

**Example 9-6**

**N**itrogen gas initially at 50 psia and 100°F is throttled through a well-insulated valve to a pressure of 15 psia. The atmospheric temperature is 60°F. Determine (a) the reversible work for the steady-flow process and (b) the irreversibility of the process, both answers in Btu/lb.

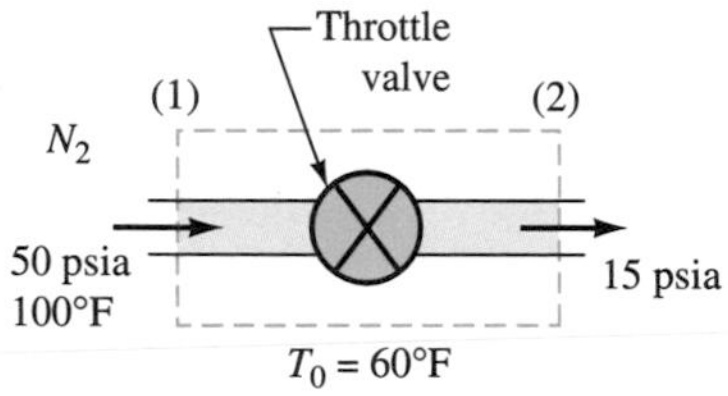

**Solution:**

**Given:**    Nitrogen gas is throttled, as shown in Fig. 9-15.

**Find:**    (a) $w_{\text{rev}}$ and (b) $i$, in Btu/lb.

**Figure 9-15**
Schematic and data for Example 9-6.

**Model:**    Steady flow, $q = w = \Delta\text{ke} = \Delta\text{pe} = 0$, ideal gas.

**Analysis:**    (a) The equation for the steady-flow, reversible work on a unit-mass basis is

$$w_{\text{sf,rev}} = \left(h_2 + \frac{V_2^2}{2} + gz_2 - T_0 s_2\right) - \left(h_1 + \frac{V_1^2}{2} + gz_1 - T_0 s_1\right) - \sum_{j=1}^{n} q_j\left(1 - \frac{T_0}{T_j}\right)$$

The steady-state energy balance for the process is

$$q + w = \Delta h + \Delta\text{ke} + \Delta\text{pe}$$

This equation reduces to $\Delta h = 0$ for a throttling process based on a model that heat and work interactions are zero and $\Delta\text{ke}$ and $\Delta\text{pe}$ are negligible. Because $h$ for an ideal-gas model is a function of the temperature only, the temperature of the nitrogen remains constant at 100°F. Since kinetic- and potential-energy changes are neglected, the equation for the reversible work becomes

$$w_{\text{sf,rev}} = \Delta h + \Delta\text{ke} + \Delta\text{pe} - T_0\Delta s = -T_0\Delta s$$

$$= -T_0\left(c_{p,\text{av}} \ln \frac{T_2}{T_1} - R \ln \frac{P_2}{P_1}\right) = RT_0 \ln \frac{P_2}{P_1}$$

$$= \frac{1.986 \text{ Btu/(lbmol·°R)} \times 520°\text{R}}{28 \text{ lb}_m/\text{lbmol}} \ln \frac{15}{50}$$

$$= -44.4 \text{ Btu/lb}_m$$

(b) For the actual process the work output is zero. Hence the irreversibility on a unit-mass basis is simply

$$i = w_{sf,act} - w_{sf,rev} = 0 - (-44.4) = 44.4 \text{ Btu/lb}_m$$

The value of $i$ can also be found from the availability balance (Eq. [9-36]),

$$\psi_2 - \psi_1 = \phi_Q + w_{act} - i$$

In this case $w = \phi_Q = 0$, so that

$$i_{cv} = \psi_1 - \psi_2 = (h_1 - T_0 s_1) - (h_2 - T_0 s_2) = T_0 \Delta s$$

This is the same result as given by the first method.

**Comment:** This example illustrates the sizable loss in work capability when a fluid undergoes a finite pressure drop through a flow restriction. Also, note that the reversible work is a measure of work *capability* that *could* be achieved. There may be *no* actual work associated with a process, but a value for the reversible work always exists.

---

An adiabatic compressor operates with air initially at 1.0 bar, 300 K, and 70 m/s. The exit conditions are 5.0 bars, 540 K, and 150 m/s. Determine (a) the actual work input, (b) the reversible (minimum) work required for the same end states, and (c) the irreversibility of the actual process, all answers in kJ/kg. Assume that $c_p$ is constant at 1.01 kJ/kg·K and that the atmosphere is at 1 bar and 17°C.

**Example 9-7**

**Solution:**

**Given:** Adiabatic air compressor operating under conditions shown in Fig. 9-16.

**Find:** (a) $w_{sf,act}$, (b) $w_{sf,rev}$, and (c) $i$, all in kJ/kg.

**Model:** Steady-state, adiabatic flow; ideal gas; $\Delta pe = 0$; $c_p = 1.01$ kJ/kg·K.

**Analysis:** (a) The actual work is found from an energy balance. On the basis of the steady-state energy equation, $q + w = \Delta h + \Delta ke + \Delta pe$, if $q = \Delta pe = 0$, and $\Delta h = c_{p,av} \Delta T$ for an ideal gas, then

$$w_{sf,act} = h_2 - h_1 + \frac{V_2^2 - V_1^2}{2}$$

$$= 1.01 \frac{\text{kJ}}{\text{kg·K}} \times (540 - 300) \text{ K} + \frac{150^2 - 70^2}{2} \frac{\text{m}^2}{\text{s}^2} \times \frac{1 \text{ N·s}^2}{\text{kg·m}} \times \frac{1 \text{ kJ}}{1000 \text{ N·m}}$$

$$= (242.4 + 8.8) \text{ kJ/kg} = 251.2 \text{ kJ/kg}$$

(b) The reversible work is calculated from the relationship:

$$w_{sf,rev} = \left(h_2 + \frac{V_2^2}{2} + gz_2 - T_0 s_2\right) - \left(h_1 + \frac{V_1^2}{2} + gz_1 - T_0 s_1\right) - \sum_{j=1}^{n} q_j \left(1 - \frac{T_0}{T_j}\right)$$

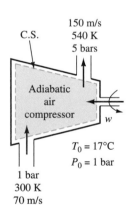

150 m/s
C.S.    540 K
5 bars

Adiabatic
air
compressor   $w$

$T_0 = 17°C$
$P_0 = 1$ bar

1 bar
300 K
70 m/s

**Figure 9-16**
Schematic and data for Example 9-7.

For the situation where $q = \Delta pe = 0$, the equation becomes

$$w_{sf,rev} = h_2 - h_1 - T_0(s_2 - s_1) + \frac{V_2^2 - V_1^2}{2} = w_{sf,act} - T_0(s_2 - s_1)$$

For an ideal-gas model,

$$\Delta s = c_{p,av} \ln(T_2/T_1) - R \ln(P_2/P_1)$$

$$= \left(1.01 \ln \frac{540}{300} - \frac{8.314}{29} \ln \frac{5}{1}\right) \frac{kJ}{kg\cdot K} = 0.1324 \text{ kJ/kg·K}$$

Therefore,

$$w_{sf,rev} = 251.2 \text{ kJ/kg} - 290 \text{ K} \times (0.1324 \text{ kJ/kg·K})$$
$$= (251.2 - 38.4) \text{ kJ/kg} = 212.8 \text{ kJ/kg}$$

This is the minimum work required for the specified end states.

(c) The irreversibility is found from the availability balance for a one-inlet, one-outlet, steady-state control volume,

$$\psi_2 - \psi_1 = \phi_Q + w_{act} - i_{cv}$$

Since the process is adiabatic, $\phi_Q = 0$ and the irreversibility is

$$i_{cv} = \psi_1 - \psi_2 + w = [(h_1 - T_0 s_1) - (h_2 - T_0 s_2)] + (h_2 - h_1) = T_0 \Delta s$$

Thus the irreversibility is

$$i_{cv} = T_0 \Delta s = 300 \text{ K} (0.1324 \text{ kJ/kg·K}) = 38.4 \text{ kJ/kg}$$

The irreversibility can also be found by $i = w_{sf,act} - w_{sf,rev} = (251.2 - 212.8) \text{ kJ/kg} = 38.4 \text{ kJ/kg}$.

**Comment:**    The presence of irreversibilities *within* the compressor has led to an increase of 38.4 kJ/kg in work input, which is an 18 percent increase.

---

**Example 9-8**

**A**n adiabatic steam turbine operates with inlet conditions of 100 bars and 520°C and exit conditions of a saturated vapor at 3 bars. Kinetic- and potential-energy changes may be neglected, and the environmental state is 300 K and 1 bar. Determine (a) the actual work output, (b) the reversible (maximum) work output, and (c) the availability of the exit stream, all answers in kJ/kg.

**Solution:**

**Given:**    An adiabatic steam turbine with input data is shown in Fig. 9-17.

**Find:**    (a) $w_{act}$, (b) $w_{sf,rev}$, and (c) $\psi_2$, in kJ/kg.

**Model:**    Adiabatic, steady-state, $\Delta ke = \Delta pe = 0$.

**Analysis:**    (a) The actual work output $W_{act}$ is found from the steady-state energy equation,

$$0 = q + w - \Delta h - \Delta ke - \Delta pe$$

Since $q = 0$ and changes in kinetic and potential energies will be neglected,

$$w_{act} = h_2 - h_1 = (2725.3 - 3425.1) \text{ kJ/kg} = -699.8 \text{ kJ/kg}$$

where the data are found in Tables A-13 and A-14.

(b) The reversible steady-flow work is given by

$$w_{sf,rev} = \left(h_2 + \frac{\mathbf{V}_2^2}{2} + gz_2 - T_0 s_2\right) - \left(h_1 + \frac{\mathbf{V}_1^2}{2} + gz_1 - T_0 s_1\right) - \sum_{j=1}^{n} q_j\left(1 - \frac{T_0}{T_j}\right)$$

Hence, for the same model,

$$w_{sf,rev} = \Delta\psi = \Delta h - T_0\Delta s$$
$$= -699.8 \text{ kJ/kg} - 300 \text{ K} \times (6.9919 - 6.6622) \text{ kJ/kg·K} = -798.7 \text{ kJ/kg}$$

(c) The availability at the exit of the turbine is given by $\psi_2 = h_2 - h_0 - T_0(s_2 - s_0)$. In the dead state of 300 K and 1 bar, water is a compressed liquid. Its properties are approximated closely by using saturated-liquid data at the given dead-state temperature of 300 K in Table A-12. Hence

$$\psi_2 = h_2 - h_0 - T_0(s_2 - s_0)$$
$$= (2725.3 - 113.24) \text{ kJ/kg} - 300 \text{ K} \times (6.9919 - 0.3954) \text{ kJ/kg·K}$$
$$= 633 \text{ kJ/kg}$$

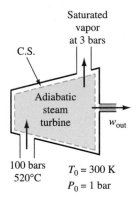

Saturated
vapor
at 3 bars

C.S.

Adiabatic
steam
turbine

$w_{out}$

100 bars
520°C

$T_0 = 300$ K
$P_0 = 1$ bar

**Figure 9-17**
Schematic and data for
Example 9-8.

**Comments:** Note that the actual work ouput is about 12 percent less than the maximum possible output, owing to the presence of irreversibilities as measured by the increase in entropy during the adiabatic process. Also, the exit availability is nearly as large as the actual work output. Although the pressure has dropped from 100 to 3 bars, a considerable amount of work output theoretically could still be obtained from the fluid.

---

## 9-5  SECOND-LAW EFFICIENCY OR EFFECTIVENESS

The *first-law* efficiencies $\eta$ of some common devices are presented in Chap. 8. These expressions are ratios of selected energy quantities. In the context of this chapter the usefulness of energy is more appropriately described by its availability. Since availability has its origin in the second law, a performance parameter for a process based on availability concepts is known as a *second-law efficiency* $\eta_{II}$, or as a *second-law effectiveness* $\varepsilon$, or simply as an *effectiveness*. (Both symbols are in current use.) A first-law efficiency gages how well energy is used when compared against an ideal process, whereas an effectiveness indicates how well availability is used.

### 9-5-1  SECOND-LAW EFFICIENCIES

First- and second-law efficiencies are different in one other important respect. The first law is a conservation principle. On the other hand, entropy and availability from a second-law viewpoint are nonconserved properties. In the presence of irreversibilities, entropy is produced and availability is destroyed. The former effect is measured by the entropy production $\sigma$, and the latter effect is measured by the irreversibility $I$. Hence second-law efficiencies measure losses in availability during a process. A general definition of a second-law effectiveness $\varepsilon$ (or $\eta_{\mathrm{II}}$) is

$$\varepsilon \equiv \frac{\text{useful availability out}}{\text{availability in}} \qquad \textbf{[9-39]}$$

$$= 1 - \frac{\text{availability destruction and losses}}{\text{availability input}}$$

where losses imply *nonuseful* transfers across the boundary. A second approach, especially useful for steady-state devices, is

$$\eta_{\mathrm{II}} = \varepsilon = \frac{\text{rate of availability output}}{\text{rate of availability input}} \qquad \textbf{[9-40]}$$

The second law stresses the fact that two forms of the same quantity of energy may have quite different availabilities. Thus energy is "weighted" according to its availability. Unlike a first-law efficiency, an effectiveness accounts for losses in work capability during a process. Note, however, that the definitions of $\varepsilon$ given above are quite general. In specific applications one must still decide what is the desired output, what is regarded as input, and what is counted as a loss. In later examples we shall demonstrate that different decisions about these items lead to different expressions for the effectiveness of a given process. Hence different $\varepsilon$ values may exist for the same data. It is not a question of which value is better, or correct. It is a matter of consistency when comparing sets of data. Fortunately, the availability balances for closed and open systems are good guidelines for establishing appropriate second-law efficiencies.

As an example of the use of the availability concept in second-law analysis, consider a heat engine which operates between two thermal reservoirs at $T_H$ and $T_L$. For an actual (irreversible) cycle $W_{\mathrm{act}} = \eta_{\mathrm{th,act}} Q_H$. However, if the cycle were reversible,

$$W_{\mathrm{rev}} = \eta_{\mathrm{Carnot}} Q_H = \left(1 - \frac{T_L}{T_H}\right) Q_H$$

Because the availability associated with shaft work is the value of the shaft work itself, we define the effectiveness of a power cycle by the availability ratio $W_{\mathrm{act}}/W_{\mathrm{rev}}$. That is,

$$\varepsilon_{\mathrm{power}} = \frac{W_{\mathrm{act}}}{W_{\mathrm{rev}}} = \frac{\eta_{\mathrm{th,act}}}{\eta_{\mathrm{Carnot}}} = \frac{\eta_{\mathrm{th,act}}}{1 - (T_L/T_H)} \qquad \textbf{[9-41]}$$

A similar type of effectiveness can be defined for refrigerators and heat pumps.

**T**wo heat engines, A and B, each have actual thermal efficiencies of 30 percent. Heat engine A receives heat transfer at 1000 K and loses energy at 300 K, while heat engine B receives heat transfer at 600 K and loses energy at 300 K. Determine the second-law effectiveness of each device, and compare to their first-law efficiencies.

**Example 9-9**

**Solution:**

**Given:** Two heat engines have a thermal efficiency of 30 percent, as shown in Fig. 9-18.

**Find:** $\varepsilon$ for each heat engine; compare to $\eta_{th}$.

**Model:** Irreversible heat engines operating between two different sets of $T_H$ and $T_L$.

**Analysis:** The first-law efficiencies are the same. This simply means that if $Q_{H,A} = Q_{H,B}$, then each device produces work in the amount of 0.30 $Q_H$. Although the work outputs are equal, we should not infer that the second-law performances are equal.

The reversible performance of either device is given by the Carnot efficiency equation $\eta_{th} = 1 - (T_L/T_H)$. Thus

$$\eta_{Carnot,A} = 1 - \frac{300}{1000} = 0.70 \quad \text{and} \quad \eta_{Carnot,B} = 1 - \frac{300}{600} = 0.50$$

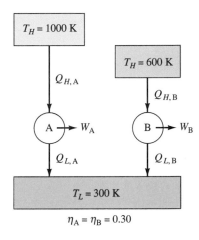

**Figure 9-18**
Schematic and data for Example 9-9.

For the same equal heat inputs $Q_{H,A}$ and $Q_{H,B}$ as before, then

$$W_{A,rev} = 0.70Q_H \quad \text{and} \quad W_{B,rev} = 0.50Q_H$$

By employing Eq. [9-41] for heat engines A and B, we find that if $W_{act} = 0.3Q_H$, then

$$\varepsilon_A = \frac{0.3Q_H}{0.7Q_H} = 0.43 \quad \text{and} \quad \varepsilon_B = \frac{0.3Q_H}{0.5Q_H} = 0.60$$

Therefore, the second-law performance of heat engine B is much better that that of A, even though in practice they have the same first-law efficiency.

## 9-5-2 THE EFFECTIVENESS FOR STEADY-STATE PROCESSES

As noted earlier, first-law efficiencies are ratios of selected energy quantities such as $w_{shaft}$, $q$, $h$, and ke. A second-law effectiveness is a ratio of selected availability quantities such as $w_{shaft}$, $\phi_Q$, and $\psi$. The latter quantities are related through the steady-flow availability balance

$$\Delta\psi = \phi_Q + w_{act} - i_{cv} \qquad \textbf{[9-36]}$$

on a unit-mass basis for a one-inlet, one-outlet control volume.

The effectiveness $\varepsilon_C$ for a compressor or $\varepsilon_P$ for a pump is defined as the change (increase) in fluid availability divided by the actual work input.

That is,

$$\varepsilon_C = \varepsilon_P \equiv \frac{\psi_e - \psi_i}{w_{\text{act,in}}} = 1 - \frac{i - \phi_Q}{w_{\text{act,in}}} \qquad \textbf{[9-42]}$$

where any heat transfer is solely with the environment. When heat transfer is negligible the above equation reduces to

$$\varepsilon_C = \varepsilon_P = \frac{\psi_e - \psi_i}{w_{\text{act,in}}} = \frac{w_{\text{act,in}} - i}{w_{\text{act,in}}} \qquad \text{(adiabatic)} \qquad \textbf{[9-43]}$$

For a turbine the definition of $\varepsilon_T$ is the inverted form for $\varepsilon_C$. That is,

$$\varepsilon_T \equiv \frac{w_{\text{act,out}}}{\psi_i - \psi_e} = 1 - \frac{i - \phi_Q}{\psi_i - \psi_e} \qquad \textbf{[9-44]}$$

where any heat transfer is with the environment. For an adiabatic turbine

$$\varepsilon_T = \frac{w_{\text{act,out}}}{-\Delta\psi} = \frac{w_{\text{act,out}}}{w_{\text{act,out}} + i} \qquad \text{(adiabatic)} \qquad \textbf{[9-45]}$$

The availability balance for an adiabatic nozzle is $\psi_1 = \psi_2 + i$. The second-law nozzle effectiveness $\varepsilon_N$ on an output/input basis is

$$\varepsilon_N = \frac{\psi_2}{\psi_1} = \frac{\psi_1 - i}{\psi_1} \qquad \text{(adiabatic)} \qquad \textbf{[9-46]}$$

The availability loss in a subsonic nozzle is usually small. This same equation applies to a throttling process, since again $q$ and $w$ are zero. Hence,

$$\varepsilon_{\text{throttle}} = \frac{\psi_2}{\psi_1} = \frac{\psi_1 - i}{\psi_1} \qquad \textbf{[9-47]}$$

In this case $i$ can be fairly large relative to $\psi_1$, which indicates a significant loss in work potential for the fluid.

As a final category, first consider heat exchange between two fluids *without mixing*. Figure 9-19 shows a steady-state heat exchanger with mass flows $\dot{m}_h$ and $\dot{m}_c$ for the hot and cold streams, respectively. Neglecting heat transfer to the environment, the availability losses which occur in such devices are: (1) loss due to heat exchange across a finite temperature difference and (2) loss due to fluid friction. The availability-rate equation for this situation is

$$0 = \dot{m}_c(\psi_2 - \psi_1) + \dot{m}_h(\psi_4 - \psi_3) + \dot{I}_{\text{cv}}$$

One measure of the effectiveness of a heat exchanger is given by the increase in availability of the cold stream divided by the decrease in availability of the hot stream on a rate basis. That is,

$$\varepsilon = \frac{\dot{m}_c(\psi_2 - \psi_1)}{-\dot{m}_h(\psi_4 - \psi_3)} \qquad \text{(heat exchanger)} \qquad \textbf{[9-48]}$$

Other forms for $\varepsilon$ appear in the literature, but this form is usually preferred.

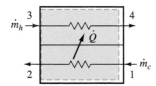

**Figure 9-19**
Schematic of the steady-state heat exchange between two fluids without mixing.

A second consideration is energy exchange by *mixing* two or more fluids in direct contact. A schematic for the overall process is shown in Fig. 9-20 for the mixing of two fluids: a cold fluid initially at state 1 and a hotter fluid initially at state 2. The availability balance has the format

$$0 = \dot{m}_3\psi_3 - \dot{m}_1\psi_1 - \dot{m}_2\psi_2 + \dot{I}_{cv}$$

where $\dot{m}_c = \dot{m}_1$ and $\dot{m}_h = \dot{m}_2$. This equation can be rewritten

$$\dot{m}_h(\psi_2 - \psi_3) = \dot{m}_c(\psi_3 - \psi_1) + \dot{I}_{cv}$$

Similar to the heat exchanger discussed above, the effectiveness for direct mixing may be defined as the gain in availability of the colder entering stream divided by the decrease in the availability of the hotter entering stream. Thus

$$\varepsilon = \frac{\dot{m}_c(\psi_3 - \psi_1)}{\dot{m}_h(\psi_2 - \psi_3)} \quad \text{(mixing)} \qquad \textbf{[9-49]}$$

Like the previous discussion for heat exchangers, alternate definitions for $\varepsilon$ are possible. The following example illustrates the use of the availability balance and an average boundary temperature.

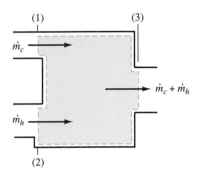

(1)                                    (3)

$\dot{m}_c$ →

→ $\dot{m}_c + \dot{m}_h$

$\dot{m}_h$ →

(2)

**Figure 9-20**
Schematic of energy exchange by mixing of two fluids.

---

**A** steam turbine operates with inlet conditions of 100 bars and 520°C and exit conditions of a saturated vapor at 3 bars. Kinetic- and potential-energy changes may be neglected, and the environmental state is 300 K and 1 bar. (*a*) Determine (1) the irreversibility within the turbine, in kJ/kg, if a heat loss of 40.0 kJ/kg occurs to the environment, and (2) the second-law effectiveness of the device. (*b*) Now find the irreversibility for a control volume that includes both the turbine and the heat-transfer region. Finally, (*c*) find the second-law effectiveness if the process is adiabatic.

**Example 9-10**

**Solution:**

**Given:** A steam turbine operating under the conditions shown in Fig. 9-21.

**Find:** (*a*) $i_{cv}$, in kJ/kg, and $\varepsilon_T$; (*b*) $i_{tot}$; (*c*) $\varepsilon_T$ for $q = 0$.

**Model:** Steady-state, $\Delta ke = \Delta pe = 0$.

**Analysis:** This problem is the same as in Example 9-8, except that now a heat loss occurs.

(*a*) (1) The specific irreversibility within the turbine will be evaluated from an availability balance on the fluid as given by

$$i_{cv} = -\Delta\psi + \sum_{j=1}^{n} q_j\left(1 - \frac{T_0}{T_j}\right) + w_{act}$$

The value of $\Delta\psi$ was found to be $-798.7$ kJ/kg in Example 9-8. A steady-state energy balance on the control volume for negligible kinetic- and potential-energy

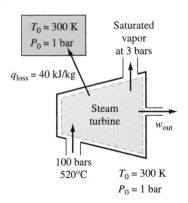

$T_0 = 300$ K
$P_0 = 1$ bar

Saturated vapor at 3 bars

$q_{loss} = 40$ kJ/kg

Steam turbine

$w_{out}$

100 bars
520°C

$T_0 = 300$ K
$P_0 = 1$ bar

**Figure 9-21**
Schematic and data for Example 9-10, part *a*.

changes shows that

$$w_{act} = \Delta h + \Delta ke + \Delta pe - q = h_2 - h_1 - q$$
$$= [2725.3 - 3425.1 - (-40.0)] \text{ kJ/kg} = -659.8 \text{ kJ/kg}$$

The remaining term in the availability balance, $\phi_Q$, is handled similarly to Eq. [9-38]. The temperature of the fluid changes from 520 to 133.6°C as it passes through the turbine. We *model* the variable temperature $T_i$ of the turbine fluid by a *constant* boundary temperature $T_b$ which is the average of the inlet- and outlet-fluid temperatures. In this case the average fluid temperature is $(133.6 + 520)/2 = 326.8$°C, or 600 K. Hence

$$\phi_Q = \sum_j q_j\left(1 - \frac{T_0}{T_j}\right) \approx q\left(1 - \frac{T_0}{T_b}\right) = -40.0 \text{ kJ/kg}\left(1 - \frac{300}{600}\right) = -20.0 \text{ kJ/kg}$$

When all the calculated terms are substituted into the availability balance we find that

$$i_{cv} = [-(-798.7) + (-20.0) + (-659.8)] \text{ kJ/kg} = 118.9 \text{ kJ/kg}$$

This is a 20 percent increase in $i_{cv}$ over the value of 98.9 kJ/kg found for adiabatic operation in Example 9-8. This increase is due solely to the availability loss $\phi_Q$ associated with the heat transfer.

(2) The second-law effectiveness of a turbine $\varepsilon_T$ generally is given by

$$\varepsilon_T = \frac{w_{act,out}}{-\Delta\psi} = \frac{659.8}{798.7} = 0.826 \qquad \text{(or 82.6 percent)}$$

This same result could be obtained from the relationship

$$\varepsilon_T = 1 - \frac{\text{availability destruction and loss}}{\text{availability input}}$$
$$= 1 - \frac{i - \phi_Q}{\Delta\psi} = 1 - \frac{118.9 - (-20.0)}{798.7} = 0.826$$

Although the quantity $\phi_Q$ associated with $q$ has the potential to produce more work, it is a loss as far as turbine performance is concerned. This loss eventually shows up as an irreversibility in the surroundings.

(b) For the enlarged control volume shown in Fig. 9-22 the availability balance becomes

$$i = -\Delta\psi + q\left(1 - \frac{T_0}{T_b}\right) + w_{act}$$
$$= [-(-798.7) + 0 + (-659.8)] \text{ kJ/kg} = 138.9 \text{ kJ/kg}$$

where $T_b = T_0$. Since the irreversibility $i$ within the turbine is 118.9 kJ/kg, then $i_Q = 138.9 - 118.9 = 20.0$ kJ/kg. This is the same value given by the expression for a heat-transfer region,

$$i_Q = T_0 q\left(\frac{1}{T_0} - \frac{1}{T_b}\right) = 300(-40.0)\left(\frac{1}{300} - \frac{1}{600}\right) = 20.0 \text{ kJ/kg}$$

where $T_b$ now is 600 K. Thus the irreversibility due to heat transfer is about 17 percent of that within the turbine due to frictional losses.

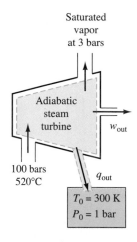

**Figure 9-22**
Schematic and data for Example 9-10, part b.

(c) From Example 9-8, $w_{act} = -699.8$ kJ/kg and $\Delta\psi = -798.7$ kJ/kg. Therefore, under adiabatic conditions,

$$\varepsilon_T = \frac{w_{act,out}}{-\Delta\psi} = \frac{699.8}{798.7} = 0.876 \quad \text{(or 87.6 percent)}$$

**Comments:** (1) For the fluid availability decrease of 798.7 kJ/kg, 82.6 percent appears as work output, 14.9 percent is destroyed, and 2.5 percent accompanies the heat transfer out.

(2) For the adiabatic turbine in Example 9-9 the effectiveness is 87.6 percent for adiabatic performance. In the current example the loss in availability due to heat transfer has substantially lowered the effectiveness of the device to 82.6 percent. Hence heat losses from turbines are highly undesirable.

(3) The irreversibility for the enlarged control volume could also be found from $i_{cv} = T_0\sigma_{cv}$, where $\sigma_{cv}$ is determined from an entropy balance.

---

**Water** enters a heat exchanger at a rate of 100 lb$_m$/s as a compressed liquid at 30 psia and 100°F and leaves at the same pressure and 200°F. Heat transfer supplied to the water stream comes from a stream of hot air which enters at 1000°R and 40 psia and leaves at 600°R at the same pressure. Determine (a) the change in stream availability of the water, (b) the change in the stream availability of the air, and (c) the overall irreversibility of the heat-exchange process, all answers in Btu/s. The environmental temperature is 520°R. Finally, determine (d) the effectiveness of the heat exchanger.

**Example 9-11**

**Solution:**

**Given:** Heat transfer between water and air streams in a steady-state heat exchanger as illustrated in Fig. 9-23.

**Find:** (a) $\Delta\dot{\psi}_{water}$, (b) $\Delta\dot{\psi}_{air}$, (c) $i_{tot}$, in Btu/s, and (d) $\varepsilon$.

**Model:** Adiabatic, steady-flow process; air an ideal gas; neglect ke and pe effects.

**Analysis:** The change in stream availability is given by

$$\Delta\psi = \psi_2 - \psi_1 = (h_2 - h_1) + (ke_2 - ke_1) + (pe_2 - pe_1) - T_0(s_2 - s_1)$$

By neglecting the changes in kinetic and potential energy, this equation reduces to

$$\Delta\psi = h_2 - h_1 - T_0(s_2 - s_1)$$

(a) For the water stream as the system, the saturation temperature at 30 psia is 250.3°F. Hence the water remains a liquid at the exit. Compressed-liquid data may be estimated by using saturated-liquid data at the given temperature. On the basis of data from Table A-12E,

$$\Delta\psi_{H_2O} = (168.07 - 68.05) \text{ Btu/lb}_m - 520°R \times (0.2940 - 0.1296) \text{ Btu/(lb}_m\cdot°R)$$
$$= 14.5 \text{ Btu/lb}_m$$

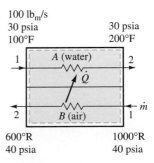

100 lb$_m$/s
30 psia
100°F

30 psia
200°F

600°R
40 psia

1000°R
40 psia

**Figure 9-23**
Schematic and data for Example 9-11

The stream availability change per unit time is found to be

$$(\dot{m}\,\Delta\psi)_{\text{water}} = 100 \text{ lb}_\text{m}/\text{s} \times 14.5 \text{ Btu/lb}_\text{m} = 1450 \text{ Btu/s}$$

(b) Calculation of the air stream rate of change of availability requires information on the mass flow rate of air. This latter quantity is found from an energy balance on the heat exchanger. For a steady-flow control volume,

$$0 = \dot{Q} + \dot{W}_{\text{shaft}} + \sum_{\text{in}} \left( h + \frac{\mathbf{V}^2}{2} + gz \right)_i \dot{m}_i - \sum_{\text{out}} \left( h + \frac{\mathbf{V}^2}{2} + gz \right)_e \dot{m}_e$$

In the absence of heat and work transfers, and neglecting kinetic- and potential-energy changes, the energy balance reduces to

$$\dot{m}_A(h_{A1} - h_{A2}) = \dot{m}_B(h_{B2} - h_{B1})$$

where $A$ and $B$ represent the water and air streams, respectively. Enthalpy data for air as an ideal gas are found in Table A-5E. Substitution of data into the energy equation yields

$$100 \text{ lb}_\text{m}/\text{s} \times (168.07 - 68.05) \text{ Btu/lb}_\text{m} = \dot{m}_{\text{air}} \times (143.47 - 240.98) \text{ Btu/lb}_\text{m}$$

$$\dot{m}_{\text{air}} = \frac{100(100.02)}{97.51} = 102.6 \text{ lb}_\text{m}/\text{s}$$

Since $\Delta P = 0$ for the air stream as the system, the entropy change for the ideal-gas model is $\Delta s = s_2^0 - s_1^0$. The specific availability change for air is given by

$$\Delta\psi_{\text{air}} = h_2 - h_1 - T_0(s_2 - s_1) = h_2 - h_1 - T_0(s_2^0 - s_1^0)$$
$$= (143.47 - 240.98) \text{ Btu/lb}_\text{m} - 520°\text{R} \times (0.62607 - 0.75042) \text{ Btu/lb}_\text{m}\cdot°\text{R}$$
$$= -32.85 \text{ Btu/lb}_\text{m}$$

The rate of availability change for the air becomes

$$(\dot{m}\,\Delta\psi)_{\text{air}} = 102.6 \text{ lb}_\text{m}/\text{s} \times (-32.85 \text{ Btu/lb}_\text{m}) = -3370 \text{ Btu/s}$$

(c) The irreversibility rate for the heat exchanger is found by using the availability balance for a steady-state control volume,

$$\sum_{\text{out}} \dot{m}_e\psi_e - \sum_{\text{in}} \dot{m}_i\psi_i = \dot{\Phi}_Q + \dot{W} - \dot{I}_{\text{cv}}$$

In terms of two fluids A and B this can be written as

$$\dot{m}_A(\psi_2 - \psi_1)_A + \dot{m}_B(\psi_2 - \psi_1)_B = \dot{\Phi}_Q + \dot{W} - \dot{I}_{\text{cv}}$$

But $\dot{W} = \dot{\Phi}_Q = 0$ for the control volume. Consequently,

$$\dot{I}_{\text{cv}} = -[(\dot{m}\,\Delta\psi)_A + (\dot{m}\,\Delta\psi)_B] = -[1450 + (-3370)] = 1920 \text{ Btu/s}$$

Another method for evaluating $\dot{I}_{\text{cv}}$ for an adiabatic process is

$$\dot{I}_{\text{cv}} = T_0\dot{\sigma}_{\text{cv}} = T_0 \sum \dot{m}\,\Delta s = T_0[(\dot{m}\,\Delta s)_{\text{water}} + (\dot{m}\,\Delta s)_{\text{air}}]$$
$$= 520°\text{R}[100(0.2940 - 0.1296) + 102.6(0.62606 - 0.75042)] \text{ Btu/(°R}\cdot\text{s})$$
$$= 1915 \text{ Btu/s}$$

The slight difference in answers is due to roundoff error.

(*d*) The effectiveness of the heat exchanger is determined from

$$\varepsilon = \frac{(\dot{m}\,\Delta\psi)_{\text{water}}}{-(\dot{m}\,\Delta\psi)_{\text{air}}} = \frac{1450}{-(-3370)} = 0.43$$

**Comment:** The effectiveness of the heat exchanger is fairly low owing to the sizable temperature difference between the two fluids.

## 9-6 SUMMARY

The *thermomechanical availability* associated with the state of a system is measured by the maximum reversible work that could be delivered by the system as it proceeds toward a state of equilibrium with the environment while any heat transfer occurs solely with the environment. This state of equilibrium is its *dead state* at $T_0$ and $P_0$, and the *standard atmosphere* or environment is recommended to be 298.15 K and 1.01325 bar (1 atm). The general equation for the actual useful work associated with a region of space with mass, heat, and work transfer to or from it is

$$\dot{W}_u = \sum_{\text{out}} \left( h + \frac{V^2}{2} + gz - T_0 s \right)_e \dot{m}_e - \sum_{\text{in}} \left( h + \frac{V^2}{2} + gz - T_0 s \right)_i \dot{m}_i$$

$$- \sum_{j=1}^{n} \dot{Q}_j \left( 1 - \frac{T_0}{T_j} \right) + \frac{d(E + P_0 V - T_0 S)_{\text{cv}}}{dt} + T_0 \dot{\sigma}_{\text{cv}}$$

The reversible work is found by setting $T_0 \dot{\sigma}_{\text{cv}} = 0$.

For a simple compressible substance within a *closed system* with heat transfer $Q_j$ crossing the boundary at $T_j$, the reversible useful work on a unit-mass basis is

$$w_{\text{rev},u} = \Delta u + P_0 \Delta v - T_0 \Delta s - \sum_{j=1}^{n} q_j \left( 1 - \frac{T_0}{T_j} \right)$$

The irreversibility $i$ is given by

$$i_{\text{cm}} = w_{\text{act},u} - w_{\text{rev},u} = T_0 \sigma_{\text{cm}}$$

The *closed-system availability function* $\phi$ is defined as

$$\phi \equiv (u + P_0 v - T_0 s) - (u_0 + P_0 v_0 - T_0 s_0)$$

where the total availability $\Phi = m\phi$. Consequently,

$$w_{\text{rev},u} = \phi_2 - \phi_1 - \sum_{j=1}^{n} q_j \left( 1 - \frac{T_0}{T_j} \right)$$

Generally, for heat transfer $Q_i$ to or from a closed system of variable temperature, the *availability transfer associated with heat transfer* at a uniform boundary temperature $T_b$ is given by

$$\Phi_Q \equiv \int_1^2 \left(1 - \frac{T_0}{T_b}\right) \delta Q$$

The availability transfer associated with heat transfer $Q$ into or out of a closed system at a constant temperature $T_R$ is given by

$$\Phi_{Q,R} = Q\left(1 - \frac{T_0}{T_R}\right)$$

Therefore, the *availability balance for a control mass* symbolically becomes

$$\Delta\Phi_{cm} = \Phi_Q + W_{act,u} - I_{cm}$$

where $I_{cm}$ measures the availability destruction within the closed system.

For a steady-state *control volume,* the reversible work transfer is found from

$$\dot{W}_{rev} = \sum_{out}\left(h + \frac{\mathbf{V}^2}{2} + gz - T_0s\right)_e \dot{m}_e - \sum_{in}\left(h + \frac{\mathbf{V}^2}{2} + gz - T_0s\right)_i \dot{m}_i$$
$$- \sum_{j=1}^n \dot{Q}_j\left(1 - \frac{T_0}{T_j}\right)$$

For a time period during which a unit mass enters the single inlet and a unit mass leaves the single exit of a steady-state control volume, the reversible work is given by

$$w_{sf,rev} = \left(h_2 + \frac{\mathbf{V}_2^2}{2} + gz_2 - T_0s_2\right) - \left(h_1 + \frac{\mathbf{V}_1^2}{2} + gz_1 - T_0s_1\right)$$
$$+ \sum_{j=1}^n q_j\left(1 - \frac{T_0}{T_j}\right)$$

Under the same circumstances the irreversibility is

$$i_{sf} = w_{sf,act} - w_{sf,rev} = T_0\sigma_{cv}$$

The *stream availability* $\psi$ for a control volume is defined as

$$\psi = (h + ke + pe - T_0s) - (h_0 + pe_0 - T_0s_0)$$
$$= h - h_0 - T_0(s - s_0) + \frac{\mathbf{V}^2}{2} + gz$$

The *rate of availability transfer* $\dot{\Phi}_Q$ associated with heat transfer across the boundary of a control volume is

$$\dot{\Phi}_Q = \sum_{j=1}^n \dot{Q}_j\left(1 - \frac{T_0}{T_j}\right)$$

As a result, the availability balance for steady-state control volume is symbolically written as

$$\sum_{\text{out}} \dot{m}_e \psi_e - \sum_{\text{in}} \dot{m}_i \psi_i = \dot{\Phi}_Q + \dot{W}_{\text{act}} - \dot{I}_{\text{cv}}$$

where $\dot{I}_{\text{cv}} = T_0 \dot{\sigma}_{\text{cv}}$.

## PROBLEMS

### REVERSIBLE WORK AND IRREVERSIBILITY IN CLOSED SYSTEMS

9-1. A rigid tank initially contains steam at 8.0 MPa and 400°C, while the environment is at 0.1 MPa and 25°C. The steam is cooled until the pressure drops to 4.0 MPa. Determine (*a*) the reversible useful work associated with the change of state and (*b*) the irreversibility of the process, both answers in kJ/kg.

9-2. Air initially at 1 bar and 27°C is contained in a well-insulated tank. An impeller inside the tank is turned by an external mechanism until the pressure is 1.2 bars. Determine (*a*) the actual work required, (*b*) the reversible useful work associated with the change of state, and (*c*) the irreversibility, all answers in kJ/kg. Let $T_0 = 27$°C and $P_0 = 1$ bar.

9-3. A piston-cylinder device contains 0.40 kg of air at 0.10 MPa and 27°C. Determine the minimum useful work input, in kilojoules, required to compress the air to 0.40 MPa and 127°C if $T_0 = 20$°C and $P_0 = 0.10$ MPa.

9-4. A closed, rigid tank contains 0.5 kg of saturated water vapor at 4 bars. Heat is added in the amount of 70 kJ from a thermal reservoir at 500 K, and some work is done by means of a paddle wheel until the fluid is at 7 bars. Determine (*a*) the paddle wheel required, (*b*) the reversible useful work associated with the change of state, and (*c*) the irreversibility of the process, all in kilojoules, if $T_0 = 25$°C.

9-5E. A rigid tank initially contains steam at 1000 psia and 800°F, while the environment is at 14.7 psia and 70°F. The steam is cooled until the pressure drops to 500 psia. Determine (*a*) the reversible useful work (or availability change) associated with the change of state and (*b*) the irreversibility of the process, both answers in Btu/lb$_m$.

9-6E. Air initially at 15 psia and 90°F is contained in a well-insulated tank. An impeller inside the tank is turned by an external mechanism until the pressure is 18 psia. Determine (*a*) the actual work required, (*b*) the reversible useful work (or availability change) associated with the change of state, and (*c*) the irreversibility, all answers in Btu/lb$_m$. Let $T_0 = 90$°F and $P_0 = 15$ psia.

9-7E. Refrigerant 134a is compressed in a piston-cylinder device from a saturated vapor at 15°F to a final pressure of 160 psia. The process is adiabatic, and the compressor efficiency is 75 percent. Determine (a) the actual work required, in Btu/lb, (b) the actual outlet temperature, in degrees Fahrenheit, (c) the minimum useful work required, in Btu/lb, for the actual final state found in part b, and (d) the irreversibility of the process, in Btu/lb$_m$. Note that $T_0 = 60°F$ and $P_0 = 1$ atm.

9-8E. Steam is contained in a piston-cylinder device. Before expansion the state is 160 psia, 500°F, and 0.10 ft$^3$. After expansion the pressure and volume are 20 psia and 0.65 ft$^3$, respectively. The heat transfer during the process is $-0.80$ Btu, and the surroundings are at 80°F and 14.7 psia. Determine (a) the actual work, (b) the reversible useful work (or availability change), and (c) the irreversibility of the process, all answers in Btu.

9-9E. A piston-cylinder device contains 0.88 lb$_m$ of water initially at 300°F and 150 psia. During an internally reversible, isothermal expansion, 811 Btu of heat is transferred to the water from a thermal reservoir at 1100°F. The state of the environment is 530°R and 14.8 psia. Determine (a) the actual work, (b) the reversible useful work, and (c) the irreversibility of the process, in Btu.

9-10E. A thermal reservoir at 1500°R transfers heat to 1 lb$_m$ of air in a piston-cylinder device maintained at a constant pressure of 100 psia. During the process the air temperature within the cylinder changes from 540 to 940°F. The surrounding atmosphere is at 14.7 psia and 540°R. Using data from Table A-5E, determine (a) the heat transfer, (b) the actual useful work, (c) the reversible useful work, and (d) the irreversibility, in Btu/lb$_m$.

9-11. Refrigerant 134a is compressed in a piston-cylinder device from a saturated vapor at $-4°C$ to a final pressure of 9 bars. The process is adiabatic and the compressor efficiency is 78 percent. Determine (a) the actual work required, in kJ/kg, and (b) the actual outlet temperature, in degrees Celsius. (c) Find the minimum useful work required, in kJ/kg, for the actual final state found in part b. (d) Find the irreversibility of the process, in kJ/kg. Note that $T_0 = 20°C$ and $P_0 = 1$ bar.

9-12. Steam is contained in a piston-cylinder device. Before expansion the state is 10 bars, 280°C, and 0.010 m$^3$. After expansion the pressure and volume are 1.5 bars and 0.060 m$^3$, respectively. The heat transfer during the process is $-0.80$ kJ, and the surroundings are at 20°C and 1 bar. Determine (a) the actual work, (b) the reversible useful work (or availability change), and (c) the irreversibility of the process, all answers in kilojoules.

9-13. A piston-cylinder device contains 0.44 kg of water initially at 160°C and 1 MPa. During an internally reversible, isothermal expansion,

988 kJ of heat is transferred to the water from a thermal reservoir at 600°C. The state of the environment is 298 K and 0.1 MPa. Determine (a) the actual work, (b) the reversible useful work, and (c) the irreversibility of the process, in kilojoules.

9-14. A thermal reservoir at 830 K transfers heat to 1 kg of air in a piston-cylinder device maintained at a constant pressure of 6.0 bars. During the process the air temperature within the cylinder changes from 287 to 507°C. The surrounding atmosphere is at 1 bar and 290 K. Using data from Table A-5, determine (a) the heat transfer, (b) the actual useful work, (c) the reversible useful work, and (d) the irreversibility, in kJ/kg.

9-15. A piston cylinder expands air from 6 bars, 77°C, and 0.060 m$^3$ to 3.5 bars and 0.150 m$^3$. During the process, 65 kJ of heat is added to the air from a source at 600 K. The atmosphere is at 1 bar and 300 K. (a) Determine the reversible useful work associated with the process, in kilojoules. (b) Determine the irreversibility of the process, in kilojoules.

9-16. Refrigerant 134a is contained in a rigid tank initially at 2 bars, a quality of 50.4 percent, and a volume of 0.10 m$^3$. Heat is added from a thermal reservoir at 100°C until the pressure reaches 5 bars. Determine (a) the quantity of heat added, in kilojoules, and (b) the reversible useful work associated with the overall process, in kilojoules, if $T_0 = 24°C$.

9-17. Carbon dioxide is contained in a frictionless piston-cylinder device at an initial condition of 2 bars and 17°C. Heat is added from a thermal reservoir at 700 K to the 0.88 kg of gas until the volume is doubled. On the basis of data in Table A-9, determine (a) the quantity of heat supplied, in kilojoules, and (b) the reversible useful work associated with the overall process, in kilojoules, if the process is constant pressure, $T_0 = 290$ K, and $P_0 = 1$ bar.

9-18E. A piston-cylinder expands air from 100 psia, 140°F, and 2.0 ft$^3$ to 60 psia and 5.0 ft$^3$. During the process, 75 Btu of heat is added to the air from a source at 1000°R. The atmosphere is at 1 atm and 500°R. Determine (a) the reversible useful work associated with the process, and (b) the irreversibility of the process, in Btu.

9-19E. Refrigerant 134a is contained in a rigid tank initially at 30 psia, a quality of 46.7 percent, and a volume of 3.0 ft$^3$. Heat is added from a thermal reservoir at 300°F until the pressure reaches 80 psia. Determine (a) the quantity of heat added, and (b) the reversible useful work associated with the overall process, in Btu, if $T_0 = 70°F$.

9-20. One-fourth kilogram of nitrogen initially at 140 kPa and 25°C is in an insulated tank. An impeller within the tank is turned by an external motor until the pressure is 180 kPa. Determine (a) the paddle-wheel work and (b) the irreversibility of the process, in kilojoules, if the atmosphere is at 96 kPa and 22°C.

9-21E. One-fourth pound of nitrogen initially at 16 psia and 70°F is in an insulated tank. An impeller within the tank is turned by an external motor until the pressure is 26 psia. Determine (a) the paddle-wheel work and (b) the irreversibility of the process, in Btu, if the atmosphere is at 14.8 psia and 70°F.

## AVAILABILITY AND IRREVERSIBILITY IN CLOSED SYSTEMS

9-22. A tank with a volume of 0.30 m³ contains air at 600 kPa and 600 K. The surrounding atmosphere is at 96 kPa and 300 K.
(a) Determine the availability of the air, in kilojoules.
(b) The air now undergoes a free expansion until its volume is doubled. Determine the change in the closed-system availability, in kilojoules.

9-23. Determine the availability, in kilojoules, associated with 50 kg of liquid water at 0°C and 0.95 bar if the surroundings are at 0.95 bar and 20°C.

9-24. Determine the availability of steam in a closed system at 8.0 MPa and 400°C, in kJ/kg, if the environment is at 0.10 MPa and 25°C.

9-25E. A tank with a volume of 10.0 ft³ contains air at 100 psia and 300°F. The surrounding atmosphere is at 14.5 psia and 70°F.
(a) Determine the availability of the air, in Btu.
(b) The air now undergoes a free expansion until its volume is doubled. Determine the change in the closed-system availability, in Btu.

9-26E. Determine the availability, in Btu, associated with 50 lb$_m$ of liquid water at 32°F and 1 atm if the surroundings are at 1 atm and 60°F.

9-27E. Determine the availability of steam in a closed system at 1000 psia and 800°F, in Btu/lb$_m$, if the environment is at 14.7 psia and 70°F.

9-28E. Carbon dioxide is contained in a frictionless piston-cylinder device at an initial condition of 20 psia and 40°F. Heat is added from a thermal reservoir at 740°F to the 0.88 lb$_m$ of gas until the volume is doubled. On the basis of data in Table A-9E, determine (a) the quantity of heat supplied, (b) the availability of the final state, and (c) the reversible useful work associated with the overall process, all answers in Btu, if the process is constant pressure, $T_0 = 60°F$, and $P_0 = 14.7$ psia.

9-29E. A tank of air at 200 psia and 360°F has a volume of 30.0 ft³. The air is cooled by heat transfer until the temperature is 80°F. The surroundings are at 1 atm and 80°F. Determine (a) the availability of the initial and final states, (b) the reversible useful work, and (c) the irreversibility of the process, in Btu.

9-30. Fifty kilojoules of heat are transferred between thermal reservoirs at 1000 and 600 K. The environmental temperature is 280 K. Determine (a) the net availability change and (b) the irreversibility of the process, in kilojoules.

9-31. A heat transfer of 100 kJ occurs between a thermal reservoir at $T_R$ and the environment at 300 K. Determine the availability transfers and the irreversibility of the heat-exchange process, in kilojoules, for $T_R$ values of (a) 1000 K and (b) 600 K.

9-32. A tank of air at 12 bars and 227°C has a volume of 0.80 m³. The air is cooled by heat transfer until the temperature is 27°C. The surroundings are at 1 bar and 27°C. Determine (a) the availability of the initial and final states, (b) the reversible useful work, and (c) the irreversibility of the process, in kilojoules.

9-33. A storage battery is capable of delivering 1 kWh of energy. Determine the volume of air stored in a tank at 27°C and 20 bars that is needed theoretically to have the same work capability, in cubic meters. The state of the environment is 27°C and 1 bar.

9-34. Fifty kilograms of water at 0°C and 0.95 bar are allowed to mix in a closed system with 30 kg of water at 80°C and 0.95 bar. Determine the reversible useful work associated with the change of state of (a) the 50 kg of water and (b) the 30 kg of water. Find (c) the change in availability for the overall adiabatic process and (d) the irreversibility, in kilojoules, if $T_0 = 20$°C and $P_0 = 1$ bar.

9-35. A 5-kg block of aluminum at 300°C is brought into thermal contact with a 10-kg block of copper initially at $-50$°C. Contact is maintained until thermal equilibrium is reached. The process is adiabatic, and the specific heats of aluminum and copper may be taken to be 0.99 and 0.38 kJ/kg·°C, respectively. Determine the change in availability (a) of the aluminum block, (b) of the copper block, and (c) for the overall process, all in kilojoules. (d) Then find the irreversibility, in kilojoules, if $T_0 = 27$°C.

9-36. Determine the availability of a unit mass of an ideal gas in a closed system at temperature $T$ (different from $T_0$ of the surroundings), but at a pressure $P$ which is the same as $P_0$ of the surroundings. By using your knowledge of property relations of ideal gases, express the answer in terms of $T_0$, $P_0$, $T$, and any required constants of the gas.

9-37. Determine the availability of a unit mass of an ideal gas in a closed system at temperature $T_0$ which is the same as that of the surroundings, but at a pressure $P$ which is different from $P_0$ of the surroundings. By using your knowledge of property relations of ideal gases, express your answer in terms of $T_0$, $P_0$, $P$, and any required constants of the gas.

9-38E. Fifty pounds of water at 40°F and 14.6 psia are allowed to mix in a closed system with 30 lb$_m$ of water at 160°F and 14.6 psia. Determine the reversible useful work associated with the change of state of (a) the 50 lb$_m$ of water and (b) the 30 lb$_m$ of water. Find (c) the change in availability for the overall adiabatic process and (d) the irreversibility, in Btu, if $T_0 = 70$°F and $P_0 = 14.6$ psia.

9-39E. A 5-lb$_m$ block of aluminum at 250°F is brought into thermal contact with a 10-lb$_m$ block of copper initially at 30°F. Contact is maintained until thermal equilibrium is reached. The process is adiabatic, and the specific heats of aluminum and copper may be taken to be 0.225 and 0.092 Btu/lb$_m$·°R, respectively. Determine the change in availability (a) of the aluminum block, (b) of the copper block, and (c) for the overall process, all in Btu. (d) Then find the irreversibility, in Btu, if $T_0 = 80$°F.

## REVERSIBLE WORK AND IRREVERSIBILITY IN STEADY FLOW

9-40. Air enters a steady-flow turbine at 300 kPa and 480 K and exhausts at 100 kPa and 380 K. The process is adiabatic, and the surroundings are at 100 kPa and 20°C. Compute (a) the actual work output, (b) the reversible shaft work output, and (c) the irreversibility, in kJ/kg.

9-41. Steam enters a turbine at 30 bars and 400°C and expands to 1 bar and 120°C in a steady-flow, adiabatic process. The ambient conditions are 1 bar and 27°C. Disregard changes in kinetic and potential energy, and determine (a) the actual work delivered, (b) the reversible shaft work, and (c) the irreversibility, in kJ/kg.

9-42. Air enters a steady-flow compressor at 1.4 bars, 17°C, and 70 m/s. It leaves the adiabatic device at 4.2 bars, 147°C, and 110 m/s. Determine (a) the actual work input, (b) the reversible work required, and (c) the irreversibility, in kJ/kg, if $T_0 = 17$°C and $P_0 = 1$ bar.

9-43. Refrigerant 134a is compressed in steady flow from a saturated vapor at $-12$°C to a final state of 8 bars and 50°C. For the adiabatic process determine (a) the actual work required, (b) the minimum work required, and (c) the irreversibility, in kJ/kg, if $T_0 = 20$°C and $P_0 = 1$ bar.

9-44E. Air enters a steady-flow turbine at 45 psia and 400°F and exhausts at 15 psia and 200°F. The process is adiabatic, and the surroundings are at 14.7 psia and 70°F. Compute (a) the actual work output, (b) the reversible shaft work output, and (c) the irreversibility, in Btu/lb$_m$.

9-45E. Steam enters a turbine at 400 psia and 700°F and expands to 14.7 psia and 250°F in a steady-flow adiabatic process. The ambient conditions are 14.7 psia and 80°F. Disregard changes in kinetic and potential energy, and determine (a) the actual work delivered, (b) the reversible shaft work, and (c) the irreversibility, in Btu/lb$_m$.

9-46E. Air enters a steady-flow compressor at 20 psia, 50°F, and 200 ft/s. It leaves the adiabatic device at 50 psia, 260°F, and 350 ft/s. Determine (a) the actual work input, (b) the reversible work required, and (c) the irreversibility, in Btu/lb$_m$, if $T_0 = 40$°F and $P_0 = 1$ atm.

9-47E. Refrigerant 134a is compressed in steady flow from a saturated vapor at $-10$°F to a final state of 100 psia and 120°F. The inlet volume flow rate is 4 ft$^3$/min. For the adiabatic process determine (a) the actual work required and (b) the minimum work required, in Btu/lb$_m$, and (c) the irreversibility, in Btu/min, if $T_0 = 60$°F and $P_0 = 1$ atm.

9-48. Saturated-liquid refrigerant 134a enters an expansion valve at 6 bars and leaves at 2 bars. Determine the irreversibility of the process, in kJ/kg if (a) it is adiabatic and (b) the fluid receives 4.0 kJ/kg of heat from the atmosphere, which is at 1 bar and 27°C.

9-49. Saturated water vapor at 30 bars is throttled to 7 bars. If the atmospheric temperature is 12°C, calculate the irreversibility of the process, in kJ/kg.

## Availability, Irreversibility, and Effectiveness in Steady Flow

9-50. Steam enters a turbine at 80 bars and 560°C at a rate of 50,000 kg/h. Partway through the turbine, 25 percent of the flow is bled off at 20 bars and 440°C. The rest of the steam leaves the turbine at 0.10 bar as a saturated vapor. Determine (a) the availability at the three states of interest, in kJ/kg, (b) the maximum power output possible, in kilowatts, and (c) the actual power output, in kilowatts, if the flow is adiabatic. The environment is at 1 bar and 20°C.

9-51. A hydrocarbon oil is to be cooled in a heat exchanger from 440 to 320 K by exchanging heat with water which enters the exchanger at 20°C at a rate of 3000 kg/h. The oil flows at a rate of 750 kg/h and has an average specific heat of 2.30 kJ/kg·°C. Compute the change in flow availability, in kJ/h, for (a) the hydrocarbon oil stream and (b) the water stream. Then find (c) the loss in availability for the overall process, and (d) the irreversibility of the process, in kJ/h, if $T_0 = 17$°C.

9-52. Refrigerant 134a with a mass flow rate of 5 kg/min enters a condenser at 14 bars, 80°C, and leaves at a state of 52°C, 13.9 bars. Determine the loss in availability, in kJ/min, if the coolant in the condenser is water which enters at 12°C and 7 bars and leaves at 24°C and 7 bars. Note that $T_0 = 15$°C.

9-53. An open feedwater heater operates at 7 bars. Compressed liquid water at 35°C enters at one section, while superheated vapor enters at another section. The fluids mix and leave the heater as a saturated liquid. Determine the change in stream availability, in kJ/min, if the flow rate of compressed liquid is 4370 kg/min and the mass flow rate of superheated vapor is 1000 kg/min. It is given that $T_0 = 20$°C and $P_0 = 1$ bar.

9-54E. Saturated water vapor at 400 psia is throttled to 100 psia. If the atmospheric temperature is 50°F, calculate the irreversibility of the process, in Btu/lb$_m$.

9-55E. Steam enters a turbine at 1000 psia and 1100°F at a rate of 100,000 lb$_m$/h. Partway through the turbine, 25 percent of the flow is bled off at 300 psia and 800°F. The rest of the steam leaves the turbine at 1 psia as a saturated vapor. Determine (a) the availability at the three states of interest, in Btu/lb$_m$, (*b*) the maximum power output possible, in horsepower, and (*c*) the actual power output, in horsepower, if the flow is adiabatic. The environment is at 14.7 psia and 70°F.

9-56E. A hydrocarbon oil is to be cooled in a heat exchanger from 260 to 120°F by exchanging heat with water which enters the exchanger at 70°F at a rate of 4000 lb$_m$/h. The oil flows at a rate of 1500 lb$_m$/h and has an average specific heat of 0.55 Btu/lb$_m$·°R. Compute the change in flow availability, in Btu/h, for (*a*) the hydrocarbon oil stream and (*b*) the water stream. Then find (*c*) the loss in availability for the overall process and (*d*) irreversibility of the process, in Btu/h, if $T_0 = 60$°F.

9-57E. Refrigerant 134a with a mass flow rate of 10 lb$_m$/min enters a condenser at 200 psia, 180°F, and leaves at a state of 120°F, 190 psia. Determine the loss in availability, in Btu/min, if the coolant in the condenser is water which enters at 55°F and 90 psia and leaves at 75°F and 90 psia. Note that $T_0 = 60$°F.

9-58E. An open feedwater heater operates at 100 psia. Compressed liquid water at 100°F enters at one section, while superheated vapor enters at another section. The fluids mix and leave the heater as a saturated liquid. Determine the change in stream availability, in Btu/min, if the flow rate of compressed liquid is 4690 lb$_m$/min and the mass flow rate of superheated vapor is 1000 lb$_m$/min. It is given that $T_0 = 60$°F and $P_0 = 1$ atm.

9-59. Air is expanded adiabatically in a steady-flow turbine from 6 bars, 560 K, to 1.0 bar. Changes in kinetic and potential energies are negligible, and the environmental conditions are 1 bar and 27°C. If the adiabatic turbine efficiency is 88.0 percent, use the air table for data to determine
(*a*) The actual work of expansion, kJ/kg.
(*b*) The actual outlet temperature, in kelvins.
(*c*) The reversible work output for the same end state as part *b*.
(*d*) The irreversibility, in kJ/kg.
(*e*) The second-law effectiveness.

9-60. Steam enters a turbine at 40 bars, 500°C, and 140 m/s and leaves as a saturated vapor at 100°C and 80 m/s. The measured work output is 746.0 kJ/kg, and the average temperature $T_b$ at the outer surface of the turbine may be taken as the average of the inlet and outlet steam temperature. (*a*) Determine the availability change and the irreversibility, both in kJ/kg, for the process within the turbine. (*b*) Now, enlarge the control volume so that the local environment at 25°C is included. Find the availability change and the irreversibility in this new situation.

9-61. Air is compressed adiabatically in steady flow from 2 bars, 27°C, to 4 bars. Neglect changes in kinetic and potential energies. The measured work input is 80.6 kJ/kg, $P_0 = 1$ bar, and $T_0 = 27$°C.

Determine (*a*) the irreversibility of the process, (*b*) the change in stream availability, in kJ/kg, and (*c*) the effectiveness of the process.

9-62. Nitrogen gas initially at 3.6 bars and 27°C is throttled through a well-insulated valve to a pressure of 1.1 bars. The atmospheric temperature is 15°C. Determine (*a*) the change in stream availability and (*b*) the irreversibility of the process, in kJ/kg.

9-63. A hand-held hair dryer has 0.015 kg/s of atmospheric air entering at 22°C, 100 kPa, and 3.6 m/s, and exiting at 87°C, 100 kPa, and 9.0 m/s. Using Table A-5 data, and assuming adiabatic operation, find (*a*) the actual power supplied to the dryer, (*b*) the minimum power necessary to transform the air from the given inlet to outlet states, and (*c*) the irreversibility rate, in kilowatts, if $T_0$ is 22°C.

9-64. Water enters a heat exchanger at a rate of 50 kg/s as a liquid at 0.20 MPa and 90°C and leaves at the same pressure and 120°C. The heat is supplied to the water stream from a stream of hot air which enters at 680 K and 0.30 MPa and leaves at 460 K at the same pressure. Determine (*a*) the change in stream availability of the water, (*b*) the change in stream availability of the air, and (*c*) the overall irreversibility of the heat-exchange process, all answers in kJ/s. The environmental temperature is 290 K.

# CHAPTER
# 10

# NONREACTIVE IDEAL-GAS MIXTURES

Hyperbolic cooling tower at Nine Mile Point nuclear power station near Oswego, NY.

The basic thermodynamic laws introduced so far are of general validity. In the application of these laws to closed and open systems, however, we have dealt primarily with systems containing a single chemical species. Analytical expressions, tables, and graphs have been presented which related such intensive, intrinsic properties as $P$, $v$, $T$, $u$, $h$, $s$, $c_v$, and $c_p$ for systems of a single component. Since many engineering applications involve multicomponent systems, it is vital to have some understanding of methods for evaluating the properties of such systems.

A complete description of a multicomponent system requires a specification not only of two properties, such as the pressure and the temperature of the mixture, but also of the composition. Thus properties such as $u$, $h$, $v$, and $s$ of a mixture are different for every different composition. Note, however, that the properties of individual components are readily available. Hence one method of evaluating mixture properties is to devise rules for averaging the properties of the individual pure components, so that the resulting value is representative of the overall composition. This approach is used in this chapter to model the behavior of ideal-gas mixtures in general and the special case of air–water vapor mixtures.

## 10-1 COMPOSITION ANALYSIS OF GAS MIXTURES

The composition of a mixture is commonly specified in terms of either the mass of each component or the number of moles of each component. An analysis of a mixture made on the basis of mass (or weight) is called a *gravimetric analysis*. For a nonreacting gas mixture, the total mass of the mixture $m_m$ is the sum of the mass of each of the $k$ components. That is,

$$m_m = m_1 + m_2 + \cdots + m_k = \sum_{i=1}^{k} m_i \qquad \text{[10-1]}$$

The *mass fraction* $mf_i$ of the $i$th component is defined as

$$mf_i \equiv \frac{m_i}{m_m} \qquad \text{[10-2]}$$

If Eq. [10-2] is divided through by $m_m$, clearly the sum of the mass fractions of all the components in a mixture is unity:

$$\sum_{i=1}^{k} mf_i = 1 \qquad \text{[10-3]}$$

If an analysis of a gas mixture is based on the amount of substance or the number of moles of each component present, the analysis is termed a *molar analysis*. The total number of moles $N_m$ for a mixture is given by

$$N_m = N_1 + N_2 + \cdots + N_k = \sum_{i=1}^{k} N_i \qquad \text{[10-4]}$$

and the mole fraction $y_i$ of any component is defined as

$$y_i = \frac{N_i}{N_m} \qquad \text{[10-5]}$$

Division of Eq. [10-4] by $N_m$ shows that the sum of the mole fractions in a gas mixture is equal to unity:

$$\sum_{i=1}^{k} y_i = 1 \qquad \text{[10-6]}$$

From the definition of the molar mass $M_i$ of a component, the mass of a component is related to the number of moles of that component by

$$m_i = N_i M_i \qquad \text{[10-7]}$$

In a similar fashion, for the gas mixture an *apparent molar mass* (or average molecular weight) $M_m$ of the mixture defined by

$$M_m = \frac{m_m}{N_m} \qquad \textbf{[10-8]}$$

The substitution of Eqs. [10-1], [10-7], and [10-5] in sequence into Eq. [10-8] yields

$$M_m = \frac{m_m}{N_m} = \frac{\sum m_i}{N_m} = \frac{\sum N_i M_i}{N_m} = \sum_{i=1}^{k} y_i M_i \qquad \textbf{[10-9]}$$

The apparent or average molar mass of a gas mixture, then, is the sum over all the components of the product of the mole fraction and the molar mass.

As an example of this latter relationship, the apparent molar mass of atmospheric air can be determined in the following way. If we neglect the presence of trace amounts of other components, dry air (excluding water vapor) is composed of roughly 78.08 percent $N_2$, 20.95 percent $O_2$, 0.93 percent argon, and 0.035 percent carbon dioxide on a mole basis. Substitution of these values into Eq. [10-9] yields

$$\begin{aligned} M_{\text{dry air}} &= 0.7808(28.01) + 0.2095(32.0) \\ &\quad + 0.0093(39.94) + 0.00035(44.01) \\ &= 28.97 \end{aligned}$$

This is the commonly quoted value for dry air, as shown in Table A-2.

A relationship between mass fractions and mole fractions is found by substituting Eqs. [10-7] and [10-8] into Eq. [10-2] and by also making use of Eq. [10-5]. The result is

$$mf_i = y_i \frac{M_i}{M_m} \qquad \textbf{[10-10]}$$

To make use of Eq. [10-10], the average molar mass $M_m$ of the mixture must be calculated first. Also recall that a specific gas constant for a pure ideal gas is related to a universal gas constant by $R = R_u/M$. In a similar fashion the **apparent gas constant** $R_m$ of a gas mixture is defined by

$$R_m = \frac{R_u}{M_m} \qquad \textbf{[10-11]}$$

The preceding relationships involving composition variables are illustrated in the following two examples.

---

**Example 10-1**

The molar analysis of a gaseous fuel is 20 percent $CH_4$, 40 percent $C_2H_6$, and 40 percent $C_3H_8$. Determine (*a*) the gravimetric analysis in terms of mass fractions, (*b*) the apparent molar mass of the mixture, and (*c*) the apparent gas constant.

**Solution:**

**Given:** The molar analysis of a gaseous hydrocarbon fuel, as shown in Fig. 10-1.

**Find:** (a) $mf_i$'s, (b) $M_m$, and (c) $R_m$.

**Model:** Ideal-gas mixture.

**Analysis:** (a) The actual moles of the individual components are not known. It is convenient to select 100 mol as a basis for the subsequent calculation. Thus the percentages given in the fuel analysis are also the number of moles per 100 mol of mixture. (It is not important whether the moles are expressed in kilomoles or pound-moles.) These values are listed in column 2 of Table 10-1. Column 3 lists the molar mass of each. Recall that the mass of a component is given by $N_i M_i$ [Eq. [10-7]]. Consequently, the mass of each component per 100 mol of mixture is found by multiplying the value in column 2 by the value in column 3 for each species. The result is shown in column 4. The number at the bottom of column 4 is the summation of the masses $m_m$. The mass analysis, then, is found by dividing the mass of each component by $m_m$. The resulting $mf_i$ values are given (in percentages) in column 5. Note that the gravimetric analysis shown in column 5 is significantly different from the molar analysis in column 2.

(b) The sum of column 4 is the mass of the mixture per 100 mol of mixture. Therefore, the mass per mole $M_m$ is $\frac{3280}{100} = 32.80$. An alternate method for determining the apparent molar mass is through the use of Eq. [10-9]. In this case

$$M_m = \sum y_i M_i = 0.20(16) + 0.40(30) + 0.40(44) = 32.8$$

which is in agreement with the preceding calculation. The solutions to parts $a$ and $b$ are independent of whether the data are in SI or USCS units.

(c) The apparent gas constant is calculated from Eq. [10-11]. Its value depends on the value of $R_u$ chosen in a particular set of units. In SI units, for example, one choice is

$$R_m = \frac{R_u}{M_m} = \frac{8.314 \text{ kJ/kmol·K}}{32.8 \text{ kg/kmol}} = 0.253 \text{ kJ/kg·K}$$

20 percent $CH_4$
40 percent $C_2H_6$
40 percent $C_3H_8$
(Percent by moles)

**Figure 10-1**
Molar analysis of the gas mixture in Example 10-1.

**Table 10-1**

| (1) Component | (2) Moles per 100 mol of Mixture $N_1$ | | (3) Molar Mass $M_i$ | | (4) Mass per 100 mol of Mixture $m_i$ | (5) Mass Analysis $mf_i(\%)$ |
|---|---|---|---|---|---|---|
| $CH_4$ | 20 | × | 16 | = | 320 | 9.76 |
| $C_2H_6$ | 40 | × | 30 | = | 1200 | 36.59 |
| $C_3H_8$ | 40 | × | 44 | = | 1760 | 53.65 |
| Total | 100 | | | | 3280 | 100.00 |

In USCS units a possible choice might be

$$R_m = \frac{R_u}{M_m} = \frac{10.73 \text{ psia·ft}^3/\text{lbmol·°R}}{32.8 \text{ lb}_m/\text{lbmol}} = 0.327 \text{ psia·ft}^3/\text{lb}_m\text{·°R}$$

**Comment:**   Although not necessary, the tabular-type calculation shown in this example is quite convenient.

**Example 10-2**

**A** mixture consists of 10 percent hydrogen, 48 percent oxygen, and 42 percent carbon monoxide by mass. Determine (a) the molar analysis, in percent, (b) the apparent molar mass, and (c) the apparent gas constant in bar·m³/kg·K and psia·ft³/lb$_m$·°R, if an ideal-gas mixture.

**Solution:**

**Given:**   The mass analysis of the gas mixture shown in Fig. 10-2.

**Find:**   (a) $y_i$'s, (b) $M_m$, and (c) $R_m$.

**Model:**   Ideal-gas mixture.

**Analysis:**   (a) The basis for the calculations is a unit mass of the mixture. In Table 10-2, the first three columns list the chemical species, mass fraction, and molar mass, respectively. Column 4 is the moles of each per unit mass of mixture, found by dividing column 2 by column 3. The mole fraction of each constituent in column 5 is then determined by dividing each value in column 4 by the total in column 4. Note that hydrogen, which is present in the smallest amount in the gravimetric analysis, has the largest percentage on a molar basis.

(b) The apparent molar mass is obtained most easily from the last value in column 4:

$$M_m = \frac{1}{\text{moles/unit mass of mixture}} = \frac{1}{0.080} = 12.5$$

---

**Figure 10-2**
Mass analysis for Example 10-2.

10 percent H$_2$
48 percent O$_2$
42 percent N$_2$
(Percent by mass)

---

**Table 10-2**

| (1) | (2) | | (3) | | (4) | (5) |
|---|---|---|---|---|---|---|
| | | | **Molar** | | **Moles per Unit Mass** | **Mole** |
| **Component** | **Mass** | | **Mass** | | **of Mixture** | **Fraction** |
| | $m_i$ | ÷ | $M_i$ | = | $N_i$ | $y_i$ |
| H$_2$ | 0.10 | ÷ | 2 | = | 0.050 | 0.6250 |
| O$_2$ | 0.48 | ÷ | 32 | = | 0.015 | 0.1875 |
| CO | 0.42 | ÷ | 28 | = | 0.015 | 0.1875 |
| Total | 1.00 | | | | 0.080 | 1.0000 |

(c) The apparent gas constant is determined from Eq. [10-11]:

SI :    $R_m = \dfrac{R_u}{M_m} = \dfrac{0.08314 \text{ bar·m}^3/\text{kmol·K}}{12.5 \text{ kg/kmol}} = 6.65 \times 10^{-3} \text{ bar·m}^3/\text{kg·K}$

USCS :    $R_m = \dfrac{R_u}{M_m} = \dfrac{10.73 \text{ psia·ft}^3/\text{lbmol·°R}}{12.5 \text{ lb}_m/\text{lbmol}} = 0.858 \text{ psia·ft}^3/\text{lb}_m\text{·°R}$

Other values of $R_m$ may be calculated by using other values of $R_u$ from Tables A-1 and A-1E.

**Comment:**    In some cases, the mass of each component will be known initially rather than the mass fractions. In this case the known masses would be placed in column 2, rather than the mass fractions. When these masses are divided by the molar masses in column 3, then column 4 would list the moles of each component in the mixture, rather than the moles per unit mass of mixture. Knowledge of the moles of each then leads to the mole fraction of each in column 5.

## 10-2    $PvT$ RELATIONSHIPS FOR IDEAL-GAS MIXTURES

The $PvT$ relationship for a mixture of gases is based on two approaches known as the Dalton and Amagat models. These models are used for both ideal- and real-gas mixtures. The application to real-gas mixtures appears in Chap. 12.

### 10-2-1    THE DALTON MODEL

One method of evaluating the $PvT$ behavior of gas mixtures is through the model known as **Dalton's law of additive pressures:**

The total pressure exerted by a mixture of gases is the sum of the component pressures $p_i$ each gas would exert if it existed alone at the mixture temperature and volume.

Hence Dalton's law can be written in the form

$$P = p_1 + p_2 + \cdots + p_k = \sum_{i=1}^{k} p_i \qquad \textbf{[10-12]}$$

where $p_i$ is the **component pressure** of the $i$th component and $p_i = f(T, V)$. A physical representation of the additive-pressure rule is shown in Fig. 10-3 for the case of two gases A and B. We would expect ideal gases to fulfill Dalton's rule exactly, since the concept of an ideal gas implies that intermolecular forces are negligible, and thus the gases act independently of one another.

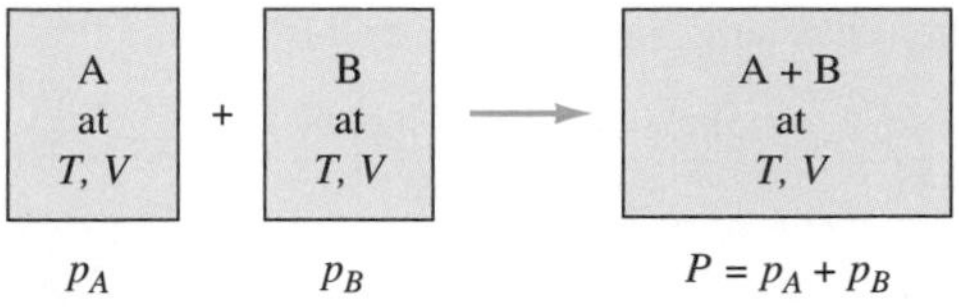

**Figure 10-3**    Schematic representation of Dalton's law of additive pressures.

The component pressure exerted by a gas in an ideal-gas mixture, in view of Dalton's rule, can be expressed as

$$p_i = \frac{N_i R_u T}{V} \qquad \text{[10-13]}$$

where $T$ and $V$ are the absolute temperature and the volume of the mixture. The total pressure of the ideal-gas mixture is given by

$$P = \frac{N_m R_u T}{V} \qquad \text{[10-14]}$$

A relationship between the component pressure $p_i$ and its mole fraction $y_i$ is found by dividing Eq. [10-13] by Eq. [10-14]. This leads to

$$\frac{p_i}{P} = \frac{N_i R_u T/V}{N_m R_u T/V} = \frac{N_i}{N_m} = y_i$$

or, in terms of $p_i$,

$$\boxed{p_i = y_i P} \qquad \text{[10-15]}$$

The product $y_i P$ is defined as the **partial pressure** $p_i'$ of a gas. The partial pressure and the component pressure are identical only for ideal-gas mixtures.

### 10-2-2   THE AMAGAT MODEL

Another *PvT* description of gas mixtures is based on **Amagat's law of additive volumes:**

The total volume of a mixture of gases is the sum of the component volumes $V_i$ that each gas would occupy if it existed alone at the mixture temperature and pressure.

This law is expressed by the relation

$$V = V_1 + V_2 + \cdots + V_k = \sum_{i=1}^{k} V_i \qquad \text{[10-16]}$$

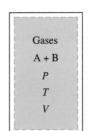

**Figure 10-4**   Schematic representation of Amagat's law of additive volumes.

where $V_i$ is the *component volume* of the $i$th component and $V_i = f(T, P)$. A schematic representation of the additive-volume rule is shown in Fig. 10-4 for the case of two gases.

A physical description of the component volume $V_i$ for ideal-gas mixtures is realized in the following way: Consider a mixture of two ideal gases in a system of volume $V$ with a total pressure $P$ and temperature $T$. Hypothetically, the gases could be separated so that one species alone occupied a certain part of the volume and the other species filled the remaining volume. The temperature and pressure of each separate gas would still be identical. The volumes that each occupies are the component volumes of the two species. This hypothetical separation of two gases (A and B) into regions representing their component or partial volumes $V_A$ and $V_B$ is shown in Fig. 10-5.

The component volume occupied by an ideal gas in an ideal-gas mixture is given by

$$V_i = \frac{N_i R_u T}{P} \qquad \textbf{[10-17]}$$

Now if Eq. [10-17] is divided by the ideal-gas equation for the mixture volume $V$, we obtain

$$\frac{V_i}{V} = \frac{N_i R_u T/P}{N_m R_u T/P} = \frac{N_i}{N_m} = y_i \qquad \textbf{[10-18]}$$

On the basis of Eqs. [10-15] and [10-18] it is evident that

$$\boxed{y_i = \frac{p_i}{P} = \frac{N_i}{N_m} = \frac{V_i}{V}} \qquad \text{(ideal gas)} \qquad \textbf{[10-19]}$$

Hence, for *ideal-gas mixtures,* the mole fraction, the volume fraction, and the ratio of component pressure to total pressure are all equal at a given state. In Example 10-2 the conversion of a mass analysis to a mole analysis revealed a gas mixture with mole fractions of 0.6250 for hydrogen, 0.1875 for oxygen, and 0.1875 for carbon monoxide. Therefore, the volumetric analysis of the mixture is 62.50 percent $H_2$, 18.75 percent $O_2$, and 18.75 percent CO. The fact that *a molar analysis is identical to a volumetric analysis of*

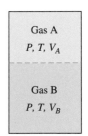

**Figure 10-5**
Hypothetical separation of gases A and B into partial volume regions.

*an ideal-gas mixture* is of major importance. It enables us to convert between volumetric analyses made in the laboratory to mole or mass analyses needed in thermodynamic calculations.

## 10-3 INTERNAL ENERGY, ENTHALPY, AND ENTROPY FOR IDEAL-GAS MIXTURES

In a mixture of ideal gases, the temperature $T$ applies to all gases within the mixture that occupies a volume $V$ at a total pressure $P$. The component pressures are given by the quantity $y_i P$. Other thermodynamic properties of the individual gases and of the mixtures may be obtained by applying the **Gibbs-Dalton law,** a generalization of Dalton's rule of additive pressures:

> In a mixture of ideal gases each component of the mixture acts as if it were alone in the system at the volume $V$ and the temperature $T$ of the mixture.

Consequently all extensive properties of the multicomponent mixture can be found by summing the contributions made by each gas component.

### 10-3-1 EVALUATION OF $\Delta U$ AND $\Delta H$ FOR IDEAL-GAS MIXTURES

On the basis of the Gibbs-Dalton law, the total internal energy of the mixture $U_m$ is given by

$$U_m = U_1 + U_2 + \cdots + U_k = \sum_{i=1}^{k} U_i \qquad \textbf{[10-20]}$$

The internal energy of each component may be expressed as $U_i = N_i \bar{u}_i = m_i u_i$, where $\bar{u}_i$ is the specific internal energy on a mole basis and $u_i$ is the specific internal energy on a mass basis. (A bar over any symbol, as in $\bar{u}$, is used to indicate a molar quantity.) As a result, Eq. [10-20] may be written in the following two ways:

$$U_m = N_m \bar{u}_m \qquad \textbf{[10-21a]}$$

$$= N_1 \bar{u}_1 + N_2 \bar{u}_2 + \cdots + N_k \bar{u}_k = \sum_{i=1}^{k} N_i \bar{u}_i$$

and

$$U_m = m_m u_m \qquad \textbf{[10-21b]}$$

$$= m_1 u_1 + m_2 u_2 + \cdots + m_k u_k = \sum_{i=1}^{k} m_i u_i$$

where $u_m$ (or $\bar{u}_m$) is the specific internal energy of the mixture. Solving these equations for the mixture specific internal energy, we find that

$$\bar{u}_m = y_1\bar{u}_1 + y_2\bar{u}_2 + \cdots + y_k\bar{u}_k = \sum_{i=1}^{k} y_i\bar{u}_i \qquad \textbf{[10-22a]}$$

and
$$u_m = mf_1u_1 + N_2u_2 + \cdots + mf_ku_k = \sum_{i=1}^{k} mf_iu_i \qquad \textbf{[10-22b]}$$

In the energy analysis of closed systems, it is the change in internal energy that is required. It follows directly from Eqs. [10-21] and [10-22] that on either a molar or mass basis

$$\Delta U_m = \sum_{i=1}^{k} N_i\Delta\bar{u}_i \qquad \text{or} \qquad \Delta U_m = \sum_{i=1}^{k} m_i\Delta u_i \qquad \textbf{[10-23]}$$

and the specific internal-energy change on a molar or mass basis is found from

$$\Delta\bar{u}_m = \sum_{i=1}^{k} y_i\Delta\bar{u}_i \qquad \text{or} \qquad \Delta u_m = \sum_{i=1}^{k} mf_i\Delta u_i \qquad \textbf{[10-24]}$$

The internal energy of an ideal-gas mixture is solely a function of the temperature of the mixture.

The enthalpy of a mixture of ideal gases is also the sum of the enthalpies of the individual components. Consequently, equations similar to Eqs. [10-21] through [10-24] may be written for the enthalpy function. The total enthalpy of a mixture $H_m$ is given by

$$H_m = N_m\bar{h}_m = \sum_{i=1}^{k} N_i\bar{h}_i \qquad \text{or} \qquad H_m = m_mh_m = \sum_{i=1}^{k} m_ih_i \qquad \textbf{[10-25]}$$

The specific enthalpies of an ideal-gas mixture on molar and mass bases are

$$\bar{h}_m = \sum_{i=1}^{k} y_i\bar{h}_i \qquad \text{or} \qquad h_m = \sum_{i=1}^{k} mf_ih_i \qquad \textbf{[10-26]}$$

The total enthalpy change on a molar and a mass basis is

$$\Delta H_m = \sum_{i=1}^{k} N_1\Delta\bar{h}_i \qquad \text{and} \qquad \Delta H_m = \sum_{i=1}^{k} m_i\Delta h_i \qquad \textbf{[10-27]}$$

The specific enthalpy change on a molar and a mass basis is

$$\Delta\bar{h}_m = \sum_{i=1}^{k} y_i\Delta\bar{h}_i \qquad \text{and} \qquad \Delta h_m = \sum_{i=1}^{k} mf_i\Delta h_i \qquad \textbf{[10-28]}$$

The enthalpy of an ideal-gas mixture is solely a function of the temperature of the mixture.

The methods of obtaining the specific internal energy and specific enthalpy data of pure ideal gases needed in the preceding equations were discussed in Chap. 4. Ideal-gas property tables contain values of $u_i$ and $h_i$ (or $\bar{u}_i$ and $\bar{h}_i$) solely as a function of temperature. The changes in internal energy and enthalpy can also be evaluated from specific-heat data. Recall that for each component

$$\Delta u_i = c_{v,i} \Delta T \qquad \text{and} \qquad \Delta h_i = c_{p,i} \Delta T$$

where $c_{v,i}$ and $c_{p,i}$ are usually taken to be constant or the arithmetic average values in the given temperature range. For an ideal-gas mixture these equations are written as

$$\Delta u_m = c_{v,m} \Delta T \qquad \text{and} \qquad \Delta h_m = c_{p,m} \Delta T \qquad \textbf{[10-29]}$$

where the average specific heats of a mixture, on a molar or a mass basis, are defined by

$$\bar{c}_{v,m} \equiv \sum_{i=1}^{k} y_i \bar{c}_{v,i} \qquad \text{or} \qquad c_{v,m} \equiv \sum_{i=1}^{k} mf_i c_{v,i} \qquad \textbf{[10-30]}$$

$$\text{and} \qquad \bar{c}_{p,m} \equiv \sum_{i=1}^{k} y_i \bar{c}_{p,i} \qquad \text{or} \qquad c_{p,m} \equiv \sum_{i=1}^{k} mf_i c_{p,i} \qquad \textbf{[10-31]}$$

Thus in general we can evaluate changes in internal energy and enthalpy either on a molar basis or a mass basis, depending on whether the available data are on a molar or mass basis.

**Example 10-3**

Consider the mixture of three ideal gases used in Example 10-2 at an initial state of 300 K and 2 bars. The gas mixture flows steadily through a heat exchanger at constant pressure until its temperature is raised to 500 K. Determine the heat transfer to the mixture, in kJ/kg, by using (a) $c_p$ data and (b) tabular $\bar{h}$ data.

**Solution:**

**Given:**   A gas mixture flows through a heat exchanger as shown in Fig. 10-6.

**Find:**   $q$ by using (a) $c_p$ data and (b) tabular $\bar{h}$ data, in kJ/kg.

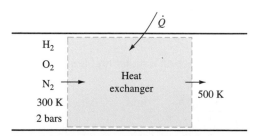

Gas mixture is 10% $H_2$, 48% $O_2$, and 42% $N_2$, by mass

**Figure 10-6**   Heat exchanger schematic and data for Example 10-3.

**Model:** Open, steady-state system; ideal-gas mixture; $\Delta ke = \Delta pe = 0$.

**Analysis:** The mass and energy balances for the fluid in the heat exchanger are

$$0 = \dot{m}_1 - \dot{m}_2 \quad \text{and} \quad 0 = \dot{Q}_{in} + \dot{m}_1 h_1 - \dot{m}_2 h_2$$

where $\dot{W} = 0$ for the steady-state control volume and kinetic- and potential-energy effects are neglected. Thus on a unit-mass basis

$$q = h_2 - h_1$$

(a) Using $c_p$ data, the specific-enthalpy change of the mixture is found by employing Eq. [10-28], namely,

$$\Delta h_m = \sum_{i=1}^{k} mf_i \Delta h_i$$

A summary of the necessary data is shown in Table 10-3. The gravimetric and molar analyses are taken from Example 10-2, and the values of $c_p$ and $\bar{h}$ for each gas are taken from Tables A-3, A-7, A-8, and A-11. Since the $c_p$ data are on a mass basis, we must use the format of Eq. [10-31] involving mass fractions to find $c_{p,m}$. In addition, the arithmetic average of the $c_p$ data between 300 and 500 K is used. Hence

$$\Delta h_m = \sum_{i=1}^{k} mf_i c_{p,i} \Delta T$$

$$= [0.10(14.420) + 0.48(0.945) + 0.42(1.051)] \text{ kJ/(kg·K)} \times (500 - 300) \text{ K}$$

$$= 2.337 \text{ kJ/kg·K} \times 200 \text{ K} = 467.4 \text{ kJ/kg mixture}$$

Thus

$$q = \Delta h_m = 467.4 \text{ kJ/kg mixture}$$

(b) Since the enthalpy data in the tables are reported on a molar basis, the mixture enthalpy is found by using the molar format of Eq. [10-28], that is,

$$\Delta \bar{h}_m = \sum_{i=1}^{k} y_i \Delta h_i = [0.6250(14,350 - 8522) + 0.1875(14,770 - 8736)$$

$$+ 0.1875(14,600 - 8723)] \text{ kJ/kmol} = 5876 \text{ kJ/kmol}$$

**Table 10-3**

| Species | $mf_i$ | $y_i$ | $c_{p,300K}$ kJ/kg·K | $c_{p,500K}$ kJ/kg·K | $h_{300K}$ kJ/kmol | $h_{500K}$ kJ/kmol |
|---------|--------|-------|----------------------|----------------------|--------------------|--------------------|
| $H_2$ | 0.10 | 0.6250 | 14.307 | 14.513 | 8,522 | 14,350 |
| $O_2$ | 0.48 | 0.1875 | 0.918 | 0.972 | 8,736 | 14,770 |
| CO | 0.42 | 0.1875 | 1.040 | 1.063 | 8,723 | 14,600 |

To convert this value to a kilogram basis,

$$\Delta h = \frac{\Delta \bar{h}}{M_m} = \frac{5876 \text{ kJ/kmol}}{12.5 \text{ kg/kmol}} = 470.1 \text{ kJ/kg}$$

where the value of $M_m$ is obtained from Example 10-2.

**Comment:**   The answer in part $b$ differs from that found in part $a$ by about 1/2 percent and is due to using average specific heats in part $a$.

---

## 10-3-2   EVALUATION OF $\Delta$S FOR IDEAL-GAS MIXTURES

The entropy of a mixture of ideal gases can also be determined on the basis of the Gibbs-Dalton rule. Since each gas behaves as if it alone occupied the volume $V$ of the system at the mixture temperature $T$, we may write

$$S_m = S_1(T, V) + S_2(T, V) + \cdots + S_k(T, V) = \sum_{i=1}^{k} S_i(T, V) \quad \text{[10-32]}$$

The total entropy change for a mixture of gases is determined from the sum of the entropy changes of the individual constituents, on either a molar or a mass basis. That is,

$$\Delta S_m = \sum_{i=1}^{k} N_i \Delta \bar{s}_i \quad \text{or} \quad \Delta S_m = \sum_{i=1}^{k} m_i \Delta s_i \quad \text{[10-33]}$$

In similar fashion, the specific-entropy change on a molar or a mass basis is

$$\Delta \bar{s}_m = \sum_{i=1}^{k} y_i \Delta \bar{s}_i \quad \text{and} \quad \Delta s_m = \sum_{i=1}^{k} mf_i \Delta s_i \quad \text{[10-34]}$$

The entropy change of an ideal gas is expressed usually as a function of temperature and pressure, rather than $T$ and $V$ of the system. Recall that for an ideal gas

$$ds = c_p \frac{dT}{T} - R \frac{dp}{p} \quad \text{[7-8]}$$

The symbol $p$ has been used for the pressure in this case because, at the temperature and volume of the mixture, the pressure of any component in an ideal-gas mixture is measured by its *component pressure* (or partial pressure) and not the *total pressure* $P$ of the system. The change in the specific entropy of an ideal gas that is a component of the mixture is given on an intensive basis by

$$\Delta s_i = c_{p,i} \ln \frac{T_2}{T_1} - R \ln \frac{p_{i2}}{p_{i1}} \quad \text{or} \quad \Delta \bar{s}_i = \bar{c}_{p,i} \ln \frac{T_2}{T_1} - R_u \ln \frac{p_{i2}}{p_{i1}}$$

$$\text{[10-35]}$$

in terms of its component or partial pressure $p_i$. Recall that in the gas tables the integral of $c_p \, dT/T$ is given by $s_2^o - s_1^o$. Consequently, the entropy change for the $i$th component in a mixture of ideal gases may also be found from

$$\Delta s_i = s_{i2}^o - s_{i1}^o - R_i \ln \frac{p_{i2}}{p_{i1}} \qquad \text{or} \qquad \Delta \bar{s}_i = \bar{s}_{i2}^o - \bar{s}_{i1}^o - R_u \ln \frac{p_{i2}}{p_{i1}}$$

**[10-36]**

where $s^o$ or $\bar{s}^o$ data are available for the component. When an ideal-gas mixture undergoes a process without a change in composition, the component pressure of each species remains the same. In this case the value of $p_{i2}/p_{i1}$ is the same as $P_2/P_1$ for the total mixture.

For isentropic processes of a pure ideal gas $T_2/T_1 = (P_2/P_1)^{(k-1)/k}$, where $k = c_p/c_v$. This same relation may be used for an ideal-gas mixture of constant composition undergoing an isentropic process. That is,

$$\frac{T_2}{T_1} \equiv \left( \frac{P_2}{P_1} \right)^{(k_m - 1)/k_m}$$

**[10-37]**

where $k_m \equiv c_{p,m}/c_{v,m}$.

---

**A** mixture of ideal gases consisting of 0.20 kg of nitrogen and 0.30 kg of carbon dioxide is compressed from 2 bars and 300 K to 6 bars adiabatically and internally reversibly in a closed system. Accounting for variable specific heats, determine (*a*) the final temperature, (*b*) the work required, in kilojoules, and (*c*) the entropy change of the nitrogen and the carbon dioxide, in kJ/K. (*d*) Determine the final temperature, using constant specific-heat data.

**Solution:**

**Given:**    A mixture of two gases is compressed adiabatically and reversibly, as shown in Fig. 10-7.

**Find:**    (*a*) $T_2$, (*b*) $W$, in kJ, (*c*) $\Delta S$ values, in kJ/K, and (*d*) $T_2$ using constant specific-heat data.

**Model:**    Adiabatic closed system, ideal-gas mixture, internally reversible process.

**Analysis:**    The energy and entropy balances for a closed system with a finite change in state are

$$\Delta E = Q + W \qquad \text{and} \qquad \Delta S = \int \frac{\partial Q}{T} + \sigma$$

These can be simplified by neglecting changes in kinetic and potential energy ($\Delta E = \Delta U$) and recalling that $Q = 0$ for an adiabatic process and that $\sigma = 0$ for an internally reversible process. Thus the energy and entropy balances reduce to

$$\Delta U = W \qquad \text{and} \qquad \Delta S = 0$$

**Example 10-4**

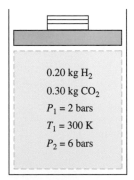

0.20 kg H$_2$

0.30 kg CO$_2$

$P_1 = 2$ bars

$T_1 = 300$ K

$P_2 = 6$ bars

**Figure 10-7**
Schematic and data for adiabatic compression in Example 10-4.

(*a*) To solve for the final temperature, we shall try the entropy balance first, since the energy equation has two unknowns $U_2$ and $W$. For the ideal-gas mixture, $\Delta S_m = \Delta S_{N_2} + \Delta S_{CO_2}$. Since we have molar data in the Appendix, we shall let $\Delta S_i = N_i \Delta \bar{s}_i$. Now the entropy change becomes

$$\Delta S_m = 0 = N_{N_2} \Delta \bar{s}_{N_2} + N_{CO_2} \Delta \bar{s}_{CO_2}$$

where $\quad N_{N_2} \Delta \bar{s}_{N_2} = \dfrac{m_{N_2}}{M_{N_2}} \left( \Delta \bar{s}^0_{N_2} - R_u \ln \dfrac{p_{N_2,2}}{p_{N_2,1}} \right)$

$$= \frac{0.20 \text{ kg}}{28 \text{ kg/kmol}} \left( \bar{s}^0_{N_2} - 191.682 - 8.314 \ln \frac{6}{4} \right) \frac{\text{kJ}}{\text{kmol·K}}$$

$$= (0.00714 \text{ kmol})\bar{s}^0_{N_2} - 1.434 \text{ kJ/K}$$

and $\quad N_{CO_2} \Delta \bar{s}_{CO_2} = \dfrac{m_{CO_2}}{M_{CO_2}} \left( \Delta \bar{s}^0_{CO_2} - R_u \ln \dfrac{p_{CO_2,2}}{p_{CO_2,1}} \right)$

$$= \frac{0.30 \text{ kg}}{44 \text{ kg/kmol}} \left( \bar{s}^0_{CO_2} - 213.915 - 8.314 \ln \frac{6}{4} \right) \frac{\text{kJ}}{\text{kmol·K}}$$

$$= (0.00682 \text{ kmol})\bar{s}^0_{CO_2} - 1.521 \text{ kJ/K}$$

Note that for each gas the term $p_{i2}/p_{i1}$ has been replaced by the ratio of mixture total pressures $P_2/P_1$. This is valid because $p_i = y_i P$, and the mole fraction of each gas remains constant during the process. Substituting the known values into the entropy balance and rearranging gives

$$(0.00714 \text{ kmol})\bar{s}^0_{N_2} + (0.00682 \text{ kmol})\bar{s}^0_{CO_2} = 2.955 \text{ kJ/K}$$

where the two $\bar{s}^0$ terms are solely a function of temperature. The solution to this equation is found by iteration. A final temperature is assumed, and the $\bar{s}^0$ values for the two gases are found in their respective tables. If these two values of $\bar{s}^0$ do not satisfy the equation, another temperature must be assumed until the correct final temperature is determined. In this case

At 390K : $\quad \sum N_i \bar{s}^0_i = 0.00714(199.331) + 0.00682(224.181)$
$$= 2.952 \text{ kJ/K}$$

At 400K : $\quad \sum N_i \bar{s}^0_i = 0.00714(200.071) + 0.00682(225.225)$
$$= 2.965 \text{ kJ/K}$$

By interpolation, the final temperature is close to 392 K.

(*b*) The work required for the adiabatic-compression process of a closed system has been shown to be

$$W = U_2 - U_1 = \sum N_i \Delta \bar{u}_i$$

Interpolating again for the final $\bar{u}_i$ we find that

$$W = 0.00714 \text{ kmol } (8146 - 6229) \text{ kJ/kmol}$$
$$+ 0.00682 \text{ kmol } (9784 - 6939) \text{ kJ/kmol}$$
$$= 33.1 \text{ kJ}$$

(*c*) The entropy changes of the individual gases in the mixture are found by applying Eq. [10-36]. Thus, after interpolating for the final $\bar{s}^0_i$ values,

$$\Delta S_{N_2} = 0.00714 \text{ kmol} \times \left(199.479 - 191.682 - 8.314 \ln \frac{6}{2}\right) \text{kJ/kmol·K}$$

$$= -0.00954 \text{ kJ/K}$$

$$\Delta S_{CO_2} = 0.00682 \text{ kmol} \times \left(224.391 - 213.915 - 8.314 \ln \frac{7}{2}\right) \text{kJ/kmol·K}$$

$$= -0.00915 \text{ kJ/K}$$

Since the overall process is isentropic, the sum of $\Delta S$ for the two gases should be zero. The slight difference here is because the final temperature is not exactly 392 K. Note, however, that each gas undergoes a definite entropy change of its own. For a mixture of two gases, this change is always equal in magnitude but opposite in sign for an isentropic process.

(d) Because $\Delta S = 0$, the final temperature can also be estimated from the relationship $T_2/T_1 = (P_2/P_1)^{(k_m-1)/k_m}$. The specific-heat ratio must be the average for the mixture, that is, $k_m = c_{p,m}/c_{v,m}$. Equations [10-30] and [10-31] are used to evaluate $c_{v,m}$ and $c_{p,m}$. Since $T_2$ is unknown, we must either use the specific-heat data at the original temperature or guess the final temperature and use the average specific-heat data in the temperature range. If we assume that $T_2$ is around 400 K and use average values at 350 K, then

$$c_{v,m} = \frac{[0.20(0.744) + 0.30(0.706)] \text{ kJ/K}}{(0.20 + 0.30) \text{ kg}} = 0.721 \text{ kJ/kg·K}$$

$$c_{p,m} = \frac{[0.20(1.041) + 0.30(0.895)] \text{ kJ/K}}{(0.20 + 0.30) \text{ kg}} = 0.953 \text{ kJ/kg·K}$$

Therefore, the specific-heat ratio $k_m$ for the mixture is $0.953/0.721 = 1.32$. The final temperature $T_2$, then, is

$$T_2 = T_1 \left(\frac{P_2}{P_1}\right)^{(k_m-1)/k_m} = 300\left(\frac{6}{2}\right)^{0.242} = 391 \text{ K}$$

Because of the relatively small temperature change, this result essentially agrees with the answer obtained in part a.

## 10-4 MIXING PROCESSES INVOLVING IDEAL GASES

As noted earlier, when an ideal-gas mixture undergoes a process without a composition change, the value of $p_{i2}/p_{i1} = P_2/P_1$. However, when two or more pure gases are mixed or two gas mixtures come into contact, the component or partial pressure of each component changes. This change must be accounted for when the entropy changes of the individual component gases are evaluated.

Consider a rigid, insulated tank divided into several compartments by partitions, as shown in Fig. 10-8. Different ideal gases fill each compartment, and the total pressure and temperature of each pure gas initially is the same. If the partitions are removed from the tank, each gas spreads into the total volume of the tank, and a new equilibrium state is reached.

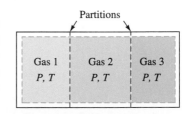

**Figure 10-8**
A rigid tank contains three pure gases initially separated by partitions and at the same pressure and temperature.

In the absence of heat and work interactions, the energy equation for the closed system reduces to $\Delta U = 0$. In terms of the individual components this equation takes the form

$$U_1 = U_2$$

or
$$\sum N_{i1}\bar{u}_{i1} = \sum N_{i2}\bar{u}_{i2} \qquad \text{[10-38]}$$

where $\bar{u}_{i1}$ and $\bar{u}_{i2}$ are the initial and final molar internal energies of the $i$th component, respectively, and they are solely a function of the temperature. Since the values of $N_i$ and $\bar{u}_{i1}$ are fixed, the above equation requires that

$$U_2 = \sum N_{i2}\bar{u}_{i2} = U_1 \qquad \text{(a constant)}$$

Although the $N_i$ values are constant, note that $\bar{u}_i$ for every component increases with increasing temperature. Thus if $T_2$ differs from $T_1$, then $U_2$ would not equal $U_1$. Therefore, the final temperature must equal the initial value for the case under consideration.

The effect of mixing several gases initially at the same pressure and temperature on the final pressure may also be found by using the ideal-gas equation. For any component $i$ we may write

$$\frac{p_{i2}V_{i2}}{p_{i1}V_{i1}} = \frac{N_i R_u T}{N_i R_u T}$$

or
$$\frac{p_{i2}}{p_{i1}} = \frac{V_{i1}}{V_{i2}} = \frac{V_{i1}}{V} = y_i$$

where $V_{i2} = V$, the total volume of the mixture. Hence $p_{i2} = y_i p_{i1} = y_i P_{\text{init}}$. If we sum over all components,

$$P_{\text{final}} = \sum p_{i2} = \sum y_i P_{\text{init}} = P_{\text{init}}$$

Therefore, the pressure also does not change upon mixing in this special case.

In general, the final state of an ideal-gas mixture after a mixing process can be determined by application of an energy balance combined with the ideal-gas property data. Mixing processes may occur within closed or open systems. The mixing of two or more gases is highly irreversible and always produces entropy. The amount or rate of entropy production can be found by applying an entropy balance to the mixing process. The following examples demonstrate this procedure.

**Example 10-5**

A rigid, insulated tank is divided into two compartments by a partition. Initially 0.02 kmol of nitrogen fills one compartment at 2 bars and 100°C. The other compartment contains 0.03 kmol of carbon dioxide at 1 bar and 20°C. The partition is removed, and the gases are allowed to mix. Determine (*a*) the final temperature and pressure of the mixture, and (*b*) the entropy production for the mixing process.

**Solution:**

**Given:** The mixing of two gases separated by a partition in a rigid, insulated tank, as shown in Fig. 10-9.

**Find:** (a) $T_2$, $P_2$, (b) $\sigma_{\text{mix}}$.

**Model:** Adiabatic, rigid, closed system; ideal gases.

**Analysis:** (a) The energy balance for the closed system shown in Fig. 10-9 is $Q + W = \Delta U$. Since both $Q$ and $W$ are zero, the energy balance dictates that $\Delta U = 0$. For the low pressures involved, the gases are assumed to behave as ideal gases, and $\Delta U = \Delta U_{N_2} + \Delta U_{CO_2} = 0$. Employing $\bar{c}_v$ data, the energy equation becomes

$$0 = \Delta U_{N_2} + \Delta U_{CO_2} = [N\bar{c}_v(T_2 - T_1)]_{N_2} + [N\bar{c}_v(T_2 - T_1)]_{CO_2}$$

The final temperature $T_2$ is the same for each gas and lies between 20 and 100°C. As a first approximation we let $c_v$ for nitrogen and carbon dioxide be constant at the values of their initial temperatures. On a mass basis (Table A-3) these values are 0.746 and 0.649 kJ/kg·K, respectively. On a molar basis these values are 20.9 and 28.6 kJ/kmol·K. Substitution of numerical values into the above energy equation yields

$$0 = 0.02 \text{ kmol} \times \frac{20.9 \text{ kJ}}{\text{kmol·K}} \times (T_2 - 100°C)$$

$$+ \; 0.03 \text{ kmol} \times \frac{28.6 \text{ kJ}}{\text{kmol·K}} \times (T_2 - 20°C)$$

Therefore, as a first estimate $T_2 = 46.0°C$. Because $\bar{c}_v$ should be evaluated at the *average* temperature, recalculate $T_2$ using new estimates of the $\bar{c}_v$ values. For $CO_2$ at $T_{av} = 33°C$, $\bar{c}_v = 29.2$ kJ/kmol·K and for $N_2$ at $T_{av} = 73°C$, $\bar{c}_v = 20.8$ kJ/kmol·K. Substituting these values back into the energy equation leads to a value for $T_2$ of 45.8°C. No further iterations are necessary.

The final pressure is determined from the ideal-gas relation $PV = NR_uT$. The total volume is the sum of the volumes of the original compartments, which may also be determined from the same equation. Hence

$$V_{N_2,1} = \frac{NR_uT}{P} = \frac{(0.02 \text{ bar})(0.08314 \text{ bar·m}^3/\text{kmol·K})(373 \text{ K})}{2 \text{ bars}} = 0.310 \text{ m}^3$$

$$V_{CO_2,1} = \frac{0.03(0.08314)(293)}{1} = 0.731 \text{ m}^3$$

$$V_{\text{tot},1} = 0.310 + 0.731 = 1.041 \text{ m}^3 = V_2$$

As a result, by employing the same set of units,

$$P_2 = \frac{N_mR_uT_2}{V_m} = \frac{0.05(0.08314)(318)}{1.041} = 1.27 \text{ bars}$$

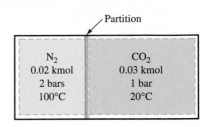

| N₂ | CO₂ |
|---|---|
| 0.02 kmol | 0.03 kmol |
| 2 bars | 1 bar |
| 100°C | 20°C |

Partition

**Figure 10-9**
Schematic and data for two gases mixing within a rigid tank in Example 10-5.

(b) The entropy production is determined from the basic entropy balance for a closed system,

$$\Delta S = \int \frac{\delta Q}{T} + \sigma$$

In the absence of heat transfer the entropy production equals the total entropy change, that is, $\sigma = \Delta S$. As a result,

$$\sigma = \Delta S = \Delta S_{N_2} + \Delta S_{CO_2}$$

Assuming ideal-gas behavior, the change in entropy for each individual component is given by $\Delta S_i = N_i \Delta \bar{s}_i$, and Eq. [10-35] indicates that

$$\Delta \bar{s}_i = \bar{c}_{p,i} \ln \frac{T_2}{T_1} - R_u \ln \frac{p_{i2}}{p_{i1}}$$

where $p_i = y_i P$. Keep in mind that each gas exerts only its component pressure at the total volume and temperature of the system. On the basis of Table A-3, the $\bar{c}_p$ value for $N_2$ at 346 K is 29.1 kJ/kmol·K and the value for $CO_2$ at 306 K is 38.2 kJ/kmol·K. Solving for the two entropy changes, we find that

$$\Delta S_{N_2} = 0.02 \text{ kmol} \left[ 29.1 \ln \frac{319}{373} - 8.314 \ln \frac{0.4(1.27)}{2} \right] \frac{\text{kJ}}{\text{kmol·K}}$$

$$= 0.02(-4.55 + 11.39) \text{ kJ/K}$$

$$= 0.137 \text{ kJ/K}$$

$$\Delta S_{CO_2} = 0.03 \text{ kmol} \left[ 37.4 \ln \frac{319}{293} - 8.314 \ln \frac{0.6(1.27)}{1} \right] \frac{\text{kJ}}{\text{kmol·K}}$$

$$= 0.03(3.18 + 2.26)$$

$$= 0.163 \text{ kJ/K}$$

Therefore, for the entire system,

$$\sigma = \Delta S = (0.137 + 0.163) \text{ kJ/K} = 0.300 \text{ kJ/K}$$

**Comment:**    (1) The entropy change for each gas could also be calculated from Eq. [10-36], which employs the $\bar{s}^o$ values from the gas tables.

(2) The entropy production in this process is due to the following factors: (1) the two gases are initially at different temperatures, (2) the gases are initially at different pressures, and (3) the gases are of different identity. These types of irreversibilities were discussed earlier in Sec. 6-4.

---

Mixing of ideal gases also occurs in steady-flow open systems. The calculations are essentially the same as for the closed-system analysis shown above, except the correct energy and entropy balances must be used. In addition, there is no general way to evaluate the final pressure unless some additional data on flow conditions entering and leaving the control volume are given.

**Example 10-6**

**A**ir at 27°C, 1 bar, and a rate of 100 kg/min is mixed adiabatically with a stream of carbon dioxide at 127°C and 1 bar to form a gas mixture at 57°C and 1 bar. Determine (a) the mass flow rate of carbon dioxide, in kg/min, (b) the mole fractions in the final gas mixture, and (c) the rate of entropy production, in kJ/(K·min).

**Solution:**

**Given:** An air stream and a carbon dioxide stream are mixed adiabatically as shown in Fig. 10-10.

**Find:** (a) $\dot{m}_{CO_2}$, kg/min, (b) $y_i$ for mixture, (c) $\dot{\sigma}$ kJ/min·K.

**Model:** Steady-state, $\dot{Q} = \dot{W} = \Delta ke = \Delta pe = 0$, ideal gases.

**Analysis:** (a) The general steady-state mass and energy balances for a control volume with two inlets and one exit are written on a rate basis as

$$0 = \dot{m}_1 + \dot{m}_2 - \dot{m}_3$$

$$0 = \dot{Q} + \dot{W} + \dot{m}_1\left(h_1 + \frac{V_1^2}{2} + gz_1\right) + \dot{m}_2\left(h_2 + \frac{V_2^2}{2} + gz_2\right) - \dot{m}_3\left(h_3 + \frac{V_3^2}{2} + gz_3\right)$$

In the absence of heat transfer and work effects and for negligible kinetic and potential energy changes, the steady-state energy balance becomes

$$0 = \dot{m}_1 h_1 + \dot{m}_2 h_2 - \dot{m}_3 h_3$$

The mass balance indicates that $\dot{m}_3 = \dot{m}_1 + \dot{m}_2$. If this relationship is substituted for $\dot{m}_3$ in the above energy balance, and the result is rearranged, we obtain

$$0 = \dot{m}_a(h_{a,3} - h_{a,1}) + \dot{m}_c(h_{c,3} - h_{c,2})$$

where $\dot{m}_a$ and $\dot{m}_c$ now represent the mass flow rates $\dot{m}_1$ of air and $\dot{m}_2$ of carbon dioxide, respectively, as shown in Fig. 10-10. The only unknown in this equation is $\dot{m}_c$. Assuming the substances are ideal gases, the enthalpy data are found in Tables A-5 and A-9. The molar data for $CO_2$ must be converted to a mass basis. Thus further rearrangement of the energy equation yields

$$\dot{m}_c = \dot{m}_a \frac{h_{a,3} - h_{a,1}}{h_{c,2} - h_{c,3}}$$

$$= 100 \text{ kg/min} \times \frac{44.01 \text{ kg}}{\text{kmol}} \times \frac{(330.34 - 300.19) \text{ kJ/kg}}{(13,372 - 10,570) \text{ kJ/kmol}} = 47.4 \text{ kg/min}$$

where 44.01 is the molar mass of carbon dioxide.

(b) The molar flow rates for the two gases are

$$\dot{N}_a = \frac{\dot{m}_a}{M_a} = \frac{100}{28.97} = 3.45 \text{ kmol/min}$$

$$\dot{N}_c = \frac{\dot{m}_c}{M_c} = \frac{47.4}{44.01} = 1.08 \text{ kmol/min}$$

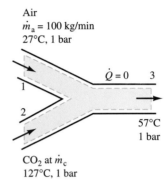

Air
$\dot{m}_a$ = 100 kg/min
27°C, 1 bar

$\dot{Q} = 0$    3

57°C
1 bar

$CO_2$ at $\dot{m}_c$
127°C, 1 bar

**Figure 10-10**
Schematic and data for adiabatic, steady-flow mixing in Example 10-6.

The total molar flow rate $\dot{N}_m$ of the mixture is 4.53 kmol/min. Therefore,

$$y_a = \frac{\dot{N}_a}{\dot{N}_m} = \frac{3.45}{4.53} = 0.762 \qquad \text{and} \qquad y_c = \frac{\dot{N}_c}{\dot{N}_m} = \frac{1.08}{4.53} = 0.238$$

(*c*) The entropy balance for the control volume at steady state is given by

$$0 = \dot{m}_1 s_1 + \dot{m}_2 s_2 - \dot{m}_3 s_3 + \sum_{j=1}^{n} \frac{\dot{Q}_j}{T_j} + \dot{\sigma}_{\text{cv}}$$

Using the mass balance with the entropy balance in a fashion similar to that used above for the energy balance, we find for the mixing process under adiabatic conditions that

$$\dot{\sigma}_{\text{cv}} = \dot{m}_a (s_{a,3} - s_{a,1}) + \dot{m}_c (s_{c,3} - s_{c,2})$$

The entropy change for each gas is found from Eq. [10-36],

$$\Delta s_i = s_{i2}^\circ - s_{i1}^\circ - R_i \ln \frac{p_{i2}}{p_{i1}}$$

In this case $p_{i2} = y_i P_2$ and $p_{i1} = P_1$. Since $P_1 = P_2$, the equation becomes

$$\Delta s_i = s_{i2}^\circ - s_{i1}^\circ - R_i \ln \frac{p_{i2}}{p_{i1}} = s_{i2}^\circ - s_{i1}^\circ - R_i \ln y_i$$

Applied to the two ideal gases, we find that

$$\Delta s_a = \left( 1.79783 - 1.70203 - \frac{8.314}{28.97} \ln 0.762 \right) \text{kJ/kg·K} = 0.1738 \text{ kJ/kg·K}$$

$$\Delta s_c = \frac{(217.534 - 225.225 - 8.314 \ \ln 0.238) \text{ kJ/kmol·K}}{44.01 \text{ kg/kmol}} = 0.0964 \text{ kJ/kg·K}$$

The use of these values in the entropy production equation leads to

$$\dot{\sigma}_{\text{cv}} = [100(0.1738) + 47.4(0.0964)] \text{ kJ/K} = 21.95 \text{ kJ/K·min}$$

**Comment:** (1) The irreversibilities leading to entropy production are: (*a*) the different temperatures and (*b*) the different composition of the inlet flow streams.

(2) It is not mere coincidence that we were able to manipulate the various balances until we had $\Delta h$ and $\Delta s$ values. This is a direct consequence of the results of Sec. 10-3. When used with ideal-gas mixtures, the approaches used in this example are very common for closed and open systems.

## 10-5 PROPERTIES OF AN IDEAL GAS–VAPOR MIXTURE

Although the relationships developed in preceding sections for the properties of ideal-gas mixtures are of general usefulness, one additional complication must be recognized in dealing with gas mixtures. There is always the possibility that one or more of the gases may exist in a state that is close to a saturation state for the given component. We have seen that each gas in an ideal-gas mixture exerts a pressure that is equal to its component pressure. But the component pressure can never be greater than the saturation pressure for that component at the mixture temperature. Any attempt to increase the component pressure beyond the saturation pressure results in partial condensing of the vapor. For example, consider increasing the total pressure of an ideal-gas mixture at constant temperature. Since the mole fractions of every component gas are fixed (at least temporarily), the component pressure of each increases in direct proportion to the increase in the total pressure, i.e., $p_i = y_i P$. However, if the component pressure of any constituent eventually exceeds its saturation pressure for that temperature, the gas will begin to condense out as the pressure is increased further. This process is illustrated on a $Pv$ diagram in Fig. 10-11$a$. The gas that condenses under this circumstance is usually spoken of as a *vapor*. Consequently, such ideal-gas mixtures are referred to as ***gas-vapor mixtures.***

A similar situation also occurs when a gas-vapor mixture is cooled at constant pressure. In this case the component pressure of the vapor remains constant (up to the point of condensation), but the temperature eventually is lowered sufficiently to equal the saturation temperature for the given component pressure. As the temperature is lowered still further, the saturation pressure corresponding to the temperature becomes less than the actual partial pressure, and hence some of the vapor must condense. The effect of lowering the temperature of a mixture containing a condensable vapor is shown on a $Ts$ diagram in Fig. 10-11$b$, where state 2 is the state where condensation begins. A well-known example of a gas mixture containing a condensable vapor is the air in the atmosphere. The condensation of water from the air as the temperature is lowered, forming dew, is a common experience.

Many systems of engineering interest involve the use of gas-vapor mixtures. One of the most common and important applications is the conditioning of atmospheric air for human comfort. In fact, an entire industry—the HVAC industry—has developed to meet the demands for heating, ventilating, and air conditioning of buildings. In the following sections, we will

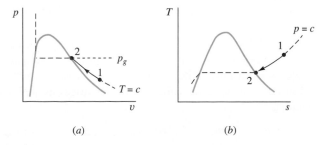

(a)　　　　　　　　　　(b)

**Figure 10-11**　　The effect of raising the total pressure or lowering the temperature on a gas mixture containing a vapor.

restrict the discussion to two-component mixtures containing dry air and water vapor. These mixtures are commonly called *moist air* or *atmospheric air,* and the study of their behavior is called *psychrometrics.*

The temperature of a gas-vapor mixture as measured by a conventional thermometer is called the *dry-bulb temperature,* such as $T_l$ in Fig. 10-11. In Fig. 10-11*b* a process was illustrated in which the temperature of a gas-vapor mixture was lowered while the total pressure of the mixture remained constant. When a mixture of dry air and water vapor is cooled at constant pressure from an unsaturated state, the temperature at which the mixture becomes saturated, or condensation begins, is called the *dew-point temperature.* Hence the temperature at state 2 on the $Ts$ diagram in Fig. 10-11*b* is the dew point of any mixture for which the partial pressure of the water vapor is represented by the constant-pressure line. The dew-point temperature of atmospheric air equals the saturation temperature of water that corresponds to the partial pressure of the water vapor actually in the mixture. A dry-air–water-vapor mixture which is saturated with water is frequently called *saturated air.*

For dry-air–water-vapor mixtures that are not saturated we need to be able to quantify the amount of water vapor present at a given state of the mixture. This can be done by using properties called the relative humidity and the specific humidity (or humidity ratio). The *relative humidity* $\phi$ for an ideal-gas mixture is defined as the ratio of the partial pressure of the vapor in a mixture to the saturation pressure of the vapor at the same dry-bulb temperature of the mixture. If $p_v$ represents the actual vapor pressure and $p_g$ represents the saturation pressure at the same temperature, then

$$\phi \equiv \frac{p_v}{p_g} \qquad \textbf{[10-39]}$$

The pressures used to define the relative humidity are shown in the $Ts$ diagram of Fig. 10-12, which is an extension of the data shown in Fig. 10-11*b*. State 1 is the initial state of the water vapor in the mixture, and its vapor pressure at this state is $p_1$. If this same vapor were present in saturated air at the same temperature, its pressure would necessarily have to be that given at state 3, which is the saturation pressure $p_g$ for that temperature. In terms of Fig. 10-12, $\phi = p_1/p_3$. (In terms of the $Pv$ diagram in Fig. 10-11*a*, $\phi = p_1/p_2$.) The relative humidity $\phi$ is always less than or equal to unity. Based on the assumption that both the dry air and the water vapor in the mixture behave as ideal gases, the equation for the relative humidity can be expressed in terms of specific volumes (or densities) as well as partial pressures. That is,

$$\phi = \frac{p_v}{p_g} = \frac{RT_v/v_v}{RT_g/v_g} = \frac{v_g}{v_v} = \frac{\rho_v}{\rho_g} \qquad \textbf{[10-40]}$$

since the temperatures and the gas constants in the relation are equal.

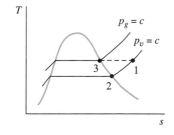

**Figure 10-12**
A $Ts$ diagram for water vapor in atmospheric air.

The *humidity ratio* (or *specific humidity*) $\omega$ describes the quantity of water vapor in a mixture relative to the amount of dry air present. It is defined as the ratio of the mass of water vapor $m_v$ to the mass of dry air $m_a$. Thus,

$$\omega \equiv \frac{m_v}{m_a} \qquad \textbf{[10-41]}$$

These masses could be on a kilogram or a pound-mass basis. It should be noted that the humidity ratio is not a measure of the mass fraction of the water vapor in the mixture.

The humidity ratio is directly related to the relative humidity. To find this relation, we first find the humidity ratio in terms of the component pressures of the two constituent gases. The mass of an ideal gas in a gas mixture is given by $m_i = p_i V M_i / R_u T$, where $p_i$ is the partial pressure and $M_i$ is the molar mass of the component. By substituting this equation into Eq. [10-41], we find that

$$\omega = \frac{p_v V M_v / R_u T}{p_a V M_a / R_u T} = \frac{M_v p_v}{M_a p_a}$$

where both gases occupy the same volume at the same temperature. The ratio of molar masses for water to dry air is $M_v / M_a = 0.622$, and the partial pressure of dry air $p_a = P - p_v$; thus

$$\omega = 0.622 \frac{p_v}{p_a} = 0.622 \frac{p_v}{P - p_v} = 0.622 \frac{\phi p_g}{P - \phi p_g} \qquad \textbf{[10-42]}$$

Finally, we are able to relate the humidity ratio and the relative humidity of a mixture by combining Eqs. [10-39] and [10-42]. This yields

$$\phi = \frac{p_v}{p_g} = \frac{p_a \omega}{0.622 p_g} \qquad \textbf{[10-43]}$$

The saturation pressure $p_g$ needed in some of the above relations is found as a function of temperature in the saturation temperature table for water. However, in design calculations this pressure-temperature relation may be needed in an equation format. A simple equation of reasonable accuracy in SI units is

$$\ln[p_g / \text{bars}] = 12.1929 - \frac{4109.1}{(T/\text{K}) - 35.50} \qquad \textbf{[10-44]}$$

where $T$ is the temperature in kelvins. In USCS units

$$\ln[p_g / (\text{lb}_f / \text{in}^2)] = 14.9716 - \frac{7490.8}{(T/^\circ\text{R}) - 61.05} \qquad \textbf{[10-45]}$$

where $T$ is the temperature in degrees Rankine.

**Example 10-7**

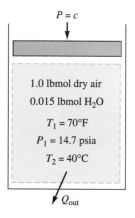

$P = c$

1.0 lbmol dry air

0.015 lbmol $H_2O$

$T_1 = 70°F$

$P_1 = 14.7$ psia

$T_2 = 40°C$

$Q_{out}$

**Figure 10-13**
Schematic and data for
the dry air-water vapor
mixture in Example 10-7.

**A** piston-cylinder device contains a mixture of 1.0 lbmol of dry air and 0.015 lbmol of water vapor at 70°F and 14.7 psia. The mixture is cooled at constant pressure to 40°F. Determine (*a*) the initial humidity ratio, (*b*) the initial relative humidity, (*c*) the initial dew-point temperature, and (*d*) the amount of mass condensed per unit mass of dry air.

**Solution:**

**Given:**    A piston cylinder contains dry air and water vapor as shown in Fig. 10-13. Cooling occurs to 40°F.

**Find:**    (*a*) $\omega_1$, (*b*) $\phi_1$, (*c*) $T_{dew}$, (*d*) $\Delta m$ condensed.

**Model:**    Closed system, ideal-gas mixture, constant pressure.

**Analysis:**    (*a*) The initial humidity ratio is found using the definition of $\omega$,

$$\omega_1 = \frac{m_v}{m_a} = \frac{N_v M_v}{N_a M_a} = \frac{0.015(18.02)}{1.0(28.97)} = 0.00933 \text{ lb}_m \text{ water/lb}_m \text{ dry air}$$

(*b*) At 70°F the value of $p_g$ from Table A-12E is 0.3632 psia. In addition,

$$p_v = y_v P = (0.015/1.015)(14.7) = 0.2172 \text{ psia}$$

Using the definition of $\phi$, Eq. [10-39],

$$\phi_1 = \frac{p_v}{p_g} = \frac{0.2172}{0.3632} = 0.598 \text{ (or 59.8 percent)}$$

(*c*) The dew point is the saturation temperature for the vapor in the mixture. For a vapor pressure of 0.2172 psia, linear interpolation in Table A-12E indicates a temperature of 55.4°F. The same value is found by using Eq. [10-45].

(*d*) The mass of water that condenses on cooling the mixture to 40°F is determined by applying a mass balance to the water component entering and leaving the system. Noting that both vapor and liquid water are present in the system at the final state, the mass balance on the water component is

$$m_{v1} = m_{v2} + m_{liq,2}$$

Note also that $m_{a1} = m_{a2} = m_a$. Division of the water balance by $m_a$, and recalling that $\omega \equiv m_v/m_a$, leads to

$$\omega_1 = \omega_2 + \frac{m_{liq,2}}{m_a} \quad \text{or} \quad \frac{m_{liq,2}}{m_a} = \omega_1 - \omega_2$$

Thus the mass of water that condenses per mass of dry air is measured by $\Delta\omega$. The value of $\omega_1$ was found in part *a* to be 0.00933 lb$_m$ water/lb$_m$ dry air. The saturation pressure for state 2 at 40°F is 0.1217 psia. Since the vapor is saturated at the exit state,

$$\omega_2 = 0.622 \frac{p_g}{P - p_g} = \frac{0.622(0.1217)}{14.7 - 0.1217} = 0.00519 \text{ lb}_m \text{ water/lb}_m \text{ dry air}$$

Therefore,

$$\frac{m_{\text{liq},2}}{m_a} = \omega_1 - \omega_2 = 0.00933 - 0.00519 = 0.00414 \text{ lb}_m \text{ water/lb}_m \text{ dry air}$$

**Comment:**    Water starts condensing from the mixture when the mixture temperature reaches the dew-point temperature of 55.4°F.

---

The enthalpy of a dry-air–water-vapor mixture is the sum of the enthalpies of the individual components. That is,

$$H_{\text{mix}} = H_{\text{dry air}} + H_{\text{water vapor}}$$

On the basis of a unit mass of dry air, the specific enthalpy becomes

$$\boxed{h_{\text{mix}} = h_a + \omega h_v}  \hspace{2cm} \textbf{[10-46]}$$

where $h_a$ is the enthalpy of the dry-air component and $h_v$ is the enthalpy of the water vapor. In atmospheric air problems, due to the low vapor pressure of water, $h_v$ may be evaluated by using $h_g$ for a saturated vapor at the given dry-bulb temperature.

As an example, we shall calculate the specific enthalpy of a moist air mixture at 27°C and 1.01 bar with a humidity ratio $\omega = 0.015$ kg water/kg dry air. Using Eq. [10-46], we need to find $h_a$ and $h_v$. The enthalpy of dry air is found from Table A-5 as $h_a = 300.19$ kJ/kg dry air. Similarly, by assuming $h_v \simeq h_g$, the enthalpy of saturated water vapor at 27°C is found in Table A-12 as $h_g = 2550.8$ kJ/kg water. Substituting these values back into Eq. [10-46] yields

$$h_{\text{mix}} = h_a + \omega h_v$$
$$= 300.19 \text{ kJ/kg dry air}$$
$$+ (0.015 \text{ kg water/kg dry air})(2550.8 \text{ kJ/kg water})$$
$$= 338.45 \text{ kJ/kg dry air}$$

It is important to note that the numerical value of $h_{\text{mix}}$ is arbitrary. The enthalpy for dry air was taken from the ideal-gas table where $h_a = 0$ when $T = 0$ K, and the enthalpy of water vapor was taken from the steam tables where the enthalpy of saturated liquid is zero at 0.01°C, the triple state of water. When enthalpy differences are calculated, the enthalpy values for the dry air must have the same reference values, that is, must be taken from the same tables. The same is true for the water component.

In performing psychrometric calculations, it is frequently useful to have simple expressions for the specific enthalpy of dry air, water vapor, moist air, and liquid water as a function of temperature. The basic equations for these ideal-gas, saturated-vapor, and compressed-liquid properties are

$$h_a = c_{p,a}(T - T_{\text{ref}}) + h_{a,\text{ref}} \quad \text{(dry air)} \qquad \textbf{[10-47a]}$$

$$h_v = h_{g,\text{ref}} + c_{p,v}(T - T_{\text{ref}}) \quad \text{(water vapor)} \qquad \textbf{[10-47b]}$$

$$h_{\text{mix}} = h_a + \omega h_v \quad \text{(moist air)} \qquad \textbf{[10-47c]}$$

$$h_f = h_{f,\text{ref}} + c_{p,\text{liq}}(T - T_{\text{ref}}) \quad \text{(liquid water)} \qquad \textbf{[10-47d]}$$

In SI units, the above expressions for $h$ in kJ/kg become

$$h_a = 1.005^{(}T_C/^\circ C) \quad \text{(dry air)} \qquad \textbf{[10-48a]}$$

$$h_v = 2501.7 + 1.82(T_C/^\circ C) \quad \text{(water vapor)} \qquad \textbf{[10-48b]}$$

$$h_{\text{mix}} = 1.005(T_C/^\circ C) + \omega[2501.7 + 1.82(T_C/^\circ C)] \quad \text{(moist air)} \quad \textbf{[10-48c]}$$

$$h_f = 4.190(T_C/^\circ C) \quad \text{(saturated liquid water)} \qquad \textbf{[10-48d]}$$

where $T_C$ is the dry-bulb temperature in degrees Celsius and $0^\circ C \leq T_C \leq 40^\circ C$. The SI equations require the reference values for both the enthalpy of dry air and for saturated liquid water to be zero at the reference temperature of $0^\circ C$. In USCS units, similar expressions for $h$ in Btu/lb are

$$h_a = 0.240(T_F/^\circ F) \quad \text{(dry air)} \qquad \textbf{[10-49a]}$$

$$h_v = 1061.5 + 0.435(T_F/^\circ F) \quad \text{(water vapor)} \qquad \textbf{[10-49b]}$$

$$h_{\text{mix}} = 0.240(T_F/^\circ F) + \omega[1061.5 + 0.435(T_F/^\circ F)] \quad \text{(moist air)}$$
$$\textbf{[10-49c]}$$

$$h_f = 1.00(T_F/^\circ F) - 32.0 \quad \text{[saturated liquid water)} \qquad \textbf{[10-49d]}$$

where $T_F$ is the dry-bulb temperature in degrees Fahrenheit and $32^\circ F \leq T_F \leq 110^\circ F$. The USCS equations assume the reference value of the enthalpy of dry air to be zero at $0^\circ F$, the reference temperature of water vapor also to be $0^\circ F$, and the enthalpy for saturated liquid water to be zero at $32^\circ F$.

Finally, it is convenient to have values of the specific volume of the gas-vapor mixture on the basis of a unit mass of dry air. Consider a sample of atmospheric air with a volume $V_{\text{mix}}$ such that it contains a unit mass of dry air plus water vapor. The volume $V_{\text{mix}}$ occupied by the mixture is the same as the volume $V_a$ occupied by the dry-air component. Consequently

$$v_{\text{mix}} = \frac{V_{\text{mix}}}{m_a} = \frac{V_a}{m_a} = v_a = \frac{R_a T}{p_a} = \frac{R_a T}{P - P_v} \qquad \textbf{[10-50]}$$

where $p_a$ is the component pressure and $R_a$ is the specific gas constant for dry air. The evaluation of the above quantities is illustrated in the following example.

**Example 10-8**

An air–water-vapor mixture at 25°C and 1 bar has a relative humidity of 50 percent. Find (a) the humidity ratio, (b) the dew point, (c) the enthalpy, in kJ/kg dry air, and (d) the specific volume, in m³/kg dry air.

**Solution:**

**Given:** The air–water-vapor mixture shown in Fig. 10-14.

**Find:** (a) $\omega$, (b) $T_{dew}$, in °C, (c) $h$, in kJ/kg, and (d) $v$, in m³/kg.

**Model:** Ideal-gas–vapor mixture.

**Analysis:** (a) The humidity ratio is evaluated from Eq. [10-45].

$$\omega = 0.622\frac{p_v}{p_a} = 0.622\frac{p_v}{P - p_v} = 0.622\frac{\phi p_g}{P - \phi p_g}$$

From Table A-12 for steam it is found that the saturation pressure $p_g$ at 25°C is 0.0317 bar. Therefore, the actual vapor pressure $p_v = \phi p_g = 0.50(0.0317) = 0.0159$ bar. Then

$$\omega_1 = \frac{0.622(0.0159)}{1.00 - 0.0159} = 0.01005 \text{ kg water/kg dry air}$$

(b) The dew point by definition is the temperature at which the actual vapor pressure becomes the saturation pressure. From Table A-12, the vapor pressure at 13°C is 0.01497 bar, and at 14°C it is 0.01598 bar. By linear interpolation the dew point is roughly 13.9°C.

(c) The enthalpy can be determined by two methods. If tabular $h$ data are used from Tables A-5 and A-12 at 25°C,

$$h_{mix} = h_a + \omega h_v$$
$$= 298.33 \text{ kJ/kg dry air} + (0.01005 \text{ kg water/kg dry air})(2547.2 \text{ kJ/kg water})$$
$$= 323.93 \text{ kJ/kg dry air}$$

A second method is to use Eq. [10-48c] as an approximation. On this basis

$$h_{mix} = 1.005T_C + \omega(2501.7 + 1.82T_C)$$
$$= 1.005(25) + 0.01005[2501.7 + 1.82(25)] = 50.72 \text{ kJ/kg dry air}$$

Note the significant difference in answers for the same state. These calculations emphasize the fact that the choice of the reference datum for the dry air and water components may greatly affect the numerical values for the enthalpy of the mixture. In calculating $\Delta h_{mix}$ in energy analyses, consistent data must be used.

(d) The specific volume of the mixture on a dry-air basis is found from Eq. [10-50]:

$$v = \frac{R_a T}{p_a} = \frac{0.08314}{29}\frac{\text{bar·m}^3}{\text{kg·K}} \times \frac{298\text{K}}{(1.0 - 0.0159)\text{bar}} = 0.868 \text{ m}^3/\text{kg dry air}$$

The specific volume of atmospheric air typically runs between 0.80 and 0.95 m³/kg.

**Figure 10-14**
Schematic and data for the dry air-water vapor mixture in Example 10-8.

## 10-6    THE ADIABATIC-SATURATION AND WET-BULB TEMPERATURES

Mixture composition is one of the properties required to calculate enthalpy values for a gas-vapor mixture. For moist air, the humidity ratio $\omega$ is a measure of the water vapor in the mixture; however, it is difficult to measure directly. Fortunately, relative humidity $\phi$ is easily measured and can be used along with the mixture pressure and temperature to calculate the humidity ratio by Eq. [10-42]. Currently, electronic sensors are available that accurately measure relative humidity over a range of pressures and temperatures within 1 percent. Another technique for determining the humidity ratio based on an adiabatic saturation process is important for at least three reasons. First, it provides a fundamental method for calibrating modern electronic sensors. Second, it provides a precise definition of the adiabatic-saturation temperature, an important psychrometric property. And finally, it provides an excellent vehicle for learning how to apply thermodynamic relations to a psychrometric process.

In the ***adiabatic-saturation process,*** an unsaturated moist air stream is brought into direct contact with the surface of liquid water in a flow channel. The experimental apparatus for a steady-state, adiabatic-saturation process is sketched in Fig. 10-15a, and the $Ts$ diagram for the water component in the process is shown in Fig. 10-15b. In operation, unsaturated moist air enters the insulated flow channel at state 1 with a dry-bulb temperature $T_1$ and a relative humidity $\phi_1$ that is less than 100 percent. Liquid makeup water enters at state 3, and the process occurs at a constant pressure. The entering moist air will pick up additional moisture as it contacts the liquid water, lose energy to the liquid water, and leave the device at state 2 with temperature $T_2$ and relative humidity $\phi_2$. If the contact time between the moist air and the liquid water is long enough, the moist air will leave the device as a saturated mixture, that is, $\phi_2 = 100$ percent. When the device is operated so that $\phi_2 = 100$ percent, and $T_3$ of the makeup water equals $T_2$, the temperature $T_2$ equals the ***adiabatic-saturation temperature*** $T_{ad,1}$

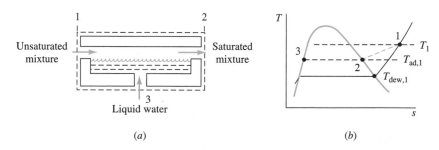

(a)                                        (b)

**Figure 10-15**    Physical description of the adiabatic-saturation process and its representation on a Ts diagram.

of the entering moist air. The process for the moist-air mixture is shown by the dashed line from state 1 to state 2 in Fig. 10-15$b$. Note that the adiabatic-saturation temperature $T_{ad,1}$ is below the dry-bulb temperature $T_1$ and above the dew-point temperature $T_{dew,1}$ of state 1.

If the adiabatic-saturation temperature $T_{ad,1}$, the dry-bulb temperature $T_1$, and the total pressure $P$ of a moist-air mixture are known at state 1, the humidity ratio $\omega_1$ can be calculated. To show this, we write a steady-state energy balance for the control volume shown in Fig. 10-15$a$. For the adiabatic-saturation process, changes in kinetic and potential energy are negligible, and there is no work or heat transfer for this process. Under these conditions the energy equation reduces to

$$0 = \dot{m}_{a1}h_{mix,1} - \dot{m}_{a2}h_{mix,2} + \dot{m}_{w3}h_{w3} \qquad \textbf{[a]}$$

where the subscript $w$ refers to liquid makeup water. Recalling the definition of mixture enthalpy for moist air, $h_{mix} = h_a + \omega h_v$, the energy equation can be written as

$$0 = \dot{m}_{a1}(h_a + \omega h_v)_1 - \dot{m}_{a2}(h_a + \omega h_v)_2 + \dot{m}_w h_{w3} \qquad \textbf{[b]}$$

Applying conservation of mass to the system for the dry air shows as expected that $\dot{m}_{a1} = \dot{m}_{a2} = \dot{m}_a$. Applying conservation of mass to the system for the water shows that

$$0 = \dot{m}_{v1} - \dot{m}_{v2} + \dot{m}_{w3} \qquad \text{or} \qquad \dot{m}_{w3} = \dot{m}_{v2} - \dot{m}_{v1} \qquad \textbf{[c]}$$

Using the definition of the humidity ratio, $\omega = \dot{m}_v/\dot{m}_a$, and the equality of the dry-air mass flow rates, the mass balance of water, Eq. [c], can be rewritten as

$$\dot{m}_{w3} = \dot{m}_a(\omega_2 - \omega_1) \qquad \textbf{[d]}$$

Substituting Eq. [d] back into Eq. [b] and dividing by $\dot{m}_a$ yields

$$0 = (h_{a1} + \omega_1 h_{v1}) - (h_{a2} + \omega_2 h_{v2}) + (\omega_2 - \omega_1)h_{w3}$$

Solving this equation for the humidity ratio $\omega_1$ of the moist air entering the adiabatic-saturation process gives the following expression:

$$\omega_1 = \frac{(h_{a2} - h_{a1}) + \omega_2(h_{v2} - h_{w3})}{h_{v1} - h_{w3}} \approx \frac{c_{p,a}(T_2 - T_1) + \omega_2(h_{g2} - h_{w3})}{h_{g1} - h_{w3}} \qquad \textbf{[e]}$$

where $\Delta h_a = c_{p,a}\Delta T$ and $h_v = h_g$. If the device is operated so that the moist air leaving the device is saturated ($\phi_2 = 100$ percent) and the makeup liquid water is supplied at the same temperature as that of the exiting moist air ($T_3 = T_2$), then the temperature $T_2$ of the saturated moist air leaving the device equals the adiabatic-saturation temperature $T_{ad,1}$ of the entering moist air stream. Note that the adiabatic-saturation temperature is solely a function of the system pressure and the humidity ratio and dry-bulb temperature or the entering moist air. Under these specific operating conditions, $h_{w3} \simeq h_{f2}$, $h_{g2} - h_{f2} = h_{fg2}$, and $T_3 = T_2 = T_{ad,1}$. Substituting these latter relations into Eq. [e] for $\omega_1$ leads to the desired result,

$$\omega_1 = \frac{c_{p,a}(T_{ad,1} - T_1) + \omega_{sat,2}h_{fg2}}{h_{g1} - h_{f2}} \qquad \textbf{[10-51]}$$

where $\omega_{sat,2}$, $h_{fg2}$, and $h_{f2}$ are measured at the adiabatic-saturation temperature $T_2 = T_{ad,1}$. The above expression relates the humidity ratio $\omega_1$ to the adiabatic-saturation temperature $T_{ad,1}$, the total pressure $P$, and the dry-bulb temperature $T_1$. Based on Eq. [10-42] with $\phi = 100$ percent, the value of $\omega_{sat,2}$ is obtained from $\omega_{sat,2} = 0.622p_{g2}/(P - p_{g2})$, where $p_{g2}$ is evaluated again at $T_{ad,1}$.

Although the technique of adiabatic saturation leads to the desired results, it is difficult in practice to attain a saturated state by this method without employing an extremely long flow channel, which is impractical. In lieu of this, a temperature equivalent to the adiabatic-saturation temperature for water-vapor–air mixtures, known as the **wet-bulb temperature,** is used. This temperature is easily measured by the following technique. The bulb of an ordinary thermometer is covered with a wick moistened with water. Unsaturated atmospheric air is then passed over the wetted wick until dynamic equilibrium is attained and the temperature of the wick (and hence the thermometer) reaches a stable value. For dry-air–water-vapor mixtures at normal temperatures and pressures, the wet-bulb temperature, the determination of which relies on heat- and mass-transfer rates, is a very close approximation to the adiabatic-saturation temperature. Thus the temperature $T_{ad}$ used in Eq. [10-51] to calculate the humidity ratio is normally the wet-bulb temperature $T_{wb}$, and this leads to an answer of sufficient accuracy. This correspondence between the wet-bulb and the adiabatic-saturation temperatures occurs only for moist-air mixtures at atmospheric pressure. It is invalid for other mixtures and other pressures.

**Example 10-9**

**A** sample of atmospheric air at 1 bar has a dry-bulb temperature of 24°C and a wet-bulb temperature of 16°C. Determine (a) the humidity ratio, (b) the relative humidity.

**Solution:**

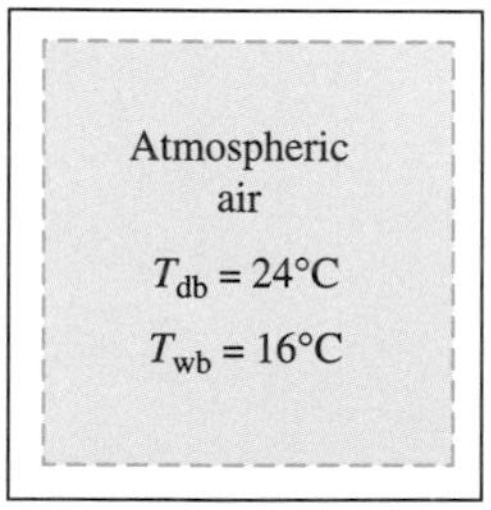

**Given:**  Atmospheric air with the properties shown in Fig. 10-16.

**Find:**  (a) $\omega$, and (b) $\phi$ in kJ/kg dry air.

**Model:**  Ideal-gas mixture.

**Analysis:**  (a) The humidity ratio, or specific humidity, is calculated from Eq. [10-51],

$$\omega_1 = \frac{c_{p,a}(T_{ad,1} - T_1) + \omega_{sat,2}h_{fg2}}{h_{g1} - h_{f2}}$$

Atmospheric
air

$T_{db} = 24°C$

$T_{wb} = 16°C$

**Figure 10-16**
A schematic and data
for the atmospheric air in
Example 10-9.

The value of $\omega_{sat,2}$ in this equation is determined from the relation

$$\omega_{sat,2} = 0.622\frac{\phi p_{g2}}{P - \phi p_{g2}}$$

In this special case $\phi = 1.0$ and $p_{g2}$ is the saturation pressure of 0.01818 bar at the wet-bulb temperature of 16°C. Therefore,

$$\omega_{sat,2} = \frac{0.622(0.01818)}{1.0 - 0.0.01818} = 0.0115 \text{ kg water/kg dry air}$$

The enthalpies of the water in the liquid and vapor phases are found at 16°C in Table A-12. Hence use of Eq. [10-51] yields

$$\omega_1 = \frac{(1.0 \text{ kJ/kg·C})(16 - 24)C + (0.0115 \text{ kg water/kg dry air})(2463.6 \text{ kJ/kg water})}{(2545.4 - 67.2) \text{ kJ/kg water}}$$

$$= 0.00820 \text{ kg water/kg dry air}$$

(b) The relative humidity is now computed from Eqs. [10-45] and [10-42]. First,

$$\omega_1 = 0.00820 = \frac{0.622 p_v}{1.0 - p_v}$$

The solution of this equation yields a value of $p_v = 0.0130$ bar. Thus the relative humidity, by Eq. [10-42], is

$$\phi = \frac{p_v}{p_g} = \frac{0.0130}{0.02985} = 0.436 \text{ (or 43.6\%)}$$

## 10-7 THE PSYCHROMETRIC CHART

To facilitate the computation and visualization of processes involving moist-air mixtures, it is useful to plot the important moist-air parameters on a diagram known as a *psychrometric chart.* An outline of a psychrometric chart is shown in Fig. 10-17. Full psychrometric charts in SI and USCS units can be found in the Appendix as Figs. A-25 and A-25E. The psychrometric chart provides a convenient method for finding the properties of moist-air mixtures.

Three major moist-air properties used to construct a psychrometric chart are the dry-bulb temperature $T_{db}$, the humidity ratio $\omega$, and the total pressure $P$. Because most psychrometric processes occur at essentially constant pressure, each psychrometric chart is constructed for only one total pressure. In this text we shall consider charts constructed only for standard atmospheric pressure (101.324 kPa or 14.696 $lb_f/in^2$). Referring to Fig. 10-17, the dry-bulb temperature appears on the horizontal axis and the humidity ratio appears on the vertical axis. Lines of constant relative humidity appear as concave upward curves that run from the lower left to the

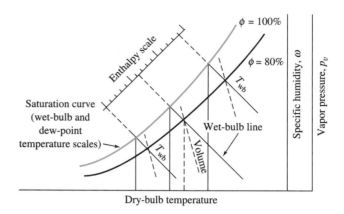

**Figure 10-17**    Outline of a psychrometric chart.

upper right of the chart. Of particular interest is the ***saturation curve.*** This line represents the locus of states where $\phi = 100$ percent and a moist-air mixture is saturated with water vapor. Additional lines of constant relative humidity are also plotted, and follow the general shape of the saturation curve. For a specified total pressure, the humidity ratio, relative humidity, and dry-bulb temperature are related mathematically through Eq. [10-42]. Some psychrometric charts also show the vapor pressure $p_v$ on a separate vertical axis.

We have already shown by Eq. [10-51] that the wet-bulb temperature is directly related to the humidity ratio and the dry-bulb temperature of the mixture. Lines of constant wet-bulb temperature run from the upper left to the lower right of the chart. The wet-bulb temperature lines begin at the saturation curve ($\phi = 100$ percent), and on the saturation curve the wet-bulb and dry-bulb temperatures have the same value. Lines of constant dew-point temperature $T_{dew}$ are parallel to lines of constant humidity ratio, and the scale for both the wet-bulb and the dew-point temperature is usually found along the saturation curve. To show how these temperatures are related, consider a mixture with $T_{db} = 25°C$ and $\phi = 50$ percent. Consulting Fig. A-25 in the Appendix, we start from the state point at the intersection of the given $T_{db}$ and $\phi$ values. Reading across to the vertical axis, we find the humidity ratio is 0.010 kg/kg dry air. The horizontal line passing through the state point is also a line of constant dew-point temperature and intersects the saturation curve at $T_{dew} = 14°C$. The wet-bulb temperature line is the oblique line passing through the state point, and it intersects the saturation curve at $T_{wb} = 18°C$. For any unsaturated moist-air mixture, $\phi < 100$ percent and $T_{dew} < T_{wb} < T_{db}$. For any state along the saturation curve, all three temperatures are equal.

In addition to the above parameters, it is useful to have information about the specific enthalpy and specific volume for the mixture. Lines of constant specific enthalpy $h_{mix}$ lie approximately parallel to lines of constant wet-bulb temperatures. During an adiabatic saturation process, the wet-bulb temperature for the mixture is constant. In addition, the specific enthalpy of the mixture remains very nearly constant, except for the slight enthalpy change due to the evaporation of water to the air stream. As a first approximation, the lines of constant wet-bulb temperature and constant mixture enthalpy are plotted as parallel lines. For increased accuracy, some charts take this slight deviation into account by providing separate scales for the two properties. Lines of constant enthalpy appear as straight lines on the chart. For the SI chart (Fig. A-25) the enthalpy values are reported in units of kJ/kg dry air and assume that the enthalpy of both dry air and saturated liquid water is zero at 0°C. For the USCS chart (Fig. A-25E) the enthalpy values are reported in Btu/lb$_m$ and are based on a zero value for the enthalpy of dry air at 0°F and a zero value for the enthalpy of liquid water at 32°F. These values are consistent with the approximate equations developed as Eqs. [10-48] and [10-49] in Sec. 10-5. In addition, SI or USCS enthalpy values for water (liquid or vapor) can also be taken directly from the steam tables in this text and used in conjunction with the mixture values from the appropriate psychrometric chart.

Finally, lines of constant specific volume are plotted on the psychrometric chart as nearly vertical lines running from upper left to lower right. The values are reported in m$^3$/kg dry air in SI and ft$^3$/lb$_m$ dry air in USCS units. Correction tables are frequently attached to the psychrometric chart that allow the user to correct for different total pressures. These corrections can also be determined through the use of Eq. [10-50]. The correction is negligible for conditions close to the total pressure value used to construct the psychrometric chart.

---

The dry-bulb and wet-bulb temperatures of atmospheric air at a local pressure of 1 bar are 23 and 16°C, respectively. From the psychrometric chart (Fig. A-25) estimate (a) the humidity ratio, (b) the relative humidity, (c) the vapor pressure, in bars, (d) the dew point, (e) the enthalpy, and (f) the specific volume, in m$^3$/kg.

**Example 10-10**

**Solution:**

**Given:** A dry-air–water-vapor mixture with properties shown in Fig. 10-18a.

**Find:** (a) $\omega$, (b) $\phi$, (c) $p_v$, in bars, (d) $T_{dew}$, in °C, (e) $h$, in kJ/kg dry air, and (f) $v$, in m$^3$/kg.

**Model:** Ideal-gas mixture.

**Analysis:** The properties are found from the chart by first finding the point at the intersection of the vertical dry-bulb temperature and the sloping wet-bulb temperature line (see Fig. 10-18b).

(a) The humidity ratio is read from the ordinate at the right to be about 0.0087 kg water per kilogram of dry air.

(b) It is found that the 50 percent relative humidity line runs through the point selected.

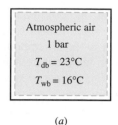

(a)

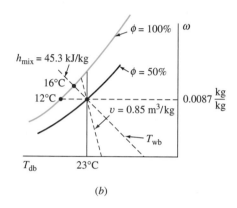

(b)

**Figure 10-18**        Illustration of the atmospheric
air conditions in Example 11-10.

(*c*) The vapor pressure is also read from the ordinate scale at the right and is roughly 0.0140 bar.

(*d*) The dew point is the temperature at which condensation would just begin if the mixture were cooled at constant pressure. Since the vapor pressure and the humidity ratio remain constant until condensation begins, the dew-point temperature is found by moving horizontally to the left from the initial state until the saturation line is reached. The temperature at this point is the dew point and is close to 12°C.

(*e*) The enthalpy is found by following the wet-bulb line (which is also an enthalpy line) from the initial state up to the enthalpy scale. The value read is approximately 45.3 kJ/kg of dry air.

(*f*) At the initial state a constant-specific-volume line of $0.85 \text{m}^3/\text{kg}$ passes just to the left of the point. The actual value is close to $0.851 \text{m}^3/\text{kg}$ of dry air.

**Comment:**    These estimated values read from the chart should be checked against calculated values. Be careful to use the correct reference state when evaluating the mixture enthalpy.

## 10-8    AIR-CONDITIONING PROCESSES

A person generally feels more comfortable when the air within a building is maintained in a fairly limited range of temperatures and relative

**Photograph 10-1**   Outdoor central-station air handling unit. (Courtesy of The Trane Company)

humidities. However, owing to mass and energy transfer between the inside of the building and the local environment and owing to internal effects such as cooking, baking, and clothes washing in the home, the temperature and the relative humidity frequently reach undesirable levels. To achieve values of $T$ and $\phi$ within the desired ranges (the comfort zone), it is usually necessary to alter the state of the air. As a result, equipment must be designed to raise or lower the temperature and the relative humidity. In addition to altering the state of a specific air stream by heating, cooling, humidifying, or dehumidifying, a change in state can also be attained by mixing the building air directly with another air stream from, for example, outside the building. Thus there are a number of basic processes to be considered for conditioning atmospheric air. Photograph 10-1 shows an air-handling unit that uses these processes to condition air supplied to a building.

## 10-8-1   BASIC MASS AND ENERGY RELATIONS FOR PROCESS ANALYSIS

The basic relations available for evaluating any air-conditioning process include an energy balance, mass balances for dry air and for water (liquid and vapor), and appropriate property information. To demonstrate the evaluation of the mass flow rates and the thermodynamic properties of the mixture, consider a steady-state mixing and humidification process with heat transfer as shown in Fig. 10-19. Moist air enters the device at both states 1 and 2 of the steady-state device, pure liquid water enters at state 3, and moist air leaves the device at state 4.

It is possible to write three different but related mass balances for this steady-state control volume. Starting with the total mass of the control

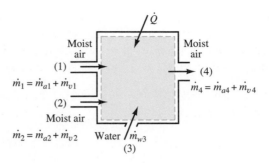

**Figure 10-19**    Illustration of a typical air-conditioning process.

volume, the conservation of mass requires that

$$0 = \dot{m}_1 + \dot{m}_2 - \dot{m}_3 - \dot{m}_4$$

where $\dot{m}$ is the total mass flow rate. Neglecting chemical reactions inside the device, the conservation of mass can also be applied to the dry air and the water separately. This leads to

$$0 = \dot{m}_{a1} + \dot{m}_{a2} - \dot{m}_{a4} \qquad 0 = \dot{m}_{v1} + \dot{m}_{v2} - \dot{m}_{w3} - \dot{m}_{v4} \quad \textbf{[10-52]}$$

where $\dot{m}_a$, $\dot{m}_v$, and $\dot{m}_w$ are the mass flow rates for dry air, water vapor, and liquid water, respectively, for the inlet and exit gas streams. For most psychrometric problems, the separate balances for water and dry air are preferred. In addition, the vapor mass flow rate can be expressed in terms of the humidity ratio and the dry air mass flow rate by $\dot{m}_v = \omega \dot{m}_a$.

Assuming that kinetic- and potential-energy effects are negligible and that there is no work transfer, the steady-state energy balance can be written as

$$0 = \dot{Q} + \dot{m}_1 h_1 + \dot{m}_2 h_2 + \dot{m}_3 h_3 - \dot{m}_4 h_4$$

where $\dot{m}h$ applies at any inlet or outlet. This general relationship can be rewritten in the following two ways. The first or *mixture approach* makes direct use of the mixture enthalpy for moist air. This is the preferred way to write the energy balance if you use enthalpy values from the psychrometric chart. The equation above takes on the format

$$0 = \dot{Q} + \dot{m}_{a1} h_{\text{mix}1} + \dot{m}_{a2} h_{\text{mix}2} + \dot{m}_{w3} h_{w3} - \dot{m}_{a4} h_{\text{mix}4} \quad \textbf{[10-53]}$$

where $h_{\text{mix}} = h_a + \omega h_v$. Numerical values for $h_{\text{mix}}$ can be obtained from the psychrometric chart or from Eqs. [10-48] and [10-49] developed earlier for SI and USCS units.

If we use a *component approach*, we recognize that in terms of the dry-air and water-vapor components, $\dot{m}h = \dot{m}_a h_a + \dot{m}_v h_v$. Thus the energy balance can be written in terms of the components as

$$0 = \dot{Q} + (\dot{m}_{a1} h_{a1} + \dot{m}_{v1} h_{v1}) + (\dot{m}_{a2} h_{a2} + \dot{m}_{v2} h_{v2}) \quad \textbf{[10-54]}$$
$$+ \dot{m}_{w3} h_{w3} - (\dot{m}_{a4} h_{a4} + \dot{m}_{v4} h_{v4})$$

Combining this equation with the mass balances for dry air and for water, it is possible to rearrange the above equation so that only differences in enthalpy for dry air and water need to be calculated. For example, if the mass balances are used to replace $\dot{m}_{a4}$ and $\dot{m}_{v4}$, we obtain the expression

$$0 = \dot{Q} + \dot{m}_{a1}(h_{a1} - h_{a4}) + \dot{m}_{a2}(h_{a2} - h_{a4}) + \dot{m}_{v1}(h_{v1} - h_{v4})$$
$$+ \dot{m}_{v2}(h_{v2} - h_{v4}) + \dot{m}_{w3}(h_{w3} - h_{v4})$$

A significant advantage of the component approach is that there is no need to compute a mixture enthalpy, and since we are calculating enthalpy differences, the problem of reference states disappears.

### 10-8-2  HEATING AND COOLING AT CONSTANT HUMIDITY RATIO

A major use of the energy and mass balances developed above is to gas streams that are heated or cooled at constant humidity ratio. This requires that the temperature of the moist-air stream not be lowered below its dew-point temperature. A schematic of the process is shown in Fig. 10-20a, and possible paths for the process are shown on the psychrometric chart outlined in Fig. 10-20b. The energy equation for this situation is

$$0 = \dot{Q} + \dot{m}_{a1}h_{a1} + \dot{m}_{v1}h_{v1} - \dot{m}_{a2}h_{a2} - \dot{m}_{v2}h_{v2}$$

where $\dot{m}_v = \omega\dot{m}_a$ for both inlet and exit streams. On the basis of Eq. [10-47], $h_a \cong c_{p,a}T$. Hence the energy equation becomes

$$\dot{Q} = \dot{m}_{a2}(h_{a2} + \omega_2 h_{v2}) - \dot{m}_{a1}(h_{a1} + \omega_1 h_{v1}) \qquad \textbf{[10-55]}$$
$$= \dot{m}_a(c_{p,a}T_2 + \omega_2 h_{g2}) - \dot{m}_a(c_{p,a}T_1 + \omega_1 h_{g1})$$

where $\omega_2 = \omega_1$. Knowledge of the initial state and the outlet temperature fixes all the data on the right-hand side of this equation.

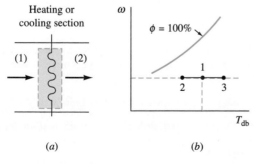

Heating or cooling section

(1)   (2)

$\omega$

$\phi = 100\%$

1

2   3

$T_{db}$

(a)

(b)

**Figure 10-20**    Schematic for heating or cooling at a constant humidity ratio.

**Example 10-11**

**A**tmospheric air at 30°C and 40 percent relative humidity enters a heat exchanger at a volumetric rate of 200 m³/min. The moist air is cooled to 18°C at a constant pressure of 1.01 bar. Determine (a) the mass flow rate of dry air, in kg/min, (b) the humidity ratio, (c) the heat-transfer rate, in kJ/min, and (d) the final relative humidity.

**Solution:**

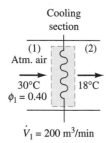

Cooling
section

(1)        (2)
Atm. air

30°C        18°C
$\phi_1 = 0.40$

$\dot{V}_1 = 200$ m³/min

**Figure 10-21**
Schematic and data for cooling of atmospheric air in Example 10-11.

**Given:**   Atmospheric air is cooled under the conditions shown in Fig. 10-21.

**Find:**   (a) $\dot{m}_a$, in kg/min, (b) $\omega$, (c) $\dot{Q}$, in kJ/min, (d) $\phi_2$.

**Model:**   Steady-state, open system; ideal-gas mixture.

**Analysis:**   (a) The mass flow rate of dry air is found from the volumetric rate by $\dot{m}_a$ = (volume rate)/$v$. The specific volume of the moist air is given by Eq. [10-50], $v_{mix} = R_a T/p_a$. The vapor pressure of water is

$$p_{v1} = \phi_1 p_{g1} = 0.40(0.04246) = 0.0170 \text{ bar}$$

and the partial pressure of dry air is

$$p_a = P - p_v = 1.01 - 0.017 = 0.993 \text{ bar}$$

Use of Eq. [10-50] now shows that

$$v = \frac{R_a T}{p_a} = \frac{0.08314(303)}{28.97(0.993)} = 0.876 \text{ m}^3/\text{kg dry air}$$

Finally, the mass flow rate of dry air is

$$\dot{m}_a = \frac{\text{vol. rate}}{v} = \frac{200\text{m}^3/\text{min}}{0.876 \text{ m}^3/\text{kg dry air}} = 228 \text{ kg dry air/min}$$

(b) The humidity ratio is given by Eq. [10-42]. In this case

$$\omega_1 = \omega_2 = \frac{0.622 p_v}{p_a} = \frac{0.622(0.017)}{0.993} = 0.01065 \text{ kg water/kg dry air}$$

(c) The heat-transfer rate is found from the steady-state energy balance,

$$0 = \dot{Q} + \dot{W} - \dot{m}(\Delta h + \Delta \text{ke} + \Delta \text{pe})$$

In the absence of shaft work and with negligible kinetic- and potential-energy changes, the equation reduces to $\dot{Q} = \dot{m}_a(h_2 - h_1)$. The inlet and outlet enthalpies, based on Eq. [10-48], are

$$h_1 = c_{p,a}T_1 + \omega h_{g1} = 1.005(30) + 0.01065(2556.3) = 57.4 \text{ kJ/kg dry air}$$

$$h_2 = 1.005(18) + 0.01065(2534.4) = 45.1 \text{ kJ/kg dry air}$$

Thus the heat-transfer rate is

$$\dot{Q} = \dot{m}_a(h_2 - h_1) = \frac{228 \text{ kg dry air}}{\text{min}} \times \frac{(45.1 - 57.4) \text{ kJ}}{\text{kg dry air}} = -2805 \text{ kJ/min}$$

The heat-transfer rate can also be found from a modified form of the energy equation, namely,

$$\dot{Q} = \dot{m}_a[c_{p,a}(T_2 - T_1) + \omega(h_{g2} - h_{g1})]$$

This format uses the same data as used above. However, in this method the individual $h$ values cannot be compared to data in the psychrometric chart.

(d) The final relative humidity occurs at 18°C, $\omega = 0.0170$, and 1.01 bars. Since the saturation vapor pressure at 18°C is 0.02064 bars, Eq. [10-42] for $\omega$ becomes

$$\omega = 0.0170 = \frac{\phi p_g}{P - \phi p_g} = \frac{\phi_2(0.02064)}{1.01 - \phi_2(0.0264)}$$

or

$$\phi_2 = 0.814 \ \ (\text{or } 81.4 \text{ percent})$$

**Comment:** The calculated values of $\omega$, $h_1$, $h_2$, $v_1$, and $\phi_2$ can be checked against the values found in the psychrometric chart.

---

## 10-8-3 DEHUMIDIFICATION WITH HEATING

A fairly common condition within industrial and residential buildings, especially in the summer, is the tendency toward high temperatures and high relative humidities. The discomfort of the human body in this situation is well known. A major method of lowering both $T$ and $\phi$ simultaneously is illustrated in Fig. 10-22a. The air to be treated is passed through a flow channel which contains cooling coils. The fluid inside the coils might be, for example, relatively cold water or a refrigerant which has been cooled in a vapor-compression refrigeration cycle. The initial state of the air stream is shown as state 1 on the sketch of a psychrometric chart in Fig. 10-22b. As the air passes through the cooling coil, its temperature decreases and its relative humidity increases, at constant specific humidity. If the air remains

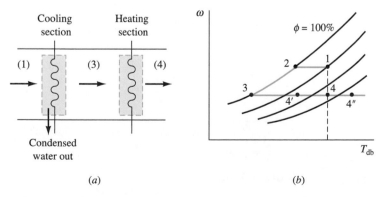

**Figure 10-22** Dehumidification process with heating. (a) Equipment; (b) process on a psychrometric chart.

in contact with the cooling coil sufficiently, the air stream will reach its dew point, indicated by state 2 in Fig. 10-22*b*. Further cooling requires the air to remain saturated, and its state follows the l00 percent relative-humidity line to the left toward state 3. During this latter process water condenses out from the air, and its specific humidity is lowered. Hence, by sufficient contact with the coil, both the temperature and the water content of the air are lowered. In many cases this conditioned air stream flows directly back and mixes with the air in the building.

However, in some cases the conditioned air may be at too low a temperature. This is overcome by then passing the air stream leaving the cooling coil section through a heating section. By proper choice of the temperature of the fluid within the heating coil, the temperature of the air stream leaving the overall equipment may be adjusted to the desired value. Three possibilities are shown as states 4, 4′, and 4″ in Fig. 10-22*b*. By proper adjustment of the amount of cooling (which controls the position of state 3) and the amount of heating, a suitable state 4 may be attained. The following example illustrates the overall process.

**Example 10-12**

Outside atmospheric air at 32°C and 70 percent relative humidity is to be conditioned so that it enters a home at 22°C and 45 percent relative humidity. First the air passes through a heat exchanger where the air is cooled below its dew point, and water condenses from the air stream until the desired humidity ratio is reached. Then the air passes over a heating coil in a heat exchanger until the temperature reaches 22°C. Determine (*a*) the amount of water removed, in kg/kg dry air, (*b*) the heat removed by the cooling system, in kJ/kg dry air, and (*c*) the quantity of heat added in the final heat exchanger, in kJ/kg dry air.

**Solution:**

**Given:**   A process to dehumidify and heat an air stream, as shown in Fig. 10-23.

**Find:**   (*a*) $\Delta\omega$, (*b*) $q_{\text{out}}$, in kJ/kg dry air, (*c*) $q_{\text{in}}$, in kJ/kg dry air.

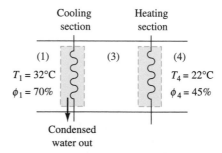

**Figure 10-23**   Illustration of the dehumidification and heating process in Example 10-12.

**Model:** Steady-state, open system; ideal-gas mixture.

**Analysis:** The heat-transfer quantities are found from the steady-state energy balance,

$$0 = \dot{Q} + \dot{W} + \sum_{\text{in}} \dot{m}(h + \text{ke} + \text{pe}) - \sum_{\text{out}} \dot{m}(h + \text{ke} + \text{pe})$$

In the absence of shaft work and with negligible kinetic- and potential-energy changes, the energy equation reduces to

$$\dot{Q} = \sum \dot{m}_{\text{out}} h_{\text{out}} - \sum \dot{m}_{\text{in}} h_{\text{in}}$$

and the water removed on a unit-mass basis is measured by $\Delta\omega$ for process 2-3 on Fig. 10-22b. Therefore, the $h$ and $\omega$ values are the important properties to evaluate for the overall process. The basic property expressions are

$$h_m = c_p T_{\text{air}} + \omega h_g \qquad \text{and} \qquad \omega = \frac{0.622 \phi p_g}{P - \phi p_g}$$

From Table A-12 the saturation vapor pressures at 32 and 22°C are 0.04759 and 0.02645 bar, respectively. We assume $P = 1$ bar, and we recall that $c_p$ for air is 1.005 kJ/kg·K. Therefore,

$$\omega_1 = \omega_2 = \frac{0.622(0.70)(0.04759)}{1.00 - 0.70(0.04759)} = 0.0214 \text{ kg water/kg dry air}$$

$$\omega_3 = \omega_4 = \frac{0.622(0.45)(0.02645)}{1.00 - 0.45(0.02645)} = 0.00749 \text{ kg water/kg dry air}$$

At states 1 and 4 the enthalpies are

$$h_1 = 1.005(32) + 0.0214(2559.9) = 86.94 \text{ kJ/kg dry air}$$

and

$$h_4 = 1.005(22) + 0.00749(2541.7) = 41.15 \text{ kJ/kg dry air}$$

To find $h_2$ and $h_3$, we need information on $T_2$ and $T_3$. Temperature $T_2$ is the dew-point temperature and is found from the fact that $p_{v2} = p_{g2} = 0.7(0.04759) = 0.0333$ bar. From Table A-12 a pressure of 0.0333 bar falls between saturation temperatures of 25 and 26°C. By linear interpolation, $T_2 = 25.8°C$ and $h_{g,2} = 2548.7$ kJ/kg. Thus

$$h_2 = 1.005(25.8) + 0.0214(2548.7) = 80.47 \text{ kJ/kg dry air}$$

Finally, $p_{v3} = \phi p_{g3} = 0.45(0.02645) = 0.01190$ bar. This pressure falls between 8 and 10°C in Table A-12. By linear interpolation, $T_3 = 9.5°C$ and $h_{g,3} = 2518.9$ kJ/kg. Hence

$$h_3 = 1.005(9.5) + 0.00749(2518.9) = 28.42 \text{ kJ/kg dry air}$$

(a) The quantity of water removed per unit mass of dry air is given by the difference in the humidity ratios between states 2 and 3. Consequently,

$$\Delta\omega = \omega_3 - \omega_2 = 0.00749 - 0.0214 = -0.01391 \text{ kg water/kg dry air}$$

The negative sign indicates water was removed from the flow.

(b) The heat transfer for the cooling-coil section is found from $\dot{Q} = \sum \dot{m}_{out} h_{out} - \sum \dot{m}_{in} h_{in}$. For the cooling-coil section this becomes

$$\dot{Q} = \dot{m}_a h_3 + \dot{m}_a(\omega_1 - \omega_3) h_{f3} - \dot{m}_a h_a$$

On a unit mass of dry air basis this becomes

$$
\begin{aligned}
q &= h_3 + (\omega_1 - \omega_3) h_{f3} - h_1 \\
&= [(28.37) + 0.01391(39.9) - (86.78)] \text{ kJ/kg dry air} \\
&= 28.37 + 0.56 - 86.78 \\
&= -57.85 \text{ kJ/kg dry air}
\end{aligned}
$$

Note that the energy removed by the condensed-liquid stream is extremely small and might be neglected as a first approximation.

(c) The heat transfer for the final section is equal to the enthalpy change of the air stream between states 3 and 4. Hence

$$q_{in} = h_4 - h_3 = 41.15 - 28.42 = 12.73 \text{ kJ/kg dry air}$$

The data for state 2 are not necessary for the solution but were listed merely to indicate the dew-point properties of the mixture. The values calculated for the enthalpies and the humidity ratios at the various states should now be checked by means of the psychrometric chart in the Appendix.

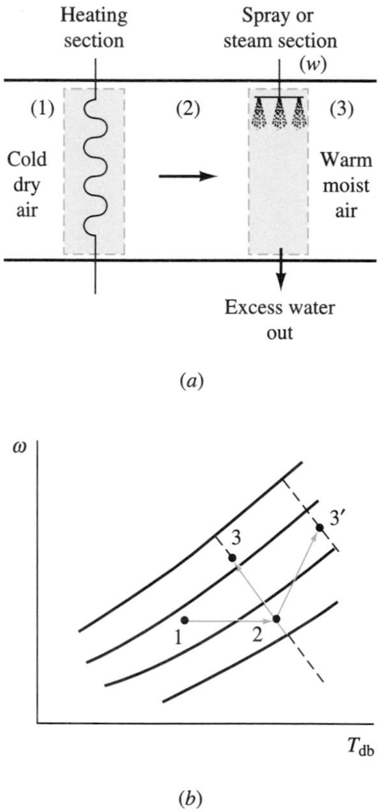

## 10-8-4   HUMIDIFICATION

In winter or at high altitudes, air in the atmosphere frequently is dry (it has low relative humidity) and cold. Thus the engineering problem is one of increasing both the water content and the dry-bulb temperature of the inlet air to a building. The process of adding moisture to an air stream is called **humidification.** An example of humidification is the adiabatic-saturation process discussed in Sec. 10-7. Another method of increasing the moisture content of an atmospheric air stream is by steam injection, as shown in Fig. 10-24a. If the temperature of the steam injected is relatively high, both the humidity ratio and the dry-bulb temperature will increase. This possibility is shown by the process line 2-3 on the psychrometric chart in Fig. 10-24b. The condition and amount of the inlet steam will determine the slope of line 2-3. In some cases it may be necessary to heat the inlet air at state 1 to state 2 before steam injection occurs. The overall process could be reversed, of course, with steam injection followed by heating.

The energy equation for the adiabatic steam humidifier is $0 = \sum(\dot{m}h)_{in} - \sum(\dot{m}h)_{out}$. Since $\dot{m}_{a3} = \dot{m}_{a2} = \dot{m}_a$, the energy equation may be written as

$$0 = \dot{m}_a h_2 + \dot{m}_s h_s - \dot{m}_a h_3$$

**Figure 10-24**
Schematic and data for heating and humidification (a) Process equipment; (b) process diagram on a psychrometric chart.

Upon rearrangement,

$$\dot{m}_s h_s = \dot{m}_a(h_3 - h_2) \qquad \text{[10-56]}$$

where the subscript $s$ represents the steam flow. The mass balance for the water (liquid and vapor) is

$$\omega_3 \dot{m}_a - \omega_2 \dot{m}_a - \dot{m}_s = 0$$

$$\dot{m}_s = (\omega_3 - \omega_2)\dot{m}_a \qquad \text{[10-57]}$$

The following example illustrates steam injection into an atmospheric air stream. In the following section the effect of liquid water injection is studied.

---

**Example 10-13**

An air stream at 8°C and 30 percent relative humidity is first heated to 32°C. Steam at 10 bars is then introduced into the humidification section until the air reaches a state of 30°C and 50 percent relative humidity at 1 bar. Determine (a) the heat transfer, in kJ/kg dry air, (b) the mass of steam required, in kg steam/kg dry air, and (c) the temperature of the steam added, in degrees Celsius.

**Solution:**

**Given:** Atmospheric air is heated and humidified as shown in Fig. 10-25.

**Find:** (a) $q_{in}$, (b) $\dot{m}_s/\dot{m}_a$, in kg water/kg dry air, and (c) $T_s$, in °C.

**Model:** Steady-state, open system; $w = \Delta ke = \Delta pe = 0$.

**Analysis:** (a) Because shaft work is zero and the kinetic- and potential-energy changes are assumed negligible, the steady-state energy balance reduces to

$$0 = \dot{Q} + \sum_{in} \dot{m}h - \sum_{out} \dot{m}h$$

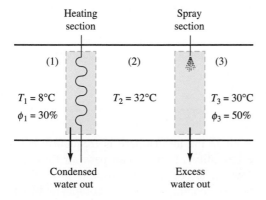

**Figure 10-25** Schematic and data for heating and humidification in Example 10-13.

This equation reduces to $q = h_2 - h_1$ on a dry-air basis for the heating section. The initial enthalpy and humidity ratio are evaluated as follows from Eqs. [10-42] and [10-48]:

$$\omega_1 = \frac{0.622\phi p_g}{P - \phi p_g} = \frac{0.622(0.30)(0.01072)}{1.0 - 0.30(0.01072)} = 0.00201 \text{ kg water/kg dry air}$$

$$h_1 = c_{p,a}T + \omega h_g = 1.005(8) + 0.00201(2516.1) = 13.10 \text{ kJ/kg dry air}$$

Because the humidity ratio is constant during heating,

$$h_2 = 1.005(32) + 0.00201(2559.9) = 37.30 \text{ kJ/kg dry air}$$

Therefore, the heat transfer to the moist air is

$$q = h_2 - h_1 = 37.30 - 13.10 = 24.20 \text{ kJ/kg dry air}$$

(b) The mass of steam required per unit mass of dry air is found from a mass balance on the water component in the humidification section: $\dot{m}_w = \dot{m}_a(\omega_3 - \omega_2)$. This relation requires a knowledge of $\omega_3$. At 30°C, $p_g = 0.04246$ bar. Hence, on the basis of Eq. [10-42]:

$$\omega_3 = \frac{0.622\phi_3 p_{g3}}{P - \phi_3 p_{g3}} = \frac{0.622(0.50)(0.4246)}{1.0 - 0.50(0.4246)} = 0.01349 \text{ kg water/kg dry air}$$

Therefore, since $\omega_2 = \omega_1$,

$$\frac{\dot{m}_w}{\dot{m}_a} = \omega_3 - \omega_2 = 0.01349 - 0.00201 = 0.01148 \text{ kg water/kg dry air}$$

(c) The temperature of the steam added is determined from its enthalpy value $h_w$. Before an energy balance is applied to the humidifier section, the value of $h_3$ must be calculated.

$$h_3 = c_p T_3 + \omega_3 h_{g3} = 1.005(30) + 0.01349(2556.3) = 64.63 \text{ kJ/kg dry air}$$

An energy balance around the humidifier reduces to $\sum \dot{m}h = 0$ or $0 = \dot{m}_a h_2 - \dot{m}_a h_3 + \dot{m}_w h_w$. Upon rearrangement,

$$h_w = \frac{\dot{m}_a}{\dot{m}_w}(h_3 - h_2) = \frac{1}{0.0115}(64.63 - 37.30) = 2376.5 \text{ kJ/kg steam}$$

At 10 bars the value of $h_g$ is 2778.1 kJ/kg. Therefore, the steam is a wet vapor as it enters with a saturation temperature of 179.9°C and a quality of roughly 80 percent.

## 10-8-5   EVAPORATIVE COOLING

In desert climates the air in the atmosphere is frequently hot and dry (it has very low relative humidity). Rather than pass air through a refrigerated cooling section, which is costly, it is possible to take advantage of the low humidity to achieve cooling. This is accomplished by passing the air stream through a water-spray section, as shown in Fig. 10-26a. (The equivalent effect may be achieved by passing the air through a filter bed

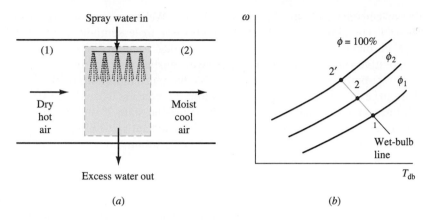

**Figure 10-26** Evaporative cooling. (a) Process equipment; (b) process on a psychrometric chart.

of some type, through which water is allowed to trickle. This provides reasonably good air-water contact.) Owing to the low relative humidity, part of the liquid-water stream evaporates. The energy for the evaporation process comes from the air stream, so it is cooled. The overall effect is a cooling and humidification of the air stream, and the process is called ***evaporative cooling.***

This process is essentially equivalent to the adiabatic-saturation process discussed in Sec. 10-6, which is another example of humidification. In both cases the energy balance on the process in the absence of work interactions and negligible kinetic- and potential-energy changes is

$$0 = \dot{Q} + \sum_{in} \dot{m}h - \sum_{out} \dot{m}h$$

For the adiabatic cooling process this can be written as

$$0 = \dot{m}_{a1}h_1 + \dot{m}_w h_w - \dot{m}_{a2}h_2$$

or, on the basis of a unit mass of dry air entering and leaving,

$$0 = h_1 + (\omega_2 - \omega_1)h_w - h_2$$

where $h_w$ is evaluated as the saturated-liquid enthalpy $h_f$ at the liquid temperature. The only difference in the use of these equations is that in an adiabatic-saturation process the liquid is brought in at temperature $T_2$, while for evaporative cooling the temperature of the entering liquid water stream takes on any reasonable value. This difference in entering water temperature is relatively insignificant to the analysis. Hence the path of the process follows closely a constant wet-bulb line on a psychrometric chart, as shown in Fig. 10-26*b*.

From another viewpoint, the energy associated with the injected liquid water stream is much less than the energy of the inlet and exit moist-air streams. Hence $h_2 \approx h_1$, and an evaporative-cooling process approaches one of constant enthalpy for the gas stream. We have already noted that constant enthalpy and constant wet-bulb lines are essentially parallel on a psychrometric chart. As a result it is quite accurate to represent an evaporative-cooling process as either one of constant enthalpy or one of constant wet-bulb temperature. Also note that there is a minimum temperature that can be achieved by evaporative cooling. This occurs when the air stream becomes saturated and is noted by the saturation state 2' on Fig. 10-26b. Because evaporative cooling is essentially a constant wet-bulb process, the wet-bulb temperature of the entering air is the minimum temperature that can be achieved by this process.

**Example 10-14**

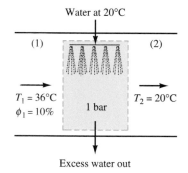

**Figure 10-27**
Evaporative cooling schematic and data for Example 10-14.

**Desert** air at 36°C, 1 bar, and 10 percent relative humidity passes through an evaporative cooler. Water is added at 20°C, and the final air temperature is 20°C. Determine (a) the amount of water added per kilogram of dry air, (b) the final relative humidity, and (c) the minimum temperature that could be achieved.

**Solution:**

**Given:** Atmospheric air passes through an evaporative cooler, as shown in Fig. 10-27.

**Find:** (a) $m_w$ added, in kg/kg dry air, (b) $\phi_2$, (c) $T_{2,\min}$, in °C.

**Model:** Steady-state, adiabatic open system; $w = \Delta\text{ke} = \Delta\text{pe} = 0$.

**Analysis:** (a) The amount of water added per unit mass of dry air is $\Delta\omega$. For the initial state, using Eq. [10-42] and Table A-12 yields

$$\omega_1 = \frac{0.622\phi p_g}{P - \phi p_g} = \frac{0.622(0.10)(0.05947)}{1.0 - 0.10(0.05947)} = 0.00372 \text{ kg water/kg dry air}$$

In the absence of heat and work interactions and $\Delta\text{ke} = \Delta\text{pe} = 0$, a steady-state energy balance leads to $\sum \dot{m}_{\text{in}}h_{\text{in}} - \sum \dot{m}_{\text{out}}h_{\text{out}} = 0$. For the process shown in Fig. 10-27 this becomes

$$h_1 + \Delta\omega h_{f,w} = h_2$$

or

$$(c_p T_1 + \omega_1 h_{g1}) + (\omega_2 - \omega_1)h_{f,w} = (c_p T_2 + \omega_2 h_{g2})$$

Substitution of appropriate data shows that

$$1.005(36) + 0.00372(2567.1) + (\omega_2 - 0.00372)(83.96) = 1.005(20) + \omega_2(2538.1)$$

$$\omega_2 = \frac{36.2 + 9.55 - 0.31 - 20.1}{2538.1 - 83.96} = 0.0103 \text{ kg water/kg dry air}$$

If the liquid-water contribution had been neglected, the final humidity ratio would be found to be 0.0101 kg water per kilogram of dry air. Thus this contribution usually is small. Finally, the amount of water added is

$$\omega_2 - \omega_1 = 0.0103 - 0.00372 = 0.00658 \text{ kg water/kg dry air}$$

(b) To determine the final relative humidity, we must first find the vapor pressure by applying Eq. [10-42] in a rearranged form:

$$p_v = \frac{\omega P}{\omega + 0.622} = \frac{0.0103(1)}{0.0103 + 0.622} = 0.0163 \text{ bar}$$

At 20°C the saturation vapor pressure is 0.02339 bar. Hence

$$\phi_2 = \frac{p_v}{p_g} = 0.697 \text{ or } 69.7\%$$

The values calculated above may be checked on the psychrometric chart.

(c) The minimum temperature is the adiabatic-saturation value, as represented by $T_{ad,1}$ in Eq. [10-51]. Unfortunately, this equation cannot be solved directly for $T_{ad,1}$. A trial-and-error solution is necessary. Based on known data, Eq. [10-51] becomes

$$0.00372 = \frac{1.005(T_{ad,1} - 36) + \omega_{sat,2}h_{fg2}}{2567.1 - h_{f2}}$$

In addition, the equation for $\omega_2$ at the final saturation state is

$$\omega_{sat,2} = \frac{0.622p_{g2}}{1.0 - p_{g2}}$$

To solve these two equations, we first assume that $T_2 = 16°C$. Then

$$\omega_{sat,2} = \frac{0.622(0.01818)}{1.0 - 0.01818} = 0.0115 \text{ kg water/kg dry air}$$

Finally, the right side of Eq. [10-51] becomes

$$\frac{1.005(16 - 36) + 0.0115(2463.6)}{2567.1 - 67.2} = 0.00330$$

This result compares to the desired value of 0.00372. When 17°C is used as an estimate, the right side of Eq. [10-51] is equal to 0.00451. Therefore, the minimum temperature lies between 16 and 17°C, and this state is denoted state 2′ on Fig. 10-26b. This minimum temperature can also be found on the psychrometric chart by following a constant wet-bulb line from the initial state to a state of 100 percent relative humidity.

**Comment:** The trial-and-error solution in part c can be avoided through use of a computer program which also includes atmospheric-air data.

---

## 10-8-6 ADIABATIC MIXING OF TWO AIR STREAMS

An important application in air conditioning is the mixing of two air streams, as shown in Fig. 10-28a. The entering air streams have different dry-bulb

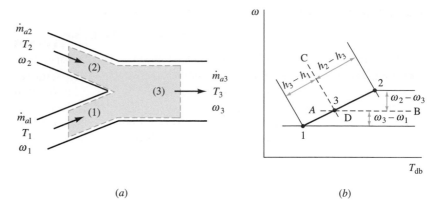

**Figure 10-28**    Adiabatic mixing of two air streams. (a) Schematic of the mixing process; (b) process on a psychrometric chart.

temperatures and humidity ratios. Heat transfer to the environment is usually negligible, hence the process is called *adiabatic mixing*. Kinetic- and potential-energy changes are usually neglected. Three basic relations can be written for the overall control volume on a rate basis:

1. Mass balance on dry air:

$$\dot{m}_{a1} + \dot{m}_{a2} = \dot{m}_{a3} \qquad\qquad \textbf{[10-58a]}$$

2. Mass balance on water vapor:

$$\dot{m}_{a1}\omega_1 + \dot{m}_{a2}\omega_2 = \dot{m}_{a3}\omega_3 \qquad\qquad \textbf{[10-58b]}$$

3. Energy balance for adiabatic mixing:

$$\dot{m}_{a1}h_1 + \dot{m}_{a2}h_2 = \dot{m}_{a3}h_3 \qquad\qquad \textbf{[10-58c]}$$

where $h = c_{p,a}T + \omega h_g$. When the properties and flow rates of two flow streams are known, Eqs. [10-58] are three equations with three unknowns. When the inlet states are known, for example, the unknowns for the exit state are $\dot{m}_{a3}$, $\omega_3$, and $h_3$. These latter two properties fix all other properties of the exit stream, such as its dry-bulb and wet-bulb temperatures and its relative humidity. The exit stream will leave with a humidity ratio and dry-bulb temperature that are between the values of the entering streams.

The mixing process also has an interesting interpretation on a psychrometric chart. For this purpose it is useful to combine Eqs. [10-58a] through [10-58c] to form two additional expressions. When Eqs. [10-58a] and [10-58b] are combined so that $\dot{m}_{a3}$ is eliminated, then

$$\frac{\dot{m}_{a1}}{\dot{m}_{a2}} = \frac{\omega_2 - \omega_3}{\omega_3 - \omega_1} \qquad\qquad \textbf{[10-59]}$$

If Eq. [10-58a] is substituted into Eq. [10-59c], the result is

$$\frac{\dot{m}_{a1}}{\dot{m}_{a2}} = \frac{h_2 - h_3}{h_3 - h_1}$$ [10-60]

Both Eqs. [10-59] and [10-60] have a geometric interpretation with respect to the psychrometric chart. Consider the horizontal dashed line $AB$ in Fig. 10-28b. Equation [10-59] requires that the vertical distances between states 2 and 3 and between states 3 and 1 on the chart are in proportion to the ratio of the mass flow rate of dry air for streams 1 and 2. The possible states which satisfy Eq. [10-59] lie on line $AB$. Likewise, Eq. [10-60] is satisfied by states on the dashed line $CD$ which runs parallel to the enthalpy lines. The only state which satisfies both of these equations geometrically is state 3 which *lies on a straight line connecting states* 1 *and* 2 *on the psychrometric chart*. For fixed states 1 and 2, the position of state 3 on process line 1-3-2 depends on the ratio of $\dot{m}_{a1}$ to $\dot{m}_{a2}$.

Generally, states 1 and 2 are situated so that state 3 lies below the 100 percent relative-humidity line. In some situations the choice of states 1 and 2 and their respective mass flow rates may put state 3 on the saturation line ($\phi = 1$). Finally, if states 1 and 2 lie close to the saturation line, then state 3 may lie above the saturation line. In this case water will condense during the mixing process. Normally this would be an undesirable condition if the flow stream goes directly into a home or business.

---

**A**n air stream (1) enters an adiabatic mixing chamber at a rate of 150 m³/min at 10°C and $\phi_1 = 0.80$. It is mixed with another stream (2) at 32°C and $\phi_2 = 0.60$ at a rate of 100 m³/min. Determine the final temperature and relative humidity of the exit stream (3), if the total pressure is 1 bar.

**Example 10-15**

**Solution:**

**Given:** An adiabatic mixing process occurs as shown with appropriate data in Fig. 10-29.

**Find:** $T_3$ and $\phi_3$.

**Model:** Steady state, adiabatic, $w = \Delta \text{ke} = \Delta \text{pe} = 0$.

**Analysis:** The mass balances on dry air and water vapor are

$$\dot{m}_{a1} + \dot{m}_{a2} = \dot{m}_{a3} \qquad \text{and} \qquad \dot{m}_{a1}\omega_1 + \dot{m}_{a2}\omega_2 = \dot{m}_{a3}\omega_3$$

and the energy balance on the adiabatic mixing process is

$$\dot{m}_{a1}h_1 + \dot{m}_{a2}h_2 = \dot{m}_{a3}h_3$$

These equations require a knowledge of the mass flow rates of dry air. The volume rates given are for the total flow, including water vapor. However, $\dot{m}_a$ can be found

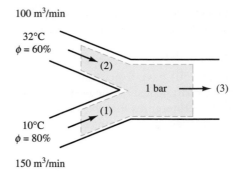

**Figure 10-29**   Schematic and data for the
adiabatic mixing process in
Example 10-15.

by dividing the total volume rate by the specific volume of the air. It was noted in
Sec. 10-5 that $v = R_aT/p_a$. Thus for state 1

$$p_{v1} = \phi_1 p_g = 0.8(0.01228) = 0.0098 \text{ bar}$$

$$p_{a1} = P - p_{v1} = 1.0 - 0.0098 = 0.9902 \text{ bar}$$

$$v_1 = \frac{R_aT}{p_a} = \frac{0.08314(283)}{29(0.9902)} = 0.819 \text{ m}^3/\text{kg dry air}$$

$$\dot{m}_{a1} = \frac{\text{volume rate}}{v} = \frac{150}{0.819} = 183 \text{ kg dry air/min}$$

$$\omega_1 = \frac{0.622p_v}{p_a} = \frac{0.622(0.0098)}{0.9902} = 0.00616 \text{ kg water/kg dry air}$$

$$h_1 = c_pT + \omega h_g = 1.005(10) + 0.00616(2519.8) = 25.5 \text{ kJ/kg dry air}$$

Similarly for state 2,

$$p_{v2} = \phi_2 p_g = 0.6(0.04759) = 0.0286 \text{ bar}$$

$$p_{a2} = P - p_{v2} = 1.0 - 0.0286 = 0.9714 \text{ bar}$$

$$v_2 = \frac{0.08314(305)}{29(0.9714)} = 0.900 \text{ m}^3/\text{kg dry air}$$

$$\dot{m}_{a2} = \frac{100}{0.900} = 111 \text{ kg dry air/min}$$

$$\omega_2 = \frac{0.622(0.0286)}{0.9714} = 0.0183 \text{ kg water/kg dry air}$$

$$h_2 = 1.005(32) + 0.0183(2559.9) = 79.0 \text{ kJ/kg dry air}$$

From these data $\omega_3$ is found by combining Eqs. [10-58] and [10-59], while $h_3$ is found by combining Eqs. [10-58] and [10-60]. The results are

$$\omega_3 = \frac{\dot{m}_{a1}\omega_1 + \dot{m}_{a2}\omega_2}{\dot{m}_{a1} + \dot{m}_{a2}}$$

$$= \frac{183(0.00616) + 111(0.0183)}{183 + 111} = 0.01074 \text{ kg water/kg dry air}$$

$$h_3 = \frac{\dot{m}_{a1}h_1 + \dot{m}_{a2}h_2}{\dot{m}_{a1} + \dot{m}_{a2}}$$

$$= \frac{183(25.5) + 111(79.0)}{183 + 111} = 45.7 \text{ kJ/kg dry air}$$

The temperature $T_3$ is found from the value of $h_3$, since $h = c_pT + \omega h_g$, or

$$h_3 = 45.7 = 1.005(T_3) + 0.01074(h_{g3})$$

One method of solving this equation is by trial and error. A temperature is guessed, the value of $h_g$ at this temperature is found from the steam tables, and $T_3$ and $h_{g3}$ are substituted into the equation. If the equation is not satisfied, another temperature must be tried. A simpler method is to employ the approximation equation for $h$ given by Eq. [10-48c]. Thus,

$$h_3 = 45.7 = 1.005T_3 + 0.01074(2501.7 + 1.82T_3)$$

$$T_3 = \frac{18.68}{1.025} = 18.2°C$$

Although an approximation technique was used, the answer is probably within 0.1°C of that found from tabular $h$ data. To determine the relative humidity at state 3, we must first find $p_{v3}$ from the value of $\omega_3$:

$$\omega_3 = 0.01074 = \frac{0.622p_{v3}}{1 - p_{v3}} \quad \text{and} \quad p_{v3} = 0.01697 \text{ bar}$$

Since the saturation pressure at 18.2 °C is around 0.0209 bar, then

$$\phi_3 = \frac{p_v}{p_g} = \frac{0.01697}{0.0209} = 0.812 \text{ (or 81.2\%)}$$

**Comment:** The initial values in this problem can be checked against values from the psychrometric chart. The final state can be located on the chart by using the inlet mass flow rates in terms of a "lever rule" on the straight line connecting the initial states.

## 10-8-7 WET COOLING TOWER

At fossil- and nuclear-fueled power plants, a considerable portion of the energy released by the fuel must be rejected to the environment. Cooling water from natural sources such as rivers and lakes is commonly used to

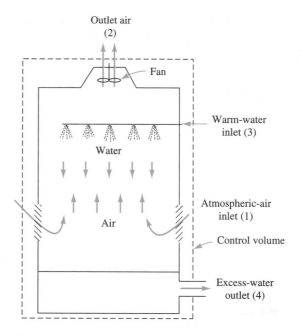

**Figure 10-30**     Schematic of a wet cooling tower.

carry off the energy rejected from the condenser in the power plant. Because of environmental concern, there is a limitation on the temperature of the cooling water from the plant which is discharged back to the natural source. In many cases the solution is to reduce the temperature of the cooling water by some means and then recirculate the water back to the condenser in the power plant. The use of a *wet cooling tower* is one method of cooling any water stream by transferring the energy to the air in the atmosphere.

A schematic of a wet cooling tower is shown in Fig. 10-30. The air passing through this tower is drawn upward by a fan; hence the device is classified as an *induced-draft* tower. A small induced-draft tower, as shown in Photograph 10-2, is often used in a cooling-water loop to serve as a heatsink for some system, e.g. a water-cooled condenser. A *natural-draft* tower operates similarly to a chimney, with the air rising because of a density difference. The warm cooling water is sprayed from the top of the tower and falls downward under gravity. Unsaturated atmospheric air is inducted at the bottom of the tower and flows upward counter to the falling droplets. The water and air streams are thus brought into intimate contact, and a small fraction of the water evaporates into the air stream. The evaporation process results in a cooling of the remaining water stream. The moisture content and the temperature of the air stream increase during the process. The cooled water is then returned to the power (or industrial) plant to pick up additional waste energy. Since a portion of the circulating water is evaporated, an equivalent amount must be added as makeup somewhere external to the tower.

**Photograph 10-2** Typical installation of an induced-draft wet cooling tower. (Courtesy of the Marley Company)

An energy and a water balance are now applied to a control volume drawn around the entire tower, as shown by the dashed line in Fig. 10-30. The process is assumed to be adiabatic, the fan work is neglected, and changes in kinetic and potential energy are negligible. Consequently, the basic energy balance is written as $\sum(\dot{m}h)_{\text{in}} = \sum(\dot{m}h)_{\text{out}}$. Therefore,

$$\dot{m}_{a1}h_{m1} + \dot{m}_{w3}h_{w3} = \dot{m}_{a2}h_{m2} + \dot{m}_{w4}h_{w4} \qquad \textbf{[10-61]}$$

or
$$\dot{m}_{a1}(h_{a1} + \omega_1 h_{g1}) + \dot{m}_{w3}h_{w3} = \dot{m}_{a2}(h_{a2} + \omega_2 h_{g2}) + \dot{m}_{w4}h_{w4}$$
$$\textbf{[10-62]}$$

where $h_m$ in Eq. [10-61] is the mixture enthalpy per unit mass of dry air. This value can be calculated directly, as shown in Eq. [10-62], or read from a psychrometric chart. The water-stream enthalpy $h_w$ may be evaluated by $h_f$ at the given temperature. Note that $\dot{m}_{a1} = \dot{m}_{a2} = \dot{m}_a$. In addition, it is necessary to relate $\dot{m}_{w3}$ and $\dot{m}_{w4}$ by means of a mass balance on the water passing through the control volume. This mass balance leads to

$$\dot{m}_{w4} = \dot{m}_{w3} - \dot{m}_a(\omega_2 - \omega_1) \qquad \textbf{[10-63]}$$

In the analysis of wet cooling towers, the value of $\dot{m}_{w3}$ is known as well as the temperature of the water entering and leaving the tower. Additional data on the inlet- and outlet-air conditions allow one to determine the mass flow rate of dry air required.

**Example 10-16**

Water enters a cooling tower at 40°C and leaves at 25°C. The tower receives atmospheric air at 20°C and 40 percent relative humidity. The air leaves the tower at 35°C and 95 percent relative humidity. Find the mass flow rate of dry air, in kg/min, passing through the tower if the water inlet rate is 12,000 kg/min.

**Solution:**

**Given:** Data for a cooling tower are given on the schematic shown in Fig. 10-31.

**Find:** $\dot{m}$ of dry air, in kg/min.

**Model:** Steady state, $\dot{Q} = \dot{W} = \Delta\text{ke} = \Delta\text{pe} = 0$.

**Analysis:** The basic energy equation involving the two inlet streams and the two outlet streams is

$$\dot{m}_{a1}(h_{a1} + \omega_1 h_{g1}) + \dot{m}_{w3}h_{w3} = \dot{m}_{a2}(h_{a2} + \omega_2 h_{g2}) + \dot{m}_{w4}h_{w4}$$

The main data needed, in addition to steam-table information, are the humidity ratios for the inlet and exit air streams. These are found by using Eq. [10-42] as follows:

$$\omega_1 = \frac{0.622(0.40)(0.02339)}{1.0 - 0.40(0.02339)} = 0.00587 \text{ kg water/kg dry air}$$

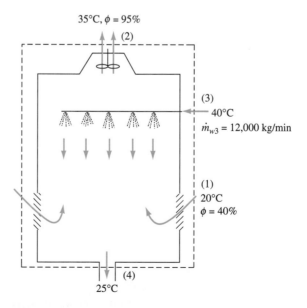

35°C, $\phi = 95\%$

(2)

(3)
40°C
$\dot{m}_{w3} = 12,000$ kg/min

(1)
20°C
$\phi = 40\%$

(4)
25°C

**Figure 10-31**    A cooling tower schematic for Example 10-16.

$$\omega_2 = \frac{0.622(0.95)(0.05628)}{1.0 - 0.95(0.05628)} = 0.0351 \text{ kg water/kg dry air}$$

Other data include

$$h_{g1} = 2538.1 \text{ kJ/kg} \qquad h_{f3} = 167.6 \text{ kJ/kg}$$
$$h_{g2} = 2565.3 \text{ kJ/kg} \qquad h_{f4} = 104.9 \text{ kJ/kg}$$

With these data, and recalling that $c_{p,a} = 1.005$ kJ/kg·°C, we find that the energy balance cited above leads to the following result:

$$\dot{m}_a[1.005(20) + 0.00587(2538.1)] + 12,000(167.6)$$
$$= \dot{m}_a[1.005(35) + 0.0351(2565.3)] + \dot{m}_{w4}(104.9)$$

In addition, a water mass balance on the system is given by

$$\dot{m}_{w4} = \dot{m}_{w3} - \dot{m}_a(\omega_2 - \omega_1)$$

Substitution of data yields

$$\dot{m}_{w4} = 12,000 - \dot{m}_a(0.0351 - 0.00587)$$

When $\dot{m}_{w4}$ from the last equation is substituted into the energy equation, then

$$35.00\dot{m}_a + 2.011 \times 10^6 = 125.22\dot{m}_a + 1.259 \times 10^6 - 3.07\dot{m}_a$$

$$\dot{m}_a = \frac{0.752 \times 10^6}{87.15} = 8630 \text{ kg dry air/min}$$

If the air enters at 1 bar, then the specific volume is

$$v = \frac{R_a T}{P - p_v} = \frac{0.08314(293)}{[1.0 - 0.40(0.02339)](28.97)} = 0.849 \text{ m}^3\text{/kg dry air}$$

Then

$$\text{Volume rate} = v_1 \dot{m}_a = 0.849(8630) = 7330 \text{ m}^3\text{/min}$$

**Comment:** Data for this problem could be obtained from the steam tables and the psychrometric chart. From the mass balance on the water streams, the water evaporated amounts to 252 kg/min, or 2.1 percent of the inlet cooling-water flow.

## 10-9 SUMMARY

The composition of a gas mixture is expressed on either a gravimetric (mass) or a mole basis. On a mass basis for $k$ components

$$m_m = \sum_{i=1}^{k} m_i \qquad mf_i \equiv \frac{m_i}{m_m} \qquad \sum_{i=1}^{k} mf_i = 1$$

and on a mole basis

$$N_m = \sum_{i=1}^{k} N_i \qquad y_i \equiv \frac{N_i}{N_m} \qquad \sum_{i=1}^{k} y_i = 1$$

An apparent (or average) molar mass $M_m$ and specific gas constant $R_m$ of a gas mixture are

$$M_m \equiv \frac{m_m}{N_m} \qquad \text{and} \qquad R_m = \frac{R_u}{M_m}$$

where $R_u$ is the universal gas constant.

Two important $PvT$ relationships are Dalton's law of additive pressures and Amagat's law of additive volumes, namely,

$$P = \sum_{i=1}^{k} p_i \qquad \text{and} \qquad V = \sum_{i=1}^{k} V_i$$

where the component pressure $p_i$ is measured at $T$ and $V$ of the mixture, while the component volume $V_i$ is measured at $P$ and $T$ of the mixture. For ideal-gas mixtures this leads to

$$y_i = \frac{p_i}{P} = \frac{N_i}{N_m} = \frac{V_i}{V}$$

at a given state. Also for ideal-gas mixtures other thermodynamic properties are additive. That is, at a specified state

$$U_m = \sum_{i=1}^{k} m_i u_i = \sum_{i=1}^{k} N_i \bar{u}_i$$

$$H_m = \sum_{i=1}^{k} m_i h_i = \sum_{i=1}^{k} N_i \bar{h}_i$$

$$S_m = \sum_{i=1}^{k} m_i s_i = \sum_{i=1}^{k} N_i \bar{s}_i$$

On a molar or mass basis these equations become

$$\bar{u}_m = \sum_{i=1}^{k} y_i \bar{u}_i \qquad \text{or} \qquad u_m = \sum_{i=1}^{k} mf\, u_i$$

$$\bar{h}_m = \sum_{i=1}^{k} y_i \bar{h}_i \qquad \text{or} \qquad h_m = \sum_{i=1}^{k} mf\, h_i$$

$$\bar{s}_m = \sum_{i=1}^{k} y_i \bar{s}_i \qquad \text{or} \qquad s_m = \sum_{i=1}^{k} mf\, s_i$$

Since $\Delta u_m \equiv c_{v,m}\Delta T$ and $\Delta h_m \equiv c_{p,m}\Delta T$, the average specific heat of a mixture is defined by

$$c_{v,m} = \sum_{i=1}^{k} y_i \bar{c}_{v,i} = \sum_{i=1}^{k} mf\, c_{v,i}$$

$$c_{p,m} = \sum_{i=1}^{k} y_i c_{p,i} = \sum_{i=1}^{k} = mf\, c_{p,i}$$

When the entropy is expressed as a function of temperature and pressure,

$$\Delta s_i = c_{p,i}\ln\frac{T_2}{T_1} - R_i\ln\frac{p_{i2}}{p_{i1}} = s_{i2}^o - s_{i1}^o - R_i\ln\frac{p_{i2}}{p_{i1}}$$

where $p_i$ again is the component pressure in the mixture. This point regarding $p_i$ is particularly important when dealing with mixing processes.

An important application of ideal-gas mixture-property relations is to atmospheric air. Two important properties in this case are the relative humidity $\phi$ and the specific humidity $\omega$, given by

$$\phi \equiv \frac{p_v}{p_g} \qquad \omega \equiv \frac{m_v}{m_a} = \frac{0.622\phi p_g}{P - \phi p_g}$$

The enthalpy of atmospheric air per unit mass of dry air is represented by

$$h_{\text{mix}} = h_a + \omega h_v = c_{p,a}T + \omega h_g$$

where each term is expressed on a mass of dry air basis. The specific volume per unit mass of dry air is

$$v_{\text{mix}} = \frac{R_u T}{P - p_v}$$

where $p_v$ is measured at $T$. Measurement of the adiabatic-saturation or wet-bulb temperature $T_2 = T_{\text{wb},1} = T_{\text{ad},1}$ of air initially at dry-bulb temperature $T_1$ and specific humidity $\omega_1$ leads to the specific humidity through the equations

$$\omega_{\text{sat},2} = \frac{0.622 p_{g2}}{P - p_{g2}}$$

and

$$\omega_1 = \frac{c_{p,a}(T_2 - T_1) + \omega_{\text{sat},2} h_{fg2}}{h_{v1} - h_{f2}}$$

Thus measurement of the dry-bulb and wet-bulb temperatures and total pressure leads to the evaluation of $\omega_1$, and hence to all other properties.

In conjunction with the conservation of mass and conservation of energy equations, the above property relations lead to the analysis of various air-conditioning processes. These include (1) heating and cooling at constant specific humidity, (2) dehumidification with heating, (3) humidification by water addition, (4) evaporative cooling, (5) adiabatic mixing of two atmospheric-air streams, and (6) wet cooling towers.

## PROBLEMS

10-1. A gas mixture contains 40 percent methane ($CH_4$) and 60 percent carbon monoxide (CO) by volume. Determine (*a*) the gravimetric analysis, (*b*) the apparent molar mass of the mixture, and (*c*) the mass, in kilograms, of $10 m^3$ of the mixture at 1.2 bars and 127°C.

10-2. The gravimetric analysis of a mixture of ideal gas is $N_2$, 56 percent; CO, 28 percent, and $O_2$, 16 percent. Determine (*a*) the volumetric analysis, (*b*) the apparent molar mass of the mixture, and (*c*) the volume, in cubic meters, of 0.20 kg of the mixture at 67°C and 140 kPa.

10-3. A gas mixture contains 0.28 kg of CO, 0.16 kg of $O_2$, and 0.66 kg of $CO_2$ at 1.4 bars and 17°C. Compute (*a*) the volumetric analysis, (*b*) the apparent gas constant, in kJ/kg·K, and (*c*) the volume occupied by the mixture, in cubic meters.

10-4. A gas mixture has the following volumetric analysis: $N_2$, 60 percent; $CO_2$, 33 percent; $O_2$, 7 percent.
(*a*) Determine the gravimetric analysis.
(*b*) Determine the apparent molar mass, in kg/kmol.
(*c*) If the volume flow rate in steady state is 10 $m^3$/s at 100 kPa and 57°C, determine the mass flow rate in kg/s.

10-5. The volumetric analysis of a mixture of ideal gases is $N_2$, 70 percent; $CO_2$, 20 percent; $O_2$, 10 percent. The volume flow rate of the mixture at 1 bar and 90°C is 12 $m^3$/s. Determine (*a*) the gravimetric analysis, (*b*) the component pressure of $CO_2$, and (*c*) the mass flow of the mixture, in kg/s.

10-6. A mixture of ideal gases has the following volumetric analysis: $CO_2$, 50 percent; $N_2$, 40 percent; $H_2O$, 10 percent.
(*a*) Calculate the gravimetric analysis of the mixture and the apparent molar mass.
(*b*) A tank of 0.0277 $m^3$ capacity contains 0.060 kg of the mixture at 77°C. Determine the gas pressure, in kilopascals.

10-7. A natural gas mixture has the following molar analysis: $CH_4$, 70 percent; $C_2H_6$, 10 percent; $N_2$, 20 percent. It enters a furnace at 1.01 bar and 27°C with a volume flow rate of 5 $m^3$/s. Determine (*a*) the mass fractions of each component and (*b*) the mass flow rate in kg/s.

10-8E. The volumetric analysis of a mixture of ideal gases is $N_2$, 60 percent; $CO_2$, 30 percent; $O_2$, 10 percent. The volume flow rate of the mixture at 15 psia and 140°F is 20 ft$^3$/s. Determine (a) the gravimetric analysis, (b) the component pressure of $CO_2$, and (c) the mass flow of the mixture, in lb$_m$/s.

10-9E. The gravimetric analysis of an ideal-gas mixture is $N_2$, 56 percent; $CO_2$, 28 percent; and $O_2$, 16 percent. Determine (a) the volumetric analysis, (b) the apparent molar mass of the mixture, and (c) the volume, in cubic feet, of 0.30 lb$_m$ of the mixture at 90°F and 20 psia.

10-10E. A gas mixture contains 0.42 lb$_m$ of CO, 0.16 lb$_m$ of $O_2$, and 0.88 lb$_m$ of $CO_2$ at 20 psia and 90°F. Compute (a) the volumetric analysis, (b) the apparent gas constant, in Btu/lb$_m$·°R, and (c) the volume occupied by the mixture, in cubic feet.

10-11E. A gas mixture has the following volumetric analysis: $N_2$, 50 percent; $CO_2$, 40 percent; $O_2$, 10 percent. Determine (a) the gravimetric analysis, (b) the apparent molar mass, in lb$_m$/lbmol, and (c) the mass, in pounds, of 1000 ft$^3$ of the gas mixture at 30 psia and 80°F.

10-12. A mixture at 300 K and 160 kpa has the following volumetric analysis: $O_2$, 60 percent; $CO_2$, 40 percent. Compute (a) the mass analysis, (b) the partial pressure of $O_2$, in kilopascals, (c) the apparent molar mass, and (d) the volume occupied by 0.13 kg of the mixture, in cubic meters.

10-13. A gas mixture at 120 kPa and 0.063 m$^3$ has the following gravimetric analysis: $O_2$, 32 percent; $CO_2$, 56 percent; He, 12 percent. Determine (a) the volumetric analysis, (b) the apparent gas constant, in kJ/kg·K, and (c) the temperature of 0.050 kg of the mixture, in kelvins.

10-14. A gas mixture is made up of 64 percent $O_2$ and 36 percent $H_2$, by mass. The total pressure is 3 bars, and the volume is 0.10 m$^3$. Determine (a) the partial pressure of hydrogen, in bars, (b) the apparent molar mass, in kg/kmol, (c) the gas constant, in kJ/kg·K, and (d) the temperature, in kelvins, if the mass of the mixture is 0.052 kg.

10-15. A 0.1-m$^3$ tank contains 0.70 kg of $N_2$ and 1.10 kg of $CO_2$ at 27°C. Compute (a) the component pressure of $N_2$, in kilopascals, (b) the component volume of each species, in cubic meters, (c) the total pressure, in kilopascals, and (d) the gas constant of the mixture, in kJ/kg·K.

10-16E. A gas mixture at 23 psia and 1.90 ft$^3$ has the following gravimetric analysis: $O_2$, 24 percent; $CO_2$, 56 percent; He, 20 percent. Determine (a) the volumetric analysis, (b) the apparent gas constant, in Btu/lb$_m$·°R, and (c) the temperature of 0.10 lb$_m$ of the mixture, in degrees Rankine.

10-17E. An ideal-gas mixture is made up of 80 percent $O_2$ and 20 percent $H_2$ by weight. The total pressure is 50 psia, and the volume is 2.0 ft$^3$. Determine (a) the partial pressure of hydrogen, in psia, (b) the apparent molar mass, in lb$_m$/lbmol, (c) the gas constant, in Btu/(lb$_m$·°R), and (*d*) the temperature, in degrees Fahrenheit, if the mass of the mixture is 0.14 lb$_m$.

10-18E. A 20.0-ft$^3$ tank contains 1.40 lb$_m$ of N$_2$ and 2.20 lb$_m$ of CO$_2$ at 90°F. Compute (*a*) the component pressure of N$_2$, in psia, (*b*) the component volume of each species, in cubic feet, (*c*) the total pressure, in psia, and (*d*) the gas constant of the mixture, in Btu/lb$_m$·°R.

10-19. A rigid tank initially contains 2.0 kg of CO and 1.0 kg of CO$_2$ at 2.4 bars and 400 K. During a process 174.5 kJ of heat is added, and a current of 6.5 A passes through a resistor within the tank for a period of 0.20 h. If the final temperature of the mixture is 500 K, determine the constant voltage applied across the resistor during the process. Employ Tables A-8 and A-9 for data.

10-20. A gas at 77°C and 100 kPa has the following volumetric analysis: H$_2$, 4 percent; CO, 12 percent; CO$_2$, 25 percent; and N$_2$, 59 percent. The initial mixture passes at 20 m$^3$/min through a steady-flow heat exchanger until the temperature reaches 227°C. Determine (*a*) the gravimetric analysis, (*b*) the heat transfer based on tabular ideal-gas tables, (*c*) the heat transfer based on average specific-heat data, in kJ/kg, and (*d*) the heat transfer rate in kJ/min.

10-21. A rigid tank contains a gas mixture at an initial state of 227°C and 200 kPa with the following volumetric analysis: N$_2$, 50 percent; CO$_2$, 30 percent; O$_2$, 20 percent. The mixture is cooled to 27°C. Determine the quantity of heat transfer required, in kJ/kmol, based on (*a*) tabular ideal-gas data and (*b*) average specific-heat data, and (*c*) the heat transfer in kJ for a volume of 2 m$^3$.

10-22. A gas mixture contains 0.28 kg of CO, 0.16 kg of O$_2$, and 0.66 kg of CO$_2$ at 1.4 bars and 17°C. The mixture is heated at constant pressure in a piston-cylinder device to a final temperature of 277°C. Determine the heat transfer required, in kilojoules, by employing (*a*) ideal-gas tabular data and (*b*) average specific-heat data, and (*c*) the heat transfer in kJ/kmol.

10-23. A gas mixture has the following volumetric analysis: N$_2$, 70 percent; CO$_2$, 23 percent; O$_2$, 7 percent. The mixture is cooled as it passes through a steady-state heat exchanger from 550 to 325 K. Compute the heat transferred, in kJ/kmol and kJ/kg, using tabular ideal-gas data.

10-24E. A gas at 100°F and 14.5 psia has the following volumetric analysis: H$_2$, 4 percent; CO, 12 percent; CO$_2$, 35 percent; and N$_2$, 49 percent. The mixture passes through a steady-flow heat exchanger until the temperature reaches 300°F. Determine (*a*) the gravimetric analysis, (*b*) the heat transfer based on tabular ideal-gas tables, and (*c*) the heat transfer based on average specific-heat data, in Btu/lb$_m$.

10-25E. A rigid tank contains a gas mixture at an initial state of 500°F and 25 psia with the following volumetric analysis: N$_2$, 70 percent; CO$_2$,

20 percent; $O_2$, 10 percent. The mixture is cooled to 100°F. Determine the quantity of heat transfer required, in Btu/lbmol, based on (a) tabular ideal-gas data and (b) average specific-heat data, and (c) the heat transfer in Btu for a volume of 5 ft³.

10-26E. A gas mixture contains 0.28 $lb_m$ of CO, 0.16 $lb_m$ of $O_2$, and 0.66 $lb_m$ of $CO_2$ at 20 psia and 90°F. The mixture is heated at constant pressure in a piston-cylinder device to a final temperature of 440°F. Determine the heat required, in Btu, by employing (a) ideal-gas tabular data and (b) average specific-heat data.

10-27E. A gas mixture has the following volumetric analysis: $N_2$, 60 percent; $CO_2$, 33 percent; $O_2$, 7 percent. The mixture is cooled as it passes through a steady-state heat exchanger from 540 to 140°F. Compute the heat transferred, in Btu/lbmol and Btu/$lb_m$, using tabular ideal-gas data.

10-28. A rigid tank contains 0.2 kg of nitrogen and 0.1 kg of carbon dioxide at 200 kPa and 37°C. During a process, 4.90 kJ of heat is added, and a current of 4.5 A passes through a resistor within the tank for a period of 6.0 min. If the final temperature of the gas mixture is 147°C, determine the constant voltage applied across the resistor. Use tabular data from the ideal-gas tables.

10-29. An ideal-gas mixture consists of 56 percent CO, 28 percent $N_2$, and 16 percent He by weight at 0.34 MPa and 327°C. It expands adiabatically through a steady-flow turbine until the exit temperature is 77°C. Determine (a) the shaft-work output, in kJ/kmol, and (b) the power, in kilowatts, for 4.0 m³/min at the inlet.

10-30. An equimolar mixture of two ideal gases expands through an adiabatic nozzle from an initial state of 400 K and 50 m/s to a final state of 350 K. Determine the exit velocity, in m/s, if the mixture is Ar and CO.

10-31. An ideal-gas mixture has the following volumetric analysis: CO, 30 percent; $CO_2$, 50.0 percent; $O_2$, 20 percent. It enters a steady-state compressor at 37°C and 60 m/s and leaves the device at 237°C and 100 m/s.
(a) Determine the shaft-work input required, in kJ/kg, if a heat loss of 4.0 kJ/kg occurs during the process.
(b) If the volume flow rate at the entrance is 6.0 m³/min and the pressure is 120 kPa, determine the power input, in kilowatts.

10-32. An equimolar mixture of CO and Ar flows steadily through a diffuser. The initial state is 300 K and 200 m/s, and the final velocity is 50 m/s. Using specific-heat data, estimate the final temperature, in kelvins, for the mixture, if the process is adiabatic.

10-33E. A rigid tank contains 1.0 $lb_m$ of nitrogen and 1.0 $lb_m$ of carbon dioxide at 30 psia and 100°F. During a process, 47.0 Btu of heat is added, and a current of 4.5 A passes through a resistor within the tank for a period of 15.0 min. If the final temperature of the gas mixture is 300°F, determine the constant voltage applied across the resistor. Use tabular data from the ideal-gas tables.

10-34E. An ideal-gas mixture consists of 56 percent CO, 32 percent $N_2$, and 12 percent He by weight at 50 psia and 620°F. It expands adiabatically through a steady-flow turbine until the exit temperature is 160°F. Determine (a) the shaft-work output, in Btu/lbmol, and (b) the power, in kilowatts, for 30 $ft^3$/min at the inlet.

10-35E. An equimolar mixture of He and CO expands through an adiabatic nozzle from an initial state of 260°F and 100 ft/s to a final state of 200°F. Determine the exit velocity, in ft/s, for the mixture.

10-36E. An ideal-gas mixture has the following volumetric analysis: CO, 33.3 percent; $CO_2$, 50.0 percent; $O_2$, 16.7 percent. It enters a steady-state compressor at 100°F and 150 ft/s and leaves the device at 440°F and 300 ft/s.

(a) Determine the shaft-work input required, in Btu/$lb_m$, if a heat loss of 4.0 Btu/$lb_m$ occurs during the process.

(b) If the volume flow rate at the entrance is 120 $ft^3$/min and the pressure 15 psia, determine the power input, in horsepower.

10-37. Carbon dioxide ($CO_2$) in the amount of 0.1 kmol and initially at 2 bars and 27°C is mixed adiabatically with 0.2 kmol of $O_2$ initially at 5 bars and 152°C. During the constant-volume mixing process, electric energy equivalent to 670 kJ/kmol of mixture is added. Determine (a) the final temperature of the mixture, in degrees Celsius, if tabular data are used and (b) the final pressure, in bars.

10-38. Carbon dioxide ($CO_2$), in the amount of 0.9 kmol and initially at 0.2 MPa and 327°C, is mixed adiabatically in a closed system at constant pressure with 2 kmol of $N_2$, initially at 0.2 MPa and 27°C. During the process 2000 kJ is added in the form of electrical work. Determine the final temperature of the mixture, in degrees Celsius.

10-39. Nitrogen in the amount of 2 kg and initially in a rigid tank at 700 kPa and 177°C is mixed with 1.2 kg of oxygen, initially in another tank at 300 kPa and 27°C, by opening a valve connecting the two tanks. If the final equilibrium temperature is 117°C, determine (a) the heat transfer, in kilojoules, and (b) the final pressure, in kilopascals.

10-40E. Carbon dioxide ($CO_2$) in the amount of 1 lbmol and initially at 30 psia and 100°F is mixed adiabatically with 2 lbmol of $O_2$ initially at 80 psia and 300°F. During the constant-volume mixing process, electric energy equivalent to 200 Btu/lbmol of mixture is added. Determine (a) the final temperature of the mixture, in degrees Fahrenheit, if tabular data are used and (b) the final pressure, in psia.

10-41E. Carbon dioxide ($CO_2$), in the amount of 1 lbmol and initially at 30 psia and 200°F, is mixed adiabatically in a closed system at constant pressure with 2 lbmol of $N_2$, initially at 30 psia and 80°F. During the process 400 Btu is added in the form of electrical work. Determine the final temperature of the mixture, in degrees Fahrenheit.

10-42E. Nitrogen in the amount of 2 $lb_m$ and initially in a rigid tank at 100 psia and 140°F is mixed with 1 $lb_m$ of oxygen, initially in another tank at 50 psia and 60°F, by opening a valve connecting the two tanks. If the final equilibrium temperature is 100°F, determine (a) the heat transfer, in Btu, and (b) the final pressure, in psia.

## ISENTROPIC PROCESSES AND ENTROPY CHANGES OF IDEAL-GAS MIXTURES

10-43. A mixture of CO, $CO_2$, and $N_2$ in a molar ratio of 1 : 2 : 5 is expanded through a nozzle from 600 K, 1000 kPa, and 10 m/s to 100 kPa. If the final velocity is 710 m/s, determine the nozzle efficiency.

10-44. A gas mixture consisting of $N_2$, $CO_2$, and $H_2O$ in a molar ratio of 4 : 1 : 1 enters a turbine at 1000 K and expands through a 6 : 1 pressure ratio. Using $s^o$ data, compute the power output in kilowatts if the turbine adiabatic efficiency is 80 percent and the mass flow rate is 2 kg/s of mixture.

10-45. A mixture of $H_2O$, CO, and $N_2$, in a molar ratio of 1 : 2 : 5 is compressed in a piston-cylinder device from 1 bar and 300 K to 4 bars. Using $s^o$ data, determine (a) the final temperature, in kelvins, and (b) the work input, kJ/kmol of mixture.

10-46. A mixture of $N_2$, $CO_2$, and $H_2O$ in a molar ratio of 4 : 1 : 1 is compressed in steady flow through a pressure ratio of 6 : 1 from an initial temperature of 300 K. Using $s^o$ data, determine the compressor efficiency if the actual outlet temperature is 520 K.

10-47. An equimolar mixture of helium and argon enters a turbine at 660 K and expands adiabatically through a 4.5:1 pressure ratio to a temperature of 400 K. Determine (a) the actual work output, in kJ/kg of mixture, and (b) the isentropic turbine efficiency.

10-48. An equimolar mixture of helium and nitrogen initially at 300 K is compressed adiabatically in steady flow through a 3:1 pressure ratio. The compressor adiabatic efficiency is 82 percent. Determine the actual work required, in kJ/kmol, using specific-heat data.

10-49E. A mixture of $H_2O$, CO, and $N_2$ with component pressures in the ratio 1 : 2 : 5 is expanded isentropically from 2000°R and 10 atm to 1 atm in a piston-cylinder device. Using $s^o$ data, determine how much work is done, in Btu/lbmole of mixture.

10-50E. A gas mixture consisting of $N_2$, $CO_2$, and $H_2O$ in a molar ratio of 4 : 1 : 1 enters a turbine at 2500°R and expands through a 6 : 1 pressure ratio. Using $s^o$ data, compute the power output in horsepower if the turbine adiabatic efficiency is 84 percent and the mass flow rate is 5 lb$_m$/s of mixture.

10-51E. A mixture of $H_2O$, CO, and $N_2$ in a molar ratio of 1 : 2 : 5 is compressed isentropically in a piston-cylinder device from 1 atm and 80°F to 4 atm. Using $s^o$ data, determine (*a*) the final temperature, in degrees Rankine, and (*b*) the work required, in Btu/lbmol of mixture.

10-52E. A mixture of $N_2$, $CO_2$, and $H_2O$ in a molar ratio of 4 : 1 : 1 is compressed adiabatically through a 6 : 1 pressure ratio in steady flow from an initial temperature of 540°R. Using $s^o$ data, determine the compressor efficiency if the actual outlet temperature is 940°R.

10-53. A rigid tank is subdivided into two compartments, one containing 0.020 kmol of CO initially at 500 K and 0.420 m$^3$ and the other containing 0.040 kmol of $CO_2$ initially at 300 K and 1.00 m$^3$. The partition separating the two gases is broken, and the gases mix adiabatically. The final pressure is found to be 123 kPa. Determine (*a*) the final temperature of the mixture, in kelvins, and (*b*) the entropy generation for the process, in kJ/K.

10-54. Carbon monoxide and argon in separate streams enter an adiabatic mixing chamber in a 2:1 mass ratio. At the inlet the carbon monoxide is at 120 kPa and 300 K, and the argon is at 120 kPa and 450 K. The mixture leaves at 110 kPa. Determine (*a*) the final temperature of the mixture in kelvins, and (*b*) the entropy production for the process, in kJ/K per kilomole of mixture.

10-55. Hydrogen and nitrogen in separate streams enter an adiabatic mixing chamber in a 3:1 molar ratio. At the inlet the hydrogen is at 2 bars and 77°C, and the nitrogen is at 2 bars and 277°C. The mixture leaves at 1.9 bars.
(*a*) Determine the final temperature of the mixture, in kelvins.
(*b*) Compute the entropy production of the overall process, in kJ/kmol·K.

10-56. An insulated 0.3-m$^3$ tank is divided into two sections by a partition. One section is 0.2 m$^3$ in volume and initially contains hydrogen gas at 0.20 MPa and 127°C. The remaining section initially holds nitrogen gas at 0.40 MPa and 27°C. The adiabatic partition is removed, and the gases are allowed to mix. Determine (*a*) the temperature of the equilibrium mixture, in kelvins, (*b*) the pressure of the mixture, in bars, and (*c*) the entropy production, in kJ/K.

10-57. A system consists of two tanks interconnected with a pipe and valve. One tank contains 2.0 kg of argon at 1.5 bars and 27°C. The other tank holds 1.6 kg of oxygen at 5 bars and 127°C. The valve is opened, and the gases are allowed to mix. The atmospheric temperature is 17°C, and the final temperature of the mixture is 77°C. Determine (*a*) the heat transfer, in kilojoules, (*b*) the final mixture pressure, (*c*) the change in entropy of the argon and of the oxygen, in kJ/K, and (*d*) the total entropy generation during the process, in kJ/K.

10-58. Two tanks are interconnected by a pipe and valve. One tank with a volume of 0.692 m$^3$ contains 2.2 kg of carbon dioxide at 27°C. The other tank has a volume of 0.33m$^3$ and holds 1.60 kg of oxygen at 127°C. The valve is opened, and the gases mix. The final temperature of the mixture is 177°C, due to heat transfer from a reservoir at 600 K. Determine (a) the heat transfer, in kilojoules, (b) the final pressure, in bars, (c) the change in entropy of the carbon dioxide and the oxygen, in kJ/K, and (d) the total entropy generation for the process, in kJ/K.

10-59. An insulated 0.06-m$^3$ tank is divided into two sections by a partition. One section is 0.02 m$^3$ in volume and initially contains 0.070 kg of carbon monoxide at 267°C. The remaining section initially contains 0.010 kg of helium at 17°C. The adiabatic partition is then removed, and the gases are allowed to mix. Determine (a) the temperature of the equilibrium mixture, in kelvins, (b) the final pressure of the mixture, in kilopascals, and (c) the entropy production for the process, in kJ/K.

10-60. At one inlet to a control volume 33 kg/min of $CO_2$ and 9 kg/min of $H_2O$ enter as a mixture at 0.20 MPa and 440 K. At another inlet 8 kg/min of $O_2$ and 14 kg/min of $N_2$ enter as a mixture at 0.20 MPa and 340 K. The mixture of four gases leaves the control volume at 0.19 MPa and 400 K. The shaft work is zero. Use the ideal-gas tables to calculate (a) the magnitude and direction of any heat transfer, in kJ/h, (b) the rate of entropy change of the $CO_2$ component, in kJ/h·K, and (c) the cross-sectional area at the control surface where $CO_2$ and $H_2O$ enter, in square centimeters, if the gas velocity at that surface is 8 m/s.

10-61E. A rigid tank is subdivided into two compartments, one containing 0.0125 lbmol of CO initially at 400°F and 40 psia and the other containing 0.0148 lbmol of $CO_2$ initially at 200°F and 20 psia. The partition separating the two gases is broken, and the gases mix adiabatically. The final pressure is found to be 26.4 psia. Determine (a) the final temperature of the mixture, in degrees Fahrenheit, and (b) the entropy generation of the overall process, in Btu/°R.

10-62E. Carbon monoxide and argon in separate streams enter an adiabatic mixing chamber in a 2:1 mass ratio. At the inlet the carbon monoxide is at 18 psia and 540°R, and the argon is at 18 psia and 840°R. The mixture leaves at 16 psia. Determine (a) the final temperature of the mixture, in degrees Rankine, and (b) the entropy production for the process, in Btu/°R per pound mole of mixture.

10-63E. Hydrogen and nitrogen in separate streams enter an adiabatic mixing chamber in a 3:1 molar ratio. At the inlet the hydrogen is at 25 psia, 140°F, and the nitrogen is at 25 psia, 540°F. The mixture leaves at 24 psia.
(a) Determine the final temperature of the mixture, in degrees Rankine.

(b) Compute the entropy change of the hydrogen gas, in Btu/lbmol·°R.

10-64E. An insulated 3-ft$^3$ tank is divided into two sections by a partition. One section is 2 ft$^3$ in volume and initially contains hydrogen gas at 30 psia and 110°F. The remaining section initially holds nitrogen gas at 50 psia and 50°F. The adiabatic partition is removed, and the gases are allowed to mix. Determine (a) the temperature of the equilibrium mixture, in degrees Fahrenheit, (b) the pressure of the mixture, in psia, and (c) the entropy production, in Btu/°R.

10-65E. A system consists of two tanks interconnected with a pipe and valve. One tank contains 5.0 lb of argon at 20 psia and 90°F. The other tank holds 4.0 lb of oxygen at 100 psia and 140°F. The valve is opened, and the gases are allowed to mix. The atmospheric temperature is 40°F, and during the process 31 Btu of heat leaves the uninsulated system. Determine (a) the final temperature, (b) the final mixture pressure, (c) the apparent molar mass of the mixture, (d) the change in entropy of the argon and of the oxygen, in Btu/°R, and (e) the total entropy generation during the process, in Btu/°R.

10-66E. Two tanks are interconnected by a pipe and valve. One tank with a volume of 37.0 ft$^3$ contains 5.0 lb$_m$ of argon at 90°F. The other tank has a volume of 8.0 ft$^3$ and holds 4.0 lb$_m$ of oxygen at 140°F. The valve is opened, and the gases mix. During the mixing process 49.0 Btu of heat is transferred to the mixture from a reservoir at 540°F. Determine (a) the final temperature, in degrees Fahrenheit, (b) the final pressure, in psia, (c) the change in entropy of the argon and oxygen, in Btu/°R, and (d) the total entropy generation for the process, in Btu/°R.

10-67E. An insulated 1.5-ft$^3$ tank is divided into two sections by a partition. One section is 0.5 ft$^3$ in volume and initially contains 0.10 lb$_m$ of carbon monoxide at 500°F. The remaining section initially contains 0.020 lb$_m$ of helium at 60°F. The adiabatic partition is then removed, and the gases are allowed to mix. Determine (a) the temperature of the equilibrium mixture, in degrees Fahrenheit, (b) the final pressure of the mixture, in psia, and (c) the entropy generation for the process, in Btu/°R.

## PROPERTIES OF IDEAL GAS–VAPOR MIXTURES

10-68. A tank contains 300 g of dry air and 50 g of saturated water vapor at 70°C. Determine (a) the volume of the tank, in liters, and (b) the total pressure, in millibars.

10-69. A rigid tank with a volume of 300 L initially contains 0.34 kg of dry air at 60°C. Water vapor is added until the gas is saturated at the same temperature. Determine (*a*) the mass of water added, in kilograms, and (*b*) the final total pressure, in kilopascals.

10-70. A rigid tank with a volume of 40 L initially contains 0.060 kg of oxygen at 20°C. Refrigerant 134a is added until the oxygen is saturated at the same temperature. Determine (*a*) the mass of refrigerant 134a added, in kilograms, and (*b*) the final total pressure, in bars.

10-71. A rigid tank contains 300 g of dry air and 70 g of saturated water vapor at a temperature of 80°C. Determine (*a*) the volume of the tank, in liters, and (*b*) the total pressure, in kilopascals.

10-72. A rigid tank with a volume of 0.2 m³ initially contains atmospheric air at 30°C and 100 kPa with a relative humidity of 10 percent. Water vapor is added to the tank until the air is saturated at 30°C. Determine (*a*) the mass of water added, in kilograms, and (*b*) the final pressure in kilopascals.

10-73. A tank contains 10 kg of dry air and 0.15 kg of water vapor at 26°C and 100 kPa. Determine (*a*) the specific humidity, (*b*) the relative humidity, and (*c*) the volume of the tank in m³.

10-74. A piston-cylinder device contains 0.10 kmol of dry air and 0.0014 kmol of water vapor at 20°C and 100 kPa. The mixture is cooled at constant pressure to 5°C. Determine (*a*) the initial humidity ratio, (*b*) the initial relative humidity in percent, (*c*) the initial dew-point temperature, and (*d*) the amount of water condensed per unit mass of dry air.

10-75. A sample of atmospheric air contains 0.012 kg of water and 1.0 kg of dry air at 22°C and 95 kPa. During the night the temperature drops to 10°C at the same pressure. Determine (*a*) the initial relative humidity, (*b*) the initial dew point, and (*c*) the mass of water condensed per unit mass of dry air.

10-76. Atmospheric air initially at 20°C, 1 bar, and 70 percent relative humidity is compressed isothermally until the pressure reaches 2 bars. (*a*) Show that condensation must occur. (*b*) Determine the amount of water condensed, in kg $H_2O$/kg dry air.

10-77. Dry air initially at −20°C and 4 bars contains refrigerant 134a, for which the "relative humidity" of R-134a is 75 percent. The ideal-gas mixture is compressed isothermally until the pressure reaches 6 bars. (*a*) Show that condensation must occur. (*b*) Determine the amount of R-134a condensed, in kg R-134a/kg dry air.

10-78E. A rigid tank with a volume of 10 ft³ initially contains 0.65 lb$_m$ of dry air at 140°F. Water vapor is added until the gas is saturated at the same temperature. Determine (*a*) the mass of water added, in pounds, and (*b*) the final total pressure, in psia.

10-79E. A rigid tank with a volume of 1.25 ft³ initially contains 0.110 lb$_m$ of oxygen at 70°F. Refrigerant 134a is added until the gas is saturated at the same temperature. Determine (a) the mass of refrigerant 134a added, in pounds, and (b) the final total pressure, in psia.

10-80E. A tank contains 0.25 $lb_m$ of dry air and 0.15 $lb_m$ of saturated water vapor at a temperature of 180°F. Determine (a) the volume of the tank and (b) the total pressure in the tank, in psia.

10-81. A 3-$m^3$ storage tank initially contains air at 5 bars and 150°F with a relative humidity of 10 percent. The air is then cooled back to the ambient temperature of 17°C. Determine (a) the dew point of the initial mixture, (b) the temperature at which condensation actually begins, (c) the amount of water condensed, and (d) the heat transferred from the tank, in kilojoules.

10-82. Two-tenths of a kilogram of atmospheric air is contained in a closed, rigid tank at the initial conditions of 28°C dry-bulb, 90 percent relative humidity, and 120 kPa. Calculate the heat required to increase the temperature of the mixture to 48°C, in kilojoules.

10-83E. A 100-$ft^3$ storage tank initially contains air at 80 psia and 300°F and a relative humidity of 10 percent. The air then cools back to the ambient temperature of 60°F. Determine (a) the dew point of the initial mixture, (b) the temperature at which condensation actually begins, (c) the amount of water condensed per pound of dry air, (d) the heat transferred to or from the tank during the process, in Btu/$lb_m$ dry air.

10-84. Two-tenths of a pound of atmospheric air is contained in a closed, rigid tank at the initial conditions of 90°F dry-bulb temperature, 90 percent relative humidity, and 16.0 psia. Calculate the heat required to increase the temperature of the mixture by 30°F, in Btu.

10-85. If the partial pressure of water vapor in atmospheric air at 1 bar is 30 mbars at 30°C, calculate (a) the relative humidity, (b) the dew-point temperature, (c) the humidity ratio, (d) the enthalpy, in kJ/kg (based on $h = 0$ at 0°C for both dry air and water), and (e) the specific volume of the mixture, in $m^3$/kg dry air. Use steam-table data where necessary.

10-86. The partial pressure of water vapor in atmospheric air at 980 mbar is 2.0 kPa at 25°C. Calculate (a) the relative humidity, (b) the dew-point temperature, (c) the humidity ratio, (d) the enthalpy, in kJ/kg dry air (based on $h = 0$ at 0°C for both dry air and water), and (e) the specific volume of the mixture, in $m^3$/kg dry air. Use steam-table data where necessary.

10-87. An atmospheric air mixture at 960 mbars contains 2.4 percent water vapor by volume. If the dry-bulb temperature is 25°C, calculate (a) the relative humidity, (b) the dew-point temperature, (c) the humidity ratio, in g/kg, and (d) the enthalpy, in kJ/kg dry air (based on $h = 0$ at 0°C for both dry air and water). Use steam-table data where necessary.

10-88. Atmospheric air with a relative humidity of 40 percent is held at 35°C and 970 mbars. Calculate (a) the specific humidity, in g/kg, (b) the dew-point temperature, (c) the enthalpy, in kJ/kg dry air, where $h = 0$ at 0°C, and (d) the specific volume, in m³/kg dry air. Use steam-table data.

10-89E. If the partial pressure of water vapor in atmospheric air at 14.7 psia is 0.400 psia at 90°F, calculate (a) the relative humidity, (b) the approximate dew-point temperature, (c) the humidity ratio, (d) the enthalpy, in Btu/lb$_m$, based on $h = 0$ at 0°F for dry air, and (e) the specific volume of the mixture. Use steam-table data where necessary.

10-90E. The partial pressure of water vapor in atmospheric air is 0.35 psia at 80°F. Calculate (a) the relative humidity, (b) the dew-point temperature, (c) the specific humidity, in lb$_m$/lb$_m$, (d) the enthalpy, in Btu/lb$_m$ dry air (based on $h = 0$ at 0°F for dry air), and (e) the specific volume, in ft³/lb$_m$ dry air, for 14.6 psia total pressure. Use the Steam Tables where necessary.

10-91E. An atmospheric air mixture at 14.5 psia contains 2 percent water by volume. If the dry-bulb temperature is 74°F, calculate (a) the relative humidity, (b) the dew-point temperature, (c) the humidity ratio, in gr/lb$_m$, and (d) the enthalpy, in Btu/lb$_m$ dry air (based on $h = 0$ at 0°F for dry air). Use steam-table data.

10-92E. Atmospheric air with a relative humidity of 40 percent is held at 94°F and 14.30 psia. Calculate (a) the specific humidity, in gr/lb$_m$, (b) the dew-point temperature, (c) the enthalpy, in Btu/lb$_m$ dry air, where $h = 0$ at 0°F for dry air, and (d) the specific volume, in ft³/lb$_m$ dry air. Use steam-table data.

10-93E. Atmospheric air at 76°F and 14.6 psia has an enthalpy of 27.0 Btu/lb$_m$ dry air (based on $h = 0$ at 0°F for dry air). Calculate (a) the specific humidity, in lb$_m$/lb$_m$, (b) the vapor pressure of water, in psia, (c) the relative humidity, and (d) the specific volume, in ft³/lb$_m$ dry air.

10-94. Atmospheric air at 22°C and 99 kPa has an enthalpy of 40.0 kJ/kg dry air (based on $h = 0$ at 0°C for dry air). Calculate (a) the specific humidity, in kg/kg, (b) the vapor pressure of water, in kilopascals, (c) the relative humidity, and (d) the specific volume, in m³/kg dry air.

10-95. Condensation on cold-water pipes often occurs in warm, humid rooms. (a) If the water temperature in the pipes may reach a minimum of 14°C and the room temperature is kept at 24°C, find the maximum limit of relative humidity in the room to avoid condensation at normal atmospheric pressure. (b) When the room relative humidity is 54 percent at 25°C, find the minimum pipe temperature if condensation is to be avoided.

10-96. On a cold winter day the inside surface of a wall in a home is found to be 16°C, and the air within the room is at 23°C.

(a) Find the maximum relative humidity that the air can have without the occurrence of condensation of water on the wall.

(b) If added insulation in the wall raises the maximum permissible relative humidity to 75 percent, find the new permissible inside-wall temperature, in degrees Celsius.

10-97E. Condensation on cold-water pipes often occurs in warm, humid rooms. (a) If the water temperature in the pipes might reach a minimum of 50°F and the room temperature is kept at 74°F, find the maximum limit of relative humidity in the room to avoid condensation at normal atmospheric pressure. (b) When the room relative humidity is 58 percent at 76°F, find the minimum pipe temperature if condensation is to be avoided.

10-98E. On a cold winter day the inside surface of a wall in a home is found to be 60°F. (a) If the air within the room is at 72°F, find the maximum relative humidity that the air can have without the occurrence of condensation of water on the wall. (b) If added insulation in the wall raises the maximum permissible relative humidity to 75 percent, find the new permissible inside-wall temperature, in degrees Fahrenheit.

10-99. Atmospheric air at 0.10 MPa has a dry-bulb temperature of 25°C and a wet-bulb temperature of 18°C. Determine, through the use of steam-table data and Eq. [10-54], (a) the humidity ratio, in g/kg, (b) the relative humidity, and (c) the enthalpy, in kJ/kg (based on $h = 0$ at 0°C for dry air).

10-100. Atmospheric air at 970 mbars has dry-bulb and wet-bulb temperatures of 29 and 20°C, respectively. Determine, through the use of steam-table data and Eq. [10-54], (a) the humidity ratio, in g/kg, (b) the relative humidity, and (c) the enthalpy, in kJ/kg (based on $h = 0$ at 0°C for dry air).

10-101. Atmospheric air at 1 bar has dry-bulb and wet-bulb temperatures of 27 and 21°C, respectively. Determine, through the use of steam-table data and Eq. [10-54], (a) the humidity ratio, in g/kg, (b) the relative humidity, and (c) the enthalpy, in kJ/kg (based on $h = 0$ at 0°C for dry air).

10-102E. Atmospheric air at 14.7 psia has a dry-bulb temperature of 80°F and a wet-bulb temperature of 68°F. Determine, through the use of steam-table data and Eq. [10-54], (a) the humidity ratio, in $lb_m/lb_m$, (b) the relative humidity, and (c) the enthalpy, in $Btu/lb_m$ (based on $h = 0$ at 0°F for dry air).

10-103E. Atmospheric air at 14.6 psia has dry-bulb and wet-bulb temperatures of 86 and 68°F, respectively. Determine, through the use of steam-table data and Eq. [10-54], (a) the humidity ratio, in $lb_m/lb_m$, (b) the relative humidity, and (c) the enthalpy, in $Btu/lb_m$ (based on $h = 0$ at 0°F for dry air).

10-104E. Atmospheric air at 14.7 psia has dry-bulb and wet-bulb temperatures of 82 and 72°F, respectively. Determine, through the use of steam-table data and Eq. [10-54], (a) the humidity ratio, in $lb_m/lb_m$, (b) the relative humidity, and (c) the enthalpy, in Btu/$lb_m$ (based on $h = 0$ at 0°F for dry air).

10-105. Atmospheric air at 1 bar has dry-bulb and wet-bulb temperature values of 25 and 20°C. By use of a psychrometric chart, estimate the values of (a) relative humidity, (b) dew point, (c) specific humidity, and (d) specific enthalpy.

10-106. Atmospheric air at 1 bar has a dry-bulb temperature of 28°C and a specific humidity of 0.0090 kg $H_2O$/kg dry air. Use a psychrometric chart to estimate (a) the wet-bulb temperature, (b) the relative humidity, (c) the dew point, and (d) the specific enthalpy.

10-107. Atmospheric air at 1 bar has a dry-bulb temperature of 25°C and a relative humidity of 50 percent. Use a psychrometric chart to estimate (a) the wet-bulb temperature, (b) the dew point, (c) the specific humidity, and (d) the specific enthalpy.

10-108. Atmospheric air at 1 bar has a dry-bulb temperature of 24°C and a specific humidity of 0.012 kg water/kg dry air. Use a psychrometric chart to estimate (a) the relative humidity, (b) the dew point, (c) the specific enthalpy, and (d) the specific volume.

10-109. Atmospheric air at 1 bar has a relative humidity of 60 percent and a specific humidity of 0.010 kg water/kg dry air. Use a psychrometric chart to estimate (a) the dry-bulb temperature, (b) the specific enthalpy, (c) the dew point, and (d) the specific volume.

10-110. Atmospheric air at 1 bar has a dry-bulb temperature of 26°C and a specific enthalpy of 65 kJ/kg dry air. Use a psychrometric chart to estimate (a) the wet-bulb temperature, (b) the relative humidity, (c) the specific humidity, and (d) the specific volume.

10-111E. Atmospheric air at 1 atm has dry-bulb and wet-bulb temperature values of 85 and 74°F. By use of a psychrometric chart, estimate the values of (a) relative humidity, (b) dew point, (c) specific humidity, (d) the specific volume, and (e) specific enthalpy.

10-112E. Atmospheric air at 1 atm has a dry-bulb temperature of 90°F and a specific humidity of 90 gr $H_2O$/$lb_m$ dry air. Use a psychrometric chart to estimate (a) the wet-bulb temperature, (b) the relative humidity, (c) the dew point, and (d) the specific enthalpy.

10-113E. Atmospheric air at 1 atm has a dry-bulb temperature of 73°F and a relative humidity of 60 percent. Use a psychrometric chart to estimate (a) the wet-bulb temperature, (b) the dew point, (c) the specific humidity, and (d) the specific enthalpy.

10-114E. Atmospheric air at 1 atm has dry-bulb and wet-bulb temperatures of 78 and 64°F. Use a psychrometric chart to estimate (a) the relative humidity, (b) the dew point, (c) the specific humidity, (d) the specific volume, and (e) the specific enthalpy.

10-115. Atmospheric air at 20°C, 105 kPa, and 80 percent relative humidity enters an insulated compressor with a volumetric flow rate of 0.2 m³/s. If the air leaves at 200 kPa and 100°C, determine (a) the relative humidity at the exit, (b) the power input required, in kilowatts, and (c) the rate of entropy production, in kW/K.

10-116. Atmospheric air with dry-bulb and wet-bulb temperatures of 30 and 20°C, respectively, is cooled to 16°C at a constant pressure of 100 kPa. The surface of the equipment is maintained at 15°C. If the mass flow rate through the cooling unit is 500 kg dry air/h, calculate (a) the dew point of the mixture, (b) the heat removed, in kJ/h, and (c) the entropy generation within the fluid, in kJ/min·K. Check calculated property data against the psychrometric chart.

## HEATING, COOLING, DEHUMIDIFICATION, AND HUMIDIFICATION

10-117. Air which originally exists at 36°C, 100 kPa, and 40 percent relative humidity is cooled at constant pressure to 24°C. Calculate (a) the relative humidity and (b) the dew-point temperature at the final state, and (c) the heat removed, in kJ/kg dry air. Check your data against the psychrometric chart.

10-118. Atmospheric air at 98 kPa and 26°C dry-bulb temperature, with a relative humidity of 70 percent, is cooled to 12°C. Calculate any required property data, using the steam table, and determine (a) the grams of water condensed per kilogram of dry air, and (b) the heat removed, in kJ/kg of dry air.

10-119. Atmospheric air with dry-bulb and wet-bulb temperatures of 28 and 20°C, respectively, is cooled to 18°C at a constant pressure of 995 mbars. The volume flow rate entering the cooling unit is 100 m³/h. Calculate any required property data, using the steam table, and determine (a) the mass flow rate of dry air, in kg/h, and (b) the heat removed, in kJ/h.

10-120. Atmospheric air at 28°C and 70 percent relative humidity flows over a cooling coil at a rate of 500 m³/min. The liquid which condenses is removed from the system at 15°C. Subsequent heating of the air results in a final temperature of 30°C and a relative humidity of 30 percent. Determine (a) the heat removed in the cooling section, in kJ/min, (b) the heat added in the heating section, in kJ/min, and (c) the amount of water vapor condensed, in kg/min. Calculate the required property data (and check against the psychrometric chart) if $P = 101$ kPa.

10-121. Atmospheric air at 101 kPa, 30°C, and 60 percent relative humidity flows over a set of cooling coils at an inlet flow rate of 1500 m³/min. The liquid which condenses is removed from the system at 17°C.

Subsequent heating of the air results in a final state of 25°C and 60 percent relative humidity. Determine (a) the heat removed in the cooling section, in kJ/min, (b) the heat added in the heating section, in kJ/min, and (c) the amount of vapor condensed, in kg/min.

10-122. Ambient air from outside a store is to be conditioned from a state of 29°C and 80 percent relative humidity to a final state of 24°C and 40 percent relative humidity. If the pressure remains constant at 1 bar, calculate (a) the amount of water removed, in kg/kg dry air, (b) the heat removed in the cooling process, and (c) the heat added in the final step of the process, in kJ/kg dry air.

10-123E. Air which originally exists at 100°F, 14.7 psia, and 40 percent relative humidity is cooled at constant pressure to 78°F. Calculate (a) the relative humidity, (b) the dew-point temperature at the final state, and (c) the heat removed, in Btu/lb_m dry air. Check your data against the psychrometric chart.

10-124E. Atmospheric air at 14.6 psia and 80°F dry-bulb temperature, with a relative humidity of 70 percent, is cooled to 60°F. Calculate any required property data, using the steam table, and determine (a) the pounds of water condensed per pound of dry air, and (b) the heat removed, in Btu/lb_m of dry air.

10-125E. Atmospheric air with dry-bulb and wet-bulb temperatures of 82 and 68°F, respectively, is cooled to 60°F at a constant pressure of 14.7 psia. The volume flow rate entering the cooling unit is 100 ft³/min. Calculate any required property data, using the steam table, and determine (a) the mass flow rate of dry air, in lb_m/min, and (b) the heat removed, in Btu/min.

10-126E. Atmospheric air at 86°F and 60 percent relative humidity flows over a cooling coil at a rate of 2000 ft³/min. The liquid which condenses is removed from the system at 60°F. Subsequent heating of the air results in a final temperature of 75°F and a relative humidity of 50 percent. Determine (a) the heat removed in the cooling section, in Btu/h, (b) the heat added in the heating section, in Btu/h, and (c) the amount of water vapor condensed, in lb_m/h. Calculate the required property data (and check against the psychrometric chart) if $P = 14.7$ psia.

10-127E. Atmospheric air at 14.7 psia, 84°F, and 70 percent relative humidity flows over a set of cooling coils at an inlet flow rate of 15,000 ft³/min. The liquid which condenses is removed from the system at 50°F. Subsequent heating of the air results in a final state of 75°F and 40 percent relative humidity. Determine (a) the heat removed in the cooling section, (b) the heat added in the heating section, in Btu/h.

10-128E. Ambient air from outside a store is to be conditioned from a state of 84°F and 80 percent relative humidity to a final state of 68°F and 40 percent relative humidity. If the pressure remains constant at 14.7 psia, calculate (a) the amount of water removed, in lb_m/lb_m dry air, (b) the heat removed in the cooling process, and (c) the heat added in the final step of the process, in Btu/lb$_m$ dry air.

10-129. Atmospheric air at 16°C, 100 kPa, and 40 percent relative humidity is injected with superheated steam at 100 kPa until the air reaches a state of 20°C and 70 percent relative humidity. For the adiabatic process, determine (a) the mass of steam required, in kg/kg dry air, and (b) the temperature of the steam used, in degree Celsius.

10-130. Liquid water at 20°C is sprayed into atmospheric air at 100 kPa with dry-bulb and wet-bulb temperatures of 50 and 30°C. The exit temperature for the adiabatic process is 38°C. Determine (a) the relative humidities at the inlet and exit and (b) the quantity of water sprayed in, in kg/kg dry air.

10-131. An air stream at 30°C and a specific humidity of 0.0150 kg/kg dry air enters an insulated spray chamber. Saturated water vapor at 110°C is sprayed into the air stream at a mass ratio of 0.0035 kg water/kg dry air. Determine the final state of the atmospheric air, in terms of (a) the dry-bulb temperature and (b) the relative humidity.

## EVAPORATIVE COOLING

10-132. Atmospheric air at 34°C and 20 percent relative humidity passes through an evaporative cooler until the air reaches 21°C. Calculate (a) the amount of liquid water added, in kg/kg dry air, (b) the final relative humidity, and (c) the minimum possible final temperature. Calculate the required property data, and check these values against the psychrometric chart.

10-133. Atmospheric air at 28°C and 10 percent relative humidity passes through an evaporative cooler until the final relative humidity is 60 percent. Calculate (a) the amount of water added to the air, in kg/kg dry air, (b) the final dry-bulb temperature, and (c) the minimum final temperature, in degrees Celsius. Use the steam table to evaluate required property data.

10-134. Atmospheric air at 36°C and 10 percent relative humidity passes through an evaporative cooler. If the dew-point temperature is 14°C, calculate (a) the amount of liquid water added in the cooler, in g/kg dry air, (b) the final temperature, and (c) the minimum possible final temperature, in degrees Celsius.

10-135. An atmospheric air stream at 10°C and 40 percent relative humidity is first heated to 33°C and then passed through an evaporative cooler until the dry-bulb temperature reaches 20°C. The volume flow rate is 40 m$^3$/min at the initial state. Calculate the required property data, using steam-table data, and determine (a) the rate of heat addition, in kJ/min, (b) the rate of water added, in kg/min, and (c) the final relative humidity. Check calculations by using the psychrometric chart.

10-136. Atmospheric air at 12°C and 30 percent relative humidity is first heated to 35°C and then passed through an evaporative cooler until the temperature reaches 22°C. The volumetric flow at the initial state is 50 m³/min. Calculate (a) the rate of heat addition, in kJ/min, (b) the rate of water added in the second section, in kg/min, and (a) the final relative humidity. Calculate the required property data using steam-table data where necessary.

10-137. An air stream at 34°C with a wet-bulb temperature of 15°C passes through an evaporative cooler until the temperature is 17°C. It is then heated to 30°C. If the volume flow rate at the initial state is 35 m³/min, determine (a) the rate of water added in kg/min in the first section, (b) the rate of heat addition in kJ/min in the second section, and (c) the final relative humidity. Calculate the required property data using the steam table where necessary.

10-138E. Atmospheric air at 94°F and 20 percent relative humidity passes through an evaporative cooler until the air reaches 68°F. Calculate (a) the final relative humidity, (b) the amount of liquid water added, in $lb_m/lb_m$ dry air, and (c) the minimum possible final temperature. Calculate the required property data, and check these values against the psychrometric chart.

10-139E. Atmospheric air at 92°F and 10 percent relative humidity passes through an evaporative cooler until the final relative humidity is 80 percent. Calculate (a) the final dry-bulb temperature, (b) the amount of water added to the air, in $lb_m/lb_m$ dry air, and (c) the minimum final temperature, in degrees Fahrenheit. Use the steam table to evaluate required property data.

10-140E. Atmospheric air at 90°F and 10 percent relative humidity passes through an evaporative cooler. If the final dew-point temperature is 54°F, calculate (a) the amount of liquid water added in the cooler, in $lb_m/lb_m$ dry air, (b) the final temperature, and (c) the minimum possible final temperature, in degrees Fahrenheit.

10-141E. An atmospheric air stream at 50°F and 40 percent relative humidity is first heated to 90°F, and then passed through an evaporative cooler until the dry-bulb temperature reaches 70°F. The volume flow rate is 1000 ft³/min at the initial state. Calculate the required property data, using steam-table data, and determine (a) the rate of heat addition, in Btu/min, (b) the rate of water added, in $lb_m/min$, and (c) the final relative humidity. Check calculations by using the psychrometric chart.

10-142E. Atmospheric air at 56°F and 30 percent relative humidity is first heated to 94°F and then passed through an evaporative cooler until the temperature reaches 68°F. The volumetric flow at the initial state is 1500 ft³/min. Calculate (a) the rate of heat addition, in Btu/min, (b) the rate of water added in the second section, in $lb_m/min$, and (*c*) the final relative humidity. Calculate the required property data using steam-table data where necessary.

## ADIABATIC MIXING

10-143.  Two atmospheric air streams (1 and 2) undergo an adiabatic mixing process in steady flow at 1 bar to form a new mixture at state 3. Stream 1 has a dry-bulb temperature of 20°C and a relative humidity of 90 percent, while stream 2 has values of 33°C and 20 percent. If the final mixture is to have a specific humidity of 0.0090 kg water/kg dry air, calculate (*a*) the mass flow rate of stream 2, in kg dry air/min, (*b*) the final mixture dry-bulb temperature, in degrees Celsius, and (*c*) the volume rate of stream 3, in m³/min, if the mass rate for stream 1 is 100 kg dry air/min. Check calculated property data against the psychrometric chart.

10-144.  Two streams of atmospheric air at 1 bar are mixed in an adiabatic, steady-flow process. One stream, at 36°C and 40 percent relative humidity, enters at a rate of 5 kg dry air/min, while the second stream, at 5°C and 100 percent relative humidity, enters at 15 kg dry air/min. Determine (*a*) the exit humidity ratio, (*b*) the exit stream enthalpy, and (*c*) the exit dry-bulb temperature, in degrees Celsius. Compute the required property data, using steam-table data where necessary.

10-145.  In an air-conditioning process 50 m³/min of outside air at 29°C and 80 percent relative humidity is mixed adiabatically with 100 m³/min of inside air initially at 20°C and 30 percent relative humidity. Determine for the resultant mixture (*a*) the humidity ratio, (*b*) the dry-bulb temperature, and (*c*) the relative humidity. Compute the initial property values required, using steam-table data where necessary. The system pressure is 0.10 MPa.

10-146.  An atmospheric air mixture at 29°C dry bulb and 21°C wet bulb mixes adiabatically with another air stream originally at 14°C dry bulb and 12°C wet bulb. The system pressure is 100 kPa. The ratio of the volume flow rate of the cold stream to that of the hot stream is 2.4. Determine for the resultant mixture (*a*) the humidity ratio, (*b*) the dry-bulb temperature, and (*c*) the relative humidity. Compute the required inlet property data, using steam-table data where necessary.

10-147.  Atmospheric air with dry-bulb and wet-bulb temperatures of 16 and 12°C, respectively, enters a steady-flow chamber and mixes with another air stream entering with dry-bulb and wet-bulb temperatures of 33 and 29°C, respectively. The volumetric flow rate of the lower-temperature stream is three times that of the other stream. For the exiting stream determine (*a*) the humidity ratio, (*b*) the dry-bulb temperature, and (*c*) the relative humidity. Compute property values where necessary from steam-table data, for $P = 1.0$ bar.

10-148. Air at 40°C and 30 percent relative humidity is mixed adiabatically at 1 bar with a second stream. A single mixed stream leaves at 30°C and 40 percent relative humidity with a mass flow rate of 4 kg/s. If the mass flow rates of the entering streams are equal, calculate for the second entering stream (*a*) its temperature, in degrees Celsius, and (*b*) its relative humidity.

10-149E. Two streams of atmospheric air at 14.7 psia are mixed in an adiabatic, steady-flow process. One stream, at 100°F and 40 percent relative humidity, enters at a rate of 5 lb$_m$ dry air/min, while the second stream, at 40°F and 100 percent relative humidity, enters at 15 lb$_m$ dry air/min. Determine (*a*) the exit humidity ratio, (*b*) the exit stream enthalpy, and (*c*) the exit dry-bulb temperature, in degrees Fahrenheit. Compute the required property data, using steam-table data where necessary.

10-150E. In an air-conditioning process 500 ft$^3$/min of outside air at 86°F and 80 percent relative humidity is mixed adiabatically with 640 ft$^3$/min of inside air initially at 70°F and 30 percent relative humidity. Determine for the resultant mixture (*a*) the dry-bulb temperature, (*b*) the humidity ratio, and (*c*) the relative humidity. Compute the initial property values required, using steam-table data where necessary. The system pressure is 14.7 psia.

10-151E. An atmospheric air mixture at 90°F dry bulb and 80°F wet bulb mixes adiabatically with another air stream originally at 60°F dry bulb and 50°F wet bulb. The system pressure is 14.7 psia. The ratio of the volume flow rate of the cold stream to that of the hot stream is 2.4. Determine for the resultant mixture (*a*) the dry-bulb temperature, (*b*) the humidity ratio, and (*c*) the relative humidity. Compute the required inlet property data, using steam-table data where necessary.

## WET COOLING TOWER

10-152. Water enters a cooling tower at 36 m$^3$/min and is cooled from 28 to 20°C. The entering atmospheric air at 1 bar has dry-bulb and wet-bulb temperatures of 21 and 15°C, respectively. The air is saturated when it leaves the tower. Determine (*a*) the volume rate of inlet air required, in m$^3$/min, and (*b*) the rate of water evaporated, in kg/h. The exit-air temperature is 28°C.

10-153. Water enters a cooling tower at 33°C and leaves at 22°C. The tower receives 10,000 m$^3$/min of atmospheric air at 1 bar, 20°C, and 40 percent relative humidity. The air-exit condition is 32°C and 90 percent relative humidity. Determine (*a*) the mass flow rate of dry air passing through the tower, in kg/min, (b) the mass flow rate of entering water, in kg/min, and (c) the rate of water evaporated, in kg/min.

10-154. It is desired to cool 1000 kg/min of water from 34 to 25°C. A cooling tower receives 700 m³/min of atmospheric air at 1 bar with dry-bulb and wet-bulb temperatures of 29 and 21°C, respectively. If the evaporation rate from the liquid water stream is 850 kg/h, determine the (a) temperature and (b) relative humidity of the exit-air stream.

10-155. It is desired to cool liquid water from 40 to 26°C. A cooling tower receives 800 m³/min of atmospheric air at 1 bar with dry-bulb and wet-bulb temperatures of 29 and 21°C, respectively. The outlet water rate is 1250 kg/min, and the exit-air relative humidity is 100 percent. Determine (a) the outlet-air temperature, in degrees Celsius, and (b) the evaporation rate of the water, in kg/min.

10-156E. Water enters a cooling tower at 9000 gal/min and is cooled from 84 to 68°F. The entering atmospheric air at 14.7 psia has dry-bulb and wet-bulb temperatures of 70 and 60°F, respectively. The air is saturated when it leaves the tower. Determine (a) the volume rate of inlet air required, in ft³/min, and (b) the rate of water evaporated, in lb$_m$/h. The exit-air temperature is 82°F.

10-157E. Water enters a cooling tower at 92°F and leaves at 72°F. The tower receives 300,000 ft³/min of atmospheric air at 14.7 psia, 68°F, and 40 percent relative humidity. The air exit condition is 90°F and 90 percent relative humidity. Determine (a) the mass flow rate of dry air passing through the tower, in lb$_m$/min, (b) the mass flow rate of entering water, in lb$_m$/min, and (c) the rate of water evaporated, in lb$_m$/min.

10-158E. It is desired to cool 1000 lb$_m$/min of water from 100 to 77°F. A cooling tower receives 24,000 ft³/min of atmospheric air at 14.7 psia with dry-bulb and wet-bulb temperatures of 85 and 70°F, respectively. If the evaporation rate from the liquid water stream is 1200 lb$_m$/h, determine (a) the temperature and (b) the relative humidity of the exit-air stream.

10-159E. It is desired to cool liquid water from 100 to 80°F. A cooling tower receives 24,000 ft³/min of atmospheric air at 14.7 psia with dry-bulb and wet-bulb temperatures of 84 and 70°F, respectively. The outlet water rate is 2800 lb$_m$/min, and the exit-air relative humidity is 95 percent. Determine (a) the outlet-air temperature, in degrees Fahrenheit, and (b) the evaporation rate of the water, in lb$_m$/min.

## PARAMETRIC STUDIES AND DESIGN

10-160. A rigid tank contains 0.2 kg of nitrogen gas and 0.1 kg of carbon dioxide at 200 kPa and 37°C. During a process, the heat transfer out is given by $Q = 0.445(T_2 - 37)$, where $Q$ is in kilojoules and $T_2$ is in degrees Celsius. During the process a current of 4.5 A passes through a resistor within the tank for a period of 6.0 min. On the basis of a computer analysis, determine the constant voltage applied across the resistor as a function of the final temperature $T_2$. The range of values for $T_2$ is from 50 to 250°C. Use tabular $u$ data in your solution.

10-161. An equimolar mixture of CO and argon flows steadily and adiabatically through a diffuser. The initial state is 300 K and 200 m/s. By means of a computer program, investigate the effect of the exit velocity on the exit-gas temperature, in kelvins. The range of the exit velocity is from 150 to 30 m/s.

10-162. An ideal-gas mixture has the following volumetric analysis: CO, 30 percent; $CO_2$, 50 percent; $O_2$, 20 percent. It enters a steady-state compressor at 37°C and 60 m/s and leaves the device at 100 m/s. At the entrance the volume flow rate is 6.0 m³/min and the pressure is 120 kPa. By means of a computer program investigate the effect of the exit temperature on the power input, in kilowatts. The range of exit temperatures is from 150 to 350°C.

10-163E. An equimolar mixture of He and CO expands through an adiabatic nozzle from an initial state of 260°F and 100 ft/s. By means of a computer analysis investigate the effect of the exit temperature on the exit velocity. The range of temperatures is from 170 to 220°F.

10-164. A gas mixture consisting of $N_2$, $CO_2$, and $H_2O$ in a molar ratio of 4:1:1 enters an adiabatic, frictionless turbine. By means of a computer program investigate the effect of the pressure ratio on the power output, in kilowatts. The mass flow rate is 2 kg/s of mixture, and the range of pressure ratios to consider is from 2:1 to 7:1.

10-165. A mixture of $H_2O$, CO, and $N_2$ in a molar ratio of 1:2:5 is compressed adiabatically and frictionlessly in a piston-cylinder device from 1 bar and 300 K to a final pressure $P_2$. By means of a computer program investigate the effect of the final pressure of the final temperature, in kelvins, and the work input, in kJ/kg of mixture. The range of final pressures to consider is from 2 to 7 bars.

10-166. An atmospheric air stream at 10°C and 40 percent relative humidity is first heated to $T_2$ and then passed through an evaporative cooler until the dry-bulb temperature reaches 20°C. The volume flow rate is 40 m³/min at the initial state. By means of a computer program investigate the effect of temperature $T_2$ on the rate of heat addition, in kJ/min, and the final relative humidity. The range of $T_2$ is from 24 to 40°C.

10-167E. A building in the desert is to be maintained at 74°F and 40 percent relative humidity at a pressure of 14.7 psia. Rather than a conventional vapor-compression air conditioner, it is proposed to cool the building with the aid of an evaporative cooler and a desiccant system. The overall process is shown in Fig. P10-167. Environmental law

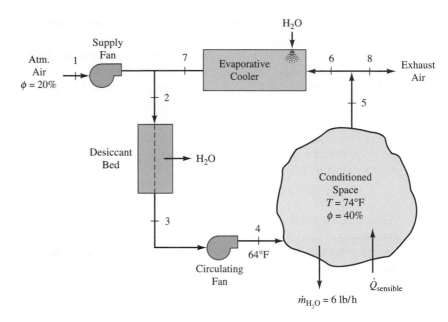

**Figure P10-167**

requires that outside ventilation air must be used to control indoor air pollution. The sensible heat gain across the walls of the building, in Btu/h, can be determined from the empirical relation:

$$\text{Sensible heat gain} = -111,000 + 1500 T_F$$

where $T_F$ is the outside temperature in degrees Farhenheit. Other information includes:

1. Water vapor diffuses through the walls at 6.0 $\text{lb}_m/\text{h}$.
2. The air returning to the living space at state 4 is required to be 64°F.
3. The air leaving the evaporative cooler is always saturated.
4. The effect of the fan on any energy analysis is negligible.
5. The water removal by the desiccant is isothermal.
6. The relative humidity of the atmospheric air is always 20 percent.
7. The relationship between the dry-bulb temperature in degrees Fahrenheit and the enthalpy in $\text{Btu/lb}_m$ dry air at 100 percent relative humidity on the psychrometric chart is

$$T_F = 2.1365 + 3.0525h - 0.038213h^2 + 0.21044 \times 10^{-3}h^3$$

8. The vapor pressure of water in psia as a function of temperature in degrees Rankine is represented at room conditions by

$$\ln P_{\text{sat}} = 14.9716 - \frac{7490.8}{T_R - 61.05}$$

For the design process, determine for outside air temperatures between 80 and 105°F in 5°F intervals:

(a) The mass flow rate of dry air through the conditioned space, in lb$_m$/h.

(b) The mass flow rate of ventilation air required, in lb$_m$/h of dry air.

(c) The quantity of water added in the evaporative cooler, in lb$_m$/h.

(d) The quantity of water removed by the desiccant, in lb$_m$/h.

(e) The volume flow rate of ventilation air, in ft$^3$/min.

Your comments should include any positive or negative features of the design.

10-168E. A shower head is to be designed for 2.5 gpm to meet new energy standards. The shower head should also address the common problem of a drop in temperature of the water from the shower head to the feet of a person taking a shower, and annoying condensation on the bathroom mirror. Hence the design problem is twofold: (1) meet the energy standard and minimize the temperature drop, and (2) determine the size of the exhaust fan in the shower area, expressed as CFM of airflow versus volume of bathroom in cubic feet. Physical conditions include:

1. Bathroom temperature during shower is 80°F.
2. Temperature of makeup air is 70°F.
3. Relative humidity of makeup air is 40 percent.
4. Mirror surface temperature is 65°F.
5. Water temperature at shower head (maximum) is 95°F.
6. Water pressure at shower head (maximum) is 60 psig.

The following empirical equations are provided:

(A) The relationship between the mass flow rate of water $\dot{m}$ through one hole in the shower head is $\dot{m} = KA(2\rho\,\Delta P)^{1/2}$, where K is 0.3 for this shower head, A is the area of the hole, $\rho$ is the density of water, and $\Delta P$ is the pressure drop across the shower head.

(B) The evaporation rate of the water into the room $\dot{m}_v$ for each small hole is $\dot{m}_v = CDL(T_s - T_r)^{0.8}$, where D is the diameter of the hole in inches, L is the length of water jet in inches, $T_s$ is the temperature at the shower head in °F, $T_r$ is the bathroom temperature in °F, and $C$ is an empirical constant equal to 0.15 lb$_m$/(h·in$^2$·°F$^{0.8}$).

10-169E. A cooling and dehumidifying coil is to be sized to meet the requirements for conditioning the air within a small building, as shown in Fig. P10-169. A flow rate of 2000 lb$_m$/h of fresh outside air at 90°F and 60 percent relative humidity is mixed with return air from the conditioned space. The design temperature for the air supplied to the building is chosen to be 55°F. A detailed heat- and mass-transfer

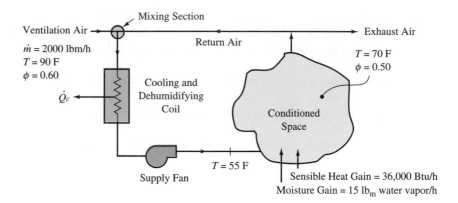

**Figure P10-169**

analysis estimates a sensible-heat gain of 36,000 Btu/h and a total moisture gain of 15 $lb_m$/h of water evaporated at room temperature. Estimate the cooling coil airflow rate and heat-transfer rate necessary to maintain design conditions of 70°F and 50 percent relative humidity for the conditioned space within the building.

Electricity is used to provide chilled water to the cooling coil. One approach for reducing the operating costs in dry climates involves the use of an evaporative cooler for precooling the outside ventilation air prior to mixing with the return air. Evaluate the change in cooling-coil heat-transfer rate associated with use of an evaporative cooler, assuming a source of 70°F water is available for the evaporative cooler. Neglect the energy supplied to the fan.

10-170. A natural gas-fired furnace is to produce a maximum of 100,000 kJ/h for a residential heating application. The furnace consists of a burner, where a mixture of natural gas and air burn, and a heat exchanger where the combustion products transfer energy to air that circulates through the house. A mixture of $CO_2$, $H_2O$, $O_2$, and $N_2$, in a $1 : (2 + 30\omega) : 2 : 15$ volumetric ratio exits the furnace at 1530 K and 1 atm, where $\omega$ is the specific humidity. These hot gases immediately enter the heat exchanger. House air flowing through the heat exchanger must exit at 35°C and 1 atm.

Provide pertinent details for the design of this furnace. In particular, you should indicate (1) the required temperature differential between the exiting air and exiting combustion products stream, (2) how you would optimize the efficiency of energy transfer from the combustion product stream to the house air stream, (3) what additional equipment or subsystems you will have to add to accomplish item 2, (4) how atmospheric conditions, through the specific humidity $\omega$, influence your design and its performance, and (5) what fraction of the available combustion product energy is transferred to the house air, on a percentage basis.

10-171. A cooling tower is to be connected to a power plant that employs a condenser to transform the 10 kPa, 73.5 percent quality steam exiting the turbine to a saturated liquid. The steam is condensed by using a cold-water stream from the cooling tower as the working fluid. The power plant is designed to generate 10 MW of electic power, and will reject energy at the rate of 24.4 MW in the condenser. A circulating pump is used to return the cooled water from the cooling tower to the condenser, and a feed pump is used to supply makeup water to the cooling-tower loop. Your task is to suggest sizes for the two pumps and the capacity of the roof-mounted blower on the tower, in kilowatts.

# 11

# *PvT* Behavior of Real Gases and Real-Gas Mixtures

Engine-driven emergency stand-by generator.

In Chap. 3 some quantitative data for the *PvT* behavior of real gases were presented in the form of tables which appear in the Appendix. However, in many cases it would be convenient to have analytical expressions for the relationships among these three properties. There is no set method for doing this; consequently, in this chapter we look at a few of the many correlations available for the equations of state of real gases. The accuracy of these *PvT* relations varies with the type of gas and the range of the properties under consideration. The best methods have not necessarily been chosen for analysis, since the purpose here is simply to provide some insight into typical correlations. When the occasion arises, a search of the literature will reveal numerous equations of state which might profitably be used for particular applications.

A general method of correlating *PvT* data, which has a reasonable degree of accuracy combined with an inherent simplicity, is based on the van der Waals *principle of corresponding states*. This principle was introduced in Sec. 4-5. The reader should review that section now.

## 11-1 THE VIRIAL EQUATION OF STATE

For simple compressible systems we have postulated that two independent properties are required to fix the equilibrium state. If $Pv$ is chosen as the dependent thermodynamic property, it may be considered to be a function of two independent properties such as pressure and temperature or specific volume and temperature. The data of Fig. 11-1 illustrate a functional relationship of the type $Pv = f(P, T)$. This relationship may be expressed to any desired accuracy by the infinite-series expansion of the type

$$Pv = a + bP + cP^2 + dP^3 + \cdots \qquad \text{[11-1a]}$$

Since all $PvT$ equations of state should reduce to $Pv = RT$ as the pressure approaches zero, the constant $a$ must equal $RT$. Hence Eq. [11-1a] may also be written in terms of the compressibility factor $Z$ as

$$Z = \frac{Pv}{RT} = 1 + BP + CP^2 + DP^3 + \cdots \qquad \text{[11-1b]}$$

If the independent variables were selected as $v$ and $T$, the following form of an infinite series would be applicable:

$$Pv = a\left(1 + \frac{b'}{v} + \frac{c'}{v^2} + \frac{d'}{v^3} + \cdots\right) \qquad \text{[11-2a]}$$

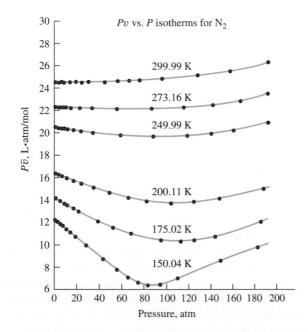

**Figure 11-1** Variation of $P\bar{v}$ of nitrogen with pressure at constant temperature. (From M. W. Zemansky, "Heat and Thermodynamics," 4th ed., McGraw-Hill, New York, 1957.)

where, again, $a = RT$. Thus, in terms of $Z$,

$$Z = \frac{Pv}{RT} = 1 + \frac{b'}{v} + \frac{c'}{v^2} + \frac{d'}{v^3} + \cdots \qquad \textbf{[11-2b]}$$

Equations of this type, which relate $P$, $v$, and $T$, are known as **virial equations of state.** The coefficients $a$, $b$, $c$, etc., are the first, second, third, etc., virial coefficients for the respective equations. These coefficients are solely a function of the temperature of the gas, and have a physical significance on a molecular scale. As a result, they can be determined, at least in theory, from the field of statistical mechanics as well as from direct macroscopic measurements illustrated in Fig. 11-1. The virial coefficients are also, of course, a function of the substance of interest.

As an example, a virial equation for nitrogen gas at 0°C and for pressures up to 200 atm has the following format:

$$P\bar{v} = 22{,}414.6 - 10.281P + 0.065189P^2 + 5.1955 \times 10^{-7}P^4 \qquad \textbf{[11-3]}$$
$$- 1.3156 \times 10^{-11}P^6 + 1.009 \times 10^{-16}P^8$$

The specific volume in this equation is expressed in $cm^3/mol$, and the pressure is in atmospheres.

When the pressure is relatively low, it is seen from Eq. [11-3] that the latter terms for $Pv$ contribute little compared with the first several terms. Thus at low pressures a virial equation of the type

$$P\bar{v} = R_u T + bP \qquad \textbf{[11-4]}$$

is often of sufficient accuracy. An empirical equation for the second virial coefficient $b$ of nitrogen has the form

$$b = 39.5 - \frac{1.00 \times 10^4}{T} - \frac{1.084 \times 10^6}{T^2} \qquad \textbf{[11-5]}$$

where $b$ is in $cm^3/gmol$ and $T$ is in kelvins.

## 11-2    TWO-CONSTANT EQUATIONS OF STATE

When very little is known about the $PvT$ behavior of a substance, equations that are cubic in the specific volume are quite useful. Such equations are a reasonable compromise between simplicity and generality. Two of these will be discussed in this section. The van der Waals equation was developed from molecular considerations, and the method of determining the two constants in the equation is of general validity for all two-constant equations of state. The more recent Redlich-Kwong equation of state is empirical in nature and has proved to be one of the more useful and widely accepted cubic equations.

## 11-2-1  THE VAN DER WAALS EQUATION OF STATE

In 1873, van der Waals proposed an equation of state which was an attempt to correct the ideal-gas equation so that it would be applicable to real gases. On the basis of simple kinetic theory, particles are assumed to be point masses, and there are no intermolecular forces between particles. However, as the pressure increases on a gaseous system, the volume occupied by the particles may become a significant part of the total volume. In addition, the intermolecular attractive forces become important under this condition. To account for the volume occupied by the particles, van der Waals proposed that the specific volume in the ideal-gas equation be replaced by the term $v - b$. Also the ideal pressure was to be replaced by the term $P + a/v^2$. The constant $b$ is the covolume of the particles, and the constant $a$ is a measure of the attractive forces. Thus the van der Waals equation is

$$\left(P + \frac{a}{v^2}\right)(v - b) = RT \qquad \textbf{[11-6]}$$

Both $a$ and $b$ have units which must be consistent with those employed for $P$, $v$, and $T$. Note that as the pressure approaches zero and the specific volume approaches infinity, the correction terms are negligible and the equation reduces to $Pv = RT$. Empirically, it is observed that as the pressure increases, the $a/v^2$ term usually becomes important sooner than the $b$ correction factor.

The van der Waals equation is moderately successful. However, for increased accuracy different sets of the constants $a$ and $b$ should be determined empirically for particular regions of pressure and temperature. To avoid this situation, a method of evaluating a single set of constants is based on the experimental fact that the critical isotherm on a $Pv$ diagram for any gas passes through a point of inflection at the critical state. This phenomenon is illustrated in Fig. 11-2, which shows some typical isotherms based on the van der Waals equation of state. The curve marked $T_c$ is the critical isotherm, and it is fairly representative of actual experimental data. This point of inflection at the critical state is expressed mathematically by the relations

$$\left(\frac{\partial P}{\partial v}\right)_{T_c} = 0 \qquad \text{and} \qquad \left(\frac{\partial^2 P}{\partial v^2}\right)_{T_c} = 0$$

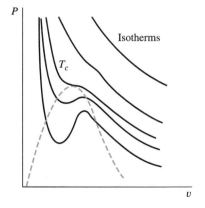

**Figure 11-2**
Isotherms for a van der Waals gas on a *Pv* diagram.

These equations permit the evaluation of the constants in any two-constant equation of state. For example, the van der Waals constants $a$ and $b$ may now be determined by this method. On the basis of Eq. [11-6],

$$\left(\frac{\partial P}{\partial v}\right)_{T_c} = \frac{-RT_c}{(v_c - b)^2} + \frac{2a}{v_c^3} = 0$$

and

$$\left(\frac{\partial^2 P}{\partial v^2}\right)_{T_c} = \frac{2RT_c}{(v_c - b)^3} - \frac{6a}{v_c^4} = 0$$

The subscript $c$ indicates that the property must be evaluated at the critical state. By combining these two expressions with the van der Waals equation of state, the following equations for $a$, $b$, and $Z_c$ are developed.

$$a = \frac{27}{64} \frac{RT_c}{P_c} \qquad b = \frac{RT_c}{8P_c} \qquad Z_c = \frac{P_c v_c}{RT_c} = \frac{3}{8}$$

The above equations for $a$ and $b$ allow the determination of these equations from $P_c$ and $T_c$, which are obtained experimentally. However, since the values of $a$ and $b$ actually need to be varied in order to relate $PvT$ data accurately, the use of critical data to evaluate the two constants will lead to an equation of state of limited accuracy. Another indication of this is the fact that $Z_c$ is 0.375 according to the van der Waals equation. Experimentally, this term is found to lie between 0.20 and 0.30 for the majority of gases. Thus the van der Waals equation is sometimes in serious error. Nevertheless, it is of historical interest as one of the first attempts to correct the ideal-gas equation so that real-gas behavior might be predicted. Tables A-22 and A-22E list some typical values for the van der Waals constants based on critical data.

### 11-2-2   THE REDLICH-KWONG EQUATION OF STATE

Although other two-constant equations of state could be presented, the Redlich-Kwong equation is of considerable interest. It has considerable accuracy over a wide range of $PvT$ conditions, especially when $T$ is greater than the critical value. On the basis of empirical considerations, Redlich and Kwong proposed in 1949 the following relationship:

$$P = \frac{RT}{v - b} - \frac{a}{T^{1/2}v(v + b)} \qquad \textbf{[11-7]}$$

Constants $a$ and $b$ again can be evaluated from critical data. It is found that $a = 0.4275R^2T_c^{2.5}/P_c$ and $b = 0.0867RT_c/P_c$. One of the primary considerations in the development of this equation was that, at high pressures, the specific volume of all gases tends to approach the limiting value of $0.26v_c$, and this value is essentially independent of the temperature. Consequently, $b$ also equals $0.26v_c$. As a result, the equation gives quite good results at high pressures. The equation also appears to be fairly accurate for temperatures above the critical value. When $T$ is less than $T_c$, the equation is found to deviate from experimental data as the temperature is lowered. However, in several instances it has been shown that the Redlich-Kwong equation is as accurate as the eight-constant Benedict-Webb-Rubin equation of state. Hence the Redlich-Kwong equation has the advantage of simplicity combined with a fair degree of accuracy, especially with respect to other two-constant equations of state. Values of $a$ and $b$ for a few common gases are presented in Tables A-22 and A-22E.

Both the van der Waals and the Redlich-Kwong equations must be solved by iteration when $v$ is the unknown, unless hand calculators or personal computers are programmed to solve cubic equations.

---

Estimate the pressure which would be exerted by 3.7 kg of CO in a 0.030-m³ container at 215 K, employing (a) the ideal-gas equation, (b) the van der Waals equation, (c) the Redlich-Kwong equation of state, and (d) the compressibility factor Z.

**Solution:**

**Given:**   Carbon monoxide in a container as shown in Fig. 11-3.

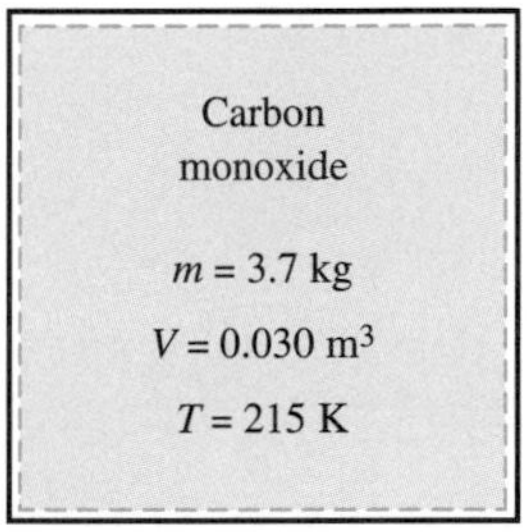

**Find:**   $P$, in bars, by (a) ideal-gas equation, (b) van der Waals equation, (c) Redlich-Kwong equation, and (d) the Z factor.

**Model:**   Gas in equilibrium.

**Analysis:**   The specific volume of the gas is

$$v = \frac{V}{m} = \frac{0.030}{3.7} = 0.00811 \text{ m}^3/\text{kg} = 0.227 \text{ m}^3/\text{kmol}$$

(a) The pressure is computed directly from the ideal-gas equation as

$$P = \frac{RT}{v} = \frac{0.08314(215)}{28(0.00811)} = 78.7 \text{ bars}$$

(b) The constants for the van der Waals equation are found in Table A-22. For CO we find that $a = 1.463$ bar·m⁶/kmol² and $b = 0.0394$ m³/kmol. Substitution of these values into Eq. [11-6] leads to

$$\left(P + \frac{a}{v^2}\right)(v - b) = RT$$

$$\left[P + \frac{1.463}{(0.227)^2}\right][0.227 - 0.0394] = 0.08314(215)$$

The solution to this equation gives the pressure to be 66.9 bars.

(c) The specific volume is 0.227 m³/kmol, and the temperature is 215 K. The constants $a$ and $b$ for the Redlich-Kwong equation are found in Table A-23 to be 17.26 bar·m⁶·K^{1/2}/kmol² and 0.02743 m³/kmol for CO. Making the proper substitutions, we find that

$$P = \frac{RT}{v - b} - \frac{a}{T^{1/2}v(v + b)}$$

$$= \frac{0.08314(215)}{0.227 - 0.0274} - \frac{17.26}{(215)^{1/2}(0.227)(0.227 + 0.0274)}$$

$$= 69.2 \text{ bars}$$

**Example 11-1**

**Figure 11-3**
Schematic and data for Example 11-1.

Carbon monoxide

$m = 3.7$ kg

$V = 0.030$ m³

$T = 215$ K

(d) Using the method introduced in Sec. 4-5, the following reduced properties are evaluated:

$$T_r = T/T_c = 215/133 = 1.62$$

$$v_r' = \frac{vP_c}{RT_c} = \frac{(0.00811)(35.0)(28.01)}{0.08314(133)} = 0.718$$

On the basis of Fig. A-28 we find that $P_r$ is roughly 1.93. Therefore

$$P = P_r P_c = 1.93(35.0) = 67.6 \text{ bars}$$

**Comment:**    The NIST table for carbon monoxide lists a value of 69.13 bars.

## 11-3    OTHER EQUATIONS OF STATE

One of the inherent limitations of two-constant equations of state is the lack of accuracy over a wide range of conditions. To achieve higher accuracy requires equations with a considerably larger number of terms, or constants. Over a hundred different equations of state have been proposed in the past fifty years. They vary in their complexity, the type of substances for which the equations were originally developed, and the appropriate range of pressure or density. Each year, new equations of state appear in the literature. The two equations discussed below are of reasonable accuracy, and the number of empirical constants required is not large. Some equations of state may require up to 50 constants to be fitted from experimental data.

### 11-3-1    BEATTIE-BRIDGEMAN EQUATION

One of the earliest equations of state of reasonable simplicity and accuracy is the Beattie-Bridgeman equation of 1928. The five-constant equation has the format

$$P = \frac{RT}{v^2}\left(1 - \frac{c}{vT^3}\right)(v + B) - \frac{A}{v^2} \qquad \textbf{[11-8]}$$

where    $A = A_0\left(1 - \frac{a}{v}\right)$    and    $B = B_0\left(1 - \frac{b}{v}\right)$

The equation is generally used when the density is not above $0.8\rho_c$. An expansion of the equation reveals it has a virial format truncated after the $1/v^3$ term. Table 11-1 lists the values of the five constants, in SI units, for two common gases.

### 11-3-2    BENEDICT-WEBB-RUBIN EQUATION

An example of a more complex equation which has been successful in predicting $PvT$ behavior, especially of hydrocarbons, is the Benedict-Webb-

**Table 11-1**  Constants for Beattie-Bridgeman equation of state, with $v$ in $m^3/kmol$, $T$ in kelvins, and $P$ in kPa

| Gas | $A_0$ | $a$ | $B_0$ | $b$ | $c \times 10^{-4}$ |
|---|---|---|---|---|---|
| Carbon dioxide | 507.2836 | 0.07132 | 0.10476 | 0.07235 | 66.00 |
| Nitrogen | 136.2315 | 0.02617 | 0.05046 | −0.00691 | 4.20 |

Rubin equation. This equation, developed in 1940, is

$$P = \frac{RT}{v} + \left(B_0 RT - A_0 - \frac{C_0}{T^2}\right)\frac{1}{v^2} + \frac{bRT - a}{v^3} + \frac{a\alpha}{v^6} \quad \text{[11-9]}$$

$$+ \frac{c}{v^3 T^2}\left(1 + \frac{\gamma}{v^2}\right)\exp\left(-\frac{\gamma}{v^2}\right)$$

The equation is an extension of the Beattie-Bridgeman equation and has eight adjustable constants for a given substance. Values of these constants for five common substances are found in Tables A-22 and A-22E. The units on these constants must be consistent with those used for $P$, $v$, $R$, and $T$. The equation generally is applied to states where the density is less than 2.5 times the critical density.

---

Estimate the pressure of CO at 215 K and a specific volume of 0.227 L/mol, on the basis of the Benedict-Webb-Rubin (BWR) equation of state.

**Example 11-2**

**Solution:**

**Given:**  Carbon monoxide at conditions shown in Fig. 11-4.

**Find:**  $P$, in bars, by BWR equation of state.

**Model:**  Gas in equilibrium.

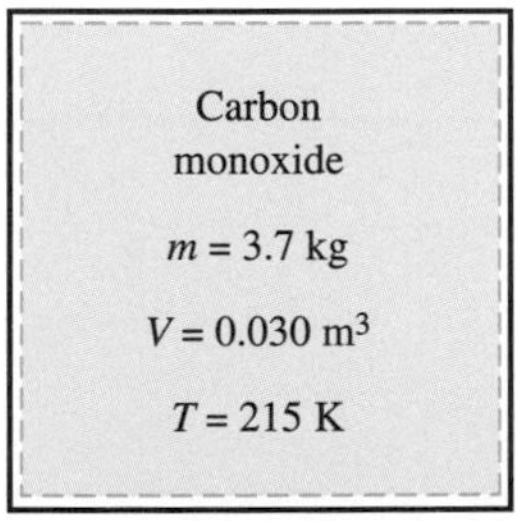

**Analysis:**  The values of the constants for the BWR equation for CO are listed in Table A-22. The specific volume is 0.227 L/mol and $R_u = 0.08314$ bar·L/mol·K. Making the proper substitutions, we find that

$$P = \frac{0.08314(215)}{0.227} + \left[0.05454(0.08314)(215) - 1.3587 - \frac{8.673 \times 10^3}{215^2}\right]\frac{1}{(0.227)^2}$$

$$+ \frac{0.002632(0.08314)(215) - 0.0371}{(0.227)^3} + \frac{0.371(1.35 \times 10^{-4})}{(0.227)^6}$$

$$+ \frac{1.054 \times 10^3}{(0.227)^3 215^2}\left[1 + \frac{0.0060}{(0.227)^2}\right]\exp\left[-\frac{0.0060}{(0.227)^2}\right]$$

$$= 78.75 - 11.09 + 0.86 + 0.04 + 1.94 = 70.50 \text{ bars}$$

| Carbon monoxide |
|---|
| $m = 3.7$ kg |
| $V = 0.030$ m$^3$ |
| $T = 215$ K |

**Figure 11-4**
Schematic and data for Example 11-2.

**Comment:**   NBS Technical Note 202 reports a pressure of 70.91 bars for the given state. Thus the BWR equation is in error less than 0.6 percent. In comparison, the results of Example 11-1 indicate that the van der Waals equation is 5 percent too low, while the Redlich-Kwong equation is 1.8 percent too low.

---

Another example of a complex equation of state is that suggested by Martin and Hou in 1955. It fits polar and nonpolar substances with densities up to 50 percent greater than the critical density. The original form of this equation in 1955 had nine adjustable constants. Since that time (1959), the equation has been modified to include eleven constants.

### 11-3-3   PENG-ROBINSON EQUATION

Many equations of state are poor predictors of wet-region data, especially for the saturated-liquid state. The Peng-Robinson equation of state was developed in 1976 primarily to improve vapor-liquid equilibrium data. It introduces a third parameter, the acentric factor $\omega$, in addition to the critical pressure and temperature. The Peng-Robinson equation is given by

$$P = \frac{RT}{v - b} - \frac{a\alpha}{v(v + b) + b(v - b)} \qquad \textbf{[11-10]}$$

where
$$\alpha = [1 + S(1 - T_r^{1/2})]^2 \qquad \textbf{[11-11]}$$

and
$$S = 0.37464 + 1.54226\omega - 0.26992\omega^2$$

Values of $\omega$ are given in Table A-2. Equations for the constants $a$ and $b$ are developed in the same way as for the van der Waals and Redlich-Kwong equations.

### 11-4   REAL-GAS MIXTURES

In Chap. 10 it was pointed out that the evaluation of the properties of mixtures is a difficult one, since any one property depends not only on two independent properties such as pressure and temperature but also on a specification of the composition of the mixture such as the mole fractions of each component. Hence the tabulation of the properties of mixtures is not too fruitful because of the tremendous quantity of data required. As in the approach for ideal gases presented earlier, one solution is to employ the properties of the individual pure constituents in some sort of mixture rule. The two most common rules are those of additive pressures and additive volumes introduced in Chap. 10. For example, the total pressure of a mixture of gases might be computed by Dalton's rule of additive pressures as the sum of the pressures exerted by the individual constituents. That is,

$$P_m = p_1 + p_2 + p_3 + \cdots]_{T,V} \qquad \textbf{[11-12]}$$

where the pressures of the components 1, 2, 3, etc., are computed from more exact methods for each individual gas. Each component pressure is evaluated at the volume and the temperature of the mixture, as noted in Eq. [11-12]. The real error in this approach is that each gas is assumed to act as if it alone occupied the entire volume and its behavior were not affected by the presence of other chemical species. As a consequence, it might be expected that the law of additive pressures will be more valid at relatively low pressures or densities, and considerable error might occur at higher pressures.

If each gas in a gas mixture were assumed to follow the van der Waals equation, the equation for component $i$ would be

$$p_i]_{T,V} = \frac{R_u T}{\bar{v}_i - b_i} - \frac{a_i}{\bar{v}_i^2}$$

where $\bar{v}_i$ is the molar volume of the component. However, the molar volume of any species is related by definition to the molar volume $\bar{v}_m$ of the mixture by $\bar{v}_m = y_i \bar{v}_i$. The substitution of this relation into the above equations yields

$$p_i]_{T,V} = \frac{y_i R_u T}{\bar{v}_m - y_i b_i} - \frac{y_i^2 a_i}{\bar{v}_m^2} \qquad \textbf{[11-13]}$$

If this general form of the equation is substituted into the additive-pressure law, we find upon rearrangement that

$$P_m = R_u T \left( \frac{y_1}{\bar{v}_m - y_1 b_1} + \frac{y_2}{\bar{v}_m - y_2 b_2} + \cdots \right) - \frac{1}{\bar{v}_m^2} (a_1 y_1^2 + a_2 y_2^2 + \cdots)$$

$$\textbf{[11-14]}$$

For any mixture of given temperature, volume, and composition the right-hand side of this equation can be computed directly, assuming that the van der Waals constant for each constituent has already been determined.

Another approach to the use of the additive-pressure rule would be to assume that the compressibility factors for the individual components could be used. If an equation of the type $p_i V = Z_i N_i R_u T$ is substituted in the additive-pressure law for each component, the result is

$$P_m = \frac{R_u T}{V} (N_1 Z_1 + N_2 Z_2 + \cdots)$$

However, if an average compressibility factor $Z_m$ for the mixture is defined in terms of the equation $P_m V = Z_m N_m R_u T$, then we see that

$$Z_m N_m = Z_1 N_1 + Z_2 N_2 + \cdots$$

or $\qquad\qquad\qquad Z_{m,V,T} = y_1 Z_1 + y_2 Z_2 + \cdots \qquad\qquad \textbf{[11-15]}$

where $y$, again, is the mole fraction for that species. The subscripts $V$ and $T$ on $Z_m$ emphasize that the compressibility factors for each component are to be evaluated at the volume and temperature of the mixture. In general, the rule of additive pressures in the form of Eq. [11-15] tends to give values of $Z_m$ that are greater than the experimental value at low pressures or densities and to give values that are too low at high pressures. To overcome this difficulty, a method, frequently called the *Bartlett rule of additive pressures,* may be employed. By this method the same form of Eq. [11-15] is used, but the individual compressibility factors are evaluated by using the temperature and the molar volume of the mixture rather than the molar volume of the component. This approach tends to give better results in the low-pressure region. Other equations of state can be used, of course, in conjunction with the law of additive pressures.

Amagat's rule of additive volumes may also be applied to gas mixtures as an approximation technique. In this case

$$V_m = V_1 + V_2 + \cdots ]_{P,T} \qquad [11\text{-}16]$$

where the individual volumes are computed from appropriate equations of state based on the pressure and the temperature of the mixture. If the compressibility-factor methods are employed, then $P_m V_i = Z_i N_i R_u T$. The use of this expression in Eq. [11-16] leads to the following relationship for the overall mixture compressibility factor $Z_m$:

$$Z_{m,P,T} = y_1 Z_1 + y_2 Z_2 + \cdots \qquad [11\text{-}17]$$

The subscripts $P$ and $T$ emphasize that the individual $Z$ factors are to be evaluated at the pressure and the temperature of the mixture. Although Eqs. [11-15] and [11-17] are identical in form for the additive-pressure and additive-volume rules, the method of evaluation is quite different. As a result, the two methods give different values for a given problem when they are applied to real gases. Since Amagat's rule takes into account the total pressure of the system, effectively it is accounting for the influence of intermolecular forces between different chemical species as well as between like molecules. It is a matter of experience that the rule of additive volumes for real-gas mixtures leads to $Z_m$ values that are too low in the low-pressure region. In general, Amagat's rule is probably superior to the additive-pressure rule except at relatively low pressures.

For preliminary engineering estimates another mixture rule known as *Kay's rule* is satisfactory to within roughly 10 percent. In this method, *pseudocritical* temperature and pressure ($T'_c$ and $P'_c$) for mixtures are defined in the following manner:

$$T'_c = y_1 T_{c1} + y_2 T_{c2} + y_3 T_{c3} + \cdots \qquad [11\text{-}18a]$$

and

$$P'_c = y_1 P_{c1} + y_2 P_{c2} + y_3 P_{c3} + \cdots \qquad [11\text{-}18b]$$

This technique is quite useful since it merely requires a knowledge of the critical temperatures and pressures of the constituent gases. Other types of correlations for real-gas mixtures are available in the literature.

**A** system contains a mixture which is 75 mole percent hydrogen and 25 mole percent nitrogen at 25°C. When the specific volume is 0.0845 m³/kmol, the pressure experimentally is found to be 405 bars. Estimate what the pressure would be on the basis of (a) ideal-gas behavior, (b) the additive-pressure rule and van der Waals gases, (c) the additive volume rule and van der Waals gases, (d) the additive-pressure rule and compressibility factors, (e) the additive-volume rule and compressibility factors, (f) the Bartlett rule of additive pressures, and (g) the pseudocritical temperature and pressure approach.

**Example 11-3**

**Solution:**

**Given:** A gas mixture containing 75 mole percent $H_2$ and 25 mole percent $N_2$ at 25°C and 0.0845 m³/kmol, as shown in Fig. 11-5.

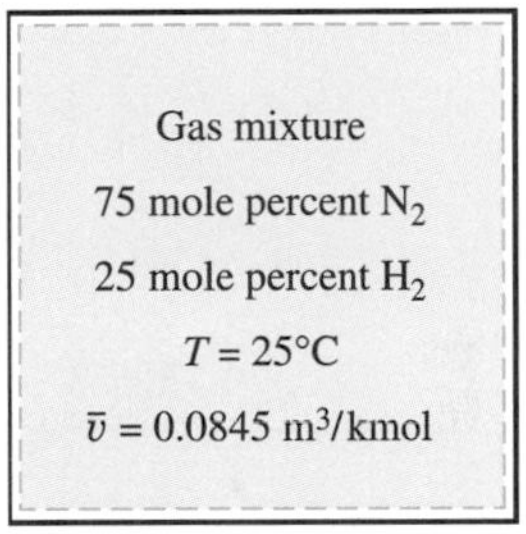

**Find:** P based on (a) ideal-gas behavior, (b) the additive-pressure rule and van der Waals gases, (c) the additive-volume rule and van der Waals gases, (d) the additive-pressure rule and compressibility factors, (e) the additive-volume rule and compressibility factors, (f) the Bartlett rule of additive pressures, and (g) the pseudocritical temperature and pressure approach.

---

Gas mixture

75 mole percent $N_2$

25 mole percent $H_2$

$T = 25°C$

$\bar{v} = 0.0845$ m³/kmol

---

**Figure 11-5**
Schematic and data for
Example 11-3.

**Analysis:** The critical data and the van der Waals constants for hydrogen and nitrogen are as follows:

$$T_{c,H_2} = 33.2 + 8.0 = 41.2 \text{ K} \qquad a_{H_2} = 0.247 \text{ bar·m}^6/\text{kmol}^2$$

$$P_{c,H_2} = 13.0 + 8.0 = 21 \text{ bars} \qquad b_{H_2} = 0.0265 \text{ m}^3/\text{kmol}$$

$$T_{c,N_2} = 126.2 \text{ K} \qquad a_{N_2} = 1.361 \text{ bar·m}^6/\text{kmol}^2$$

$$P_{c,N_2} = 33.9 \text{ bars} \qquad b_{N_2} = 0.0385 \text{ m}^3/\text{kmol}$$

For use with the generalized compressibility charts, the critical temperature and pressure of hydrogen have been corrected by the factor 8.0, as indicated in the main text, for greater accuracy.

(a) The pressure based on the ideal-gas equation is

$$P = \frac{NR_uT}{V} = \frac{0.08314(298)}{0.0845} = 293 \text{ bars}$$

This value is in error by over 25 percent.

(b) For the additive-pressure rule with the van der Waals equation, we may employ Eq. [11-14]. Thus

$$P = 0.08314(298)\left[\frac{0.75}{0.0845 - 0.75(0.0265)} + \frac{0.25}{0.0845 - 0.25(0.0385)}\right]$$

$$- \frac{1}{(0.0845)^2}[0.247(0.75)^2 + 1.361(0.25)^2]$$

$$= 24.8(14.95) - 140.1(0.224) = 339 \text{ bars}$$

The result is better than the ideal-gas relation but still quite low compared with the experimental value.

(c) According to the additive-volume rule, the total volume is the sum of the volumes of the components, measured at the pressure and temperature of the mixture. The pressure is unknown; so we write the following three equations:

$$0.0845 = 0.75\bar{v}_{H_2} + 0.25\bar{v}_{N_2}$$

$$P_{H_2} = \frac{R_u T}{\bar{v} - b} - \frac{a}{\bar{v}^2} = \frac{0.08314(298)}{\bar{v} - 0.0265} - \frac{0.247}{\bar{v}^2} = \frac{24.78}{\bar{v} - 0.0265} - \frac{0.247}{\bar{v}^2}$$

$$P_{N_2} = \frac{24.78}{\bar{v} - 0.0385} - \frac{1.361}{\bar{v}^2}$$

For a satisfactory solution the pressures of the two gases must be equal. These three equations can be solved by means of a computer program. If we employ an iteration process, we first assume that $\bar{v}$ for hydrogen is 0.0862 m$^3$/kmol, for example. Then, from the first equation above,

$$\bar{v}_{N_2} = \frac{0.0845 - 0.75(0.0862)}{0.25} = 0.0794 \text{ m}^3/\text{kmol}$$

Substituting the two values of $\bar{v}$ into the last two equations, we find that

$$P_{H_2} = \frac{24.78}{0.0862 - 0.0265} - \frac{0.247}{(0.0862)^2} = 382 \text{ bars}$$

$$P_{N_2} = \frac{24.78}{0.0794 - 0.0385} - \frac{1.361}{(0.0794)^2} = 390 \text{ bars}$$

These two answers are sufficiently close not to warrant another trial. This solution is roughly 5 percent lower than the measured value.

(d) The additive-pressure rule and the compressibility chart can be combined into an approximation method if pseudoreduced-volume lines are available. Recall that $v'_{r,H_2} = \bar{v} P_c / R_u T_c$. On this basis the pseudoreduced volumes and the reduced temperatures for the two gases become

$$v'_{r,H_2} = \frac{0.0845(21)}{0.75(41.2)(0.08314)} = 0.69 \qquad T_{r,H_2} = \frac{298}{41.2} = 7.23$$

$$v'_{r,N_2} = \frac{0.0845(33.9)}{0.25(126.2)(0.08314)} = 1.09 \qquad T_{r,N_2} = \frac{298}{126.2} = 2.36$$

From Figs. A-28 and A-29, the $Z$ factors for hydrogen and nitrogen are found to be 1.06 and 0.99, respectively. Hence

$$Z_m = 0.75(1.06) + 0.25(0.99) = 1.05$$

and                         $$P = Z_m P_{\text{ideal}} = 1.05(293) = 308 \text{ bars}$$

This method leads to considerable error in predicting the pressure.

(e) According to the additive-volume rule, each gas exists at the mixture pressure and temperature. Since the pressure is unknown, it must be assumed under an iteration process. If we first assume that the pressure is 350 bars, then

$$P_{r,H_2} = \frac{350}{21} = 16.7 \qquad T_{r,H_2} = 7.23$$

$$P_{r,N_2} = \frac{350}{33.9} = 10.3 \qquad T_{r,N_2} = 2.36$$

The individual compressibility factors based on these reduced properties are 1.20 and 1.21 for hydrogen and nitrogen, respectively. Therefore

$$Z_m = 0.75(1.20) + 0.25(1.21) = 1.20$$

and

$$P = Z_m P_{ideal} = 1.20(293) = 352 \text{ bars}$$

This value agrees substantially with the assumed value; so no further trial is necessary. In this case the additive-volume rule is a better approximation than the additive-pressure rule, although both give too low a value.

(*f*) The Bartlett rule of additive pressures is based on using the molar specific volume of the mixture for the specific volume of any component. If the given value of 0.0845 m³/kmol is used in the pseudoreduced-volume definition, we find that, from the calculation made in part *d*,

$$v'_{r,H_2} = 0.75(0.69) = 0.52 \qquad T_{r,H_2} = 7.23$$

$$v'_{r,N_2} = 0.25(1.07) = 0.27 \qquad T_{r,N_2} = 2.36$$

The compressibility factors for hydrogen and nitrogen under these conditions are approximately 1.25 and 1.17, respectively. Thus

$$Z_m = 0.75(1.25) + 0.25(1.17) = 1.23$$

and

$$P = Z_m P_{ideal} = 1.23(293) = 360 \text{ bars}$$

Note that the answer is much improved over that obtained by the unmodified rule of additive pressures.

(*g*) If, finally, the definitions of the pseudocritical temperature and pressure defined in the main text are employed, then

$$T'_c = y_1 T_{c1} + y_2 T_{c2} = 0.75(41.2) + 0.25(126.2) = 62.5 \text{ K}$$

and

$$P'_c = y_1 P_{c1} + y_2 P_{c2} = 0.75(21) + 0.25(33.9) = 24.2 \text{ bars}$$

Consequently,

$$v'_r = \frac{\bar{v} P'_c}{R_u T'_c} = \frac{0.0845(24.2)}{0.08314(62.5)} = 0.394$$

$$T'_r = \frac{T}{T'_c} = \frac{298}{62.5} = 4.77$$

From Fig. A-29, the *Z* factor is 1.31, and the corresponding pressure of the system is $P = 1.31(293) = 384$ bars.

**Comment:** All the methods employed above gave answers that were too low, although some were much better than others.

## 11-5 SUMMARY

It is convenient to have analytical expressions for the *PvT* behavior of real gases. This chapter presented a few of the many correlations available for the equations of state of real gases. The accuracy of these *PvT* relations varies with the type of gas and the range of the properties under consideration. Data of the type $Pv = f(P, T)$ may be represented to any desired accuracy by infinite-series expansions known as *virial equations:*

$$P_V = RT + bP + cP^2 + dP^3 + \cdots$$

and

$$Pv = RT\left(1 + \frac{b'}{v} + \frac{c'}{v^2} + \frac{d'}{v^3} + \cdots\right)$$

These series expansions can also be written in terms of the compressibility factor $Z(\equiv Pv/RT)$.

Two-constant equations of state include the van der Waals and the Redlich-Kwong formats. The van der Waals equation is

$$\left(P + \frac{a}{v^2}\right)(v - b) = RT$$

Both constants $a$ and $b$ have units that must be consistent with those employed for $P$, $v$, and $T$. Evaluation of the constants is based on an experimental fact expressed mathematically by the relations

$$\left(\frac{\partial P}{\partial v}\right)_{T_c} = 0 \quad \text{and} \quad \left(\frac{\partial^2 P}{\partial v^2}\right)_{T_c} = 0$$

These equations permit the evaluation of the constants in any two-constant equation of state. The Redlich-Kwong equation is

$$P = \frac{RT}{v - b} - \frac{a}{T^{1/2}v(v + b)}$$

Constants $a$ and $b$ again can be evaluated from critical data. This equation appears to be fairly accurate for temperatures above the critical value.

One of the inherent limitations of two-constant equations of state is the lack of accuracy over a wide range of conditions. Higher accuracy requires equations with a considerably larger number of terms, or constants. One of the earliest equations of state of reasonable simplicity and accuracy is the Beattie-Bridgeman equation. A more complex equation that has been successful in predicting $PvT$ behavior, especially of hydrocarbons, is the Benedict-Webb-Rubin equation. This equation is an extension of the Beattie-Bridgeman equation and has eight adjustable constants for a given substance.

Many equations of state are poor predictors of wet-region data, especially for the saturated-liquid state. The Peng-Robinson equation of state was developed primarily to improve vapor-liquid equilibrium data. It introduces a third parameter, the acentric factor $\omega$, in addition to the critical pressure and temperature.

The $PvT$ relationship for a real-gas mixture is often estimated using a mixture rule and the properties of the individual pure components. The two most common mixing rules are *Dalton's rule of additive pressures* and

*Amagat's rule of additive volumes*. For example, each gas could be modeled by the van der Waals equation, and the mixture assumed to fulfill Dalton's rule. If the behavior of the individual component is described in terms of its compressibility factor, it is possible to estimate the compressibility factor for the mixture. Two empirical mixture rules that are commonly used are *Bartlett's rule of additive pressures* and *Kay's rule* for calculating the pseudocritical temperature and pressure for the mixture. Care must be taken to check the validity of any mixture rule against experimental data for any given mixture.

## Suggested Readings and References

Benedict, O., G. B. Webb, and L. C. Rubin: *J. Chem. Phys.,* **8**(4): 334–345 (1940).

Lewis, G. N., and M. Randall: "Thermodynamics," 2d ed., McGraw-Hill, New York, 1961.

Martin, J. J., and Y. C. Hou: *A.I. Ch. E. J.,* **1**(2): 142–151 (1955).

Martin, J. J., R. M. Kapoor, and N. deNevers: *A.I. Ch. E. J.,* **5**(2): 159–164 (1959).

Obert, E. F.: "Concepts of Thermodynamics," McGraw-Hill, New York, 1960.

Otto, J., A. Michels, and H. Wouters: *Physik. Z.,* **35**(3): 97–100 (1934).

Redlich, O., and J. N. S. Kwong: *Chem. Rev.,* **44**(1): 233–244 (1949).

Wark, K.: "Advanced Thermodynamics for Engineers," McGraw-Hill, New York, 1995.

## Problems

11-1. The constants for the virial equation of state in the form $Pv = A(1 + b'/v + c'/v^2 + \cdots)$ have been determined experimentally for nitrogen at $-100°C$. The values are $A = 14.39$, $b' = -0.05185$ m³/kmol, $c' = 0.002125$ m⁶/kmol². Also, $P$ is in bars and $v$ is in m³/kmol.

(a) Determine the compressibility factor at 68 bars and $-100°C$ from the above equation.

(b) Compare the above result to that obtained from a generalized compressibility chart.

11-2E. The constants for the virial equation of state in the form $Pv = A(1 + b'/v^2 + c'/v^2 + \cdots)$ have been determined experimentally for nitrogen at $-148°F$. The values are $A = 0.634$, $b' = -2.3146 \times 10^{-3}$, and $c' = 4.235 \times 10^{-6}$, with $P$ in atmospheres.

(a) Determine the compressibility factor at 67 atm and $-148°F$ from the above equation.

(b) Compare the above result to that obtained from a generalized compressibility chart.

11-3. Calculate the specific volume, in $m^3/kmol$, of nitrogen at 102 bars and $-45°C$ by (a) the ideal-gas equation, (b) the compressibility chart, and (c) the virial equation as given by Eq. (11-1a). The virial coefficients for nitrogen at this temperature are $b = -2.34 \times 10^{-2}$, $c = 3.61 \times 10^{-5}$, and $d = 5.18 \times 10^{-7}$, where $P$ is in bars, $T$ is in kelvins, and $v$ is in $m^3/kmol$.

11-4. Consider nitrogen gas at 150 bars and $0°C$.
   (a) Determine the specific volume, in $m^3/kmol$, by means of Eq. (11-3).
   (b) Compare with the answer found by using Eqs. (11-4) and (11-5).
   (c) Check against the tabulated value, in $m^3/kmol$.

11-5. Determine the specific volume of nitrogen at 150 K and 64 bars (a) by means of Eqs. (11-4) and (11-5), and (b) compare with the result based on the generalized compressibility chart in $m^3/kmol$.

11-6E. Calculate the specific volume, in $ft^3/lbmol$, of nitrogen at 2000 psia and $-50°F$ by (a) the ideal-gas law, (b) the compressibility-factor method, and (c) the virial equation as given by Eq. (11-1a). (The virial coefficients for nitrogen at this temperature are $b = -3.75 \times 10^{-1}$, $c = 5.86 \times 10^{-4}$, and $d = 8.53 \times 10^{-6}$, where $P$ is in atmospheres and $T$ is in degrees Rankine.) (d) Then, estimate a value from Table A-20E.

11-7. Determine the specific volume of nitrogen at 204 atm and $0°C$, by means of Eq. (11-3). Compare with the answer found by using Eqs. (11-4) and (11-5), in $cm^3/mol$. Check against Table A-20E.

11-8. Derive the equations for $a$ and $b$ in the Redlich-Kwong equation of state in terms of the critical values of $P$, $v$, and $T$.

11-9E. Steam at 1600 psia and $740°F$ is a real gas. Determine the specific volume by using (a) the ideal-gas equation, (b) the compressibility factor, (c) the van der Waals equation, and (d) the tabulated value, in $ft^3/lb_m$.

11-10E. Carbon dioxide at $576°R$ and 69.1 atm has an observed specific volume of 3.94 $ft^3/lbmol$. Calculate the value based on (a) the ideal-gas equation, (b) the principle of corresponding states, and (c) the van der Waals equation.

11-11. Steam at 100 bars and $360°C$ is a real gas. Determine the specific volume, in $m^3/kg$, by using (a) the ideal-gas equation, (b) the compressibility factor, (c) the van der Waals equation, (d) the Redlich-Kwong equation, and (e) the tabulated value.

11-12. Carbon dioxide at 350 K and 7.0 MPa has an observed specific volume of 0.321 $m^3/kmol$. Calculate the value based on (a) the ideal-gas equation, (b) the Redlich-Kwong equation, (c) the van der Waals equation, and (d) the compressibility factor.

11-13. Compute the specific volume of steam at 140 bars and 400°C by means of (*a*) the ideal-gas equation, (*b*) the van der Waals equation, (*c*) the Redlich-Kwong equation, (*d*) the compressibility factor, and (*e*) the tabulated value, in m³/kg.

11-14. Determine the pressure, in bars, of steam at 360°C and 0.0309 m³/kg by means of (*a*) the ideal-gas equation, (*b*) the van der Waals equation, (*c*) the Redlich-Kwong equation, (*d*) the compressibility-factor method, and (*e*) the tabulated superheat value.

11-15E. Compute the specific value of steam at 2000 psia and 800°F by means of (*a*) the ideal-gas equation, (*b*) the van der Waals equation, (*c*) the Redlich-Kwong equation, (*d*) the compressibility chart, and (*e*) the tabulated superheat data, in ft³/lbm.

11-16E. Determine the temperature, in degrees Fahrenheit, for steam at 1600 psia and 0.342 ft³/lbm by means of (*a*) the principle of corresponding states, (*b*) the van der Waals equation, (*c*) the Redlich-Kwong equation, and (*d*) the tabulated superheat value.

11-17. Compute the temperature, in degrees Celsius, for steam at 100 bars and 0.02331 m³/kg by means of (*a*) the compressibility factor, (*b*) the van der Waals equation, (*c*) the Redlich-Kwong equation, and (*d*) the tabulated superheat value.

11-18. The specific volume of R-134a at 60°C is 0.02301 m³/kg. Determine the pressure in bars by means of (*a*) the ideal-gas equation, (*b*) the van der Waals equation, (*c*) the Redlich-Kwong equation, (*d*) the compressibility chart, and (*e*) the tabulated superheat value.

11-19. Determine the pressure in bars of R-134a at 80°C and 0.01435 m³/kg by means of (*a*) the ideal-gas equation, (*b*) the van der Waals equation, (*c*) the Redlich-Kwong equation, (*d*) the compressibility chart, and (*e*) the tabulated superheat value.

11-20. Determine the temperature in degrees Celsius of R-134a at 9 bars and 0.02609 m³/kg by means of (*a*) the ideal-gas equation, (*b*) the van der Waals equation, (*c*) the Redlich-Kwong equation, (*d*) the compressibility chart, and (*e*) the tabulated superheat value.

11-21E. The specific volume of refrigerant 134a at 160°F is 0.3011 ft³/lbm. Determine the pressure in psia by means of (*a*) the ideal-gas equation, (*b*) the van der Waals equation, (*c*) the Redlich-Kwong equation, (*d*) the compressibility chart, and (*e*) the tabulated superheat value.

11-22E. The specific volume of refrigerant 134a at 120 psia is 0.461 ft³/lbm. Determine the temperature of the fluid, in degrees Fahrenheit, by means of (*a*) the ideal-gas equation, (*b*) the van der Waals equation, (*c*) the Redlich-Kwong equation, (*d*) the compressibility chart, and (*e*) the tabulated superheat value.

11-23E. The specific volume of nitrogen at −110°F is 0.115 ft³/lbm. Determine the pressure of the substance by means of (*a*) the ideal-gas equation, (b) the compressibility factor, (c) the van der Waals equation, (d) the Redlich-Kwong equation, and (e) the tabulated value.

11-24. The specific volume of saturated $CO_2$ vapor at 25°C is 0.00419 $m^3$/kg. Estimate the pressure, in bars, on the basis of (a) the ideal-gas equation, (b) the Redlich-Kwong equation, (c) the Benedict-Webb-Rubin equation, (d) the van der Waals equation, and (e) the compressibility factor. The tabulated value is 64.0 bars.

11-25. Determine the pressure of $N_2$ at −123°C if the specific volume is 0.00239 $m^3$/kg, employing (a) the Redlich-Kwong equation, (b) the Benedict-Webb-Rubin equation, (c) the Beattie-Bridgeman equation, and (d) the principle of corresponding states. (e) Compare to tabulated superheat value, in bars.

11-26E. Determine the pressure of $N_2$ at −160°F if the specific volume is 0.0399 $ft^3$/$lb_m$, employing (a) the Benedict-Webb-Rubin equation, (b) the Redlich-Kwong equation, and (c) the principle of corresponding states. (d) Compare to tabulated value, in psia.

11-27. NBS Technical Note 202 (1963) lists the specific volume of carbon monoxide at 200 K and 60 atm to be 0.2356 $m^3$/kmol. Estimate the value of $P$ at the given $v$ and $T$ by means of (a) the ideal-gas equation, (b) the generalized compressibility chart, (c) the Redlich-Kwong equation, and (d) the Benedict-Webb-Rubin equation, in atm.

11-28. Carbon dioxide at 75°C has a specific volume of 0.014 $m^3$/kg. Estimate the pressure in MPa by means of (a) the Beattie-Bridgeman equation and (b) the generalized compressibility chart.

11-29. Nitrogen gas at −123°C has a specific volume of 0.00239 $m^3$/kg. Estimate the pressure in MPa by means of (a) the generalized compressibility chart, (b) the Beattie-Bridgeman equation, and (c) the superheat table value.

11-30. A mixture consists of 1 kg of $CO_2$ (an ideal gas) and 1 kg of water vapor (a real gas) at 20 bars and 200°C. Calculate (a) the partial pressure of each component, (b) the component pressure of $CO_2$ based on the additive-pressure rule, and (c) the volume of the mixture, in cubic meters.

11-31E. A mixture consists of 1.0 $lb_m$ of $CO_2$ (an ideal gas) and 1.0 $lb_m$ of water vapor (a real gas) at 300 psia and 400°F. Calculate (a) the partial pressure of each component, (b) the component pressure of $CO_2$ based on the additive-pressure rule, and (c) the volume of the mixture, in cubic feet.

11-32. Assume air is a mixture of two gases, and the composition is 78 mole percent nitrogen and 22 mole percent oxygen. What is the pressure of the mixture if 14 $m^3$ of it at 20°C and 1 bar is compressed to 0.028 $m^3$ and 37°C? Assume the following: (a) ideal-gas law; (b) van der Waals gases, using the additive-pressure law: (c) van der Waals gases, using the additive-volume law; and (d) corresponding states and the additive-volume law.

11-33. Two gases $A$ and $B$, each in the amount of 1 kmol, exist in a mixture at 260 K and a total pressure of 50 bars. Find the total volume, in cubic meters, using compressibility data and (a) the additive-pressure rule and (b) the additive-volume rule. The critical pressures and temperatures of $A$ and $B$ are $P_{cA}$ = 6 bars, $P_{cB}$ = 8 bars, $T_{cA} = T_{cB} =$ 200 K.

11-34E. Air is a mixture of two gases, and the composition is 79 mole percent nitrogen and 21 mole percent oxygen. What is the pressure of the mixture if 500 ft$^3$ of it at 70°F is compressed to 1.0 ft$^3$? The original pressure was 1 atm, and the final temperature is 100°F. Assume the following: (a) ideal gas; (b) van der Waals gases, using the additive-pressure law; (c) van der Waals gases, using the additive-volume law; and (d) corresponding states and the additive-volume law.

11-35E. Two gases $A$ and $B$, each in the amount of 1 lbmol, exist in a mixture at 420°R and a total pressure of 60 atm. Find the total volume, in cubic feet, using $Z$ data and (a) the additive-pressure rule and (b) the additive-volume rule. The critical pressures and temperatures of $A$ and $B$ are $P_{cA}$ = 6 atm, $P_{cB}$ = 7 atm, $T_{cA} = T_{cB} = 300°R$.

11-36. One wishes to prepare a mixture of 60 mole percent acetylene ($C_2H_2$) and 40 mole percent $CO_2$ at 47°C and 100 bars, in a 1.0-m$^3$ tank. The tank initially contains acetylene at 47°C and pressure $P_1$. Carbon dioxide is then bled into the tank from a line containing $CO_2$ at 47°C and 100 bars until the tank pressure reaches 100 bars. What is the value of $P_1$ such that when the tank pressure reaches 100 bars, the composition within the tank will be 60 mole percent acetylene? Assume the validity of Kay's rule for the mixture.

11-37. Calculate the pressure exerted by a mixture of 0.5 kmol of methane ($CH_4$) and 0.5 kmol of propane ($C_3H_8$) for a temperature of 90°C and a volume of 0.48 m$^3$. Use (a) the ideal-gas equation, (b) van der Waals equation and the additive-pressure law, (c) the compressibility chart and additive volumes, and (d) the compressibility chart and Kay's rule. The observed value is 50.6 bars.

11-38E. One wishes to prepare a mixture of 60 mole percent acetylene ($C_2H_2$) and 40 percent $CO_2$ at 120°F, 1500 psia, in a 1.0-ft$^3$ tank. The tank initially contains acetylene at 120°F and pressure $P_1$. Carbon dioxide is then bled into the tank from a line containing $CO_2$ at 120°F and 1500 psia until the tank pressure reaches 1500 psia. What is the value of $P_1$ such that when the tank pressure reaches 1500 psia, the composition within the tank will be 60 percent acetylene? Assume the validity of Kay's rule for the mixture.

11-39. A gaseous mixture of 60 percent propane and 40 percent ethane on a mole basis occupies a volume of 0.070 m$^3$ at 127°C. If the total mass is 15.0 kg, determine the mixture pressure, in bars, based on (a) an ideal-gas model, (b) the additive-pressure rule and compressibility factors, and (c) Kay's rule and compressibility factors.

# CHAPTER
# 12

# GENERALIZED THERMODYNAMIC RELATIONSHIPS

Nuclear power plant at Indian Point in New York. Each dome contains a pressurized-water reactor that provides thermal energy to drive a Rankine cycle power plant.

In the solution of engineering and scientific problems, it is essential to be able to determine the values of thermodynamic properties. Since classical thermodynamics is an experimental science, a great deal of empirical data have been obtained over the past decades. The reader must recognize, however, that only a relatively few properties can be evaluated by direct experimentation. Of these, the correlation among pressure, specific volume, and temperature and the relationship between the specific heats and temperature at low pressures are most easily measured. The evaluation of such properties as the internal energy, the enthalpy, and the entropy is made on the basis of calculations involving the directly measurable data mentioned above. Consequently, one of the major tasks of thermodynamics is to provide basic equations to evaluate properties such as $u$, $h$, and $s$, and others, from measurable property data.

A second point to recognize is that, in many cases, insufficient data are available to calculate properties, even if the basic mathematical relations have been developed. Hence approximation techniques are needed, since in numerous situations insufficient time or money may require that alternative methods of evaluating properties be available. A case in point is the use of the generalized compressibility factor $Z$, discussed in Chaps. 4 and 11, which provides a means of correlating $PvT$ data in the absence of sufficient direct experimental information. This particular method is extended to other properties in this chapter. Before we investigate some general thermodynamic relations for the qualitative and quantitative insight they provide, a brief review of some rules of partial differential calculus is in order.

## 12-1 FUNDAMENTALS OF PARTIAL DERIVATIVES

In this chapter we are primarily interested in property relationships for simple compressible substances. Therefore, on the basis of the state postulate, most of the expressions developed in this chapter involve a dependent variable expressed as a function of two independent variables. With this in mind, consider three thermodynamic variables (such as $P$, $v$, and $T$) represented in general by $x$, $y$, and $z$. Their functional relationship is expressed in the form $x = x(y, z)$. We could also write that $y = y(x, z)$ or that $z = z(x, y)$. The *total differential* when $x$ is the dependent variable is given by the equation

$$dx = \left(\frac{\partial x}{\partial y}\right)_z dy + \left(\frac{\partial x}{\partial z}\right)_y dz \qquad \text{[12-1a]}$$

This expression can also be written in the notation

$$dx = M\, dy + N\, dz \qquad \text{[12-1b]}$$

where $M = (\partial x/\partial y)_z$

= the partial derivative of $x$ with respect to $y$ at constant $z$

$N = (\partial x/\partial z)_y$

= the partial derivative of $x$ with respect to $z$ at constant $y$

Similar expressions for $dy$ and $dz$ may also be written, depending upon which variable is selected as the dependent variable.

The physical significance of a partial derivative is easily seen by recalling that the equilibrium states of a simple compressible substance can be represented by a three-dimensional surface. Such a surface is shown in Fig. 12-1 for a single-phase region, where a set of three properties is again symbolized by $x$, $y$, and $z$. As an example, this surface could represent the superheat region of a $PvT$ diagram similar to that shown in Fig. 4-1. Sections of the surface in Fig. 12-1 have been cut away around the equilibrium state $D$, so that the curvature of the surface is more clearly seen. Consider a plane of constant $z$ which intersects the surface. The curve of intersection is marked with the points $C$, $D$, and $E$ on the diagram. The quantity $(\partial x/\partial y)_z$ is the slope of the surface at any state along this curve. In particular, the value of this partial derivative at state $D$ is the slope of tangent line $AB$. Similar interpretations of quantities $(\partial x/\partial z)_y$ and $(\partial y/\partial z)_x$ can be made when planes of constant $y$ and $x$, respectively, intersect the surface of equilibrium states.

In the following sections we wish to formulate the total differentials for $u$, $h$, and $s$, and others, in the format of Eq. [12-1a]. Integration of these differential expressions will then provide us with general equations for evaluating property changes such as $\Delta u$, $\Delta h$, and $\Delta s$ for solids, liquids, and gases, as well as for phase changes. These total differentials are derived from the basic laws of thermodynamics and definitions. Unfortunately, the expressions initially will contain independent variables on the right side of

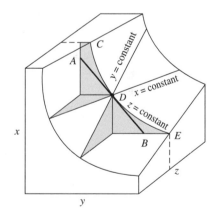

**Figure 12-1**
A representation of a partial derivative for a thermodynamic surface.

Eq. [12-1a] that are not directly measurable. Hence some manipulation of partial derivatives is necessary in order to convert the initial total differentials into a useful format. Three relationships among partial derivatives that are important to these transformations are discussed below.

The first relationship we seek is based on Eq. [12-1b]. Partial differentiation of $M$ and $N$ with respect to $z$ and $y$, respectively, leads to

$$\frac{\partial M}{\partial z} = \frac{\partial^2 x}{\partial y \, \partial z} \quad \text{and} \quad \frac{\partial N}{\partial y} = \frac{\partial^2 x}{\partial z \, \partial y}$$

If the variable $x$ is continuous, then it is known from calculus that the order of differentiation is immaterial, so that

$$\left(\frac{\partial M}{\partial z}\right)_y = \left(\frac{\partial N}{\partial y}\right)_z \qquad \textbf{[12-2]}$$

Equation [12-2] is known as the **test for exactness.** Its usefulness in deriving the Maxwell relations is demonstrated in the next section.

The two remaining relationships among partial derivatives are derived in the following manner. Equation (a) is written with $y$ and $z$ as the independent variables. For the situation where $y = y(x, z)$, the total differential is written as

$$dy = \left(\frac{\partial y}{\partial x}\right)_z dx + \left(\frac{\partial y}{\partial z}\right)_x dz \qquad \textbf{[12-3a]}$$

When Eq. [12-1a] is substituted into Eq. [12-3a] in order to eliminate $dx$, and the subsequent result is rearranged, we find that

$$\left[1 - \left(\frac{\partial y}{\partial x}\right)_z \left(\frac{\partial x}{\partial y}\right)_z\right] dy = \left[\left(\frac{\partial y}{\partial x}\right)_z \left(\frac{\partial x}{\partial z}\right)_y + \left(\frac{\partial y}{\partial z}\right)_x\right] dz \quad \textbf{[12-3b]}$$

We select $y$ and $z$ to be the independent variables. When $dz = 0$ and $dy \neq 0$, then the coefficient of $dy = 0$ in Eq. [12-3b]. As a result,

$$\left(\frac{\partial y}{\partial x}\right)_z \left(\frac{\partial x}{\partial y}\right)_z = 1 \quad \text{and} \quad \left(\frac{\partial y}{\partial x}\right)_z = \frac{1}{(\partial x / \partial y)_z} \qquad \textbf{[12-4]}$$

This latter equation, known as the **reciprocity relation**, shows that the inverted form of a partial derivative equals its reciprocal. In a similar fashion, when $dy = 0$ and $dz \neq 0$, then the coefficient of $dz = 0$ in Eq. [12-3b]. Thus, after rearrangement and use of Eq. [12-4],

$$\left(\frac{\partial x}{\partial y}\right)_z \left(\frac{\partial y}{\partial z}\right)_x \left(\frac{\partial z}{\partial x}\right)_y = -1 \qquad \textbf{[12-5]}$$

The above equation, known as the **cyclic relation**, is used frequently in developing property relations. Equations [12-2], [12-4], and [12-5] are important for manipulating property relations.

## 12-2 SOME FUNDAMENTAL PROPERTY RELATIONS

In Chap. 7 several important equations were developed for the entropy change of simple compressible systems of fixed chemical composition by combining equations which evolve from the first and second laws of thermodynamics. The first-law format used was $\delta q + \delta w = du$. For an internally reversible process, the heat and work interactions are represented by $T\,ds$ and $-P\,dv$. Substitution of these two expressions into the energy equation yields the *first $T\,ds$ equation*

$$T\,ds - P\,dv = du \qquad\qquad \textbf{[12-6]}$$

Since $h \equiv u + Pv$, then $du = dh - P\,dv - v\,dP$. Replacement of $du$ in Eq. [12-6] by this relationship leads to the *second $T\,ds$ equation*, namely,

$$T\,ds + v\,dP = dh \qquad\qquad \textbf{[12-7]}$$

Two additional equations of interest may be formed by defining two other properties of matter. The **Helmholtz function $a$** is defined by the equation

$$a \equiv u - Ts \qquad\qquad \textbf{[12-8]}$$

Hence

$$da = du - T\,ds - s\,dT \qquad\qquad \textbf{[12-9]}$$

If Eq. [12-6] is substituted into Eq. [12-9], then we obtain a third important property relation

$$da = -P\,dv - s\,dT \qquad\qquad \textbf{[12-10]}$$

Finally, a fourth relation is derived by first defining the **Gibbs function $g$** by

$$g \equiv h - Ts \qquad\qquad \textbf{[12-11]}$$

Therefore, the differential change in $g$ is

$$dg = dh - T\,ds - s\,dT \qquad\qquad \textbf{[12-12]}$$

Substitution of $dh$ from Eq. [12-7] into Eq. [12-12] yields

$$dg = v\,dP - s\,dT \qquad\qquad \textbf{[12-13]}$$

To summarize these four important relationships among properties of simple compressible substances, they are collected together as a set and presented here:

$$du = T\,ds - P\,dv \qquad\qquad \textbf{[12-6]}$$

$$dh = T\,ds + v\,dP \qquad\qquad \textbf{[12-7]}$$

$$da = -P\,dv - s\,dT \qquad\qquad \textbf{[12-10]}$$

$$dg = v\,dP - s\,dT \qquad\qquad \textbf{[12-13]}$$

Note that the variables on the right-hand sides of these equations include only $P$, $v$, $T$, and $s$. These equations are referred to as the **Gibbsian equations**.

One of the most important sets of thermodynamic relations that arise from the Gibbsian, or $T\,ds$, equations is obtained by applying Eq. [12-2] to the expressions for $du$, $dh$, $da$, and $dg$. For example, consider the total differential written in the format of Eq. [12-1b] and the total differential for $du$ given by Eq. [12-6].

$$dx = M\,dy + N\,dz \qquad\qquad \textbf{[12-1b]}$$

$$du = T\,ds - P\,dv \qquad\qquad \textbf{[12-6]}$$

In comparison, $M$ is replaced by $T$, $N$ is replaced by $-P$ , $y$ is replaced by $s$, and $z$ is replaced by $v$. Making these replacements in Eq. [12-2], $(\partial M/\partial z)_y = (\partial N/\partial y)_z$, leads to the relation

$$\boxed{\left(\frac{\partial T}{\partial v}\right)_s = \left(\frac{\partial P}{\partial s}\right)_v} \qquad\qquad \textbf{[12-14]}$$

Application of Eq. [12-2] to the three Gibbsian equations for $dh$, $da$, and $dg$ yields the following set of relations among partial derivatives:

$$\boxed{\left(\frac{\partial T}{\partial P}\right)_s = \left(\frac{\partial v}{\partial s}\right)_P} \qquad\qquad \textbf{[12-15]}$$

$$\left(\frac{\partial P}{\partial T}\right)_v = \left(\frac{\partial s}{\partial v}\right)_T \qquad\qquad \textbf{[12-16]}$$

$$\left(\frac{\partial v}{\partial T}\right)_P = -\left(\frac{\partial s}{\partial P}\right)_T \qquad\qquad \textbf{[12-17]}$$

This set of four equations is referred to as the **Maxwell relations**. Their importance is not apparent at this point, but a simple example may help illustrate their usefulness. Consider a system at a given equilibrium state. For a differential change of state, it is desirable to know the rate of change of entropy of the system as the volume is altered isothermally. This could apply, for example, to a gas contained within a piston-cylinder assembly that is expanded isothermally. It is not possible to evaluate $(\partial s/\partial v)_T$ directly for the process, since entropy variations cannot be measured directly. However, Eq. [12-16] states that it is necessary only to measure the rate of change of pressure with temperature at constant volume, since $(\partial P/\partial T)_v = (\partial s/\partial v)_T$. From an experimental viewpoint it is relatively easy to measure pressure and temperature variations.

**Example 12-1**

**E**stimate the value of $(\partial s/\partial P)_T$ for refrigerant 134a at 20°C and 0.28 MPa by employing $PvT$ data from the superheat table. Check the answer against entropy data from the table.

**Solution:**

**Given:** Refrigerant 134a at 20°C and 0.28 MPa.

**Find:** $(\partial s/\partial P)_T$, in kJ/kg·K·MPa, using $PvT$ data; check against tabular entropy data.

**Model:** Validity of a Maxwell relation.

**Analysis:** Equation [12-17] shows that $(\partial s/\partial P)_T = (\partial v/\partial T)_P$. To estimate this second partial derivative from tabular $PvT$ data, we shall assume that a linear, finite approximation in the vicinity of the state is reasonably accurate. As shown on the $vT$ diagram in Fig. 12-2, the slope of the tangent to the constant pressure line at the actual state will be approximated by the slope of the line connecting states at 10 and 30°C for a pressure of 0.28 MPa. That is,

$$\left(\frac{\partial v}{\partial T}\right)_P \approx \left(\frac{\Delta v}{\Delta T}\right)_P = \frac{(v_{30} - v_{10})}{(T_{30} - T_{10})} = \frac{0.08320 - 0.07613}{30 - 10}$$

$$= 3.535 \times 10^{-4} \text{ m}^3/\text{kg·K} = 0.3535 \text{ kJ/kg·K·MPa}$$

since $1 \text{ kJ} = 10^{-3} \text{ m}^3 \cdot \text{MPa}$. The value calculated above also is an estimated value of $(\partial s/\partial P)_T$.

As a check on the approximation, we can use entropy and pressure data from the refrigerant table. Choosing data at 0.24 and 0.32 MPa and 20°C,

$$\left(\frac{\partial s}{\partial P}\right)_T \approx \left(\frac{\Delta s}{\Delta P}\right)_T = \frac{s_{0.32} - s_{0.24}}{P_{0.32} - P_{0.24}} = \frac{0.9749 - 1.0034}{0.32 - 0.24}$$

$$= 0.3563 \text{ kJ/kg·K·MPa}$$

These two estimates of the partial derivatives differ by only 0.8 percent. Some difference is to be expected, since we have used finite approximations to differential quantities.

**Comment:** In the absence of entropy data for a substance, $(\partial s/\partial P)_T$ could be found with reasonable accuracy from $PvT$ data alone.

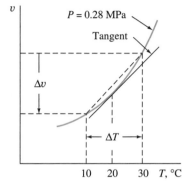

**Figure 12-2**
Illustration for estimating $(\partial v/\partial T)_P$ from superheat data.

## 12-3 GENERALIZED RELATIONS FOR CHANGES IN ENTROPY, INTERNAL ENERGY, AND ENTHALPY

One of the most important functions of thermodynamics is to provide fundamental equations for the evaluation of properties or the change in properties under the most general considerations. These equations, for example, should be independent of the type or phase of a substance. Once these "generalized" equations have been developed, then it will be necessary to provide specific information on a substance if further numerical evaluation of property data is desired. We again restrict our analysis to simple compressible substances. The generalized equations for entropy will be developed first, followed by equations for the internal energy and the enthalpy. This procedure is a matter of choice. It is no more difficult to develop expressions for internal energy and enthalpy first, followed by equations for the entropy change. The independent properties are selected to be any pair of the group $(P, v, T)$. For practical reasons we are usually concerned only with the two pairs of variables $(T, v)$ and $(T, P)$.

Before we proceed, it is necessary to develop two special expressions for $c_v$ and $c_p$. Recall that these two specific heats are defined in terms of partial derivatives, namely,

$$c_v \equiv \left(\frac{\partial u}{\partial T}\right)_v \qquad \text{and} \qquad c_p \equiv \left(\frac{\partial h}{\partial T}\right)_P$$

If we differentiate Eq. [12-6] with respect to temperature and at constant volume, we find that

$$\left(\frac{\partial u}{\partial T}\right)_v = T\left(\frac{\partial s}{\partial T}\right)_v \qquad \text{or} \qquad \left(\frac{\partial s}{\partial T}\right)_v = \frac{c_v}{T} \qquad \textbf{[12-18]}$$

In a similar manner the differentiation of Eq. [12-7] with respect to temperature and at constant pressure leads to

$$\left(\frac{\partial h}{\partial T}\right)_P = T\left(\frac{\partial s}{\partial T}\right)_P \qquad \text{or} \qquad \left(\frac{\partial s}{\partial T}\right)_P = \frac{c_p}{T} \qquad \textbf{[12-19]}$$

These two relations are used in the development of entropy relations in the following subsection. Generalized expressions for the specific heats appear in Sec. 12-4.

### 12-3-1    GENERALIZED ENTROPY RELATIONS

If the entropy is chosen to be a function of $T$ and $v$, we may write that the total differential of $s$ is given by

$$ds = \left(\frac{\partial s}{\partial T}\right)_v dT + \left(\frac{\partial s}{\partial v}\right)_T dv \qquad \textbf{[12-20]}$$

It is now desirable to express $ds$ solely in terms of measurable quantities. This requires replacing the partial derivatives in Eq. [12-20] with other terms that include only the variables $P, v, T$ and the specific heats $c_v$ and $c_p$. The first partial derivative can be replaced by Eq. [12-18]. The second partial derivative may be replaced by the Maxwell relation given by Eq. [12-16]. The result is

$$ds = \frac{c_v\, dT}{T} + \left(\frac{\partial P}{\partial T}\right)_v dv \qquad \textbf{[12-21]}$$

This is the result we sought, since the right-hand side of the equation is now expressed solely in terms of measurable quantities.

An equivalent equation for $ds$ in terms of the variables $T$ and $P$ may be developed by starting with the relation

$$ds = \left(\frac{\partial s}{\partial T}\right)_P dT + \left(\frac{\partial s}{\partial P}\right)_T dP \qquad \textbf{[12-22]}$$

Again, the second partial derivative may be replaced by the Maxwell relation given by Eq. [12-17]. In addition, the first partial derivative may be replaced by Eq. [12-19]. The substitution of these two equations into Eq. [12-22] yields the desired result:

$$ds = \frac{c_p \, dT}{T} - \left(\frac{\partial v}{\partial T}\right)_P dP \qquad \textbf{[12-23]}$$

Integration of Eqs. [12-21] and [12-23] between the same two states should lead to the same values for $\Delta s$, since the change in a property is independent of the method used to calculate it. These equations are *generalized* relations, because they are not restricted to any particular substance or to any particular phase of a substance. They are restricted, however, to simple compressible systems.

Estimate the entropy change for refrigerant 134a which is compressed isothermally at 20°C from 0.24 to 0.32 MPa, in kJ/kg·K, on the basis of tabular $PvT$ data.

Example 12-2

**Solution:**

**Given:** Refrigerant 134a is compressed under the conditions shown in Fig. 12-3.

**Find:** $\Delta s$, in kJ/kg·K.

**Model:** Use of generalized entropy equation and tabular data.

**Analysis:** Under the isothermal conditions of the problem Eq. [12-23] reduces to

$$ds = -\left(\frac{\partial v}{\partial T}\right)_P dP$$

Integration of this expression requires information on the variation of the partial derivative with pressure. Because the pressure change is small, as a first approximation we shall assume that the partial derivative remains essentially constant between 0.24 and 0.32 MPa. In addition, the partial derivative will be evaluated by the finite-difference method discussed in Example 12-1. In this case the above equation becomes

$$\Delta s \approx -\left(\frac{\Delta v}{\Delta T}\right)\Delta P$$

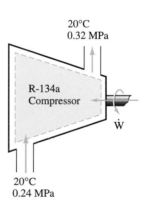

**Figure 12-3**
Schematic and data for Example 12-2.

The value of $\Delta v/\Delta T$ at the intermediate pressure of 0.28 MPa was determined in Example 12-1 to be 0.3535 kJ/kg·K·MPa. Use of this value in the above equation leads to the result

$$\Delta s \approx -0.3535(0.32 - 0.24) = -0.0283 \text{ kJ/kg·K}$$

From the superheat table for R-134a the entropy change is found to be

$$(s_2 - s_1)_T = (0.9749 - 1.0034) = -0.0285 \text{ kJ/kg·K}$$

The superheat values of $s$ have been determined from an accurate $PvT$ equation of state.

**Comment:**   In spite of the approximations involved, the answer is underestimated by only roughly 1 percent. The last two examples illustrate some mathematical methods which are frequently good estimating tools in the absence of an accurate $PvT$ equation of state.

### 12-3-2   GENERALIZED INTERNAL ENERGY AND ENTHALPY RELATIONS

A generalized equation for the internal-energy change is now possible by recalling the first $T\,ds$ equation developed in Sec. 12-2:

$$du = T\,ds - P\,dv \qquad \textbf{[12-6]}$$

The quantity $ds$ is eliminated from this equation by substituting Eq. [12-21] for it. After separation of variables,

$$du = c_v\,dT + \left[T\left(\frac{\partial P}{\partial T}\right)_v - P\right]dv \qquad \textbf{[12-24]}$$

The generalized equation for the change in enthalpy is found by employing the second $T\,ds$ equation, namely,

$$dh = T\,ds + v\,dP \qquad \textbf{[12-7]}$$

Substitution of Eq. [12-23] for $ds$ in Eq. [12-7] and subsequent rearrangement lead to

$$dh = c_p\,dT + \left[v - T\left(\frac{\partial v}{\partial T}\right)_P\right]dP \qquad \textbf{[12-25]}$$

Integration of Eqs. [12-21], [12-23], [12-24], and [12-25] requires experimental knowledge of the $PvT$ behavior of the substance in the region of interest plus experimental information on the relationship between the specific heats and temperature.

The change in enthalpy, for example, is found by integration of Eq. [12-25]. As a result,

$$h_2 - h_1 = \int_1^2 c_p \, dT + \int_1^2 \left[ v - T \left( \frac{\partial v}{\partial T} \right)_P \right] dP \qquad \textbf{[12-26]}$$

To integrate the first term, information is required about the variation of $c_p$ with temperature at a fixed pressure. Frequently this pressure is chosen to be essentially zero, so that ideal-gas specific-heat data $(c_{p,0})$ are used. The integration of the second term requires knowledge of the $PvT$ behavior of the substance for the range of pressure desired at a given temperature. Figure 12-4 shows one possible path of integration between two real-gas states, 1 and 2, on a $Ts$ diagram. For this particular path the first term of Eq. [12-26] is integrated at zero pressure between states $x$ and $y$. The second term in Eq. [12-26] must be integrated twice in this case. One integration is at constant temperature $T_1$ between states 1 and $x$, and the other integration is at $T_2$ between states $y$ and 2.

Owing to the format of the coefficient of $dv$ in Eq. [12-24] for $du$, it is helpful if the $PvT$ equation of state is explicit in $P$. However, an equation explicit in $v$ would be advantageous when Eq. [12-25] for $dh$ is integrated. One can use the first of these equations to find $\Delta h$ by first calculating $\Delta u$ and then finding $\Delta h$ from the definition

$$h_2 - h_1 = u_2 - u_1 + (P_2 v_2 - P_1 v_1) \qquad \textbf{[12-27]}$$

Similarly, if an equation of state explicit in $v$ is to be used, one first calculates $\Delta h$ by Eq. [12-25] and then finds $\Delta u$ from Eq. [12-27].

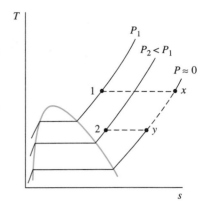

**Figure 12-4**
Possible path of integration between two real-gas states.

---

In Chap. 3 it is pointed out that the internal energy of gases at low pressures can be approximated quite successfully by the relation $du = c_v \, dT$. Demonstrate that this relation holds exactly for ideal gases.

**Example 12-3**

**Solution:**

**Given:**  An ideal gas.

**Find:**  Validate $du = c_v \, dT$.

**Model:**  Generalized equation for internal energy, ideal gas.

**Analysis:**  On the basis of Eq. [12-24], the generalized equation for $du$ reduces to $du = c_v \, dT$ if the coefficient on the $dv$ term is zero. From the ideal-gas relationship, $Pv = RT$, we find that $(\partial P/\partial T)_v = R/v$ for an ideal gas. Substituting this last ideal-gas relation into the coefficient of $dv$ in Eq. [12-24] yields

$$T \left( \frac{\partial P}{\partial T} \right)_v - P = T \left( \frac{R}{v} \right) - P = P - P = 0$$

Thus an ideal gas is always a substance such that the internal energy is truly only a function of temperature, as given by $du = c_v \, dT$. Because many gases at low pressures approximately fulfill the ideal-gas equation, their internal energies also are only a function of temperature, to a high degree of accuracy.

**Comment:** If a similar treatment is carried out for the enthalpy function, it may be shown that the coefficient of $dP$ in Eq. [12-25] is always zero for ideal gases. This proves that, for ideal gases, the enthalpy change is always given by $dh = c_p \, dT$.

**Example 12-4**

Find the change in enthalpy and entropy for a real gas along an isothermal path between pressures $P_1$ and $P_2$. Assume for the range of pressures involved that the $PvT$ behavior of the gas is represented by the relation $v = RT/P + b - a/RT$, where $a$ and $b$ are constants.

**Solution:**

**Given:** The equation of state $v = RT/P + b - a/RT$.

**Find:** Equations for $\Delta h$ and $\Delta s$.

**Model:** Use of generalized equations.

**Analysis:** For an isothermal process the change in enthalpy is given by the second term of Eq. [12-26], namely,

$$(h_2 - h_1)_T = \int_1^2 \left[ v - T \left( \frac{\partial v}{\partial T} \right)_P \right] dP$$

The equation of state is explicit in $v$. Hence $(\partial v/\partial T)_P = R/P + a/RT^2$, and

$$v - T \left( \frac{\partial v}{\partial T} \right)_P = \frac{RT}{P} + b - \frac{a}{RT} - T \left( \frac{R}{P} + \frac{a}{RT^2} \right) = b - \frac{2a}{RT}$$

Therefore,

$$(h_2 - h_1)_T = \left( b - \frac{2a}{RT} \right)(P_2 - P_1)$$

To evaluate the entropy change, we choose Eq. [12-23], rather than Eq. [12-21], because it requires an equation of state explicit in $v$. Hence

$$(s_2 - s_1) = -\int_1^2 \left( \frac{\partial v}{\partial T} \right)_P dP = -\int_1^2 \left( \frac{R}{P} + \frac{a}{RT^2} \right) dP$$

$$= -R \ln \frac{P_2}{P_1} - \frac{a(P_2 - P_1)}{RT^2}$$

## 12-4 GENERALIZED RELATIONS FOR $c_p$ AND $c_v$

Two expressions for the specific heats have already been developed in this chapter:

$$c_v = T \left( \frac{\partial s}{\partial T} \right)_v \qquad \text{[12-18]}$$

$$c_p = T\left(\frac{\partial s}{\partial T}\right)_P \qquad \text{[12-19]}$$

These expressions are generalized relations for $c_v$ and $c_p$, and they may be used in any single-phase region where $PvT$ data are available. An alternate method of evaluating specific-heat data is based on the experimental fact that it is relatively easier to measure specific-heat data at low pressure than at elevated pressures. For example, we have already seen from Chap. 3 that much is known about the specific heats of common gases as a function of temperature at low pressures. Such data were called *zero-pressure* specific heats. Consequently, we are primarily concerned with determining in what manner specific-heat values vary with increasing pressure (or decreasing specific volume) at constant temperature. Such an evaluation must again be based solely on the use of measured $PvT$ data in the desired range of equilibrium states. In a mathematical sense what we seek are expressions for the two terms $(\partial c_v/\partial v)_T$ and $(\partial c_p/\partial P)_T$. Generalized equations for these two expressions are obtained by starting with Eqs. [12-21] and [12-23]:

$$ds = \frac{c_v\,dT}{T} + \left(\frac{\partial P}{\partial T}\right)_v dv \qquad \text{[12-21]}$$

$$ds = \frac{c_p\,dT}{T} - \left(\frac{\partial v}{\partial T}\right)_P dP \qquad \text{[12-23]}$$

These equations are of the form $dx = M\,dy + N\,dz$. Therefore Eq. [12-2] applies, that is, $\partial M/\partial z = \partial N/\partial y$. When applied to Eq. [12-21], the following relation results:

$$\left(\frac{\partial c_v}{\partial v}\right)_T = T\left(\frac{\partial^2 P}{\partial T^2}\right)_v \qquad \text{[12-28]}$$

This is the desired relation. If we start with Eq. [12-23], it can be shown by an analogous procedure that

$$\left(\frac{\partial c_p}{\partial P}\right)_T = -T\left(\frac{\partial^2 v}{\partial T^2}\right)_P \qquad \text{[12-29]}$$

To obtain the value of $c_p$ at an elevated pressure, for example, Eq. [12-29] is integrated from zero pressure to the desired value. Hence

$$c_p - c_{p,0} = -T\int_0^P \left(\frac{\partial^2 v}{\partial T^2}\right)_P dP \qquad \text{[12-30]}$$

where $c_{p,0}$ again is the zero-pressure, or ideal-gas, specific heat. The integration of the right-hand side requires a knowledge of the $PvT$ behavior of the substance in either tabular or analytical form.

Another thermodynamic relation of interest is the difference between the constant-pressure and constant-volume specific heats, that is, $c_p - c_v$. One reason for this interest is that $c_p$ values usually are easier to measure than $c_v$ values. Hence $c_v$ can be evaluated solely from $c_p$ and $PvT$ data. Since the change in any property is not dependent on the method of evaluation, we may equate the two equations for $ds$ previously presented. Equating Eqs. [12-21] and [12-23], we find that

$$\frac{c_v\, dT}{T} + \left(\frac{\partial P}{\partial T}\right)_v dv = \frac{c_p\, dT}{T} - \left(\frac{\partial v}{\partial T}\right)_P dP$$

or

$$\frac{c_p - c_v}{T} = \left(\frac{\partial P}{\partial T}\right)_v dv + \left(\frac{\partial v}{\partial T}\right)_P dP$$

Differentiation with respect to pressure at constant volume yields

$$\frac{c_p - c_v}{T}\left(\frac{\partial T}{\partial P}\right)_v = \left(\frac{\partial v}{\partial T}\right)_P$$

or upon rearrangement,

$$\boxed{c_p - c_v = T\left(\frac{\partial v}{\partial T}\right)_P\left(\frac{\partial P}{\partial T}\right)_v} \qquad \textbf{[12-31]}$$

An equivalent form for $c_p - c_v$ can be found by replacing $(\partial P/\partial T)_v$ in terms of the cyclic rule $(\partial P/\partial T)_v = -(\partial v/\partial T)_P(\partial P/\partial v)_T$. Use of this expression in Eq. [12-31] leads to the relation

$$\boxed{c_p - c_v = -T\left(\frac{\partial v}{\partial T}\right)_P^2\left(\frac{\partial P}{\partial v}\right)_T} \qquad \textbf{[12-32]}$$

A number of important qualitative results stem from Eq. [12-32].

1. On the basis of experimental data it is known that $(\partial P/\partial v)_T$ is always negative for all substances in all phases. Since the first partial derivative in Eq. [12-32] is a squared term, $c_p - c_v$ must always be positive, or zero. This quantity becomes zero on two occasions:

   (a) The first is when $T$ is absolute zero on the thermodynamic scale, if the remaining terms remain finite at this state. Consequently, the specific heats at constant pressure and constant volume are identical at the absolute zero of temperature.

   (b) The specific heats will also be equal if the value of $(\partial v/\partial T)_P$ is ever zero. This occurs, for example, in the case of liquid water at 4°C, where the fluid is at its state of maximum density.

2. Even at temperatures above absolute zero, the difference between $c_v$ and $c_p$ will generally be small for liquids and solids. This is true because the value of $(\partial v/\partial T)_P$ is very small for these two phases. Hence one frequently speaks of the specific heat of a liquid or a solid without specifying the type.

3. The fact that $c_p > c_v$ for the gas phase also leads to the generalization that constant-volume lines always have steeper slopes than constant-pressure lines at the same point on a $Ts$ diagram.

The following example illustrates one use of the generalized specific-heat equations.

---

**Example 12-5**

**D**etermine the isothermal change in $c_p$ with pressure for the gas studied in Example 12-4, for which $v = RT/P + b - a/RT$, and $a$ and $b$ are constants.

**Solution:**

**Given:**   A gas with the equation of state: $v = RT/P + b - a/RT$.

**Find:**   $(c_{p,2} - c_{p,1})_T$.

**Model:**   Use of generalized specific-heat equations.

**Analysis:**   The isothermal change in $c_p$ with pressure is given by Eq. [12-29], namely, $(\partial c_p/\partial P)_T = -T(\partial^2 v/\partial T^2)_P$. The equation of state is $v = RT/P + b - a/RT$, and in Example 13-4 the first derivative was found to be

$$\left(\frac{\partial v}{\partial T}\right)_P = \frac{R}{P} + \frac{a}{RT^2}$$

The second partial derivative, then, is

$$\left(\frac{\partial^2 v}{\partial T^2}\right)_P = -\frac{2a}{RT^3}$$

Consequently,

$$(c_{p2} - c_{p1})_T = -\int_1^2 T\left(-\frac{2a}{RT^3}\right)dP = \frac{2a(P_2 - P_1)}{RT^2}$$

**Comment:**   This example and the preceding one illustrate the importance of an accurate $PvT$ equation of state when evaluating property data from theoretical relationships.

---

The property variation of solids and liquids is often expressed in terms of the **volumetric expansion coefficient** $\beta$ and the **isothermal coefficient of compressibility** $K_T$. These two quantities are defined as

$$\beta = \frac{1}{v}\left(\frac{\partial v}{\partial T}\right)_P \qquad\qquad \textbf{[12-33]}$$

$$K_T = -\frac{1}{v}\left(\frac{\partial v}{\partial P}\right)_T \qquad\qquad \textbf{[12-34]}$$

Substitution of these two equations in Eq. [12-32] leads to

$$c_p - c_v = \frac{vT\beta^2}{K_T} \qquad\qquad \textbf{[12-35]}$$

Note from the defining equations that $\beta$ is expressed in reciprocal absolute temperature units and $K_T$ is expressed in reciprocal pressure units.

The use of $\beta$ and $K_T$ is quite helpful in many calculations since their values may often be assumed to be constant during a given process. The slow variation of these two properties with temperature is illustrated by the data of Table 12-1, which shows data for solid copper and liquid water. As an example, consider the isothermal compression of a solid or liquid. The change in entropy generally is given by Eq. [12-23]. For an isothermal process this reduces to $ds_T = -(\partial v/\partial T)_P\, dP$. If we substitute Eq. [12-33] into this expression, we find that

$$ds_T = -v\beta\, dP$$

This result is easily integrated if we assume that $v$ and $\beta$ are constant over the pressure range. As a result,

$$(s_2 - s_1)_T = -v\beta(P_2 - P_1) \qquad\qquad \textbf{[12-36a]}$$

In addition, if the isothermal compression is reversible, then the heat transfer during the process is found from $q = T\,\Delta s$. Another example is the isothermal work of compression. Rearrangement of Eq. [12-34] leads to

$$dv_T = -K_T v\, dP$$

When this equation is substituted into the integral of $-P\,dv$, then the isothermal work is

$$w_T = -\int_1^2 P\, dv = \int_1^2 v K_T P\, dP$$

**Table 12-1**     $\beta$ and $K_T$ for copper and liquid water as a function of temperature

| (a) Copper | | | (b) Water | | |
|---|---|---|---|---|---|
| $T$, K | $\beta \times 10^6$, K$^{-1}$ | $K_T \times 10^6$, bar$^{-1}$ | $T$, °C | $\beta \times 10^6$, K$^{-1}$ | $K_T \times 10^6$, bar$^{-1}$ |
| 100 | 31.5 | 0.721 | 0 | −68.1 | 50.89 |
| 150 | 41.0 | 0.733 | 10 | 87.9 | 47.81 |
| 200 | 45.6 | 0.748 | 20 | 206.6 | 45.90 |
| 250 | 48.0 | 0.762 | 30 | 303.1 | 44.77 |
| 300 | 49.2 | 0.776 | 40 | 385.4 | 44.24 |
| 500 | 54.2 | 0.837 | 50 | 457.8 | 44.18 |
| 800 | 60.7 | 0.922 | 60 | | 44.50 |

In addition to $v$, the coefficient $K_T$ is nearly constant for liquids and solids over wide ranges of pressure at a given temperature. If we designate the specific volume by its initial value $v_1$, then integration of the above equation yields

$$w_T = \frac{v_1 K_T}{2}(P_2^2 - P_1^2) \qquad \textbf{[12-36b]}$$

This equation provides the means for estimating fairly accurately the work required to compress a solid or liquid under isothermal conditions.

---

**A**t 500 K the specific volume of copper is 7.115 cm³/mol. Other data are given in Table 12-1.

Example 12-6

    (a) Determine the value of $\bar{c}_p - \bar{c}_v$ in J/mol· K.

    (b) If the value of $\bar{c}_p$ is 26.15 J/mol·K at this temperature, what percentage error would be made in $\bar{c}_v$ if we assume that $\bar{c}_p = \bar{c}_v$?

**Solution:**

**Given:** Solid copper at 500 K with $\bar{v} = 7.115$ cm³/mol.

**Find:** (a) $\bar{c}_p - \bar{c}_v$, (b) percentage error in $\bar{c}_v$ if $\bar{c}_p = \bar{c}_v$.

**Model:** Generalized specific-heat relations.

**Analysis:** (a) The difference between $\bar{c}_p$ and $\bar{c}_v$ is found directly by substituting the appropriate values into Eq. [12-35]. Thus

$$\bar{c}_p - \bar{c}_v = \frac{\bar{v}T\beta^2}{K_T} = \frac{7.115(500)(54.2 \times 10^{-6})^2}{0.837 \times 10^{-7}} \frac{\text{N·cm}}{\text{mol·K}} \times \frac{\text{m}}{10^2 \text{ cm}}$$

$$= 1.249 \text{ J/mol·K}$$

    (b) If $\bar{c}_p = 26.15$ J/mol·K, then $\bar{c}_v = 26.15 - 1.25 = 24.90$ J/mol·K. Therefore, the percentage of error in assuming that $\bar{c}_v$ equals $\bar{c}_p$ is 1.249/24.90 = 0.050, or 5 percent. As a result, one must be careful not to assume that $\bar{c}_p$ and $\bar{c}_v$ are equal for solid materials if the temperature is sufficiently high.

---

## 12-5    VAPOR PRESSURE AND THE CLAPEYRON EQUATION

The dependency of the vapor pressure on the saturation temperature will now be developed from theoretical considerations. The generalized relationship that results is also valid for solid-gas and solid-liquid phase changes. We begin by calculating the entropy change during a phase change. This entropy change in terms of the variables $v$ and $T$ has already been presented in the form of Eq. [12-20], namely,

$$ds = \left(\frac{\partial s}{\partial T}\right)_v dT + \left(\frac{\partial s}{\partial v}\right)_T dv \qquad \textbf{[12-20]}$$

However, the temperature is constant for any process involving a phase change. Therefore, the $dT$ term in Eq. [12-20] above is zero. In addition, the Maxwell relation given by Eq. [12-16] shows that $(\partial s/\partial v)_T = (\partial P/\partial T)_v$. Hence Eq. [12-20] for a phase change becomes

$$ds = \left(\frac{\partial P}{\partial T}\right)_v dv \qquad \text{or} \qquad \Delta s = \int \left(\frac{\partial P}{\partial T}\right)_v dv$$

The term $(\partial P/\partial T)_v$ is the slope of the saturation curve at a given saturation state, as shown in Fig. 12-5. This quantity is independent of the volume during a change of phase. Consequently, the partial derivative may be written as a total derivative $(dP/dT)_{\text{sat}}$, and it may be moved outside the integral sign during the integration of the above equation. Integration leads to

$$s_2 - s_1 = \left(\frac{dP}{dT}\right)_{\text{sat}} (v_2 - v_1)$$

$$\left(\frac{dP}{dT}\right)_{\text{sat}} = \frac{s_2 - s_1}{v_2 - v_1} \qquad \textbf{[12-37]}$$

where the subscripts 1 and 2 represent the saturation phases for the process. For example, they may represent the saturated-vapor and saturated-liquid phases during a vaporization process.

An alternative form of Eq. [12-37] can be written in terms of the enthalpy change during a phase change. Equation [12-7] states that

$$dh = T \, ds + v \, dP \qquad \textbf{[12-7]}$$

For a phase change $dP$ is zero and $T$ is a constant. Thus $ds = dh/T$ and $s_2 - s_1 = (h_2 - h_1)/T$. Equation [12-37] then becomes

$$\left(\frac{dP}{dT}\right)_{\text{sat}} = \frac{h_2 - h_1}{T(v_2 - v_1)} = \frac{\Delta h}{T \, \Delta v} \qquad \textbf{[12-38a]}$$

Equation [12-38a] is called the ***Clapeyron equation***. In general, $\Delta h$ and $\Delta v$ are the enthalpy and volume changes between the two saturation states at the given pressure or temperature. For a liquid-vapor phase change this equation is written as

$$\left(\frac{dP}{dT}\right)_{\text{sat}} = \frac{h_{fg}}{Tv_{fg}} \qquad \textbf{[12-38b]}$$

This equation allows us to evaluate $h_{fg}$ from a knowledge of only $PvT$ data, that is, from the slope of the saturation curve on a $PT$ diagram and $v_f$ and $v_g$ at the given temperature.

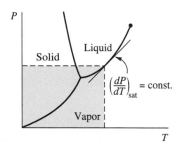

**Figure 12-5**
Illustration of the slope of the liquid-vapor saturation line.

**Example 12-7**

Estimate the enthalpy of vaporization of water at 200°C, using $PvT$ data from Table A-12.

**Solution:**

**Given:** A wet mixture of water at 200°C.

**Find:**    $h_{fg}$, in kJ/kg.

**Model:**    The Clapeyron equation.

**Analysis:**    On the basis of Eq. [12-38b],

$$h_{fg} = T \, \Delta v \left( \frac{dP}{dT} \right)_{\text{sat}}$$

The value of $dP/dT$ may be approximated to a reasonable degree by $\Delta P/\Delta T$ in the region of interest. The saturation data at a 10°temperature interval on either side of 200°C are as follows: At 190°C the saturation pressure is 12.54 bars, and at 210°C the pressure is 19.06 bars. Hence

$$\left( \frac{dP}{dT} \right)_{\text{sat}} \approx \left( \frac{\Delta P}{\Delta T} \right)_{\text{sat}} = \frac{(19.06 - 12.54)\text{bars}}{(210 - 190)°\text{C}} = 0.326 \text{ bar/K}$$

The change in volume $v_{fg}$ at 200°C is given as 0.1262 m³/kg. Substitution of these values into the Clapeyron equation yields

$$h_{fg} = 473(0.1262)(0.326)(100) = 1946 \text{ kJ/kg}$$

where the factor 100 converts units of m³·bar to kJ.

**Comments:**

1. Table A-12 lists the value of $h_{fg}$ as 1941 kJ/kg. An error of roughly 0.26 percent results from this technique.

2. The slope of the saturation line also could be found by a polynomial curve-fit of the saturation data in the vicinity of 200°C. Differentiation of the curve-fit equation would lead to an improved value for $(dP/dT)_{\text{sat}}$.

3. The entropy of vaporization would be found from the relation $s_{fg} = h_{fg}/T$.

---

For liquid-vapor and solid-vapor phase changes Eq. [12-38] can be further modified by introducing several approximations. For purposes of discussion we consider only the first of these, but the results are equally applicable to solid-vapor phase changes. For liquid-vapor phase changes at relatively low pressures, the value of $v_g$ is many times the size of $v_f$. Thus a good approximation is to replace $v_{fg}$ by $v_g$ in the above equations. Also, at these low pressures, the $PvT$ relation for the vapor closely follows that for an ideal gas; that is, $v_g = RT/P$. By making these two successive approximations in Eq. [12-38a], we find that

$$\frac{dP}{dT} = \frac{Ph_{fg}}{RT^2}$$

or

$$\frac{dP}{P} = \frac{h_{fg} \, dT}{RT^2} \qquad\qquad \textbf{[12-39]}$$

Equation [12-39] is frequently called the ***Clausius-Clapeyron equation***. Integration of this equation depends on the variation of $h_{fg}$ with temperature. If a small variation of pressure (or temperature) is chosen so that the change in $h_{fg}$ over the interval of integration is small, then integration yields

$$\ln P = -\frac{h_{fg}}{R}\left(\frac{1}{T}\right) + C \qquad \textbf{[12-40]}$$

where $C$ is a constant of integration. This equation indicates that the vapor pressure of a liquid is very closely an exponential function of the saturation temperature. The general form of the equation is also valid for saturation data below the triple state in the sublimation region.

Remember that Eq. [12-40] is only an approximation. A more accurate analytical expression for the variation of the saturation pressure with respect to temperature requires that additional terms be added. For example, a better approximation might be given by an equation of the form

$$\ln P_{\text{sat}} = A + \frac{B}{T} + C \ln T + DT + ET^2 + \cdots \qquad \textbf{[12-41]}$$

The constants $A$, $B$, $C$, etc., are adjusted to obtain the best fit with the experimental data. Nevertheless, the simple exponential form given by Eq. [12-40] is fairly accurate in many cases.

An equation like Eq. [12-41] is important for the following reason. If the equation represents a precise fit to experimental data, then the derivative of the equation will give an accurate value of $dP/dT$. Substitution of this value into the Clapeyron equation, (Eq. [12-38]), will lead to an accurate evaluation of $\Delta h$ for a liquid-gas or a solid-gas phase transformation.

The relationship given by Eq. [12-38] is quite useful to demonstrate one other point: The slope of a saturation line on a $PT$ diagram apparently depends on the signs of $\Delta h$ and $\Delta v$. In most cases, when heat is added to a closed system to bring about a phase change, the volume also increases. Hence $dP/dT$ is usually positive. However, in the case of melting of water and a few other substances, the volume decreases. The slope of the melting curve on a phase diagram for these few substances must then be negative. This was pointed out in Chap. 3 during the discussion of phase diagrams, but now the Clapeyron equation theoretically substantiates what is empirically observed. The freezing temperature of any substance that expands on freezing is lowered when the pressure is increased.

---

**Example 12-8**

The saturation pressure and the enthalpy of vaporization of refrigerant 134a at 20°C are found to be 5.716 bars and 181.09 kJ/kg, respectively. Without any additional experimental data, estimate the saturation pressure at 0°C.

**Solution:**

**Given:**   Refrigerant 134a at 20°C with $P_{\text{sat}}$ of 5.716 bars and $h_{fg}$ of 181.09 kJ/kg.

**Find:**    $P_{sat}$ at 0°C.

**Model:**    Clausius-Clapeyron equation, $h_{fg}$ is constant.

**Analysis:**    If the value of $h_{fg}$ is assumed constant, then definite integration of Eq. [12-39] leads to

$$\ln \frac{P_2}{P_1} = \frac{h_{fg}(T_2 - T_1)}{RT_1T_2}$$

Although it would be better to use an average value for $h_{fg}$ between the temperatures of interest, we must use the value at $T_1$ for lack of information. Substituting in values, we find that

$$\ln \frac{P_2}{P_1} = \frac{181.09(102.03)(0 - 20)}{8.314(293)(273)} = -0.556$$

Therefore,

$$P_2/P_1 = 0.574 \quad \text{and} \quad P_2 = 0.574(5.716) = 3.28 \text{ bars}$$

**Comment:**    The tabulated value based on a more accurate method of evaluation is 2.93 bars. Although the estimate is 12 percent too high, it is a fair approximation in the absence of further experimental data, since $h_{fg}$ does vary by 5 percent in the given temperature range.

## 12-6    THE JOULE-THOMSON COEFFICIENT

The adiabatic throttling of fluids through some type of flow restriction has been discussed in Sec. 5-6-3. A schematic of the equipment is shown in Fig. 12-6a. The effect of the restriction is a significant pressure drop. For

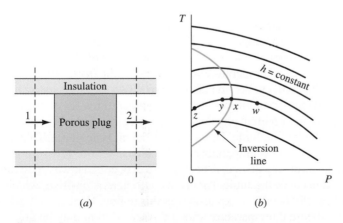

(a)                                    (b)

**Figure 12-6**    (a) Equipment for Joule-Thomson experiment. (b) A plot of data resulting from typical throttling measurements.

such processes the work is zero and changes in kinetic and potential energies are made negligible. As a result, the steady-state energy equation reduces to

$$h_1 = h_2$$

The throttling process has a number of important scientific and engineering applications.

One of the important effects of throttling is the change in temperature which accompanies the drop in pressure. It is an experimental fact that throttling of a fluid leads to a final temperature which may be higher or lower than the initial value, depending on the values of $P_1$, $T_1$, and $P_2$. The mathematical measure of this effect is given by the *Joule-Thomson coefficient,* which is defined as

$$\mu_{JT} = \left(\frac{\partial T}{\partial P}\right)_h \qquad\qquad \textbf{[12-42]}$$

The Joule-Thomson (JT) coefficient may be readily determined at various states by plotting experimental data in terms of a family of constant-enthalpy lines on a $TP$ diagram. To obtain this plot, the values of $P_1$ and $T_1$ upstream from the restriction are held fixed, and the pressure $P_2$ downstream is varied experimentally. For each setting of $P_2$ the downstream temperature $T_2$ is measured. Under throttling conditions the state for each measurement made downstream has the same enthalpy as the initial state upstream. After a sufficient number of measurements are made downstream for a given state upstream, a line of constant enthalpy can be drawn on the $TP$ diagram. Then either the initial pressure or temperature is altered, and the measurement procedure is repeated for this new value of the enthalpy. In this manner a whole family of constant-enthalpy lines on a $TP$ plot may be obtained. A typical result is shown in Fig. 12-6*b*. The slope of a constant-enthalpy line at any state is a measure of the Joule-Thomson coefficient at that state, i.e., a measure of $(\partial T/\partial P)_h$.

Figure 12-6*b* shows that a number of the constant-enthalpy lines have a state of maximum temperature. The line shown in the figure which passes through these states of maximum temperature is called the *inversion line*, and the value of the temperature for that state is the *inversion temperature*. A pressure line will cut the inversion curve at two different states; hence one speaks of the upper and lower inversion temperatures for a given pressure. This line has an important physical significance. To the right of the inversion line on a $TP$ plot, the Joule-Thomson coefficient is negative. That is, in this particular region the temperature will increase as the pressure decreases through the throttling device. A heating effect occurs. However, to the left of the inversion curve the Joule-Thomson coefficient is positive, which means that cooling will occur for expansions in this region. Hence, on the throttling of a fluid, the final temperature after a flow restriction may be greater than, equal to, or less than the initial temperature, depending on the final pressure for any given set of initial conditions. For example, in Fig. 12-6*b* a typical initial state might be point $w$. Expansion to the inversion curve (point $x$) results in heating of the fluid. If further expansion to point $y$ is permitted, some cooling will occur, but this is not sufficient to lower the temperature to that of the initial state. However, if expansion to point $z$ is possible, enough cooling will occur to bring the final temperature to a lower value than that for the initial state.

Also note that for some initial states a cooling process is impossible. The upper part of the inversion curve passes through zero pressure at some finite temperature for all substances. Consequently, many enthalpy lines at high temperatures never pass through the inversion line, as seen in Fig. 12-6b. For these enthalpy lines the Joule-Thomson coefficient is always negative throughout the range of pressures. Examples are hydrogen and helium, which have negative coefficients at ordinary temperatures and low pressures. Hence, for these two gases the temperature must be artificially lowered considerably before throttling can be employed for an additional cooling effect. For most substances, however, at ordinary temperatures the Joule-Thomson coefficient is negative at high pressures, and it becomes positive at low pressures. For a given pressure drop the maximum cooling effect is attained only if the initial state lies on the inversion line. If the initial state lies to the right of the inversion curve, part of the expansion results in heating, which counters the desired effect.

The importance of the Joule-Thomson coefficient is severalfold. First, other intrinsic properties may be related or evaluated from measurements of the coefficient. These include, for example, specific volumes, specific heats, and enthalpies. Second, the result of throttling in some situations is a reduction in the temperature of the fluid. Thus lower temperatures may be achieved with a device that has no moving parts. In fact, under proper conditions, it is possible that one or more of the components of a gas flow stream might pass into the liquid phase during a throttling process. Such liquefaction provides a means of separating components of a gas mixture. Solidification of a gas, such as carbon dioxide, is also possible by throttling.

It is useful to examine the Joule-Thomson coefficient in terms of a generalized equation, that is, its relation to the variables $P$, $v$, and $T$ and the specific heats. This is easily obtained by recalling the generalized equation for the enthalpy, namely,

$$dh = c_p \, dT + \left[ v - T \left( \frac{\partial v}{\partial T} \right)_P \right] dP \qquad \textbf{[12-25]}$$

By differentiating this equation with respect to the pressure at constant enthalpy, one obtains the following result:

$$\left( \frac{\partial T}{\partial P} \right)_h = \frac{1}{c_p} \left[ T \left( \frac{\partial v}{\partial T} \right)_P - v \right] = \mu_{JT} \qquad \textbf{[12-43]}$$

Thus the Joule-Thomson coefficient may be calculated from a knowledge of the $PvT$ relationship of the fluid and the specific heat at constant pressure for that state. In practice, one could use the Joule-Thomson coefficient, which is easily measured, to evaluate specific-heat data at elevated pressures. If this generalized relationship is applied to an ideal gas, an interesting result occurs. We find that since $(\partial v/\partial T)_P$ for an ideal gas is simply $R/P$,

$$\mu_{\text{JT, ideal gas}} = \frac{1}{c_p}\left(\frac{RT}{P} - v\right) = 0$$

Hence an ideal gas undergoes no change in temperature upon throttling. This is not surprising since it has already been pointed out in Chap. 4 that the enthalpy of an ideal gas is solely a function of its temperature. If the initial and final enthalpies of a fluid are equal for a throttling process by definition, an ideal gas under this condition would also have the same initial and final temperatures.

**Example 12-9**

**E**stimate the value of the Joule-Thomson coefficient for steam at 360°C and 30 bars, in °C/bar, on the basis of superheat data.

**Solution:**

**Given:**   Steam at 360°C and 30 bars.

**Find:**   $\mu_{\text{JT}}$, in °C/bar.

**Model:**   $\mu_{\text{JT}} = (\partial T/\partial P)_h$.

**Analysis:**   To estimate the Joule-Thomson coefficient from superheat data, we write the definition of $\mu_{\text{JT}}$ in finite-difference form.

$$\mu_{\text{JT}}\left(\frac{\partial T}{\partial P}\right)_h \approx \left(\frac{\Delta T}{\Delta P}\right)_h$$

The value of $h$ at the given state is 3138.7 kJ/kg. As shown in Fig. 12-7, the pressures of 20 and 40 bars are chosen for evaluating $(\Delta T/\Delta P)_h$. The temperatures at these pressures are found by linear interpolation in the superheat table. Thus

At $P = 20$ bars and $h = 3138.7$ kJ/kg:      $T = 350.8°C$
At $P = 40$ bars and $h = 3138.7$ kJ/kg:      $T = 368.9°C$

Therefore, at 30 bars and 360°C,

$$\mu_{\text{JT}} \approx \frac{(368.9 - 350.8)°C}{(40 - 20) \text{ bars}} = 0.906°C/\text{bar}$$

Because the coefficient is positive, the temperature will decrease as the pressure decreases during the throttling process.

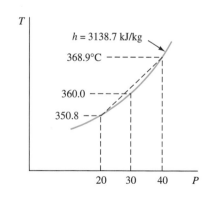

**Figure 12-7**
A *TP* plot for estimating Joule-Thomson coefficient.

**Comment:**   The Joule-Thomson coefficient could also be found from Eq. [12-43]. This would require that $(\partial v/\partial T)_p$ be evaluated in a manner like that used to find $(\partial T/\partial P)_h$. Information on $c_p$ is also required, which could be estimated from $(\partial h/\partial T)_p$.

## 12-7   GENERALIZED THERMODYNAMIC CHARTS

The principle of corresponding states discussed in Sec. 4-8 is extremely useful in predicting property values other than $P$, $v$, and $T$. These three values were previously correlated through the compressibility factor $Z$ and the reduced properties $P_r$, $v_r'$, and $T_r$. The compressibility factor and reduced coordinates may be used to evaluate such properties as the enthalpy, the entropy, and the specific heat at constant pressure for gases at elevated pressures. The usefulness of such a method is that only the critical pressure and temperature are required for any given substance. The correlations for these properties again are presented in graphical form. The method of evaluation involves the generalized equations previously developed in this chapter.

### 12-7-1   GENERALIZED ENTHALPY CHART

Recall from Sec. 12-3 that the enthalpy of simple compressible substances may be evaluated from the generalized equation

$$dh = c_p \, dT + \left[ v - T \left( \frac{\partial v}{\partial T} \right)_P \right] dP \qquad \textbf{[12-25]}$$

The first term on the right of the above equation is not too difficult to evaluate, since it requires a knowledge only of the variation of $c_p$ with temperature at the desired pressure. However, the variation of $h$ with pressure is not so straightforward because it requires a knowledge of the $PvT$ behavior of each substance of interest. Since detailed data for many compounds will be lacking, a more general method must be employed.

At constant temperature the enthalpy change is given by

$$dh_T = \left[ v - T \left( \frac{\partial v}{\partial T} \right)_P \right] dP$$

If the compressibility relation $Pv = ZRT$ is used, one finds that

$$dh_T = \left[ \frac{ZRT}{P} - \frac{ZRT}{P} - \frac{RT^2}{P} \left( \frac{\partial Z}{\partial T} \right)_P \right] dP = -\frac{RT^2}{P} \left( \frac{\partial Z}{\partial T} \right)_P dP$$

Before we integrate this expression, it must be transformed to reduced coordinates so that the result will be of general validity. By definition, $T = T_c T_r$ and $P = P_c P_r$. Hence

$$dT = T_c dT_r \qquad \text{and} \qquad dP = P_c dP_r$$

Substitution of these expressions into the equation for $dh_T$ yields

$$dh_T = -\frac{RT_c^2 T_r^2}{P_c P_r} \left( \frac{\partial Z}{T_c \partial T_r} \right)_{P_r} P_c dP_r = -RT_c T_r^2 \left( \frac{\partial Z}{\partial T_r} \right)_{P_r} d \ln P_r$$

Upon integrating at constant temperature, we obtain

$$\frac{\Delta h_T}{RT_c} = -\int_i^f T_r^2 \left(\frac{\partial Z}{\partial T_r}\right)_{P_r} d \ln P_r \qquad \textbf{[12-44]}$$

where the symbols $i$ and $f$ signify the initial and final reduced-pressure limits. For convenience, the enthalpy is evaluated from the ideal-gas to a real-gas state at the same temperature. The lower limit on the right, then, is zero pressure, for which state $P_r$ is likewise zero. The enthalpy of an ideal gas is indicated by an asterisk, that is, $h^*$. The upper limit is the actual real-gas enthalpy $h$ at some elevated pressure $P$. Hence

$$\frac{h^* - h}{RT_c} = \int_0^P T_r^2 \left(\frac{\partial Z}{\partial T_r}\right)_{P_r} d \ln P_r \qquad \textbf{[12-45]}$$

The value of the integral is obtained by graphical integration, employing data from the generalized compressibility chart. The result of integration leads to values of the **departure function** $(h^* - h)/RT_c$ as a function of $P_r$ and $T_r$. A plot of these data is called a **generalized enthalpy chart**, and a typical chart is shown in Fig. A-30. The use of this chart is illustrated in the following examples.

**Example 12-10**

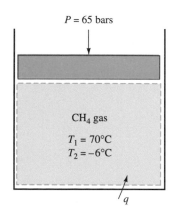

$P = 65$ bars

CH$_4$ gas

$T_1 = 70°C$
$T_2 = -6°C$

$q$

**Figure 12-8**
Schematic and data for Example 12-10.

**M**ethane gas (CH$_4$) is cooled from 70 to $-6°C$ in a closed system maintained at 65 bars. The zero-pressure specific heat is given by $\bar{c}_{p,0} = 18.9 + 0.0555T$, where $T$ is in kelvins and the specific heat is in kJ/kmol·K. Calculate the heat transfer, in kJ/kg, (a) using the generalized chart (Fig. A-30) and (b) assuming ideal-gas behavior.

**Solution:**

**Given:**   Methane gas is cooled under the conditions shown in Fig. 12-8.

**Find:**   $q$, in kJ/kg, by using (a) the generalized chart, and (b) ideal-gas relations.

**Model:**   Closed, constant pressure, $\bar{c}_{p,0} = 18.9 + 0.0555T$.

**Analysis:**   A sketch of the process from state 1 to state 2 on a $Ts$ diagram is shown in Fig. 12-9. The actual path follows the constant-pressure line. The energy balance for a closed system is $q + w = \Delta u$. At constant pressure $w = -P \Delta v$. Therefore, the energy equation reduces to

$$q = \Delta u + P \Delta v = \Delta h$$

However, the enthalpy change $h_2 - h_1$ cannot be obtained directly from the generalized enthalpy chart, since the chart always gives the enthalpy difference between the actual state and the ideal-gas state at the same temperature. Hence the sketch also shows a line of approximately zero pressure, which is the ideal-gas state. The generalized enthalpy chart (Fig. A-30) gives the enthalpy change between such states as 1 and 1* or 2 and 2*. Since the enthalpy is a state function, we may calculate the change of this property by the path 1-1*-2*-2 as well as by the direct path 1-2.

Consequently

$$q = \Delta h = h_2 - h_1 = (h_1^* - h_1) + (h_2^* - h_1^*) - (h_2^* - h_2)$$

The first and third terms on the right are obtained from Fig. A-30, the generalized enthalpy chart.

(a) From the statement of the problem,

$$P_{r1} = \frac{65}{46.4} = 1.40 \qquad P_{r2} = 1.40$$

$$T_{r1} = \frac{343}{191} = 1.80 \qquad T_{r2} = \frac{267}{191} = 1.40$$

At these values we find from the generalized enthalpy chart that $(h* - h)/RT_c$ at states 1 and 2 has values of 0.38 and 0.80, respectively. Hence

$$h_1^* - h_1 = 0.38(8.314/16.04)(191) = 37.6 \text{ kJ/kg}$$

$$h_2^* - h_2 = 0.80(8.314/16.04)(191) = 79.2 \text{ kJ/kg}$$

The value of $h_2^* - h_1^*$ is found from the integration of the zero-pressure specific-heat data; that is,

$$h_2^* - h_1^* = \int_{343}^{267} \frac{(18.9 + 0.0555T)}{16.04} dT = \frac{-1436 - 1286}{16.04} = -169.7 \text{ kJ/kg}$$

Substitution of these values into the energy balance yields

$$q = (h_1^* - h_1) + (h_2^* - h_1^*) - (h_2^* - h_2)$$
$$= 37.6 + (-169.7) - 79.2$$
$$= -211.3 \text{ kJ/kg}$$

(b) If the effect of pressure is neglected, the change in enthalpy for the process is the same as for an ideal gas; that is,

$$q = h_2^* - h_1^*$$

This value has already been found to be $-169.7$ kJ/kg. Thus a 20 percent error will result if the effect of pressure on the enthalpy is neglected.

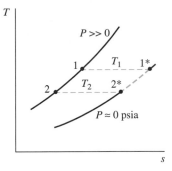

**Figure 12-9**
A $Ts$ process diagram for Example 12-10.

## 12-7-2 GENERALIZED ENTROPY CHART

For a number of different processes, including those with heat transfer, it is important to have a generalized entropy chart as well as the enthalpy chart. The entropy chart is based on the generalized equation for the entropy change of a simple compressible substance, namely,

$$ds = \frac{c_p \, dT}{T} - \left(\frac{\partial v}{\partial T}\right)_P dP \qquad \textbf{[12-23]}$$

As in the situation for the enthalpy function, we note that the first term on the right requires only specific-heat data for a substance at the required pressure. The term in Eq. [12-23] that is difficult to evaluate in some cases is the second one, simply because sufficient $PvT$ data may not be available for the substance of interest. Therefore, a generalized approach is necessary in these cases. Following the procedure for the enthalpy function, we integrate Eq. [12-23] from essentially zero pressure to the desired pressure at constant temperature. This is denoted mathematically by

$$(s_P - s_0^*)_T = -\int_0^P \left(\frac{\partial v}{\partial T}\right)_P dP \qquad \textbf{[12-46]}$$

Normally, the next procedure would be to insert the definition of the compressibility factor and the reduced pressure and temperature into this expression. However, Eq. [12-46] is not directly useful because the entropy at the ideal-gas state of zero pressure is infinite in value. This dilemma is circumvented in the following way: We apply Eq. [12-23] to an isothermal change between zero pressure and the given pressure $P$, but we assume that the gas behaves as an ideal gas at all times. Since $(\partial v/\partial T)_P = R/P$ for an ideal gas, then

$$(s_P^* - s_0^*)_T = -\int_0^P \left(\frac{\partial v}{\partial T}\right)_P dP = -R\int_0^P \frac{dP}{P} \qquad \textbf{[12-47]}$$

The state represented by $s_P^*$ is a fictional state since an ideal gas exists only at zero pressure. However, we may still assign values to this state even if it is nonexistent. If Eq. [12-46] is now subtracted from Eq. [12-47], we obtain

$$(s_P^* - s_P)_T = -\int_0^P \left[\frac{R}{P} - \left(\frac{\partial v}{\partial T}\right)_P\right] dP \qquad \textbf{[12-48]}$$

From the definition of the compressibility factor, $Z = Pv/RT$,

$$\left(\frac{\partial v}{\partial T}\right)_P = \frac{RZ}{P} + \frac{RT}{P}\left(\frac{\partial Z}{\partial T}\right)_P$$

The use of this equation permits Eq. [12-48] to be written as

$$(s_P^* - s_P)_T = -R\int_0^P \left[\frac{1 - Z}{P} - \frac{T}{P}\left(\frac{\partial Z}{\partial T}\right)_P\right] dP$$

This latter result can now be expressed in terms of reduced properties as

$$(s_P^* - s_P)_T = -R\int_0^{P_r} \frac{1 - Z}{P_r} dP_r + RT_r\int_0^{P_r} \left(\frac{\partial Z}{\partial T_r}\right)_{P_r} \frac{dP_r}{P_r}$$

By comparing the last term of this equation with Eq. [12-45], one finds that this term can be written as a function of $h^* - h$. The final result is

$$\frac{(s_P^* - s_P)_T}{R} = \frac{h^* - h}{RT_rT_c} - \int_0^{P_r} (1 - Z)\frac{dP_r}{P_r} \qquad \textbf{[12-49]}$$

The value of the first term on the right is available from the generalized enthalpy chart. The last term on the right must be evaluated by graphical integration of compressibility-factor data. Equation (12-49) permits the evaluation of the departure of the entropy value from the ideal-gas value at the same pressure and temperature. A graphical presentation of the **departure function** $(s_P^* - s_P)_T/R$ versus the reduced pressure and temperature is shown in Fig. A-31 in the form of a generalized entropy chart.

The departure function presented graphically in Fig. A-31 is used in the following manner. Since entropy is a property, its change is independent of the path chosen to evaluate it. Thus, between two real-gas states we may write

$$s_2 - s_1 = (s_{P_1}^* - s_{P_1})_{T_1} + (s_2^* - s_1^*) - (s_{P_2}^* - s_{P_2})_{T_2} \quad \textbf{[12-50]}$$

The first and third terms on the right are found from the generalized entropy chart for the initial and final states. The middle term is simply the entropy change of an ideal gas between the initial and final states. But this latter term is given by either

$$s_2^* - s_1^* = c_{p,\mathrm{av}} \ln \frac{T_2}{T_1} - R \ln \frac{P_2}{P_1} \quad \textbf{[12-51]}$$

or

$$s_2^* - s_1^* = s_2^0 - s_1^0 - R \ln \frac{P_2}{P_1} \quad \textbf{[12-52]}$$

Substitution of Eq. [12-52] into Eq. [12-50], for example, shows that

$$s_2 - s_1 = (s_{P_1}^* - s_{P_1})_{T_1} + s_2^0 - s_1^0 - R \ln \frac{P_2}{P_1} - (s_{P_2}^* - s_{P_2})_{T_2} \quad \textbf{[12-53]}$$

Of course, Eq. [12-51] could be used in place of Eq. [12-52] when Eq. [12-53] is developed. Besides the evaluation of entropy changes for real gases, Eq. [12-53] is also useful for the evaluation of isentropic processes of real gases.

---

**C**arbon dioxide is compressed reversibly and adiabatically from 1 bar and 220 K to 40 bars in a steady-flow process. Determine the final temperature with the aid of a generalized entropy chart if the gas at 1 bar is ideal.

**Example 12-11**

**Solution:**

**Given:** Carbon dioxide is compressed reversibly and adiabatically in steady flow as shown in Fig.12-10.

**Find:** $T_2$, in kelvins.

**Model:** Steady, reversible, adiabatic flow; generalized $s$ chart.

**Analysis:** The process is reversible and adiabatic, hence isentropic. Thus we seek the value of $T_2$ for which $s_2 = s_1$. When Eq. [12-53] is employed, we find

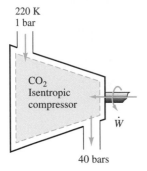

220 K
1 bar

CO$_2$
Isentropic
compressor

$\dot{W}$

40 bars

**Figure 12-10**
Schematic and data for
Example 12-11.

that the first term on the right is zero, since the initial state is one of an ideal gas. Consequently the equation on a molar basis reduces to

$$\bar{s}_1 - \bar{s}_2 = \bar{s}_1^0 - \bar{s}_2^0 + R_u \ln \frac{P_2}{P_1} + (\bar{s}_2^* - \bar{s}_2)_{\text{chart}} = 0$$

The values of $s_2^0$ and the chart correction are unknown, however, since the final temperature is not known. Hence the solution becomes one of iteration. A trial temperature is chosen repeatedly until the above equation is satisfied.

The following data are available:

$$P_{r2} = \frac{40}{73.9} = 0.541 \qquad \bar{s}_1^0 = 202.966 \text{ kJ/kmol·K}$$

$$R_u \ln \frac{P_2}{P_1} = 8.314 \ln 40 = 30.67 \text{ kJ/kmol·K}$$

As a first trial we assume that $T_2$ is 500 K. On this basis,

$$T_{r2} = \frac{500}{304} = 1.64 \qquad \frac{s_2^* - s_2}{R_u} = 0.073 \qquad \bar{s}_2^0 = 234.814 \text{ kJ/kmol·K}$$

Substitution of all the data into the equation for $\bar{s}_1 - \bar{s}_2$ yields

$$\bar{s}_1 - \bar{s}_2 = 202.97 - 234.81 + 30.67 + 0.073(8.314) = -0.56 \text{ kJ/kmol·K}$$

The assumption of $T_2$ equal to 500 K gives a total entropy change close to zero. If a further assumption of $T_2 = 490$ K is made, we find that

$$\bar{s}_1 - \bar{s}_2 = 202.97 - 233.92 + 30.67 + 0.078(8.314) = 0.37 \text{ kJ/kmol·K}$$

The first temperature guess leads to a small negative $\Delta s$ value, while the second evaluation yields a small positive $\Delta s$. Hence the best approximation to the final temperature lies between 490 and 500 K, and is probably around 494 K.

---

Although only two generalized charts have been presented here, clearly any number of these might be devised, once a generalized equation is available for a property in terms of the variables $P$ and $T$. For example, a generalized chart that permits the estimation of $c_p$ values at high pressures is available. A generalized chart for a property called the fugacity is also found to be quite useful. In the absence of an abundance of $PvT$ data for a substance, generalized charts are powerful tools for predicting the properties of a fluid—gas or liquid.

## 12-7-3 GENERALIZED CHARTS APPLIED TO GAS MIXTURES

In Chap. 10 the internal energy, enthalpy, and entropy of an ideal-gas mixture are determined by adding the contributions from the individual components. That is, on a mole basis,

$$u_m = \sum y_i u_i \qquad h_m = \sum y_i h_i \qquad s_m = \sum y_i s_i$$

This same additivity rule can be applied to real-gas mixtures, with several words of caution. First, the component properties $u_i$, $h_i$, and $s_i$ must be evaluated at the mixture pressure and temperature, and not at the component pressures. If the mixture volume and temperature are given data, one must first use Dalton's rule of additive pressure [see Eq. [12-12]] to find the appropriate mixture pressure. Second, this additive rule for $u_m$, $h_m$, and $s_m$ gives *approximate* results, similar to applying Dalton's or Amogat's rule for $PvT$ data to real-gas mixtures.

The values of $u_i$, $h_i$, and $s_i$ are determined from the generalized property relations developed in Sec. 12-3. As an alternative, property data from generalized charts may be used. In this case the reduced pressure $P_r$ of each component must be evaluated at the *mixture* pressure. The example below illustrates the use of the generalized enthalpy chart for a real-gas mixture.

---

**A** gas mixture consisting of 70 mole percent methane and 30 mole percent nitrogen is compressed reversibly and isothermally in a steady-flow process from 10 bars to 100 bars at 250 K. Calculate the enthalpy change in kJ/kmol of mixture by using (*a*) the generalized *h* chart and (*b*) tabular *h* data.

**Example 12-12**

**Solution:**

**Given:** A 70 mole percent methane and 30 mole percent nitrogen mixture is compressed reversibly and isothermally as shown in Fig. 12-11.

**Find:** $\Delta \bar{h}$, in kJ/kmol, by (*a*) the generalized *h* chart and (*b*) tabular data.

**Model:** Steady, reversible, isothermal flow.

**Analysis:** The enthalpy change on a mole of mixture basis is given by

$$\Delta \bar{h}_m = \sum y_i(\bar{h}_{i,2} - \bar{h}_{i,1})$$

The evaluation of $\bar{h}_i$ depends on the chosen source.

(*a*) To use a generalized enthalpy chart the reduced properties must be calculated. These are

Methane: $\quad T_{r1} = T_{r2} = \dfrac{T_1}{T_c} = \dfrac{250}{190.7} = 1.31 \qquad P_{r1} = \dfrac{P_1}{P_c} = \dfrac{10}{46.4} = 0.22$

$$P_{r2} = P_2/P_c = \frac{100}{46.4} = 2.2$$

Nitrogen: $\quad T_{r1} = T_{r2} = \dfrac{T_1}{T_c} = \dfrac{250}{126.2} = 2.0 \qquad P_{r1} = \dfrac{P_1}{P_c} = \dfrac{10}{33.9} = 0.30$

$$P_{r2} = \frac{P_2}{P_c} = \frac{100}{33.9} = 3.0$$

From Fig. A-30 we find the following approximate values:

Methane: $\qquad \dfrac{h_1^* - h_1}{RT_c} = 0.12 \qquad \dfrac{h_2^* - h_2}{RT_c} = 1.65$

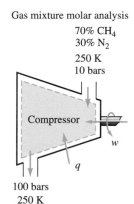

Gas mixture molar analysis

70% $CH_4$
30% $N_2$
250 K
10 bars

Compressor

$w$

$q$

100 bars
250 K

**Figure 12-11**
Schematic and data for Example 12-12.

Nitrogen:
$$\frac{h_1^* - h_1}{RT_c} = 0.03 \qquad \frac{h_2^* - h_2}{RT_c} = 0.48$$

In terms of the format of the generalized $h$ chart,

$$(h_2 - h_1)_i = RT_{c,i}\left[\frac{(h_1^* - h_1)}{RT_c} - \frac{(h_2^* - h_2)}{RT_c}\right]_i + (h_2^* - h_1^*)_i$$

The last term on the right side is the enthalpy change for an ideal gas. In this example the term is zero because the process is isothermal. Therefore, for methane on a molar basis:

$$(\bar{h}_2 - \bar{h}_1)_{CH_4} = 8.314(190.7)(0.12\text{-}1.65) = -2425 \text{ kJ/kmol}$$

Likewise, for nitrogen:

$$(\bar{h}_2 - \bar{h}_1)_{N_2} = 8.314(126.2)(0.03 - 0.48) = -472 \text{ kJ/kmol}$$

The enthalpy change per mole of mixture is

$$\Delta\bar{h}_m = \sum y_i(\bar{h}_{i,2} - \bar{h}_{i,1}) = 0.70(-2425) + 0.30(-472) = -1840 \text{ kJ/kmol}$$

(b) The tabular superheat data for methane, taken from the literature, are $h_1 = 1077.9$ kJ/kg and $h_2 = 928.5$ kJ/kg. The nitrogen data, taken from Table A-20 in the Appendix, are $h_1 = 356.3$ kJ/kg and $h_2 = 330.4$ kJ/kg. These data are determined by employing an accurate $PvT$ equation in the generalized enthalpy equation. In this case the enthalpy change becomes

$$\Delta\bar{h}_m = \sum y_i\Delta\bar{h}_i = 0.7(928.5 - 1077.9)(16.04) + 0.3(330.3 - 356.3)(28.01)$$
$$= -1895 \text{ kJ/kmol}$$

where the molar masses of 16.04 and 28.01 are used to correct the units. These answers for $\Delta\bar{h}_m$ differ by roughly 3 percent.

**Comments:**

1. An ideal-gas model would give an answer of zero.

2. The two methods are in substantial agreement. However, whether they match reality depends on the accuracy of the mixing rule in this temperature and pressure range.

---

## 12-8    DEVELOPMENT OF PROPERTY TABLES

The presentation of saturation and superheat tables first appeared in Chap. 3. With information provided in this current chapter, the general method for developing tables that contain $v$, $u$, $h$, and $s$ data as a function of $P$ and $T$ is presented below. The method requires three sets of experimental data that are represented analytically by suitable equations. These three items are:

1. An accurate $PvT$ equation of state for the superheat and saturation regions based on experimental data (see Chap. 11)

2. A vapor-pressure equation similar to Eq. [12-41] based on experimental vapor-pressure data

3. An equation for ideal-gas $c_{p,0}$ data over the desired temperature range must be developed from experimental measurements

The first item above permits the evaluation of $v$ data at selected states.

To illustrate the general method, the values of $h$ and $s$ are assigned arbitrary values at a specified *reference* state. These reference values could be zero, for example. For the purpose of illustration, the reference state is chosen to be the saturated-liquid state at a specified temperature. This reference state is shown as state 1 on the $Ts$ diagram in Fig. 12-12. Relative to this state, we seek to determine property values for the arbitrary saturation states 2, 5, and 6, and for the superheat states 3 and 4. Values at other states would be determined similarly.

Property data for state 2 are found from the Clapeyron equation. Differentiation of the vapor-pressure equation yields data for $(dP/dT)_{\text{sat}}$. Substitution of this quantity into the Clapeyron equation, written in the format

$$h_2 - h_1 = h_{fg} = -T(v_g - v_f)\left(\frac{dP}{dT}\right)_{\text{sat}}$$

yields a value for $h_2$. Values for $v_f$ and $v_g$ for states 1 and 2 come from the equation of state. The entropy and internal energy at state 2 are found from

$$s_2 = s_1 + \frac{h_{fg}}{T_1} \qquad \text{and} \qquad u_2 = u_1 + h_{fg} - P_1(v_2 - v_1)$$

This same type of calculation would lead to $\Delta h$, $\Delta s$, and $\Delta u$ between states 5 and 6.

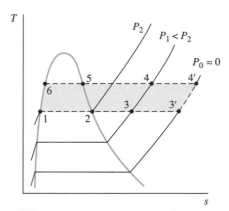

**Figure 12-12**    A $Ts$ diagram illustrating evaluation of data for saturation and superheat tables.

State 3 is at the same temperature as state 2 but is at a different pressure. Isothermal calculations of this type are most easily done through the concept of the *departure function* discussed in Sec. 12-7. Recall that a departure function $y^R$ is defined as

$$y^R \equiv y^* - y$$

where $y$ is the desired value of $y$ at $(T, P)$, and $y^*$ is the property value the fluid would have at $(T, P)$ if the fluid were an ideal gas. Since equations of state typically are explicit in pressure, we start from the general Helmholtz relation $da_T = -P \, dv$. By suitable manipulation, this equation leads to the residual Helmholtz function in the form

$$a^* - a = \int_\infty^v \left( P - \frac{RT}{v} \right) dv + RT \ln Z \qquad \textbf{[12-54]}$$

Since $da = -P \, dv - s \, dT$, then $s = -(\partial a / \partial T)_v$, or

$$s^* - s = -\frac{\partial}{\partial T} (a^* - a)_v \qquad \textbf{[12-55]}$$

Substitution of Eq. [12-54] into Eq. [12-55] and subsequent differentiation leads to

$$s^* - s = -\int_\infty^v \left[ \left( \frac{\partial P}{\partial T} \right)_v - \frac{R}{v} \right] dv - \ln Z \qquad \textbf{[12-56]}$$

Noting that $h = a + Ts + Pv$, Eqs. (12-54) and (12-56) can be used to show that

$$h^* - h = -\int_\infty^v \left[ T \left( \frac{\partial P}{\partial T} \right)_v - P \right] dv + RT(1 - Z) \qquad \textbf{[12-57]}$$

Finally, by definition, $u^* - u = h^* - h - P^*v^* - Pv$. Therefore,

$$u^* - u = h^* - h + RT(Z - 1) \qquad \textbf{[12-58]}$$

Equations [12-56], [12-57], and [12-58], in conjunction with the $PvT$ equation of state, lead to the evaluation of the residual values at a given state for $s$, $h$, and $u$. For the change in state in the superheat region, for example,

$$y_3 - y_2 = (y_2^* - y_2) - (y_3^* - y_3) + (y_3^* - y_2^*) \qquad \textbf{[12-59]}$$

where $y$ is any property of interest. The third term on the right of Eq. [12-59] is the property change if the gas were an ideal gas at the two given states. Recall for an ideal gas

$$h_y^* - h_x^* = \int_x^y c_{p,0} \, dT \qquad \textbf{[12-60]}$$

$$s_y - s_x = \int \frac{c_{p,0} \, dT}{T} - R \ln \frac{P_y}{P_x} \qquad \textbf{[12-61]}$$

where $x$ and $y$ are two arbitrary states. To determine property values at state 4 the calculation must proceed by path $3 - 3' - 4' - 4$ in Fig. 12-12. This is necessary because $c_{p,0}$ data are known only along pressure line $P_0$, which is low enough for the gas to behave as an ideal gas. That is,

$$h_4 - h_3 = (h_4^* - h_4) - (h_3^* + h_3) + \int_3^4 c_{p,0}\, dT \qquad \textbf{[12-62]}$$

$$s_4 - s_3 = (s_4^* - s_4) - (s_3^* - s_3) + \int_3^4 \frac{c_{p,0}\, dT}{T} - R \ln \frac{P_4}{P_3} \qquad \textbf{[12-63]}$$

Once data at state 4 are known, data at states 5 and 6 are determined by the reverse procedure for states 1, 2, and 3. Thus by the series of calculations discussed above, property values at any desired state can be evaluated relative to reference values for $h$ and $s$.

## 12-9 SUMMARY

On the basis of concepts developed from the first and second laws, and the definitions of the Helmholtz function $a$ and Gibbs function $g$, four important property relationships among properties of simple compressible substances can be developed. These are

$$du = T\, ds - P\, dv$$
$$dh = T\, ds + v\, dP$$
$$da = -P\, dv - s\, dT$$
$$dg = v\, dP - s\, dT$$

From these relationships the four additional equations shown below can be derived.

$$\left(\frac{\partial T}{\partial v}\right)_s = -\left(\frac{\partial P}{\partial s}\right)_v \qquad \left(\frac{\partial T}{\partial P}\right)_s = \left(\frac{\partial v}{\partial s}\right)_P$$

$$\left(\frac{\partial P}{\partial T}\right)_v = \left(\frac{\partial s}{\partial v}\right)_T \qquad \left(\frac{\partial v}{\partial T}\right)_P = -\left(\frac{\partial s}{\partial P}\right)_T$$

This latter set of equations is referred to as the *Maxwell relations.*

Two important specific-heat relations are

$$\left(\frac{\partial s}{\partial T}\right)_v = \frac{c_v}{T} \qquad \text{and} \qquad \left(\frac{\partial s}{\partial T}\right)_P = \frac{c_p}{T}$$

When these expressions and the Maxwell relations are substituted into the total differentials for $du$, $dh$, and $ds$, the following generalized relations are produced:

$$du = c_v\, dT + \left[ T\left(\frac{\partial P}{\partial T}\right)_v - P \right] dv$$

$$dh = c_p\, dT + \left[ v - T \left( \frac{\partial v}{\partial T} \right)_P \right] dP$$

$$ds = \frac{c_v\, dT}{T} + \left( \frac{\partial P}{\partial T} \right)_v dv = \frac{c_p\, dT}{T} - \left( \frac{\partial v}{\partial T} \right)_P dP$$

These are *generalized relations* because they are not restricted to any particular substance or to any particular phase. They are restricted, however, to simple compressible substances. The generalized relations for $c_p$ and $c_v$ are as follows:

$$\left( \frac{\partial c_v}{\partial v} \right)_T = T \left( \frac{\partial^2 P}{\partial T^2} \right)_v$$

$$\left( \frac{\partial c_p}{\partial P} \right)_T = -T \left( \frac{\partial^2 v}{\partial T^2} \right)_P$$

$$c_p - c_v = -T \left( \frac{\partial v}{\partial T} \right)_P^2 \left( \frac{\partial P}{\partial v} \right)_T$$

Application of experimental data, in conjunction with the last equation, shows that $c_p - c_v \geq 0$.

The slope of the vapor-pressure curve on a *PT* diagram is given theoretically by the *Clapeyron equation,* namely,

$$\left( \frac{dP}{dT} \right)_{\text{sat}} = \frac{h_{fg}}{T v_{fg}}$$

An approximation to this equation, which does not contain the specific volume, is

$$\ln \left( \frac{P_2}{P_1} \right)_{\text{sat}} = -\frac{h_{fg}}{R} \left( \frac{1}{T_2} - \frac{1}{T_1} \right)$$

This equation is one form of the *Clausius-Clapeyron equation.* It predicts that $\ln P_{\text{sat}}$ is a linear function of $1/T$.

The generalized relation for the Joule-Thomson coefficient is

$$\mu_{\text{JT}} \equiv \left( \frac{\partial T}{\partial P} \right)_h = \frac{1}{c_p} \left[ T \left( \frac{\partial v}{\partial T} \right)_P - v \right]$$

This equation is useful in predicting when the temperature of a gas will decrease during a throttling process.

Based on the generalized equations for *dh* and *ds,* generalized charts and tables have been developed on the basis of generalized $Z$ data as a function of reduced pressure $P_R$ and reduced temperature $T$. Usually, values of the departure functions, such as $(h^* - h)/RT_c$ and $(s_P^* - s_P)_T/R$, are presented for selected values of $P_r$ and $T_r$. This permits estimation of $\Delta h$ and $\Delta s$ solely from the initial and final pressures and temperatures, and the critical data for the substance. This concept is also applicable to real-gas mixtures.

## PROBLEMS

12-1. Prove that the constant-pressure lines in the wet region of an $hs$ diagram are straight and not parallel and that the slope of a constant-pressure line in the superheat region increases with temperature.

12-2. Starting with any of the four Maxwell relations, derive the other three, making use of Eq. [12-5].

12-3. Derive expressions for (a) $(\partial u/\partial P)_T$ and (b) $(\partial u/\partial v)_T$ that involve only $P$, $T$, and $v$.

12-4. Derive the relation $c_p = T(\partial s/\partial T)_P$.

12-5. From Eqs. [12-18] and [12-19], show that the slope of a constant-volume line is greater than that for a constant-pressure line through the same state point in the gas region of a $Ts$ diagram.

12-6. Derive the expression $c_p = T(\partial P/\partial T)_s(\partial v/\partial T)_P$.

12-7. Derive the expression $c_v = -T(\partial v/\partial T)_s(\partial P/\partial T)_v$.

12-8. Approximate the value of $c_p$ for steam at 120 bars and $480°C$, (a) by employing the equation derived in Prob. 12-6, and (b) by using the definition $c_p = (\partial h/\partial T)_P$.

12-9. Approximate the value of $c_p$ for refrigerant 134a at 12 bars and $80°C$, by employing the equation derived in Prob. 12-6, and (b) by using the definition $c_p = (\partial h/\partial T)_P$.

12-10E. Approximate the value of $c_p$ for steam at 800 psia and $650°F$ by employing the equation derived in Prob. 12-6. Compare with the value obtained directly from the definition of this property.

12-11E. Approximate the value of $c_v$ for steam at 600 psia and $600°F$ by employing the equation derived in Prob. 12-7, in $Btu/(lb_m \cdot °F)$.

12-12. Show that the slope of a constant-pressure line in the vapor region of a $Ts$ diagram normally increases with increasing temperature.

12-13. Derive the relation $(\partial^2 g/\partial T^2)_P = -c_p/T$.

12-14. Derive the relations $(\partial^2 a/\partial T^2)_v = -c_v/T$.

12-15. What can be concluded qualitatively about the change in the enthalpy of a fluid during an isentropic compression?

12-16. Prove that, for any homogeneous system, $c_p/c_v = (\partial P/\partial v)_s(\partial P/\partial v)_T$.

12-17. Check the validity of Eq. [12-17] by finding approximate values of the derivatives on both sides of the equation for steam at 80 bars and $400°C$, in $m^3/kg \cdot °C$.

12-18. Check the validity of Eq. [12-17] by finding approximate values of the derivatives on both sides of the equation for refrigerant 134a at 7 bars and $50°C$, in $m^3/kg \cdot °C$.

12-19E. Find the approximate values of the derivatives on both sides of Eq. [12-17] for steam at 200 psia and $500°F$, in units of $ft^3/lb_m \cdot °F$.

12-20. Develop the general relation $(\partial h/\partial v)_T = T(\partial P/\partial T)_v + v(\partial P/\partial v)_T$, and evaluate $(\partial h/\partial v)_T$ for an ideal gas.

12-21. On the basis of the van der Waals equation of state and the generalized relations:
(a) Show that $(h_2 - h_1)_T = (P_2 v_2 - P_1 v_1) + a(1/v_1 - 1/v_2)$.
(b) Show that $(s_2 - s_1)_T = R \ln [(v_2 - b)/(v_1 - b)]$.
(c) Evaluate $(h_2 - h_1)_T$, in kJ/kmol, for oxygen for 27°C when it is compressed from 1 to 100 bars. Compare to an ideal-gas solution to the problem.

12-22E. On the basis of the van der Waals equation of state and the generalized relations:
(a) Show that $(h_2 - h_1)_T = (P_2 v_2 - P_1 v_1) + a(1/v_1 - 1/v_2)$.
(b) Show that $(s_2 - s_1)_T = R \ln [(v_2 - b)/(v_1 - b)]$.
(c) Evaluate $(h_2 - h_1)_T$, in Btu/lbmol, for oxygen at 80°F which is compressed isothermally from 1 to 100 atm. Compare to an ideal-gas solution to the problem.

12-23. For a substance which is modeled by the Redlich-Kwong equation of state, (a) develop expressions for (1) the isothermal change in the specific internal energy and (2) the specific entropy. The results should be given in terms of $T$, $v$, $a$, $b$, and a constant. Then (b) evaluate $\Delta u$ and $\Delta s$ for steam as it is compressed isothermally at 360°C from $23.31 \times 10^{-3}$ to $11.05 \times 10^{-3}$ m³/kg. (c) Check the answers against tabular superheat data.

12-24. For a substance which is modeled by the Redlich-Kwong equation of state, (a) develop expressions for the isothermal change in (1) the specific internal energy and (2) the specific entropy. The results should be given in terms of $T$, $v$, $a$, $b$, and a constant. Then (b) evaluate $\Delta u$ and $\Delta s$ for refrigerant 134a as it is compressed isothermally at 80°C from 0.03264 to 0.01435 m³/kg. (c) Check the answers against tabular superheat data.

12-25E. For a substance which is modeled by the Redlich-Kwong equation of state, (a) develop expressions for the isothermal change in (1) the specific internal energy and (2) the specific entropy. The results should be given in terms of $T$, $v$, $a$, $b$, and a constant. Then (b) evaluate the internal-energy and entropy changes for steam as it is compressed isothermally at 650°F from 0.564 to 0.206 ft³/lb$_m$. (c) Check the answers against tabular superheat data in Btu/lb.

12-26E. For a substance which is modeled by the Redlich-Kwong equation of state, (a) develop expressions for the isothermal change in (1) the specific internal energy and (2) the specific entropy. The results should be given in terms of $T$, $v$, $a$, $b$, and a constant. Then (b) evaluate $\Delta u$ and $\Delta s$ for refrigerant 134a as it is compressed isothermally at 160°F from 0.4852 to 0.2636 ft³ /lb$_m$. (c) Check the answers against tabular superheat data.

12-27 Derive expressions for the isothermal change (a) of enthalpy and (b) of entropy for a substance which fulfills the Berthelot equation of state, which is of the form $P = RT(v - b) - a/Tv^2$.

12-28E. Nitrogen gas exists at $350°R$ and 1000 psia. (a) Use Eqs. [12-26], [11-4], and [11-5] to predict the $c_p$ value, in Btu/lbmol·°F. (b) Compare your answer to the result obtained from the definition of $c_p$, in conjunction with superheat tabular data.

12-29. Predict the $c_p$ value of nitrogen, in kJ/kmol·K, at 200 K and 50 bars by means of (a) Eqs. [12-27], [11-4], and [11-5] and (b) the definition of $c_p$ in conjunction with tabular superheat data.

12-30. A gas has a compressibility factor $Z$ given by $Z = 1 + AP/T^2$. Derive an expression for the change in enthalpy between two states at the same temperature but different pressures.

12-31. Derive an expression for the change in enthalpy in terms of temperature and pressure changes from a state of $T_1$ and $P_1$ to a state of higher values of $T_2$ and $P_2$ for a gas which is modeled by the equation $Pv/RT = 1 + AP/T$ and whose specific heat at a low pressure is given by $c_{p,0} = B + CT$; $A$, $B$, and $C$ are constants, and $P_0$ is less than $P_1$ and $P_2$.

12-32. Show that the change in volume of a substance can be related to $\beta$ and $K_T$ such that $dv/v = \beta \, dT - K_T \, dP$.

12-33. On the basis of the result shown in Prob. 12-32, estimate the percentage change in the volume of copper when it changes state from 200 K and 1 bar to 300 K and 1000 bars.

12-34 A 0.1-kg mass of copper is heated from 250 to 500 K and is compressed from 1 to 500 bars. On the basis of Prob. 12-32, estimate the change in volume, in cubic centimeters, if the value of $v$ at 500 K is 7.115 cm³/mol.

12-35. At 800 K the value of $v$ for solid copper is $7.215 \times 10^{-3}$ m³/kmol. (a) Determine the value of $c_p - c_v$ for copper, in kJ/kmol·K. (b) If $c_p$ is 28.48 kJ/kmol·K, find the percentage error that would be made in $c_v$ if we assumed that $c_p = c_v$ at this temperature.

12-36. At 300 K the value for $v$ for solid copper is $7.062 \times 10^{-3}$ m³/kmol. Determine the value of $c_p - c_v$, in kJ/kmol·K.

12-37. At 0°C the values of $v$, $\beta$, and $K_T$ for liquid mercury are $14.67 \times 10^{-3}$ m³/kmol, $174 \times 10^{-6}$ K⁻¹, and $3.79 \times 10^{-7}$ cm²/N, respectively. (a) Determine the value of $c_p - c_v$, in kJ/kmol·K. (b) If $c_p$ is 28.0 kJ/kmol·K, find the percentage error that would be made in $c_v$ if we assumed $c_p = c_v$.

12-38E. On the basis of the result shown in Prob. 12-32, estimate the percentage change in the volume of copper when it changes state from $360°R$ and 1 atm to $540°R$ and 1000 atm, on the basis of data in Table 12-1.

12-39E. A 0.1 $lb_m$ mass of copper changes state from 540°R and 1 atm to 900°R and 500 atm. On the basis of Prob. 12-32 estimate the change in volume, in cubic inches, if the value of $v$ at 900°R is 0.114 $ft^3$/lbmol.

12-40E. At 900°R the values of $v$, $\beta$, and $K_T$ for solid copper are 0.114 $ft^3$/lbmol, $30.1 \times 10^{-6}$°$R^{-1}$, and $400.3 \times 10^{-12}$ $ft^2$/$lb_f$, respectively.
   (a) Determine the value of $c_p - c_v$, in Btu/lbmol·°F.
   (b) If $c_p$ is 6.25 Btu/lbmol·°F, what percentage error would be made in $c_v$ if we assumed that $c_p = c_v$?

12-41E. At 540°R the values of $v$, $\beta$, and $K_T$ for solid copper are 0.113 $ft^3$/lbmol, $27.3 \times 10^{-6}$°$R^{-1}$, and $5.35 \times 10^{-8}$ $in^2$/$lb_f$, respectively. Determine the value of $c_p - c_v$, in Btu/lbmol·°R.

12-42E. At 32°F and 14,225 psia the values of $v$, $\beta$, and $K_T$ for liquid mercury are 0.235 $ft^3$/lbmol, $96.7 \times 10^{-6}$°$R^{-1}$, and $1686 \times 10^{-12}$ $ft^2$/$lb_f$, respectively.
   (a) Determine the value of $c_p - c_v$, in Btu/lbmol·°F.
   (b) If $c_p$ is 6.69 Btu/lbmol·°F, what percentage of error would be made in $c_v$ if we assumed that $c_p = c_v$?

12-43. The isothermal compressibility of ethyl alcohol at 20°C and 1 bar is $110 \times 10^{-6}$ $bar^{-1}$ and the specific gravity is 0.789. Find the work required to compress the fluid isothermally from 1 to 100 bars at 20°C, if $v$ is assumed constant.

12-44. Copper, with a density of 8900 kg/$m^3$, is compressed isothermally and reversibly at 250 K from 1 to 400 bars. Find (a) the work required and (b) the heat transfer, in kJ/kg.

12-45. The pressure on water is increased isothermally and reversibly at 20°C from 1 to 500 bars. Find (a) the work required and (b) the change in internal energy, both in kJ/kg.

12-46. The isothermal compressibility for liquid water at 50°C is given by $K_T = 0.125 \times 10^{-3}/v(P + 2740)$, where $K_T$ is in $bar^{-1}$, $P$ is in bars, and $v$ is in $m^3$/kg. A mass of 2.5 kg of water is compressed isothermally and reversibly at 50°C from 1 to 600 bars. Determine (a) the work required and (b) the heat transfer, both answers in kilojoules.

12-47. Determine the work required, in joules, to compress 20 $cm^3$ of liquid mercury at a constant temperature of 0°C from a pressure of 1 bar to (a) 500 bars and (b) 1000 bars. The isothermal compressibility of mercury at the given temperature is given by $K_T = 3.9 \times 10^{-6} - 1.0 \times 10^{-10}P$, where $K_T$ is in $bar^{-1}$ and $P$ is in bars. The density of mercury may be taken as 13,600 kg/$m^3$.

12-48. Copper, with a density of 8900 kg/$m^3$, is compressed isothermally and reversibly at 500 K from 1 to 300 bars. Compute (a) the work required and (b) the change in internal energy, both in kJ/kg.

## VAPOR PRESSURE AND ENTHALPY OF VAPORIZATION

12-49. The vapor pressure of carbon tetrachloride at several temperatures is as follows:

| $T$, °C | 25 | 35 | 45 | 55 |
|---|---|---|---|---|
| $P$, mbars | 151.7 | 232.5 | 345.1 | 498.0 |

Plot $\ln P$ versus $1/T$, and from the slope evaluate the mean enthalpy of vaporization, in kJ/kg.

12-50. Estimate from tabular $PvT$ data the enthalpy of vaporization of water at 150°C by employing (*a*) the Clapeyron equation and (*b*) the Clausius-Clapeyron equation. (*c*) Compare with data in Table A-12.

12-51. Estimate from tabular $PvT$ data the enthalpy of vaporization of refrigerant 134a at (*a*) 8°C and (*b*) 44°C by employing (1) the Clapeyron equation, (2) the Clausius-Clapeyron equation, and then (3) comparing with data in Table A-16.

12-52. The vapor pressure of carbon tetrachloride at several temperatures is as follows:

| $T$, °F | 77 | 95 | 113 | 131 |
|---|---|---|---|---|
| $P$, mmHg | 113.8 | 174.4 | 258.9 | 373.6 |

Plot $\ln P$ versus $1/T$, and from the slope evaluate the mean enthalpy of vaporization, in Btu/lb$_m$.

12-53E. Estimate the enthalpy of vaporization of water at 400°F by employing (*a*) the Clapeyron equation and (*b*) the Clausius-Clapeyron equation, using tabular $PvT$ saturation data. (*c*) Compare the answers with the tabulated value, in Btu/lb$_m$.

12-54E. From tabular $PvT$ saturation data estimate the enthalpy of vaporization of refrigerant 134a at 70°F using (*a*) the Clapeyron equation and (*b*) the Clausius-Clapeyron equation, in Btu/lb$_m$. (*c*) Compare with the tabulated value.

12-55. The vapor-pressure data for nitrogen over the range from the triple state to the normal (1 atm) boiling point is closely represented by $\ln P = A - BT - (C/T)$, where $P$ is in bars, $T$ is in kelvins, $A = 11.302$, $B = 0.01443$, and $C = 786.7$.
   (*a*) Derive an expression for the enthalpy of vaporization in terms of the constants A, B, and C, the temperature, and the change in specific volume $v_{fg}$.
   (*b*) Evaluate $h_{fg}$, in kJ/kg, at 80 K if the value of $v_{fg}$ is $162.7 \times 10^{-3}$ m$^3$/kg.

12-56E. The vapor pressure of ammonia ($NH_3$) is given by the relation $\log_{10} P = 25.574 - 3.295 \times 10^3/T - 6.401 \log_{10} T - 4.148 \times 10^{-4}T + 1.476 \times 10^{-6}T^2$, where $T$ is in degrees Rankine and $P$ is in psia. Evaluate $h_{fg}$ at 40°F and compare to the tabulated value of 536.2 Btu/lb$_m$.

12-57. The specific volumes of liquid water and ice at 0°C are $1.0002 \times 10^{-3}$ and $1.0911 \times 10^{-3}$ m$^3$/kg, respectively. Estimate the melting-point temperature of ice at (a) 300 bars and (b) 500 bars if the enthalpy of fusion at 0°F is 333.4 kJ/kg.

12-58. The vapor pressure of water between 5 and 50 bars can be represented by the equation $\ln P = -4692/T + 0.0124 \ln T + 12.58$, with $P$ in bars and $T$ in kelvins. Compute the enthalpy of vaporization at (a) 160°C and (b) 200°C, using values of $v_f$ and $v_g$ from a saturation table. Compare to tabulated $h_{fg}$ values.

12-59. The triple state of carbon dioxide is $-56.6°C$ and 5.178 bars. Predict the slopes of the three saturation lines in the vicinity of the triple state on a $PT$ diagram on the basis of the following triple-state data:

| Phase | Solid | Liquid | Vapor |
|---|---|---|---|
| $h$, kJ/kg | 181.3 | 380.2 | 728.5 |
| $v$, cm$^3$/g | 0.661 | 0.849 | 72.22 |

12-60. Helium 4 at 1 bar boils at 4.22 K, and the enthalpy of vaporization is 83.3 kJ/kmol. By producing a vacuum over the liquid phase, the fluid boils at a lower temperature. Estimate the pressure, in millibars, necessary to produce a boiling temperature of (a) 2 K and (b) 1 K.

12-61. At the triple state of water the pressure and temperature are 6.12 mbars and 0.010°C, respectively. The enthalpy of melting is 333.4 kJ/kg, and the specific volumes of the liquid and solid phases are $1.0002 \times 10^{-3}$ and $1.0911 \times 10^{-3}$ m$^3$/kg, respectively. A person skates on ice at $-2°C$ on blades with a contact area of 0.32 cm$^2$. Determine the weight and the mass, in newtons and kilograms, respectively, of a person so that the ice beneath the blades just melts.

12-62. At the triple state of water the pressure and temperature are 6.12 mbars and 0.010°C, respectively. The enthalpy of melting is 333.4 kJ/kg, and the specific volumes of the liquid and solid phases are $1.0002 \times 10^{-3}$ and $1.0911 \times 10^{-3}$ m$^3$/kg, respectively. A person with a mass of 80 kg is skating on blades that have a total surface area of 0.25 cm$^2$ in contact with the ice. Consider the temperature of the ice to be (a) $-2°C$ and (b) $-3°C$. Determine whether the ice will melt under the blades at the two given temperatures.

12-63E. The specific volumes of liquid water and ice at 32°F are 0.01602 and 0.01747 ft$^3$/lb$_m$, respectively. Estimate the melting-point temperature of ice at 500 atm if the enthalpy of fusion at 32°F is 143.32 Btu/lb$_m$.

12-64E. The vapor pressure of water can be represented by the equation $\ln P = -8445T + 0.0124 \ln T + 15.252$, with $P$ in psia and $T$ in degrees Rankine. Compute the enthalpy of vaporization at 400°F, using values of $v_f$ and $v_g$ from a saturation table. Compare to the tabulated $h_{fg}$ value, in Btu/lb$_m$.

12-65. The effect of pressure on the melting point of sodium has been measured by Bridgman (*Phys. Rev.,* vol. 3, p. 127, 1914). Some of the data include these:

| $P$, atm | $T$,°F | $\Delta v \times 10^5$, ft$^3$/lb$_m$ |
|---|---|---|
| 5810 | 288.5 | 30.06 |
| 7740 | 310.6 | 27.46 |
| 9680 | 332.0 | 24.97 |

Determine the approximate enthalpy of fusion at 310.6°F, and compare with the tabulated value of 51.6 Btu/lb$_m$.

12-66E. The enthalpy of fusion of water is practically constant at 143.8 Btu/lb$_m$. Estimate the freezing-point temperature at 10,000 psia, if the specific volumes of liquid water and ice at 32°F are 0.01602 and 0.01747 ft$^3$/lb$_m$, respectively.

## JOULE-THOMSON COEFFICIENT

12-67. Estimate the Joule-Thomson coefficient for water at (a) 60 bars and 320°C and (b) 100 bars and 400°C, in °C/bar, on the basis of the definition of the coefficient.

12-68. A gas fulfills the relation $P(v - b) = RT$, where $b$ is a positive constant. (a) Derive an expression for the Joule-Thomson coefficient of the gas. (b) Could this gas be cooled effectively by throttling?

12-69. Calculate the Joule-Thomson coefficient for a gas that is modeled by the Dieterici equation: $P(v - b) = RT \exp(-a/RTv)$. Prove that the inversion temperature for such a gas is $2a(v - b)/(Rbv)$.

12-70. By employing the generalized equation for the Joule-Thomson coefficient and the tabular superheat data for refrigerant 134a, estimate the coefficient, in °C/bar, at 12 bars and 80°C.

12-71. The Joule-Thomson coefficient for nitrogen at 40 bars and $-73$°C is approximately 0.40°C/bar. Determine the value of $c_p$, in kJ/kmol·K, if Eqs. [11-4] and [11-5] represent the $PvT$ behavior of nitrogen at this state.

12-72E. Using superheat data estimate (a) the value of $c_p = (\partial h/\partial T)_p$, and (b) the Joule-Thomson coefficient for steam at 100 psia and 700°F.

12-73. Using tabular superheat data, estimate the Joule-Thomson coefficient for R-134a, in K/bar, at 80°C and 14 bars on the basis of Eq. [12-43] and the definition of $c_p$.

12-74E. Determine the Joule-Thomson coefficient for water substance at 500 psia and 550°F, in °F/psi, by means of the generalized equation for $\mu_{JT}$ and the definition of $c_p$.

12-75E. By employing the generalized equation for the Joule-Thomson coefficient and the tabular superheat data for refrigerant 134a, estimate (a) $c_p$ from its definition and (b) the coefficient, in °F/atm, at 300 psia and 200°F.

12-76. The equation of state for dry air may be expressed by the truncated virial equation $Pv = RT + bP$, and $b = 37.28 - 8.43 \times 10^3/T - 1.75 \times 10^6/T^2 + 7.35 \times 10^7/T^3$, where $b$ is in $cm^3$/mol and $T$ is in kelvins. If the measured Joule-Thomson coefficient at 1 atm and 0°C is 0.275°C/atm, (a) calculate the value of $c_p$ at this state, in kJ/kg·K. (b) Compare to $c_{p,0}$ data at this temperature.

12-77. An acceptable equation of state for helium gas is given by $Pv = RT - aP/T + bP$, where $a = 386.7$ K·$cm^3$/gmol and $b = 15.29$ $cm^3$/gmol.
   (a) Compute the Joule-Thomson coefficient for helium at 150 and 15K, in K/bar.
   (b) Find the inversion temperature, in kelvins.
   (c) Estimate the temperature reached in an ideal throttling process from 25 bars and 15 K to 1 bar.

12-78. On the basis of Eq. [11-5], determine the Joule-Thomson coefficient of nitrogen at 1 bar and (a) 300 K, (b) 500 K, and (c) 700 K.

## Use of Generalized Charts

12-79. Methane gas in a closed system is cooled at a constant pressure of 55 bars from 100 to 30°C. The equation for $c_{p,0}$ is given in Table A-3. Calculate the heat transferred, in kJ/kg, using generalized data.

12-80. Nitrogen at 50 bars and 200 K is contained in a tank of 0.3 $m^3$. Heat is transferred to the nitrogen until the temperature is 300 K. Determine the heat transferred, in kJ/kmol, and the final pressure, (a) using generalized charts, (b) assuming ideal-gas behavior, and (c) using data from the superheat table.

12-81E. Nitrogen at a pressure of 1500 psia and $-100$°F is contained in a tank of 10 $ft^3$. Heat is transferred to the nitrogen until the temperature is 200°F. Determine the heat transfer, in Btu/$lb_m$, and the final pressure, (a) using generalized charts, (b) assuming ideal-gas behavior.

12-82E. Methane gas in a closed system is cooled at a constant pressure of 800 psia from 200 to 100°F. The molar specific heat at zero pressure is given by $\bar{c}_{p,0} = 4.52 + 0.00737T$, where $T$ is in degrees Rankine. Calculate the heat transferred, in Btu/lb$_m$.

12-83E. Ethane is compressed from 90°F and 15 atm to 145°F and 96 atm in a steady-flow process. The molar specific heat at zero pressure is given by $\bar{c}_{p,0} = 4.01 + 0.001636T$, where $T$ is in degrees Rankine. Determine the enthalpy change of the gas, in Btu/lbmol, if (a) an ideal-gas model is used and (b) a generalized chart is used.

12-84. Ethane ($C_2H_6$) is compressed from 30°C and 15 bars to 124°C and 98 bars in a steady-flow process. The molar specific heat, in kJ/kmol·K, is given by $\bar{c}_{p,0} = 16.8 + 0.123T$, where $T$ is in kelvins. Determine the enthalpy change, in kJ/kmol, if (a) an ideal-gas model is used and (b) a generalized chart is used.

12-85. Methane ($CH_4$) is compressed in a steady-flow process from 13.9 bars and 51°C to 186 bars and 108°C. Using generalized charts, determine (a) the change in enthalpy, in kJ/kmol, and (b) the change in entropy, in kJ/kmol·K.

12-86E. Methane ($CH_4$) is compressed in a steady-flow process from 168 psia and 90°F to 2860 psia and 228°F. Using generalized charts, determine (a) the change in enthalpy, in Btu/lbmol, and (b) the change in entropy, in Btu/lbmol·°R.

12-87. Ethylene is being expanded in a gas turbine from 180°C and 307 bars to 67°C and 61 bars. The inlet volume flow rate is 1 m$^3$/min. A mean $\bar{c}_{p,0}$ value of 50.0 kJ/kmol·K may be assumed. Estimate the shaft power delivered, in kilowatts, for the adiabatic process.

12-88E. Propane is expanded from 340°F and 1235 psia to 140°F and 185 psia. The mean $c_{p,0}$ value is 0.44 Btu/lb$_m$·°F. The turbine inlet volume flow rate is 100 ft$^3$/min. Determine the horsepower output for the expansion process, if adiabatic.

12-89. Ethylene at 67°C and 1 bar is compressed isothermally in a reversible, nonflow process to 255 bars. On the basis of generalized charts determine (a) the change in entropy, in kJ/kmol·K, (b) the change in internal energy, in kJ/kmol, (c) the heat transferred, and (d) the work required, in kJ/kmol.

12-90. Calculate (a) the heat transfer and (b) the work required, in kJ/kg, when ethane is compressed reversibly and isothermally from 5 to 98 bars at 30°C in a steady-flow process, using generalized charts.

12-91E. Calculate (a) the heat transfer and (b) the work required, in Btu/lb$_m$, when ethane ($C_2H_6$) is compressed reversibly and isothermally from 70 to 1420 psia at 90°F in a steady-flow process, using generalized charts.

12-92E. Ethylene ($C_2H_4$) at 100°F and 1 atm is compressed isothermally to 101 atm in a reversible nonflow process. Using the principle of corresponding states, determine (a) the change in specific internal energy,

(b) the change in entropy in Btu/lbmol·°R, and (c) the work done, in Btu/lbmol.

12-93E. Ethylene ($C_2H_4$) at 127°F and 4080 psia is expanded isothermally to 90 psia in a reversible nonflow process. Using the principle of corresponding states, determine (a) the change in entropy in Btu/lb$_m$·°R, (b) the change in specific internal energy, (c) the heat added, and (d) the work done, in Btu/lb$_m$.

12-94. Refrigerant 134a is throttled from 16 bars and 100°C to 4 bars. Determine the final temperature based (a) on a generalized chart and (b) on tabulated superheat data. The value of $c_{p,0}$ in this temperature range may be taken to be 0.85 kJ/kg·K.

12-95. Carbon dioxide is compressed in a reversible, adiabatic, steady-flow process from 10 bars, 30°C, to 74 bars. Determine the work of compression, in kJ/kg, using generalized charts.

12-96. Steam at 280 bars and 520°C is expanded adiabatically and reversibly in a steady-flow process to 100 bars. Determine the final temperature, in degrees Celsius, and the change in enthalpy, in kJ/kg, using (a) generalized charts and (b) tabular superheat data.

12-97. Oxygen initially at 40 bars and −41°C is compressed adiabatically in a steady-flow process to 100 bars. Determine the minimum work of compression, in kJ/kmol, if (a) the generalized charts are used and (b) the gas is assumed to be ideal.

12-98E. Carbon dioxide is compressed in a reversible, adiabatic, steady-flow process from 10 atm, 460°R, to 40 atm. Determine the work of compression, in Btu/lb$_m$, if based on (a) generalized charts and (b) an ideal-gas model.

12-99E. Steam at 4500 psia and 1000°F is expanded adiabatically and reversibly in a steady-flow process to 1600 psia. Determine the final temperature and the change in enthalpy, in Btu/lb$_m$, employing (a) the principle of corresponding states and (b) the superheat table.

12-100. Propane ($C_3H_8$) enters a constant-area pipe at 34.2 bars, 100°C, and a velocity of 40 m/s. The gas flows adiabatically through the pipe until the pressure reaches 10.7 bars. If $c_{p,0}$ is relatively constant at 1.76 kJ/kg·K, estimate (a) the exit temperature, in degrees Celsius, and (b) the exit velocity, in m/s.

# CHAPTER

# 13

# CHEMICAL REACTIONS

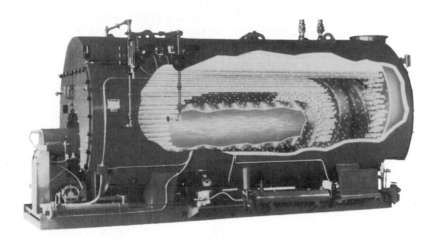

Cutaway of a Classic III Scotch design firetube packaged boiler. (Courtesy of Kewaunee Boiler Mfg. Co., Inc.)

In the preceding chapters attention was focused on the thermodynamic analysis of nonreacting systems. There are many practical examples, however, of engineering systems in which chemical reactions play a major role. In this chapter the basic concepts of the conservation of mass, the conservation of energy, and the entropy balance are applied to systems in which a change in chemical composition occurs. Although the equations representing these principles remain the same in format as before, it is necessary to modify the method of evaluating certain thermodynamic properties. For example, in preceding chapters we only considered methods for evaluating the change in *sensible* internal energy due to temperature and pressure changes and the change in *latent* internal energy due to a phase change. In this chapter the method for evaluating the change in *chemical* internal energy due to a change in composition will be introduced.

## 13-1     STOICHIOMETRY OF REACTIONS

Although chemical reactions may involve many different types of reactants, our interest is directed toward combustion reactions. A **combustion reaction** generally is one involving a **fuel** and an **oxidizer**, and energy is released. In this chapter the combustion studies involve the reaction of fuels containing primarily carbon and hydrogen. Most of the reactions to be studied are for gaseous fuels. The oxidizer is either oxygen or air, and carbon dioxide ($CO_2$) and water ($H_2O$) are the primary products. **Complete combustion** requires sufficient oxygen to be present that all the carbon and all the hydrogen in the fuel are converted to $CO_2$ and $H_2O$. During **incomplete combustion** other products will appear, the major one being carbon monoxide (CO).

The combustion of fuels in industrial practice is normally accomplished by employing air as the oxidizer. The major components of dry air are considered to be approximately 21 percent oxygen, 78 percent nitrogen, and 1 percent argon by volume. Small amounts of carbon dioxide and other gases are present, of course. It is convenient to assume that dry air is composed of 21 percent oxygen and 79 percent nitrogen by volume for preliminary calculations. Hence there are 21 mol of oxygen to every 79 mol of nitrogen in our assumed composition of atmospheric air. Therefore, we may write

$$1 \text{ mol } O_2 + 3.76 \text{ mol } N_2 = 4.76 \text{ mol air}$$

$$1 \text{ kg } O_2 + 3.31 \text{ kg } N_2 = 4.31 \text{ kg air}$$

$$1 \text{ lb } O_2 + 3.31 \text{ lb } N_2 = 4.31 \text{ lb air}$$

Recall that the average molar mass of dry air is 28.97, which we round off to 29.0 in many calculations. When air is used as a source of oxygen, a major fraction of the oxidizer source is diatomic nitrogen ($N_2$). Nitrogen is generally assumed to be a stable element that does not react chemically with other species. While this is true at normal temperatures, nitrogen does react with oxygen to form nitric oxide (NO) at the temperatures produced with a hot flame. At lower temperatures the NO will react with oxygen to form nitrogen dioxide ($NO_2$). Unless specifically stated otherwise, the reaction between nitrogen and oxygen is neglected in this chapter. In addition, when *moist* air is used in a combustion process, the mass of water vapor must be considered when writing a mass balance.

One basic consideration in the analysis of combustion processes is the theoretical or stoichiometric reaction for a given fuel. By definition, a **theoretical** or **stoichiometric reaction** occurs when the reaction goes to *completion* and there is *no excess oxygen* in the products of combustion. The amount of air (or oxygen) required in a stoichiometric reaction is very important in combustion studies and is determined for methane ($CH_4$), for example, in the following manner. On the basis of 1 mol of fuel and an unknown number of moles $w$ of air, we can write that

$$CH_4 + w(O_2 + 3.76N_2) \rightarrow xCO_2 + yH_2 + zN_2$$

where $w$, $x$, $y$, and $z$ represent the unknown number of moles of air, carbon dioxide, water, and nitrogen, respectively. Atomic species are conserved (in the absence of nuclear reactions). Thus the conservation of mass principle may be applied to the four chemical species present on the left and right sides of the equation. Hence,

C balance: $\quad\quad\quad\quad\quad\quad 1 = x$

H balance: $\quad\quad\quad\quad\quad\quad 4 = 2y$

O balance: $\quad\quad\quad\quad\quad\quad 2w = 2x + y$

$N_2$ balance $\quad\quad\quad\quad 3.76w = z$

Thus $w$, $x$, $y$, and $z$ are 2, 1, 2, and 7.52, respectively. Consequently the **stoichiometric equation** for the combustion of methane with air is

$$CH_4 + 2O_2 + 7.52N_2 \rightarrow CO_2 + 2H_2O + 7.52N_2$$

By a similar procedure the 100 percent theoretical or chemically correct combustion of propane ($C_3H_8$) with air is given by the stoichiometric equation

$$C_3H_8 + 5O_2 + 5(3.76)N_2 \rightarrow 3CO_2 + 4H_2O + 18.80N_2$$

Note that for this reaction, the number of moles of reactants and products are not equal. However, on a mass basis the above equation can be interpreted as

44 kg $C_3H_8$ + 160 kg $O_2$ + 526.4 kg $N_2 \rightarrow$
$$132 \text{ kg } CO_2 + 72 \text{ kg } H_2O + 526.4 \text{ kg } N_2$$

Because mass must be conserved, the mass of reactants equals the mass of products.

In each case no oxygen appears in the products of combustion, and the nitrogen has been assumed to undergo no chemical change. The numbers that appear before each species in a stoichiometric equation are known as the **stoichiometric coefficients** of the various species, and usually they are represented by the symbol $\nu_i$. One might consider the general chemical reaction

$$\nu_A A + \nu_B B + \cdots \rightarrow \nu_L L + \nu_M M + \cdots \quad\quad \textbf{[13-1]}$$

where the uppercase letters $A$, $B$, etc., represent the chemical species of the reactants and $L$, $M$, etc., represent the chemical species of the products. In general, any number of reactants or products might need to be considered.

For the complete combustion of carbon and hydrogen to $CO_2$ and $H_2O$, we may use the term *theoretical* or *stoichiometric* oxygen, or air, requirements. When this quantity is not used in a process, we speak of the *percentage of theoretical* air, or oxygen, actually used. The stoichiometric quantity is the 100 percent theoretical requirement. When a deficiency is used, the percentage of theoretical oxygen is somewhere between 0 and 100 percent, and an excess of oxygen (of air) means that some value greater than the

100 percent theoretical value was employed. Thus 200 percent theoretical air means that twice as much air is supplied as is necessary for complete combustion. In such a case oxygen will necessarily appear in the product gases. Other terms in frequent use are the *percentage excess* and the *percentage deficiency* of air (of oxygen). As examples, 150 percent theoretical air is equivalent to 50 percent excess air, and 80 percent theoretical air is a 20 percent deficiency of air. When 150 percent theoretical air or 50 percent excess air is supplied, the combustion reaction for propane becomes

$$C_3H_8 + 7.5\,O_2 + 28.20\,N_2 \rightarrow 3\,CO_2 + 4\,H_2O + 2.5\,O_2 + 28.20\,N_2$$

This reaction is again for complete combustion. In practice, other products such as carbon monoxide may appear in small quantities. In writing chemical equations for complete combustion like the one above, no information about the products of the actual process is necessary.

In addition to the preceding nomenclature, the relationship between the fuel and air supplied to a combustion process is frequently given in terms of the **air-fuel** or **fuel-air** ratio. The air-fuel ratio (AF) is defined as the mass of air supplied per unit mass of fuel supplied. The fuel-air ratio (FA) is the reciprocal of the above definition. For the 100 percent theoretical combustion of propane, for example, the chemical equation shows that 23.80 mol of air (5 mol of $O_2$ + 18.80 mol of $N_2$) is required per mole of fuel. Consequently, the air-fuel ratio for theoretical combustion of this fuel is

$$AF = \frac{23.80\text{ kmol air}}{\text{kmol fuel}} \times \frac{29\text{ kg air}}{\text{kmol air}} \times \frac{\text{kmol fuel}}{44\text{ kg fuel}} = 15.7\,\frac{\text{kg air}}{\text{kg fuel}}$$

Since the air-fuel ratio is expressed as mass of air per mass of fuel, its value is the same whether expressed as kilograms per kilogram, as above, or as pounds per pound. The fuel-air ratio for the same combustion process would be 0.0637, in units of kg fuel/kg air or lb fuel/lb air. Many hydrocarbon fuels from oil or natural gas require an air-fuel ratio in the neighborhood of 15 to 16 for stoichiometric combustion.

Finally, the relationship between the amounts of fuel and air supplied to a combustion process is also given by the **equivalence ratio** $\phi$. By definition,

$$\phi \equiv \frac{FA_{act}}{FA_{stoich}}$$

where FA in the numerator represents the fuel-air ratio used under actual (act) combustion conditions and FA in the denominator is the stoichiometric or chemically correct value. The value of $\phi$ is less than 1 when an excess of oxidant (such as air or oxygen) is used. This is also called a *lean* mixture. A *rich* mixture is one where $\phi$ is greater than unity, and the fuel is in excess of the stoichiometric requirement. The term equivalency ratio is used frequently with respect to spark ignition and compression ignition engine operation and to gas turbine combustor analysis.

**A** gaseous fuel contains the following components on a volumetric basis: hydrogen, 2 percent; methane, 64 percent; and ethane, 34 percent. Calculate (*a*) the air-fuel ratio required, kg air/kg fuel, (*b*) the equivalence ratio used, and (*c*) the volume of dry air required per kilogram and per kilomole of fuel, if 20 percent excess air is used and the air conditions are 27°C and 0.98 bar (98 kPa).

**Example 13-1**

**Solution:**

**Given:** A gas mixture with composition and state as shown in Fig. 13-1.

**Find:** (*a*) AF ratio, in kg/kg, (*b*) $\phi$, (*c*) *V*, in m³ per kg fuel and m³ per kmol fuel.

**Model:** Ideal-gas mixture, complete combustion, 20 percent excess air.

**Analysis:** (*a*) The first step is to write the chemical reactions for the combustion of the fuel per mole of fuel. In terms of the theoretical or chemically correct oxygen requirements,

$$0.02H_2 + 0.01O_2 \rightarrow 0.2H_2O$$

$$0.64CH_4 + 1.28O_2 \rightarrow 0.64CO_2 + 1.28H_2O$$

$$0.34C_2H_6 + 1.19O_2 \rightarrow 0.68CO_2 + 1.02H_2O$$

As a result we find that 2.48 mol of oxygen per mole of fuel is required for complete combustion. The theoretical air-fuel ratio is

$$AF = \frac{2.48(4.76)(29)}{0.02(2) + 0.64(16) + 0.34(30)} = \frac{342.3}{20.48} = 16.7 \text{ kg air/kg fuel}$$

For 20 percent excess air the required value is

$$AF = 16.7(1.20) = 20.0 \text{ kg air/kg fuel}$$

(*b*) The equivalence ratio is defined as $FA_{act}/FA_{stoich}$. If this is written in terms of the AF ratio, then since 20 percent excess air is used,

$$\phi = \frac{AF_{stoich}}{AF_{act}} = \frac{1}{1.2} = 0.83$$

(*c*) Under the stated conditions, the air is assumed to behave as an ideal gas. The volume occupied by each kilogram of air is found from the ideal-gas equation:

$$v = \frac{RT}{P} = \frac{0.08314 \text{ bar·m}^3}{\text{kmol·K}} \times \frac{300 \text{ K}}{0.98 \text{ bar}} \times \frac{\text{kmol}}{29 \text{ kg}} = 0.878 \text{ m}^3/\text{kg}$$

In part *a* the air-fuel ratio was found to be 20.0 for the 20 percent excess air supplied. Therefore, the volume of air required per kilogram of fuel is

$$V = 0.878 \text{ m}^3/\text{kg air} \times 20.0 \text{ kg air/kg fuel} = 17.6 \text{ m}^3 \text{ per kg fuel}$$

Finally, in part *a* it was determined that there is 20.48 kg fuel per kilomole of fuel. Therefore, the volume of air per kilomole of fuel is

$$V = 17.6 \text{ m}^3/\text{kg fuel} \times 20.48 \text{ kg fuel/kmol fuel} = 360 \text{ m}^3 \text{ per kmol fuel}$$

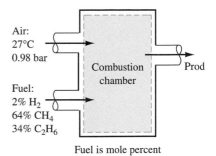

Air:
27°C
0.98 bar

Combustion chamber

Prod

Fuel:
2% $H_2$
64% $CH_4$
34% $C_2H_6$

Fuel is mole percent

**Figure 13-1**
Schematic and data for Example 13-1.

**Comment:**    The theoretical equation for the overall reaction with complete combustion and 20 percent excess air is

$$0.02H_2 + 0.64CH_4 + 0.34C_2H_6 + 2.98\,O_2 + 11.19N_2 \rightarrow 1.32CO_2 + 2.32H_2O$$
$$+ 11.19N_2 + 0.50O_2$$

In the incomplete combustion of carbon in a fuel, the carbon reacts according to the reaction $C + \frac{1}{2}O_2 \rightarrow CO$. Since oxygen has a greater affinity for combining with hydrogen than it does with carbon, all the hydrogen in a fuel normally is converted to water. If there is insufficient oxygen to ensure complete combustion, it is always the carbon that is not completely reacted. In actual practice, there is usually CO in the products, even though an excess of oxygen was supplied. This may be attributed either to incomplete mixing during the process or to insufficient time for complete combustion. The following example involves a combustion process carried out with a deficiency of air. The example again illustrates the use of the principle of the conservation of mass of the elements present in the overall reaction. By this technique we can systematically determine additional information on the initial reactant state or the state of the products.

**Example 13-2**

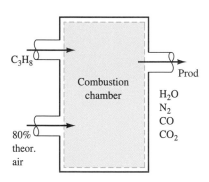

**Figure 13-2**
Schematic for Example 13-2.

**Propane** gas is allowed to react with 80 percent theoretical air. Determine the chemical equation for the reaction if both CO and $CO_2$, but no $O_2$, appear in the products.

**Solution:**

**Given:**    Propane reacts with 80 percent theoretical air, as shown in Fig. 13-2.

**Find:**    Chemical equation for the reaction.

**Model:**    Both CO and $CO_2$ in the products, but no $O_2$.

**Analysis:**    With a deficiency of air it is assumed that all the hydrogen is converted to water, but the carbon is converted to both CO and $CO_2$. The equation for 100 percent theoretical air has been shown to be

$$C_3H_8 + 5\,O_2 + 18.80\,N_2 \rightarrow 3\,CO_2 + 4\,H_2O + 18.80\,N_2$$

For 80 percent theoretical air we may write, in general, that

$$C_3H_8 + 4\,O_2 + 15.04\,N_2 \rightarrow aCO + bCO_2 + 4\,H_2O + 15.04\,N_2$$

The problem is to predict theoretically the quantities of CO and $CO_2$ in the products, as given by the unknowns $a$ and $b$. One of the basic premises for reacting systems of this sort is that all atomic species present are conserved. Therefore, we can write mass balances for each of the atomic species present in a reaction. Balances on hydrogen and nitrogen are not informative in this particular case. However, a carbon and an oxygen balance lead to the following equations:

C balance: $$3 = a + b$$

O balance: $$8 = a + 2b + 4$$

The solution to these two equations is that $a$ equals 2 and $b$ equals 1. Hence the correct theoretical equation for the combustion of propane with 80 percent theoretical air is

$$C_3H_8 + 4O_2 + 15.05N_2 \rightarrow 2CO + 1CO_2 + 4H_2O + 15.04N_2$$

The values of $a$ and $b$ would vary, depending on the percentage of theoretical air used. This equation would not be applicable to the actual combustion process, since there may be $O_2$ in the products as well as other species in small amounts. However, the chemical equation can be used as a first approximation of what should be expected in the way of products for this particular reaction under the given conditions.

---

Up to this point we have not designated the phases of the reactants and the products for a reaction, since we have been interested only in reviewing the chemistry of combustion processes and in introducing terminology pertinent to the subject. Nevertheless, acknowledgment of the fact that different phases might be present during the reaction is important for several reasons. One is the problem of the dew point of the product gases. In the combustion of hydrocarbon fuels, one of the main products is water. In a mixture of ideal product gases, this water vapor has a certain partial pressure. If the partial pressure ever becomes greater than the saturation pressure of water at a given temperature, some of the water will condense out as the temperature is lowered further. The presence of liquid-water droplets in the combustion gases may lead to corrosion problems, for one thing. Consequently, it is useful to be able to predict the dew point of a given product gas. This requires knowledge of the partial pressure of the water vapor in a gas, which in turn is a function of the mole fraction of the water vapor. Example 13-3 illustrates a typical calculation for the dew point of a combustion gas when dry air is used.

---

**Example 13-3**

Gaseous propane is burned with 150 percent theoretical air at a pressure of 970 mbars (97 kPa). If the entering air is dry, determine (*a*) the mole analysis of the product gas assuming complete combustion, (*b*) the dew point of the gas mixture, in degrees Celsius, and (*c*) the percentage of the water formed that is condensed if the product gases are cooled to 20°C.

**Solution:**

**Given:** Propane is burned with 150 percent theoretical air at 970 mbars, as shown in Fig. 13-3.

**Find:** (*a*) Mole analysis of products, (*b*) dew point of products, in degrees Celsius, (*c*) percent of water condensed when products are cooled to 20°C.

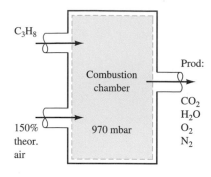

**Figure 13-3**
Schematic for Example 13-3.

**Model:**   Complete combustion, ideal gases.

**Analysis:**   (a) The chemical equation for the reaction of propane ($C_3H_8$) with 150 percent theoretical air has been given previously as

$$C_3H_8 + 7.5O_2 + 28.20\,N_2 \rightarrow 3CO_2 + 4H_2O + 2.5O_2 + 28.20N_2$$

The number of total moles of products is 37.7; therefore, the molar analysis of the products is simply

$$\text{Mole fraction } CO_2 = \frac{3}{37.7} = 0.0796\ (7.96\%)$$

$$\text{Mole fraction } H_2O = \frac{4}{37.7} = 0.1061\ (10.61\%)$$

$$\text{Mole fraction } O_2 = \frac{2.5}{37.7} = 0.0663\ (6.63\%)$$

$$\text{Mole fraction } N_2 = \frac{28.20}{37.7} = 0.7480\ (74.80\%)$$

(b) The partial pressure of the water vapor in the product gas is $0.1061(970) = 103$ mbars $= 0.103$ bar. From Table A-12, the saturation temperature corresponding to this pressure is roughly 46°C. When the gas mixture is cooled to 46°C at constant pressure, the dew point is reached. For temperatures below this value, water will condense out of the product gas.

(c) Since the dew point of the mixture is 46°C, water will condense out as the gases are cooled to 20°C. At 20°C, the gas mixture will be saturated with water, and the partial pressure of the water will be the saturation pressure at 20°C of 0.02339 bar. Assuming an ideal-gas mixture, this partial pressure of water must equal its mole fraction in the gas phase times the total pressure, that is, $p_i = y_i P = (N_i/N_m)P$. The number of moles of dry products is $3 + 2.5 + 28.2$, or 33.7. Representing the unknown moles of water still in the vapor phase by the symbol $W$ gives

$$p_g = 23.39 \text{ mbar} = \frac{W(970 \text{ mbar})}{33.7 + W} \qquad \text{or} \qquad W = 0.83 \text{ mol}$$

Because the number of moles of water formed is 4, the percentage of water vapor that condenses is

$$\text{Percentage condensed} = \frac{4 - 0.83}{4}(100) = 79\%$$

Thus roughly 20 percent of the water formed remains in the gas phase as a saturated vapor.

## 13-2   ACTUAL COMBUSTION PROCESSES

In the preceding section the discussion and examples were based on the premise that complete information was available on the reactants entering into a combustion process. In addition, it was necessary to assume that, in the presence of excess air, all carbon within a fuel would be converted completely to carbon dioxide. However, it is a common experience based on measurements of product gases that carbon monoxide is frequently present in significant quantities, even if an excess of air has been used. As a further point, there are a number of applications for which actual measurement of the air-fuel ratio is difficult to obtain. The flow of fuel, whether solid, liquid, or gas, into the reaction chamber is normally fairly well known; however, the air flow might be hard to measure accurately. To overcome these problems, an analysis of the gaseous products may be made with a good degree of accuracy. From this analysis a significant amount of information about the overall combustion process is learned.

Numerous experimental methods can be used to determine the concentration of various components in the actual gaseous products of combustion. The need for more accurate techniques increased with the passage of strict air-pollution emission standards by the state and federal governments, beginning in the late 1960s. Today there are a large number of manufacturers of suitable equipment. The analysis of combustion gases is usually reported on either a "dry" or a "wet" basis. On a dry basis, the percentage of water vapor in the gas stream is not reported. This lack of water vapor data does not limit the usefulness of a "dry" analysis. The well-established Orsat analyzer is a typical piece of equipment which reports the overall analysis on a dry basis. Although the mole fraction of water vapor in the original gas sample is not reported by such a measurement, this does not limit the usefulness of the technique. Another important gas analyzer is the gas chromatograph, used in conjunction with a thermal-conductivity or a flame-ionization detector. Typical calculations based on representative "dry" analyses from such analyzers are given below. The general method is based on the atomic species present in the reacting mixture.

---

Propane is reacted with dry air in such a ratio that an analysis of the dry products of combustion gives $CO_2$, 11.5 percent; $O_2$, 2.7 percent; and CO, 0.7 percent. Determine the percentage of theoretical air used during the process.

**Example 13-4**

**Solution:**

**Given:**   Propane reacts with dry air as shown in Fig. 13-4.

**Find:**   Percent theoretical air used.

**Model:**   Incomplete combustion; nitrogen is inert.

**Analysis:**   The basis for the analysis is chosen to be 100 mol of dry product gases. The given product analysis of $CO_2$ (11.5 percent), $O_2$ (2.7 percent), and CO (0.7 percent) adds up to 14.9 mol. If the remaining gas is assumed to be nitrogen, the amount of $N_2$ is 85.1 mol. Hence the chemical equation for the reaction takes on the format

$$x C_3H_8 + a O_2 + 3.76a N_2 \rightarrow 11.5 CO_2 + 0.7 CO + 2.7 O_2 + 85.1 N_2 + b H_2O$$

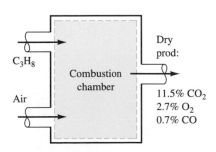

**Figure 13-4**
Schematic and data for Example 13-4.

A mass balance on each element enables us to evaluate the unknown coefficients. For nitrogen,

$N_2$ balance:                              $3.76a = 85.1$

$$a = 22.65$$

A carbon balance leads directly to the quantity of fuel used:

C balance:                              $3x = 11.5 + 0.7 = 12.2$

$$x = 4.07$$

The only unknown at this point is the value of $b$, the number of moles of water in the products. However, two mass balances have not been used, namely, those for oxygen and hydrogen. The extra mass balance provides in this case, where the composition of the fuel is known, an independent check on the preceding calculations. From an oxygen balance:

O balance:             $2(22.65) = 2(11.5) + 0.7 + 2(2.7) + b$

$$b = 16.2$$

Then, employing a hydrogen balance, we check the computation:

H balance:                              $8x = 2b$

Using the values of $x$ and $b$ calculated above, the left and right sides of the hydrogen balance equal 32.56 and 32.4, respectively. This check is fairly good, given the limited significant figures used for the calculations. The actual reaction, then, is represented by the chemical equation

$4.07C_3H_8 + 22.65O_2 + 85.1N_2 \rightarrow$

$$11.5CO_2 + 0.7CO + 2.7O_2 + 85.1N_2 + 16.2H_2O$$

By dividing through by 4.07, the equation can be placed on the basis of 1 mol of fuel. The stoichiometric equation for propane has been shown to be

$$C_3H_8 + 5O_2 + 18.80N_2 \rightarrow 3CO_2 + 4H_2O + 18.80N_2$$

The number of moles of oxygen used theoretically per mole of fuel is 5. For the actual combustion, this ratio is $22.65/4.07 = 5.57$. Therefore, the percentage of theoretical air (or oxygen) used in the actual combustion process was

$$\text{Percentage of theoretical air} = \frac{5.57(100)}{5.0} = 111\%$$

or the percentage of excess air was 11 percent.

---

**Example 13-5**

**A**n unknown hydrocarbon fuel $C_xH_y$ reacts with dry air. An Orsat analysis is made on the product gases, with the following result: $CO_2$, 12.1 percent; $O_2$, 3.8 percent; and CO, 0.9 percent. Determine the chemical equation for the actual reaction and the composition of the fuel, the air-fuel ratio used during the test, and the percent excess or deficiency of air used.

**Solution:**

**Given:**   Hydrocarbon fuel $C_xH_y$ reacts with dry air to give product analysis shown in Fig. 13-5.

**Find:** The equation for the reaction and fuel composition, the air-fuel ratio used, and the percent excess or deficiency of air used.

**Model:** Incomplete combustion; nitrogen is inert.

**Analysis:** The percentages of the three gases in the Orsat analysis add up to 16.8 percent. If it is assumed that the remaining gas in the sample on a dry basis is nitrogen, the volumetric percent of nitrogen must be 83.2 percent. On the basis of 100 mol of dry product gases, we write the general chemical equation in the form

$$C_xH_y + aO_2 + 3.76aN_2 \rightarrow 12.1CO_2 + 3.8O_2 + 0.9CO + 83.2N_2 + bH_2O$$

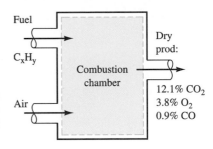

**Figure 13-5**
Schematic and data for Example 13-5.

The equation actually reverses the technique used in the preceding section. Basically, we ask ourselves what initial composition of fuel and what air-fuel ratio would be required to produce certain known percentages of product gases. The unknown values of $x$ and $y$ determine the fuel composition, and the value of $a$ will establish the air-fuel ratio. In addition, the value of $b$, the moles of water vapor formed, will need to be found if the dew point is desired. A nitrogen balance determines the value of $a$:

N₂ balance:

$$3.76a = 83.2$$
$$a = 22.1$$

An oxygen balance enables one to determine the number of moles of water formed during the reaction:

O balance:

$$2(22.1) = 2(12.1) + 2(3.8) + 0.9 + b$$
$$b = 11.5$$

Now carbon and hydrogen balances may be used to find the values of $x$ and $y$:

C balance: $\quad x = 12.1 + 0.9 = 13.0$

H balance: $\quad y = 2b = 2(11.5) = 23.0$

With a knowledge of the values of the initially unknown quantities, the chemical equation may now be written as

$$C_{13}H_{23} + 22.1O_2 + 83.2N_2 \rightarrow 12.1CO_2 + 3.8O_2 + 0.9CO + 83.2N_2 + 11.5H_2O$$

The fact that the values of $x$ and $y$ came out to be whole numbers has no significance, since three-figure accuracy is involved. Usually, the values of $x$ and $y$ for similar problems of this type will not be whole numbers. Also, the formula $C_{13}H_{23}$ should not be thought of as belonging to a single chemical species but as the average formula for a fuel which probably contains a large number of different compounds.

From the balanced chemical equation above, the air-fuel ratio can be computed. Since 105.3 mol of air was used per mole of fuel,

$$AF = \frac{105.3 \text{ kmol air}}{1 \text{ kmol fuel}} \times \frac{29 \text{ kg air}}{1 \text{ kmol air}} \times \frac{1 \text{ kmol fuel}}{179 \text{ kg fuel}}$$
$$= 17.1 \text{ kg air/kg fuel (or lb}_m \text{ air/lb}_m \text{ fuel)}$$

(Note that the air-fuel ratio is the same whether the analysis is made in SI or USCS units.) Finally, the excess or deficiency of air can be found by first computing the stoichiometric AF value for this fuel. The chemical equation for combustion of $C_{13}H_{23}$

with stoichiometric air is

$$C_{13}H_{23} + 18.75O_2 + 3.76(18.75)N_2 \rightarrow 13CO_2 + 11.5H_2O + 18.75(3.76)N_2$$

Since 18.75 mol of $O_2$ is required and 22.1 mol of $O_2$ was actually used, the percentage of excess air used during the test is

$$\text{Percentage excess} = \frac{22.1 - 18.75}{18.75}(100) = 18\%$$

---

The preceding discussion has dealt with the chemistry of reactive systems in terms of theoretical and actual considerations. Terminology used in engineering studies of reactive systems has been reviewed. The presentation has been based primarily on the principle of conservation of atomic species during chemical reactions. We are now in a position to consider reactive systems in light of the conservation of energy principle.

## 13-3    THE ENTHALPY OF FORMATION

The application of the conservation of energy principle to reacting systems requires an extension of the ideas presented earlier for systems of fixed composition. Engineering studies of reacting systems are generally directed toward steady-state, steady-flow processes. For a simple compressible system the energy equation for a chemically reactive flow system is

$$\dot{Q} + \dot{W} = \sum_{\text{prod}} \dot{N}_i(\bar{h} + \text{ke} + \text{pe})_i - \sum_{\text{reac}} \dot{N}_i(\bar{h} + \text{ke} + \text{pe})_i \quad \textbf{[13-2]}$$

where $\bar{h}_i$ is the molar enthalpy at a specified the temperature and pressure, and $\dot{N}_i$ is the molar flow rate of any product or reactant.

The evaluation of the $\bar{h}_i$ quantities in Eq. [13-2] introduces a difficulty unique to reactive systems. In Chaps. 3 and 4, methods for evaluating the enthalpy of an ideal or real gas are introduced. A common way of presenting such data is in tabular format. The Appendix contains a number of tables of thermodynamic data for ideal and real gases. Recall, however, that the values of $\bar{h}$ in these tables depend on the choice of a reference state. The ideal-gas data, for example, are based arbitrarily on a zero value of the enthalpy at absolute zero of temperature. The steam data, however, are based on a zero reference value for the saturated liquid at the triple state. In the literature we may find tables of these same substances with reference states different from those discussed above, since it is merely an arbitrary choice of the author of a given table. Because reference states are completely arbitrary, different values of $\bar{h}_i$ in Eq. [13-2] will result when tables based on different reference states are employed. We must seek another technique for evaluating the enthalpy function which avoids this difficulty.

A consistent accounting of the enthalpy change for a chemically reactive system is accomplished by introducing the concept of the enthalpy of formation $\Delta \bar{h}_f$ of a pure substance:

> The **enthalpy of formation** is defined as the enthalpy change that occurs when a compound is formed from its stable elements at a given temperature and specified standard reference pressure.

This enthalpy change is equal to the energy released or absorbed during the formation process. On a molar basis, this formation process is denoted symbolically by

$$\Delta \bar{h}_f = \bar{h}_{\text{compound}} - \sum_i (\nu_i \bar{h}_i)_{\text{stable elements}} \qquad \textbf{[13-3]}$$

where $\nu_i$ again is the stoichiometric coefficient for the given element. The enthalpy of formation $\Delta \bar{h}_f$ is determined by laboratory measurements or by methods of statistical thermodynamics, which employs spectroscopic data for the species of interest.

Enthalpy of formation data for a number of common substances are given in Tables A-23 and A-23E in the Appendix. To standardize $\Delta \bar{h}_f$ data, they are reported for a state of 25°C (298.15 K) or 77°F (536.7°R) and unit pressure. The unit pressure chosen for the above tables is 1 atm. (Data based on 1 bar also appear in the literature.) This choice of temperature and pressure is a *standard reference state* for thermochemical data. Properties at the standard pressure are symbolized by the superscript °. At this standard reference pressure a gas is also taken to be an ideal gas. Hence the molar values of the enthalpy of formation in Tables A-23 and A-23E are listed as $\Delta \bar{h}_f^\circ$. The sign convention for $\Delta \bar{h}_f^\circ$ values is the same as that for $Q$ and $W$. When energy released as a compound is formed from its stable elements, the value of $\Delta \bar{h}_f^\circ$ is negative. For example, with reference to Fig. 13-6, if $H_2(g)$ and $O_2(g)$ are reacted in a steady-flow system at 25°C, it would be necessary to remove 241,820 kJ/kmol of gaseous water formed to keep the temperature constant. Thus the enthalpy of formation of gaseous water at 25°C is listed as −241,820 kJ/kmol.

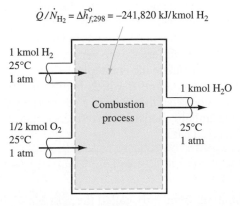

$$\dot{Q}/\dot{N}_{H_2} = \Delta \bar{h}^\circ_{f,298} = -241{,}820 \text{ kJ/kmol } H_2$$

1 kmol $H_2$
25°C
1 atm

Combustion process

1 kmol $H_2O$
25°C
1 atm

1/2 kmol $O_2$
25°C
1 atm

**Figure 13-6**  Illustration for the enthalpy of formation for water.

If Eq. [13-3] is rearranged, then at the standard reference state

$$\bar{h}_{\text{compound,ref}} = \Delta\bar{h}^{\circ}_{f,298} + \sum_{i}(\nu_i\bar{h}_i)_{\text{stable elements,ref}} \qquad \textbf{[13-4]}$$

This equation relates the enthalpy of a compound to its enthalpy of formation and the enthalpy of the elements from which it is formed, all in the standard reference state. An enthalpy datum for reacting species is established in the following manner.

> By *convention*, the enthalpy of all *stable elements* in the standard reference state of 1 atm and 25°C is assigned a value of *zero*.

The term "stable element" must be strictly followed. The stable forms of elements such as hydrogen, nitrogen, and oxygen, for example, at 1 atm and 25°C are $H_2$, $N_2$, and $O_2$, respectively. However, the stable form of carbon under these same conditions is solid graphite, $C(s)$, and not diamond. Note also that the enthalpy of formation of any *stable element* at any temperature, by concept, is zero in value. For example, the enthalpy of formation of the stable form of oxygen, $O_2$, is listed as zero in Tables A-23 and A-23E. However, note in these tables that the unstable form of oxygen (the monatomic species, O) has a nonzero value for its enthalpy of formation.

On the basis of the above discussion, Eq. [13-4] may now be written as

$$\bar{h}_{\text{compound}}(\text{at 25°C and 1 atm}) = \Delta\bar{h}^{\circ}_{f,298} = \Delta\bar{h}^{\circ}_{f,537} \qquad \textbf{[13-5]}$$

Hence the enthalpy of any compound (or unstable element) at 298 K (537°R) and 1 atm is equal to its enthalpy of formation of that state, while the enthalpy of stable elements at this same state is zero. Note that the tables in the Appendix present $\Delta\bar{h}^{\circ}_f$ data for a particular phase of a substance. The difference between the enthalpy of formation for the gas phase and the liquid phase can be approximated quite closely by the enthalpy of vaporization $\bar{h}_{fg}$ at that temperature. Thus, to convert data, one applies the relation

$$\Delta\bar{h}^{\circ}_f \text{ (gas phase)} \approx \Delta\bar{h}^{\circ}_f \text{ (liquid phase)} + \bar{h}_{fg} \qquad \textbf{[13-6]}$$

(Note that when a substance vaporizes at the standard reference temperature of 298 K, the saturation pressure will not normally be the standard pressure of 1 atm. However, the additional corrections in the vapor- and liquid-phase enthalpies for this pressure difference are very small and can be neglected.)

The final step is to evaluate the enthalpy of a pure substance at a specified temperature $T$ and pressure $P$ which is different from the standard reference state. To accomplish this, we must add to the value given by Eq. [13-5] the change in enthalpy between the reference state at 25°C and 1 atm and the specified state. That is,

$$\boxed{\bar{h}_{i,T,P} = \Delta\bar{h}^{\circ}_{f,298,i} + (\bar{h}_{T,P} - \bar{h}_{298\text{ K,1atm}})_i} \qquad \textbf{[13-7a]}$$

This latter contribution, frequently called the *sensible* enthalpy change, is found from methods discussed in Chaps. 3 and 4. If we restrict ourselves to reacting ideal-gas mixtures, then the enthalpy of each gas is independent of pressure. In this case the total enthalpy $h_{i,T}$ of an ideal gas at temperature $T$ is given by

$$\overline{h}_{i,T} = \Delta \overline{h}^{\circ}_{f,298,i} + (\overline{h}_T - \overline{h}_{298})_i \qquad \text{ideal gas} \qquad \textbf{[13-7b]}$$

where $\overline{h}_T$ is the enthalpy at the specified temperature $T$ and $\overline{h}_{298}$ is the enthalpy at the reference temperature of 298 K (537°R). If tabular data are not available to evaluate the last term in Eq. [13-7b], the enthalpy change $(\overline{h}_T - \overline{h}_{298})_i$ for an ideal gas must be computed from the integral of $\overline{c}_{p,0} dT$. Thus the enthalpy of any pure substance is composed of two parts—that associated with its formation from stable elements at a standard reference state and that associated with a change of state at constant composition.

## 13-4 STEADY-FLOW ENERGY ANALYSIS OF REACTING MIXTURES

If the enthalpy defined in Eq. (13-7a) is combined with Eq. (13-2), the steady-state energy balance for chemically reactive systems on a rate basis becomes

$$\dot{Q} + \dot{W} = \sum_{\text{prod}} \dot{N}_i \underbrace{(\Delta \overline{h}^{\circ}_{f,298} + \Delta \overline{h})_i}_{\overline{h}_{i,T,P}} - \sum_{\text{reac}} \dot{N}_i \underbrace{(\Delta \overline{h}^{\circ}_{f,298} + \Delta \overline{h})_i}_{\overline{h}_{i,T,P}} \qquad \textbf{[13-8]}$$

where $\Delta \overline{h}^{\circ}_{f,298}$ is the enthalpy of formation at standard temperature and pressure and $\Delta \overline{h} = \overline{h}_{T,P} - \overline{h}_{298,1\,\text{atm}}$ is the sensible enthalpy change. For many applications, the kinetic- and potential-energy changes of the flow streams can be neglected as done in the equation above. In general, the numerical value of $\dot{Q}$ and $\dot{W}$ can be either positive or negative depending on the direction of energy flow on the system boundary. For many combustion processes, there is no work and $\dot{W} = 0$. Under these conditions, isothermal chemical reactions that release energy in the form of heat transfer are called **exothermic** and those that absorb energy are called **endothermic**.

For studying combustion processes, it is frequently useful to have Eq. [13-8] expressed on a molar basis. Since combustion calculations are typically based on the amount of fuel present, it is common to rewrite Eq. [13-8] in terms of the molar flow rate of fuel as follows:

$$\dot{Q} + \dot{W} = \dot{N}_{\text{fuel}} \left[ \sum_{\text{prod}} \left( \frac{\dot{N}_i}{\dot{N}_{\text{fuel}}} \right) (\Delta \overline{h}^{\circ}_{f,298} + \Delta \overline{h})_i - \sum_{\text{reac}} \left( \frac{\dot{N}_i}{\dot{N}_{\text{fuel}}} \right) (\Delta \overline{h}^{\circ}_{f,298} + \Delta \overline{h})_i \right]$$

$$\textbf{[13-9a]}$$

This equation can be written more compactly by defining a new symbol for the molar flow rate ratios and using the standard symbol for the enthalpy of a compound. Once this is done Eq. [13-9a] becomes

$$\dot{Q} + \dot{W} = \dot{N}_{\text{fuel}}\left[\sum_{\text{prod}} n_i \bar{h}_i - \sum_{\text{reac}} n_i \bar{h}_i\right] \qquad \text{[13-9b]}$$

where

> $n_i = \dot{N}_i/\dot{N}_{\text{fuel}}$, the ratio of the molar flow rate of compound $i$ entering or leaving the system to the molar flow rate of the fuel, and

> $\bar{h}_i = (\Delta\bar{h}^{\circ}_{f,298} + \Delta\bar{h})_i$, the enthalpy of compound $i$ evaluated with respect to the enthalpy of the stable elements at the standard reference state (25°C and 1 atm).

*The numerical values of $n_i$ are the coefficients from a balanced chemical equation that is written on the basis of one mole of fuel and includes <u>all</u> compounds entering and leaving the system;* as a consequence $n_{\text{fuel}}$ is always unity. It is sometimes useful to talk about the enthalpy of the reactants and the products. This can be done by rewriting Eq. [13-9b] as

$$\dot{Q} + \dot{W} = \dot{N}_{\text{fuel}}[\bar{h}_{\text{prod}} - \bar{h}_{\text{reac}}] \qquad \text{[13-9c]}$$

where $\bar{h}_{\text{prod}} = \sum_{\text{prod}} n_i \bar{h}_i$, the enthalpy of the products (compounds leaving the system) per mole of fuel, and $\bar{h}_{\text{reac}} = \sum_{\text{reac}} n_i \bar{h}_i$, the enthalpy of the reactants (compounds entering the system) per mole of fuel.

For *gases at low pressures,* the effect of pressure on the enthalpy may frequently be neglected and the gases modeled as ideal gases. Under these conditions, Eq. [13-9] can be written as

$$\dot{Q} + \dot{W} = \dot{N}_{\text{fuel}}\left[\sum_{\text{prod}} n_i \left(\Delta\bar{h}^{\circ}_{f,298} + \bar{h}_T - \bar{h}_{298}\right)_i - \sum_{\text{reac}} n_i \left(\Delta\bar{h}^{\circ}_{f,298} + \bar{h}_T - \bar{h}_{298}\right)_i\right]$$

$$\text{[13-10]}$$

where $\bar{h}_T$ is the ideal-gas enthalpy at the specified temperature $T$ and $\bar{h}_{298}$ is the ideal-gas enthalpy at the reference temperature of 298 K (537°R).

---

**Example 13-6**

The combustion of methane with the stoichiometric amount of oxygen is given by the chemical equation

$$CH_4(g) + 2O_2(g) \rightarrow CO_2(g) + 2H_2O(g)$$

where the symbol $(g)$ after the chemical formula indicates the gas phase. Determine the heat transfer per kmole of fuel if this reaction occurs at 1 atm and 25°C.

**Solution:**

**Given:**    Combustion of methane with stoichiometric air, as shown in Fig. 13-7.

**Find:**    $\dot{Q}/\dot{N}_{CH_4}$, in kJ per kmol $CH_4$.

**Model:**    Steady-state, open system; complete combustion; ideal gases; $\dot{W} = 0$.

**Analysis:** For a steady-state combustion process, the energy equation is

$$\dot{Q} = \sum_{\text{prod}} \dot{N}_i \bar{h}_i - \sum_{\text{reac}} \dot{N}_i \bar{h}_i$$

where $\bar{h}_i = \Delta \bar{h}^{\,o}_{f,298} + \bar{h}_T - \bar{h}_{298}$ for an ideal gas. In this particular case, the sensible enthalpy change $\bar{h}_T - \bar{h}_{298}$ is zero for reactants and products, because both the initial and final temperatures represent the standard reference temperature. Dividing the energy balance through by the molar flow rate of the methane and recognizing that $\bar{h}_i = \Delta \bar{h}^{\,o}_{f,298,i}$ for this problem, we have

$$\dot{Q}/\dot{N}_{CH_4} = n_{CO_2}\Delta\bar{h}^{\,o}_{f,CO_2} + n_{H_2O}\Delta\bar{h}^{\,o}_{f,H_2O} - n_{CH_4}\Delta\bar{h}^{\,o}_{f,CH_4} - n_{O_2}\Delta\bar{h}^{\,o}_{f,O_2}$$

where $n$ is the ratio of the molar flow rate to the molar flow rate of fuel. The numerical values for $n$ are the coefficients from the balanced chemical reaction for the combustion process written per mole of fuel: $CH_4 + 2O_2 \rightarrow CO_2 + 2H_2O$. Substituting in these values and evaluating the enthalpy of formation data from Table A-23 at 25°C yields

$$\dot{Q}/\dot{N}_{CH_4} = 1(-393{,}520) + 2(-241{,}820) - 1(-74{,}850) - 2(0)$$
$$= -802{,}310 \text{ kJ/kmol } CH_4$$

Note that the above answer is independent of the amount of oxidant supplied to the reaction. If air is supplied, even in an excess amount, the oxygen and nitrogen enter and leave at the standard reference temperature. Hence the value of $\bar{h}_T - \bar{h}_{298}$ is zero for these elements regardless of the quantity of each.

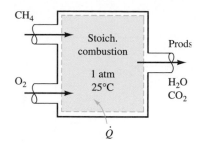

**Figure 13-7**

Schematic for Example 13-6.

---

The answer to the preceding example illustrates a term called the enthalpy of reaction. The **enthalpy of reaction** $\Delta\bar{h}_R$ is defined as the enthalpy change when a reaction occurs at a specified temperature and pressure. Values of the enthalpy of reaction at the standard reference state are designated by $\Delta\bar{h}^{\,o}_R$. For example, the value of $\Delta\bar{h}^{\,o}_R$ for methane is $-802{,}319$ kJ/kmol, as found in Example 4-6. The enthalpy of reaction for a combustion process is called the **enthalpy of combustion** $\Delta\bar{h}_c$. The answer in Example 13-6 above is also the enthalpy of combustion for gaseous methane at 25°C. Such values can be measured experimentally when the enthalpy of formation data of a fuel is not known. This technique is discussed in more detail in Sec. 13-7.

---

**M**ethane gas initially at 400 K is burned with 50 percent excess air which enters the combustion chamber at 500 K. The reaction, which occurs at 1 atm, goes to completion, and the temperature of the product gases is 1800 K. Determine the heat transfer to or from the combustion chamber, in kJ per kmol of fuel.

**Example 13-7**

**Solution:**

**Given:** Methane gas reacts with air in the combustion chamber shown in Fig. 13-8.

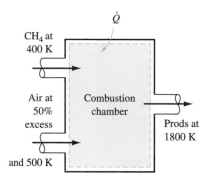

$\dot{Q}$

CH$_4$ at
400 K

Air at
50%
excess

Combustion
chamber

Prods at
1800 K

and 500 K

**Figure 13-8**
Schematic and data for Example 13-7.

**Find:**  $\dot{Q}/\dot{N}_{CH_4}$, in kJ/kmol of fuel.

**Model:**  Steady-state, open system; complete combustion; ideal gases; $\dot{W} = 0$.

**Analysis:**  The heat transfer is found from the steady-state energy balance for a reacting mixture

$$\dot{Q} = \sum_{prod} \dot{N}_i \bar{h}_i - \sum_{reac} \dot{N}_i \bar{h}_i$$

where $\bar{h}_i = \Delta\bar{h}^{\circ}_{f,298} + \bar{h}_T - \bar{h}_{298}$ for an ideal gas. Dividing through by the molar flow rate of fuel gives

$$\dot{Q}/\dot{N}_{Fuel} = \sum_{prod} n_i \bar{h}_i - \sum_{reac} n_i \bar{h}_i$$

where $n_i$ is the ratio of the molar flow rate of component $i$ to the molar flow rate of fuel. First, the values of $n_i$ must be established. The chemical equation for the complete combustion of methane with 50 percent excess air is

$$CH_4(g) + 3O_2(g) + 3(3.76)N_2(g) \rightarrow CO_2(g) + 2H_2O(g) + 11.28N_2(g) + O_2(g)$$

The state of the water in the products is indicated as a gas, since the final temperature is well above the dew point of water. Also, since the partial pressure of the water vapor is only 132 mbars $[p_{water} = (1.01325\ bar)(2/15.8)]$, the vapor may be assumed to be an ideal gas along with the other product gases.

Now that the reactants and products have been identified along with the values of $n$, the energy balance becomes

$$\dot{Q}/\dot{N}_{CH_4} = [1\bar{h}_{CO_2,3} + 2\bar{h}_{H_2O,3} + 11.28\bar{h}_{N_2,3} + 1\bar{h}_{CO_2,3}]$$
$$- [1\bar{h}_{CH_4,1} + 3\bar{h}_{O_2,2} + 11.28\bar{h}_{N_2,2}]$$

The sensible component of the enthalpy, $\Delta\bar{h}_i$, for all the gases except methane are obtained from Tables A-6 through A-10, and the enthalpy of formation $\Delta\bar{h}^{\circ}_{f,298}$ data are listed in Table A-23. No tabular data are available in this text that permit the calculation of the enthalpy change for methane from 298 to 400 K. This quantity must be determined by integrating $\bar{c}_p$ data. An equation for $\bar{c}_p$ of methane, found in Table A-3 in the Appendix, is

$$\frac{\bar{c}_p}{R_u} = 3.826 - 3.979 \times 10^{-3}T + 24.558 \times 10^{-6}T^2$$
$$- 22.733 \times 10^{-9}T^3 + 6.963 \times 10^{-12}T^4$$

where $T$ is expressed in kelvins. The appropriate value for $R_u$ is 8.314 kJ/kmol·K. The enthalpy change between 298 and 400 K for methane, using the above $c_p$ equation and omitting the integration steps, is found to be

$$\Delta\bar{h}(CH_4) = \int_{298}^{400} c_p\, dT$$
$$= 8.314(390.25 - 141.64 + 307.27 - 100.67 + 10.99)$$
$$= 3876\ kJ/kmol\ CH_4$$

Using the available data to evaluate the enthalpy of each gas gives the following values:

| Compound $i$ | $n_i$ | $T$ (K) | $\Delta\bar{h}_{f,298}^{o}$ (kJ/kmol) | $\bar{h}_T - \bar{h}_{298}$ (kJ/kmol) | $\bar{h}_i$ (kJ/kmol) |
|---|---|---|---|---|---|
| $CH_4(g)$ | 1 | 400 | $-74{,}870$ | 3876 | $-70{,}994$ |
| $O_2(g)$ | 3 | 500 | 0 | 14,770–8682 | 6088 |
| $N_2(g)$ | 11.28 | 500 | 0 | 14,581–8669 | 5912 |
| $CO_2(g)$ | 1 | 1800 | $-393{,}520$ | 88,806–9364 | $-314{,}078$ |
| $H_2O(g)$ | 2 | 1800 | $-241{,}820$ | 72,513–9904 | $-179{,}211$ |
| $N_2(g)$ | 11.28 | 1800 | 0 | 57,651–8669 | 48,982 |
| $O_2(g)$ | 1 | 1800 | 0 | 60,371–8682 | 51,689 |

Substitution of these values into the energy balance gives

$$\dot{Q}/\dot{N}_{CH_4} = [1(-314{,}078) + 2(-179{,}211) + 11.28(48.982) + 1(51{,}689)]$$
$$- [1(-70{,}994) + 3(6088) + 11.28(5912)]$$
$$= \underbrace{[-68{,}294]}_{\bar{h}_{\text{prod}}} - \underbrace{[13{,}957]}_{\bar{h}_{\text{reac}}} = -82{,}251 \text{ kJ/kmol } CH_4$$

**Comment:** In Example 13-6 for the stoichiometric combustion of methane at 25°C, it was found that the heat transferred out was 802,290 kJ/kmol. When excess air is present and the products are heated to 1800 K, as in this example, roughly 90 percent of the energy released at 25°C is used to heat the products to 1800 K.

---

**L**iquid butane ($C_4H_{10}$) at 25°C is sprayed into a combustion chamber. In addition, 400 percent theoretical air is supplied at an inlet temperature of 600 K. The gaseous products of combustion leave the chamber at 1100 K. If complete combustion is assumed, determine the heat transfer in kJ per kmol of fuel.

**Solution:**

**Given:** Combustion of liquid butane with 400 percent theoretical air as shown in Fig. 13-9.

**Find:** $\dot{Q}/\dot{N}_{\text{fuel}}$, in kJ/kmol of fuel.

**Model:** Steady-state, open system; complete combustion; gases are ideal; $\dot{W} = 0$.

**Analysis:** For the fuel under consideration, the stoichiometric chemical equation is

$$C_4H_{10}(l) + 6.5O_2(g) + 24.4N_2(g) \rightarrow 4CO_2(g) + 5H_2O(g) + 24.4N_2(g)$$

For 400 percent theoretical air the chemical equation becomes

$$C_4H_{10}(l) + 26.0O_2(g) + 97.6N_2(g)$$
$$\rightarrow 4CO_2(g) + 5H_2O(g) + 19.5O_2 + 97.6N_2(g)$$

**Example 13-8**

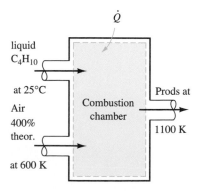

**Figure 13-9**
Schematic and data for Example 13-8.

The heat transfer is found by applying the steady-state energy balance to the combustion chamber neglecting kinetic and potential energy effects:

$$\dot{Q} = \sum_{\text{prod}} \dot{N}_i \bar{h}_i - \sum_{\text{reac}} \dot{N}_i \bar{h}_1 = \dot{N}_{\text{fuel}} \left[ \sum_{\text{prod}} n_i \bar{h}_i - \sum_{\text{reac}} n_i \bar{h}_1 \right]$$

where $\bar{h} = \Delta \bar{h}^{\circ}_{f,298} + \bar{h}_T - \bar{h}_{298}$ for ideal gases. The tabulated value of $\Delta \bar{h}^{\circ}_{f,298}$ for $n$-butane in Table A-23 is listed for the gas phase. This value must be corrected by $\bar{h}_{fg}$, as shown in Eq.[13-6], when it is substituted into the above energy balance. The other required enthalpy values are obtained from Tables A-6 through A-10 in the Appendix. Use of these data and the corrected enthalpy of formation of $n$-butane yields

| Compound $i$ | $T$ (K) | $n_i$ | $\Delta \bar{h}^{\circ}_{f,298}$ (kJ/kmol) | $\bar{h}_T - \bar{h}_{298}$ (kJ/kmol) | $\bar{h}_i$ (kJ/kmol) |
|---|---|---|---|---|---|
| $C_4H_{10}(l)$ | 298 | 1 | $-126{,}150-21{,}060$ | 0 | $-147{,}210$ |
| $O_2(g)$ | 600 | 26.0 | 0 | $17{,}929-8682$ | 9247 |
| $N_2(g)$ | 600 | 97.6 | 0 | $17{,}563-8669$ | 8894 |
| $CO_2(g)$ | 1100 | 4 | $-393{,}520$ | $48{,}258-9364$ | $-354{,}626$ |
| $H_2O(g)$ | 1100 | 5 | $-241{,}820$ | $40{,}071-9904$ | $-211{,}643$ |
| $O_2(g)$ | 1100 | 19.5 | 0 | $34{,}899-8682$ | $26{,}217$ |
| $N_2(g)$ | 1100 | 97.6 | 0 | $33{,}426-8669$ | $24{,}757$ |

Substituting this information back into the energy balance gives

$$\begin{aligned}
\dot{Q}/\dot{N}_{C_4H_{10}} &= [4\bar{h}_{CO_2,3} + 5\bar{h}_{H_2O,3} + 19.5\bar{h}_{O_2,3} + 97.6\bar{h}_{N_2,3}] \\
&\quad - [1\bar{h}_{C_4H_{10},1} + 26.0\bar{h}_{O_2,2} + 97.6\bar{h}_{N_2,2}] \\
&= [4(-354{,}626) + 5(-211{,}643) + 19.5(26.217) + 97.6(24{,}757)] \\
&\quad - [1(-147{,}210) + 26.0(9247) + 97.6(8894)] \\
&= [450{,}796] - [961{,}266] \\
&= -510{,}470 \text{ kJ/kmol } C_4H_{10}
\end{aligned}$$

## 13-5   ADIABATIC FLAME TEMPERATURE

In the absence of work effects and any appreciable kinetic-energy change of the flow stream, the energy released by a chemical reaction in a steady-flow reactor appears in two forms: heat transfer to the surroundings and a temperature rise of the product gases. The smaller the heat transfer, the larger the temperature rise becomes. In the limit of adiabatic operation of the reactor, the maximum temperature rise will occur. In a number of engineering applications of reacting systems, such as rocket-propulsion and gas-turbine cycles, it is desirable to be able to predict the maximum temperature attainable by the product gases. This maximum temperature is referred to as the ***adiabatic-flame*** or ***adiabatic-combustion temperature*** $T_{af}$ of the reacting mixture. On the basis of Eq. [13-10] the energy balance for a reacting mixture in steady flow and under adiabatic conditions becomes

$$\sum_{prod} n_i(\Delta \bar{h}^o_{f,298} + \bar{h}_T - \bar{h}_{298})_i = \sum_{reac} n_i(\Delta \bar{h}^o_{f,298} + \bar{h}_T - \bar{h}_{298})_i \quad \textbf{[13-11]}$$

Since the initial temperature and composition of the reactants normally is known, the right-hand side of Eq. [13-11] can be evaluated directly.

To obtain the maximum temperature rise of the products a reaction must go to completion. Hence the coefficient $n_i$ of the products are also known from the chemistry of the reaction. In addition, the values of $\Delta \bar{h}^o_f$ and $\bar{h}_{298}$ for each of the products are obtainable from tables of thermochemical data. Therefore, the only unknowns in Eq. [13-11] are the $\bar{h}_T$ values for each of the product gases at the unknown adiabatic-combustion temperature. Since the $\bar{h}_T$ values are tabulated against temperature in the Appendix, the solution to Eq. [13-11] is one of iteration. That is, a temperature must be guessed, and then the enthalpy values of the product gases at that temperature are found from their respective tables. If the guessed temperature is correct, then the numerical values of the left- and right-hand sides of Eq. [13-11] are equal. If not, then another estimate of the final temperature must be tried. (This type of calculation is quickly and accurately evaluated if data for the reacting species as a function of temperature are accessible on the computer.) Calculated temperatures in the range of 2300 to 2500 K or 4200 to 4500°R are fairly typical for the adiabatic combustion of many hydrocarbons with the stoichiometric quantity of air when the reactants are at 25°C.

An estimate of the maximum combustion temperature made on this basis normally is conservative. That is, the calculated value frequently will be several hundred degrees higher than a measured value. In actual practice, the calculated combustion temperature is not attained because of several effects. First, combustion is seldom complete. Second, heat losses may be minimized but not eliminated. Last, some of the products of combustion will dissociate into other chemical species as a result of the higher temperatures present. These dissociation reactions normally are endothermic and consume some of the energy released by the overall reaction. The next two examples illustrate adiabatic-combustion-temperature calculations.

---

**L**iquid butane ($C_4 H_{10}$) at 25°C and 400 percent theoretical air at 600 K react in a steady-flow process. Determine the adiabatic-flame temperature, in kelvins, assuming complete combustion.

**Example 13-9**

**Solution:**

**Given:**   Liquid butane and 400 percent theoretical air react according to the steady-flow system shown in Fig. 13-10.

**Find:**   Adiabatic-flame temperature, in kelvins.

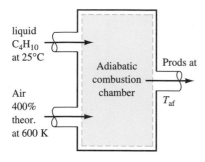

liquid
$C_4H_{10}$
at 25°C

Adiabatic
combustion
chamber

Prods at

$T_{af}$

Air
400%
theor.
at 600 K

**Figure 13-10**
Schematic and data for Example 13-9.

**Model:**   Steady-flow, adiabatic, complete combustion, ideal gases.

**Analysis:**   The equation for the reaction under consideration is

$$C_4H_{10}(l) + 26\,O_2(g) + 97.6\,N_2(g)$$
$$\rightarrow 4\,CO_2(g) + 5H_2O(g) + 19.5O_2(g) + 97.6N_2(g)$$

Writing the steady-state energy balance for this adiabatic combustion process and then dividing through by the molar flow rate of the fuel gives

$$\sum_{\text{reac}} n_i \bar{h}_i = \sum_{\text{prod}} n_i \bar{h}_i$$

$$1\bar{h}_{C_4H_{10},1} + 26\bar{h}_{O_2,2} + 97.6\bar{h}_{N_2,2} = 4\bar{h}_{CO_2,3} + 5\bar{h}_{H_2O,3} + 19.5\bar{h}_{O_2,3} + 97.6\bar{h}_{N_2,3}$$

where $\bar{h} = \Delta\bar{h}^{\circ}_{f,298} + \bar{h}_T - \bar{h}_{298}$ for an ideal gas. As noted in the discussion that precedes this example, all the values of the terms in this equation are known, except the $\bar{h}_T$ values for the products. These values are unknown because the final temperature of the products is not known. Substitutions of the known values from tables in the Appendix yields

$$1(-126,150 - 21,060) + 26(0 + 17,929 - 8682) + 97.6(0 + 17,563 - 8669)$$
$$= 4(-393,520 + \bar{h}_{T,CO_2} - 9364) + 5(-241,820 + \bar{h}_{T,H_2O} - 9904)$$
$$+ 19.5(0 + \bar{h}_{T,O_2} - 8682) + 97.6(0 + \bar{h}_{T,N_2} - 8669)$$

In the above equation the value 21,060 in the first term is the enthalpy of vaporization of $n$-butane. Solving for the unknown enthalpy values at the adiabatic-combustion temperature, we find that

$$4\bar{h}_{T,CO_2} + 5\bar{h}_{T,H_2O} + 19.5\bar{h}_{T,O_2} + 97.6\bar{h}_{T,N_2} = 4,846,800 \text{ kJ/kmol}$$

We are now involved in an iteration, or trial-and-error, type of solution. The value of the adiabatic-flame temperature is guessed, which enables us to look up the corresponding $\bar{h}_T$ values for each of the products. These values permit the evaluation of the left-hand side of the above equation. Several estimates may be required before the approximate answer is obtained.

A reasonably fast convergence on the correct temperature is achieved in the following manner. The chemical equation for the overall reaction shows that most of the combustion product is nitrogen (roughly 77 percent on a mole basis). Hence as a first approximation, we assume the product gas is entirely nitrogen. In this case the energy balance reduces to

$$126.1\bar{h}_{T,N_2} = 4,846,800 \text{ kJ/kmol}$$

or

$$\bar{h}_{T,N_2} = 38,400 \text{ kJ/kmol}$$

From Table A-6 this enthalpy value corresponds roughly to a temperature of 1240 K. This value will be found to be too high. The actual product gases contain $CO_2$ $H_2O$, and $O_2$ as well as $N_2$. The triatomic gases, $CO_2$ and $H_2O$, have a larger average specific heat $c_p$ than that of nitrogen in the given temperature range. As a result, they absorb more energy per mole for a given temperature change than does nitrogen. The effect of the presence of these triatomic gases, then, is to lower the final temperature from that predicted solely on the basis of nitrogen. The table below shows a summary of the iteration process in the vicinity of the final answer.

| | 1200 K | 1220 K |
|---|---|---|
| $4\bar{h}_T(CO_2)$ | 215,400 | 219,900 |
| $5\bar{h}_T(H_2O)$ | 221,900 | 226,300 |
| $19.5\bar{h}_T(O_2)$ | 749,700 | 763,700 |
| $97.6\bar{h}_T(N_2)$ | 3,589,400 | 3,655,300 |
| $\sum_{prod} n_i\bar{h}_{Ti}$ | 4,776,400 | 4,865,200 |

The sum of the enthalpy values for the products of combustion should be 4,846,800 kJ/kmol. The data in the table above indicate that the adiabatic-combustion temperature $T_{af}$ is close to 1215 K.

**Determine** the adiabatic-flame temperature for the reaction of liquid butane ($C_4H_{10}$) with the theoretical-air requirements if all the reactants enter the combustion chamber at 25°C in steady flow. Assume complete combustion.

**Example 13-10**

**Solution:**

**Given:** Adiabatic combustion of liquid butane with stoichiometric air as illustrated in Fig. 13-11.

**Find:** Adiabatic-flame temperature, in kelvins.

**Model:** Steady, adiabatic flow; stoichiometric, complete combustion; ideal gases.

**Analysis:** The chemical equation for the reaction is

$$C_4H_{10}(l) + 6.5O_2(g) + 24.4N_2(g) \rightarrow 4CO_2(g) + 5H_2O(g) + 24.4N_2(g)$$

Writing the steady-state energy balance for this adiabatic combustion process and then dividing through by the molar flow rate of the fuel gives

$$\sum_{reac} n_i\bar{h}_i = \sum_{prod} n_i\bar{h}_i$$

$$1\bar{h}_{C_4H_{10}} + 6.5\bar{h}_{O_2} + 24.4\bar{h}_{N_2} = 4\bar{h}_{CO_2} + 5\bar{h}_{H_2O} + 24.4\bar{h}_{N_2}$$

where $\bar{h} = \Delta\bar{h}_{f,298}^{\circ} + \bar{h}_T - \bar{h}_{298}$ for an ideal gas. The only reactant terms in this case are $\Delta\bar{h}_f^{\circ}$ and the enthalpy of vaporization for butane. Substitution of tabular data yields

$$1(-126,150 - 21,060)$$
$$= 4(-393,520 + \bar{h}_{T,CO_2} - 9364) + 5(-241,820 + \bar{h}_{T,H_2O} - 9904)$$
$$+ 24.4(\bar{h}_{T,N_2} - 8669)$$

Therefore,

$$4\bar{h}_{T,CO_2} + 5\bar{h}_{T,H_2O} + 24.4\bar{h}_{T,N_2} = 2,934,500 \text{ kJ/kmol}$$

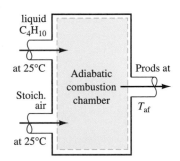

**Figure 13-11**
Schematic and data for Example 13-10.

To shorten the iteration process, we again assume that all the products are nitrogen. This leads to the approximation

$$\bar{h}_{T,\,N_2} = \frac{2,934,500}{33.4} = 87,860\,\text{kJ/kmol}$$

From the nitrogen table, A-6, the temperature corresponding to this enthalpy is between 2600 and 2650 K. The actual temperature must lie below this value. A summary of several trial calculations in the vicinity of the final answer is shown here:

|  | 2350 K | 2400 K |
|---|---|---|
| $4\bar{h}_T$ ($CO_2$) | 488,400 | 500,600 |
| $5\bar{h}_T$ ($H_2O$) | 504,200 | 517,500 |
| $24.4\bar{h}_T$ ($O_2$) | 1,890,900 | 1,935,400 |
| $\sum_{\text{prod}} n_i\bar{h}_{Ti}$ | 2,883,500 | 2,935,500 |

Since the summation of enthalpies for the products should equal 2,934,500 kJ, the preceding table indicates that the adiabatic-flame temperature is very nearly 2400 K.

**Comment:**    It is interesting to compare the above result with that obtained in Example 13-9. Even though the air in that case is preheated to 600 K, the adiabatic-combustion temperature is only around 1220 K when 400 percent theoretical air is used. This is some 1200 degrees less than the value for stoichiometric combustion. In the original example, the excess nitrogen and oxygen molecules are a large sink for the energy released by the reaction. The presence of this excess gas greatly reduces the combustion temperature.

## 13-6   CONSTANT-VOLUME THERMOCHEMICAL ANALYSIS

If a chemical reaction occurs in a closed system maintained at constant volume, then the basic energy balance is

$$Q = \Delta U = U_{\text{prod}} - U_{\text{reac}} = \sum_{\text{prod}} N_i\bar{u}_i - \sum_{\text{reac}} N_i\bar{u}_i \qquad \textbf{[13-12]}$$

The specific internal energy $\bar{u}_i$ of any component is related to its specific enthalpy $\bar{h}_i$ by recalling that $\bar{u} \equiv \bar{h} - P\bar{v}$. Using this definition and Eq. (13-8b) for $\bar{h}_i$, we find that

$$\bar{u}_{i,T} = \bar{h}_{i,T} - (P\bar{v})_{i,T} = [\Delta\bar{h}_{f,298}^{\circ} + \bar{h}_T - \bar{h}_{298} - (P\bar{v})_T]_i \qquad \textbf{[13-13]}$$

Furthermore, the sensible enthalpy change given by $h_T - h_{298}$ can be replaced by the definition relating $\bar{h}$ to $\bar{u}$. This leads to a second format for $u_i$, namely,

$$\bar{u}_{i,T} = [\Delta \bar{h}^{\circ}_{f,298} + \bar{u}_T - \bar{u}_{298} + (P\bar{v})_T - (P\bar{v})_{298} - (P\bar{v})_T]_i \qquad \textbf{[13-14]}$$

$$= [\Delta \bar{h}^{\circ}_{f,298} + \bar{u}_T - \bar{u}_{298} - (P\bar{v})_{298}]_i$$

The substitution of Eq. [13-14] into Eq. [13-12] leads to an equation for the heat transfer under constant-volume conditions:

$$Q = \sum_{\text{prod}} N_i \Delta \bar{h}^{\circ}_{f,298,i} - \sum_{\text{reac}} N_i \Delta \bar{h}^{\circ}_{f,298,i} + \sum_{\text{prod}} N_i(\bar{u}_T - \bar{u}_{298}) \qquad \textbf{[13-15]}$$

$$- \sum_{\text{reac}} N_i(\bar{u}_T - \bar{u}_{298})_i - \sum_{\text{prod}} N_i(P\bar{v})_{i,298} + \sum_{\text{reac}} N_i(P\bar{v})_{i,298}$$

Furthermore, the specific volume of solid and liquid components in a chemical reaction is usually negligible in comparison to the specific volume occupied by the gaseous components. Hence in the summations above which involve $P\bar{v}$ terms, it is necessary to account for only the gaseous species. In addition, if we assume that all gaseous components are ideal gases, then $Pv$ can be replaced by $R_uT$. Consequently the above energy balance for the *constant-volume* process can be written in the following form:

$$Q = \sum_{\text{prod}} N_i(\Delta \bar{h}^{\circ}_{f,298} + \bar{u}_T - \bar{u}_{298})_i \qquad \textbf{[13-16]}$$

$$- \sum_{\text{reac}} N_i(\Delta \bar{h}^{\circ}_{f,298} + \bar{u}_T - \bar{u}_{298})_i - \Delta N R_u T_{298}$$

where $\Delta N$ in the last equation represents the moles of gaseous products minus the moles of gaseous reactants. In general, $\Delta N$ may be positive, negative, or zero. The above two equations are equally valid in USCS units, if data at 298 K are replaced by values at 537°R.

---

**Example 13-11**

Carbon monoxide with 200 percent theoretical air at 25°C reacts completely in a constant-volume reaction chamber. After a given time the gas temperature is measured to be 1200 K. Determine the heat transfer for the chamber, in kJ per kmol of carbon monoxide.

**Solution:**

**Given:** Carbon monoxide reacts with 200 percent theoretical air in the constant-volume chamber shown in Fig. 13-12.

**Find:** $Q/N_{CO}$, in kJ per kmol of CO, by Eq. [13-16].

**Model:** Constant-volume, closed system; ideal gases.

**Analysis:** The overall chemical reaction is given by

$$CO(g) + O_2(g) + 3.76N_2(g) \rightarrow CO_2(g) + \tfrac{1}{2}O_2(g) + 3.76N_2(g)$$

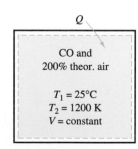

**Figure 13-12**
Schematic and data for Example 13-11.

The heat transfer is determined by applying the energy balance for a reaction in a constant-volume, closed system.

$$Q = \sum_{\text{prod}} N_i(\Delta\bar{h}^{\circ}_{f,298} + \bar{u}_T - \bar{u}_{298})_i - \sum_{\text{reac}} N_i(\Delta\bar{h}^{\circ}_{f,298} + \bar{u}_T - \bar{u}_{298})_i - \Delta N R_u T_{298}$$

Assuming 1 kmol of CO and substituting in the required data from the Appendix, we find that

$$
\begin{aligned}
Q &= 1(-393{,}520 + 43{,}871 - 6885) + \tfrac{1}{2}(0 + 28{,}469 - 6203) \\
&\quad + 3.76(0 + 26{,}799 - 6190) - 1(-110{,}520 + 0) - 1(0) - 3.76(0) \\
&\quad - (-0.5)(8.314)(298) \\
&= -156{,}110 \text{ kJ per kmol CO}
\end{aligned}
$$

When chemical reactions occur under constant-volume adiabatic conditions, not only elevated temperatures but also higher pressures usually result. It is of practical interest to estimate the maximum possible pressures that might develop. This would occur theoretically under adiabatic conditions, with the simultaneous achievement of the adiabatic-flame temperature. From a more practical viewpoint, the heat release would have to be nearly instantaneous, and the conditions of internal equilibrium would have to be met. For calculation purposes these criteria will be assumed to be valid. This will lead to an upper limit, which is of real interest and importance. Dissociation effects at these high temperatures are neglected at this time.

In general, two methods of calculation present themselves. One requires a knowledge of the initial volume; the other requires information about the initial system pressure. To simplify both methods, we assume that the gases behave ideally. If one applies the ideal-gas relation to the product gases after a chemical reaction at a known volume $V$, the expression for the final pressure $P_f$ is

$$P_f = \frac{N_f R_u T_f}{V} \qquad \text{[13-17]}$$

where $N_f$, the total number of gaseous moles of products, may be determined from the basic chemical equation representing the reaction. The adiabatic-flame temperature $T_f$ may be calculated by methods previously outlined.

The second method applies the ideal-gas relation to the initial and final states, such that

$$P_f = P_i \frac{N_f}{N_i} \frac{T_f}{T_i} \qquad \text{[13-18]}$$

In this case the $N$ values come from the stoichiometry of the reaction, while $T_f$ is the adiabatic-flame temperature. The initial pressure $P_i$ must also be measured within the test chamber.

**Example 13-12**

Liquid benzene ($C_6H_6$) in the amount of 2.5 g is placed in a constant-volume test chamber. The initial temperature of the fuel and the oxidizer is 25°C. If the fuel is burned with 20 percent excess air and the reaction goes to completion, determine the maximum pressure in the chamber, in bars, if (a) the chamber volume is 0.030 m$^3$, and (b) the initial pressure $P_i$ is 1.0 bar.

**Solution:**

**Given:** Liquid benzene and 20 percent excess air is burned in a constant-volume chamber, as shown in Fig. 13-13.

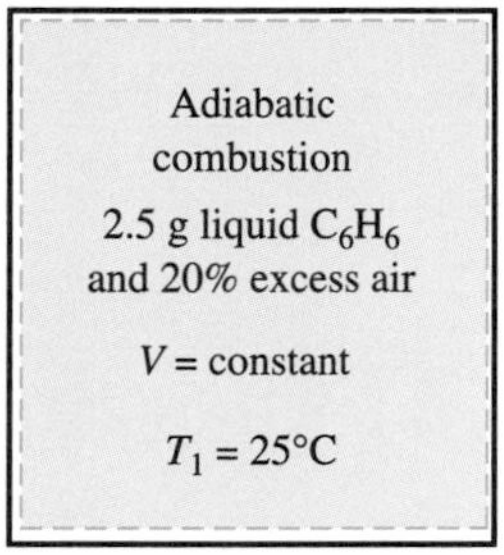

**Find:** Maximum pressure, in bars, if (a) $V = 0.030$ m$^3$ and (b) $P_i = 1.0$ bar.

**Model:** Constant-volume, closed system; adiabatic process; ideal gases.

**Analysis:** The chemical equation for the reaction is

$$C_6H_6(l) + 9O_2(g) + 33.8N_2(g) \rightarrow 6CO_2(g) + 3H_2O(g) + 1.5O_2(g) + 33.8N_2(g)$$

Before determining the maximum pressure, we must calculate the maximum final temperature expected. In this case we set $Q$ in the constant-volume, closed-system energy balance,

$$Q = \sum_{prod} N_i(\Delta\bar{h}_{f,298}^\circ + \bar{u}_T - \bar{u}_{298})_i - \sum_{reac} N_i(\Delta\bar{h}_{f,298}^\circ + \bar{u}_T - \bar{u}_{298})_i - \Delta NR_uT_{298}$$

equal to zero and solve by iteration for the unknown temperature. Before we substitute the appropriate data into the equation, one correction must be made. Table A-23 lists the enthalpy of formation for benzene in the gaseous state, but it enters the reaction in the liquid state. The enthalpy of vaporization is given in the same table as 33,830 kJ/kmol. The enthalpy of formation must be decreased by this amount in the energy balance. For purposes of the energy balance, we will assume 1 kmol of benzene. Other appropriate data from the Appendix are now substituted into the energy balance, so that

$$0 = 6(-393,520 + \bar{u}_{T,CO_2} - 6885) + 3(-241,810 - \bar{u}_{T,H_2O} - 7425)$$
$$+ 1.5(0 + \bar{u}_{T,H_2O} - 6203) + 33.8(0 + \bar{u}_{T,N_2} - 6190)$$
$$- 1(82,930 - 33,830 + 0) - 9(0) - 33.8(0) - 1.5(8.314)(298)$$

Solving for the unknown quantities, we write the energy balance as

$$6\bar{u}_{T,CO_2} + 3\bar{u}_{T,H_2O} + 1.5\bar{u}_{T,O_2} + 33.8\bar{u}_{T,N_2} = 3,421,500 \text{ kJ}$$

A temperature must now be found which satisfies the above equation. Results of the iteration process are summarized here:

| | 2650 K | 2700 K |
|---|---|---|
| $3\bar{u}(H_2O)$ | 284,870 | 291,800 |
| $6\bar{u}(CO_2)$ | 711,000 | 727,030 |
| $1.5\bar{u}(O_2)$ | 106,320 | 108,650 |
| $33.8\bar{u}(N_2)$ | 2,246,180 | 2,294,340 |
| $\sum N_i\bar{u}$ (prod) | 3,348,370 | 3,421,820 |

---

Adiabatic combustion

2.5 g liquid $C_6H_6$ and 20% excess air

$V$ = constant

$T_1 = 25°C$

(a) $V = 0.030$ m$^3$
(b) $P_1 = 1$ bar

**Figure 13-13**
Schematic and data for Example 13-12.

Results of the table indicate that the maximum combustion temperature is essentially 2700 K.

(*a*) Finally, there remains the determination of $N_f$, the number of kilomoles of gaseous products at equilibrium. From the chemical equation it is noted that 44.3 mol of gaseous products is formed for each mole of benzene. The initial quantity of 2.5 g of benzene is equivalent to $3.21 \times 10^{-5}$ kmol. The number of kilomoles of gaseous products then is $44.3(3.21 \times 10^{-5}) = 0.00142$ kmol. Therefore, the maximum combustion pressure is

$$P_f = \frac{N_f R_u T_f}{V} = \frac{0.00142(0.08314)(2700)}{0.030} = 10.6 \text{ bars}$$

(*b*) The following data are known from part *a*: $N_i/N_f = 42.8/44.3$, $T_i = 298$ K, $T_f = 2700$ K. In addition, $P_i = 1.0$ bar. Therefore

$$P_f = P_i \frac{N_f}{N_i} \frac{T_f}{T_i} = \frac{1.0(44.3)(2700)}{42.8(298)} = 9.38 \text{ bars}$$

**Comment:**   The answers to parts *a* and *b* are different because some of the input *PvT* information is different for the two cases.

## 13-7   ENTHALPY OF REACTION AND HEATING VALUES

In earlier sections of this chapter the energy analysis of reacting substances in constant-volume or steady-flow systems was presented. In either type of system the only thermodynamic information required was the enthalpy-of-formation data and sensible-enthalpy or internal-energy data for each substance. There are practical situations, however, where the enthalpy of formation is not known for the fuel. For example, a fuel oil may contain dozens of individual compounds, and the exact mole analysis of the fuel may not be known. Similarly, solid fuels such as coal have a variable composition, depending on which mine supplied the fuel. In addition, coal is not considered to be a mixture of various compounds, but rather a material which is represented by some overall chemical composition. This composition is usually expressed in terms of a gravimetric analysis known as the **ultimate analysis**. Such an analysis reports the carbon, hydrogen, oxygen, nitrogen, and sulfur content of a particular sample. In addition, the moisture (water) and ash content are also noted for coal. In Table 13-1 the ultimate analysis of some typical coals in the United States is listed.

Owing to the wide range of chemical compositions that fuel oil and coal generally have, the enthalpy of formation is not a useful concept in such cases. A different approach must be used when an energy analysis is made for the combustion of these fuels. To replace enthalpy-of-formation data, experiments are carried out with the fuels to determine the enthalpy of reaction or the heating value of the fuel. Solid and liquid fuels are commonly

**Table 13-1** Ultimate analysis of representative coals in the United States on a moisture-and-ash basis, percent by mass

| Type (rank) | State | Ash | S | H | C | N | O | $H_2O$ | Heating Value* Btu/lb$_m$ | Heating Value* kJ/kg |
|---|---|---|---|---|---|---|---|---|---|---|
| Anthracite | PA | 10.7 | 0.5 | 1.7 | 81.6 | 0.6 | 0.8 | 4.1 | 12,590 | 29,290 |
| Bituminous | WV | 5.1 | 0.7 | 5.2 | 78.8 | 1.6 | 5.8 | 2.8 | 13,980 | 32,520 |
|  | IL | 8.1 | 0.9 | 4.0 | 68.5 | 1.1 | 7.6 | 9.8 | 12,015 | 27,950 |
|  | IN | 5.6 | 1.1 | 4.4 | 66.0 | 1.5 | 7.9 | 13.5 | 11,788 | 27,420 |
| Subbituminous | WY | 6.2 | 0.4 | 4.4 | 60.6 | 1.0 | 15.4 | 12.0 | 10,640 | 24,750 |
| Lignite | SD | 8.2 | 2.2 | 2.2 | 38.0 | 0.5 | 9.6 | 39.2 | 6,307 | 14,670 |

| *Heating value listed is the higher heating value at 25°C (77°F).

tested in a constant-volume system known as the bomb calorimeter, while gaseous fuels are tested in a steady-flow system known as a steady-flow calorimeter. In either case, a fuel is burned completely in such a manner that the products are returned to the same temperature as the initial reactants. The required heat transfer per unit mass of fuel is measured accurately. This heat transfer per mole of fuel for the steady-flow calorimeter is the *enthalpy of reaction* $\Delta \bar{h}_{R,T}$ introduced in Sec. 13-4. As noted earlier, it is commonly reported at the standard reference state of 298 K (537°R) and 1 atm, and thus is symbolized by $\Delta \bar{h}^o_{R,298}$ or $\Delta \bar{h}^o_{R,537}$. If we apply Eq. [13-9] to a steady-state reaction in a calorimeter at 298 K (537°R) and apply the definition of the enthalpy of reaction at the standard reference state, we have

$$\dot{Q} = \dot{N}_{\text{fuel}} \left[ \sum_{\text{prod}} n_i \Delta \bar{h}^o_{f,298,i} - \sum_{\text{reac}} n_i \Delta \bar{h}^o_{f,298,i} \right] = \dot{N}_{\text{fuel}} \Delta \bar{h}^o_{R,298}$$

where $n_i = \nu_i$ because the reaction is complete. Solving for the enthalpy of reaction at 298 K (537°R) and 1 atm, we now have

$$\Delta \bar{h}^o_{R,298} = \frac{\dot{Q}}{\dot{N}_{\text{fuel}}} = \sum_{\text{prod}} \nu_i \Delta \bar{h}^o_{f,298,i} - \sum_{\text{reac}} \nu_i \Delta \bar{h}^o_{f,298,i} \quad \textbf{[13-19]}$$

where the $\nu_i$-terms are defined as before for Eq. [13-10]. As a result, by measuring the overall enthalpy change $\Delta \bar{h}_{R,298}$ at the given reference temperature experimentally, we bypass the problem of not knowing the value of the enthalpy of formation $\Delta \bar{h}^o_{f,298}$ for the fuel.

If Eq. [13-19] is now substituted into Eq. [13-10] for the general situation where both reactants and products could be at temperatures other than 298 K then

$$\dot{Q} = \dot{N}_{\text{fuel}} \left[ \Delta \bar{h}^o_{R,298} + \sum_{\text{prod}} n_i \left( \bar{h}_T - \bar{h}_{298} \right)_i - \sum_{\text{reac}} n_i \left( \bar{h}_T - \bar{h}_{298} \right)_i \right] \quad \textbf{[13-20]}$$

Values at 298 K are replaced by values at 537°R when USCS units are employed. Similar to the use of Eq. [13-10], Eq. [13-20] can be used to find heat transfer quantities or to calculate adiabatic-combustion temperatures in those situations where enthalpy-of-reaction data are available for a fuel, rather than enthalpy-of-formation data.

The enthalpy of reaction at a reference temperature of 298 K (537°R) is determined by means of Eq. [13-19]. The enthalpy of reaction at any temperature $T$ is defined as the term in brackets in Eq. [13-20]:

$$\Delta \bar{h}_{R,T} = \Delta \bar{h}_{R,298}^{\circ} + \sum_{\text{prod}} \nu_i \left( \bar{h}_T - \bar{h}_{298} \right)_i - \sum_{\text{reac}} \nu_i \left( \bar{h}_T - \bar{h}_{298} \right)_i \quad \textbf{[13-21]}$$

where $h_T$ is measured at the same temperature for both the reactant and product terms. (The subscript 298 again becomes 537 when the temperature is measured in degrees Rankine.)

Although we have specifically been discussing the enthalpy of reaction in terms of fuel-oxidant systems, the term is a general one. That is, it is a measure of the isothermal enthalpy change for any type of reaction. However, many engineering applications of chemical reactions involve the combustion of hydrocarbon fuels with oxygen or air. As noted earlier in Sec. 13-4, the enthalpy of reaction for a combustion process is frequently called the *enthalpy of combustion* $\Delta \bar{h}_c$. The enthalpy of combustion of a fuel is also referred to as the **heating value of the fuel.** The heating value is defined as energy released from the combustion process, and it is always positive in value. Therefore, the enthalpy of combustion $\Delta \bar{h}_c$ and the heating value for a given fuel are opposite is sign. Another difference is that $\Delta \bar{h}_c$ is usually quoted on a mole basis, while heating values are nearly always listed on a mass (kilogram or pound) basis.

Two different heating values are quoted in the literature. The value commonly cited is the **higher heating value,** which is based on liquid water in the products of combustion. When water in the product gases is in the vapor state, the energy released is called the **lower heating value.** This value is of particular importance in engineering calculations, because combustion gases normally leave equipment before the dew point of the water is reached. When a higher heating value $q_H$ is listed at 25°C or 77°F, the lower heating value $q_L$ may be calculated by

$$q_L = q_H - 2442 \frac{m_w}{m_f} \qquad \text{kJ/kg fuel} \qquad \textbf{[13-22a]}$$

or
$$q_L = q_H - 1050 \frac{m_w}{m_f} \qquad \text{Btu/lb}_m \text{ fuel} \qquad \textbf{[13-22b]}$$

where $m_w/m_f$ is the mass of water formed per mass of fuel. The numerical constants in the above equations are the enthalpy-of-vaporization values at the given temperature. The examples below illustrate the difference between lower and higher heating values of a given fuel. The notation of whether $H_2O(l)$ or $H_2O(g)$ appears in a chemical equation is quite important from an energy standpoint.

**Example 13-13**

In Example 13-6 the enthalpy of reaction (or combustion) of gaseous methane ($CH_4$) at 25°C was found to be −802,290 kJ/kmol when water appears as a gas in the products. Determine the lower and higher heating values for the fuel at the same temperature.

**Solution:**

**Given:** $\Delta \bar{h}_R$ of −802,290 kJ/kmol at 25°C.

**Find:** Higher and lower heating values at 25°C.

**Model:** Steady-state, open system; complete combustion.

**Analysis:** The lower heating value of methane gas is found by dividing the enthalpy of combustion by the molar mass of methane and changing the sign on the answer. Thus

$$q_L = \frac{\Delta \bar{h}_c}{M} = \frac{802,290 \text{ kJ/kmol}}{16.04 \text{ kg/kmol}} = 50,020 \text{ kJ/kg fuel}$$

The higher heating value is found by using Eq. (13-22a) and the fact that 2 mol of water is formed per mole of methane. Hence,

$$q_H = q_L + 2442 \frac{m_w}{m_f} = 50,020 + 2442 \left(\frac{2}{1}\right)\left(\frac{18.02}{16.04}\right) = 55,510 \text{ kJ/kg fuel}$$

**Comment:** In this case the higher heating value is approximately 11 percent higher that the lower heating value at 25°C.

**Example 13-14**

Consider an Indiana coal with the ultimate analysis on a wet basis taken from Table 13-1. The values are sulfur, 1.1; hydrogen, 4.4; carbon, 66.0; nitrogen, 1.5; oxygen, 7.9; ash, 5.6; water, 13.5, in percent by mass. (a) Determine the reaction equation and the air-fuel ratio used for complete combustion with 20 percent excess air. (b) If the reactants enter the steady-flow combustor at 25°C and the products are cooled to 500 K, determine the heat transfer, in kJ/kg fuel.

**Solution:**

**Given:** An Indiana coal reacts with 20 percent excess air in a steady-flow combustor as shown in Fig. 13-14.

**Find:** (a) Reaction equation and AF, and (b) $\dot{Q}/\dot{m}_{\text{fuel}}$, in kJ per kg fuel.

**Model:** Steady-flow, open system; complete combustion.

**Analysis:** (a) The basis for the calculation is 1 kg of moist fuel. To write a suitable reaction equation, the gravimetric analysis of the fuel must be converted to a mole basis, by using the appropriate molar masses of the atomic reactants. For example, the number of moles of carbon per kilogram of fuel is 0.66/12 = 0.055 kmol. Note that any oxygen in the fuel reduces the oxygen that needs to be supplied

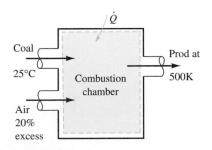

**Figure 13-14**
Schematic and data for Example 13-14.

in the air. Also it is assumed that any nitrogen in the fuel ends up as gaseous $N_2$. On this basis the equation for the chemically correct amount of air for complete combustion is found to be

$$0.055C + 0.044H + 0.00034S + 0.00107N + 0.004940 + 0.06390_2$$
$$+0.2402N_2 \rightarrow 0.055CO_2 + 0.022H_2O + 0.00034SO_2 + 0.2407N_2$$

The numerical coefficients in this equation each represent the kilomoles of a component per kilogram of fuel. The reaction equation for 20 percent excess air would be

$$0.055C + 0.044H + 0.00034S + 0.00107N + 0.004940 + 0.07670_2$$
$$+ 0.2883N_2 \rightarrow 0.055CO_2 + 0.022H_2O + 0.00034SO_2 + 0.2888N_2 + 0.01280_2$$

The air-fuel ratio for the 20 percent excess air case is

$$AF = \frac{0.0767(4.76)(29)}{1.0} = 10.59 \text{ kg air/kg fuel}$$

(*b*)  Because the mole fraction of $SO_2$ in the product gas is less than 0.1 percent, its contribution to the energy analysis is neglected. The heating value of 27,420 kJ/kg in Table 13-1 for the coal is the higher heating value. Because the products of combustion are cooled only to 500 K, we need the lower heating value, in kJ/kg. In part *a* we found that 0.022 mol of water, or 0.396 kg of water, is formed per kilogram of fuel. Thus Eq. (13-22*a*) shows that

$$q_L = q_H - 2442\frac{m_w}{m_f} = 27,420 - 2442\frac{0.396}{1} = 26,450 \text{ kJ/kg fuel}$$

The energy analysis is based on Eq. [13-20], where the reactant term is zero in this case. Note from Table 13-1 that the Indiana coal contains 13.5 percent moisture. This is equivalent to 0.0075 kmol $H_2O$ per kilogram of fuel. The heating of this additional water to 500 K must be accounted for in the energy analysis below. Therefore, the steady-flow energy balance becomes

$$\dot{Q} = \dot{N}_{\text{fuel}}\Delta\bar{h}^o_{R,298} + \sum_{\text{prod}}\dot{N}_i(\bar{h}_T - \bar{h}_{298})_i - \sum_{\text{reac}}\dot{N}_i(\bar{h}_T - \bar{h}_{298})_i$$

Dividing through by the mass flow rate of fuel gives

$$\frac{\dot{Q}}{\dot{m}_{\text{fuel}}} = \frac{\dot{N}_{\text{fuel}}}{\dot{m}_{\text{fuel}}}\Delta\bar{h}^o_{R,298} + \sum_{\text{prod}}\frac{\dot{N}_i}{\dot{m}_{\text{fuel}}}(\bar{h}_T - \bar{h}_{298})_i - \sum_{\text{reac}}\frac{\dot{N}_i}{\dot{m}_{\text{fuel}}}(\bar{h}_T - \bar{h}_{298})_i$$

Now the first term on the right-hand side equals $-q_L$. The remaining terms are known or available in the appendix. Thus

$$\dot{Q}/\dot{m}_{\text{fuel}} = -26,450 + 0.055(17,678 - 9364) + 0.022(16,828 - 9904)$$
$$+ 0.2888(14,581 - 8669) + 0.0128(14,770 - 8682)$$
$$+ 0.0075(16,828 - 9904 + 44,010)$$
$$= -24,055 \text{ kJ/kg fuel}$$

**Comment:**     Note that the sign on the heating value is changed when it is substituted into the energy equation. The calculation indicates that roughly 10.5 percent of the energy released by the fuel is used to heat the products to 500 K. The remaining energy appears as a heat loss from the system.

## 13-8 SECOND LAW ANALYSIS OF REACTIONS

For a chemical reaction in a closed system, the change in entropy is given by

$$\Delta S = \sum_{\text{prod}} N_i \bar{s}_i - \sum_{\text{reac}} N_i \bar{s}_i \qquad \textbf{[13-23a]}$$

where $\bar{s}_i$ is some appropriate entropy value for the $i$th species. As with applications of the energy balance, it is frequently useful to express the change in entropy per mole of fuel as follows

$$\Delta S = N_{\text{fuel}} \left[ \sum_{\text{prod}} n_i \bar{s}_i - \sum_{\text{reac}} n_i \bar{s}_i \right] = N_{\text{fuel}} [\bar{s}_{\text{prod}} - \bar{s}_{\text{reac}}] \qquad \textbf{[13-23b]}$$

where $n_i = N_i / N_{\text{fuel}}$, the ratio of the moles of species $i$ to the moles of fuel initially in the system, $\bar{s}_{\text{prod}} = \sum_{\text{prod}} n_i \bar{s}_i$, the entropy of the products per mole of fuel, and $\bar{s}_{\text{reac}} = \sum_{\text{reac}} n_i \bar{s}_i$, the entropy of the reactants per mole of fuel. As discussed earlier, the numerical values of $n_i$ are the coefficients from a balanced chemical equation, like Eq. [13-1], that is written on the basis of one mole of fuel and includes all compounds within the system before and after the reaction.

In the application of the conservation of energy to chemical reactions, it became necessary to assign to the enthalpy of stable elements at 25°C (77°F) and 1 atm the value of zero, by convention. Only in this manner could the change in the enthalpy of a reacting system be found in a consistent manner. The same problem of a reference state or datum occurs in evaluating the entropy change for a reaction. In this case, however, we rely on a more fundamental approach based on what is known as the third law of thermodynamics.

The ***third law of thermodynamics,*** based on experimental evidence, states that the entropy change for isothermal processes at absolute zero of temperature is zero. Further insight has led to the following statement:

> By convention, the entropy of a pure crystalline substance is taken to be zero at the absolute zero of temperature, 0 K or 0°R.

An entropy value based on this datum is known as an ***absolute entropy.*** The third law itself also leads to the conclusion that the entropy of crystalline substances at absolute zero temperature is not a function of pressure. At temperatures above absolute zero, however, the entropy is a function of pressure. Because of this, the tabulated values of the absolute entropy are usually given at the standard reference pressure of 1 atm (or 1 bar). Similar to the enthalpy function, the entropy at this reference pressure is denoted by $\bar{s}_{i,T}^{o}$.

The absolute entropy values at the standard reference state of 25°C (77°F) and 1 atm are given in Tables A-23 and A-23E. The ideal-gas tables (Tables A-5 through A-11 and A-5E through A-11E) list the values of $\bar{s}^o_{i,T}$ versus temperature for selected gases at 1 atm.

The evaluation of entropy changes for reactive mixtures requires a means of determining the entropy of a substance, either pure or in a mixture, at pressures other than that used in reference tables. For ideal gases, for example, recall from Chap. 11 that, based on the Gibbs-Dalton rule, each constituent of a mixture of gases behaves as if it existed alone at the temperature and volume of the mixture. Thus each ideal gas exerts a pressure equal to its component or partial pressure $p_i$. The entropy of a pure ideal gas at a pressure other than that found in a table is determined by employing Eq. [7-8] on a molar basis, namely,

$$d\bar{s} = \bar{c}_p \frac{dT}{T} - R_u \frac{dP}{P} \qquad \text{[7-8]}$$

This equation is integrated at constant temperature from the standard reference pressure $P_{\text{ref}}$ to a mixture state where its pressure is its partial pressure $p_i$:

$$\bar{s}_{i,T,P} - \bar{s}^o_{i,T} = -R_u \ln \frac{p_i}{P_{\text{ref}}}$$

Solving for the absolute entropy of an ideal gas at any pressure $P$ and temperature $T$, we have

$$\boxed{\bar{s}_{i,T,P} = \bar{s}^o_{i,T} - R_u \ln \frac{p_i}{P_{\text{ref}}} = \bar{s}^o_{i,T} - R_u \ln \frac{y_i P}{P_{\text{ref}}}} \qquad \text{Ideal gas}$$

$$\text{[13-24]}$$

where the partial pressure is given by $p_i = y_i P$ and the reference pressure is $P_{\text{ref}} = 1$ atm. Substituting the absolute entropy values from Eq. [13-24] back into Eq. [13-23a] gives the entropy change for a chemical reaction in a closed system involving *ideal gases* as

$$\Delta S = \sum_{\text{prod}} N_i \left( \bar{s}^o_{i,T} - R_u \ln \frac{y_i P}{P_{\text{ref}}} \right) - \sum_{\text{reac}} N_i \left( \bar{s}^o_{i,T} - R_u \ln \frac{y_i P}{P_{\text{ref}}} \right) \qquad \text{[13-25]}$$

A similar expression can be obtained for the entropy change per mole of fuel as in Eq. [13-23b], $\Delta S / N_{\text{fuel}} = \bar{s}_{\text{prod}} - \bar{s}_{\text{reac}}$, by substituting the coefficients $n_i$ from a balanced chemical equation for the number of moles $N_i$ in Eq. [13-25]. These expressions are required in any second-law analysis.

When a chemical reaction occurs in a steady-state, open system, the entropy balance can be written as

$$0 = \sum_{\text{reac}} \dot{N}_i \bar{s}_i - \sum_{\text{prod}} \dot{N}_i \bar{s}_i + \sum_j \frac{\dot{Q}_j}{T_j} + \dot{\sigma}_{\text{cv}} \qquad \text{[13-26a]}$$

where the heat transfer $\dot{Q}_j$ occurs at a location on the control surface with uniform temperature $T_j$. As was done with the energy balance earlier, it is useful to rewrite the entropy balance in terms of the molar flow rate of fuel as follows:

$$0 = \dot{N}_{\text{fuel}}\left[\sum_{\text{reac}} n_i \bar{s}_i - \sum_{\text{prod}} n_i \bar{s}_i\right] + \sum_j \frac{\dot{Q}_j}{T_j} + \dot{\sigma}_{\text{cv}} \qquad \textbf{[13-26b]}$$

where $n_i = \dot{N}_i/\dot{N}_{\text{fuel}}$, the ratio of the molar flow rate of species $i$ to the molar flow rate of the fuel as defined earlier. It is sometimes useful to talk about the entropy of the reactants and the products. This can be done by rewriting Eq. [13-26b] as

$$0 = \dot{N}_{\text{fuel}}[\bar{s}_{\text{reac}} - \bar{s}_{\text{prod}}] + \sum_j \frac{\dot{Q}_j}{T_j} + \dot{\sigma}_{\text{cv}} \qquad \textbf{[13-26c]}$$

where $\bar{s}_{\text{reac}} = \sum_{\text{reac}} n_i \bar{s}_i$, the entropy of the reactants per mole of fuel, and $\bar{s}_{\text{prod}} = \sum_{\text{reac}} n_i \bar{s}_i$, the entropy of the products per mole of fuel.

The term $\bar{s}_{\text{prod}} - \bar{s}_{\text{reac}}$ is often referred to as the entropy change for the chemical reaction in a steady flow reactor. Note that the units on this entropy change are per mole of fuel. Solving Eq. [13-26c] for this entropy change gives

$$\bar{s}_{\text{prod}} - \bar{s}_{\text{reac}} = \frac{1}{\dot{N}_{\text{fuel}}}\left[\sum_j \frac{\dot{Q}_j}{T_j} + \dot{\sigma}_{\text{cv}}\right] \qquad \textbf{[13-27a]}$$

Thus, the entropy change for a chemical reaction can in general take on any numerical value. However, if the steady-state, open system is operated adiabatically, Eq. [13-27a] reduces to

$$\bar{s}_{\text{prod}} - \bar{s}_{\text{reac}} = \frac{\dot{\sigma}_{\text{cv}}}{\dot{N}_{\text{fuel}}} \geq 0 \qquad \textbf{[13-27b]}$$

where the inequality sign results from the second law limitations on the rate of entropy production $\dot{\sigma}_{\text{cv}}$. Looking at this another way, Eq. [13-27] demonstrates that when a chemical reaction occurs in a steady-state, adiabatic, open system, the entropy of the products must always be greater than or equal to the entropy of the reactants. Another way of saying this is that the entropy of the products leaving the device can never be less than the entropy of reactants entering the device. The procedure for evaluating the entropy change for a chemical reaction is illustrated in the following two examples.

---

**Example 13-15**

Carbon monoxide and oxygen in a stoichiometric ratio at 25°C and 1 atm react to form carbon dioxide at the same pressure and temperature. If the reaction goes to completion, evaluate (a) the entropy change for the reaction and (b) the entropy production for the process, in kJ/K per kilomole of CO.

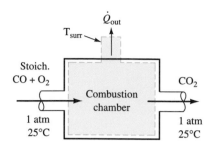

**Figure 13-15**
Schematic and data for Example 13-15.

**Solution:**

**Given:** The stoichiometric reaction of CO and $O_2$ to form $CO_2$, as shown in Fig. 13-15.

**Find:** (a) $\Delta \bar{s}_{reaction}$ and (b) $\sigma_{tot}$, in kJ/K·kmol of CO.

**Model:** Steady-state, isothermal, open system; ideal gases; complete reaction.

**Analysis:** The basic reaction is $CO + \frac{1}{2}O_2 \rightarrow CO_2$ in the gas phase.
(a) The entropy change for the steady-state reaction is given by

$$\bar{s}_{prod} - \bar{s}_{reac} = \sum_{prod} n_i \bar{s}_i - \sum_{reac} n_i \bar{s}_i$$

The absolute entropies of the three reacting species are calculated below on the basis of the relation $\bar{s}_{i,T,P} = \bar{s}^o_{i,T} - R_u \ln y_i$ since $P = P_{ref}$. The $s^o_i$ values are from Tables A-7, A-8, and A-9. In units of kJ/kmol·K,

$$\bar{s}(CO_2) = 213.685 - 8.314 \ln 1.0 = 213.685$$

$$\bar{s}(CO) = 197.543 - 8.314 \ln \tfrac{2}{3} = 200.914$$

$$\bar{s}(O_2) = 205.033 - 8.314 \ln \tfrac{1}{3} = 214.168$$

Therefore, the total change in entropy for the complete reaction is

$$\bar{s}_{prod} - \bar{s}_{reac} = 213.685 - 200.914 - \tfrac{1}{2}(214.168) = -94.313 \text{ kJ/K·kmol CO}$$

(b) The entropy production for the control volume that includes the heat-transfer region is given by the entropy balance

$$\dot{\sigma}_{cv} = \sum_{prod} \dot{N}_i \bar{s}_i - \sum_{reac} \dot{N}_i \bar{s}_i - \frac{\dot{Q}}{T_{surr}} \geq 0$$

The calculation of the term $\dot{Q}/T_{surr}$ requires information on $\dot{Q}$ for the reaction. To keep the reacting mixture at 25°C, the energy released by the reaction appears as a heat transfer to the surroundings. This heat transfer is found by applying the energy balance. Using the appropriate data, we find that

$$\frac{\dot{Q}}{\dot{N}_{CO}} = \sum_{prod} n_i \Delta \bar{h}_{f,298,i} - \sum_{reac} n_i \Delta \bar{h}_{f,298,i}$$

$$= 1(-393,520) - 1(-110,530) - \tfrac{1}{2}(0) = -282,990 \text{ kJ/kmol CO}$$

On a molar basis, the entropy balance becomes

$$\frac{\dot{\sigma}_{cv}}{\dot{N}_{CO}} = \bar{s}_{prod} - \bar{s}_{reac} - \frac{\dot{Q}/\dot{N}_{CO}}{T_{surr}} = -94.313 - \frac{-282,900}{298} = 855.3 \text{ kJ/K·kmol}$$

Thus a large entropy production occurs per kilomole of CO.

---

**Example 13-16**

A 1:1 molar ratio of gaseous carbon monoxide and water vapor enters a steady-state adiabatic reactor at 400 K and 1 atm, and the products are carbon dioxide and hydrogen. If the pressure is constant and the reaction goes to completion, determine the entropy production for the reaction, in kJ/K per kilomole of CO.

**Solution:**

**Given:** A 1:1 molar ratio of CO and $H_2O$ reacts as shown in Fig. 13-16.

**Find:** $\dot{\sigma}_{CV}/\dot{N}_{CO}$, in kJ/K·kmol of CO.

**Model:** Steady-state, adiabatic, open system; ideal gases; complete reaction.

**Analysis:** The reaction under consideration is

$$CO(g) + H_2O(g) \rightarrow CO_2(g) + H_2(g)$$

To evaluate the entropy change, we must first determine the final temperature. The steady-flow energy balance reduces to

$$0 = \sum_{prod} n_i(\Delta\bar{h}_{f,298}^{0} + \bar{h}_T - \bar{h}_{f,298})_i - \sum_{reac} n_i(\Delta\bar{h}_{f,298}^{0} + \bar{h}_T - \bar{h}_{298})_i$$

Using $\Delta\bar{h}_f^{0}$ data from Table A-23 and sensible-enthalpy data for ideal gases from Tables A-8 through A-11, we find that

$$0 = 1(-393,520 + \bar{h}_{T,CO_2} - 9364) + 1(0 + \bar{h}_{T,H_2} - 8468)$$
$$- 1(-110,530 + 11,644 - 8669) - 1(-241,820 + 13,356 - 9904)$$

Upon rearrangement the equation becomes

$$\bar{h}_{T,CO_2} + \bar{h}_{T,H_2} = 65,430 \text{ kJ/kmol}$$

This equation can be solved directly with a general equation-solving program that contains the enthalpy data. If the data from Tables A-9 and A-11 in the Appendix are used, then at 920 K

$$\bar{h}_{T,CO_2} + \bar{h}_{T,H_2} = 38,467 + 26,747 = 65,210 \text{ kJ/kmol}$$

Similarly, at 960 K,

$$\bar{h}_{T,CO_2} + \bar{h}_{T,H_2} = 40,607 + 27,947 = 68,555 \text{ kJ/kmol}$$

Therefore, the final temperature of the products is very close to 920 K. The entropy change is found by evaluating $\bar{s}_{i,T}^{0} - R_u \ln(p_i/P_{ref})$ for each chemical species in the reaction. Since the mole fraction of each reactant and each product is $\frac{1}{2}$ and the total pressure is constant, the partial pressure of each constituent is the same. Hence the value of $R_u \ln(p_i/P_{ref})$ is the same for each, and the terms will cancel when the change in entropy is calculated. The steady-state entropy balance for the conditions of this problem is

$$\frac{\dot{\sigma}}{\dot{N}_{CO}} = \bar{s}_{prod} - \bar{s}_{reac} = (\bar{s}_{CO_2}^{0} + \bar{s}_{H_2}^{0})_{920 \text{ K}} - (\bar{s}_{CO}^{0} + \bar{s}_{H_2O}^{0})_{400 \text{ K}}$$
$$= 264.728 + 163.607 - 206.125 - 198.673$$
$$= 23.537 \text{ kJ/K per kmol CO}$$

The positive answer is in accord with the second law.

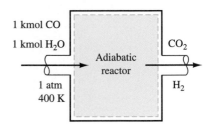

**Figure 13-16**
Schematic for Example 13-16.

---

The calculations in Example 13-16 illustrate entropy generation within an adiabatic steady-flow combustor. A similar second-law analysis of the adiabatic combustion of hydrocarbon fuels with air may be used to show the influence of air-fuel ratios on entropy generation. For example, consider the reaction of ethylene ($C_2H_4$) with air. The fuel and air are unmixed and each enters the reactor at 298 K and 1 atm. The adiabatic combustion temperatures for 100, 200, and 400 percent theoretical air are close to 2570, 1605, and 1015 K, respectively.

For the case of 200 percent theoretical air the overall gas-phase reaction is

$$C_2H_4 + 6O_2 + 22.56N_2 \rightarrow 2CO_2 + 2H_2O + 3O_2 + 22.56N_2$$

For all the reactants, in all cases, the specific entropy of each reactant is

$$\bar{s}_{C_2H_4} = 219.83 \text{ kJ/kmol·K}$$

$$\bar{s}_{O_2} = 205.03 - 8.314\ln(6/28.56) = 218.00 \text{ kJ/kmol·K}$$

$$\bar{s}_{N_2} = 191.50 - 8.314\ln(22.56/28.56) = 193.46 \text{ kJ/kmol·K}$$

The entropy of the reactants per kilomole of fuel for 200 percent theoretical is

$$\bar{s}_{reac} = 1\bar{s}_{C_2H_4} + 6\bar{s}_{O_2} + 22.56\bar{s}_{N_2} = 592.31 \text{ kJ/K per kilomole of fuel.}$$

The specific entropy of each product for the 200 percent case is

$$\bar{s}_{CO_2} = 295.901 - 8.314\ln(2/29.56) = 318.30 \text{ kJ/kmol·K}$$

$$\bar{s}_{H_2O} = 253.513 - 8.314\ln(2/29.56) = 275.91 \text{ kJ/kmol·K}$$

$$\bar{s}_{O_2} = 260.333 - 8.314\ln(3/29.56) = 279.35 \text{ kJ/kmol·K}$$

$$\bar{s}_{N_2} = 244.028 - 8.314\ln(22.56/29.56) = 246.27 \text{ kJ/kmol·K}$$

and the entropy of the products per kilomole of fuel is

$$\bar{s}_{prod} = 2\bar{s}_{CO_2} + 2\bar{s}_{H_2O} + 3\bar{s}_{O_2} + 22.56\bar{s}_{N_2}$$
$$= 7582.42 \text{ kJ/K per kilomole of fuel.}$$

Therefore for the overall reaction at 200 percent theoretical air

$$\frac{\dot{\sigma}_{cv}}{\dot{N}_{fuel}} = \bar{s}_{prod} - \bar{s}_{reac} = 1690 \text{ kJ/K per kilomole of fuel.}$$

Similar calculations at 100 and 400 percent theoretical air lead to entropy-generation values of 1190 and 2270 kJ/K per kilomole of fuel. The results indicate a steadily increasing entropy generation as the amount of excess air increases. Generally, the irreversibility of an adiabatic combustion reaction is minimized by operating close to the stoichiometric ratio of air to fuel.

## 13-9  AVAILABILITY ANALYSIS OF REACTING SYSTEMS

Thermodynamics, through the use of the second law, provides a means of measuring the quality of energy as well as its quantity. Toward this goal the concepts of reversible work, availability, and irreversibility were introduced in Chap 9. In this section we apply these concepts to chemically reacting systems. This has a practical importance, for example, in the study of heat engines. The chemical energy released by a combustion reaction increases the temperature of the product gases to a high value. This high-temperature gas then is a source of thermal energy for the heat engine. The performance of the heat engine is characterized by its thermal efficiency, the ratio of net work output to heat input. Rarely does the actual thermal efficiency exceed 40 percent. However, regardless of the system used to produce the work, it is important to determine the maximum potential work output associated with a given chemical reaction for specified end states of the reaction.

If we restrict ourselves to steady-state, open systems, the reversible work associated with a chemical reaction is given by the *change in the stream availability* for the specified end states and the state of the surroundings. In the absence of significant changes in kinetic and potential energy,

$$\dot{W}_{rev,sf} = \sum_{prod} \dot{N}_i \overline{\psi}_i - \sum_{reac} \dot{N}_i \overline{\psi}_i = \dot{N}_{fuel}(\overline{\psi}_{prod} - \overline{\psi}_{reac})$$

where $\overline{\psi}_i = \overline{h}_i - T_0 \overline{s}_i$. For a mixture of ideal gases this becomes

$$\dot{W}_{rev,sf} = \sum_{prod} \dot{N}_i(\overline{h}_i - T_0\overline{s}_i) - \sum_{reac} \dot{N}_i(\overline{h}_i - T_0\overline{s}_i) \qquad \textbf{[13-28]}$$

$$= \sum_{prod} \dot{N}_i \left[ (\Delta\overline{h}^{\circ}_{f,298} + \overline{h}_T - \overline{h}_{298})_i - T_0 \left( \overline{s}^{\circ}_{i,T} - R_u \ln \frac{p_i}{P_{ref}} \right) \right]$$

$$- \sum_{reac} \dot{N}_i \left[ (\Delta\overline{h}^{\circ}_{f,298} + \overline{h}_T - \overline{h}_{298})_i - T_0 \left( \overline{s}^{\circ}_{i,T} - R_u \ln \frac{p_i}{P_{ref}} \right) \right]$$

Values at 298 K are replaced by values at 537°R when USCS units are used.

The irreversibility of a steady-state chemical reaction is still defined, as in Chap. 9, by the two equivalent relations

$$\dot{I} = \dot{W}_{act} - \dot{W}_{rev} = T_0 \dot{\sigma}_{cv} \qquad \textbf{[13-29]}$$

where $\dot{\sigma}_{cv}$ is the rate of entropy generation within the control volume. On the basis of the second law for a steady-state reacting system, Eq. [13-29] may also be written in the form

$$\dot{I} = T_0 \dot{N}_{fuel}(\overline{s}_{prod} - \overline{s}_{reac}) - \dot{Q} \qquad \textbf{[13-30]}$$

The following two examples illustrate the calculation procedures. Generally, the reactants and products of a combustion reaction are in a mixture state. However, in some cases the reactants enter and the products leave the system essentially in the pure state. These two possibilities are noted in the examples.

Acetylene gas ($C_2H_2$) and 80 percent excess air enter a steady-state combustion chamber at 25°C. Assume that the reaction is complete and that the products leave

**Example 13-17**

at 25°C. As a further simplification, we assume that each species in the reaction either enters or leaves the reaction chamber separately at 1 atm. In addition, for comparative purposes with later examples, the water in the products is assumed to be in a hypothetical vapor state. Determine (*a*) the heat transfer for the chamber, (*b*) the change in stream availability, and (*c*) the irreversibilty of the reaction, all in kJ/kmol of fuel, if the gases are ideal gases.

**Solution:**

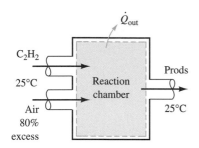

**Figure 13-17**
Schematic and data for Example 13-17.

**Given:** Acetylene gas reacts with 80 percent excess air as shown in Fig. 13-17.

**Find:** (*a*) $\dot{Q}/\dot{N}_{fuel}$, (*b*) $\overline{\psi}_{prod} - \overline{\psi}_{reac}$, and (*c*) $\dot{I}/\dot{N}_{fuel}$, all in kJ/kmol fuel.

**Model:** Steady-flow, separate ideal gases enter or leave, complete reaction.

**Analysis:** The equation for the gas-phase reaction is

$$C_2H_2 + 4.50O_2 + 16.92N_2 \rightarrow 2CO_2 + H_2O + 2.00O_2 + 16.92N_2$$

(*a*) The heat transfer is found from the steady-state energy balance

$$\dot{Q}/\dot{N}_{fuel} = \sum_{prod} n_i(\Delta\overline{h}^{\circ}_{f,298} + \overline{h}_T - \overline{h}_{f,298})_i - \sum_{reac} n_i(\Delta\overline{h}^{\circ}_{f,298} + \overline{h}_T - \overline{h}_{298})_i$$

Employing data from Table A-23, we find that

$$\dot{Q}/\dot{N}_{fuel} = 2(-393,520) + 1(-241,810) + 2(0) + 16.92(0) - 1(226,730)$$
$$- 4.5(0) - 16.92(0)$$
$$= -1,255,580 \text{ kJ/kmol } C_2H_2$$

(*b*) The maximum work is calculated by means of the stream availability change, namely,

$$\dot{W}_{rev,sf} = \dot{N}_{fuel}(\overline{\psi}_{prod} - \overline{\psi}_{reac}) = \dot{N}_{fuel}[\overline{h}_{prod} - \overline{h}_{reac} - T_0(\overline{s}_{prod} - \overline{s}_{reac})]$$

The enthalpy change for the reaction has already been calculated, since this quantity is simply the heat transfer found in part *a*. Hence the only additional information required is the entropy change for the reaction. Using the absolute-entropy values found in Table A-23, we find that

$$\overline{s}_{prod} - \overline{s}_{reac} = \sum_i (n_i s_i)_{prod} - \sum_i (n_i s_i)_{reac}$$
$$= 2(213.64) + 1(188.72) + 2(205.03) + 16.92(191.50)$$
$$- 1(200.85) - 4.50(205.03) - 16.92(191.50)$$
$$= -97.425 \text{ kJ/K per kmol } C_2H_2$$

The use of the $\Delta\overline{h}$ and $\Delta\overline{s}$ values leads to

$$\overline{\psi}_{prod} - \overline{\psi}_{reac} = -1,255,580 - 298.15(-97.425) = -1,226,530 \text{ kJ/kmol } C_2H_2$$

Thus the maximum possible work output from the chemical reaction is 1,226,530 kJ/kmol of fuel.

(*c*) The irreversibility is most easily determined by the difference between the actual work and the reversible work for the given end states. The actual process involves no work because the total energy released by the reaction leaves the control volume by heat transfer. Therefore, the irreversibility of the process is

$$\frac{\dot{I}}{\dot{N}_{\text{fuel}}} = \frac{\dot{W}_{\text{act}} - \dot{W}_{\text{rev}}}{\dot{N}_{\text{fuel}}} = 0 - (-1,226,530) = 1,226,530 \text{ kJ/kmol C}_2\text{H}_2$$

Thus the opportunity to produce a large amount of work is totally lost.

---

**Example 13-18**

**R**econsider Example 13-17 in the following manner. Rather than having each reactant enter and each product leave separately at 1 atm, the reactants (acetylene and 80 percent excess air) now enter as a *mixture* at 1 atm and the products of combustion leave as a *mixture* at 1 atm. Again, determine (*a*) the heat transfer and (*b*) the availability change, both in kJ/kmol of fuel.

**Solution:**

**Given:** Acetylene gas reacts with 80 percent excess air, as shown in Fig. 13-18.

**Find:** (*a*) $\dot{Q}/\dot{N}_{\text{fuel}}$ and (*b*) $\bar{\psi}_{\text{prod}} - \bar{\psi}_{\text{reac}}$, both in kJ/kmol fuel.

**Model:** Steady flow, ideal gases; reactants enter and products leave as mixtures.

**Analysis:** (*a*) On the basis of Example 13-17, the heat transfer out is still 1,255,580 kJ/kmol of $C_2H_2$, because the enthalpies of ideal gases are not a function of pressure.

(*b*) The value of the availability change or the maximum work is altered from the original value, because the entropies of ideal gases are a function of the component pressures. The correction for each gas is given by $n_i(-R_u \ln p_i/P_{\text{ref}})$. The overall chemical reaction is

$$C_2H_2 + 4.50O_2 + 16.92N_2 \rightarrow 2CO_2 + H_2O + 2.00O_2 + 16.92N_2$$

On this basis the total correction to the $\Delta\bar{s}$ term is evaluated for products minus reactants as:

$$\sum_i n_i \ln p_i = 2\ln\frac{2}{21.92} + \ln\frac{1}{21.92} + 2.00\ln\frac{2}{21.92} + 16.92\ln\frac{16.92}{21.92}$$

$$- \ln\frac{1}{22.42} - 4.50\ln\frac{4.50}{22.42} - 16.92\ln\frac{16.92}{22.42}$$

$$= -1.946$$

Therefore,

$$\Delta\bar{s}_{\text{corr}} = -R_u \sum_i n_i \ln\frac{p_i}{P_{\text{ref}}} = -8.314 \text{ kJ/kmol·K}(-1.946 \text{ kmol}) = 16.18 \text{ kJ/K}$$

When this result is used to correct the $\bar{\psi}_{\text{prod}} - \bar{\psi}_{\text{reac}}$ value of $-1,226,530$ kJ/kmol found in Example 13-17, then

$$\bar{\psi}_{\text{prod}} - \bar{\psi}_{\text{reac}} = -1,226,530 - 298.15(16.18) = -1,231,300 \text{ kJ/kmol}$$

**Comment:** In this case the availability change or the maximum work output is little affected by whether the reactants and products are separate or in a mixture at 1 atm.

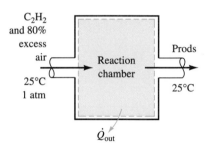

**Figure 13-18**
Schematic and data for Example 13-18.

The preceding examples illustrate that availability change associated with combustion reactions is quite large. The engineer must develop methods of utilizing this vast work potential. The conventional way is to carry out the combustion process adiabatically. The hot product gases are then used as the energy source for some type of heat engine. By comparing the work output of the engine to the reversible work output of the chemical reaction, we have some measure of the effectiveness of the process. There is, however, an even more basic question. Combustion processes by nature are highly irreversible. We would expect the availability of the product gases, even at the high adiabatic-combustion temperature, to be lower than the availability of the initial reactants. What is the typical loss in availability, then, of a combustion process, even before the energy is used in a heat engine? The following example illustrates the point.

**Example 13-19**

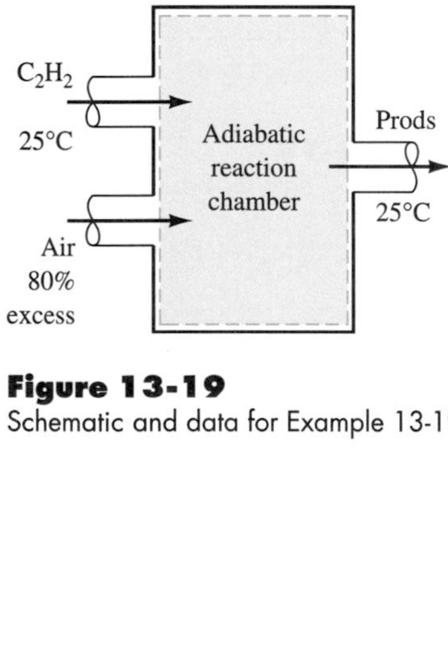

**Figure 13-19**
Schematic and data for Example 13-19.

**R**econsider Example 13-17 for the combustion of acetylene gas with 80 percent excess air, both initially at 25°C. The steady-flow process is now carried out adiabatically. Determine (*a*) the adiabatic-flame temperature, in kelvins, (*b*) the availability change, and (*c*) the irreversibility associated with the process, the last two answers in kJ/kmol of fuel.

**Solution:**

**Given:**    Acetylene gas and 80 percent excess air, both initially at 25°C, reacted adiabatically in steady flow, as shown in Fig. 13-19.

**Find:**    (*a*) $T_{af}$, in kelvins, (*b*) $\bar{\psi}_{prod} - \bar{\psi}_{reac}$, and (*c*) $\dot{I}/\dot{N}_{fuel}$, in kJ/kmol of fuel.

**Model:**    Steady flow, separate ideal gases enter and leave, complete combustion.

**Analysis:**    (*a*) To find the adiabatic-combustion temperature, one equates the enthalpy of the reactants to that of the products. Since we have already determined the heat transfer at 25°C in Example 13-17, we can also find the final temperature by equating $-\dot{Q}$ to the enthalpy change of the products gases above 25°C. That is,

$$-\dot{Q} = \sum_{prod} \dot{N}_i \Delta h_i = \dot{N}_{fuel} \sum_{prod} n_i (h_T - h_{298})_i$$

or

$$1{,}255{,}580 = 2(\bar{h}_{CO_2} - 9364) + 1(\bar{h}_{H_2O} - 9904)$$
$$+ 2.00(\bar{h}_{O_2} - 8682) + 16.92(\bar{h}_{N_2} - 8669)$$

Upon rearrangement,

$$2\bar{h}_{CO_2} + \bar{h}_{H_2O} + 2.0\bar{h}_{O_2} + 16.92\bar{h}_{N_2} = 1{,}448{,}260 \text{ kJ/kmol}$$

This equation can be solved by a computer program, or by iteration in the tables in the Appendix. By iteration, using 1920 K as a trial,

$$\sum_i n_i \bar{h}_i = 2(95{,}995) + 78{,}527 + 2(64{,}868) + 16.92(61{,}936) = 1{,}448{,}210 \text{ kJ}$$

Thus the adiabatic-combustion temperature is essentially 1920 K.

(b) The change in the stream availability for the adiabatic process is given by the relation

$$\dot{W}_{\text{rev}} = \dot{N}_{\text{fuel}}(\bar{\psi}_{\text{prod}} - \bar{\psi}_{\text{reac}}) = \dot{N}_{\text{fuel}}[\bar{h}_{\text{prod}} - \bar{h}_{\text{reac}} - T_0(\bar{s}_{\text{prod}} - \bar{s}_{\text{reac}})]$$

But $\Delta\bar{h}$ for the overall adiabatic process is zero. Therefore, the above equation reduces to

$$
\begin{aligned}
\bar{\psi}_{\text{prod}} - \bar{\psi}_{\text{reac}} &= T_0\left[\sum(n_i\bar{s}_i^0)_{\text{reac}} - \sum(n_i\bar{s}_i^0)_{\text{prod}}\right] \\
&= 298.15[1(200.85) + 4.50(205.03) + 16.92(191.50) \\
&\quad - 2(306.751) - 1(262.497) - 2(267.115) - 16.96(250.502)] \\
&= 298.15(4363.7 - 5648.7) \\
&= -383,120 \text{ kJ/kmol } C_2H_2
\end{aligned}
$$

Note that this value applies only to the situation where the products are separate and each at 1 atm at the adiabatic-combustion temperature. In practice, the product gases are in a mixture at 1 atm, and this should be taken into account. Example 13-19 has shown, however, that the contributions of the $R_u \ln p_i/P_{\text{ref}}$ terms are fairly small. Hence the value we have obtained for the reversible work is reasonably close to that which we would find for a mixture of product gases.

(c) Similar to Example 13-18, the irreversibility is simply the reversible work calculated, because the actual work output is zero. That is,

$$\frac{\dot{I}}{\dot{N}_{\text{fuel}}} = \frac{\dot{W}_{\text{act}} - \dot{W}_{\text{rev}}}{\dot{N}_{\text{fuel}}} = 0 - (-383,120) = 383,120 \text{ kJ/kmol } C_2H_2$$

This is also the value of $T_0\dot{\sigma}_{\text{cv}}/\dot{N}_{\text{fuel}}$.

---

Example 13-19 illustrates a significant point with respect to adiabatic-combustion processes. The primary objective of creating a high-temperature gas (1920 K or 3460°R in this case) for use as a thermal source for a heat engine is achieved. However, the theoretical work potential of the hot gas is only 69 percent of the work potential of the reactants, for the specified initial and final states. That is, 31 percent of the work potential of the reactants is already lost, owing to the irreversibility of the adiabatic-combustion process. If the heat-engine process were totally reversible, only 69 percent of the work potential of the reactants would be delivered. Since actual heat engines rarely reach 40 percent actual thermal efficiency, the actual work output will be less than 28 percent (0.69 × 0.40 × 100) of the theoretical capability of the reactants in this given case. When the air supplied is closer to the stoichiometric requirements, the work potential of the hot gas will be somewhat higher than 69 percent. If the excess air used with a hydrocarbon fuel is several hundred percent, the work potential of the product gases could approach only 20 percent of the original reactants. Hence irreversibilities associated with the adiabatic combustion of hydrocarbon fuels lead to a significant reduction in the work potential, even though the quantity of energy is still the same. Irreversibilities associated with the heat engine itself will further reduce the effectiveness of this process.

To make full use of the work potential of a reaction, the irreversibilities which degrade the process must be reduced. As an ultimate goal, the chemical reaction should be reversible. The galvanic or reversible chemical cell discussed in Sec. 2-7 is a device for converting chemical energy to electric energy by means of a controlled chemical reaction. Theoretically, the electrical work output from the cell should approach the change in availability of the chemical reaction. In practice, this direct conversion of chemical to electrical energy occurs in a device known as a fuel cell. The first modern fuel cells used hydrogen and oxygen as the fuel and oxidizer. Current research and development is aimed at the use of gaseous hydrocarbons and air and at improving the conversion efficiency. A more detailed discussion of fuel cells appears in the following section.

## 13-10   FUEL CELLS

The maximum work obtainable from a heat engine is limited to the Carnot efficiency. Because of the presence of irreversibilities in actual practice, modern cyclic heat-to-work converters seldom achieve thermal efficiencies greater than about 40 percent. It would be highly desirable if other methods of energy conversion were available that did not rely on the conversion of heat to work, since such methods would not be limited to the Carnot efficiency. One well-known device that bypasses the heat-work conversion step is the conventional battery. A battery converts chemical energy stored in the chemical bonds of the reactants to electrical work. Usually a small quantity of energy is discharged as heat transfer so that the process is essentially *isothermal*. A *fuel cell* is a form of a battery with several important changes. The electrode materials are not consumed during its operation but remain invariant. As a result, fuel and oxidizer must be supplied continuously from outside sources (see Fig. 13-20). In addition, a means of eliminating the

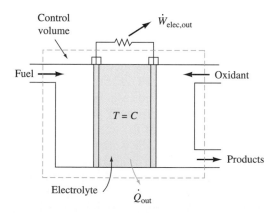

**Figure 13-20**     Schematic of a fuel cell.

products of the reaction must be provided. The conventional battery is a closed system, whereas the fuel cell operates as an open system.

The electrical work of an internally reversible fuel cell is determined from a thermodynamic analysis of an open system. Consider the fuel cell shown in Fig. 13-20. We assume that heat transfer to or from the fuel cell results in an isothermal operation. If changes in kinetic and potential energy are neglected, the steady-state energy equation for the fuel-cell volume reduces to

$$\dot{W}_{\text{elec}} = \dot{N}_{\text{fuel}}\left(\sum_{\text{prod}} n_i \bar{h}_i - \sum_{\text{reac}} n_i \bar{h}_i\right) - \dot{Q}$$

$$= \dot{N}_{\text{fuel}}\left(\bar{h}_{\text{prod}} - \bar{h}_{\text{reac}}\right) - \dot{Q}$$

Similarly the steady-state entropy balance for an internally reversible, isothermal control volume is

$$0 = \frac{\dot{Q}}{T} + \dot{N}_{\text{fuel}}\left(\sum_{\text{reac}} n_i \bar{s}_i - \sum_{\text{prod}} n_i \bar{s}_i\right)$$

$$= \frac{\dot{Q}}{T} + \dot{N}_{\text{fuel}}\left(\bar{s}_{\text{reac}} - \bar{s}_{\text{prod}}\right)$$

Solving for the entropy balance for $\dot{Q}$ and substituting the result back into the energy balance gives an expression for the reversible electric power for the fuel cell

$$\dot{W}_{\text{elec,rev}} = \dot{N}_{\text{fuel}}\left(\bar{h}_{\text{prod}} - \bar{h}_{\text{reac}}\right) - \left[-\dot{N}_{\text{fuel}}T\left(\bar{s}_{\text{reac}} - \bar{s}_{\text{prod}}\right)\right]$$

$$= \dot{N}_{\text{fuel}}\left[\bar{h}_{\text{prod}} - \bar{h}_{\text{reac}} - T\left(\bar{s}_{\text{prod}} - \bar{s}_{\text{reac}}\right)\right]$$

$$= \dot{N}_{\text{fuel}}\left[\Delta\bar{h}_R - T\Delta\bar{s}_R\right]$$

where $\Delta\bar{h}_R$ and $\Delta\bar{s}_R$ are respectively the enthalpy and entropy of reaction. Because the Gibbs function is defined as $g \equiv h - Ts$, under isothermal conditions $\Delta g = \Delta h - T\Delta s$. Comparing this relation with the above equation for the reversible electric power, we conclude that

$$\dot{W}_{\text{elec,rev}} = \dot{N}_{\text{fuel}}\,\Delta\bar{g}_R \qquad \text{isothermal} \qquad \textbf{[13-31]}$$

Therefore, the electrical work per mole of fuel for an isothermal, internally reversible fuel cell is measured by the change in the Gibbs function for the reaction.

There are several different efficiencies frequently defined for the operation of batteries and fuel cells. One possible standard of performance is the ratio of the maximum useful-work output to the energy input. We have seen that the maximum useful-work output per mole of fuel is given by $\Delta\bar{g}_R$. The energy input per mole of fuel is the enthalpy of reaction $\Delta\bar{h}_R$ released by the overall chemical reaction. The *ideal efficiency* of a fuel cell, therefore, may be defined by the relation

$$\eta_i = \frac{\Delta \bar{g}_R}{\Delta \bar{h}_R} \qquad \text{[13-32]}$$

where both $\Delta \bar{g}_R$ and $\Delta \bar{h}_R$ normally have negative values. For isothermal operation $\Delta g = \Delta h - T \Delta s$. Therefore, the ideal efficiency frequently is seen in the form

$$\eta_i = 1 - \frac{T \Delta s_R}{\Delta h_R} \qquad \text{[13-33]}$$

For isothermal, reversible processes the quantity $T \Delta \bar{s}_R$ represents the heat transfer for the system. When the reversible heat transfer is out of the system ($\Delta \bar{s}_R$ is negative), the ideal efficiency is less than unity. An ideal efficiency greater than unity occurs when $\Delta \bar{s}_R$ is positive. This is possible, and simply indicates that heat transfer into the fuel cell occurs to keep the process isothermal.

The evaluation of the ideal efficiency requires information about the enthalpy of reaction $\Delta \bar{h}_R$, the Gibbs-function change for the reaction $\Delta \bar{g}_R$, and the entropy change for the reaction $\Delta \bar{s}_R$ at the specified temperature. Recall that the value of $\Delta h_R$ for a system of reacting ideal gases is

$$\Delta \bar{h}_{R,T} = \Delta \bar{h}^\circ_{R,298} + \sum_{\text{prod}} \nu_i \left( \bar{h}_T - \bar{h}_{298} \right)_i - \sum_{\text{reac}} \nu_i \left( \bar{h}_T - \bar{h}_{298} \right)_i \qquad \text{[13-21]}$$

Similarly $\Delta \bar{s}_{R,T}$ is given by

$$\Delta \bar{s}_{R,T} = \sum_{\text{prod}} \nu_i \left( s^\circ_{i,T} - R \ln \frac{p_i}{P_{\text{ref}}} \right) - \sum_{\text{reac}} \nu_i \left( s^\circ_{i,T} - R \ln \frac{p_i}{P_{\text{ref}}} \right) \qquad \text{[13-34]}$$

where the subscript $T$ in this case represents the same temperature for the reactants and products, since the process is isothermal. For reactions that occur at the standard reference temperature of 298 K, the equation for the enthalpy of reaction reduces to

$$\Delta \bar{h}^\circ_{R,298} = \sum_{\text{prod}} \nu_i \Delta \bar{h}^\circ_{f,298,i} - \sum_{\text{reac}} \nu_i \Delta \bar{h}^\circ_{f,298,i} \qquad \text{[13-19]}$$

A similar equation holds for the Gibbs function:

$$\Delta \bar{g}^\circ_{R,298} = \sum_{\text{prod}} \nu_i \Delta \bar{g}^\circ_{f,298,i} - \sum_{\text{reac}} \nu_i \Delta \bar{g}^\circ_{f,298,i} \qquad \text{[13-35]}$$

Typical values of the Gibbs function of formation of compounds are found in Table A-23 for conditions of 1 atm and 25°C (77°F).

In addition to the ideal efficiency of a battery or fuel cell, another important parameter is the ideal open-circuit voltage developed by the cell. We have already seen that the Gibbs-function change for the reversible case is a measure of the electrical work produced by the cell. The electrical work is the product of the amount of charge $Q_c$ that passes from the cell per mole of reacting fuel and the ideal electrostatic potential $V_i$ developed. That is, $w_{elec} = -Q_c V_i$. The quantity of charge $Q_c$ is equal to the number of moles of electrons $N_c$ produced by the cell reaction per mole of reacting fuel multiplied by the number of coulombs (C) per mole of electrons, $\mathscr{F}$. Hence $Q_c = N_c \mathscr{F}$. Therefore,

$$\dot{W}_{elec,rev}/\dot{N}_{fuel} = -N_c \mathscr{F} V_i = \Delta \bar{g}_R$$

or

$$V_i = \frac{-\Delta \bar{g}_R}{N_c \mathscr{F}} \qquad \text{ideal} \qquad \textbf{[13-36]}$$

The quantity $\Delta \bar{g}_R$ is negative for battery and fuel-cell reactions, so that the ideal voltage $V_i$ is a positive value.

The quantity of charge $\mathscr{F}$ is called a *faraday*. Its value is

$$\mathscr{F} = \frac{6.023 \times 10^{23} \text{ electrons}}{\text{mol electrons}} \times \frac{1.602 \times 10^{-19} \text{ C}}{\text{electron}}$$

$$= 96{,}487 \text{ C/mol electrons}$$

If this value is multiplied by the identity 1 J $=$ 1 V·C, then

$$\mathscr{F} = 96{,}487 \text{ kJ/V·kmol electrons} \qquad \textbf{[13-37]}$$

Therefore, Eq. [13-35] becomes

$$V_i = \frac{-\Delta \bar{g}_R}{96{,}487 N_c} \qquad \text{ideal} \qquad \textbf{[13-38]}$$

where $\Delta \bar{g}_R$ is expressed in kilojoules per kilomole and $V_i$ is in volts.

The value of $N_c$ is determined from a knowledge of the reactions that occur at the two electrodes. The *anode* is the negative electrode where the fuel is ionized and electrons are supplied to the external circuit. The *cathode* is the positive electrode where the oxidizer is ionized and electrons are received from the external circuit. An *electrolyte* between the electrodes transports ions to complete the electric circuit. When an aqueous solution is used that is acidic, positive $H^+$ ions are transported through the electrolyte from anode to cathode. If the electrolyte is a basic solution, negative $OH^-$ ions move from cathode to anode. Nonaqueous liquids and solids may also act as electrolytes.

The method of evaluation of the quantity $N_c$ in Eq. [13-38] is made more clear by examining two specific examples. First consider a fuel cell operating on hydrogen and oxygen such as has been used on space missions. The overall reaction at room temperature is

$$H_2(g) + \tfrac{1}{2} O_2(g) \rightarrow H_2O(l) \qquad \textbf{[13-39]}$$

A schematic of the cell operating with an acidic electrolyte is shown in Fig. 13-21a. It is convenient to express the overall chemical reaction in

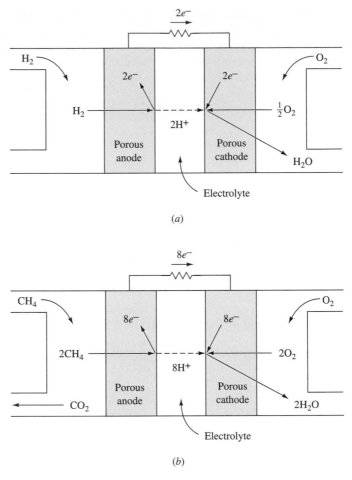

**Figure 13-21**     Schematic of (a) a hydrogen-oxygen fuel cell and (b) a methane-oxygen fuel cell.

terms of the individual reactions which occur at each electrode. The reactions that occur at sites on the anode and cathode are called ***half-cell reactions***. For the $H_2$–$O_2$ fuel cell the anode half-cell reaction is

Anode:     $$H_2 \rightarrow 2e^- + 2H^+$$

The electrons released at the anode now pass through an external circuit in order to complete the reaction. By placing an external load in the circuit, electrical work can be produced, even though the process is isothermal. Since chemical energy is converted directly to electric energy, there is no Carnot limitation on the efficiency of the conversion process. At the cathode, the half-cell reaction is

Cathode:     $$2H^+ + 2e^- + \tfrac{1}{2}O_2 \rightarrow H_2O(l)$$

Thus during the process hydrogen ions migrate through the electrolyte from the anode to the cathode, while electrons pass through the external circuit. In this case $N_c$ is 2, since two electrons are liberated at the anode for every molecule of hydrogen consumed.

---

Consider a hydrogen-oxygen fuel cell that contains an acidic electrolyte. For a temperature of 25°C and assuming that the water formed as a product at the cathode is all in the liquid state, determine (a) the maximum work output, in kJ/kmol of fuel, (b) the ideal efficiency, and (c) the ideal voltage.

**Example 13-20**

**Solution:**

**Given:** A hydrogen-oxygen fuel cell as shown in Fig. 13-21a.

**Find:** (a) $\dot{W}_{elec,rev}/\dot{N}_{fuel}$, (b) $\eta_{ideal}$, and (c) $V_i$, in volts.

**Model:** Steady-state, internally reversible, isothermal cell.

**Analysis:** (a) In this particular fuel cell, all reactants and products may be assumed to be pure, and the operating pressure is taken to be 1 atm. Consequently, the maximum work output as given by the Gibbs-function change for the reaction is determined from the standard-state value $\Delta \bar{g}^o_{R,298}$. Using Eq. [13-35] and the data from Table A-23, we find that

$$\dot{W}_{max}/\dot{N}_{fuel} = \Delta \bar{g}^o_{R,298} = \sum_i (\nu_i \Delta \bar{g}^o_{f,i})_{prod} - \sum_i (\nu_i \Delta \bar{g}^o_{f,i})_{reac}$$

$$= 1(-237,180) - 1(0) - \tfrac{1}{2}(0)$$

$$= -237,180 \text{ kJ/kmol}$$

(b) The ideal efficiency requires the evaluation of the enthalpy change for the reaction, which in this case of pure substances each at 1 atm is given by $\Delta \bar{h}^o_{298}$. Hence, using Eq. [13-19] and data from Table A-23,

$$\Delta \bar{h}^o_{R,298} = \sum_{prod} \nu_i \Delta \bar{h}^o_{f,298,i} - \sum_{reac} \nu_i \Delta \bar{h}^o_{f,298,i}$$

$$= 1(-285,830) - 1(0) - \tfrac{1}{2}(0) = -285,830 \text{ kJ/kmol}$$

On the basis of Eq. [13-32], then, the ideal efficiency at 298 K becomes

$$\eta_i = \frac{\Delta \bar{g}_R}{\Delta \bar{h}_R} = \frac{-237,180}{-285,830} = 0.830 \text{ (or 83.0\%)}$$

Note that the energy output $\Delta \bar{g}_R$ is less than the energy input $\Delta \bar{h}_R$ in this case. The difference between these values, $\Delta \bar{h}_R - \Delta \bar{g}_R$, is a measure of the heat transfer necessary to the environment if this cell is to be kept at constant temperature.

(c) The ideal voltage of the cell is determined by using Eq. [13-38]. In this case the value of $N_c$ is 2, as shown by the anode or cathode half-cell reaction discussed above. Therefore,

$$V_i = \frac{-\Delta \bar{g}_R}{96,487 N_c} = \frac{-(-237,180)}{96,487(2)} = 1.23 \text{ V}$$

**Comment:**    When other substances are used in batteries and fuel cells, the ideal voltage may typically range from 1 to 2 V per cell. Higher overall voltages are obtained by placing several cells in series.

---

As a second example, the number of moles of electrons $N_c$ produced per mole of fuel can be ascertained for a hydrocarbon such as methane ($CH_4$) in the following manner. A schematic of the process is shown in Fig. 13-21$b$. The overall chemical reaction may be written as

$$CH_4 + 2O_2 \rightarrow CO_2 + 2H_2O$$

At the anode the reactants are $CH_4$ and water from the electrolyte, and the products are gaseous carbon dioxide, hydrogen ions, and electrons. The half-cell reaction at the anode is

Anode:          $CH_4 + 2H_2O \rightarrow CO_2 + 8H^+ + 8e^-$

At the cathode the electrons and hydrogen ions react with oxygen introduced to the cell to form water. In this case,

Cathode:          $2O_2 + 8H^+ + 8e^- \rightarrow 4\,H_2O$

For any hydrocarbon that does not contain oxygen in its formula, the number of electrons $N_c$ released per molecule of fuel $C_xH_y$ is $4x + y$.

One important consideration in the analysis of any fuel cell is the effect of temperature on performance. First, the ideal efficiency may be affected by temperature. This depends on the variation of $\Delta \bar{g}_R$ and $\Delta \bar{h}_R$ with temperature. These properties are fairly sensitive to temperature for the hydrogen-oxygen cell, for example. Second, the influence of temperature on the rate of reaction is also important. Cells employing carbon or hydrocarbons as the fuel frequently must be operated at elevated temperatures in order to achieve sufficient reaction rates.

The large-scale fuel-cell power plant has a number of distinct advantages over conventional fossil-fuel plants. A fuel cell operates as a direct-energy converter, and not as a heat engine. Therefore, it is not affected by the Carnot-efficiency limitation of the second law. Thus it has the potential for greater energy-conversion efficiencies than conventional heat-engine devices. First-generation fuel cells have efficiencies around 40 percent, while second-generation cell efficiency is expected to approach 45 to 50 percent. This compares to 30 to 40 percent for fossil-fuel plants currently in use. In addition, the efficiency of a fuel-cell system is directly related to that of the individual cell. Hence plant size will not affect efficiency. In comparison, the thermal efficiency of fossil-fuel plants decreases with decreasing size. For a given size, a conventional plant will be much less efficient at part load. However, the efficiency of fuel-cell power plants will remain roughly the same when they are operating anywhere between 25 and 100 percent of rated load. Another major advantage of the fuel-cell power plant is its ability to respond quickly to load variations. This response can be less than a minute. Large conventional power plants have very long response times. Therefore, large load variations are handled by some other means. Other desirable features of the fuel-cell power plant include extremely small amounts of air pollution and large quantities of waste energy that are not rejected to the environment.

## 13-11 SUMMARY

Chemical reactions involving hydrocarbon fuels with dry air are usually based on a dry air composition of 21 percent oxygen and 79 percent nitrogen. A stoichiometric reaction occurs when the reaction goes to completion and there is no excess oxygen in the products. When the fuel and air are not in a stoichiometric ratio, the reactant composition is expressed in terms of the percent theoretical or percent excess air used. Also employed is the air-fuel ratio (on a mass basis) or the equivalence ratio. When the composition of the product gases is measured, the reactant composition can be determined from atomic species balances.

In applying the conservation of energy equation to processes with chemical reactions care must be taken to include the energy stored in the chemical bonds that change during a chemical reaction. This is done by first defining the *enthalpy of formation* $\Delta \bar{h}_f$ as the change in enthalpy that occurs when a compound is formed from its stable elements at a specified temperature and pressure. By convention it is also assumed that the enthalpy of the stable form of the elements at the *standard reference state* (25°C and 1 atm) is assigned a zero value. Now the enthalpy of any compound $i$ can be defined as

$$\bar{h}_{i,T,P} = \Delta \bar{h}^{\circ}_{f,298,i} + (\bar{h}_{T,P} - \bar{h}_{298\text{ K,1 atm}})_i$$

where $\Delta \bar{h}^{\circ}_{f,298}$ is the enthalpy of formation at the standard reference state and $\bar{h}_{T,P} - \bar{h}_{298\text{ K,1 atm}}$ is the sensible enthalpy change of the compound between the standard reference state and the desired state. For an ideal gas the enthalpy is only a function of temperature and this equation simplifies to

$$\bar{h}_{i,T} = \Delta \bar{h}^{\circ}_{f,298,i} + (\bar{h}_T - \bar{h}_{298})_i \qquad \text{ideal gas}$$

where $\bar{h}_T$ and $\bar{h}_{298}$ are the idea-gas enthalpy at temperature $T$ and 298 K, respectively.

Now the energy equation neglecting changes in kinetic and potential energy can be written for any steady-state, open system with chemical reactions as

$$\dot{Q} + \dot{W} = \sum_{\text{prod}} \dot{N}_i \bar{h}_i - \sum_{\text{reac}} \dot{N}_i \bar{h}_i$$

where $\dot{N}_i$ is the molar flow rate of compound $i$. Because most combustion calculations are done on the basis of fuel flow rates, a more useful form of the energy balance is

$$\dot{Q} + \dot{W} = \dot{N}_{fuel}\left[\sum_{prod} n_i \bar{h}_i - \sum_{reac} n_i \bar{h}_i\right] = \dot{N}_{fuel}[\bar{h}_{prod} - \bar{h}_{reac}]$$

where $n_i = \dot{N}_i/\dot{N}_{fuel}$ and $\bar{h}_{prod}$ and $\bar{h}_{reac}$ are the enthalpy per mole of fuel of the products and the reactants, respectively. Typically, the work term is zero when the energy balance is applied to a combustion process. When the energy balance is applied to a combustion process for which the initial and final temperatures for the reaction are equal, the heat transfer per mole of fuel, $\dot{Q}/\dot{N}_{fuel}$, is called the *enthalpy of reaction* $\Delta \bar{h}_R$ or the *enthalpy of combustion* $\Delta \bar{h}_c$ and the numerical values are negative. When the absolute value of the enthalpy of combustion is reported on a mass basis, it is called the *heating value of a fuel*.

An important use of the energy balance is to predict the *adiabatic flame (or combustion) temperature* for specified reactants. In this case, the energy balance gives an equation where the final enthalpy of the products depends implicitly on the final temperature of the process. This type of calculation often requires an iterative solution when only tabular enthalpy data is available.

Calculations for combustion processes may also be carried out for constant-volume systems. In this case

$$Q = \sum_{prod} N_i(\Delta \bar{h}^o_{f,298} + \bar{u}_T - \bar{u}_{298})_i$$
$$- \sum_{reac} N_i(\Delta \bar{h}^o_{f,298} + \bar{u}_T - \bar{u}_{298})_i - \Delta N R_u T_{298}$$

In this latter case the evaluation of the final (explosion) pressure is important for safety considerations.

The entropy balance can be applied to a steady-state chemical reaction, and in a manner similar to the energy balance it is useful to write the balance in terms of the molar flow rate of fuel:

$$0 = \dot{N}_{fuel}\left[\sum_{reac} n_i \bar{s}_i - \sum_{prod} n_i \bar{s}_i\right] + \sum_j \frac{\dot{Q}_j}{T_j} + \dot{\sigma}_{cv}$$

where $\bar{s}_{i,T,P}$ is the *absolute entropy* of a compound. For an ideal gas, the absolute entropy is

$$\bar{s}_{i,T,P} = \bar{s}^o_{i,T} - R_u \ln(p_i/P_{ref}) \qquad \text{ideal gas}$$

where $\bar{s}^o_{i,T}$ values are based on zero entropy at absolute zero of temperature and $p_i = y_i P$, the partial pressure of the component. The change in entropy for a reaction is defined in terms of the entropy of the reactants and of the products per mole of fuel as

$$\bar{s}_{prod} - \bar{s}_{reac} = \sum_{prod} n_i \bar{s}_i - \sum_{reac} n_i \bar{s}_i$$

When the energy balance is written for a steady-state, adiabatic open system with chemical reactions, we find that $\bar{s}_{prod} - \bar{s}_{reac} = \dot{\sigma}_{cv}/\dot{N}_{fuel} \geq 0$.

In terms of availability concepts, the reversible work associated with a chemical reaction in a steady-state open system is given by

$$\dot{W}_{rev,sf} = \dot{N}_{fuel}\left[\sum_{prod} n_i \bar{\psi}_i - \sum_{reac} n_i \bar{\psi}_i\right] = \dot{N}_{fuel}[\bar{\psi}_{prod} - \bar{\psi}_{reac}]$$

where $\bar{\psi}_i = \bar{h}_i - T_0 \bar{s}_i$, the flow availability, is calculated using the appropriate enthalpy and entropy to account for the chemical contributions. The irreversibility of a combustion process in in a steady-state, open system can be written as

$$\dot{I}_{cv} = T_0 \dot{\sigma}_{cv} = T_0 \dot{N}_{fuel}(\bar{s}_{prod} - \bar{s}_{reac}) - \dot{Q}$$

The maximum electrical work obtainable for an isothermal fuel cell is

$$\dot{W}_{elec,rev} = \dot{N}_{fuel} \Delta \bar{g}_T$$

the ideal efficiency is

$$\eta_i = \frac{\Delta \bar{g}_R}{\Delta \bar{h}_R} = 1 - \frac{T \Delta \bar{s}_R}{\Delta \bar{h}_R}$$

and the ideal voltage is found by

$$V_i = \frac{-\Delta \bar{g}_R}{96,487 N_c}$$

where $\Delta \bar{g}_R$ is expressed in kJ/kmol and $V_i$ is in volts. The value of $N_c$ is determined from the half-cell reactions which occur at the anode and cathode.

## PROBLEMS

13-1. Ethane ($C_2H_6$) at a fuel rate of 1 kg/min is burned with dry air in a fuel-to-air mass ratio of 1:17. Assume complete combustion and a total pressure of 1 bar. Compute (*a*) the percent of excess air used, (*b*) the equivalence ratio, (*c*) the percentage of $CO_2$ by volume in the product gases, (*d*) the dew-point temperature, in degrees Celsius, (*e*) the percentage of water vapor condensed if the product gases are cooled to 20°C, and (*f*) the required volume flow rate of dry air at 1 bar and 27°C, in m³/min. If the actual air supplied has a humidity ratio of 16 g $H_2O$ per kilogram of dry air, (*g*) determine the dew-point temperature for the products of the reaction.

13-2. Ethylene ($C_2H_4$) at a fuel rate of 1 kg/min is burned with dry air in a fuel-to-air mass ratio of 1:17. Assume complete combustion and a total pressure of 95 kPa. Compute (*a*) the percent of excess air used, (*b*) the equivalence ratio, (*c*) the percentage of $CO_2$ by volume in the product gases, ($d$) the dew-point temperature, in degrees Celsius, ($e$) the percentage of water vapor condensed if the product gases are cooled to 20°C, and ($f$) the required volume flow rate of dry air at 95 kPa and 27°C, in m$^3$/min. If the actual air supplied has a humidity ratio of 15 g $H_2O$ per kilogram of dry air, ($g$) determine the dew-point temperature for the products of the reaction.

13-3. Acetylene ($C_2H_2$) at a fuel rate of 1 kg/min is burned with dry air in an air-to-fuel mass ratio of 17:1. Assume complete combustion and a total pressure of 110 kPa. Compute ($a$) the percent of excess air used, ($b$) the equivalence ratio, ($c$) the percentage of $CO_2$ by volume in the product gases, ($d$) the dew-point temperature, in degrees Celsius, ($e$) the percentage of water vapor condensed if the product gases are cooled to 20°C, and ($f$) the required volume flow rate of dry air at 110 kPa and 22°C, in m$^3$/min. If the actual air supplied has a humidity ratio of 16 g $H_2O$ per kilogram of dry air, ($g$) determine the dew-point temperature for the products of the reaction.

13-4. Propylene ($C_3H_6$) at a fuel rate of 1 kg/min is burned with dry air in an air-to-fuel mass ratio of 17:1. Assume complete combustion and a total pressure of 1.05 bars. Compute ($a$) the percent of excess air used, ($b$) the equivalence ratio, ($c$) the percentage of $CO_2$ by volume in the product gases, ($d$) the dew-point temperature, in degrees Celsius, ($e$) the percentage of water vapor condensed if the product gases are cooled to 20°C, and ($f$) the required volume flow rate of dry air at 1.05 bar and 32°C, in m$^3$/min. If the actual air supplied has a humidity ratio of 17 g $H_2O$ per kilogram of dry air, ($g$) determine the dew-point temperature for the products of the reaction.

13-5. Ethane ($C_2H_6$) at a fuel rate of 1 kg/min is burned with 20 percent excess dry air. Assume complete combustion and a total pressure of 100 kPa. Compute ($a$) the air-fuel ratio used, ($b$) the equivalence ratio, ($c$) the mole percent of $N_2$ in the product gases, ($d$) the dew-point temperature, in degrees Celsius, ($e$) the percentage of water vapor condensed if the product gases are cooled to 25°C, and ($f$) the required volume flow rate of dry air at 100 kPa and 22°C, in m$^3$/min. If the actual air supplied has a humidity ratio of 17 g $H_2O$ per kilogram of dry air, ($g$) determine the dew-point temperature for the products of the reaction.

13-6. Ethylene ($C_2H_4$) at a fuel rate of 1 kg/min is burned with 20 percent excess dry air. Assume complete combustion and a total pressure of 1.05 bars. Compute ($a$) the air-fuel ratio used, ($b$) the equivalence ratio, ($c$) the mole percent of $N_2$ in the product gases, ($d$) the dew-point temperature, in degrees Celsius, ($e$) the percentage of water vapor condensed if the product gases are cooled to 25°C, and ($f$) the required volume flow rate of dry air at 1.05 bars and 27°C, in m$^3$/min. If the actual air supplied has a humidity ratio of 17 g $H_2O$ per kilogram of dry air, ($g$) determine the dew-point temperature for the products of the reaction.

13-7. Acetylene ($C_2H_2$) at a fuel rate of 1 kg/min is burned with 20 percent excess dry air. Assume complete combustion and a total pressure of 98 kPa. Compute (*a*) the air-fuel ratio used, (*b*) the equivalence ratio, (*c*) the mole percent of $N_2$ in the product gases, (*d*) the dew-point temperature, in degrees Celsius, (*e*) the percentage of water vapor condensed if the product gases are cooled to 25°C, and (*f*) the required volume flow rate of dry air at 98 kPa and 22°C, in $m^3$/min. If the actual air supplied has a humidity ratio of 16 g $H_2O$ per kilogram of dry air, (*g*) determine the dew-point temperature for the products of the reaction.

13-8. Propylene ($C_3H_6$) at a fuel rate of 1 kg/min is burned with 20 percent excess dry air. Assume complete combustion and a total pressure of 1 bar. Compute (*a*) the air-fuel ratio used, (*b*) the equivalence ratio, (*c*) the mole percent of $N_2$ in the product gases, (*d*) the dew-point temperature, in degrees Celsius, (*e*) the percentage of water vapor condensed if the product gases are cooled to 25°C, and (*f*) the required volume flow rate of dry air at 1 bar and 27°C, in $m^3$/min. If the actual air supplied has a humidity ratio of 15 g $H_2O$ per kilogram of dry air, (*g*) determine the dew-point temperature for the products of the reaction.

13-9E. Ethane ($C_2H_6$) at a fuel rate of 1 $lb_m$/min is burned with dry air in an air-to-fuel mass ratio of 17:1. Assume complete combustion and a total pressure of 14.5 psia. Compute (*a*) the percent of excess air used, (*b*) the equivalence ratio, (*c*) the percentage of $CO_2$ by volume in the product gases, (*d*) the dew-point temperature, in degrees Fahrenheit, (*e*) the percentage of water vapor condensed if the product gases are cooled to 70°F, and (*f*) the required volume flow rate of dry air at 14.5 psia and 80°F, in $ft^3$/min. If the actual air supplied has a humidity ratio of 0.015 $lb_m$ $H_2O$/$lb_m$ dry air, (*g*) determine the dew-point temperature for the products of the reaction.

13-10E. Propylene ($C_3H_6$) at a fuel rate of 1 $lb_m$/min is burned with dry air in a fuel-to-air mass ratio of 1:17. Assume complete combustion and a total pressure of 14.5 psia. Compute (*a*) the percent of excess air used, (*b*) the equivalence ratio, (*c*) the percentage of $CO_2$ by volume in the product gases, (*d*) the dew-point temperature, in degrees Fahrenheit, (*e*) the percentage of water vapor condensed if the product gases are cooled to 70°F, and (*f*) the required volume flow rate of dry air at 14.7 psia and 80°F, in $ft^3$/min. If the actual air supplied has a humidity ratio of 0.015 $lb_m$ $H_2O$ per pound of dry air, (*g*) determine the dew-point temperature for the products of the reaction.

13-11E. Ethane ($C_2H_6$) at a fuel rate of 1 lb/min is burned with 20 percent excess dry air. Assume complete combustion and a total pressure of 14.8 psia. Compute (*a*) the air-fuel ratio used, (*b*) the equivalence ratio, (*c*) the mole percent of $N_2$ in the product gases, (*d*) the dew-point temperature, in degrees Fahrenheit, (*e*) the percentage of water vapor condensed if the product gases are cooled to 80°F, and (*f*) the required volume flow rate of dry air at 14.8 psia and 80°F, in ft$^3$/min. If the actual air supplied has a humidity ratio of 0.017 lb$_m$ H$_2$O per pound of dry air, (g) determine the dew-point temperature for the products of the reaction.

13-12E. Acetylene ($C_2H_2$) at a fuel rate of 1 lb$_m$/min is burned with 20 percent excess dry air. Assume complete combustion and a total pressure of 14.6 psia. Compute (a) the air-fuel ratio used, (b) the equivalence ratio, (c) the mole percent of $N_2$ in the product gases, (d) the dew-point temperature, in degrees Fahrenheit, (e) the percentage of water vapor condensed if the product gases are cooled to 80°F, and (f) the required volume flow rate of dry air at 14.6 psia and 80°F, in ft$^3$/min. If the actual air supplied has a humidity ratio of 0.014 lb$_m$ H$_2$O per pound of dry air, (g) determine the dew-point temperature for the products of the reaction.

13-13. Propane ($C_3H_8$) is burned with 20 percent excess air. Owing to incomplete mixing, only 96 percent of the carbon in the fuel is converted to $CO_2$, the rest appearing as CO. Determine (a) the air-fuel ratio used, (b) the mole fraction of $CO_2$ in the total products, and (c) the dew point, in degrees Celsius, if the pressure is 108 kPa.

13-14. Liquid benzene ($C_6H_6$) is burned with 120 percent theoretical-air requirements. Compute (a) the air-fuel ratio used, (b) the mole fraction of $N_2$ in the total products, and (c) the dew point, in degrees Celsius, if the pressure is 110 kPa.

13-15. Pentane gas ($C_5H_{12}$) is burned to completion with 25 percent excess air at a pressure of 1.03 bars.
(a) If the air supplied is dry, find (1) the dew point, in degrees Celsius, and (2) the number of moles of water per mole of fuel burned which would condense if the product gases were cooled to 22°C.
(b) If the air supplied has a specific humidity of 14 g H$_2$O per kilogram of dry air, find the answers to the same questions asked in part a.

13-16. Methane gas ($CH_4$) is burned with 50 percent excess air at 0.95 bar.
(a) If the air supplied is dry, find (1) the dew point, in degrees Celsius, and (2) the number of moles of water per mole of fuel which would condense if the product gases were cooled to 25°C.
(b) If the air supplied has a specific humidity of 16 g H$_2$ per kilogram of dry air, find the answers to the same questions in part a.

13-17E. Ethylene gas ($C_2H_4$) is burned with 20 percent excess air. Owing to incomplete mixing, only 96 percent of the carbon in the fuel is converted to $CO_2$, the rest appearing as CO. Determine (a) the air-fuel ratio used, (b) the mole fraction of $CO_2$ in the total products, and (c) the dew point, in degrees Fahrenheit, if the pressure is 14.8 psia.

13-18E. Liquid benzene ($C_6H_6$) is burned with 140 percent theoretical-air requirements. Compute (*a*) the air-fuel ratio used, (*b*) the mole fraction of $N_2$ in the total products, and (*c*) the dew point, in degrees Fahrenheit, if the pressure is 15.0 psia.

13-19E. Pentane gas ($C_5H_{12}$) is burned to completion with the stoichiometric amount of air at a pressure of 14.8 psia.

    (*a*) If the air supplied is dry, find (1) the dew point, in degrees Fahrenheit, and (2) the number of moles of water per mole of fuel burned which would condense if the product gases were cooled to 75°F.

    (*b*) If the air supplied has a specific humidity of 0.014 $lb_m$ $H_2O$ per pound of dry air, find the answers to the same questions asked in part *a*.

13-20. A fuel gas has a volumetric composition of $CH_4$, 60 percent; $C_2H_6$, 30 percent; and $N_2$, 10 percent. It is burned to completion, using 20 percent excess of dry air. Determine (*a*) the mole percent of the various products on a dry basis, (*b*) the air-fuel ratio used, (*c*) the dew-point temperature, in degrees Celsius, if the total pressure is 100 kPa, and (*d*) the mole percent of $CO_2$ in the product gas on a wet basis if the gas is cooled to 20°C.

13-21. A gaseous fuel having a volumetric analysis of 65 percent $CH_4$, 25 percent $C_2H_6$, 5 percent $CO_2$, and 5 percent $N_2$ is burned with 30 percent excess of dry air. Determine (*a*) the air-fuel ratio used, (*b*) the mole percent $CO_2$ in the total products, (*c*) the dew-point temperature if the pressure is 1.0 bar, and (*d*) the mole percent $N_2$ in the total gaseous products if the gas is cooled to 22°C.

13-22. A gaseous fuel having a volumetric analysis of 80 percent $CH_4$ and 20 percent $C_3H_8$ is burned with 20 percent excess of dry air. Determine (*a*) the air-fuel ratio used, (*b*) the mole percent $N_2$ in the total gaseous products, (*c*) the dew-point temperature if the pressure is 105 kPa, and (*d*) the mole percent $N_2$ in the gaseous products if the gas is cooled to 30°C.

13-23. A gaseous fuel which is 60 mole percent $CH_4$ and 40 mole percent $C_2H_4$ is burned with 10 percent excess dry air. Determine (*a*) the air-fuel ratio used, (*b*) the mole percent of $CO_2$ in the total products, (*c*) the dew-point temperature if the pressure is 1.05 bars, and (*d*) the mole percent $CO_2$ in the gaseous products if the mixture is cooled to 25°C.

13-24E. A fuel gas has a volumetric composition of $CH_4$, 60 percent; $C_2H_6$, 30 percent; and $N_2$, 10 percent. It is burned to completion, using 20 percent excess of dry air. Determine (*a*) the mole percent of the various products on a dry basis, (*b*) the air-fuel ratio used, (*c*) the dew-point temperature, in degrees Fahrenheit, if the total pressure is 14.7 psia, and (*d*) the mole percent of $CO_2$ in the product gas on a wet basis if the gas is cooled to 70°F.

13-25E. A gaseous fuel having a volumetric analysis of 70 percent $CH_4$, 25 percent $C_2H_6$, and 5 percent $CO_2$ is burned with 30 percent excess of dry air. Determine (a) the air-fuel ratio used, (b) the mole percent $CO_2$ in the total products, (c) the dew-point temperature if the pressure is 14.7 psia, and (d) the mole percent $N_2$ in the total gaseous products if the gas is cooled to 80°F.

## ACTUAL COMBUSTION PROCESSES

13-26. Octane ($C_8H_{18}$) is burned with dry air, and a volumetric analysis on a dry basis of the products reveals 10.39 percent $CO_2$, 4.45 percent $O_2$, and 1.48 percent CO. Compute (a) the actual air-fuel ratio used and (b) the percentage of theoretical air used.

13-27. The combustion of methane gas ($CH_4$) with dry air leads to the following volumetric analysis during a test: 9.7 percent $CO_2$, 0.5 percent CO, and 3.0 percent $O_2$. Determine (a) the number of moles of air used per mole of fuel, (b) the percentage of theoretical air used, and (c) the air-fuel ratio.

13-28. Solid carbon is burned with dry air in a combustion test. A volumetric analysis reveals that the products of combustion include 3.5 percent CO, 13.8 percent $CO_2$, and 5.2 percent $O_2$ on a dry basis. Determine (a) the air-fuel ratio used and (b) the percentage of theoretical air used.

13-29. Ethylene ($C_2H_4$) is burned with 33 percent excess air. An analysis of the products of combustion on a dry basis reveals 6.06 percent $O_2$ by volume. The remaining data for the product analysis are missing. What percentage of the carbon in the fuel was converted to CO instead of $CO_2$?

13-30. Propylene ($C_3H_6$) is burned with 20 percent excess air. An analysis of the products of combustion on a dry basis reveals 4.31 percent $O_2$ by volume. What percentage of the carbon in the fuel was converted to CO instead of $CO_2$?

13-31. The volumetric analysis of the dry products of combustion of a hydrocarbon fuel described by the general formula $C_xH_y$ is $CO_2$, 13.6 percent; $O_2$, 0.4 percent; CO, 0.8 percent; $CH_4$, 0.4 percent; and $N_2$, 84.8 percent. There is 13.6 mol of $CO_2$ formed per mole of fuel. Determine (a) the values of $x$ and $y$ and (b) the air-fuel ratio used.

13-32. The volumetric analysis of the dry products of combustion of a hydrocarbon described by the general formula $C_xH_y$ is $CO_2$, 12.37 percent; CO, 0.87 percent; $O_2$, 2.47 percent; and $N_2$, 84.29 percent. There is 1.42 mol of $O_2$ in the products of combustion per mole of fuel. Determine (a) the values of $x$ and $y$ and (b) the air-fuel ratio used.

13-33. A fuel gas with a volumetric analysis of 60 percent CO, 20 percent $H_2$, and 20 percent $N_2$ is burned with dry air. A volumetric analysis of the product gases gives 20.0 percent $CO_2$, 5.0 percent CO, and

2.8 percent $O_2$ on a dry basis. Calculate (*a*) the percentage of theoretical air used and (*b*) the air-fuel ratio used.

13-34. A gaseous fuel is composed of 20 percent $CH_4$, 40 percent $C_2H_6$, and 40 percent $C_3H_8$, where all percentages are by volume. The volumetric analysis of the dry combustion gases gives 11.4 percent $CO_2$, 1.7 percent $O_2$, and 1.2 percent CO. Determine (*a*) the theoretically correct air-fuel ratio required, and (*b*) the percent excess air used.

## ENERGY ANALYSIS

13-35. Liquid octane (*n*-$C_8H_{18}$) at 25°C and air at 500 K enter a combustion chamber in steady flow at 1 bar. The volume flow rate of air is 500 $m^3$/h. Determine (*a*) the mass flow rate of fuel required, in kmol/h, and (*b*) the rate of heat transfer that occurs, in kJ/min, if 150 percent theoretical air is used and the temperature of the product gases is 1000 K.

13-36. Ethane gas ($C_2H_6$) at 25°C and air at 227°C enter a combustion chamber in steady flow at 100 kPa. The products of combustion leave at 1100 K. If 25 percent excess air is used, combustion is complete, and the fuel rate is 1 kg/min, determine the rate of heat loss, in kJ/min.

13-37. Propane ($C_3H_8$) at 25°C is burned in steady flow with 100 percent excess air initially at 25°C and 1 bar. The reaction goes to completion and the products are at 600 K. For a volume flow rate of air of 20 $m^3$/min, determine (*a*) the mass flow rate of fuel required, in kg/min, and (*b*) the rate of heat transfer required, in kJ/min.

13-38. Carbon monoxide is burned with air, each entering at 25°C. The final temperature of the products is 1100 K, and during the steady-flow process 185,000 kJ of heat per kilomole of CO is removed. Determine the percentage of excess air used if combustion is complete.

13-39. Hydrogen gas is burned with air in steady flow with each at an initial temperature of 400 K. The final temperature is 800 K. During the process, 190,000 kJ of heat is removed per kilomole of $H_2$. Determine the percentage of excess air used if combustion is complete.

13-40. Ethylene gas ($C_2H_4$) is burned with 10 percent excess air. Owing to incomplete combustion, only 96 percent of the carbon in the fuel is converted to $CO_2$, the rest appearing as CO. The fuel enters at 25°C, while the air enters at 400 K, 120 kPa, and 30 $m^3$/min. If the final temperature of the products of combustion is 1500 K, determine the magnitude and direction of the heat-transfer rate, in kJ/min.

13-41. Methane ($CH_4$) and air enter a steady-flow combustor at 1 bar and 25°C in a ratio such that 90 percent of the carbon in the fuel is converted to $CO_2$ and the rest is converted to CO. There is no oxygen present in the products. If the products leave the combustor at 500 K, calculate the heat transfer, in kJ/kmol of fuel.

13-42. Butane ($C_4H_{10}$) at 25°C and 60 percent excess air at 500 K enter a steady-flow combustor. Find the heat transfer, in kJ/kmol of fuel, when the final temperature after complete combustion is (a) 1500 K and (b) 2200 K. Then, on the basis of parts a and b, (c) estimate by linear extrapolation the temperature, in kelvins, of the products of combustion when the net heat transfer is zero.

13-43. Butane ($C_4H_{10}$) at 25°C and air at 500 K enter a steady-flow combustor. The complete products of combustion leave at 1600 K, and a heat loss of 944,000 kJ/kmol of fuel occurs during combustion. Determine the percentage of theoretical air used.

13-44. Pentane gas ($C_5H_{12}$) at 25°C and 40 percent excess air at 600 K enter a steady-flow combustor. Find the heat transfer, in kJ/kmol of fuel, when the final temperature after complete combustion is (a) 1400 K and (b) 1900 K. Then, on the basis of parts a and b, (c) estimate by linear extrapolation the temperature, in kelvins, of the products of combustion when the net heat transfer is zero.

13-45. Propane gas ($C_3H_8$) with 20 percent excess air enters a combustion chamber in steady flow at 1 bar and 25°C. Owing to incomplete combustion, only 94 percent of the carbon in the fuel is converted to $CO_2$, the rest being converted to CO. If the rate of air flow is 10 kg/min, determine the rate of heat transfer required, in kJ/min, if the temperature of the products gases is 900 K.

13-46. Hydrogen gas is burned with 100 percent excess air, both initially at 25°C and 100 kPa. During the process a heat-transfer of 35,000 kJ/kmol of fuel occurs. Find the exit-gas temperature for the steady-flow process, in kelvins.

13-47E. Ethane gas ($C_2H_6$) at 77°F and air at 540°F enter a combustion chamber in steady flow at 14.7 psia. The products of combustion leave at 2000°R. If 15 percent excess air is used, combustion is complete, and the air rate is 10 $lb_m$/min, determine the rate of heat transfer, in Btu/min.

13-48E. Propane ($C_3H_8$) is burned in steady flow with 50 percent excess air, both initially at 77°F and 1 atm. The reaction goes to completion and the products are at 940°F. For a volume flow rate of air of 300 $ft^3$/min, determine (a) the mass flow rate of fuel required, in $lb_m$/min, and (b) the rate of heat transfer required, in Btu/min.

13-49E. Hydrogen gas is burned with air in steady flow with each at an initial temperature of 720°R. The final temperature is 1800°R. During the process, 76,000 Btu of energy is removed by heat transfer per pound-mole of $H_2$. Determine the percentage of excess air used if combustion is complete.

13-50E. Butane ($C_4H_{10}$) at 77°F and 60 percent excess air at 900°R enter a steady-flow combustor. Find the heat transfer, in Btu/lbmol of fuel, when the final temperature after complete combustion is (a) 2700°R and (b) 3700°R. Then, on the basis of parts a and b, (c) estimate by linear extrapolation the temperature, in degrees Rankine, of the products of combustion when the net heat transfer is zero.

13-51E. Butane ($C_4H_{10}$) at 77°F and air at 900°R enter a steady-flow combustor. The complete products of combustion leave at 2900°R, and a heat transfer out of 450,000 Btu/lbmol of fuel occurs during combustion. Determine the percentage of theoretical air used.

13-52E. Pentane gas ($C_5H_{12}$) at 77°F and 40 percent excess air at 1080°R enter a steady-flow combustor. Find the heat transfer, in Btu/lbmol of fuel, when the final temperature after complete combustion is (a) 2500°R and (b) 3500°R. Then, on the basis of parts a and b, (c) estimate by linear extrapolation the temperature, in degrees Rankine, of the products of combustion when the net heat transfer is zero.

13-53. A gas-turbine combustion chamber is supplied with air at 400 K and liquid octane at 25°C. The products of combustion leave at 1400 K. Calculate the air-fuel ratio used if the flow is steady, the combustion is complete, and the heat transfer is negligible.

13-54. Ethylene ($C_2H_4$) at 25°C is burned with excess air which enters the steady-flow combustor at 400 K. If the final temperature of the products of combustion is 1700 K, determine the percentage of excess air used under adiabatic conditions.

## ADIABATIC COMBUSTION

13-55. Ethylene ($C_2H_4$) at 25°C is burned with 200 percent excess air supplied at 400 K, and the reaction occurs in steady flow in the gas phase. Find the maximum combustion temperature, in kelvins.

13-56. Hydrogen gas at 25°C is reacted in steady flow with 500 percent theoretical oxygen requirements which enter at 500 K. Find the maximum combustion temperature, in kelvins.

13-57. Propane gas ($C_3H_8$) is burned in steady flow with 20 percent excess air starting at 25°C. Determine the maximum combustion temperature, in kelvins.

13-58. Determine the maximum theoretical combustion temperature for the reaction of ethane with 30 percent excess air in a steady-flow process. The reactants enter at 25°C, and the reaction goes to completion.

13-59. Determine the maximum temperature under adiabatic, steady-flow conditions when methane ($CH_4$) is burned with stoichiometric air, both entering at 25°C. Assume that 10 percent of the carbon is burned only to CO.

13-60. Carbon monoxide undergoes adiabatic combustion with 50 percent excess air in steady flow. Determine the maximum combustion temperature, in kelvins, if the initial reactants are at 25°C.

13-61. Determine the adiabatic-combustion temperature for the reaction of liquid methanol ($CH_3OH$) with the stoichiometric-air requirements at 1 bar. The methanol enters at 25°C, and the air enters at 400 K. Assume complete combustion and steady flow.

13-62. Consider the decomposition of tetranitromethane according to the reaction $C(NO_2)_4 \rightarrow CO_2 + 2N_2 + 3O_2$. If the reactants enter at 25°C, and the enthalpy of formation of $C(NO_2)_4$ is $-141,920$ kJ/kmol, estimate the adiabatic-combustion temperature of the product gases, in kelvins.

13-63. Ethylene gas ($C_2H_4$) at 25°C is burned with 10 percent excess air at 400 K in a steady flow process. Owing to incomplete combustion, only 96 percent of the carbon in the fuel is converted to $CO_2$, the rest appearing as CO. If the process is adiabatic, determine the final temperature of the product gases, in kelvins.

13-64E. Ethylene ($C_2H_4$) at 77°F is burned with 300 percent excess air supplied at 260°F and the reaction occurs in steady flow in the gas phase. Find the maximum combustion temperature, in degrees Rankine.

13-65E. Hydrogen gas at 77°F is reacted in steady flow with 600 percent theoretical oxygen requirements which enter at 1000°R. Find the maximum combustion temperature, in degrees Rankine.

13-66E. Propane gas ($C_3H_8$) is burned in steady flow with 20 percent excess air starting at 77°F. Determine the maximum combustion temperature, in degrees Rankine.

13-67E. Determine the maximum theoretical combustion temperature for the reaction of ethane with 50 percent excess air in a steady-flow process. The reactants both enter at 77°F, and the reaction goes to completion.

13-68E. Determine the maximum temperature under adiabatic, steady-flow conditions when methane ($CH_4$) is burned with 20 percent excess air, both entering at 77°F. Assume that 10 percent of the carbon is burned only to CO, the rest to $CO_2$.

13-69E. Carbon monoxide undergoes adiabatic combustion with the stoichiometric amount of air in steady flow. Determine the maximum combustion temperature, in degrees Rankine, if the initial reactants are at 77°F.

13-70. Determine the adiabatic-combustion temperature, in kelvins, and the explosion pressure, in megapascals, for the constant-volume combustion of CO with 100 percent excess air. Initial conditions are 0.1 MPa and 27°C.

13-71. An equimolar mixture of hydrogen and carbon monoxide, together with the theoretical amount of air for complete combustion, is ignited in a constant-volume bomb. The initial conditions are 2 bars and 25°C. Estimate the maximum temperature, in kelvins, and pressure, in bars, that would be attained, assuming complete combustion.

13-72. Two cubic centimeters of liquid benzene ($C_6H_6$) are placed in a 28.3-L constant-volume bomb at 25°C. If the fuel is burned with the stoichiometric amount of air and the specific gravity of the fuel is 0.879, determine (*a*) the maximum combustion temperature, in kelvins, and (*b*) the explosion pressure, in megapascals.

13-73. Two cubic centimeters of liquid methyl alcohol ($CH_3OH$) are placed in a 5.80-L constant-volume test cell at 25°C. If the fuel is burned with the stoichiometric amount of air and the specific gravity of the fuel is 0.795, determine (*a*) the maximum combustion temperature, in kelvins, and (*b*) the explosion pressure, in bars, for complete combustion.

13-74E. Determine the adiabatic-combustion temperature, in degrees Rankine, and the explosion pressure, in atmospheres, for the constant-volume combustion of CO with 100 percent excess air. Initial conditions are 1.2 atm and 80°F.

13-75E. An equimolar mixture of hydrogen and carbon monoxide, together with the theoretical amount of air for complete combustion, is ignited in a constant-volume chamber. The initial conditions are 1.1 atm and 77°F. Estimate (*a*) the maximum temperature, in degrees Rankine, and (*b*) the pressure, in psia, that would be attained, assuming complete combustion.

13-76E. One-tenth of a cubic inch of liquid benzene ($C_6H_6$) is placed in a 1.0-$ft^3$ constant-volume bomb at 77°F. If the fuel is burned with the stoichiometric amount of air and the specific gravity of the fuel is 0.879, determine (*a*) the maximum combustion temperature, in degrees Rankine, and (*b*) the explosion pressure, in psia, for complete combustion.

## ENTHALPY OF REACTION, HEATING VALUES

13-77. The enthalpy of reaction for the combustion of hydrazine ($N_2H_4$) with oxygen is found to be $-624,800$ kJ/kmol at 25°C for the specific reaction $N_2H_4(l) + O_2(g) \rightarrow N_2(g) + 2H_2O(l)$. Determine the enthalpy of formation of liquid hydrazine at 25°C, in kJ/kmol.

13-78. Consider the use of hydrogen peroxide ($H_2O_2$) and methane as the oxidant and fuel in a propulsion system, where the overall reaction is $4H_2O_2(l) + CH_4(g) \rightarrow 6H_2O(g) + CO_2(g)$. Determine the enthalpy of reaction for the above reaction at 25°C.

13-79. The enthalpy of combustion of isobutane ($i$-$C_4H_{10}$) is $-2,872,600$ kJ/kmol at 25°C. Determine at 25°C the enthalpy of formation of this isomer of *n*-butane, in kJ/kmol.

13-80. The lower heating value of gaseous butene ($C_4H_8$) at 25°C is 45,220 kJ/kg. Determine (*a*) the enthalpy of formation of butene, in kJ/kmol, and (*b*) the enthalpy of reaction, in kJ/kmol, for the reaction $C_4H_8 + H_2 \rightarrow C_4H_{10}$.

13-81E. The enthalpy of reaction for the combustion of hydrazine ($N_2H_4$) with oxygen is found to be $-268,600$ Btu/lbmol at 77°F for the specific reaction $N_2H_4(l) + O_2(g) \rightarrow N_2(g) + 2H_2O(l)$. Determine the enthalpy of formation of liquid hydrazine at 77°F, in Btu/lbmol.

13-82E. Consider the use of hydrogen peroxide ($H_2O_2$) and methane as the oxidant and fuel in a propulsion system, where the overall reaction is $4H_2O_2(l) + CH_4(g) \rightarrow 6H_2O(g) + CO_2(g)$. Determine the enthalpy of reaction for the above reaction at 77°F.

13-83E. The enthalpy of combustion of isobutane ($i$-$C_4H_{10}$) is $-1,234,850$ Btu/lbmol at 77°F. Determine at 77°F the enthalpy of formation of this isomer of $n$-butane, in Btu/lbmol.

13-84. Consider a Pennsylvania coal with the ultimate analysis given in Table 13-1.
  (*a*) Determine (1) the number of moles of $CO_2$, $N_2$, and $O_2$ present in the products of combustion per kilogram of fuel when 20 percent excess air is used, and (2) the air-fuel ratio used.
  (*b*) Neglecting the effect of $SO_2$, determine the heat transfer, in kJ/kg fuel, if the reactants enter a steady-flow combustor at 25°C and the product gases leave at 460 K.

13-85. Consider the Illinois coal with the ultimate analysis given in Table 13-1.
  (*a*) Determine (1) the number of moles of $CO_2$, $N_2$, and $O_2$ present in the products of combustion per kilogram of fuel when 20 percent excess air is used, and (2) the air-fuel ratio used.
  (*b*) Neglecting the effect of $SO_2$, determine the heat transfer, in kJ/kg fuel, if the reactants enter a steady-flow combustor at 25°C and the product gases leave at 440 K.

13-86. Consider the Wyoming coal with the ultimate analysis given in Table 13-1.
  (*a*) Determine (1) the number of moles of $CO_2$, $N_2$, and $O_2$ present in the products of combustion per kilogram of fuel when 20 percent excess air is used, and (2) the air-fuel ratio used.
  (*b*) Neglecting the effect of $SO_2$, determine the heat transfer, in kJ/kg fuel, if the reactants enter a steady-flow combustor at 25°C and the product gases leave at 480 K.

13-87. Consider the South Dakota coal with the ultimate analysis given in Table 13-1.
  (*a*) Determine (1) the number of moles of $CO_2$, $N_2$, and $O_2$ present in the products of combustion per kilogram of fuel when 20 percent excess air is used, and (2) the air-fuel ratio used.
  (*b*) Neglecting the effect of $SO_2$, determine the heat transfer, in kJ/kg fuel, if the reactants enter a steady-flow combustor at 25°C and the product gases leave at 500 K.

13-88E. Consider a Pennsylvania coal with the ultimate analysis given in Table 13-1.
  (*a*) Determine (1) the number of moles of $CO_2$, $N_2$, and $O_2$ present in the products of combustion per pound of

fuel when 20 percent excess air is used, and (2) the air-fuel ratio used.

(b) Neglecting the effect of $SO_2$, determine the heat transfer, in $Btu/lb_m$ fuel, if the reactants enter a steady-flow combustor at 77°F and the product gases leave at 340°F.

13-89E. Consider the Illinois coal with the ultimate analysis given in Table 13-1.

(a) Determine (1) the number of moles of $CO_2$, $N_2$, and $O_2$ present in the products of combustion per pound of fuel when 20 percent excess air is used, and (2) the air-fuel ratio used.

(b) Neglecting the effect of $SO_2$, determine the heat transfer, in $Btu/lb_m$ fuel, if the reactants enter a steady-flow combustor at 77°F and the product gases leave at 300°F.

13-90E. Consider the Wyoming coal with the ultimate analysis given in Table 13-1.

(a) Determine (1) the number of moles of $CO_2$, $N_2$, and $O_2$ present in the products of combustion per pound of fuel when 20 percent excess air is used, and (2) the air-fuel ratio used.

(b) Neglecting the effect of $SO_2$, determine the heat transfer, in $Btu/lb_m$ fuel, if the reactants enter a steady-flow combustor at 77°F and the product gases leave at 380°F.

## ENTROPY CHANGES FOR REACTING MIXTURES

13-91. Methane gas ($CH_4$) is burned with the stoichiometric amount of air, both initially at 25°C, in a steady-flow process. The products are at 25°C, and water appears as a liquid. The environmental temperature is 25°C, combustion is complete, and both fuel and air enter separately at 1 atm. Determine (a) the entropy change for the reaction and (b) the entropy generation for the overall process, in kJ/K per kilomole of fuel. Water appears as a liquid in the products, which are at 25°C and 1 atm.

13-92. Ethane gas ($C_2H_6$) is burned with the stoichiometric amount of air, both initially at 25°C, in a steady-flow process. The products are at 25°C, and water appears as a liquid. The environmental temperature is 25°C, combustion is complete, and both fuel and air enter separately at 1 atm. Determine (a) the entropy change for the reaction and (b) the entropy generation for the overall process, in kJ/K per kilomole of fuel. Water appears as a liquid in the products, which are at 25°C and 1 atm.

13-93. Propane gas ($C_3H_8$) is burned with the stoichiometric amount of air, both initially at 25°C, in a steady-flow process. The products are at 25°C, and water appears as a liquid. The environmental temperature is 25°C, combustion is complete, and both fuel and air enter separately at 1 atm. Determine (a) the entropy change for the reaction and (b) the entropy generation for the overall process, in kJ/K per kilomole of fuel. Water appears as a liquid in the products, which are at 25°C and 1 atm.

13-94. Liquid methyl alcohol ($CH_3OH$) is burned with the stoichiometric amount of air, both initially at 25°C, in a steady-flow process. The complete products are at 25°C, and water appears as a liquid. The environmental temperature is 25°C, and both fuel and air enter separately at 1 atm. Determine (a) the entropy change for the reaction and (b) the entropy generation for the overall process, in kJ/K per kilomole of fuel. Water appears as a liquid in the products, which are at 25°C and 1 atm.

13-95E. Methane gas ($CH_4$) is burned with the stoichiometric amount of air, both initially at 77°F, in a steady-flow process. The environmental temperature is 60°F, combustion is complete, and both fuel and air enter separately at 1 atm. Water appears as a liquid in the products, which also are at 77°F. Determine (a) the entropy change for the reaction and (b) the entropy generation for the overall process, in Btu/°R per pound-mole of fuel. Water appears as a liquid in the products, which are at 77°F and 1 atm.

13-96E. Propane gas ($C_3H_8$) is burned with the stoichiometric amount of air, both initially at 77°F, in a steady-flow process. The environmental temperature is 60°F, combustion is complete, and both fuel and air enter separately at 1 atm. Water appears as a liquid in the products, which also are at 77°F. Determine (a) the entropy change for the reaction and (b) the entropy generation for the overall process, in Btu/°R per pound-mole of fuel. Water appears as a liquid in the products, which are at 77°F and 1 atm.

13-97E. Liquid methyl alcohol ($CH_3OH$) is burned with the stoichiometric amount of air, both initially at 77°F, in a steady-flow process. The environmental temperature is 60°F, combustion is complete, and both fuel and air enter separately at 1 atm. Water appears as a liquid in the products, which also are at 77°F. Determine (a) the entropy change for the reaction and (b) the entropy generation for the overall process, in Btu/°R per pound-mole of fuel. Water appears as a liquid in the products, which are at 77°F and 1 atm.

13-98. Propane ($C_3H_8$) reacts adiabatically in steady flow with 20 percent excess air. When both are initially at 25°C, the adiabatic combustion temperature is 2125 K. Evaluate the entropy generation for the reaction, in kJ/K per kilomole of fuel, if the fuel and air initially are separate and each is at 1 atm.

13-99. Ethane ($C_2H_6$) reacts adiabatically in steady flow with 30 percent excess air, both initially at 25°C. The adiabatic combustion temperature is 2005 K. Evaluate the entropy generation for the reaction, in kJ/K per kilomole of fuel, if the fuel and air initially are separate and each is at 1 atm.

13-100. Carbon monoxide undergoes adiabatic combustion with 50 percent excess air in steady flow, both reactants initially at 25°C. The final flame temperature is 2107 K. Evaluate the entropy generation for the reaction, in kJ/K per kilomole of fuel, if the fuel and air initially are separate and each is at 1 atm.

13-101E. Propane ($C_3H_8$) reacts adiabatically in steady flow with 20 percent excess air. When both are initially at 77°F, the adiabatic combustion temperature is 3820°R. Evaluate the entropy generation for the reaction, in Btu/°R per pound-mole of fuel, if the fuel and air initially are separate and each is at 1 atm.

13-102E. Carbon monoxide undergoes adiabatic combustion with stoichiometric air in steady flow, both reactants initially at 77°F. The final adiabatic temperature is 4800°R. Evaluate the entropy generation for the reaction, in Btu/°R per pound-mole of fuel, if the fuel and air initially are separate and each is at 1 atm.

13-103. Butane gas ($C_4H_{10}$) at 25°C and 130 percent excess air at 500 K enter a steady-flow combustor. The complete products of combustion leave at 1600 K, and a heat loss of 944,000 kJ/kmol of fuel occurs during combustion to the environment at 25°C. Determine (*a*) the entropy change for the reaction and (*b*) the entropy generation for the overall process, in kJ/K per kilomole of fuel.

13-104. Hydrogen gas is burned with 54 percent excess air with each initially at 400 K. The complete products of combustion in the steady-flow process leave at 800 K. During the process, 190,000 kJ of heat per kilomole of $H_2$ is transferred to the environment at 25°C. Determine (*a*) the entropy change for the reaction and (*b*) the entropy generation for the overall process, in kJ/K per kilomole of fuel.

## AVAILABILITY AND IRREVERSIBILITY OF REACTING MIXTURES

13-105. Propane gas ($C_3H_8$) is burned in steady flow at 1 atm with 40 percent excess air, both fuel and air entering at 25°C. Assume the reaction is complete and that the products leave at the environmental temperature of 25°C. The fuel and the oxidant each enter separately at 1 atm, and the water is in the liquid state in the products. Determine (*a*) the heat released by the reaction, (*b*) the reversible work (or availability change) for the process, and (*c*) the irreversibility of the reaction, all in kJ/kmol of fuel.

13-106. Reconsider Prob. 13-105. The steady-flow combustion process is now carried out adiabatically. For 40 percent excess air initially at the environmental temperature of 25°C, the exit temperature is 1920 K. Determine (*a*) the availability of the hot product gases, (*b*) the availability change (reversible work) for the adiabatic process, (*c*) the irreversibility for the process, all in kJ/kmol, and (*d*) the ratio of the availability of the hot gases to that of the original reactants.

13-107. Ethane gas ($C_2H_6$) reacts with 50 percent excess air in a steady-flow process, both fuel and air entering at 25°C. The reaction is complete, and the products leave as a mixture at 1 atm at the environmental temperature of 25°C. The fuel and air enter separately at 1 atm, and water in the products is in the liquid state. Determine (a) the heat transfer, (b) the reversible work (or availability change), and (c) the irreversibility of the process, all in kJ/kmol of fuel.

13-108. Reconsider Prob. 13-107. The steady-flow combustion process is now carried out adiabatically. The fuel and 50 percent excess air still enter separately at 1 atm at the environmental temperature of 25°C, and the products are a mixture at 1 atm and 1825 K. Determine (a) the availability of the hot product gases, (b) the availability change (or the reversible work) for the adiabatic process, (c) the irreversibility for the process, all three answers in kJ/kmol of fuel, and (d) the ratio of the availability of the product gases to that of the original reactants.

13-109. Carbon monoxide (CO) reacts with the stoichiometric quantity of air in a steady-flow process, both fuel and air entering at 25°C. The reaction is complete, and the products leave as a mixture at 1 atm at an environmental temperature of 25°C. The carbon monoxide and air enter separately at 1 atm. Determine (a) the heat transfer, (b) the reversible work (or availability change), and (c) the irreversibility of the process, all in kJ/kmol of fuel.

13-110. Reconsider Prob. 13-109. The steady-flow combustion process is now carried out adiabatically. The carbon monoxide and stiochiometric air still enter separately at 1 atm at the environmental temperature of 25°C, and the products are a mixture at 1 atm and 2665 K. Determine (a) the availability of the hot product gases, (b) the availability change (or the reversible work) for the adiabatic process, (c) the irreversibility for the process, all three answers in kJ/kmol of fuel, and (d) the ratio of the availability of the hot product gases to that of the original reactants.

13-111. Liquid methyl alcohol ($CH_3OH$) reacts with the stoichiometric quantity of air in a steady-flow process, both fuel and air entering at 25°C. The reaction is complete, and the products leave as a mixture at 1 atm and the environmental temperature of 25°C. The fuel and air enter separately at 1 atm, and the water produced is liquid. Determine (a) the heat transfer, (b) the reversible work (or availability change), and (c) the irreversibility of the process, all in kJ/kmol of fuel.

13-112. Reconsider Prob. 13-111. The steady-flow combustion process is now carried out adiabatically. The methyl alcohol and stoichiometric air still enter separately at 1 atm at the environmental temperature of 25°C, and the products are a mixture at 1 atm and 2230 K. Determine (a) the availability of the hot product gases, (b) the availability change (or the reversible work) for the adiabatic process, (c) the irreversibility for the process, all three answers in kJ/kmol of fuel, and (*d*) the ratio of the availability of the hot product gases to that of the original reactants.

13-113E. Propane gas ($C_3H_8$) is burned in steady flow at 1 atm with 20 percent excess air, both entering at 77°F. Assume the reaction is complete and that the products leave at the environmental temperature of 77°F. The fuel and the oxidant each enter separately at 1 atm, and the water is in the liquid state in the products. Determine (*a*) the heat transfer, (*b*) the reversible work (or availability change) for the process, and (*c*) the irreversibility of the reaction, all in Btu/lbmol of fuel.

13-114E. Reconsider Prob. 13-113. The steady-flow combustion process is now carried out adiabatically at 1 atm. For propane and 20 percent excess air initially at 77°F, the adiabatic combustion temperature is 3820°R. Determine (*a*) the availability of the hot product gases, (*b*) the availability change (or the reversible work) for the adiabatic process, (*c*) the irreversibility for the process, all in Btu/lbmol, and (*d*) the ratio of the availability of the hot gases to that of the original reactants.

13-115E. Ethane gas ($C_2H_6$) reacts with 50 percent excess air in a steady-flow process at 1 atm, both fuel and air entering at 77°F. The reaction is complete, and the products leave as a mixture at the environmental temperature of 77°F. The fuel and air enter separately at 1 atm, and water in the products is in the liquid state. Determine (*a*) the heat transfer, (*b*) the reversible work (or availability change), and (*c*) the irreversibility of the process, all in Btu/lbmol of fuel.

13-116E. Reconsider Prob. 13-115. The steady-flow combustion process is now carried out adiabatically. The ethane and 50 percent excess air still enter separately at 1 atm and 77°F, and the products are a mixture at 1 atm and 3280°R. Determine (*a*) the availability of the hot product gases, (*b*) the availability change (or the reversible work) for the adiabatic process, (*c*) the irreversibility for the process, all three answers in Btu/lbmol of fuel, and (*d*) the ratio of the availability of the product gases to that of the original reactants.

13-117E. Carbon monoxide (CO) reacts with the stoichiometric quantity of air in a steady-flow process at 1 atm, both fuel and air entering at 77°F. The reaction is complete, and the products leave as a mixture at the environmental temperature of 77°F. The carbon monoxide and air enter separately at 1 atm. Determine (*a*) the heat transfer, (*b*) the reversible work (or availability change), and (*c*) the irreversibility of the process, all in Btu/lbmol of fuel.

13-118E. Reconsider Prob. 13-117. The steady-flow combustion process is now carried out adiabatically. The carbon monoxide and stoichiometric air still enter separately at 1 atm and the environmental temperature of 77°F, and the products are a mixture at 1 atm and 4800°R. Determine (*a*) the availability of the hot product gases, (*b*) the availability change (the reversible work) for the adiabatic process, (*c*) the irreversibility for the process, all three answers in Btu/lbmol of fuel, and (*d*) the ratio of the availability of the hot product gases to that of the original reactants.

## FUEL CELLS

13-119. A fuel cell operates on an equimolar mixture of CO and $H_2$ as the fuel. The fuel and the stoichiometric amount of oxygen enter at the anode and cathode, respectively, at the environmental conditions of 25°C and 1 atm. Assume that all the water formed at the cathode leaves as a liquid and that the $CO_2$ formed at the anode leaves at 25°C and 1 atm as a pure gas. Determine (*a*) the maximum work output for complete reaction in a steady-flow process, in kJ/kmol of mixture, (*b*) the ideal efficiency of the cell, and (*c*) the ideal voltage.

13-120. Methyl alcohol and stoichiometric oxygen in the gas phase enter a fuel cell each at 1 atm and 25°C, and they form complete products also at 25°C. All the water formed at the cathode leaves at 1 atm as a liquid, and the $CO_2$ formed at the anode leaves at 1 atm as a pure gas. Determine (*a*) the maximum work output for the steady-flow process, in kJ/kmol of fuel, (*b*) the ideal cell efficiency, and (*c*) the ideal voltage.

13-121. Propane ($C_3H_8$) and stoichiometric oxygen in the gas phase enter a fuel cell each at 1 atm and 25°C, and they form complete products also at 25°C. All the water formed at the cathode leaves at 1 atm as a liquid, and the $CO_2$ formed at the anode leaves at 1 atm as a pure gas. Determine (*a*) the maximum work output for the steady-flow process, in kJ/kmol of fuel, (*b*) the ideal cell efficiency, and (*c*) the ideal voltage.

13-122. Ethylene ($C_2H_4$) and stoichiometric oxygen in the gas phase enter a fuel cell each at 1 atm and 25°C, and they form complete products also at 25°C. All the water formed at the cathode leaves at 1 atm as a liquid, and the $CO_2$ formed at the anode leaves at 1 atm as a pure gas. Determine (*a*) the maximum work output for the steady-flow process, in kJ/kmol of fuel, (*b*) the ideal cell efficiency, and (*c*) the ideal voltage.

13-123. Ethyl alcohol ($C_2H_5OH$) and stoichiometric oxygen in the gas phase enter a fuel cell each at 1 atm and 25°C, and they form complete products also at 25°C. All the water formed at the cathode leaves at 1 atm as a liquid, and the $CO_2$ formed at the anode leaves at 1 atm as a pure gas. Determine (*a*) the maximum work output for the steady-flow process, in kJ/kmol of fuel, (*b*) the ideal cell efficiency, and (*c*) the ideal voltage.

13-124. Hydrogen ($H_2$) and stoichiometric oxygen in the gas phase each enter a fuel cell at 1 atm. The reactants enter and the products leave at (*a*) 400 K, (*b*) 600 K, and (*c*) 800 K. Note that water is now a gaseous product at 1 atm. Determine (1) the maximum work output for the steady-flow process, in kJ/kmol of fuel, (2) the ideal cell efficiency, and (3) the ideal voltage.

## PARAMETRIC STUDIES AND DESIGN

13-125. Propane gas is reacted in steady flow with air starting at 25°C. By means of a computer analysis investigate the effect of excess air on the maximum combustion temperature. The excess air ranges from 0 to 500 percent. Plot the temperature versus percent excess air.

13-126. Carbon monoxide undergoes adiabatic combustion with excess air in steady flow. The initial reactants are at 25°C. By means of a computer analysis investigate the effect of excess air on the maximum combustion temperature. The excess air ranges from 0 to 500 percent. Plot the temperature versus percent excess air.

13-127. Methane gas initially at 25°C reacts in steady flow with 20 percent excess air. By means of a computer analysis investigate the effect of preheating of the air on the adiabatic combustion temperature. The initial temperature of the air ranges from 300 to 700 K. Plot the adiabatic flame temperature versus the initial air temperature.

13-128. Ethylene gas initially at 25°C is burned with excess air which enters the steady-flow combustor at 400 K. On the basis of a computer analysis determine the excess air used as a function of the temperature of the products of combustion under adiabatic conditions. The final temperature ranges from 1200 to 1800 K. Plot the percent excess air used versus the adiabatic combustion temperature.

13-129. Ethane gas at 25°C and air at 227°C enter a combustion chamber in steady flow. The complete products leave at 1100 K, and the fuel rate is 1 kg/min. By means of a computer analysis determine the effect of excess air on the rate of heat loss, in kJ/min. The range of excess air used is from 0 to 200 percent. Plot the rate of heat loss versus the percent excess air supplied.

13-130. A new combustor must be designed for metallurgical heat treatment in a small foundry. EPA source regulations require that the combustor meet emission indices for CO and NO of 8.6 and 1.0, respectively. The emission indices for carbon monoxide (COEI) and nitric oxide (NOEI) are both defined as grams of pollutant per kilogram of fuel. An atmospheric combustor is recommended, with a well-insulated first stage used for primary combustion, followed by either heat treatment or secondary combustion. The fuel is gaseous propane ($C_3H_8$) which enters the combustor at 298 K. The inlet-air temperature also is 298 K. Several designs are to be considered.

  1. Investigate the feasibility of simply using a single-stage combustor followed by heat treatment without secondary

combustion for an equivalence-ratio range of $0.5 \leq \phi \leq 1.5$. Provide COEI and NOEI curves versus $\phi$ on a single plot. What do the curves tell you about the feasibility of simultaneously meeting the emission standards for NO and CO in a single-stage combustor?

Based on the above analysis, a two-stage combustor must be investigated in an attempt to meet EPA regulations. In this case a primary region operates fuel-rich to minimize NO formation, while a second region operates fuel-lean at a lower temperature to complete CO oxidation while simultaneously controlling NO emissions. Two alternative designs are to be considered for the second stage.

2. First, completion of combustion in the second stage occurs at 5 percent overall excess air, with an exit temperature equal to 70 percent of the primary-zone exhaust temperature. This strategy utilizes heat recovery to reduce NOEI but delays heat treatment until after completion of secondary combustion.

3. Second, air addition and heat treatment reduce the temperature at the inlet of the second stage to 20 percent of the primary zone exhaust temperature. Secondary combustion then occurs adiabatically with an overall stoichiometry corresponding to 100 percent excess air. In this case, heat treatment occurs between combustor stages and NOEI is reduced by diluent addition.

For the latter two designs the primary-zone equivalence ratio $\phi_p$ occurs in the range $1.6 \leq \phi \leq 2.0$ so as to avoid both NO and soot emissions in the primary combustor. Since the exhaust CO level is less than 1 ppm when using two-stage combustion, the design procedure is completed by evaluating and plotting NOEI as a function of $\phi_p$ for each design option. Which of the latter two designs is preferable from a combined energy and environmental point of view? List limitations and advantages of each design.

13-131. High-temperature heat treatment of materials requires gas temperatures in the range of 1300 to 2100 K. To reduce fuel costs the process of recuperation is employed. A recuperator is essentially a heat exchanger that recovers energy from the exhaust gases leaving an industrial furnace and uses this energy to preheat the combustion air entering the furnace. In essence, the heat-treatment process is modeled as an adiabatic furnace followed by heat treatment and recuperation. Currently, material properties limit air-preheat temperatures to 800 K. However, new technology has the potential to raise preheat temperatures to 1600 K. The cost savings of this new recuperator technology is to be investigated for preheat temperatures of 1200 and 1600 K, compared to a baseline preheat temperature of 800 K for current technology.

The fuels of interest are gaseous methane, gaseous propane, and liquid octane. Previous research indicates that the most direct procedure for estimating fuel saved by heat recovery is the "temperature-matching" method. In this method, the adiabatic flame temperature is first calculated at a chosen baseline condition. The reduction in fuel use at higher air preheat temperatures is then determined by maintaining the same adiabatic flame temperature as for the baseline case. The following design analysis is required.

1. As an initial step, determine the adiabatic flame temperatures with a preheat air temperaure of 800 K and without preheat (an oxidizer/diluent temperature of 298 K). Use excess air of 5, 25, and 50 percent. Compare the various temperatures and note any trends. The presence of equilibrium species CO and NO should be taken into consideration.

2. Again taking into account equilibrium species, determine the percent fuel saved with respect to the baseline air preheat temperature of 800 K at air preheat temperatures of 1200 and 1600 K and for excess air values of 3, 25, and 50 percent. Discuss the trends.

3. The energy analyses above are incomplete without environmental analyses. The emission index is defined as the grams of pollutant emitted per kilogram of fuel burned in the combustor. Calculate the emission indices for carbon monoxide (COEI) and nitric oxide (NOEI) at the same conditions used to evaluate fuel savings from recuperation. Discuss the relevant trends.

13-132. Incineration is one method used for the reduction of waste material normally sent to landfills. Some categories of waste, such as food waste, may require supplemental energy to achieve complete combustion. An existing food waste incinerator is not performing up to specifications, mainly owing to the low combustion temperature. A supplementary energy source is necessary to increase the temperature to 1200 K. You have been contacted to analyze the performance of the incinerator and propose a solution. The following information is already available during previous tests. The waste-food composition, on a wet basis, is $C:H:O:N:S:NC = 45:6.5:29:3:0.5:16$, where NC represents noncombustibles. The noncombustibles consist of 90 percent water and 10 percent ash. The higher heating value is 5510 kJ per kilogram of wet waste. The atmospheric-pressure incinerator receives waste and air at 25°C and operates at stoichiometric conditions. Heat loss from the combustion section is known to be 5 percent of the enthalpy of reaction, and the incinerator must handle one ton of waste per hour.

You are asked to investigate three possible energy supplements: (1) injection of natural gas, (2) injection of No. 2 fuel oil, and (3) enrichment of the incoming air with oxygen. Relevant properties of these supplements are listed below:

(a) Properties of natural gas:
   Volume composition—$CH_4:C_2H_6:N_2 = 83.4:15.8:0.8$
   Price—$4.82 per 1000 cubic feet

(b) Properties of No. 2 fuel oil:
   Volume composition—C:H:S = 87.3:12.5:0.2; HHV = 44,150 kJ/kg
   Price—$0.66 per gallon; specific gravity—0.865
(c) Properties of oxygen:
   Volume composition—O:N = 0.996:0.004; price—$7.50 per 1000 cubic feet

There are two key aspects of this analysis. First, you must propose a specific supplemental fuel system. Second, you must specify the criteria used when choosing the supplement.

13-133. The adiabatic flame temperature for a given fuel depends on the amount of air supplied to the reaction. For (a) methane, (b) butane, and (c) acetylene in the gas phase, and (d) methyl alcohol and (e) octane in the liquid phase initially at 1 bar and 298 K, determine the minimum percent excess air required so that the adiabatic combustion temperature will not exceed (1) 1100 K, (2) 1400 K, (3) 1700 K, and (4) 2000 K. Plot the percent excess air versus temperature for all five fuels. Discuss the trends versus the molecular structure and phase of the fuels.

13-134. A natural gas-fired furnace is to produce a maximum of 100,000 kJ/h for a residential-heating application. The furnace consists of a burner, where a mixture of natural gas and air burn, and a counterflow heat exchanger where the combustion products transfer energy to air that circulates through the house. A mixture of $CO_2$, $H_2O$, $O_2$, and $N_2$, in a $1:(2 + 30\omega):2:15$ volumetric ratio exits the furnace at 1530 K and 1 atm, where $\omega$ is the specific humidity. The combustion gases exit the heat exchanger at 25°C; house air enters the heat exchanger at 15°C and exits at 35°C. The atmospheric air enters the furnace at 10°C with a relative humidity of 80 percent.

The choices for a fuel are natural gas and liquid propane. You are to recommend one of these based on furnace efficiency, operating cost, and other issues you deem important. Provide arguments why you chose the issues you did and why you assigned the rankings you chose when deciding upon your design.

# CHEMICAL EQUILIBRIUM

Ballard fuel cell engine installed in a transit bus. The Ballard fuel cell combines hydrogen and air to produce thermal energy, water, and electricity. (Courtesy of Ballard Power Systems, Inc.)

In the preceding chapter, the basic concepts of the conservation of mass, the conservation of energy, and the entropy balance were applied to systems in which a chemical reaction occurs. However, in all cases the composition of the reaction products was specified by either assuming that the reaction went to completion or that an independent measurement of the composition of the products was available. Either of these restrictions can be removed by applying the second law in the form of the entropy balance and considering the limitations imposed on the products by the entropy production in the reaction. In this chapter, a more general method for evaluating composition changes due to chemical reactions will be introduced, and its subsequent effect on energy analysis will be discussed.

## 14-1    INTRODUCTION

The change in entropy for the adiabatic gas-phase reaction $CO + H_2O \rightarrow CO_2 + H_2$ is evaluated in Example 13-16. For an initial temperature of 400 K and a final temperature of 920 K, the entropy change for the stoichiometric process to complete products is $\bar{s}_{prod} - \bar{s}_{reac} = 23.54$ kJ/K. This result appears to be in accord with the increase in entropy principle for an adiabatic system. However, the principle is expressed correctly by Eq. [13-27b], namely, $\bar{s}_{prod} - \bar{s}_{reac} \geq 0$ for an adiabatic process. That is, the value of the entropy must increase *continuously* for all spontaneous, or irreversible, processes. In view of this requirement, the system will reach an equilibrium state whenever the entropy reaches a *maximum* value. If the entropy attains a maximum value at some concentration of the four species other than that for completion, the reaction will stop at that point. Whatever composition of reactants and products leads to the highest total entropy value is the composition at the equilibrium state. Although the results in Example 13-16 are numerically correct, the reaction in actuality stops far short of completion.

To determine the equilibrium composition of a reacting mixture of CO and $H_2O$ in a 1:1 molar ratio initially at 400 K (260°F) in an adiabatic process at 1 atm, we need to evaluate the total *absolute* entropy of the system for different degrees of completion. Arbitrarily, we shall evaluate the total absolute entropy of the gases mixture under those conditions for which 0, 0.2, 0.4, 0.6, 0.7, 0.8, 0.9, and 1.0 kmol of $CO_2$ is formed under adiabatic conditions. Only one actual calculation is shown, since the same method is used throughout. For example, if 0.8 kmol of $CO_2$ is formed, the chemical reaction proceeds accordingly as

$$CO(g) + H_2O(g) \rightarrow 0.8CO_2(g) + 0.8H_2(g) + 0.2CO(g) + 0.2H_2O(g)$$

As in Example 13-16, we must employ the conservation of energy principle to determine the final temperature that would be reached for the above reaction. By equating the enthalpy of the products with that of the reactants, it is found that the final temperature of the four product gases is 834 K, or roughly 1040°F. This is approximately 30°C, or 60°F, less than for the reaction to completion. This is to be expected, because the total enthalpy of reaction is not released.

The absolute entropy of each gas is now found from

$$S_i = N_i(\bar{s}_{i,T}^{\circ} - R_u \ln p_i)$$

where $p_i$ is the partial pressure measured in atmospheres. Recall that $\bar{s}_{i,T}^{\circ}$ is the value of the absolute entropy at one atmosphere and the desired temperature. The values of $\bar{s}_{i,T}^{\circ}$ used in the calculations below are based on data from the ideal-gas tables in the Appendix. The following calculations are in terms of SI data, with the USCS results shown in parentheses at the right of each line. Hence

$$S_{CO_2} = 0.8(259.560 - 8.314 \ln 0.4) = 213.743 \text{ kJ/K} \qquad (51.035 \text{ Btu/}°\text{R})$$

$$S_{H_2} = 0.8(160.673 - 8.314 \ln 0.4) = 134.634 \text{ kJ/K} \qquad (32.154 \text{ Btu/}°\text{R})$$

$$S_{CO} = 0.2(228.493 - 8.314 \ln 0.1) = 49.528 \text{ kJ/K} \qquad (11.846 \text{ Btu/}°\text{R})$$

$$S_{H_2O} = 0.2(225.311 - 8.314 \ln 0.1) = 48.891 \text{ kJ/K} \qquad (11.676 \text{ Btu/}°\text{R})$$

$$S_{tot} = 446.796 \text{ kJ/K} \qquad (106.711 \text{ Btu°R})$$

Thus the total entropy, when the reaction goes to 80 percent completion, is 446.796 kJ/K, or 106.711 Btu/°R, per mole of initial carbon monoxide. The total entropy for the system for no reaction and for complete (comp) reaction can be found from SI data used in Example 13-16. These values are:

$$S_{init} = 206.125 + 198.673 - 2(8.314)(\ln 0.5) = 416.325 \text{ kJ/K}$$

$$S_{comp} = 264.728 + 163.607 - 2(8.314)(\ln 0.5) = 439.862 \text{ kJ/K}$$

For the same calculation in USCS units

$$S_{init} = 49.317 + 47.450 - 2(1.986)(\ln 0.5) = 99.520 \text{ Btu/}°\text{R}$$

$$S_{comp} = 63.250 + 39.090 - 2(1.986)(\ln 0.5) = 105.093 \text{ Btu/}°\text{R}$$

For these data, in either SI or USCS units, the entropy for 80 percent completion is greater than that for 100 percent completion, and both values are greater than the initial value. This indicates that the entropy function is a maximum before the reaction reaches completion. To ascertain the actual state of equilibrium, one needs to carry out similar calculations at other degrees of completion. A summary of these computations appears in Table 14-1. The system temperature of each degree of completion is also indicated. The data from this table are plotted in Fig. 14-1. It is found that the state of maximum entropy occurs when approximately 70 percent of the CO and $H_2O$ has reacted. This, then, is the state of equilibrium.

**Table 14-1**

| Moles CO$_2$ Formed | SI | | USCS | |
|---|---|---|---|---|
| | $S_{tot}$, kJ/K | $T_{final}$, K | $S_{tot}$, Btu/°R | $T_{final}$, °R |
| 0 | 416.325 | 400 | 99.520 | 720 |
| 0.2 | 433.870 | 520 | 103.759 | 940 |
| 0.4 | 442.525 | 630 | 105.828 | 1140 |
| 0.6 | 447.218 | 740 | 106.815 | 1330 |
| 0.7 | 447.752 | 790 | 106.935 | 1420 |
| 0.8 | 446.796 | 834 | 106.711 | 1500 |
| 1.0 | 439.862 | 920 | 105.093 | 1660 |

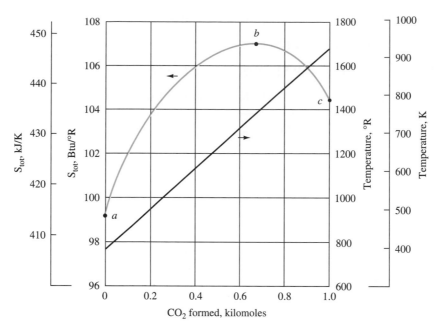

**Figure 14-1**   The total entropy and the combustion temperature for the adiabatic reaction of an equimolar mixture of CO and $H_2O$ initially at 400 K (720°R) as a function of the extent of the reaction, for the reaction $CO(g) + H_2O(g) \rightleftharpoons CO_2(g) + H_2(g)$.

The foregoing discussion illustrates the fact that a second-law analysis is frequently as important as a first-law analysis in dealing with reactive systems. In the absence of second-law principles, it is possible to arrive at incorrect conclusions. Although the entropy balance can be used for the analysis of reactive systems under adiabatic conditions, the determination of the extent of a reaction for a given temperature and pressure is of major engineering interest. Our first goal in this chapter is to establish a suitable criterion for determining the equilibrium state of a reactive mixture at a specified temperature and pressure. Although we restrict ourselves to ideal-gas mixtures, the major points presented are of general usefulness.

## 14-2   THE GIBBS CRITERION

Consider a closed system that contains an arbitrary number of components and phases. The temperature and pressure are uniform throughout the system, and the system initially is in nonequilibrium. The conservation of energy for a simple compressible system with compression/expansion work is

$$dU = \delta Q + \delta W = \delta Q - P\,dV$$

or solving for the heat transfer

$$\delta Q = dU + P\,dV \qquad \text{[a]}$$

In addition, an entropy balance for a closed system in the form of Eq. [6-34] states that

$$dS = \frac{\delta Q}{T} + \delta\sigma$$

or, upon rearrangement,

$$\delta Q = T\,dS - T\,\delta\sigma \qquad \text{[b]}$$

Substitution of Eq. [b] into Eq. [a] leads to

$$T\,dS - (dU + P\,dV) = T\,\delta\sigma \geq 0 \qquad \text{[14-1]}$$

where $\delta\sigma \geq 0$ is always valid. The variables—$T$, $S$, $U$, $P$, and $V$—in Eq. [14-1] are all properties solely of the system. Hence it applies to *any* change in state of a closed system of uniform pressure and temperature. The inequality applies to an incremental change in the closed system between nonequilibrium states. Of major importance, the direction of the inequality dictates the direction of change that leads toward equilibrium. The equality applies to changes between equilibrium states, that is, to reversible processes.

Equation [14-1] is not particularly useful in this general format. However, restricted forms of the equation are of great utility. For example, consider any closed system held at constant volume and internal energy. Under this condition, Eq. [14-1] dictates that

$$dS_{U,V} \geq 0 \qquad \text{[14-2]}$$

This result is simply another application of the entropy balance, namely, $dS_{U,V} = \delta\sigma \geq 0$. The above equation may also be written in the notation

$$(dS)_{U,V} > 0 \qquad \text{spontaneous process} \qquad \text{[14-3a]}$$

$$(dS)_{U,V} = 0 \qquad \text{reversible process} \qquad \text{[14-3b]}$$

$$(dS)_{U,V} < 0 \qquad \text{unnatural process} \qquad \text{[14-3c]}$$

Equations [14-2] and [14-3] provide no new information. A modification of Eq. [14-1] does, however, lead to a practical equilibrium criterion in terms of temperature and pressure.

Under the conditions that temperature and pressure remain constant during a process, Eq. [14-1] can be written as

$$dU_{T,P} + d(PV)_{T,P} - d(TS)_{T,P} \leq 0$$

or $d(U + PV - TS)_{T,P} \leq 0$. But the Gibbs function $G \equiv U + PV - TS$. Thus

$$dG_{T,P} \leq 0 \qquad \text{[14-4]}$$

This equation states that the Gibbs function always decreases for a spontaneous (irreversible) change of a closed system under isothermal, isobaric

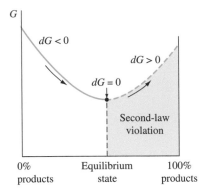

**Figure 14-2**

Illustration of the minimization of the Gibbs function as an equilibrium criterion.

conditions. That is, only $dG_{T,P} < 0$ is permissible for spontaneous processes. As a process approaches equilibrium, the Gibbs function attains a minimum value, and in the limiting case of equilibrium, $dG_{T,P} = 0$. These mathematical results are illustrated in Fig. 14-2. Equation [14-5] may also be expressed in the format

| | | |
|---|---|---|
| $dG_{T,P} < 0$ | spontaneous process | **[14-5a]** |
| $dG_{T,P} = 0$ | reversible process | **[14-5b]** |
| $dG_{T,P} > 0$ | unnatural process | **[14-5c]** |

This set of equations would be directly applicable, for example, to a chemically reactive system at a given pressure and temperature.

A graphical interpretation of Eq. [14-4] may be obtained in a manner similar to that used in Sec. 14-1 for the entropy function. Consider the gas-phase reaction $CO + H_2O \rightarrow CO_2 + H_2$. The temperature is chosen to be 1000 K, and the pressure is low enough so that the gases are essentially ideal. On the basis of 1 kmol each of the initial reactants, the reaction might proceed until 1 kmol each of $CO_2$ and $H_2$ is formed. As CO and $H_2O$ are consumed, the composition of the system continually changes. Consequently, the Gibbs function for the total system also changes. Employing the concepts of total-enthalpy and absolute-entropy values for each component introduced in Chap. 13, one may calculate the value of $G_{tot}$ at the given temperature and pressure for various compositions of the system. These compositions depend, of course, on the extent of the reaction. The values of $G_{tot}$ (in kJ/kmol) for various values of the number of moles of $CO_2$ formed are summarized for a temperature of 1000 K:

| $CO_2$, kmol | 0 | 0.2 | 0.4 | 0.5 | 0.6 | 0.7 | 0.8 | 1.0 |
|---|---|---|---|---|---|---|---|---|
| $G_{tot}$, kJ/kmol | −783.2 | −792.2 | −795.6 | −796.3 | −796.3 | −795.5 | −793.8 | −786.3 |

These data are plotted in Fig. 14-3. When an equimolar mixture of CO and $H_2O$ is allowed to react, the Gibbs function of the system decreases until approximately 0.55 mol of $CO_2$ is formed, for a temperature of 1000 K. A further change in composition represented by the process from state $b$ to state $c$ is impossible, since this requires that the Gibbs function increase for a process at constant temperature and pressure. Although state $c$ has a lower value of $G$ than state $a$, it is not the minimum value based on the composition of the original system. Consequently, only state $b$ will eventually prevail. Also if an equimolar mixture of $CO_2$ and $H_2$ is heated to 1000 K, it will react, and the quantity of $CO_2$ from state $c$ to state $b$ is also possible, since this is not a violation of the criterion established by Eq. [14-5]. It must be kept in mind that the application of the Gibbs function as a criterion for equilibrium is equivalent to applying the entropy balance for these

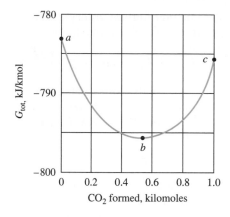

**Figure 14-3**
The total Gibbs function for the reaction of an equimolar mixture of CO and $H_2O$ at 1000 K as a function of the extent of the reaction, for the reaction $CO(g) + H_2O(g) \rightleftharpoons CO_2(g) + H_2(g)$.

conditions. The use of the Gibbs function, however, has the advantage that easily controlled properties such as the temperature and the pressure are involved in its application to reactive systems.

## 14-3 EQUILIBRIUM AND THE CHEMICAL POTENTIAL

The equilibrium criterion $dG_{T,P} = 0$ will now be used to find the equilibrium composition of a reacting mixture. On the basis of the state postulate for simple compressible substances, any extensive property is a function of two independent properties, such as $T$ and $P$, and the number of moles $N_i$ of each of the substances present at a given time. This is expressed mathematically by

$$G = G(T, P, N_1, N_2, \ldots, N_i) \qquad \textbf{[14-6]}$$

where the quantities $N_i$ represent the number of moles of each chemical species within the system. The overall change in $G$ at constant temperature and pressure is given by the total differential of $G$, that is,

$$dG_{T,P} = \left(\frac{\partial G}{\partial N_1}\right)_{P,T,N_j} dN_1 + \left(\frac{\partial G}{\partial N_2}\right)_{P,T,N_j} dN_2 + \cdots + \left(\frac{\partial G}{\partial N_i}\right)_{P,T,N_j} dN_i$$

$$= \sum_i \left(\frac{\partial G}{\partial N_i}\right)_{P,T,N_j} dN_i \qquad \textbf{[14-7]}$$

In this equation the subscript $N_j$ on the partial derivative indicates that the numbers of moles of every component are held constant except one, along with fixed values of $P$ and $T$.

The partial derivative that appears in the summation in Eq. [14-7] is defined as the **chemical potential** $\mu_i$. That is,

$$\mu_i \equiv \left(\frac{\partial G}{\partial N_i}\right)_{T,P,N_j} \qquad \textbf{[14-8]}$$

On the basis of Eqs. [14-4] and [14-8], the equilibrium criterion and the definition of $\mu_i$, Eq. [14-7] can now be written as

$$dG_{T,P} = \sum_i \mu_i \, dN_i = 0 \qquad \textbf{[14-9]}$$

This relationship is the key to determining the equilibrium composition.

Equation [14-9] can be simplified by noting that the $dN_i$ terms are related to the stoichiometric coefficients of a chemical reaction. Consider a generalized chemical reaction represented by

$$\nu_A A + \nu_B B \rightleftarrows \nu_E E + \nu_F F \qquad \textbf{[14-10]}$$

The $\nu$ symbols represent the stoichiometric coefficients for the balanced chemical reaction equation, and the uppercase letters stand for the chemical species involved in the reaction. Although the reaction we have chosen has two reactants and two products, it must be realized that the format of the resulting equations will be valid for any number of reactants and products. For the reaction represented by Eq. [14-10], Eq. [14-9] is written as

$$\sum_i \mu_i \, dN_i = \mu_A \, dN_A + \mu_B \, dN_B + \mu_E \, dN_E + \mu_F \, dN_F = 0 \quad \textbf{[14-11]}$$

The $dN_i$ terms may be positive or negative, depending on whether they represent reactants or products. In addition, these terms are not independent of each other, because the change in the numbers of moles of reactants and products is always in proportion to the stoichiometric coefficients. Thus we may write for the forward direction of the reaction:

$$\frac{-dN_A}{\nu_A} = \frac{-dN_B}{\nu_B} = \frac{dN_E}{\nu_E} = \frac{dN_F}{\nu_F} = d\varepsilon \qquad \textbf{[14-12]}$$

where $\varepsilon$ is known as the *extent of reaction* (or "degree of advancement"). The negative signs are necessary for the two reactant terms since the stoichiometric coefficients such as $\nu_A$ and $\nu_B$ are considered positive. Thus, in general, $dN_i = \pm \nu_i \, d\varepsilon$. Substitution of the relations contained within Eq. [14-12] into Eq. [14-11] yields the criterion of equilibrium for the reaction expressed by Eq. [14-10], namely,

$$\nu_E \mu_E + \nu_F \mu_F - \nu_A \mu_A - \nu_B \mu_B = 0 \qquad \textbf{[14-13]}$$

The term $d\varepsilon$ has dropped out since it appears in each term of the equation as a multiplying factor. When Eq. [14-13] is satisfied, the Gibbs function of the system at a given temperature and pressure will be a minimum, in line with the general criterion set up earlier.

A more general statement of Eq. [14-13] is

$$\sum_{\text{prod}} \nu_i \mu_i - \sum_{\text{reac}} \nu_i \mu_i = 0 \qquad \textbf{[14-14]}$$

where the summations are taken over all products and all reactants in the overall reaction. Equation [14-14] is known as the **equation of reaction**

**equilibrium,** and it relates intensive properties of the reactants and products. The equation is valid for any chemical reaction, regardless of the phases of the reacting species. To establish the equilibrium composition of a reacting mixture, we must derive expressions for the chemical potential $\mu_i$ of a given component as a function of the temperature, pressure, and composition. In the next section, this is done for a mixture of ideal gases.

## 14-4 THE CHEMICAL POTENTIAL OF AN IDEAL GAS

According to the Gibbs-Dalton rule introduced in Chap. 10, an ideal gas in a mixture of gases behaves as if it alone occupies the volume of the system at the given temperature. Under this circumstance, the gas exerts a pressure equal to its component or partial pressure $p_i$. Therefore, the chemical potential of an ideal gas in a mixture is determined from knowledge of its temperature and component pressure $p_i$. The Gibbs-Dalton rule also provides us with the method of evaluating the chemical potential of an ideal gas under the above conditions. Because such gases do behave independently, at the temperature and total pressure of an ideal-gas mixture,

$$G_{\text{mix}} = \sum_i N_i \bar{g}_i = N_1 \bar{g}_1 + N_2 \bar{g}_2 + \cdots + N_i g_i \qquad \textbf{[14-15]}$$

where each $\bar{g}_i$ term is measured at the temperature and component pressure of that species. If we take the partial derivative of $G_{\text{mix}}$ with respect to $N_i$, holding temperature, pressure, and the remaining $N$'s constant, then

$$\left( \frac{\partial G}{\partial N_i} \right)_{T,P,N_j} = \bar{g}_{i,\text{pure},p_i}$$

Since the term on the left is the definition of the chemical potential of the $i$th species, for an ideal gas in an ideal-gas mixture

$$\mu_{i,\text{mix},P} = \mu_{i,\text{pure},p_i} = g_{i,\text{pure},p_i} \qquad \textbf{[14-16]}$$

Hence the chemical potential of an ideal gas in a mixture may be evaluated in terms of the specific Gibbs function of the pure component at a pressure $p_i$ and temperature $T$. The specific Gibbs function of a substance at temperature $T$ is, by definition,

$$\bar{g}_{i,T} = \bar{h}_{i,T} - T\bar{s}_{i,T} \qquad \textbf{[a]}$$

The enthalpy and the entropy of an ideal gas must be evaluated in terms of the standard reference state discussed in Chap. 14. Recall that the standard state of an ideal gas is taken as 1 atm, and it is symbolized by the superscript $^\circ$. Since the enthalpy of an ideal gas is not a function of pressure, we may write

$$\bar{h}_{i,T} = \bar{h}^\circ_{i,T} \qquad \textbf{[b]}$$

where $h_{i,T}^o$ accounts for the *enthalpy of formation* as well as the *enthalpy difference* between 298 K (537°R) and the specified temperature $T$. Also, the absolute entropy at temperature $T$ and pressure $p_i$ is given by Eq. [13-24], namely,

$$\bar{s}_{i,T} = \bar{s}_{i,T}^o - R_u \ln p_i \qquad \text{[c]}$$

where $p_i$ is measured in atmospheres. The chemical potential of a component in an ideal-gas mixture is then found by combining Eqs. [a], [b], and [c] with Eq. [14-16]. The result is

$$\mu_{i,T} = \bar{g}_{i,T} = \bar{h}_{i,T}^o - T\bar{s}_{i,T}^o + R_u T \ln p_i$$

With reference to Eq. [a], the sum of the first two terms of the right-hand side of the equation above is the standard-state Gibbs function $\bar{g}_{i,T}^o$. Therefore, the chemical potential of an ideal gas at temperature $T$ and pressure $p_i$ is given by

$$\mu_{i,\text{ideal},T} = \bar{g}_{i,T}^o + R_u T \ln p_i \qquad \text{[14-17]}$$

where the quantity $p_i$ must be measured in atmospheres. This relation, in conjunction with the equation of reaction equilibrium, enables us to determine the equilibrium composition of a reacting ideal-gas mixture.

## 14-5    The Equilibrium Constants $K_o$ and $K_p$

The equilibrium composition of a reacting ideal-gas mixture is determined by means of (1) the equation of reaction equilibrium and (2) the equation for the chemical potential of an ideal gas. In review, these equations are

$$\nu_E \mu_E + \nu_F \mu_F - \nu_A \mu_A - \nu_B \mu_B = 0 \qquad \text{[14-13]}$$

and

$$\mu_i = \bar{g}_i^o + R_u T \ln p_i \qquad \text{[14-17]}$$

where $A$ and $B$ represent reactants and $E$ and $F$ are products. By using Eq. [14-17] to replace the $\mu_i$ terms in Eq. [14-13], we find that

$$\nu_E(\bar{g}_E^o + R_u T \ln p_E) + \nu_F(\bar{g}_F^o + R_u T \ln p_F)$$
$$- \nu_A(\bar{g}_A^o + R_u T \ln p_A) - \nu_B(\bar{g}_B^o + R_u T \ln p_B) = 0 \qquad \text{[14-18]}$$

At this point it is convenient to rearrange Eq. [14-18] and collect common terms. This leads to

$$(\nu_E \bar{g}_E^o + \nu_F \bar{g}_F^o - \nu_A \bar{g}_A^o - \nu_B \bar{g}_B^o) + \nu_E R_u T \ln p_E + \nu_F R_u T \ln p_F$$
$$- \nu_A R_u T \ln p_A - \nu_B R_u T \ln p_B = 0 \qquad \text{[14-19]}$$

The quantity in the parentheses is called the **standard-state Gibbs-function change** for a reaction and is given the symbol $\Delta G_T^o$. Thus in this particular case

$$(\nu_E \bar{g}_E^0 + \nu_F \bar{g}_F^0 - \nu_A \bar{g}_A^0 - \nu_B \bar{g}_B^0) = \Delta G_T^0$$

or, more generally for any reaction, since $G = H - TS$,

$$\Delta G_T^0 = \Delta H_T^0 - T \Delta S_T^0 \equiv \sum_{\text{prod}} \nu_i \bar{g}_{i,T}^0 - \sum_{\text{reac}} \nu_i \bar{g}_{i,T}^0 \qquad \textbf{[14-20]}$$

Recall that the standard state for ideal gases is defined as 1 atm pressure. Equation [14-20] then indicates that the standard-state Gibbs-function change can be evaluated from a knowledge of the enthalpy of reaction and the entropy change for the stoichiometric reaction occurring at 1 atm pressure and temperature $T$. Methods for computing these quantities were introduced in Chap. 13. Consequently, the value of $\Delta G_T^0$ is known once the stoichiometric chemical equation and the temperature are specified.

We now substitute this definition of $\Delta G_T^0$ into Eq. [14-19]. At the same time the terms containing logarithms can be combined into a single term. Thus we find that

$$
\begin{aligned}
-\Delta G_T^0 &= \nu_E R_u T \ln p_E + \nu_F R_u T \ln p_F - \nu_A R_u T \ln p_A - \nu_B R_u T \ln p_B \\
&= R_u T [\ln(p_E)^{\nu_E} + \ln(p_F)^{\nu_F} - \ln(p_A)^{\nu_A} - \ln(p_B)^{\nu_B}] \\
&= R_u T \ln \frac{(p_E)^{\nu_E} (p_F)^{\nu_F}}{(p_A)^{\nu_A} (p_B)^{\nu_B}} \qquad \textbf{[14-21]}
\end{aligned}
$$

where the $p_i$ values are the *actual component pressures of the reacting gases at equilibrium* in atmospheres. The exponents of the component pressures are the stoichiometric coefficients based on the balanced theoretical chemical equation.

It is convenient to rewrite Eq. [14-21] in the form

$$\exp \frac{-\Delta G_T^0}{R_u T} = \frac{(p_E)^{\nu_E} (p_F)^{\nu_F}}{(p_A)^{\nu_A} (p_B)^{\nu_B}} \qquad \textbf{[14-22]}$$

Since $G_T^0$ is solely a function of temperature, the left-hand side of Eq. [14-22] is only a function of temperature. Hence its value can be determined for any reaction once a temperature is selected. This type of computation is formalized by defining this term as the ***standard-state equilibrium constant*** $K_0$. That is,

$$K_0 \equiv \exp \frac{-\Delta G_T^0}{R_u T} \qquad \text{or} \qquad \ln K_0 = \frac{-\Delta G_T^0}{R_u T} \qquad \textbf{[14-23]}$$

The first step in an equilibrium calculation, once the reaction equation is chosen, is to use Eq. [14-23] to evaluate $K_0$. The general method is discussed in the following section. For a few specific reactions the values of $K_0$ or $\log_{10} K_0$ are tabulated against the temperature in thermodynamic tables. The use of base 10 on the logarithm, rather than base $e$, is for convenience. A tabulation of $\log_{10} K_o$ values over a range of temperatures for some common ideal-gas reactions is presented in Table A-24.

As a second step, the right-hand side of Eq. [14-22] is defined as $K_p$, an *equilibrium constant based on partial pressure,* for ideal-gas reactions. That is,

$$K_p \equiv \frac{(p_E)^{\nu_E}(p_F)^{\nu_F}}{(p_A)^{\nu_A}(p_B)^{\nu_B}} \qquad \text{[14-24]}$$

In general, of course, the number of terms in the numerator and denominator depends on the number of product and reactant species in the reaction equation. Equation [14-24] is more meaningful and useful if it is written in terms of the number of moles of each constituent present at equilibrium, rather than the partial pressures. Since the component pressure of any ideal gas is defined by

$$p_i = y_i P = \frac{N_i}{N_m} P$$

the expression for $K_p$ may be modified to the following form for a reaction with two reactants and two products:

$$K_p = \frac{(N_E)^{\nu_E}(N_F)^{\nu_F}}{(N_A)^{\nu_A}(N_B)^{\nu_B}} \left(\frac{P}{N_m}\right)^{\Delta\nu} \qquad \text{[14-25]}$$

where $\Delta\nu = \nu_E + \nu_F - \nu_A - \nu_B$ and $P$ is measured in atmospheres. More generally,

$$\Delta\nu \equiv \sum_{\text{prod}} \nu_i - \sum_{\text{reac}} \nu_i \qquad \text{[14-26]}$$

In addition, $N_m$ in Eq. [14-25] is the sum of the *total* number of moles of mixture present at equilibrium.

In summary, the equilibrium composition of an ideal-gas mixture is determined from the relation

$$K_o = K_p \qquad \text{[14-27]}$$

Equation [14-23] is used to evaluate $K_o$. This type of calculation is discussed in Sec. 14-6. To avoid repetitious calculations, the tabulation of $\log_{10} K_o$ data in Table A-24 is helpful. However, in practice such data are usually not available and $K_o$ must be determined from fundamental property data. Once $K_o$ is known, from whatever source, then Eq. [14-25] is used to determine the $N_i$ values at equilibrium. This final step is discussed in detail in Sec. 14-7. The overall process is illustrated in Fig. 14-4.

Although the definition of $K_o$ appears straightforward, the following items need to be emphasized when using data from Table A-24:

1. The value of $K_o$ is independent of pressure, and depends solely on temperature. The reason is that $\Delta G_T^o$ is defined as a value in the

---

**Equilibrium Analysis of Ideal-gas Mixtures**

(1) Equilibrium criterion:
$K_o = K_p$

(2) $K_o$ in terms of $\Delta G_T^o$:

$$K_o = \exp \frac{-\Delta G_T^o}{R_u T}$$

(3) $K_p$ in terms of the equilibrium composition:

$$K_p = \frac{(N_E)^{\nu_E}(N_F)^{\nu_F}}{(N_A)^{\nu_A}(N_B)^{\nu_B}} \left[\frac{P}{N_m}\right]^{\Delta\nu}$$

where

$$\Delta\nu = \nu_E + \nu_F - \nu_A - \nu_B$$

**Figure 14-4**

Summary of steps in the equilibrium analysis of chemical reactions.

standard state of 1 atm, regardless of the actual pressure. Its value also is not influenced by the presence of inert gases in the gas mixture.

2. The value of $K_0$ depends on the method of writing a chemical equation. For example, consider the following three reactions at a given temperature:

$$CO + \tfrac{1}{2}O_2 \rightarrow CO_2 \qquad\qquad \textbf{[a]}$$

$$2CO + O_2 \rightarrow 2CO_2 \qquad\qquad \textbf{[b]}$$

$$CO_2 \rightarrow CO + \tfrac{1}{2}O_2 \qquad\qquad \textbf{[c]}$$

The standard-state Gibbs-function change for reaction $b$ is twice that for reaction $a$. Consequently, doubling the stoichiometric equation will square the value of $K_0$; that is, $(K_0)_b = (K_0)_a^2$. But reversal of the direction of a chemical equation reverses the sign on $\Delta G_T^0$. Thus the $K_0$ value for reaction $c$ is the reciprocal of the value of reaction $a$. These results stem from the exponential relationship between $K_0$ and $\Delta G_T^0$. One can conclude from these results that tabulated values of $K_0$ as a function of temperature are meaningless unless the chemical equation to which they refer is also cited.

3. The larger the value of $K_0$, the more a reaction approaches completion. In general, when $\log_{10} K_0$ is greater than $+4$ or $+5$, the reaction is essentially complete. On the other hand, when it is less than $-4$ or $-5$, the extent of the reaction toward products is negligible. It is in the range between these values that equilibrium calculations are essential to determine the equilibrium composition.

The problem discussed in item 2 above is important when $K_0$ data are read from a table. When these data are calculated from basic property data, as illustrated in the next section, the correct $\Delta G_T^0$ value for the selected chemical reaction is found directly.

## 14-6 CALCULATION OF $K_0$ VALUES

As demonstrated in Sec. 14-5, the equilibrium constant $K_0$ is directly related to the standard-state Gibbs function change $\Delta G_T^0$. For the chemical reaction represented by the equation

$$\nu_A A + \nu_B B \rightleftharpoons \nu_E E + \nu_F F \qquad\qquad \textbf{[14-11]}$$

the standard-state Gibbs function change is determined by

$$\Delta G_T^0 = \nu_E \bar{g}_E^0 + \nu_F \bar{g}_F^0 - \nu_A \bar{g}_A^0 - \nu_B \bar{g}_B^0$$

Any term $\bar{g}_i^0$ for the pure $i$th species is found by evaluating the quantity $\bar{h}_i^0 - T\bar{s}_i^0$. By employing Eq. [14-8b] for the enthalpy $\bar{h}_i^0$ of any reacting species, the standard-state Gibbs function for any species is

$$\bar{g}_i^0 = (\Delta \bar{h}_{f,298}^0 + \bar{h}_T - \bar{h}_{298} - T\bar{s}_T^0)_i$$

The standard-state Gibbs-function change for any reaction, then, is given by

$$\Delta G_T^o = \sum_{prod} \nu_i (\Delta \bar{h}_{f,298}^o + \bar{h}_T - \bar{h}_{298} - T\bar{s}_T^o)_i \qquad \textbf{[14-28]}$$

$$- \sum_{reac} \nu_i (\Delta \bar{h}_{f,298}^o + \bar{h}_T - \bar{h}_{298} - T\bar{s}_T^o)_i$$

For evaluation purposes, Table A-23 provides $\Delta \bar{h}_{f,298}^o$ data, and Tables A-6 through A-11 are sources of data for the values of $\bar{h}$ and $\bar{s}$ at the required temperature. Once $\Delta G_T^o$ has been determined from Eq. [14-27], then Eq. [14-23] is used to evaluate the equilibrium constant $K_o$. That is,

$$K_o = \exp\left(\frac{-\Delta G_T^o}{R_u T}\right) \qquad \textbf{[14-23]}$$

The evaluation technique is illustrated below in Example 14-1.

**Example 14-1**

Evaluate the equilibrium constant $K_o$ at (a) 298 K and (b) 1000 K for the gas-phase reaction $CO + H_2O \rightleftharpoons CO_2 + H_2$.

**Solution:**

**Given:**   The gas-phase reaction $CO + H_2O \rightleftharpoons CO_2 + H_2$.

**Find:**   $K_o$ at (a) 298 K and (b) 1000 K.

**Model:**   Ideal-gas reaction.

**Analysis:**   (a) First the standard-state Gibbs-function change is evaluated for the reaction, making use of Eq. [14-28]:

$$\Delta G_T^o = \sum_{prod} \nu_i (\Delta \bar{h}_{f,298}^o + \bar{h}_T - \bar{h}_{298} - T\bar{s}_T^o)_i$$

$$- \sum_{reac} \nu_i (\Delta \bar{h}_{f,298}^o + \bar{h}_T - \bar{h}_{298} - T\bar{s}_T^o)_i$$

At 298 K the $(\bar{h}_T - \bar{h}_{298})$ terms are zero. The remaining enthalpy of formation and absolute entropy data are read from Table A-23. In this case

$$\Delta G_T^o = \sum_i \nu_i (\Delta \bar{h}_{f,298} - T\bar{s}_T^o)_i$$

$$= 1[-393,520 - 298(213.685)] + 1[0 - 298(130.574)]$$

$$- 1[-110,530 - 298(197.543)] - 1[-241,810 - 298(188.720)]$$

$$= -28,663 \text{ kJ/kmol CO}$$

Substitution of this value into Eq. [14-23] yields

$$K_o = \exp\frac{-\Delta G_T^o}{R_u T} = \exp\frac{28,663}{8.314(298)} = \exp 11.57 = 105,760$$

The value of $\log_{10} K_o$ is 5.02, which is in good agreement with data from Table A-24, except for the sign. In Table A-24 the chemical equation is written in the reverse direction. Hence the calculated value of $\log K_o$ is the negative of the tabulated value.

(b) At a temperature of 1000 K the calculation is

$$\Delta G_T^o = 1[-393,520 + 42,769 - 9364 - 1000(269.215)]$$
$$+ 1[0 + 29,154 - 8468 - 1000(166.114)]$$
$$- 1[-110,530 + 30,355 - 8669 - 1000(234.421)]$$
$$- 1[-241,810 + 35,882 - 9904 - 1000(232.597)]$$
$$= -3064 \text{ kJ/kmol CO}$$

This value is now substituted into Eq. [14-23]. At 1000 K

$$K_o = \exp \frac{3064}{8.314(1000)} = \exp 0.369 = 1.45$$

This result is in substantial agreement with the value found in Table A-24, except for the sign difference.

**Comment:** By continuing this type of calculation for other temperatures and tabulating the results, the column of data in Table A-24 for this reaction could be developed. The same procedure is applicable for all the reactions shown in the complete table.

## 14-7 CALCULATION OF EQUILIBRIUM COMPOSITIONS

As shown in the preceding section, the value of $\Delta G_T^o$ can be computed for any reaction if enthalpy and entropy data are available. Therefore, the value of $K_o$ for any reaction involving ideal gases is also known, once the temperature has been specified. For reacting ideal-gas mixtures $K_o = K_p$. Moreover, the equilibrium constant $K_p$ is related to the moles of the gases at chemical equilibrium by Eq. [14-25], namely,

$$K_p = \frac{(N_E)^{\nu_E}(N_F)^{\nu_F}}{(N_A)^{\nu_A}(N_B)^{\nu_B}} \left(\frac{P}{N_m}\right)^{\Delta\nu} \qquad \textbf{[14-25]}$$

where $P$ is measured in atmospheres. In Eq. [14-25], $N_m$ is the sum of the total number of moles of mixture present at equilibrium. It is related to the $N_i$ values by

$$N_m = N_A + N_B + \cdots + N_E + N_F + \cdots + N_{\text{inert}} \qquad \textbf{[14-29]}$$

It is important not to omit the last term in this expression. The presence of inert gases affects the component pressure of each reacting species. If we again restrict ourselves for the moment to a reaction with two reactants and two products, Eqs. [14-25] and [14-29] constitute two equations with five unknowns. The remaining equations necessary for the solution are based on the principle of conservation of atomic species previously used in Chap. 14. In practice, there are sufficient mass balances to make a solution possible.

Before proceeding with example calculations, several points need emphasis.

1. The mixture pressure does affect the equilibrium composition, even though it does not affect the value of $K_0$. Equation [14-25] involves the term $P^{\Delta\nu}$. The value and sign on $\Delta\nu$ will influence the equilibrium composition. Only when $\Delta\nu$ is zero will the pressure have no effect.

2. The presence of inert gases in the reacting mixture affects the composition, because of the term $N_m^{-\Delta\nu}$ in Eq. [14-25]. The moles of inert gases must be included in $N_m$, as noted in Eq. [14-28]. When $\Delta\nu$ is zero, the presence of inert gases will have no effect.

3. The relative amounts of the initial reactants will affect the final composition. This occurs because the quantities of the reactants influence the mass balances on atomic species.

4. The temperature of the reacting mixture affects the final composition, for the obvious reason that a temperature change can greatly alter the value of $K_0$.

Thus temperature, pressure, inert gases, and initial reactant composition all generally affect the equilibrium composition.

In the examples below, the equilibrium constant $K_0$ is read from Table A-24. In practice this type of table is not available for other reactions. In addition, the values in Table A-24 are quoted for only a limited number of temperatures. Hence the general procedure would be to first calculate $K_0$ by the method in Sec. 14-6. Then, since $K_0 = K_p$ for ideal-gas mixtures, Eq. [14-25] is used, in conjunction with atomic species balances, to determine the equilibrium composition. When many such computations are required, a computer program such as EES that is capable of solving nonlinear equations and contains the required property data for computing $K_0$ is very useful. The general method for evaluating the equilibrium composition of a reacting ideal-gas mixture is illustrated below by several examples.

**Example 14-2**

Carbon monoxide (CO) and oxygen ($O_2$) initially in an equimolar ratio are allowed to reach equilibrium at 1 atm and 3000 K. Determine the composition of the equilibrium mixture.

**Solution:**

**Given:**   Initial equimolar mixture of CO and $O_2$ attains equilibrium at 1 atm and 3000 K, as shown in Fig. 14-5.

**Find:**   Equilibrium composition.

**Model:**   Ideal-gas reaction; only reacting species are CO, $O_2$, and $CO_2$.

**Analysis:**   The single reaction to consider is

$$CO(g) + \tfrac{1}{2}O_2(g) \rightleftharpoons CO_2(g)$$

Another possible reaction might be the further dissociation of $O_2$ into monatomic oxygen (O). In Table A-24 the value of $K_o$ for dissociation of $O_2$ is roughly 0.013. Thus the formation of O will be quite small. We shall neglect the effect in this calculation. Consequently, we assume that the only chemical species present at equilibrium are CO, $O_2$, and $CO_2$. The equilibrium constant $K_o$ at 3000 K is listed as 3.06 in Table A-24. Therefore,

$$K_p = \frac{(N_{CO_2})^1}{(N_{O_2})^{1/2}(N_{CO})^1}\left(\frac{P}{N_m}\right)^{1-1-1/2}$$

or

$$3.06 = \frac{(N_{CO_2})}{(N_{O_2})^{1/2}(N_{CO})}\left(\frac{1}{N_m}\right)^{-1/2}$$

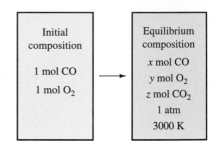

**Figure 14-5**
Schematic for Example 14-2.

The actual chemical reaction itself, which does not go to completion, may be written as

$$1CO + 1O_2 \rightarrow xCO + yO_2 + zCO_2$$

where $x$, $y$, and $z$ represent the number of moles of CO, $O_2$, and $CO_2$ present in the mixture at equilibrium. Since $N_m = x + y + z$, the expression for $K_p$ becomes, upon rearrangement,

$$3.06 = \frac{z(x + y + z)^{1/2}}{x(y)^{1/2}}$$

In addition to this relationship, which contains three unknowns, two balances may be made on the carbon and oxygen atoms in the actual reaction above. Hence

C balance                   $1 = x + z$

O balance                   $3 = x + 2y + 2z$

Solving for $y$ and $z$, we find that

$$z = 1 - x$$

and

$$y = \tfrac{1}{2}(3 - x - 2z) = \tfrac{1}{2}(1 + x)$$

Also,

$$N_m = x + y + z = x + (1 - x) + \tfrac{1}{2}(1 + x) + \tfrac{1}{2}(3 + x)$$

We have chosen to evaluate the numbers of moles of $O_2$ and $CO_2$ in terms of CO. This is an arbitrary choice, and either $y$ or $z$ could have been selected as the remaining unknown. Substitution of the equations for $y$, $z$, and $N_m$ into the expression for $K_p$ yields

$$3.06 = \frac{(1 - x)[(3 + x)/2]^{1/2}}{x[(1 + x)/2]^{1/2}} = \frac{(1 - x)(3 + x)^{1/2}}{x(1 + x)^{1/2}}$$

This equation for $x$ may be solved by iteration, synthetic division, Newton's method, or any other suitable technique. A value of 0.34 satisfies the equation within the desired accuracy. Therefore, the correct chemical equation for the reaction becomes

$$1CO + 1O_2 \rightarrow 0.34CO + 0.67O_2 + 0.66CO_2$$

This compares with the theoretical reaction for complete combustion as

$$1CO + 1O_2 \rightarrow CO_2 + \tfrac{1}{2}O_2$$

For the particular reactant mixture and final pressure and temperature, the $CO_2$ formed is roughly two-thirds that expected for complete combustion.

### 14-7-1   EFFECT OF PRESSURE ON EQUILIBRIUM

The total pressure frequently is an important design variable. The effect of pressure on the equilibrium composition is illustrated by the example below.

**Example 14-3**

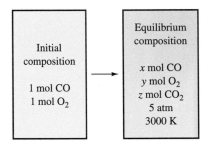

**Figure 14-6**
Schematic and data for Example 14-3.

**An** initial equimolar mixture of carbon monoxide and oxygen reaches equilibrium at 3000 K and 5 atm. Determine the composition of the equilibrium mixture.

**Solution:**

**Given:**   Initial equimolar mixture of CO and $O_2$ attains equilibrium at 5 atm and 3000 K, as shown in Fig. 14-6.

**Find:**   Equilibrium composition.

**Model:**   Ideal-gas reaction; only reacting species are CO, $O_2$, and $CO_2$.

**Analysis:**   On the basis of the analysis made in Example 14-2, we may write the following set of equations:

$$1CO + 1O_2 \rightarrow xCO + yO_2 + zCO_2$$

$$z = 1 - x \qquad y = \tfrac{1}{2}(1 + x) \qquad N_m = \tfrac{1}{2}(3 + x)$$

$$K_p = \frac{z(5)^{-1/2}}{x(y)^{1/2}(x + y + z)^{-1/2}} = 3.06$$

or, upon rearrangement,

$$3.06(5)^{1/2} = \frac{(1 - x)(3 + x)^{1/2}}{x(1 + x)^{1/2}}$$

The method of solution is exactly the same as in the preceding example. However, the change in pressure does affect the numerical answer. The right-hand side of the equation is the same as before, but the numerical value of the left-hand side has increased by the square root of the pressure, in atmospheres. The value of $x$ that satisfies the above relation is approximately 0.193 mol. Thus 0.807 mol of $CO_2$ is formed at equilibrium.

**Comment:**   When the pressure is 1 atm, the $CO_2$ formed is 0.66 mole per mole of initial CO. Apparently, the effect of pressure on the equilibrium composition is quite pronounced for this reaction.

As a general rule, any time the exponent of the pressure term in the $K_p$ expression is negative (i.e., the sum $\nu_E + \nu_F - \nu_A - \nu_B$ is negative), an increase in the pressure will always increase the number of moles of products formed at a given temperature and decrease the number of moles of reactants present at equilibrium. The opposite result is true when the exponent on the pressure is positive. When the exponent is zero, the pressure has no effect on the equilibrium composition.

### 14-7-2 EFFECT OF INERT GASES ON EQUILIBRIUM

Like the pressure effect discussed above, the effect of inert gases on the equilibrium composition depends on the sign and value of $\Delta\nu$ in the $K_p$ expression. For a combustion reaction, for example, the oxidant could be pure oxygen or air. In the latter case we assume that the nitrogen in air is nonreactive. Thus the use of these two oxidants should lead to different results. This point is illustrated in the following example.

**Example 14-4**

**T**o determine the effect of the presence of inert gases on the equilibrium composition, compute the mixture composition at 3000 K (5400°R) and 1 atm for a mixture initially composed of 1 mol of carbon monoxide and 4.76 mol of air.

**Solution:**

**Given:** Initial mixture of 1 mol of CO and 4.76 mol of air at 3000 K, as shown in Fig. 14-7.

**Find:** Equilibrium composition.

**Model:** Ideal-gas reaction; only reacting species are CO, $O_2$, $CO_2$, and $N_2$.

**Analysis:** The approach is basically the same as in the two preceding examples, except that the equation for the total number of moles $N_m$ must be revised. The chemical equation of the ideal-gas reaction now becomes

$$1CO + 1O_2 + 3.76N_2 \rightarrow xCO + yO_2 + zCO_2 + 3.76N_2$$

Again, from the carbon and oxygen balances $z = 1 - x$ and $y = \frac{1}{2}(1 + x)$. However, the total number of moles of mixture at equilibrium is now

$$N_m = x + y + z + 3.76 = \tfrac{1}{2}(10.52 + x)$$

The expression for $K_p$ is then

$$3.06 = \frac{(1 - x)(1)^{-1/2}}{x[(1 + x)/2]^{1/2}[(10.52 + x)/2]^{-1/2}} = \frac{(1 - x)(10.52 + x)^{1/2}}{x(1 + x)^{1/2}}$$

A suitable solution for $x$ from this equation is 0.47 mol of CO at the equilibrium state of 3000 K and 1 atm.

**Comment:** Compared with the original example, we find that the presence of inert nitrogen has decreased the $CO_2$ formed from 0.66 to 0.53 mol for the same pressure and temperature. Clearly the pressure of the system and the presence of

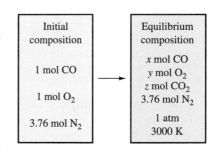

| Initial composition | Equilibrium composition |
|---|---|
| 1 mol CO | x mol CO |
| 1 mol $O_2$ | y mol $O_2$ |
| 3.76 mol $N_2$ | z mol $CO_2$ |
| | 3.76 mol $N_2$ |
| | 1 atm |
| | 3000 K |

**Figure 14-7**
Schematic and data for Example 14-4.

inert gases must be taken into account in evaluating the equilibrium composition of a reacting ideal-gas mixture at a given temperature.

The preceding examples have illustrated the general method for evaluating the equilibrium composition of an ideal-gas mixture when the system pressure and temperature are known. To make any computations for a reacting mixture, the identity of the chemical species expected in the equilibrium mixture must be assumed. This is necessary, because the presence of various species determines which equilibrium reactions must be considered. To date, we have examined mixtures for which only one equilibrium reaction was significant. Other reactions undoubtedly would occur, such as the dissociation of $O_2$ into atomic oxygen (O) or $N_2$ into atomic nitrogen (N), but the extent of these reactions was assumed to be negligible. Experience enables one to predict which reactions can be safely neglected and which reactions should be taken into account.

As mentioned in Sec. 14-5, the magnitude of $K_o$ is often a good guide to the importance of a reaction. As a reasonable rule, when $K_o$ is less than $10^{-3}$ or $10^{-4}$ (or log $K_o$ is less than $-3.0$ or $-4.0$), the extent of the reaction is usually not significant. Likewise, when $K_o$ is greater than $10^3$ or $10^4$ (or log $K_o$ is greater than 3.0 or 4.0), the reaction probably proceeds fairly close to completion. For example, consider the reaction $CO + \frac{1}{2}O_2 \rightleftharpoons CO_2$. The equilibrium constant $K_o$ at 2000 K (3600°R) is roughly 760. For an equimolar ratio of CO and $O_2$ at 1 atm, the CO present at equilibrium is approximately 0.0024 mol per initial mole of CO. The reaction essentially goes to completion. Consequently, for temperatures below roughly 2200 K or 4000°R, the dissociation of $CO_2$ into CO and $O_2$ may usually be neglected, except at pressures very much below atmospheric. At higher temperatures the effects of dissociation must be considered, as shown by previous examples, since the value of $K_o$ falls rapidly with increasing temperatures above 2200 K or 4000°R. Figure 14-8 shows the decrease in $K_o$ with increasing temperature for the exothermic combustion of CO to $CO_2$. The moles of CO at equilibrium per 1 mol initially of CO are also listed. We prove later that the value of $K_o$ *always* decreases with increasing temperatures for all exothermic reactions.

| $CO + \frac{1}{2}O_2 \rightleftharpoons CO_2$ | | |
|---|---|---|
| $T$, K | $K_o$ | $N_{CO,eq}$ |
| 1500 | $2 \times 10^5$ | $10^{-5}$ |
| 2000 | 766 | 0.0024 |
| 2500 | 27.5 | 0.058 |
| 3000 | 3.05 | 0.34 |

**Figure 14-8**
Illustration of change in $K_o$ and CO concentration with temperature for the $CO$–$CO_2$ reaction.

### 14-7-3 IONIZATION REACTIONS

The concept of the dissociation of a compound into two or more smaller particles can be extended to ionization effects. Elements, for example, will ionize into a positively charged ion and an electron. At elevated temperatures diatomic nitrogen dissociates into its monatomic form according to the equation

$$N_2 \rightleftharpoons 2N$$

As the gas is heated to higher temperatures, the following reaction also occurs:

$$N \rightleftharpoons N^+ + e^-$$

It is reasonable to assume in many situations that the positive ions and the electrons behave as ideal-gas particles. Consequently the $K_p$ expression (Eq. [14-25]) is equally valid for a mixture of neutral particles, ions, and electrons at a given temperature and pressure. (In the presence of electric fields, the temperature of the electrons may not necessarily be the same as the temperature of the ions and neutral particles. In the presence of moderate fields, one can assume that the temperature of all particles is the same.) In general, the degree of ionization increases as the temperature is raised and the pressure is lowered.

---

**A** hypothetical monatomic gas species $A$ ionizes according to the relation $A \rightleftharpoons A^+ + e^-$. If at some temperature $T$ the value of $K_o$ is 0.1 and the pressure is 0.1 atm, determine the percentage of ionization.

**Example 14-5**

**Solution:**

**Given:** The ionization reaction $A \rightleftharpoons A^+ + e^-$ at 0.1 atm, for which $K_o = 0.1$.

**Find:** Percent ionization.

**Model:** Ideal-gas reaction.

**Analysis:** The $K_p$ expression for the reaction is

$$K_p = \frac{N_{A^+} N_{e^-}}{N_A} \frac{P}{N_m}$$

where $N_m = N_{A^+} + N_{e^-} + N_A$. The conservation of mass and conservation of charge equations are

Positive charge balance: $\quad 0 = N_{A^+} + N_{e^-}$

Species $A$ balance: $\quad 1 = N_A + N_{A^+}$

The first balance states that since charge must be conserved, $N_{A^+} = N_{e^-}$. The second balance states that every ion must come originally from a neutral particle. Therefore, on the basis of 1 initial mole of species $A$, $1 - N_A = N_{A^+}$. If these latter two equalities are substituted into the $K_p$ expression, we find that

$$K_p = 0.1 = \frac{(1 - N_A)(1 - N_A)}{N_A} \frac{0.1}{2 - N_A}$$

or $\quad N_A(2 - N_A) = (1 - N_A)^2$

The solution to this equation is

$$N_A = 0.293$$

Therefore, the degree of ionization is close to 70 percent.

---

## 14-8    FIRST-LAW ANALYSIS OF EQUILIBRIUM IDEAL-GAS MIXTURES

In Chap. 13 the conservation of energy principle was applied to chemical reactions that were assumed to proceed to completion. Two types of processes were analyzed. In the first, the initial and final states of the reactants and products were known. On the basis of this information, the heat interactions could be evaluated. In the second, the process was assumed to be adiabatic. For a knowledge of the initial state of the reactants, the maximum reaction temperature (the adiabatic-flame temperature) was computed. These two types of calculations may now be repeated in the light of the second-law restrictions on chemical reactions. We realize now that in actual combustion processes two things may occur. Either the principal reaction itself may not go to completion, or products of the principal reaction may dissociate into other chemical species not initially present. Hence computations for chemical reactions should take into account these two factors if it is thought that they might be important. If sufficient heat transfer occurs, for example, during a combustion process, it is quite possible that neither effect need be considered. The two examples that follow illustrate the use of both the first and second laws in the analysis of chemical reactions of ideal-gas mixtures.

**Example 14-6**

**A**n equimolar mixture of carbon monoxide and water vapor enters a steady-flow device at 400 K (260°F) and 1 atm. The final products, which include $CO_2$ and $H_2$, leave at 1000 K (1340°F). Compute the quantity and direction of the heat transferred per mole of CO initially in the reactants.

**Solution:**

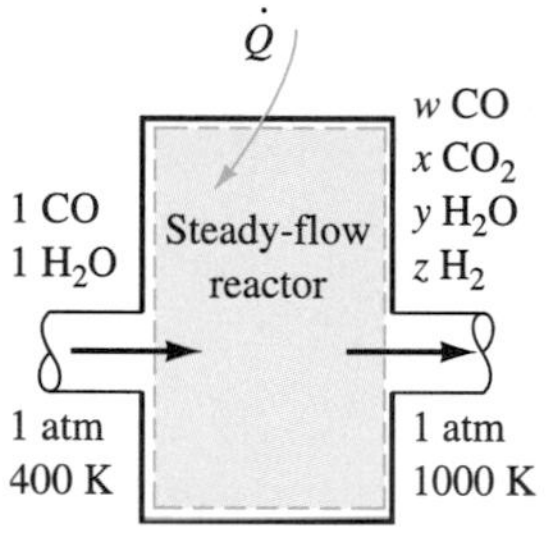

**Given:**    Equimolar mixture of CO and $H_2O$ at 400 K reacts in steady flow to form $CO_2$ and $H_2$ at 1000 K, as shown in Fig. 14-9.

**Find:**    $\dot{Q}/\dot{N}_{CO}$ in kJ/kmol of CO.

**Model:**    Steady flow, ideal-gas mixture.

**Analysis:**    The chemical reaction is given by

$$CO(g) + H_2O(g) \rightleftharpoons CO_2(g) + H_2(g)$$

In lieu of the expected equilibrium reaction among the constituents, the actual reaction must be written as

$$CO + H_2O \rightleftharpoons wCO + xCO_2 + yH_2O + zH_2$$

The use of mass balances leads to relationships among $w$, $x$, $y$, and $z$:

| | | |
|---|---|---|
| O balance: | $2 = w + 2x + y$ |
| C balance: | $1 = w + x$ |
| H balance: | $2 = 2y + 2z$ |

**Figure 14-9**
Schematic and data for Example 14-6.

A further relationship among the four variables is provided by the equilibrium-constant expression for the water-gas reaction:

$$K_p = \frac{(N_{CO_2})(N_{H_2})}{(N_{CO})(N_{H_2O})} \left(\frac{P}{N_m}\right)^0$$

At 1000 K the value of $\log K_0$ for this reaction is listed in Table A-24 as 0.159, or $K_0 = 1.44$. Therefore, in terms of the unknown quantities:

$$1.44 = \frac{xz}{wy}$$

The three mass balances and the equation for $K_p$ can be solved simultaneously by computer or by hand calculation. The solution in terms of the actual chemical equation is

$$1CO + 1H_2O \rightarrow 0.455CO + 0.545CO_2 + 0.455H_2O + 0.545H_2$$

Thus the reaction goes to 54.5 percent of completion. With this information the value of the heat interaction may be calculated.

The conservation of energy principle for this steady-flow process is

$$\dot{Q} = \sum_{prod} \dot{N}_i h_i - \sum_{reac} \dot{N}_i h_i$$

$$\dot{Q}/\dot{N}_{CO} = \sum_{prod} n_i(\Delta \bar{h}^\circ_{f,298} + \bar{h}_T - \bar{h}_{298})_i - \sum_{reac} n_i(\Delta \bar{h}^\circ_{f,298} + \bar{h}_T - \bar{h}_{298})_i$$

Substituting $\Delta \bar{h}_f$ data from Table A-23 and $\bar{h}$ data from Tables A-8 to A-11, we find that

$$\dot{Q}/\dot{N}_{CO} = 0.455(-110,530 + 30,355 - 8669) + 0.545(-393,520 + 42,769 - 9364)$$

$$+ 0.455(-241,820 + 35,882 - 9904) + 0.545(0 + 29,154 - 8468)$$

$$- 1(-110,530 + 11,644 - 8669) - 1(-241,820 + 13,356 - 9904)$$

$$= 22,300 \text{ kJ/kmol CO}$$

**Comment:** The positive value of $Q$ indicates that 22,300 kJ of heat would have to be added per initial kilomole of CO in order for the products to reach 1000 K. At this temperature, dissociation of $CO_2$ and $H_2O$ into CO, $H_2$, and $O_2$ may be neglected.

---

**Example 14-7**

One mole of carbon monoxide and 220 percent theoretical oxygen requirements, both initially at 25°C, undergo a reaction to form $CO_2$ in a steady-flow process at 1 atm. Neglecting dissociation of $O_2$, determine the final equilibrium composition and the final temperature if the process is adiabatic.

**Solution:**

**Given:** CO with 220 percent theoretical $O_2$ at 25°C react adiabatically in steady flow at 1 atm to form $CO_2$, as shown in Fig. 14-10.

**Find:** Final equilibrium composition and temperature.

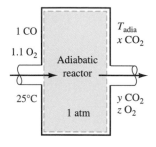

**Figure 14-10**
Schematic and data for Example 14-7.

**Model:** Adiabatic, steady-flow; ideal gas mixture.

**Analysis:** The theoretical equation for complete combustion would be

$$CO + 2.2(0.5)O_2 \rightarrow CO_2 + 0.6O_2$$

In this particular case. however, when the effects of chemical equilibrium are to be considered, we must write

$$CO + 1.1O_2 \rightleftharpoons xCO_2 + yCO + zO_2$$

The mass balances on the equation are

O balance: $\qquad\qquad 1 + 2.2 = 2x + y + 2z$

C balance: $\qquad\qquad 1 = x + y$

The expression for $K_p$ for the reaction $CO + \frac{1}{2}O_2 \rightleftharpoons CO_2$ is

$$K_p = \frac{N_{CO_2}}{(N_{CO})(N_{O_2})^{1/2}} \left(\frac{P}{N_m}\right)^{-1/2} = \frac{x(x + y + z)^{1/2}}{y(z)^{1/2}}$$

At this point we have three equations and four unknowns. The unknowns are $x$, $y$, $z$, and $K_p$. The value of $K_p$ ($= K_o$), of course, is solely a function of the temperature. The additional required equation is the conservation of energy principle, which in this case reduces to

$$\bar{h}_{reac} = \bar{h}_{prod}$$

The above energy balance can be written more explicitly as

$$\sum_{reac} n_i(\Delta\bar{h}^\circ_{f,298} + \bar{h}_T - \bar{h}_{298})_i = \sum_{prod} n_i(\Delta\bar{h}^\circ_{f,298} + \bar{h}_T - \bar{h}_{298})_i$$

In terms of the above reaction, this can be written as

$$1(-110,530 + 0) + 1.1(0)$$
$$= x(-393,520 + \bar{h}_{T,CO_2} - 9364) + y(-110,530 + \bar{h}_{T,CO} - 8669)$$
$$+ z(0 + \bar{h}_{T,O_2} - 8682)$$

where the enthalpy terms are to be evaluated at the final but as yet unknown temperature. This equation introduces no more unknowns; hence we now have a sufficient number of equations for a solution. The method of solution, however, requires an iteration, or trial-and-error, technique by hand or a computerized solution such as EES. Before starting the iteration process, the mass balances and the $K_p$ expression may be combined into the form

$$K_p = \frac{x(2.1 - 0.5x)^{1/2}}{(1 - x)(1.1 - 0.5x)^{1/2}}$$

At the same time a combination of the mass balances and the energy equation leads to

$$-110,530 = x(-402,884 + \bar{h}_{T,CO_2}) + (1 - x)(-119,199 + \bar{h}_{T,CO})$$
$$+ (1.1 - 0.5x)(\bar{h}_{T,O_2} - 8682)$$

Thus we have two equations with two unknowns, primarily $x$ and $T$. As a first approximation we let $T$ equal 2900 K. The enthalpy values in the energy balance may now be obtained from gas tables A-7 to A-9, and the energy equation can be solved for $x$. This yields $x = 0.707$. This value of $x$ is now substituted into the expression for $K_p$. Thus

$$K_p = \frac{0.707(2.1 - 0.354)^{1/2}}{0.293(1.1 - 0.354)^{1/2}} = 3.69$$

At 2900 K, however, the equilibrium constant $K_0$ is found to be 4.46 by evaluation in Table A-24. As a second guess, we let $T$ equal 3000 K. The energy balance then gives $x = 0.739$. Substitution of this quantity into the relation for $K_p$ yields a value of 4.36. At 3000 K the value of $K_0$ given in Table A-24 is 3.06. Since this latter value is smaller than the value calculated from the value of $x$, we have now guessed too high a value for the temperature. The correct answer must lie between 2900 and 3000 K where $K_p = K_0$. We need not refine the method further, since the general approach has been sufficiently demonstrated.

The foregoing example is fairly simple because only one equilibrium reaction is involved. In practice, there may be a number of equilibrium reactions involved for a given set of initial reactants. Nevertheless, the method of solution is exactly the same, whether the process is adiabatic or nonadiabatic. Mass balances based on the conservation of atomic species, the conservation of energy principle, and the increase in entropy principle in the form of $K_p$ expressions must be employed. Care must be taken to ensure that all the $K_p$ expressions are independent of each other. In addition, it is important to keep in mind that our study of chemical reactions involving the use of $K_p$ requires ideal-gas behavior.

## 14-9 THE VAN'T HOFF EQUATION RELATING $K_O$ AND $\Delta h_R$

It is informative to derive a general expression for the variation of $\ln K_0$ with temperature. In Sec. 14-5 the equilibrium constant $K_0$ was related to the standard-state Gibbs-function change $\Delta G_T^0$ by the definition $-R_u \ln K_0 \equiv \Delta G_T^0/T$. When the relation $\Delta G_T = \Delta H_T - T \Delta S_T$ is substituted and the resulting expression differentiated with respect to temperature, we find that

$$-R\frac{d \ln K_0}{dT} = -\frac{\Delta H_T}{T^2} + \frac{d(\Delta H_T)}{T\, dT} - \frac{d(\Delta S_T)}{dT}$$

However, if one applies the basic equation $T\, dS = dH - V\, dP$ to a chemical reaction at constant pressure, then $T\, d(\Delta S)/dT = d(\Delta H)/dT$. Hence the second and third terms on the right of the above equation cancel, and it reduces to

$$\frac{d \ln K_0}{dT} = \frac{\Delta H_T^0}{R_u T^2} = \frac{\Delta \bar{h}_R^0}{R_u T^2} \qquad \textbf{[14-30]}$$

where $\Delta \bar{h}_R^{\circ}$ is the enthalpy of reaction introduced in Sec. 13-7. The expression for the change in $\ln K_o$ with temperature may be written also as

$$\frac{d \ln K_o}{d(1/T)} = \frac{-\Delta \bar{h}_R^{\circ}}{R} \qquad \textbf{[14-31]}$$

Either of these equations is referred to as the **van't Hoff isobar equation.**

To integrate these equations, the functional relationship between $\Delta \bar{h}_R^{\circ}$ and $T$ must be known. The enthalpy of reaction for some reactions is nearly independent of temperature. Hence it is often possible to assume that $\Delta \bar{h}_R^{\circ}$ is constant over the range of temperatures of interest. If some average or initial value for the enthalpy of reaction is chosen, integration of Eq. [14-31] leads directly to

$$\ln \frac{K_{o2}}{K_{o1}} \approx -\frac{\Delta \bar{h}_R^{\circ}}{R_u}\left(\frac{1}{T_2} - \frac{1}{T_1}\right) \qquad \textbf{[14-32]}$$

This approximation is generally quite good for small temperature intervals.

Equation [14-32] leads to an interesting qualitative result. If a reaction is exothermic (heat transfer out), $\Delta \bar{h}_R^{\circ}$ is negative by convention. In addition, when $T_2 > T_1$ for such a reaction, the right-hand side of Eq. [14-32] is negative. Hence $K_{o2}$ must be less than $K_{o1}$. Thus for *exothermic reactions, the equilibrium constant $K_o$ decreases with increasing temperature.* As $K_o$ decreases, the tendency for the reaction to proceed to completion is reduced. Under adiabatic conditions, then, the energy released by an exothermic reaction is diminished since the reaction does not proceed to completion.

---

**Example 14-8**

For the gas-phase reaction $CO + \frac{1}{2}O_2 \rightleftharpoons CO_2$ the equilibrium constant $K_o$ is found to be 3.055 at 3000 K. Estimate the value at 2000 K, based on the van't Hoff isobar equation.

**Solution:**

**Given:** $K_o = 3.055$ at 3000 K for reaction: $CO + 1/2O_2 \rightleftharpoons CO_2$.

**Find:** $K_o$ value at 2000 K.

**Model:** van't Hoff isobar equation, ideal-gas mixture.

**Analysis:** The $\Delta \bar{h}_R^{\circ}$ values at 2000 and 3000 K are $-277,950$ and $-272,690$ kJ/kmol, respectively. Since the change in $\Delta \bar{h}_R^{\circ}$ is relatively small, Eq. [14-32] should lead to a fairly good evaluation of $K_o$ at the lower temperature. The average value of $\Delta \bar{h}_R^{\circ}$ is $-275,320$ kJ/kmol. Therefore,

$$\ln \frac{K_{o2}}{K_{o1}} = \frac{275,320}{8.314}\left(\frac{1}{2000} - \frac{1}{3000}\right) = 33,115(1.667 \times 10^{-4}) = 5.52$$

Hence, $K_{o2}/K_{o1} = 252$, or $K_{o2} = 770$.

**Comment:** The tabulated value found in Table A-24 is 766. The error is 0.5 percent, but this is for a temperature change of 1000 K. The calculation does indicate that reasonable accuracy is obtained if $\Delta \bar{h}_R^{\circ}$ is relatively constant.

---

**At** high temperature the potassium atom is ionized according to the equation $K \rightleftharpoons K^+ + e^-$. The values of the equilibrium constant $K_o$ for this gas-phase reaction at 3000 and 3500 K are $8.33 \times 10^{-6}$ and $1.33 \times 10^{-4}$, respectively. Estimate the average enthalpy of reaction in the given temperature range, in J/mol and electron-volts per molecule.

**Example 14-9**

**Solution:**

**Given:** For the reaction $K \rightleftharpoons K^+ + e^-$, $K_o$ equals $8.33 \times 10^{-6}$ at 3000 K and $1.33 \times 10^{-4}$ at 3500 K.

**Find:** Average $\Delta \bar{h}_R$ in given temperature range.

**Model:** van't Hoff isobar equation, ideal-gas mixture.

**Analysis:** The average enthalpy of reaction can be determined from the van't Hoff isobar equation. Use of Eq. [14-32] yields

$$\ln \frac{1.33 \times 10^{-4}}{8.33 \times 10^{-6}} = \frac{\Delta h_R^{\circ}}{8.314} \left( \frac{1}{3500} - \frac{1}{3000} \right)$$

or

$$\Delta \bar{h}_R^{\circ} = \frac{-2.778(8.314)}{-4.76 \times 10^{-5}} = 483{,}700 \text{ kJ/kmol}$$

The conversion of this quantity to electron-volts per molecule is carried out as follows:

$$\Delta \bar{h}_R^{\circ} = 483{,}700 \text{ kJ/kmol} \times \frac{\text{kmol}}{6.023 \times 10^{26} \text{ molecule}} \times \frac{\text{eV}}{1.60 \times 10^{-22} \text{ kJ}}$$

$$= 5.02 \text{ eV/molecule}$$

**Comment:** This value of the enthalpy of reaction is also known as the *ionization potential* of the potassium atom. Note that the reaction is highly endothermic.

---

## 14-10 SIMULTANEOUS REACTIONS

In the preceding discussion we focused on determining the equilibrium state for a single reaction at a given temperature and pressure. To attain equilibrium, the total Gibbs function of all the reacting species must be minimized. It is quite common, however, for two or more reactions to occur simultaneously in a reacting mixture. In addition, some of the reacting species will appear in several of the competing reactions. As might be anticipated, the correct analysis of this more complex process requires the use of the equation of reaction equilibrium for every independent reaction that occurs in the mixture.

As a generalization, consider a system for which there are $R$ *independent* reactions. The solution to this problem of simultaneous reactions requires that $R$ equations of the form

$$\sum_{\text{prod}} \nu_i \mu_i - \sum_{\text{reac}} \nu_i \mu_i = 0 \qquad \text{[14-14]}$$

be written. By substituting the chemical potential of an ideal gas for $\mu_i$ in these $R$ equations, there will arise $R$ different expressions for $K_p$. Each $K_p$ is evaluated from the general relationship

$$K_p = K_0 = \exp\left(\frac{-\Delta G_T^0}{R_u T}\right)$$

where a different $\Delta G_T^0$ value exists for each reaction at a given temperature. In addition, atomic balances are written in terms of the initial state of the reacting mixture. The $K_p$ expressions plus the atomic balances will provide sufficient information to determine the equilibrium composition of the ideal-gas mixture.

As an example, consider the adiabatic combustion of a hydrocarbon fuel $C_x H_y$ with a small excess of air. To calculate the adiabatic-combustion temperature, a number of competing reactions may have to be considered in addition to the main combustion reaction. Among the possible reactions at a high temperature are

$$CO_2 \rightleftharpoons CO + \tfrac{1}{2}O_2 \qquad\qquad H_2O \rightleftharpoons H_2 + \tfrac{1}{2}O_2$$

$$O_2 \rightleftharpoons 2O \qquad\qquad H_2O \rightleftharpoons H + OH$$

$$N_2 \rightleftharpoons 2N \qquad\qquad N_2 + O_2 \rightleftharpoons 2NO$$

$$H_2 \rightleftharpoons 2H \qquad\qquad 2NO + O_2 \rightleftharpoons 2NO_2$$

When a number of reactions are involved, the numerical calculations are difficult by hand. However, this type of problem is easily solved with a computer program such as EES.

It has been stressed that the $R$ reactions must be independent, in order to achieve a valid solution. As an example, consider the situation where the reacting species include CO, $CO_2$, $H_2$, $H_2O$, and $O_2$. In this case we might write

$$CO + \tfrac{1}{2}O_2 \rightleftharpoons CO_2 \qquad\qquad \text{[a]}$$

$$CO + H_2O \rightleftharpoons CO_2 + H_2 \qquad\qquad \text{[b]}$$

A third reaction we also might have written is

$$H_2 + \tfrac{1}{2}O_2 \rightleftharpoons H_2O \qquad\qquad \text{[c]}$$

These three reactions are not independent, however, since Eq. [c] can be formed by subtracting Eq. [b], algebraically, from Eq. [a]. Hence only two

$K_p$ expressions should be written in this case, although which two is a matter of choice.

Example 14-10

One mole of CO and 1 mol of $H_2O$ are heated to 2500 K (4500°R) and 1 atm. Determine the equilibrium composition, if it is assumed that only CO, $CO_2$, $H_2$, $H_2O$, and $O_2$ are present and that all these gases are ideal gases.

**Solution:**

**Given:**   One mole of CO and 1 mol of $H_2O$ react at 2500 K and 1 atm. The products include only CO, $CO_2$, $H_2$, $H_2O$, and $O_2$, as shown in Fig. 14-11.

**Find:**   The equilibrium composition.

**Model:**   Ideal gases.

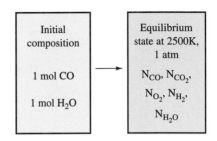

**Figure 14-11**
Schematic and data for Example 14-10.

**Analysis:**   The theoretical equations we choose are

$$CO + \tfrac{1}{2}O_2 \rightleftarrows CO_2 \qquad [a]$$

$$CO + H_2O \rightleftarrows CO_2 + H_2 \qquad [b]$$

From Table A-24 we find that the $K_o$ values for these reactions as written are 27.5 and 0.164, respectively. Thus

$$K_{p,a} = 27.5 = \frac{N_{CO_2}}{N_{CO}(N_{O_2})^{1/2}}\left(\frac{P}{N_T}\right)^{-1/2}$$

and

$$K_{p,b} = 0.164 = \frac{(N_{CO_2})(N_{H_2})}{(N_{CO})(N_{H_2O})}\left(\frac{P}{N_T}\right)^{0}$$

where

$$N_T = N_{CO} + N_{CO_2} + N_{O_2} + N_{H_2} + N_{H_2O}$$

The overall chemical equation relating the reactants and the products is

$$1CO + 1H_2O \rightarrow N_{CO} + N_{CO_2} + N_{O_2} + N_{H_2} + N_{H_2O}$$

This relationship can be used to write mass balances on the atomic species present in the overall reaction. These are

C balance:          $1 = N_{CO} + N_{CO_2}$

O balance:          $2 = N_{CO} + 2N_{CO_2} + N_{H_2O} + 2N_{O_2}$

$H_2$ balance:        $1 = N_{H_2} + N_{H_2O}$

The two equilibrium-constant expressions and the above three species balances constitute five equations with five unknowns. Hence a unique numerical solution exists. First, we solve the three atomic balances for three of the $N_i$ quantities in terms of the remaining two independent $N_i$ values. As an arbitrary choice, let $N_{CO_2}$ and $N_{H_2O}$ be the two independent variables. Then the expressions for the three dependent variables and $N_T$ are

$$N_{CO} = 1 - N_{CO_2} \qquad\qquad N_{H_2} = 1 - N_{H_2O}$$

$$N_{O_2} = \tfrac{1}{2}(1 - N_{CO_2} - N_{H_2O}) \qquad N_T = \tfrac{1}{2}(5 - N_{CO_2} - N_{H_2O})$$

Substitution of these four relations into the $K_p$ relations yields

$$27.5 = \frac{N_{CO_2}(5 - N_{CO_2} - N_{H_2O})^{1/2}}{(1 - N_{CO_2})(1 - N_{CO_2} - N_{H_2O})^{1/2}}$$

and

$$0.164 = \frac{N_{CO_2}(1 - N_{H_2O})}{(1 - N_{CO_2})(N_{H_2O})}$$

By suitable numerical techniques (including computer analysis) the simultaneous solution of the two preceding equations may be carried out. The table below summarizes the results.

| Species | Moles Initially | Moles Finally |
|---------|-----------------|---------------|
| CO | 1 | 0.71 |
| $H_2O$ | 1 | 0.71 |
| $CO_2$ | 0 | 0.29 |
| $H_2$ | 0 | 0.29 |
| $O_2$ | 0 | 0.005 |

**Comment:**    From these data, nearly 30 percent of both CO and $H_2O$ have reacted at the state of equilibrium.

## 14-11    SUMMARY

For chemical reactions occurring at a given temperature and pressure, the general criterion for equilibrium is

$$dG_{T,P} \leq 0$$

Thus a spontaneous chemical reaction proceeds in the direction that decreases the Gibbs function at constant $T$ and $P$. As a reaction reaches equilibrium, the Gibbs function attains a minimum value. A more useful format of the criterion for equilibrium is

$$\sum_{prod} \nu_i \mu_i - \sum_{reac} \nu_i \mu_i = 0$$

where $\nu_i$ is the stoichiometric coefficient and $\mu_i$ is the chemical potential.

The chemical potential $\mu_{i,T,P}$ of a gas in an ideal-gas mixture is

$$\mu_{i,T,P} = \bar{g}_{i,T}^0 + R_u T \ln p_i$$

where $p_i$ is measured in atmospheres. As a result, the equilibrium criterion for an ideal-gas mixture can be expressed as $K_0 = K_p$, where

$$K_0 \equiv \exp \frac{-\Delta G_T^0}{RT} \quad \text{and} \quad K_p \equiv \frac{(p_E)^{\nu_E}(p_F)^{\nu_F}}{(p_A)^{\nu_A}(p_B)^{\nu_B}}$$

$K_o$ is the *standard-state equilibrium constant,* and $K_p$ is defined as an *equilibrium constant based on partial pressure* for ideal-gas reactions. Also,

$$\Delta G_T^o = \Delta H_T^o - T\Delta S_T^o = \sum_{prod} \nu_i \bar{g}_{i,T}^o - \sum_{reac} \nu_i \bar{g}_{i,T}^o$$

Since $\Delta G_T^o$ is solely a function of temperature, $K_o$ can be determined for any reaction once a temperature is selected.

The $p_i$ values in $K_p$ are the *actual component pressures of the reacting gases at equilibrium.* Since $p_i = y_i P = (N_i/N_m)P$, the expression for $K_p$ may be modified to the following form:

$$K_p = \frac{(N_E)^{\nu_E}(N_F)^{\nu_F}}{(N_A)^{\nu_A}(N_B)^{\nu_B}} \left(\frac{P}{N_m}\right)^{\Delta\nu}$$

where $\Delta\nu = \nu_E + \nu_F - \nu_A - \nu_B$ and $P$ is measured in atmospheres. To find the equilibrium composition $N_i$ of an ideal-gas mixture, one uses the relation $K_o = K_p$. The equilibrium composition is affected by (1) the temperature, (2) the pressure when $\Delta\nu \neq 0$, (3) the presence of inert gases when $\Delta\nu \neq 0$, and (4) the composition of the initial reactants. Finally, when the first and second laws are combined for a reacting mixture, we find

$$\frac{d\ln K_p}{d(1/T)} = \frac{-\Delta\bar{h}_R^o}{R_u}$$

This expression, known as the *van't Hoff isobar equation,* relates the enthalpy of reaction to the equilibrium constant. An important result is that for exothermic reactions, the equilibrium constant $K_o$ decreases with increasing temperature. Thus the equilibrium composition may vary significantly with temperature.

## PROBLEMS

14-1. Carbon monoxide and oxygen in an equimolar ratio are at 1000 K and 5.0 bars. Calculate the value of the chemical potential of carbon monoxide, in kJ/kmol, assuming ideal-gas behavior.

14-2. An ideal-gas mixture at 2 atm and 800 K contains the following gases with the indicated mole fractions: $CO_2$, 11.0 percent; $N_2$, 79.0 percent; $O_2$, 9.0 percent, and $CO$, 1.0 percent. Calculate the value of the chemical potential of (*a*) carbon dioxide, and (*b*) oxygen, in kJ/kmol.

14-3. An ideal-gas mixture at 3 bars and 500 K contains a mixture of hydrogen and oxygen in a 2:1 molar ratio. Calculate the value of the chemical potential of hydrogen, in kJ/kmol.

14-4. The products of a combustion process contain the following gases and their mole fractions: $CO_2$, 8.5 percent; $O_2$, 6.7 percent; $H_2O$, 9.6 percent; and $N_2$, 75.2 percent. For a state of 1.5 bars and 700 K, determine the chemical potential of (*a*) the water and (*b*) the nitrogen in the mixture, in kJ/kmol.

14-5. Consider the gaseous reaction $H_2 + \frac{1}{2}O_2 \rightarrow H_2O$ at (a) 298 K and (b) 2000 K.
   (1) Calculate the value of $\bar{g}^\circ$ for each gas in the reaction, in kJ/kmol.
   (2) Evaluate $\Delta G_T^\circ$ for the stoichiometric reaction, in kilojoules.
   (3) On the basis of part 2 and Eq. [14-23], calculate $\log_{10} K_0$. Compare this value to that given in Table A-24.

14-6. Consider the gaseous reaction $CO + \frac{1}{2}O_2 \rightarrow CO_2$ at (a) 298 K and (b) 2000 K.
   (1) Calculate the value of $\bar{g}^\circ$ for each gas in the reaction, in kJ/kmol.
   (2) Evaluate $\Delta G_T^\circ$ for the stoichiometric reaction, in kilojoules.
   (3) On the basis of part 2 and Eq. [14-23], calculate $\log_{10} K_0$. Compare this value to that given in Table A-24.

14-7. Consider the gaseous reaction $O_2 \rightarrow 2O$ at (a) 298 K and (b) 2500 K.
   (1) Calculate the value of $\bar{g}^\circ$ for each gas in the reaction, in kJ/kmol.
   (2) Evaluate $\Delta G_T^\circ$ for the stoichiometric reaction, in kilojoules.
   (3) On the basis of part 2 and Eq. [14-23], calculate $\log_{10} K_0$. Compare this value to that given in Table A-24.

14-8. Consider the gaseous reaction $CO_2 + H_2 \rightarrow CO + H_2O$ at (a) 298 K and (b) 1200 K.
   (1) Calculate the value of $\bar{g}^\circ$ for each gas in the reaction, in kJ/kmol.
   (2) Evaluate $\Delta G_T^\circ$ for the stoichiometric reaction, in kilojoules.
   (3) On the basis of part 2 and Eq. [14-23], calculate $\log_{10} K_0$. Compare this value to that given in Table A-24.

14-9. One mole of $O_2$ and one mole of argon, an inert gas, are heated to an unknown temperature at a pressure of 1 atm. The $O_2$ dissociates according to the reaction $O_2 \rightleftarrows 2O$, and the resultant equilibrium mixture consists of 0.92 mol of $O_2$, 0.16 mol of O, and 1 mol of argon. Making use of Table A-24, find the unknown temperature, in degrees kelvin.

14-10E. Carbon monoxide and oxygen in an equimolar ratio are at 1000°R and 5.0 atm. Calculate the value of the chemical potential of (a) the carbon monoxide, and (b) the oxygen in Btu/lbmol, assuming ideal-gas behavior.

14-11E. An ideal-gas mixture at 2 atm and 880°R contains the following gases with the indicated mole fractions: $CO_2$, 12.0 percent; $N_2$, 78.0 percent; $O_2$, 9.0 percent, and CO, 1.0 percent. Calculate the value of the chemical potential of (a) the carbon dioxide and (b) the carbon monoxide, in Btu/lbmol.

14-12E. Consider the gaseous reaction $H_2 + \frac{1}{2}O_2 \rightarrow H_2O$ at (a) 537°R and (b) 3060°R.
   (1) Calculate the value of $\bar{g}^\circ$ for each gas in the reaction, in Btu/lbmol.

(2) Evaluate $\Delta G_T^0$ for the stoichiometric reaction, in Btu.

(3) On the basis of part 2 and Eq. [14-23], calculate the value of $\log_{10} K_0$. Compare this value to that given in Table A-24.

14-13E. Consider the gaseous reaction $CO + \frac{1}{2}O_2 \rightarrow CO_2$ at (a) 537°R and (b) 3780°R.

(1) Calculate the value of $\bar{g}^0$ for each gas in the reaction, in Btu/lbmol.

(2) Evaluate $\Delta G_T^0$ for the stoichiometric reaction, in Btu.

(3) On the basis of part 2 and Eq. [14-23], calculate $\log_{10} K_0$. Compare this value to that given in Table A-24.

14-14E. Consider the gaseous reaction $CO_2 + H_2 \rightarrow CO + H_2O$ at (a) 537°R and (b) 1800°R.

(1) Calculate the value of $\bar{g}^0$ for each gas in the reaction, in Btu/lbmol.

(2) Evaluate $\Delta G_T^0$ for the stoichiometric reaction, in Btu.

(3) On the basis of part 2 and Eq. [14-23], calculate $\log_{10} K_0$. Compare this value to that given in Table A-24.

14-15. Consider the gas-phase reaction $CO + H_2O \rightarrow CO_2 + H_2$. Tests at 2 atm indicate an equilibrium composition of 0.501 mol of $CO_2$, 0.499 mol of CO, 1.499 mol of $H_2O$, and 0.501 mol of $H_2$. Determine (a) the value of $K_p$ and (b) the temperature of the equilibrium mixture, in kelvins, using data from Table A-24.

14-16. A gas mixture at 10 atm consists of 0.1 mol of CO, 0.9 mol of $CO_2$, 0.3 mol of $O_2$, and 3 mol of $N_2$. The mixture is in equilibrium for the reaction $CO + \frac{1}{2}O_2 \rightleftarrows CO_2$. Determine (a) the value of $K_p$ and (b) the system temperature, in kelvins, using data from Table A-24.

14-17. Consider the reaction $O_2 + N_2 \rightleftarrows 2NO$. Measurement of the equilibrium composition indicates the presence of 0.942 mol of $O_2$, 2.942 mol of $N_2$, and 0.116 mol of NO. Determine (a) the value of $K_p$ and (b) the temperature of the equilibrium mixture, in kelvins, using data from Table A-24.

14-18. Consider the gas-phase reaction $CO + 3H_2 \rightleftarrows CH_4 + H_2O$. Initially a reaction vessel is filled with 2 mol of CO, 5 mol of $H_2$, and 2 mol of nitrogen (an inert gas). For the reaction given above, the products at equilibrium include 0.6 mol CO at 9 atm and temperature $T$. Determine the equilibrium constant $K_p$ at the given temperature and pressure.

## EQUILIBRIUM COMPOSITION CALCULATIONS

14-19. Using the data of Table A-24, find the temperature, in kelvins, at which (a) 10 percent and (b) 7 percent of $CO_2$ dissociates to CO and $O_2$ at 2 atm.

14-20. One mole of water vapor is heated at 2 bars until it is (a) 10 percent and (b) 7 percent dissociated into $H_2$ and $O_2$. Using data from

Table A-24, determine the required final temperature in each case, in kelvins.

14-21. Find the temperature, in kelvins, at which CO will be (a) 10 percent and (b) 7 percent of the total moles at equilibrium if CO is reacted with the stoichiometric amount of $O_2$ at 2 atm to form $CO_2$.

14-22. Find (a) the temperature, in kelvins, necessary to dissociate diatomic oxygen to a state where the monatomic species comprises 20 percent of the total moles present at equilibrium for a pressure of 0.20 atm. (b) For the same amount of O at equilibrium and a pressure of 0.40 atm, would the required temperature be higher or lower then that in part a? Discuss.

14-23. The equilibrium constant $K_0$ for the reaction $I_2 \rightleftharpoons 2I$ at 1500 K is 1.22.
(a) If 1 mol of diatomic iodine is heated to 1500 K at 0.5 atm, determine the number of moles of $I_2$ at equilibrium.
(b) Determine the number of moles of $I_2$ at equilibrium at 1500 K and 0.5 atm if the initial reactants include 1 mol of $I_2$ and 2 mol of argon (an inert gas).
(c) Determine the moles of $I_2$ at equilibrium at 1500 K and 0.3 atm if the initial reactants include 1 mol of $I_2$ and 2 mol of argon.

14-24. The standard-state Gibbs-function change for the dissociation of diatomic fluorine ($F_2$) into its monatomic species at 1200 K is 12,550 kJ/kmol.
(a) Compute the percent dissociation of pure $F_2$ at this temperature and (1) 1 atm and (2) 0.2 atm.
(b) If the initial reactants include 1 mol of $F_2$ and 2 mol of $N_2$ (an inert gas) at 1200 K, determine the moles of $F_2$ present at (1) 1 atm and (2) 0.2 atm.

14-25. At 300 K the equilibrium constant for the reaction $N_2O_4 \rightleftharpoons 2NO_2$ is $K_0 = 0.18$. (a) Find the system pressure required for 20 percent of the initially pure $N_2O_4$ to dissociate into $NO_2$ at this temperature. (b) Will lowering the pressure at constant temperature increase or decrease the amount of $NO_2$ at equilibrium? Discuss.

14-26. Consider the reaction $NO \rightleftharpoons \frac{1}{2}O_2 + \frac{1}{2}N_2$ at 2500 K.
(a) Determine the percentage of initially pure NO present at equilibrium at 2 atm.
(b) Determine the number of moles of NO present at 1 atm if the initial reactants include 2 mol of NO and 1 mol of He (an inert gas).
(c) Repeat part b if the initial reactants are 2 mol of NO and 3 mol of $O_2$.

14-27. Carbon dioxide at 1 atm is heated to 2500 K. For the dissociation reaction $CO_2 \rightleftharpoons CO + \frac{1}{2}O_2$, (a) determine the number of moles of $N_2$ which must be added per mole of original $CO_2$ so that the equilibrium mixture contains just 75 percent of the initial $CO_2$. (b) For the dissociation reaction in general, discuss the effect of raising the pressure on the amount of $CO_2$ at equilibrium.

14-28. A mixture of CO and $H_2O$ reacts at 1000 K according to the reaction $CO + H_2O \rightleftarrows CO_2 + H_2$. Determine the moles of $CO_2$ formed per initial mole of CO if the initial $CO/H_2O$ mole ratio is (a) 1:1, (b) 1:2, (c) 1:5.

14-29. The hydrogenation of iso-octene ($C_8H_{16}$) to iso-octane ($C_8H_{18}$) follows the reaction $C_8H_{16} + H_2 \rightleftarrows C_8H_{18}$. At 900 K and 2 atm the equilibrium conversion is 45 percent for a one-to-one ratio of initial reactants. If the pressure is changed to 10 atm at 900 K, and the initial reactant ratio $C_8H_{16}/H_2$ is 1:2, calculate the percent conversion at equilibrium, assuming the gases are ideal gases.

14-30. A reaction vessel initially contains 3 moles of $H_2$ and 6 moles of helium, a nonreacting gas. The gases are heated to 3500 K and 2 atm. Based on the reaction $H_2 \rightleftarrows 2H$, (a) determine the number of moles of H present at equilibrium. (b) Discuss qualitatively whether the pressure should be raised or lowered to increase the amount of H formed at the given temperature.

14-31. One mole of water at 3000 K dissociates according to the reaction $H_2O \rightleftarrows H_2 + \frac{1}{2}O_2$.
   (a) Determine the percentage of dissociation at this temperature and 1 atm pressure.
   (b) Repeat the calculation for the same temperature and pressure if the initial reactants include 1 mol of water and 1 mol of nitrogen (an inert gas).
   (c) Repeat the calculation for the same temperature and reactants as part a, but with the pressure 0.5 atm.

14-32. The equilibrium constant $K_o$ for the reaction $\frac{3}{2}H_2 + N_2 \rightleftarrows NH_3$ is 0.0068 at 450°C. Determine the equilibrium composition at 450°C if 3 mol of $H_2$ and 1 mol of $N_2$ are mixed at a pressure of (a) 20 atm and (b) 80 atm.

14-33. A mixture consisting of 1 mol of $H_2$, 0.6 mol of $O_2$, and 1 mol of $N_2$ reacts at 3000 K and 0.5 atm. For the reaction $H_2 + \frac{1}{2}O_2 \rightleftarrows H_2O$, determine the number of moles of $H_2O$ present at equilibrium.

14-34. At 1000 K the value of $K_o$ for the reaction $NO + \frac{1}{2}O_2 \rightleftarrows NO_2$ is 0.110.
   (a) Determine the number of moles of NO present at 1000 K and 1.1 atm if the initial reactant is 1 mol of pure $NO_2$.
   (b) Determine the number of moles of NO present if the initial reactants include 1 mole of NO and 2 mol of $NO_2$ at 1000 K and 1.1 atm.
   (c) Repart part b if the pressure changes to 3.5 atm.

14-35. One mol of $CO_2$ is mixed with 1 mol of $H_2$. Determine the equilibrium composition for the reaction $CO_2 + H_2 \rightleftarrows CO + H_2O$ if the temperature is (a) 1000 K and (b) 2000 K.

14-36. Consider the *water-gas* reaction $CO + H_2O \rightleftharpoons CO_2 + H_2$. Determine the moles of $H_2$ formed at 1000 K when the initial mixture contains only 1 mol of CO and (*a*) 1 mol of $H_2O$, (*b*) 3 mol of $H_2O$, and (*c*) 6 mol of $H_2O$.

14-37. Determine the mole percent of diatomic hydrogen that dissociates into monatomic hydrogen by the reaction $H_2 \rightleftharpoons 2H$ at 3500 K and (*a*) 1 atm and (*b*) 3 atm.

14-38. Hydrogen cyanide is produced in a reactor at 1 atm by the gas-phase reaction $N_2 + C_2H_2 \rightleftharpoons 2HCN$. The standard-state Gibbs function change at 300°C is 30,050 kJ/kmol of HCN. Determine the moles of HCN formed per mole of initial $N_2$ under the following conditions.
(*a*) The initial reactants are an equimolar mixture of $N_2$ and $C_2H_2$.
(*b*) The initial reactants include 1 mol of $N_2$ and 2 mol of $C_2H_2$.
(*c*) The initial reactants are 1 mol of $N_2$, 1 mol of $C_2H_2$, and 0.040 mol of HCN.

14-39. Sodium vapor ionizes according to the reaction $Na \rightleftharpoons Na^+ + e^-$. At 2000 K the value of $K_0$ is 0.668. Find the system pressure necessary for 75 percent of the sodium to have ionized, in atmospheres.

14-40. The equilibrium constant for the ionization reaction $Ar \rightleftharpoons Ar^+ + e^-$ for argon at 10,000 K is 0.00042. Determine the mole fraction of ionized argon atoms at (*a*) 0.01 atm and (*b*) 0.05 atm.

14-41. Cesium vapor ionizes according to the reaction $Cs \rightleftharpoons Cs^+ + e^-$ at elevated temperatures. Find the percentage of ionization at (*a*) 1400 K, (*b*) 1800 K, and (*c*) 2000 K, if the corresponding $\log_{10} K_0$ values are $-1.01$, 0.609, and 1.19, respectively, and the pressure is 1.5 atm.

14-42. For the ionization reaction $Na \rightleftharpoons Na^+ + e^-$ the value of $\log_{10} K_0$ at 1800 K is $-0.913$. Determine the percentage of ionization at this temperature at (*a*) 5 atm, (*b*) 1 atm, and (*c*) 0.5 atm.

14-43. The ionization equilibrium constant for the reaction $N \rightleftharpoons N^+ + e^-$ is (*a*)$6.26 \times 10^{-4}$ at 10,000 K, (*b*)$1.51 \times 10^{-2}$ at 12,000 K, and (*c*) 0.151 at 14,000 K. Find the system pressure necessary for 5 percent of N to ionize, assuming this is the sole species present initially, for the three temperatures.

## ENERGY ANALYSIS OF EQUILIBRIUM MIXTURES

14-44. Equal moles of $O_2$ and $H_2$ enter a combustion chamber separately at 1 atm and 298 K. For the reaction $H_2 + O_2 \rightleftharpoons 2OH$, the products ($H_2$, $O_2$, andOH) leave at 1 atm and 2000 K. If the equilibrium constant is 0.41, find (*a*) the moles of OH and $O_2$ at equilibrium per mole of initial $O_2$, (*b*) the heat transfer, in kJ per kilomole of $O_2$, and (*c*) the entropy change for the reaction itself, in kJ/K per kilomole of oxygen.

14-45. Pure diatomic hydrogen enters a steady-flow reactor at 298 K and 2 atm. In the reactor it is heated to 3000 K, where it dissociates according to the equation $H_2 \rightleftarrows 2H$. Determine (a) the moles of H present at equilibrium, per mole of initial $H_2$, and (b) the heat transfer, in kJ/kmol of $H_2$.

14-46. A mixture of 1 kmol of $H_2O$ and 1 kmol of $O_2$ enters a steady-flow reaction chamber at 1.2 atm and 1000 K. An equilibrium mixture consisting of $O_2$, $H_2$, and $H_2O$ leaves the chamber at 3200 K and 1.2 atm. Determine the heat transfer in kJ/kmol of $H_2O$ entering the chamber.

14-47. Diatomic oxygen is heated at a constant pressure of 0.1 atm from 298 to 2800 K. Determine the heat transferred to the reaction vessel, in kJ/kmol of initial $O_2$, if the reaction $O_2 \rightleftarrows 2O$ is in steady flow.

14-48. Diatomic hydrogen is heated at a constant pressure of 0.2 atm from 298 to 2800 K. Determine the heat transfer to or from the reaction vessel, in kJ/kmol of initial $H_2$, if the reaction $H_2 \rightleftarrows 2H$ is considered and the flow is steady.

14-49. One mole of carbon monoxide and 80 percent stoichiometric air, both initially at 25°C, react adiabatically at 1 atm. Compute the maximum combustion temperature, in kelvins, accounting for the dissociation reaction $CO_2 \rightleftarrows CO + \frac{1}{2}O_2$ in steady flow.

14-50. Carbon monoxide is reacted with a 10 percent excess of stoichiometric air, both initially at 500 K.
   (a) Determine the adiabatic-flame temperature without dissociation.
   (b) Determine the equilibrium flame temperature and the percentage of dissociation for the same initial conditions if the pressure is 2 atm.

14-51. Carbon monoxide is burned at constant volume, using the stoichiometric amount of air for combustion to $CO_2$. The initial reactant pressure and temperature are 1 atm and 500 K, respectively. Determine the maximum combustion temperature, in kelvins, if dissociation is considered, and the percent of initial CO left at equilibrium.

14-52. Diatomic oxygen is heated at a constant pressure of 0.1 atm from 537 to 5040°R. Determine the heat transfer, in Btu/lbmol of initial $O_2$, if the reaction $O_2 \rightleftarrows 2O$ occurs in steady flow.

14-53E. Diatomic hydrogen is heated at a constant pressure of 0.2 atm from 537 to 5040°R. Determine the heat transfer to or from the reaction vessel, in Btu/lbmol of initial $H_2$, if the reaction $H_2 \rightleftarrows 2H$ is considered and the flow is steady.

14-54E. One mole of carbon monoxide and 80 percent stoichiometric air, both initially at 77°F, react adiabatically at 1 atm in steady flow.
   (a) Determine the adiabatic-flame temperature without dissociation.
   (b) Compute the maximum flame temperature, accounting for the dissociation reaction $CO_2 \rightleftarrows CO + \frac{1}{2}O_2$, in degrees Rankine.

14-55E. Carbon monoxide is reacted with a 10 percent excess of stoichiometric air, both initially at 440°F and in steady flow.
  (a) Determine the adiabatic-flame temperature without dissociation.
  (b) Determine the equilibrium flame temperature and the percentage of dissociation for the same initial conditions if the pressure is 2 atm.

14-56E. Carbon monoxide is burned at constant volume, using the stoichiometric amount of air for combustion to $CO_2$. The initial reactant pressure and temperature are 1 atm and 900°R, respectively. Determine the maximum combustion temperature, in degrees Rankine, if dissociation is considered, and the percent of the initial CO left at equilibrium.

14-57. One mole of carbon monoxide is burned with 20 percent excess air at a pressure of 0.5 atm. If the original temperature of the reactants is 25°C, determine the adiabatic-combustion temperature with dissociation, in kelvins for the steady-flow process.

14-58. Solid carbon at 25°C and the stoichiometric quantity of air at 600 K are burned adiabatically at a constant pressure of 1 atm. Determine the number of moles of $CO_2$ formed per mole of carbon, assuming that only $CO_2$, CO, $O_2$, and $N_2$ are present in the products.

14-59. Carbon monoxide is burned with 120 percent of the theoretical-air requirement, both being supplied at 25°C and burned at a pressure of 0.1 atm. It is desired to supply the products of the reaction at 2500 K, assuming that only $CO_2$, CO, $O_2$, and $N_2$ are present in the products.
  (a) Determine the equilibrium composition at 2500 K.
  (b) Determine the heat transfer to or from the product gases, in kJ/kmol of initial CO, in order to achieve the required final temperature in the steady-flow process.
  (c) Assume that CO initially is at 25°C, and the products found in part a again are at 2500 K and 0.1 atm. To what temperature must the air be preheated before the reaction so that the reaction itself is adiabatic, in kelvins?

14-60. Carbon dioxide and hydrogen in a 1:2 molar ratio and at 400 K enter a reaction vessel maintained at 1 atm. The mixture is heated to 1000 K. For the reaction $CO_2 + H_2 \rightleftharpoons CO + H_2O$, determine the magnitude and direction of any heat transfer, in kJ/kmol of initial $CO_2$.

14-61 A mixture consisting of 1 mol of $H_2$, 0.7 mol of $O_2$, and 1 mol of $N_2$ initially at 25°C reacts at 0.5 atm to form products at 3000 K. For the reaction $H_2 + \frac{1}{2}O_2 \rightleftharpoons H_2O$, determine the magnitude and direction of the heat transfer, in kJ/kmol of initial $H_2$ in steady flow.

14-62E. Solid carbon at 77°F and the stoichiometric quantity of air at 620°R are burned adiabatically at a constant pressure of 1 atm. Determine the number of moles of $CO_2$ formed per mole of carbon, assuming that only $CO_2$, $CO$, $O_2$, and $N_2$ are present in the products.

14-63E. Carbon monoxide is reacted with 120 percent of the theoretical-air requirement, both being supplied at 77°F. Combustion occurs at a pressure of 0.1 atm. It is desired to supply the products of the reaction at 4500°R, assuming that only $CO_2$, $CO$, $O_2$, and $N_2$ are present in the products in steady flow.
(a) Determine the equilibrium composition at 4500°R.
(b) Determine the heat transfer to or from the gases, in Btu/lbmol of initial $CO$, in order to achieve the required final temperature.
(c) Assume that $CO$ initially is at 77°F, and the products again are at 4500°R and 0.1 atm. To what temperature must the air be preheated before the reaction so that the reaction itself is adiabatic, in degrees Rankine?

14-64E. Carbon dioxide and hydrogen in a 1:2 molar ratio and at 700°R enter a reaction vessel. The reaction $CO_2 + H_2 \rightleftharpoons CO + H_2O$ occurs as the mixture is heated at 1 atm to 1800°R. Determine the magnitude and direction of any heat transfer, in Btu/lbmol of initial $CO_2$.

14-65E. A mixture consisting of 1 mol of $H_2$, 0.7 mol of $O_2$, and 1 mol of $N_2$ initially at 77°F reacts at 0.5 atm to form products at 5400°R. For the reaction $H_2 + \frac{1}{2}O_2 \rightleftharpoons H_2O$, determine the magnitude and direction of the heat transfer, in Btu/lbmol of initial $H_2$, for steady flow.

### SIMULTANEOUS REACTIONS

14-66. A mixture of 1 mol of $CO_2$, 1 mol of $H_2$, and 0.5 mol of $O_2$ is heated to 3000 K and 1 atm. Assume that the equilibrium mixture consists of $CO_2$, $H_2$, $O_2$, $CO$, and $H_2O$. Determine the equilibrium composition, assuming H, O, and OH are absent.

14-67. The reaction of 1 mol of $CO_2$, 1 mol of $H_2$, and 0.5 mol of $O_2$ yields an equilibrium mixture of 0.5 mol of $CO_2$, 0.11 mol of $H_2$, 0.305 mol of $O_2$, 0.5 mol of $CO$, and 0.89 mol of $H_2O$. Determine the temperature, in kelvins, at which this condition exists if the pressure is 30.3 atm.

14-68. One mole of $H_2O$ vapor is heated to 2800 K and 1 atm. Determine the equilibrium composition assuming that $H_2O$, $H_2$, $O_2$, and OH are present.

14-69. A mixture containing 1 mol of $H_2O$ and 2 mol of $O_2$ is heated to 4000 K at 1 atm. The equilibrium constant for the reaction $H_2O \rightleftharpoons H_2 + \frac{1}{2}O_2$ is 0.582 and for the reaction $H_2O \rightleftharpoons \frac{1}{2}H_2 + OH$ the value of $K_o$ is 0.957. On the basis of these two reactions, determine the equilibrium number of moles of $H_2O$, $H_2$, $O_2$, and OH.

14-70. A system initially contains 1 mol of $CO_2$, 0.5 mol of $O_2$, and 0.5 mol of $N_2$. Upon heating to 3000 K and 1 atm, the mixture contains only

$CO_2$, $CO$, $O_2$, $N_2$, and $NO$. Determine the moles of each at equilibrium.

## RELATIONSHIP BETWEEN $K_0$ AND $\Delta \bar{h}_R^0$

14-71. Consider the formation of $O_2$ from monatomic oxygen (O) by the reaction $2O \rightleftharpoons O_2$.
  (a) On the basis of $K_0$ data and the van't Hoff isobar equation, estimate at (1) 2000 K and (2) 3000 K the enthalpy of formation of $O_2$, in kJ/kmol of $O_2$.
  (b) For a comparison, calculate the $\Delta \bar{h}_R^0$ value based on $\Delta \bar{h}_{f,298}^0$ and sensible enthalpy data, for the same two temperatures.
  (c) Will the amount of $O_2$ formed increase or decrease with an increase in temperature at constant pressure? Discuss.

14-72. Consider the dissociation of diatomic hydrogen into monatomic hydrogen according to the reaction $H_2 \rightleftharpoons 2H$.
  (a) On the basis of $K_0$ data and the van't Hoff isobar equation, estimate the energy required to break $H_2$ down to H at (1) 2000 K and (2) 2500 K and 1 atm, in kJ/kmol of $H_2$.
  (b) For a comparison, calculate the $\Delta \bar{h}_R^0$ value at the given temperatures by using $\Delta \bar{h}_{f,298}^0$ and by noting that the sensible-enthalpy change of monatomic hydrogen for parts 1 and 2 is 35,375 and 45,770 kJ/kmol, respectively.
  (c) Will the amount of H formed increase or decrease with an increase in temperature at constant pressure? Discuss.

14-73. Consider the reaction $CO + \frac{1}{2}O_2 \rightleftharpoons CO_2$.
  (a) Estimate $\Delta \bar{h}_R^0$ at (1) 2200 K and (2) 2500 K for this reaction by using $K_0$ data and the van't Hoff equation, in kJ/kmol of CO.
  (b) For a comparison, calculate the $\Delta \bar{h}_R^0$ value based on $\Delta \bar{h}_{f,298}^0$ and sensible enthalpy data, for the same two temperatures.
  (c) Will the amount of $CO_2$ formed increase or decrease with an increase in temperature at constant pressure? Discuss.

14-74. Consider the reaction $H_2O \rightleftharpoons H_2 + \frac{1}{2}O_2$.
  (a) Estimate $\Delta \bar{h}_R^0$ at (1) 1800 K and (2) 2600 K for this reaction by using $K_0$ data and the van't Hoff equation, in kJ/kmol of $H_2O$.
  (b) For a comparison, calculate the $\Delta \bar{h}_R^0$ value based on $\Delta \bar{h}_{f,298}^0$ and sensible enthalpy data, for the same two temperatures.
  (c) Will the amount of $H_2O$ undergoing dissociation increase or decrease with an increase in temperature at constant pressure? Discuss.

14-75. Consider the reaction $CO + \frac{1}{2}O_2 \rightleftharpoons CO_2$. On the basis of $\Delta \bar{h}_R$ and equilibrium-constant data at 2000 K, and assuming the enthalpy of reaction is constant, (a) estimate from the van't Hoff equation the value of $\log_{10} K_0$ at 2500 K. (b) Compare the calculated $\log_{10} K_0$ value with that shown in Table A-24 for 2500 K.

14-76. On the basis of data given in Prob. 14-43, estimate the enthalpy of ionization for the reaction $N \rightleftharpoons N^+ + e^-$ at 12,000 K, in kJ/kmol and eV/molecule of N.

14-77. For the reaction $N_2O_4 \rightleftharpoons 2NO_2$, the enthalpy of reaction at 300 K is 57,930 kJ/kmol. Discuss whether the amount of $N_2O_4$ in the equilibrium mixture will increase or decrease with (*a*) an increase in temperature at constant pressure and (*b*) an increase in pressure at constant temperature.

14-78. For the dissociation of $SO_3$ according to the reaction $SO_3 \rightleftharpoons SO_2 + \frac{1}{2}O_2$, the equilibrium constant is given by the relation $\ln K_0 = -11,400/T + 10.75$, where $T$ is in kelvins.
   (*a*) Estimate the enthalpy of reaction, in kJ/kmol of $SO_3$.
   (*b*) If 0.01 mol of $SO_3$ is sealed in a rigid but uninsulated vessel initially at 1 atm and 300 K, determine the temperature to which the system must be heated so that 5 percent of the $SO_3$ dissociates.
   (*c*) For the final conditions of part *b*, determine the final pressure in the vessel, in atmospheres.

14-79. The equilibrium-constant data for the reaction $Cs \rightleftharpoons Cs^+ + e^-$ are given in Prob. 14-41. In addition, the value of $\log_{10} K_0$ is $-0.108$ at 1600 K. Estimate the enthalpy of ionization for the above reaction at 1800 K, in kJ/kmol and eV/molecule.

14-80. The equilibrium-constant data for the reaction $Na \rightleftharpoons Na^+ + e^-$ are

| $T$, K | 1600 | 1800 | 2000 | 2200 |
|---|---|---|---|---|
| $\log_{10} K_0$ | $-1.819$ | $-0.913$ | $-0.175$ | 0.438 |

Estimate the enthalpy of ionization for the above reaction at (*a*) 1800 K and (*b*) 2000 K, in kJ/kmol and eV/molecule.

14-81E. Consider the formation of $O_2$ from monatomic oxygen (O) by the reaction $2O \rightleftharpoons O_2$.
   (*a*) On the basis of $K_0$ data and the van't Hoff isobar equation, estimate at (1) 2700°R and (2) 4500°R the enthalpy of formation of $O_2$, in Btu/lbmol of $O_2$.
   (*b*) For a comparison, calculate the $\Delta \bar{h}_R^0$ value based on $\Delta \bar{h}_{f,537}^0$ and sensible enthalpy data, for the same two temperatures.
   (*c*) Will the amount of $O_2$ formed increase or decrease with an increase in temperature at constant pressure? Discuss.

14-82E. Consider the dissociation of diatomic hydrogen into monatomic hydrogen according to the reaction $H_2 \rightleftharpoons 2H$.
   (*a*) On the basis of $K_0$ data and the van't Hoff isobar equation, estimate the energy required to break $H_2$ down to H at (1) 2700°R and (2) 4500°R and 1 atm, in Btu/lbmol of $H_2$.

(b) For a comparison, calculate the $\Delta \bar{h}_R^o$ value at the given
temperatures by using $\Delta \bar{h}_{f,537}^o$ and by noting that the sensible-
enthalpy change of monatomic hydrogen for parts 1 and 2 is
10,740 and 19,670 Btu/lbmol, respectively.

(c) Will the amount of H formed increase or decrease with an
increase in temperature at constant pressure? Discuss.

14-83E. Consider the reaction $CO + \frac{1}{2}O_2 \rightleftharpoons CO_2$.

(a) Estimate $\Delta \bar{h}_R^o$ at (1) 2700°R and (2) 4500°R for this reaction by
using $K_o$ data and the van't Hoff equation, in Btu/lbmol of CO.

(b) For a comparison, calculate the $\Delta \bar{h}_R^o$ value based on $\Delta \bar{h}_{f,537}^o$ and
sensible enthalpy data, for the same two temperatures.

(c) Will the amount of $CO_2$ formed increase or decrease with an
increase in temperature at constant pressure? Discuss.

14-84E. The equilibrium-constant data for the reaction $Cs \rightleftharpoons Cs^+ + e^-$ in
terms of $\log_{10} K_o$ are $-1.01$ at 2520°R and $-0.108$ at 2880°R. Estimate the enthalpy of ionization for the above reaction at 2700°R, in
Btu/lbmol and eV/molecule.

14-85E. The values of $\log_{10} K_o$ for the reaction $Na \rightleftharpoons Na^+ + e^-$ are $-1.819$
at 2880°R and $-0.175$ at 3600°R. Estimate the enthalpy of ionization
for the above reaction at 3240°R, in Btu/lbmol and eV/molecule.

## Parametric Studies and Design

14-86. Nitric oxide (NO) is formed from oxygen and nitrogen in atmospheric
air by the reaction $N_2 + O_2 \rightleftharpoons 2NO$. By means of a computer analysis
determine at 200-degree intervals from 2000 to 3400 K the moles of
NO formed per mole of initial $O_2$ if the initial reactants are (a) 1 mol
of $N_2$ and 1 mol of $O_2$ and (b) 5 mols of $N_2$ and 1 mol of $O_2$. Plot the
results.

14-87. Consider the dissociation reaction $CO_2 \rightleftharpoons CO + \frac{1}{2}O_2$ at 3000 K.
By means of a computer analysis determine the effect of pressure
on the equilibrium amount of $CO_2$. Determine and plot the percent
dissociation of $CO_2$ at equilibrium as the pressure varies from 0.5 to
10 atm.

14-88. Consider the dissociation reaction $H_2O \rightleftharpoons \frac{1}{2}H_2 + \frac{1}{2}O_2$. By means
of a computer analysis determine the percent dissociation of $H_2O$ at
equilibrium at 3000 K as the pressure varies from 0.5 to 10 atm. Plot
the results.

14-89. Consider the gas-phase reaction $CO + \frac{1}{2}O_2 \rightleftharpoons CO_2$. By means of a
computer analysis (a) evaluate $\log_{10} K_o$ as a function of temperature
from 1200 to 3200 K in intervals of 400 K. Also, (b) determine the
percent dissociation of initially pure $CO_2$ as a function of the same
temperatures.

14-90. Consider the gas-phase reaction $O_2 \rightleftharpoons 2O_2$. By means of a computer analysis (a) evaluate $\log_{10} K_o$ as a function of temperature from

2400 to 3400 K in intervals of 200 K. Also, (*b*) determine the percent dissociation of initially pure $O_2$ as a function of the same temperatures.

14-91. Consider the gas-phase reaction $H_2 + \frac{1}{2}O_2 \rightleftarrows H_2O$. By means of a computer analysis (*a*) evaluate $\log_{10} K_o$ as a function of temperature from 2200 to 3400 K in intervals of 200 K. Also, (*b*) determine the percent dissociation of initially pure $H_2O$ as a function of the same temperatures.

14-92. Consider the reaction $CO + H_2O \rightleftarrows CO_2 + H_2$ at 1000 K. By means of a computer analysis determine the moles of $CO_2$ formed per initial mole of CO for various amounts of $H_2O$ present initially with the CO. The range of moles of $H_2O$ per mole of CO is from 1 to 10.

14-93. A reaction vessel initially contains 3 moles of $H_2$ and 6 moles of helium. The gases are heated to 3500 K. Based on the reaction $H_2 \rightleftarrows 2H$, use a computer analysis to determine the effect of pressure on the number of moles of H present at equilibrium. The range of pressures is from 0.5 to 10 atm.

14-94. The energy value of gas streams containing a sizable portion of carbon monoxide can be upgraded by adding steam to the mixture to produce hydrogen in the gas mixture by the reaction $CO + H_2O \rightleftarrows CO_2 + H_2$. The amount of hydrogen formed can be increased by increasing the $H_2O/CO$ ratio in the initial mixture. However, even as the amount of hydrogen increases, the mole fraction of hydrogen eventually will decrease as the ratio increases. As an initial estimate of this trend, consider an initial mixture of only CO and $H_2O$. At 1 atm and 1600 K, determine the initial ratio of $H_2O/CO$ that will maximize the mole fraction of hydrogen in the equilibrium mixture.

14-95. A heat-treatment process requires temperatures ranging from 1300 to 2100 K. Recuperation is a common strategy to reduce fuel costs. In this case recuperation involves a heat exchanger that recovers energy from the exhaust gases and uses it to preheat the combustion air. Currently, air preheat is limited to temperatures to 800 K, but new ceramic technologies have the potential to increase air preheat temperatures to 1600 K. The designer is to provide an assessment of the additional energy savings made possible by using the new recuperator technology.

An estimate is required of the amount of fuel saved at preheat temperatures of 1200 and 1600 K compared to the baseline preheat temperature of 800 K. The fuels of interest are gaseous methane and propane, and liquid octane. The required range of excess-air values is 5 to 50 percent. To estimate the fuel saved, the adiabatic flame temperature is first calculated at the chosen baseline condition. The reduction in fuel use at the higher air preheat temperature is then determined by maintaining the same adiabatic flame temperature as for the baseline flame temperature.

1. As an initial step, determine the adiabatic flame temperatures without preheat and with 800 K preheat at percent excess air values of 5, 25, and 50 percent, assuming the fuel and oxidizer enter at 298 K.
2. Next, determine the percent fuel saved with respect to the baseline air preheat temperature of 800 K, at air preheat temperatures of 1200 and 1600 K for excess air values of 5, 25, and 50 percent. Discuss trends and implications of your results.
3. An energy analysis is incomplete without an accompanying pollutant analysis. The emission index is defined as the mass of pollutant emitted per mass of fuel burned in the combustor, in g/kg. Determine emission indices for carbon monoxide (COEI) and nitric oxide (NOEI) at the same conditions used to evaluate fuel savings from recuperation.

# CHAPTER
# 15

# GAS POWER CYCLES

Mechanics work on a hi-bypass turbofan jet engine.

**A**n important use of thermodynamics is in the study of cyclic devices for power production. In this chapter we restrict ourselves to devices that employ a gas as the working fluid. Modern automotive, truck, and gas-turbine engines are examples of the extremely fruitful application of thermodynamic analysis. In the next two chapters the study is directed toward power and refrigeration processes that involve the presence of two phases during the course of the cycle.

## 15-1    THE AIR-STANDARD CYCLE

Because of the complexities of the actual processes, it is profitable in the initial study of gas power and refrigeration cycles to examine the general characteristics of each cycle without going into a detailed analysis. The advantage of a simple model is that the main parameters which govern the cycle are made more apparent. By stripping the actual process of all its complications and retaining only a bare minimum of detail, we are able to examine the influence of major operating variables on the performance of the device. However, numerical values calculated from such models may not be strictly representative of the actual process. Thus modeling is an important tool in engineering analysis, but at times it is highly qualitative.

Gas cycles are those in which the working fluid remains a gas throughout the cycle. In actual gas power cycles the fluid consists mainly of air, plus the products of combustion such as carbon dioxide and water vapor. Since the gas is predominantly air, especially in gas-turbine cycles, it is convenient to examine gas power cycles in terms of an air-standard cycle. An *air-standard cycle* is an idealized cycle based on the following approximations:

1. Throughout the cycle the working fluid is taken to be air, modeled as an ideal gas.

2. The combustion process is replaced by heat transfer from an external source.

3. A heat-transfer process to the surroundings is used to restore the fluid to its initial state.

In applying the air-standard cycle restrictions to various processes, it is sometimes customary to place additional constraints on the property values of air. In the *cold-air-standard cycle* the specific heats $c_v$ and $c_p$ and the specific-heat ratio $k$ are assumed to have constant values, and these are measured at room temperature. This approach leads to simple relations for parameters such as thermal efficiency. Although they are only semiquantitative in value, these cold-air-standard relations indicate the operating variables of major importance for a particular cycle. However, the numerical results from a cold-air-standard study may be considerably different from those obtained by accounting for variable specific heats. This is due to the wide temperature variation in most gas power cycles, which severely alters the values of $c_v$ and $c_p$ throughout the cycle. Because accurate quantitative results are readily achieved by using the air tables in the Appendix, the cold-air-standard approach is used in this chapter primarily to indicate trends. In practice, of course, it is desirable to use information about the actual gases which result from the combustion of hydrocarbon fuels mixed with air.

## 15-2 THE AIR-STANDARD CARNOT CYCLE

In Chap. 6 it is pointed out that the maximum thermal efficiency of any heat engine operating between two fixed temperature levels is the Carnot efficiency, which is given by the relation

$$\eta_{\text{Carnot}} = 1 - \frac{T_L}{T_H} \qquad \textbf{[6-58]}$$

where $T_L$ and $T_H$ represent the sink temperature and the source temperature, respectively. A Carnot heat engine, as described in Sec. 8-3, undergoes a cyclic process composed of two isothermal, reversible processes and two adiabatic, reversible processes. The $PV$ and $TS$ diagrams for air undergoing a Carnot cycle are shown in Fig. 15-1. This cycle may take place in a closed system, such as a reciprocating piston-cylinder device, or in a steady-flow device. The equipment required for a steady-flow Carnot cycle using air as the working fluid is shown in Fig. 15-2. Steady-flow turbines are required for the isothermal, reversible and adiabatic, reversible expansion processes (steps 1-2 and 2-3). Similarly, steady-flow compressors are needed for the isothermal, reversible and adiabatic, reversible compression

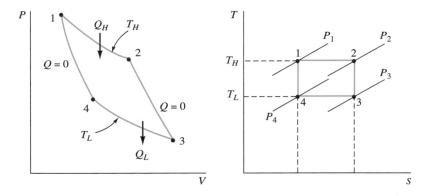

**Figure 15-1**    The *PV* and *Ts* diagrams for a Carnot heat-engine cycle.

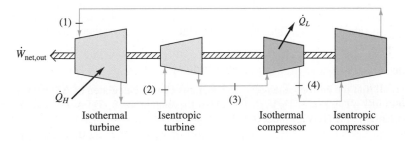

**Figure 15-2**    Steady-flow Carnot heat engine.

processes (steps 3-4 and 4-1). Heat addition $\dot{Q}_H$ occurs to the fluid in the isothermal turbine, while heat removal $\dot{Q}_L$ from the air occurs in the isothermal compressor.

To approach the Carnot efficiency given by Eq. (6-58), an actual engine must be relatively free of dissipative effects, such as friction. In addition, the fluid temperature should be constant during the heat-addition and heat-removal processes. In practice, these restrictions (among others) are impossible to meet. The construction of a heat engine that approximates the Carnot cycle is not practical. Hence the thermal efficiency of an actual engine is always considerably less than that for a Carnot engine operating between the same maximum and minimum temperatures. Nevertheless the performance of a Carnot engine operating on an air-standard cycle is an important standard against which actual engines may be compared.

**Example 15-1**

The pressure, volume, and temperature in a Carnot heat engine using air as the working medium are, at the beginning of the isothermal expansion, 5 bars, 550 cm³, and 260°C, respectively. During the isothermal expansion, 0.30 kJ of energy as heat transfer is added, and the maximum volume for the cycle is 5300 cm³. Determine (a) the volume after isothermal expansion, in cm³ (b) the sink temperature, in degrees Celsius, (c) the volume after isothermal compression, in cm³, (d) the heat-transfer output per cycle, and (e) the thermal efficiency of the closed cycle.

**Solution:**

**Given:** Carnot heat engine using air operates under the conditions shown in Fig. 15-3.

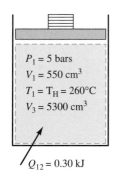

$P_1 = 5$ bars
$V_1 = 550$ cm³
$T_1 = T_H = 260°C$
$V_3 = 5300$ cm³

$Q_{12} = 0.30$ kJ

**Figure 15-3**
Schematic and data for Example 15-1.

**Find:** (a) Volume after isothermal expansion, in cm³ (b) $T_L$, in °C, (c) volume after isothermal compression, in cm³, (d) $Q_L$, per cycle, and (e) $\eta_{th}$.

**Model:** Reversible cycle, closed system, ideal gas.

**Analysis:** (a) The final volume after the isothemal expansion, process 1-2, is set by the heat transfer that occurs during the process. The heat transfer can be related to changes in system properties through either the energy equation or the entropy balance. Writing the entropy balance for a closed system, we have $\Delta S = \int_1^2 (\delta Q/T) + \sigma$. However, the process is isothermal ($T = T_H$) and, since a Carnot cycle is internally reversible, then $\sigma = 0$. Thus, the entropy balance for process 1-2 reduces to

$$Q_{12} = T_H \Delta S.$$

Recall that the entropy change of an ideal gas can be calculated as $\Delta S = m\,\Delta s = m[c_v \ln(T_2/T_1) + R\ln(V_2/V_1)]$. For an isothermal process, $T_2 = T_1$, and $\Delta S = mR\ln(V_2/V_1)$. Thus for process 1-2

$$Q_{12} = T_H\left(mR\ln\frac{V_2}{V_1}\right) = mRT_H\ln\frac{V_2}{V_1} = P_1V_1\ln\frac{V_2}{V_1}$$

where we recognized that $T_H = T_1$ and $PV = mRT$ for an ideal gas. Solving for $V_2$ and substituting in the data gives

$$V_2 = V_1 \exp\left(\frac{Q_{1-2}}{P_1 V_1}\right) = (550 \text{ cm}^3)\exp\left[\frac{0.30\text{kJ}}{(5 \text{ bars})(550 \text{ cm}^3)}\left(\frac{10^4 \text{ bar·cm}^3}{\text{kJ}}\right)\right]$$

$$= (550 \text{ cm}^3)(2.977) = 1637 \text{ cm}^3$$

This result could also be obtained by applying the energy equation to a closed system with an isothermal reversible expansion. For an isothermal expansion of an ideal gas, $\Delta U = 0$ and the energy equation becomes $Q_{12} = -W_{12} = \int P\,dV$.

(b) The sink temperature $T_L(= T_3)$ is found by examining process 2-3, the isentropic expansion. For this process $T_2(= T_1 = T_H)$, $V_2$, and $V_3(= V_{max})$ are known. In addition, since it is isentropic $\Delta S = 0$, and the equation in part a for $\Delta S$ can be written as $0 = c_v \ln(T_3/T_2) + R\ln(V_3/V_2)$. As shown in Sec. 8-1 this equation can be rewritten as $V_3/V_2 = (T_2/T_3)^{1/(k-1)}$ where $k = c_p/c_v$ and is evaluated at an appropriate average value. Assuming a cold-air-standard model, $k = 1.4$. Now solving for $T_3$ and substituting in the data, we find that

$$T_3 = T_2\left(\frac{V_2}{V_3}\right)^{(k-1)} = (533 \text{ K})\left(\frac{1637 \text{ cm}^3}{5300 \text{ cm}^3}\right)^{1.4-1} = (533 \text{ K})(0.625) = 333 \text{ K or } 60°\text{C}$$

(c) The volume after the isothermal compression, process 3-4, could be found in a manner analogous to part a if the heat transfer was known. However, since $Q_{34}$ is unknown, an alternate approach is to examine process 4-1, an isentropic compression. Using the approach of part b, we see that $V_4/V_1 = (T_1/T_4)^{1/(k-1)}$. For a Carnot cycle, we know that $T_H = T_1 = T_2$ and $T_L = T_3 = T_4$. Consequently, $T_4/T_1 = T_3/T_2$ and so $V_4/V_1 = V_3/V_2$. Solving for $V_4$ we have

$$V_4 = V_1\frac{V_3}{V_2} = (550 \text{ cm}^3)\left(\frac{5300 \text{ cm}^3}{1637 \text{ cm}^3}\right) = 1780 \text{ cm}^3$$

(d) The heat transfer for the isothermal compression process, process 3-4, could be found by applying the energy equation to the process. It can also be found by applying the entropy balance to the entire cycle. Recalling that the change in any property for a cycle is zero and $\sigma = 0$ for an internally reversible (Carnot) cycle, the entropy balance can be written as

$$0 = \frac{Q_{12}}{T_H} + \frac{Q_{34}}{T_L}$$

Solving for $Q_{34}$ we have

$$Q_{3-4} = -Q_{1-2}\frac{T_L}{T_H} = -(0.30 \text{ kJ})\left(\frac{333 \text{ K}}{533 \text{ K}}\right) = -0.187 \text{ kJ}$$

Thus the heat transfer *out* of the system during process 3-4 is 0.187 kJ.

(e) The thermal efficiency of this Carnot cycle can be calculated from the simple relation

$$\eta_{\text{Carnot}} = 1 - \frac{T_L}{T_H} = 1 - \frac{333 \text{ K}}{533 \text{ K}} = 0.375 \text{ (or 37.5\%)}$$

A similar result would be obtained using the definition of thermal efficiency, $\eta = W_{\text{net,out}}/Q_{\text{in}}$

**Comment:**   The efficiency is relatively low, because of the low temperature of supply.

## 15-3   SOME INTRODUCTORY NOMENCLATURE FOR RECIPROCATING DEVICES

A number of applications make use of a piston-cylinder arrangement in which the piston is observed to undergo cycles, or revolutions. The *bore* of the piston is its diameter, and the distance the piston moves in one direction is known as the *stroke*. When the piston has moved to a position such that a minimum volume of fluid is left in the cylinder, the piston is said to be at the top dead center (TDC) (see Fig. 15-4*a*). This minimum volume is called the *clearance volume*. When the piston has moved the distance of the stroke so that the fluid now occupies the maximum volume, the piston is in the bottom-dead-center (BDC) position. The volume displaced by the piston as it moves the distance of the stroke between TDC and BDC is the *displacement volume*. The clearance volume is frequently cited in terms of the *percentage of clearance*, which is the percentage of the piston displacement equal to the clearance volume. The **compression ratio r** of a reciprocating device is defined as the volume of the fluid at BDC divided by the volume of the fluid at TDC:

$$r = \frac{V_{\text{BDC}}}{V_{\text{TDC}}} = \frac{\text{clearance volume} + \text{displacement volume}}{\text{clearance volume}} \qquad \textbf{[15-1]}$$

The compression ratio is always expressed in terms of a volume ratio.

The **mean effective pressure** (MEP) is a useful parameter in the study of reciprocating devices used for power production. It is defined as the average pressure that would act on the piston during the entire power or outward stroke and produce the same work output as the net work output for the actual cyclic process. The net work output per cycle is given by

$$W_{\text{cycle,out}} = (\text{MEP})(\text{piston area})(\text{stroke}) \qquad \textbf{[15-2]}$$
$$= (\text{MEP})(\text{displacement volume})$$

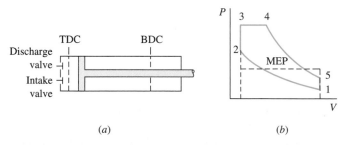

(a)                              (b)

**Figure 15-4**     Interpretation of mean effective pressure on a *PV* diagram.

For reciprocating engines of comparable size, a larger mean effective pressure is an indication of better performance in terms of power produced at the same rated speed. For the purpose of illustration, consider the hypothetical clockwise cycle 1-2-3-4-5-1 on the *PV* diagram in Fig. 15-4*b*. The net work produced is represented by the enclosed area on the diagram. The mean effective pressure of the cycle is shown by the horizontal line, and the area under this line equals the enclosed area of the actual cycle.

## 15-4 THE AIR-STANDARD OTTO CYCLE

The four-stroke spark-ignition (SI) engine is an important component in the operation of a technology to meet the modern needs of society. Although it has undergone some modifications to meet pollution standards for mobile equipment, this engine will undoubtedly continue to play a significant role as a device for producing relatively small quantities of power. A typical *PV* diagram for such an engine at wide-open throttle is shown in Fig. 15-5. The series of events includes the intake stroke *ab*, the compression stroke *bc*, the expansion or power stroke *cd*, and finally the exhaust stroke *da*. The intake and exhaust strokes occur essentially at atmospheric pressure. The process lines *ab* and *da* do not lie on top of each other; but since Fig. 15-5 is drawn to scale, it is difficult to show the separation between the intake- and exhaust-process lines, except near the BDC position. Normally the point of ignition occurs on the compression stroke before the TDC position, since flame propagation across the combustion chamber takes a finite time. For a given engine, the point of ignition can be altered until the setting for maximum power is determined. Note also that the exhaust valve is opened before the piston reaches BDC. This allows the exhaust gases to nearly reach atmospheric pressure before the exhaust stroke begins.

As discussed in Sec. 15-1, the initial step in analyzing the performance of a four-stroke reciprocating spark-ignition engine is to prepare a simple model of the overall process. Although such a model is of value only in a qualitative sense, it does provide some information on the influence of major operating variables on performance.

A theoretical cycle of interest in analyzing the behavior of reciprocating spark-ignition engines is the **Otto cycle.** It is named for Nicholas Otto, a German engineer, who produced a successful four-stroke engine in 1876. A four-stroke Otto cycle is composed of four internally reversible processes, plus an intake and exhaust portion of the cycle. Both *PV* and *TS* diagrams for the theoretical cycle are shown in Fig. 15-6. Consider a piston-cylinder assembly containing air and the piston situated at the bottom-dead-center position. This is shown as point 1 on the diagram. As the piston moves to the top-dead-center position, compression of the air occurs adiabatically. Since the processes are reversible, the compression process is isentropic, ending at state 2. Then heat transfer to the air occurs instantaneously, so that both the pressure and the temperature rise to high values during a constant-volume process 2-3. As the piston now moves toward the BDC position once more,

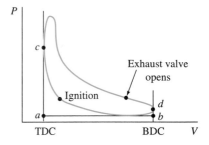

**Figure 15-5**
Typical *PV* diagram for wide-open throttle for a four-stroke spark-ignition engine.

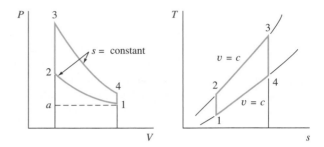

**Figure 15-6**    The *PV* and *Ts* diagrams for an air-standard Otto cycle.

the expansion is carried out adiabatically and internally reversibly, i.e., isentropically, to state 4. Now with the piston in its BDC position, heat transfer occurs from the air at constant volume until the initial state is achieved.

At this point the fluid, theoretically, could begin to go through another cycle. To make the cycle somewhat more realistic, the following sequence might be considered before resuming the cyclic pattern: In practice, the gases would contain the products of hydrocarbon combustion, so that an exhaust stroke would be necessary. Consequently, the exhaust valve now opens, and the piston moves from BDC to TDC, expelling the gases to the ambient surroundings. Then the exhaust valve closes and the intake valve opens, while the piston returns to the BDC position. During this suction stroke the cylinder is filled with fresh air for the next cycle. In the air-standard cycle this recharging of the cylinder is not necessary, since the same fluid continually undergoes the cyclic variations. Note that the work required to push the charge from the cylinder is the same in magnitude but opposite in sign to that required to suck in the new charge. Hence these two portions of the theoretical cycle do not affect the net work done by the cycle, and only cycle 1-2-3-4-1 in Fig. 15-6 is important in the thermodynamic analysis. In review, the theoretical ***Otto cycle*** is composed of the following internally reversible processes:

1-2.   Adiabatic compression

2-3.   Constant-volume heat addition

3-4.   Adiabatic expansion

4-1.   Constant-volume heat rejection

In addition, one may consider an exhaust stroke, 1-*a*, and an intake stroke, *a*-1, for completeness, although this is not necessary.

Since the air acts as a closed system, the conservation of energy principle on a unit mass basis is

$$q + w = \Delta u$$

where kinetic- and potential-energy effects are neglected. For the adiabatic compression and expansion process, $q = 0$ and

$$w = \Delta u$$

For the constant-volume heat-input and heat-rejection processes $w = 0$ and

$$q = \Delta u$$

At this point it is informative to analyze the Otto cycle on the basis of the cold-air-standard cycle, since it provides some insight as to the important parameters that affect the thermal efficiency of the cycle. On the basis of the cold-air-standard cycle, we find that

$$q_{in} = q_{23} = u_3 - u_2 = c_v(T_3 - T_2)$$

and
$$q_{out} = -q_{41} = u_4 - u_1 = c_v(T_4 - T_1)$$

Since the net work out is $q_{in} - q_{out}$, the thermal efficiency is given by

$$\eta_{th} = \frac{w_{net,out}}{q_{in}} = \frac{c_v(T_3 - T_2) - c_v(T_4 - T_1)}{c_v(T_3 - T_2)}$$

$$= 1 - \frac{T_4 - T_1}{T_3 - T_2} = 1 - \left(\frac{T_1}{T_2}\right)\frac{T_4/T_1 - 1}{T_3/T_2 - 1}$$

Note that $V_2 = V_3$ and $V_1 = V_4$. Since the isentropic relations show that

$$\frac{T_2}{T_1} = \left(\frac{V_1}{V_2}\right)^{k-1} \qquad \text{and} \qquad \frac{T_3}{T_4} = \left(\frac{V_4}{V_3}\right)^{k-1} = \left(\frac{V_1}{V_2}\right)^{k-1}$$

then $T_2/T_1 = T_3/T_4$, or $T_4/T_1 = T_3/T_2$. When this result is substituted into the equation for the thermal efficiency of a cold-air-standard Otto cycle,

$$\eta_{th,Otto} = 1 - \frac{T_1}{T_2} = 1 - \left(\frac{V_2}{V_1}\right)^{k-1} = 1 - \frac{1}{r^{k-1}} \qquad \textbf{[15-3]}$$

where $r$ is the *compression ratio* for the engine, defined by

$$r \equiv \frac{V_1}{V_2} = \frac{v_1}{v_2} \qquad \textbf{[15-4]}$$

Equation [15-3] indicates that the major parameters governing the thermal efficiency of an Otto cycle are the compression ratio and the specific-heat ratio. The influence of these factors on the thermal efficiency is demonstrated in Fig. 15-7. For a given specific-heat ratio, the value of the thermal efficiency increases with increasing compression ratio. Note, however, that the curves flatten out at compression ratios above 10 or so. Hence the advantage of operating at high compression ratios lessens rapidly. From a practical viewpoint, the compression ratio is limited by the occurrence of preignition or engine knock when the compression ratio rises much above 10 for common hydrocarbon fuels. Figure 15-7 also shows that the thermal efficiency increases with increasing specific-heat ratio. This means that higher values of the thermal efficiency are obtained when simple monatomic or diatomic molecules are present in the working fluid. Unfortunately, the presence of carbon dioxide, water vapor, and other heavier molecules in actual practice makes the attainment of high specific-heat-ratio values not practical.

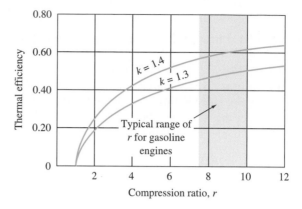

**Figure 15-7**    Thermal efficiency of the air-standard Otto cycle as a function of compression ratio and specific-heat ratio.

To account for variable specific-heat values, the thermal efficiency must be determined by the relation

$$\eta_{th,Otto} = 1 - \frac{u_4 - u_1}{u_3 - u_2} \qquad \textbf{[15-5]}$$

In addition, the $u$ values must be read from the air table in the Appendix at the corresponding temperatures around the cycle. The temperatures at states 2 and 4 are found from the isentropic relations in terms of relative volume $v_r$ data:

$$v_{r2} = v_{r1}\frac{V_2}{V_1} = \frac{v_{r1}}{r} \qquad \text{and} \qquad v_{r4} = v_{r3}\frac{V_4}{V_3} = rv_{r3}$$

Recall that the $v_r$ data are solely a function of temperature and account for the effect of variation of specific heats with temperature.

The following example is based on an air-standard Otto cycle. The presence of combustion products, especially in the power stroke of a real engine, greatly affects the analysis and the performance values.

**Example 15-2**

The initial conditions for an air-standard Otto cycle operating with a compression ratio of 8:1 are 0.95 bar and 17°C. At the beginning of the compression stroke, the cylinder volume is 2.20 L, and 3.60 kJ of energy is added as heat transfer during the constant-volume heating process. (*a*) Calculate the pressure and temperature at the end of each process of the cycle, and (*b*) determine the thermal efficiency and mean effective pressure of the cycle.

**Solution:**

**Given:**    An air-standard Otto cycle operates under the conditions shown in Fig. 15-8.

**Find:** (a) The pressure and temperature at the end of each process, and (b) $\eta_{\text{th}}$ and MEP of the cycle.

**Model:** Closed system, air-standard Otto cycle.

**Analysis:** (a) By denoting the states in the same manner as in Fig. 15-6, we let $P_1 = 0.95$ bar and $T_1 = 290$ K. Assuming that air is an ideal gas,

$$v_1 = \frac{RT}{P} = \frac{0.08314(290)}{29(0.95)} = 0.875 \text{ m}^3/\text{kg}$$

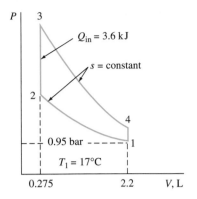

**Figure 15-8**
A PV diagram showing data for Example 15-2.

The state after isentropic compression is determined by use of relative-volume $v_r$ data, which account for variable specific heats. On the basis of data from Table A-5

$$v_{r2} = v_{r1}\frac{v_2}{v_1} = \frac{v_{r1}}{r} = \frac{676.1}{8.0} = 84.5$$

Linear interpolation in the table shows that $T_{2s} = 652$ K, $p_{r2} = 22.17$, and $u_{2s} = 475.1$ kJ/kg. One method of evaluating $P_{2s}$ is to use the isentropic relation involving relative pressure $p_r$ data,

$$P_{2s} = P_1\frac{p_{r2}}{p_{r1}} = 0.95 \text{ bars} \times \frac{22.17}{1.2311} = 17.1 \text{ bars}$$

Note that $P_{2s}$ can also be calculated from the ideal-gas relation $P_2 = P_1(V_1/V_2) \times (T_2/T_1)$. To determine state 3, the amount of heat transfer to the air per kilogram must be found from

$$q_{23} = \frac{Q_{23}}{m} = \frac{Q_{23}v_1}{V_1} = \frac{3.60 \text{ kJ}(0.875 \text{ m}^3/\text{kg})}{2.2 \times 10^{-3} \text{ m}^3} = 1432 \text{ kJ/kg}$$

Since $q_{\text{in}} = q_{23} = u_3 - u_2$, the internal energy at state 3 is found by

$$u_3 = u_2 + q_{23} = 475.1 + 1432 = 1907.1 \text{ kJ/kg}$$

Again, by linear interpolation in the air table, we find that $T_3 = 2235$ K, $p_{r3} = 3369$, and $v_{r3} = 1.907$. The pressure after constant-volume heat addition is simply $P_3 = P_2(T_3/T_2) = 17.1(2234/652) = 58.6$ bars. State 4 is found again from the isentropic relation:

$$v_{r4} = v_{r3}\frac{v_4}{v_3} = 1.907(8) = 15.26$$

Linear interpolation in the air table reveals that $T_{4s} = 1180$ K, $p_{r4} = 222.2$, and $u_{4s} = 915.6$ kJ/kg. Thus

$$P_4 = P_3(p_{r4}/p_{r3}) = 58.6(222.2/3369) = 3.9 \text{ bars}$$

(b) The thermal efficiency is found from the heat transfers for the cycle. The heat transfer out of the cycle is

$$q_{\text{out}} = -q_{41} = u_4 - u_1 = 915.6 - 206.9 = 708.7 \text{ kJ/kg}$$

Consequently,

$$\eta_{\text{th}} = 1 - \frac{q_{\text{out}}}{q_{\text{in}}} = 1 - \frac{708.7}{1432} = 0.505 \text{ (or 50.5\%)}$$

Equation [15-3] was not used to calculate the thermal efficiency, because it is based on constant-specific-heat data. Finally, the mean effective pressure of the cycle is found by applying the relation, MEP $= w_{net,out}/(v_1 - v_2)$. First, the net work is

$$w_{net,out} = q_{in} - q_{out} = 1432 - 709 = 723 \text{ kJ/kg}$$

Then Eq. [15-2] yields

$$\text{MEP} = \frac{w_{net,out}}{v_1 - v_2} = \frac{723 \text{ kJ/kg}}{(0.875 - 0.109) \text{ m}^3/\text{kg}} \times \frac{\text{bar m}^2}{10^5 \text{ N}} \times 10^3 \text{ g/kg} = 9.44 \text{ bars}$$

where $v_2$ is one-eighth of $v_1$.

## 15-5   THE AIR-STANDARD DIESEL CYCLE AND THE DUAL CYCLE

In a spark-ignition engine, the fuel is ignited by energy supplied from an external source. An alternative method for initiating the combustion process in a reciprocating engine is to raise the fuel-air mixture above its autoignition temperature. An engine built on this principle is called a *compression-ignition* (CI) *engine*. By using compression ratios in the range of 14:1 to 24:1 and using diesel fuel instead of gasoline, the temperature of the air within the cylinder will exceed the ignition temperature at the end of the compression stroke. If the fuel were premixed with the air, as in a spark-ignition engine, combustion would begin throughout the mixture when the ignition temperature was reached. As a result, we would have no control on the timing of the combustion process. To overcome this difficulty, the fuel is injected into the cylinder in a separate operation. Injection begins when the piston is near the TDC position. Thus the CI engine differs from the spark-ignition (SI) engine primarily in the method of achieving combustion and in the adjustment of the timing of the combustion process. The rest of the four-stroke CI cycle is similar to that of the SI cycle discussed in Sec. 15-4.

### 15-5-1   THE DIESEL CYCLE

A typical *PV* diagram for a compression-ignition engine is shown in Fig. 15-9. Note that the modern CI engine, frequently called a *diesel engine*, has a PV diagram very similar to that of a spark-ignition engine at full load. In the earlier history of this engine, the combustion part of the cycle was somewhat flatter, so that the initial section of the expansion process was closer to a constant-pressure process. As a result, the compression-ignition engine was modeled early in its history by a theoretical cycle known as a **Diesel cycle**. The cycle is named for Rudolf Diesel, who obtained a patent on a compression-ignition engine in the 1890s. The modern compression-ignition engine, shown in cross section in Fig. 15-10, is better modeled by the Otto cycle discussed in Sec. 15-4 or the dual cycle discussed later in

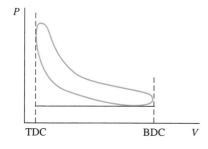

**Figure 15-9**

Typical *PV* diagram for a compression-ignition engine at rated load.

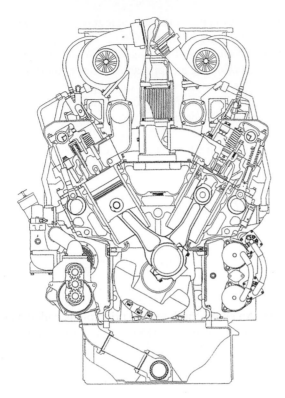

**Figure 15-10**    A cross-section of a Caterpillar 3500 Series B electronically controlled diesel engine. (Permission for use from the Lafayette Large Engine Center, Caterpillar Inc.)

this section. Only the basic characteristics of the theoretical Diesel cycle are outlined below.

The theoretical Diesel cycle for a reciprocating engine is shown in Fig. 15-11 on both $PV$ and $Ts$ diagrams. This cycle, like the Otto cycle, is composed of four internally reversible processes. The only difference

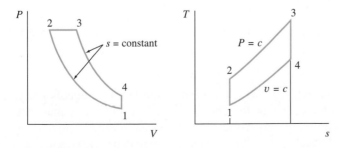

**Figure 15-11**    The $PV$ and $Ts$ diagrams for the air-standard Diesel cycle.

between the two cycles is that a Diesel cycle models the combustion as occurring at constant pressure, while the Otto cycle assumes constant-volume combustion. A useful analysis of the Diesel cycle is made possible by cold-air-standard cycle. Following an approach similar to that for the Otto cycle, the heat transfers for the cycle can be developed by applying the general energy balance $q + w = \Delta u$. For the constant-pressure heat addition process 2-3, the work for the process is $w_{23} = -\int P\,dv = -P\,\Delta v$ and the heat transfer is $q_{23} = \Delta u - w_{23} = \Delta u - (-P\,\Delta v) = \Delta h$. The heat transfer for the constant-volume heat rejection process 4-1 is calculated as it was for the Otto cycle, $q_{41} = \Delta u$. Using these results the heat transfers for the cold-air-standard Diesel cycle are

$$q_{\text{in}} = q_{23} = \Delta h = c_p(T_3 - T_2) \quad \text{and} \quad q_{\text{out}} = -q_{41} = -\Delta u = c_v(T_4 - T_1)$$

Consequently,

$$\eta_{\text{th,Diesel}} = \frac{q_{\text{in}} - q_{\text{out}}}{q_{\text{in}}} = \frac{c_p(T_3 - T_2) - c_v(T_4 - T_1)}{c_p(T_3 - T_2)} \qquad \textbf{[15-6]}$$
$$= 1 - \frac{T_4 - T_1}{k(T_3 - T_2)}$$

At this point the concept of the ***cutoff ratio*** $r_c$ is introduced. On the basis of Fig. 15-11, it is defined as the ratio of the volume $V_3$ after heat addition to the volume $V_2$ before heat addition. That is,

$$r_c \equiv \frac{V_3}{V_2} = \frac{v_3}{v_2} \qquad \textbf{[15-7]}$$

By recalling that the compression ratio $r$ is defined as $V_1/V_2$, the preceding equation for the thermal efficiency of a cold air-standard Diesel cycle containing temperatures can be expressed in terms of volumes. The result is

$$\eta_{\text{th,Diesel}} = 1 - \frac{1}{r^{k-1}}\left[\frac{r_c^k - 1}{k(r_c - 1)}\right] \qquad \textbf{[15-8]}$$

This equation indicates that the efficiency of the theoretical Diesel cycle is primarily a function of the compression ratio $r$, the cutoff ratio $r_c$, and the specific-heat ratio $k$.

The term in the brackets of Eq. [15-8] is always equal to or greater than unity. Hence for the same compression ratio and when $r_c$ is greater than unity,

$$\eta_{\text{th,Otto}} > \eta_{\text{th,Diesel}}$$

In the limit when $r_c = 1$, the term in the brackets equals unity, and the thermal efficiencies of the Otto and Diesel cycles are equal. An increase in the cutoff ratio has a drastic effect on the thermal efficiency of an air-standard Diesel cycle. This effect is shown in two diagrams in Fig. 15-12. In Fig. 15-12a the thermal efficiency is plotted against the compression ratio with lines of constant cutoff ratio. When the cutoff ratio is unity, the line

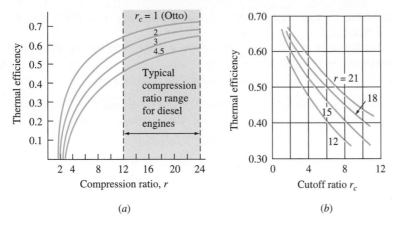

**Figure 15-12** The thermal efficiency versus cutoff ratio and compression ratio (for k = 1.40) for an air-standard Diesel cycle.

represents an Otto cycle as well as a Diesel cycle. In Fig. 15-12b $\eta_t h$ is plotted versus the cutoff ratio with lines of constant compression ratio. The rapid decrease in thermal efficiency with increasing cutoff ratio at a given compression ratio is clearly shown in this latter plot.

In the case where variable specific-heat values are to be considered, the equation for the thermal efficiency of the Diesel cycle becomes

$$\eta_{th,Diesel} = 1 - \frac{u_4 - u_1}{h_3 - h_2} \qquad \textbf{[15-9]}$$

To use Eq. [15-9], it is necessary to evaluate the $u$ and $h$ data from the air table in the Appendix. In this case the temperatures at states 2 and 4 are found from the isentropic relations

$$v_{r2} = v_{r1}\frac{V_2}{V_1} = \frac{v_{r1}}{r} \qquad \text{and} \qquad v_{r4} = v_{r3}\frac{V_4}{V_3} = \frac{rv_{r3}}{r_c} \qquad \textbf{[15-10]}$$

Recall that the $v_r$ data in the air table are solely a function of temperature.

---

**Example 15-3**

**T**he intake conditions for a theoretical Diesel cycle operating with a compression ratio of 18 are 14.4 psia and 60°F. The cutoff ratio is 2. (*a*) Calculate the pressure and temperature at the end of each process of the cycle, and (*b*) determine the thermal efficiency and mean effective pressure of the cycle.

**Solution:**

**Given:** A Diesel cycle with conditions shown in Fig. 15-13.

**Find:** (*a*) Pressures and temperatures around the cycle, and (*b*) $\eta_{th}$ and MEP.

**Model:** Air-standard Diesel cycle, variable specific heat.

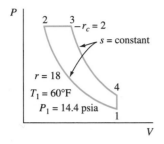

**Figure 15-13**
A PV diagram showing data for
Example 15-3.

**Analysis:** From Table A-5E, $v_{r,1} = 158.58$, $p_{r,1} = 1.2147$, and $u_1 = 88.62$ Btu/lb$_m$.

(a) Based on an ideal-gas model, state 2 is found by the isentropic relation based on relative volume data,

$$v_{r2} = v_{r1}(v_2/v_1) = 158.58(1/18) = 8.81$$

Therefore, from Table A-5E, $T_2 = 1564°R$, $p_{r2} = 65.74$, and $h_2 = 386.10$ Btu/lb$_m$. Then from an isentropic relation involving relative-pressure data,

$$P_2 = P_1(p_{r2}/p_{r1}) = 14.4(65.74/1.2147) = 779 \text{ psia}$$

Since $P_3 = P_2$, then state 3 is found from the ideal-gas relation $T_3/T_2 = v_3/v_2 = r_c$. Thus $T_3 = 2(1564) = 3128°R$, and from the air table, $v_{r3} = 1.030$, $p_{r3} = 1127$, and $h_3 = 828.27$ Btu/lb$_m$. Finally, state 4 is found from the isentropic relation involving relative-volume data,

$$v_{r4} = v_{r3}\frac{V_4}{V_3} = v_{r3}\frac{V_4}{V_2}\frac{V_2}{V_3} = v_{r3}\frac{r}{r_c}$$
$$= 1.030(18/2) = 9.27$$

Therefore, from Table A-5E, $T_4 = 1538°R$, $p_{r4} = 67.8$, and $u_4 = 273.09$ Btu/lb$_m$. The pressure after isentropic expansion is then

$$p_4 = p_3(p_{r4}/p_{r3}) = 779(67.8/1126) = 46.9 \text{ psia}$$

(b) The thermal efficiency is determined from the net work output and the heat transfer supplied. For a closed system, $q + w = \Delta u$. For a cycle, since $\Delta u = 0$, then $w_{net,out} = q_{in} - q_{out}$. Thus

$$q_{in} = q_{23} = \Delta u + P\,\Delta v = h_3 - h_2 = 828.27 - 386.10 = 442.17 \text{ Btu/lb}_m$$

$$q_{out} = -q_{41} = u_4 - u_1 = 273.09 - 88.62 = 184.47 \text{ Btu/lb}_m$$

$$w_{net,out} = 442.17 - 184.47 = 257.70 \text{ Btu/lb}_m$$

Hence

$$\eta_{th} = \frac{w_{net,out}}{q_{in}} = \frac{257.70}{442.17} = 0.583 \qquad \text{(or 58.3 percent)}$$

The mean effective pressure requires the value of $v_1$.

$$v_1 = \frac{RT_1}{P_1} = \frac{10.73(520)}{28.97\,(14.4)} = 13.37 \text{ ft}^3/\text{lb}_m$$

and $v_2 = v_1/r = 13.37/18 = 0.74 \text{ ft}^3/\text{lb}_m$. Therefore,

$$\text{MEP} = \frac{w_{net,out}}{v_1 - v_2} = \frac{257.7}{13.37 - 0.74} \times \frac{778}{144} = 110 \text{ psia}$$

where 778 and 144 are unit conversion factors.

## 15-5-2  THE DUAL CYCLE

A theoretical cycle that comes closer than the Diesel cycle to matching the actual performance of modern compression-ignition engines is the **dual cycle.** As shown in Fig. 15-14, a short heat-addition process $2 - x$ at constant volume is followed by a second heat-addition process $x - 3$ at constant pres-

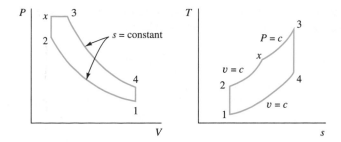

**Figure 15-14**   The *PV* and *Ts* diagrams for the air-standard dual cycle.

sure. The other three parts of the cycle are similar to those found in the Otto and Diesel cycles. The dual cycle is also called a *mixed,* or *limited-pressure, cycle.* Note that the use of a two-step heat-addition process allows the theoretical dual cycle to model fairly closely the upper left-hand portion of the actual performance curve shown in Fig. 15-9 for a compression-ignition engine.

The thermal efficiency of the air-standard dual cycle is a function of the heat-input and heat-output quantities. These are found from the general closed system energy balance $q + w = \Delta u$. The heat transfer in during the constant-volume process 2-*x* is

$$q_{2x} = u_x - u_2 = c_v(T_x - T_2) \qquad \text{(constant volume)}$$

if we assume that the specific heats are constant. For the constant-pressure path *x*-3 the heat transfer in is

$$q_{x3} = \Delta u + P\,\Delta v = h_3 - h_x = c_p(T_3 - T_x) \qquad \text{(constant pressure)}$$

Therefore, the total heat transfer in for the dual cycle is

$$q_{\text{in,tot}} = q_{2x} + q_{x3} = c_v(T_x - T_2) + c_p(T_3 - T_x)$$

The heat transfer out during the constant-volume path is

$$q_{\text{out}} = -q_{41} = u_4 - u_1 = c_v(T_4 - T_1)$$

Consequently, the thermal efficiency of an air-standard dual cycle for constant-specific-heat data is

$$\eta_{\text{th,dual}} = \frac{q_{\text{in}} - q_{\text{out}}}{q_{\text{in}}} = 1 - \frac{c_v(T_4 - T_1)}{c_v(T_x - T_2) + c_p(T_3 - T_x)} \qquad \textbf{[15-11a]}$$

$$= 1 - \frac{T_4 - T_1}{(T_x - T_2) + k(T_3 - T_x)}$$

where $k$ is the specific-heat ratio. Recall that the compression ratio $r = V_1/V_2$ and the cutoff ratio $r_c = V_3/V_x = V_3/V_2$. If, in addition, we define the *pressure ratio* $r_{p,v}$ during the constant-volume combustion process as

$$r_{p,v} = \frac{P_x}{P_2} = \frac{P_3}{P_2}$$

then the thermal efficiency as evaluated by Eq. [15-8a] may be expressed solely as a function of $r$, $r_c$, $r_{p,v}$, and $k$. The result is

$$\eta_{\text{th,dual}} = 1 - \frac{1}{r^{k-1}} \left[ \frac{r_{p,v} r_c^k - 1}{k r_{p,v}(r_c - 1) + r_{p,v} - 1} \right] \quad \textbf{[15-11b]}$$

When $r_{p,v}$ is unity, Eq. [15-11b] reduces to the Diesel efficiency equation, [Eq. [15-5]]. Also, Eq. [15-11b] reduces to Eq. [15-3] for the Otto cycle when $r_c$ is unity.

On the basis of the same heat input and compression ratio, the thermal efficiency of the three theoretical cycles decreases in the following order: Otto cycle, dual cycle, Diesel cycle. This is a major reason that modern CI engines are designed to operate closer to the Otto- or dual-cycle models, rather than the Diesel cycle.

**Example 15-4**

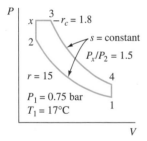

**Figure 15-15**
A *PV* diagram showing data for Example 15-4.

**T**he intake conditions for an air-standard dual cycle operating with a compression ratio of 15:1 are 0.95 bar and 17°C. The pressure ratio during constant-volume heating is 1.5:1, and the volume ratio during the constant-pressure part of the heating process is 1.8:1. Calculate (*a*) the temperatures and pressures around the cycle, (*b*) the heat transfers for the cycle, in kJ/kg, and (*c*) the thermal efficiency, using air-table data.

**Solution:**

**Given:**   An air-standard dual cycle operates under the conditions shown in Fig. 15-15.

**Find:**   (*a*) The temperatures and pressures around the cycle, (*b*) $q_{\text{in}}$ and $q_{\text{out}}$, and (*c*) $\eta_{\text{th}}$.

**Model:**   Air-standard dual cycle, variable specific heat.

**Analysis:**   (*a*) We use the notation of Fig. 15-15. Air is assumed to be an ideal gas. On this basis the pressure and temperature after isentropic compression are found from $v_r$ data:

$$v_{r2} = v_{r1} \frac{v_2}{v_1} = \frac{v_{r1}}{r} = 676.1 \frac{1}{15} = 45.07$$

By linear interpolation in Table A-5, $T_{2s} = 819$ K, $u_{2s} = 607.4$ kJ/kg, and $p_{r2} = 52.15$. Using the last value, we find that

$$P_2 = P_1(p_{r2}/p_{r1}) = 0.95(52.15/1.2311) = 40.24 \text{ bars}$$

The pressure after the constant-volume heat addition is simply

$$P_x = \left( \frac{P_x}{P_2} \right) P_2 = 1.5(40.24) = 60.4 \text{ bars}$$

On the basis of the ideal-gas equation at constant volume,

$$T_x = T_2 \frac{P_x}{P_2} = 819(1.5) = 1229 \text{ K}$$

The pressure at state 3 is the same as that at state $x$, namely, 60.4 bars. Also at state 3, based on the ideal-gas equation at constant pressure,

$$T_3 = T_x \frac{V_3}{V_x} = 1229(1.8) = 2212 \text{ K}$$

State 4 is determined from isentropic relations. First, however, we need to determine $V_3/V_4$. This ratio is $V_3/V_4 = V_3/V_1 = (V_3/V_2)(V_2/V_1) = r_c/r$. Therefore, since $v_{r3} = 1.98$ at 2212 K,

$$v_{r4} = v_{r3} \frac{V_4}{V_3} = v_{r3} \frac{r_c}{r} = 1.98 \frac{15}{1.8} = 16.5$$

By linear interpolation in Table A-5, $T_{4s} = 1150$ K, $p_{r4} = 200.2$, and $u_{4s} = 889.1$ kJ/kg. Finally, based on $p_{r3} = 3216$ at 2212 K for the isentropic process 3-4,

$$P_4 = P_3 \frac{p_{r4}}{p_{r3}} = 60.4 \frac{200.2}{3216} = 3.76 \text{ bars}$$

(b) The heat transfer to the cycle is the sum of terms of processes 2-$x$ and $x$-3. Hence

$$q_{\text{in}} = q_{2x} + q_{x3} = u_x - u_2 + h_3 - h_x$$
$$= 959.1 - 607.4 + 2518.4 - 1311.9 = 1558.2 \text{ kJ/kg}$$

The heat transfer from the cycle is

$$q_{\text{out}} = q_{41} = u_1 - u_4 = 206.9 - 889.1 = -682.2 \text{ kJ/kg}$$

(c) The thermal efficiency for this dual cycle is

$$\eta_{\text{th}} = 1 - \frac{q_{\text{out}}}{q_{\text{in}}} = 1 - \frac{682.2}{1558.2} = 0.562 \text{ (or 56.2\%)}$$

**Comment:** If $k = 1.4$ for a cold-air cycle, Eq. [15-11$b$] predicts a thermal efficiency of 62.5 percent.

## 15-6 THE AIR-STANDARD BRAYTON CYCLE

In a simple gas-turbine power cycle, separate equipment is used for the various processes of the cycle. Initially, air is compressed adiabatically in a rotating axial or centrifugal compressor. At the end of this process, the air enters a combustion chamber where fuel is injected and burned at essentially constant pressure. The products of combustion are then expanded through a turbine until they reach the ambient pressure of the surroundings. A cycle composed of these three steps is called an *open cycle* because the cycle is not actually completed. Actual gas-turbine cycles are open cycles, since air must continually be introduced into the compressor. A typical open cycle gas turbine designed to power an electric generator is shown in Photograph 15-1. The shaft power output is on the left end. Air enters the engine to the left of the compressor and travels axially through the compressor, the combustor, and the turbine. The hot exhaust gases leave the turbine on the right end. If one wishes to

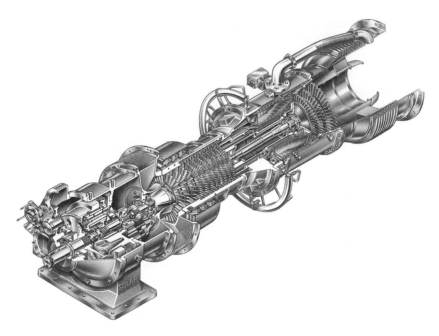

**Photograph 15-1**     4.6-MW Centaur® 50 gas turbine for generator applications. (Courtesy of Solar Turbines Incorporated)

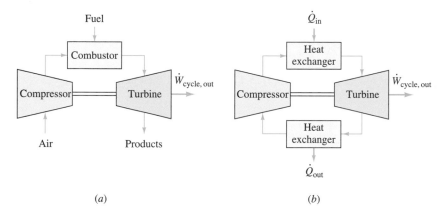

(a)                                          (b)

**Figure 15-16**     Gas turbine operating on the (a) open and (b) closed Brayton cycle.

examine a closed cycle, the products of combustion which have expanded through the turbine must be sent through a heat exchanger, where heat transfer occurs from the gas until the initial temperature is attained. The open and closed gas-turbine cycles are shown in Fig. 15-16.

In the analysis of gas-turbine cycles, it is helpful initially to employ an idealized air-standard cycle. The ideal cycle is called a **_Brayton cycle,_** in honor of the work in the 1870s of George Brayton, a U.S. engineer. Compared with the Otto and Diesel cycles, the Brayton cycle operates over a wider range of volumes but a smaller range of pressures and temperatures.

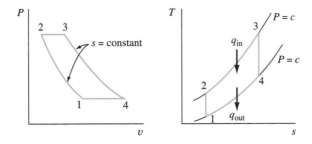

**Figure 15-17**     Typical $Pv$ and $Ts$ diagrams of the air-standard Brayton cycle.

This feature makes the Brayton cycle unadaptable for use with reciprocating machinery. The air-standard closed gas-turbine cycle is composed of four internally-reversible processes: adiabatic compression, heat-transfer input at constant pressure, adiabatic expansion, and heat-transfer output at constant pressure. Thus the compression and expansion processes are isentropic, and the actual combustion process is replaced by a heat-transfer process. The use of air as the working medium throughout the cycle is a fairly good model of actual practice because an air-fuel ratio of roughly 50:1 on a mass basis is quite common. The $Pv$ and $Ts$ diagrams for the ideal cycle are shown in Fig. 15-17. The steady-state energy balance for each device in the cycle is $q + w = \Delta h + \Delta \text{ke} + \Delta \text{pe}$. If $\Delta \text{ke}$ and $\Delta \text{pe}$ are negligible, then for isentropic work processes 1-2 and 3-4,

$$w = \Delta h$$

Similarly, for the heat exchangers used in processes 2-3 and 4-1

$$q = \Delta h$$

The thermal efficiency of the ideal Brayton cycle is given by

$$\eta_{\text{th,Brayton}} = 1 - \frac{q_{\text{out}}}{q_{\text{in}}} = 1 - \frac{h_{4s} - h_1}{h_3 - h_{2s}} \qquad \textbf{[15-12]}$$

where the subscript $s$ denotes an isentropic outlet state. For a cold air-standard Brayton cycle with constant specific-heat values, Eq. [15-12] becomes

$$\eta_{\text{th,Brayton}} = 1 - \frac{q_{\text{out}}}{q_{\text{in}}} = 1 - \frac{c_p(T_4 - T_1)}{c_p(T_3 - T_2)} = 1 - \frac{T_1(T_4/T_1 - 1)}{T_2(T_3/T_2 - 1)}$$

To simplify further, note that processes 1-2 and 3-4 are isentropic. Also, $P_2 = P_3$, and $P_1 = P_4$, or $P_2/P_1 = P_3/P_4$. Thus we can equate the isentropic relations for processes 1-2 and 3-4 to form the temperature relationship

$$\frac{T_2}{T_1} = \left(\frac{P_2}{P_1}\right)^{(k-1)/k} = \left(\frac{P_3}{P_4}\right)^{(k-1)/k} = \frac{T_3}{T_4}$$

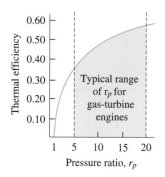

**Figure 15-18**
Thermal efficiency versus pressure ratio of the air-standard Brayton cycle ($k = 1.4$).

As a result, $T_4/T_1 = T_3/T_2$. This relation allows us to rewrite the above relationship for $\eta_{\text{th,Brayton}}$ as

$$\eta_{\text{th,Brayton}} = 1 - \frac{T_2}{T_1} = 1 - \frac{1}{r_p^{(k-1)/k}} \qquad \textbf{[15-13]}$$

where

$$r_p \equiv P_2/P_1 = P_3/P_4$$

Thus the thermal efficiency of a cold air-standard Brayton cycle is primarily a function of the overall pressure ratio. A plot of the thermal efficiency versus $r_p$ of a cold air-standard Brayton cycle for $k = 1.4$ is shown in Fig. 15-18.

The use of constant-specific-heat data, which led to Eq. [15-13], is quite useful in the initial modeling of a gas-turbine power cycle. However, to obtain reasonable values for the heat and work terms in a cycle analysis, it is necessary to account for the variation of $c_p$ with temperature. This requires the use of the air tables in the Appendix. For the isentropic compression and expansion processes in the Brayton cycle, the final state can be found by employing either $s^o$ or $p_r$ data, which were first introduced in Sec. 8-1.

**Example 15-5**

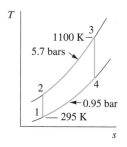

**Figure 15-19**
A $Ts$ diagram showing data for Example 15-5.

**A**n ideal air-standard Brayton cycle operates with air entering the compressor at 0.95 bar and 22°C. The pressure ratio $r_p$ is 6:1, and the air leaves the combustion chamber at 1100 K. Compute the compressor work input, the turbine work output, and the thermal efficiency of the cycle, based on the data of Table A-5.

**Solution:**

**Given:**    Ideal Brayton cycle operates under the conditions shown in Fig. 15-19.

**Find:**    $w_{C,\text{in}}$, $w_{T,\text{out}}$, and $\eta_{\text{th}}$.

**Model:**    Steady flow, air-standard cycle, variable specific heat.

**Analysis:**    Air is assumed to be an ideal gas. The compressor outlet condition can be determined from isentropic relationships. For the given inlet state, from Table A-5, $h_1 = 295.17$ kJ/kg and $p_{r1} = 1.3068$. The relative pressure at the compressor outlet is

$$p_{r2} = p_{r1}\frac{P_2}{P_1} = P_{r1}r_p = 1.3068(6) = 7.841$$

The enthalpy and temperature at state 2 are then approximated by interpolation of the $p_{r2}$ value in Table A-5. We find that $T_{2s} = 490$ K and $h_{2s} = 492.7$ kJ/kg. In addition, since $T_3$ is given as 1100 K, the value of $h_3 = 1161.1$ kJ/kg and the relative pressure $p_{r3} = 167.1$. Hence, for state 4 we find that

$$p_{r4} = p_{r3}\frac{P_4}{P_3} = 167.1\left(\frac{1}{6}\right) = 27.85$$

Again interpolating in Table A-5, we estimate that $T_{4s} = 694$ K and $h_{4s} = 706.5$ kJ/kg. On the basis of these data, and the steady-state energy balance $q + w =$

$\Delta h + \Delta \text{ke} + \Delta \text{pe}$, the following quantities are determined if $\Delta \text{ke}$ and $\Delta \text{pe}$ are neglected:

$$w_{C,\text{in}} = w_{12} = h_{2s} - h_1 = 492.7 - 295.2 = 197.5 \text{ kJ/kg}$$

$$w_{T,\text{out}} = -w_{34} = h_3 - h_{4s} = 1161.1 - 706.5 = 454.6 \text{ kJ/kg}$$

$$q_{\text{in}} = q_{23} = h_3 - h_{2s} = 1161.1 - 492.7 = 668.4 \text{ kJ/kg}$$

$$\eta_{\text{th}} = \frac{w_{T,\text{out}} - w_{C,\text{in}}}{q_{\text{in}}} = \frac{454.6 - 197.5}{668.4} = \frac{257.1}{668.4} = 0.385 \text{ (or 38.5\%)}$$

Note that states 2s and 4s also can be determined from $s^o$ data. At state 1 the value of $s^o$ is 1.68515, and at state 3 the value is 3.07732. Hence

$$s_2^o = s_1^o + R \ln r_p = 1.68515 + \frac{8.314}{29} \ln 6 = 2.1988 \text{ kJ/kg·K}$$

and $\quad s_4^o = s_3^o + R \ln \frac{1}{r_p} = 3.07732 + \frac{8.314}{29} \ln \frac{1}{6} = 2.5636 \text{ kJ/kg·K}$

From Table A-5, $T_{2s}$ is 490 K and $T_{4s}$ is 694 K. These temperatures are in agreement with the values obtained above using $p_r$ data.

---

Example 15-5 illustrates an important feature of the gas-turbine power cycle. In the ideal Brayton cycle analyzed above, the ratio of the compressor work to the turbine work is roughly 0.44, or 44 percent. This quantity, $w_{C,\text{in}}/w_{T,\text{out}}$, is known as the **back work ratio** of a power cycle. In practice, this ratio typically ranges from 40 to 80 percent. As the next several sections show, the effect of irreversibilities in the compressor and turbine is to greatly increase the back work ratio. Obviously, the ratio cannot exceed unity, for in that limiting case the net work output becomes zero. In the case of vapor power cycles, discussed in Chap. 16, the back work ratio of pump work to turbine work is several percent or less.

## 15-6-1 MAXIMUM WORK RELATIONS

Figure 15-20 illustrates the effect of the pressure ratio $r_p$ and the combustor-outlet temperature $T_3$ on the Brayton cycle. Equation [15-13] shows that the thermal efficiency increases with an increasing pressure ratio; but as $r_p$ increases, $T_3$ increases for the same heat rejection. For example, increasing the compressor-outlet pressure from state 2 to state 5 on the figure requires that the combustor-outlet temperature increase from state 3 to state 6. The heat rejections for cycles 1-2-3-4-1 and 1-5-6-4-1 are exactly the same, as given by the area under curve 4-1. Increasing $T_3$ to $T_6$ may be disadvantageous, since the temperature may now exceed the maximum allowable at the inlet to the turbine. To overcome this difficulty, it might be proposed that $r_p$ be increased but that the combustor-outlet temperature be restricted to a value not exceeding $T_3$. Such a cycle would be represented by 1-5-7-8-1. The efficiency of this cycle would be greater than that for the basic cycle 1-2-3-4-1, but the figure easily demonstrates (on the basis of the enclosed

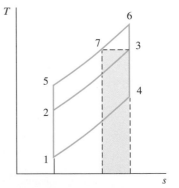

**Figure 15-20**

Illustration of the effect of pressure ratio and the combustor-outlet temperature on the thermal efficiency of the Brayton cycle.

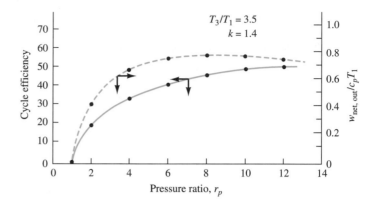

**Figure 15-21**     Ideal Brayton-cycle efficiency and net work as a function of pressure ratio.

area for the cycle), that the net work output would be much less. To achieve the same power output as the basic cycle, a larger mass flow rate would be required, which in turn means a larger physical system would be required. This may be undesirable.

In the limit, as $r_p$ increases and the combustor-outlet temperature is restricted to a value equivalent to $T_3$, the net work approaches zero. Also as $r_p$ decreases for a fixed value of $T_3$, the net work decreases. Consequently, there must be a value of $r_p$ (and hence $P_2$ and $T_2$) that leads to a maximum net work output. This maximum output for a cold air-standard Brayton cycle occurs when $T_2 = (T_1 T_3)^{1/2}$, based on constant specific heats. (The derivation of this relation and others for the occurrence of maximum work output is shown in Example 15-6.) The solid curve in Fig. 15-21 shows the variation of thermal efficiency with pressure ratio for $k = 1.4$ and a turbine-inlet to compressor-inlet temperature ratio of 3.5:1. The net work output is shown as the dashed line. Plotted as a dimensionless parameter $w_{net,out}/(c_p T_1)$, the net work output maximizes at a pressure ratio around 9:1, and then falls off very slowly at higher pressure ratios.

**Example 15-6**

**D**etermine (*a*) the pressure ratio across the compressor and the compressor-outlet temperature for the maximum net work output from a Brayton cold air-standard cycle. Then find an expression for (*b*) the thermal efficiency and (*c*) the maximum net work output in terms of the fixed compressor-inlet and turbine-inlet temperatures.

**Solution:**

**Given:**   A Brayton cold air-standard cycle with fixed compressor-inlet and turbine-inlet temperatures $T_1$ and $T_3$, and variable values of $T_2$ and $P_2$, is shown in Fig. 15-22.

**Find:**   Equations for (*a*) $r_p$ and $T_2$, (*b*) $\eta_{th}$, and (*c*) $w_{net,out,max}$ in terms of $T_1$ and $T_3$.

**Model:** Ideal gas, isentropic compressor and turbine, constant $c_p$ and $k$.

**Analysis:** The solution involves the maximization of $w_{\text{net,out}}$ with respect to the pressure ratio $r_p$ at constant temperatures $T_1$ and $T_3$. The basic energy equation is $q + w = \Delta h + \Delta \text{ke} + \Delta \text{pe}$. Neglecting $\Delta \text{ke}$ and $\Delta \text{pe}$, and assuming air is an ideal gas, the equations for the net work out, the thermal efficiency, and the heat transfer in are

$$w_{\text{net,out}} = \eta_{\text{th}} q_{\text{in}}$$

$$\eta_{\text{th}} = 1 - r_p^{(1-k)/k}$$

$$q_{\text{in}} = c_p(T_3 - T_2) = c_p(T_3 - r_p^{(k-1)/k}T_1)$$

where $T_2 = T_1 r_p^{(k-1)/k}$ for the isentropic compression process. The substitution of the last two equations into the first equation yields

$$w_{\text{net,out}}/c_p = (1 - r_p^{(1-k)/k})(T_3 - r_p^{(k-1)/k}T_1) = [(1 - r_p^{(1-k)/k})T_3 - (r_p^{(k-1)/k} - 1)T_1]$$

(a) To find the maximum, this last equation is now differentiated with respect to $r_p$ and the partial derivative is set equal to zero. The important result is

$$r_p^* = \frac{P_2}{P_1} = \left(\frac{T_3}{T_1}\right)^{1/2(k-1)} \qquad \text{(maximum work)}$$

where $r_p^*$ is the pressure ratio that maximizes the net work output. For the isentropic compression process we noted above that $r_p = (T_2/T_1)^{k/(k-1)}$. When this expression for $r_p$ is equated to the $r_p^*$ equation derived above, we find the compressor outlet temperature $T_2^*$ for maximum work to be

$$T_2^* = (T_1 T_3)^{1/2} \qquad \text{(maximum work)}$$

Thus the maximum work occurs when the compressor exit temperature is the geometric mean of the minimum and maximum temperatures for the cycle based on $k = \text{constant}$.

(b) The cycle thermal efficiency at maximum net work $\eta_{\text{th}}^*$ is found by substitution of $r_p^*$ into the equation given above for the thermal efficiency. Thus

$$\eta_{\text{th}}^* = 1 - r_p^{(1-k)/k} = 1 - (T_1/T_3)^{1/2} \qquad \text{(maximum work)}$$

(c) Finally, substitution of $r_p^*$ from part a into the equation for $w_{\text{net,out}}/c_p$ yields

$$w_{\text{net,out,max}}/c_p = (T_3^{1/2} - T_1^{1/2})^2$$

**Comment:** When $k = 1.4$ for cold air, the equation for $r_p^*$ becomes $r_p^* = (T_3/T_1)^{5/4}$.

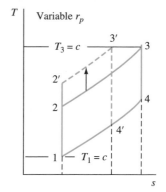

**Figure 15-22**
A $Ts$ diagram for developing expressions for thermal efficiency and maximum net work in Example 15-6.

## 15-6-2 AIR-FUEL REQUIREMENTS IN A COMBUSTOR

The maximum allowable temperature of combustion gases in a gas turbine, due to the metallurgical limit of the turbine blades, is around 1200 to 1500 K (2200 to 2700°R). However, the combustion temperature of typical hydrocarbon fuels with stoichiometric air is around 2200 to 2500 K (4000 to 4500°R). The actual combustion-gas temperature is reduced to allowable limits by using a relatively high air-fuel ratio, compared to stoichiometric, in the combustor. The required air-fuel ratio in a given situation may be predicted from an energy analysis of the combustor.

By writing a steady-state energy balance for adiabatic combustion (and negligible kinetic-energy change) on a mass basis, we find that

$$0 = \dot{m}_{fuel}\Delta h^{o}_{C,298} + [\dot{m}(h_T - h_{298})]_{prod}$$
$$- [\dot{m}(h_T - h_{298})]_{air} - [\dot{m}(h_T - h_{298})]_{fuel}$$

where $\Delta h_{C,298}$ is the enthalpy of combustion of the fuel on a *mass* basis. The combustor inlet air and outlet product gases will now be designated as states 2 and 3, as in Fig. 15-17. In addition, owing to the small amount of fuel added per unit mass of air, it is assumed that $\dot{m}_{prod} = \dot{m}_{air}$ and that $h_{prod} = h_{air}$. As a result, the above equation becomes

$$0 = \dot{m}_{fuel}\Delta h^{o}_{C,298} + \dot{m}_{air}(h_3 - h_{298})_{air}$$
$$- \dot{m}_{air}(h_2 - h_{298})_{air} - \dot{m}_{fuel}(h_T - h_{298})_{fuel}$$
$$= \dot{m}_{fuel}(\Delta h^{o}_{C,298} - h_T + h_{298})_{fuel} + \dot{m}_{air}(h_3 - h_2)_{air}$$

where $h_T$ is the enthalpy of the entering fuel at temperature $T$. Rearrangement of the above equation shows that

$$\frac{\dot{m}_{air}}{\dot{m}_{fuel}} = \frac{[-\Delta h^{o}_{C,298} - (h_T - h_{298})]_{fuel}}{(h_3 - h_2)_{air}}$$

When the fuel enters the combustor at temperatures close to 25° C, the term $(h_T - h_{298})_{fuel}$ is negligible. In this case the equation for the required air-fuel ratio reduces to

$$\frac{\dot{m}_{air}}{\dot{m}_{fuel}} = \frac{\Delta h^{o}_{C,298}}{h_3 - h_2}$$

Thus a knowledge of the enthalpy of combustion per unit mass (or heating value) of the fuel and the enthalpy change of the air across the combustor is sufficient to estimate the air-fuel ratio required to maintain the combustion gas outlet temperature at a prescribed value.

Keep in mind that irreversibilities occur in the compressor and turbine in actual operation. The effect of these irreversibilities is to decrease the energy input in the combustor and to increase the heat transfer output in the heat-rejection part of the cycle. The overall result is a decrease in the thermal efficiency of the cycle.

## 15-7 EFFECT OF COMPRESSOR AND TURBINE IRREVERSIBILITIES

For compressors and turbines that are close to being adiabatic, the isentropic work associated with these devices is a standard, or model, against which all actual equipment may be compared. The actual performance of work-absorbing or work-producing devices which are essentially adiabatic is described by an adiabatic efficiency (sometimes called a *first-law* efficiency). From Sec. 8-2-3 the *adiabatic compressor efficiency* $\eta_C$ is defined by Eq. [8-30] as

$$\eta_C \equiv \frac{\text{isentropic work input}}{\text{actual work input}} = \frac{w_s}{w_a} = \frac{\dot{W}_s}{\dot{W}_a} \qquad \textbf{[15-14]}$$

When the kinetic- and potential-energy changes are negligible, so that the steady-state energy balance reduces to $w = \Delta h$, then Eq. [15-14] may be written as

$$\eta_C = \frac{h_{2s} - h_1}{h_{2a} - h_1} \qquad \textbf{[15-15]}$$

An *adiabatic turbine efficiency* $\eta_T$ is defined in Sec. 8-2-1 by Eq. [8-19] as

$$\eta_T = \frac{\text{actual work output}}{\text{isentropic work output}} = \frac{w_a}{w_s} = \frac{\dot{W}_a}{\dot{W}_s} \qquad \textbf{[15-16]}$$

If the changes in kinetic and potential energies are neglected, the work terms above may be evaluated as a function of the enthalpy changes. Use of the notation from Fig. 15-17 allows the adiabatic efficiency of a turbine in a simple gas-turbine cycle to be written as

$$\eta_T = \frac{h_3 - h_{4a}}{h_3 - h_{4s}} \qquad \textbf{[15-17]}$$

where state $4a$ is the actual outlet state for the turbine and state $4s$ is the outlet state under isentropic expansion.

The effect of irreversibilities in both the compressor and the turbine is shown on the *Ts-hs* diagram in Fig. 15-23. The effect of the irreversibilities is to require a larger compressor-work input, which is delivered from a smaller turbine output. Hence the compressor in actual practice may consume 40 to 70 percent of the turbine output. In addition, the presence of a pressure drop in the heat exchanger (combustor) where energy is added has a negative influence on performance.

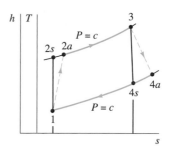

**Figure 15-23**

Illustrations of the effect of compressor and turbine irreversibilities on *Ts* and *hs* diagrams.

---

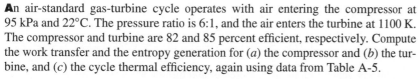

An air-standard gas-turbine cycle operates with air entering the compressor at 95 kPa and 22°C. The pressure ratio is 6:1, and the air enters the turbine at 1100 K. The compressor and turbine are 82 and 85 percent efficient, respectively. Compute the work transfer and the entropy generation for (*a*) the compressor and (*b*) the turbine, and (*c*) the cycle thermal efficiency, again using data from Table A-5.

**Example 15-7**

**Solution:**

**Given:** A gas-turbine power cycle operates under the conditions shown in Fig. 15-24.

**Find:** (*a*) $w_{C,\text{in}}$, $\sigma_{m,C}$; (*b*) $w_{T,\text{out}}$, $\sigma_{m,T}$; and (*c*) $\eta_{\text{th}}$.

**Model:** Steady flow, air-standard cycle, irreversible compressor and turbine.

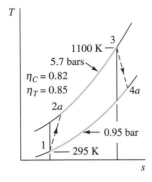

**Figure 15-24**
A $Ts$ diagram showing data for Example 15-7.

The cycle operates under the conditions of Example 15-5, except the compressor and turbine are now irreversible. The ideal-gas data for the compressor- and turbine-inlet states remain as $h_1 = 295.2$ kJ/kg at 22°C and $h_3 = 1161.1$ kJ/kg at 1100 K, respectively, and the pressure ratio is 6:1. The isentropic work values previously calculated are $w_{C,in,s} = 197.5$ kJ/kg and $w_{T,out,s} = 454.6$ kJ/kg. These isentropic values are employed to find the actual states, in conjunction with the definitions of compressor and turbine efficiencies.

(a) For the compressor, in the absence of kinetic- and potential-energy effects,

$$\Delta h_{C,a} = \frac{\Delta h_{C,s}}{\eta_C} = \frac{197.5}{0.82} = 240.9 \text{ kJ/kg}$$

Since $\Delta h_{C,a} = h_{2a} - h_1$, the actual compressor-exit enthalpy is

$$h_{2,a} = h_1 + \Delta h_{C,a} = 240.9 + 295.2 = 536.1 \text{ kJ/kg}$$

From Table A-5, this corresponds to a compressor-exit temperature of 532 K. Hence the irreversibilities within the compressor have resulted in a temperature rise of 42 K over that for the isentropic compression. The entropy production is found from an entropy balance on a control volume with one inlet and one exit. This takes the form

$$s_2 - s_1 = \frac{1}{\dot{m}}\left(\sum_j \frac{\dot{Q}_j}{T_j}\right) + \sigma_{m,cv}$$

For the compressor, in the absence of heat transfer across the control surface,

$$\sigma_{m,C} = s_2 - s_1 = s_2^\circ - s_1^\circ - R\ln\frac{P_2}{P_1}$$

$$= 2.28355 - 1.68515 - \frac{8.314}{29}\ln 6 = 0.0847 \text{ kJ/kg·K}$$

(b) For the turbine

$$\Delta h_{T,a} = \eta_T \Delta h_{T,s} = 0.85(-454.6) = -386.4 \text{ kJ/kg}$$

Since $\Delta h_{T,a} = h_{4a} - h_3$, the actual turbine-outlet enthalpy is

$$h_{4a} = h_3 + \Delta h_{T,a} = 1161.1 - 386.4 = 774.7 \text{ kJ/kg}$$

This value of $h_{4,a}$ corresponds to a temperature of 757 K, which is 63 K above that for the isentropic expansion. The entropy generation in the adiabatic turbine is

$$\sigma_{m,T} = s_4 - s_3 = s_4^\circ - s_3^\circ - R\ln\frac{P_4}{P_3}$$

$$= 2.65744 - 3.07732 - \frac{8.314}{29}\ln\frac{1}{6} = 0.0938 \text{ kJ/kg·K}$$

The entropy generation in the compressor and turbine are nearly equal.

(c) The thermal efficiency becomes

$$\eta_{th} = \frac{w_{net,out}}{q_{in}} = \frac{w_{T,out,a} - w_{C,in,a}}{h_3 - h_{2a}} = \frac{386.4 - 240.9}{1161.1 - 536.1} = 0.233 \text{ (or 23.3\%)}$$

The presence of irreversibilities within the compressor and turbine has led to a decrease in the thermal efficiency from 38.5 to 23.3 percent, which is quite significant. In addition, irreversibilities have increased the ratio of compressor work to turbine work (back work ratio) to 0.62 from the value of 0.44 for isentropic flow. Relatively small changes in adiabatic efficiencies greatly influence the net work output and the back work ratio of the device.

The general effect of irreversibilities in the compressor and turbine is more clearly shown in Fig. 15-25. This figure is based on a cold-air-standard cycle with $k = 1.40$, and the compressor and turbine efficiencies are 0.82 and 0.85, respectively. Figure 15-25$a$ shows the thermal efficiency of the irreversible cycle as a function of pressure ratio and selected ratios of the turbine-inlet temperature to the compressor-inlet temperature, $T_3/T_1$. Unlike the curve in Fig. 15-21 for the ideal Brayton cycle, these efficiency curves exhibit maximum points. Note also that the thermal efficiency maximizes at higher values of the pressure ratio as the temperature ratio $T_3/T_1$ increases. Figure 15-25$b$ illustrates the effect of pressure ratio and temperature ratio on the net work output. As in Fig. 15-21, the net work is presented as a dimensionless parameter $w_{net,out}/(c_p T_1)$. Note again that the net work output maximizes at larger values of $r_p$ as the temperature ratio increases. For comparative purposes, the net work curve for an ideal cycle with a $T_3/T_1$ of 3.0 is also shown. In this case two effects are clear when the two curves for $T_3/T_1$ of 3.0 are compared in Fig. 15-25$b$. First, the net work maximizes at a lower pressure ratio when irreversibilities are taken into account. Second, the net work is drastically reduced over the range of pressure ratios, even though the compressor and turbine efficiencies chosen are relatively high.

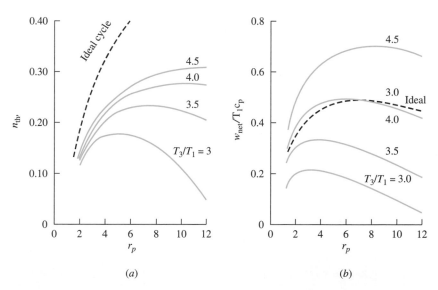

**Figure 15-25**    The effect of compressor and turbine irreversiblities on (a) the thermal efficiency and (b) the net work output of a gas-turbine power cycle as a function of pressure ratio and the ratio of turbine-inlet temperature to compressor-inlet temperature. (Data are based on $k = 1.40$, $\eta_C = 0.82$, and $\eta_T = 0.85$).

## 15-8   THE REGENERATIVE GAS-TURBINE CYCLE

One way to modify the basic gas-turbine cycle to increase the thermal efficiency is based on the concept of regeneration. In many cases the turbine exit temperature is considerably higher than the compressor exit temperature. (This temperature difference decreases as the pressure ratio $r_p$ is increased.) It is possible, then, to preheat the air leaving the compressor with energy taken from the turbine exhaust gases. Because the air entering the combustor is now at a higher temperature than without the preheating process, less fuel must be injected into the combustion section. The heat transfer between the two flow streams takes place in a heat exchanger called a *regenerator* or *recuperator*. A flow diagram for the regenerative gas-turbine cycle is shown in Fig. 15-26a. Note that $T_4$ is greater than $T_2$. In the ideal situation it is assumed that the flow through the regenerator is frictionless, and therefore occurs at constant pressure. In this internally reversible situation the heat transfer between the two flow streams may be represented as areas on a $Ts$ diagram. Since the heat transfer from the turbine exit stream must equal the heat transfer into the compressor exit stream, the two areas in Fig. 15-26b must be equal in magnitude; that is, $\dot{Q}_{2x} = -\dot{Q}_{45}$.

If the heat-transfer process in the regenerator were ideal, the temperature difference between the two fluids would approach zero at all locations within the heat exchanger. In this situation it would be possible to preheat the compressor exit stream to the temperature of the turbine exit stream, and state $x$ on the $Ts$ diagram would lie horizontally across from state 4. This is impractical, though, because a very large surface area is required for heat transfer as the temperature difference approaches zero. As a measure of the approach of an actual regenerator to the performance at this limiting condition, the *regenerator effectiveness* $\varepsilon_{\text{regen}}$ is defined in terms of the actual heat transfer to the air flowing on the compressor side of the regenerator to

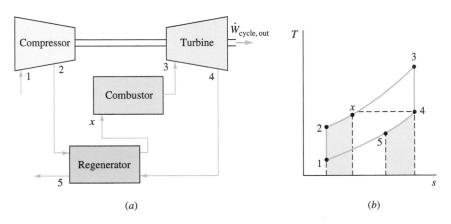

**Figure 15-26**     Flow diagram and $Ts$ diagrams for a regenerative gas-turbine cycle.

the maximum possible heat transfer. That is,

$$\varepsilon_{regen} \equiv \frac{q_{regen,act}}{q_{regen,max}} = \frac{h_x - h_2}{h_4 - h_2}$$ **[15-18]**

where the temperature corresponding to $h_x$ is somewhat lower than the temperature at state 4. In the presence of a regenerator, $q_{in} = q_{x3} = h_3 - h_x$, and $q_{out} = -q_{51} = h_5 - h_1$. Consequently, the thermal efficiency becomes

$$\eta_{th,regen} = 1 - \frac{h_5 - h_1}{h_3 - h_x}$$ **[15-19]**

For a regenerative cycle operating under the assumptions of a cold-air-standard cycle, the above equation becomes

$$\eta_{th,regen,cold} = 1 - \frac{T_1}{T_3} r_p^{(k-1)/k}$$ **[15-20]**

and the effectiveness becomes

$$\varepsilon_{regen,cold} = \frac{T_x - T_2}{T_4 - T_2}$$ **[15-21]**

Thus the thermal efficiency of a regenerative gas-turbine cycle is a function not only of the pressure ratio but also of the ratio of the minimum to maximum temperatures occurring in the cycle.

Figure 15-27 illustrates the variation of thermal efficiency with pressure ratio and temperature ratio ($T_3/T_1$) for an *ideal* regenerative cycle on a cold air standard. The efficiency curve for the ideal Brayton cycle is also shown for comparison. Unlike the Brayton cycle, the thermal efficiency of the regenerative cycle *decreases* with increasing pressure ratio for a fixed value of $T_3/T_1$. This trend is predicted by Eq. [15-20]. This equation also indicates

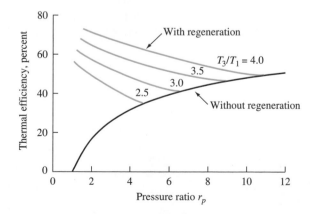

**Figure 15-27** Gas-turbine-cycle thermal efficiency as a function of pressure ratio, for $k = 1.40$, with and without ideal regeneration.

that regeneration is more effective at high maximum-to-minimum temperature ratios, as shown in Fig. 15-27. To increase the thermal efficiency, the value of $h_x$, and hence $T_x$, should be as high as possible, because the heat transfer in is reduced but the net work out remains the same. The typical value of the effectiveness lies between 0.6 and 0.8 for gas-turbine stationary power plant applications. To increase it beyond this range usually leads to higher equipment costs which negate any advantage of the higher thermal efficiency. Also, a higher effectiveness requires a greater heat-transfer area. However, this leads to an increased pressure drop through the regenerator, which is a loss in terms of cycle efficiency. Pressure drops due to the regenerator are an important factor in deciding whether a regenerator should be added.

**Example 15-8**

**A**n ideal air-standard gas-turbine cycle operates with air entering the compressor at 95 kPa and 22°C. The pressure ratio is 6:1, and the air enters the turbine at 1100 K. Accounting for variable specific heats, determine the effect on the thermal efficiency of the cycle if an ideal regenerator is inserted into the cycle.

**Solution:**

**Given:**    An ideal gas-turbine power cycle with an ideal regenerator is shown in Fig. 15-28.

**Find:**    Thermal efficiency compared to an ideal Brayton cycle.

**Model:**    Steady state, isentropic compressor and turbine, air is an ideal gas, no pressure drop in combustor.

**Analysis:**    The cycle operates under the conditions of Example 15-5, except an ideal regenerator has been added to the cycle. The data for the enthalpies at the inlet and exit of the compressor and turbine remain the same. Employing the symbols shown in Fig. 15-27, we note that $h_1 = 295.2$, $h_{2s} = 492.7$, $h_3 = 1161.1$, and $h_{4s} = 706.5$, all values in kJ/kg. From these data $w_{C,s} = 197.5$ kJ/kg,

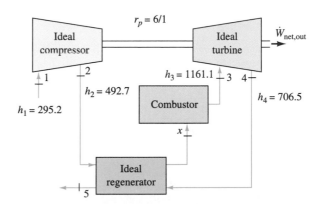

**Figure 15-28**    Schematic and data for Example 15-8.

$w_{T,s} = 454.6$ kJ/kg, and $q_{in} = 668.4$ kJ/kg. For an ideal regenerator the enthalpy of the fluid $h_x$ leaving the regenerator must be equal to $h_{4s}$. The heat-transfer input saved because of the regenerator is

$$q_{saved} = h_x - h_{2s} = h_{4s} - h_{2s} = 706.5 - 492.7 = 213.8 \text{ kJ/kg}$$

The required heat-transfer input now becomes

$$q_{in,regen} = h_3 - h_x = h_3 - h_4 = 1161.1 - 706.5 = 454.6 \text{ kJ/kg}$$

The percentage saved over the case without regeneration is

$$\text{Percent of } q_{saved} = \frac{213.8}{668.4} 100 = 32.0\%$$

Since the turbine and compressor work are exactly the same as in Example 15-5,

$$\eta_{th} = \frac{w_{net,out}}{q_{in}} = \frac{454.6 - 197.5}{454.6} = \frac{257.1}{454.6} = 0.565 \text{ (or 56.5\%)}$$

**Comment:** This compares with an efficiency of 38.5 percent without regeneration. Thus the effect of regeneration on the thermal efficiency is considerable, although in actual cases the effect would not be quite as large, because ideal regeneration is not possible. In addition, internal irreversibilities and heat-transfer losses reduce the efficiency further.

---

**In** Example 15-7 an air-standard gas-turbine cycle operates with air entering the compressor at 0.95 bar and 22°C. The pressure ratio is 6:1, and the air enters the turbine at 1100 K. The compressor and turbine are 82 and 85 percent efficient, respectively. Determine the effect on the thermal efficiency if a regenerator of 70 percent effectiveness is now inserted into the cycle.

**Example 15-9**

**Solution:**

**Given:** An air-standard gas-turbine cycle with irreversible compressor and turbine and a regenerator is shown with applicable data on an equipment schematic in Fig. 15-29.

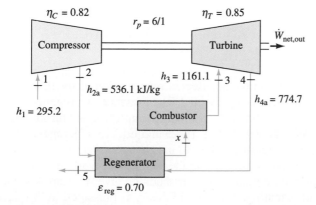

**Figure 15-29** Schematic and data for Example 15-9.

**Find:**   Effect on thermal efficiency compared to a cycle without regeneration.

**Model:**   Steady-state, irreversible compressor and turbine; air is an ideal gas; no pressure drop in heat exchangers.

**Analysis:**   In Example 15-7 the compressor and turbine efficiencies are 82 and 85 percent, respectively. The data for the enthalpies at the inlet and exit of the compressor and turbine remain the same. Using the symbols shown in Fig. 15-29, we note that $h_1 = 295.2$, $h_{2a} = 536.1$, $h_3 = 1611.1$, and $h_{4a} = 774.7$, all values in kJ/kg. Use of Eq. [15-18] for the effectiveness enables us to determine the heat transfer required between states 2 and $x$:

$$0.70 = \frac{h_x - h_2}{h_4 - h_2} = \frac{h_x - h_2}{774.7 - 536.1}$$

$$h_x - h_2 = 0.7(238.6) = 167.0 \text{ kJ/kg}$$

This quantity represents the heat transfer input saved by regeneration. Without regeneration the heat transfer in would need to be

$$q_{\text{without regen}} = h_3 - h_{2a} = 1161.1 - 536.1 = 625.0 \text{ kJ/kg}$$

The percentage of heat transfer input saved over the case without regeneration is

$$\text{Percent of } q_{\text{saved}} = \frac{167.0}{625.0} 100 = 26.7\%$$

Since the turbine work input and the compressor work output are exactly the same as in Example 15-7,

$$\eta_{\text{th}} = \frac{w_{\text{net,out}}}{q_{\text{in}}} = \frac{386.4 - 240.9}{625.0 - 167.0} = \frac{145.5}{458.0} = 0.318 \text{ (or 31.8\%)}$$

**Comment:**   This compares to an efficiency of 23.3 percent without regeneration. It is also a more realistic value than the 56.5 percent found in Example 15-8, because reasonable turbine and compressor efficiencies and regenerator effectiveness have been assumed.

## 15-9   THE PROCESSES OF INTERCOOLING AND REHEATING

In the preceding section the effect of regeneration on the thermal efficiency of a gas-turbine cycle is presented. The presence of a regenerator decreases the heat transfer to the cycle, while the compressor and turbine work remain the same. As indicated by Fig. 15-25 and Eq. [15-17], regeneration is more effective at low pressure ratios. To improve the cycle even further, we need to examine methods of lowering the compressor work input and increasing the turbine work output. These methods involve what are known as intercooling and reheating. The general theory is presented in this section, and the application to a gas-turbine cycle is presented in the following section.

### 15-9-1 ADIABATIC AND ISOTHERMAL COMPRESSION

In the preceding sections on the gas-turbine cycle the compression process is assumed to be adiabatic, although irreversibilities might be present. The work of compression in reversible, steady-flow processes is given by the integral of $v\,dP$. When the compression is adiabatic and internally reversible, and specific heats are constant, then

$$Pv^k = \text{constant} \qquad \text{and} \qquad T_2/T_1 = (P_2/P_1)^{(k-1)/k} \quad \textbf{[15-22]}$$

Under these circumstances the steady-flow compressor work for a cold-air-standard process becomes

$$w_{\text{sf}} = c_p RT_1[(P_2/P_1)^{(k-1)/k} - 1] \qquad\qquad \textbf{[15-23]}$$

$$= \frac{kRT_1[(P_2/P_1)^{(k-1)/k} - 1]}{k-1} \qquad \text{(isentropic)}$$

Another method of compression involves heat transfer out in sufficient quantity so that the process approaches an isothermal one. An expression for the steady-flow shaft work of a frictionless process in this case is

$$w_{\text{sf}} = \int v\,dP = \int \frac{RT}{P}\,dP = RT\ln\frac{P_2}{P_1} \qquad \text{(isothermal)} \quad \textbf{[15-24]}$$

Even if the heat transfer is not negligible during a compression process, the format of Eqs. [15-22] and [15-23] can be retained by writing the $Pv$ relation in the form

$$Pv^n = \text{constant} \qquad\qquad \textbf{[15-25]}$$

which is known as a *polytropic process* (see Sec. 2-5, Example 2-6, and Sec. 7-6-2). Analogous to the $TP$ relation in Eq. [15-22] for an isentropic process, we find for polytropic processes of ideal gases that

$$TP^{-(n-1)/n} = \text{constant} \qquad \text{or} \qquad T_2/T_1 = (P_2/P_1)^{(n-1)/n} \quad \textbf{[15-26]}$$

For compression processes with heat transfer out, the value of $n$ lies between the values for isothermal and isentropic behavior, that is, $1 < n < 1.4$. The polytropic compressor work is given by Eq. [15-23], except that $k$ is replaced by $n$.

A comparison of the frictionless steady-flow shaft work for these three processes is shown in Fig. 15-30 for a fixed pressure ratio $P_2/P_1$. Since the area to the left of the process curve is a measure of the steady flow work, the maximum frictionless work is required for an isentropic compression, where $n = k$. As the value of $n$ is decreased by heat transfer out during the compression process, the work input also decreases. Figure 15-30 illustrates the simple fact that the removal of energy by heat transfer during a compression process is advantageous with regard to the work-input requirements. This result is equally true for both reciprocating and rotating (axial and centrifugal) compressors.

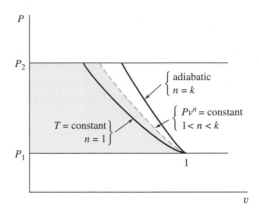

**Figure 15-30**   The effect of cooling on the required work of compression for a steady-flow process between fixed pressure limits.

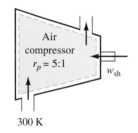

**Figure 15-31**
Schematic and data for Example 15-10.

**Example 15-10**

**A**ir initially at 300 K is compressed with a pressure ratio of 5:1. For a cold-air-standard process, determine the work required if the process is (a) isentropic with $k = 1.4$, (b) isothermal, and (c) polytropic with $n = 1.3$, in kJ/kg.

**Solution:**

**Given:**   A steady-flow compression process as shown in Fig. 15-31.

**Find:**   $w_{shaft}$ for (a) isentropic, (b) isothermal, and (c) polytropic processes.

**Model:**   Steady-flow, cold air-standard process; ideal gas, $k = 1.4$, $n = 1.3$.

**Analysis:**   (a) The isentropic work for an ideal-gas model is given by integrating $v\,dP$, where $Pv^k = $ constant. The result is Eq. [15-23],

$$w_{sf} = \frac{kRT_1[(P_2/P_1)^{(k-1)/k} - 1]}{k - 1}$$

Substitution of data yields

$$w_{isen} = \frac{1.4(8.314)(300)(5^{0.286} - 1)}{28.97(1.4 - 1)} = 176.0 \text{ kJ/k}$$

(b) The isothermal work is determined from $\int v\,dP$, where $Pv = RT$. For this case

$$w_{isoth} = RT \ln \frac{P_2}{P_1} = \frac{8.314(300)}{28.97} \ln 5 = 138.6 \text{ kJ/kg}$$

(c) The polytropic work is found by using the expression for isentropic processes, except $k$ is replaced by $n$. Hence

$$w_{poly} = \frac{nRT_1[(P_2/P_1)^{(n-1)/n} - 1]}{n - 1} = \frac{1.3(8.314)(300)(5^{0.231-1})}{28.97(1.3 - 1)} = 168.0 \text{ kJ/kg}$$

**Comment:** The isentropic work required is 27 percent greater than the isothermal work for the same initial state and overall pressure ratio. When heat transfer drops the polytropic constant $n$ to 1.3, the work required is still 21 percent greater than the isothermal case.

## 15-9-2 MULTISTAGE COMPRESSION WITH INTERCOOLING

Although cooling the gas as it passes through a compressor has its theoretical advantages, in many cases the heat-transfer rate through the compressor casing is not sufficient to significantly lower the exit temperature. By external cooling the value of $n$ may be reduced only to the neighborhood of 1.35. To achieve the maximum benefits of cooling, another physical arrangement is used called *multistage compression with intercooling*. This approach is especially effective when a comparatively large pressure change is desired. The method involves the separation of the work and heat-transfer processes into a sequence of processes. As shown in Fig. 15-32 for a two-stage compression, the fluid is first compressed to some intermediate pressure $P_x$ at state $a$, which lies between $P_1$ and $P_2$. The fluid is then passed through a heat exchanger called an *intercooler*. Here it is cooled down by heat transfer at constant pressure $P_x$ (if friction is neglected) to a lower temperature at state $b$ in the figure. In some instances this lower temperature could reach the initial temperature $T_1$. Next the fluid passes through a second stage of the compressor, where the pressure is further raised to $P_2$. Although only two stages of compression are shown in Fig. 15-32, the fluid could pass through another intercooler process, and then another stage of the compressor, until the final pressure is reached. The overall result is a lowering of the net work required for a given overall pressure ratio. Photograph 15-2 shows an application of two-stage compression with intercooling for producing process air for industrial use.

The effect of intercooling on a two-stage compressor is shown on the $Pv$ diagram in Fig. 15-33. For this figure it has been assumed that the intercooler cools the fluid to the initial temperature before the fluid enters the second stage. The compression processes between pressures $P_1$ to $P_x$ and $P_x$ to $P_2$ are shown as the type $Pv^n = $ constant, where $n$ is probably close to $k$. The shaded area represents the work input saved by the process of intercooling. Figure 15-34 shows a $Ts$ diagram for the same two-stage compression process shown in Figs. 15-32 and 15-33. Note that the final exit temperature $T_2$ is less than the temperature at state $f$ which would occur without intercooling. As a consequence of this lower temperature, a gas-turbine cycle with intercooling is especially adaptable to regeneration. In fact, intercooling is promising only if a regenerator is used at the same time; otherwise, a considerable additional amount of heat transfer must be supplied to the cycle at a relatively low temperature. It must also be kept in mind that a considerably larger regenerator will be required when intercooling is employed.

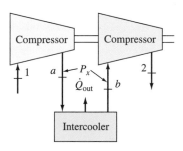

**Figure 15-32**
Schematic of a two-stage compressor with intercooling.

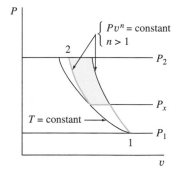

**Figure 15-33**
A $Pv$ diagram illustrating two-stage, polytropic compression with intercooling.

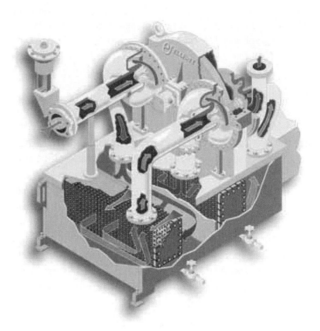

**Photograph 15-2**    Elliott Plant Air Package utilizes two-stage compression with intercooling to supply industrial process air. (Courtesy of the Elliott Company)

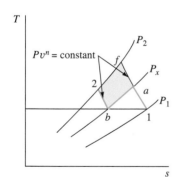

**Figure 15-34**

Two-stage, polytopic compression with intercooling shown on a *Ts* diagram.

The size of the area representing the reduction in work input depends on two variables: (1) the temperature $T_b$ of the fluid leaving the intercooler, and (2) the intermediate pressure $P_x$. According to Eq. [15-23], compressor work varies directly with the compressor inlet temperature. When heat transfer from the intercooler occurs only to the surroundings at $T_1$, then $T_1$ is the lowest possible temperature of the fluid leaving the intercooler at state $b$. An intercooler that reduces the fluid temperature to $T_1$ before the fluid enters the second stage is called an *ideal intercooler*. On the basis of an ideal intercooler, the total work for two stages of compression is found by employing Eq. [15-23] for the general case of a polytropic process. The expression is

$$ w = \frac{nRT_1[(P_x/P_1)^{(n-1)/n} - 1]}{n - 1} + \frac{nRT_1[(P_2/P_x)^{(n-1)/n} - 1]}{n - 1} \qquad \textbf{[15-27]} $$

To find the minimum total work of compression, the above equation for $w$ is differentiated with respect to the variable pressure $P_x$, and the resulting equation is set equal to zero. The minimum work input occurs when the pressures are related by the expression

$$ P_x = (P_1 P_2)^{1/2} \qquad \text{or} \qquad \frac{P_x}{P_1} = \frac{P_2}{P_x} \qquad \textbf{[15-28]} $$

If this result is substituted into Eq. [15-27], the work done by each stage is equal; that is, $w_{1x} = w_{x2}$.

## 15-9-3 TURBINE STAGING WITH REHEAT

In conjunction with intercooling, it is often effective to use turbine staging as well. Instead of expanding directly through a single turbine, the gases are allowed to expand only partially before they are returned to another combustion chamber marked "reheat combustor" in Fig. 15-35. In the reheat combustor heat transfer ideally occurs at constant pressure until the limiting temperature is reached at state 5. Then further expansion occurs until the ambient pressure is attained at state 6 in the figure. A $Ts$ diagram showing isentropic expansions for the two stages is also illustrated in Fig. 15-35.

The energy equation for an adiabatic turbine gives the turbine work out as $w_{T,\text{out}} = -\Delta h \approx -c_{p,\text{av}} \Delta T$. If $c_p$ were not a function of temperature, turbine staging would not increase the work output, because $c_{p,\text{av}} \Delta T$ would not change with staging. However, $c_p$ is a fairly strong function of temperature when the temperature range is large. For example, Table A-4 for air indicates that $c_p$ increases about 14 percent from 300 to 1000 K. Thus more turbine work is possible if the average temperature level of the gas is kept high. This is achieved by the reheating process. From this discussion it is also apparent that the reheat temperature $T_5$ ideally should reach the limiting value of $T_3$ for maximum benefit from turbine staging, as shown in Fig. 15-35.

Note that the use of reheating requires a substantial increase in the heat transfer into the basic cycle. However, the final turbine-exhaust temperature $T_6$ shown in the figure is somewhat above the outlet-turbine temperature without reheat, $T_7$. Consequently, reheating is quite effective when it is used in conjunction with regeneration, since the quantity of heat exchanged in the regenerator can be greatly increased.

Under the condition of ideal reheating ($T_3$ and $T_5$ are equal in Fig. 15-35), the total work output from the two stages depends on the pressure ratio across each stage. Analogous to two-stage compression with ideal intercooling, the maximum work output is found by differentiation of a general equation for the work of the two states. Equation [15-27] again

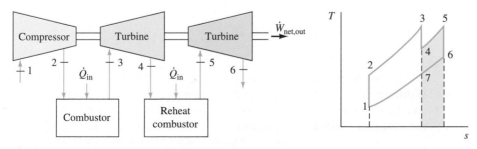

**Figure 15-35**    The air-standard gas turbine with reheating.

applies, except $T_1$ in this equation becomes $T_3$ and the pressure ratios in the equation now represent those across the two turbine stages. Since the derivation is the same, the answer for maximum work output is the same as minimum work input for a two-stage compressor. That is, the intermediate pressure is the geometric mean between the inlet pressure and the outlet pressure, or the pressure ratio across each stage is the square root of the overall pressure ratio. In terms of Fig. 15-35, the maximum work output would occur when $P_4 = P_5 = (P_3 P_6)^{1/2}$ for two-stage expansion with ideal reheating. As a result, the work output of each stage will be equal for the case of ideal reheating.

## 15-10    GAS-TURBINE CYCLES WITH INTERCOOLING AND REHEATING

From the discussion in the preceding section, clearly a gas-turbine power cycle is improved most when a combination of intercooling and reheating is employed with regeneration, as shown in Fig. 15-36a. Under any

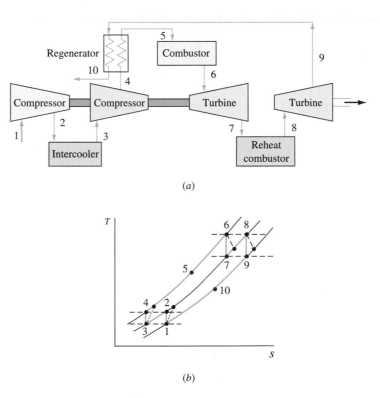

(a)

(b)

**Figure 15-36**    (a) Schematic equipment diagram; (b) $Ts$ process diagram for a gas-turbine power cycle with intercooling, reheating, and regeneration.

circumstance, the effects of irreversibilities in the turbine and compressor, and pressure losses in the combustor, etc., must be considered in predicting the actual performance of a gas-turbine cycle. One should not conclude, however, that intercooling and reheating without regeneration improve the thermal efficiency. In fact, the effect without regeneration is always to decrease the thermal efficiency, although the net work will increase. The reason is that intercooling and reheating used alone decrease the average temperature of the heat transfer input and increase the average temperature of heat transfer out. This argument can be seen qualitatively from the *Ts* diagram in Fig. 15-36*b*. The major purpose in using intercooling and reheating is to increase the effective use of a regenerator.

Example 15-11

In Example 15-9 a regenerative air-standard gas-turbine cycle operates with air entering the compressor at 0.95 bar and 22°C. The pressure ratio is 6:1, and the air enters the turbine at 1100 K. The compressor and turbine are 82 and 85 percent efficient, respectively, and the regenerator effectiveness is 70 percent. Ideal intercooling and reheating are now added to this cycle. In addition, the pressure ratios across the two-stage compressor and two-stage turbine are set equal to provide the minimum work input and the maximum work output, respectively. Determine (*a*) the effect of this additional equipment on the thermal efficiency obtained with regeneration alone and (*b*) the entropy generation in the compressor, turbine, and regenerator.

**Solution:**

**Given:** A gas-turbine cycle with intercooling, reheating, and regeneration is shown on a *Ts* diagram in Fig. 15-37.

**Find:** (*a*) Effect on $\eta_{th}$ due to addition of intercooling and reheating and (*b*) $\sigma_{m,C}$, $\sigma_{m,T}$, and $\sigma_{m,reg}$.

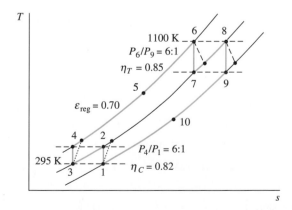

**Figure 15-37** A *Ts* diagram with data for Example 15-11.

**Model:**   Compressors and turbines are adiabatic, no pressure drops in heat exchangers, air is an ideal gas, variable specific heats.

**Analysis:**   Data for the problem in Example 15-9 include:

$$T_1 = T_3 = 295 \text{ K} \qquad T_6 = T_8 = 1100 \text{ K} \qquad P_4/P_1 = P_6/P_9 = 6/1$$

$$\eta_C = 0.82 \qquad \eta_T = 0.85 \qquad \varepsilon_{\text{regen}} = 0.70 \qquad \eta_{\text{th,regen}} = 0.318$$

where the subscript numbers refer to states in Fig. 15-37.

(a) To minimize the compressor work and to maximize the turbine work, the pressure ratios across each stage are set equal. This requires the pressure ratio across each stage to be the square root of the overall pressure ratio. Since this latter value is 6, the ratio for each stage (compressor or turbine) is 2.45. At the inlet to the compressor's first stage, $h_1 = 295.2$ kJ/kg and $p_{r1} = 1.3068$. The relative pressure at the outlet of this stage for isentropic flow would be

$$p_{r2} = p_{r1}\frac{P_2}{P_1} = 1.3068(2.45) = 3.202$$

Interpolation in the abridged air table A-5 indicates that $h_{2s} = 381.8$ kJ/kg and $T_{2s} = 381$ K. The actual compressor-enthalpy rise is

$$\Delta h_{C,a} = \frac{\Delta h_{C,s}}{\eta_C} = \frac{381.8 - 295.2}{0.82} = 105.6 \text{ kJ/kg}$$

For two-stage compression the actual compressor work is twice this value, that is,

$$w_{C,\text{in}} = 2(105.6) = 211.2 \text{ kJ/kg}$$

The actual outlet enthalpy at state 4 (see dotted lines in Fig. 15-35a) for the compressor is

$$h_{4a} = h_3 + \Delta h_{C,a} = h_1 + \Delta h_{C,a}$$
$$= 295.2 + 105.6 = 400.8 \text{ kJ/kg}$$

For the first stage of the turbine, $h_6 = 1161.1$ kJ/kg, $p_{r6} = 167.1$, and

$$p_{r7} = p_{r6}\left(\frac{P_7}{P_6}\right) = 167.1\left(\frac{1}{2.45}\right) = 68.2$$

From the air table A-5, $h_{7s} = 907.6$ kJ/kg and $T_{7s} = 877$ K. The actual turbine-enthalpy drop is

$$-\Delta h_{T,a} = -\Delta h_{T,s}\eta_T = (1161.1 - 907.6)(0.85)$$
$$= 253.5(0.85) = 215.5 \text{ kJ/kg}$$

The total turbine output is twice this value. Thus

$$w_{T,\text{out}} = 2(215.5) = 431.0 \text{ kJ/kg}$$

The actual enthalpy at state 9 for the turbine outlet is

$$h_{9a} = h_8 + \Delta h_{T,a} = h_6 + \Delta h_{T,a}$$
$$= 1161.1 - 215.5 = 945.6 \text{ kJ/kg}$$

For the regenerator, $\varepsilon_{\text{eff}} = (h_5 - h_{4a})/(h_{9a} - h_{4a})$. Hence

$$0.70 = \frac{h_5 - h_{4a}}{945.6 - 400.8}$$

or $\qquad h_5 - h_{4a} = q_{saved} = 0.7(544.8) = 381.4$ kJ/kg

The heat transfer required in the combustor becomes

$$q_{comb} = (h_6 - h_5) = (h_6 - h_{4a}) - (h_5 - h_{4a})$$
$$= (1161.1 - 400.8) - 381.4 = 378.9 \text{ kJ/kg}$$

The total heat transfer into the cycle, combustor plus reheat, is

$$q_{in} = q_{comb} + h_8 - h_{7a} = q_{comb} + h_6 - h_{7a}$$
$$= q_{comb} + w_{T,a} = 378.9 + 215.5 = 594.4 \text{ kJ/kg}$$

The thermal efficiency then is

$$\eta_{th} = \frac{w_{T,out} - w_{C,in}}{q_{in}} = \frac{431.0 - 211.1}{594.4} = 0.370 \text{ (or 37.0\%)}$$

(b) The entropy generation in either stage of the compressor or turbine is found from the basic rate equation for $\dot{\sigma}_{cv}$ given by Eq. [7-26],

$$\dot{\sigma}_{cv} = \sum_{out} \dot{m}_e s_e - \sum_{in} \dot{m}_i s_i - \sum_{j=1}^{n} \frac{\dot{Q}_j}{T_j}$$

But both devices are assumed adiabatic, and $\dot{m}$ is the same through any stage. Therefore, the rate equation reduces to $\sigma_{m,cv} = \Delta s$ on a unit-mass basis. In addition, air is assumed to be an ideal gas. Therefore, the entropy generation through any stage is given by

$$\sigma_{m,cv} = \Delta s = s_{out}^o - s_{in}^o - R \ln \frac{P_{out}}{P_{in}}$$

Hence the evaluation of $\sigma_m$ values depends on knowledge of $s^o$ data. This in turn requires that we find values for the enthalpy and temperature at states 4a, 5, 9a, and 10.

Recall that $h_{4a} = 400.8$ kJ/kg at the outlet of either compressor stage. For the turbine $\Delta h_{T,a} = -215.5$ kJ/kg per each stage. Therefore,

$$h_{7a} = h_{9a} = h_6 + \Delta h_{T,a} = 1161.1 - 215.5 = 945.6 \text{ kJ/kg}$$

Also recall that $h_5 - h_{4a} = 381.4$ kJ/kg. Hence $h_5 = 381.4 + 400.8 = 782.2$ kJ/kg. Finally, state 10 is found from an energy balance on the regenerator. Since heat transfer and work transfer are zero, and $\Delta$ke and $\Delta$pe are neglected, the steady-state energy equation reduces to $\sum_{in} \dot{m}_i h_i - \sum_{out} \dot{m}_e h_e = 0$. For the regenerator, this becomes

$$h_{9a} - h_{10} = h_5 - h_{4a} = 381.4 \text{ kJ/kg}$$

Since $h_{9a} = 945.6$ kJ/kg, $h_{10} = 945.6 - 381.4 = 564.2$ kJ/kg.

With this information on enthalpy values, the values of $T$ and $s^o$ are read directly from Table A-5. The results are:

$T_{2a} = T_{4a} = 400$ K, $\qquad s_{2a}^o = s_{4a}^o = 1.99194$ kJ/kg·K

$T_5 = 763.7$ K, $\qquad s_5^o = 2.66697$ kJ/kg·K

$T_{7a} = T_{9a} = 911.3$ K, $\qquad s_{7a}^o = s_{9a}^o = 2.86249$ kJ/kg·K

$T_{10} = 559.1$ K, $\qquad s_{10}^o = 2.33511$ kJ/kg·K

In addition, $s_1^0 = 1.68515$ kJ/kg·K at 295 K and $s_6^0 = 3.07732$ kJ/kg·K at 1100 K. On this basis the total entropy generation for the compressor and for the turbine is

$$\sigma_{m,C} = 2\left(1.99194 - 1.68515 - \frac{8.314}{29}\ln 2.45\right) = 0.0998 \text{ kJ/kg·K}$$

$$\sigma_{m,T} = 2\left(2.86249 - 3.07732 - \frac{8.314}{29}\ln\frac{1}{2.45}\right) = 0.0842 \text{ kJ/kg·K}$$

Finally, there is irreversible heat transfer within the regenerator. The entropy generation in the regenerator is also found from an entropy balance. The device is adiabatic and the mass flow rates are equal, thus the entropy balance can be written as

$$\sigma_{m,\text{reg}} = \dot{\sigma}/\dot{m} = (s_5^0 - s_{4a}^0) + (s_{10}^0 - s_{9a}^0)$$

The pressure terms in the ideal-gas equation for $\Delta s$ drop out, because it is assumed that no pressure drop occurs within the regenerator. Substitution of appropriate data yields

$$\sigma_{m,\text{reg}} = (2.66697 - 1.99194) + (2.33511 - 2.86249) = 0.1477 \text{ kJ/kg·K}$$

**Comment:**   In comparison to Example 15-9, the addition of intercooling and reheating to the cycle with regeneration increases the cycle thermal efficiency from 31.8 to 37.0 percent. Also note that the entropy generation in both the compressor and the turbine is roughly 55 to 65 percent of that occurring in the regenerator.

## 15-11   AVAILABILITY ANALYSIS OF A GAS TURBINE CYCLE

An important function of availability analyses is to reveal the magnitudes of the losses which occur in cycles. On this basis one can make improvements in existing or proposed cycles. In this section we shall examine a stationary gas-turbine power cycle, first with turbine and compressor irreversibilities, and second with regeneration.

### 15-11-1   IRREVERSIBLE COMPRESSOR AND TURBINE

The equipment schematic and $Ts$ diagram for an open gas-turbine cycle are shown in Fig. 15-38. Irreversible compressor and turbine performance is shown as processes 1-2$a$ and 3-4$a$ on the $Ts$ plot. The basic energy relation for the steady-flow process is

$$q + w = \Delta h + \Delta \text{ke} + \Delta \text{pe}$$

For the compressor, combustor, and turbine this equation reduces to

$$w_{C,\text{in}} = w_{12} = h_{2a} - h_1 \quad q_{\text{comb}} = q_{23} = h_3 - h_{2a} \quad w_{T,\text{out}} = -w_{34} = h_3 - h_{4a}$$

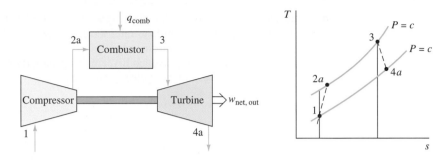

**Figure 15-38**   Equipment schematic and $Ts$ process diagram for availability analysis of a gas-turbine cycle.

where kinetic and potential energy are neglected. The compressor and turbine first-law adiabatic efficiencies are

$$\eta_C = \frac{h_{2s} - h_1}{h_{2a} - h_1} \quad \text{and} \quad \eta_T = \frac{h_3 - h_{4a}}{h_3 - h_{4s}}$$

where subscripts $a$ and $s$ represent actual and isentropic states. Recall that the stream availability is given by

$$\psi = h - h_0 - T_0(s - s_0) + \frac{V^2}{2} + gz \qquad \textbf{[9-33]}$$

In the absence of kinetic- and potential-energy effects, $\Delta\psi = \Delta h - T_0 \Delta s$. The general availability balance on a unit-mass basis for a steady-state control volume is

$$\psi_2 - \psi_1 = \phi_Q + w_{\text{act}} - i_{\text{cv}} \qquad \textbf{[9-36]}$$

where $\phi_Q = \sum q_j(1 - T_0/T_j)$.

On the basis of the above relations, the availability balances for the compressor, combustor, and turbine and the second-law efficiency are

$$w_{C,\text{in}} = w_{12} = (\psi_{2a} - \psi_1) + i_C \qquad \text{(compressor)} \quad \textbf{[15-29a]}$$

$$\sum_j q_j \left(1 - \frac{T_0}{T_j}\right) = \psi_3 - \psi_{2a} + i_{\text{comb}} \qquad \text{(combustor)} \quad \textbf{[15-29b]}$$

$$w_{T,\text{out}} = -w_{34} = \psi_3 - \psi_{4a} - i_T \qquad \text{(turbine)} \quad \textbf{[15-29c]}$$

$$\varepsilon_{\text{cycle}} = \frac{w_{\text{net,out}}}{\Delta\psi_{\text{comb}}} = \frac{w_{T,\text{out}} - w_{C,\text{in}}}{\psi_3 - \psi_2} \qquad \textbf{[15-29d]}$$

The combustion of fuel and air is modeled as heat transfer through the combustor walls, and the irreversibility within the combustor is neglected as a first approximation. The use of these relations is illustrated in the example below.

**Example 15-12**

**An** irreversible gas-turbine power plant operates between pressures of 1.0 and 6.0 bars with compressor and turbine inlet temperatures of 295 and 1100 K, respectively. The compressor and turbine adiabatic efficiencies are 82 and 85 percent, respectively. Determine energy, availability, and irreversibility data for the devices within the cycle, in kJ/kg, and the cycle effectiveness.

**Solution:**

**Given:**   A gas turbine power cycle operates with an irreversible compressor and turbine. The schematic is shown in Fig. 15-39.

**Find:**   Energy, availability, and irreversibility data for the cycle.

**Model:**   Steady state; $\eta_C = 0.82$, $\eta_T = 0.85$; neglect $\Delta$ke and $\Delta$pe; $i_{\text{comb}} \simeq 0$.

**Analysis:**   Table 15-1 summarizes the major property values at the four states of interest, for $T_0 = 25°C$ and $P_0 = 1$ bar. The entropy of an ideal gas and the availability function $\psi$ are calculated from the relations

$$s_e - s_i = s_e^o - s_i^o - R\ln(P_e/P_i)$$

$$\psi_e - \psi_i = h_e - h_i - T_0(s_e - s_i)$$

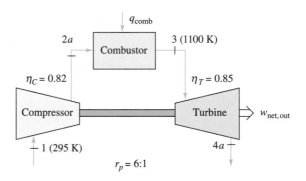

**Figure 15-39**   Schematic and data for Example 15-12.

**Table 15-1**   Property data for an irreversible gas-turbine power cycle

| State | $T_a$, K | $P$, bars | $h$, kJ/kg | $s$, kJ/kg·K | $\psi$, kJ/kg |
|-------|----------|-----------|------------|--------------|---------------|
| 1     | 295      | 1.0       | 295.2      | 1.6852       | 0.0           |
| 2     | 532      | 6.0       | 536.1      | 1.7693       | 216.1         |
| 3     | 1100     | 6.0       | 1161.1     | 2.5631       | 606.9         |
| 4     | 757      | 1.0       | 774.7      | 2.6574       | 192.7         |

where $h$ and $s^o$ data come directly from the Air Table (or see Example 15-7) and $\psi_1 = 0$. An energy accounting of inputs and outputs is shown in part $A$ of Table 15-2. The irreversibility within the compressor and turbine are found from $i = T_0 \Delta s$ for the adiabatic devices. The value of $i$ for the combustor is zero since the flow is assumed to be frictionless.

The energy analysis in part $A$ of Table 15-2 accounts for inputs and outputs for the cycle. It indicates that significant energy leaves with the turbine exhaust, and that the first-law efficiency is 23.3 percent. However, it gives no information about the effect of irreversibilities or availability losses on performance. An availability accounting for the cycle is shown in part $B$ of Table 15-2. The data are more meaningful if they are applied to the steady-state availability balance. In an input/output format

$$\sum_j q_j \left(1 - \frac{T_0}{T_j}\right) + \psi_1 = \psi_{4a} + (w_{T,\text{out}} - w_{C,\text{in}}) + i_C + i_T$$

Substitution of values yields

$$390.8 + 0 = 192.7 + 145.5 + 24.8 + 27.8$$

The availability analysis is much more revealing than the energy balance. The availability input of 390.8 kJ/kg to the cycle appears primarily as 192.7 kJ/kg in the turbine exhaust stream, compared to 145.5 kJ/kg of net work output. The destruction of availability in the compressor and turbine is fairly insignficant, totaling 52.6 kJ/kg. The increase in availability of the exhaust stream over the inlet is 49 percent, and the irreversibilities are 13 percent of the availability added in the combustor. Thus

**Table 15-2**    Energy and availability accounting for an irreversible gas-turbine power cycle

| A. Energy Accounting (in kJ/kg) | | | |
|---|---|---|---|
| **Energy in** | | **Energy out** | |
| Air in at state 1 ($h_1$) | 295.2 | Air out at state 4 ($h_4$) | 774.7 |
| Compressor work ($w_C$) | 240.9 | Turbine work ($w_T$) | 386.4 |
| Heat transfer in ($q_{23}$) | 625.0 | | |
| Total | 1161.1 | Total | 1161.1 |

| B. Availability Accounting (in kJ/kg) | | | | |
|---|---|---|---|---|
| **Process** | $q$ | $w$ | $\Delta\psi$ or $\psi$ | $i$ |
| Compressor | — | 240.9 | 216.1 | 24.8 |
| Combustor | 625.0 | — | 390.8 | — |
| Turbine | — | −386.4 | −414.2 | 27.8 |
| Inlet fluid | — | — | 0.0 | — |
| Outlet fluid | — | — | −192.7 | — |
| | 625.0 | −145.5 | 0.0 | 52.6 |

far too much work capability is lost in the turbine exhaust stream. The second-law efficiency for the cycle, on a desired output/required input basis, is

$$\varepsilon_{\text{cycle}} = \frac{w_{\text{net,out}}}{\Delta\psi_{\text{comb}}} = \frac{145.5}{390.8} = 0.372$$

Only about one-third of the availability input with the heat transfer is converted into useful work output.

### 15-11-2    REGENERATIVE CYCLE WITH IRREVERSIBILITIES

It is well known that the thermal efficiency of a gas-turbine cycle can be enhanced by use of a regenerator within the cycle. A schematic of this process and the $Ts$ plot are shown in Fig. 15-40. The "waste" energy in the turbine exhaust is used to increase the energy and temperature of the compressor exhaust stream before it enters the combustor. The subsequent decrease in heat input in the combustor, coupled with no change in net work output, could lead to a significant increase in thermal efficiency and second-law efficiency. The only additional energy relations required are the effectiveness and first-law analysis for the regenerator. In terms of the states on the $Ts$ plot, these are

$$\eta_{\text{reg}} = \frac{h_x - h_2}{h_4 - h_2} \qquad \text{and} \qquad h_x - h_2 = h_5 - h_6$$

Because $\phi_Q$ and $w_{\text{act}}$ are zero and $\dot{m}_2 = \dot{m}_5$, the availability balance for the regenerator is

$$0 = (\psi_2 - \psi_x) + (\psi_4 - \psi_5) - i_{\text{reg}} \qquad \textbf{[15-30]}$$

These relations are now applied to the preceding example.

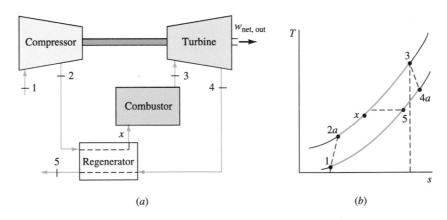

(a)    (b)

**Figure 15-40**    Equipment schematic and $Ts$ plot for regenerative gas-turbine cycle.

**Example 15-13**

**A** regenerator with 70 percent effectiveness is added to the gas-turbine cycle examined in Example 15-12. Complete the energy and availability analyses for this cycle.

**Solution:**

**Given:** A gas turbine cycle has an irreversible compressor and turbine and a regenerator with 70 percent effectiveness, as shown in Fig. 15-41.

**Find:** Energy and availability quantities for the cycle, and compare results with Example 15-12.

**Model:** Steady state, $\eta_C = 0.82$, $\eta_T = 0.85$, $\varepsilon_{regen} = 0.70$; neglect $\Delta$ke and $\Delta$pe.

**Analysis:** Table 15-3 summarizes property values for the cycle. The energy and availability accountings are shown in Table 15-4. The thermal efficiency is now 0.318, versus 0.233 without a regenerator. Also, the exhaust enthalpy has been

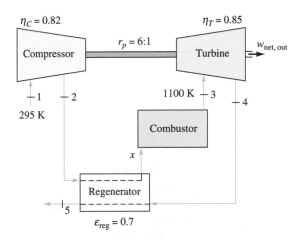

**Figure 15-41** Schematic and data for Example 15-13.

**Table 15-3** Property data for a regenerative gas-turbine power cycle

| State | $T_a$, K | $P$, bars | $h$, kJ/kg | $s$, kJ/kg·K | $\psi$, kJ/kg |
|-------|----------|-----------|-----------|--------------|----------------|
| 1 | 295 | 1.0 | 295.2 | 1.6852 | 0.0 |
| 2 | 532 | 6.0 | 536.1 | 1.7693 | 216.1 |
| x | 690 | 6.0 | 703.1 | 2.0439 | 302.1 |
| 3 | 1100 | 6.0 | 1161.1 | 2.5631 | 606.9 |
| 4 | 757 | 1.0 | 774.7 | 2.6574 | 192.7 |
| 5 | 601 | 1.0 | 607.7 | 2.4101 | 98.6 |

**Table 15-4**　Energy and availability accounting for a regenerative gas-turbine power cycle

| A. Energy Accounting (in kJ/kg) | | | |
|---|---|---|---|
| **Energy in** | | **Energy out** | |
| Air in at state 1 | 295.2 | Air out at state 5 | 607.7 |
| Compressor work | 240.9 | Turbine work | 386.4 |
| Heat in | 458.0 | | |
| Total | 994.1 | Total | 994.1 |

| B. Availability Accounting (in kJ/kg) | | | | |
|---|---|---|---|---|
| **Process** | $q$ | $w$ | $\Delta\psi$ or $\psi$ | $i$ |
| Compressor | — | 240.9 | 216.1 | 24.8 |
| Combustor | 458.0 | — | 304.8 | — |
| Turbine | — | −386.4 | −414.2 | 27.8 |
| Inlet fluid | — | — | 0.0 | — |
| Outlet fluid | — | — | −98.6 | — |
| Regenerator | — | — | −8.1 | 8.1 |
| | 458.0 | −145.5 | 0.0 | 60.7 |

reduced from 774.7 to 607.7 kJ/kg as the exhaust temperature is lowered from 757 to 601 K. The availability analysis is more revealing. Writing the availability balance as input = output,

$$\sum_j q_j \left(1 - \frac{T_0}{T_j}\right) + \psi_1 = \psi_5 + (w_{T,\text{out}} - w_{C,\text{in}}) + i_C + i_T + i_{\text{reg}}$$

Substitution of data leads to

$$304.8 + 0 = 98.6 + 145.5 + 24.8 + 27.8 + 8.1$$

In this case, the increase in the exhaust stream availability over the inlet stream is only 32 percent of the availability added in the combustor. This is down from 49 percent without a regenerator. The irreversibilities of the cycle are 20 percent and the net work is 48 percent of the availability input in the combustor. This latter value also is the effectiveness $\varepsilon$ for the cycle. Thus a sizable increase in $\varepsilon$ has occurred with the installation of the regenerator, since the effectiveness without a regenerator is 37 percent.

## 15-12　GAS TURBINES FOR JET PROPULSION

Before we discuss the application of the gas-turbine power cycle to aircraft propulsion, it is appropriate to review briefly the performance criteria for diffusers and nozzles as expressed in terms of their reversible and irreversible operation.

## 15-12-1   DIFFUSER AND NOZZLE PERFORMANCE

A diffuser decelerates flow and increases the pressure. If the heat transfer is negligible, the steady-flow energy equation $q + w = \Delta h + \Delta ke + \Delta pe$ reduces to

$$\Delta h + \Delta ke = 0$$

To simplify the analysis in this subsection, a cold air-standard basis is used. In this case, the energy equation simplifies to

$$c_p \Delta T = -\Delta ke \quad \text{or} \quad c_p(T_2 - T_1) = -\frac{\mathbf{V}_2^2 - \mathbf{V}_1^2}{2} \quad \textbf{[a]}$$

where 1 and 2 represent the inlet and outlet states, respectively. If the final velocity is small compared to the initial value, then $\mathbf{V}_1$ fixes the kinetic-energy change, which in turn fixes $\Delta T$. Hence the temperature change and $T_2$ are independent of whether irreversibilities are present for a given value of $T_1$. The property that is affected by irreversibilities is the pressure. This is clearly seen from a second-law analysis of the flow through the diffuser.

The entropy balance for an adiabatic diffuser can be written as

$$\sigma_{m,\text{diff}} = \frac{\dot{\sigma}_{\text{cv}}}{\dot{m}} = s_2 - s_1 \quad \textbf{[b]}$$

Recall that for an ideal gas, the entropy change can be calculated using the relation $\Delta s = s_2 - s_1 = c_p \ln(T_2/T_1) - R \ln(P_2/P_1)$ where $c_p$ is a constant for the cold-air-standard model. Combining the results from these two equations and solving for the actual pressure ratio across the diffuser, $P_{2a}/P_1$, we have the following relation

$$\ln \frac{P_{2a}}{P_1} = \frac{c_p}{R} \ln \frac{T_2}{T_1} - \frac{\sigma_{m,\text{diff}}}{R} \quad \textbf{[c]}$$

Since the temperature ratio $T_2/T_1$ is essentially fixed for the diffuser by the energy balance, this equation clearly shows that any entropy production (irreversibility) inside the diffuser will reduce the exit pressure. When the process is internally reversible, the pressure ratio becomes $\ln(P_{2s}/P_1) = (c_p/R) \ln(T_2/T_1)$. Using this result to eliminate $T_2/T_1$ in Eq. [c] above and solving for the actual pressure ratio gives a simple relation between the actual and the ideal pressure ratios across a diffuser when the inlet and outlet temperatures are specified:

$$\frac{P_{2a}}{P_1} = \frac{P_{2s}}{P_1} \exp\left(-\frac{\sigma_{m,\text{diff}}}{R}\right) \quad \textbf{[d]}$$

This effect is illustrated on a $Ts$ diagram in Fig. 15-42. Note that higher values of $s_2$, caused by increased $\sigma_m$, lead to lower values of $P_{2a}$ for a fixed value of $T_2$.

One measure of the performance of a diffuser in terms of pressure data is the *pressure coefficient* $K_P$. For a given value of $\Delta ke$ for a process

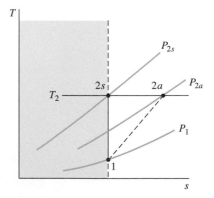

**Figure 15-42**
Effect of irreversibilities on diffuser-outlet pressure.

$$K_P = \frac{\text{actual pressure rise}}{\text{isentropic pressure rise}} = \frac{P_{2a} - P_1}{P_{2s} - P_1} \qquad \textbf{[15-31]}$$

(Note that $K_P$ is sometimes defined in a slightly different way. The definition may require $P_{2s}$ to be the pressure reached isentropically when the final kinetic energy is zero. This limiting pressure, which lies vertically above $P_{2s}$ in Fig. 15-42 if the final kinetic energy is not zero, is known as the stagnation pressure for the ideal diffuser.)

An appropriate measure of the performance of a nozzle is the adiabatic efficiency $\eta_N$ of the device. This is defined in Sec. 8-2-2 as

$$\eta_N = \frac{\mathbf{V}_{2a}^2/2 - \mathbf{V}_1^2/2}{\mathbf{V}_{2s}^2/2 - \mathbf{V}_1^2/2} \qquad \textbf{[8-25]}$$

If the inlet velocity is negligible compared to $\mathbf{V}_{2a}$ and $\mathbf{V}_{2s}$ at the exit, then Eq. [8-25] above may be expressed also by

$$\eta_N = \frac{h_1 - h_{2a}}{h_1 - h_{2s}} \qquad \textbf{[8-26]}$$

These performance criteria for diffusers and nozzles are used in the analysis of jet-propulsion cycles.

## 15-12-2   AIRCRAFT GAS TURBINE ENGINES

One of the most effective adaptations of the gas-turbine power cycle has been for the propulsion of aircraft. This is due to the favorable power-to-weight and power-to-volume ratios for a gas-turbine unit. An example of an aircraft gas turbine engine is shown in Photograph 15-3. The stationary gas-turbine cycle studied earlier and the jet-engine cycle do have some major differences, however. One lies in the operation of the compressor and turbine. In a turbojet unit, no net work output is required from the turbine beyond that needed to drive the compressor and auxiliary equipment. Hence there is no output shaft from the turbine. This is seen in Fig. 15-43, which is a schematic of a turbojet engine. The center section of the engine contains the three major components of a gas-turbine unit–the compressor, combustor, and turbine. Since the work requirement of the turbine is less, the gas does not expand to ambient pressure in the turbine. The final expansion occurs in the nozzle which follows the turbine. Here the fluid is accelerated to a relatively high velocity. The pressure ratio $P_4/P_5$ in the nozzle may be 2 or more. A third difference in operation is the placement of a diffuser in front of the compressor. Its purpose is to slow down the entering fluid and increase the pressure. A small pressure rise of a few decibars (or psi) accompanies the decrease in kinetic energy. This pressure rise is referred to as the *ram effect*.

The general thermodynamic characteristics of an ideal turbojet engine are shown on the $Ts$ diagram of Fig. 15-44. Process $y$-1 shows a pressure rise in the diffuser due to a decrease in kinetic energy. The following three

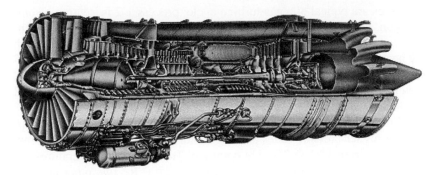

**Photograph 15-3**   Pratt & Whitney JT80 Engine (14,000 to 17,000 pounds-force thrust). (Courtesy of Pratt & Whitney)

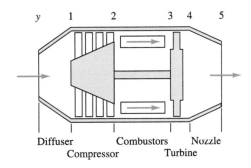

**Figure 15-43**   Turbojet engine schematic.

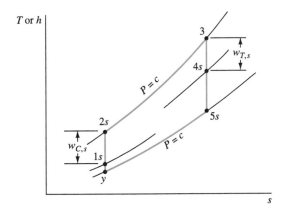

**Figure 15-44**   The $Ts$ diagram for a turbojet engine showing ideal performance.

processes are typical of a gas-turbine cycle: process 1-2 is isentropic compression, process 2-3 is constant-pressure heating, and process 3-4 is isentropic expansion. Finally, process 4-5 shows isentropic expansion through the nozzle, where the pressure decrease is accompanied by a significant increase in kinetic energy. If one considers Fig. 15-44 also as an $hs$ diagram, then the vertical distance between states $1s$ and $2s$ must equal the vertical distance between states $3$ and $4s$, because $w_{C,in,s} = w_{T,out,s}$.

Compared to stationary gas-turbine power plants, turbojet engines operate at higher pressure ratios and higher turbine-inlet temperatures. Pressure ratios from 10:1 to 25:1 are common. Turbine-inlet temperatures are in the range of 1100 to 1500 K, or 2000 to 2700°R. Two other applications of the gas-turbine cycle to aircraft propulsion include the *turbofan* and the *turbo-prop* (or propjet). In the turbofan engine, the turbine drives not only the compressor but also a set of large-diameter fan blades in front of the engine. A large portion of the total air flow to the engine bypasses the compressor and flows solely through the fan blades. This portion of the flow augments the thrust developed by the nozzle in the main engine. In a propjet engine there is no cowling on the outside of the engine through which the bypass air must flow. Thus the fan blades become a propeller. It requires a larger turbine to drive the compressor and propeller, and this reduces the thrust provided by the nozzle. However, propellers are more efficient than pure-jet engines, and most of the air passes through the propellers. Originally limited to low-speed aircraft and low altitudes, new designs with lower fuel consumption than jets and competitive speeds are under study. The following discussion covers the basic energy analysis of a turbojet cycle.

### 15-12-3    PERFORMANCE PARAMETERS FOR A JET-ENGINE CYCLE

For an aircraft gas-turbine cycle we need to examine the effect of inlet- and exit-air velocity (relative to the aircraft) on thrust and propulsion efficiency. The following discussion is based on an air-standard cycle, so that the effect of fuel addition on mass flow rates is neglected. The *thrust* developed by the engine is the unbalanced force which acts to overcome the resistance to motion known as fluid drag. On the basis of Newton's second law, the thrust is equal to the rate of change of momentum of the air passing through the engine. When the inlet and exit pressures are the same, the thrust $F$ developed by the engine is simply

$$F = \dot{m}(\mathbf{V}_{out} - \mathbf{V}_{in}) = \dot{m}(\mathbf{V}_5 - \mathbf{V}_y) \qquad \textbf{[15-32]}$$

where $\mathbf{V}_5$ and $\mathbf{V}_y$ are the air velocities measured relative to the engine.

The *propulsive efficiency* $\eta_P$ is a measure of how well the energy provided to the cycle is converted to thrust power to overcome drag forces. The thrust power $\dot{W}_{thrust}$ is related to thrust by the relation

$$\dot{W}_{thrust} = F\mathbf{V}_y = \dot{m}\mathbf{V}_y(\mathbf{V}_5 - \mathbf{V}_y)$$

The energy added to the cycle in the combustor is $\dot{Q}_{in} = \dot{m}(h_3 - h_2)$. Therefore, one measure of the propulsive efficiency is

$$\eta_P \equiv \frac{\dot{W}_{thrust}}{\dot{Q}_{in}} = \frac{\mathbf{V}_y(\mathbf{V}_5 - \mathbf{V}_y)}{h_3 - h_2} \qquad \textbf{[15-33]}$$

This is just one of many possible performance measures for a jet-engine cycle.

### 15-12-4 ENERGY ANALYSIS OF IDEAL AIRCRAFT GAS-TURBINE CYCLE

The energy analysis of a turbojet engine is based on a knowledge of the following data: the air-inlet temperature $T_y$, pressure $P_y$, velocity $\mathbf{V}_y$, the pressure ratio across the compressor $r_p$, and the limiting turbine-inlet temperature $T_3$. Subscripts for the preceding symbols are based on the notation of Fig. 15-43. An air-standard cycle is used as a model, and the variation of specific heat with temperature is accounted for by using air-table data. As a further model, the exit velocity from the diffuser is assumed to be very small in comparison to the inlet value, and the gas velocity remains small until the nozzle exit is reached. The fluid expands through the nozzle to the ambient pressure.

The actual numerical analysis of a gas-turbine cycle using tabular data follows the general pattern illustrated in preceding examples in this chapter. With the inclusion of the diffuser and nozzle in the cycle, however, three comments are pertinent. First, the diffuser analysis begins with an energy balance. Because the diffuser outlet velocity $\mathbf{V}_1$ is assumed to be negligible, the energy balance shows that the inlet conditions fix the value of $h_1$ at the outlet, and hence $T_1$, $p_{r1}$, and $s_1^0$ at the diffuser outlet. The outlet pressure $P_1$ is then found by either $P_1 = P_y(p_{r1}/p_{ry})$ or $s_1^0 - s_y^0 = -R \ln (P_1/P_y)$. Second, the turbine analysis is altered from previous examples, because the pressure ratio $P_3/P_4$ across it is unknown. However, for the overall cycle, $w_{T,out,s} = h_3 - h_{4s} = w_{C,in,s}$. From this relation $h_{4s}$ is calculated, and the corresponding values of $T_{4s}$, $p_{r4}$, and $s_4^0$ are found in the air table. Since $T_3$ (the turbine-inlet temperature) is input information, $p_{r3}$ and $s_3^0$ are also known. Therefore, the turbine-outlet pressure $P_4$ can now be found by the use of $p_r$ or $s^0$ data for states 3 and 4, as in the diffuser calculation outlined above. Finally, because the nozzle-outlet pressure is assumed to be the ambient value, the value of $p_{r5}$ or $s_5^0$ can be determined for the nozzle outlet. This in turn fixes $T_{5s}$ and $h_{5s}$. Since the enthalpy change across the nozzle under isentropic conditions is now known, the kinetic-energy change can be found from the basic energy relation $\Delta ke_s = -\Delta h_s$. This in turn leads to the value of $\mathbf{V}_5$, since $\mathbf{V}_4$ is assumed to be negligible.

Thus the method of determining values in an aircraft gas-turbine cycle does not follow the same pattern as that for stationary gas-turbine units. The following examples illustrate the numerical calculations outlined above for a reversible gas-turbine propulsion cycle. Although either $p_r$ or $s^\circ$ data could be used for the isentropic calculations, only $p_r$ data are employed in the examples below.

**Example 15-14**

**A** turbojet aircraft flies at 260 m/s at an altitude of 5000 m, where the atmospheric pressure is 0.60 bar and the temperature is 250 K. The compressor pressure ratio is 8:1, and the temperature at the turbine inlet is 1300 K. Assuming ideal performance of the various components of the engine, determine (*a*) the compressor-work input, the pressures and temperatures throughout the cycle, and the exit-jet velocity, and (*b*) the thrust and propulsive efficiency if the mass flow rate is 60 kg/s.

**Solution:**

**Given:**    An ideal turbojet gas-turbine cycle, as shown in Fig. 15-45.

**Find:**    (*a*) Compressor work, pressures and temperatures throughout cycle, and exit-jet velocity, and (*b*) thrust and propulsion efficiency.

**Model:**    Air-standard cycle; diffuser, compressor, turbine, and nozzle are isentropic; no pressure drop in combustor; air is an ideal gas.

**Analysis:**    (*a*) The values calculated correspond to the states shown in Fig. 15-45. For the diffuser process 1-*y*, we assume that the outlet velocity is negligible compared to the inlet value. At 250 K the enthalpy is 250.1 kJ/kg, and $p_{ry} = 0.7329$ from Table A-5. For isentropic flow through the diffuser, we find that

$$h_y + \frac{\mathbf{V}_y^2}{2} = h_1 + \frac{\mathbf{V}_1^2}{2}$$

$$250.1 + \frac{260^2}{2(1000)} = h_1 + 0$$

$$h_1 = 250.1 + 33.8 = 283.9 \text{ kJ/kg}$$

From the air table, $T_1 = 284$ K and $p_{r1} = 1.141$. Hence

$$P_1 = P_y \frac{p_{r1}}{p_{ry}} = 0.60 \frac{1.141}{0.7329} = 0.934 \text{ bar}$$

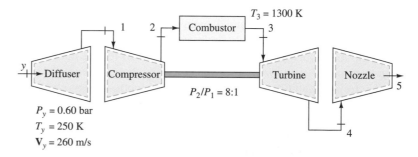

**Figure 15-45**    Schematic and data for Example 15-14.

Thus there is a 0.33-bar pressure rise in the diffuser. For the compressor, with $r_p = 8$, $P_2 = 8(0.934) = 7.47$ bars. Therefore,

$$p_{r2} = p_{r1}\frac{P_2}{P_1} = 1.141(8) = 9.13$$

From the air table A-5, $T_{2s} = 511$ K and $h_{2s} = 514.9$ kJ/kg. The isentropic compressor work is

$$w_{C,\text{in},s} = h_{2s} - h_1 = 514.9 - 283.9 = 231.0 \text{ kJ/kg}$$

At the turbine inlet, where $T_3 = 1300$ K, we find that $h_3 = 1396.0$ kJ/kg and $p_{r3} = 330.9$. Since the turbine work out must equal the compressor work in, $w_{T,\text{out}} = h_3 - h_{4s} = w_{C,\text{in}}$ and

$$h_{4s} = h_3 - w_{c,\text{in}} = 1396.0 - 231.0 = 1165.0 \text{ kJ/kg}$$

For this value of the enthalpy, Table A-5 shows that $T_{4s} = 1102$ K and $p_{r4} = 169.3$. Hence

$$P_4 = P_3\frac{p_{r4}}{p_{r3}} = 7.47\frac{169.3}{330.9} = 3.82 \text{ bars}$$

Note that the turbine-outlet pressure is considerably above the atmospheric value, and hence a reasonably large pressure drop still exists across the nozzle. The nozzle-outlet state is found from

$$p_{r5} = p_{r4}\frac{P_5}{P_4} = 169.3\frac{0.60}{3.82} = 26.6$$

At this condition, $T_{5s} = 685$ K and $h_{5s} = 697.4$ kJ/kg. An energy balance on the nozzle yields, if we neglect the inlet velocity,

$$h_4 + \frac{\mathbf{V}_4^2}{2} = h_5 + \frac{\mathbf{V}_5^2}{2}$$

$$(1165.0 + 0) \text{ kJ/kg} = 697.4 \text{ kJ/kg} + \frac{\mathbf{V}_5^2}{2} \times \frac{1 \text{ N·s}^2}{1 \text{ kg·m}} \times \frac{1 \text{ kJ}}{1000 \text{ N·m}}$$

$$\mathbf{V}_5 = [2(1000)(1165.0 - 687.4)]^{1/2} = 967 \text{ m/s}$$

This velocity is supersonic, with a Mach number around 1.8. A more meaningful calculation would include compressor, turbine, and nozzle efficiencies and a diffuser pressure coefficient.

(b) The thrust, as given by Eq. [15-32], is

$$F = \dot{m}(\mathbf{V}_5 - \mathbf{V}_y) = 60\ \frac{\text{kg}}{\text{s}} \times (967 - 260)\ \frac{\text{m}}{\text{s}} \times \frac{1 \text{ N·s}^2}{\text{kg·m}} = 42{,}420 \text{ N}$$

The propulsion efficiency, in terms of Eq. [15-33], is

$$\eta_P = \frac{\mathbf{V}_y(\mathbf{V}_5 - \mathbf{V}_y)}{h_3 - h_2} = \frac{260(967 - 260) \text{ m}^2/\text{s}^2}{(1396.0 - 514.9) \text{ kJ/kg}} \times \frac{1 \text{ kJ}}{1000 \text{ N·m}} \times \frac{1 \text{ N·s}^2}{\text{kg·m}} = 0.21$$

This is the fraction of the energy input used to overcome drag forces.

## 15-12-5 ENERGY ANALYSIS FOR A NONIDEAL CYCLE

In practice, irreversibilities occur within a turbojet engine. The diffuser, compressor, turbine, and nozzle are not isentropic. These irreversibilities modify the basic cycle shown in Fig. 15-40, and these modifications are

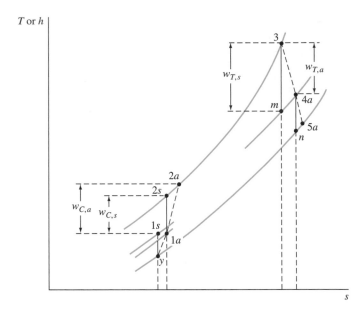

**Figure 15-46**    The $Ts$ and $hs$ diagrams for a turbojet engine showing irreversible performance.

shown in the $Ts$ diagram of Fig. 15-46. Because irreversibilities do not affect the energy analysis of the diffuser, states $1a$ and $1s$ have the same temperature and enthalpy. However, state $1a$ lies at a lower pressure than $1s$, owing to the irreversibility. The pressure coefficient $K_P$ is used to determine the actual pressure rise across the diffuser and hence $P_{1a}$. The turbine analysis is based on the requirement that $w_{T,\text{out},a} = h_{3a} - h_4 = w_{C,\text{in},a} = h_{2a} - h_1$ in the irreversible case. From this relation $h_{4a}$ is found as well as the corresponding value of $T_{4a}$. At the same time, $w_{T,\text{out},s} = w_{T,\text{out},a}/\eta_T = h_3 - h_m$. The only unknown in this expression is $h_m$, and its value leads to values of $T_m$, $p_{rm}$, and $s_m^0$. Since $T_3$ (the turbine-inlet temperature) is input data, $p_{r3}$ and $s_3^0$ are known. Therefore, the isentropic turbine-outlet pressure $P_m (= P_{4a})$ can now be found by use of $p_r$ or $s^0$ data for states 3 and 4, as in the diffuser calculation outlined above. Finally, from a knowledge of $T_{4a}$, $P_{4a}$, and $P_5$, an isentropic calculation using $p_r$ or $s^0$ data leads to $T_n$ and hence $h_n$. Since the enthalpy change across the nozzle under isentropic condition is now known, the kinetic-energy change across the nozzle is calculated from the energy balance $\Delta ke_s = -\Delta h_s$. The nozzle efficiency may then be used to evaluate the actual kinetic-energy increase and the actual exit-jet velocity.

The following example illustrates the numerical calculations outlined above for an irreversible gas-turbine jet engine. Again, $p_r$ data are used for all isentropic calculations.

**Example 15-15**

Rework Example 15-14 under the following conditions: The actual pressure rise in the diffuser is 92 percent of theoretical, the compressor efficiency is 82 percent, the

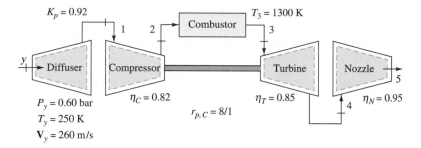

**Figure 15-47**    Schematic and data for Example 15-15.

turbine efficiency is 86 percent, and the nozzle efficiency is 95 percent. Determine the pressure and temperature throughout the cycle, the compressor work, and the exit-jet velocity, in m/s.

**Solution:**

**Given:**  A nonideal turbojet aircraft engine with data for the diffuser, compressor, turbine, and nozzle performance indicated in Fig. 15-47.

**Find:**  Compressor work, pressures and temperatures throughout cycle, and exit-jet velocity.

**Model:**  Air-standard cycle; diffuser, compressor, turbine, and nozzle are irreversible, no pressure drop in combustor, air is an ideal gas.

**Analysis:**  The effect of irreversibilities on the cycle is shown by the dashed lines in Fig. 15-46. The energy analysis on the diffuser remains the same, so $h_{1a} = h_{1s} = 283.9$ kJ/kg, $T_{1a} = T_{1s} = 284$ K, $p_{r1} = 1.141$, $P_{1s} = 0.934$ bar, and $P_{1s} - P_y = 0.334$ bar, as in Example 15-14. Because the pressure coefficient is now 0.92, Eq. [15-31] leads to

$$P_{1a} = K_P\,\Delta P_s + P_y = 0.92(0.334) + 0.6 = 0.907 \text{ bar}$$

To find state $2a$, we first must calculate the isentropic compressor work and then use $\eta_C$ to correct the value. Properties of state $2s$ are found by noting that

$$p_{r2} = p_{r1}(r_p) = 1.141(8) = 9.13$$

From Table A-5, $T_{2s} = 511$ K and $h_{2s} = 514.9$ kJ/kg. Therefore, the actual compressor work becomes

$$w_{C,\text{in},a} = \frac{w_{C,\text{in},s}}{\eta_C} = \frac{514.9 - 283.9}{0.82} = 281.7 \text{ kJ/kg}$$

and         $$h_{2a} = h_1 + w_{C,a} = 283.9 + 281.7 = 565.6 \text{ kJ/kg}$$

For this enthalpy value, $T_{2a} = 560$ K. The pressure at this state is

$$P_{2a} = P_{1a}r_p = 0.907(8) = 7.26 \text{ bars}$$

The turbine-inlet conditions are $T_3 = 1300$ K, $h_3 = 1396.9$ kJ/kg, and $P_3 = 7.26$ bars. State $4a$ is fixed by the requirement that $w_{T,\text{out},a} = w_{C,\text{in},a} = 281.7$ kJ/kg. Hence

$$h_{4a} = h_3 - w_{T,\text{out},a} = 1396.0 - 281.7 = 1114.3 \text{ kJ/kg}$$

For this enthalpy value, $T_{4a} = 1060$ K. The pressure $P_{4a}$ is found indirectly by finding $P_m$. (See Fig. 15-46 for the position of state $m$.) The properties associated with state $m$ are easily found by noting that state $m$ is the turbine-outlet state if the expansion is isentropic. Hence

$$h_m = h_3 - \frac{w_{T,\text{out},a}}{\eta_T} = 1396.0 - \frac{281.7}{0.86} = 1068.4 \text{ kJ/kg}$$

For this enthalpy $T_{4s} = 1020$ K and $p_{rm} = 123.4$. Thus for the isentropic process 3-$m$,

$$P_m = P_3 \frac{p_{rm}}{p_{r3}} = 7.26 \frac{123.4}{330.9} = 2.71 \text{ bars}$$

But states $m$ and $4a$ lie on the same pressure line. Thus $P_{4a} = P_m = 2.71$ bars.

Finally, if the nozzle is isentropic and expansion occurs to a $P_n(= P_{5a})$ value of 0.6 bar (ambient pressure), then

$$p_{rn} = p_{r4} \frac{P_5}{P_4} = 143.9 \frac{0.6}{2.71} = 31.86$$

For this value of $p_{rn}$, $T_n = 719$ K and $h_n = 734$ kJ/kg. Using a nozzle efficiency of 95 percent, we find

$$\Delta h_{N,a} = \eta_N \, \Delta h_{N,s} = 0.95(1114 - 734) = 361 \text{ kJ/kg}$$

This actual enthalpy change equals the kinetic-energy change of the fluid as it passes through the nozzle. Assuming a negligible inlet velocity $\mathbf{V}_{4a}$ to the nozzle, we find that

$$\mathbf{V}_{\text{jet}} = \left[ 2(361 \text{ kJ/kg}) \times \frac{1 \text{ kg·m}}{\text{N·s}^2} \times \frac{1000 \text{ N·m}}{1 \text{ kJ}} \right]^{1/2} = 850 \text{ m/s}$$

The exit temperature $T_{5a}$ is found from $h_{5a}$:

$$h_{5a} = h_{4a} - \Delta h_{N,a} = 1114 - 361 = 753 \text{ kJ/kg}$$

From the air table, $T_{5a} = 737$ K.

## 15-13   CLOSED-LOOP GAS-TURBINE CYCLES

The modern gas-turbine engine has two major applications: as a power source for the production of electric power when coupled with an electric generator and as a power source for the propulsion of aircraft. The device is also applicable as a power source for ground-level propulsion equipment, such as passenger cars, trucks, and trains. The thermodynamic fundamentals of the basic air-breathing gas-turbine cycle are discussed in Secs. 15-6 through 15-8. The Brayton cycle may also be used as a power system for orbiting vehicles in the space environment outside our atmosphere. In this case the space power system would be used to generate electric power aboard the spacecraft, and not to propel the vehicle. The major features of the cycle described previously remain the same for space applications. Typical turbine expansion ratios are around 2:1 or less, and cycle temperature ratios (minimum to maximum temperature) are around 0.25:0.30. However, three major modifications must be made to permit operation in a space environment:

1. The device must operate as a closed cycle, since the working fluid must be conserved for reuse. Air is not available as in terrestrial applications of open gas-turbine cycles.

2. The energy source must be of a different nature than the conventional hydrocarbons which undergo combustion directly in the cycle. This is necessary since the cycle loop is closed, and hence accumulation of products of combustion cannot be permitted.

3. The temperature of the gas prior to entering the compressor in the cycle must be reduced to its initial value. In an open cycle relatively hot gases are simply thrown out into the atmosphere, even if a regenerator (recuperator) has been employed to preheat the gases between the compressor and the heat source.

The first modification to a closed cycle increases the physical complexity of the system. However, the closed-loop operation is advantageous, since it permits the use of working fluids which have properties more desirable than those of air. The use of inert gases, for example, reduces the problem of corrosion. Fundamental studies of turbomachinery indicate that the number of turbine blade rows decreases rapidly as the molecular weight increases. An application requiring 10 stages using helium (molar mass of 4) requires two stages with neon (molar mass of 20) and one stage with argon (molar mass of 40). For gases of even higher molecular weight, a single stage could be used at lower speeds than for argon, and hence at lower stresses. The turbomachinery diameter also increases as the molecular weight increases. For the relatively low-power turbomachinery used in space application, this also is an advantage. For power levels in the range of 10 to 50 kW, gases having molecular weights in the range of 40 to 80 could be used. However, high-molecular-weight gases do have disadvantages. A major one is discussed when the third modification of the basic system is presented.

The energy sources of principal interest are solar radiation and nuclear reactors, although radioisotopes are also a possibility. Nuclear reactors are compact and require no special orientation with respect to the space environment, but they could be hazardous. A solar heat source is not dangerous, but it does require special means of orientation with respect to the sun. In addition, a relatively large solar concentrator (mirror) with good collection efficiency is required. For example, for a 10-kW Brayton power cycle with a maximum temperature of 1100 K (2000°R), a 25-ft-diameter concentrator may be required. Whichever energy source is used, the heat generated must be transferred to the working fluid by means of a heat exchanger. This heat exchanger is shown as the "heat-source heat exchanger" in Fig. 15-48. The regenerator is, of course, a second heat exchanger in the loop. Regenerators working with argon have been built with an effectiveness of 90 percent and a 2 percent pressure drop.

The third modification for a Brayton space-power system involves the "heat-sink heat exchanger" shown in Fig. 15-48. After passing through the regenerator (recuperator), the working fluid is cooled further by passing through this third heat exchanger. The energy discharged from the working

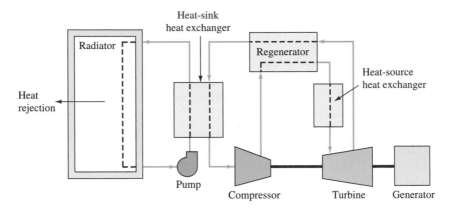

**Figure 15-48**    Gas-turbine space-power system.

fluid to a secondary liquid in the radiator loop is then radiated into space. The radiator size is significantly reduced by increasing the turbine-inlet temperature. This third heat-exchange process reduces the temperature of the primary working fluid to the required value at the compressor inlet.

It has been mentioned that high-molecular-weight gases (molecular weight greater than air) have properties favorable to turbomachinery design. Unfortunately, the thermal conductivity of a gas falls rapidly with increasing molecular weight. The conductivity of krypton (molecular weight of 83) is roughly one-fifteenth of that of helium (molecular weight of 4). This implies a severe penalty in heat-exchanger size. Recall from Fig. 15-48 that the working fluid passes through three different heat exchangers. A heat exchanger designed for krypton would be about 10 times the size of one designed for helium. The increased weight and volume suggest that the choice of a working fluid is a compromise between turbomachinery needs and heat-exchanger requirements. An interesting engineering-design feature is that the benefits of high molecular weight to turbomachinery design can be met in a subtle way without sacrificing heat-exchanger size. This is done by using a mixture of inert gases, rather than a pure gas. For example, helium and krypton could be mixed to give a gas that has the same molecular weight as argon. However, the average thermal conductivity of the gas mixture is greatly increased by the presence of helium. The result is a reduction in heat-exchanger size of about one-half of that of pure argon.

## 15-14    THE ERICSSON AND STIRLING CYCLES

The maximum thermal efficiency of any heat engine operating between two thermal reservoirs at $T_H$ and $T_L$ is the Carnot efficiency given by $\eta_{\text{Carnot}} = 1 - (T_L/T_H)$. A Carnot heat engine operating under these temperature conditions is discussed in Sec. 8-3. There are two other theoretical cycles that fulfill the Carnot efficiency but operate physically in different manners from a Carnot heat engine.

## 15-14-1 THE ERICSSON CYCLE

In Sec. 15-10 we demonstrated that the composite effect of intercooling, reheating, and regeneration is an increase in the thermal efficiency of a gas-turbine power cycle. It is interesting to examine the situation where the number of stages of both intercooling and reheating become infinitely large. In this situation the isentropic compression and expansion processes become isothermal ones. That is, for an infinitely large number of stages, the cycle can be represented by two constant-temperature processes and two constant-pressure processes with regeneration. Such a cycle is called an ***Ericsson cycle***.

Figure 15-49 shows the $Pv$ and $Ts$ diagrams for the cycle and a schematic diagram of an Ericsson engine operating as a steady-flow device. The fluid expands isothermally through the turbine from state 1 to 2. Work is produced, and heat transfer occurs reversibly from a reservoir at $T_H$. The fluid is then cooled at constant pressure in a regenerator (heat exchanger). From state 3 to state 4 the fluid is compressed isothermally. This requires work input and reversible heat transfer to a reservoir at a temperature $T_L$. Finally, the fluid temperature is raised at constant pressure to the initial state by passing the fluid in a countercurrent direction through the regenerator and picking up the energy rejected to the regenerator during the constant-pressure cooling process. In this manner the temperature difference across

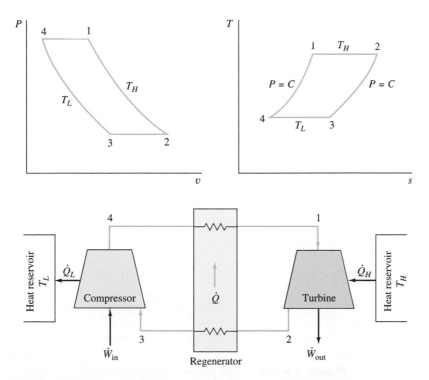

**Figure 15-49** Typical $Pv$, $Ts$, and schematic diagrams of the Ericsson cycle.

the heat exchanger is always infinitesimal throughout the length of the exchanger, as required for reversible heat transfer. The regenerator again acts as an energy-storage unit within the system. For any complete cycle the net energy storage is zero. This requires that the area under curve 4-1 on the $Ts$ diagram in Fig. 15-49 equals the area under curve 2-3.

Since the only external heat transfer is between heat reservoirs, and since all processes are described as reversible, the thermal efficiency of the Ericsson cycle equals that of the Carnot cycle, given by $1 - T_L/T_H$. The original engine constructed by Ericsson (1803–1889) was a nonflow device, but the thermodynamic cycle is the same as for the steady-flow device described above. Such a design is impractical, however, in terms of intercooling and reheating, because the cost and size requirements are prohibitive.

### 15-14-2  THE STIRLING CYCLE

Although the Ericsson cycle is impractical, the cycle does demonstrate how a regenerator might be placed in a cycle to increase the thermal efficiency. Another theoretical cycle of more practical importance that incorporates a regenerator in the basic scheme is the ***Stirling cycle***. Proposed by Robert Stirling (1790–1878), the cycle is composed of two isothermal reversible processes and two constant-volume reversible processes. The $Pv$ and $Ts$ diagrams for the cycle are shown in Fig. 15-50.

From an initial state 1 the gas is expanded isothermally to state 2, and heat transfer occurs reversibly from a thermal reservoir at $T_H$. From state 2 to state 3 energy is removed by heat transfer at constant volume until the temperature of the fluid reaches $T_L$. The volume is then reduced to its original value isothermally, and heat transfer occurs reversibly to a second thermal reservoir at $T_L$. Finally, energy is added by heat transfer at constant volume from state 4 to state 1. The cycle would operate between two fixed-temperature reservoirs if the heat transfers for the processes 2-3 and 4-1 could be kept within the system. Application of an energy balance on the closed system for these two processes shows that the two heat-transfer quantities are equal in magnitude. This is illustrated by the crosshatched areas under process curves 4-1 and 2-3 on the $Ts$ plot. What is needed is simply a means of storing the heat given up by process 2-3 and then

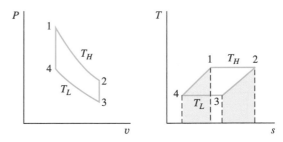

**Figure 15-50**    The $Pv$ and $Ts$ diagrams for the Stirling cycle.

supplying this same energy to the working medium during the process 4-1. This requirement of energy storage within the system necessitates the use of a regenerator. Thus the only heat effect external to the system during each cycle of a Stirling cycle is the heat exchange between two fixed-temperature reservoirs. As a result, the thermal efficiency of a Stirling cycle will equal that of a Carnot cycle operating between the same temperatures.

For many years the Stirling cycle has been of only theoretical interest. However, since the early 1950s considerable work has been done to develop a practical engine working on the Stirling cycle. Although it matches the Carnot cycle in thermal efficiency, it is difficult to build a Stirling engine without introducing some inherent disadvantages. For example, the Stirling engine operates at very high pressures, and the most suitable working fluids are helium and hydrogen. Its weight-to-horsepower ratio is not too favorable, except possibly for large vehicles such as trucks and buses. The high temperature in the cycle also leads to problems, since the pistons are not lubricated to avoid fouling of the regenerator. However, one major advantage of the Stirling engine is its excellent emission quality. This engine is an "external-combustion engine," as opposed to the common internal-combustion type for automotive use. Hence the combustion process is much more complete than in an internal-combustion type in terms of carbon monoxide, hydrocarbon, and nitrogen oxide content in the exhaust. Other advantages of Stirling engines are their relatively silent operation, reliability and long life, and multifuel capability.

Research and development work on a practical Stirling engine continues around the world. The Stirling cycle has some inherent advantages when it is operated in the reverse direction, so that it produces a refrigeration effect. Stirling refrigeration devices are particularly effective in achieving temperatures in the range of $-100$ to $-200°C$ ($-150$ to $-330°F$). A discussion of the Stirling cycle as a refrigeration technique appears in Sec. 18-7.

## 15-15   SUMMARY

An air-standard cycle is frequently used in the introductory analysis of cyclic devices used for power production. A cold-air-standard cycle uses constant values of $c_v$, $c_p$, and $k$, and these are measured at room temperature. In any air-standard cycle the working fluid is taken to be only air, and the combustion process is replaced by heat addition from an external source.

For reciprocating power devices the *compression ratio r* is

$$r = \frac{V_{BDC}}{V_{TDC}} = \frac{\text{clearance volume} + \text{displacement volume}}{\text{clearance volume}}$$

The *mean effective pressure* (MEP) is defined by the relation

$$W_{cycle} = (MEP)(\text{displacement volume})$$

where $W_{cycle}$ is the net work output.

The theoretical cycle used to analyze spark-ignition engines is the *Otto cycle*. Compression and expansion are isentropic, and heat transfers occur at constant volume. For a cold air-standard cycle,

$$\eta_{th,Otto} = 1 - \frac{1}{r^{k-1}}$$

while for variable specific heat,

$$\eta_{th,Otto} = 1 - \frac{q_{out}}{q_{in}} = 1 - \frac{u_4 - u_1}{u_3 - u_2}$$

where $u_2$ and $u_4$ are found from $v_r$ data. For compression-ignition engines the model is the *Diesel cycle*. For the cold air-standard cycle,

$$\eta_{th,Diesel} = 1 - \frac{1}{r^{k-1}} \left[ \frac{r_c^{k-1}}{k(r_c - 1)} \right]$$

where $r_c$ is the *cutoff ratio*. When variable specific heats are considered,

$$\eta_{th,Diesel} = 1 - \frac{q_{out}}{q_{in}} = 1 - \frac{u_4 - u_1}{h_3 - h_2}$$

since in this case heat addition is at constant pressure and heat transfer is at constant volume. The *dual cycle* is another model for a compression-ignition engine. This model divides the heat-addition process into a constant-volume process followed by a constant-pressure process.

A simple gas-turbine power cycle is modeled by the *Brayton cycle*. The compressor and turbine are modeled as isentropic, and heat transfer is at constant pressure. The thermal efficiency of a cold-air-standard Brayton cycle is

$$\eta_{th,Brayton} = 1 - \frac{1}{r_p^{(k-1)/k}}$$

where $r_p$ is the overall pressure ratio. The maximum work output of an ideal Brayton cycle occurs when $T_2 = (T_1 T_3)^{1/2}$. The thermal efficiency of the basic gas-turbine cycle is increased by regeneration. For a cold-air-cycle with regeneration

$$\eta_{th,regen,cold} = 1 - \frac{T_1}{T_3} r_p^{(k-1)/k}$$

where $T_1$ and $T_3$ are the compressor and turbine inlet temperatures, respectively. The compression and expansion processes are improved by intercooling and reheating with staging. For two-stage compression with ideal intercooling, and two-stage expansion with ideal reheating, the optimum intermediate pressure is the geometric mean of the inlet and outlet pressures. Intercooling and reheating are only effective when used in conjunction with regeneration.

When a diffuser and nozzle are added to a simple gas-turbine cycle, the device becomes a turbojet engine for aircraft propulsion. The diffuser slows the air flow and increases the pressure at the compressor inlet. The turbine work essentially equals the compressor work in a turbojet engine. The engine thrust is produced by expanding the turbine exhaust gases through a nozzle to produce a high exit velocity.

## PROBLEMS

15-1. An air-standard Carnot cycle rejects 100 kJ/kg as heat transfer to a sink at 300 K. The minimum and maximum pressures in the closed cycle are 0.10 and 17.4 MPa, respectively. On the basis of the air table, find (a) the pressure after isothermal compression, (b) the temperature of the heat-supply reservoir, in kelvins, (c) the specific volume after isothermal compression and after isentropic compression, both in m$^3$/kg, (d) the thermal efficiency, (e) the compression ratio, and (f) the mean effective pressure for the reciprocating device.

15-2. An air-standard Carnot cycle for a closed system is supplied with 150 kJ/kg as heat transfer from a source at 960 K. The minimum and maximum pressures in the cycle are 1 and 69.3 bars, respectively. On the basis of the air table, determine (a) the pressure after isothermal heat addition, (b) the temperature of heat rejection, in kelvins, (c) the specific volume, in m$^3$/kg, after isothermal heat addition and after isentropic expansion, (d) the thermal efficiency, (e) the compression ratio, and (f) the mean effective pressure, in bars.

15-3. The heat transfer supplied to an air-standard Carnot cycle operating between 300 and 1100 K is 150 kJ/kg. The minimum pressure in the cycle is 1 bar. Determine (a) the maximum pressure in the cycle, (b) the pressure after isothermal compression in the closed system, in bars, (c) the compression ratio, and (d) the mean effective pressure, in bars.

15-4. A Carnot heat engine which produces 10 kJ of work for one cycle has a thermal efficiency of 50 percent. The working fluid is 0.40 kg of air, and the pressure and volume at the beginning of isothermal expansion are 8 bars and 0.119 m$^3$, respectively. Find (a) the maximum and minimum temperatures for the cycle, in kelvins, (b) the heat and work transfers for each of the four processes, in kJ/cycle, (c) the volume at the end of the isothermal expansion process and at the end of the isentropic expansion process, in cubic meters, (d) the compression ratio, and (e) the mean effective pressure for the reciprocating device, in bars.

15-5. A Carnot heat engine contains 0.010 kg of nitrogen. During the isothermal heat-rejection process at 350 K the volume decreases by one-half. The high-temperature thermal reservoir is at 800 K, the minimum pressure in the cycle is 1 bar, and the volume after isothermal heat addition is 0.50 L. Determine (a) the thermal efficiency, (b) the heat rejection, in kilojoules, (c) the maximum pressure in the cycle, in bars, (*d*) the compression ratio, and (*e*) the mean effective pressure, in bars.

15-6E. An air-standard Carnot cycle for a closed system is supplied with 100 Btu/lb$_m$ of heat transfer from a source at 1200°R. The minimum and maximum pressures in the cycle are 1 and 88 atm, respectively. On the basis of the air table, determine (*a*) the pressure after isothermal heat addition, (*b*) the temperature of heat rejection, in degrees Rankine, (*c*) the specific volume, in ft$^3$/lb$_m$, after isothermal heat addition and after isentropic expansion, (*d*) the thermal efficiency, (*e*) the compression ratio, and (*f*) the mean effective pressure, in psi.

15-7E. The heat transfer supplied to an air-standard Carnot cycle operating between 540 and 2000°R is 70 Btu/lb$_m$. The minimum pressure in the cycle is 1 atm. Determine (*a*) the maximum pressure in the cycle, (*b*) the pressure after isothermal compression in the closed system, (*c*) the compression ratio, and (*d*) the mean effective pressure, in psi.

15-8E. A Carnot heat engine which produces 10 Btu of work for one cycle has a thermal efficiency of 50 percent. The working fluid is 1.0 lb$_m$ of air, and the pressure and temperature at the beginning of isothermal expansion are 100 psia and 4.0 ft$^3$, respectively. Find (*a*) the maximum and minimum temperatures for the cycle, in degrees Rankine, (*b*) the heat and work transfers for each of the four processes, in Btu/cycle, (*c*) the volume at the end of the isothermal expansion process and at the end of the isentropic expansion process, in cubic feet, (*d*) the compression ratio, and (*e*) the mean effective pressure for the reciprocating device, in psi.

## OTTO, DIESEL, AND DUAL CYCLES

15-9. The compression ratio of an Otto cycle is 8:1. Before the compression stroke of the cycle begins the pressure is 0.98 bar and the temperature is 27°C. The heat transfer to the air per cycle is 1430 kJ/kg. Employing data from Table A-5, determine (*a*) the pressure and temperature at the end of each process of the cycle, (*b*) the thermal efficiency, (*c*) the mean effective pressure, in bars, and (*d*) the volume flow rate of air, measured at conditions at the beginning of compression, needed to produce 120 kW, in m$^3$/min.

15-10. Solve Prob. 15-9 using the cold-air standard cycle with specific heats evaluated at 300 K.

15-11. The compression ratio of an Otto cycle is 8:1. Before the compression stroke of the cycle begins, the pressure is 0.98 bar and the temperature is 27°C. The value of $P_3/P_2$ is 2.90 during heat addition. Employing data from Table A-5, determine (*a*) the pressure and the temperature at the end of each process of the cycle, (*b*) the thermal efficiency, (*c*) the mean effective pressure, in bars, and (*d*) the volume flow rate of air, measured at conditions at the beginning of compression, needed to produce 120 kW, in $m^3$/min.

15-12. The air at the beginning of the compression stroke of an air-standard Otto cycle is at 0.095 MPa and 22°C, and the cylinder volume is 2800 $cm^3$. The compression ratio is 9, and 4.30 kJ is added during the heat-addition process. On the basis of Table A-5, determine (a) the temperature and pressure after the heat-addition and expansion processes, (b) the thermal efficiency, (c) the mean effective pressure, and (d) the volume flow rate of air, measured at conditions at the beginning of compression, needed to produce 120 kW of power, in $m^3$/min.

15-13. Solve Prob. 15-12 using the cold-air standard cycle with specific heats evaluated at 300 K.

15-14. The air at the beginning of the compression stroke of an air-standard Otto cycle is at 0.095 MPa and 22°C, and the cylinder volume is 2800 $cm^3$. The compression ratio is 9, and 3.54 kJ is added during the heat-addition process. On the basis of Table A-5, determine (a) the temperature and pressure after the heat-addition and expansion processes, (b) the thermal efficiency, (c) the mean effective pressure, and (d) the volume flow rate of air, measured at conditions at the beginning of compression, needed to produce 110 kW of net power, in $m^3$/min.

15-15. An air-standard Otto cycle operates with a compression ratio of 8.55, and at the beginning of compression the air is at 98 kPa and 32°C. During the heat-addition process the pressure is tripled. On the basis of Table A-5 data, determine (a) the temperatures around the cycle, in kelvins, (b) the thermal efficiency, (c) the mean effective pressure, in kPa, and (d) the thermal efficiency of a Carnot engine operating between the same overall temperature limits.

15-16. Consider an air-standard Otto cycle that has a compression ratio of 8.3 and a heat addition of 1456 kJ/kg. If the pressure and temperature at the beginning of the compression process are 0.095 MPa and 7°C, determine on the basis of Table A-5 data (a) the maximum pressure and temperature for the cycle, (b) the net work output, in kJ/kg, (c) the thermal efficiency, and (d) the mean effective pressure, in megapascals.

15-17. Consider an air-standard Otto cycle that has a compression ratio of 8.0 and a heat addition of 1188 kJ/kg. If the pressure and temperature at the beginning of the compression process are 0.098 MPa and 27°C, determine on the basis of Table A-5 data (a) the pressures and temperatures around for the cycle, (b) the net work output, in kJ/kg, (c) the thermal efficiency, and (d) the mean effective pressure, in megapascals.

15-18. Reconsider the Otto cycle in Prob. 15-9. Determine (a) the closed-system availability of the air at the end of the isentropic expansion, in kJ/kg, if $T_0 = 27°C$ and $P_0 = 0.98$ bar, and (b) the ratio of this availability quantity to the net work output of the cycle.

15-19. Reconsider the Otto cycle in Prob. 15-12. Determine (a) the closed-system availability of the air at the end of the isentropic expansion, in kJ/kg, if $T_0 = 22°C$ and $P_0 = 0.095$ MPa, and (b) the ratio of this availability quantity to the net work output of the cycle.

15-20. Reconsider the Otto cycle in Prob. 15-16. Determine (a) the closed-system availability of the air at the end of the isentropic expansion, in kJ/kg, if $T_0 = 7°C$ and $P_0 = 0.095$ MPa and (b) the ratio of this availability quantity to the net work output of the cycle.

15-21E. The compression ratio of an Otto cycle is 8:1. Before the compression stroke of the cycle begins, the pressure is 14.5 psia and the temperature is 80°F. The heat transfer to the air per cycle is 888 Btu/lb$_m$. Employing data from Table A-5E, determine (a) the pressure and temperature at the end of each process of the cycle, (b) the thermal efficiency, (c) the mean effective pressure, in psi, and (d) the volume flow rate of air, measured at conditions at the beginning of compression, needed to produce 110 hp, in ft$^3$/min.

15-22E. The air at the beginning of the compression stroke of an air-standard Otto cycle is at 14.0 psia and 80°F, and the cylinder volume is 0.20 ft$^3$. The compression ratio is 9, and 9.20 Btu is added during the heat-addition process. On the basis of Table A-5E, determine (a) the temperature and pressure after the heat-addition and expansion processes, (b) the thermal efficiency, (c) the mean effective pressure, in psi, and (d) the volume flow rate of air, measured at conditions at the beginning of compression, needed to produce 120 hp of power, in ft$^3$/min.

15-23E. An air-standard Otto cycle operates with a compression ratio of 8.50, and at the beginning of compression the air is at 14.5 psia and 90°F. During the heat-addition process the pressure is tripled. On the basis of Table A-5E data, determine (a) the temperatures around the cycle, in degrees Rankine, (b) the thermal efficiency, (c) the mean effective pressure, in psi, and (d) the thermal efficiency of a Carnot engine operating between the same overall temperature limits.

15-24E. Consider an air-standard Otto cycle that has a compression ratio of 9.0 and a heat addition of 821 Btu/lb$_m$. If the pressure and temperature at the beginning of the compression process are 14.0 psia and 40°F, determine on the basis of Table A-5E data (a) the maximum pressure and temperature for the cycle, (b) the net work output, in Btu/lb$_m$, (c) the thermal efficiency, and (d) the mean effective pressure, in psi.

15-25E. Reconsider the Otto cycle in Prob. 15-21. Determine (a) the closed-system availability of the air at the end of the isentropic expansion, in Btu/lb$_m$, if $T_0 = 80°F$ and $P_0 = 14.5$ psia, and (b) the ratio of this availability quantity to the net work output of the cycle.

15-26E. Reconsider the Otto cycle in Prob. 15-22. Determine (a) the closed-system availability of the air at the end of the isentropic expansion, in Btu/lb$_m$, if $T_0$ = 80°F and $P_0$ = 14.0 psia, and (b) the ratio of this availability quantity to the net work output of the cycle.

15-27E. Reconsider the Otto cycle in Prob. 15-24. Determine (a) the closed-system availability of the air at the end of the isentropic expansion, in Btu/lb$_m$, if $T_0$ = 40°F and $P_0$ = 14.0 psia, and (b) the ratio of this availability quantity to the net work output of the cycle.

15-28. An air-standard Diesel cycle operates with a compression ratio of 16.7 and a cutoff ratio of 2. At the beginning of compression the air temperature and pressure are 37°C and 0.10 MPa, respectively. Determine (a) the maximum temperature in the cycle, in kelvins, (b) the pressure after isentropic expansion, in megapascals, (c) the heat input per cycle, in kJ/kg, and (d) the volume flow rate of air, measured at conditions at the beginning of compression, needed to produce 200 kW. Use data from Table A-5.

15-29. An engine operates on the Diesel cycle with a compression ratio of 15:1, and fuel is injected for 10 percent of the stroke. The pressure and temperature of the air entering the cylinder are 98 kPa and 17°C, respectively. Determine (a) the cutoff ratio, (b) the temperature at the end of the compression process, in kelvins, (c) the pressure after isentropic expansion, in kilopascals, (d) the heat input, in kJ/kg, and (e) the volume flow rate of air in m$^3$/min, measured at conditions at the beginning of compression, needed to produce 200 kW. Use data from Table A-5.

15-30. The intake conditions for an air-standard Diesel cycle operating with a compression ratio of 15:1 are 0.95 bar and 17°C. At the beginning of compression the cylinder volume is 3.80 L, and 7.5 kJ of heat transfer into the system occurs during the constant-pressure heating process. Determine (a) the pressure and temperature at the end of each process of the cycle and (b) the thermal efficiency and mean effective pressure.

15-31. An air-standard Diesel cycle is supplied with 1659 kJ/kg of heat transfer per cycle. The pressure and temperature at the beginning of compression are 0.095 MPa and 27°C, respectively, and the pressure after compression is 3.60 MPa. Determine (a) the compression ratio, (b) the maximum temperature in the cycle, in kelvins, (c) the cutoff ratio, (d) the pressure after isentropic expansion, in megapascals, and (e) the net work, in kJ/kg. Use Table A-5 for data.

15-32. An air-standard Diesel cycle has a compression ratio of 15.08:1. The pressure and temperature at the beginning of compression are 1 bar and 17°C, respectively. If the maximum temperature of the cycle is 2250 K, determine (a) the cutoff ratio, (b) the maximum pressure, in bars, (c) the thermal efficiency, and (d) the mean effective pressure, in bars.

15-33. Reconsider the Diesel cycle in Prob. 15-28. Determine (a) the closed-system availability of the air at the end of the isentropic expansion, in kJ/kg, if $T_0 = 17°C$ and $P_0 = 0.10$ MPa and (b) the ratio of this availability quantity to the net work output of the cycle.

15-34. Reconsider the Diesel cycle in Prob. 15-30. Determine (a) the closed-system availability of the air at the end of the isentropic expansion, in kJ/kg, if $T_0 = 17°C$ and $P_0 = 0.95$ bar, and (b) the ratio of this availability quantity to the net work output of the cycle.

15-35E. An air-standard Diesel cycle operates with a compression ratio of 14.8 and a cutoff ratio of 2. At the beginning of compression the air temperature and pressure are 100°F and 14.5 psia, respectively. Determine (a) the maximum temperature in the cycle, in degrees Rankine, (b) the pressure after isentropic expansion, in psia, and (c) the heat input per cycle, in Btu/lb$_m$. Use data from Table A-5E.

15-36E. The intake conditions for an air-standard Diesel cycle operating with a compression ratio of 15 are 14.4 psia and 60°F. The cutoff ratio is (a) 2.84 and (b) 2.0. Determine (1) the pressure and temperature at the end of each process of the cycle and (2) the thermal efficiency and mean effective pressure. Use Table A-5E.

15-37E. An air-standard Diesel cycle is supplied with 724 Btu/lb$_m$ of heat transfer per cycle. The pressure and temperature at the beginning of compression are 14.0 psia and 80°F, respectively, and the pressure after compression is 540 psia. Determine (a) the compression ratio, (b) the maximum temperature in the cycle, in degrees Rankine, (c) the cutoff ratio, (d) the pressure after isentropic expansion, in psia, and (e) the volume flow rate of air, measured at the beginning of compression, needed to produce 150 hp. Use Table A-5E for data.

15-38E. Reconsider the Diesel cycle in Prob. 15-35. Determine (a) the closed-system availability of the air at the end of the isentropic expansion, in Btu/lb$_m$, if $T_0 = 60°F$ and $P_0 = 14.5$ psia, and (b) the ratio of this availability quantity to the net work output of the cycle.

15-39E. Reconsider the Diesel cycle in Prob. 15-36. Determine (1) the closed-system availability of the air at the end of the isentropic expansion, in Btu/lb$_m$, if $T_0 = 60°F$ and $P_0 = 14.4$ psia, and (2) the ratio of this availability quantity to the net work output of the cycle, for parts a and b.

15-40. An air-standard dual cycle operates with a compression ratio of 15:1. At the beginning of compression the conditions are 17°C, 0.95 bar, and 3.80 L. The heat transfer into the cycle is 6.60 kJ, of which one-third is added at constant volume and the remainder at constant pressure. Determine (a) the pressure after the constant-volume heat-addition process, in bars, (b) the temperature before and after the constant-pressure heat-addition process, in kelvins, (c) the temperature after isentropic expansion, and (d) the thermal efficiency.

15-41. An air-standard dual cycle operates with a compression ratio of 14:1. At the beginning of the isentropic compression the conditions are 27°C and 96 kPa. The total heat addition is 1480 kJ/kg, of which one-fourth is added at constant volume and the remainder at constant pressure. Determine (a) the temperatures at the end of each process around the cycle, in kelvins, (b) the thermal efficiency, and (c) the mean effective pressure, in bars.

15-42. An air-standard dual cycle operates with a compression ratio of 15:1. At the beginning of compression the conditions are 17°C, 0.95 bar, and 3.80 L. The heat transfer into the cycle is 6.0 kJ, of which 30 percent is added at constant volume and the remainder at constant pressure. Determine (a) the pressure after the constant-volume heat-addition process, in bars, (b) the temperature before and after the constant-pressure heat-addition process, in kelvins, (c) the temperature after isentropic expansion, and (d) the thermal efficiency.

15-43. An air-standard dual cycle operates with a compression ratio of 14:1. At the beginning of the isentropic compression the conditions are 27°C and 96 kPa. The total heat addition is 1470 kJ/kg, of which one-third is added at constant volume and the remainder at constant pressure. Determine (a) the temperatures at the end of each process around the cycle, in kelvins, (b) the thermal efficiency, and (c) the mean effective pressure, in bars.

15-44E. An air-standard dual cycle operates with a compression ratio of 15:1. At the beginning of compression the conditions are 60°F, 14.6 psia, and 230 in$^3$. The amount of heat transfer into the system is 7.5 Btu, of which one-third is added at constant volume and the remainder at constant pressure. Determine (a) the pressure after the constant-volume heat-addition process, (b) the temperature before and after the constant-pressure heat-addition process, in degrees Rankine, (c) the temperature after isentropic expansion, and (d) the thermal efficiency.

15-45E. An air-standard dual cycle operates with a compression ratio of 14:1. At the beginning of the isentropic compression the conditions are 80°F and 14.5 psia. The total heat addition is 800 Btu/lb, of which one-fourth is added at constant volume and the remainder at constant pressure. Determine (a) the temperatures at the end of each process around the cycle, in degrees Rankine, (b) the thermal efficiency, and (c) the mean effective pressure.

## IDEAL AND NONIDEAL OPEN GAS-TURBINE CYCLE

15-46. A gas-turbine power plant operates on an air-standard cycle between pressure limits of 0.1 and 0.60 MPa. The inlet-air temperature is 22°C, and the temperature limitation on the turbine is 747°C. Using the data of Table A-5, determine (a) the net work output and the heat input, in kJ/kg, (b) the thermal efficiency if the cycle is ideal, and (c) the net power output, in kilowatts, if the volume flow rate at the compressor inlet is 3.8 m³/s.

15-47. A gas-turbine power plant operates on an air-standard cycle between pressure limits of 1 and 6.4 bars. The inlet-air temperature is 22°C, and the temperature limitation on the turbine is 807°C. Calculate (a) the net work output, in kJ/kg, (b) the thermal efficiency if the cycle is ideal, and (c) the net power output, in kilowatts, if the volume flow rate at the compressor inlet is 4.4 m³/s. Use the air table for data.

15-48. A gas-turbine power plant operates on an air-standard cycle between pressure limits of 0.10 and 0.68 MPa. The inlet-air temperature to the compressor is 17°C, and the inlet-air temperature to the turbine is 1180 K. Calculate (a) the net work output, in kJ/kg, (b) the heat input, and (c) the thermal efficiency for the ideal cycle. Use Table A-5 for data.

15-49. A stationary gas-turbine power plant has maximum and minimum cycle temperatures of 827 and 27°C and a pressure ratio of 5.2:1. Find (a) the ratio of compressor work to turbine work, (b) the thermal efficiency, and (c) the mass flow rate of air required, in kg/min, for a net power output of 1000 kW. Use air-table data for the ideal-cycle analysis.

15-50. The pressure ratio of an air-standard Brayton cycle is 4.5, and the inlet conditions to the compressor are 0.10 MPa and 27°C. The turbine is limited to a temperature of 827°C, and the mass flow rate is 4 kg/s. Determine (a) the compressor and turbine work, in kJ/kg, (b) the thermal efficiency, (c) the net power output, in kilowatts, and (d) the volume flow rate at the compressor inlet, in m³/min. Use Table A-5 for data.

15-51. The pressure ratio of an air-standard Brayton cycle is 7.2:1, and the inlet conditions are 1.0 bar and 17°C. The turbine is limited to a temperature of 1260 K, and the mass flow rate is 3.5 kg/s. Determine (a) the compressor and turbine work, in kJ/kg, (b) the thermal efficiency, (c) the net power output, in kilowatts, and (d) the volume flow rate at the compressor inlet, in m³/min. Use Table A-5 for data.

15-52. An air-standard Brayton cycle operates with a pressure ratio of 7.0, an inlet pressure of 96 kPa, and minimum and maximum temperatures of 300 and 1200 K. For an inlet volume flow rate of air of 30 m³/min, determine (a) the back work ratio, (b) the thermal efficiency, and (c) the net power developed, in kilowatts.

15-53. The air entering a Brayton cycle has a compressor inlet condition of 300 K and 100 kPa, and the pressure ratio is 7:1. For turbine inlet temperatures from 900 to 1500 K, in 100-degree intervals, calculate from a computer program and plot (a) the thermal efficiency, (b) the back work ratio, and (c) the net work output, in kJ/kg.

15-54. An ideal Brayton cycle operates between temperature limits of 290 and 1200 K. Calculate from a computer program and plot (*a*) the net work, in kJ/kg, and (*b*) the thermal efficiency for pressure ratios ranging from 4 to 14 at intervals of 2.

15-55. An ideal gas-turbine power cycle operates on air between temperature limits of 22°C and 747°C. Calculate from a computer program and plot (*a*) the thermal efficiency and (*b*) the net work, in kJ/kg, as a function of the pressure ratio, for pressure ratios of 4.78, 6.00, 9.07, 11.72, and 14.05.

15-56. An ideal Brayton cycle has minimum and maximum temperature limits of 295 and 1080 K, and an inlet pressure of 1 bar. Calculate from a computer program and plot (*a*) the net work output, in kJ/kg, and (*b*) the thermal efficiency as a function of pressure ratio, for pressure ratios of 2.43, 4.78, 6.40, 9.69, and 12.46.

15-57. If the sink temperature $T_0$ is the same as the compressor-inlet temperature, then determine (1) the steady-flow availability of the turbine exhaust stream, in kJ/kg, and (2) the percentage increase in net work output if this availability could be completely converted to work output, for the Brayton cycle described in (*a*) Prob. 15-46, (*b*) Prob. 15-47, (*c*) Prob. 15-48, (*d*) Prob. 15-49.

15-58. Prove that the maximum net work for a simple Brayton cycle, for fixed compressor- and turbine-inlet temperatures, occurs when $T_2 = (T_1 T_3)^{1/2}$ if the specific heats are constant.

15-59. Show that the pressure ratio which yields the largest net work for a simple Brayton cycle with fixed compressor- and turbine-inlet temperatures is given by $P_2/P_1 = (T_3/T_1)^n$, where $n = k/[2(k-1)]$ and $k$ is a constant.

15-60E. A gas-turbine power plant operates on an ideal air-standard cycle between pressure limits of 14.5 and 87.0 psia. The inlet-air temperature is 80°F, and the air enters the turbine at 1290°F. Using the data of Table A-5E, determine (*a*) the net work output and the heat input, in Btu/lb$_m$, (*b*) the thermal efficiency, and (*c*) the net power output, in horsepower, if the volume flow rate at the compressor inlet is 120 ft$^3$/s.

15-61E. A gas-turbine power plant operates on an ideal air-standard cycle between pressure limits of 1 and 4 atm. The inlet-air temperature is 60°F, and the temperature limitation on the turbine is 1540°F. Calculate (*a*) the net work output, in Btu/lb$_m$, (*b*) the thermal efficiency, and (*c*) the net power output, in kilowatts, if the volume flow rate at the compressor inlet is 80 ft$^3$/s. Use the air table for data.

15-62E. A gas-turbine power plant operates on an air-standard cycle between pressure limits of 14.7 and 50.7 psia. The inlet-air temperature to the compressor is 60°F, and the inlet-air temperature to the turbine is 1120°F. Calculate (*a*) the net work output, in Btu/lb$_m$, (*b*) the heat input, and (c) the thermal efficiency for the ideal cycle. Use Table A-5E for data.

15-63E. The pressure ratio of an air-standard Brayton cycle is 6.0, and the inlet conditions to the compressor are 15 psia and 40°F. The turbine is limited to a temperature of 1440°F, and the mass flow rate is 10 lb/s. Determine (a) the compressor and turbine work, in Btu/lb$_m$, (b) the thermal efficiency, (c) the net power output, in horsepower, and (d) the volume flow rate at the compressor inlet, in ft$^3$/min. Use Table A-5E for data.

15-64E. An ideal Brayton cycle has minimum and maximum temperature limits of 520 and 2000°R. Calculate from a computer program and plot (a) the net work output, in Btu/lb$_m$, and (b) the thermal efficiency as a function of pressure ratio, for pressure ratios of 4.0, 7.0, 10.0, and 15.0.

15-65E. An ideal gas-turbine power cycle operates on air between temperature limits of 80°F and 1290°C, and a compressor inlet pressure of 14.5 psia. Calculate from a computer program and plot (a) the thermal efficiency and (b) the net work, in Btu/lb$_m$, as a function of the pressure ratio, for compressor outlet pressures of 87, 115, 145, and 174 psia.

15-66E. If the sink temperature $T_0$ is the same as the compressor-inlet temperature, then determine (1) the steady-flow availability of the turbine exhaust stream, in Btu/lb$_m$, and (2) the percentage increase in net work output if this availability could be completely converted to work output, for the gas-turbine cycle described in (a) Prob. 15-60, (b) Prob. 15-61, (c) Prob. 15-62.

15-67. A gas-turbine power plant operates on an air-standard cycle between pressure limits of 0.1 and 0.60 MPa. The inlet-air temperature is 22°C, and the temperature limitation on the turbine is 747°C. The compressor and turbine adiabatic efficiencies are 84 and 87 percent, respectively. Using the data of Table A-5, determine (a) the heat input, in kJ/kg, (b) the thermal efficiency, and (c) the net power output, in kilowatts, if the volume flow rate at the compressor inlet is 3.8 m$^3$/s. (See Prob. 15-46.)

15-68. A gas-turbine power plant operates on an air-standard cycle between pressure limits of 1 and 6.4 bars. The inlet-air temperature is 22°C, and the temperature limitation on the turbine is 807°C. The adiabatic efficiencies of the compressor and turbine are 82 and 85 percent, respectively. Calculate (a) the net work output, in kJ/kg, (b) the thermal efficiency, and (c) the net power output, in kilowatts, if the volume flow rate at the compressor inlet is 4.4 m$^3$/s. Use Table A-5 for data. (See Prob. 15-47.)

15-69. A gas-turbine power plant operates on an air-standard cycle between pressure limits of 0.10 and 0.68 MPa. The inlet-air temperature to the compressor is 17°C, and the inlet-air temperature to the turbine is 1180 K. The compressor and turbine adiabatic efficiencies are 82 and 85 percent, respectively. Calculate (a) the net work output, in kJ/kg, (b) the heat input, and (c) the thermal efficiency. Use Table A-5 for data. (See Prob. 15-48.)

15-70. A stationary gas-turbine power plant has maximum and minimum cycle temperatures of 827 and 27°C and a pressure ratio of 5.2:1. The adiabatic efficiencies of the compressor and turbine are 81 and 86 percent, respectively. Find (a) the ratio of compressor work to turbine work, (b) the thermal efficiency, and (c) the mass flow rate of air required, in kg/min, for a net power output of 1000 kW. Use air-table data for the cycle analysis. (See Prob. 15-49.)

15-71. The pressure ratio of an air-standard Brayton cycle is 4.5, and the inlet conditions to the compressor are 0.10 MPa and 27°C. The turbine is limited to a temperature of 827°C, and the mass flow rate is 4 kg/s. The compressor and turbine adiabatic efficiencies are 83 and 86 percent, respectively. Determine (a) the compressor and turbine work, in kJ/kg, (b) the thermal efficiency, (c) the net power output, in kilowatts, and (d) the volume flow rate at the compressor inlet, in m³/min. Use Table A-5 for data. (See Prob. 15-50.)

15-72. The pressure ratio of an air-standard Brayton cycle is 7.2:1, and the inlet conditions are 1.0 bar and 17°C. The turbine is limited to a temperature of 1260 K, and the mass flow rate is 3.5 kg/s. The compressor and turbine adiabatic efficiencies are 84 and 88 percent, respectively. Determine (a) the compressor and turbine work, in kJ/kg, (b) the thermal efficiency, (c) the net power output, in kilowatts, and (d) the volume flow rate at the compressor inlet, in m³/min. Use Table A-5 for data. (See Prob. 15-51.)

15-73. A gas-turbine power cycle operates on air between temperature limits of 22 and 747°C. The compressor and turbine adiabatic efficiencies are 84 and 87 percent, respectively. Calculate from a computer program and plot (a) the thermal efficiency and (b) the net work, in kJ/kg, as a function of the pressure ratio, for pressures ratios of 4.78, 6.00, 9.07, 11.72, and 14.05. (See Prob. 15-55.)

15-74. A gas-turbine cycle has minimum and maximum temperature limits of 295 and 1080 K, and an inlet pressure of 1 bar. The compressor and turbine adiabatic efficiencies are 82 and 85 percent. Calculate from a computer program and plot (a) the net work output, in kJ/kg, and (b) the thermal efficiency as a function of pressure ratio, for pressure ratios of 2.43, 4.78, 6.40, 9.69, and 12.46. (See Prob. 15-56.)

15-75. An air-standard Brayton cycle operates with a pressure ratio of 7, and the minimum and maximum temperatures are 300 and 1200 K. For compressor and turbine adiabatic efficiencies of 81 and 86 percent, respectively, and for an inlet volume flow rate of air of 30 m³/min, determine (a) the back work ratio, (b) the thermal efficiency, and (c) the net power developed, in kilowatts. (See Prob. 15-52.)

15-76. A gas-turbine power cycle operates with a pressure ratio of 12:1. The compressor and turbine adiabatic efficiencies are 85 and 90 percent, respectively. The compressor inlet temperature is 22°C, and the turbine inlet temperature is 1027°C. For a mass flow rate of 1 kg/s, find the power output from the cycle. Now, double the pressure ratio and find the power output in this latter case, in kilowatts.

15-77. If the sink temperature $T_0$ is the same as the compressor-inlet temperature, then determine (1) the steady-flow availability of the turbine exhaust stream, in kJ/kg, and (2) the percentage increase in net work output if this availability could be completely converted to work output, for the irreversible cycle described in (a) Prob. 15-67, (b) Prob. 15-68, (c) Prob. 15-69, (d) Prob. 15-70.

15-78E. A gas-turbine power plant operates on an air-standard cycle between pressure limits of 14.5 and 87.0 psia. The inlet-air temperature is 80°F, and the air enters the turbine at 1290°F. Using the data of Table A-5E, and noting that the compressor and turbine adiabatic efficiencies are 83 and 86 percent, respectively, determine (a) the net work output and the heat input, in Btu/lb$_m$, (b) the thermal efficiency, and (c) the net power output, in horsepower, if the volume flow rate at the compressor inlet is 120 ft³/s. (See Prob. 15-60.)

15-79E. A gas-turbine power plant operates on an air-standard cycle between pressure limits of 1 and 4 atm. The inlet-air temperature is 60°F, and the temperature limitation on the turbine is 1540°F. The adiabatic efficiencies of the compressor and turbine are 78 and 84 percent, respectively. Calculate (a) the net work output, in Btu/lb$_m$, (b) the thermal efficiency, and (c) the net power output, in horsepower, if the volume flow rate at the compressor inlet is 80 ft³/s. Use the air table for data. (See Prob. 15-61.)

15-80E. A gas-turbine power plant operates on an air-standard cycle between pressure limits of 14.7 and 50.7 psia. The inlet-air temperature to the compressor is 60°F, and the inlet-air temperature to the turbine is 1120°F. The compressor and turbine adiabatic efficiencies are 82 and 85 percent, respectively. Calculate (a) the net work output, in Btu/lb$_m$, (b) the heat input, and (c) the thermal efficiency for the cycle. Use Table A-5E for data. (See Prob. 15-62.)

15-81E. The pressure ratio of an air-standard Brayton cycle is 6.0, and the inlet conditions to the compressor are 15 psia and 40°F. The turbine is limited to a temperature of 1440°F, and the mass flow rate is 10 lb/s. The compressor and turbine adiabatic efficiencies are 80 and 85 percent, respectively. Determine (a) the compressor and turbine work, in Btu/lb$_m$, (b) the thermal efficiency, (c) the net power output, in horsepower, and (d) the volume flow rate at the compressor inlet, in ft³/min. Use Table A-5E for data. (See Prob. 15-63.)

15-82E. A gas turbine cycle has minimum and maximum temperature limits of 520 and 2000°R. The compressor and turbine adiabatic efficiencies are 78 and 84 percent, respectively. Calculate from a computer program and plot (*a*) the net work output, in Btu/lb$_m$, and (*b*) the thermal efficiency as a function of pressure ratio, for pressure ratios of 4.0, 7.0, 10.0, and 14.0. (See Prob. 15-64.)

15-83E. If the sink temperature $T_0$ is the same as the compressor-inlet temperature, then determine (1) the steady-flow availability of the turbine exhaust stream, in Btu/lb$_m$, and (2) the percentage increase in net work output if this availability could be completely converted to work output, for the irreversible cycle described in (*a*) Prob. 15-78E, (*b*) Prob. 15-79E, (*c*) Prob. 15-80E.

## REGENERATIVE GAS-TURBINE CYCLE

15-84. An air-standard Brayton cycle operates with a pressure ratio of 7, and the minimum and maximum temperatures are 300 and 1200 K. The compressor and turbine adiabatic efficiencies are 81 and 86 percent, respectively. If a regenerator with an effectiveness of 75 percent is installed, determine (*a*) the cycle thermal efficiency and (*b*) the percentage of fuel saved. Use Table A-5 for data. (See Probs. 15-52 and 15-75.)

15-85. A gas-turbine power plant operates on an air-standard cycle between pressure limits of 0.1 and 0.60 MPa. The inlet-air temperature is 22°C, and the temperature limitation on the turbine is 747°C. The compressor and turbine adiabatic efficiencies are 84 and 87 percent, respectively. If a regenerator is installed with an effectiveness of (*a*) 80 percent and (*b*) 60 percent, determine (1) the thermal efficiency and (2) the percentage of fuel saved. Use Table A-5 for data. (See Probs. 15-46 and 15-67.)

15-86. A gas-turbine power plant operates on an air-standard cycle between pressure limits of 1 and 6.4 bars. The inlet-air temperature is 22°C, and the temperature limitation on the turbine is 807°C. The adiabatic efficiencies of the compressor and turbine are 82 and 85 percent, respectively. If a regenerator were installed with an effectiveness of (*a*) 60 percent and (*b*) 80 percent, compute (1) the thermal efficiency and (2) the percentage of fuel saved. Use Table A-5 for data. (See Probs. 15-47 and 15-68.)

15-87. A gas-turbine power plant operates on an air-standard cycle between pressure limits of 0.10 and 0.68 MPa. The inlet-air temperature to the compressor is 17°C, and the inlet-air temperature to the turbine is 1180 K. The compressor and turbine adiabatic efficiencies are 82 and 85 percent, respectively. If a regenerator is installed with an effectiveness of (*a*) 60 percent and (*b*) 75 percent, determine (1) the thermal efficiency and (2) the percentage of fuel saved. Use Table A-5 for data. (See Probs. 15-48 and 15-69.)

**Table P15-89**

| System | Entering | Leaving |
|---|---|---|
| Compressor | 290.2 | 505.0 |
| Regenerator | 505.0 | 629.4 |
| Combustor | 629.4 | 1046.0 |
| Turbine | 1046.0 | 713.7 |
| Regenerator | 713.7 | 590.1 |

**Table P15-90**

| System | Entering | Leaving |
|---|---|---|
| Compressor | 300 | 525 |
| Regenerator | 525 | 715 |
| Combustor | 715 | 1161 |
| Turbine | 1161 | 800 |
| Regenerator | 800 | 610 |

15-88. A stationary gas-turbine power plant has maximum and minimum cycle temperatures of 827 and 27°C and a pressure ratio of 5.2:1. The adiabatic efficiencies of the compressor and turbine are 81 and 86 percent, respectively. If a regenerator is installed with an effectiveness of (a) 75 percent and (b) 55 percent, determine (1) the thermal efficiency, and (2) the percentage of fuel saved. Use air-table data. (See Probs. 15-49 and 15-70.)

15-89. The enthalpy data shown in Table P15-89 are based on test data taken during the operation of a regenerative gas turbine with a pressure ratio of 5.41:1. Determine (a) the thermal efficiency, (b) the effectiveness of the regenerator, (c) the adiabatic efficiency of the compressor, and (d) the adiabatic efficiency of the turbine. Data are in kJ/kg.

15-90. The enthalpy data shown in Table P15-90 are based on test data taken during the operation of a regenerative gas turbine with a pressure ratio of 5.20:1. Determine (a) the thermal efficiency, (b) the effectiveness of the regenerator, (c) the adiabatic efficiency of the compressor, and (d) the adiabatic efficiency of the turbine. Data are in kJ/kg.

15-91. An automotive manufacturer is considering the use of a regenerative-type gas turbine for a power source for a new model. Outside air enters the compressor at 22°C and 1 bar, with a flow rate of 1 kg/s, and is compressed at a ratio of 4:1. The air from the compressor enters the regenerator, where it is heated until the temperature is 537°C. It then passes to the combustion chamber, where it is heated to 927°C. After expanding through the turbine, the hot gas passes through the other half of the regenerator and then is exhausted to the atmosphere. Assuming compressor and turbine efficiencies of 90 percent and negligible pressure losses through the burner and regenerator, determine (a) the thermal efficiency, (b) the net power output, in kilowatts, (c) the regenerator effectiveness, and (d) the exhaust-stream temperature, in degrees Celsius.

15-92. If the sink temperature $T_0$ is the same as the compressor-inlet temperature, then determine (1) the steady-flow availability of the stream leaving the regenerator, in kJ/kg, and (2) the irreversibility within the compressor, turbine, and regenerator for the cycle described in (a) Prob. 15-84, (b) Prob. 15-85, (c) Prob. 15-86, (d) Prob. 15-87, (e) Prob. 15-88.

15-93E. A gas-turbine power plant operates on an air-standard cycle between pressure limits of 14.5 and 87.0 psia. The inlet-air temperature is 80°F, and the air enters the turbine is 1290°F. The compressor and turbine adiabatic efficiencies are 83 and 86 percent, respectively. If a regenerator is installed with an effectiveness of (a) 80 percent and (b) 65 percent, determine (1) the thermal efficiency and (2) the percentage of fuel saved. Use Table A-5E for data. (See Probs. 15-60 and 15-78.)

15-94E. A gas-turbine power plant operates on an air-standard cycle between pressure limits of 1 and 4 atm. The inlet-air temperature is 60°F, and the temperature limitation on the turbine is 1540°F. The adiabatic efficiencies of the compressor and turbine are 78 and 84 percent, respectively. If a regenerator is installed with an effectiveness of (a) 80 percent and (b) 60 percent, determine (1) the thermal efficiency and (2) the percentage of fuel saved. Use the air table for data. (See Probs. 15-61 and 15-79.)

15-95E. A gas-turbine power plant operates on an air-standard cycle between pressure limits of 14.7 and 50.7 psia. The inlet-air temperature to the compressor is 60°F, and the inlet-air temperature to the turbine is 1120°F. The compressor and turbine adiabatic efficiencies are 82 and 85 percent, respectively. If a regenerator is installed with an effectiveness of 70 percent, determine (a) the thermal efficiency and (b) the percentage of fuel saved. Use Table A-5E for data. (See Probs. 15-62 and 15-80.)

15-96E. Air enters a gas-turbine cycle with ideal regeneration at 1 atm and 80°F, and the turbine inlet temperature is 1750°R. With respect to the states shown in Fig. 15-25, $h_{2a} = 239.2$ Btu/lb$_m$, $h_x = 273.6$ Btu/lb$_m$, and $h_{4a} = 288.3$ Btu/lb$_m$. If the pressure ratio for the cycle is 6.26:1, determine (a) the regenerator effectiveness, (b) the compressor efficiency, (c) the turbine adiabatic efficiency, (d) the percent increase in fuel required if the regenerator is omitted from the cycle, and (e) the regenerator outlet temperature, in degrees Rankine.

15-97E. An automotive manufacturer is considering the use of a regenerative-type gas turbine for a power source for a new model. Outside air enters the compressor at 60°F and 1 atm, with a flow rate of 1.8 lb$_m$/s, and is compressed at a ratio of 4:1. The air from the compressor enters a regenerator, where it is heated until the temperature is 940°F. It then passes to the combustion chamber, where it is heated to 1740°F. After expanding through the turbine, the hot gas passes through the other half of the regenerator and then is exhausted to the atmosphere. Assuming ideal compression and expansion and negligible pressure losses through the burner and regenerator, determine (a) the thermal efficiency, (b) the net power output, in horsepower, (c) the regenerator effectiveness, and (d) the exhaust-stream temperature, in degrees Fahrenheit.

15-98E. If the sink temperature $T_0$ is the same as the compressor-inlet temperature, then determine (1) the steady-flow availability of the stream leaving the regenerator, in Btu/lb$_m$, and (2) the irreversibility within the compressor, turbine, and regenerator for the cycle described in (a) Prob. 15-93(a), (b) Prob. 15-94(a), (c) Prob. 15-95.

## POLYTROPIC PROCESS AND COMPRESSOR STAGING WITH INTERCOOLING

15-99. Air is compressed reversibly in a piston-cylinder device from 0.10 MPa, 7°C, to 0.50 MPa. Calculate (1) the work required and (2) the heat transferred, in kJ/kg, if the process is (a) polytropic with $n = 1.30$, (b) adiabatic, and (c) isothermal.

15-100. Air is compressed reversibly in a piston-cylinder device from 1 bar, 7°C, to 7 bars. Calculate (1) the work required and (2) the heat transferred, in kJ/kg, if the process is (a) polytropic with $n = 1.30$, (b) adiabatic, and (c) isothermal.

15-101. A steady-flow air compressor operates between an inlet condition of 1 bar and 37°C and an outlet pressure of 5 bars. Determine (1) the work input and (2) the heat transferred, in kJ/kg, for the following reversible processes: (a) isothermal, (b) $n = 1.30$, and (c) adiabatic.

15-102. A steady-flow air compressor operates between an inlet condition of 0.10 MPa and 37°C and an outlet pressure of 0.60 MPa. Determine (1) the work input and (2) the heat transferred, in kJ/kg, for the following reversible processes: (a) isothermal, (b) $n = 1.30$, and (c) adiabatic.

15-103. The inlet conditions of a two-stage steady-flow compressor are 0.95 bar and 27°C. The outlet pressure is 7.5 bars, and the pressure ratio across each stage is the same. If an intercooler cools the air to the initial temperature and the stages are isentropic, (a) determine the total work input, in kJ/kg, and (b) compare the result of part a with the work required for a single-stage compressor. Use Table A-5 for data.

15-104. Reconsider Prob. 15-103 and include in the analysis that each compressor stage has an adiabatic efficiency of 84 percent.

15-105. The inlet conditions of a two-stage steady-flow compressor are 1.05 bars and 37°C. The outlet pressure is 6.3 bars, and the pressure ratio across each stage is the same. If an intercooler cools the air to the initial temperature and the stages are isentropic, (a) determine the total work input, in kJ/kg, and (b) compare the result of part a with the work required for a single-stage compressor. Use Table A-5 for data.

15-106. Reconsider Prob. 15-105 and include in the analysis that each compressor stage has an adiabatic efficiency of 82 percent.

15-107. The inlet conditions of a two-stage steady-flow compressor are 0.10 MPa and 27°C. The outlet pressure is 0.70 MPa, the stages are isentropic, and the intercooler cools the air to the initial inlet temperature. Determine the total work input, in kJ/kg, if (a) the pressure ratio across each stage is the same, and (b) the pressure ratio across the first stage is twice that across the second stage. (c) Determine the work input for a single-stage compressor operating between the same overall pressure limits, in kJ/kg.

15-108. The inlet conditions of a two-stage steady-flow compressor are 1 bar and 17°C. The outlet pressure is 4.5 bars. The stages are isentropic, and the intercooler cools the air to the initial inlet temperature. Determine the total work input, in kJ/kg, if (a) the pressure ratio across each stage is the same and (b) the pressure ratio across the first stage is twice that across the second stage. (c) Determine the work input for a single-stage compressor operating between the same overall pressure limits, in kJ/kg.

15-109. Reconsider Prob. 15-108 and include in the analysis that each compressor stage has an adiabatic efficiency of 82 percent.

15-110. Consider a three-stage steady-flow compressor with intercoolers which cool the gas to the initial inlet temperature. The outlet pressures of the first and second stages of the isentropic compressor are denoted by $P_a$ and $P_b$, respectively. Show that the minimum work of compression is attained when $P_a = (P_1^2 P_2)^{1/3}$ and $P_b = (P_1 P_2^2)^{1/3}$.

15-111. A three-stage steady-flow compressor has an inlet condition of 0.95 bar and 27°C and an outlet pressure of 7.5 bars. Based on the results of Prob. 15-110, determine the minimum work of compression, in kJ/kg, using Table A-5 for data. (See Prob. 15-103 for the two-stage analysis.)

15-112. A three-stage steady-flow compressor has an inlet condition of 1.05 bar and 37°C and an outlet pressure of 6.3 bars. Based on the results of Prob. 15-110, determine the minimum work of compression, in kJ/kg, using Table A-5 for data. (See Prob. 15-105 for the two-stage analysis.)

15-113E. A steady-flow compressor operates between an inlet condition of 15 psia and 80°F and an outlet pressure of 75 psia. Determine (1) the work input and (2) the heat transferred, in Btu/lb$_m$, for the following reversible processes: (a) isothermal, (b) $n = 1.30$, and (c) adiabatic.

15-114E. A steady-flow compressor operates between an inlet condition of 15 psia and 100°F and an outlet pressure of 90 psia. Determine (1) the work input and (2) the heat transferred, in Btu/lb$_m$, for the following reversible processes: (a) isothermal, (b) $n = 1.30$, and (c) adiabatic.

15-115E. The inlet conditions of a two-stage steady-flow compressor are 14 psia and 80°F. The outlet pressure is 112 psia, and the pressure ratio across each stage is the same. If an intercooler cools the air to the initial temperature and the stages are isentropic, (a) determine the total work input, in Btu/lb$_m$, and (b) compare the result of part a with the work required for a single-stage compressor. Use Table A-5E for data.

15-116. Reconsider Prob. 15-115 and include in the analysis that each compressor stage has an adiabatic efficiency of 84 percent.

15-117E. The inlet conditions of a two-stage steady-flow compressor are 16 psia and 100°F. The outlet pressure is 96 psia, and the pressure ratio across each stage is the same. If an intercooler cools the air to the initial temperature and the stages are isentropic, (a) determine the total work input, in Btu/lb$_m$, and (b) compare the result of part a with the work required for a single-stage compressor. Use Table A-5E for data.

15-118E. A three-stage steady-flow compressor has an inlet condition of 14 psia and 80°F and an outlet pressure of 112 psia. Based on the results of Prob. 15-110, determine the minimum work of compression, in Btu/lb$_m$, using Table A-5E for data. (See Prob. 15-115 for the two-stage analysis.)

15-119E. A three-stage steady-flow compressor has an inlet condition of 16 psia and 100°F and an outlet pressure of 96 psia. Based on the results of Prob. 15-110, determine the minimum work of compression, in Btu/lb$_m$, using Table A-5E for data. (See Prob. 15-117 for the two-stage analysis.)

## TURBINE STAGING WITH REHEATING

15-120. The inlet conditions of a two-stage steady-flow turbine are 4 bars and 1000 K. The outlet pressure is 1 bar, and the pressure ratio across each stage is the same. If a combustor reheats the air to the initial temperature and the stages are isentropic, (a) determine the total work output, in kJ/kg, and (b) compare the result of part a with the work required for a single-stage turbine. Use Table A-5 for data.

15-121. Reconsider Prob. 15-120 and include in the analysis that each turbine stage has an adiabatic efficiency of 84 percent.

15-122. The inlet conditions of a two-stage steady-flow turbine are 1.20 MPa and 1200 K. The outlet pressure is 0.10 MPa, and the pressure ratio across each stage is the same. If a combustor reheats the air to the initial temperature and the stages are isentropic, (a) determine the total work output, in kJ/kg, and (b) compare the result of part a with the work required for a single-stage compressor. Use Table A-5 for data.

15-123. Reconsider Prob. 15-122 and include in the analysis that each turbine stage has an adiabatic efficiency of 85 percent.

15-124. The inlet conditions of a two-stage steady-flow turbine are 900 kPa and 1100 K. The outlet pressure is 100 kPa, and the pressure ratio across each stage is the same. If a combustor reheats the air to the initial temperature and the stages are isentropic, (a) determine the total work output, in kJ/kg, and (b) compare the result of part a with the work required for a single-stage turbine. Use Table A-5 for data.

15-125. Reconsider Prob. 15-124 and include in the analysis that each turbine stage has an adiabatic efficiency of 86 percent.

## GAS-TURBINE CYCLE WITH INTERCOOLING, REHEATING, AND REGENERATION

15-126. A gas-turbine power plant operates on an air-standard cycle between pressure limits of 0.1 and 0.60 MPa. The inlet-air temperature is 22°C, and the temperature limitation on the turbine is 747°C. The compressor and turbine adiabatic efficiencies are 84 and 87 percent, respectively. If a regenerator is installed with an effectiveness of (a) 80 percent and (b) 60 percent, and two-stage compression and expansion are used, determine (1) the net work and (2) the thermal efficiency. Use Table A-5 for data. (See Probs. 15-67 and 15-85.)

15-127. A gas-turbine power plant operates on an air-standard cycle between pressure limits of 1 and 6.4 bars. The inlet-air temperature is 22°C, and the tcmperature limitation on the turbine is 807°C. The adiabatic efficiencies of the compressor and turbine are 82 and 85 percent, respectively. If a regenerator were installed with an effectiveness of (a) 60 percent and (b) 80 percent, and ideal two-stage compression and expansion are used, find (1) the net work output and (2) the thermal efficiency. Use Table A-5 for data. (See Probs. 15-68 and 15-86.)

15-128. A gas-turbine power plant operates on an air-standard cycle between pressure limits of 0.10 and 0.68 MPa. The inlet-air temperature to the compressor is 17°C, and the inlet-air temperature to the turbine is 1180 K. The compressor and turbine adiabatic efficiencies are 82 and 85 percent, respectively. If a regenerator is installed with an effectiveness of (a) 60 percent and (b) 75 percent, and ideal two-stage compression and expansion are used, determine (1) the net work output and (2) the thermal efficiency. Use Table A-5 for data. (See Probs. 15-69 and 15-87.)

15-129. A stationary gas-turbine power plant has maximum and minimum cycle temperatures of 827 and 27°C and a pressure ratio of 5.2:1. The adiabatic efficiencies of the compressor and turbine are 81 and 86 percent, respectively. If a regenerator is installed with an effectiveness of (a) 75 percent and (b) 55 percent, and ideal two-stage compression and expansion are used, determine (1) the net work output and (2) the thermal efficiency. Use air-table data. (See Probs. 15-70 and 15-88.)

15-130. An air-standard Brayton cycle operates with a pressure ratio of 7, and the minimum and maximum temperatures are 300 and 1200 K. The compressor and turbine adiabatic efficiencies are 81 and 86 percent, respectively. If a regenerator with an effectiveness of 75 percent is installed, and the two-stage intercooling and reheating are ideal, determine (*a*) the net work output and (*b*) the thermal efficiency. Use Table A-5 for data. (See Probs. 15-75 and 15-84.)

15-131. If the sink temperature $T_0$ is the same as the compressor-inlet temperature, then determine (1) the steady-flow availability of the stream leaving the regenerator, in kJ/kg, and (2) the irreversibility within the compressor, turbine, and regenerator for the cycle described in (*a*) Prob. 15-126*a*, (*b*) Prob. 15-127*a*, (*c*) Prob. 15-128*a*, (*d*) Prob. 15-129*a*, and (*e*) Prob. 15-130.

15-132. A gas-turbine cycle operates with two stages of compression and expansion. The pressure ratio across each stage is 2. The inlet temperature is 22°C to each stage of compression and 827°C to each stage of expansion. The compressor and turbine efficiencies are 81 and 86 percent, respectively, and the regenerator has an effectiveness of 75 percent. Determine (*a*) the total compressor and turbine work, in kJ/kg, (*b*) the thermal efficiency, and (*c*) the temperature of the air stream leaving the regenerator and entering the atmosphere, in degrees Celsius. Use air-table data.

15-133. A gas-turbine cycle operates with two-stage compression and expansion. Across each stage of compression the pressure ratio is 2.0, and the efficiency is 81 percent. The compressor-inlet temperature is 22°C, but intercooling reduces the temperature only to 37°C before the air enters the second stage. The inlet temperature to each stage of expansion is 827°C, but a pressure drop between the compressor and turbine reduces the expansion pressure ratio to 1.9:1 across each turbine stage, which has an adiabatic efficiency of 86 percent. The regenerator effectiveness is 75 percent. Determine (*a*) the compressor work, in kJ/kg (*b*) the turbine work, (*c*) the heat removed in the intercooler, (*d*) the thermal efficiency, (*e*) the temperature of the air leaving the regenerator and entering the combustor, in kelvins, and (*f*) the stream availability of the air leaving the regenerator and entering the environment, which is at 22°C, in kJ/kg. Use air-table data.

15-134E. A gas-turbine power plant operates on an air-standard cycle between pressure limits of 14.5 and 87.0 psia. The inlet-air temperature is 80°F, and the air enters the turbine at 1290°F. The compressor and turbine adiabatic efficiencies are 83 and 86 percent, respectively. If a regenerator is installed with an effectiveness of (*a*) 80 percent and (*b*) 65 percent and ideal two-stage compression and expansion are used, determine (1) the net work output and (2) the thermal efficiency. Use Table A-5E for data. (See Probs. 15-78 and 15-93.)

15-135E. A gas-turbine power plant operates on an air-standard cycle between pressure limits of 1 and 4 atm. The inlet-air temperature is 60°F, and the temperature limitation on the turbine is 1540°F. The adiabatic efficiencies of the compressor and turbine are 78 and 84 percent, respectively. If a regenerator is installed with an effectiveness of (*a*) 80 percent and (*b*) 60 percent, and ideal two-stage compression and expansion are used, determine (1) the net work output and (2) the thermal efficiency. Use the air table for data. (See Probs. 15-79 and 15-94.)

15-136E. A gas-turbine power plant operates on an air-standard cycle between pressure limits of 14.7 and 50.7 psia. The inlet-air temperature to the compressor is 60°F, and the inlet-air temperature to the turbine is 1120°F. The compressor and turbine adiabatic efficiencies are 82 and 85 percent, respectively. If a regenerator is installed with an effectiveness of 70 percent, and ideal two-stage compression and expansion are used, determine (*a*) the net work output and (*b*) the thermal efficiency. Use Table A-5E for data. (See Probs. 15-80 and 15-95.)

15-137E. If the sink temperature $T_0$ is the same as the compressor-inlet temperature, then determine (1) the steady-flow availability of the stream leaving the regenerator, in Btu/lb, and (2) the irreversibility within the compressor, turbine, and regenerator for the cycle described in (*a*) Prob. 15-134E, (*b*) Prob. 15-135E, (*c*) Prob. 15-136E.

15-138. A gas-turbine cycle operates with two stages of compression and expansion. The pressure ratio across each stage is 2. The inlet temperature is 60°F to each stage of compression and 1540°F to each stage of expansion. The compressor and turbine efficiencies are 81 and 86 percent, respectively, and the regenerator has an effectiveness of 75 percent. Determine (*a*) the total compressor and turbine work, in Btu/lb$_m$, (*b*) the thermal efficiency, and (*c*) the temperature of the air stream leaving the regenerator and entering the atmosphere, in degrees Fahrenheit. Use air-table data.

15-139E. A gas-turbine cycle operates with two-stage compression and expansion. Across each stage of compression the pressure ratio is 2.0, and the efficiency is 81 percent. The compressor-inlet temperature is 60°F, but intercooling reduces the temperature only to 100°F before the air enters the second stage. The inlet temperature to each stage of expansion is 1540°F, but a pressure drop between the compressor and turbine reduces the expansion pressure ratio to 1.9:1 across each turbine stage, which has an adiabatic efficiency of 86 percent. The regenerator effectiveness is 75 percent. Determine (*a*) the compressor work, in Btu/lb$_m$, (*b*) the turbine work, (*c*) the heat removed in the intercooler, (*d*) the thermal efficiency, (*e*) the temperature of the air leaving the regenerator and entering the combustor, in degrees Fahrenheit, and (*f*) the stream availability of the air leaving the regenerator and entering the environment, which is at 60°F, in Btu/lb$_m$. Use air-table data.

## TURBOJET AND CLOSED GAS-TURBINE CYCLES

15-140. Air enters the diffuser of a jet engine at 270 m/s, 0.25 bar, and 220 K. Determine (*a*) the adiabatic outlet temperature, in kelvins,

(b) the isentropic outlet pressure, in bars, and (c) the actual outlet pressure if the pressure coefficient for the device is 0.87, in bars.

15-141. Air enters the diffuser of a jet engine at 280 m/s, 0.045 MPa, and 240 K. Determine (a) the adiabatic outlet temperature, in kelvins, (b) the isentropic outlet pressure, in bars, and (c) the actual outlet pressure if the pressure coefficient for the device is 0.85, in bars.

15-142. Air enters the diffuser of a jet engine at 300 m/s, 0.065 MPa, and 260 K. Determine (a) the adiabatic outlet temperature, in kelvins, (b) the isentropic outlet pressure, in bars, and (c) the actual outlet pressure if the pressure coefficient for the device is 0.88, in bars.

15-143. The airspeed of a turbojet aircraft is 300 m/s in still air at 0.25 bar and 220 K. The compressor pressure ratio is 9, and the maximum temperature in the cycle is 1320 K. Assume ideal performance of the various components. Determine (a) the temperatures and pressures throughout the cycle, (b) the compressor work required, in kJ/kg, (c) the exit-jet velocity, in m/s, (d) the total thrust, in newtons, and (e) the propulsive efficiency, if the mass flow rate is 50 kg/s. Use Table A-5 for data.

15-144. Solve Prob. 15-143 for the specified quantities under the following additional conditions. The actual pressure rise in the diffuser is 90 percent of the isentropic value, the compressor and turbine adiabatic efficiencies are 83 and 86 percent, respectively, and the nozzle efficiency is 94 percent.

15-145. The airspeed of a turbojet aircraft is 280 m/s in still air at 0.050 MPa and 250 K. The compressor pressure ratio is 11, and the maximum temperature in the cycle is 1400 K. Assume ideal performance of the various components. Determine (a) the compressor work, in kJ/kg, (b) the pressure at the turbine outlet, in megapascals, (c) the gas velocity at the exit, in m/s, (d) the total thrust, in newtons, and (e) the propulsive efficiency, if the mass flow rate is 55 kg/s. Use Table A-5 for data.

15-146. Solve Prob. 15-143 for the specified quantities under the following additional conditions. The actual pressure rise in the diffuser is 92 percent of the isentropic value, the compressor and turbine efficiencies are 84 and 87 percent, respectively, and the nozzle efficiency is 94 percent.

15-147. The airspeed of a turbojet aircraft is 260 m/s in still air at 0.060 MPa and 240 K. The compressor pressure ratio is 12, and the maximum temperature in the cycle is 1440 K. Assume ideal performance of the various components. Determine (a) the compressor work, in kJ/kg, (b) the pressure at the turbine outlet, in megapascals, (c) the gas velocity at the exit, in m/s, (d) the total thrust, in newtons, and (e) the propulsive efficiency, if the mass flow rate is 95 kg/s. Use Table A-5 for data.

15-148. Solve Prob. 15-147 for the specified quantities under the following additional conditions. The actual pressure rise in the diffuser is 88 percent of the isentropic value, the compressor and turbine efficiencies are 82 and 86 percent, respectively, and the nozzle efficiency is 92 percent.

15-149. A closed-cycle gas-turbine power plant operates with helium as the working fluid. The compressor and turbine adiabatic efficiencies are each 85 percent. The compressor-inlet state is 5 bars and 47°C, while the turbine-inlet state is 12 bars and 980 K. For a mass flow rate of 5 kg/s, determine (a) the net power, in kilowatts, and (b) the rate of heat removed from the cycle, in kJ/min.

15-150. A closed-cycle gas-turbine power plant operates with argon as the working fluid. The compressor and turbine adiabatic efficiencies are 83 and 86 percent, respectively. The compressor-inlet conditions are 8 bars and 340 K, while the turbine-inlet state is 15 bars and 1000 K. For a mass flow rate of 7 kg/s, determine (a) the net power output, in kilowatts, and (b) the rate of heat removed from the cycle, in kJ/min.

15-151. A closed-cycle gas-turbine power plant operates with argon as the working fluid. The compressor and turbine adiabatic efficiencies are 84 and 86 percent, respectively. Compressor-inlet conditions are 6 bars and 37°C, while the turbine-inlet conditions are 21 bars and 652°C. If the net power output is 3500 kW, determine (a) the required mass flow rate, in kg/s, (b) the gross kilowatt output of the turbine, and (c) the rate of heat removed from the cycle and radiated to outer space, in kJ/min.

15-152E. The airspeed of a turbojet aircraft is 900 ft/s in still air at 4.0 psia and −40°F. The compressor pressure ratio is 9, and the maximum temperature in the cycle is 1740°F. Assume ideal performance of the various components. Determine (a) the temperatures and pressures throughout the cycle, (b) the compressor work required, in Btu/lb$_m$, (c) the exit-jet velocity, in ft/s, (d) the total thrust, in pound-force, and (e) the propulsive efficiency, if the mass flow rate is 100 lb/s. Use Table A-5 for data.

15-153E. The airspeed of a turbojet aircraft is 900 ft/s in still air at 4.0 psia and −40°F. The compressor pressure ratio is 9, and the maximum temperature in the cycle is 1740°F. The actual pressure rise in the diffuser is 90 percent of the isentropic value, the compressor and turbine adiabatic efficiencies are 83 and 86 percent, respectively, and the nozzle efficiency is 94 percent. Determine (a) the temperatures and pressures throughout the cycle, (b) the compressor work required, in Btu/lb$_m$, (c) the exit-jet velocity, in ft/s, (d) the total thrust, in pound-force, and (e) the propulsive efficiency, if the mass flow rate is 100 lb$_m$/s. Use Table A-5 for data.

15-154E. The airspeed of a turbojet aircraft is 800 ft/s in still air at 8.0 psia and 0°F. The compressor pressure ratio is 11, and the maximum temperature in the cycle is 1940°F. Assume ideal performance of the various components. Determine (a) the compressor work, in Btu/lb$_m$, (b) the pressure at the turbine outlet, in psia, (c) the gas velocity at the exit, in ft/s, (d) the total thrust, in pound-force, and (e) the propulsive efficiency, if the mass flow rate is 110 lb$_m$/s.

15-155E. The airspeed of a turbojet aircraft is 800 ft/s in still air at 8.0 psia and 0°F. The compressor pressure ratio is 11, and the maximum temperature in the cycle is 1940°F. The actual pressure rise in the diffuser is 92 percent of the isentropic value, the compressor and turbine efficiencies are 84 and 87 percent, respectively, and the nozzle efficiency is 94 percent. Determine (a) the compressor work, in Btu/lb$_m$, (b) the pressure at the turbine outlet, in psia, (c) the total thrust, in pound-force, and (d) the propulsive efficiency, if the mass flow rate is 110 lb$_m$/s. Use Table A-5 for data.

15-156E. A closed-cycle gas-turbine power plant operates with helium as the working fluid. The compressor and turbine adiabatic efficiencies are each 84 percent. The compressor-inlet state is 80 psia and 120°F, while the turbine-inlet state is 180 psia and 1300°F. For a mass flow rate of 10 lb$_m$/s, determine (a) the net power output, in horsepower, and (b) the rate of heat transfer from the cycle, in Btu/min.

15-157E. A closed-cycle gas-turbine power plant operates with argon as the working fluid. The compressor and turbine adiabatic efficiencies are 83 and 87 percent, respectively. The compressor-inlet conditions are 120 psia and 150°F, while the turbine-inlet state is 250 psia and 1400°F. For a mass flow rate of 15 lb$_m$/s, determine (a) the net power output, in horsepower, and (b) the rate of heat transfer from the cycle, in Btu/min.

15-158E. A closed-cycle gas-turbine power plant operates with argon as the working fluid. The compressor and turbine adiabatic efficiencies are 84 and 86 percent, respectively. Compressor-inlet conditions are 100 psia and 100°F, while the turbine-inlet conditions are 350 psia and 1220°F. If the net power output is 2500 hp, determine (a) the required mass flow rate, in lb/s, (b) the gross output of the turbine, in horsepower, and (c) the rate of heat transfer from the cycle and radiated to outer space, in Btu/min.

ERICSSON AND STIRLING CYCLES

15-159. An Ericsson cycle operates on air, and the minimum volume of the cycle is 0.01 m$^3$. The maximum pressure is 6 bars, and at the end of the constant-pressure expansion the volume is 0.02 m$^3$. Draw a PV diagram for the cycle and compute, for each 20 kJ of heat added, (a) the heat transfer out, and (b) the pressure at the end of the isothermal heating process if the mass is 0.0465 kg.

15-160. An Ericsson cycle operates on air, and the minimum volume of the cycle is 5 L. The maximum pressure is 5 bars, and at the end of constant-pressure expansion the volume is 12 L. Draw a *PV* diagram for the cycle, and compute, for each 9 kJ of heat added, (*a*) the heat transfer out, and (*b*) the pressure at the end of the isothermal heating process if the mass is 0.0299 kg.

15-161. A Stirling cycle operates with air, and at the beginning of isothermal expansion the state is 510°C and 9 bars. The minimum pressure in the cycle is 2 bars, and at the end of isothermal compression the volume is 60 percent of the maximum volume. Determine (*a*) the thermal efficiency of the cycle, and (*b*) the mean effective pressure.

15-162. A Stirling cycle operates with air, and at the beginning of isothermal compression the state is 77°C and 2 bars. The maximum pressure in the cycle is 6 bars, and during isothermal expansion the volume increases by 40 percent. Determine (*a*) the thermal efficiency and (*b*) the mean effective pressure of the cycle, in bars.

15-163. A Stirling cycle operates with air, and at the beginning of isothermal expansion the state is 257°C and 5 bars. The thermal efficiency is 44 percent, and at the end of isothermal compression the volume is two-thirds of the maximum volume. Determine the mean effective pressure of the cycle.

15-164E. An Ericsson cycle operates on air, and the minimum volume of the cycle is 1 ft$^3$. The maximum pressure is 100 psia, and at the end of constant-pressure expansion the volume is 2 ft$^3$. Draw a *PV* diagram for the cycle, and compute, for each 66.6 Btu of heat added, (*a*) the heat transfer out, and (*b*) the pressure at the end of the isothermal heating process if the mass is 0.333 lb$_m$.

15-165E. An Ericsson cycle operates on air, and the minimum volume of the cycle is 0.20 ft$^3$. The maximum pressure is 120 psia, and at the end of constant-pressure expansion the volume is 0.44 ft$^3$. Draw a *PV* diagram for the cycle, and compute for each 20 Btu of heat transfer input (*a*) the heat transfer out, and (*b*) the pressure at the end of the isothermal heating process if the mass is 0.113 lb$_m$.

15-166E. A Stirling cycle operates with air, and at the beginning of isothermal expansion the state is 740°F and 80 psia. The minimum pressure in the cycle is 20 psia, and at the end of isothermal compression the volume is 60 percent of the maximum volume. Determine (*a*) the thermal efficiency, and (*b*) the mean effective pressure of the cycle.

15-167E. A Stirling cycle operates with air, and at the beginning of isothermal compression the state is 240°F and 25 psia. The maximum pressure in the cycle is 65 psia, and during isothermal expansion the volume increases by 40 percent. Determine (*a*) the thermal efficiency, and (*b*) the mean effective pressure of the cycle.

**PARAMETRIC STUDIES AND DESIGN**

15-168. By means of a computer program investigate the effect of compression ratio and maximum cycle temperature on the thermal efficiency and net specific work output of an air-standard Otto cycle. The volume ratio should range from 8 to 12, and the maximum cycle temperature should vary from 1400 to 2200 K. The air enters at 1 bar and 290 K, and the specific heats vary with temperature.

15-169. By means of a computer program investigate the effect of compression ratio and cutoff ratio on the net specific work output of a cold-air-standard Diesel cycle. Compression ratios of 15 and 20 should be considered, and the cutoff ratio ranges from 1.5 to 4.0. The air enters at 100 kPa and 290 K.

15-170. By means of a computer program investigate the effect of pressure ratio and maximum cycle temperature on the thermal efficiency and net specific work output of an ideal Brayton cycle. The pressure ratio should range between 6 and 18, and the maximum cycle temperature should range from 900 to 1500 K. Assume that the working fluid is air at an initial state of 100 kPa and 300 K, and that the specific heats are constant at their 300 K values.

15-171. Reconsider Prob. 15-170 by accounting for variable specific-heat data.

15-172. By means of a computer program investigate the effect of compressor and turbine adiabatic efficiencies on the thermal efficiency and net specific work output of a simple gas-turbine cycle. The efficiencies are the same for a given case, and range from 75 to 100 percent. The pressure ratios to consider are from 6 to 12, and the maximum cycle temperatures are 900 and 1200 K. Assume that the working fluid is air initially at 100 kPa and 300 K, and that the specific heats are constant at their 300 K values.

15-173. Reconsider Prob. 15-172 by accounting for variable specific-heat data.

15-174. By means of a computer program investigate the effect of regeneration on the thermal efficiency of an ideal gas-turbine cycle. The regenerator effectiveness should range from 60 to 90 percent. The air enters the compressor at 100 kPa and 300 K, and the specific heats are considered constant at their 300 K values. Consider the following parameters: (*a*) pressure ratio of 6 and 12, (*b*) maximum cycle temperature of 900 and 1200 K.

15-175. Reconsider Prob. 15-174 by accounting for variable specific-heat data.

15-176. Reconsider Prob. 15-174 by accounting for compressor and turbine adiabatic efficiencies each of 80 or 90 percent.

15-177. By means of a computer program investigate the effect of staging on the thermal efficiency and net specific work output for an ideal regenerative gas-turbine cycle with multistage compression and expansion. Assume the overall pressure ratio is 10, and each stage has an equal pressure ratio. The air enters each stage of the compressor at 300 K, and the specific heats are constant at their 300 K values. The temperature at the inlet to each stage of the turbine is 1000 K and the number of stages should range from 1 to 5.

15-178. Reconsider Prob. 15-177 by substituting argon as the working fluid.

15-179. Reconsider Prob. 15-177 by accounting for variable specific-heat data.

15-180E. It is desired to compress 6 lb/min of air from 14.7 psia and 70°F to 588 psia. This may be done by a single compressor or a series of compressors with intercooling between each compressor stage. When two or more stages are used, the pressure ratio (PR) for each stage is the same. Design the lowest life-cycle cost system, based on the following information:
1. The maximum exit temperature for any stage cannot exceed 450°F.
2. The adiabatic efficiency $\eta_C$ of any stage is $\eta_C = 1 - 0.2PR^{0.25}$.
3. Each intercooler used increases the cost of the basic unit by $650.
4. The cost of electrical power to drive the compressors is $0.11 per kW·h.
5. The cost of removing energy in the intercoolers is $0.01 per kBtu.
6. The compressor operates an average of 30 hours per week.

The following assumptions may be made: (1) specific heats are constant over the entire temperature range, (2) pressure drops in intercoolers can be neglected, (3) cost of borrowing money can be neglected, and (4) maintenance cost does not increase when an intercooler is added. In the design process the following questions should be addressed:
1. Should each intercooler be designed to give the minimum outlet temperature?
2. Would a doubling of the cost of electricity and the cost of removing energy from an intercooler in the future affect your design?

For a given compressor design, the results should include the inlet and outlet conditions for each stage and each intercooler.

15-181. The inlet conditions to an open-cycle gas-turbine plant with regeneration are the ambient values of 1.013 bars and 300 K. The turbine inlet temperature is 1100 K, and the adiabatic efficiencies of the compressor and turbine are 85 and 88 percent. The pressure loss in the combustion chamber is 3 percent of the inlet pressure. In the regenerator the pressure loss is 2 percent of the inlet pressure on both sides of the heat exchanger, while its effectiveness is 50 percent. Either air-table data or constant specific heats at room temperature may be used for parts $a$ through $d$ below.

(a) Estimate the compressor pressure ratio $r_p$ for maximum work output of a nonregenerative cycle, and the corresponding thermal efficiency.

(b) Repeat part $a$ for a regenerative cycle.

(c) Estimate the compressor pressure ratio for maximum thermal efficiency of a nonregenerative cycle, and the corresponding net work output.

(d) Repeat part $c$ for a regenerative cycle.

(e) Your business office has prepared cost figures for your final design. They assume that for $r_p = 6$ a turbine plant costs one million dollars, with the cost increasing as $r_p^2$. For such a plant, the operating cost is initially \$50,000 per year, varying directly proportional to the compression ratio. Inflation is estimated at 6 percent a year. Determine the compression ratio for the cycle that, when operated over 20 years, would have a total economic cost exactly double that of the 10-year lifetime device operated at $r_p = 6$. [Note that $\sum ar^{n-1} = a(1 - r^n)/(1 - r)$ for $n = 1, 2, 3, \ldots$.]

15-182. Investigate the effects of turbine inlet temperature and adiabatic efficiencies of the compressor and turbine on the thermal efficiency and specific net work output of a gas-turbine power cycle. The range of inlet temperatures to consider is 900 to 1600 K, while the compressor and turbine isentropic efficiencies are each 100, 94, 88, and 84 percent. Show suitable plots of the results, and discuss.

15-183. A simple gas-turbine cycle is to produce 75 kW of net power for automotive service. Neglect heat losses from piping and main components, and neglect pressure drops in piping and combustor. The working fluid is assumed to be air throughout, but do not assume constant specific heats. The air enters the compressor at 1 bar and 290 K and the turbine at 1140 K. The turbine and compressor adiabatic efficiencies are 0.84 and 0.82, respectively. Determine (a) the compressor work, (b) the turbine work, (c) the net work, and (d) the heat input, all in kJ/kg, (e) the thermal efficiency, (f) the mass flow rate, in kg/min, (g) the volume flow rate at the compressor inlet, and (h) the turbine outlet, both in m³/min. Evaluate at sufficient pressure ratios between 4 and 13.5 so that the variation of the required quantities with pressure ratio can be noted.

15-184. Consider the addition of a regenerator to the gas-turbine cycle in Prob. 15-183. Determine items $a$ through $f$ for the same data, if the regenerator effectiveness is 70 percent.

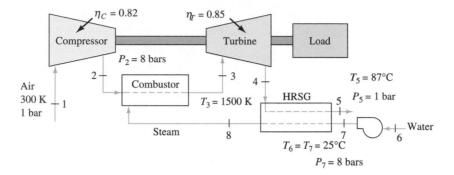

**Figure P15-185**

15-185. In order for utilities to meet varying power needs quickly, steam injection into gas turbines is under consideration. To accommodate flow rates in excess of design ratings, gas turbines from the aircraft industry are chosen. High-pressure steam is injected into the air entering the combustor, as shown in Fig. P15-185. Energy from the hot gases leaving the turbine is used to generate steam in the heat-recovery steam generator (HRSG) at a pressure $P_8$ which equals $P_2$ for the air leaving the compressor. The performance of the overall cycle is primarily a function of the pressure ratio and the mass flow rate ratio used, all other factors being the same. As shown in Fig. P15-l85, air enters the compressor at 1 bar and 300 K, the turbine inlet temperature is 1400 K, water leaves the pump and enters the HRSG at state 7, and the air-steam mixture leaves the HRSG at 1 bar and 87°C. The compressor and turbine are assumed to be isentropic; pressure drops in connecting pipes and in the heat exchanger are neglected.

For a pressure ratio of 8 across the compressor and turbine, and for a mass flow rate ratio of steam to air of 0.02 to 0.12, determine the variation of thermal efficiency and power (in kW) with mass flow rate ratio. Do the data suggest an optimum mass flow rate ratio? Then increase the pressure ratio to 12, and discuss how the thermal efficiency and power are affected, with minimum supporting calculations.

15-186. A closed-cycle gas-turbine power plant, such as might be used to provide power for a remote experimental station, is shown in Fig. P15-186. Typical component performance factors include $\eta_C = 0.80$, $\eta_T = 0.82$, and for the regenerator $T_3 - T_2 = 0.87(T_5 - T_2)$. The compressor pressure ratio might lie between 2.5:1 and 5:1. Owing to pressure drops in the cycle, the turbine inlet pressure is 90 percent of the compressor outlet pressure and the pressure ratio for the turbine is 85 percent of that for the compressor. The function of the bypass valve is load control; that is, it controls the turbine gross output and hence the turbine net power output. The design study should include at least three cases: (1) bypass valve closed with the net

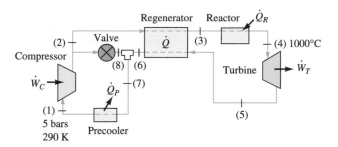

**Figure P15-186**

output power of 500 kW, (2) bypass valve opened so that the system is self-sustaining (no net output power), and (3) bypass valve set so that the net output power is 50 percent of case 1. Note that states 1 and 4 are given, and remain the same for the three cases. Design parameters to be determined for the three cases and two pressure ratios might include (*a*) compressor and turbine powers, in kW, (*b*) the flow rate in kg/h, (*c*) the reactor heat-transfer rate, in kW, (*d*) the precooler load in kW, and (*e*) the overall energy-conversion efficiency.

# CHAPTER
# 16

# VAPOR POWER CYCLES

With a 750-megawatt output from 14 units, The Geysers is the largest producer of geothermal power in the world.

One of the major applied areas of thermodynamics is electric power generation by means of steam power plants. A brief introduction to the subject was presented in Chaps. 5 and 6, where some basic concepts, definitions, and second-law restrictions on power cycles were presented. Steam power plants operate essentially on the same basic cycle, whether the input energy is from the combustion of fossil fuels (coal, gas, or oil), or from a fission process in a nuclear reactor. The steam cycle differs from the gas cycles presented in Chap. 15 in that both vapor and liquid phases are presented during portions of the cycle. A modern large-scale electric power cycle is fairly complicated in terms of mass and energy flows. To simplify the nature of these cycles, they are discussed largely in terms of simple models which involve idealized processes. The advantage of such models is that we gain some important qualitative information on the major parameters that influence the overall performance of the cycle. More comprehensive analyses of vapor power cycles may be obtained from standard textbooks on the subject.

## 16-1    THE RANKINE CYCLE

The thermal efficiency of a power cycle is maximized if all the energy supplied from a source occurs at the maximum possible temperature and all the
energy rejected to a sink occurs at the lowest possible temperature. For a
reversible cycle operating under these conditions, the thermal efficiency is
the Carnot efficiency, given by $(T_H - T_L)/T_H$. One theoretical cycle that fulfills these conditions is the Carnot heat-engine cycle introduced in Sec. 8-3.
To review, a Carnot cycle is composed of two reversible, isothermal processes and two reversible, adiabatic (isentropic) processes. If the working
fluid appears in both the liquid and the vapor phases during various parts of
the cycle, then the $Ts$ diagram for the cycle will be similar to that shown in
Fig. 16-1.

The Carnot heat-engine cycle described by Fig. 16-1 can be summarized in the following statements:

1. Wet steam at state 1 is compressed isentropically to saturated liquid
   at state 2.

2. At the elevated pressure for state 2, energy is added by heat transfer
   at constant pressure (and constant temperature) until the water is a
   saturated vapor at state 3.

3. The fluid is then expanded isentropically through a turbine to state 4.

4. The wet steam leaving the turbine is then partially condensed at constant pressure (and constant temperature) back to state 1, with the
   energy removed by heat transfer.

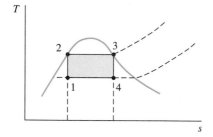

**Figure 16-1**
A $Ts$ diagram of a Carnot vapor power
cycle.

The thermal efficiency of the cycle is, of course, the highest for any heat engine operating between temperatures $T_1$ and $T_2$ and is given by $(T_2 - T_1)/T_2$.
The Carnot heat engine model is impractical to use, however, with fluids that
undergo phase changes. It is difficult, for example, to compress a two-phase
mixture isentropically, as required by process 1-2. Second, the condensing
process 4-1 would have to be controlled very accurately to end up with the
desired quality of state 1. Third, the isentropic expansion produces a fluid
with a high moisture content at state 4. This would cause erosion of the
turbine blades by liquid droplets. Fourth, the efficiency of a Carnot cycle
is greatly affected by temperature $T_2$ at which energy is added. For steam,
the critical temperature is only 374°C (705°F). Therefore, if the cycle is to
be operated within the wet region, the maximum possible temperature is
severely limited.

The objections listed in the preceding paragraph for the basic Carnot
vapor power cycle can be eliminated by two modifications. First, process
4-1 in Fig. 16-1 is carried out so that the wet steam leaving the turbine
is completely condensed to saturated liquid at the turbine-outlet pressure.
The compression process 1-2 is now handled by a liquid pump, which isentropically raises the pressure of the liquid leaving the condenser to that desired in process 2-3. Second, the fluid is superheated in process 2-3 to a

temperature $T_3$ frequently greater than the critical temperature. This model for a steam power cycle is called the **Rankine cycle**. The basic cycle is presented schematically and on a $Ts$ diagram in Fig. 16-2. The Rankine cycle then consists of

1-2. Isentropic compression in a pump

2-3. Constant-pressure energy addition by heat transfer in a boiler-superheater

3-4. Isentropic expansion in a turbine

4-1. Constant-pressure energy removal by heat transfer from a condenser

The heat transfer in at constant temperature in the Carnot cycle has been replaced by heat transfer at constant pressure and increasing temperature in the Rankine cycle.

After the saturated vapor leaves the boiler at state $3'$, the fluid passes through another energy-input section called a superheater. The heat-transfer process of *superheating* leads to a higher temperature at the turbine inlet without increasing the maximum pressure in the cycle. Temperatures at state 3 in the range of 540 to 600°C (1000 to 1100°F) are generally permissible. Note that the average temperature at which heat transfer occurs during process $3'$-3 is greater than that during the heat-transfer process in the boiler section, whereas the temperature of rejection (process 4-$4'$) is still the same. On the basis of a Carnot-engine analysis, then, superheating would be expected to increase the thermal efficiency. The other important point is that the quality at state 4 is considerably higher than that at state $4'$. Hence superheating has also alleviated the moisture problem in the turbine.

If the heat-transfer processes are reversible, the heat transfer to the fluid in the boiler-superheater is represented on the $Ts$ diagram in Fig. 16-2 by the area enclosed by states $2$-$d$-$3'$-$3$-$b$-$a$-$2$. The area enclosed by $1$-$4$-$b$-$a$-$1$ then represents the heat transfer from the fluid in the condenser. The energy

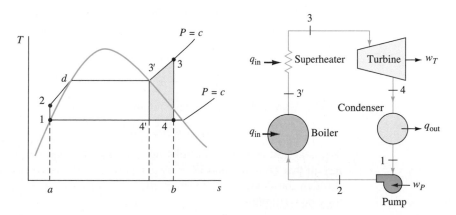

**Figure 16-2**    The *Ts* diagram and equipment schematic for a Rankine cycle with super-heating.

equation for a cyclic steady-flow process requires that the net heat transfer in equals the net work transfer out. Hence the net work input for the cycle is represented by the difference in the areas for the heat input and heat rejection, that is, area 1-2-$d$-3'-3-4-1. The thermal efficiency is defined as $w_{\text{net,out}}/q_{\text{in}}$.

Expressions for the work and heat interactions in the Rankine cycle are found by applying the steady-flow energy equation on a unit-mass basis:

$$q + w = \Delta h + \Delta \text{ke} + \Delta \text{pe}$$

to each separate piece of equipment. If we neglect kinetic- and potential-energy changes, the basic steady-state energy equation for each process reduces to $q + w = h_{\text{out}} - h_{\text{in}}$. The isentropic pump work is given by

$$w_P = h_2 - h_1$$

The value of $h_2$ may be found by interpolation in the compressed-liquid table for water, or from a computer program of steam data. An alternate method is to compute the isentropic pump work from the steady-flow work equation

$$w_{\text{sf}} = \int v \, dP \qquad \textbf{[7-32]}$$

Since the change in the specific volume of liquid water from the saturation state to pressures in the compressed-liquid state normally encountered in steam power plants is less than 1 percent, the fluid in the pump may be considered incompressible. Consequently, the pump work is frequently determined within the desired accuracy from the relation

$$w_{P,\text{in}} = v_{f,1}(P_2 - P_1) \qquad s_1 = s_2 \qquad \textbf{[16-1]}$$

where $v_{f,1}$ is the saturated-liquid specific volume at state 1. Note in Fig. 16-2 that the length of line 1-2 is greatly exaggerated. The temperature rise due to isentropic compression is actually quite small.

On a unit-mass basis the heat input is

$$q_{\text{in}} = q_{23} = h_3 - h_2 \qquad P_3 = P_2 \qquad \textbf{[16-2]}$$

the isentropic turbine work output is

$$w_{T,\text{out}} = h_3 - h_4 \qquad s_3 = s_4 \qquad \textbf{[16-3]}$$

and the heat transfer from the condenser is

$$q_{\text{cond,out}} = h_4 - h_1 \qquad P_4 = P_1 \qquad \textbf{[16-4]}$$

The heat and work transfer relations may also be expressed on a rate basis. Recall that $\dot{Q} = \dot{m}q$ and $\dot{W} = \dot{m}w$, where $\dot{m}$ is the mass flow rate of steam through the device. The thermal efficiency of an ideal Rankine cycle then may be written as

$$\eta_{\text{th}} = \frac{w_{T,\text{out}} - w_{P,\text{in}}}{q_{\text{in}}} = \frac{h_3 - h_4 - v_{f,1}(P_2 - P_1)}{h_3 - h_2} \qquad \textbf{[16-5a]}$$

The thermal efficiency can also be expressed in the alternative form

$$\eta_{th} = 1 - \frac{q_{out}}{q_{in}} = 1 - \frac{h_4 - h_1}{h_3 - h_2} \qquad \textbf{[16-5b]}$$

This latter equation is based on the fact that the net work output equals the net heat input.

The energy removed from the steam in the condenser is transferred to a liquid stream of water provided from a river, lake, or pond, as shown in Fig. 16-3. This "cooling water" (cw) undergoes a small temperature rise as it passes through the condenser. An energy balance on a control volume around the condenser reduces to

$$\dot{m}(h_1 - h_4) + \dot{m}_{cw}(h_{out} - h_{in})_{cw} = 0 \qquad \textbf{[16-6]}$$

where heat and work transfers as well as kinetic- and potential energy changes are negligible. The enthalpy for the cooling water may be approximated by using the saturated liquid enthalpy at the given temperature. As an alternative, if we assume an incompressible fluid at constant pressure, then $\Delta h_{inc} = c\,\Delta T$. Either method is acceptable. For a given steam cycle, the above energy balance relates the required mass flow rate of cooling water to its temperature rise. Typical examples are given below.

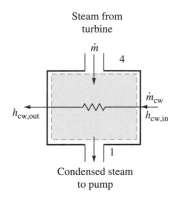

Steam from turbine

$\dot{m}$

$\dot{m}_{cw}$

$h_{cw,out}$

$h_{cw,in}$

Condensed steam to pump

**Figure 16-3**
Schematic for a condenser of a vapor power cycle.

**Determine** (a) the quality of the steam leaving the turbine and (b) the thermal efficiency of an ideal Rankine cycle without superheating for which the turbine-inlet conditions are a saturated vapor at 30 bars and the condenser pressure is 0.1 bar.

**Example 16-1**

**Solution:**

**Given:** An ideal Rankine cycle without superheating is shown in Fig. 16-4.

**Find:** (a) $x_4$ and (b) $\eta_{th}$.

**Model:** Steady-state, internally reversible processes; neglect $\Delta$ke and $\Delta$pe.

**Analysis:** (a) The basic steady-state energy balance on a unit-mass basis is

$$q + w = \Delta h + \Delta ke + \Delta pe$$

If $\Delta$ke and $\Delta$pe are neglected, then $q + w = \Delta h$. Using the notation of Fig. 16-2, we find the following data from the steam tables:

$$h_3 = h_g \text{ at 30 bars} = 2804.2 \text{ kJ/kg}$$

$$h_1 = h_f \text{ at 0.1 bar} = 191.8 \text{ kJ/kg}$$

The pump work is

$$w_{P,in} = v_f\,\Delta P = 1.01 \times 10^{-3} \text{ m}^3/\text{kg} \times 29.9 \text{ bars} \times \frac{10^2 \text{ kJ}}{\text{bar}\cdot\text{m}^2} = 3.0 \text{ kJ/kg}$$

Using the energy balance for the pump, $h_2$ can be found as

$$h_2 = h_1 + w_{P,in} = (191.8 + 3.0) \text{ kJ/kg} = 194.8 \text{ kJ/kg}$$

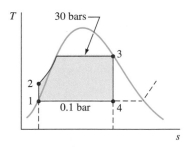

**Figure 16-4**
A Ts process diagram with data for Example 16-1.

The enthalpy at state 4 is the value for 0.1 bar and for $s_4 = s_3 = s_g$ at 30 bars. Since $s_3 = 6.1869$ kJ/kg·K, then

$$s_3 = 6.1869 = (s_f + x s_{fg})_{\text{at 0.1 bar}} = 0.6493 + x_4(8.1502 - 0.6493)$$

$$x_4 = \frac{5.5376}{7.5009} = 0.738 \text{ (or 73.8\%)}$$

$$h_4 = h_f + x h_{fg} = [191.8 + 0.738(2393.8)] \text{ kJ/kg} = 1958.3 \text{ kJ/kg}$$

(b) The evaluation of the thermal efficiency requires calculations of the turbine and pump work and the heat-transfer input. On the basis of the energy balance discussed above,

$$w_{T,\text{out}} = h_3 - h_4 = 2804.2 - 1958.3 = 845.9 \text{ kJ/kg}$$

$$q_{\text{in}} = h_3 - h_2 = 2804.2 - 194.8 = 2609.4 \text{ kJ/kg}$$

The thermal efficiency is

$$\eta_{\text{th}} = \frac{w_{T,\text{out}} - w_{P,\text{in}}}{q_{\text{in}}} = \frac{845.9 - 3.0}{2609.4} = 0.323 \text{ (or 32.3\%)}$$

---

**Example 16-2**

**S**uperheated steam enters the turbine of an ideal Rankine cycle at 30 bars and 500°C and exits the condenser as a saturated liquid at 0.1 bar. Determine (a) the thermal efficiency, (b) the mass flow rate of steam required, in kg/h, (c) the rate of heat transfer to the cycle, in megawatts, and (d) the mass flow rate of condenser cooling water, in kg/h, if the cooling water undergoes a temperature rise from 18 to 28°C. The net power output is 100 MW.

**Solution:**

**Given:**   An ideal Rankine cycle with superheating is shown in Fig. 16-5.

**Find:**   (a) $\eta_{\text{th}}$, (b) $\dot{m}$, in kg/h, (c) $\dot{Q}_{\text{in}}$, in MW, and (d) $\dot{m}_{\text{cw}}$, in kg/h.

**Model:**   Steady-state, internally reversible processes; neglect ke and pe effects.

**Analysis:**   The basic steady-state energy balance on a unit-mass basis is

$$q + w = \Delta h + \Delta \text{ke} + \Delta \text{pe}$$

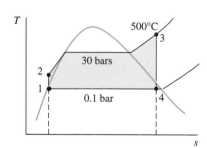

**Figure 16-5**
A $Ts$ diagram with data for Example 16-2.

If $\Delta$ke and $\Delta$pe are neglected, then $q + w = \Delta h$.

(a) This example contains the same pressure conditions as Example 16-1. However, the steam entering the turbine has been superheated from the saturation state of 234 to 500°C. The pump work will remain the same, that is, 3.0 kJ/kg, but all other heat and work quantities are altered. The turbine-inlet enthalpy $h_3$ is now 3456.5 kJ/kg, and the entropy at this state is 7.2338 kJ/kg·K. Hence, using $h_2 = 194.8$ kJ/kg from the preceding example, we find that

$$q_{\text{in}} = h_3 - h_2 = 3456.5 - 194.8 = 3261.7 \text{ kJ/kg}$$

where the notation of Fig. 16-5 is being used. Calculation of the turbine work requires that the final isentropic state 4 be found first. Since $s_3 = s_4 = 7.2338$ kJ/kg·K,

$$7.2338 = (s_f + x s_{fg})_{\text{at 0.1 bar}} = 0.6493 + x_4(7.5009)$$

or $\qquad\qquad x_4 = \dfrac{6.5845}{7.5009} = 0.878 \text{ (or 87.8\%)}$

$\qquad h_4 = (h_f + x h_{fg})_{\text{at 0.1 bar}} = 191.8 + 0.878(2392.8) = 2292.7 \text{ kJ/kg}$

Therefore,

$\qquad\qquad w_{T,\text{out}} = h_3 - h_4 = 3456.5 - 2292.7 = 1163.8 \text{ kJ/kg}$

The thermal efficiency becomes

$\qquad\qquad \eta_{\text{th}} = \dfrac{w_{T,\text{out}} - w_{P,\text{in}}}{q_{\text{in}}} = \dfrac{1163.8 - 3.0}{3261.7} = 0.356 \text{ (or 35.6\%)}$

$\quad$ (*b*) The mass flow rate of steam is found from the basic relation between work and power, $\dot{W} = \dot{m}w$. Therefore,

$$\dot{m} = \dfrac{\dot{W}_{\text{net,out}}}{w_{\text{net,out}}} = \dfrac{\dot{W}_{\text{net,out}}}{(h_3 - h_4) - (h_2 - h_1)}$$

$$= \dfrac{(100 \text{ MW})(10^3 \text{ kW/MW})}{(1163.8 - 3.0) \text{ kJ/kg}} \times \dfrac{1 \text{ kJ}}{\text{kW·s}} \times \dfrac{3600 \text{ s}}{\text{h}}$$

$$= 3.10 \times 10^5 \text{ kg/h}$$

$\quad$ (*c*) The rate of heat transfer into the cycle is found from $\dot{Q}_{\text{in}} = \dot{m} q_{\text{in}}$. Hence

$$\dot{Q}_{\text{in}} = (3.10 \times 10^5 \text{ kg/h})(3261.7 \text{ kJ/kg})(\text{h}/3600 \text{ s})$$

$$= 281{,}000 \text{ kW}$$

$$= 281 \text{ MW}$$

As an alternative method, $\dot{Q}_{\text{in}} = \dot{W}_{\text{net,out}}/\eta_{\text{th}} = 100 \text{ MW}/0.356 = 281 \text{ MW}$.

$\quad$ (*d*) On the basis of an energy balance for a control volume around the condenser, Eq. [16-6] gives an equation for $\dot{m}_{\text{cv}}$ as

$$\dot{m}_{\text{cw}} = \dfrac{\dot{m}(h_4 - h_1)}{(h_{\text{out}} - h_{\text{in}})_{\text{cw}}}$$

$$= \dfrac{3.10 \times 10^5 \text{ kg/h}(2292.7 - 191.8) \text{ kJ/kg}}{(117.43 - 75.58) \text{ kJ/kg}}$$

$$= 15.56 \times 10^6 \text{ kg/h}$$

where the enthalpies of the cooling water are saturated-liquid values at the given temperature. On the basis of a specific-heat value of 4.184 kJ/kg·K for liquid water from Table A-4, the enthalpy change of the cooling water would be 41.84 kJ/kg rather than the value of 42.15 kJ/kg used above. This would alter the value of $\dot{m}_{\text{cw}}$ by less than 1 percent.

**Comment:** The result of superheating is an increase in the thermal efficiency from 31.4 to 35.6 percent, and an increase in the quality of the steam leaving the turbine from 74.8 to 87.8 percent. Although this latter value of the quality is still objectionable, it must be remembered that the above calculations are based on isentropic flow through the turbine. The presence of irreversibilities in the flow, although decreasing the work output of the turbine, will be found to increase the quality at the outlet. As another point, the Carnot efficiency based on the maximum and minimum temperatures of the superheat cycle (500 and 45.8°C) is now 58.7 percent. This is considerably above the 35.6 percent for the actual ideal cycle. Because heat is added to the fluid at 500°C only at the very end of the heat-addition process, the average temperature of heat addition is well below 500°C. Consequently, the cycle efficiency begins to deviate considerably from that predicted for the Carnot cycle.

It is interesting to note from the preceding examples that the pump work is a very small fraction of the turbine work in steam power cycles. The ratio of work input to work output is known as the *back work ratio*, or bwr, and for the Rankine cycle is given by bwr $= w_{P,in}/w_{T,out}$. For the ideal Rankine cycle in the preceding example this ratio is less than 0.01. In the analysis of the gas-turbine cycle in Chap. 15, it is found that the compressor consumes a large fraction of the turbine output. This major difference is due to the difference between the specific volume of a liquid and that of a gas. Since the work required to compress a fluid reversibly over a given increment of pressure is given by $w = \int v\, dP$ [see Eq. [7-32]], it takes considerably less work to compress a liquid than to compress a gas for the same pressure change.

## 16-1-1    EFFECT OF CONDENSER PRESSURE ON THE RANKINE CYCLE

The efficiency of the simple Rankine cycle discussed above may be increased, on the basis of the theoretical Carnot cycle, either by decreasing the temperature at which energy is rejected as heat transfer or by increasing the average temperature at which energy is added as heat transfer. The first effect is accomplished by lowering the exhaust pressure, which in turn lowers the value of $T_4$ (and $T_1$), shown in Fig. 16-2. This point is illustrated by Fig. 16-6. The crosshatched area enclosed by states 1-2-2''-1''-4''-4-1 is a measure of the decrease in heat transfer out of the cycle when the exhaust pressure is lowered from $P_4''$ to $P_4$. This area is also a measure of the increase in net work output. The increase in heat input is measured by the total area under curve 2-2''. Hence, as a result of the lower exhaust pressure,

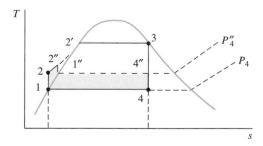

**Figure 16-6**    Effect of turbine-exhaust pressure on the thermal efficiency of an ideal Rankine cycle.

both the quantity of heat input required and the net work output increase. The overall effect of these two changes is an increase in thermal efficiency. The following example illustrates this.

---

**D**etermine (*a*) the thermal efficiency and (*b*) the back work ratio of an ideal Rankine cycle with superheating, for which the turbine-inlet conditions are 30 bars and 500° C and the condenser pressure is 0.06 bar.

**Solution:**

**Given:**   An ideal Rankine cycle with superheating is shown in Fig. 16-7.

**Find:**   (*a*) $\eta_{\text{th}}$ and (*b*) the back work ratio

**Model:**   Steady-state, internally reversible processes; neglect ke and pe effects.

**Analysis:**   This problem is the same as that in Example 16-2, except the condenser pressure has been lowered from 0.1 to 0.06 bar. Therefore, the values of $h_1$, $h_2$, and $h_3$ remain the same as before, while $w_P$ and $q_{\text{in}}$ change insignificantly. These values are:

$$h_1 = 191.8 \text{ kJ/kg} \qquad h_2 = 194.8 \text{ kJ/kg} \qquad h_3 = 3456.5 \text{ kJ/kg}$$

$$w_{P,\text{in}} = 3.0 \text{ kJ/kg} \qquad q_{\text{in}} = 3261.7 \text{ kJ/kg}$$

(*a*) The only value that changes is $h_4$. Since $s_3 = s_4 = 7.2338$ kJ/kg·K,

$$7.2338 = (s_f + s_{fg})_{\text{at 0.06 bar}} = 0.5210 + x_4(8.3304 - 0.5210)$$

or
$$x_4 = \frac{6.7128}{7.8094} = 0.860 \text{ (or 86.0\%)}$$

$$h_4 = (h_f + h_{fg})_{\text{at 0.06 bar}} = 191.8 + 0.860(2415.9) = 2268.4 \text{ kJ/kg}$$

Therefore, on the basis of the restrictive energy balance, $q + w = \Delta h$,

$$w_{T,\text{out}} = h_3 - h_4 = 3456.5 - 2268.4 = 1188.1 \text{ kJ/kg}$$

The thermal efficiency becomes

$$\eta_{\text{th}} = \frac{w_{T,\text{out}} - w_{P,\text{in}}}{q_{\text{in}}} = \frac{1188.1 - 3.0}{3261.7} = 0.363 \text{ (or 36.3\%)}$$

(*b*) The back work ratio (bwr) for the cycle is

$$\text{bwr} \equiv \frac{w_{P,\text{in}}}{w_{T,\text{out}}} = \frac{3.0}{1188.1} = 0.0025$$

Thus a negligible amount of the turbine output is used to drive the pump.

**Comment:**   In comparing Examples 16-2 and 16-3, we find that lowering the condenser pressure from 0.1 to 0.06 bar increases the thermal efficiency from 35.6 to 36.3 percent.

**Example 16-3**

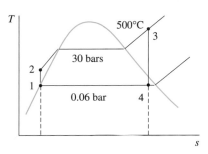

**Figure 16-7**
A *Ts* process diagram with data for Example 16-3.

The preceding example illustrates the effect of the turbine-exhaust pressure on the work output and thermal efficiency of a simple Rankine cycle. There is a limit, however, to the minimum pressure in the condenser. Heat transfer occurs from the condensing steam to cooling water or atmospheric air. The temperature of the cooling water or air normally would only vary over a very narrow range. Typically this might be 15 to 30°C (60 to 90°F). A temperature differential of 10 to 15°C (15 to 25°F) must exist across the heat-transfer surface in order to maintain adequate heat-transfer rates. Hence the minimum condensing temperature for the steam would range from 25 to 45°C (75 to 115°F). From the saturation-steam tables, the saturation pressures corresponding to these ranges of temperature are roughly from 0.03 to 0.10 bar (0.5 to 1.5 psia). Thus, fairly low pressures, well below atmospheric values, are possible in the condensers of modern steam power plants. However, this pressure varies over a fairly small range.

Although the effect of lowering the exhaust pressure is advantageous from the standpoint of increasing the thermal efficiency, it has the great disadvantage of increasing the moisture content (decreasing the quality) of the fluid leaving the turbine. This increased moisture content through the turbine decreases the efficiency of an actual turbine. In addition, the impingement of liquid droplets on the turbine blades may lead to a serious erosion problem. In practice, it is desirable to keep the moisture content less than 10 percent at the low-pressure end of the turbine. Also, increasing the average temperature at which heat is supplied will increase the thermal efficiency of the Rankine cycle. However, the maximum temperature in the cycle at the turbine inlet is usually limited by metallurgical considerations.

## 16-1-2 EFFECT OF IRREVERSIBILITIES ON VAPOR-POWER-CYCLE PERFORMANCE

The various discussions of the Rankine cycle presented earlier in this section all involve ideal cycles. Irreversibilities have not been considered. In all cases there are frictional losses in the piping which lead to pressure drops. Heat-transfer losses occur in equipment modeled as adiabatic. Irreversibilities caused by frictional flow are especially important in the turbine and pump. If heat losses are assumed negligible in a turbine or pump, the adiabatic efficiencies of this equipment must be considered, to bring the ideal analysis more in line with the actual performance. The concept of adiabatic efficiency for various flow devices is introduced in Sec. 8-2. To review, for turbines and pumps

$$\eta_T \equiv \frac{w_{a,\text{out}}}{w_{s,\text{out}}} = \frac{h_3 - h_{4a}}{h_3 - h_{4s}} \qquad \text{[8-17]}$$

and

$$\eta_P \equiv \frac{w_{s,\text{in}}}{w_{a,\text{in}}} = \frac{h_{2s} - h_1}{h_{2a} - h_1} \qquad \text{[8-29]}$$

Figure 16-8 shows the change in the position of the exit states for the turbine and pump when irreversibilities are present.

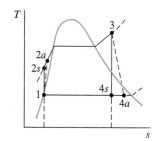

**Figure 16-8**

The $Ts$ diagram of a simple steam power cycle with irreversible turbine and pump performance.

The turbine-inlet conditions in a Rankine cycle are 30 bars and 500°C, and the turbine-exhaust condition is 0.1 bar. The turbine operates adiabatically with an efficiency of 82 percent, and the efficiency of the pump is 78 percent. Compute (a) the thermal efficiency of the cycle, (b) the back work ratio, (c) the mass flow rate of steam required, in kg/h, if the net power output is 100 MW, (d) the mass flow rate of cooling water required, in kg/h, if the water undergoes a temperature rise from 18 to 28°C, and (e) the entropy generation in the turbine and pump, in kJ/kg·K.

Example 16-4

**Solution:**

**Given:** A Rankine cycle operating with irreversible turbine and pump is shown in Fig. 16-9.

**Find:** (a) $\eta_{\text{th}}$, (b) bwr, (c) $\dot{m}_{\text{steam}}$, (d) $\dot{m}_{\text{cw}}$, (e) $\sigma_{m,T}$, $\sigma_{m,P}$.

**Model:** Steady state, irreversible turbine and pump.

**Analysis:** Losses other than those in the turbine and pump are neglected. The steady-state energy balance $q + w = \Delta h$ is appropriate if $\Delta\text{ke}$ and $\Delta\text{pe}$ are neglected.

(a) From Example 16-3, the turbine-inlet enthalpy $h_3$ and entropy $s_3$ are 3456.5 kJ/kg and 7.2338 kJ/kg·K, respectively. Also, isentropic expansion to 0.1 bar gives a turbine-outlet enthalpy $h_{4s}$ of 2292.7 kJ/kg. The actual work output is the isentropic enthalpy change times the turbine adiabatic efficiency. Hence

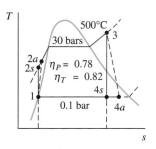

**Figure 16-9**
A $Ts$ process diagram with data for Example 16-4.

$$w_{T,a,\text{out}} = \eta_T(-\Delta h_{T,s}) = 0.82(3456.5 - 2292.7) = 954.3 \text{ kJ/kg}$$
$$= h_3 - h_{4a}$$

Using $h_3 = 3456.5$ kJ/kg, $h_{4a} = 3456.5 - 954.3 = 2502.2$ kJ/kg. Since $h_g$ at 0.1 bar is 2584.7 kJ/kg, the steam leaving the turbine is in the wet region. The exit quality is roughly 96.6 percent. The pump work for the isentropic process was previously determined to be 3.0 kJ/kg from $w_P = v \Delta p$. For the irreversible process this becomes

$$w_{P,a,\text{in}} = \frac{w_{P,s,\text{in}}}{\eta_P} = \frac{3.0}{0.78} = 3.8 \text{ kJ/kg}$$

Therefore, the enthalpy of the water leaving the pump is

$$h_{2a} = h_1 + w_{P,a,\text{in}} = 191.8 + 3.8 = 195.6 \text{ kJ/kg}$$

The thermal efficiency then becomes

$$\eta_{\text{th}} = \frac{w_{T,\text{out}} - w_{P,\text{in}}}{h_3 - h_2} = \frac{954.3 - 3.8}{3456.5 - 195.6} = 0.29 \text{ (or 29.1\%)}$$

(b) The back work ratio (bwr) for the cycle is

$$\text{bwr} \equiv \frac{w_{P,\text{in}}}{w_{T,\text{out}}} = \frac{3.8}{954.3} = 0.0040$$

Thus the back work ratio has changed from 0.0025 to 0.0040 owing to turbine and pump irreversibilities. The value is still very small.

(c) The mass flow rate of steam is found from the basic relation between work and power, $\dot{W} = \dot{m}w$. Therefore,

$$\dot{m} = \frac{\dot{W}_{\text{net,out}}}{w_{\text{net,out}}} = \frac{\dot{W}_{\text{net,out}}}{(h_3 - h_4) - (h_2 - h_1)}$$

$$= \frac{(100 \text{ MW})(10^3 \text{ kW/MW})}{(954.3 - 3.8) \text{ kJ/kg}} \times \frac{1 \text{ kJ}}{\text{kW·s}} \times \frac{3600 \text{ s}}{\text{h}}$$

$$= 3.79 \times 10^5 \text{ kg/h}$$

This is an increase of 22 percent compared to the ideal Rankine cycle.

(d) An energy balance for a control volume around the condenser with $\dot{Q} = 0$ and $\dot{W} = 0$ leads to an expression for the cooling water flow rate

$$\dot{m}_{\text{cw}} = \frac{\dot{m}(h_4 - h_1)}{(h_{\text{out}} - h_{\text{in}})_{\text{cw}}} = \frac{3.79 \times 10^5 \text{ kg/h} (2502.2 - 191.8)}{117.43 - 75.58} = 20.92 \times 10^6 \text{ kg/h}$$

where the enthalpies of the cooling water are saturated liquid values at the given temperature.

(e) The basic equation for steady-state entropy generation is

$$\dot{\sigma}_{\text{cv}} = \sum_{\text{out}} \dot{m}_e s_e - \sum_{\text{in}} \dot{m}_i s_i - \sum_{j=1}^{n} \frac{\dot{Q}_j}{T_j}$$

For an adiabatic device with one inlet and one outlet, this equation on a unit-mass basis reduces to $\sigma_{m,\text{cv}} = s_e - s_i$. For the turbine,

$$s_{4a} = (1 - x)s_f + xs_g = 0.034(0.6493) + 0.966(8.1502) = 7.8952 \text{ kJ/kg·K}$$

For state 3 at 30 bars and 500°C, $s_3 = 7.2338 \text{ kJ/kg·K}$. Hence

$$\sigma_{m,T} = s_{4a} - s_3 = 7.8952 - 7.2338 = 0.6614 \text{ kJ/kg·K}$$

The entropy generation within the pump is given by $\sigma_{m,P} = s_{2a} - s_1$. State $2a$ occurs at 30 bars and $h_{2a} = 195.6 \text{ kJ/kg}$. Use of a computerized table of water data (or interpolation in Table A-15) yields $s_{2a} = 0.6516 \text{ kJ/kg·K}$, and $s_1 = s_f$ at 0.1 bars. Therefore,

$$\sigma_{m,P} = 0.6516 - 0.6493 = 0.0023 \text{ kJ/kg·K}$$

A more approximate answer for the pump is found by assuming the fluid is incompressible. In this situation an energy balance shows that $\Delta u = c \Delta T = w_{P,a} - w_{P,s} = 0.8 \text{ kJ/kg}$. From Table A-4, $c = 4.18 \text{ kJ/kg·K}$. Thus

$$\Delta T = \frac{0.8 \text{ kJ/kg}}{4.18 \text{ kJ/kg·K}} = 0.19°C \quad T_1 = 45.81°C = 318.96 \text{ K} \quad T_2 = 319.15 \text{ K}$$

Then using the incompressible equation for $\Delta s$,

$$\sigma_{m,P} = \Delta s = c \ln \frac{T_2}{T_1} = 4.18 \ln \frac{319.15}{318.96} = 0.0025 \text{ kJ/kg·K}$$

This value agrees reasonably well with the value based on tabular data.

**Comment:**   Irreversibilities lead to a decrease in thermal efficiency from 35.6 percent for the ideal Rankine cycle to 29.1 percent. The cooling-water requirement increases by 34 percent over that for the ideal cycle. In addition, the entropy generation in the turbine is several hundred times that in the pump.

## 16-2 THE REHEAT CYCLE

In the ideal Rankine cycle, the efficiency may be increased by the use of a superheater section. The process of superheating raises the average temperature at which the heat transfer occurs into the cycle, thus theoretically raising the efficiency. An equivalent gain in the average temperature during the heat-transfer process may be accomplished by raising the boiler pressure in the cycle. This may result in a higher initial cost of the steam generator (boiler plus superheater), because of the higher pressure that must be contained; but over the years the higher efficiency of the overall unit may more than compensate for this additional cost. However, for a given maximum temperature in the steam generator, an increase in the generator pressure results in a decrease in the quality of the steam leaving the turbine. To avoid a potential turbine-blade erosion problem and still take advantage of the higher temperature made available by increasing the boiler pressure, the *reheat cycle* was developed.

In the reheat cycle, the steam is not allowed to expand completely to the condenser pressure in a single stage. After partial expansion the steam is withdrawn from the turbine and reheated at constant pressure. Then it is returned to the turbine for further expansion to the exhaust pressure. The turbine may be considered to consist of two stages, a high-pressure one and a low-pressure one. Figure 16-10 illustrates the reheat cycle on a $Ts$ diagram and the equipment schematic. The position of state 4 after the first stage of expansion is usually close to the saturation line. The temperature upon reheating to state 5 in Fig. 16-10 is usually equal to or slightly less than the inlet temperature to the first stage of the turbine.

Extreme care must be used in selecting path 4-5 for reheating, because the average temperature for the reheat process may turn out to be less than the average temperature for the heat-addition process 2-3. Thus reheating does not necessarily increase the thermal efficiency of the basic Rankine cycle. Correct use of reheating will, however, remove the objectionable feature of high moisture content at the turbine exhaust as well as increase the

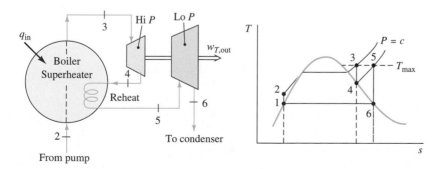

**Figure 16-10**  Equipment schematic and *Ts* diagram for an ideal reheat vapor power cycle.

thermal efficiency. There is a reheat pressure $P_4$ which will maximize the thermal efficiency for the given values of $P_3$, $T_3$, $T_5$, and $P_6$ in Fig. 16-10. For conventional values of these parameters, the maximum efficiency of an ideal reheat cycle typically occurs when the ratio $P_4/P_3$ is in the range of 0.15 to 0.35.

In computing the thermal efficiency of a reheat cycle, one must remember to account for the work output of both the high and low pressure stages of the turbine ($w_{T,\text{hi}P,\text{out}}$ and $w_{T,\text{lo}P,\text{out}}$) as well as the heat transfers into both the boiler-superheater section ($q_{\text{boil}}$) and the reheat section ($q_{\text{reheat}}$). In terms of notation of Fig. 16-10, the thermal efficiency is

$$\eta_{\text{th}} = \frac{w_{T,\text{hi}P,\text{out}} + w_{T,\text{lo}P,\text{out}} - w_{P,\text{in}}}{q_{\text{boil}} + q_{\text{reheat}}} \qquad \textbf{[16-7]}$$

$$= \frac{(h_3 - h_4) + (h_5 - h_6) - w_P}{(h_3 - h_2) + (h_5 - h_4)} \qquad \text{reheat cycle}$$

The pump work is found from $v\,\Delta P$.

---

**Example 16-5**

In an ideal reheat cycle the turbine-inlet state is 30 bars and 500°C. After expansion to 5 bars, the steam is reheated to 500°C and then expanded to the condenser pressure of 0.1 bar. The net power output is 100 MW. Determine (*a*) the cycle thermal efficiency, (*b*) the mass flow rate of steam, in kg/h, and (*c*) the rate of heat transfer from the condenser, in MW.

**Solution:**

**Given:**   An ideal Rankine cycle with reheat, as shown in Fig. 16-11.

**Find:**   (*a*) $\eta_{\text{th}}$, (*b*) $\dot{m}$, in kg/h, and (*c*) $\dot{Q}_{\text{cond,out}}$, in MW.

**Model:**   Steady state, reversible flow; adiabatic turbine and pump, neglect $\Delta$ke and $\Delta$pe.

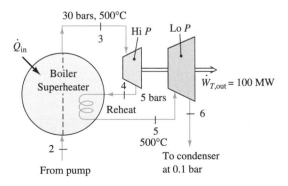

**Figure 16-11**   Schematic and data for Example 16-5.

**Analysis:** (a) The problem is based on the data of Example 16-2, except for the addition of the reheat section. From Example 16-2 the following data are still applicable. The subscripts on the property values are consistent with the notation in Fig. 16-11.

$$h_1 = 191.8 \text{ kJ/kg} \qquad h_2 = 194.8 \text{ kJ/kg} \qquad h_3 = 3456.5 \text{ kJ/kg}$$

$$w_{P,\text{in}} = h_2 - h_1 = 3.0 \text{ kJ/kg} \qquad q_{\text{boil-super}} = h_3 - h_2 = 3261.7 \text{ kJ/kg}$$

$$s_3 = 7.2338 \text{ kJ/(kg·K)} = s_4$$

From the last value we can estimate the enthalpy at state 4, where the pressure is 5 bars. The entropy of saturated vapor at 5 bars is 6.8213 kJ/ kg·K. Therefore, state 4 is still in the superheat region. From superheat table A-14 it is found by linear interpolation that $T_4$ is close to 241°C and that $h_4$ is 2941.6 kJ/kg·K. State 5 is at 5 bars and 500°C, for which $h_5 = 3483.9$ kJ/kg and $s_5 = 8.0873$ kJ/(kg·K). Finally, state 6 is the result of isentropic expansion from state 5 to 0.1 bar. Using data from the saturation table at 0.1 bar

$$s_6 = 8.0873 = (s_f + x s_{fg})_{\text{at 0.1 bar}} = 0.6493 + x_6(8.5009)$$

or
$$x_6 = \frac{8.0873 - 0.6493}{7.5009} = 0.992 \text{ (or 99.2\%)}$$

Hence

$$h_6 = h_f + x_6 h_{fg} = 191.8 + 0.992(2392.8) + 2565.5 \text{ kJ/kg}$$

For the steady-state processes, $q + w = \Delta h + \Delta \text{ke} + \Delta \text{pe}$ and neglecting $\Delta$ke and $\Delta$pe, we find that

$$w_{T,\text{hi}P,\text{out}} = h_3 - h_4 = 3456.5 - 2941.6 = 514.9 \text{ kJ/kg}$$

$$w_{T,\text{lo}P,\text{out}} = h_5 - h_6 = 3483.9 - 2565.5 = 918.4 \text{ kJ/kg}$$

$$q_{\text{reheat}} = h_5 - h_4 = 3483.9 - 2941.6 = 542.3 \text{ kJ/kg}$$

The thermal efficiency, then, is

$$\eta_{\text{th}} = \frac{w_{T,\text{hi}P,\text{out}} + w_{T,\text{lo}P,\text{out}} - w_{P,\text{in}}}{q_{\text{boil}} + q_{\text{reheat}}}$$

$$= \frac{514.9 + 918.4 - 3.0}{3261.7 + 542.3} = 0.376 \text{ (or 37.6\%)}$$

Compared with Example 16-2 without reheating, the thermal efficiency has increased only from 35.6 to 37.6 percent. However, the outlet quality of the steam at the turbine exhaust has increased from 87.8 to 99.2 percent.

(b) In determining the mass flow rate from the net work and power, we must take into consideration the work output of both turbine stages, as well as the pump work.

$$\dot{m} = \frac{\dot{W}_{\text{net,out}}}{w_{\text{net,out}}} = \frac{\dot{W}_{\text{net,out}}}{w_{T,\text{hi,out}} + w_{T,\text{lo,out}} - w_{P,\text{in}}}$$

$$= \frac{(100 \text{ MW})(10^3 \text{ kW/MW})}{514.9 + 918.4 - 3.0} \times \frac{3600 \text{ s}}{\text{h}} = 2.52 \times 10^5 \text{ kg/h}$$

(c) The rate of heat loss from the condenser to the cooling water is

$$\dot{Q}_{\text{cond,out}} = \dot{m}(h_6 - h_1) = (2.52 \times 10^5 \text{ kg/h})(2565.5 - 191.8) \text{ kJ/kg} \times (1 \text{ h}/3600 \text{ s})$$

$$= 1.66 \times 10^5 \text{ kJ/s} = 166 \text{ MW}$$

**Comment:** The mass flow rate without reheat from Example 16-2 is $3.10 \times 10^5$ kg/h. The increase in thermal efficiency due to reheat has led to a 19 percent reduction in the required mass flow rate. Equally important, the quality of the steam leaving the turbine has increased from 87.8 percent without reheat to 99.9 percent with reheat. Thus a major purpose of the reheat cycle is to reduce the moisture content in the steam leaving the turbine.

The preceding example illustrated an ideal reheat steam power cycle. By accounting for turbine and pump efficiencies, more realistic values for the thermal efficiency, mass flow rate, turbine exit quality, and condenser cooling-water requirements could be obtained.

## 16-3    THE REGENERATIVE CYCLE

Referring to Fig. 16-2 for the simple Rankine cycle with superheating reveals a serious disadvantage of the basic cycle. For the portion of the heat-addition process 2-$d$, the average temperature is much below the temperature of the vaporization and superheating process $d$-3-3′. From the viewpoint of the second law, the cycle efficiency is greatly reduced as a result of this relatively low-temperature heat-addition process. If the average temperature for this portion of the heat-addition process were raised, the efficiency of the cycle would more nearly approach that of the Carnot cycle. One practical method of accomplishing this is by the use of a *regeneration* process internal to the overall cycle.

### 16-3-1    OPEN FEEDWATER HEATER

The ideal regeneration vapor power cycle, shown in Fig. 16-12, is accomplished as described here. Part of the superheated steam which enters the turbine at state 3 is bled or extracted from the turbine at state 4, which is an

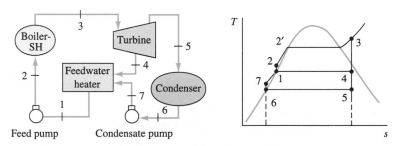

**Figure 16-12**     Equipment schematic and *Ts* diagram for an ideal regenerative vapor power cycle with one open feedwater heater.

intermediate state in the turbine expansion process. The extracted steam is directed into a heat exchanger known as a feedwater heater. The portion of the steam that is not extracted expands completely to the condenser pressure (state 5), and it is then condensed to saturated liquid at state 6. A pump increases the pressure of the liquid leaving the condenser isentropically to the same pressure as that of the extracted steam. The compressed liquid at state 7 then enters the feedwater heater, where it mixes directly with the flow stream extracted from the turbine. Because of this direct-mixing process, the feedwater heater in Fig. 16-12 is called an *open,* or *direct-contact,* type of heater. In the ideal situation, the mass flow rates for the two streams entering the heater are adjusted so that the state of the mixture leaving the heater is a saturated liquid at the heater pressure (state 1). A second pump then isentropically raises the pressure of the liquid to state 2, which corresponds to the steam-generator inlet pressure.

The theoretical analysis of an ideal regenerative cycle employs the conservation of mass and energy principles to a control volume which lies just inside the open feedwater heater. The conservation of mass requires that $\sum \dot{m}_i = \sum \dot{m}_e$. On the basis of Fig. 16-12 we may write this as

$$\dot{m}_1 = \dot{m}_4 + \dot{m}_7$$

Likewise, the energy balance with $\dot{Q} = 0$ and $\dot{W} = 0$ becomes $\sum \dot{m}_i h_i = \sum \dot{m}_e h_e$. For the open heater this reduces to

$$\dot{m}_1 h_1 = \dot{m}_4 h_4 + \dot{m}_7 h_7$$

Combining the two equations so that $\dot{m}_7$ is eliminated, we observe that

$$\dot{m}_1 h_1 = \dot{m}_4 h_4 + (\dot{m}_1 - \dot{m}_4) h_7$$

or

$$h_1 = \frac{\dot{m}_4}{\dot{m}_1} h_4 + \left(1 - \frac{\dot{m}_4}{\dot{m}_1}\right) h_7$$

If the fraction of the steam extracted from the turbine $\dot{m}_4/\dot{m}_1$ at state 4 is represented by $y_4$, then

$$1(h_1) = y_4 h_4 + (1 - y_4) h_7 \qquad \text{open heater} \qquad \textbf{[16-8]}$$

Since the enthalpies of the flow streams entering and leaving the open feedwater heater are known for the ideal cycle, the above relation permits the evaluation of the quantity of steam which must be extracted from the turbine at state 4 to ensure that the flow stream leaving the heater will be a saturated liquid.

Note that both the work output of the turbine and the work input to the (condensate) pump between states 6 and 7 in Fig. 16-12 are affected by the fraction of the steam bled from the turbine at state 4. On the basis of a unit mass passing through the boiler-superheater section, the total turbine work output is

$$w_{T,\text{out}} = \frac{\dot{W}_{T,\text{out}}}{\dot{m}_1} = 1(h_3 - h_4) + (1 - y_4)(h_4 - h_5) \qquad \textbf{[16-9]}$$

On the same basis the condensate pump work under isentropic conditions is

$$w_{P,in} = v(P_7 - P_6)(1 - y_4) \qquad \text{[16-10]}$$

The work input of the feed pump between states 1 and 2 is still given by $v \, \Delta P$, because the total flow passes through this pump.

---

**Example 16-6**

**A** regenerative steam power cycle operates so that steam enters the turbine at 30 bars and 500°C and exhausts at 0.1 bar. A single open feedwater heater is employed which operates at 5 bars. The turbine adiabatic efficiency is 82 percent and both pump efficiencies are 78 percent. Find (a) the thermal efficiency of the cycle, (b) the mass flow rate of steam required for a net power output of 100 MW, (c) the heat-transfer rate from the condenser, in MW, and (d) the entropy generation in the turbine and the heater, in kJ/kg·K.

**Solution:**

**Given:** A regenerative steam power cycle is shown with appropriate data in Fig. 16-13.

**Find:** (a) $\eta_{th}$, (b) $\dot{m}$, in kg/h, (c) $\dot{Q}_{cond,out}$, in MW, and (d) $\sigma_{m,T}$ and $\sigma_{m,heater}$, in kJ/kg·K.

**Model:** Steady state; adiabatic, irreversible turbine and pump; neglect $\Delta ke$ and $\Delta pe$.

**Analysis:** From the preceding examples the following data are still valid:

$P_3 = 30$ bars $\qquad T_3 = 500°C \qquad h_3 = 3456.5$ kJ/kg $\qquad s_3 = 7.2338$ kJ/kg·K

$P_4 = 5$ bars $\qquad h_{4s} = 2941.6$ kJ/kg

$P_5 = 0.1$ bar $\qquad h_{5a} = 2502.2$ kJ/kg $\qquad s_{5a} = 7.8952$ kJ/kg·K

$P_6 = 0.1$ bar, saturated liquid $\qquad h_6 = 191.8$ kJ/kg $\qquad s_6 = 0.6493$ kJ/kg·K

$P_1 = 5$ bars $\qquad h_1 = h_f = 640.2$ kJ/kg $\qquad s_1 = s_f = 1.8607$ kJ/kg·K

The enthalpies at states 2 and 7 can be found by adding the pump work $w_P$ to the enthalpies at states 1 and 6, respectively. The actual pump work is approximated by the value of $v \, \Delta P / \eta_P$. Hence

1–2: $\qquad w_{P,a,in} = 1.09 \times 10^{-3}(25)(10^2)/0.78 = 3.5$ kJ/kg

$\qquad\qquad h_{2a} = h_1 + w_{P,a,in} = 640.2 + 3.5 = 643.7$ kJ/kg

6–7: $\qquad w_{P,a,in} = 1.01 \times 10^{-3}(4.9)(10^2)/0.78 = 0.6$ kJ/kg

$\qquad\qquad h_{7a} = h_6 + w_{P,a,in} = 191.8 + 0.6 = 192.4$ kJ/kg

At $h_{7a} = 192.4$ kJ/kg and 5 bars, $s_{7a} = 0.6495$ kJ/kg·K. Finally, state 4 is found similarly by

3–4: $\qquad w_{T,a,out} = \eta_T(h_3 - h_{4s}) = 0.82(3456.6 - 2941.6) = 422.2$ kJ/kg

$\qquad\qquad h_{4a} = h_3 - w_{T,a,out} = 3456.5 - 422.2 = 3034.3$ kJ/kg

At $h_{4a} = 3034.3$ kJ/kg and 5 bars, $s_{4a} = 7.4071$ kJ/kg·K.

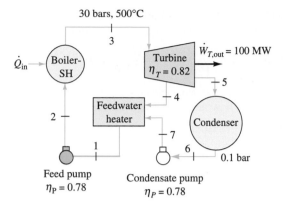

30 bars, 500°C

**Figure 16-13**     Equipment schematic with data for Example 16-6.

(a) To determine the turbine work output, it is necessary to first find the fraction of the steam which is extracted at 5 bars. An energy balance on the adiabatic heater reduces to $\sum \dot{m}_i h_i = \sum \dot{m}_e h_e$ when $\Delta$ke and $\Delta$pe are neglected. In terms of Fig. 16-13, this becomes

$$\dot{m}_1 h_1 = \dot{m}_4 h_4 + \dot{m}_7 h_7$$

In combination with the mass balance, $\dot{m}_1 = \dot{m}_4 + \dot{m}_7$, we find that

$$1(h_1) = y_4 h_4 + (1 - y_4)h_7$$

Rearrangement of this equation and substitution of enthalpy values yields

$$y_4 = \frac{\dot{m}_4}{\dot{m}_1} = \frac{h_1 - h_7}{h_4 - h_7} = \frac{640.2 - 192.4}{3034.3 - 192.4} = 0.158$$

Therefore, 0.158 kg of steam is extracted for every kilogram that enters the turbine. The work output of the turbine is $w_{T,\text{out}} = w_{T,\text{hi}P} + (1 - y_4)w_{T,\text{lo}P}$. Hence

$$w_{T,\text{out}} = h_3 - h_4 + (1 - y_4)(h_4 - h_5)$$
$$= 3456.5 - 3034.3 + 0.842(3034.3 - 2502.2)$$
$$= 422.2 + 448.0 = 870.2 \text{ kJ/kg}$$

On the basis of a unit mass passing through the boiler, the total pump work is

$$w_{P,\text{in}} = w_{P,\text{feed,in}} + (1 - y_4)w_{P,\text{cond,in}}$$
$$= 3.5 + 0.842(0.6) = 4.0 \text{ kJ/kg}$$

and the heat-transfer input is

$$q_{\text{in}} = h_3 - h_2$$
$$= 3456.5 - 643.7 = 2812.8 \text{ kJ/kg}$$

Consequently, the thermal efficiency for the cycle is

$$\eta_{\text{th}} = \frac{w_{T,\text{out}} - w_{P,\text{in}}}{q_{\text{in}}} \equiv \frac{870.2 - 4.0}{2812.8} = 0.308 \text{ (or 30.8\%)}$$

(b) The mass flow rate is found from $\dot{W}_{net,out} = \dot{m}_3(w_{T,out} - w_{P,in})$, where $w_{T,out}$ is the total turbine ouput per unit mass entering the turbine. Using data from part a,

$$\dot{m}_3 = \frac{\dot{W}_{T,net,out}}{w_{T,out} - w_{P,in}} = \frac{100 \text{ MW}(10^3 \text{ kW/MW})}{(870.2 - 4.0) \text{ kJ/kg}} \times \frac{3600 \text{ s}}{\text{h}} = 4.16 \times 10^5 \text{ kg/h}$$

(c) The heat-transfer rate from the condensing steam is given by $\dot{Q}_{cond,out} = \dot{m}_5(h_5 - h_6)$. The mass flow rate $\dot{m}_5$ through the condenser is

$$\dot{m}_5 = (1 - y)\dot{m}_3 = 0.842(4.16 \times 10^5) = 3.50 \times 10^5 \text{ kg/h}$$

Thus

$$\dot{Q}_{cond,out} = (3.50 \times 10^5 \text{ kg/h})(2502.2 - 191.8) \text{ kJ/kg} \times (\text{h}/3600 \text{ s})$$
$$= 224{,}600 \text{ kJ/s} = 224.6 \text{ MW}$$

Note that the mass flow rate bled to the open heater is $0.66 \times 10^5$ kg/h.

(d) The entropy generation in a steady-state control volume is found from

$$\dot{\sigma}_{cv} = \sum_{out} \dot{m}_e s_e - \sum_{in} \dot{m}_i s_i - \sum_{j=1}^{n} \frac{\dot{Q}_j}{T_j}$$

Under adiabatic conditions and for a unit mass entering the turbine,

$$\sigma_{m,T} = (s_4 - s_3) + (1 - y_4)(s_5 - s_4)$$
$$= (7.4071 - 7.2338) + 0.842(7.8952 - 7.4071)$$
$$= 0.584 \text{ kJ/kg·K}$$

In a similar manner, the entropy generation within the closed heater is

$$\sigma_{m,heater} = s_1 - y_4 s_4 - (1 - y_4)s_7$$
$$= 1.8607 - 0.158(7.4071) - 0.842(0.6495)$$
$$= 0.1435 \text{ kJ/kg·K}$$

Thus the entropy generation within the heater is roughly 25 percent of that in the turbine. The total entropy generation in the pumps would be similar to the small value of 0.0025 kJ/kg·K found for a single pump in Example 16-4.

**Comment:** In Example 16-4 the conditions were the same as for the above cycle, except that no regeneration was used. In that case the thermal efficiency was 29.1 percent. Thus the inclusion of one open feedwater heater in the ideal cycle has increased the cycle efficiency by some 5.8 percent. Note, however, that the work output of the turbine has been lowered by nearly 11 percent and the total mass flow rate has increased 10 percent as compared to the case without regeneration.

## 16-3-2  CLOSED FEEDWATER HEATER

A modification of the regenerative cycle described above occurs when *closed feedwater heaters* are used to preheat the water being returned from the condenser to the boiler. In a closed feedwater heater, the two entering flow streams do not mix. The feedwater leaving the condenser flows inside tubes that pass through the heater. The steam extracted from the turbine

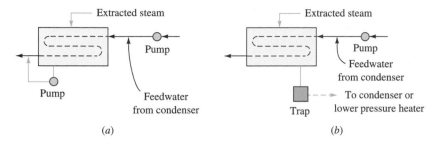

**Figure 16-14** Schematic of a closed feedwater heater with condensate (a) pumped directly back into feedwater line and (b) trapped (throttled) back to a lower pressure in the feedwater line.

enters the heater and condenses on the outsides of the tubes which carry the feedwater. (This equipment is called a *shell-and-tube heat exchanger*.) A schematic diagram of a closed feedwater heater with two alternatives for removal of the extracted steam is shown in Fig. 16-14. In Fig. 16-14*a* a pump is used to return the extracted steam which has condensed directly to the feedwater line to the boiler. As a second alternative, Fig. 16-14*b* shows the condensate being collected in a trap which permits only liquid to flow to a region of lower feedwater pressure. This lower-pressure region may be the condenser itself or another feedwater heater. On the basis of either diagram in Fig. 16-14, the steady-flow energy balance on the closed feedwater heater in the absence of heat and work transfer, and with negligible kinetic- and potential-energy changes, can be written as

$$0 = (\dot{m}\,\Delta h)_{\text{extract}} + (\dot{m}\,\Delta h)_{\text{feed}} \qquad \textbf{[16-11]}$$

The heater is assumed to be well insulated, and kinetic- and potential-energy changes of the flows are assumed negligible. Note that the $\dot{m}$ values in this equation are not the same.

In the ideal closed feedwater heater, the extracted steam condenses and leaves the feedwater heater as a *saturated liquid* at the turbine extraction pressure. In the ideal case the feedwater from the condenser is assumed to leave as a compressed liquid at the *same temperature* as the condensed extracted steam. One advantage of a closed feedwater heater is that the pressures of the extracted steam and the feedwater can be significantly different. Owing to the type of construction, closed feedwater heaters usually operate at higher pressures than the open type. However, since open feedwater heaters operate at fairly low pressures, they are less expensive. Another advantage of the open heater is that it brings the feedwater to the boiler up to its saturation temperature at the heater pressure. The following example illustrates the use of closed heaters.

**A**n ideal regenerative steam power cycle operates with steam that enters the turbine at 30 bars and 500°C and leaves the condenser as a saturated liquid at 0.1 bar.

**Example 16-7**

A single closed feedwater heater is used with steam entering at 5 bars. Determine (a) the thermal efficiency of the cycle and (b) the mass flow rate through the closed heater and the condenser, in kg/h, for a net power output of 100 MW.

**Solution:**

**Given:**  An ideal regenerative steam power cycle with a closed feedwater heater, as shown in Fig. 16-15.

**Find:**  (a) $\eta_{\text{th}}$ and (b) $\dot{m}_4$ and $\dot{m}_5$.

**Model:**  Steady state, internally reversible processes; adiabatic turbine and pump, negligible $\Delta$ke and $\Delta$pe.

**Analysis:**  The following data for the open heater in Example 16-6 are still valid:

$$P_3 = 30 \text{ bars} \qquad T_3 = 500°C \qquad h_3 = 3456.5 \text{ kJ/kg}$$

$$P_4 = 5 \text{ bars} \qquad s_4 = s_3 \qquad h_{4s} = 2941.6 \text{ kJ/kg}$$

$$P_5 = 0.1 \text{ bar} \qquad s_5 = s_3 \qquad h_{5s} = 2292.7 \text{ kJ/kg}$$

$$P_6 = 0.1 \text{ bar, saturated liquid} \qquad h_6 = 191.8 \text{ kJ/kg}$$

In addition, state 9 is a saturated liquid at 5 bars; thus $h_9 = h_f$ at 5 bars = 640.2 kJ/kg. The enthalpies at states 1 and 7 can be found by adding the pump work $w_P$ to the enthalpies at states 9 and 6, respectively. The pump work is approximated by the value of $v \, \Delta P$. Hence

9–1:    $$w_{P,\text{in}} = 1.09 \times 10^{-3}(25)(10^2) = 2.7 \text{ kJ/kg}$$

$$h_1 = h_9 + w_{P,\text{in}} = 640.2 + 2.7 = 642.9 \text{ kJ/kg}$$

6–7:    $$w_{P,\text{in}} = 1.01 \times 10^{-3}(29.9)(10^2) = 3.0 \text{ kJ/kg}$$

$$h_7 = h_6 + w_{P,\text{in}} = 191.8 + 3.0 = 194.8 \text{ kJ/kg}$$

The fluid at state 8 is assumed to be a compressed liquid at the *same temperature* as state 9. The saturation temperature at 5 bars is 151.9°C. Thus state 8 is at 30 bars and 151.9°C.

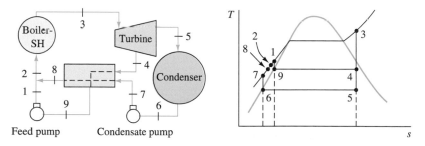

**Figure 16-15**    Equipment schematic and *T*s diagram for an ideal regenerative vapor power cycle with one closed feedwater heater in Example 16-7.

The enthalpy at state 8 is found by double interpolation in the printed compressed-liquid table, or from a computerized table. The value of $h_8$ is close to 642.4 kJ/kg if linear interpolation is used.

(a) Further analysis of the cycle requires that we first find the fraction of the steam which is extracted (bled) from the turbine at 5 bars. The basic energy balance on the closed heater in the absence of heat and work transfer and with negligible kinetic- and potential-energy changes becomes

$$0 = (\dot{m}\,\Delta h)_{\text{extract}} + (\dot{m}\,\Delta h)_{\text{feed}}$$

or

$$\dot{m}_{\text{extract}}(h_4 - h_9) = \dot{m}_{\text{feed}}(h_8 - h_7)$$

Substitution of the enthalpy values yields

$$\frac{\dot{m}_{\text{extract}}}{\dot{m}_{\text{feed}}} = \frac{642.4 - 194.8}{2941.6 - 640.2} = 0.1945$$

The fraction $y$ of the steam extracted at state 4 then is

$$y_4 = \frac{\dot{m}_{\text{extract}}}{\dot{m}_{\text{boiler}}} = \frac{0.1945}{1.1945} = 0.163$$

Therefore, 0.163 kg of steam is extracted at state 4 for every kilogram which enters the turbine at state 3. This information is necessary to find the enthalpy at state 2. For the mixing of the two flow streams after the heater, an energy balance around the mixing point yields

$$\dot{m}_8 h_8 + \dot{m}_1 h_1 = \dot{m}_2 h_2$$

Hence

$$(1 - 0.163)(642.4) + 0.163(642.9) = 1(h_2)$$

and

$$h_2 = 642.5 \text{ kJ/kg}$$

State 2 lies between states 1 and 8 on the 30-bar pressure line on the $Ts$ diagram.

The work output of the turbine becomes

$$
\begin{aligned}
w_{T,\text{out}} &= w_{T,\text{hi}P,\text{out}} + (1 - y_4)w_{T,\text{lo}P,\text{out}} \\
&= h_3 - h_4 + (1 - y_4)(h_4 - h_5) \\
&= 3456.5 - 2941.6 + 0.837(2941.6 - 2292.7) \\
&= 514.9 + 543.1 \\
&= 1058.0 \text{ kJ/kg}
\end{aligned}
$$

On the basis of a unit mass passing through the boiler, the total pump work is

$$
\begin{aligned}
w_{P,\text{in}} &= w_{P,\text{feed},\text{in}} + (1 - y_4)w_{P,\text{cond},\text{in}} \\
&= 0.163(2.7) + 0.837(3.0) = 3.0 \text{ kJ/kg}
\end{aligned}
$$

and the heat-transfer input is

$$q_{\text{in}} = h_3 - h_2 = 3456.5 - 642.5 = 2814.0 \text{ kJ/kg}$$

Consequently, the thermal efficiency for the cycle is

$$\eta_{\text{th}} = \frac{w_{T,\text{out}} - w_{P,\text{in}}}{q_{\text{in}}} = \frac{1058.0 - 3.0}{2814.0} = 0.375 \text{ (or 37.5\%)}$$

The result in this case is the same as that for an open heater operating at the same extraction pressure. However, closed heaters usually are employed at higher pressures than open heaters.

(b) The mass flow rates in the cycle are found by starting with the steam entering the turbine. The net power $\dot{W}_{net,out} = \dot{m}_3(w_{T,out} - w_{P,in})$. Using data from part a,

$$\dot{m}_3 = \frac{100 \text{ MW}(10^3 \text{ kW/MW})}{(1058.0 - 3.0) \text{ kJ/kg}} \times \frac{3600 \text{ s}}{\text{h}} = 3.41 \times 10^5 \text{ kg/h}$$

This is the same calculation as in Example 16-6 for the open heater. Replacing an open heater with a closed heater does not affect the turbine performance. The mass flow rate through the condenser is

$$\dot{m}_5 = (1 - y)\dot{m}_3 = 0.837(3.41 \times 10^5) = 2.85 \times 10^5 \text{ kg/h}$$

The mass flow rate through the open heater $\dot{m}_4$ is the difference between $\dot{m}_3$ and $\dot{m}_5$, or $0.56 \times 10^5$ kg/h.

**Comment:**   The percent of the flow extracted from the turbine to the closed heater is the same as that for an open heater in Example 16-6. In most cases these values are slightly different.

### 16-3-3   OPTIMUM EXTRACTION PRESSURE

When one considers the open-heater regenerative cycle shown in Fig. 16-12, the question arises as to the appropriate extraction pressure $P_4$ to be used. The extraction pressure must lie between the boiler pressure $P_3$ and the condenser pressure $P_5$. If the steam is extracted at the limiting conditions of either before the turbine inlet at $P_3$ or after the turbine at $P_5$, then the thermal efficiency is unaffected by the presence of the heater. Since regeneration does increase the thermal efficiency, there must exist an optimum pressure which will maximize the thermal efficiency for given values of $P_3$, $T_3$, and $P_5$. For the case of one open heater it can be shown that the feedwater exit temperature at state 1 should be close to halfway between the saturation temperatures in the boiler and the condenser.

**Example 16-8**

**R**econsider the ideal regenerative cycle with an open feedwater heater in Example 16-6. Estimate the optimum extraction pressure for the cycle.

**Solution:**

**Given:**   An ideal regenerative cycle with an inlet turbine state of 30 bars and 500°C and a condenser pressure of 0.1 bar.

**Find:**   Optimum extraction pressure.

**Model:**   Approximation rule for best extraction pressure.

**Analysis:** The optimum extraction pressure is based on the temperature which is halfway between the boiler and condenser saturation temperatures. At 30 and 0.1 bar the saturation temperatures are 234 and 46°C, respectively. The average of these values is 140°C. The saturation pressure corresponding to this temperature is 3.6 bars, and this pressure would be a good estimate of the extraction pressure to use to optimize the thermal efficiency.

### 16-3-4 THE SUPERCRITICAL CYCLE

In the preceding discussions of the vapor power cycle, the final pressure at the outlet of the pump preceding the heat-addition section of the cycle was always chosen to be below the critical pressure of the fluid. That is, the maximum pressure was less than 221 bars, or 3200 psia, when the working fluid was water. A number of modern steam power plants operate on a cycle for which the turbine-inlet pressure is supercritical. A typical value might be 250 to 325 bars, or 3500 to 5000 psia.

A Rankine cycle with a supercritical turbine-inlet pressure is shown in Fig. 16-16a. Note that during the heat-addition process (4-1) a phase change does not occur. The pressurized fluid enters the heat-exchanger tubes in the furnace and gradually expands in volume with no phase change occurring as it passes through the tubes. Inlet-turbine temperatures may approach 620°C or 1150°F. For large power units working under these conditions, two reheat sections are often employed.

One of the major thermodynamic considerations in the analysis of power cycles is the average temperature level of the fluid during heat addition with reheat. For comparative purposes a subcritical cycle is also sketched on Fig. 16-16a as a dashed line. During the heat-transfer process 4'-1' for the subcritical cycle, a considerable portion of the heating occurs at a constant temperature which is less than the critical temperature of the fluid. For the supercritical cycle, the temperature continually increases

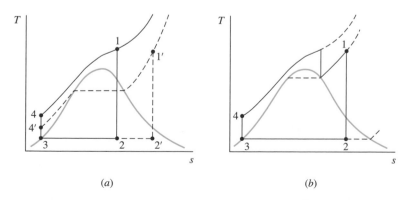

(a)                                                            (b)

**Figure 16-16**     Supercritical Rankine cycle (a) with heat rejection in the wet region and (b) with reheat.

during the heat-transfer process 4-1. For the same limiting temperature at the turbine inlet for both types of cycles, the average temperature level during heat addition for the supercritical cycle will be greater. Theoretically, this should lead to a higher thermal efficiency, all other factors being the same. The actual thermal efficiency of supercritical units is around 40 percent. Another possibility is shown in Fig. 16-16b. In this figure the upper pressure is supercritical, and two stages of expansion with reheat have been added.

### 16-3-5   MULTIPLE USE OF FEEDWATER HEATERS

Modern high-pressure steam power plants usually employ several open and closed feedwater heaters in a given cycle. The maximum number is around six to eight. Although the thermal efficiency does increase as the number of heaters is increased, so does the capital cost. Therefore, beyond a certain point the use of additional heaters cannot be economically justified.

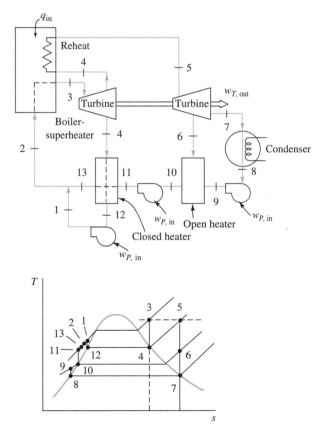

**Figure 16-17**   Schematic of equipment and *Ts* diagram for a regenerative-reheat cycle with one open and one closed feedwater heater.

Methods of optimizing a regenerative cycle with two or more heaters with respect to using the proper extraction pressures are well known but beyond the scope of this text. The examples below illustrate a cycle with one closed and one open heater. The method of analyzing a closed heater is examined in detail. A schematic of the equipment and a $Ts$ diagram for the cycle are shown in Fig. 16-17. Note that the process also includes reheating after the first extraction point. Recall that reheating helps prevent excessive moisture content in the turbine exhaust stream.

**Example 16-9**

The steam in an ideal regenerative power cycle enters the turbine at 30 bars and 500°C and exhausts at 0.1 bar. Steam is bled from the turbine at 10 bars for a closed feedwater heater and at 5 bars for an open heater. Condensate from the closed heater is pumped to 30 bars and joins the feedwater stream after it leaves the closed heater. Find the thermal efficiency of the cycle if the stream is reheated to 500°C at the 10-bar extraction point.

**Solution:**

**Given:** The ideal regenerative steam power cycle shown in Fig. 16-17 has the following properties: $P_3 = 30$ bars, $T_3 = 500°C$, $P_7 = 0.1$ bar, $P_4 = 10$ bars, $P_6 = 5$ bars, $P_1 = P_{11} = 30$ bars, $T_5 = 500°C$.

**Find:** $\eta_{th}$.

**Model:** Steady state, internally reversible processes; negligible ke and pe effects.

**Analysis:** The following data are still valid from Example 16-6, except the numbering system has been altered. Based on the notation of Fig. 16-17,

$$P_3 = 30 \text{ bars} \qquad T_3 = 500°C \qquad h_3 = 3456.5 \text{ kJ/kg}$$

$$P_8 = 0.1 \text{ bar, saturated liquid} \qquad h_8 = 191.8 \text{ kJ/kg}$$

$$P_{10} = 5 \text{ bars, saturated liquid} \qquad h_{10} = 640.2 \text{ kJ/kg}$$

8–9:  $w_{P,in} = v_8(P_9 - P_8) = 1.01 \times 10^{-3}(5 - 0.1)(10^2) = 0.5 \text{ kJ/kg}$

$$h_9 = h_8 + w_{P,in} = 191.8 + 0.5 = 192.3 \text{ kJ/kg}$$

10–11:  $w_{P,in} = v_{10}(P_{11} - P_{10}) = 1.09 \times 10^{-3}(30 - 5)(10^2) = 2.7 \text{ kJ/kg}$

$$h_{11} = h_{10} + w_{P,in} = 640.2 + 2.7 = 642.9 \text{ kJ/kg}$$

In addition, the enthalpy values for states 4, 5, 6, 12, 13, and 1 may be found from tables or by direct calculations:

$$P_4 = 10 \text{ bars} \qquad s_4 = s_3 = 7.2338 \text{ kJ/kg·K} \qquad h_4 = 3116.9 \text{ kJ/kg}$$

$$P_5 = 10 \text{ bars} \qquad T_5 = 500°C \qquad h_5 = 3478.5 \text{ kJ/kg}$$

$$P_6 = 5 \text{ bars} \qquad s_6 = s_5 = 7.7622 \text{ kJ/kg·K} \qquad h_6 = 3251.3 \text{ kJ/kg}$$

$$P_7 = 0.1 \text{ bar} \qquad s_7 = s_5 \qquad h_7 = 2460.9 \text{ kJ/kg}$$

$$P_{12} = 10 \text{ bars, saturated liquid} \qquad h_{12} = 762.8 \text{ kJ/kg}$$

$P_{13} = 30$ bars    $T_{13} = T_{12} = 179.9°C$, compressed liquid    $h_{13} = 765.0$ kJ/kg

12–1:    $w_{P,\text{in}} = v_{12}(P_1 - P_{12}) = 1.13 \times 10^{-3}(30 - 10)(10^2) = 2.3$ kJ/kg

$h_1 = h_{12} + w_{P,\text{in}} = 762.8 + 2.3 = 765.1$ kJ/kg

At this point every enthalpy value is known except that for state 2. Generally this is found from energy and mass balances around the point where streams 1 and 13 enter and stream 2 leaves. That is, in the absence of heat and work effects and negligible kinetic- and potential-energy changes, $\sum \dot{m}_i h_i = \sum \dot{m}_e h_e$ and $\sum \dot{m}_i = \sum \dot{m}_e$, or

$$\dot{m}_1 h_1 + \dot{m}_{13} h_{13} = \dot{m}_2 h_2 \qquad \text{and} \qquad \dot{m}_1 + \dot{m}_{13} = \dot{m}_2$$

However, the split in total flow between streams 1 and 13 has not yet been determined. Once this split is found, state 2 may be evaluated. In this particular case a numerical solution is available without determining the fraction bled at state 4. This occurs because the enthalpies at states 1 and 13 are essentially identical, that is, 765.0 and 765.1 kJ/kg. Thus regardless of the split in flow, $h_2$ after mixing will also be 765.0 kJ/kg.

The fractional flow extracted to the two heaters is important, however, for the evaluation of the turbine and pump work and the heat input. The fraction of flow through the boiler-superheater section that is extracted at state 4 is found from energy and mass balances around the closed heater. In the absence of heat and work effects and with negligible kinetic- and potential-energy changes, the steady-state energy equation reduces to $\sum \dot{m}_i h_i = \sum \dot{m}_e h_e$ and $\sum \dot{m}_i = \sum \dot{m}_e$. For the closed heater:

$$\dot{m}_1(h_4 - h_{12}) = \dot{m}_{13}(h_{13} - h_{11}) \qquad \text{and} \qquad \dot{m}_1 + \dot{m}_{13} = \dot{m}_3$$

If we let $\dot{m}_4/\dot{m}_3 = \dot{m}_1/\dot{m}_3 = y_4$ and $\dot{m}_{13}/\dot{m}_3 = 1 - y_4$, then

$$y_4(h_4 - h_{12}) = (1 - y_4)(h_{13} - h_{11})$$

$$y_4(3116.9 - 762.8) = (1 - y_4)(765.0 - 642.9)$$

$$y_4 = 0.0493$$

Hence 4.93 percent of the total flow around the cycle is bled to the closed heater. An energy balance and a mass balance around the open heater, similar to the balances for the closed heater, will lead to the fraction of the flow bled to the open heater:

$$\dot{m}_6 h_6 + \dot{m}_9 h_9 = \dot{m}_{10} h_{10} \qquad \text{and} \qquad \dot{m}_6 + \dot{m}_9 = \dot{m}_{10}$$

If we let $\dot{m}_6/\dot{m}_{10} = z_6$ and $\dot{m}_9/\dot{m}_{10} = 1 - z_6$, then

$$z_6(3251.3) + (1 - z_6)(192.3) = 1(640.2)$$

$$z_6 - 0.1464$$

Thus 14.64 percent of the flow reheated and passing through the second stage of the turbine is extracted to the open heater. Note that the symbol $z_6$ has been used rather than $y_6$, because we wish to let $y_6$ represent the fraction of the *total flow* bled to the open heater. It can be easily shown that $y_6 = 0.1392$ on the basis of the values of $y_4$ and $z_6$.

Finally we may evaluate the turbine and pump work and the heat input:

$$w_{T,\text{out}} = 1(h_3 - h_4) + (1 - y_4)(h_5 - h_6) + (1 - y_4 - y_6)(h_6 - h_7)$$

$$= 3456.5 - 3116.9 + 0.9507(3478.5 - 3251.3) + 0.8115(3251.3 - 2460.9)$$

$$= 339.6 + 216.0 + 641.4 = 1197.0 \text{ kJ/kg}$$

$$w_{P,\text{in}} = w_{P,8-9}(1 - y_4 - y_6) + w_{P,10-11}(1 - y_4) + w_{P,12-1}y_4$$
$$= 0.5(0.8115) + 2.7(0.9507) + 2.3(0.0493) = 3.1 \text{ kJ/kg}$$

$$q_{\text{in}} = 1(h_3 - h_2) + (1 - y_4)(h_5 - h_4)$$
$$= 3456.5 - 765.0 + 0.9507(3478.5 - 3116.9) = 3035.3 \text{ kJ/kg}$$

As a result of the above calculations, the thermal efficiency becomes

$$\eta_{\text{th}} = \frac{w_{T,\text{out}} - w_{P,\text{in}}}{q_{\text{in}}} = \frac{1197.0 - 3.1}{3035.3} = 0.393 \text{ (or } 39.3\%)$$

**Comment:** In Example 16-6 the efficiency was 37.5 percent for a single open heater without reheating. The increase to 39.3 percent in this example is due both to a second heater in the cycle and to the reheat section.

## 16-4 COGENERATION SYSTEMS

The various descriptions of vapor cycles in the preceding sections deal with systems for which the sole purpose is power production. It must be recognized that there are industrial and commercial situations where thermal energy is also required. For example, a power plant on a college campus may provide steam for heating buildings as well as for electric power. Consider also the following five major energy-intensive industries: chemicals, oil refining, steel making, food processing, and pulp and paper production. Large plants in these basic industries require steam for the operation of various processes in addition to their electrical needs. Although the multiple use of steam generation has been in practice for a number of decades, there continues to be an increased interest in such systems. This is due to potential cost savings and more effective use of our energy resources.

For a large power plant which produces only electricity, the thermal efficiency typically ranges from 0.30 to 0.40, or 30 to 40 percent. For energy conservation, it is important that a larger fraction of the energy from basic sources be utilized. One method of achieving this is to integrate the use of steam for heating or industrial purposes with the general production of electric power. This technique of sequential production of energy (usually electric and thermal) from a single energy source is called **cogeneration.** The performance of a cogeneration system is frequently measured in terms of its effectiveness $\varepsilon$, which is defined as

$$\varepsilon = \frac{\text{electric energy delivered} + \text{thermal energy delivered}}{\text{combustion energy input}} \qquad \textbf{[16-12]}$$

Compared to the thermal-efficiency range quoted earlier, with cogeneration it is estimated that an effectiveness as high as 55 to 70 percent is possible. A schematic of one possible cogeneration system, as might be used on a college campus, is shown in Fig. 16-18. Steam is bled from an intermediate point in the turbine to provide energy for the thermal load. In practice the

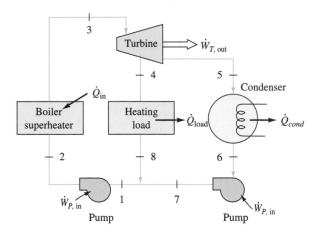

**Figure 16-18**   Schematic of a cogeneration plant employing steam bled from the turbine for the thermal load.

heating load might be integrated into a cycle which includes reheat and regeneration.

Note that the thermal energy from a steam power plant can be delivered two ways—as steam bled from the turbine, as noted above, or as waste heat removed from the turbine exhaust stream. In this latter case the energy removed from the exhaust stream is used to heat water, which is circulated separately. The technique used depends on the end use of the thermal energy. Process heat supplied by steam in the temperature range of 150 to 200°C (300 to 400°F) is a common industrial need. Extraction of steam commonly occurs at pressures of 5 to 7 bars (75 to 100 psia). For conventional turbine inlet and exit pressures the steam will have produced over 60 percent of the total work possible before extraction occurs. The steam that is not extracted, of course, continues to produce work as it expands to the condenser pressure. Although the total work output is decreased for a given mass flow rate, a larger fraction of the total heat input is used for a useful purpose when cogeneration of this type is employed. Hot water for residential heating obtained by removing heat from the turbine exhaust stream would be supplied at temperatures considerably below that for extracted steam.

Cogeneration is especially attractive when a power plant can be integrated into a community so that it provides also for residential and commercial heating. The distribution of thermal energy from a central source to industrial, commercial, and residential customers for space heating, domestic hot water, and process needs is called *district heating*. Although cogeneration is an attractive approach to energy conservation, a plant operating in this mode must be carefully designed. Such a plant must be able to follow the power and heating loads on a local basis and possibly integrate its electrical load with the large electrical utilities which produce the base electrical needs of a state or region. This situation is fairly complex and requires careful evaluation.

Example 16-10

**Steam** expands through a turbine in a cogeneration power cycle from 30 bars and 500°C to 0.1 bar. The turbine has an adiabatic efficiency of 82 percent and delivers 40 MW of power. Steam is extracted at 5 bars for a heating load of $3 \times 10^6$ kJ/min. The steam leaves the heating load as a saturated liquid at 3 bars and mixes with the steam leaving the condensate pump at 3 bars. The pump efficiencies are both 78 percent. Determine (a) the mass flow rates, in kg/min, through the heating load and through the condenser, (b) the rate of heat input, in kJ/min, and (c) the effectiveness of the cycle.

**Solution:**

**Given:** A cogeneration power cycle as shown in Fig. 16-19.

**Find:** (a) $\dot{m}_{\text{heating}}$ and $\dot{m}_{\text{cond}}$, (b) $\dot{Q}_{\text{in}}$, in kJ/min, and (c) $\varepsilon_{\text{cycle}}$.

**Model:** Steady state; irreversible, adiabatic turbine and pump.

**Analysis:** A number of the states shown for the system in Fig. 16-19 have been used in earlier examples. On this basis, data around the cycle are as follows:

$h_3 = 3456.3$ kJ/kg, $s_3 = 7.2338$ kJ/kg·K, $T_{4s} = 240.8°C$, $h_{4s} = 2941.6$ kJ/kg

$h_{4a} = h_3 - \eta_T(h_3 - h_{4s}) = 3456.3 - 0.82(3456.3 - 2941.6) = 3034.2$ kJ/kg

$$x_{5s} = \frac{s_3 - s_{f,5}}{s_{g,5} - s_{f,5}} = \frac{7.2338 - 0.6493}{8.1502 - 0.6493} = 0.878$$

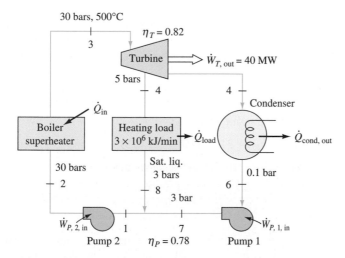

**Figure 16-19** Equipment schematic and data for Example 16-10.

$$h_{5s} = h_f + xh_{fg} = 191.83 + 0.878(2392.8) = 2292.3 \text{ kJ/kg}$$

$$h_{5a} = h_3 - \eta_T(h_3 - h_{5s}) = 3456.3 - 0.82(3456.3 - 2292.3) = 2501.8 \text{ kJ/kg}$$

$$h_6 = h_f \text{ at } 0.1 \text{ bar} = 191.8 \text{ kJ/kg}, \quad h_8 = h_f \text{ at } 3 \text{ bars} = 561.5 \text{ kJ/kg}$$

$$w_{P,1,\text{in}} = v_{f,6}(P_7 - P_6)/\eta_P = 1.01 \times 10^{-3}(2.9)(100)/0.78 = 0.4 \text{ kJ/kg}$$

$$h_7 = h_6 + w_{P,1a} = 191.8 + 0.4 = 192.2 \text{ kJ/kg}$$

(a) The energy removed for heating is $q_{\text{load}} = h_{4a} - h_8$. Hence the mass flow rate through the heating load is

$$\dot{m}_{\text{load}} = \frac{\dot{Q}_{\text{load}}}{q_{\text{load}}} = \frac{3 \times 10^6}{3034.2 - 561.2} = 1213 \text{ kg/min}$$

The total mass flow rate $\dot{m}_3$ must be found first before the rate through the condenser is determined. An energy balance on the turbine is given by

$$\dot{W}_{T,\text{out}} = \dot{m}_3(h_3 - h_{4a}) + (\dot{m}_3 - \dot{m}_{\text{load}})(h_{4a} - h_{5a})$$

Substitution of data leads to

$$40{,}000(60) = \dot{m}_3(3456.3 - 3034.2) + (\dot{m}_3 - 1213)(3034.2 - 2501.8)$$

or
$$\dot{m}_3 = \frac{2.4 \times 10^6 + 1213(532.4)}{422.1 + 532.4} = 3191 \text{ kg/min}$$

Therefore, the mass flow rate through the condenser is

$$\dot{m}_{\text{cond}} = \dot{m}_3 - \dot{m}_{\text{load}} = 3191 - 1213 = 1978 \text{ kg/min}$$

(b) The evaluation of the rate of heat input to the cycle requires information on $h_2$. First we need to find $h_1$. For the adiabatic mixing of streams 7 and 8, the energy balance reduces to

$$\dot{m}_3 h_1 = \dot{m}_8 h_8 + \dot{m}_7 h_7$$

Thus,

$$h_1 = \frac{\dot{m}_8 h_8 + \dot{m}_7 h_7}{\dot{m}_3} = \frac{1213(561.6) + 1978(192.2)}{3191} = 332.6 \text{ kJ/kg}$$

and $\quad w_{P,2,\text{in}} = v_{f,1}(P_2 - P_1)/\eta_P = 1.073 \times 10^{-3}(27)(10^2)/0.78 = 3.7 \text{ kJ/kg}$

As a result,

$$h_2 = h_1 + w_{P,2,\text{in}} = 332.6 + 3.7 = 336.3 \text{ kJ/kg}$$

Finally, the rate of heat transfer addition is

$$\dot{Q}_{\text{in}} = \dot{m}_3(h_3 - h_2) = 3191(3456.3 - 336.8) = 9.956 \times 10^6 \text{ kJ/min}$$

(c) On the basis of Eq. [16-12] and neglecting pump work,

$$\varepsilon = \frac{\dot{W}_{T,\text{out}} + \dot{Q}_{\text{load}}}{\dot{Q}_{\text{in}}} = \frac{2.4 \times 10^6 + 3 \times 10^6}{9.956 \times 10^6} = 0.542$$

**Comment:**  Based on the turbine output alone, the energy-conversion efficiency is only 24 percent. However, 54 percent of the energy input appears as a *useful* output.

## 16-5   THE COMBINED CYCLE

The thermal efficiency of the gas-turbine power cycle discussed in Chap. 15 and of the vapor power cycles discussed in the preceding sections of this chapter is typically less than 40 percent. Although techniques such as reheating and regeneration do improve the cycle performance, the rejected or waste energy is still a large fraction of the energy input. One possible way of achieving further improvement is by an arrangement called a combined or coupled cycle. A ***combined power cycle*** is one based on the coupling of two different power cycles such that the waste (rejected) heat from one cycle is used partially or totally as the heat source for the other cycle. This basic concept is also used in binary vapor cycles, which are discussed in Sec. 16-6. One important combined cycle involves the use of a gas-turbine (Brayton) topping cycle with a steam-turbine (Rankine) cycle. A *topping cycle* is the one which operates with the temperature level of its waste heat higher than that of the maximum temperature of the second cycle.

In a gas-turbine cycle, the exhaust stream leaving the turbine is relatively hot. As discussed in Sec. 16-8, one method of utilizing this high-temperature energy uses the concept of regeneration. A portion of the energy is returned to the air as it passes from the compressor outlet to the combustor inlet. An alternative possibility is shown in Fig. 16-20. In place of regeneration, the gas-turbine exhaust stream is used as the energy source

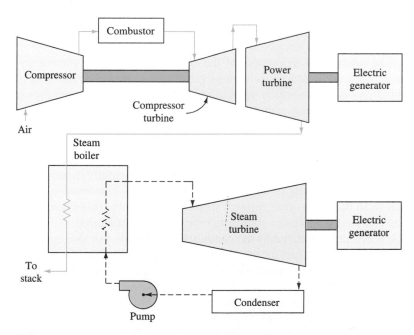

**Figure 16-20**     Schematic of a combined cycle composed of a basic Rankine steam cycle and a gas-turbine topping cycle.

in the boiler of a conventional steam power cycle. Although this is not illustrated, the steam cycle would probably employ feedwater heaters. However, reheating in the turbine section of the steam cycle would not normally be possible. The energy in the hot gas-turbine exhaust is needed primarily for the boiling and superheating of the steam. The thermodynamic analysis of this combined cycle would follow procedures established in the sections on the Brayton and Rankine cycles.

The thermal efficiency $\eta_{th,comb}$ of the combined cycle shown in Fig. 16-20 is the ratio of the sum of the two power outputs to the rate of heat input to the gas-turbine cycle. That is,

$$\eta_{th,comb} = \frac{\dot{W}_{gas,out} + \dot{W}_{steam,out}}{\dot{Q}_{gas,in}} = \frac{\dot{m}_{gas} w_{gas,out} + \dot{m}_{steam} w_{steam,out}}{\dot{m}_{gas} q_{gas,in}} \qquad \textbf{[16-13]}$$

Note that the mass flow rates for the two cycles must be included, because there is not a common fluid passing through the equipment. These two mass flow rates are not independent but must satisfy the energy balance for the heat exchanger labeled as a steam boiler in Fig. 16-20. In the absence of heat and work interactions and for negligible kinetic- and potential-energy effects, the heat-exchanger energy balance is $\sum \dot{m}_i h_i = \sum \dot{m}_e h_e$. In terms of Fig. 16-20, this becomes

$$0 = \dot{m}_{gas}(h_{in} - h_{out})_{gas} + \dot{m}_{steam}(h_{in} - h_{out})_{steam} \qquad \textbf{[16-14]}$$

The waste energy in the gas-turbine exhaust cannot be completely used to heat the steam, because a finite temperature must exist between the flow streams for an adequate rate of heat transfer. Once the temperatures at the inlets and exits of the steam boiler are set, the mass flow rate ratio $\dot{m}_{gas}/\dot{m}_{steam}$ is fixed.

The practical development of the combined gas-steam-turbine cycle was delayed until modern technology provided the means of building gas-turbine power plants which operate at relatively high pressure ratios and with turbine-inlet temperatures that exceed 1350 K (2400°R). Pressure ratios of 10:1 to 13:1 are typical for this temperature. When turbine-inlet temperatures around 1600 K (2900°R) are possible on commercially available units, pressure ratios in the vicinity of 20:1 may be used. A limited number of commercial gas-steam-turbine power plants are currently in use. As gas-turbine technology improves, the advantages of this combined cycle will improve in relation to conventional steam power plants. This cycle appears to be extremely useful in conjunction with coal gasification. Not only is cycle efficiency improved by the combined cycle itself, but also coal gasification as a source of fuel for the gas-turbine combustor offers the additional advantage of removing potential air pollutants before they enter the combustion process.

There are some power cycles, such as a diesel engine, for which the temperature of rejection may be a few hundred degrees above that of the environment. In such cases another power cycle could operate between the low exhaust temperature of the cycle and the atmospheric temperature. These low-temperature power cycles used in conjunction with another power cycle are called *bottoming cycles*. Only a very few fluids have the reasonably high critical temperature and large enthalpy of vaporization required to be a successful working medium for a bottoming cycle. Fluorocarbons are only fair in meeting these constraints and usually are restricted in their use to temperatures below 150°C (300°F) or slightly higher. Some relatively low molecular weight hydrocarbons, such as isobutane and ammonia, have potential as working fluids at low temperatures.

**Example 16-11**

The gas-turbine cycle in a combined gas-steam cycle has a pressure ratio of 6, a compressor inlet temperature of 295 K, and a turbine inlet temperature of 1200 K. The compressor and turbine adiabatic efficiencies are 82 and 85 percent, respectively. The bottoming cycle is a simple Rankine cycle for which the turbine-inlet conditions are 30 bars and 500°C, and the condenser pressure is 0.1 bar. The adiabatic efficiencies of the steam turbine and pump are 82 and 78 percent, respectively. The exhaust gases leave the heat exchanger at 440 K. Determine (*a*) the mass-flow-rate ratio of air to steam, (*b*) the overall thermal efficiency, and (*c*) the entropy generation in the air compressor, air turbine, steam turbine, water pump, and air-steam heat exchanger, all in kJ/K per kilogram of air.

**Solution:**

**Given:** A combined air-steam cycle is shown in Fig. 16-21.

**Find:** (*a*) $\dot{m}_{air}/\dot{m}_{steam}$, (*b*) $\eta_{th,comb}$, and (*c*) $\sigma_m$ for various components per kilogram of air.

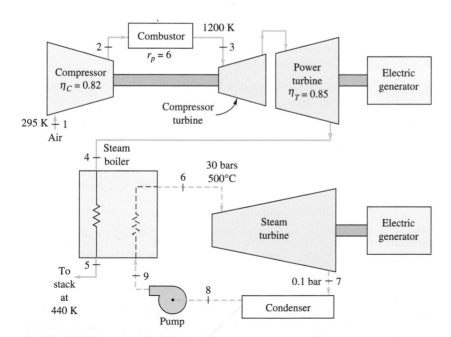

**Figure 16-21** Schematic and data for Example 16-11.

**Model:**    Steady state; adiabatic, irreversible turbines, compressor, and pump; $\Delta \text{ke} = \Delta \text{pe} = 0$.

**Analysis:**    The gas-turbine cycle is similar to one analyzed in Example 15-7 except that the turbine inlet temperature has increased to 1200 K from 1100 K. The steam cycle alone was analyzed in Example 16-4. The following data using the notation in Fig. 16-21 are still valid:

*Steam cycle:*

$$h_6 = 3456.5 \text{ kJ/kg}, \ s_6 = 7.2338 \text{ kJ/kg·K},$$

$$h_9 = 195.6 \text{ kJ/kg}, \ s_9 = 0.6516 \text{ kJ/kg·K}, \ T_9 = 46°C,$$

$$\sigma_{m,T} = 0.6614 \text{ kJ/(kg steam)(K)}, \ \sigma_{m,P} = 0.0023 \text{ kJ/(kg steam)(K)}$$

$$w_{\text{net,out}} = 950.5 \text{ kJ/kg}, \ \eta_{th} = 29.1\%$$

*Gas cycle:*

$$h_1 = 295.2 \text{ kJ/kg}, \ h_2 = 536.1 \text{ kJ/kg}$$

$$w_C = 240.9 \text{ kJ/kg}, \ \sigma_{m,C} = 0.0847 \text{ kJ/(kg air)(K)}$$

To finish the analysis of the gas cycle and solve for state 4 we must examine the performance of the combustor and the turbine in the gas cycle. Knowing state 3, the turbine inlet conditions, we find that $h_3 = 1277.79 \text{ kJ/kg}$ and $s° = 3.17888 \text{ kJ/kg·K}$. Solving first for the heat transfer into the combustor: $q_{\text{comb}} = h_3 - h_2 = 741.7 \text{ kJ/kg}$. Next for the turbine work, we find that for an isentropic expansion of an ideal gas through the turbine $0 = s_{4s}° - s_3° - R\ln(P_4/P_3)$. Solving for the isentropic exit state gives

$$s_{4s}° = s_3° + R\ln\left(\frac{P_4}{P_3}\right) = 3.17888 + \frac{8.314}{29}\ln\left(\frac{1}{6}\right) = 2.6652 \text{ kJ/kg·K}$$

And interpolating in Table A-5, we find that $T_{4s} = 762.4 \text{ K}$ and $h_{4s} = 780.8 \text{ kJ/kg}$. An energy balance on the turbine gives the adiabatic turbine work out as $w_{T,\text{out}} = h_3 - h_4$. Using the turbine efficiency, we have

$$w_{T,a,\text{out}} = \eta_T w_{T,s,\text{out}} = \eta_T(h_3 - h_{4s}) = 0.85(1277.79 - 780.8) = 422.4 \text{ kJ/kg}$$

$$h_{4a} = h_3 - w_{T,a,\text{out}} = 1277.79 - 422.4 = 855.4 \text{ kJ/kg}$$

Again from Table A-5, we find that $T_{5a} = 830.3 \text{ K}$. To find the entropy production within the irreversible turbine, we apply the entropy balance to give $\sigma_{m,T} = s_4 - s_3$. Now for an ideal gas

$$\sigma_{m,T} = s_{4a}° - s_3° - R\ln\left(\frac{P_4}{P_3}\right) = 2.75882 - 3.17888 - \frac{8.314}{29}\ln\left(\frac{1}{6}\right)$$

$$= 0.09362 \text{ kJ/(kg air)(K)}$$

The net work output from the gas cycle and the cycle efficiency are calculated as

$$w_{\text{net,out}} = w_{T,\text{out}} - w_{C,\text{in}} = 422.4 - 240.9 = 181.5 \text{ kJ/kg}$$

$$\eta_{th} = \frac{w_{\text{net,out}}}{q_{\text{comb}}} = \frac{181.5}{741.7} = 24.5\%$$

In addition, $h_5 = 441.6 \text{ kJ/kg}$, and $s_5° = 2.08870 \text{ kJ/kg·K}$ at 440 K.

(a) The ratio of mass flow rates is found from an energy balance on the air-steam heat exchanger, Eq. [16-14],

$$\frac{\dot{m}_{air}}{\dot{m}_{steam}} = \frac{h_6 - h_9}{h_4 - h_5} = \frac{3456.5 - 195.6}{855.4 - 441.6} = 7.88$$

Thus it takes 7.88 kg of air to heat 1 kg of steam.

(b) The overall thermal efficiency is given by Eq. [16-13]. Dividing the numerator and denominator by $\dot{m}_{steam}$ and substituting values, we find that

$$\eta_{th,comb} = \frac{\dot{m}_{gas}w_{gas,out} + \dot{m}_{steam}w_{steam,out}}{\dot{m}_{gas}q_{gas,in}} = \frac{7.88(181.5) + 950.5}{7.88(741.7)} = 0.407$$

(c) The entropy generation values found for the gas-turbine cycle are

$$\sigma_{m,C} = 0.0847 \text{ kJ/(kg air)(K)} \qquad \sigma_{m,T} = 0.0936 \text{ kJ/(kg air)(K)}$$

The entropy-generation values found in Example 16-4 for the steam power cycle are reported in units of kJ/K per kilogram of steam. Using the mass flow rate ratio found in part a to convert the units to kJ/K per kilogram of air,

$$\sigma_{m,T} = \frac{1 \text{ kg steam}}{7.88 \text{ kg air}} \frac{0.6614 \text{ kJ}}{(kg \text{ steam})(K)} = 0.0839 \text{ kJ/(kg air)(K)}$$

$$\sigma_{m,P} = \frac{1 \text{ kg steam}}{7.88 \text{ kg air}} \frac{0.0023 \text{ kJ}}{(kg \text{ steam})(K)} = 0.0003 \text{ kJ/(kg air)(K)}$$

Finally, the entropy generation within the adiabatic heat exchanger is found from

$$\dot{\sigma}_{cv} = \sum_{out} \dot{m}_e s_e - \sum_{in} \dot{m}_i s_i$$

On the basis of a unit mass of air, the above equation becomes

$$\begin{aligned}
\sigma_{m,exch} &= (\dot{m}_{steam}/\dot{m}_{air})(s_6 - s_9) + (s_5 - s_4) \\
&= (1/7.88)(7.2338 - 0.6516) + (2.08870 - 2.75882) \\
&= 0.8353 + (-0.6701) = 0.1652 \text{ kJ/(kg air)(K)}
\end{aligned}$$

where $s_5 - s_4 = s_5^0 - s_4^0$ because $P_5 = P_4$. This latter value is the largest of the five answers and is due to irreversible heat transfer within the heat exchanger. The pump value is the only insignificant contribution.

**Comment:** The overall thermal efficiency of 40.7 percent is considerably higher than the values of 24.5 and 29.1 percent when the gas and steam cycles, respectively, operate alone. Keep in mind that the entropy-generation calculations did not include pressure drops within the heat exchanger or in piping connecting the pieces of equipment.

## 16-6 APPLICATIONS OF VAPOR CYCLES

This section introduces three applications of vapor power cycles of interest.

### 16-6-1 BINARY VAPOR CYCLES

On the basis of the Carnot efficiency equation for the conversion of thermal to mechanical energy, it is apparent that heat addition to the working fluid should occur at the highest possible temperature. However, the use of higher temperatures is restricted by the availability of materials which can withstand such conditions. Materials technology has limited the steam power cycle to around 560°C (1050°F) and the industrial gas-turbine cycle to around 1000 to 1200°C (1800 to 2200°F). As technology develops materials to withstand higher temperatures, it may become possible to operate a Rankine-type cycle at temperatures now reserved for the gas-turbine cycle. In such a case it would be highly desirable to use some other fluid than water. As noted in the discussion in Sec. 16-2-2, the great disadvantage of water is its critical temperature. When a boiler is used, the heat is added to the fluid at a constant and relatively low temperature. This low-temperature heat addition could be partially overcome by employing a substance with a much higher critical temperature.

The properties of alkali metals are fairly suitable for high-temperature Rankine cycles. Most promising are potassium, sodium, and mixtures of these two elements. In addition to the appropriateness of their thermodynamic properties, such as vapor pressure and heat capacity, these substances have reasonably high heat-transfer coefficients. As a result, heat-exchanger size is significantly reduced. Such a cycle has been proposed for the generation of power aboard a spacecraft.

The maximum temperature for a Rankine cycle operating with potassium, for example, may be as high as 1200°C (2200°F). The degree of superheating at the turbine inlet typically might be 30 to 80°C (50 to 150°F). The vapor pressure of potassium at 1150°C is roughly 13.5 bars, which is equivalent to 200 psia at 2100°F. The condensing temperature of potassium may be as low as 600°C, with a vapor pressure of 0.17 bar. This is equivalent to 1100°F, with a vapor pressure of 2.5 psia. These temperature limits give a turbine expansion ratio of 80, which is relatively high when only two stages of expansion are used. Hence the expansion ratio places a practical lower limit on the cycle-temperature ratio (minimum to maximum temperature). Typical space power Rankine cycles are constrained to operate with cycle-temperature ratios on the order of 0.65 to 0.75. Consequently the cycle efficiency is relatively low, probably 25 percent or less.

The low cycle-temperature ratio does have one advantage in terms of space applications. Since the condensing temperature is high, the temperature of the fluid in the heat-rejection loop is also high. A high radiator rejection temperature leads to a small radiator size, since the radiant heat transfer per unit area varies as the fourth power of the temperature. Tabular data for potassium in the saturation and superheat regions are presented in Tables A-21 and A-21E.

The use of alkali metals in a high-temperature Rankine cycle for space applications has led to a renewed interest in binary vapor power cycles. A *binary cycle* is one in which the heat removed during the heat-rejection process of one power cycle is used as the heat input for another power cycle. It was noted above that the condensing temperature of a potassium cycle may be around 600°C (1100°F). Heat removed at this temperature may be used to supply a Rankine cycle operating on steam and rejecting at atmospheric

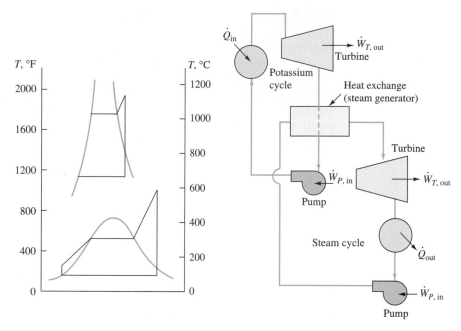

**Figure 16-22**    Equipment schematic and *Ts* diagram for a binary vapor power cycle involving steam and potassium.

temperature. Because one cycle operates at temperatures above that in the other cycle, the high-temperature cycle is frequently called a *topping cycle*. An equipment schematic and a *Ts* diagram for a binary cycle using potassium and water are shown in Fig. 16-22. Only a portion of the saturation curve for potassium is shown on the *Ts* diagram, and the processes are idealized. In fact, the pump work is neglected on the potassium cycle. In practice, irreversible performance must be considered as well as the possible presence of a reheat section and feedwater heaters for the steam cycle. Also heat addition in the steam cycle might be supercritical. Note from Fig. 16-22 that a finite difference must exist between the potassium condensing temperature and the boiler-superheat temperatures of the steam cycle.

For reasonable temperature values and turbine and pump efficiencies the predicted thermal efficiency of a potassium cycle runs around 20 to 30 percent. When a potassium cycle is used as a topping cycle for a steam cycle, the thermal efficiency of the overall binary cycle may approach 50 to 60 percent. A disadvantage of the binary cycle is a high capital cost. However, this becomes less important as fuel costs rise.

## 16-6-2   GEOTHERMAL ENERGY CONVERSION

Geothermal energy is the natural energy stored in the earth at depths close enough to the surface to be tapped and used either for electric generation or for thermal energy. This type of energy is stored in the form of dry steam, hot water, hot water with dissolved methane gas, and dry hot rock. The first use of geothermal energy in the United States began in 1960, when dry steam from The Geysers, an area north of San Francisco, was used to produce 11 MW of commercial electric power. The steam conditions at The Geysers are around 0.76 MPa (110 psi) and 180°C (350°F). By 1980, geothermal plants in this area produced about 800 MW of power, and by the end of the 1980s this capacity had nearly doubled. Unfortunately, natural dry steam occurs in very few places in the United States, and much of this is within national park boundaries. It is estimated that dry-steam resources constitute only about 0.5 percent of the total U.S. geothermal resources.

Geothermal hot water, however, underlies a considerable portion of the continental United States. Twenty-four states have known hot-water resources with temperatures above 90°C (195°F) and located no deeper than 3 km (2 mi) below the surface. The presence of underground hot-water resources in the western half of the country is well established. It is significant that 60°C (140°F) water was discovered in 1979 in Maryland in the Atlantic coastal plain. Hence geothermal sources span the continent. The recoverable energy from hot-water reservoirs has been estimated to be as high as 200 to 400 quadrillion ($10^{15}$) kJ (or Btu).

When water is above 150°C (300°F), it is said to be of electrical grade. All hot water of this type lies in the western states. For those water sources which have a temperature above this lower limit, two distinct technologies are used to produce electric power. When the hydrothermal fluid is above 210°C (410°F), the fluid can be throttled to a lower pressure so that part of the liquid will flash or vaporize into steam. The dry steam produced is then passed through a steam turbine of the same type used in the dry-steam geothermal plants discussed above, such as at The Geysers and other sites around the world. Energy-conversion efficiency based on this method is around 15 percent. When the fluid temperature is below 210°C (410°F), the available temperature difference between turbine inlet and outlet is too small for economic operation by the above method. Instead, in this latter case of moderate temperatures one must rely on the operation of a cycle where two separate fluids are involved in the overall energy-conversion system. However, only one of these fluids is responsible for the actual power production cycle.

In a two-fluid cycle, the geothermal fluid transfers thermal energy across a heat-exchanger surface to a second fluid that vaporizes at a much lower temperature than water for a given pressure. This vaporized secondary fluid then passes through a turbine that drives an electric generator, as shown in Fig. 16-23. The fluid passes through a condenser and feedwater pump as in any conventional Rankine vapor power cycle. Both fluids in this system pass through closed loops. That is, the geothermal water is returned to the hot-water reservoir through injection wells after it passes through the heat exchanger. It is significant that to produce the same amount of power as a direct-flash cycle, the two-fluid cycle requires only about two-thirds the water flow rate. This type of cycle is designed to use a low-molecular-weight hydrocarbon (three to six carbon atoms per molecule) as

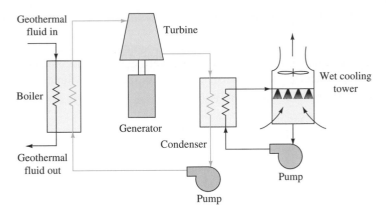

**Figure 16-23**    Schematic of a geothermal power cycle.

the secondary working fluid in the Rankine cycle. Propane, isobutane, and isopentane are typical examples. One engineering problem in the design of geothermal binary cycles is the scale-up of turbines to the large size required with hydrocarbon fluids. Scaling and corrosion are two other problem areas in working with geothermal fluids, especially with respect to the heat-exchanger operation. Another major problem is the heat-rejection process. Because the thermal efficiency of geothermal plants is so low, a much higher fraction of the input energy must be rejected, in comparison to conventional power plants based on fossil fuels. In direct-flash power plants, the turbine condensate passes through an evaporative cooling tower and then back through the condenser. Two-fluid cycles, as shown in Fig. 16-23, may have to use conventional wet cooling towers with water as the cooling medium for the condenser.

A third source of geothermal energy is called *geopressured water,* and it accounts for roughly 15 percent of the accessible energy in geothermal form in the United States. Most of this energy lies beneath the Gulf Coast. Water and organic material were deposited in past geological times beneath shale caps, and this organic material has been transformed to natural gas, or methane. These geopressure sources exhibit temperatures around 100 to 180°C (220 to 350°F) at depths of 2 to 6 km (1 to 4 mi). Interestingly, the pressures in these deposits are far greater than those based solely on hydrostatic effects at these depths, since they run as high as 70 MPa (10,000 psi). Although research is being done in this area, the use of geopressured energy on a large scale is not expected in the near future.

Finally, one must consider petrothermal or magma energy. This is the thermal energy found in vast regions of molten rock or magma or the hot dry rock formed when magma is cooled. Magma itself can produce thermal energy at temperatures around 1000 to 1110°C (1800 to 2000°F). Principal magma-type locations are in the western United States. As might be expected, there currently are no known methods or materials for drilling into magma. For hot-rock regions, it is expected that water would be injected into a region that has been fractured, and after heating, the warm water would be pumped to the surface through another well in the vicinity. Current research deals with drilling methods, fracture techniques, and water circulation, among others. Although petrothermal energy accounts for over 80 percent of accessible geothermal energy in the United States, its commercial development lies in the future.

### 16-6-3    Ocean Thermal-Energy Conversion

Heat engines are cyclic devices which receive heat from a high-temperature source, produce a net work output, and reject heat to a low-temperature sink. The larger the temperature difference between heat source and sink, the greater the theoretical energy-conversion efficiency. When fuels (such as fossil sources) become more costly and less available, efforts are made to increase the thermal efficiency by operating at a higher temperature of supply. Other methods include the use of combined or cogeneration cycles (as discussed in preceding sections) which transform a larger fraction of the energy input to useful output. Nevertheless, if an energy source were relatively free in cost and of nearly unlimited availability, then heat engines operating between heat reservoirs with a very small difference in temperature might be attractive, even though the thermal efficiency would be extremely low. It is this latter situation that is encouraging research and development in ocean thermal energy conversion (OTEC).

An OTEC device is simply a heat engine designed to operate between the relatively warm temperature at the surface of an ocean and the cooler water temperature found deep below the surface. This approach was first suggested by the French physicist d'Arsonval in 1881. In practice, for economical operation the temperature difference required is around 20°C (36°F). At the 600- to 900-m (2000- to 3000-ft) level below the ocean surface the temperature is around 5°C (40°F). To achieve the desired temperature difference quoted above, one must seek geographically those regions of the ocean's sun-warmed surface where the temperature averages 25 to 26°C (78 to 80°F) at least. As might be anticipated, such regions exist only at latitudes close to the equator. The greatest temperature differences are found in the western Pacific Ocean. However, regions east and west of Central America are also satisfactory, and areas off the coast of southern United States and east of Florida are marginally good.

An OTEC system is simply a Rankine vapor power cycle operating under some rather special conditions. Because the temperatures in the evaporator and condenser are low, a working fluid must be chosen so that its vapor pressure is fairly high for these temperatures. For example propane ($C_3H_8$) has a vapor pressure of about 0.55 MPa (80 psia) at 5°C (40°F), and the value is around 0.95 MPa (140 psia) at 25°C (78°F). Similarly, ammonia ($NH_3$) at these cited temperatures has vapor pressures of 0.48 Mpa (70 psia) and 1.03 MPa (150 psia), respectively. Other fluids are possible choices. Warm water is drawn from near the surface of the ocean into a heat

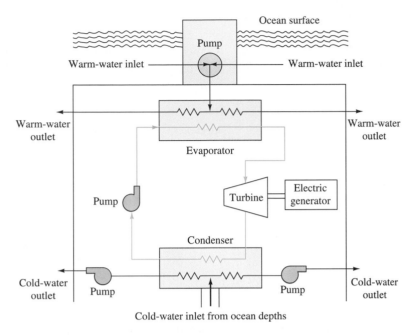

**Figure 16-24** Equipment layout for a floating platform containing an OTEC system.

exchanger, or evaporator, where the liquid working fluid of the Rankine cycle is evaporated (see Fig. 16-24). The vapor then passes through a turbine, which powers an electric generator and passes through the condenser. Cold water pumped up from the depths of the ocean through a long pipe is used to cool the working medium, which condenses back to a liquid. Next the liquid is pumped to the evaporator for another loop around the cycle. In addition to the vapor-pressure requirements, another special condition of an OTEC plant is the obvious enormous dimensions of the circulating systems for the warm- and cold-water streams. Note also that the cold water is pumped from depths of 600 to 900 m (2000 to 3000 ft).

## 16-7 AVAILABILITY ANALYSIS OF A SIMPLE STEAM POWER CYCLE

An availability analysis of vapor power systems is used to reveal the magnitudes of the irreversibilties which occur in the cycles. Improvements in any cycle then can be made by starting with the largest existing irreversibility. As an example, we shall examine a Rankine steam power cycle with irreversible turbine and pump. The equipment schematic and $Ts$ diagram are shown in Fig. 16-25. On the basis of the numbered states on the $Ts$ diagram, the basic energy relations for the steady-state boiler, turbine, condenser,

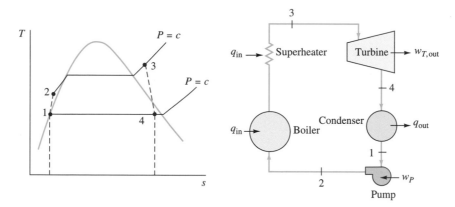

**Figure 16-25**     The $Ts$ and equipment schematic for a Rankine cycle with superheating.

and pump are

$$q_{in} = h_3 - h_2 \quad w_{T,out} = h_4 - h_3 \quad q_{cond,out} = h_1 - h_4 \quad w_P = h_2 - h_1$$

where kinetic and potential energy are neglected. The turbine and pump adiabatic efficiencies are

$$\eta_T = \frac{h_3 - h_{4a}}{h_3 - h_{4s}} \quad \text{and} \quad \eta_P = \frac{h_{2s} - h_1}{h_{2a} - h_1}$$

where subscripts $a$ and $s$ represent actual and isentropic states. Recall that for steady-state devices the stream availability is given by

$$\psi = h - h_0 - T_0(s - s_0) + \frac{V^2}{2} + gz \qquad \textbf{[9-47]}$$

where $z$ is measured relative to $z_0$. In the absence of kinetic- and potential-energy effects, $\Delta\psi = \Delta h - T_0\,\Delta s$. The availability balance on a unit mass basis for a steady-state control volume with one inlet and one outlet is

$$\psi_e - \psi_i = \phi_Q + w_{act} - i_{cv} \qquad \textbf{[9-54]}$$

where $\phi_Q = \sum q_j(1 - T_0/T_j)$.

On the basis of the above review, the availability balances for the four devices may be written on an input/output form as

$$\sum_j q_j \left(1 - \frac{T_0}{T_j}\right) + \psi_{2a} = \psi_3 + i_{boiler} \qquad \text{(boiler-superheater)}$$

$$\textbf{[16-15a]}$$

$$\psi_3 = \psi_{4a} - w_{T,out} + i_T \qquad \text{(turbine)} \qquad \textbf{[16-15b]}$$

$$\psi_{4a} = \psi_1 - \sum_j q_j \left(1 - \frac{T_0}{T_j}\right) + i_{cond} \qquad \text{(condenser)} \qquad \textbf{[16-15c]}$$

$$w_P + \psi_1 = \psi_{2a} + i_P \qquad \text{(pump)} \qquad \textbf{[16-15d]}$$

In addition, the effectiveness of the cycle is given by

$$\varepsilon = \frac{w_{\text{net,out}}}{\Delta\psi_{\text{boiler}}}$$  **[16-15e]**

The irreversibilities due to frictional flow in the boiler and the condenser will be neglected. The use of these relations is illustrated in the example below.

Example 16-12

**A** simple steam power cycle generates steam in the boiler-superheater at 30 bars and 500°C and condenses steam at 0.10 bar. The cooling water required in the condenser experiences a temperature rise from 18 to 28°C. The adiabatic efficiency of the turbine is 82 percent and that of the pump is 78 percent. Determine energy and availability quantities for the devices within the cycle, and the cycle effectiveness.

**Solution:**

**Given:** A simple Rankine power cycle with irreversible turbine and pump is shown in Fig. 16-26.

**Find:** Energy, availability, and irreversibility data around the cycle, and the effectiveness.

**Model:** Steady state, irreversible turbine and pump, negligible kinetic- and potential-energy effects.

**Analysis:** Table 16-1 summarizes the major property values at the four states of interest for the dead state at $T_0 = 25°C$ and $P_0 = 1.0$ bar. These data, partially evaluated from a computerized steam table, are taken primarily from Example 16-4. The states are numbered according to the diagram for the cycle. An

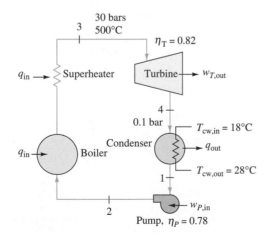

**Figure 16-26** Schematic and data for Example 16-12.

**Table 16-1**  Property data for a steam power cycle

| State | $T,°C$ | $P,$ bars | $h,$ kJ/kg | $s,$ kJ/kg·K | $\psi,$ kJ/kg |
|-------|--------|-----------|------------|--------------|---------------|
| 1     | 45.8   | 0.10      | 191.8      | 0.6493       | 2.8           |
| 2     | 46.1   | 30        | 195.6      | 0.6516       | 5.9           |
| 3     | 500.0  | 30        | 3456.5     | 7.2338       | 1304.3        |
| 4     | 45.8   | 0.10      | 2502.2     | 7.8918       | 153.8         |
| 0     | 25.0   | 1.00      | 105.0      | 0.3674       | 0.0           |

energy accounting is shown in part *A* of Table 16-2. The major items based on these data are the back work ratio, bwr $= w_P/w_T = 0.003$, and the thermal efficiency, $\eta_{th} = w_{net,out}/q_{in} = 950.5/3260.9 = 0.291$. The first item indicates that a very small amount of turbine work is used to drive the pump. The second item reveals that 71 percent of the heat added is removed in the condenser cooling stream.

An availability analysis reveals more information on the condenser. With reference to part *B* of Table 16-2, the effectiveness of the cycle in terms of useful output/input is

$$\varepsilon = \frac{w_{T,out} - w_{P,in}}{\Delta\psi_{boiler}} = \frac{950.5}{1298.4} = 0.732$$

The decrease in fluid availability in the condenser is minimal, amounting to 12 percent of the availability input in the boiler. A tremendous energy loss occurs, but only a small availability change occurs to the fluid in the condenser. This is due to the low temperature of the water (45.8°C) as it condenses, relative to the environment at 25°C.

**Table 16-2**  An energy and availability accounting for a steam power cycle

| **A. Energy Accounting (in kJ/kg)** | | | |
|---|---|---|---|
| **Energy in** | | **Energy out** | |
| Heat added in boiler | 3260.9 | Heat out in condenser | 2310.4 |
| Pump work | 3.8 | Turbine work | 954.3 |
| Total | 3264.7 | Total | 3264.7 |

| **B. Availability Accounting (in kJ/kg)** | | | |
|---|---|---|---|
| **Process** | $q$ | $w_a$ | $\Delta\psi$ | $i$ |
| Boiler | 3260.9 | — | 1298.4 | — |
| Turbine | — | −954.3 | −1150.5 | 196.2 |
| Condenser | −2310.4 | — | −151.0 | — |
| Pump | — | 3.8 | 3.1 | 0.7 |
| Total | 950.5 | −950.5 | 0.0 | 196.9 |

The irreversibilities in the boiler and condenser are taken as zero in Table 16-2. The *internal* losses are zero, because frictional losses are neglected. External irreversibilities exist in both cases, however, owing to heat transfer across a finite temperature gradient. The irreversibility for the condenser can be evaluated since the inlet and outlet temperatures for the cooling water (cw) have been specified as 18 and 28°C. An energy balance around the condenser shows that

$$\frac{\dot{m}_{cw}}{\dot{m}_{steam}} = \frac{-\Delta h_{steam}}{\Delta h_{cw}} = \frac{-(191.8 - 2502.2)}{(117.43 - 75.58)} = 55.2$$

where the water-coolant enthalpies are the saturated liquid values $h_f$ at the given temperatures. The irreversibility of the adiabatic condenser on a kilogram of steam basis is

$$i_{cond} = \frac{\dot{I}}{\dot{m}_{steam}} = T_0 \frac{\dot{\sigma}_{cond}}{\dot{m}_{steam}}$$

$$= T_0 \left[ (s_1 - s_4) + \left( \frac{\dot{m}_{cw}}{\dot{m}_{steam}} \right)(s_{out} - s_{in})_{cw} \right]$$

$$= 298.15[(0.6493 - 7.8918) + 55.2(0.4093 - 0.2679)]$$

$$= 167.8 \text{ kJ/kg steam}$$

where the coolant entropy values are $s_f$ data from the saturation table. On the basis of part *B* of Table 16-2, the condenser irreversibility (loss of work capability) is about 85 percent of the loss in the turbine. Hence this loss is worth reducing through the use of feedwater heaters.

The boiler loss of availability is more difficult to ascertain since the end states of the combustion gases during the heat-transfer process may not be well known. This irreversibility might be approximated by assigning an "average" combustion-gas temperature for the heat-transfer process. For this model the irreversibility for the heat-exchange process is given by Eq. [16-15a]. If we assume an average combustion-gas temperature of 1500 K, then

$$i_{q,comb} = \sum_j q_j \left( 1 - \frac{T_0}{T_j} \right) + \psi_2 - \psi_3$$

$$= 3260.9 \left( 1 - \frac{298.15}{1500} \right) + 5.9 - 1304.3$$

$$= 1314.3 \text{ kJ/kg}$$

Regardless of the choice of an average combustion temperature, the irreversibility associated with the heat-transfer process from the hot gases is far larger than other losses in the cycle. In addition, another irreversibility is associated with the combustion process itself. This latter topic is discussed in Chap. 13.

## 16-8 SUMMARY

An ideal Rankine power cycle consists of an isentropic turbine and an isentropic pump, with heat exchange in a boiler-superheater and a condenser at constant pressure. The pump work is frequently approximated by

$w_{P,\text{in}} = v_f(P_2 - P_1)$. In terms of Fig. 16-2,

$$q_{\text{in}} = h_3 - h_2 \qquad w_{T,\text{out}} = h_3 - h_4 \qquad q_{\text{cond,out}} = h_4 - h_1$$

and the thermal efficiency is

$$\eta_{\text{th}} = \frac{w_{T,\text{out}} - w_{P,\text{in}}}{q_{\text{in}}} = \frac{h_3 - h_4 - v_{f,1}(P_2 - P_1)}{h_3 - h_2}$$

The effect of lowering the condenser pressure is to increase heat input, work output, thermal efficiency, and moisture content at the turbine outlet. Irreversibilities in the turbine and pump decrease the net work output and the thermal efficiency but increase the quality at the turbine outlet.

A correctly designed reheat cycle increases the average temperature during heat addition and thus increases the thermal efficiency. The regenerative cycle uses open or closed feedwater heaters to increase the average temperature of heat addition. Although the net power out will be lowered for a given total mass flow rate, the thermal efficiency is improved for a regenerative cycle.

The effective use of energy is increased by the simultaneous production of power and thermal energy for heating. Such plants are called *cogeneration* systems. Another means of improving the efficiency of power generation is through the use of a *combined cycle,* where a gas-turbine cycle is used in tandem with a steam power cycle. The gas-cycle exhaust stream acts as an energy source for the steam cycle. Other vapor power cycles of interest include binary-vapor cycles, geothermal energy conversion, and ocean-thermal energy conversion.

## PROBLEMS

16-1. A Carnot cycle operates in steady flow with steam. At the end of adiabatic compression the pressure is 1.5 MPa, and the quality is 10 percent. Heat is added during the isothermal expansion until the steam is a saturated vapor. The fluid then expands adiabatically until the pressure is 0.10 MPa. Plot a $Ts$ diagram for the cycle. Then determine (*a*) the quality at the end of adiabatic expansion, (*b*) the thermal efficiency, (*c*) the heat input, in kJ/kg, (*d*) the work during adiabatic compression, and (*e*) the net work, in kJ/kg.

16-2. A Carnot heat engine operates in steady flow with steam. At the end of adiabatic compression the pressure is 30 bars, and the quality is 20 percent. At the end of isothermal heat addition the steam is a saturated vapor, and during isothermal heat rejection the pressure is 0.3 bar. Plot a $Ts$ diagram for the cycle. Then determine (*a*) the quality at the end of adiabatic expansion, (*b*) the thermal efficiency, (*c*) the heat rejected, in kJ/kg, (*d*) the work during adiabatic expansion, and (*e*) the net work output per cycle, in kJ/kg.

16-3. A Carnot heat engine operates in steady flow with water. At the beginning of the heat-addition process the fluid is a saturated liquid, and at the end of this process it is a saturated vapor. Heat addition occurs at 12.0 MPa, and heat rejection is at 0.030 MPa. Plot a $Ts$ diagram for the cycle. Then determine (a) the quality at the end of adiabatic expansion and at the end of isothermal heat rejection, (b) the thermal efficiency, (c) the heat added per cycle, in kJ/kg, (d) the work during adiabatic compression, and (e) the net work per cycle.

16-4. Water enters the boiler of a steady-flow Carnot engine as a saturated liquid at 2.5 MPa and leaves with a quality of 95 percent. The steam pressure at the exit of the turbine is 0.050 MPa. Plot a $Ts$ diagram for the cycle. Then determine (a) the thermal efficiency, (b) the quality at the end of the isothermal heat-rejection process, (c) the heat input, (d) the work during adiabatic compression, and (e) the net work output, in kJ/kg.

16-5E. A Carnot heat engine operates in steady flow with steam. At the end of adiabatic compression the pressure is 400 psia and the quality is 20 percent. At the end of isothermal heat addition the steam is a saturated vapor, and during isothermal heat rejection the pressure is 20 psia. Plot a $Ts$ diagram for the cycle. Then determine (a) the quality at the end of adiabatic expansion, (b) the thermal efficiency, (c) the heat rejected, in Btu/$lb_m$, (d) the work during adiabatic expansion, and (e) the net work output per cycle, in Btu/$lb_m$.

16-6E. A Carnot heat engine operates in steady flow with water. At the beginning of the heat-addition process the fluid is a saturated liquid, and at the end of this process it is a saturated vapor. Heat addition occurs at 640°F, and heat rejection is at 40°F. Plot a $Ts$ diagram for the cycle. Then determine (a) the quality at the end of adiabatic expansion and at the end of isothermal heat rejection, (b) the thermal efficiency, (c) the heat added per cycle, in Btu/$lb_m$, (d) the work during adiabatic compression, and (e) the net work per cycle.

## RANKINE SIMPLE AND NONIDEAL CYCLES

16-7. An ideal Rankine cycle has steam entering the turbine at 40 bars and 440°C. If the turbine output is 10,000 kW, determine (1) the quality at the turbine outlet, (2) the thermal efficiency, and (3) the mass flow rate of steam, in kg/h, for condenser pressures of (a) 0.08 bar and (b) 0.04 bar.

16-8. A Rankine cycle has an exhaust pressure of 0.008 MPa and a turbine-inlet pressure of 6.0 MPa. Determine (1) the quality at the turbine outlet, (2) the thermal efficiency, and (3) the mass flow rate of steam, in kg/h, for a net power output of 10 MW for turbine-inlet temperatures of (a) 540°C and (b) 440°C.

16-9. An ideal Rankine cycle has a turbine-exhaust pressure of 0.08 bar and a turbine-inlet temperature of 600°C. Determine (1) the quality at the turbine outlet and (2) the thermal efficiency of the cycle for a turbine-inlet pressure of (*a*) 120 bars and (*b*) 60 bars.

16-10. An ideal Rankine cycle generates steam at 140 bars and 560°C and condenses the fluid at 0.06 bar. The net power output is 20 MW. Determine (*a*) the heat input, in kJ/kg, (*b*) the thermal efficiency, (*c*) the mass flow rate of steam, in kg/h, and (*d*) the mass flow rate of cooling water required in the condenser if the water experiencies a 10°C temperature rise.

16-11. An ideal Rankine steam power cycle which produces 125 MW of gross turbine power has the following operating conditions: turbine inlet, 80 bars and 560°C; condenser pressure, 0.06 bar. Determine (*a*) the heat input, in kJ/kg, (*b*) the thermal efficiency, (*c*) the mass flow rate of steam, in kg/h, and (*d*) the mass flow rate of cooling water required in the condenser, in kg/h, if the cooling water experiences an 8°C temperature rise.

16-12. An ideal Rankine cycle which produces 150 MW of net power has a turbine inlet condition for the steam of 80 bars and 440°C, and a condenser pressure of 0.080 bar. The cooling water for the condenser experiences a 9.4°C temperature rise. Determine (*a*) the quality of the steam at the turbine outlet, (*b*) the thermal efficiency, (*c*) the required mass flow rate of steam, in kg/s, and (*d*) the volume flow rate of the cooling water required, in m$^3$/s, if the water enters at 18°C.

16-13. A Rankine cycle has steam entering the turbine at 40 bars and 440°C. Use a computer analysis to determine (*a*) the quality at the turbine outlet and (*b*) the thermal efficiency for condenser pressures of 0.04, 0.08, 0.20, 0.40, 0.60, 0.80, and 1.0 bar. Plot the two desired quantities as a function of the condenser pressure.

16-14. An ideal Rankine cycle has a turbine-inlet and turbine-exhaust pressure of 6.0 and 0.008 MPa, respectively. Use a computer analysis to determine (*a*) the quality at the turbine outlet and (*b*) the thermal efficiency for turbine-inlet temperatures of 400, 440, 480, 520, 560, and 600°C. Plot the two desired quantities versus the turbine-inlet temperature.

16-15. A steam power unit operating on the Rankine cycle has a mass flow rate of 23,740 kg/h through it. Water enters the boiler at 100 bars, and it leaves the condenser at 0.10 bar and 45.0°C. Cooling water in the condenser is circulated at a rate of $1.31 \times 10^6$ kg/h, and it experiences a temperature rise of 8.50°C. Determine (*a*) the enthalpy and entropy at the turbine outlet, (*b*) the enthalpy at the turbine inlet, (*c*) the heat added, in kJ/kg, (*d*) the thermal efficiency, and (*e*) the power output of the turbine, in kilowatts.

16-16. A Rankine cycle uses solar energy for the heat input and refrigerant 134a as the working fluid. The fluid enters the pump as a saturated liquid at 9 bars and is pumped to 16 bars. The turbine-inlet temperature is 160°C, and the mass flow rate is 1000 kg/h. Determine (*a*) the net work output, in kJ/kg, (b) the thermal efficiency, and (c) the area, in square meters, of the solar collector needed if the collector picks up 650 J/m²·s.

16-17. A Rankine power cycle uses a small solar pond at 90°C for the heat input and refrigerant 134a as the working fluid. The fluid enters the pump as a saturated liquid at 7 bars and leaves the pump at 16 bars. The turbine-inlet temperature is 80°C, and the mass flow rate is 1200 kg/h. Determine (a) the net work output, in kJ/kg, (b) the thermal efficiency, and (c) the mass flow rate of water needed from the pond, in kg/h, if the water undergoes a 5°C temperature drop as it passes through the evaporator of the cycle.

16-18. Reconsider the Rankine cycle of Prob. 16-7, but include in the analysis that the adiabatic efficiency of the turbine is 85 percent and that of the pump is 70 percent. Determine (1) the quality at the turbine outlet, (2) the thermal efficiency for the modified cycle, and (3) the entropy generation in the turbine and pump, in kJ/kg·K.

16-19. Reconsider the Rankine cycle of Prob. 16-8, but include in the analysis that the adiabatic efficiency of the turbine is 85 percent and that of the pump is 70 percent. Determine (1) the quality at the turbine outlet, (2) the thermal efficiency for the modified cycle, and (3) the entropy generation in the turbine and pump, in kJ/kg·K.

16-20. An ideal Rankine cycle has an exhaust pressure of 0.08 bar and a turbine-inlet temperature of 600°C. The adiabatic efficiency of the turbine is 85 percent, and that of the pump is 70 percent. Determine (1) the quality at the turbine outlet and (2) the thermal efficiency for the cycle for a turbine-inlet pressure of (a) 120 bars and (b) 60 bars. (See Prob. 16-9.)

16-21. A Rankine cycle generates steam at 140 bars and 560°C and the turbine-exhaust pressure is 0.06 bar. The adiabatic efficiency of the turbine is 85 percent, and that of the pump is 70 percent. Determine (a) the net work, in kJ/kg, (b) the thermal efficiency for the cycle, and (c) the entropy generation for the turbine and pump, in kJ/kg·K. (See Prob. 16-10.)

16-22. A Rankine steam power cycle which produces 125 MW of gross turbine power has the following operating conditions: turbine inlet, 80 bars and 560°C; condenser pressure, 0.06 bar. The adiabatic efficiency of the turbine is 85 percent, and that of the pump is 70 percent. Determine (a) the thermal efficiency, (b) the mass flow rate of steam, in kg/h, and (c) the mass flow rate of cooling water required in the condenser, in kg/h, if the cooling water experiences an 8°C temperature rise. (See Prob. 16-11.)

16-23. Reconsider the Rankine cycle of Prob. 16-12, but include in the analysis that the adiabatic efficiency of the turbine is 84 percent, and that of the pump is 75 percent. Determine (a) the quality of the steam at the turbine outlet, (b) the thermal efficiency, (c) the required mass flow rate of steam, in kg/s, (*d*) the volume flow rate of the cooling water required, in m³/s, if the water enters at 18°C, and (*e*) the entropy generation in the turbine and pump, in kJ/kg·K, for the modified cycle.

16-24E. An ideal Rankine cycle has steam entering the turbine at 1200 psia and 800°F. If the turbine output is 10,000 kW, determine (1) the quality at the turbine outlet, (2) the thermal efficiency, and (3) the mass flow rate of steam, in lb$_m$/h, for condenser pressures of (*a*) 1 psia and (*b*) 0.60 psia.

16-25E. A Rankine cycle has an exhaust pressure of 1 psia and a turbine-inlet pressure of 800 psia. Determine (1) the quality at the turbine outlet, (2) the thermal efficiency, and (3) the mass flow rate of steam, in lb$_m$/h, for a net power output of 10 MW for turbine-inlet temperatures of (*a*) 1000°F and (*b*) 800°F.

16-26E. A Rankine cycle generates steam at 2000 psia and 1000°F and the turbine-exhaust pressure is 0.80 psia. The net work output is 100 MW. Determine (*a*) the heat input, in Btu/lb$_m$, and (*b*) the thermal efficiency for the cycle, (*c*) the mass flow rate of steam, in lb$_m$/h, and (*d*) the mass flow rate of cooling water required in the condenser, in lb$_m$/h, if the water experiences a 14°F rise.

16-27E. An ideal Rankine steam power cycle which produces 125 MW of gross turbine power has the following operating conditions: turbine inlet, 1600 psia and 1000°F; condenser pressure, 1.0 psia. Determine (*a*) the heat input, in Btu/lb$_m$, (*b*) the thermal efficiency, (*c*) the mass flow rate of steam, in lb$_m$/h, and (*d*) the mass flow rate of cooling water required in the condenser, in lb$_m$/h, if the cooling water experiences a 12°F temperature rise.

16-28E. A Rankine cycle uses solar energy for the heat input and refrigerant 134a as the working fluid. The fluid enters the pump as a saturated liquid at 100 psia and is pumped to 300 psia. The turbine-inlet temperature is 240°F, and the mass flow rate is 1200 lb$_m$/h. Determine (*a*) the net work output, in Btu/lb$_m$, (*b*) the thermal efficiency, and (*c*) the area, in square feet, of the solar collector needed if the collector picks up 200 Btu/ft²·h.

16-29E. Reconsider the Rankine cycle of Prob. 16-24*a*, but include in the analysis that the adiabatic efficiency of the turbine is 85 percent and that of the pump is 70 percent. Determine (*a*) the quality at the turbine outlet and (*b*) the thermal efficiency for the modified cycle.

16-30E. Reconsider the Rankine cycle of Prob. 16-25, but include in the analysis that the adiabatic efficiency of the turbine is 85 percent and that of the pump is 70 percent. Determine (1) the quality at the turbine outlet and (2) the thermal efficiency for the modified cycle.

16-31E. A Rankine cycle generates steam at 2000 psia and 1000°F and the turbine-exhaust pressure is 0.80 psia. The adiabatic efficiency of the turbine is 85 percent, and that of the pump is 70 percent. Determine (*a*) the net work, in Btu/lb$_m$, (*b*) the thermal efficiency for the cycle, and (*c*) the entropy generation for the turbine and pump, in Btu/lb$_m$·°R. (See Prob. 16-26.)

16-32E. A Rankine steam power cycle which produces 125 MW of gross turbine power has the following operating conditions: turbine inlet, 1600 psia and 1000°F; condenser pressure, 1.0 psia. The adiabatic efficiency of the turbine is 85 percent, and that of the pump is 70 percent. Determine (*a*) the thermal efficiency, (*b*) the mass flow rate of steam, in lb$_m$/h, (*c*) the mass flow rate of cooling water required in the condenser, in lb$_m$/h, if the cooling water experiences a 12°F temperature rise, and (*d*) the entropy generation for the turbine and the pump, in Btu/lb$_m$·°R. (See Prob. 16-27.)

## VAPOR POWER CYCLES WITH REHEAT

16-33. Modify the data of Prob. 16-7 in the following manner. The steam in the turbine is expanded to 7 bars, reheated to 440°C, and then expanded to (*a*) 0.08 bar and (*b*) 0.04 bar. Determine (1) the quality at the turbine outlet and (2) the thermal efficiency.

16-34. Modify the data of Prob. 16-8 in the following manner. (*a*) Steam at 6.0 MPa and 540°C is expanded to 0.50 MPa, reheated to 500°C, and expanded to 0.008 MPa. (*b*) Steam at 6.0 MPa and 440°C is expanded to 1.0 MPa, reheated to 440°C, and expanded to 0.008 MPa. Determine (1) the quality at the condenser inlet and (2) the thermal efficiency.

16-35. An ideal vapor-power cycle is operated under the two following conditions. Steam (*a*) at 120 bars and 600°C is expanded to 10 bars, reheated to 540°C, and expanded to 0.08 bar and (*b*) at 60 bars and 600°C is expanded to 10 bars, reheated to 540°C, and expanded to 0.08 bar. Calculate (1) the quality at the turbine outlet and (2) the thermal efficiency of the cycle. (See Prob. 16-9.)

16-36. Steam generated in an ideal vapor power cycle at 14.0 MPa and 560°C is expanded to 1.5 MPa, reheated to 540°C, and expanded further to 0.006 MPa. The net power output is 20 MW. Determine (*a*) the heat input, in kJ/kg, (*b*) the thermal efficiency, (*c*) the mass flow rate of steam, in kg/h, and (*d*) the mass flow rate of cooling water required in the condenser if the water experiences a 10°C temperature rise. (See Prob. 16-10.)

16-37. An ideal steam power cycle which produces 125 MW of gross turbine power has the following operating conditions: turbine inlet, 8.0 MPa and 560°C; expansion to 0.5 MPa; reheat to 500°C, expansion to a condenser pressure of 0.006 MPa. Determine (*a*) the heat input, in kJ/kg, (*b*) the thermal efficiency, (*c*) the mass flow rate of steam, in kg/h, and (*d*) the mass flow rate of cooling water required in the condenser, in kg/h, if the cooling water experiences an 8°C temperature rise. (See Prob. 16-11.)

16-38. An ideal Rankine cycle which produces 150 MW of net power has a turbine inlet condition for the steam of 8.0 MPa and 440°C. Expansion occurs to 1.5 MPa, followed by reheat to 440°C, and then further turbine expansion to a condenser pressure of 0.008 bar. The cooling water for the condenser experiences a 9.4°C temperature rise. Determine (a) the quality of the steam at the turbine outlet, (b) the thermal efficiency, (c) the required mass flow rate of steam, in kg/s, and (d) the volume flow rate of the cooling water required, in m³/s, if the water enters at 18°C. (See Prob. 16-12.)

16-39. An ideal steam vapor power cycle with reheat operates with steam entering the turbine at 40 bars and 440°C, a reheat temperature of 440°C, and a condenser pressure of 0.08 bar. Determine the thermal efficiency for reheat pressures of 40, 20, 10, 7, and 5 bars, and plot the results versus reheat pressure. (See Prob. 16-33a.)

16-40. An ideal steam vapor power cycle with reheat operates with steam entering the turbine at 140 bars and 560°C, a reheat temperature of 540°C, and a condenser pressure of 0.06 bar. Determine the thermal efficiency for reheat pressures of 140, 80, 40, and 15 bars, and plot the results versus reheat pressure. (See Probs. 16-10 and 16-36.)

16-41. The adiabatic efficiency of both stages of the turbine is 85 percent and that of the pump is 70 percent in a reheat cycle. On this basis determine the thermal efficiency for the data given in (a) Prob. 16-33a, (b) Prob. 16-33b, (c) Prob. 16-34a, and (d) Prob. 16-34b.

16-42. Recall the data given in Prob. 16-35a. Determine (a) the thermal efficiency and (b) the entropy generation in the two turbine stages if the reheat cycle analysis takes into account that the adiabatic efficiency of both stages of the turbine is 85 percent and that of the pump is 70 percent.

16-43. Modify the reheat cycle analysis of Prob. 16-36 to take into account that the adiabatic efficiency of both stages of the turbine is 85 percent and that of the pump is 70 percent. Determine (a) the heat input, in kJ/kg, (b) the thermal efficiency, (c) the mass flow rate of steam, in kg/h, (d) the mass flow rate of cooling water required in the condenser if the water undergoes a 10°C temperature rise, and (e) the entropy generation in each turbine stage, in kJ/kg·K.

16-44. Modify the reheat cycle analysis for Prob. 16-37 so that the adiabatic efficiency of both stages of the turbine is 85 percent and that of the pump is 70 percent. Determine (a) the heat input, in kJ/kg, (b) the thermal efficiency, (c) the mass flow rate of steam, in kg/h, (d) the mass flow rate of cooling water required in the condenser, in kg/h, if the cooling water undergoes an 8°C temperature rise, and (e) the entropy generation in each turbine stage, in kJ/kg·K.

16-45. Modify the reheat cycle analysis of Prob. 16-38 so that the adiabatic efficiency of the turbine is 85 percent and that of the pump is 70 percent. Determine (*a*) the quality of the steam at the turbine outlet, (*b*) the thermal efficiency, (*c*) the required mass flow rate of steam, in kg/s, (*d*) the volume flow rate of the cooling water required, in m³/s, if the water enters at 18°C, and (*e*) the entropy generation in the two turbine stages, in kJ/kg·K.

16-46E. Modify the data of Prob. 16-24 in the following manner. The steam in the turbine is expanded to 100 psia, reheated to 800°F, and then expanded to (*a*) 1 psia and (*b*) 0.60 psia. Determine (1) the quality at the turbine outlet and (2) the thermal efficiency.

16-47E. Modify the data of Prob. 16-25 in the following manner. (*a*) Steam at 800 psia and 1000°F is expanded to 60 psia, reheated to 1000°F, and expanded to 1 psia. (*b*) Steam at 800 psia and 800°F is expanded to 120 psia, reheated to 800°F, and expanded to 1 psia. Determine (1) the quality at the condenser inlet and (2) the thermal efficiency.

16-48E. Steam generated in an ideal vapor power cycle at 2000 psia and 1000°F is expanded to 160 psia, reheated to 900°F, and expanded further to 0.80 psia. The net power output is 100 MW. Determine (*a*) the heat input, in Btu/lb$_m$, (*b*) the thermal efficiency, (*c*) the mass flow rate of steam, in lb$_m$/h, and (*d*) the mass flow rate of cooling water required in the condenser, in lb$_m$/h, if the water undergoes a 14°F temperature rise. (See Prob. 16-26.)

16-49E. An ideal steam power cycle which produces 125 MW of gross turbine power has the following operating conditions: turbine inlet, 1600 psia and 1000°F; expansion to 80 psia; reheat to 900°F, expansion to a condenser pressure of 1 psia. Determine (*a*) the heat input, in Btu/lb$_m$, (*b*) the thermal efficiency, (*c*) the mass flow rate of steam, in lb$_m$/h, and (*d*) the mass flow rate of cooling water required in the condenser, in lb$_m$/h, if the cooling water experiences a 12°F temperature rise. (See Prob. 16-27.)

16-50E. An ideal steam vapor power cycle with reheat operates with steam entering the turbine at 1200 psia and 800°F, a reheat temperature of 800°F, and a condenser pressure of 1 psia. Determine the thermal efficiency for reheat pressures of 1200, 400, 200, and 100 psia, and plot the results versus reheat pressure. (See Probs. 16-24*a* and 16-46*a*.)

16-51E. An ideal steam vapor power cycle with reheat operates with steam entering the turbine at 2000 psia and 1000°F, a reheat temperature of 900°F, and a condenser pressure of 0.8 psia. Determine the thermal efficiency for reheat pressures of 2000, 600, 400, 250, and 160 psia, and plot the results versus reheat pressure. (See Probs. 16-26 and 16-48.)

16-52E. The adiabatic efficiency of the turbine is 85 percent and that of the pump is 70 percent in a reheat cycle. On this basis determine (1) the thermal efficiency and (2) the entropy generation in the two turbine stages, in Btu/lb$_m$·°R, for the cycle data given in (*a*) Prob. 16-46*a*, (*b*) Prob. 16-46*b*, (*c*) Prob. 16-47*a*, (*d*) Prob. 16-47*b*, (*e*) Prob. 16-48, (*f*) Prob. 16-49.

## REGENERATIVE VAPOR CYCLES

16-53. Modify the data of Prob. 16-7 in the following manner. The steam in the turbine is expanded to 7 bars, where a portion is bled to a single open feedwater heater operating at the same pressure. The remaining fluid in the turbine is expanded to (*a*) 0.08 bar and (*b*) 0.04 bar. Determine (1) the fraction of the total flow bled to the heater, (2) the thermal efficiency, and (3) the total mass flow of steam, in kg/h. (Also see Prob. 16-33.)

16-54. Modify the data of Prob. 16-8 in the following manner. (*a*) Steam at 6.0 MPa and 540°C is expanded to 0.50 MPa, where a portion is bled to a single open heater and the remainder is expanded to 0.008 MPa. (*b*) Steam at 6.0 MPa and 440°C is expanded to 1.0 MPa, where a portion is bled to a single open heater and the remainder is expanded to 0.008 MPa. Determine (1) the fraction of the total flow bled to the heater and (2) the thermal efficiency. (Also see Prob. 16-34.)

16-55. An ideal vapor power cycle is operated under the two following conditions. (*a*) Steam at 120 bars and 600°C is expanded to 10 bars, where a portion is bled to a single open heater and the remainder is expanded to 0.08 bar. (*b*) Steam at 60 bars and 600°C is expanded to 10 bars, where a portion is bled to a single open heater and the remainder is expanded to 0.08 bar. Calculate (1) the fraction of the total flow bled to the heater and (2) the thermal efficiency of the cycle. (Also see Prob. 16-35.)

16-56. Steam generated in an ideal vapor power cycle at 14.0 MPa and 560°C is expanded to 1.5 MPa, where a portion is bled to (*a*) a single open feedwater heater and (*b*) a single closed feedwater heater followed by a pump. The remainder of the fluid is expanded further in the turbine to 0.006 MPa. The net power output is 20 MW. Determine (1) the fraction of the total flow bled to the heater, (2) the total mass flow rate of steam required, in kg/h, and (3) the thermal efficiency. (Also see Prob. 16-36.)

16-57. An ideal steam power cycle which produces 125 MW of gross turbine power has the following operating conditions: turbine inlet, 8.0 MPa and 560°C; expansion to 0.5 MPa to (*a*) a single open heater and (*b*) a single closed heater, with the remaining expansion to a condenser pressure of 0.006 MPa. Determine (1) the fraction of the total flow bled to the heater, (2) the total mass flow rate of steam, in kg/h, and (3) the thermal efficiency. (Also see Prob. 16-37.)

16-58. An ideal Rankine cycle which produces 150 MW of net power has a turbine inlet condition for the steam of 80 bars and 440°C. Expansion occurs to 15 bars, where a portion is bled to a open feedwater heater. The remainder of the turbine expansion occurs to a condenser pressure of 0.080 bar. Determine (*a*) the fraction of the total flow bled to the heater, (*b*) the thermal efficiency, and (*c*) the total mass flow rate of steam required, in kg/h. (Also see Prob. 16-38.)

16-59. The boiler-superheater of an ideal regenerative steam cycle produces steam at 120 bars and 600°C. A closed feedwater heater receives steam from the turbine at 30 bars, and an open heater operates at 10 bars. The condenser operates at 0.08 bar, and the liquid condensate from the closed heater is throttled back into the open heater. There is a pump after the condenser and after the open heater. Determine (*a*) the fraction of the total flow which goes to the closed heater and to the open heater, (*b*) the work output of the turbine and the total pump work, in kJ/kg of total flow, and (*c*) the thermal efficiency.

16-60. Steam enters the turbine of an ideal regenerative cycle at 4.0 MPa and 500°C. Steam is extracted at 0.7 and 0.3 MPa and introduced into two open feedwater heaters that are in series. Appropriate pumps are used after the condenser, which operates at 0.006 MPa, and after each heater. Determine (*a*) the fraction of the total flow which goes to the 0.7-MPa heater and the 0.3-MPa heater, (*b*) the work output of the turbine and the total pump work, in kJ/kg of total flow, and (*c*) the thermal efficiency.

16-61. Steam is generated at 140 bars and 520°C. The steam is expanded through the first stage of the turbine to 40 bars. Part of the exhaust stream from the first stage is supplied to a closed feedwater heater, and the remainder is reheated to 520°C. The reheated steam is then expanded in a second turbine stage to an exhaust pressure of 0.08 bar. Some steam is bled from the second stage at 7 bars for use in an open feedwater heater. The steam bled from the first stage condenses as it passes through the closed heater and is throttled back into the open heater at 7 bars. The saturated liquid leaving the open heater passes first through a pump to 140 bars and then through the closed heater to the boiler. If we assume an ideal cycle and neglect pump work, determine (*a*) the percentage of the steam entering the first stage which goes to the closed heater, (*b*) the percentage of the steam entering the second stage which is bled to the open heater, and (*c*) the thermal efficiency of the cycle.

16-62. The adiabatic efficiencies of the turbine and pump in a regenerative cycle are 85 and 70 percent, respectively. On this basis determine (1) the thermal efficiency of the cycle and (2) the entropy generation in the turbine and in the heater, in kJ/kg·K, for the data given in (*a*) Prob. 16-53*a*, (*b*) Prob. 16-53*b*, (*c*) Prob. 16-54*a*, (*d*) Prob. 16-54*b*, (*e*) Prob. 16-55*a*, (*f*) Prob. 16-55*b*, (*g*) Prob. 16-56*a*, (*h*) Prob. 16-57*a*, and (*i*) Prob. 16-59.

16-63E.  Modify the data of Prob. 16-24 in the following manner. The steam in the turbine is expanded to 100 psia, where a portion is bled to a single open feedwater heater operating at the same pressure. The remaining fluid in the turbine is expanded to (*a*) 1 psia and (*b*) 0.60 psia. Determine (1) the fraction of the flow bled to the heater, (2) the thermal efficiency, and (3) the total mass flow rate of the steam, in $lb_m$/h. (Also see Prob. 16-46.)

16-64E.  Modify the data of Prob. 16-25 in the following manner. (*a*) Steam at 800 psia and 1000°F is expanded to 60 psia, where a portion is bled to a single open heater and the remainder is expanded to 1 psia. (*b*) Steam at 800 bars and 800°F is expanded to 120 psia, where a portion is bled to a single open heater and the remainder is expanded to 1 psia. Determine (1) the fraction of the total flow bled to the heater, and (2) the thermal efficiency. (See also Prob. 16-47.)

16-65E.  Steam generated in an ideal vapor power cycle at 2000 psia and 1000°F is expanded to 160 psia, where a portion is bled to (*a*) a single open feedwater heater and (*b*) a single closed feedwater heater followed by a pump. The remainder of the fluid is expanded further in the turbine to 0.80 psia. The net power output is 100 MW. Determine (1) the fraction of the total flow bled to the heater, (2) the total mass flow rate of steam required, in $lb_m$/h, and (3) the thermal efficiency. (See Probs. 16-26 and 16-48.)

16-66E.  An ideal steam power cycle which produces 125 MW of gross turbine power has the following operating conditions: turbine inlet, 1600 psia and 1000°F; expansion to 80 psia to (*a*) a single open heater and (*b*) a single closed heater followed by a pump, with the remaining expansion to a condenser pressure of 1 psia. Determine (1) the fraction of the total flow bled to the heater, (2) the total mass flow rate of steam, in $lb_m$/h, and (3) the thermal efficiency. (Also see Prob. 16-49.)

16-67E.  The boiler-superheater of an ideal regenerative steam cycle produces steam at 1800 psia and 1000°F. A closed feedwater heater receives steam from the turbine at 400 psia, and an open heater operates at 140 psia. The condenser operates at 0.80 psia, and the liquid condensate from the closed heater is throttled back into the open heater. There is a pump after the condenser and after the open heater. Determine (*a*) the fraction of the total flow which goes to the closed heater and to the open heater, (*b*) the work output of the turbine and the total pump work, in Btu/$lb_m$ of total flow, and (*c*) the thermal efficiency.

16-68E.  Steam enters the turbine of an ideal regenerative cycle at 500 psia and 900°F. Steam is extracted at 100 and 40 psia and introduced into two open feedwater heaters that are in series. Appropriate pumps are used after the condenser, which operates at 1 psia, and after each heater. Determine (a) the fraction of the total flow which goes to the 100-psia heater and the 40-psia heater, (b) the work output of the turbine and the total pump work, in Btu/lb$_m$ of total flow, and (c) the thermal efficiency.

16-69E. The adiabatic efficiencies of the turbine and pump in a regenerative cycle are 85 and 70 percent, respectively. On this basis determine the thermal efficiency of the cycle for the data given in (a) Prob. 16-63a, (b) Prob. 16-63b, (c) Prob. 16-64a, (d) Prob. 16-64b, (e) Prob. 16-65a, (f) Prob. 16-65b, (g) Prob. 16-66a, (h) Prob. 16-67.

## REHEAT-REGENERATIVE VAPOR CYCLES

16-70. Consider the following modification of Probs. 16-33 and 16-53. The steam in the turbine is expanded to 7 bars, where a portion is bled to an open feedwater heater operating at this pressure. The remaining portion is reheated to 440°C and then expanded to (a) 0.08 bar and (b) 0.04 bar. Determine the thermal efficiency.

16-71. Consider the following modification of Probs. 16-34 and 16-54. (a) Steam at 6.0 MPa and 540°C is expanded to 0.5 MPa, where a portion is bled to an open feedwater heater. The remaining portion is reheated to 500°C and then expanded to 0.008 MPa. (b) Steam at 6.0 MPa and 440°C is expanded to 1.0 MPa, where a portion is bled to an open feedwater heater. The remaining portion is reheated to 440°C and then expanded to 0.008 MPa. Find the thermal efficiency.

16-72. An ideal vapor power cycle is operated under the two following conditions. (a) Steam at 120 bars and 600°C is expanded to 10 bars, where a portion is bled to a single open heater. The remaining portion is reheated to 540°C and is expanded to 0.08 bar. (b) Steam at 60 bars and 600°C is expanded to 10 bars, where a portion is bled to a single open heater. The remaining portion is reheated to 540°C and expanded to 0.08 bar. Calculate (1) the fraction of the total flow bled to the heater and (2) the thermal efficiency of the cycle. (Also see Probs. 16-35 and 16-55.)

16-73. Steam generated in an ideal vapor power cycle at 14.0 MPa and 560°C is expanded to 1.5 MPa, where a portion is bled to a single closed feedwater heater followed by a pump. The remainder of the fluid is reheated to 540°C before it is expanded further in the turbine to 0.006 MPa. The net power output is 20 MW. Determine (a) the fraction of the total flow bled to the heater, (b) the total mass flow rate of steam required, in kg/h, and (c) the thermal efficiency. (Also see Probs. 16-36 and 16-56b.)

16-74. Consider the following modification of Prob. 16-59. The portion of the flow which does not go to the closed heater is reheated to 500°C before it enters the second stage of the turbine. Determine (a) the fraction of the total flow which goes to the open heater, (b) the heat input, in kJ/kg, and (c) the thermal efficiency.

16-75. Consider the following modification of Prob. 16-60. The portion of the flow which does not go to the open heater at 0.7 MPa is reheated to 500°C before it enters the second stage of the turbine. Determine (a) the fraction of the total flow which goes to the open heater at 0.3 MPa, (b) the heat input, in kJ/kg, and (c) the thermal efficiency.

16-76. An ideal reheat-regenerative cycle operates with conditions at the turbine inlet of 140 bars and 600°C, and reheating is at 7 bars and 500°C. A closed feedwater heater operates at 15 bars, and the drain from the closed heater is trapped back into an open feedwater heater which operates at 3 bars. The condenser pressure is 0.06 bar. Determine (a) the thermal efficiency and (b) the mass flow rate through the steam generator required for a turbine output of 100 MW, in kg/h.

16-77. The adiabatic efficiencies of the turbine and pump in a regenerative-reheat cycle are 85 and 70 percent, respectively. On this basis determine the thermal efficiency of the cycle for the data given in (a) Prob. 16-70a, (b) Prob. 16-70b, (c) Prob. 16-71a, (d) Prob. 16-71b, (e) Prob. 16-72a, (f) Prob. 16-72b, (g) Prob. 16-74, and (h) Prob. 16-75.

16-78E. Consider the following modification of Probs. 16-46 and 16-63. The steam in the turbine is expanded to 100 psia, where a portion is bled to an open feedwater heater operating at this pressure. The remaining portion is reheated to 800°F and then expanded to (a) 1 psia and (b) 0.60 psia. Determine the thermal efficiency.

16-79E. Consider the following modification of Probs. 16-47 and 16-64. (a) Steam at 800 psia and 1000°F is expanded to 60 psia, where a portion is bled to an open feedwater heater. The remaining portion is reheated to 1000°F and then expanded to 1 psia. (b) Steam at 800 psia and 800°F is expanded to 120 psia, where a portion is bled to an open feedwater heater. The remaining portion is reheated to 800°F and then expanded to 1 psia. Find the thermal efficiency.

16-80. Consider the following modification of Prob. 16-67. The portion of the flow which does not go to the closed heater at 400 psia is reheated to 1000°F before it enters the second stage of the turbine. Determine (a) the fraction of the total flow which goes to the open heater, (b) the heat input, in Btu/lb$_m$, and (c) the thermal efficiency.

16-81. Consider the following modification of Prob. 16-68. The portion of the flow which is not extracted to the open heater at 100 psia is reheated to 900°F before it enters the second stage of the turbine. Determine (a) the fraction of the total flow which goes to the open heater at 40 psia, (b) the heat input, in Btu/lb$_m$, and (c) the thermal efficiency.

16-82E. The adiabatic efficiencies of the turbine and pump in a regenerative-reheat cycle are 85 and 70 percent, respectively. On this basis determine the thermal efficiency of the cycle for the data given in (*a*) Prob. 16-78*a*, (*b*) Prob. 16-78*b*, (*c*) Prob. 16-79*a*, (*d*) Prob. 16-79*b*, (*e*) Prob. 16-80, (*f*) Prob. 16-81.

## SUPERCRITICAL POWER CYCLE

16-83. A supercritical steam power cycle operates with reheat and regeneration. The steam enters the turbine at 240 bars and 600°C and expands to 0.06 bar. Steam leaves the first stage at 60 bars, where part of it enters a closed heater, while the rest is reheated to 540°C. Both sections of the turbine have an adiabatic efficiency of 88 percent. There is a condensate pump between the condenser and the heater. Another pump for the extracted steam that is condensed lies between the heater and the condensate line from the heater. The feedwater enters the steam generator at 280 bars. Determine (*a*) the enthalpies at the 10 states around the cycle, (*b*) the fraction of the total mass flow rate which is extracted to the heater, and (*c*) the thermal efficiency of the cycle.

16-84. In a supercritical steam power cycle, steam expands ideally from 320 bars and 640°C to 240 bars. It is then cooled at constant pressure to 400°C, followed by isentropic compression to 320 bars. Heat is then added until the temperature reaches 640°C. Determine (*a*) the heat added, in kJ/kg, (*b*) the heat rejected, and (*c*) the thermal efficiency.

16-85. A supercritical steam power cycle operates with reheat and regeneration. The steam enters the turbine at 240 bars and 560°C and expands to 0.08 bar. Steam leaves the first stage at 40 bars, where part of it enters a closed heater, while the rest is reheated to 540°C. Both sections of the turbine have an adiabatic efficiency of 87 percent. There is a condensate pump between the condenser and the heater. Another pump for the extracted steam that is condensed lies between the heater and the condensate outlet line from the heater. Determine (*a*) the four enthalpies at the inlets and outlets of the closed heater, (*b*) the enthalpy at the condenser inlet and the enthalpy at the inlet to the heat-addition section, (*c*) the fraction of the total mass flow which is extracted to the heater, and (*d*) the thermal efficiency of the cycle.

16-86E. A supercritical steam power cycle operates with reheat and regeneration. The steam enters the turbine at 3500 psia and 1200°F and expands to 1 psia. Steam leaves the first stage at 200 psia, where part of it enters a closed heater, while the rest is reheated to 1000°F. Both sections of the turbine have an adiabatic efficiency of 89 percent. There is a condensate pump between the condenser and the heater. Another pump for the extracted steam that is condensed lies between the heater and the condensate line from the heater. The total feedwater enters the steam generator at 4000 psia. Determine (*a*) the enthalpies at the 10 states around the cycle, (*b*) the fraction of the total mass flow rate which is extracted to the heater, and (*c*) the thermal efficiency of the cycle.

16-87E. A supercritical steam power cycle operates with reheat and regeneration. The steam enters the turbine at 4400 psia and 1100°F and expands to 1 psia. Steam leaves the first stage at 400 psia, where part of it enters a closed heater, while the rest is reheated to 1000°F. Both sections of the turbine have an adiabatic efficiency of 88 percent. There is a condensate pump between the condenser and the heater. Another pump for the extracted steam that is condensed lies between the heater and the condensate-outlet line from the heater. Determine (*a*) the four enthalpies at the inlets and outlets of the closed heater, (*b*) the enthalpy at the condenser inlet and the enthalpy at the inlet to the heat-addition section, (*c*) the fraction of the total mass flow which is extracted to the heater, and (*d*) the thermal efficiency of the cycle.

## COGENERATION CYCLE

16-88. Steam expands through a turbine in a cogeneration power cycle from 60 bars and 440°C to 0.08 bar. The turbine has an efficiency of 85 percent and delivers 50 MW of power. Steam is extracted at 5 bars for a heating load of $5 \times 10^6$ kJ/min. The steam leaves the heating load as a saturated liquid at 4 bars and mixes with the steam leaving the condensate pump at 4 bars (see Fig. 16-18). The pump efficiencies are both 70 percent. Determine (*a*) the mass flow rates, in kg/min, through the heating load and through the condenser, (*b*) the heat input to the boiler-superheater section, in kJ/min, (*c*) the dimensionless ratio of the heating load plus the turbine output to the heat input, and (*d*) the dimensionless ratio of the turbine output to the heat input if the cycle were operated without the heating load, but at the same total mass flow rate.

16-89. A turbine in a cogeneration cycle expands steam from 100 bars and 480°C to 0.06 bar with an efficiency of 84 percent. The total mass flow rate through the boiler-superheater is $0.30 \times 10^6$ kg/h, and 45 percent of the flow is extracted at 7 bars for the heating load (see Fig. 16-18). Assume the fluid leaves the heating load at 5 bars and 150°C and mixes with the water leaving the condensate pump at 5 bars. The pump efficiencies are both 65 percent. Determine (*a*) the heating load, in kJ/min, (*b*) the turbine output, in megawatts, (*c*) the dimensionless ratio of the heating load plus the turbine output to the heat input, and (*d*) the ratio of the turbine output to the heat input if the cycle were operated without the heating load, but at the same total mass flow rate.

heat input, and (*d*) the dimensionless ratio of the turbine output to the heat input if the cycle were operated without the heating load, but at the same total mass flow rate.

16-94E.  A turbine in a cogeneration cycle expands steam from 1600 psia and 900°F to 0.80 psia with an efficiency of 84 percent. The total mass flow rate through the boiler-superheater is $0.30 \times 10^6$ lb$_m$/h, and 45 percent of the flow is extracted at 100 psia for the heating load (see Fig. 16-18). Assume the fluid leaving the heating load at 80 psia and 310°F and mixes with the water leaving the condensate pump at 80 psia. The pump efficiencies are both 65 percent. Determine (*a*) the heating load, in Btu/min, (*b*) the turbine output, in megawatts, (*c*) the dimensionless ratio of the heating load plus the turbine output to the heat input, and (*d*) the ratio of the turbine output to the heat input if the cycle were operated without the heating load, but at the same total mass flow rate.

16-95E.  Steam expands through a turbine in a cogeneration power cycle from 1800 psia and 800°F to 1 psia. The turbine has an efficiency of 87 percent and delivers 60 MW of power. Steam leaves the first stage of the turbine at 100 psia, where a portion is used for a heating load of $5 \times 10^6$ Btu/min. The remaining portion of the steam is reheated to 700°F before expansion to the condenser pressure. The steam leaves the heating load as a saturated liquid at 60 psia and then mixes with the steam leaving the condensate pump at 60 psia (see Fig. 16-18). The pump efficiencies are both 60 percent. Determine (*a*) the mass flow rates, in lb$_m$/min, through the heating load and through the condenser, (*b*) the heat input to the boiler-superheater section, in Btu/min, (*c*) the dimensionless ratio of the heating load plus the turbine output to the heat input, and (*d*) the dimensionless ratio of the turbine output to the heat input if the cycle were operated without the heating load, but at the same total mass flow rate.

## COMBINED GAS-STEAM POWER CYCLE

16-96.  A gas-turbine power plant operates on a 12:1 pressure ratio with compressor- and turbine-inlet temperatures of 300 and 1400 K, respectively. The adiabatic efficiencies of the compressor and turbine are 85 and 87 percent, respectively. The turbine exhaust, used as the energy source for a steam cycle, leaves the heat exchanger at 480 K. The inlet conditions to the 85 percent efficient turbine in the steam cycle are 140 bars and 520°C. The condenser pressure is 0.10 bar, and the pump is 75 percent efficient. For the combined cycle, determine (*a*) the heat input to the air, in kJ/kg of air, (*b*) the net work output of the gas-turbine cycle, in kJ/kg of air, (*c*) the net work output of the steam cycle, in kJ/kg of steam, (*d*) the mass flow rates of the air and the water, in kJ/s, (*e*) the overall thermal efficiency of the combined cycle, and (*f*) the entropy generation in the compressor, both turbines, and the heat exchanger, in kJ/(kg air)(K).

16-97. Determine the quantities that are required in Prob. 16-96 if, in addition, an open feedwater heater is added to the steam cycle with steam bled from the turbine to the heater at 7 bars.

16-98. A gas-turbine power plant operates on a 13:1 pressure ratio with compressor- and turbine-inlet temperatures of 290 and 1440 K, respectively. The adiabatic efficiencies of the compressor and turbine are 84 and 88 percent, respectively. The turbine exhaust, used as the energy source for a steam cycle, leaves the heat exchanger at 500 K. The inlet conditions to the 86 percent efficient turbine in the steam cycle are 160 bars and 560°C. The condenser pressure is 0.08 bar, and the pump is 70 percent efficient. Determine (*a*) the heat input to the air, in kJ/kg of air, (*b*) the dimensionless mass-flow-rate ratio of air to steam, (*c*) the net work output of the gas-turbine cycle, in kJ/kg of air, (*d*) the net work output of the steam cycle, in kJ/kg of steam, (*e*) the overall thermal efficiency of the combined cycle, and (*f*) the entropy generation in the compressor, both turbines, and the heat exchanger, in kJ/(kg air)(K).

16-99. Determine the quantities that are required in Prob. 16-98 if, in addition, an open feedwater heater is added to the steam cycle with steam bled from the turbine to the heater at 10 bars. In part *f*, include calculating $\sigma$ for the heater.

16-100. A gas-turbine power plant operates on a 10:1 pressure ratio with compressor- and turbine-inlet temperatures of 300 and 1340 K, respectively. The adiabatic efficiencies of the compressor and turbine are 84 and 87 percent, respectively. The turbine exhaust, used as the energy source for a steam cycle, leaves the boiler at 460 K. The inlet conditions to the 86 percent efficient turbine in the steam cycle are 100 bars and 480°C. The condenser pressure is 0.08 bar, and the pump is 70 percent efficient. Determine (*a*) the heat input to the air, in kJ/kg of air, (*b*) the dimensionless mass-flow-rate ratio of air to steam, (*c*) the net work output of the gas-turbine cycle, in kJ/kg of air, (*d*) the net work output of the steam cycle, in kJ/kg of steam, (*e*) the overall thermal efficiency of the combined cycle, and (*f*) the entropy generation in the compressor, both turbines, and the heat exchanger, in kJ/(kg air)(K).

16-101. Determine the quantities that are required in Prob. 16-100 if, in addition, an open feedwater heater is added to the steam cycle with steam bled from the turbine to the heater at 7 bars. In part *f*, include calculating $\sigma$ for the heater.

16-102. A gas-turbine power plant operates on a 12:1 pressure ratio with compressor- and turbine-inlet temperatures of 300 and 1400 K, respectively. The adiabatic efficiencies of the compressor and turbine are 84 and 88 percent, respectively. The turbine exhaust, used as the energy source for a steam cycle, leaves the heat exchanger at 480 K. The inlet conditions to the 90 percent efficient turbine in the steam cycle are 8.0 MPa and 400°C. The condenser pressure is 8 kPa, and the pump is 80 percent efficient. For a net power output of 20 MW for the cycle, determine (*a*) the mass flow rates of air and of water, in kg/s, (*b*) the rate of heat input to the air, in kJ/s, (*c*) the net work output of the overall cycle, in kJ/kg of air, (*d*) the overall thermal efficiency of the combined cycle, and (*e*) the entropy generation in the compressor, both turbines, and the heat exchanger, in kJ/(kg air)(K).

16-103E.  A gas-turbine power plant operates on a 12:1 pressure ratio with compressor- and turbine-inlet temperatures of 80 and 2040°F, respectively. The adiabatic efficiencies of the compressor and turbine are 85 and 87 percent, respectively. The turbine exhaust, used as the energy source for a steam cycle, leaves the heat exchanger at 400°F. The inlet conditions to the 85 percent efficient turbine in the steam cycle are 2000 psia and 900°F. The condenser pressure is 1 psia, and the pump is 75 percent efficient. Determine (*a*) the heat input to the air, in Btu/lb$_m$ of air, (*b*) the dimensionless mass-flow-rate ratio of air to steam, (*c*) the net work output of the gas-turbine cycle, in Btu/lb$_m$ of air, (*d*) the net work output of the steam cycle, in Btu/lb$_m$ of steam, (*e*) the overall thermal efficiency of the combined cycle, and (*f*) the entropy generation in the compressor, both turbines, and the heat exchanger, in Btu/(lb$_m$ air)(°R).

16-104E.  Determine the quantities required in Prob. 16-103 if, in addition, an open feedwater heater is added to the steam cycle with steam bled from the turbine to the heater at 100 psia.

16-105E.  A gas-turbine power plant operates on a 10:1 pressure ratio with compressor- and turbine-inlet temperatures of 80 and 1940°F, respectively. The adiabatic efficiencies of the compressor and turbine are 84 and 87 percent, respectively. The turbine exhaust, used as the energy source for a steam cycle, leaves the heat exchanger at 380°F. The inlet conditions to the 86 percent efficient turbine in the steam cycle are 1600 psia and 900°F. The condenser pressure is 1 psia, and the pump is 70 percent efficient. Determine (*a*) the heat input to the air, in Btu/lb$_m$ of air, (*b*) the dimensionless mass-flow-rate ratio of air to steam, (*c*) the net work output of the gas-turbine cycle, in Btu/lb$_m$ of air, (*d*) the net work output of the steam cycle, in Btu/lb$_m$ of steam, and (*e*) the overall thermal efficiency of the combined cycle.

## BINARY VAPOR CYCLE

16-106.  A Rankine cycle with potassium as the working fluid has a turbine-inlet state of 1325 K and 6 atm, and the temperature of the condensing fluid is 950 K.

(*a*) Determine the thermal efficiency for an ideal cycle.

(*b*) Determine the thermal efficiency if the turbine- and pump-adiabatic efficiencies are 87 and 60 percent, respectively.

16-107. A potassium Rankine cycle operates with a turbine-inlet state of 1450 K and 10 atm, and the condenser temperature is 900 K.

(*a*) Determine the thermal efficiency for an ideal cycle.

(*b*) Repeat part *a* if the turbine and pump efficiencies are 86 and 65 percent, respectively.

16-108. A Rankine cycle with potassium as the working fluid has a turbine-inlet state of 1200 K and 3 atm, and the temperature of the condensing fluid is 900 K.

(*a*) Determine the thermal efficiency for an ideal cycle.

(*b*) Determine the thermal efficiency if the turbine- and pump-adiabatic efficiencies are 86 and 65 percent, respectively.

16-109. A binary vapor power cycle operates with the potassium cycle described in Prob. 16-106*b*. The heat rejected from this cycle enters the boiler-superheater section of a reheat-regenerative steam cycle. This cycle is described by Prob. 16-77*c*. Assume that the heat for the reheat section of the steam cycle is provided from another source, and not from the potassium cycle. Determine (*a*) the ratio of the mass flow rate of potassium to that of steam and (*b*) the overall thermal efficiency.

16-110. A binary vapor power cycle operates with the potassium cycle described in Prob. 16-107*b*. The heat rejected from this cycle enters the boiler-superheater section of a reheat-regenerative steam cycle. This cycle is described by Prob. 16-77*e*. Assume that the heat of the reheat section of the steam cycle is provided from another source, and not from the potassium cycle. Determine (*a*) the ratio of the mass flow rate of potassium to that of steam and (*b*) the overall thermal efficiency.

16-111. A binary vapor power cycle operates with the potassium cycle described in Prob. 16-108*b*. The heat rejected from this cycle enters the boiler-superheater section of a reheat-regenerative steam cycle. This cycle is described by Prob. 16-77*a*. Assume that the heat of the reheat section of the steam cycle is provided from another source, and not from the potassium cycle. Determine (*a*) the ratio of the mass flow rate of potassium to that of steam and (*b*) the overall thermal efficiency.

16-112E. A Rankine cycle with potassium as the working fluid has a turbine-inlet state of 2400°R and 90 psia, and the temperature of the condensing fluid is 1600°R.

(*a*) Determine the thermal efficiency for an ideal cycle.

(*b*) Determine the thermal efficiency if the turbine- and pump-adiabatic efficiencies are 87 and 60 percent, respectively.

16-113E. A potassium Rankine cycle operates with a turbine-inlet state of 2200°R and 50 psia, and the condenser temperature is 1600°R.
(a) Determine the thermal efficiency for an ideal cycle.
(b) Repeat part a if the turbine and pump efficiencies are 86 and 65 percent, respectively.

16-114E. A binary vapor power cycle operates with the potassium cycle described in Prob. 16-112b. The heat rejected from this cycle enters the boiler-superheater section of a reheat-regenerative steam cycle. This cycle is described by Prob. 16-82c. Assume that the heat for the reheat section of the steam cycle is provided from another source, and not from the potassium cycle. Determine (a) the ratio of the mass flow rate of potassium to that of steam and (b) the overall thermal efficiency.

16-115E. A binary vapor power cycle operates with the potassium cycle described in Prob. 16-113b. The heat rejected from this cycle enters the boiler-superheater section of a reheat-regenerative steam cycle. This cycle is described by Prob. 16-82a. Assume that the heat of the reheat section of the steam cycle is provided from another source, and not from the potassium cycle. Determine (a) the ratio of the mass flow rate of potassium to that of steam and (b) the overall thermal efficiency.

## OCEAN THERMAL-ENERGY CONVERSION

16-116. An OTEC system operates in a region of the ocean where the surface temperature is 25°C and the cold-water temperature is 7°C. Assume that both water streams experience a 3°C temperature change in the evaporator and condenser and that the minimum temperature between either water stream and the refrigerant 134a used as the working fluid is 2°C. The power requirement for the device is 10 MW. Determine (a) the refrigerant 134a flow rate, in kg/min, (b) the evaporator heat-transfer rate, in kJ/min, (c) the condenser heat-transfer rate, in kJ/min, (d) the refrigerant 134a pump power input, in kilowatts, and (e) the required mass flow rate for the cold-water pump which draws water from a depth of 800 m, in kg/min. Assume that all the pumps and the turbine are isentropic in operation.

16-117E. An OTEC system operates in a region of the ocean where the surface temperature is 80°F and the cold-water temperature is 40°F. Assume that both water streams experience a 5°F temperature change in the evaporator and condenser and that the minimum temperature between either water stream and the refrigerant 134a used as the working fluid is 5°F. The power requirement for the device is 10,000 hp. Determine (a) the refrigerant 134a flow rate, in Btu/min, (b) the evaporator heat-transfer rate, in Btu/min, (c) the condenser heat-transfer rate, in Btu/min, (*d*) the refrigerant 134a pump power input, in horse-power, and (*e*) the required mass flow rate for the cold-water pump which draws water from a depth of 2000 ft, in $lb_m$/min. Assume that all the pumps and the turbine are isentropic in operation.

## AVAILABILITY ANALYSIS OF STEAM POWER CYCLES

16-118. A simple steam power cycle has a turbine inlet state of 140 bars and 560°C and an exhaust pressure of 0.06 bar. The cooling water in the condenser undergoes a temperature rise from 18 to 28°C and $T_0$ is 25°C. The adiabatic efficiencies of the turbine and pump are 85 and 70 percent, respectively. Determine, in kJ/kg of steam,
   (*a*) The availability changes across the boiler, turbine, condenser, and pump.
   (*b*) The irreversibilities in the turbine, pump, and condenser.
   (*c*) Then find the effectiveness of the cycle.

16-119. A simple steam power cycle has a turbine inlet state of 6.0 MPa and 540°C and an exhaust pressure of 0.008 MPa. The cooling water in the condenser undergoes a temperature rise from 20 to 27°C and $T_0$ is 20°C. The adiabatic efficiencies of the turbine and pump are 85 and 70 percent, respectively. Determine, in kJ/kg of steam,
   (*a*) The availability changes across the boiler, turbine, condenser, and pump.
   (*b*) The irreversibilities in the turbine, pump, and condenser.
   (*c*) Then find the effectiveness of the cycle.

16-120. The simple steam power cycle in Prob. 16-119 is modified in the fol-lowing manner. Steam entering the turbine at 6.0 MPa and 540°C is expanded to 0.50 MPa, reheated reversibly to 500°C, and expanded to 0.008 MPa. Determine, in kJ/kg,
   (*a*) The availability changes across the two turbine stages, condenser, reheat section, boiler, and pump.
   (*b*) The irreversibilities in the two turbine stages, condenser, and pump.
   (*c*) Then find the effectiveness of the cycle.

16-121. The simple steam power cycle in Prob. 16-119 is modified in the fol-lowing manner. Steam entering the turbine at 6.0 MPa and 540°C is expanded to 0.50 MPa, where a portion is bled to a single open heater and the remainder is expanded to 0.008 MPa. Determine, in kJ/kg of steam through the boiler,
   (*a*) The availability changes across the two turbine stages, condenser, open heater, boiler, and pumps.
   (*b*) The irreversibilities in the two turbine stages, condenser, open heater, and pumps.
   (*c*) Then find the effectiveness of the cycle.

16-122. A simple steam power cycle has a turbine inlet state of 120 bars and 600°C and an exhaust pressure of 0.08 bar. The cooling water in the condenser undergoes a temperature rise from 22 to 28°C and $T_0$ is 22°C. The adiabatic efficiencies of the turbine and pump are 85 and 70 percent, respectively. Determine, in kJ/kg of steam,

(a) The availability changes across the boiler, turbine, condenser, and pump.
(b) The irreversibilities in the turbine, pump, and condenser.
(c) Then find the effectiveness of the cycle.

16-123. The simple steam power cycle in Prob. 16-122 is modified in the following manner. Steam entering the turbine at 120 bars and 600°C is expanded to 10 bars, reheated reversibly to 540°C, and expanded to 0.08 bar. Determine, in kJ/kg,

(a) The availability changes across the two turbine stages, condenser, reheat section, boiler, and pump.
(b) The irreversibilities in the two turbine stages, condenser, and pump.
(c) Then find the effectiveness of the cycle.

16-124. The simple steam power cycle in Prob. 16-122 is modified in the following manner. Steam entering the turbine at 120 bars and 600°C is expanded to 10 bars, where a portion is bled to a single open heater and the remainder is expanded to 0.08 bar. Determine, in kJ/kg of steam through the boiler,

(a) The availability changes across the two turbine stages, condenser, open heater, boiler, and pumps.
(b) The irreversibilities in the two turbine stages, condenser, open heater, and pumps.
(c) Then find the effectiveness of the cycle.

## PARAMETRIC AND DESIGN STUDIES

16-125. By means of a computer program investigate the effect of condenser pressure and boiler pressure on the thermal efficiency and specific work output of a simple Rankine power cycle. The condenser pressure ranges from 5 to 50 kPa and the boiler pressure ranges from 0.1 to 10 MPa. The turbine inlet temperature is constant at 500°C. Neglect the pump work.

16-126. By means of a computer program investigate the effect of superheating on the thermal efficiency and specific work output of a simple Rankine power cycle. The cycle operates between 5 kPa and 14 MPa; the range of turbine inlet temperatures is from 350 to 600°C. Neglect the pump work.

16-127. By means of a computer program investigate the effect of reheat on the thermal efficiency and turbine exit steam quality of an ideal steam power cycle operating between 8 kPa and 10 MPa. The turbine inlet temperature is 500°C and the range of reheat pressures to consider is from 0.1 to 5 MPa. Neglect the pump work. Discuss the pressure required to maximize the thermal efficiency.

16-128. Reconsider Prob. 16-127 if the adiabatic efficiency of any stage of the turbine is 82 percent.

16-129. By means of a computer program investigate the effect of the extraction pressure of a single open feedwater heater on the thermal efficiency and specific work output of an ideal steam power cycle without reheat. The turbine inlet state is 520°C and 140 bars, and the condenser pressure is 0.080 bar. The range of extraction pressures is from 1 to 10 bars. Neglect pump work. Discuss whether an optimum extraction pressure exists.

16-130. Reconsider Prob. 16-129 for the case where the adiabatic efficiency of each stage of the turbine is 85 percent.

16-131. By means of a computer program investigate the effect of open feedwater heaters on the thermal efficiency and work output of an ideal steam power cycle without reheat. Consider the cases of (a) no feedwater heater, (b) a single heater, and (c) two heaters. The turbine inlet state is 120 bars and 520°C, and the condenser pressure is 0.10 bar.

16-132. By means of a computer program investigate the effect of changing the heating load of the cogeneration plant in Prob. 16-88 on the mass flow rates and the boiler heat input rate. The heating load ranges from $10^6$ to $10^7$ kJ/min, with all other data remaining the same.

16-133. Three options are proposed for the design of a 100-MW power plant. Option 1 is a gas turbine cycle with an overall pressure ratio of 10 and a maximum cycle temperature of 1400 K. The compressor and turbine each have three stages for which the pressure ratios are equal. Two ideal intercoolers and two ideal reheaters are used, and the compressor and turbine adiabatic efficiencies are 80 and 85 percent, respectively. The inlet air from the surroundings is at 300 K and 1 bar. Option 2 includes the same gas-turbine cycle as option 1, but a regenerator with an effectiveness of 70 percent is added to the cycle. Option 3 includes the same gas-turbine cycle as option 2, but the exhaust stream leaving the regenerator is used as the energy source for a Rankine steam cycle. Thus option 3 operates as a combined cycle. The steam cycle operates between 20 and 0.1 bar, with a turbine inlet temperature of 320°C and a turbine exit quality of 90 percent. The exhaust gases leaving the steam generator are required to be at 490 K for satisfactory operation of the stack. In addition, the minimum temperature difference in the steam generator heat exchanger must be equal to or greater than 40°C (the $\Delta T$ at the pinch point).

In order to make a feasiblility and economic analysis of the three options based on life cycle costs over a period of 40 years, the following information regarding fuel and operating cost and capital cost are available:

(1) Fuel and operating cost for all three options: $0.01 per kWh of heat input.

(2) The capital cost of the gas-turbine cycle without regenerations is 20 million dollars.

(3) The capital cost of the gas-turbine cycle with regeneration is 25 million dollars.

(4) The capital cost of the combined cycle is 35 million dollars.

Any economic analysis should consider the life cycle cost (LLC) as

$$\text{LCC} = (\text{fuel and operating costs added over 40 years}) + P(1 + R)^N$$

where $P$ is the capital cost, $R$ is the interest rate, assumed to be 5 percent, and $N$ is the number of years of operation. Inflation and salvage value are to be neglected, as well as changes and fluctuations in (*a*) interest rates, and (*b*) cost of fuel and operation. The results should include a table of the thermal efficiency and heat rate and a table of life-cycle costs for the three options.

16-134. A company has a new requirement for both electrical power and process heat. One method for providing these needs is the combination of a gas-turbine power system to drive an electrical generator and a steam-generation plant for process heat. The energy in the exhaust of a conventional gas-turbine power cycle is to be used as an energy source for the production of process steam. A schematic of the steam-cycle portion of the overall design is shown in Fig. P16-134, where state 4 represents the exhaust condition from the gas-turbine cycle. Suggested pressure ratios ($r_p$) for the gas-turbine cycle are 3.46, 4.78, 6.0, 7.41, and 9.07. The compressor and turbine inlet temperatures have been set at 295 and 1020 K, and the compressor and turbine

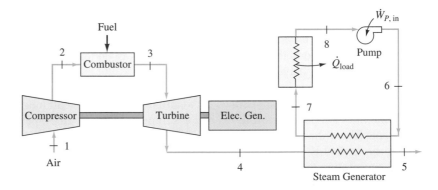

**Figure P16-134**

adiabatic efficiencies are 84 and 87 percent, respectively. For the process steam loop, the steam pressure in the steam generator is constant at 10 bars. The exit conditions for the heating load are 3 bars and 80°C at state 8, and the adiabatic efficiency of the pump is 70 percent. Pressure drops through all piping are neglected. The heat-exchanger group has specified that $T_7 = T_4 - 30$ and $T_5 = T_6 + 30$, where $T$ is in degrees Celsius, and $T_6 = T_8$. Design results should include (a) $\dot{m}_7$ for an $\dot{m}_4$ of 10 kg/s in the gas turbine, in kg/s, (b) $\dot{Q}_{load}$, in kJ/s, and (c) $\dot{W}_{net,out}$, in kJ/s. To account for differences in the cost of the heating and power equipment, the economics section of the company has estimated that overall costs are related by

$$\text{Cost} = K(0.50\dot{Q}_{load}\dot{m}_{load}^2 + 0.05\dot{W}_{net,out}r_p^2)$$

Determine the variation of costs (in terms of $K$, a cost factor) with pressure ratio, and whether there is a pressure ratio which leads to an optimum cost.

16-135. An important variable in the selection of a regenerative steam power cycle is the pressures chosen for the extraction of steam from the turbine. You are asked to evaluate such a system with a single open heater. The turbine inlet conditions are 100 bars and 360°C, and the condenser pressure is 1 bar. The turbine and pump are assumed to be isentropic, and pressure drops in piping are neglected.

(a) Estimate the extraction pressure ($P_4$) which leads to the maximum thermal efficiency, using the "rule of thumb" discussed in this chapter.

(b) Determine (1) the extraction pressure, (2) the mass flow rate split, $y$, (3) the net work output, and (4) the thermal efficiency for the pressure which maximizes the thermal efficiency. (Note, these calculations are quite sensitive to the significant figures used.)

(c) Economic considerations involve the following capital cost of equipment. The boiler costs $100q_h$ dollars, the condenser costs $50q_{cond}$ dollars, a feedwater heater costs $1000P_4$ dollars, the feedwater pump costs $1000w_P$ dollars, and the turbine costs $1000w_T$ dollars, where $q$ and $w$ terms are measured in kJ/kg, and $P_4$ is measured in bars. It is suggested that the total cost be determined per unit of net work output for selected $P_4$ values.

16-136. A cogeneration system is being considered for an industrial plant that requires new sources of both high-temperature heating and electricity. Currently, electricity is bought from the local utility at a rate of 13 cents per kilowatt-hour and heating is provided by a high-temperature furnace at a cost of $5.00 per million Btu. The boiler and condenser of the proposed cogeneration system will operate at 440 and 50°C, respectively. Assume a boiler energy input of $5 \times 10^6$ kJ/h, a steam quality of 0.90 at the turbine exit, isentropic turbines and pumps, and no inefficiencies associated with the generation of electricity.

(*a*) Determine and plot the overall efficiency of the cogeneration system for providing useful work and process heating as a function of the fraction of flow diverted to the process-heat heat exchanger, using temperatures of 120 and 180°C for the steam condensing in the process-heat heat exchanger. Assume the fluid leaves the heat exchanger as a saturated liquid.

(*b*) Determine and plot the operating cost savings in dollars per hour associated with the cogeneration system as compared to the current situation as a function of flow diverted to the process heat exchanger. Do plots for temperatures of 120 and 180°C for the steam condensing in the heat exchanger. Assume that energy delivered to the boiler costs $5.00 per million Btu.

What do you recommend for the flow split to the process-heat heat exchanger? Should the second-stage turbine and heat-rejection condenser be removed for this application?

16-137. A Rankine cycle operating between 1 bar and 540°C, 100 bars, is to be modified with the addition of open feedwater heaters. Each feedwater heater is to be operated at the optimum extraction pressure from the turbine to maximize the cycle efficiency. The optimum pressure for a single heater is that pressure which corresponds to a saturation temperature one-half of that of the boiler plus condenser saturation temperatures. For two heaters, the procedure is repeated for the boiler to second heater and the first heater to the condenser, and these two equations can be solved for the two unknown saturation temperatures. A similar method is used for a greater number of feedwater heaters. Each turbine stage is assumed to be isentropic. The designer is required to determine:

(1) The bleed-off pressures, the fraction of flow to be bled off, and net work output, in kJ/kg, and the thermal efficiency for the number of open heaters *n* varying from 0 to 4.

(2) The number of open heaters which maximize the thermal efficiency. If the maximization occurs after $n = 4$, do your data enable an estimate of any such maximum, or do they show that no maximum will occur?

# CHAPTER

# 17

# REFRIGERATION SYSTEMS

An air-cooled condenser (20 to 120 tons capacity) is used to reject thermal energy from a vapor-compression refrigeration cycle. (Courtesy of the Trane Company)

**A** refrigeration system is used to maintain a region of space at a temperature below that of the environment. The working fluid employed in the system may remain in a single phase (gas refrigeration) or may change phases (vapor-compression refrigeration). It is common to associate refrigeration with the preservation of foods and the air conditioning of buildings. However, there are many other needs for refrigeration techniques. The use of liquid fuels for rocket propulsion, liquefied oxygen for steel making, liquid nitrogen for low-temperature (cryogenic) research and surgery techniques, and liquefied natural gas for shipment between continents are a few examples where refrigeration is essential. The heat pump, which is capable of providing both a cooling and a heating effect with the same equipment, continues its popular use in residential and commercial buildings. In this chapter we examine some of the basic thermodynamic cycles used for the maintenance of low temperatures.

## 17-1    THE REVERSED CARNOT CYCLE

In Sec. 8-3 the Carnot heat-engine cycle is introduced as a theoretical cycle which achieves the Carnot efficiency $\eta_{th,Carnot} = 1 - T_L/T_H$. Since the cycle is reversible, the four processes which comprise the Carnot heat-engine cycle may be reversed in direction. A schematic of a reversed Carnot heat-engine cycle, now operating as a refrigerator or heat pump, is shown in Fig. 17-1$a$. Heat transfer $\dot{Q}_L$ occurs reversibly in a heat exchanger from a low-temperature source of temperature $T_L$ to the fluid in the reversed heat engine. The fluid undergoes an isothermal change of state at $T_L$ from state 4 to state 1. The fluid is then compressed isentropically from state 1 to state 2 by work transfer until the temperature reaches $T_H$. Then the fluid undergoes an isothermal, reversible change from state 2 to state 3. This is accomplished by reversible heat transfer $\dot{Q}_H$ in another heat exchanger to a high-temperature sink at $T_H$. Finally the fluid is expanded isentropically in a turbine from state 3 to the initial state 4 of the fluid.

In order for processes 4-1 and 2-3 to be isothermal, the fluid undergoes evaporation during the heat addition $\dot{Q}_L$ and condensation during the heat removal $\dot{Q}_H$. The steady-flow Carnot refrigerator or heat-pump cycle is shown on a $Ts$ diagram in Fig. 17-1$b$. The conservation of energy for the steady-state cyclic process taken as a closed system is $\dot{Q}_{L,in} - \dot{Q}_{H,out} + \dot{W}_{net,in} = 0$. Similarly the entropy balance reduces to $\sum \dot{Q}_j/T_j = 0$, or $\dot{Q}_{H,out}/T_H = \dot{Q}_{L,in}/T_L$ because the system is steady-state and reversible. The Carnot refrigeration cycle is a standard of performance because it requires the minimum work input for a given refrigeration effect between two regions of fixed temperature.

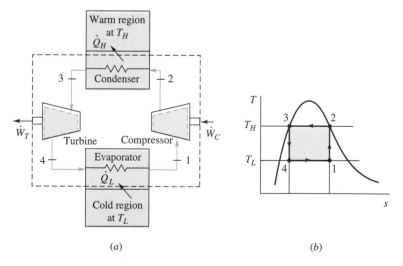

(a)                                                (b)

**Figure 17-1**    A reversed Carnot heat engine operating in the wet region.

The energy-efficiency standard for refrigeration processes is the coefficient of performance (COP) which is defined as

$$\text{COP}_{\text{Ref}} \equiv \frac{\dot{Q}_{L,\text{in}}}{\dot{W}_{\text{in}}} \qquad \qquad \textbf{[17-1]}$$

as noted earlier in Sec. 6-2-2. Recall that the areas under the $T_H$ line between states 2 and 3 and under the $T_L$ line between states 1 and 4 in Fig. 17-1$b$ represent $\dot{Q}_H$ and $\dot{Q}_L$, respectively. Therefore, for a Carnot refrigerator,

$$\text{COP}_{\text{Ref,Carnot}} = \frac{T_L}{T_H - T_L} \qquad \qquad \textbf{[17-2]}$$

Note that the value of the COP can exceed unity and, in fact, does so for a well-designed unit. Note also that the main variable that controls the COP of a Carnot refrigerator is the temperature difference $T_H - T_L$. For a high COP, $T_H$ should be as low as possible and $T_L$ should be as high as possible. However, $T_H$ cannot be less than the temperature of the environment to which heat is rejected, and $T_L$ cannot be greater than the temperature of the cold region from which heat is removed.

The disadvantages of a Carnot refrigeration cycle as a model for practical devices lie in the compression and expansion processes. The compression of a wet mixture generally is to be avoided, owing to possible damage to the compressor by the presence of liquid droplets. For a turbine, expansion involving a fluid with a high moisture content also is highly undesirable owing to the liquid droplets as well as the low turbine efficiency. Modifications to overcome these two inherent problem areas in the Carnot refrigeration cycle lead to the practical vapor-compression refrigeration cycle discussed in the next section.

## 17-2   THE VAPOR-COMPRESSION REFRIGERATION CYCLE

Although the reversed Carnot cycle is a standard to which all actual cycles may be compared, it is not a practical cycle for refrigeration purposes. It is highly desirable to retain, as far as possible, the constant-temperature heat-addition and heat-rejection processes. However, the compression process should involve only the vapor phase. In addition, the turbine is replaced by a throttling device, which is much less expensive than a work expander and is nearly maintenance-free. When these changes are made, the cycle is known as a *vapor-compression refrigeration cycle.* The schematic of the equipment for the cycle, along with $Ts$ and $Ph$ diagrams of the *ideal* cycle, is shown in Fig.17-2. The specific processes for the fluid in an ideal cycle are:

1-2. Isentropic compression from saturated to superheated vapor

2-3. Internally reversible, constant pressure change with heat transfer from the fluid until a saturated liquid

3-4. Throttling through a valve or capillary tube to $P_4 = P_1$, where $h_4 = h_3$

4-1. Internally reversible, constant pressure change with heat transfer to a low-quality wet mixture until a saturated vapor

Note on the $Ts$ and $Ph$ diagrams that all of process 4-1 and a large portion of process 2-3 occur at constant temperature. The assumptions of saturated vapor at state 1 and saturated liquid at state 3 are part of the definition of the *ideal* cycle. Unlike many other "ideal" cycles, the vapor-compression cycle modeled in Fig. 17-2 contains an irreversible process within it, the throttling process. All other portions of the cycle are assumed to be internally reversible. The effects of internal and external irreversibilities are discussed in later sections. Typically for residential air conditioning, processes 1-2, 2-3, and 3-4 occur in an outdoor unit as shown in Photograph 17-1. Process 1-4 occurs in an evaporator coil located in the air ducts.

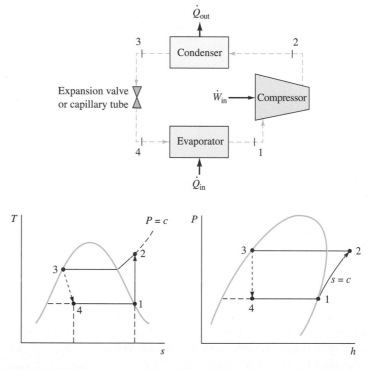

**Figure 17-2**   A schematic of the equipment and *Ts* and *Ph* diagrams for an ideal vapor-compression refrigeration cycle.

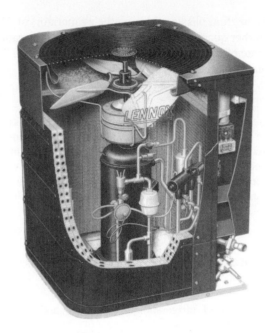

**Photograph 17-1**   Outdoor unit for a vapor-
compression air conditioner
(heat pump) showing the compressor and condenser (evaporator). (Courtesy of Lennox)

For the four processes described above for the ideal vapor-compression refrigeration cycle, the kinetic- and potential-energy changes are usually negligible. Therefore, the general steady-state energy equation on a unit-mass basis for each process reduces to

$$q + w = h_e - h_i \qquad \textbf{[17-3]}$$

where subscripts $i$ and $e$ again represent inlet and exit states. Heat transfer is absent from the compressor and throttling processes, while a work interaction is present only in the compressor process. The coefficient of performance for the cycle in Fig. 17-2 is

$$\text{COP}_{\text{Ref}} = \frac{q_{\text{evap}}}{w_C} = \frac{h_1 - h_4}{h_2 - h_1} \qquad \text{(vapor compression)} \quad \textbf{[17-4]}$$

The energy and COP equations are equally valid for an idealized cycle as well as for an actual cycle where irreversibilities are present in the compressor, condenser, and evaporator.

The rating of refrigeration systems is frequently given on the basis of the **refrigeration capacity,** which is the rate of heat transfer $\dot{Q}_{\text{evap}}$ in the evaporator. On the basis of Fig. 17-2, $\dot{Q}_{\text{evap}} = \dot{m}(h_1 - h_4)$. Frequently the capacity is expressed in terms of tons of refrigeration. A **ton of refrigeration** is defined as a heat-transfer rate from the cold region (or the heat-transfer rate to the fluid passing through the evaporator) of 211 kJ/min or 200 Btu/min. Another quantity frequently cited for a refrigeration device is the volume flow rate of the refrigerant at the compressor inlet. This is the **effective displacement** of the compressor.

**Example 17-1**

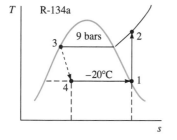

**Figure 17-3**
A *Ts* process diagram with data for Example 17-1.

An ideal vapor-compression refrigeration cycle with refrigerant 134a as the working fluid operates with an evaporator temperature of $-20°C$ and a condenser pressure of 9.0 bars. The mass flow rate of the refrigerant is 3 kg/min. Compute (*a*) the coefficient of performance, (*b*) the refrigeration capacity in tons of refrigeration, (*c*) the coefficient of performance of a Carnot reversed heat engine operating under the same evaporation and condensation temperatures as the actual cycle, and (*d*) the effective displacement, in $m^3/min$.

**Solution:**

**Given:** An ideal vapor-compression refrigeration cycle operating with R-134a under the conditions shown on the *Ts* diagram in Fig. 17-3.

**Find:** (*a*) $COP_{Ref}$, (*b*) tons of refrigeration, (*c*) $COP_{Carnot}$, and (*d*) $\dot{V}_1$, $m^3/min$.

**Model:** Steady state, internally reversible processes except for throttling; adiabatic compression; negligible $\Delta ke$ and $\Delta pe$; saturated vapor enters compressor, saturated liquid leaves condenser.

**Analysis:** The evaluation of the COP of a vapor-compression refrigeration cycle requires enthalpy data. Tables A-16 to A-18 provide data for the states illustrated in Fig. 17-3. State 1 is defined as a saturated vapor; therefore, at $-20°C$

$$h_1 = h_g = 235.31 \text{ kJ/kg} \qquad s_1 = s_g = 0.9332 \text{ kJ/kg·K}$$

In addition, the fluid pressure in the evaporator is the saturation value of roughly 1.33 bars. At 9 bars the entropy of saturated vapor is 0.9054 kJ/kg·K, which is less than the compressor inlet entropy of 0.9332 kJ/kg·K. Hence isentropic compression to state 2 leads to a superheated vapor. Interpolation in the superheat table at 9 bars indicates that $T_{2s}$ is roughly 43.3°C and $h_{2s} = 274.90$ kJ/kg. At the end of the condensation process where the fluid is a saturated liquid,

$$h_3 = h_f \text{ at 9 bars} = 99.56 \text{ kJ/kg} \qquad T_3 = T_{sat} = 35.53°C$$

The enthalpy at state 3 is also the enthalpy at state 4, after the fluid passes through the throttling device. Consequently,

(*a*) The COP is given by $q_{evap}/w_C$. Expressing $q_{evap}$ and $w_c$ in terms of enthalpy data, we find that

$$COP = \frac{q_{evap}}{w_C} = \frac{h_1 - h_4}{h_2 - h_1} = \frac{235.31 - 99.56}{274.90 - 235.31} = 3.43$$

(*b*) The refrigeration capacity is

$$\dot{Q}_{refrig} = \dot{m}q_{evap} = \frac{3 \text{ kg/min} \times (235.31 - 99.56) \text{ kJ/kg}}{211 \text{ kJ/min·ton}} = 1.93 \text{ tons}$$

(c) The Carnot COP based on evaporation and condensation temperatures is

$$\text{COP}_{\text{Carnot}} = \frac{T_1}{T_3 - T_1} = \frac{253}{55.5} = 4.56$$

(d) At state 1, $v_1 = v_g = 0.1464 \text{ m}^3/\text{kg}$ at $-20°C$. Thus the effective displacement is given by

(Volume flow rate)$_1$ = $\dot{m}v_1$ = 3 kg/min (0.1464 m³/kg) = 0.439 m³/min

**Comment:** The somewhat larger COP of the Carnot cycle is not unexpected. The vapor-compression cycle has a higher average temperature during condensation, and it contains an irreversibility in the throttling process.

## 17-2-1 EFFECT OF COMPRESSOR IRREVERSIBILITIES

Irreversibilities in the flow through the compressor lead to an increase in the entropy of the fluid during an adiabatic process. This is also accompanied by an increase in the final temperature over that found in the ideal case. This is shown as path 1-2a in Fig. 17-4. For this adiabatic, irreversible case the actual exit state can be predicted from the compressor adiabatic efficiency:

$$\eta_C = \frac{h_{2s} - h_1}{h_{2a} - h_1} \qquad \textbf{[8-28]}$$

If heat transfer from the compressor is great enough, though, the actual entropy of the fluid at the compressor exit can be less than that at the inlet. This case is shown by path 1-2' in Fig. 17-4. Whether the entropy increases or decreases in an actual compressor depends on the relative contributions of friction versus heat transfer. A net heat transfer from the compressor is beneficial, since this effect lowers the average specific volume of the fluid during compression. The smaller specific volume leads to a reduced work input. State 2' is usually found from experimental measurement or from a polytropic model of the process. The use of Eq. [8-28] for predicting compressor performance is not appropriate when heat transfer occurs. The effect of frictional flow through an adiabatic compressor on cycle performance is illustrated below.

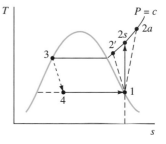

**Figure 17-4**
Effect of irreversibilities and heat losses on compressor performance.

A vapor-compression refrigeration cycle with R-134a as the working fluid operates with an evaporator temperature of $-20°C$, a condenser pressure of 9.0 bars, and a compressor efficiency of 84 percent. The mass flow rate of refrigerant is 3 kg/min. Find (a) the power input to the compressor, in kilowatts, (b) the coefficient of performance, and (c) the entropy generation in the compressor and throttling processes, in kJ/min·K.

**Example 17-2**

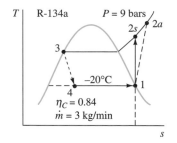

**Figure 17-5**

A $Ts$ process diagram with data for Example 17-2.

**Given:** A vapor-compression refrigeration cycle with irreversible compressor as shown in Fig. 17-5.

**Find:** (a) $\dot{W}_C$, in kW; (b) COP, and (c) $\dot{\sigma}_C$ and $\dot{\sigma}_{\text{throt}}$, in kJ/min·K.

**Model:** Steady state, irreversible compressor; neglect $\Delta$ke and $\Delta$pe.

**Analysis:** This problem is the same as that discussed in Example 17-1, except the ideal compressor is replaced by one with an efficiency of 84 percent. Data from Example 17-1 that are still applicable include:

$$T_1 = -20°C \quad P_1 = 1.33 \text{ bars} \quad h_1 = 235.31 \text{ kJ/kg} \quad s_1 = 0.9332 \text{ kJ/kg·K}$$

$$h_{2s} = 274.90 \text{ kJ/kg} \quad T_{2s} = 43.3°C \quad T_3 = 35.53°C \quad h_3 = h_4 = 99.56 \text{ kJ/kg}$$

(a) The compressor adiabatic efficiency is given by $\eta_C = (h_{2s} - h_1)/(h_{2a} - h_1)$. Therefore, the actual outlet enthalpy is

$$h_{2a} = h_1 + \frac{h_{2s} - h_1}{\eta_C} = 235.31 + \frac{274.90 - 235.31}{0.84} = 282.44 \text{ kJ/kg}$$

From Table A-18 at 9.0 bars, $h = 282.34$ kJ/kg at 50°C. Thus $T_{2a}$ is very close to 50°C. As a result, the actual compressor work is

$$w_C = h_{2a} - h_1 = 282.44 - 235.31 = 47.13 \text{ kJ/kg}$$

The required power input is the product of the mass flow rate and the compressor work. Hence

$$\dot{W}_C = \dot{m}w_C = 3 \text{ kg/min } (47.13 \text{ kJ/kg}) = 141.4 \text{ kJ/min} = 2.36 \text{ kW}$$

(b) The coefficient of performance for the cycle becomes

$$\text{COP} = \frac{q_{\text{evap}}}{w_C} = \frac{h_1 - h_4}{h_{2a} - h_1} = \frac{135.75}{47.13} = 2.88$$

(c) The steady-state rate equation for entropy generation is

$$\dot{\sigma}_{\text{cv}} = \sum_{\text{out}} \dot{m}_e s_e - \sum_{\text{in}} \dot{m}_i s_i - \sum_{j=1}^{n} \frac{\dot{Q}_j}{T_j}$$

For the adiabatic compressor this equation reduces to $\dot{\sigma}_C = \dot{m}(s_{2a} - s_1)$. The value of $s_{2,a} = 0.9566$ kJ/kg·K, found in Table A-18 at 50°C and 9.0 bars. Hence the entropy generation in the compressor is

$$\dot{\sigma}_C = 3 \text{ kg/min} \times (0.9566 - 0.9332) \text{ kJ/kg·K} = 0.0702 \text{ kJ/min·K}$$

For the adiabatic throttling process the same equation applies, that is, $\dot{\sigma}_{\text{throt}} = \dot{m}(s_4 - s_3)$. From Table A-17 at 9 bars, $s_3 = s_f = 0.3656$ kJ/kg·K. The quality at state 4 is found from saturation data in Table A-16 at $-20°C$.

$$x_4 = \frac{h_4 - h_f}{h_{fg}} = \frac{99.56 - 24.26}{211.05} = 0.357$$

As a result, the entropy at state 4 is

$$s_4 = (1 - x_4)s_f + x_4 s_g$$
$$= 0.643(0.0996) + 0.357(0.9332) = 0.3972 \text{ kJ/kg·K}$$

Therefore, the rate of entropy generation during throttling is

$$\dot{\sigma}_{\text{throt}} = 3(0.3972 - 0.3656) = 0.0948 \text{ kJ/min·K}$$

**Comment:** A compressor adiabatic efficiency of 84 percent has led to an additional temperature rise of roughly 7°C in the compressor due to irreversibilities and a decrease of 16 percent in the COP compared to the ideal-compressor case in Example 17-1. The rate of entropy generation during throttling is somewhat greater than that in the irreversible compressor.

## 17-2-2  EFFECT OF IRREVERSIBLE HEAT TRANSFER ON CYCLE PERFORMANCE

The evaluation of refrigeration cycle performance in the previous examples has been based on the assignment of saturation temperatures or pressures of the refrigerant in the evaporator and condenser. Nevertheless, the operating temperatures of the working fluid in an actual cycle are really established by the temperature to be maintained in the cold region and the temperature of the cooling water or air that is available for use in the condenser. In addition, to obtain sufficient heat-transfer rates, a finite temperature difference must be maintained in the two heat exchangers.

Figure 17-6 illustrates this point. In the evaporator, heat transfer occurs from the cold region to the refrigerant, which undergoes a phase change at constant temperature. For effective heat transfer, the saturation temperature of the refrigerant must be lower than the temperature of the cold region ($T_{\text{cr}}$ on the figure). In the condenser the refrigerant is being condensed by heat transfer to a coolant flow stream external to the cycle. Cooling water and atmospheric air are two typical coolants which might be passed over the condenser tubes. To accomplish this heat transfer, the saturation temperature of the refrigerant in the condenser must be greater than the temperature of the atmospheric coolants ($T_{\text{cw}}$ in the figure).

The finite temperature differences required for the heat-transfer processes constitute irreversibilities which are not included in the Carnot refrigeration model. On the basis of Eq. [17-2], the coefficient of performance decreases as the average temperature in the evaporator is lowered and the average temperature of the refrigerant in the condenser is raised. This effect is examined in the example below. These irreversibilities in the heat-transfer processes are also measured by entropy generation external to the fluid cycle. The evaluation of the entropy generation external to the condenser requires an idealization, as noted in the following example.

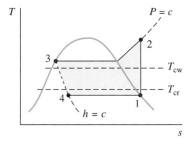

**Figure 17-6**
Effect of irreversible heat transfer on the performance of a vapor-compression refrigeration cycle.

**Example 17-3**

**R**econsider the ideal vapor-compression refrigeration cycle in Example 17-1. The cold region is to be maintained at $-20°C$. To allow for a finite temperature difference between the cold region and the evaporator fluid, the evaporator fluid is maintained at $-26°C$. The air which passes over the condenser coils in Example 17-1 is at the saturation temperature corresponding to 9 bars, or $35.53°C$. Again, to allow for a finite temperature difference, the condenser pressure is raised to 12 bars, for which the saturation temperature is $46.32°C$. Determine (*a*) the coefficient of performance of the cycle under the new conditions, and compare with the result of Example 17-1, and (*b*) the rate of entropy generation in the heat transfer region where a finite temperature difference occurs in the evaporator and in the condenser, in kJ/min·K.

**Solution:**

**Given:**    An ideal vapor-compression refrigeration cycle with finite temperature differences between the circulating fluid and the cold and hot regions, as shown in Fig. 17-7.

**Find:**    (*a*) COP, and compare to Example 17-1, and (*b*) $\dot{\sigma}_{Q,\text{evap}}$ and $\dot{\sigma}_{Q,\text{cond}}$.

**Model:**    Steady state, internally reversible compressor; neglect $\Delta$ke and $\Delta$pe.

**Analysis:**    From Table A-16 at the evaporator exit temperature of $-26°C$,

$$P_1 = p_g = 1.02 \text{ bars} \quad h_1 = h_g = 231.62 \text{ kJ/kg} \quad s_1 = s_g = 0.9390 \text{ kJ/kg·K}$$

(*a*) State 2 at the compressor outlet is found from knowledge of $P_2$ and the fact that $s_2 = s_1$. By interpolation at 12 bars in the superheat table A-18, we find that $T_{2s}$ is roughly $56.2°C$ and $h_{2s} = 282.94$ kJ/kg. At the end of the condensation process at 12 bars, from Table A-17,

$$h_3 = h_f = 115.76 \text{ kJ/kg}$$

The enthalpy at state 3 is also the enthalpy at state 4, after the fluid passes through the throttling process. Consequently,

$$\text{COP} = \frac{q_{\text{evap}}}{w_C} = \frac{h_1 - h_4}{h_2 - h_1} = \frac{231.62 - 115.76}{282.94 - 231.62} = 2.26$$

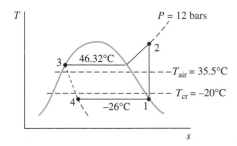

**Figure 17-7**    A $Ts$ diagram with data for Example 17-3.

(b) The entropy generation due to irreversible heat transfer across a finite temperature difference can be found from an entropy balance on the region. For a region with a steady-state heat transfer rate of $\dot{Q}$ from a high temperature $T_{high}$ to a low temperature $T_{low}$, the entropy balance reduces to

$$\dot{\sigma}_Q = \dot{Q}\left(\frac{1}{T_{low}} - \frac{1}{T_{high}}\right)$$

For the evaporator, $T_{high} = 253.15$ K, $T_{low} = 247.15$ K, and $q_{evap} = 115.86$ kJ/kg. Substitution of data into the above equation yields

$$\dot{\sigma}_{Q,evap} = 3(115.86) \text{ kJ/min}\left(\frac{1}{247.15 \text{ K}} - \frac{1}{253.15 \text{ K}}\right) = 0.0333 \text{ kJ/min·K}$$

In the condenser the R-134a begins in the superheat region at 56.2°C with an enthalpy of 282.94 kJ/kg. It is cooled to a saturated vapor at 46.32°C, for which $h_g$ is 270.99 kJ/kg, and then to a saturated liquid, for which $h_f$ is 115.76 kJ/kg. The heat is rejected to air at a constant temperature of 35.53°C, or 308.7 K. The rate of entropy generation due to irreversible heat transfer is determined by $\dot{\sigma}_Q = \dot{Q}(1/T_{low} - 1/T_{high})$. In order to use this equation for evaluating $\dot{\sigma}_Q$, we *model* the desuperheating process which occurs over a range from 56.2 to 46.3°C as occurring at an arithmetic average temperature of 51.2°C, or 324.4 K. (This model assumes that the pressure line in the superheat region just outside the saturated-vapor line on a $Ts$ diagram is straight, which is essentially correct.) Hence the entropy generation for the condenser heat-loss process in this case is calculated in two steps, namely,

$$\dot{\sigma}_{Q,cond} = \dot{m}(h_{2s} - h_g)\left(\frac{1}{T_{air}} - \frac{1}{T_{avg}}\right) + \dot{m}h_{fg}\left(\frac{1}{T_{air}} - \frac{1}{T_{sat}}\right)$$

$$= 3(282.94 - 270.99)\left(\frac{1}{308.7} - \frac{1}{324.4}\right) + 3(155.23)\left(\frac{1}{308.7} - \frac{1}{319.5}\right)$$

$$= 0.0056 + 0.0510 = 0.0566 \text{ kJ/min·K}$$

Note that the contribution of the first term is so small that use of an average temperature to evaluate $\dot{\sigma}_{Q,cond}$ has a reasonably small effect. These overall values for $\dot{\sigma}_{Q,evap}$ and $\dot{\sigma}_{Q,cond}$ are slightly smaller than the values found in Example 17-2 for entropy generation in the compressor and throttling devices.

**Comment:**    By allowing for a 6°C differential at the evaporator end of the cycle and a 9°C minimum difference along the condenser, the COP has decreased from 3.43 to 2.26. Although these calculations are based on an idealized cycle, the trend in the actual case would be similar. The COP is severely affected by the need for finite temperature differences across the condenser and evaporator heat-transfer surfaces.

---

## 17-2-3  EFFECT OF SUPERHEATING AND SUBCOOLING ON CYCLE PERFORMANCE

Since it is not possible to control exactly the state of the fluid leaving the evaporator, the system is usually designed so that the fluid leaves as a slightly superheated vapor rather than as the saturated vapor found in the ideal cycle. This ensures that the compressor is always handling the vapor phase. This overdesign leads to an increase in the specific volume of the vapor in comparison to the saturated vapor at the same pressure. This unfortunately leads to an increase in the power input to the compressor, because steady-flow work is proportional to the specific volume (recall that $w_{sf,rev} = \int v \, dP$). In the ideal cycle the refrigerant is assumed to leave the condenser as a saturated liquid. In practice, even if pressure losses in the condenser are small, the fluid usually leaves the condenser as a subcooled liquid. This is a beneficial effect, since the low enthalpy which results from the subcooling effect permits a larger quantity of energy to be added per unit mass during the evaporation process. Note, however, that subcooling and superheating increase the size of the condenser and evaporator, which increases the initial cost of the heat exchangers.

**Example 17-4**

**A** vapor-compression refrigeration cycle operates with refrigerant 134a and a pressure of 0.10 MPa in the evaporator and 1.20 MPa in the condenser. The fluid leaves the evaporator superheated by 6.43°C and leaves the condenser subcooled by 4.32°C. Calculate the coefficient of performance of the cycle if the adiabatic compressor has an efficiency of (a) 100 and (b) 84 percent.

**Solution:**

**Given:**   A vapor-compression refrigeration cycle with superheated vapor leaving the evaporator and subcooled liquid leaving the condenser, as shown in Fig. 17-8.

**Find:**   COP for (a) $\eta_C = 1.0$ and (b) $\eta_C = 0.84$.

**Model:**   Steady-state; neglect $\Delta$ke and $\Delta$pe.

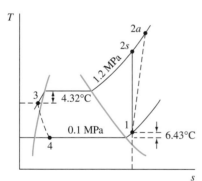

**Figure 17-8**
A $Ts$ diagram with data for Example 17-4.

**Analysis:**   Calculation of the COP requires a reevaluation of the enthalpies around the cycle. The saturation temperatures for pressures of 0.10 and 1.20 MPa are −26.43 and 46.32°C, respectively. Thus

$$T_1 = -26.43 + 6.43 = 20.0°C \qquad \text{and} \qquad T_3 = 46.32 - 4.32 = 42.0°C$$

Since the fluid at the condenser exit is subcooled, its enthalpy at state 3 is approximated by the saturated liquid enthalpy at 42°C. Hence from the superheat table A-18 for R-134a at 1 bar and −20.0°C and from the saturation table A-16 at 42°C we find that

$$h_1 = 236.54 \text{ kJ/kg} \qquad s_1 = 0.9602 \text{ kJ/kg·K} \qquad h_3 = 109.19 \text{ kJ/kg} = h_4$$

(a) The enthalpy at the compressor outlet pressure of 1.2 MPa is found by interpolation in the superheat table where $s_2 = s_1$. At this state $h_{2s}$ is 289.97 kJ/kg at roughly 62.2°C. The COP for the cycle with an isentropic compressor is

$$\text{COP} = \frac{q_{evap}}{w_C} = \frac{h_1 - h_4}{h_{2s} - h_1} = \frac{236.44 - 109.19}{289.97 - 236.44} = 2.38$$

(b) In the case of a compressor with an efficiency of 84 percent, the actual outlet enthalpy is

$$h_{2a} = h_1 + \frac{h_{2s} - h_1}{\eta_C} = 236.44 + \frac{289.97 - 236.44}{0.84} = 300.17 \text{ kJ/kg}$$

This corresponds to a temperature of 71.1°C. Thus the fluid irreversibility has raised the outlet temperature by 8.9°C over the isentropic case. The COP now is

$$\text{COP} = \frac{h_1 - h_4}{h_{2a} - h_1} = \frac{236.44 - 109.19}{300.17 - 236.44} = 2.00$$

The presence of an irreversible compressor has reduced the COP by 16 percent, similar to the same effect in Example 17-2.

**Comment:** In Example 17-3, where the evaporator pressure was 1.02 bars rather than 1.0 bar (0.1 MPa), the COP was 2.26 for an ideal compressor without superheating at the inlet. Thus superheating in the evaporator and subcooling in the condenser for part *a* above has raised the COP to 2.38, or 5.3 percent. Although the compressor work increased 4.3 percent, the cooling effect increased 9.8 percent. Similar comparisons would be valid for part *b*.

---

The effects of compressor irreversibilities, finite temperature differences in the heat exchangers, and superheating the fluid at the evaporator exit and subcooling at the condenser exit have been discussed in the previous three subsections. Other irreversibilities also should be considered. Primarily, there are pressure drops in the condenser and evaporator. Additional pressure drops due to fluid friction occur in the piping that connects the various components. The largest pressure drop usually occurs in the tubing between the evaporator and the compressor. Heat transfer from the surroundings to the refrigerant can be significant in the connecting lines where the refrigerant temperature is well below the ambient. We have already spoken of heat transfer during the compression process. Thus many other factors need to be taken into account in designing a refrigeration cycle.

## 17-2-4 SELECTION OF REFRIGERANTS

In a vapor-compression refrigeration cycle, the two desired saturation temperatures for the evaporation and condensation processes determine the operating pressures of the cycle for a given refrigerant. Consequently, the choice of a refrigerant partially depends on the saturation pressure-temperature relationship in the range of interest. Normally, the minimum pressure in the cycle should be above 1 atm to avoid leakage into the equipment from the environment. Also, maximum pressures over 10 to 20 bars (150 to 200 psia) would be undesirable from safety and equipment cost considerations. In addition, the fluid needs to be nontoxic, stable, noncorrosive, nonflammable, low-cost, and have a relatively high enthalpy of vaporization. These restrictions limit the number of compounds which are suitable for refrigerants. In fact, because of the range of applicability of refrigeration cycles, no one substance is suitable for all cases. Even with a proper choice of the refrigerant, a number of modifications may be made on the basic cycle to improve its performance.

Although chemicals such as sulfur dioxide, ammonia, and ethyl ether were used in the early development of refrigeration processes, the field has been dominated by the general class of substances called chlorofluorocarbons (CFCs) since the 1930s. The most important ones of these are designated as R-11, R-12, R-22, and R-502 (a blend of R-22 and R-115). In the late 1980s measures were taken internationally to restrict the use of certain CFCs which were found to deplete the protective ozone layer in the atmosphere and also contribute to the greenhouse effect which causes global warming. Thus the 1990s and beyond is a time period where new refrigerants such as hydrofluorocarbons (HFCs) are investigated and old favorites are banned. The problem is complicated by the fact that performance is diminished when a new refrigerant is simply "dropped in" into existing equipment. Thus compressor redesign may be necessary when a new refrigerant is selected. In this edition of the text, R-134a tables have replaced those for R-12. R-134a is environmentally friendly because the molecule is chlorine-free.

## 17-3　Heat Pumps

A refrigerator removes energy as heat transfer $\dot{Q}_L$ from a cold region and rejects energy as heat transfer $\dot{Q}_H$ to the environment. Its major purpose is energy removal from the cold region. However, the same basic cycle could have as its major purpose the supply of energy by heat transfer $\dot{Q}_H$ to a living space, such as a home or commercial building. In this latter case the energy removed as $\dot{Q}_L$ comes from the cooler environment. In fact, the modern *heat pump* combines both heating and cooling of a region of space in the same unit. When cooling is required, the heat-pump system operates as an air conditioner. In this cooling mode, energy as heat transfer $\dot{Q}_{L,\text{in}}$ is removed from a living space, and energy as heat transfer $\dot{Q}_{H,\text{out}}$ is rejected outside the building to the environment. In the *air-conditioning* mode the COP is given by Eq. [17-1]. In terms of the ideal refrigeration cycle shown in Fig. 17-2, the COP for cooling is

$$\text{COP}_{\text{HP,cool}} = \frac{\dot{Q}_{L,\text{in}}}{\dot{W}_{\text{in}}} = \frac{h_1 - h_4}{h_2 - h_1} \qquad \textbf{[17-5]}$$

In the *heating mode*, the heat-pump system in the wintertime removes energy as heat transfer $\dot{Q}_{L,\text{in}}$ from the environment and rejects energy as heat transfer $\dot{Q}_{H,\text{out}}$ to the living space. The coefficient of performance for a heat pump in the heating mode is defined as

$$\text{COP}_{\text{HP,heat}} = \frac{\dot{Q}_{H,\text{out}}}{\dot{W}_{\text{in}}} = \frac{h_2 - h_3}{h_2 - h_1} \qquad \textbf{[17-6]}$$

Because the areas under the $T_H$ and $T_L$ lines on the $Ts$ diagram in Fig. 17-1$b$ represent $\dot{Q}_H$ and $\dot{Q}_L$, respectively, the COP of a Carnot heat pump operating in the heating mode is

$$\text{COP}_{\text{HP,heat,Carnot}} = \frac{T_H}{T_H - T_L} \qquad \textbf{[17-7]}$$

Thus the performance of a heat pump in the heating mode is improved by decreasing the temperature difference $T_H - T_L$, similar to that of a refrigerator or air conditioner.

One method of decreasing $T_H - T_L$ is to increase $T_L$ through the use of a solar collector. Such units are called *solar-assisted* heat pumps. Solar energy is collected by a fluid circulating through solar panels. The relatively hot fluid is stored in a large insulated tank. By proper regulation of the flow, the temperature of the fluid within the tank may rise to a range of 20 to 30°C (70 to 90°F), for example, even though the outside temperature may be below the freezing point of water. The evaporator of the heat pump is placed internally within the tank, so that the fluid within the heat-pump system evaporates at a higher temperature (and pressure) than normal. Although more equipment is required, the solar-assisted heat pump will have a much higher COP in the heating mode compared to operation of the evaporator in the cold environment. For the same heating load, the electrical work input will be reduced sizably when solar energy assists the operation.

When a heat pump is used to air-condition a building, the evaporator is within the building and the condenser is outside the building. However, in a heating mode, the evaporator is outside the building and the condenser inside. It would be impractical to have two sets of equipment, so each heat exchanger (one inside and one outside the building) must act as both a condenser and an evaporator, depending on the mode of operation. One method of accomplishing this is to add a reversing valve in the cycle. A schematic of this arrangement is shown in Fig. 17-9. The solid flow line is the flow

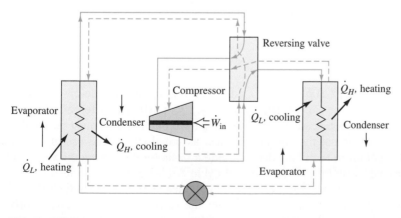

**Figure 17-9**    Equipment flow diagram for a heat pump in a heating mode (solid line) and in a cooling mode (dashed line).

direction for the heating mode of operation, and the dashed line is for the cooling mode. Note that the direction of flow through the compressor is always the same, regardless of the mode of operation.

## 17-4 CASCADE AND MULTISTAGED VAPOR-COMPRESSION SYSTEMS

This section is devoted to two variations of the basic vapor-compression refrigeration cycle. The first is the cascade cycle, which permits the use of a vapor-compression cycle when the temperature difference between the evaporator and condenser is quite large. The second variation involves the use of multistaged compression with intercooling, which reduces the required compressor-work input.

### 17-4-1 THE CASCADE CYCLE

There are industrial applications where low temperatures in the range of $-25$ to $-75°C$ ($-10$ to $-100°F$) are required. Unfortunately, a single vapor-compression cycle usually cannot be used to achieve these moderately low temperatures. The temperature difference between the condenser and evaporator is now quite large. Consequently, the vapor–pressure–saturation-temperature variation for any single refrigerant would not fit the desired values for the evaporator and condenser. To overcome this difficulty and still rely on vapor compression, a cascade system may be employed. A *cascade cycle* is simply an arrangement of simple vapor-compression cycles, in series, such that the condenser of a lower-temperature cycle provides the heat input to the evaporator of a higher-temperature cycle, as shown in Fig. 17-10a. Although only two units are shown, the use of three or four units in series is practical, if needed. Normally a different refrigerant would be used in each separate cycle, in order to match the desired ranges of temperature and pressure. In choosing the two refrigerants in Fig. 17-10a, for example, it is important that the triple-state temperature of the fluid in cycle B be lower than the critical temperature of the fluid in cycle A.

Figure 17-10b shows a $Ts$ diagram for an ideal double-cascade system employing the same refrigerant in each loop. (If two different refrigerants are used in a cascade system, two separate $Ts$ diagrams must be used.) Although it is not the normal practice, as noted above, the use of the same refrigerant in each loop does provide a basis for discussing the virtues of a cascade system. The positions of cycle A (1-2-3-4) and cycle B (5-6-7-8) are shown in the figure. The mass flow rates of refrigerant in the two cycles are typically not the same, whether the refrigerant is the same or different. The mass flow rate $\dot{m}_A$ is fixed by the refrigeration capacity required in the evaporator of cycle A. In addition, the rate of heat transfer from the

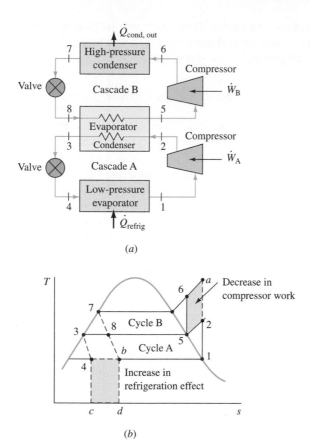

**Figure 17-10**    Equipment schematic and $Ts$ diagram for a cascade refrigeration cycle.

condenser of cycle A must equal the rate of heat transfer to the fluid in the evaporator of cycle B if the overall heat exchanger is well insulated. Neglecting kinetic- and potential-energy changes, a steady-state energy balance on the condenser-evaporator heat exchanger has the form $\sum \dot{m}_i h_i - \sum \dot{m}_e h_e = 0$. In terms of Fig. 17-10a, the energy balance becomes

$$\dot{m}_A(h_2 - h_3) = \dot{m}_B(h_5 - h_8) \qquad \text{cascade system} \qquad \textbf{[17-8]}$$

Thus the ratio of the mass flow rates through each cycle is fixed by the enthalpy changes of each fluid as it passes through the heat exchanger.

If a single refrigeration cycle could be used for the overall temperature range, this would be represented by the cycle 1-a-7-b-1 on Fig. 17-10b. Two significant effects are apparent from the $Ts$ diagram. First, for the single cycle the compressor work is increased by area 2-a-6-5, in comparison to the cascade system. Second, there is a decrease in the refrigeration capacity when a single unit is used, for the same mass flow rate through the low-temperature evaporator. This loss is represented by area 4-*b*-*d*-*c* on the *Ts* diagram. These two effects would lead to a higher COP for the cascade system in comparison to the single unit.

### 17-4-2   MULTISTAGE VAPOR COMPRESSION

Another modification of the vapor-compression refrigeration cycle involves multistage compression with intercooling to decrease the work input. For gas power cycles (see Sec. 15-10), the heat transfer from the intercooler is usually to the environment. In a refrigeration cycle, the sink for the energy can be the circulating refrigerant itself, because in many sections of the cycle the temperature of the refrigerant is below the environment temperature. Hence the intercooler heat exchanger becomes a regenerative heat exchanger, since the heat transfer now occurs internal to the system. One

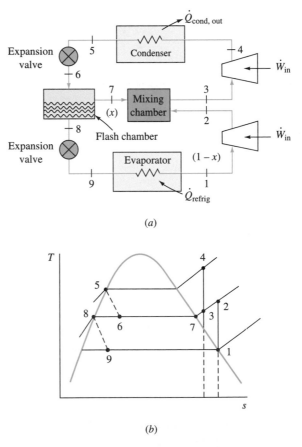

(*a*)

(*b*)

**Figure 17-11**      Equipment and *Ts* diagram for a two-stage, vapor-compression refrigeration cycle with regenerative intercooling.

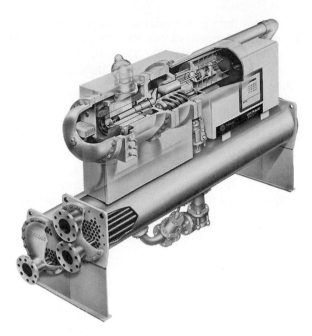

**Photograph 17-2** Vapor-Compression liquid chiller (100–300 tons) with a two-stage helical rotary compressor, an economizer (intercooling), and a water-cooled condenser. (Courtesy of The Trane Company)

scheme for two-stage compression with regenerative intercooling is shown in Fig. 17-11a. The liquid leaving the condenser is throttled (process 5-6) into a flash chamber maintained at a pressure between the evaporator and condenser pressures. All the vapor separated from the liquid in the flash chamber is transferred to a direct-mixing chamber, where it mixes with the vapor leaving the low-pressure compressor at state 2. The mixing chamber acts as a regenerative intercooler in that it cools the vapor leaving the low-pressure compressor before the total mixture enters the high-pressure stage of the compressor at state 3. The saturated liquid from the flash chamber is throttled to the evaporator pressure at state 9. A real refrigeration system with two-stage compression and intercooling is shown in Photograph 17-2.

The two-stage compression process with regenerative intercooling is shown on the $Ts$ diagram in Fig. 17-11b. Isentropic compression has been assumed. Although the same refrigerant circulates through both loops of the overall system, the mass flow rates through the loops are not the same. For the purpose of analysis of the system, it is convenient to assume that a unit mass circulates in one of the loops, but the choice is arbitrary. Let us assume that a *unit mass* passes through states 3-4-5-6 in the high-pressure loop. The fraction of vapor formed in the flash chamber is the quality $x$ of the fluid at state 6 in Fig. 17-11b, and this is the fraction of the flow through the condenser which passes through the mixing chamber from the flash chamber. The fraction of liquid formed is $1 - x$, and this is the fraction of the total flow which passed through the evaporator. The value of the enthalpy at state 3 may be ascertained from an energy balance on the mixing chamber under adiabatic conditions. In terms of the mixing chamber,

$$xh_7 + (1 - x)h_2 = 1(h_3) \qquad \textbf{[17-9]}$$

where $h_3$ is the only unknown. The refrigeration effect per unit mass through the condenser is

$$q_{\text{refrig}} = (1 - x)(h_1 - h_9) \qquad \textbf{[17-10]}$$

The total work input to the compressor per unit mass through the condenser is the sum for the two stages, namely,

$$w_C = (1 - x)(h_2 - h_1) + 1(h_4 - h_3) \qquad \textbf{[17-11]}$$

The COP of the two-stage vapor-compression cycle with regenerative intercooling is still defined as $q_{\text{refrig}}/w_C$.

**Example 17-5**

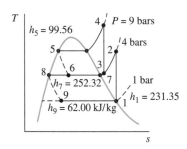

**Figure 17-12**
A $Ts$ diagram with data for Example 17-5.

**A** two-stage ideal vapor-compression refrigeration system with regenerative intercooling operates with refrigerant 134a and has pressures of 1.0, 4.0, and 9.0 bars in the evaporator, flash chamber and mixing chamber, and condenser, respectively. If the refrigeration load is 5 tons, determine (a) the power input to the low-pressure compressor, in kilowatts, (b) the mass flow rate leaving the flash chamber and entering the mixing chamber, in kg/min, (c) the power input to the high-pressure compressor, in kilowatts, and (d) the COP of the overall cycle.

**Solution:**

**Given:**    A two-stage ideal vapor-compression refrigeration system as shown in Fig. 17-12.

**Find:**    (a) $\dot{W}_{C,\text{low}}$, kW; (b) $\dot{m}_7$, kg/min; (c) $\dot{W}_{C,\text{high}}$, in kW; and (d) $COP_{\text{overall}}$.

**Model:**    Steady-state, reversible compressors and heat exchangers; neglect $\Delta\text{ke}$ and $\Delta\text{pe}$.

**Analysis:**    The evaluation of power and COP require enthalpy data. Using the notation of Fig. 17-12, data from Tables A-16 to A-18 indicate that

$h_1 = h_g$ at 1 bar $= 231.35$ kJ/kg      $s_1 = s_g = 0.9395$ kJ/kg·k

$h_5 = h_f$ at 9 bars $= 99.56$ kJ/kg $= h_6$

$h_7 = h_g$ at 4 bars $= 252.32$ kJ/kg      $h_8 = h_f$ at 4 bars $= 62.00$ kJ/kg $= h_9$

The enthalpy $h_{2s}$ depends on $P_2 = 4$ bars and $s_2 = s_1$. By interpolation in Table A-18 at 4 bars, $h_{2s} = 259.50$ kJ/kg and $T_{2s} = 16.4°C$.

(a) The power to the low-pressure compressor is found from $\dot{W}_C = \dot{m}_1 w_C$. The mass flow rate $\dot{m}_1$ is found from an energy balance on the evaporator, namely,

$\dot{Q}_{evap} = \dot{m}_1 q_{evap}$. Since $\dot{Q}_{evap}$ is equivalent to 5 tons of refrigeration,

$$\dot{m}_1 = \frac{\dot{Q}_{evap}}{h_1 - h_9} = \frac{5(211) \text{ kJ/min}}{(231.35 - 62.00) \text{ kJ/kg}} = 6.23 \text{ kg/min}$$

Therefore, for the low-pressure compressor,

$$\dot{W}_{C,low} = \dot{m}_1(h_{2s} - h_1) = 6.23 \text{ kg/min} (259.50 - 231.35) \text{ kJ/kg}$$
$$= 175.4 \text{ kJ/min} = 2.92 \text{ kW}$$

(b) In order to determine the mass flow rate $\dot{m}_7$, an energy balance is made on the flash chamber. The result, in terms of Fig. 17-11a, is

$$\dot{m}_6 h_6 = \dot{m}_7 h_7 + \dot{m}_8 h_8$$

where $\dot{m}_8 = \dot{m}_1$ and $\dot{m}_6 = \dot{m}_8 + \dot{m}_7$. Since all enthalpy values are known, the energy balance becomes

$$(\dot{m}_7 + \dot{m}_1)(99.56) = \dot{m}_7(252.32) + \dot{m}_1(62.00)$$

Hence

$$\frac{\dot{m}_7}{\dot{m}_1} = x_6 = \frac{99.56 - 62.00}{252.32 - 99.56} = 0.246$$

or

$$\dot{m}_7 = 0.246(6.23) = 1.53 \text{ kg/min}$$

(c) In order to determine the power to the high-pressure compressor we must find $h_3$ and $h_{4s}$. An energy balance on the mixing chamber leads to

$$\dot{m}_7 h_7 + \dot{m}_2 h_2 = \dot{m}_3 h_3$$

where $\dot{m}_2 = \dot{m}_1$ and $\dot{m}_3 = \dot{m}_2 + \dot{m}_7 = 7.76$ kg/min. Substitution of data yields

$$1.53(252.32) + 6.23(259.50) = 7.76 h_3$$

or

$$h_3 = 258.08 \text{ kJ/kg}$$

Also, $P_3 = 4$ bars, so $s_3$ can now be found by interpolation in the superheat table A-18.

$$s_3 = 0.9182 + (0.9515 - 0.9182)\frac{258.08 - 253.35}{262.96 - 253.35} = 0.9346 \text{ kJ/kg·K}$$

Finally, knowledge of $s_4 = s_3$ and $P_4 = 9$ bars leads to $h_{4s}$, again by interpolation. The result is $h_{4s} = 275.35$ kJ/kg. With this information

$$\dot{W}_{C,high} = \dot{m}_3(h_{4s} - h_3) = 7.76 \text{ kg/min} (275.35 - 258.08) \text{ kJ/kg}$$
$$= 134.0 \text{ kJ/min} = 2.23 \text{ kW}$$

(d) The COP is given by

$$\text{COP} = \frac{\dot{Q}_{evap}}{\dot{W}_{C,low} + \dot{W}_{C,high}} = \frac{5(211)}{175.4 + 134.0} = 3.41$$

where all energy terms are expressed in kJ/min.

**Comment:** The COP for a single-stage compressor operating between 1 and 9 bars is 2.89. Thus the COP increases by 18 percent when two-stage compression is used.

### 17-4-3   A REFRIGERATOR-FREEZER WITH DUAL EVAPORATORS

In the modern household refrigerator-freezer, the refrigerator section is maintained around 2 to 3°C (35 to 38°F), while the freezer section is held around −18°C (0°F). This is achieved through the use of a single evaporator placed in the freezer section. Cold air from the freezer is transferred into the refrigerator to keep that region at the desired temperature. While the method works well, it has an undesirable energy inefficiency. The unit is providing cooling at the freezer temperature, but a sizable fraction of the heat removal is needed only at the refrigerator temperature. But the COP of a refrigeration cycle is inversely proportional to $(T_H - T_L)$. The lower the value of $T_L$ used, the lower the COP, or energy efficiency. Therefore, removing all the heat leakage into the system at the level of the freezer temperature constitutes an energy penalty for the overall unit.

One method of improving the energy efficiency is to provide an evaporator for both the refrigerator region and the freezer region but retain a single compressor. One system for achieving this is shown in Fig. 17-13$a$. Expansion occurs from the condenser outlet at state 3 to an intermediate pressure $P_4$. This is followed by heat transfer $\dot{Q}_{L,R}$ from the refrigerator region into the refrigerant in the intermediate-temperature evaporator. It is assumed that $P_5 = P_4$. Further expansion from state 5 to state 6 lowers the refrigerant temperature to that desired in the freezer section. The heat transfer rate $\dot{Q}_{L,F}$ from the freezer region to the refrigerant occurs in the low-temperature evaporator. This is followed by a single compressor and condenser. The path of the refrigerant cycle is shown on the $Ts$ diagram in Fig. 17-13$b$. The cycle has been idealized with saturated vapor entering the isentropic compressor, saturated liquid leaving the condenser, and frictional losses neglected.

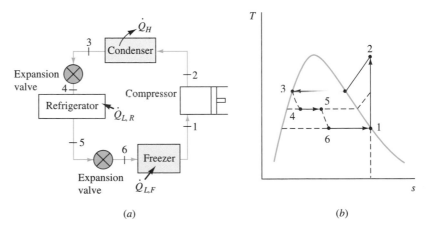

(a)                                    (b)

**Figure 17-13**     Schematic of a refrigeration cycle with dual evaporators.

## 17-5   LIQUEFACTION AND SOLIDIFICATION OF GASES

In modern technology the preparation of liquids at temperatures below $-75°C$ ($-100°F$) is quite important. In the study of properties and the behavior of substances at low temperatures, the materials are placed in baths composed of liquefied gases. Mixtures of gases may be separated by liquefaction techniques. For example, liquid oxygen and nitrogen are separated from air in this manner, and in the same manner rarefied gases such as helium may be obtained. Liquefied gases are also successful rocket propellants. Liquid helium and hydrogen are especially useful in research studies in the temperature range from 2 to 30 K, such as those of superconductivity and superfluidity. The following discussion offers a brief introduction to the thermodynamic cycles employed in the liquefaction and solidification of gases.

A simple cycle for liquefying gases is illustrated in Fig. 17-14. Makeup gas enters the steady-flow system at state 6, and after mixing the gas is compressed to an elevated pressure and temperature (state 1). Note that the compression is shown not to be isentropic on the $Ts$ diagram. In fact, in practice, multistaging with intercooling is used, so that the compression process is closer to isothermal than isentropic. Before it undergoes a throttling process, the gas is cooled to state 2 by passing through an efficient counterflow heat exchanger. (The use of the aftercooler shown in the figure is discussed in a subsequent paragraph of this section.) The gas is now throttled to state $a$, which must lie within the wet region (or solid-gas region) if a phase change is to occur. The liquid is collected in the bottom of the separation chamber as a saturated liquid at state 3, and the remaining gas is passed back through the heat exchanger. Since the gas leaving the separator as saturated vapor at state 4 may be considerably colder than the gas at state 1, it may be used effectively to cool the gas stream passing from the compressor to the

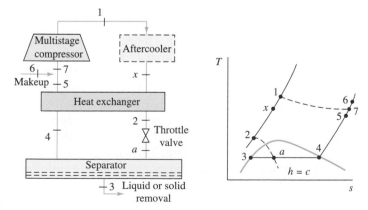

**Figure 17-14**    Equipment schematic and $Ts$ diagram of a basic cycle for liquefying gases.

throttling valve. The gas now at state 5 is mixed with makeup gas, and the cycle is repeated. The process is traced on the $Ts$ diagram in Fig. 17-14.

Consider a control volume drawn around the heat exchanger, throttling valve, and separator shown in Fig. 17-14. It is assumed that these three pieces of equipment are adiabatic and that the changes in kinetic and potential energies are negligible. The steady-state energy equation for this control volume consequently reduces to $\sum \dot{m}_i h_i = \sum \dot{m}_e h_e$. Assume a unit mass enters the control volume at state 1, and let $y$ equal the fraction of liquid or solid formed in the separator per unit mass entering. Therefore, on the basis of the symbols shown in Fig. 17-14, the energy balance becomes

$$yh_3 + (1 - y)h_5 = 1(h_1)$$

Solving for $y$, we find that

$$y = \frac{h_5 - h_1}{h_5 - h_3} \qquad\qquad \textbf{[17-12]}$$

The value of $h_3$ is set by the pressure $P_3$ in the separator. At the same time $h_5$ is set by the design of the heat exchanger. The enthalpy $h_1$ is fixed by $T_1$ and $P_1$; however, $T_1$ is set by the design of the separator and heat exchanger. Consequently, the compressor-outlet pressure $P_1$ is the main variable controlling the fraction of the gas which is liquefied or solidified. Note that $y$ is made larger by decreasing the value of $h_1$. It may be shown that the maximum degree of liquefaction occurs when the initial value of $P_1$ for a given temperature of the gas entering the heat exchanger is the inversion pressure at that temperature.

An interesting application of the apparatus described in Fig. 17-14 is in the production of solid carbon dioxide, or dry ice. A $Ts$ diagram for carbon dioxide appears as Fig. A-24E. Figure 17-15 is a representation of the approximate path of the process on a $Ts$ diagram. Carbon dioxide at atmospheric conditions (state 5) is compressed by multistaging to a high pressure (roughly 70 bars, or 1000 psia), represented by state 1. It then passes through the heat exchanger, where it is cooled and condensed to a saturated liquid at state 2. The corresponding temperature in this liquid state is room temperature. The liquid is expanded through a throttling valve to atmospheric pressure in the separator. The resulting mixture is solid and gas at state $b$. The corresponding saturation states are states 3 and 4 on the figure. Solid is formed because the triple state of carbon dioxide is roughly 5 atm and $-57°C$ $(-70°F)$. Consequently, liquid carbon dioxide cannot exist at 1 atm. The sublimation temperature of carbon dioxide at atmospheric pressure is approximately $-80°C$ $(-110°F)$. Solid carbon dioxide will be formed by cooling to this temperature by throttling. The solid phase is pressed into bars, and the gas phase at state 4 and the required makeup gas return to the compressor at state 5.

A number of modifications of the basic cycle lead to improved operation and efficiency. One is the use of an aftercooler or precooler, which appears between the compressor and the heat exchanger. This is indicated by the

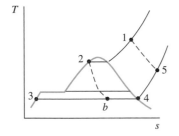

**Figure 17-15**

The $Ts$ diagram of the cycle for the production of solid carbon dioxide.

dashed box in Fig. 17-14. In practice, a separate refrigerating system is used to precool the gas as it passes through the aftercooler. Any auxiliary refrigeration device such as this is inherently more efficient in cooling the gas, compared with a throttling process, because the refrigeration device can be made more reversible than a throttling process.

## 17-6   GAS REFRIGERATION CYCLES

The adiabatic expansion of gases can be used to produce a refrigeration effect. In its simplest form, this is accomplished by using a reversed Brayton cycle. Figure 17-16 is a $Ts$ diagram for the ideal reversed Brayton cycle. The gas is first compressed isentropically from state 1 to state 2, which is at a relatively high temperature compared with the ambient temperature $T_0$. It then passes through a heat exchanger, where in the limit the gas can be cooled to the ambient temperature. (In practice, state 3 may be 10 degrees or so above the ambient value.) The gas then enters an adiabatic expansion device, such as a turbine, where work is produced and the gas is cooled to a temperature $T_4$, which is considerably below the temperature $T_L$ of the cold region. Consequently, heat transfer occurs from the cold region to the gas through a heat exchanger. Ideally the gas temperature will increase to $T_L$, which is also $T_1$ in the cycle. The fluid in the ideal gas-refrigeration cycle is assumed to pass through the cycle in an internally reversible manner. For steady-state operation, and in the absence of kinetic- and potential-energy changes, the steady-state energy balance for each component reduces to

$$q + w = \Delta h$$

Hence the heat and work interactions for each piece of equipment may be evaluated in terms of the enthalpy change during the process. On the $Ts$ diagram in Fig. 17-16, the area under curve 4-1 represents the energy removed from the cold region by heat transfer, the area enclosed by cycle 1-2-3-4-1 represents the net work input, and the ratio of these areas is a measure of the COP of the device. In this case

$$\text{COP} = \frac{q_{\text{refrig}}}{w_{\text{net,in}}} = \frac{q_{\text{in}}}{w_{C,\text{in}} - w_{T,\text{out}}} = \frac{h_1 - h_4}{h_2 - h_1 - (h_3 - h_4)} \quad \textbf{[17-13]}$$

The reversed Brayton cycle has the disadvantage of large temperature variations of the fluid during heat-addition and heat-removal processes. Hence the COP of this cycle is considerably below that for a reversed Carnot engine operating between the temperatures $T_0$ and $T_L$.

Modifications of the reversed Brayton cycle lead to some useful applications of gas-refrigeration cycles. For example, in Fig. 17-16, the temperature $T_1$ of the fluid after receiving heat transfer from the cold region is below that of state 3, where the fluid enters the expansion engine. If the gas at state 1 could be used to cool the gas further below the temperature of

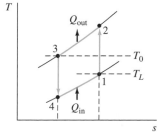

**Figure 17-16**
Gas-refrigeration cycle using a reversed Brayton cycle.

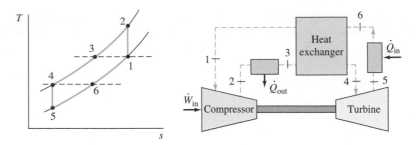

**Figure 17-17**    Gas-refrigeration cycle using a heat exchanger (regenerator) internal to the cycle.

state 3, subsequent expansion would lead to a temperature below that of state 4. In this manner extremely low temperatures might be achieved. This may be accomplished by inserting a heat exchanger internally into the cycle, as shown in Fig. 17-17. Heat transfer external to the cycle results in the usual temperature drop from state 2 to state 3 in the figure. The additional regenerator, however, allows the gas to be cooled further to state 4 in the ideal case. The gas, after expansion, receives energy as heat transfer from the cold region from state 5 to state 6 and receives a further quantity of energy in the heat exchanger from state 6 to state 1. Such use of heat exchangers internal to the cycle is important in processes for the liquefaction of gases. The gas-refrigeration cycle in the form of an open cycle is employed for the purpose of aircraft cooling, since it has a definite weight advantage over vapor-compression refrigeration. There are a number of modifications of the technique, but in general, air is compressed, cooled by rejecting energy to the ambient atmosphere, and then expands through a turbine. The cool air leaving the turbine goes directly into the cabin of the aircraft.

**Example 17-6**

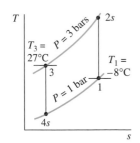

**Figure 17-18**
A $Ts$ diagram with data for Example 17-6.

**A** reversed Brayton cycle is proposed to operate between $T_0 = 27°C$ and $T_L = -8°C$. The compression and expansion occur between 1 and 3 bars. Determine (a) the maximum and minimum temperatures, and (b) the COP.

**Solution:**

**Given:**    A reversed Brayton cycle operates under the conditions shown on the $Ts$ diagram in Fig. 17-18.

**Find:**    (a) $T_{max}$ and $T_{min}$, and (b) COP.

**Model:**    Steady-state, internally reversible processes.

**Analysis:**    Evaluation of minimum and maximum temperatures and the COP requires knowledge of enthalpies around the cycle. From the problem statement we conclude that $T_3 = T_0$ and $T_1 = T_L$.

(*a*) From the air table A-5 for $T_1 = -8°C$, we find that $h_1 = 265.1$ kJ/kg and $p_{r1} = 0.900$. State 2 is found by noting that for an isentropic process of an ideal gas,

$$p_{r2} = p_{r1} \frac{P_2}{P_1} = 0.900(3) = 2.70$$

For this $p_r$ value, the air table indicates that $T_{2s} = 363$ K (90°C) and $h_{2s} = 363.6$ kJ/kg. A similar calculation is now carried out for process 3-4. At 27°C, $h_3 = 300.2$ kJ/kg and $p_{r3} = 1.386$. Consequently, $p_{r4} = 1.386/3 = 0.462$, $T_{4s} = 219$ K ($-54°C$), and $h_{4s} = 219.0$ kJ/kg. Therefore, the maximum and minimum temperatures in the cycle are 90 and $-54°C$.

(*b*) Neglecting kinetic- and potential-energy changes, the compressor work, the turbine work, and the heat removed from the cold region, respectively, are found from the energy balance, $q + w = \Delta h$. Therefore,

$$w_{C,\text{in}} = h_{2s} - h_1 = 363.6 - 265.1 = 98.5 \text{ kJ/kg}$$

$$w_{T,\text{out}} = h_3 - h_{4s} = 300.2 - 219.0 = 81.2 \text{ kJ/kg}$$

$$q_{\text{in}} = h_1 - h_{4s} = 265.1 - 219.0 = 46.1 \text{ kJ/kg}$$

Therefore,

$$\text{COP} = \frac{q_{\text{in}}}{w_{C,\text{in}} - w_{T,\text{out}}} = \frac{46.1}{98.5 - 81.2} = 2.66$$

**Comment:** This COP is somewhat high, since we have not accounted for the irreversibilities in the turbine and compressor, or elsewhere.

## 17-7 STIRLING REFRIGERATION CYCLE

The attainment of extremely low temperatures (less than 200 K, or $-100°F$) is usually achieved by three well-established methods:

1. Vaporization of a liquid
2. Joule-Thomson effect by isenthalpic expansion
3. Adiabatic expansion in an engine with production of work

Since the early 1950s considerable work has been done to develop a practical refrigeration device based on the Stirling cycle. (See Sec. 15-14-2 for a discussion of the Stirling heat engine.) Devices arising from this work have proved useful in the temperature range of 100 to 200 K. Other methods for maintaining temperatures in all this range are not too plentiful. As noted in Sec. 15-14-2, the Stirling cycle is composed of two constant-temperature processes and two constant-volume processes. Figure 17-19, similar to Fig. 15-50, shows the *PV* and *TS* diagrams for the cycle. If the cycle is reversible, then the heat quantities for the cycle are represented by areas on the *TS* diagram. In the presence of an ideal regenerator in the cycle, the heat quantities $Q_{14}$ and $Q_{32}$, which are equal in magnitude but opposite in sign, are exchanged between fluid streams within the device. Hence the

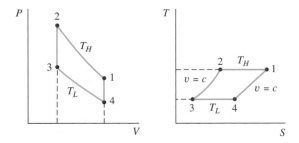

**Figure 17-19**        The *PV* and *TS* diagrams of a Stirling refrigeration cycle using a gaseous working fluid.

only external heat transfer occurs in processes 3-4 and 1-2 at constant temperatures $T_L$ and $T_H$. Consequently the COP of the Stirling refrigerator theoretically equals that of the Carnot refrigerator, namely, $T_L/(T_H - T_L)$. The COP will be quite small if $T_L$ is small, since $T_H - T_L$ will also be quite large. For example, if $T_L$ is 100 K and $T_H$ is 300 K, a COP of 0.5 results. Thus considerable work must be performed per unit of heat removal, compared to a household refrigerator or air-conditioning unit.

An ideal reciprocating Stirling refrigeration unit is shown in Fig. 17-20. A cold temperature is produced by a reversible expansion of a gas in region *E*, while the gas is heated by compression in region *D*. Particles of the gas oscillate between the two spaces connected by a regenerator *F*. The regenerator must be composed of a material with a high heat capacity. Cylinder *A* encloses the regular piston *B* and a displacer piston *C*. The shaft of the displacer piston passes through piston *B*. Consider the piston initially in position 1 shown in Fig.17-20: Four distinct ideal processes take place in the cycle:

1-2: The piston *B* moves upward, compressing the gas in region *D*. To keep the process isothermal at temperature $T_H$, heat is removed through the cylinder walls during the compression process. This is equivalent to process 1-2 in Fig. 17-20.

2-3: The displacer piston *C* now moves downward, forcing part of the gas through the regenerator to space *E*. The gas is cooled as it passes through the regenerator, the energy being stored in the regenerator material. This creates a temperature gradient in the regenerator, the temperature increasing from region *E* to region *D*. Since piston *B* does not move, this transfer of part of the gas from *E* to *D*, with energy storage in *F*, occurs at constant volume. This corresponds to process 2-3.

3-4: Both pistons now move downward, with the resulting expansion of the gas in region *E*. The expansion process would tend to cool the gas, but heat added from an outside source keeps the temperature at

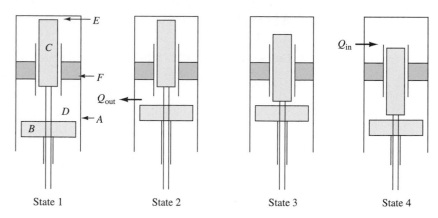

State 1          State 2          State 3          State 4

**Figure 17-20**    Relative positions of regular and displacer piston in an ideal reciprocating Stirling refrigeration cycle.

$T_L$. This heat removal at very low $T_L$ values constitutes the refrigeration effect. For example, a gas circulated around the outside of region $E$ could be liquefied. This process is equivalent to process 3-4 in
Fig. 17-20.

4-1:  Finally, the displacer piston $C$ moves upward to its initial position,
forcing gas from region $E$ to region $D$. As the gas passes through the
regenerator from the cold to the hot side, it is reheated. This internal
heat-addition process corresponds to process 4-1. Note that the temperature gradient in the regenerator makes the internal heat transfer
between the gas and the refrigerator material reversible.

Several discrepancies between theory and practice exist. First, the
entire gas within the system does not follow the process path shown in
Fig. 17-19. A considerable fraction of the gas remains in either region $D$
or region $E$. Hence the $TS$ and $PV$ diagrams for the actual device do not
have their usual clear-cut representation of the process. The system at any
instant is nonhomogeneous and so cannot be represented by a single state.
Different gas particles describe different cycles. Second, the simple piston
displacements described above would be difficult to match, since they occur discontinuously. In an actual device the reciprocating motions of the
pistons would be harmonic. To match more nearly the theoretical cycle, the
displacer and regular pistons are placed out of phase by some angle $\phi$ (on
the same drive shaft), so that the compression region $D$ lags in phase with
respect to the expansion space $E$. This compromise tends to blur even further the four steps of the theoretical cycle.

A number of applications for a Stirling refrigeration cycle include

1. Means for cooling electronic equipment and superconducting research magnets

2. Freeze drying of materials

3. A precooler for production of liquid hydrogen and helium

4. An air liquefier

5. Gas separation, e.g., as a liquid nitrogen generator from air

## 17-8   ABSORPTION REFRIGERATION

In any refrigeration process, the energy removed from the cold region eventually must be rejected to another region which is at a considerably higher temperature. This second region is usually the surrounding environment. To carry out the heat-rejection process, the temperature of the fluid within the refrigeration cycle must be raised to a value above that of the environment. In a vapor-compression refrigeration cycle, discussed in Sec. 17-2, the temperature of the vapor leaving the evaporator is raised by a compression process. The work input required in an ideal steady-flow compression process is given by the integral of $v\,dP$. The pressure limits on the integral are set by the saturation temperatures required in the evaporator and condenser of the cycle. Once the pressure range is determined for a given refrigerant, the main variable which controls the amount of work input is the specific volume of the fluid. In a vapor-compression refrigeration cycle, the value of $v$ is relatively large, since the fluid is in the superheat region throughout the compression process. Therefore, the work input is also relatively large. One method of overcoming this disadvantage is to design a refrigeration cycle in which the fluid is a liquid during the compression process. Then the work input will be significantly smaller.

The technique of absorption refrigeration is based on this approach. To accomplish this, however, the overall cycle becomes physically more complex. In addition, a two-component mixture, such as ammonia and water or lithium bromide and water, must be used as the circulating fluid in part of the cycle, rather than the single component typically used in a vapor-compression cycle. Two-component fluids have an important characteristic which must be recognized. When two phases are present in equilibrium, the composition of a given component is not the same in the two phases. The vapor phase will contain more of the component that is more volatile at the given temperature. For example, consider an ammonia-water mixture. At 43°C (110°F) the saturation pressure of ammonia is 17 bars (247 psia), while that of water is 0.09 bar (1.3 psia). Therefore, ammonia has a much greater tendency to vaporize at a given temperature than does water. Hence for an ammonia-water solution, the vapor phase contains much more ammonia (is richer in ammonia) than the liquid phase in equilibrium with it. This fact is extremely important in making mass and energy balances on equipment used in absorption refrigeration.

A schematic diagram of a simple absorption-refrigeration cycle is shown in Fig. 17-21. A condenser, throttle valve, and evaporator are shown

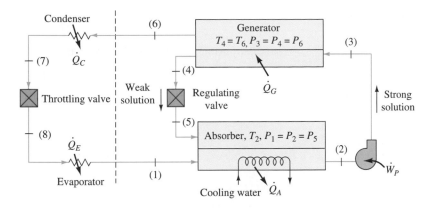

**Figure 17-21**   Schematic diagram of a simple absorption-refrigeration cycle.

on the left-hand side of the diagram. These three pieces of equipment also are used in a conventional vapor-compression cycle, as shown earlier in Fig. 17-2. The compressor in that cycle, however, is now replaced by four pieces of equipment: an absorber, a pump, a generator, and a valve. For the purpose of discussion, we consider ammonia and water as the two components in the cycle. Essentially pure ammonia vapor leaves the evaporator at state 1 and enters the absorber. The absorbing medium is a weak solution (low ammonia concentration) of ammonia and water which continually enters at state 5. The process of absorption releases energy; hence cooling water must be circulated through the absorber, to keep the solution at a constant temperature. The temperature of the absorbing fluid must be kept at as low a temperature as possible, since the amount of pure refrigerant (ammonia) that can be absorbed decreases as the temperature increases. However, the absorber must operate at 10 to 20 degrees above the cooling-water temperature to allow for adequate heat-transfer rates. The liquid that leaves the absorber at state 2 is a rich or strong solution (high ammonia concentration). This binary liquid mixture is now compressed by a pump to state 3, which is at the desired condenser pressure.

The temperature rise of the binary mixture due to the pump work usually is quite small. Thus the strong solution is subcooled liquid as it enters the generator shown in Fig. 17-21. Energy must now be added to the solution in the generator by heat transfer $\dot{Q}_G$ to warm the incoming liquid to the saturation temperature and to drive out of solution some of the ammonia. This nearly pure ammonia passes to the condenser at state 6 and eventually returns to the absorber at state 1. The weak solution left in the generator (state 4) now flows through a regulating valve, which drops the pressure of the solution to that in the absorber. It mixes with the solution remaining in the absorber, and cold vapor coming from the evaporator is added to the overall liquid solution. The strong solution is cooled, as before, and the cycle is repeated. Hence the ammonia-water solution cycles through the absorber, pump, generators, and valve and merely serves as a transport medium for

**Photograph 17-3**    A single-stage, absorption liquid chiller (100–1600 tons) using steam as the thermal energy source and $H_2O$-LiBr as the refrigerant-absorbent pair. (Courtesy of The Trane Company)

the ammonia refrigerant. Note that any absorption-refrigerator device requires an external heat source for the generation of refrigerant vapor. Thus absorption refrigeration is especially attractive if a low-temperature (100 to 200°C, or 200 to 400°F) source of thermal energy is readily available. Absorption refrigeration is often used to provide air conditioning in conjunction with a district heating system. When low-pressure steam is available an absorption chiller as shown in Photograph 17-3 can be used to provide chilled water.

In practice, absorption-refrigeration units have at least two modifications not shown in Fig. 17-21. First, the cold, strong solution at state 3 needs warming beforc it enters the generator, and the warm, weak solution at state 4 needs cooling before it enters the absorber. Consequently, a heat exchanger is placed between the absorber and the generator, which permits heat transfer from the weak solution to the strong solution. Second, a major requirement is that the ammonia from the generator should be essentially free of water as it passes through the condenser-throttling-valve-evaporator loop. Any traces of water will freeze in the expansion valve and evaporator. Hence the vapor leaving the generator passes through a device called a rectifier before it enters the condenser. The rectifier separates any remaining water vapor from the vapor stream leaving the generator and returns the water to the generator.

# 17-9 AVAILABILITY ANALYSIS OF A VAPOR-COMPRESSION REFRIGERATION CYCLE

Availability analyses are used to reveal the magnitudes of the irreversibilities in processes. On this basis improvements can be made in existing or proposed designs. In this section the internal and external irreversibilities of a vapor-compression refrigeration cycle will be examined. The equipment schematic and $Ts$ diagram for a vapor-compression refrigeration cycle with an irreversible compressor are shown in Fig. 17-22. The steady-state energy balance on a unit-mass basis is

$$q + w = \Delta h + \Delta \text{ke} + \Delta \text{pe}$$

Using the state notation found in Fig. 17-22, we can write the following steady-state energy relations for the compressor, condenser, throttling valve, and evaporator, respectively.

$$q_C + w_C = h_{2a} - h_1 \quad q_{\text{cond,out}} = h_3 - h_{2a} \quad h_3 = h_4 \quad q_{\text{evap}} = h_1 - h_4$$

where kinetic- and potential-energy changes are neglected. The steady-state availability balances for the four processes on an input/product basis are

$$\psi_1 + w_C = \psi_2 - \sum_j q_j \left(1 - \frac{T_0}{T_j}\right) + i_C \qquad \text{(compressor)} \qquad \textbf{[17-14a]}$$

$$\psi_2 = \psi_3 - \sum_j q_j \left(1 - \frac{T_0}{T_j}\right) + i_{\text{cond}} \qquad \text{(condenser)} \qquad \textbf{[17-14b]}$$

$$\psi_3 = \psi_4 + i_{\text{valve}} \qquad \text{(throttle)} \qquad \textbf{[17-14c]}$$

$$\psi_4 = \psi_1 - \sum_j q_j \left(1 - \frac{T_0}{T_j}\right) + i_{\text{evap}} \qquad \text{(evaporator)} \qquad \textbf{[17-14d]}$$

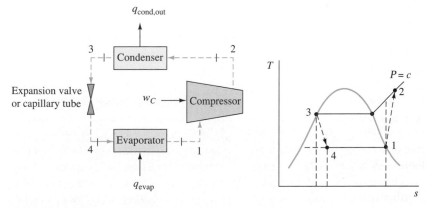

**Figure 17-22** A schematic of the equipment and a $Ts$ diagram for a vapor-compression refrigeration cycle.

The second-law effectiveness for the cycle is given by

$$\varepsilon_{\text{refrig}} = \frac{-\phi_{Q,\text{evap}}}{w_C} \qquad \text{[17-15]}$$

A $q_j(1 - T_0/T_j)$ term has been included in the compressor balance to account for heat transfer, if present. The flow through the lines connecting the equipment is assumed to be adiabatic and frictionless. The example below illustrates energy and availability analysis for a nonideal vapor-compression refrigeration cycle.

**Example 17-7**

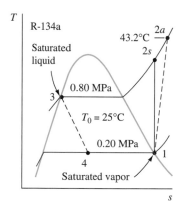

**Figure 17-23**
A *Ts* diagram
with data for Example 17-7.

**Saturated** refrigerant 134a vapor enters the adiabatic compressor of a vapor-compression refrigeration cycle at 0.20 MPa and leaves the condenser as a saturated liquid at 0.80 MPa. The temperature of the vapor leaving the irreversible compressor is 43.2°C, and the ambient temperature is 25°C. Determine the energy and availability changes, and the irreversibilities around the cycle, as well as the cycle effectiveness.

**Solution:**

**Given:** A vapor-compression refrigeration cycle with appropriate data is shown in Fig. 17-23.

**Find:** Energy, availability, and irreversibility data around the cycle, and the cycle effectiveness.

**Model:** Steady-state, irreversible compressor, negligible kinetic- and potential-energy changes

**Analysis:** Table 17-1 summarizes the major property data from Tables A-16 to A-18 for the four states of the cycle and the dead state. The $\psi$ data are calculated from the $h$ and $s$ data. Neglecting kinetic- and potential-energy changes, the steady-state energy balance is $q + w = \Delta h$. In terms of the compressor, condenser, and evaporator,

$$w_C = h_{2a} - h_1 = 277.08 - 241.30 = 35.78 \text{ kJ/kg}$$

$$q_{\text{cond,out}} = h_3 - h_{2a} = 93.42 - 277.08 = -183.66 \text{ kJ/kg}$$

$$q_{\text{evap}} = h_1 - h_4 = 241.30 - 93.42 = 147.88 \text{ kJ/kg}$$

**Table 17-1**    Property data for a vapor-compression refrigeration cycle

| State | $T$, °C | $P$, MPa | $h$, kJ/kg | $s$, kJ/kg·K | $\psi$, kJ/kg |
|-------|---------|----------|------------|--------------|---------------|
| 1 | −10.09 | 0.20 | 241.30 | 0.9253 | 18.28 |
| 2 | 43.18 | 0.80 | 277.08 | 0.9481 | 47.26 |
| 3 | 31.33 | 0.80 | 93.42 | 0.3459 | 43.15 |
| 4 | −10.09 | 0.20 | 93.42 | 0.3632 | 37.99 |
| 0 | 25.00 | 0.10 | 274.39 | 1.0976 | 0.00 |

**Table 17-2** An availability accounting for a vapor-compression refrigeration cycle

| Process | $q$ | $w$ | $\Delta\psi$ | $i$ |
|---------|------|-------|--------|------|
| Compressor | — | 35.78 | 28.98 | 6.80 |
| Condenser | −183.66 | — | −4.11 | — |
| Throttle valve | — | — | −5.16 | 5.16 |
| Evaporator | 147.88 | — | −19.71 | — |
| Total | −35.78 | 35.78 | 0.00 | 11.96 |

| All values in kJ/kg.

The compressor operation is assumed to be adiabatic, and the sum of these three terms is zero for the cycle. The only other important information to be gained is that

$$\text{COP} = \frac{q_{\text{evap}}}{w_C} = \frac{147.88}{35.78} = 4.13$$

The availability accounting shown in Table 17-2 leads to several interesting points. First, if the flow in the condenser is assumed frictionless so $i_{\text{cond}} = 0$, then the availability balance given by Eq. [17-14$b$] shows that

$$\sum_j q_i \left(1 - \frac{T_0}{T_j}\right) = \phi_{Q,j} = \psi_3 - \psi_2 = -4.11 \text{ kJ/kg}$$

Thus the direction of both $q$ and $\phi_{Q,j}$ is out of the condenser. In the case of a frictionless evaporator,

$$\sum_j q_j \left(1 - \frac{T_0}{T_j}\right) = \phi_{Q,j} = \psi_1 - \psi_4 = -19.71 \text{ kJ/kg}$$

Thus the availability transfer is negative, while $q$ is positive. Hence $q$ and $\phi_{Q,j}$ pass in the opposite direction at the boundary of the evaporator. This is not unusual if one recalls the discussion in Sec. 9-3-3 on availability transfer associated with heat transfer. When the system temperature $T$ is less than $T_0$, the system *loses* availability when heat is added.

Table 17-2 indicates that the irreversibilities in the compressor and in the throttle device are of the same magnitude. If the flow through the condenser were irreversible, then $\Delta\psi$ would be a measure of both $i$ and $\phi_{Q,j}$ in the condenser. Likewise for irreversible flow through the evaporator, $\Delta\psi$ measures the sum of availability destruction and transfer. Finally, the second-law efficiency in this example is

$$\epsilon = \frac{-\phi_{Q,\text{evap}}}{w_C} = \frac{19.71}{35.78} = 0.55$$

The preceding discussion and numerical example for a vapor-compression refrigeration cycle consider irreversibilities solely within the equipment itself. To simplify the analysis we have neglected flow losses in the evaporator and the condenser. Even if we accept this model, irreversibilities external to these two devices due to irreversible heat transfer should be considered. Hence we shall assume the source of heat transfer to the evaporator fluid to be the cold refrigerated region, at a constant temperature $T_{\text{cold}}$. First, the general steady-state availability balance is

$$i_{\text{tot}} = w_{\text{act}} + \psi_{\text{in}} - \psi_{\text{out}} + q_R\left(1 - \frac{T_0}{T_R}\right) \qquad \textbf{[17-16]}$$

where $i_{\text{tot}}$ includes both internal irreversibilities and external irreversibilities due to heat transfer to or from an external region of constant temperature $T_R$. When applied to the evaporator heat-transfer process with frictionless flow,

$$i_{Q,\text{evap}} = \psi_{\text{in}} - \psi_{\text{out}} + q_{\text{evap}}\left(1 - \frac{T_0}{T_{\text{cold}}}\right) \qquad \textbf{[17-17]}$$

This equation is equivalent to that for $i_Q$ for the heat-transfer process itself between the refrigerant and the cold region.

When Eq. [17-16] is applied to the heat transfer process in the condenser, it reduces to

$$i_{Q,\text{cond}} = \psi_{\text{in}} - \psi_{\text{out}} = h_{\text{in}} - h_{\text{out}} - T_0(s_{\text{in}} - s_{\text{out}}) \qquad \textbf{[17-18]}$$

There is no thermal reservoir term, as for the evaporator, because heat is rejected to the atmosphere at $T_0$.

---

**Example 17-8**

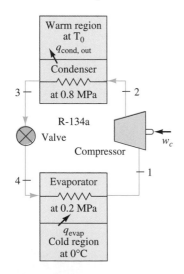

**Figure 17-24**
Equipment schematic and data for Example 17-8.

**R**econsider the vapor-compression refrigeration cycle analyzed in Example 17-7. The evaporator and condenser pressures are 0.20 and 0.80 MPa, respectively. If the cold region is maintained at 0°C, determine the irreversibilities for the heat-transfer processes in the evaporator and condenser. Compare these values to those for the compressor and the throttle device found in Example 17-7.

**Solution:**

**Given:** A vapor-compression refrigeration cycle with irreversible heat transfers is shown in Fig. 17-24.

**Find:** $i_Q$ for the evaporator and condenser.

**Model:** Steady-state, negligible kinetic- and potential-energy changes.

**Analysis:** Using data from Tables 17-1 and 17-2 the irreversibility of the evaporator heat transfer process, based on Eq. [17-17], is

$$i_{Q,\text{evap}} = \psi_4 - \psi_1 + q_{\text{evap}}\left(1 - \frac{T_0}{T_{\text{cold}}}\right)$$

$$= 37.99 - 18.28 + 147.88\left(1 - \frac{298.15}{273.15}\right) = 19.71 - 13.53 = 6.18 \text{ kJ/kg}$$

For the heat transfer from the condenser, Eq. [17-18] leads to

$$i_{Q,\text{cond}} = \psi_2 - \psi_3 = 47.26 - 43.15 = 4.11 \text{ kJ/kg}$$

**Comment:** These irreversibility values of 6.18 and 4.11 kJ/kg for the evaporator and condenser compare to 6.80 and 5.16 kJ/kg for the compressor and throttle valve, respectively. Thus the irreversibilities of all four processes are of similar magnitude. A decrease in the irreversibility of the two heat-transfer processes requires smaller temperature differences in each case.

## 17-10 SUMMARY

An idealized vapor-compression refrigeration cycle consists of four processes: (1) isentropic compression from a saturated-vapor state, (2) internally reversible, constant-pressure heat removal until a saturated-liquid state is reached, (3) throttling to the initial pressure, and (4) internally reversible, constant-pressure evaporation. The rate of heat transfer from the refrigerated space is called the *refrigeration capacity* of the cycle. Compressor irreversibilities increase the required work input and decrease the coefficient of performance.

The necessity of maintaining a finite temperature difference for the heat exchange in the evaporator and condenser constitutes another form of irreversibilities which lowers the COP. Superheating of the fluid before it enters the compressor will slightly increase the compressor work. Subcooling of the refrigerant as it leaves the condenser has a positive effect on the refrigeration capacity for a given mass flow rate. A refrigeration cycle may also be used to provide energy to a building when operating as a heat pump.

Variations of the basic compression-refrigeration cycle include the cascade and multistaged compression systems. A cascade system employing two or more cycles in series is used when a single refrigerant does not fulfill the required range of saturation temperatures and pressures. Multistaged compression is a method of decreasing the work input to the cycle by staging the compression process with intercooling between stages. Another variation of the basic cycle is the use of separate evaporators for the fresh-food space and the freezer space.

Gas-cycle and absorption refrigeration are two alternatives to mechanical vapor-compression refrigeration. Two important gas-refrigeration cycles include the use of a reversed gas-turbine (Brayton) cycle and a Stirling cycle. The absorption refrigeration cycle has been devised as a scheme to greatly reduce or eliminate the compressor work requirement. This is accomplished by using a heat input to drive the refrigeration cycle.

## PROBLEMS

17-1. Refrigerant 134a is used as the working fluid in a refrigeration cycle which follows a reversed Carnot cycle and operates between 2 and 8 bars. In the condenser the refrigerant changes from saturated vapor to saturated liquid. Determine (*a*) the coefficient of performance, (*b*) the quality of the fluid leaving the evaporator, (*c*) the work input to the compressor, in kJ/kg, and (*d*) the mass flow rate for a heat input in the evaporator of 120 kJ/min.

17-2. A refrigerator operates on a reversed Carnot cycle between evaporator and condenser temperatures of −4 and 32°C, respectively. The working fluid is refrigerant 134a, which changes from saturated vapor to saturated liquid as it flows through the condenser. Determine (*a*) the COP, (*b*) the quality of the fluid leaving the expansion process, (*c*) the work input to the compressor, in kJ/kg, and (*d*) the mass flow rate required, in kg/min, when the compressor power input is 2 kW.

17-3. A refrigerator operates on a reversed Carnot cycle between evaporator and condenser temperatures of 0 and 28°C, respectively. The working fluid is refrigerant 134a, which changes from saturated vapor to saturated liquid in the condenser. Determine (*a*) the COP, (*b*) the quality of the fluid leaving the expansion process, (*c*) the work input to the compressor, in kJ/kg, and (*d*) the mass flow rate required, in kg/min, when the heat transfer to the fluid in the evaporator is 150 kJ/min.

17-4E. Refrigerant 134a is used as the working fluid in a refrigeration cycle which follows a reversed Carnot cycle and operates between 40 and 160 psia. In the condenser the refrigerant changes from saturated vapor to saturated liquid. Determine (*a*) the coefficient of performance, (*b*) the quality of the fluid leaving the evaporator, (*c*) the work input to the compressor, in Btu/lb$_m$, and (*d*) the mass flow rate for a heat input in the evaporator of 160 Btu/min.

17-5E. A refrigerator operates on a reversed Carnot cycle between evaporator and condenser temperatures of 0 and 100°F, respectively. The working fluid is refrigerant 134a, which changes from saturated vapor to saturated liquid in the condenser. Determine (*a*) the COP, (*b*) the quality of the fluid leaving the expansion process, (*c*) the work input to the compressor, in Btu/lb$_m$, and (*d*) the mass flow rate required, in lb$_m$/min, when the heat transfer to the fluid in the evaporator is 150 Btu/min.

### VAPOR-COMPRESSION REFRIGERATION CYCLE

17-6. Saturated refrigerant 134a vapor enters the compressor of an ideal vapor-compression refrigeration cycle at 1.6 bars; saturated liquid enters the expansion valve at 7 bars. For the 5-ton unit, determine

(a) the temperature of the fluid leaving the compressor, (b) the coefficient of performance, and (c) the power input to the compressor, in kilowatts.

17-7. Modify the isentropic compressor in Prob. 17-6 so that its adiabatic efficiency is 80.2 percent. For the 5-ton unit, determine (a) the temperature of the fluid leaving the compressor, (b) the coefficient of performance, (c) the power input to the compressor, in kilowatts, and the entropy generation (d) within the compressor and (e) within the throttling process, in kJ/kg·K.

17-8. An ideal vapor-compression refrigeration cycle uses refrigerant 134a and operates between 1.8 and 9 bars. Entering the compressor, the fluid is a saturated vapor. Determine (a) the temperature of the fluid leaving the compressor, in degrees Celsius, (b) the coefficient of performance, (c) the effective displacement of the compressor, in L/min, for a 7-ton refrigeration system, and (d) the power input to the compressor, in kilowatts.

17-9. Modify the isentropic compressor in Prob. 17-8 so that its adiabatic efficiency is 78.2 percent. Determine (a) the temperature of the fluid leaving the compressor, in degrees Celsius, (b) the coefficient of performance, (c) the power input to the compressor, in kilowatts, and the entropy generation (d) within the compressor and (e) within the throttling process, in kJ/kg·K.

17-10. The pressures in the evaporator and condenser of a 4-ton refrigeration plant operating on refrigerant 134a are 0.20 and 0.80 MPa. For the ideal cycle the fluid enters the compressor as a saturated vapor, and no subcooling occurs in the condenser. Determine (a) the temperature of the fluid leaving the compressor, in degrees Celsius, (b) the coefficient of performance, (c) the effective displacement, in L/min, and (d) the power input, in kilowatts.

17-11. Modify the isentropic compressor in Prob. 17-10 so that its adiabatic efficiency is 75.9 percent. Determine (a) the temperature of the fluid leaving the compressor, in degrees Celsius, (b) the coefficient of performance, (c) the power input, in kilowatts, and the entropy generation (d) within the compressor and (e) within the throttling process, in kJ/kg· K.

17-12. The pressures in the evaporator and condenser of a refrigeration plant operating on refrigerant 134a are 0.14 and 0.90 MPa, respectively, and the mass flow rate is 5 kg/min. For the ideal cycle the fluid enters the compressor as a saturated vapor, and no subcooling occurs in the condenser. Determine (a) the temperature of the fluid leaving the compressor, in degrees Celsius, (b) the coefficient of performance, (c) the effective displacement, in L/min, (d) the power input, in kilowatts, and (e) the tons of refrigeration.

17-13. Modify the isentropic compressor in Prob. 17-12 so that its adiabatic efficiency is 83.2 percent. Determine (a) the temperature of the fluid leaving the compressor, in degrees Celsius, (b) the coefficient of performance, (c) the power input, in kilowatts, and the entropy generation (d) within the compressor and (e) within the throttling process, in kJ/kg·K.

17-14. Refrigerant 134a enters the compressor of an ideal vapor-compression refrigeration system as a saturated vapor at −16°C with a volumetric flow rate of 1.2 m³/min. The refrigerant leaves the condenser at 9 bars as a saturated liquid. Determine (a) the compressor power, in kilowatts, (b) the refrigeration capacity, in tons, and (c) the coefficient of performance.

17-15. Modify the compressor in Prob. 17-14 so that the actual exit temperature is 50°C. Determine (a) the compressor power, in kilowatts, (b) the coefficient of performance, (c) the compressor adiabatic efficiency, and the entropy generation (d) within the compressor and (e) within the throttling process, in kJ/kg·K.

17-16E. Saturated refrigerant 134a vapor enters the compressor of an ideal vapor-compression refrigeration cycle at 20 psia; saturated liquid enters the expansion valve at 160 psia. For the 5-ton unit, determine (a) the temperature of the fluid leaving the compressor, in degrees Fahrenheit, (b) the coefficient of performance, and (c) the power input to the compressor, in horsepower.

17-17E. Modify the isentropic compressor in Prob. 17-16 so that its adiabatic efficiency is 78 percent. Determine (a) the compressor fluid exit temperature, in degrees Celsius, (b) the coefficient of performance, (c) the power required to drive the compressor, in horsepower, and the entropy generation (d) within the compressor and (e) within the throttling process, in Btu/lb$_m$·°R.

17-18E. An ideal vapor-compression refrigeration cycle uses refrigerant 134a and operates between 40 and 180 psia. Entering the compressor, the fluid is a saturated vapor. Determine (a) the temperature of the fluid leaving the compressor, in degrees Fahrenheit, (b) the coefficient of performance, (c) the effective displacement of the compressor, in ft³/min, for a 6-ton refrigeration system, and (d) the power input to the compressor, in horsepower.

17-19E. Modify the isentropic compressor in Prob. 17-18E so that its adiabatic efficiency is 83 percent. Determine (a) the temperature of the fluid leaving the compressor, in degrees Fahrenheit, (b) the coefficient of performance, (c) the power input to the compressor, in horsepower, and the entropy generation (d) within the compressor and (e) within the throttling process, in Btu/lb$_m$·°R.

EFFECT OF FINITE TEMPERATURE DIFFERENCES

17-20. Reconsider Prob. 17-7 under the following conditions. To allow for a finite temperature difference between the cold region and the evaporator fluid, the evaporator is set to operate at 1.2 bar. Likewise, the pressure of the fluid in the condenser is raised to 8 bars. The cold-region temperature corresponds to a refrigerant saturation pressure of 1.6 bars. Determine, relative to Prob. 17-7, the percentage change in (a) the compressor work, (b) the refrigeration capacity, and (c) the COP. Then, (d) find the entropy generation in the compressor, the throttle device, and associated with the heat transfers to the evaporator and from the condenser, in kJ/kg·K.

17-21. Reconsider Prob. 17-9 under the following conditions. To allow for a finite temperature difference between the cold region and the evaporator fluid, the evaporator is set to operate at 1.4 bars. Likewise, the pressure of the fluid in the condenser is raised to 10 bars. The cold-region temperature corresponds to a refrigerant saturation pressure of 1.8 bars. Determine, relative to Prob. 17-9, the percentage change in (a) the compressor work, (b) the refrigeration capacity, and (c) the COP. Then, (d) find the entropy generation in the compressor, the throttle device, and associated with the heat transfers to the evaporator and from the condenser, in kJ/kg·K.

17-22. Reconsider Prob. 17-11 under the following conditions. To allow for a finite temperature difference between the cold region and the evaporator fluid, the evaporator is set to operate at 0.14 MPa. Likewise, the pressure of the fluid in the condenser is raised to 0.90 MPa. The cold-region temperature corresponds to a refrigerant saturation pressure of 0.20 MPa. Determine, relative to Prob. 17-11, the percentage change in (a) the compressor work, (b) the refrigeration capacity, and (c) the COP. Then, (d) find the entropy generation in the compressor, the throttle device, and associated with the heat transfers to the evaporator and from the condenser, in kJ/kg·K.

17-23. Reconsider Prob. 17-13 under the following conditions. To allow for a finite temperature difference between the cold region and the evaporator fluid, the evaporator is set to operate at 0.1 MPa. Likewise, the pressure of the fluid in the condenser is raised to 1.2 MPa. The cold-region temperature corresponds to a refrigerant saturation pressure of 0.14 MPa. Determine, relative to Prob. 17-13, the percentage change in (a) the compressor work, (b) the refrigeration capacity, and (c) the COP. Then, (d) find the entropy generation in the compressor, the throttle device, and associated with the heat transfers to the evaporator and from the condenser, in kJ/kg·K.

17-24. Modify Prob. 17-20 with the condition that the compressor adiabatic efficiency is 78 percent. Determine under this new constraint (a) the compressor-outlet temperature, in degrees Celsius, and (b) the percentage change in the COP compared to Prob. 17-20, and (c) the entropy generation within the compressor and within the throttling process, in kJ/kg·K.

17-25. Modify Prob. 17-21 with the condition that the compressor adiabatic efficiency is 76 percent. Determine under this new constraint (a) the compressor-outlet temperature, in degrees Celsius, and (b) the percentage change in the COP compared to Prob. 17-21, and (c) the entropy generation within the compressor and within the throttling process, in kJ/kg·K.

17-26E. Reconsider Prob. 17-17E under the following conditions. To allow for a finite temperature difference between the cold region and the evaporator fluid, the evaporator is set to operate at 15 psia. Likewise, the pressure of the fluid in the condenser is raised to 200 psia. Other information remains the same. Determine, relative to Prob. 17-17, the percentage change in (a) the compressor work, (b) the refrigeration capacity, and (c) the COP. Then, (d) find the entropy generation in the compressor, throttle device, and associated with the heat transfers to the evaporator and from the condenser, in Btu/lb$_m$·°R.

17-27E. Reconsider Prob. 17-19E under the following conditions. To allow for a finite temperature difference between the cold region and the evaporator fluid, the evaporator is set to operate at 30 psia. Likewise, the pressure of the fluid in the condenser is raised to 200 psia. Other information remains the same. Determine, relative to Prob. 17-19, the percentage change in (a) the compressor work, (b) the refrigeration capacity, and (c) the COP. Then, (d) find the entropy generation in the compressor, throttle device, and associated with the heat transfers to the evaporator and from the condenser, in Btu/lb$_m$·°R.

### EFFECT OF SUPERHEATING AND SUBCOOLING

17-28. Reconsider Prob. 17-6 under the following conditions. The refrigerant leaves the evaporator at 1.6 bars and 0°C and leaves the condenser at 7 bars and 20°C. Determine, relative to the cycle in Prob. 17-6, the percentage change in (a) the compressor work and (b) the refrigeration capacity, for the same mass flow rate.

17-29. Reconsider Prob. 17-8 under the following conditions. The refrigerant leaves the evaporator at 1.8 bars and 0°C and leaves the condenser at 9 bars and 28°C. Determine, relative to the cycle in Prob. 17-8, the percentage change in (a) the compressor work and (b) the refrigeration capacity, for the same mass flow rate.

17-30. Reconsider Prob. 17-10 under the following conditions. (a) The refrigerant leaves the condenser as a saturated liquid at 0.80 MPa, and the evaporator pressure is 0.20 MPa. Relative to Prob. 17-10, determine the percentage change in compressor work for compressor inlet temperatures of (1) 0°C and (2) 10°C. (b) The refrigerant enters the compressor as a saturated vapor at 0.20 MPa, and the condenser pressure is 0.80 MPa. Relative to Prob. 17-10, for the same mass flow rate, determine the percentage change in refrigeration capacity if the fluid enters the throttling device at (1) 26°C, and (2) 24°C.

17-31. Reconsider Prob. 17-12 under the following conditions. (*a*) The refrigerant leaves the condenser as a saturated liquid at 9 bars, and the evaporator pressure is 1.4 bars. Relative to Prob. 17-12, determine the percentage change in compressor work for compressor inlet temperatures of (1) $-10°C$ (2) $0°C$, and (3) $10°C$. (*b*) The refrigerant enters the compressor as a saturated vapor at 1.4 bars, and the condenser pressure is 9 bars. Relative to Prob. 17-12, for the same mass flow rate, determine the percentage change in refrigeration capacity if the fluid enters the throttling device at (1) $34°C$, (2) $32°C$, and (3) $28°C$.

17-32E. Reconsider Prob. 17-16E under the following conditions. The refrigerant leaves the evaporator at 20 psia and $10°F$ and leaves the condenser at 160 psia and $100°F$. Determine, relative to the cycle in Prob. 17-16, (*a*) the percentage change in the compressor work and (*b*) the percentage change in refrigeration capacity, for the same mass flow rate.

17-33E. Reconsider Prob. 17-18E under the following conditions. The refrigerant leaves the evaporator at 40 psia and $40°F$ and leaves the condenser at 180 psia and $110°F$. Determine, relative to the cycle in Prob. 17-18, (*a*) the percentage change in the compressor work and (*b*) the percentage change in refrigeration capacity, for the same mass flow rate.

17-34. Refrigerant 134a leaves the evaporator of a vapor-compression cycle at 1 bar and $-20°C$ and is adiabatically compressed to 8 bars and $60°C$. The temperature of the fluid leaving the condenser is $26°C$. Neglect any heat transfer and pressure drops in the connecting lines. The volume flow rate at the compressor inlet is 2 $m^3$/min. Determine (*a*) the rate of heat removal from the refrigerated space, (*b*) the compressor adiabatic efficiency, (*c*) the power input to the compressor, in kilowatts, (*d*) the COP of the cycle, and (*e*) the entropy generation in the compressor and in the throttling process, in kJ/kg·K.

17-35. In a vapor-compression refrigeration cycle refrigerant 134a enters the compressor superheated $8.80°C$ at 1.4 bars. The compressor adiabatic efficiency is 72 percent, and its outlet pressure is 10 bars. The inlet temperature to the throttling device is $36°C$. If the refrigeration capacity is 4 tons, determine (*a*) the mass flow rate required, in kg/min, (*b*) the power input to the compressor, in kilowatts, (*c*) the COP of the cycle, and (*d*) the entropy generation in the compressor and in the throttling process, in kJ/kg·K.

17-36. Refrigerant 134a leaves the evaporator of a vapor-compression refrigeration cycle at 1.4 bars and $-10°C$ and is compressed to 7 bars and $50°C$. The temperature of the fluid leaving the condenser is $20°C$. Determine (*a*) the coefficient of performance, (*b*) the effective displacement of the compressor, in L/min per ton of refrigeration, (*c*) the compressor efficiency, and (*d*) the entropy generation in the compressor and in the throttling process, in kJ/kg·K.

17-37. In a vapor-compression refrigeration cycle, refrigerant 134a leaves the evaporator as a saturated vapor. The evaporator and condenser pressures are 1.4 and 10 bars, respectively. The fluid entering the condenser is at 60°C, and the refrigeration capacity is 10 tons. Determine (a) the mass flow rate required, in kg/min, (b) the power input, in kilowatts, and (c) the compressor adiabatic efficiency, if the condenser-outlet temperature is 34°C.

17-38. In a vapor-compression refrigeration cycle which circulates refrigerant 134a at a rate of 6 kg/min, the refrigerant enters the compressor at 1.4 bars and −10°C and leaves at 8 bars. The fluid leaves the condenser at 28°C, and the compressor adiabatic efficiency is 73.0 percent. Determine (a) the compressor-outlet temperature, in degrees Celsius, (b) the COP, (c) the tons of refrigeration produced by the cycle, and the entropy generation (d) within the compressor and (e) within the throttling process, in kJ/kg·K.

17-39E. Refrigerant 134a leaves the evaporator of a vapor-compression refrigeration cycle at 20 psia and 10°F and is compressed to 120 psia and 140°F. The temperature of the fluid leaving the condenser is 80°F. Determine (a) the coefficient of performance, (b) the effective displacement of the compressor, in ft³/min per ton of refrigeration, and (c) the compressor efficiency.

17-40E. In a vapor-compression refrigeration cycle, refrigerant 134a leaves the evaporator at 0°F. The evaporator and condenser pressures are 20 and 180 psia, respectively. The fluid entering the condenser is at 160°F, and the refrigeration capacity is 10 tons. If the condenser-outlet temperature is 115°F, determine (a) the mass flow rate required, in $lb_m$/min, (b) the power input, in horsepower, (c) the compressor adiabatic efficiency, (d) the COP of the cycle, and (e) the entropy generation in the compressor and in the throttling process, in Btu/$lb_m$·°R.

## HEAT-PUMP CYCLES

17-41. A building is to be maintained at 22°C by means of a Carnot heat pump which removes energy from the atmosphere at −10°C. The heat-loss rate through the walls of the building is estimated to be 0.64 kW for each kelvin of temperature difference between the inside and outside of the building. Find (a) the power required to drive the heat pump, in kilowatts, and (b) the COP of the device. The same heat pump is now used to cool the building in the summer. For the same building temperature, heat-gain rate of 0.64 kW per degree temperature difference, and power input, find (c) the maximum atmospheric temperature permissible for the heat sink, in degrees Celsius, and (d) the COP of the cooling device when it is operating 10°C below the maximum permissible temperature.

17-42. A building is equipped with a Carnot heat pump which has a 2-kW power input. Measurements on the building indicate that the rate of heat gain will be 1.0 kW per kelvin of temperature difference between the interior and exterior of the building. Find (*a*) the maximum permissible summer temperature, in degrees Celsius, that the heat pump (acting as an air conditioner) can handle if the interior is kept at 20°C and (*b*) the COP of the air conditioner when it it is operating at the maximum possible value. The heat pump is then used in the winter to maintain the interior temperature again at 20°C. (*c*) For the same heat-transfer rate across the walls and the same power input, determine the rate of heat transfer from the atmosphere to the heat pump, in kJ/min, if the outside temperature is −10°C.

17-43. An ideal vapor-compression heat pump operates between an evaporator temperature of 0°C and a condenser pressure of 9 bars. Refrigerant 134a leaves the evaporator as a saturated vapor and enters the expansion valve as a saturated liquid. If the heat pump supplies 1000 kJ/min to a high-temperature region, determine (*a*) the temperature at the exit of the isentropic compressor, in degrees Celsius, (*b*) the COP, (*c*) the effective displacement of the compressor, in L/min, (*d*) the power input to the compressor, in kilowatts, and (*e*) the power input required if electrical resistance heating is used, in kilowatts.

17-44. An actual vapor-compression heat pump operates with refrigerant 134a. Underground water pumped into the building at 12°C is the source of energy, which is used to maintain the inside air supply at 30°C. The evaporator is designed to operate at a temperature 8 degrees below the cold-water inlet temperature, and the condenser operates at 9.0 bars. The fluid leaving the evaporator is saturated vapor, and it is subcooled 1.53°C as it leaves the condenser. The compressor has an adiabatic efficiency of 72 percent. Determine (*a*) the pressure in the evapoator, in bars, (*b*) the minimum temperature difference in the condenser between the refrigerant and the heat-air supply, (*c*) the temperature at the compressor outlet, in degrees Celsius, (*d*) the COP of the device, and (*e*) the volume flow rate of water required, in m³/min, if the supply has a temperature change of 2.0°C as it passes through the evaporator and the heat is supplied to the air at a rate of 1200 kJ/min.

17-45. In a heat-pump cycle using refrigerant 134a, the evaporator and condenser pressures are 2.4 and 9 bars, respectively. Vapor enters the compressor at 0°C and leaves at 53°C. Liquid leaves the condenser at 32°C. If the cycle is to supply 80,000 kJ/h to a building, calculate (*a*) the refrigerant flow rate, in kg/min, (*b*) the power input to the compressor, in kilowatts, (*c*) the compressor adiabatic efficiency, (*d*) the coefficient of performance, and (*e*) the power input, in kilowatts, if electric resistance heating is employed instead of a heat pump. Finally, if the price of electricity is $0.09 per kilowatthour, fuel oil costs $1.20 per gallon, and the heating value of oil is 130,000 kJ/gal, (f) determine the heating cost in dollars per hour if the heating source is (1) a heat pump, (2) direct electric resistance heating, and (3) an oil burner which operates with 80 percent efficiency.

17-46. A building requires 200,000 kJ/h of heat to maintain the interior air supply at 35°C when the outside temperature is −10°C. The heat is supplied by a heat pump using refrigerant 134a. The evaporator operates at a temperature 10 degrees below the outside air temperature, and the condenser operates at 10 bars. The compressor has an adiabatic efficiency of 75 percent. The fluid leaving the evaporator is a saturated vapor, and it is a saturated liquid leaving the condenser. Determine (a) the pressure in the evaporator, in bars, (b) the minimum temperature difference in the condenser between the refrigerant and the heated air supply, (c) the temperature at the compressor outlet, in degrees Celsius, (d) the quality of the fluid leaving the throttle device, and (e) the percentage increase in input energy if direct electric resistance heating were used instead of the heat pump. Finally, if the price of electricity is $0.09 per kilowatthour, fuel oil costs $1.20 per gallon, and the heating value of oil is 130,000 kJ/gal, (f) determine the heating cost in dollars per 8-hour heating day if the heating source is (1) a heat pump, (2) direct electric resistance heating, and (3) an oil burner which operates with 80 percent efficiency.

17-47E. An ideal vapor-compression heat pump operates with refrigerant 134a. The fluid leaves the evaporator as a saturated vapor at 30°F and enters the throttling device as a saturated liquid at 140 psia. If the heat pump supplies 1200 Btu/min to the high-temperature region, determine (a) the temperature of the fluid at the compressor outlet, in degrees Fahrenheit, (b) the coefficient of performance, (c) the effective displacement of the compressor, in ft³/min, (d) the power input to the compressor, in horsepower and kilowatts, and (e) the power input required, in kilowatts, if the energy source is direct electric resistance heating.

17-48E. In a heat-pump cycle using refrigerant 134a, the evaporator and condenser pressures are 40 and 180 psia, respectively. Vapor enters the compressor at 40°F and leaves at 155°F. Liquid leaves the condenser at 110°F. If the cycle is to supply 80,000 Btu/h to a building, calculate (a) the refrigerant flow rate, in lb/min, (b) the power input to the compressor, in horsepower, (c) the compressor adiabatic efficiency, (d) the coefficient of performance, and (e) the power input, in kilowatts, if electric resistance heating is employed instead of a heat pump. Finally, if the price of electricity is $0.09 per kilowatthour, fuel oil costs $1.25 per gallon, and the heating value of oil is 125,000 Btu/gal, (f) determine the heating cost in dollars per hour if the heating source is (1) a heat pump, (2) direct electric resistance heating, and (3) an oil burner which operates with 80 percent efficiency.

17-49E. A building requires 200,000 Btu/h of heat to maintain the interior air supply at 95°F when the outside temperature is 10°F. The heat is supplied by a heat pump using refrigerant 134a. The evaporator operates at a temperature 20 degrees below the outside air temperature, and the condenser operates at 160 psia. The compressor has an adiabatic efficiency of 75 percent. The fluid leaving the evaporator is a saturated vapor, and it is a saturated liquid leaving the condenser. Determine (a) the pressure in the evaporator, in psia, (b) the minimum temperature difference in the condenser between the refrigerant and the heated air supply, (c) the temperature at the compressor outlet, in degrees Fahrenheit, (d) the quality of the fluid leaving the throttle device, and (e) the percentage increase in input energy if direct electric resistance heating were used instead of the heat pump. Finally, if the price of electricity is $0.09 per kilowatthour, fuel oil costs $1.25 per gallon, and the heating value of oil is 125,000 Btu/gal, (f) determine the heating cost in dollars per hour if the heating source is (1) a heat pump, (2) direct electric resistance heating, and (3) an oil burner which operates with 80 percent efficiency.

## CASCADE AND MULTISTAGE VAPOR-COMPRESSION CYCLES

17-50. A cascade refrigeration system employs refrigerant 13 at the low-temperature end of the system and refrigerant 134a for the higher-temperature loop. The refrigerant 13 loop operates between 3.73 and 14.0 bars. The enthalpies for the saturated vapor leaving the evaporator at 220 K, the superheated vapor leaving the isentropic compressor at 275 K, and the saturated liquid entering the throttle device at 260 K are 194.9, 217.0, and 103.0 kJ/kg, respectively. The idealized cycle for the refrigerant 134a loop operates between 1.4 and 7.0 bars. The refrigeration capacity for the low-temperature evaporator is 5 tons. Determine (a) the mass flow rates, in kg/min, in both loops, (b) the power input to both compressors, in kilowatts, and (c) the COP of the cascade system.

17-51. Reconsider Prob. 17-50. Repeat the required calculations if the compressor adiabatic efficiencies are both 75 percent. Other losses are neglected.

17-52. A cascade refrigeration system employs refrigerant-22 at the low-temperature end of the system and refrigerant 134a for the higher-temperature loop. The refrigerant-22 loop operates between 1.1 and 5.0 bars. The enthalpies for the saturated vapor leaving the evaporator at −39°C, the superheated vapor leaving the isentropic compressor at 29°C, and the saturated liquid entering the throttle device at 0°C are 267.37, 304.16, and 78.47 kJ/kg, respectively. The idealized cycle for the refrigerant 134a loop operates between −12°C and 9.0 bars. The refrigeration capacity for the low-temperature evaporator is 5 tons.

16-90. Steam expands through a turbine in a cogeneration power cycle from 12 MPa and 440°C to 0.008 MPa. The turbine has an efficiency of 87 percent and delivers 60 MW of power. Steam leaves the first stage of the turbine at 0.70 MPa, where a portion is used for a heating load of $6 \times 10^6$ kJ/min. The remaining portion of the steam is reheated to 400°C before expansion to the condenser pressure. The steam leaves the heating load as a saturated liquid at 0.50 MPa and then mixes with the steam leaving the condensate pump at 0.50 MPa (see Fig. 16-18). The pump efficiencies are both 60 percent. Determine (*a*) the mass flow rates, in kg/min, through the heating load and through the condenser, (*b*) the heat input to the boiler-superheater section, in kJ/kg, (*c*) the dimensionless ratio of the heating load plus the turbine output to the heat input, and (*d*) the dimensionless ratio of the turbine output to the heat input if the cycle were operated without the heating load, but at the same total mass flow rate.

16-91. A steam plant is used for the supply of both heat and power. Steam enters the turbine at 30 bars and 320°C and expands to 0.06 bar. After expansion to 2 bars, steam is removed for heating purposes, but 8000 kg/h must always go through the low-pressure end of the turbine to keep the blades from overheating. The adiabatic efficiency of the turbine is 80 percent, and its output is 4000 kW. Determine the maximum kilograms per hour of steam which can be supplied for heating.

16-92. The power plant on a college campus is to be designed to provide both electrical power and steam heating. Relative to the schematic shown in Fig. 16-18, turbine inlet conditions are 140 bars and 560°C. Steam is bled for heating from the 82 percent efficient turbine at 5 bars, and the turbine exhaust pressure is 0.10 bar. The heating load is projected to be $5 \times 10^7$ kJ/h, and the condenser cooling requirement is expected to be $10.7 \times 10^7$ kJ/h. The steam leaves the heating load as a saturated liquid at 4 bars, and the adiabatic efficiency of the pumps is 75 percent. Determine (*a*) the gross power output of the turbine, in kilowatts, and (*b*) the dimensionless ratio of the turbine output to the heating load.

16-93E. Steam expands through a turbine in a cogeneration power cycle from 1000 psia and 800°F to 0.80 psia. The turbine has an efficiency of 85 percent and delivers 50 MW of power. Steam is extracted at 80 psia for a heating load of $6 \times 10^6$ Btu/min. The steam leaves the heating load as a saturated liquid at 60 psia and mixes with the steam leaving the condensate pump at 60 psia (see Fig. 16-18). The pump efficiencies are both 70 percent. Determine (*a*) the mass flow rates, in $lb_m$/min, through the heating load and through the condenser, (*b*) the heat input to the boiler-superheater section, in Btu/min, (*c*) the dimensionless ratio of the heating load plus the turbine output to the

Determine (*a*) the mass flow rates, in kg/min, in both loops, (*b*) the power input to both compressors, in kilowatts, and (*c*) the COP of the cascade system. (*d*) Determine the COP of a single loop cycle using refrigerant 134a between −40°C and 9 bars.

17-53. Reconsider Prob. 17-52. Repeat the required calculations if the compressor adiabatic efficiencies are both 77 percent. Other losses are neglected.

17-54. A two-stage refrigeration system with regenerative intercooling operates with refrigerant 134a and has pressures of 2.0, 4.0, and 10.0 bars in the evaporator, flash chamber and mixing chamber, and condenser, respectively. Assume isentropic compression and isenthalpic throttling and no other losses. If the refrigeration load is 5 tons, determine (*a*) the mass flow rate through the evaporator, in kg/min, (*b*) the power input to the low-pressure compressor, in kilowatts, (*c*) the mass flow rate leaving the flash chamber and entering the mixing chamber, in kg/min, (*d*) the power input to the high-pressure compressor, in kilowatts, and (*e*) the COP of the cycle.

17-55. Reconsider Prob. 17-54. Repeat the required calculation if the adiabatic efficiencies of both compressors is 77 percent. All other conditions remain the same.

17-56. A two-stage refrigeration system with regenerative intercooling operates with refrigerant 134a and has pressures of 1.6, 2.8, and 8.0 bars in the evaporator, flash chamber and mixing chamber, and condenser, respectively. Assume isentropic compression and isenthalpic throttling and no other losses. If the refrigeration load is 5 tons, determine (*a*) the mass flow rate through the evaporator, in kg/min, (*b*) the power input to the low-pressure compressor, in kilowatts, (*c*) the mass flow rate leaving the flash chamber and entering the mixing chamber, in kg/min, (*d*) the power input to the high-pressure compressor, in kilowatts, and (*e*) the COP of the cycle.

17-57. Reconsider Prob. 17-56. Repeat the required calculation if the adiabatic efficiencies of both compressors is 75 percent. All other conditions remain the same.

17-58. A two-stage refrigeration system with regenerative intercooling operates with refrigerant 134a and has pressures of 1.8, 3.2, and 12.0 bars in the evaporator, flash chamber and mixing chamber, and condenser, respectively. Assume adiabatic compression with 78 percent efficiency, isenthalpic throttling, but no other losses. If the refrigeration load is 10 tons, determine (*a*) the mass flow rate through the evaporator, in kg/min, (*b*) the power input to the low-pressure compressor, in kilowatts, (*c*) the mass flow rate leaving the flash chamber and entering the mixing chamber, in kg/min, (*d*) the power input to the high-pressure compressor, in kilowatts, and (*e*) the COP of the cycle.

17-59E. A cascade refrigeration system employs refrigerant 13 at the low-temperature end of the system and refrigerant 134a for the higher-temperature loop. The refrigerant 13 loop operates between 58.2 and 240 psia. The enthalpies for the saturated vapor leaving the evaporator at $-60°F$, the superheated vapor leaving the isentropic compressor at $48°F$, and the saturated liquid entering the throttle device at $20°F$ are 52.39, 62.50, and 15.44 $Btu/lb_m$, respectively. The idealized cycle for the refrigerant 134a loop operates between 20 and 160 psia. The refrigeration capacity for the low-temperature evaporator is 5 tons. Determine (a) the mass flow rates, in $lb_m/min$, in both loops, (b) the power input to both compressors, in horsepower, and (c) the COP of the cascade system.

17-60E. Reconsider Prob. 17-59. Repeat the required calculations if the compressor adiabatic efficiencies are both 75 percent. Other losses are neglected.

17-61E. A two-stage refrigeration system with regenerative intercooling operates with refrigerant 134a and has pressures of 30, 60, and 140 psia in the evaporator, flash chamber and mixing chamber, and condenser, respectively. Assume isentropic compression and isenthalpic throttling and no other losses. If the refrigeration load is 5 tons, determine (a) the mass flow rate through the evaporator, in $lb_m/min$, (b) the power input to the low-pressure compressor, in horsepower, (c) the mass flow rate leaving the flash chamber and entering the mixing chamber, in $lb_m/min$, (d) the power input to the high-pressure compressor, in horsepower, and (e) the COP of the cycle.

17-62. Reconsider Prob. 17-61. Repeat the required calculation if the adiabatic efficiencies of both compressors is 77 percent. All other conditions remain the same.

17-63. A simple modification of a single-evaporator refrigerator is one with two evaporators, as shown in Fig. P17-63. Evaporator 1 handles the cooling load for the freezer compartment, and evaporator 2 keeps the fresh-food compartment at the desired temperature. Consider the low-temperature evaporator operating at 1.4 bars and handling a load of 3 tons. The high-temperature evaporator produces saturated vapor at 3.2 bars and has a refrigerating capacity of 2 tons. Assume a compressor efficiency of 79 percent, a condenser pressure of 10 bars, and that saturated liquid leaves the condenser. Neglecting other losses, determine (a) the mass flow rates of R-134a through each evaporator, in kg/min, (b) the compressor power input, in kilowatts, and (c) the condenser heat loss, in kJ/min.

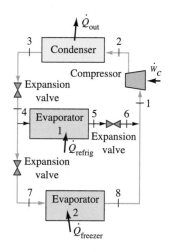

**Figure P17-63**

## LIQUEFACTION AND SOLIDIFICATION CYCLES

17-64. Consider the basic process for liquefying gases shown in Fig. 17-14, but ignore the compressor and aftercooler. Refrigerant 134a enters the heat exchanger at 16 bars and 80°C and is cooled by saturated vapor which is withdrawn from the heavily insulated separator. The high-pressure cooled gas leaves the heat exchanger and passes through a throttling valve into the separator. The separator contains liquid and vapor in equilibrium at 1.0 bar. The vapor leaving at low pressure from the heat exchanger has a temperature of 60°C. As a first approximation, neglect all pressure losses except that across the throttling valve. Determine (a) the fraction of the inlet high-pressure gas which is liquefied and (b) the temperature (if superheated) or the quality (if saturated) of the flow stream entering the valve.

17-65. Consider the basic process for liquefying gases shown in Fig. 17-14, but ignore the compressor and aftercooler. Refrigerant 134a enters the heat exchanger at 14 bars and 60°C and is cooled by saturated vapor which is withdrawn from the heavily insulated separator. The high-pressure cooled gas leaves the heat exchanger and passes through a throttling valve into the separator. The separator contains liquid and vapor in equilibrium at 1.8 bars. The vapor leaving at low pressure from the heat exchanger has a temperature of 50°C. As a first approximation, neglect all pressure losses except that across the throttling valve. Determine (a) the fraction of the inlet high-pressure gas which is liquefied and (b) the temperature (if superheated) or the quality (if saturated) of the flow stream entering the valve.

17-66. Consider the basic process for liquefying gases shown in Fig. 17-14, but ignore the compressor and aftercooler. Refrigerant 134a enters the heat exchanger at 12 bars and 60°C and is cooled by saturated vapor which is withdrawn from the heavily insulated separator. The high-pressure cooled gas leaves the heat exchanger and passes through a throttling valve into the separator. The separator contains liquid and vapor in equilibrium at 1.4 bars. The vapor leaving at low pressure from the heat exchanger has a temperature of 50°C. As a first approximation, neglect all pressure losses except that across the throttling valve. Determine (a) the fraction of the inlet high-pressure gas which is liquefied and (b) the temperature (if superheated) or the quality (if saturated) of the flow stream entering the valve.

17-67E. Consider the basic process for liquefying gases shown in Fig. 17-14, but ignore the compressor and aftercooler. Refrigerant 134a enters the heat exchanger at 300 psia and 160°F and is cooled by saturated vapor which is withdrawn from the heavily insulated separator. The high-pressure cooled gas leaves the heat exchanger and passes through a throttling valve into the separator. The separator contains liquid and vapor in equilibrium at 30 psia. The vapor leaving at low pressure from the heat exchanger has a temperature of 140°F. As a first approximation, neglect all pressure losses except that across the throttling valve. Determine (a) the fraction of the inlet high-pressure gas which is liquefied and (b) the temperature (if superheated) or the quality (if saturated) of the flow stream entering the valve.

17-68E. Consider the equipment diagram shown in Fig. 17-14 and the $Ts$ diagram for the solidification of $CO_2$ shown in Fig. 17-15. Omitting the aftercooler, state 1 has a pressure of 1000 psia and a temperature of

90°F. After passing through the heat exchanger, the fluid is throttled to 50 psia. Employing data from Fig. A-24E, determine (*a*) the percentage of solid in the separation process and (*b*) the temperature at state 5, in degrees Fahrenheit. The fluid at state 2 is 50 percent liquid at 1000 psia.

17-69E. Solid carbon dioxide is to be produced by throttling the fluid under the following conditions:

(*a*) The gas is compressed to 600 psia and precooled to 100°F. A pressure drop of 10 psi is incurred in the precooler. The fluid is passed through a heat exchanger and then throttled to 20 psia into a separator. The gas phase in the separator is passed back through the heat exchanger and discharged at 15 psia and 80°F. The solid $CO_2$ is removed at 20 psia. Determine (1) the mass fraction of solid $CO_2$ produced and (2) the temperature and condition of the $CO_2$ entering the throttle device.

(*b*) Now the same process is carried out, except that the fluid is precooled another 60 degrees to 40°F before it enters the heat exchanger. The outlet temperature at 15 psia is also 60 degrees lower than before. Repeat parts 1 and 2.

## GAS-REFRIGERATION CYCLE

17-70. A reversed Brayton cycle with a pressure ratio of 3 is used to produce a refrigeration effect. The compressor-inlet temperature is 27°C, and the turbine-inlet temperature is 7°C. If the compressor and turbine perform ideally, determine (*a*) the maximum and minimum temperatures in the cycle, in degrees Celsius, and (*b*) the coefficient of performance. If the compressor and turbine adiabatic efficiencies are 83 and 86 percent, respectively, determine (*c*) the maximum and minimum temperatures in the cycle, in degrees Celsius, and (*d*) the COP.

17-71. The compressor of an air-standard Brayton refrigeration cycle operates between 0.10 MPA, 280 K, and 0.50 MPa. The turbine-inlet temperature is 360 K. If the compressor and turbine perform ideally, determine (*a*) the maximum and minimum temperatures in the cycle, in degrees Celsius, and (*b*) the COP. If the compressor and turbine adiabatic efficiencies are 82 and 85 percent, respectively, determine (*c*) the maximum and minimum temperatures in the cycle, in degrees Celsius, (*d*) the modified COP, and (*e*) the mass flow rate of air required, in kg/min, to remove 211 kJ/min of heat from the cold region.

17-72. A reversed Brayton cycle operates with air entering the compressor at 1 bar and 7°C and leaving at 4 bars. Air enters the turbine at 27°C. If the compressor and turbine perform ideally, determine (*a*) the minimum temperature in the cycle, in degrees Celsius, (*b*) the COP, and (*c*) the mass flow rate of air required, in kg/min, to remove 211 kJ/min from the cold region. Now, if the compressor and turbine efficiencies are 84 and 88 percent, respectively, determine (d) the minimum temperature, (e) the COP, and (f) the mass flow rate for 211 kJ/min of cooling.

17-73. Using the data of Prob. 17-72, consider a regenerator placed in the cycle similar to that shown in Fig. 17-17. Assume that the air entering the turbine is precooled by the regenerator to the compressor-inlet temperature. For the ideal cycle, determine (a) the minimum temperature in the cycle, in degrees Celsius, and (b) the COP.

17-74. An ideal reversed Brayton cycle operates with a pressure ratio of 3. Air enters the compressor at 7°C and enters the turbine normally at 27°C. The cycle is now modified by considering a regenerator in the cycle similar to that shown in Fig. 17-17. The heat exchanger allows the air to be precooled to the compressor-inlet temperature before it enters the turbine. Determine (a) the old and new minimum cycle temperature, in degrees Celsius, and (b) the original and new COP values.

17-75. A regenerative-type gas refrigerator similar to that shown in Fig. 17-17 operates with helium between pressure limits of 1 and 6 bars. Compression and expansion are isentropic, and pressure drops are negligible. The compressor-inlet temperature (state 1) and the regenerator-inlet temperature (state 3) are both at 20°C, and the outlet temperature from the refrigerated region is −30°C. Determine (a) the minimum temperature achieved by the cycle, in degrees Celsius, (b) the amount of refrigeration possible, in kJ/kg, and (c) the net amount of work required, in kJ/kg.

17-76. Air at 1 bar and −13°C enters an adiabatic compressor, which discharges the fluid at 5 bars to a heat exchanger cooled by the ambient surroundings. The compressed air leaves the heat exchanger at 4.9 bars and 42°C and enters an adiabatic turbine which expands the fluid to 1.1 bars. The expanded and cold air now passes through a second heat exchanger where it picks up heat from a low-temperature region and exits at 1 bar and −13°C. The turbine and compressor are both 80 percent efficient, and the turbine helps drive the compressor. On the basis of air-table data, determine (a) the net work input per unit of heat removed from the low-temperature region, and (b) the heat removed to the ambient surroundings per unit of heat removed from the low-temperature region. (c) Sketch a Ts diagram of the process.

17-77E. The compressor of an air-standard Brayton refrigeration cycle operates between 14 psia, 40°F, and 63 psia. The turbine-inlet temperature is 190°F. If the compressor and turbine perform ideally, determine (a) the maximum and minimum temperatures in the cycle, in degrees Fahrenheit, and (b) the COP. If the compressor and turbine adiabatic efficiencies are 82 and 85 percent, respectively, determine (c) the maximum and minimum temperatures in the cycle, in degrees

Fahrenheit, (d) the modified COP, and (e) the mass flow rate of air required, in $lb_m$/min, to remove 200 Btu/min of heat from the cold region.

17-78E. A reversed Brayton cycle operates with air entering the compressor at 14.5 psia and 80°F and leaving at 58 psia. Air enters the turbine at 140°F. If the compressor and turbine perform ideally, determine (a) the minimum temperature in the cycle, in degrees Fahrenheit, (b) the COP, and (c) the mass flow rate of air required, in $lb_m$/min, to remove 200 Btu/min from the cold region. Now, if the compressor and turbine efficiencies are 84 and 88 percent, respectively, determine (d) the minimum temperature, (e) the COP, and (f) the mass flow rate for 200 Btu/min of cooling.

17-79E. Using the data of Prob. 17-78E, consider a regenerator placed in the cycle similar to that shown in Fig. 17-17. Assume that the air entering the turbine is precooled by the regenerator to the compressor-inlet temperature. For the ideal cycle, determine (a) the minimum temperature in the cycle, in degrees Fahrenheit, and (b) the COP.

17-80E. An ideal reversed Brayton cycle operates with a pressure ratio of 3. Air enters the compressor at 45°F and enters the turbine normally at 80°F. The cycle is now modified by considering a regenerator in the cycle similar to that shown in Fig. 17-17. The heat exchanger allows the air to be precooled to the compressor-inlet temperature before it enters thc turbine. Determine (a) the old and new minimum cycle temperature, in degrees Fahrenheit, and (b) the original and new COP values.

## STIRLING REFRIGERATION CYCLE

17-81. A Stirling cycle is used for refrigeration purposes. Air at 230 K and 10 bars is expanded isothermally to 1 bar. It is then heated to 315 K at constant volume. Compression at a constant temperature of 315 K follows, and the cycle is completed by constant-volume heat removal. Compute (a) the coefficient of performance for the cycle, (b) the quantity $q_L$, in kJ/kg, (c) $P_{max}$, in bars, and (d) the mass in kilograms if the maximum volume during the cycle is 1 liter.

17-82. Work Prob. 17-81 if the working fluid is carbon dioxide and it is assumed to behave as an ideal gas.

17-83. Air enters a Stirling cycle used for refrigeration purposes at 240 K and 8 bars and expands isothermally to 1.2 bars. It is then heated to 320 K at constant volume. Compression at a constant temperature of 320 K follows, and the cycle is completed by constant-volume heat removal. Compute (a) the coefficient of performance for the cycle, (b) the quantity $q_L$, in kJ/kg, (c) $P_{max}$, in bars, and (d) the mass in kilograms if the minimum volume during the cycle is 0.20 liter.

17-84. Work Prob. 17-83 if the working fluid is argon.

17-85E. A Stirling cycle is used for refrigeration purposes. Air at $-40°F$ and 10 atm is expanded isothermally to 1 atm. It is then heated to $100°F$ at constant volume. Compression at a constant temperature of $100°F$ follows, and the cycle is completed by constant-volume heat removal. Compute (a) the coefficient of performance for the cycle, (b) the quantity $q_L$, in Btu/lb$_m$, (c) $P_{max}$, in atm, and (d) the mass if the maximum volume during the cycle is 300 in$^3$.

17-86E. A Stirling cycle is used for refrigeration purposes. Air at $-20°F$ and 8 atm is expanded isothermally to 1.2 atm. It is then heated to $120°F$ at constant volume. Compression at a constant temperature of $120°F$ follows, and the cycle is completed by constant-volume heat removal. Compute (a) the coefficient of performance for the cycle, (b) the quantity $q_L$, in Btu/lb$_m$, (c) $P_{max}$, in atm, and (d) the mass if the minimum volume during the cycle is 50 in$^3$.

## AVAILABILITY ANALYSIS OF REFRIGERATION CYCLE

17-87. Saturated R-134a vapor enters the compressor of a vapor-compression refrigeration cycle at 1.4 bars; saturated liquid enters the expansion valve at 9 bars. $T_0$ is $25°C$. If the compressor outlet temperature is $50°C$ and the cold region is at $-10°C$, determine, in kJ/kg, (a) the availability changes across the compressor, condenser, expansion valve, and evaporator, (b) the irreversibilities associated with the compressor, expansion valve, evaporator, and condenser, and (c) the effectiveness of the cycle.

17-88. Saturated R-134a vapor enters the compressor of a vapor-compression refrigeration cycle at 2.8 bars; saturated liquid enters the expansion valve at 10 bars. $T_0$ is $25°C$. If the compressor outlet temperature is $50°C$ and the cold region is at $5°C$, determine, in kJ/kg, (a) the availability changes across the compressor, condenser, expansion valve, and evaporator, (b) the irreversibilities associated with the compressor, expansion valve, evaporator, and condenser, and (c) the effectiveness of the cycle.

## PARAMETRIC AND DESIGN STUDIES

17-89. By means of a computer program investigate the effect of condenser pressure on the COP of an ideal vapor-compression refrigeration cycle. The refrigerant is R-134a and the evaporator pressure is constant at 10 bars. The condenser pressure ranges from 1.0 to 3.0 bars.

17-90. By means of a computer program investigate the effect of evaporator pressure on the COP of an ideal vapor-compression refrigeration cycle. The refrigerant is R-134a and the condenser pressure is constant at 100 kPa. The evaporator pressure ranges from 700 to 1400 kPa.

17-91. Reconsider Prob. 17-88 for the situation where the adiabatic efficiency of the compressor is 82 percent.

17-92. By means of a computer program investigate the effect of superheating at the compressor outlet on the COP of a vapor-compression refrigeration cycle. The refrigerant is R-134a and the condenser pressure is 9.0 bars. The compressor outlet temperature ranges from 36 to 50°C, and the compressor inlet state is a saturated vapor at 2 bars.

17-93. By means of a computer program investigate the effect of altering the temperatures in the heat exchanger of a cascade refrigeration cycle on the overall COP and total compressor specific work input. For convenience, operate both cycles with R-134a, and let the evaporator pressure at the lowest temperature be 1.0 bar and condenser pressure on the high-temperature cycle be 12 bars. The condenser saturation temperature ranges from 10 to 26°C, and these temperatures are always 6°C higher than the evaporator temperature of the high-temperature cycle.

17-94. By means of a computer program investigate the effect of the intermediate pressure of a two-stage ideal vapor-compression refrigeration cycle on the COP and total specific-work input. The evaporator and compressor pressures are chosen to be 140 and 1000 kPa, and the range of intermediate pressures is from 300 to 800 kPa.

17-95. By means of a computer program investigate the effect of the pressure ratio on the COP and net work input of a regenerative air-standard gas refrigeration cycle as shown in Fig. 17-17. The compressor inlet state is 100 kPa and 20°C, and the pressure ratio across the compressor and expander ranges from 1.5 to 4.

17-96. An investigation of a refrigeration cycle with multistage compression and a flash chamber is to be made. The equipment in Fig. 17-11 has been modified by placing a heat exchanger after state 2 and before the mixing chamber, so that the fluid leaving the inserted heat exchanger is cooled to a saturated vapor at the mixing-chamber pressure. Thus the fluids entering and leaving the mixing chamber are all at the same state, namely, a saturated vapor at the intermediate pressure $P_i$. The evaporator pressure is 1 bar and the condenser pressure is 10 bars. Both compressors have an adiabatic efficiency of 90 percent. It is suggested that calculations be based on 1.0 kg of fluid circulating through the evaporator. You are asked to evaluate the following for your company:

(a) Determine whether there is an optimum intermediate pressure $P_i$ which maximizes the coefficient of performance, in bars.

(b) Determine the trends in the cooling requirement for the heat exchanger as a function of $P_i$. Report qualitatively whether this has a detrimental effect on the overall design.

(*c*) Provide plots of (1) the COP, (2) the total compressor work, and (3) the heat removal in the heat exchanger after the low-pressure compressor, versus $P_i$ to support your conclusions.

17-97. You have been asked by company management to consider the design and production of vapor-compression refrigeration systems which use a two-stage compressor. Such systems do not require that the vapor from the throttling process be passed through the evaporative heat exchanger, and hence may have a competitive advantage. An equipment schematic is shown in Fig. 17-11. For design purposes, the evaporator and condenser pressures are set at 2.0 and 10.0 bars, respectively. The adiabatic efficiencies of both stages are 77 percent. Pressure losses in piping and other equipment are neglected. The refrigeration capacity is set at 5 tons, and the fluids to consider are (1) R-12, or (2) R-134, or (3) both refrigerants. The task is to determine the best intermediate pressure based solely on compressor cost. The accounting office predicts that the cost of a stage, in dollars per kilowatt input, is a function of the pressure difference across the stage, the maximum pressure for that stage, and the mass flow rate through the stage. Their overall formula is

$$\text{Cost of stage (\$/kW)} = 0.5[1.5(P_{\text{Hi}} - P_{\text{Lo}})_{\text{stage}} + P^{1.2}_{\text{Hi,stage}}]\dot{m}^{0.5}_{\text{stage}}$$

where $\dot{m}$ is in kg/min and $P$ is in bars. The cost of other equipment is to be neglected in the initial investigation.

Although cost is the primary objective, performance also plays an important role. In your opinion, will the considerations of total work input and the coefficient of performance influence your selection based on costs? In fact, how well does a single-stage unit compete in cost, work input, and COP?

17-98. An industrial heat pump is to be applied to the sterilization of milk, as shown in Fig. P17-98. Milk at room temperature $T_1$ is heated by hot water which circulates around a loop that passes through the condenser of a heat pump operating with R-123a. After reaching the proper sterilization temperature $T_2$, the milk is cooled to a temperature $T_3$ of 5°C by a fluid which circulates around a loop that passes through the evaporator of the heat pump. Assume this latter fluid has the properties of water, except it freezes at −20°C. The volume flow rate for the milk is 10 L/min. The design of the system should include selection of the refrigerant pressures in the evaporator and condenser and the mass flow rates required for the fluids in the hot and cold loops.

17-99. The heat-pump concept is to be applied to remove the latent heat from the moisture that comes from a drying kiln and to transfer that energy to the air going into the drier, as shown in Fig. P17-99. The inlet room-temperature air is first heated by energy from the refrigerant in the

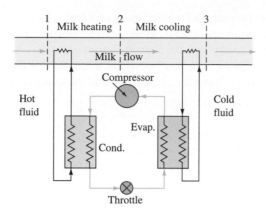

**Figure P17-98**

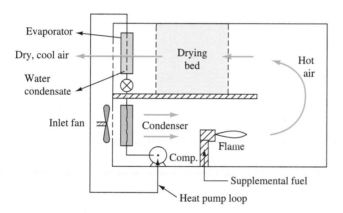

**Figure P17-99**

condenser of the heat pump, and then heated further by combustion of natural gas until it reaches 200°C. The hot air then passes through a drying bed where it removes water from a specified material, and exits the bed at 120°C and a relative humidity of 90 percent. Finally, the air passes through the evaporator of the heat pump, where it is cooled to 40°C. Investigate the design of such a unit, including the proper pressures of R-134a in the heat-pump cycle, the variation of fuel rate required, the mass flow rate of inlet air required, and the fan size required, if the energy removed from the drying bed amounts to 10,000 kJ/min.

17-100. An advanced home refrigerator has separate evaporators for the freezer and fresh-food compartments, and a heat exchanger (regenerator) placed between the relatively cold and hot refrigerant streams. The regenerator will increase the refrigeration capacity of the device but will also increase the compressor power required. Using R-134a as the refrigerant, and temperatures of 3 and $-18°C$ for the fresh food and freezer compartments temperatures, determine the effects on the COP by varying (*a*) the heat-exchanger effectiveness, and (*b*) the condenser and evaporator pressures, allowing for finite temperature differences between the refrigerant and the compartments.

.

# BIBLIOGRAPHY

Anderson, E. E.: *Thermodynamics,* PWS Publishing, Boston, 1994.

Balzhiser, R. E., and M. R. Samuels: *Engineering Thermodynamics,* Prentice-Hall, Englewood Cliffs, NJ, 1977.

Black, W. Z., and J. G. Hartley: *Thermodynamics,* Harper & Row, New York, 1985.

Burghardt, M. D.: *Engineering Thermodynamics with Applications,* 3rd ed., Harper & Row, New York, 1986.

Callen, H. B.: *Thermodynamics,* 2nd ed.,Wiley, New York, 1985.

Cravalho, E. G., and J. L. Smith, Jr.: *Engineering Thermodynamics,* Pitman, Boston, 1981.

Dixon, J. R.: *Thermodynamics I: An Introduction to Energy,* Prentice-Hall, Englewood Cliffs, NJ, 1976.

Hatsopoulos, G. N., and J. H. Keenan: *Principles of General Thermodynamics,* Wiley, New York, 1965.

Holman, J. P.: *Thermodynamics,* 3rd ed., McGraw-Hill, New York, 1980.

Howell, J. R., and R. O. Buckius: *Fundamentals of Engineering Thermodynamics,* 2nd ed., McGraw-Hill, New York, 1992.

Huang, F. F.: *Engineering Thermodynamics Fundamentals and Applications,* Macmillan, New York, 1976.

Jones, J. B., and R. E. Dugan: *Engineering Thermodynamics,* Prentice-Hall, Englewood Cliffs, NJ, 1995.

Karlekar, B. V.: *Thermodynamics for Engineers,* Prentice-Hall, Englewood Cliffs, NJ, 1983.

Keenan, J. H.: *Thermodynamics,* MIT Press, Cambridge, MA, 1970.

Kestin, J.: *A Course in Thermodynamics,* vol. 1, Blaisdell, Waltham, MA, 1966.

Lewis, G. N., and M. Randall, *Thermodynamics,* 2nd ed. (revised by K. S. Pitzer and L. Brewer), McGraw-Hill, New York, 1961.

Look, D. C., Jr., and H. S. Sauer, Jr.: *Thermodynamics,* Brooks/Cole, Monterey, CA, 1982.

Moran, M. J., and H. N. Shapiro: *Fundamentals of Engineering Thermodynamics,* 3rd ed., Wiley, New York, 1995.

Myers, G. E.: *Engineering Thermodynamics,* Prentice-Hall, Englewood Cliffs, NJ, 1989.

Obert, E. F.: *Concepts of Thermodynamics,* McGraw-Hill, New York, 1960.

——— and R. A. Gaggioli: *Thermodynamics,* McGraw-Hill, New York, 1963.

Reynolds, W. C., and H. C. Perkins: *Engineering Thermodynamics,* 2nd ed., McGraw-Hill, New York, 1977.

Rolle, K. C.: *Introduction to Thermodynamics,* 2nd ed., Charles E. Merrill, Columbus, OH, 1980.

Van Wylen, G. J., R. E. Sonntag, and C. Borgnakke: *Fundamentals of Classical Thermodynamics,* 4th ed., Wiley, New York, 1994.

Wood, B. D.: *Applications of Thermodynamics,* 2nd ed., Addison-Wesley, Reading, MA, 1982.

Zemansky, M. W., M. M. Abbott, and H. C. VanNess: *Basic Engineering Thermodynamics,* 2nd ed., McGraw-Hill, New York, 1975.

# SUPPLEMENTARY TABLES AND FIGURES (SI UNITS)

**Table A-1**     **Physical constants and conversion factors**

| Physical Constants | |
|---|---|
| Avogadro's number | $N_A = 6.023 \times 10^{26}$ atoms/kgmol |
| Universal gas constant | $R_u = 0.08205$ L·atm/(gmol·K) |
| | $= 8.314$ kJ/(kgmol·K) |
| | $= 0.08314$ bar·m$^3$/(kgmol·K) |
| | $= 8.314$ kPa·m$^3$/(kgmol·K) |
| Planck's constant | $h = 6.626 \times 10^{-34}$ J·s/molecule |
| Boltzmann's constant | $k = 1.380 \times 10^{-23}$ J/(K·molecule) |
| Speed of light | $c = 2.988 \times 10^{10}$ cm/s |
| Standard gravity | $g = 9.80665$ m/s$^2$ |

| Conversion Factors | |
|---|---|
| Length | 1 cm $= 0.3937$ in $= 10^4\ \mu$m $= 10^8$Å |
| | 1 km $= 0.6215$ mi $= 3281$ ft |
| Mass | 1 kg $= 2.205$ lb$_m$ |
| Force | 1 N $= 1$ kg·m/s$^2 = 0.2248$ lb$_f$ |
| Pressure | 1 bar $= 10^5$ N/m$^2 = 0.9869$ atm |
| | $= 100$ kPa $= 14.504$ psia |
| | 1 torr $= 1$ mmHg at 0°C $= 1.333$ mbar |
| | $= 1.933 \times 10^{-2}$ psi |
| | 1 mbar $= 0.402$ inH$_2$O |
| Volume | 1 L $= 0.0353$ ft$^3 = 0.2642$ gal $= 61.025$ in$^3 = 10^{-3}$ m$^3$ |
| Density | 1 g/cm$^3 = 1$ kg/L $= 62.4$ lb$_m$/ft$^3 = 10^3$ kg/m$^3$ |
| Energy | 1 J $= 1$ N·m $= 1$V·C |
| | $= 0.7375$ ft·lb$_f = 10$ bar·cm$^3 = 0.624 \times 10^{19}$ eV |
| | 1 kJ $= 0.948$ Btu $= 737.6$ ft·lb$_f = 10^{-2}$ bar·m$^3$ |
| | 1 kJ/kg $= 0.431$ Btu/lb |
| Power | 1 W $= 1$ J/s $= 3.413$ Btu/h |
| | 1 kW $= 1.3405$ hp $= 737.3$ ft·lb$_f$/s |
| Velocity | 1 m/s $= 2.237$ mi/h $= 3.60$ km/h $= 3.281$ ft/s |
| Specific heat | 1 kJ/(kg·K) $= 0.2389$ Btu/(lb$_m$·°F) |
| Temperature | $T(\text{K}) = \frac{5}{9}[T(°\text{F}) + 459.67] = T(°\text{C}) + 273.15 = T(°\text{R})/1.8$ |

**Table A-2** Molar mass, critical constants, and gas-phase specific heats at 25°C and 1 atm for some common substances

$(T_c, \text{ K}; P_c, \text{ Kbars}; \bar{v}_c, \text{ m}^3/\text{kmol}; c_v \text{ and } c_p, \text{ kJ/kg·K})$

| Substance | Molar mass | $T_c$ | $P_c$ | $\bar{v}_c$ | $Z_c$ | $c_v$ | $c_p$ | $\omega$ |
|---|---|---|---|---|---|---|---|---|
| Acetylene ($C_2H_2$) | 26.04 | 309 | 62.4 | 0.112 | 0.272 | 1.37 | 1.69 | 0.190 |
| Air (equivalent) | 28.97 | 133 | 37.7 | 0.0829 | 0.284 | 0.718 | 1.005 | |
| Ammonia ($NH_3$) | 17.04 | 406 | 112.8 | 0.0723 | 0.242 | 1.66 | 2.15 | 0.250 |
| Benzene ($C_6H_6$) | 78.11 | 562 | 48.3 | 0.256 | 0.274 | 0.67 | 0.775 | 0.212 |
| $n$-Butane ($C_4H_{10}$) | 58.12 | 425.2 | 37.9 | 0.257 | 0.274 | 1.56 | 1.71 | 0.199 |
| Carbon dioxide ($CO_2$) | 44.01 | 304.2 | 73.9 | 0.0941 | 0.276 | 0.657 | 0.846 | 0.224 |
| Carbon monoxide (CO) | 28.01 | 133 | 35.0 | 0.0928 | 0.294 | 0.744 | 1.04 | 0.049 |
| Refrigerant 134a ($C_2F_4H_2$) | 102.03 | 374.3 | 40.6 | 0.200 | 0.262 | 0.76 | 0.85 | 0.326 |
| Ethane ($C_2H_6$) | 30.07 | 305.4 | 48.8 | 0.148 | 0.285 | 1.48 | 1.75 | 0.099 |
| Ethylene ($C_2H_4$) | 28.05 | 283 | 51.2 | 0.128 | 0.279 | 1.23 | 1.53 | 0.089 |
| Helium (He) | 4.003 | 5.2 | 2.3 | 0.0579 | 0.300 | 3.12 | 5.19 | −0.365 |
| Hydrogen ($H_2$) | 2.016 | 33.2 | 13.0 | 0.0648 | 0.304 | 10.2 | 14.3 | −0.218 |
| Methane ($CH_4$) | 16.04 | 190.7 | 46.4 | 0.0991 | 0.290 | 1.70 | 2.22 | 0.008 |
| Nitrogen ($N_2$) | 28.01 | 126.2 | 33.9 | 0.0897 | 0.291 | 0.743 | 1.04 | 0.039 |
| Oxygen ($O_2$) | 32.00 | 154.4 | 50.5 | 0.0741 | 0.290 | 0.658 | 0.918 | 0.025 |
| Propane ($C_3H_8$) | 44.09 | 370 | 42.5 | 0.200 | 0.278 | 1.48 | 1.67 | 0.153 |
| Sulfur dioxide ($SO_2$) | 64.06 | 431 | 78.7 | 0.124 | 0.268 | 0.471 | 0.601 | 0.256 |
| Water ($H_2O$) | 18.02 | 647.3 | 220.9 | 0.0558 | 0.230 | 1.40 | 1.86 | 0.344 |

*SOURCES:* Principally from D. Ambrose, "Vapor-Liquid Critical Properties," National Physical Laboratory Report 107, 1980, and R. C. Reid et al., *The Properties of Gases and Liquids*, 4th ed., McGraw-Hill, Inc., New York, 1987.

**Additional Formulas and Values of the Molar Mass of Some Common Elements and Compounds**

| Substance | Formula | Molar mass | Substance | Formula | Molar mass |
|---|---|---|---|---|---|
| Argon | Ar | 39.94 | Silver | Ag | 107.88 |
| Aluminum | Al | 26.97 | Sodium | Na | 22.997 |
| Carbon | C | 12.01 | Hydrogen peroxide | $H_2O_2$ | 34.02 |
| Copper | Cu | 63.54 | Propylene | $C_3H_6$ | 42.08 |
| Iron | Fe | 55.85 | $n$-Pentane | $C_5H_{12}$ | 72.15 |
| Lead | Pb | 207.2 | $n$-Octane | $C_8H_{18}$ | 114.22 |
| Mercury | Hg | 200.6 | Methyl alcohol | $CH_3OH$ | 32.05 |
| Potassium | K | 39.096 | Ethyl alcohol | $C_2H_5OH$ | 46.07 |

**Table A-3** Ideal-gas specific-heat data for selected gases, kJ/kg·K

| 1. Zero-Pressure Specific Heats for Six Common Gases, Where $k = c_p/c_v$ ||||||||||
| Temp., K | $c_p$ | $c_v$ | $k$ | $c_p$ | $c_v$ | $k$ | $c_p$ | $c_v$ | $k$ | Temp., K |
| | Air ||| Carbon dioxide ($CO_2$) ||| Carbon monoxide (CO) ||| |
| 250 | 1.003 | 0.716 | 1.401 | 0.791 | 0.602 | 1.314 | 1.039 | 0.743 | 1.400 | 250 |
| 300 | 1.005 | 0.718 | 1.400 | 0.846 | 0.657 | 1.288 | 1.040 | 0.744 | 1.399 | 300 |
| 350 | 1.008 | 0.721 | 1.398 | 0.895 | 0.706 | 1.268 | 1.043 | 0.746 | 1.398 | 350 |
| 400 | 1.013 | 0.726 | 1.395 | 0.939 | 0.750 | 1.252 | 1.047 | 0.751 | 1.395 | 400 |
| 450 | 1.020 | 0.733 | 1.391 | 0.978 | 0.790 | 1.239 | 1.054 | 0.757 | 1.392 | 450 |
| 500 | 1.029 | 0.742 | 1.387 | 1.014 | 0.825 | 1.229 | 1.063 | 0.767 | 1.387 | 500 |
| 550 | 1.040 | 0.753 | 1.381 | 1.046 | 0.857 | 1.220 | 1.075 | 0.778 | 1.382 | 550 |
| 600 | 1.051 | 0.764 | 1.376 | 1.075 | 0.886 | 1.213 | 1.087 | 0.790 | 1.376 | 600 |
| 650 | 1.063 | 0.776 | 1.370 | 1.102 | 0.913 | 1.207 | 1.100 | 0.803 | 1.370 | 650 |
| 700 | 1.075 | 0.788 | 1.364 | 1.126 | 0.937 | 1.202 | 1.113 | 0.816 | 1.364 | 700 |
| 750 | 1.087 | 0.800 | 1.359 | 1.148 | 0.959 | 1.197 | 1.126 | 0.829 | 1.358 | 750 |
| 800 | 1.099 | 0.812 | 1.354 | 1.169 | 0.980 | 1.193 | 1.139 | 0.842 | 1.353 | 800 |
| 900 | 1.121 | 0.834 | 1.344 | 1.204 | 1.015 | 1.186 | 1.163 | 0.866 | 1.343 | 900 |
| 1000 | 1.142 | 0.855 | 1.336 | 1.234 | 1.045 | 1.181 | 1.185 | 0.888 | 1.335 | 1000 |
| | Hydrogen ($H_2$) ||| Nitrogen ($N_2$) ||| Oxygen ($O_2$) ||| |
| 250 | 14.051 | 9.927 | 1.416 | 1.039 | 0.742 | 1.400 | 0.913 | 0.653 | 1.398 | 250 |
| 300 | 14.307 | 10.183 | 1.405 | 1.039 | 0.743 | 1.400 | 0.918 | 0.658 | 1.395 | 300 |
| 350 | 14.427 | 10.302 | 1.400 | 1.041 | 0.744 | 1.399 | 0.928 | 0.668 | 1.389 | 350 |
| 400 | 14.476 | 10.352 | 1.398 | 1.044 | 0.747 | 1.397 | 0.941 | 0.681 | 1.382 | 400 |
| 450 | 14.501 | 10.377 | 1.398 | 1.049 | 0.752 | 1.395 | 0.956 | 0.696 | 1.373 | 450 |
| 500 | 14.513 | 10.389 | 1.397 | 1.056 | 0.759 | 1.391 | 0.972 | 0.712 | 1.365 | 500 |
| 550 | 14.530 | 10.405 | 1.396 | 1.065 | 0.768 | 1.387 | 0.988 | 0.728 | 1.358 | 550 |
| 600 | 14.546 | 10.422 | 1.396 | 1.075 | 0.778 | 1.382 | 1.003 | 0.743 | 1.350 | 600 |
| 650 | 14.571 | 10.447 | 1.395 | 1.086 | 0.789 | 1.376 | 1.017 | 0.758 | 1.343 | 650 |
| 700 | 14.604 | 10.480 | 1.394 | 1.098 | 0.801 | 1.371 | 1.031 | 0.771 | 1.337 | 700 |
| 750 | 14.645 | 10.521 | 1.392 | 1.110 | 0.813 | 1.365 | 1.043 | 0.783 | 1.332 | 750 |
| 800 | 14.695 | 10.570 | 1.390 | 1.121 | 0.825 | 1.360 | 1.054 | 0.794 | 1.327 | 800 |
| 900 | 14.822 | 10.698 | 1.385 | 1.145 | 0.849 | 1.349 | 1.074 | 0.814 | 1.319 | 900 |
| 1000 | 14.983 | 10.859 | 1.380 | 1.167 | 0.870 | 1.341 | 1.090 | 0.830 | 1.313 | 1000 |

| *SOURCE:* Data adapted from *Tables of Thermal Properties of Gases,* NBS Circular 564, 1955.

**Table A-3** (*Continued*)

---

### 2. Specific-Heat Data for Monatomic Gases

Over a wide range of temperatures at low pressures, the specific heats $c_v$ and $c_p$ of all monatomic gases are essentially independent of temperature and pressure. In addition, on a molar basis all monatomic gases have the same value for either $c_v$ or $c_p$ in a given set of units. One set of values is

$$\bar{c}_v = 12.5 \text{ kJ/(kmol·K)} \quad \text{and} \quad \bar{c}_p = 20.8 \text{ kJ/(kmol·K)}$$

*SOURCE:* Data adapted from *Tables of Thermal Properties of Gases*, NBS Circular 564, 1955.

---

### 3. Constant-Pressure Specific-Heat Equations for Various Gases at Zero Pressure (SI Units)

$$\frac{\bar{c}_p}{R_u} = a + bT + cT^2 + dT^3 + eT^4$$

**where $T$ is in kelvins, equation valid from 300 to 1000 K**

| Gas | $a$ | $b \times 10^3$ | $c \times 10^6$ | $d \times 10^9$ | $e \times 10^{12}$ |
|---|---|---|---|---|---|
| CO | 3.710 | −1.619 | 3.692 | −2.032 | 0.240 |
| $CO_2$ | 2.401 | 8.735 | −6.607 | 2.002 | |
| $H_2$ | 3.057 | 2.677 | −5.180 | 5.521 | −1.812 |
| $H_2O$ | 4.070 | −1.108 | 4.152 | −2.964 | 0.807 |
| $O_2$ | 3.626 | −1.878 | 7.056 | −6.764 | 2.156 |
| $N_2$ | 3.675 | −1.208 | 2.324 | −0.632 | −0.226 |
| Air (dry) | 3.653 | 1.334 | 3.291 | −1.910 | 0.275 |
| $NH_3$ | 3.591 | 0.494 | 8.345 | −8.383 | 2.730 |
| NO | 4.046 | −3.418 | 7.982 | −6.114 | 1.592 |
| $NO_2$ | 3.459 | 2.065 | 6.687 | −9.556 | 3.620 |
| $SO_2$ | 3.267 | 5.324 | 0.684 | −5.281 | 2.559 |
| $SO_3$ | 2.578 | 14.556 | −9.176 | −0.792 | 1.971 |
| $CH_4$ | 3.826 | −3.979 | 24.558 | −22.733 | 6.963 |
| $C_2H_2$ | 1.410 | 19.057 | −24.501 | 16.391 | −4.135 |
| $C_2H_4$ | 1.426 | 11.383 | 7.989 | −16.254 | 6.749 |

*SOURCE:* Adapted from the data in NASA SP-273, Government Printing Office, Washington, 1971.

**Table A-4**     Specific heats of some common liquids and solids

$(c_p, \text{kJ/kg} \cdot {}^\circ\text{C})$

| A. Liquids | | | | | |
|---|---|---|---|---|---|
| **Substance** | **State** | $c_p$ | **Substance** | **State** | $c_p$ |
| Water | 1 atm, 273 K | 4.217 | Benzene | 1 atm, 15°C | 1.80 |
| | 1 atm, 280 K | 4.198 | | 1 atm, 65°C | 1.92 |
| | 1 atm, 290 K | 4.186 | Glycerin | 1 atm, 10°C | 2.32 |
| | 1 atm, 300 K | 4.179 | | 1 atm, 50°C | 2.58 |
| | 1 atm, 320 K | 4.180 | Mercury | 1 atm, 10°C | 0.138 |
| | 1 atm, 340 K | 4.188 | | 1 atm, 315°C | 0.134 |
| | 1 atm, 360 K | 4.203 | Sodium | 1 atm, 95°C | 1.38 |
| | 1 atm, 373 K | 4.218 | | 1 atm, 540°C | 1.26 |
| Ammonia | sat., −20°C | 4.52 | Propane | 1 atm, 0°C | 2.41 |
| | sat., 50°C | 5.10 | Bismuth | 1 atm, 425°C | 0.144 |
| Refrigerant 134a | sat., −20°C | 1.227 | | 1 atm, 760°C | 0.164 |
| | sat., 20°C | 1.411 | Ethyl alcohol | 1 atm, 25°C | 2.43 |
| | sat., 50°C | 1.581 | | 1 atm, 50°C | 2.81 |

| B. Solids | | | | | |
|---|---|---|---|---|---|
| **Substance** | **Temp.** | $c_p$ | **Substance** | **Temp.** | $c_p$ |
| Ice | 200 K | 1.56 | Silver | 20°C | 0.233 |
| | 220 K | 1.71 | | 200°C | 0.243 |
| | 240 K | 1.86 | Lead | −173°C | 0.118 |
| | 260 K | 2.01 | | −50°C | 0.126 |
| | 270 K | 2.08 | | 27°C | 0.129 |
| | 273 K | 2.11 | | 100°C | 0.131 |
| Aluminum | 200 K | 0.797 | | 200°C | 0.136 |
| | 250 K | 0.859 | Copper | −173°C | 0.254 |
| | 300 K | 0.902 | | −100°C | 0.342 |
| | 350 K | 0.929 | | −50°C | 0.367 |
| | 400 K | 0.949 | | 0°C | 0.381 |
| | 450 K | 0.973 | | 27°C | 0.386 |
| | 500 K | 0.997 | | 100°C | 0.393 |
| Iron | 293 K | 0.448 | | 200°C | 0.403 |

**Table A-5** Ideal-gas properties of air

(*T*, K; *h* and *u*, kJ/kg; *s°*, kJ/kg·K; std state is 1 atm.)

| T | h | $p_r$ | u | $v_r$ | $s°$ | T | h | $p_r$ | u | $v_r$ | $s°$ |
|---|---|---|---|---|---|---|---|---|---|---|---|
| 200 | 199.97 | 0.3363 | 142.56 | 1707. | 1.29559 | 460 | 462.02 | 6.245 | 329.97 | 211.4 | 2.13407 |
| 210 | 209.97 | 0.3987 | 149.69 | 1512. | 1.34444 | 470 | 472.24 | 6.742 | 337.32 | 200.1 | 2.15604 |
| 220 | 219.97 | 0.4690 | 156.82 | 1346. | 1.39105 | 480 | 482.49 | 7.268 | 344.70 | 189.5 | 2.17760 |
| 230 | 230.02 | 0.5477 | 164.00 | 1205. | 1.43557 | 490 | 492.74 | 7.824 | 352.08 | 179.7 | 2.19876 |
| 240 | 240.02 | 0.6355 | 171.13 | 1084. | 1.47824 | 500 | 503.02 | 8.411 | 359.49 | 170.6 | 2.21952 |
| 250 | 250.05 | 0.7329 | 178.28 | 979. | 1.51917 | 510 | 513.32 | 9.031 | 366.92 | 162.1 | 2.23993 |
| 260 | 260.09 | 0.8405 | 185.45 | 887.8 | 1.55848 | 520 | 523.63 | 9.684 | 374.36 | 154.1 | 2.25997 |
| 270 | 270.11 | 0.9590 | 192.60 | 808.0 | 1.59634 | 530 | 533.98 | 10.37 | 381.84 | 146.7 | 2.27967 |
| 280 | 280.13 | 1.0889 | 199.75 | 738.0 | 1.63279 | 540 | 544.35 | 11.10 | 389.34 | 139.7 | 2.29906 |
| 285 | 285.14 | 1.1584 | 203.33 | 706.1 | 1.65055 | 550 | 554.74 | 11.86 | 396.86 | 133.1 | 2.31809 |
| 290 | 290.16 | 1.2311 | 206.91 | 676.1 | 1.66802 | 560 | 565.17 | 12.66 | 404.42 | 127.0 | 2.33685 |
| 295 | 295.17 | 1.0368 | 210.49 | 647.9 | 1.68515 | 570 | 575.59 | 13.50 | 411.97 | 121.2 | 2.35531 |
| 300 | 300.19 | 1.3860 | 214.07 | 621.2 | 1.70203 | 580 | 586.04 | 14.38 | 419.55 | 115.7 | 2.37348 |
| 305 | 305.22 | 1.4686 | 217.67 | 596.0 | 1.71865 | 590 | 596.52 | 15.31 | 427.15 | 110.6 | 2.39140 |
| 310 | 310.24 | 1.5546 | 221.25 | 572.3 | 1.73498 | 600 | 607.02 | 16.28 | 434.78 | 105.8 | 2.40902 |
| 315 | 315.27 | 1.6442 | 224.85 | 549.8 | 1.75106 | 610 | 617.53 | 17.30 | 442.42 | 101.2 | 2.42644 |
| 320 | 320.29 | 1.7375 | 228.42 | 528.6 | 1.76690 | 620 | 628.07 | 18.36 | 450.09 | 96.92 | 2.44356 |
| 325 | 325.31 | 1.8345 | 232.02 | 508.4 | 1.78249 | 630 | 638.63 | 19.84 | 457.78 | 92.84 | 2.46048 |
| 330 | 330.34 | 1.9352 | 235.61 | 489.4 | 1.79783 | 640 | 649.22 | 20.64 | 465.50 | 88.99 | 2.47716 |
| 340 | 340.42 | 2.149 | 242.82 | 454.1 | 1.82790 | 650 | 659.84 | 21.86 | 473.25 | 85.34 | 2.49364 |
| 350 | 350.49 | 2.379 | 250.02 | 422.2 | 1.85708 | 660 | 670.47 | 23.13 | 481.01 | 81.89 | 2.50985 |
| 360 | 360.58 | 2.626 | 257.24 | 393.4 | 1.88543 | 670 | 681.14 | 24.46 | 488.81 | 78.61 | 2.52589 |
| 370 | 370.67 | 2.892 | 264.46 | 367.2 | 1.91313 | 680 | 691.82 | 25.85 | 496.62 | 75.50 | 2.54175 |
| 380 | 380.77 | 3.176 | 271.69 | 343.4 | 1.94001 | 690 | 702.52 | 27.29 | 504.45 | 72.56 | 2.55731 |
| 390 | 390.88 | 3.481 | 278.93 | 321.5 | 1.96633 | 700 | 713.27 | 28.80 | 512.33 | 69.76 | 2.57277 |
| 400 | 400.98 | 3.806 | 286.16 | 301.6 | 1.99194 | 710 | 724.04 | 30.38 | 520.23 | 67.07 | 2.58810 |
| 410 | 411.12 | 4.153 | 293.43 | 283.3 | 2.01699 | 720 | 734.82 | 32.02 | 528.14 | 64.53 | 2.60319 |
| 420 | 421.26 | 4.522 | 300.69 | 266.6 | 2.04142 | 730 | 745.62 | 33.72 | 536.07 | 62.13 | 2.61803 |
| 430 | 431.43 | 4.915 | 307.99 | 251.1 | 2.06533 | 740 | 756.44 | 35.50 | 544.02 | 59.82 | 2.63280 |
| 440 | 441.61 | 5.332 | 315.30 | 236.8 | 2.08870 | 750 | 767.29 | 37.35 | 551.99 | 57.63 | 2.64737 |
| 450 | 451.80 | 5.775 | 322.62 | 223.6 | 2.11161 | 760 | 778.18 | 39.27 | 560.01 | 55.54 | 2.66176 |

**Table A-5**    *(Continued)*

| $T$ | $h$ | $p_r$ | $u$ | $v_r$ | $s^o$ | $T$ | $h$ | $p_r$ | $u$ | $v_r$ | $s^o$ |
|---|---|---|---|---|---|---|---|---|---|---|---|
| 780 | 800.03 | 43.35 | 576.12 | 51.64 | 2.69013 | 1360 | 1467.49 | 399.1 | 1077.10 | 9.780 | 3.32724 |
| 800 | 821.95 | 47.75 | 592.30 | 48.08 | 2.71787 | 1380 | 1491.44 | 424.2 | 1095.26 | 9.337 | 3.34474 |
| 820 | 843.98 | 52.59 | 608.59 | 44.84 | 2.74504 | 1400 | 1515.42 | 450.5 | 1113.52 | 8.919 | 3.36200 |
| 840 | 866.08 | 57.60 | 624.95 | 41.85 | 2.77170 | 1420 | 1539.44 | 478.0 | 1131.77 | 8.526 | 3.37901 |
| 860 | 888.27 | 63.09 | 641.40 | 39.12 | 2.79783 | 1440 | 1563.51 | 506.9 | 1150.13 | 8.153 | 3.39586 |
| 880 | 910.56 | 68.98 | 657.95 | 36.61 | 2.82344 | 1460 | 1587.63 | 537.1 | 1168.49 | 7.801 | 3.41247 |
| 900 | 932.93 | 75.29 | 674.58 | 34.31 | 2.84856 | 1480 | 1611.79 | 568.8 | 1186.95 | 7.468 | 3.42892 |
| 920 | 955.38 | 82.05 | 691.28 | 32.18 | 2.87324 | 1500 | 1635.97 | 601.9 | 1205.41 | 7.152 | 3.44516 |
| 940 | 977.92 | 89.28 | 708.08 | 30.22 | 2.89748 | 1520 | 1660.23 | 636.5 | 1223.87 | 6.854 | 3.46120 |
| 960 | 1000.55 | 97.00 | 725.02 | 28.40 | 2.92128 | 1540 | 1684.51 | 672.8 | 1242.43 | 6.569 | 3.47712 |
| 980 | 1023.25 | 105.2 | 741.98 | 26.73 | 2.94468 | 1560 | 1708.82 | 710.5 | 1260.99 | 6.301 | 3.49276 |
| 1000 | 1046.04 | 114.0 | 758.94 | 25.17 | 2.96770 | 1580 | 1733.17 | 750.0 | 1279.65 | 6.046 | 3.50829 |
| 1020 | 1068.89 | 123.4 | 776.10 | 23.72 | 2.99034 | 1600 | 1757.57 | 791.2 | 1298.30 | 5.804 | 3.52364 |
| 1040 | 1091.85 | 133.3 | 793.36 | 22.39 | 3.01260 | 1620 | 1782.00 | 834.1 | 1316.96 | 5.574 | 3.53879 |
| 1060 | 1114.86 | 143.9 | 810.62 | 21.14 | 3.03449 | 1640 | 1806.46 | 878.9 | 1335.72 | 5.355 | 3.55381 |
| 1080 | 1137.89 | 155.2 | 827.88 | 19.98 | 3.05608 | 1660 | 1830.96 | 925.6 | 1354.48 | 5.147 | 3.56867 |
| 1100 | 1161.07 | 167.1 | 845.33 | 18.896 | 3.07732 | 1680 | 1855.50 | 974.2 | 1373.24 | 4.949 | 3.58335 |
| 1120 | 1184.28 | 179.7 | 862.79 | 17.886 | 3.09825 | 1700 | 1880.1 | 1025 | 1392.7 | 4.761 | 3.5979 |
| 1140 | 1207.57 | 193.1 | 880.35 | 16.946 | 3.11883 | 1750 | 1941.6 | 1161 | 1439.8 | 4.328 | 3.6336 |
| 1160 | 1230.92 | 207.2 | 897.91 | 16.064 | 3.13916 | 1800 | 2003.3 | 1310 | 1487.2 | 3.944 | 3.6684 |
| 1180 | 1254.34 | 222.2 | 915.57 | 15.241 | 3.15916 | 1850 | 2065.3 | 1475 | 1534.9 | 3.601 | 3.7023 |
| 1200 | 1277.79 | 238.0 | 933.33 | 14.470 | 3.17888 | 1900 | 2127.4 | 1655 | 1582.6 | 3.295 | 3.7354 |
| 1220 | 1301.13 | 254.7 | 951.09 | 13.747 | 3.19834 | 1950 | 2189.7 | 1852 | 1630.6 | 3.022 | 3.7677 |
| 1240 | 1324.93 | 272.3 | 968.95 | 13.069 | 3.21751 | 2000 | 2252.1 | 2068 | 1678.7 | 2.776 | 3.7994 |
| 1260 | 1348.55 | 290.8 | 986.90 | 12.435 | 3.23638 | 2050 | 2314.6 | 2303 | 1726.8 | 2.555 | 3.8303 |
| 1280 | 1372.24 | 310.4 | 1004.76 | 11.835 | 3.25510 | 2100 | 2377.4 | 2559 | 1775.3 | 2.356 | 3.8605 |
| 1300 | 1395.97 | 330.9 | 1022.82 | 11.275 | 3.27345 | 2150 | 2440.3 | 2837 | 1823.8 | 2.175 | 3.8901 |
| 1320 | 1419.76 | 352.5 | 1040.88 | 10.747 | 3.29160 | 2200 | 2503.2 | 3138 | 1872.4 | 2.012 | 3.9191 |
| 1340 | 1443.60 | 375.3 | 1058.94 | 10.247 | 3.30959 | 2250 | 2566.4 | 3464 | 1921.3 | 1.864 | 3.9474 |

| *SOURCE:* Adapted from J. H. Keenan and J. Kaye, "Gas Tables," Wiley, New York, 1945. (Original tables were revised in 1983.)

**Table A-6** Ideal-gas enthalpy, internal energy, and absolute entropy of diatomic nitrogen ($N_2$)

($\Delta \bar{h}_f^\circ = 0$ kJ/kmol; $T$, K; $\bar{h}$ and $\bar{u}$, kJ/kmol; $\bar{s}^\circ$, kJ/(kmol·K); std. state is 1 atm.)

| $T$ | $\bar{h}$ | $\bar{u}$ | $\bar{s}^\circ$ | $T$ | $\bar{h}$ | $\bar{u}$ | $\bar{s}^\circ$ |
|---|---|---|---|---|---|---|---|
| 0 | 0 | 0 | 0 | 600 | 17,563 | 12,574 | 212.066 |
| 220 | 6,391 | 4,562 | 182.639 | 610 | 17,864 | 12,792 | 212.564 |
| 230 | 6,683 | 4,770 | 183.938 | 620 | 18,166 | 13,011 | 213.055 |
| 240 | 6,975 | 4,979 | 185.180 | 630 | 18,468 | 13,230 | 213.541 |
| 250 | 7,266 | 5,188 | 186.370 | 640 | 18,772 | 13,450 | 214.018 |
| 260 | 7,558 | 5,396 | 187.514 | 650 | 19,075 | 13,671 | 214.489 |
| 270 | 7,849 | 5,604 | 188.614 | 660 | 19,380 | 13,892 | 214.954 |
| 280 | 8,141 | 5,813 | 189.673 | 670 | 19,685 | 14,114 | 215.413 |
| 290 | 8,432 | 6,021 | 190.695 | 680 | 19,991 | 14,337 | 215.866 |
| 298 | 8,669 | 6,190 | 191.502 | 690 | 20,297 | 14,560 | 216.314 |
| 300 | 8,723 | 6,229 | 191.682 | 700 | 20,604 | 14,784 | 216.756 |
| 310 | 9,014 | 6,437 | 192.638 | 710 | 20,912 | 15,008 | 217.192 |
| 320 | 9,306 | 6,645 | 193.562 | 720 | 21,220 | 15,234 | 217.624 |
| 330 | 9,597 | 6,853 | 194.459 | 730 | 21,529 | 15,460 | 218.059 |
| 340 | 9,888 | 7,061 | 195.328 | 740 | 21,839 | 15,686 | 218.472 |
| 350 | 10,180 | 7,270 | 196.173 | 750 | 22,149 | 15,913 | 218.889 |
| 360 | 10,471 | 7,478 | 196.995 | 760 | 22,460 | 16,141 | 219.301 |
| 370 | 10,763 | 7,687 | 197.794 | 770 | 22,772 | 16,370 | 219.709 |
| 380 | 11,055 | 7,895 | 198.572 | 780 | 23,085 | 16,599 | 220.113 |
| 390 | 11,347 | 8,104 | 199.331 | 790 | 23,398 | 16,830 | 220.512 |
| 400 | 11,640 | 8,314 | 200.071 | 800 | 23,714 | 17,061 | 220.907 |
| 410 | 11,932 | 8,523 | 200.794 | 810 | 24,027 | 17,292 | 221.298 |
| 420 | 12,225 | 8,733 | 201.499 | 820 | 23,342 | 17,524 | 221.684 |
| 430 | 12,518 | 8,943 | 202.189 | 830 | 24,658 | 17,757 | 222.067 |
| 440 | 12,811 | 9,153 | 202.863 | 840 | 24,974 | 17,990 | 222.447 |
| 450 | 13,105 | 9,363 | 203.523 | 850 | 25,292 | 18,224 | 222.822 |
| 460 | 13,399 | 9,574 | 204.170 | 860 | 25,610 | 18,459 | 223.194 |
| 470 | 13,693 | 9,786 | 204.803 | 870 | 25,928 | 18,695 | 223.562 |
| 480 | 13,988 | 9,997 | 205.424 | 880 | 26,248 | 18,931 | 223.927 |
| 490 | 14,285 | 10,210 | 206.033 | 890 | 26,568 | 19,168 | 224.288 |
| 500 | 14,581 | 10,423 | 206.630 | 900 | 26,890 | 19,407 | 224.647 |
| 510 | 14,876 | 10,635 | 207.216 | 910 | 27,210 | 19,644 | 225.002 |
| 520 | 15,172 | 10,848 | 207.792 | 920 | 27,532 | 19,883 | 225.353 |
| 530 | 15,469 | 11,062 | 208.358 | 930 | 27,854 | 20,122 | 225.701 |
| 540 | 15,766 | 11,277 | 208.914 | 940 | 28,178 | 20,362 | 226.047 |
| 550 | 16,064 | 11,492 | 209.461 | 950 | 28,501 | 20,603 | 226.389 |
| 560 | 16,363 | 11,707 | 209.999 | 960 | 28,826 | 20,844 | 226.728 |
| 570 | 16,662 | 11,923 | 210.528 | 970 | 29,151 | 21,086 | 227.064 |
| 580 | 16,962 | 12,139 | 211.049 | 980 | 29,476 | 21,328 | 227.398 |
| 590 | 17,262 | 12,356 | 211.562 | 990 | 29,803 | 21,571 | 227.728 |

**Table A-6**     *(Continued)*

| $T$ | $\bar{h}$ | $\bar{u}$ | $\bar{s}^{\,o}$ | $T$ | $\bar{h}$ | $\bar{u}$ | $\bar{s}^{\,o}$ |
|---|---|---|---|---|---|---|---|
| 1000 | 30,129 | 21,815 | 228.057 | 1760 | 56,227 | 41,594 | 247.396 |
| 1020 | 30,784 | 22,304 | 228.706 | 1780 | 56,938 | 42,139 | 247.798 |
| 1040 | 31,442 | 22,795 | 229.344 | 1800 | 57,651 | 42,685 | 248.195 |
| 1060 | 32,101 | 23,288 | 229.973 | 1820 | 58,363 | 43,231 | 248.589 |
| 1080 | 32,762 | 23,782 | 230.591 | 1840 | 59,075 | 43,777 | 248.979 |
| 1100 | 33,426 | 24,280 | 231.199 | 1860 | 59,790 | 44,324 | 249.365 |
| 1120 | 34,092 | 24,780 | 231.799 | 1880 | 60,504 | 44,873 | 249.748 |
| 1140 | 34,760 | 25,582 | 232.391 | 1900 | 61,220 | 45,423 | 250.128 |
| 1160 | 35,430 | 25,786 | 232.973 | 1920 | 61,936 | 45,973 | 250.502 |
| 1180 | 36,104 | 26,291 | 233.549 | 1940 | 62,654 | 46,524 | 250.874 |
| 1200 | 36,777 | 26,799 | 234.115 | 1960 | 63,381 | 47,075 | 251.242 |
| 1220 | 37,452 | 27,308 | 234.673 | 1980 | 64,090 | 47,627 | 251.607 |
| 1240 | 38,129 | 27,819 | 235.223 | 2000 | 64,810 | 48,181 | 251.969 |
| 1260 | 38,807 | 28,331 | 235.766 | 2050 | 66,612 | 49,567 | 252.858 |
| 1280 | 39,488 | 28,845 | 236.302 | 2100 | 68,417 | 50,957 | 253.726 |
| 1300 | 40,170 | 29,361 | 236.831 | 2150 | 70,226 | 52,351 | 254.578 |
| 1320 | 40,853 | 29,878 | 237.353 | 2200 | 72,040 | 53,749 | 255.412 |
| 1340 | 41,539 | 30,398 | 237.867 | 2250 | 73,856 | 55,149 | 256.227 |
| 1360 | 42,227 | 30,919 | 238.376 | 2300 | 75,676 | 56,553 | 257.027 |
| 1380 | 42,915 | 31,441 | 238.878 | 2350 | 77,496 | 57,958 | 257.810 |
| 1400 | 43,605 | 31,964 | 239.375 | 2400 | 79,320 | 59,366 | 258.580 |
| 1420 | 44,295 | 32,489 | 239.865 | 2450 | 81,149 | 60,779 | 259.332 |
| 1440 | 44,988 | 33,014 | 240.350 | 2500 | 82,981 | 62,195 | 260.073 |
| 1460 | 45,682 | 33,543 | 240.827 | 2550 | 84,814 | 63,613 | 260.799 |
| 1480 | 46,377 | 34,071 | 241.301 | 2600 | 86,650 | 65,033 | 261.512 |
| 1500 | 47,073 | 34,601 | 241.768 | 2650 | 88,488 | 66,455 | 262.213 |
| 1520 | 47,771 | 35,133 | 242.228 | 2700 | 90,328 | 67,880 | 262.902 |
| 1540 | 48,470 | 35,665 | 242.685 | 2750 | 92,171 | 69,306 | 263.577 |
| 1560 | 49,168 | 36,197 | 243.137 | 2800 | 94,014 | 70,734 | 264.241 |
| 1580 | 49,869 | 36,732 | 243.585 | 2850 | 95,859 | 72,163 | 264.895 |
| 1600 | 50,571 | 37,268 | 244.028 | 2900 | 97,705 | 73,593 | 265.538 |
| 1620 | 51,275 | 37,806 | 244.464 | 2950 | 99,556 | 75,028 | 266.170 |
| 1640 | 51,980 | 38,344 | 244.896 | 3000 | 101,407 | 76,464 | 266.793 |
| 1660 | 52,686 | 38,884 | 245.324 | 3050 | 103,260 | 77,902 | 267.404 |
| 1680 | 53,393 | 39,424 | 245.747 | 3100 | 105,115 | 79,341 | 268.007 |
| 1700 | 54,099 | 39,965 | 246.166 | 3150 | 106,972 | 80,782 | 268.601 |
| 1720 | 54,807 | 40,507 | 246.580 | 3200 | 108,830 | 82,224 | 269.186 |
| 1740 | 55,516 | 41,049 | 246.990 | 3250 | 110,690 | 83,668 | 269.763 |

| *SOURCE:* Based on data from the JANAF Thermochemical Tables, NSRDS-NBS-37, 1971.

**Table A-7**  Ideal-gas enthalpy, internal energy, and absolute entropy of diatomic oxygen ($O_2$)

($\Delta \bar{h}_f^\circ$ = 0 kJ/kmol; $T$, K; $\bar{h}$ and $\bar{u}$, kJ/kmol; $\bar{s}^\circ$, kJ/(kmol·K); std state is 1 atm.)

| $T$ | $\bar{h}$ | $\bar{u}$ | $\bar{s}^\circ$ | $T$ | $\bar{h}$ | $\bar{u}$ | $\bar{s}^\circ$ |
|---|---|---|---|---|---|---|---|
| 0 | 0 | 0 | 0 | 600 | 17,929 | 12,940 | 226.346 |
| 220 | 6,404 | 4,575 | 196.171 | 610 | 18,250 | 13,178 | 226.877 |
| 230 | 6,694 | 4,782 | 197.461 | 620 | 18,572 | 13,417 | 227.400 |
| 240 | 6,984 | 4,989 | 198.696 | 630 | 18,895 | 13,657 | 227.918 |
| 250 | 7,275 | 5,197 | 199.885 | 640 | 19,219 | 13,898 | 228.429 |
| 260 | 7,566 | 5,405 | 201.027 | 650 | 19,544 | 14,140 | 228.932 |
| 270 | 7,858 | 5,613 | 202.128 | 660 | 19,870 | 14,383 | 229.430 |
| 280 | 8,150 | 5,822 | 203.191 | 670 | 20,197 | 14,626 | 229.920 |
| 290 | 8,443 | 6,032 | 204.218 | 680 | 20,524 | 14,871 | 230.405 |
| 298 | 8,682 | 6,203 | 205.033 | 690 | 20,854 | 15,116 | 230.885 |
| 300 | 8,736 | 6,242 | 205.213 | 700 | 21,184 | 15,364 | 231.358 |
| 310 | 9,030 | 6,453 | 206.177 | 710 | 21,514 | 15,611 | 231.827 |
| 320 | 9,325 | 6,664 | 207.112 | 720 | 21,845 | 15,859 | 232.291 |
| 330 | 9,620 | 6,877 | 208.020 | 730 | 22,177 | 16,107 | 232.748 |
| 340 | 9,916 | 7,090 | 208.904 | 740 | 22,510 | 16,357 | 233.201 |
| 350 | 10,213 | 7,303 | 209.765 | 750 | 22,844 | 16,607 | 233.649 |
| 360 | 10,511 | 7,518 | 210.604 | 760 | 23,178 | 16,859 | 234.091 |
| 370 | 10,809 | 7,733 | 211.423 | 770 | 23,513 | 17,111 | 234.528 |
| 380 | 11,109 | 7,949 | 212.222 | 780 | 23,850 | 17,364 | 234.960 |
| 390 | 11,409 | 8,166 | 213.002 | 790 | 24,186 | 17,618 | 235.387 |
| 400 | 11,711 | 8,384 | 213.765 | 800 | 24,523 | 17,872 | 235.810 |
| 410 | 12,012 | 8,603 | 214.510 | 810 | 24,861 | 18,126 | 236.230 |
| 420 | 12,314 | 8,822 | 215.241 | 820 | 25,199 | 18,382 | 236.644 |
| 430 | 12,618 | 9,043 | 215.955 | 830 | 25,537 | 18,637 | 237.055 |
| 440 | 12,923 | 9,264 | 216.656 | 840 | 25,877 | 18,893 | 237.462 |
| 450 | 13,228 | 9,487 | 217.342 | 850 | 26,218 | 19,150 | 237.864 |
| 460 | 13,535 | 9,710 | 218.016 | 860 | 26,559 | 19,408 | 238.264 |
| 470 | 13,842 | 9,935 | 218.676 | 870 | 26,899 | 19,666 | 238.660 |
| 480 | 14,151 | 10,160 | 219.326 | 880 | 27,242 | 19,925 | 239.051 |
| 490 | 14,460 | 10,386 | 219.963 | 890 | 27,584 | 20,185 | 239.439 |
| 500 | 14,770 | 10,614 | 220.589 | 900 | 27,928 | 20,445 | 239.823 |
| 510 | 15,082 | 10,842 | 221.206 | 910 | 28,272 | 20,706 | 240.203 |
| 520 | 15,395 | 11,071 | 221.812 | 920 | 28,616 | 20,967 | 240.580 |
| 530 | 15,708 | 11,301 | 222.409 | 930 | 28,960 | 21,228 | 240.953 |
| 540 | 16,022 | 11,533 | 222.997 | 940 | 29,306 | 21,491 | 241.323 |
| 550 | 16,338 | 11,765 | 223.576 | 950 | 29,652 | 21,754 | 241.689 |
| 560 | 16,654 | 11,998 | 224.146 | 960 | 29,999 | 22,017 | 242.052 |
| 570 | 16,971 | 12,232 | 224.708 | 970 | 30,345 | 22,280 | 242.411 |
| 580 | 17,290 | 12,467 | 225.262 | 980 | 30,692 | 22,544 | 242.768 |
| 590 | 17,609 | 12,703 | 225.808 | 990 | 31,041 | 22,809 | 243.120 |

**Table A-7**    (*Continued*)

| $T$ | $\bar{h}$ | $\bar{u}$ | $\bar{s}^o$ | $T$ | $\bar{h}$ | $\bar{u}$ | $\bar{s}^o$ |
|---|---|---|---|---|---|---|---|
| 1000 | 31,389 | 23,075 | 243.471 | 1760 | 58,880 | 44,247 | 263.861 |
| 1020 | 32,088 | 23,607 | 244.164 | 1780 | 59,624 | 44,825 | 264.283 |
| 1040 | 32,789 | 24,142 | 244.844 | 1800 | 60,371 | 45,405 | 264.701 |
| 1060 | 33,490 | 24,677 | 245.513 | 1820 | 61,118 | 45,986 | 265.113 |
| 1080 | 34,194 | 25,214 | 246.171 | 1840 | 61,866 | 46,568 | 265.521 |
| 1100 | 34,899 | 25,753 | 246.818 | 1860 | 62,616 | 47,151 | 265.925 |
| 1120 | 35,606 | 26,294 | 247.454 | 1880 | 63,365 | 47,734 | 266.326 |
| 1140 | 36,314 | 26,836 | 248.081 | 1900 | 64,116 | 48,319 | 266.722 |
| 1160 | 37,023 | 27,379 | 248.698 | 1920 | 64,868 | 48,904 | 267.115 |
| 1180 | 37,734 | 27,923 | 249.307 | 1940 | 65,620 | 49,490 | 267.505 |
| 1200 | 38,447 | 28,469 | 249.906 | 1960 | 66,374 | 50,078 | 267.891 |
| 1220 | 39,162 | 29,018 | 250.497 | 1980 | 67,127 | 50,665 | 268.275 |
| 1240 | 39,877 | 29,568 | 251.079 | 2000 | 67,881 | 51,253 | 268.655 |
| 1260 | 40,594 | 30,118 | 251.653 | 2050 | 69,772 | 52,727 | 269.588 |
| 1280 | 41,312 | 30,670 | 252.219 | 2100 | 71,668 | 54,208 | 270.504 |
| 1300 | 42,033 | 31,224 | 252.776 | 2150 | 73,573 | 55,697 | 271.399 |
| 1320 | 42,753 | 31,778 | 253.325 | 2200 | 75,484 | 57,192 | 272.278 |
| 1340 | 43,475 | 32,334 | 253.868 | 2250 | 77,397 | 58,690 | 273.136 |
| 1360 | 44,198 | 32,891 | 254.404 | 2300 | 79,316 | 60,193 | 273.981 |
| 1380 | 44,923 | 33,449 | 254.932 | 2350 | 81,243 | 61,704 | 274.809 |
| 1400 | 45,648 | 34,008 | 255.454 | 2400 | 83,174 | 63,219 | 275.625 |
| 1420 | 46,374 | 34,567 | 255.968 | 2450 | 85,112 | 64,742 | 276.424 |
| 1440 | 47,102 | 35,129 | 256.475 | 2500 | 87,057 | 66,271 | 277.207 |
| 1460 | 47,831 | 35,692 | 256.978 | 2550 | 89,004 | 67,802 | 277.979 |
| 1480 | 48,561 | 36,256 | 257.474 | 2600 | 90,956 | 69,339 | 278.738 |
| 1500 | 49,292 | 36,821 | 257.965 | 2650 | 92,916 | 70,883 | 279.485 |
| 1520 | 50,024 | 37,387 | 258.450 | 2700 | 94,881 | 72,433 | 280.219 |
| 1540 | 50,756 | 37,952 | 258.928 | 2750 | 96,852 | 73,987 | 280.942 |
| 1560 | 51,490 | 38,520 | 259.402 | 2800 | 98,826 | 75,546 | 281.654 |
| 1580 | 52,224 | 39,088 | 259.870 | 2850 | 100,808 | 77,112 | 282.357 |
| 1600 | 52,961 | 39,658 | 260.333 | 2900 | 102,793 | 78,682 | 283.048 |
| 1620 | 53,696 | 40,227 | 260.791 | 2950 | 104,785 | 80,258 | 283.728 |
| 1640 | 54,434 | 40,799 | 261.242 | 3000 | 106,780 | 81,837 | 284.399 |
| 1660 | 55,172 | 41,370 | 261.690 | 3050 | 108,778 | 83,419 | 285.060 |
| 1680 | 55,912 | 41,944 | 262.132 | 3100 | 110,784 | 85,009 | 285.713 |
| 1700 | 56,652 | 42,517 | 262.571 | 3150 | 112,795 | 86,601 | 286.355 |
| 1720 | 57,394 | 43,093 | 263.005 | 3200 | 114,809 | 88,203 | 286.989 |
| 1740 | 58,136 | 43,669 | 263.435 | 3250 | 116,827 | 89,804 | 287.614 |

| *SOURCE:* Based on data from the JANAF Thermochemical Tables, NSRDS-NBS-37, 1971.

**Table A-8** Ideal-gas enthalpy, internal energy, and absolute entropy of carbon monoxide (CO)

($\Delta \bar{h}_f^\circ = -110{,}530$ kJ/kmol; $T$, K; $\bar{h}$ and $\bar{u}$, kJ/kmol; $\bar{s}^\circ$, kJ/(kmol·K); std state is 1 atm.)

| $T$ | $\bar{h}$ | $\bar{u}$ | $\bar{s}^\circ$ | $T$ | $\bar{h}$ | $\bar{u}$ | $\bar{s}^\circ$ |
|---|---|---|---|---|---|---|---|
| 0 | 0 | 0 | 0 | 600 | 17,611 | 12,622 | 218.204 |
| 220 | 6,391 | 4,562 | 188.683 | 610 | 17,915 | 12,843 | 218.708 |
| 230 | 6,683 | 4,771 | 189.980 | 620 | 18,221 | 13,066 | 219.205 |
| 240 | 6,975 | 4,979 | 191.221 | 630 | 18,527 | 13,289 | 219.695 |
| 250 | 7,266 | 5,188 | 192.411 | 640 | 18,833 | 13,512 | 220.179 |
| 260 | 7,558 | 5,396 | 193.554 | 650 | 19,141 | 13,736 | 220.656 |
| 270 | 7,849 | 5,604 | 194.654 | 660 | 19,449 | 13,962 | 221.127 |
| 280 | 8,140 | 5,812 | 195.173 | 670 | 19,758 | 14,187 | 221.592 |
| 290 | 8,432 | 6,020 | 196.735 | 680 | 20,068 | 14,414 | 222.052 |
| 298 | 8,669 | 6,190 | 197.543 | 690 | 20,378 | 14,641 | 222.505 |
| 300 | 8,723 | 6,229 | 197.723 | 700 | 20,690 | 14,870 | 222.953 |
| 310 | 9,014 | 6,437 | 198.678 | 710 | 21,002 | 15,099 | 223.396 |
| 320 | 9,306 | 6,645 | 199.603 | 720 | 21,315 | 15,328 | 223.833 |
| 330 | 9,597 | 6,854 | 200.500 | 730 | 21,628 | 15,558 | 224.265 |
| 340 | 9,889 | 7,062 | 201.371 | 740 | 21,943 | 15,789 | 224.692 |
| 350 | 10,181 | 7,271 | 202.217 | 750 | 22,258 | 16,022 | 225.115 |
| 360 | 10,473 | 7,480 | 203.040 | 760 | 22,573 | 16,255 | 225.533 |
| 370 | 10,765 | 7,689 | 203.842 | 770 | 22,890 | 16,488 | 225.947 |
| 380 | 11,058 | 7,899 | 204.622 | 780 | 23,208 | 16,723 | 226.357 |
| 390 | 11,351 | 8,108 | 205.383 | 790 | 23,526 | 16,957 | 226.762 |
| 400 | 11,644 | 8,319 | 206.125 | 800 | 23,844 | 17,193 | 227.162 |
| 410 | 11,938 | 8,529 | 206.850 | 810 | 24,164 | 17,429 | 227.559 |
| 420 | 12,232 | 8,740 | 207.549 | 820 | 24,483 | 17,665 | 227.952 |
| 430 | 12,526 | 8,951 | 208.252 | 830 | 24,803 | 17,902 | 228.339 |
| 440 | 12,821 | 9,163 | 208.929 | 840 | 25,124 | 18,140 | 228.724 |
| 450 | 13,116 | 9,375 | 209.593 | 850 | 25,446 | 18,379 | 229.106 |
| 460 | 13,412 | 9,587 | 210.243 | 860 | 25,768 | 18,617 | 229.482 |
| 470 | 13,708 | 9,800 | 210.880 | 870 | 26,091 | 18,858 | 229.856 |
| 480 | 14,005 | 10,014 | 211.504 | 880 | 26,415 | 19,099 | 230.227 |
| 490 | 14,302 | 10,228 | 212.117 | 890 | 26,740 | 19,341 | 230.593 |
| 500 | 14,600 | 10,443 | 212.719 | 900 | 27,066 | 19,583 | 230.957 |
| 510 | 14,898 | 10,658 | 213.310 | 910 | 27,392 | 19,826 | 231.317 |
| 520 | 15,197 | 10,874 | 213.890 | 920 | 27,719 | 20,070 | 231.674 |
| 530 | 15,497 | 11,090 | 214.460 | 930 | 28,046 | 20,314 | 232.028 |
| 540 | 15,797 | 11,307 | 215.020 | 940 | 28,375 | 20,559 | 232.379 |
| 550 | 16,097 | 11,524 | 215.572 | 950 | 28,703 | 20,805 | 232.727 |
| 560 | 16,399 | 11,743 | 216.115 | 960 | 29,033 | 21,051 | 233.072 |
| 570 | 16,701 | 11,961 | 216.649 | 970 | 29,362 | 21,298 | 233.413 |
| 580 | 17,003 | 12,181 | 217.175 | 980 | 29,693 | 21,545 | 233.752 |
| 590 | 17,307 | 12,401 | 217.693 | 990 | 30,024 | 21,793 | 234.088 |

**Table A-8** (*Continued*)

| $T$ | $\bar{h}$ | $\bar{u}$ | $\bar{s}^o$ | $T$ | $\bar{h}$ | $\bar{u}$ | $\bar{s}^o$ |
|---|---|---|---|---|---|---|---|
| 1000 | 30,355 | 22,041 | 234.421 | 1760 | 56,756 | 42,123 | 253.991 |
| 1020 | 31,020 | 22,540 | 235.079 | 1780 | 57,473 | 42,673 | 254.398 |
| 1040 | 31,688 | 23,041 | 235.728 | 1800 | 58,191 | 43,225 | 254.797 |
| 1060 | 32,357 | 23,544 | 236.364 | 1820 | 58,910 | 43,778 | 255.194 |
| 1080 | 33,029 | 24,049 | 236.992 | 1840 | 59,629 | 44,331 | 255.587 |
| 1100 | 33,702 | 24,557 | 237.609 | 1860 | 60,351 | 44,886 | 255.976 |
| 1120 | 34,377 | 25,065 | 238.217 | 1880 | 61,072 | 45,441 | 256.361 |
| 1140 | 35,054 | 25,575 | 238.817 | 1900 | 61,794 | 45,997 | 256.743 |
| 1160 | 35,733 | 26,088 | 239.407 | 1920 | 62,516 | 46,552 | 257.122 |
| 1180 | 36,406 | 26,602 | 239.989 | 1940 | 63,238 | 47,108 | 257.497 |
| 1200 | 37,095 | 27,118 | 240.663 | 1960 | 63,961 | 47,665 | 257.868 |
| 1220 | 37,780 | 27,637 | 241.128 | 1980 | 64,684 | 48,221 | 258.236 |
| 1240 | 38,446 | 28,426 | 241.686 | 2000 | 65,408 | 48,780 | 258.600 |
| 1260 | 39,154 | 28,678 | 242.236 | 2050 | 67,224 | 50,179 | 259.494 |
| 1280 | 39,884 | 29,201 | 242.780 | 2100 | 69,044 | 51,584 | 260.370 |
| 1300 | 40,534 | 29,725 | 243.316 | 2150 | 70,864 | 52,998 | 261.226 |
| 1320 | 41,266 | 30,251 | 243.844 | 2200 | 72,688 | 54,396 | 262.065 |
| 1340 | 41,919 | 30,778 | 244.366 | 2250 | 74,516 | 55,809 | 262.887 |
| 1360 | 42,613 | 31,306 | 244.880 | 2300 | 76,345 | 57,222 | 263.692 |
| 1380 | 43,309 | 31,836 | 245.388 | 2350 | 78,178 | 58,640 | 264.480 |
| 1400 | 44,007 | 32,367 | 245.889 | 2400 | 80,015 | 60,060 | 265.253 |
| 1420 | 44,707 | 32,900 | 246.385 | 2450 | 81,852 | 61,482 | 266.012 |
| 1440 | 45,408 | 33,434 | 246.876 | 2500 | 83,692 | 62,906 | 266.755 |
| 1460 | 46,110 | 33,971 | 247.360 | 2550 | 85,537 | 64,335 | 267.485 |
| 1480 | 46,813 | 34,508 | 247.839 | 2600 | 87,383 | 65,766 | 268.202 |
| 1500 | 47,517 | 35,046 | 248.312 | 2650 | 89,230 | 67,197 | 268.905 |
| 1520 | 48,222 | 35,584 | 248.778 | 2700 | 91,077 | 68,628 | 269.596 |
| 1540 | 48,928 | 36,124 | 249.240 | 2750 | 92,930 | 70,066 | 270.285 |
| 1560 | 19,635 | 36,665 | 249.695 | 2800 | 94,784 | 71,504 | 270.943 |
| 1580 | 50,344 | 37,207 | 250.147 | 2850 | 96,639 | 72,945 | 271.602 |
| 1600 | 51,053 | 37,750 | 250.592 | 2900 | 98,495 | 74,383 | 272.249 |
| 1620 | 51,763 | 38,293 | 251.033 | 2950 | 100,352 | 75,825 | 272.884 |
| 1640 | 52,472 | 38,837 | 251.470 | 3000 | 102,210 | 77,267 | 273.508 |
| 1660 | 53,184 | 39,382 | 251.901 | 3050 | 104,073 | 78,715 | 274.123 |
| 1680 | 53,895 | 39,927 | 252.329 | 3100 | 105,939 | 80,164 | 274.730 |
| 1700 | 54,609 | 40,474 | 252.751 | 3150 | 107,802 | 81,612 | 275.326 |
| 1720 | 55,323 | 41,023 | 253.169 | 3200 | 109,667 | 83,061 | 275.914 |
| 1740 | 56,039 | 41,572 | 253.582 | 3250 | 111,534 | 84,513 | 276.494 |

| *SOURCE:* Based on data from the JANAF Thermochemical Tables, NSRDS-NBS-37, 1971.

**Table A-9** Ideal-gas enthalpy, internal energy, and absolute entropy of carbon dioxide ($CO_2$)

($\Delta \bar{h}_f^\circ = -393{,}520$ kJ/kmol; $T$, K; $\bar{h}$ and $\bar{u}$, kJ/kmol; $\bar{s}^\circ$, kJ/(kmol·K); std state is 1 atm.)

| $T$ | $\bar{h}$ | $\bar{u}$ | $\bar{s}^\circ$ | $T$ | $\bar{h}$ | $\bar{u}$ | $\bar{s}^\circ$ |
|---|---|---|---|---|---|---|---|
| 0 | 0 | 0 | 0 | 600 | 22,280 | 17,291 | 243.199 |
| 220 | 6,601 | 4,772 | 202.966 | 610 | 22,754 | 17,683 | 243.983 |
| 230 | 6,938 | 5,026 | 204.464 | 620 | 23,321 | 18,076 | 244.758 |
| 240 | 7,280 | 5,285 | 205.920 | 630 | 23,709 | 18,471 | 245.524 |
| 250 | 7,627 | 5,548 | 207.337 | 640 | 24,190 | 18,869 | 246.282 |
| 260 | 7,979 | 5,817 | 208.717 | 650 | 24,674 | 19,270 | 247.032 |
| 270 | 8,335 | 6,091 | 210.062 | 660 | 25,160 | 19,672 | 247.773 |
| 280 | 8,697 | 6,369 | 211.376 | 670 | 25,648 | 20,078 | 248.507 |
| 290 | 9,063 | 6,651 | 212.660 | 680 | 26,138 | 20,484 | 249.233 |
| 298 | 9,364 | 6,885 | 213.685 | 690 | 26,631 | 20,894 | 249.952 |
| 300 | 9,431 | 6,939 | 213.915 | 700 | 27,125 | 21,305 | 250.663 |
| 310 | 9,807 | 7,230 | 215.146 | 710 | 27,622 | 21,719 | 251.368 |
| 320 | 10,186 | 7,526 | 216.351 | 720 | 28,121 | 22,134 | 252.065 |
| 330 | 10,570 | 7,826 | 217.534 | 730 | 28,622 | 22,552 | 252.755 |
| 340 | 10,959 | 8,131 | 218.694 | 740 | 29,124 | 22,972 | 253.439 |
| 350 | 11,351 | 8,439 | 219.831 | 750 | 29,629 | 23,393 | 254.117 |
| 360 | 11,748 | 8,752 | 220.948 | 760 | 30,135 | 23,817 | 254.787 |
| 370 | 12,148 | 9,068 | 222.044 | 770 | 30,644 | 24,242 | 255.452 |
| 380 | 12,552 | 9,392 | 223.122 | 780 | 31,154 | 24,669 | 256.110 |
| 390 | 12,960 | 9,718 | 224.182 | 790 | 31,665 | 25,097 | 256.762 |
| 400 | 13,372 | 10,046 | 225.225 | 800 | 32,179 | 25,527 | 257.408 |
| 410 | 13,787 | 10,378 | 226.250 | 810 | 32,694 | 25,959 | 258.048 |
| 420 | 14,206 | 10,714 | 227.258 | 820 | 33,212 | 26,394 | 258.682 |
| 430 | 14,628 | 11,053 | 228.252 | 830 | 33,730 | 26,829 | 259.311 |
| 440 | 15,054 | 11,393 | 229.230 | 840 | 34,251 | 27,267 | 259.934 |
| 450 | 15,483 | 11,742 | 230.194 | 850 | 34,773 | 27,706 | 260.551 |
| 460 | 15,916 | 12,091 | 231.144 | 860 | 35,296 | 28,125 | 261.164 |
| 470 | 16,351 | 12,444 | 232.080 | 870 | 35,821 | 28,588 | 261.770 |
| 480 | 16,791 | 12,800 | 233.004 | 880 | 26,347 | 29,031 | 262.371 |
| 490 | 17,232 | 13,158 | 233.916 | 890 | 36,876 | 29,476 | 262.968 |
| 500 | 17,678 | 13,521 | 234.814 | 900 | 37,405 | 29,922 | 263.559 |
| 510 | 18,126 | 13,885 | 235.700 | 910 | 37,935 | 30,369 | 264.146 |
| 520 | 18,576 | 14,253 | 236.575 | 920 | 38,467 | 30,818 | 264.728 |
| 530 | 19,029 | 14,622 | 237.439 | 930 | 39,000 | 31,268 | 265.304 |
| 540 | 19,485 | 14,996 | 238.292 | 940 | 39,535 | 31,719 | 265.877 |
| 550 | 19,945 | 15,372 | 239.135 | 950 | 40,070 | 32,171 | 266.444 |
| 560 | 20,407 | 15,751 | 239.962 | 960 | 40,607 | 32,625 | 267.007 |
| 570 | 20,870 | 16,131 | 240.789 | 970 | 41,145 | 33,081 | 267.566 |
| 580 | 21,337 | 16,515 | 241.602 | 980 | 41,685 | 33,537 | 268.119 |
| 590 | 21,807 | 16,902 | 242.405 | 990 | 42,226 | 33,995 | 268.670 |

**Table A-9** *(Continued)*

| $T$ | $\bar{h}$ | $\bar{u}$ | $\bar{s}^o$ | $T$ | $\bar{h}$ | $\bar{u}$ | $\bar{s}^o$ |
|------|---------|---------|---------|------|----------|---------|---------|
| 1000 | 42,769 | 34,445 | 269.215 | 1760 | 86,420 | 71,787 | 301.543 |
| 1020 | 43,859 | 35,378 | 270.293 | 1780 | 87,612 | 72,812 | 302.271 |
| 1040 | 44,953 | 36,306 | 271.354 | 1800 | 88,806 | 73,840 | 302.884 |
| 1060 | 46,051 | 37,238 | 272.400 | 1820 | 90,000 | 74,868 | 303.544 |
| 1080 | 47,153 | 38,174 | 273.430 | 1840 | 91,196 | 75,897 | 304.198 |
| 1100 | 48,258 | 39,112 | 274.445 | 1860 | 92,394 | 76,929 | 304.845 |
| 1120 | 49,369 | 40,057 | 275.444 | 1880 | 93,593 | 77,962 | 305.487 |
| 1140 | 50,484 | 41,006 | 276.430 | 1900 | 94,793 | 78,996 | 306.122 |
| 1160 | 51,602 | 41,957 | 277.403 | 1920 | 95,995 | 80,031 | 306.751 |
| 1180 | 52,724 | 42,913 | 278.362 | 1940 | 97,197 | 81,067 | 307.374 |
| 1200 | 53,848 | 43,871 | 279.307 | 1960 | 98,401 | 82,105 | 307.992 |
| 1220 | 54,977 | 44,834 | 280.238 | 1980 | 99,606 | 83,144 | 308.604 |
| 1240 | 56,108 | 45,779 | 281.158 | 2000 | 100,804 | 84,185 | 309.210 |
| 1260 | 57,244 | 46,768 | 282.066 | 2050 | 103,835 | 86,791 | 310.701 |
| 1280 | 58,381 | 47,739 | 282.962 | 2100 | 106,864 | 89,404 | 312.160 |
| 1300 | 59,522 | 48,713 | 283.847 | 2150 | 109,898 | 92,023 | 313.589 |
| 1320 | 60,666 | 49,691 | 284.722 | 2200 | 112,939 | 94,648 | 314.988 |
| 1340 | 61,813 | 50,672 | 285.586 | 2250 | 115,984 | 97,277 | 316.356 |
| 1360 | 62,963 | 51,656 | 286.439 | 2300 | 119,035 | 99,912 | 317.965 |
| 1380 | 64,116 | 52,643 | 287.283 | 2350 | 122,091 | 102,552 | 319.011 |
| 1400 | 65,271 | 53,631 | 288.106 | 2400 | 125,152 | 105,197 | 320.302 |
| 1420 | 66,427 | 54,621 | 288.934 | 2450 | 128,219 | 107,849 | 321.566 |
| 1440 | 67,586 | 55,614 | 289.743 | 2500 | 131,290 | 110,504 | 322.808 |
| 1460 | 68,748 | 56,609 | 290.542 | 2550 | 134,368 | 113,166 | 324.026 |
| 1480 | 69,911 | 57,606 | 291.333 | 2600 | 137,449 | 115,832 | 325.222 |
| 1500 | 71,078 | 58,606 | 292.114 | 2650 | 140,533 | 118,500 | 326.396 |
| 1520 | 72,246 | 59,609 | 292.888 | 2700 | 143,620 | 121,172 | 327.549 |
| 1540 | 73,417 | 60,613 | 293.654 | 2750 | 146,713 | 123,849 | 328.684 |
| 1560 | 74,590 | 61,620 | 294.411 | 2800 | 149,808 | 126,528 | 329.800 |
| 1580 | 75,767 | 62,630 | 295.161 | 2850 | 152,908 | 129,212 | 330.896 |
| 1600 | 76,944 | 63,741 | 295.901 | 2900 | 156,009 | 131,898 | 331.975 |
| 1620 | 78,123 | 64,653 | 296.632 | 2950 | 159,117 | 134,589 | 333.037 |
| 1640 | 79,303 | 65,668 | 297.356 | 3000 | 162,226 | 137,283 | 334.084 |
| 1660 | 80,486 | 66,592 | 298.072 | 3050 | 165,341 | 139,982 | 335.114 |
| 1680 | 81,670 | 67,702 | 298.781 | 3100 | 168,456 | 142,681 | 336.126 |
| 1700 | 82,856 | 68,721 | 299.482 | 3150 | 171,576 | 145,385 | 337.124 |
| 1720 | 84,043 | 69,742 | 300.177 | 3200 | 174,695 | 148,089 | 338.109 |
| 1740 | 85,231 | 70,764 | 300.863 | 3250 | 177,822 | 150,801 | 339.069 |

*SOURCE:* Based on data from the JANAF Thermochemical Tables, NSRDS-NBS-37, 1971.

**Table A-10**     Ideal-gas enthalpy, internal energy, and absolute entropy of water ($H_2O$)

($\Delta \bar{h}_f^\circ = -241{,}820$ kJ/kmol; $T$, K; $\bar{h}$ and $\bar{u}$, kJ/kmol; $\bar{s}^\circ$, kJ/(kmol·K); std state is 1 atm.)

| $T$ | $\bar{h}$ | $\bar{u}$ | $\bar{s}^\circ$ | $T$ | $\bar{h}$ | $\bar{u}$ | $\bar{s}^\circ$ |
|---|---|---|---|---|---|---|---|
| 0 | 0 | 0 | 0 | 600 | 20,402 | 15,413 | 212.920 |
| 220 | 7,295 | 5,466 | 178.576 | 610 | 20,765 | 15,639 | 213.529 |
| 230 | 7,628 | 5,715 | 180.054 | 620 | 21,130 | 15,975 | 214.122 |
| 240 | 7,961 | 5,965 | 181.471 | 630 | 21,495 | 16,257 | 214.707 |
| 250 | 8,294 | 6,215 | 182.831 | 640 | 21,862 | 16,541 | 215.285 |
| 260 | 8,627 | 6,466 | 184.139 | 650 | 22,230 | 16,826 | 215.856 |
| 270 | 8,961 | 6,716 | 185.399 | 660 | 22,600 | 17,112 | 216.419 |
| 280 | 9,296 | 6,968 | 186.616 | 670 | 22,970 | 17,399 | 216.976 |
| 290 | 9,631 | 7,219 | 187.791 | 680 | 23,342 | 17,688 | 217.527 |
| 298 | 9,904 | 7,425 | 188.720 | 690 | 23,714 | 17,978 | 218.071 |
| 300 | 9,966 | 7,472 | 188.928 | 700 | 24,088 | 18,268 | 218.610 |
| 310 | 10,302 | 7,725 | 190.030 | 710 | 24,464 | 18,561 | 219.142 |
| 320 | 10,639 | 7,978 | 191.098 | 720 | 24,840 | 18,854 | 219.668 |
| 330 | 10,976 | 8,232 | 192.136 | 730 | 25,218 | 19,148 | 220.189 |
| 340 | 11,314 | 8,487 | 193.144 | 740 | 25,597 | 19,444 | 220.707 |
| 350 | 11,652 | 8,742 | 194.125 | 750 | 25,977 | 19,741 | 221.215 |
| 360 | 11,992 | 8,998 | 195.081 | 760 | 26,358 | 20,039 | 221.720 |
| 370 | 12,331 | 9,255 | 196.012 | 770 | 26,741 | 20,339 | 222.221 |
| 380 | 12,672 | 9,513 | 196.920 | 780 | 27,125 | 20,639 | 222.717 |
| 390 | 13,014 | 9,771 | 197.807 | 790 | 27,510 | 20,941 | 223.207 |
| 400 | 13,356 | 10,030 | 198.673 | 800 | 27,896 | 21,245 | 223.693 |
| 410 | 13,699 | 10,290 | 199.521 | 810 | 28,284 | 21,549 | 224.174 |
| 420 | 14,043 | 10,551 | 200.350 | 820 | 28,672 | 21,855 | 224.651 |
| 430 | 14,388 | 10,813 | 201.160 | 830 | 29,062 | 22,162 | 225.123 |
| 440 | 14,734 | 11,075 | 201.955 | 840 | 29,454 | 22,470 | 225.592 |
| 450 | 15,080 | 11,339 | 202.734 | 850 | 29,846 | 22,779 | 226.057 |
| 460 | 15,428 | 11,603 | 203.497 | 860 | 30,240 | 23,090 | 226.517 |
| 470 | 15,777 | 11,869 | 204.247 | 870 | 30,635 | 23,402 | 226.973 |
| 480 | 16,126 | 12,135 | 204.982 | 880 | 31,032 | 23,715 | 227.426 |
| 490 | 16,477 | 12,403 | 205.705 | 890 | 31,429 | 24,029 | 227.875 |
| 500 | 16,828 | 12,671 | 206.413 | 900 | 31,828 | 24,345 | 228.321 |
| 510 | 17,181 | 12,940 | 207.112 | 910 | 32,228 | 24,662 | 228.763 |
| 520 | 17,534 | 13,211 | 207.799 | 920 | 32,629 | 24,980 | 229.202 |
| 530 | 17,889 | 13,482 | 208.475 | 930 | 33,032 | 25,300 | 229.637 |
| 540 | 18,245 | 13,755 | 209.139 | 940 | 33,436 | 25,621 | 230.070 |
| 550 | 18,601 | 14,028 | 209.795 | 950 | 33,841 | 25,943 | 230.499 |
| 560 | 18,959 | 14,303 | 210.440 | 960 | 34,247 | 26,265 | 230.924 |
| 570 | 19,318 | 14,579 | 211.075 | 970 | 34,653 | 26,588 | 231.347 |
| 580 | 19,678 | 14,856 | 211.702 | 980 | 35,061 | 26,913 | 231.767 |
| 590 | 20,039 | 15,134 | 212.320 | 990 | 35,472 | 27,240 | 232.184 |

**Table A-10**     (*Continued*)

| $T$ | $\bar{h}$ | $\bar{u}$ | $\bar{s}^{o}$ | $T$ | $\bar{h}$ | $\bar{u}$ | $\bar{s}^{o}$ |
|---|---|---|---|---|---|---|---|
| 1000 | 35,882 | 27,568 | 232.597 | 1760 | 70,535 | 55,902 | 258.151 |
| 1020 | 36,709 | 28,228 | 233.415 | 1780 | 71,523 | 56,723 | 258.708 |
| 1040 | 37,542 | 28,895 | 234.223 | 1800 | 72,513 | 57,547 | 259.262 |
| 1060 | 38,380 | 29,567 | 235.020 | 1820 | 73,507 | 58,375 | 259.811 |
| 1080 | 39,223 | 30,243 | 235.806 | 1840 | 74,506 | 59,207 | 260.357 |
| 1100 | 40,071 | 30,925 | 236.584 | 1860 | 75,506 | 60,042 | 260.898 |
| 1120 | 40,923 | 31,611 | 237.352 | 1880 | 76,511 | 60,880 | 261.436 |
| 1140 | 41,780 | 32,301 | 238.110 | 1900 | 77,517 | 61,720 | 261.969 |
| 1160 | 42,642 | 32,997 | 238.859 | 1920 | 78,527 | 62,564 | 262.497 |
| 1180 | 43,509 | 33,698 | 239.600 | 1940 | 79,540 | 63,411 | 263.022 |
| 1200 | 44,380 | 34,403 | 240.333 | 1960 | 80,555 | 64,259 | 263.542 |
| 1220 | 45,256 | 35,112 | 241.057 | 1980 | 81,573 | 65,111 | 264.059 |
| 1240 | 46,137 | 35,827 | 241.773 | 2000 | 82,593 | 65,965 | 264.571 |
| 1260 | 47,022 | 36,546 | 242.482 | 2050 | 85,156 | 68,111 | 265.838 |
| 1280 | 47,912 | 37,270 | 243.183 | 2100 | 87,735 | 70,275 | 267.081 |
| 1300 | 48,807 | 38,000 | 243.877 | 2150 | 90,330 | 72,454 | 268.301 |
| 1320 | 49,707 | 38,732 | 244.564 | 2200 | 92,940 | 74,649 | 269.500 |
| 1340 | 50,612 | 39,470 | 245.243 | 2250 | 95,562 | 76,855 | 270.679 |
| 1360 | 51,521 | 40,213 | 245.915 | 2300 | 98,199 | 79,076 | 271.839 |
| 1380 | 52,434 | 40,960 | 246.582 | 2350 | 100,846 | 81,308 | 272.978 |
| 1400 | 53,351 | 41,711 | 247.241 | 2400 | 103,508 | 83,553 | 274.098 |
| 1420 | 54,273 | 42,466 | 247.895 | 2450 | 106,183 | 85,811 | 275.201 |
| 1440 | 55,198 | 43,226 | 248.543 | 2500 | 108,868 | 88,082 | 276.286 |
| 1460 | 56,128 | 43,989 | 249.185 | 2550 | 111,565 | 90,364 | 277.354 |
| 1480 | 57,062 | 44,756 | 249.820 | 2600 | 114,273 | 92,656 | 278.407 |
| 1500 | 57,999 | 45,528 | 250.450 | 2650 | 116,991 | 94,958 | 279.441 |
| 1520 | 58,942 | 46,304 | 251.074 | 2700 | 119,717 | 97,269 | 280.462 |
| 1540 | 59,888 | 47,084 | 251.693 | 2750 | 122,453 | 99,588 | 281.464 |
| 1560 | 60,838 | 47,868 | 252.305 | 2800 | 125,198 | 101,917 | 282.453 |
| 1580 | 61,792 | 48,655 | 252.912 | 2850 | 127,952 | 104,256 | 283.429 |
| 1600 | 62,748 | 49,445 | 253.513 | 2900 | 130,717 | 106,605 | 284.390 |
| 1620 | 63,709 | 50,240 | 254.111 | 2950 | 133,486 | 108,959 | 285.338 |
| 1640 | 64,675 | 51,039 | 254.703 | 3000 | 136,264 | 111,321 | 286.273 |
| 1660 | 65,643 | 51,841 | 255.290 | 3050 | 139,051 | 113,692 | 287.194 |
| 1680 | 66,614 | 52,646 | 255.873 | 3100 | 141,846 | 116,072 | 288.102 |
| 1700 | 67,589 | 53,455 | 256.450 | 3150 | 144,648 | 118,458 | 288.999 |
| 1720 | 68,567 | 54,267 | 257.002 | 3200 | 147,457 | 120,851 | 289.884 |
| 1740 | 69,550 | 55,083 | 257.589 | 3250 | 150,272 | 123,250 | 290.756 |

| *SOURCE:* Based on data from the JANAF Thermochemical Tables, NSRDS-NBS-37, 1971.

**Table A-11**  Ideal-gas enthalpy, internal energy, and absolute entropy of diatomic hydrogen ($H_2$)

($\Delta \bar{h}_f^\circ = 0$ kJ/kmol; $T$, K; $\bar{h}$ and $\bar{u}$, kJ/kmol; $\bar{s}^\circ$, kJ/(kmol·K); std state is 1 atm.)

| $T$ | $\bar{h}$ | $\bar{u}$ | $\bar{s}^\circ$ | $T$ | $\bar{h}$ | $\bar{u}$ | $\bar{s}^\circ$ |
|---|---|---|---|---|---|---|---|
| 0 | 0 | 0 | 0 | 1440 | 42,808 | 30,835 | 177.410 |
| 260 | 7,370 | 5,209 | 126.636 | 1480 | 44,091 | 31,786 | 178.291 |
| 270 | 7,657 | 5,412 | 127.719 | 1520 | 45,384 | 32,746 | 179.153 |
| 280 | 7,945 | 5,617 | 128.765 | 1560 | 46,683 | 33,713 | 179.995 |
| 290 | 8,233 | 5,822 | 129.775 | 1600 | 47,990 | 34,687 | 180.820 |
| 298 | 8,468 | 5,989 | 130.574 | 1640 | 49,303 | 35,668 | 181.632 |
| 300 | 8,522 | 6,027 | 130.754 | 1680 | 50,662 | 36,654 | 182.428 |
| 320 | 9,100 | 6,440 | 132.621 | 1720 | 51,947 | 37,646 | 183.208 |
| 340 | 9,680 | 6,853 | 134.378 | 1760 | 53,279 | 38,645 | 183.973 |
| 360 | 10,262 | 7,268 | 136.039 | 1800 | 54,618 | 39,652 | 184.724 |
| 380 | 10,843 | 7,684 | 137.612 | 1840 | 55,962 | 40,663 | 185.463 |
| 400 | 11,426 | 8,100 | 139.106 | 1880 | 57,311 | 41,680 | 186.190 |
| 420 | 12,010 | 8,518 | 140.529 | 1920 | 58,668 | 42,705 | 186.904 |
| 440 | 12,594 | 8,936 | 141.888 | 1960 | 60,031 | 43,735 | 187.607 |
| 460 | 13,179 | 9,355 | 143.187 | 2000 | 61,400 | 44,771 | 188.297 |
| 480 | 13,764 | 9,773 | 144.432 | 2050 | 63,119 | 46,074 | 189.148 |
| 500 | 14,350 | 10,193 | 145.628 | 2100 | 64,847 | 47,386 | 189.979 |
| 520 | 14,935 | 10,611 | 146.775 | 2150 | 66,584 | 48,708 | 190.796 |
| 560 | 16,107 | 11,451 | 148.945 | 2200 | 68,328 | 50,037 | 191.598 |
| 600 | 17,280 | 12,291 | 150.698 | 2250 | 70,080 | 51,373 | 192.385 |
| 640 | 18,453 | 13,133 | 152.863 | 2300 | 71,839 | 52,716 | 193.159 |
| 680 | 19,630 | 13,976 | 154.645 | 2350 | 73,608 | 54,069 | 193.921 |
| 720 | 20,807 | 14,821 | 156.328 | 2400 | 75,383 | 55,429 | 194.669 |
| 760 | 21,988 | 15,669 | 157.923 | 2450 | 77,168 | 56,798 | 195.403 |
| 800 | 23,171 | 16,520 | 159.440 | 2500 | 78,960 | 58,175 | 196.125 |
| 840 | 24,359 | 17,375 | 160.891 | 2550 | 80,755 | 59,554 | 196.837 |
| 880 | 25,551 | 18,235 | 162.277 | 2600 | 82,558 | 60,941 | 197.539 |
| 920 | 26,747 | 19,098 | 163.607 | 2650 | 84,386 | 62,335 | 198.229 |
| 960 | 27,948 | 19,966 | 164.884 | 2700 | 86,186 | 63,737 | 198.907 |
| 1000 | 29,154 | 20,839 | 166.114 | 2750 | 88,008 | 65,144 | 199.575 |
| 1040 | 30,364 | 21,717 | 167.300 | 2800 | 89,838 | 66,558 | 200.234 |
| 1080 | 31,580 | 22,601 | 168.449 | 2850 | 91,671 | 67,976 | 200.885 |
| 1120 | 32,802 | 23,490 | 169.560 | 2900 | 93,512 | 69,401 | 201.527 |
| 1160 | 34,028 | 24,384 | 170.636 | 2950 | 95,358 | 70,831 | 202.157 |
| 1200 | 35,262 | 25,284 | 171.682 | 3000 | 97,211 | 72,268 | 202.778 |
| 1240 | 36,502 | 26,192 | 172.698 | 3050 | 99,065 | 73,707 | 203.391 |
| 1280 | 37,749 | 27,106 | 173.687 | 3100 | 100,926 | 75,152 | 203.995 |
| 1320 | 39,002 | 28,027 | 174.652 | 3150 | 102,793 | 76,604 | 204.592 |
| 1360 | 40,263 | 28,995 | 175.593 | 3200 | 104,667 | 78,061 | 205.181 |
| 1400 | 41,530 | 29,889 | 176.510 | 3250 | 106,545 | 79,523 | 205.765 |

**Table A-11**     (*Continued*)
Ideal-gas enthalpy, internal energy, and absolute entropy of monatomic oxygen (O)

($\Delta \bar{h}_f^\circ$ = 249,190 kJ/kmol; std state is 1 atm.)

| $T$ | $\bar{h}$ | $\bar{u}$ | $\bar{s}^\circ$ | $T$ | $\bar{h}$ | $\bar{u}$ | $\bar{s}^\circ$ |
|---|---|---|---|---|---|---|---|
| 0 | 0 | 0 | 0 | 2400 | 50,894 | 30,940 | 204.932 |
| 298 | 6,852 | 4,373 | 160.944 | 2450 | 51,936 | 31,566 | 205.362 |
| 300 | 6,892 | 4,398 | 161.079 | 2500 | 52,979 | 32,193 | 205.783 |
| 500 | 11,197 | 7,040 | 172.088 | 2550 | 54,021 | 32,820 | 206.196 |
| 1000 | 21,713 | 13,398 | 186.678 | 2600 | 55,064 | 33,447 | 206.601 |
| 1500 | 32,150 | 19,679 | 195.143 | 2650 | 56,108 | 34,075 | 206.999 |
| 1600 | 34,234 | 20,931 | 196.488 | 2700 | 57,152 | 34,703 | 207.389 |
| 1700 | 36,317 | 22,183 | 197.751 | 2750 | 58,196 | 35,332 | 207.772 |
| 1800 | 38,400 | 23,434 | 198.941 | 2800 | 59,241 | 35,961 | 208.148 |
| 1900 | 40,482 | 24,685 | 200.067 | 2850 | 60,286 | 36,590 | 208.518 |
| 2000 | 42,564 | 25,935 | 201.135 | 2900 | 61,332 | 37,220 | 208.882 |
| 2050 | 43,605 | 26,560 | 201.649 | 2950 | 62,378 | 37,851 | 209.240 |
| 2100 | 44,646 | 27,186 | 202.151 | 3000 | 63,425 | 38,482 | 209.592 |
| 2150 | 45,687 | 27,811 | 202.641 | 3100 | 65,520 | 39,746 | 210.279 |
| 2200 | 46,728 | 28,436 | 203.119 | 3200 | 67,619 | 41,013 | 210.945 |
| 2250 | 47,769 | 29,062 | 203.588 | 3300 | 69,720 | 42,283 | 211.592 |
| 2300 | 48,811 | 29,688 | 204.045 | 3400 | 71,824 | 43,556 | 212.220 |
| 2350 | 49,852 | 30,314 | 204.493 | 3500 | 73,932 | 44,832 | 212.831 |

Ideal-gas enthalpy, internal energy, and absolute entropy of hydroxyl (OH)

($\Delta \bar{h}_f^\circ$ = 39,040 kJ/kmol; std state is 1 atm.)

| $T$ | $\bar{h}$ | $\bar{u}$ | $\bar{s}^\circ$ | $T$ | $\bar{h}$ | $\bar{u}$ | $\bar{s}^\circ$ |
|---|---|---|---|---|---|---|---|
| 0 | 0 | 0 | 0 | 2400 | 77,015 | 57,061 | 248.628 |
| 298 | 9,188 | 6,709 | 183.594 | 2450 | 78,801 | 58,431 | 249.364 |
| 300 | 9,244 | 6,749 | 183.779 | 2500 | 80,592 | 59,806 | 250.088 |
| 500 | 15,181 | 11,024 | 198.955 | 2550 | 82,338 | 61,186 | 250.799 |
| 1000 | 30,123 | 21,809 | 219.624 | 2600 | 84,189 | 62,572 | 251.499 |
| 1500 | 46,046 | 33,575 | 232.506 | 2650 | 85,995 | 63,962 | 252.187 |
| 1600 | 49,358 | 36,055 | 234.642 | 2700 | 87,806 | 65,358 | 252.864 |
| 1700 | 52,706 | 38,571 | 236.672 | 2750 | 89,622 | 66,757 | 253.530 |
| 1800 | 56,089 | 41,123 | 238.606 | 2800 | 91,442 | 68,162 | 254.186 |
| 1900 | 59,505 | 43,708 | 240.453 | 2850 | 93,266 | 69,570 | 254.832 |
| 2000 | 62,952 | 46,323 | 242.221 | 2900 | 95,095 | 70,983 | 255.468 |
| 2050 | 64,687 | 47,642 | 243.077 | 2950 | 96,927 | 72,400 | 256.094 |
| 2100 | 66,428 | 48,968 | 243.917 | 3000 | 98,763 | 73,820 | 256.712 |
| 2150 | 68,177 | 50,301 | 244.740 | 3100 | 102,447 | 76,673 | 257.919 |
| 2200 | 69,932 | 51,641 | 245.547 | 3200 | 106,145 | 79,539 | 259.093 |
| 2250 | 71,694 | 52,987 | 246.338 | 3300 | 109,855 | 82,418 | 260.235 |
| 2300 | 73,462 | 54,339 | 247.116 | 3400 | 113,578 | 85,309 | 261.347 |
| 2350 | 75,236 | 55,697 | 247.879 | 3500 | 117,312 | 88,212 | 262.429 |

| *SOURCE:* Based on data from the JANAF Thermochemical Tables, NSRDS-NBS-37, 1971.

**Table A-12** Properties of saturated water: Temperature table

$(v, m^3/kg; u, kJ/kg; h, kJ/kg; s, kJ/(kg \cdot K); 1 \text{ bar} = 0.1 \text{ MPa})$

| Temp., °C $T$ | Press., bars $P$ | Specific Volume | | Internal Energy | | Enthalpy | | | Entropy | |
|---|---|---|---|---|---|---|---|---|---|---|
| | | Sat. Liquid $v_f \times 10^3$ | Sat. Vapor $v_g$ | Sat. Liquid $u_f$ | Sat. Vapor $u_g$ | Sat. Liquid $h_f$ | Evap. $h_{fg}$ | Sat. Vapor $h_g$ | Sat. Liquid $s_f$ | Sat. Vapor $s_g$ |
| 0 | 0.00611 | 1.0002 | 206.278 | −0.03 | 2375.4 | −0.02 | 2501.4 | 2501.3 | 0.0001 | 9.1565 |
| 4 | 0.00813 | 1.0001 | 157.232 | 16.77 | 2380.9 | 16.78 | 2491.9 | 2508.7 | 0.0610 | 9.0514 |
| 5 | 0.00872 | 1.0001 | 147.120 | 20.97 | 2382.3 | 20.98 | 2489.6 | 2510.6 | 0.0761 | 9.0257 |
| 6 | 0.00935 | 1.0001 | 137.734 | 25.19 | 2383.6 | 25.20 | 2487.2 | 2512.4 | 0.0912 | 9.0003 |
| 8 | 0.01072 | 1.0002 | 120.917 | 33.59 | 2386.4 | 33.60 | 2482.5 | 2516.1 | 0.1212 | 8.9501 |
| 10 | 0.01228 | 1.0004 | 106.379 | 42.00 | 2389.2 | 42.01 | 2477.7 | 2519.8 | 0.1510 | 8.9008 |
| 11 | 0.01312 | 1.0004 | 99.857 | 46.20 | 2390.5 | 46.20 | 2475.4 | 2521.6 | 0.1658 | 8.8765 |
| 12 | 0.01402 | 1.0005 | 93.784 | 50.41 | 2391.9 | 50.41 | 2473.0 | 2523.4 | 0.1806 | 8.8524 |
| 13 | 0.01497 | 1.0007 | 88.124 | 54.60 | 2393.3 | 54.60 | 2470.7 | 2525.3 | 0.1953 | 8.8285 |
| 14 | 0.01598 | 1.0008 | 82.848 | 58.79 | 2394.7 | 58.80 | 2468.3 | 2527.1 | 0.2099 | 8.8048 |
| 15 | 0.01705 | 1.0009 | 77.926 | 62.99 | 2396.1 | 62.99 | 2465.9 | 2528.9 | 0.2245 | 8.7814 |
| 16 | 0.01818 | 1.0011 | 73.333 | 67.18 | 2397.4 | 67.19 | 2463.6 | 2530.8 | 0.2390 | 8.7582 |
| 17 | 0.01938 | 1.0012 | 69.044 | 71.38 | 2398.8 | 71.38 | 2461.2 | 2532.6 | 0.2535 | 8.7351 |
| 18 | 0.02064 | 1.0014 | 65.038 | 75.57 | 2400.2 | 75.58 | 2458.8 | 2534.4 | 0.2679 | 8.7123 |
| 19 | 0.02198 | 1.0016 | 61.293 | 79.76 | 2401.6 | 79.77 | 2456.5 | 2536.2 | 0.2823 | 8.6897 |
| 20 | 0.02339 | 1.0018 | 57.791 | 83.95 | 2402.9 | 83.96 | 2454.1 | 2538.1 | 0.2966 | 8.6672 |
| 21 | 0.02487 | 1.0020 | 54.514 | 88.14 | 2404.3 | 88.14 | 2451.8 | 2539.9 | 0.3109 | 8.6450 |
| 22 | 0.02645 | 1.0022 | 51.447 | 92.32 | 2405.7 | 92.33 | 2449.4 | 2541.7 | 0.3251 | 8.6229 |
| 23 | 0.02810 | 1.0024 | 48.574 | 96.51 | 2407.0 | 96.52 | 2447.0 | 2543.5 | 0.3393 | 8.6011 |
| 24 | 0.02985 | 1.0027 | 45.883 | 100.70 | 2408.4 | 100.70 | 2444.7 | 2545.4 | 0.3534 | 8.5794 |
| 25 | 0.03169 | 1.0029 | 43.360 | 104.88 | 2409.8 | 104.89 | 2442.3 | 2547.2 | 0.3674 | 8.5580 |
| 26 | 0.03363 | 1.0032 | 40.994 | 109.06 | 2411.1 | 109.07 | 2439.9 | 2549.0 | 0.3814 | 8.5367 |
| 27 | 0.03567 | 1.0035 | 38.774 | 113.25 | 2412.5 | 113.25 | 2437.6 | 2550.8 | 0.3954 | 8.5156 |
| 28 | 0.03782 | 1.0037 | 36.690 | 117.42 | 2413.9 | 117.43 | 2435.2 | 2552.6 | 0.4093 | 8.4946 |
| 29 | 0.04008 | 1.0040 | 34.733 | 121.60 | 2415.2 | 121.61 | 2432.8 | 2554.5 | 0.4231 | 8.4739 |
| 30 | 0.04246 | 1.0043 | 32.894 | 125.78 | 2416.6 | 125.79 | 2430.5 | 2556.3 | 0.4369 | 8.4533 |
| 31 | 0.04496 | 1.0046 | 31.165 | 129.96 | 2418.0 | 129.97 | 2428.1 | 2558.1 | 0.4507 | 8.4329 |
| 32 | 0.04759 | 1.0050 | 29.540 | 134.14 | 2419.3 | 134.15 | 2425.7 | 2559.9 | 0.4644 | 8.4127 |
| 33 | 0.05034 | 1.0053 | 28.011 | 138.32 | 2420.7 | 138.33 | 2423.4 | 2561.7 | 0.4781 | 8.3927 |
| 34 | 0.05324 | 1.0056 | 26.571 | 142.50 | 2422.0 | 142.50 | 2421.0 | 2563.5 | 0.4917 | 8.3728 |
| 35 | 0.05628 | 1.0060 | 25.216 | 146.67 | 2423.4 | 146.68 | 2418.6 | 2565.3 | 0.5053 | 8.3531 |
| 36 | 0.05947 | 1.0063 | 23.940 | 150.85 | 2424.7 | 150.86 | 2416.2 | 2567.1 | 0.5188 | 8.3336 |
| 38 | 0.06632 | 1.0071 | 21.602 | 159.20 | 2427.4 | 159.21 | 2411.5 | 2570.7 | 0.5458 | 8.2950 |
| 40 | 0.07384 | 1.0078 | 19.523 | 167.56 | 2430.1 | 167.57 | 2406.7 | 2574.3 | 0.5725 | 8.2570 |
| 45 | 0.09593 | 1.0099 | 15.258 | 188.44 | 2436.8 | 188.45 | 2394.8 | 2583.2 | 0.6387 | 8.1648 |

**Table A-12**     (*Continued*)

| Temp., °C $T$ | Press., bars $P$ | Specific Volume | | Internal Energy | | Enthalpy | | | Entropy | |
|---|---|---|---|---|---|---|---|---|---|---|
| | | Sat. Liquid $v_f \times 10^3$ | Sat. Vapor $v_g$ | Sat. Liquid $u_f$ | Sat. Vapor $u_g$ | Sat. Liquid $h_f$ | Evap. $h_{fg}$ | Sat. Vapor $h_g$ | Sat. Liquid $s_f$ | Sat. Vapor $s_g$ |
| 50 | .1235 | 1.0121 | 12.032 | 209.32 | 2443.5 | 209.33 | 2382.7 | 2592.1 | .7038 | 8.0763 |
| 55 | .1576 | 1.0146 | 9.568 | 230.21 | 2450.1 | 230.23 | 2370.7 | 2600.9 | .7679 | 7.9913 |
| 60 | .1994 | 1.0172 | 7.671 | 251.11 | 2456.6 | 251.13 | 2358.5 | 2609.6 | .8312 | 7.9096 |
| 65 | .2503 | 1.0199 | 6.197 | 272.02 | 2463.1 | 272.06 | 2346.2 | 2618.3 | .8935 | 7.8310 |
| 70 | .3119 | 1.0228 | 5.042 | 292.95 | 2469.6 | 292.98 | 2333.8 | 2626.8 | .9549 | 7.7553 |
| 75 | .3858 | 1.0259 | 4.131 | 313.90 | 2475.9 | 313.93 | 2321.4 | 2635.3 | 1.0155 | 7.6824 |
| 80 | .4739 | 1.0291 | 3.407 | 334.86 | 2482.2 | 334.91 | 2308.8 | 2643.7 | 1.0753 | 7.6122 |
| 85 | .5783 | 1.0325 | 2.828 | 355.84 | 2488.4 | 355.90 | 2296.0 | 2651.9 | 1.1343 | 7.5445 |
| 90 | .7014 | 1.0360 | 2.361 | 376.85 | 2494.5 | 376.92 | 2283.2 | 2660.1 | 1.1925 | 7.4791 |
| 95 | .8455 | 1.0397 | 1.982 | 397.88 | 2500.6 | 397.96 | 2270.2 | 2668.1 | 1.2500 | 7.4159 |
| 100 | 1.0133 | 1.0435 | 1.673 | 418.94 | 2506.5 | 419.04 | 2257.0 | 2676.1 | 1.3069 | 7.3549 |
| 110 | 1.433 | 1.0516 | 1.210 | 461.14 | 2518.1 | 461.30 | 2230.2 | 2691.5 | 1.4185 | 7.2387 |
| 120 | 1.985 | 1.0603 | 0.8919 | 503.50 | 2529.3 | 503.71 | 2202.6 | 2706.3 | 1.5276 | 7.1296 |
| 130 | 2.701 | 1.0697 | 0.6685 | 546.02 | 2539.9 | 546.31 | 2174.2 | 2720.5 | 1.6344 | 7.0269 |
| 140 | 3.613 | 1.0797 | 0.5089 | 588.74 | 2550.0 | 589.13 | 2144.7 | 2733.9 | 1.7391 | 6.9299 |
| 150 | 4.758 | 1.0905 | 0.3928 | 631.68 | 2559.5 | 632.20 | 2114.3 | 2746.5 | 1.8418 | 6.8379 |
| 160 | 6.178 | 1.1020 | 0.3071 | 674.86 | 2568.4 | 675.55 | 2082.6 | 2758.1 | 1.9427 | 6.7502 |
| 170 | 7.917 | 1.1143 | 0.2428 | 718.33 | 2576.5 | 719.21 | 2049.5 | 2768.7 | 2.0419 | 6.6663 |
| 180 | 10.02 | 1.1274 | 0.1941 | 762.09 | 2583.7 | 763.22 | 2015.0 | 2778.2 | 2.1396 | 6.5857 |
| 190 | 12.54 | 1.1414 | 0.1565 | 806.19 | 2590.0 | 807.62 | 1978.8 | 2786.4 | 2.2359 | 6.5079 |
| 200 | 15.54 | 1.1565 | 0.1274 | 850.65 | 2595.3 | 852.45 | 1940.7 | 2793.2 | 2.3309 | 6.4323 |
| 210 | 19.06 | 1.1726 | 0.1044 | 895.53 | 2599.5 | 897.76 | 1900.7 | 2798.5 | 2.4248 | 6.3585 |
| 220 | 23.18 | 1.1900 | 0.08619 | 940.87 | 2602.4 | 943.62 | 1858.5 | 2802.1 | 2.5178 | 6.2861 |
| 230 | 27.95 | 1.2088 | 0.07158 | 986.74 | 2603.9 | 990.12 | 1813.8 | 2804.0 | 2.6099 | 6.2146 |
| 240 | 33.44 | 1.2291 | 0.05976 | 1033.2 | 2604.0 | 1037.3 | 1766.5 | 2803.8 | 2.7015 | 6.1437 |
| 250 | 39.73 | 1.2512 | 0.05013 | 1080.4 | 2602.4 | 1085.4 | 1716.2 | 2801.5 | 2.7927 | 6.0730 |
| 260 | 46.88 | 1.2755 | 0.04221 | 1128.4 | 2599.0 | 1134.4 | 1662.5 | 2796.6 | 2.8838 | 6.0019 |
| 270 | 54.99 | 1.3023 | 0.03564 | 1177.4 | 2593.7 | 1184.5 | 1605.2 | 2789.7 | 2.9751 | 5.9301 |
| 280 | 64.12 | 1.3321 | 0.03017 | 1227.5 | 2586.1 | 1236.0 | 1543.6 | 2779.6 | 3.0668 | 5.8571 |
| 290 | 74.36 | 1.3656 | 0.02557 | 1278.9 | 2576.0 | 1289.1 | 1477.1 | 2766.2 | 3.1594 | 5.7821 |
| 300 | 85.81 | 1.4036 | 0.02167 | 1332.0 | 2563.0 | 1344.0 | 1404.9 | 2749.0 | 3.2534 | 5.7045 |
| 320 | 112.7 | 1.4988 | 0.01549 | 1444.6 | 2525.5 | 1461.5 | 1238.6 | 2700.1 | 3.4480 | 5.5362 |
| 340 | 145.9 | 1.6379 | 0.01080 | 1570.3 | 2464.6 | 1594.2 | 1027.9 | 2622.0 | 3.6594 | 5.3357 |
| 360 | 186.5 | 1.8925 | 0.006945 | 1725.2 | 2351.5 | 1760.5 | 720.5 | 2481.0 | 3.9147 | 5.0526 |
| 374.14 | 220.9 | 3.155 | 0.003155 | 2029.6 | 2029.6 | 2099.3 | 0 | 2099.3 | 4.4298 | 4.4298 |

| *SOURCE:* J. H. Keenan, F. G. Keyes, P. G. Hill, and J. G. Moore, "Steam Tables," Wiley, New York, 1969.

**Table A-13** Properties of saturated water: Pressure Table

$(v, m^3/kg; u, kJ/kg; h, kJ/kg; s, kJ/(kg \cdot K); 1 \text{ bar} = 0.1 \text{ MPa})$

| Press., bars $P$ | Temp., °C $T$ | Specific Volume | | Internal Energy | | Enthalpy | | | Entropy | |
| | | Sat. Liquid $v_f \times 10^3$ | Sat. Vapor $v_g$ | Sat. Liquid $u_f$ | Sat. Vapor $u_g$ | Sat. Liquid $h_f$ | Evap. $h_{fg}$ | Sat. Vapor $h_g$ | Sat. Liquid $s_f$ | Sat. Vapor $s_g$ |
|---|---|---|---|---|---|---|---|---|---|---|
| .040 | 28.96 | 1.0040 | 34.800 | 121.45 | 2415.2 | 121.46 | 2432.9 | 2554.4 | .4226 | 8.4746 |
| .060 | 36.16 | 1.0064 | 23.739 | 151.53 | 2425.0 | 151.53 | 2415.9 | 2567.4 | .5210 | 8.3304 |
| .080 | 41.51 | 1.0084 | 18.103 | 173.87 | 2432.2 | 173.88 | 2403.1 | 2577.0 | .5926 | 8.2287 |
| 0.10 | 45.81 | 1.0102 | 14.674 | 191.82 | 2437.9 | 191.83 | 2392.8 | 2584.7 | .6493 | 8.1502 |
| 0.20 | 60.06 | 1.0172 | 7.649 | 251.38 | 2456.7 | 251.40 | 2358.3 | 2609.7 | .8320 | 7.9085 |
| 0.30 | 69.10 | 1.0223 | 5.229 | 289.20 | 2468.4 | 289.23 | 2336.1 | 2625.3 | .9439 | 7.7686 |
| 0.40 | 75.87 | 1.0265 | 3.993 | 317.53 | 2477.0 | 317.58 | 2319.2 | 2636.8 | 1.0259 | 7.6700 |
| 0.50 | 81.33 | 1.0300 | 3.240 | 340.44 | 2483.9 | 340.49 | 2305.4 | 2645.9 | 1.0910 | 7.5939 |
| 0.60 | 85.94 | 1.0331 | 2.732 | 359.79 | 2489.6 | 359.86 | 2293.6 | 2653.5 | 1.1453 | 7.5320 |
| 0.70 | 89.95 | 1.0360 | 2.365 | 376.63 | 2494.5 | 376.70 | 2283.3 | 2660.0 | 1.1919 | 7.4797 |
| 0.80 | 93.50 | 1.0380 | 2.087 | 391.58 | 2498.8 | 391.66 | 2274.1 | 2665.8 | 1.2329 | 7.4346 |
| 0.90 | 96.71 | 1.0410 | 1.869 | 405.06 | 2502.6 | 405.15 | 2265.7 | 2670.9 | 1.2695 | 7.3949 |
| 1.00 | 99.63 | 1.0432 | 1.694 | 417.36 | 2506.1 | 417.46 | 2258.0 | 2675.5 | 1.3026 | 7.3594 |
| 1.50 | 111.4 | 1.0528 | 1.159 | 466.94 | 2519.7 | 467.11 | 2226.5 | 2693.6 | 1.4336 | 7.2233 |
| 2.00 | 120.2 | 1.0605 | 0.8857 | 504.49 | 2529.5 | 504.70 | 2201.9 | 2706.7 | 1.5301 | 7.1271 |
| 2.50 | 127.4 | 1.0672 | 0.7187 | 535.10 | 2537.2 | 535.37 | 2181.5 | 2716.9 | 1.6072 | 7.0527 |
| 3.00 | 133.6 | 1.0732 | 0.6058 | 561.15 | 2543.6 | 561.47 | 2163.8 | 2725.3 | 1.6718 | 6.9919 |
| 3.50 | 138.9 | 1.0786 | 0.5243 | 583.95 | 2546.9 | 584.33 | 2148.1 | 2732.4 | 1.7275 | 6.9405 |
| 4.00 | 143.6 | 1.0836 | 0.4625 | 604.31 | 2553.6 | 604.74 | 2133.8 | 2738.6 | 1.7766 | 6.8959 |
| 4.50 | 147.9 | 1.0882 | 0.4140 | 622.25 | 2557.6 | 623.25 | 2120.7 | 2743.9 | 1.8207 | 6.8565 |
| 5.00 | 151.9 | 1.0926 | 0.3749 | 639.68 | 2561.2 | 640.23 | 2108.5 | 2748.7 | 1.8607 | 6.8212 |
| 6.00 | 158.9 | 1.1006 | 0.3157 | 669.90 | 2567.4 | 670.56 | 2086.3 | 2756.8 | 1.9312 | 6.7600 |
| 7.00 | 165.0 | 1.1080 | 0.2729 | 696.44 | 2572.5 | 697.22 | 2066.3 | 2763.5 | 1.9922 | 6.7080 |
| 8.00 | 170.4 | 1.1148 | 0.2404 | 720.22 | 2576.8 | 721.11 | 2048.0 | 2769.1 | 2.0462 | 6.6628 |
| 9.00 | 175.4 | 1.1212 | 0.2150 | 741.83 | 2580.5 | 742.83 | 2031.1 | 2773.9 | 2.0946 | 6.6226 |

**Table A-13** (Continued)

| Press., bars $P$ | Temp., °C $T$ | Specific Volume | | Internal Energy | | Enthalpy | | | Entropy | |
|---|---|---|---|---|---|---|---|---|---|---|
| | | Sat. Liquid $v_f \times 10^3$ | Sat. Vapor $v_g$ | Sat. Liquid $u_f$ | Sat. Vapor $u_g$ | Sat. Liquid $h_f$ | Evap. $h_{fg}$ | Sat. Vapor $h_g$ | Sat. Liquid $s_f$ | Sat. Vapor $s_g$ |
| 10.0 | 179.9 | 1.1273 | 0.1944 | 761.68 | 2583.6 | 762.81 | 2015.3 | 2778.1 | 2.1387 | 6.5863 |
| 15.0 | 198.3 | 1.1539 | 0.1318 | 843.16 | 2594.5 | 844.84 | 1947.3 | 2792.2 | 2.3150 | 6.4448 |
| 20.0 | 212.4 | 1.1767 | 0.09963 | 906.44 | 2600.3 | 908.79 | 1890.7 | 2799.5 | 2.4474 | 6.3409 |
| 25.0 | 224.0 | 1.1973 | 0.07998 | 959.11 | 2603.1 | 962.11 | 1841.0 | 2803.1 | 2.5547 | 6.2575 |
| 30.0 | 233.9 | 1.2165 | 0.06668 | 1004.8 | 2604.1 | 1008.4 | 1795.7 | 2804.2 | 2.6457 | 6.1869 |
| 35.0 | 242.6 | 1.2347 | 0.05707 | 1045.4 | 2603.7 | 1049.8 | 1753.7 | 2803.4 | 2.7253 | 6.1253 |
| 40.0 | 250.4 | 1.2522 | 0.04978 | 1082.3 | 2602.3 | 1087.3 | 1714.1 | 2801.4 | 2.7964 | 6.0701 |
| 45.0 | 257.5 | 1.2692 | 0.04406 | 1116.2 | 2600.1 | 1121.9 | 1676.4 | 2798.3 | 2.8610 | 6.0199 |
| 50.0 | 264.0 | 1.2859 | 0.03944 | 1147.8 | 2597.1 | 1154.2 | 1640.1 | 2794.3 | 2.9202 | 5.9734 |
| 60.0 | 275.6 | 1.3187 | 0.03244 | 1205.4 | 2589.7 | 1213.4 | 1571.0 | 2784.3 | 3.0267 | 5.8892 |
| 70.0 | 285.9 | 1.3513 | 0.02737 | 1257.6 | 2580.5 | 1267.0 | 1505.1 | 2772.1 | 3.1211 | 5.8133 |
| 80.0 | 295.1 | 1.3842 | 0.02352 | 1305.6 | 2569.8 | 1316.6 | 1441.3 | 2758.0 | 3.2068 | 5.7432 |
| 90.0 | 303.4 | 1.4178 | 0.02048 | 1350.5 | 2557.8 | 1363.3 | 1378.9 | 2742.1 | 3.2858 | 5.6772 |
| 100. | 311.1 | 1.4524 | 0.01803 | 1393.0 | 2544.4 | 1407.6 | 1317.1 | 2724.7 | 3.3596 | 5.6141 |
| 110. | 318.2 | 1.4886 | 0.01599 | 1433.7 | 2529.8 | 1450.1 | 1255.5 | 2705.6 | 3.4295 | 5.5527 |
| 120. | 324.8 | 1.5267 | 0.01426 | 1473.0 | 2513.7 | 1491.3 | 1193.6 | 2684.9 | 3.4962 | 5.4924 |
| 130. | 330.9 | 1.5671 | 0.01278 | 1511.1 | 2496.1 | 1531.5 | 1130.7 | 2662.2 | 3.5606 | 5.4323 |
| 140. | 336.8 | 1.6107 | 0.01149 | 1548.6 | 2476.8 | 1571.1 | 1066.5 | 2637.6 | 3.6232 | 5.3717 |
| 150. | 342.2 | 1.6581 | 0.01034 | 1585.6 | 2455.5 | 1610.5 | 1000.0 | 2610.5 | 3.6848 | 5.3098 |
| 160. | 347.4 | 1.7107 | 0.009306 | 1622.7 | 2431.7 | 1650.1 | 930.6 | 2580.6 | 3.7461 | 5.2455 |
| 170. | 352.4 | 1.7702 | 0.008364 | 1660.2 | 2405.0 | 1690.3 | 856.9 | 2547.2 | 3.8079 | 5.1777 |
| 180. | 357.1 | 1.8397 | 0.007489 | 1698.9 | 2374.3 | 1732.0 | 777.1 | 2509.1 | 3.8715 | 5.1044 |
| 190. | 361.5 | 1.9243 | 0.006657 | 1739.9 | 2338.1 | 1776.5 | 688.0 | 2464.5 | 3.9388 | 5.0228 |
| 200. | 365.8 | 2.036 | 0.005834 | 1785.6 | 2293.0 | 1826.3 | 583.4 | 2409.7 | 4.0139 | 4.9269 |
| 220.9 | 374.1 | 3.155 | 0.003155 | 2029.6 | 2029.6 | 2099.3 | 0 | 2099.3 | 4.4298 | 4.4298 |

SOURCE: J. H. Keenan, F. Keyes, P. Hill, J. Moore, "Steam Tables," Wiley, New York, 1969.

**Table A-14**    Properties of water: Superheated-vapor table

**(v, m³/kg; u, kJ/kg; h, kJ/kg; s,kJ/kg·K)**

| Temp., °C | v | u | h | s | v | u | h | s |
|---|---|---|---|---|---|---|---|---|
| | **0.06 bar (0.006MPa)** ($T_{sat}$ = 36.16°C) | | | | **0.35 bar (0.035MPa)** ($T_{sat}$ = 72.69°C) | | | |
| Sat. | 23.739 | 2425.0 | 2567.4 | 8.3304 | 4.526 | 2473.0 | 2631.4 | 7.7158 |
| 80 | 27.132 | 2487.3 | 2650.1 | 8.5804 | 4.625 | 2483.7 | 2645.6 | 7.7564 |
| 120 | 30.219 | 2544.7 | 2726.0 | 8.7840 | 5.163 | 2542.4 | 2723.1 | 7.9644 |
| 160 | 33.302 | 2602.7 | 2802.5 | 8.9693 | 5.696 | 2601.2 | 2800.6 | 8.1519 |
| 200 | 36.383 | 2661.4 | 2879.7 | 9.1398 | 6.228 | 2660.4 | 2878.4 | 8.3237 |
| 240 | 39.462 | 2721.0 | 2957.8 | 9.2982 | 6.758 | 2720.3 | 2956.8 | 8.4828 |
| 280 | 42.540 | 2781.5 | 3036.8 | 9.4464 | 7.287 | 2780.9 | 3036.0 | 8.6314 |
| 320 | 45.618 | 2843.0 | 3116.7 | 9.5859 | 7.815 | 2842.5 | 3116.1 | 8.7712 |
| 360 | 48.696 | 2905.5 | 3197.7 | 9.7180 | 8.344 | 2905.1 | 3197.1 | 8.9034 |
| 400 | 51.774 | 2969.0 | 3279.6 | 9.8435 | 8.872 | 2968.6 | 3279.2 | 9.0291 |
| 440 | 54.851 | 3033.5 | 3362.6 | 9.9633 | 9.400 | 3033.2 | 3362.2 | 9.1490 |
| 500 | 59.467 | 3132.3 | 3489.1 | 10.134 | 10.192 | 3132.1 | 3488.8 | 9.3194 |
| | **0.70 bar (0.07 MPa)** ($T_{sat}$ = 89.95°C) | | | | **1.0 bar (0.10 MPa)** ($T_{sat}$ = 99.63°C) | | | |
| Sat. | 2.365 | 2494.5 | 2660.0 | 7.4797 | 1.694 | 2506.1 | 2675.5 | 7.3594 |
| 100 | 2.434 | 2509.7 | 2680.0 | 7.5341 | 1.696 | 2506.7 | 2676.2 | 7.3614 |
| 120 | 2.571 | 2539.7 | 2719.6 | 7.6375 | 1.793 | 2537.3 | 2716.6 | 7.4668 |
| 160 | 2.841 | 2599.4 | 2798.2 | 7.8279 | 1.984 | 2597.8 | 2796.2 | 7.6597 |
| 200 | 3.108 | 2659.1 | 2876.7 | 8.0012 | 2.172 | 2658.1 | 2875.3 | 7.8343 |
| 240 | 3.374 | 2719.3 | 2955.5 | 8.1611 | 2.359 | 2718.5 | 2954.5 | 7.9949 |
| 280 | 3.640 | 2780.2 | 3035.0 | 8.3162 | 2.546 | 2779.6 | 3034.2 | 8.1445 |
| 320 | 3.905 | 2842.0 | 3115.3 | 8.4504 | 2.732 | 2841.5 | 3114.6 | 8.2849 |
| 360 | 4.170 | 2904.6 | 3196.5 | 8.5828 | 2.917 | 2904.2 | 3195.9 | 8.4175 |
| 400 | 4.434 | 2968.2 | 3278.6 | 8.7086 | 3.103 | 2967.9 | 3278.2 | 8.5435 |
| 440 | 4.698 | 3032.9 | 3361.8 | 8.8286 | 3.288 | 3032.6 | 3361.4 | 8.6636 |
| 500 | 5.095 | 3131.8 | 3488.5 | 8.9991 | 3.565 | 3131.6 | 3488.1 | 8.8342 |
| | **1.5 bars (0.15 MPa)** ($T_{sat}$ = 111.37°C) | | | | **3.0 bars (0.30 MPa)** ($T_{sat}$ = 133.55°C) | | | |
| Sat. | 1.159 | 2519.7 | 2693.6 | 7.2233 | 0.606 | 2543.6 | 2725.3 | 6.9919 |
| 120 | 1.188 | 2533.3 | 2711.4 | 7.2693 | | | | |
| 160 | 1.317 | 2595.2 | 2792.8 | 7.4665 | 0.651 | 2587.1 | 2782.3 | 7.1276 |
| 200 | 1.444 | 2656.2 | 2872.9 | 7.6433 | 0.716 | 2650.7 | 2865.5 | 7.3115 |
| 240 | 1.570 | 2717.2 | 2952.7 | 7.8052 | 0.781 | 2713.1 | 2947.3 | 7.4774 |
| 280 | 1.695 | 2778.6 | 3032.8 | 7.9555 | 0.844 | 2775.4 | 3028.6 | 7.6299 |
| 320 | 1.819 | 2840.6 | 3113.5 | 8.0964 | 0.907 | 2838.1 | 3110.1 | 7.7722 |
| 360 | 1.943 | 2903.5 | 3195.0 | 8.2293 | 0.969 | 2901.4 | 3192.2 | 7.9061 |
| 400 | 2.067 | 2967.3 | 3277.4 | 8.3555 | 1.032 | 2965.6 | 3275.0 | 8.0330 |
| 440 | 2.191 | 3032.1 | 3360.7 | 8.4757 | 1.094 | 3030.6 | 3358.7 | 8.1538 |
| 500 | 2.376 | 3131.2 | 3487.6 | 8.6466 | 1.187 | 3130.0 | 3486.0 | 8.3251 |
| 600 | 2.685 | 3301.7 | 3704.3 | 8.9101 | 1.341 | 3300.8 | 3703.2 | 8.5892 |

**Table A-14**     *(Continued)*

| Temp., °C | $v$ | $u$ | $h$ | $s$ | $v$ | $u$ | $h$ | $s$ |
|---|---|---|---|---|---|---|---|---|
| | **5.0 bars (0.50 MPa) ($T_{sat}$ = 151.86°C)** | | | | **7.0 bars (0.70 MPa) ($T_{sat}$ = 164.97°C)** | | | |
| Sat. | 0.3749 | 2561.2 | 2748.7 | 6.8213 | 0.2729 | 2572.5 | 2763.5 | 6.7080 |
| 180 | 0.4045 | 2609.7 | 2812.0 | 6.9656 | 0.2847 | 2599.8 | 2799.1 | 6.7880 |
| 200 | 0.4249 | 2642.9 | 2855.4 | 7.0592 | 0.2999 | 2634.8 | 2844.8 | 6.8865 |
| 240 | 0.4646 | 2707.6 | 2939.6 | 7.2307 | 0.3292 | 2701.8 | 2932.2 | 7.0641 |
| 280 | 0.5034 | 2771.2 | 3022.9 | 7.3865 | 0.3574 | 2766.9 | 3017.1 | 7.2233 |
| 320 | 0.5416 | 2834.7 | 3105.6 | 7.5308 | 0.3852 | 2831.3 | 3100.9 | 7.3697 |
| 360 | 0.5796 | 2898.7 | 3188.4 | 7.6660 | 0.4126 | 2895.8 | 3184.7 | 7.5063 |
| 400 | 0.6173 | 2963.2 | 3271.9 | 7.7938 | 0.4397 | 2960.9 | 3268.7 | 7.6350 |
| 440 | 0.6548 | 3028.6 | 3356.0 | 7.9152 | 0.4667 | 3026.6 | 3353.3 | 7.7571 |
| 500 | 0.7109 | 3128.4 | 3483.9 | 8.0873 | 0.5070 | 3126.8 | 3481.7 | 7.9299 |
| 600 | 0.8041 | 3299.6 | 3701.7 | 8.3522 | 0.5738 | 3298.5 | 3700.2 | 8.1956 |
| 700 | 0.8969 | 3477.5 | 3925.9 | 8.5952 | 0.6403 | 3476.6 | 3924.8 | 8.4391 |
| | **10.0 bars (1.0 MPa) ($T_{sat}$ = 179.91°C)** | | | | **15.0 bars (1.5 MPa) ($T_{sat}$ = 198.32°C)** | | | |
| Sat. | 0.1944 | 2583.6 | 2778.1 | 6.5865 | 0.1318 | 2594.5 | 2792.2 | 6.4448 |
| 200 | 0.2060 | 2621.9 | 2827.9 | 6.6940 | 0.1325 | 2598.1 | 2796.8 | 6.4546 |
| 240 | 0.2275 | 2692.9 | 2920.4 | 6.8817 | 0.1483 | 2676.9 | 2899.3 | 6.6628 |
| 280 | 0.2480 | 2760.2 | 3008.2 | 7.0465 | 0.1627 | 2748.6 | 2992.7 | 6.8381 |
| 320 | 0.2678 | 2826.1 | 3093.9 | 7.1962 | 0.1765 | 2817.1 | 3081.9 | 6.9938 |
| 360 | 0.2873 | 2891.6 | 3178.9 | 7.3349 | 0.1899 | 2884.4 | 3169.2 | 7.1363 |
| 400 | 0.3066 | 2957.3 | 3263.9 | 7.4651 | 0.2030 | 2951.3 | 3255.8 | 7.2690 |
| 440 | 0.3257 | 3023.6 | 3349.3 | 7.5883 | 0.2160 | 3018.5 | 3342.5 | 7.3940 |
| 500 | 0.3541 | 3124.4 | 3478.5 | 7.7622 | 0.2352 | 3120.3 | 3473.1 | 7.5698 |
| 540 | 0.3729 | 3192.6 | 3565.6 | 7.8720 | 0.2478 | 3189.1 | 3560.9 | 7.6805 |
| 600 | 0.4011 | 3296.8 | 3697.9 | 8.0290 | 0.2668 | 3293.9 | 3694.0 | 7.8385 |
| 640 | 0.4198 | 3367.4 | 3787.2 | 8.1290 | 0.2793 | 3364.8 | 3783.8 | 7.9391 |
| | **20.0 bars (2.0 MPa) ($T_{sat}$ = 212.42°C)** | | | | **30.0 bars (3.0 MPa) ($T_{sat}$ = 233.90°C)** | | | |
| Sat. | 0.0996 | 2600.3 | 2799.5 | 6.3409 | 0.0667 | 2604.1 | 2804.2 | 6.1869 |
| 240 | 0.1085 | 2659.6 | 2876.5 | 6.4952 | 0.0682 | 2619.7 | 2824.3 | 6.2265 |
| 280 | 0.1200 | 2736.4 | 2976.4 | 6.6828 | 0.0771 | 2709.6 | 2941.3 | 6.4462 |
| 320 | 0.1308 | 2807.9 | 3069.5 | 6.8452 | 0.0850 | 2788.4 | 3043.4 | 6.6245 |
| 360 | 0.1411 | 2877.0 | 3159.3 | 6.9917 | 0.0923 | 2861.7 | 3138.7 | 6.7801 |
| 400 | 0.1512 | 2945.2 | 3247.6 | 7.1271 | 0.0994 | 2932.8 | 3230.9 | 6.9212 |
| 440 | 0.1611 | 3013.4 | 3335.5 | 7.2540 | 0.1062 | 3002.9 | 3321.5 | 7.0520 |
| 500 | 0.1757 | 3116.2 | 3467.6 | 7.4317 | 0.1162 | 3108.0 | 3456.5 | 7.2338 |
| 540 | 0.1853 | 3185.6 | 3556.1 | 7.5434 | 0.1227 | 3178.4 | 3546.6 | 7.3474 |
| 600 | 0.1996 | 3290.9 | 3690.1 | 7.7024 | 0.1324 | 3285.0 | 3682.3 | 7.5085 |
| 640 | 0.2091 | 3362.2 | 3780.4 | 7.8035 | 0.1388 | 3357.0 | 3773.5 | 7.6106 |
| 700 | 0.2232 | 3470.9 | 3917.4 | 7.9487 | 0.1484 | 3466.5 | 3911.7 | 7.7571 |

**Table A-14**     (*Continued*)

| Temp., °C | *v* | *u* | *h* | *s* | *v* | *u* | *h* | *s* |
|---|---|---|---|---|---|---|---|---|
| | **40 bars (4.0 MPa)** ($T_{sat}$ = 250.40°C) | | | | **60 bars (6.0 MPa)** ($T_{sat}$ = 275.64°C) | | | |
| Sat. | 0.04978 | 2602.3 | 2801.4 | 6.0701 | 0.03244 | 2589.7 | 2784.3 | 5.8892 |
| 280 | 0.05546 | 2680.0 | 2901.8 | 6.2568 | 0.03317 | 2605.2 | 2804.2 | 5.9252 |
| 320 | 0.06199 | 2767.4 | 3015.4 | 6.4553 | 0.03876 | 2720.0 | 2952.6 | 6.1846 |
| 360 | 0.06788 | 2845.7 | 3117.2 | 6.6215 | 0.04331 | 2811.2 | 3071.1 | 6.3782 |
| 400 | 0.07341 | 2919.9 | 3213.6 | 6.7690 | 0.04739 | 2892.9 | 3177.2 | 6.5408 |
| 440 | 0.07872 | 2992.2 | 3307.1 | 6.9041 | 0.05122 | 2970.0 | 3277.3 | 6.6853 |
| 500 | 0.08643 | 3099.5 | 3445.3 | 7.0901 | 0.05665 | 3082.2 | 3422.2 | 6.8803 |
| 540 | 0.09145 | 3171.1 | 3536.9 | 7.2056 | 0.06015 | 3156.1 | 3517.0 | 6.9999 |
| 600 | 0.09885 | 3279.1 | 3674.4 | 7.3688 | 0.06525 | 3266.9 | 3658.4 | 7.1677 |
| 640 | 0.1037 | 3351.8 | 3766.6 | 7.4720 | 0.06859 | 3341.0 | 3752.6 | 7.2731 |
| 700 | 0.1110 | 3462.1 | 3905.9 | 7.6198 | 0.07352 | 3453.1 | 3894.1 | 7.4234 |
| 740 | 0.1157 | 3536.6 | 3999.6 | 7.7141 | 0.07677 | 3528.3 | 3989.2 | 7.5190 |
| | **80 bars (8.0 MPa)** ($T_{sat}$ = 295.06°C) | | | | **100 bars (10.0 MPa)** ($T_{sat}$ = 311.06°C) | | | |
| Sat. | 0.02352 | 2569.8 | 2758.0 | 5.7432 | 0.01803 | 2544.4 | 2724.7 | 5.6141 |
| 320 | 0.02682 | 2662.7 | 2877.2 | 5.9489 | 0.01925 | 2588.8 | 2781.3 | 5.7103 |
| 360 | 0.03089 | 2772.7 | 3019.8 | 6.1819 | 0.02331 | 2729.1 | 2962.1 | 6.0060 |
| 400 | 0.03432 | 2863.8 | 3138.3 | 6.3634 | 0.02641 | 2832.4 | 3096.5 | 6.2120 |
| 440 | 0.03742 | 2946.7 | 3246.1 | 6.5190 | 0.02911 | 2922.1 | 3213.2 | 6.3805 |
| 480 | 0.04034 | 3025.7 | 3348.4 | 6.6586 | 0.03160 | 3005.4 | 3321.4 | 6.5282 |
| 520 | 0.04313 | 3102.7 | 3447.7 | 6.7871 | 0.03394 | 3085.6 | 3425.1 | 6.6622 |
| 560 | 0.04582 | 3178.7 | 3545.3 | 6.9072 | 0.03619 | 3164.1 | 3526.0 | 6.7864 |
| 600 | 0.04845 | 3254.4 | 3642.0 | 7.0206 | 0.03837 | 3241.7 | 3625.3 | 6.9029 |
| 640 | 0.05102 | 3330.1 | 3738.3 | 7.1283 | 0.04048 | 3318.9 | 3723.7 | 7.0131 |
| 700 | 0.05481 | 3443.9 | 3882.4 | 7.2812 | 0.04358 | 3434.7 | 3870.5 | 7.1687 |
| 740 | 0.05729 | 3520.4 | 3978.7 | 7.3782 | 0.04560 | 3512.1 | 3968.1 | 7.2670 |
| | **120 bars (12.0 MPa)** ($T_{sat}$ = 324.75°C) | | | | **140 bars (14.0 MPa)** ($T_{sat}$ = 336.75°C) | | | |
| Sat. | 0.01426 | 2513.7 | 2684.9 | 5.4924 | 0.01149 | 2476.8 | 2637.6 | 5.3717 |
| 360 | 0.01811 | 2678.4 | 2895.7 | 5.8361 | 0.01422 | 2617.4 | 2816.5 | 5.6602 |
| 400 | 0.02108 | 2798.3 | 3051.3 | 6.0747 | 0.01722 | 2760.9 | 3001.9 | 5.9448 |
| 440 | 0.02355 | 2896.1 | 3178.7 | 6.2586 | 0.01954 | 2868.6 | 3142.2 | 6.1474 |
| 480 | 0.02576 | 2984.4 | 3293.5 | 6.4154 | 0.02157 | 2962.5 | 3264.5 | 6.3143 |
| 520 | 0.02781 | 3068.0 | 3401.8 | 6.5555 | 0.02343 | 3049.8 | 3377.8 | 6.4610 |
| 560 | 0.02977 | 3149.0 | 3506.2 | 6.6840 | 0.02517 | 3133.6 | 3486.0 | 6.5941 |
| 600 | 0.03164 | 3228.7 | 3608.3 | 6.8037 | 0.02683 | 3215.4 | 3591.1 | 6.7172 |
| 640 | 0.03345 | 3307.5 | 3709.0 | 6.9164 | 0.02843 | 3296.0 | 3694.1 | 6.8326 |
| 700 | 0.03610 | 3425.2 | 3858.4 | 7.0749 | 0.03075 | 3415.7 | 3846.2 | 6.9939 |
| 740 | 0.03781 | 3503.7 | 3957.4 | 7.1746 | 0.03225 | 3495.2 | 3946.7 | 7.0952 |

## Table A-14    (Continued)

| Temp., °C | v | u | h | s | v | u | h | s |
|---|---|---|---|---|---|---|---|---|
| | **160 bars (16.0 MPa) ($T_{sat.}$ = 347.44°C)** | | | | **180 bars (18.0 MPa) ($T_{sat}$ = 357.06°C)** | | | |
| Sat. | 0.00931 | 2431.7 | 2580.6 | 5.2455 | 0.00749 | 2374.3 | 2509.1 | 5.1044 |
| 360 | 0.01105 | 2539.0 | 2715.8 | 5.4614 | 0.00809 | 2418.9 | 2564.5 | 5.1922 |
| 400 | 0.01426 | 2719.4 | 2947.6 | 5.8175 | 0.01190 | 2672.8 | 2887.0 | 5.6887 |
| 440 | 0.01652 | 2839.4 | 3103.7 | 6.0429 | 0.01414 | 2808.2 | 3062.8 | 5.9428 |
| 480 | 0.01842 | 2939.7 | 3234.4 | 6.2215 | 0.01596 | 2915.9 | 3203.2 | 6.1345 |
| 520 | 0.02013 | 3031.1 | 3353.3 | 6.3752 | 0.01757 | 3011.8 | 3378.0 | 6.2960 |
| 560 | 0.02172 | 3117.8 | 3465.4 | 6.5132 | 0.01904 | 3101.7 | 3444.4 | 6.4392 |
| 600 | 0.02323 | 3201.8 | 3573.5 | 6.6399 | 0.02042 | 3188.0 | 3555.6 | 6.5696 |
| 640 | 0.02467 | 3284.2 | 3678.9 | 6.7580 | 0.02174 | 3272.3 | 3663.6 | 6.6905 |
| 700 | 0.02674 | 3406.0 | 3833.9 | 6.9224 | 0.02362 | 3396.3 | 3821.5 | 6.8580 |
| 740 | 0.02808 | 3486.7 | 3935.9 | 7.0251 | 0.02483 | 3478.0 | 3925.0 | 6.9623 |
| | **200 bars (20.0 MPa) ($T_{sat}$ = 365.81°C)** | | | | **240 bars (24.0 MPa)** | | | |
| Sat. | 0.00583 | 2293.0 | 2409.7 | 4.9269 | | | | |
| 400 | 0.00994 | 2619.3 | 2818.1 | 5.5540 | 0.00673 | 2477.8 | 2639.4 | 5.2393 |
| 440 | 0.01222 | 2774.9 | 3019.4 | 5.8450 | 0.00929 | 2700.6 | 2923.4 | 5.6506 |
| 480 | 0.01399 | 2891.2 | 3170.8 | 6.0518 | 0.01100 | 2838.3 | 3102.3 | 5.8950 |
| 520 | 0.01551 | 2992.0 | 3302.2 | 6.2218 | 0.01241 | 2950.5 | 3248.5 | 6.0842 |
| 560 | 0.01689 | 3085.2 | 3423.0 | 6.3705 | 0.01366 | 3051.1 | 3379.0 | 6.2448 |
| 600 | 0.01818 | 3174.0 | 3537.6 | 6.5048 | 0.01481 | 3145.2 | 3500.7 | 6.3875 |
| 640 | 0.01940 | 3260.2 | 3648.1 | 6.6286 | 0.01588 | 3235.5 | 3616.7 | 6.5174 |
| 700 | 0.02113 | 3386.4 | 3809.0 | 6.7993 | 0.01739 | 3366.4 | 3783.8 | 6.6947 |
| 740 | 0.02224 | 3469.3 | 3914.1 | 6.9052 | 0.01835 | 3451.7 | 3892.1 | 6.8038 |
| 800 | 0.02385 | 3592.7 | 4069.7 | 7.0544 | 0.01974 | 3578.0 | 4051.6 | 6.9567 |
| | **280 bars (28.0 MPa)** | | | | **320 bars (32.0 MPa)** | | | |
| 400 | 0.00383 | 2223.5 | 2330.7 | 4.7494 | 0.00236 | 1980.4 | 2055.9 | 4.3239 |
| 440 | 0.00712 | 2613.2 | 2812.6 | 5.4494 | 0.00544 | 2509.0 | 2683.0 | 5.2327 |
| 480 | 0.00885 | 2780.8 | 3028.5 | 5.7446 | 0.00722 | 2718.1 | 3949.2 | 5.5968 |
| 520 | 0.01020 | 3906.8 | 3192.3 | 5.9566 | 0.00853 | 2860.7 | 3133.7 | 5.8357 |
| 560 | 0.01136 | 3015.7 | 3333.7 | 6.1307 | 0.00963 | 2979.0 | 3287.2 | 6.0246 |
| 600 | 0.01241 | 3115.6 | 3463.0 | 6.2823 | 0.01061 | 3085.3 | 3424.6 | 6.1858 |
| 640 | 0.01338 | 3210.3 | 3584.8 | 6.4187 | 0.01150 | 3184.5 | 3552.5 | 6.3290 |
| 700 | 0.01473 | 3346.1 | 3758.4 | 6.6029 | 0.01273 | 3325.4 | 3732.8 | 6.5203 |
| 740 | 0.01558 | 3433.9 | 3870.0 | 6.7153 | 0.01350 | 3415.9 | 3847.8 | 6.6361 |
| 800 | 0.01680 | 3563.1 | 4033.4 | 6.8720 | 0.01460 | 3548.0 | 4015.1 | 6.7966 |
| 900 | 0.01873 | 3774.3 | 4298.8 | 7.1084 | 0.01633 | 3762.7 | 4285.1 | 7.0372 |

SOURCE: J. H. Keenan, F. G. Keyes, P. G. Hill, and J. G. Moore, "Steam Tables," Wiley, New York, 1969.

**Table A-15**  Properties of water: Compressed-liquid table

($v$, m³/g; $u$, kJ/kg; $h$, kJ/kg; $s$, kJ/(kg·K))

| Temp., °C | $v \times 10^3$ | $u$ | $h$ | $s$ | $v \times 10^3$ | $u$ | $h$ | $s$ |
|---|---|---|---|---|---|---|---|---|
| | **25 bars (2.5 MPa) (223.99°C)** | | | | **50 bars (5.0 MPa) (263.99°C)** | | | |
| 20 | 1.0006 | 83.80 | 86.30 | .2961 | .9995 | 83.65 | 88.65 | .2956 |
| 40 | 1.0067 | 167.25 | 169.77 | .5715 | 1.0056 | 166.95 | 171.97 | .5705 |
| 80 | 1.0280 | 334.29 | 336.86 | 1.0737 | 1.0268 | 333.72 | 338.85 | 1.0720 |
| 100 | 1.0423 | 418.24 | 420.85 | 1.3050 | 1.0410 | 417.52 | 422.72 | 1.3030 |
| 140 | 1.0784 | 587.82 | 590.52 | 1.7369 | 1.0768 | 586.76 | 592.15 | 1.7343 |
| 180 | 1.1261 | 761.16 | 763.97 | 2.1375 | 1.1240 | 759.63 | 765.25 | 2.1341 |
| 200 | 1.1555 | 849.9 | 852.8 | 2.3294 | 1.1530 | 848.1 | 848.1 | 2.3255 |
| 220 | 1.1898 | 940.7 | 943.7 | 2.5174 | 1.1866 | 938.4 | 944.4 | 2.5128 |
| Sat. | 1.1973 | 959.1 | 962.1 | 2.5546 | 1.2859 | 1147.8 | 1154.2 | 2.9202 |
| | **75 bars (7.5 MPa) (290.59°C)** | | | | **100 bars (10 MPa) (311.06°C)** | | | |
| 20 | .9984 | 83.50 | 90.99 | .2950 | .9972 | 83.36 | 93.33 | .2945 |
| 40 | 1.0045 | 166.64 | 174.18 | .5696 | 1.0034 | 166.35 | 176.38 | .5686 |
| 80 | 1.0256 | 333.15 | 340.84 | 1.0704 | 1.0245 | 332.59 | 342.83 | 1.0688 |
| 100 | 1.0397 | 416.81 | 424.62 | 1.3011 | 1.0385 | 416.12 | 426.5 | 1.2992 |
| 140 | 1.0752 | 585.72 | 593.78 | 1.7317 | 1.0737 | 584.68 | 595.42 | 1.7292 |
| 180 | 1.1219 | 758.13 | 766.55 | 2.1308 | 1.1199 | 756.65 | 767.84 | 2.1275 |
| 220 | 1.1835 | 936.2 | 945.1 | 2.5083 | 1.1805 | 934.1 | 945.9 | 2.5039 |
| 260 | 1.2696 | 1124.4 | 1134.0 | 2.8763 | 1.2645 | 1121.1 | 1133.7 | 2.8699 |
| Sat. | 1.3677 | 1282.0 | 1292.2 | 3.1649 | 1.4524 | 1393.0 | 1407.6 | 3.3596 |
| | **150 bars (15 MPa) (342.24°C)** | | | | **200 bars (20 MPa) (365.81°C)** | | | |
| 20 | .9950 | 83.06 | 97.99 | .2934 | .9928 | 82.77 | 102.62 | .2923 |
| 40 | 1.0013 | 165.76 | 180.78 | .5666 | .9992 | 165.17 | 185.16 | .5646 |
| 80 | 1.0222 | 331.48 | 346.81 | 1.0656 | 1.0199 | 330.40 | 350.80 | 1.0624 |
| 100 | 1.0361 | 414.74 | 430.28 | 1.2955 | 1.0337 | 413.39 | 434.06 | 1.2917 |
| 140 | 1.0707 | 582.66 | 598.72 | 1.7242 | 1.0678 | 580.69 | 602.04 | 1.7193 |
| 180 | 1.1159 | 753.76 | 770.50 | 2.1210 | 1.1120 | 750.95 | 773.20 | 2.1147 |
| 220 | 1.1748 | 929.9 | 947.5 | 2.4953 | 1.1693 | 925.9 | 949.3 | 2.4870 |
| 260 | 1.2550 | 1114.6 | 1133.4 | 2.8576 | 1.2462 | 1108.6 | 1133.5 | 2.8459 |
| 300 | 1.3770 | 1316.6 | 1337.3 | 3.2260 | 1.3596 | 1306.1 | 1333.3 | 3.2071 |
| Sat. | 1.6581 | 1585.6 | 1610.5 | 3.6848 | 2.036 | 1785.6 | 1826.3 | 4.0139 |
| | **250 bars (25 MPa)** | | | | **300 bars (30 MPa)** | | | |
| 20 | .9907 | 82.47 | 107.24 | .2911 | .9886 | 82.17 | 111.84 | .2899 |
| 40 | .9971 | 164.60 | 189.52 | .5626 | .9951 | 164.04 | 193.89 | .5607 |
| 100 | 1.0313 | 412.08 | 437.85 | 1.2881 | 1.0290 | 410.78 | 441.66 | 1.2844 |
| 200 | 1.1344 | 834.5 | 862.8 | 2.2961 | 1.1302 | 831.4 | 865.3 | 2.2893 |
| 300 | 1.3442 | 1296.6 | 1330.2 | 3.1900 | 1.3304 | 1287.9 | 1327.8 | 3.1741 |

SOURCE: J. H Keenan, F. G. Keyes, P. G. Hill, and J. G. Moore, "Steam Tables," Wiley, New York, 1969.

**Table A-16**  Properties of saturated refrigerant 134a ($CF_4H_2$): Temperature table

($v$, m³/kg; $u$, kJ/kg; $h$, kJ/kg; $s$, kJ/kg·K)

| Temp., °C $T$ | Press., bars $P$ | Specific Volume | | Internal Energy | | Enthalpy | | | Entropy | |
|---|---|---|---|---|---|---|---|---|---|---|
| | | Sat. Liquid $v_f \times 10^3$ | Sat. Vapor $v_g$ | Sat. Liquid $u_f$ | Sat. Vapor $u_g$ | Sat. Liquid $h_f$ | Evap. $h_{fg}$ | Sat. Vapor $h_g$ | Sat. Liquid $s_f$ | Sat. Vapor $s_g$ |
| −40 | 0.5164 | 0.7055 | 0.3569 | −0.04 | 204.45 | 0.00 | 222.88 | 222.88 | 0.0000 | 0.9560 |
| −36 | 0.6332 | 0.7113 | 0.2947 | 4.68 | 206.73 | 4.73 | 220.67 | 225.40 | 0.0201 | 09506 |
| −32 | 0.7704 | 0.7172 | 0.2451 | 9.47 | 209.01 | 9.52 | 218.37 | 227.90 | 0.0401 | 0.9456 |
| −28 | 0.9305 | 0.7233 | 0.2052 | 14.31 | 211.29 | 14.37 | 216.01 | 230.38 | 0.0600 | 0.9411 |
| −26 | 1.0199 | 0.7265 | 0.1882 | 16.75 | 212.43 | 16.82 | 214.80 | 231.62 | 0.0699 | 0.9390 |
| −24 | 1.1160 | 0.7296 | 0.1728 | 19.21 | 213.57 | 19.29 | 213.57 | 232.85 | 0.0798 | 0.9370 |
| −22 | 1.2192 | 0.7328 | 0.1590 | 21.68 | 214.70 | 21.77 | 212.32 | 234.08 | 0.0897 | 0.9351 |
| −20 | 1.3299 | 0.7361 | 0.1464 | 24.17 | 215.84 | 24.26 | 211.05 | 235.31 | 0.0996 | 0.9332 |
| −18 | 1.4483 | 0.7395 | 0.1350 | 26.67 | 216.97 | 26.77 | 209.76 | 236.53 | 0.1094 | 0.9315 |
| −16 | 1.5748 | 0.7428 | 0.1247 | 29.18 | 218.10 | 29.30 | 208.45 | 237.74 | 0.1192 | 0.9298 |
| −12 | 1.8540 | 0.7498 | 0.1068 | 34.25 | 220.36 | 34.39 | 205.77 | 240.15 | 0.1388 | 0.9267 |
| −8 | 2.1704 | 0.7569 | 0.0919 | 39.38 | 222.60 | 39.54 | 203.00 | 242.54 | 0.1583 | 0.9239 |
| −4 | 2.5274 | 0.7644 | 0.0794 | 44.56 | 224.84 | 44.75 | 200.15 | 244.90 | 0.1777 | 0.9213 |
| 0 | 2.9282 | 0.7721 | 0.0689 | 49.79 | 227.06 | 50.02 | 197.21 | 247.23 | 0.1970 | 0.9190 |
| 4 | 3.3765 | 0.7801 | 0.0600 | 55.08 | 229.27 | 55.35 | 194.19 | 249.53 | 0.2162 | 0.9169 |
| 8 | 3.8756 | 0.7884 | 0.0525 | 60.43 | 231.46 | 60.73 | 191.07 | 251.80 | 0.2354 | 0.9150 |
| 12 | 4.4294 | 0.7971 | 0.0460 | 65.83 | 233.63 | 66.18 | 187.85 | 254.03 | 0.2545 | 0.9132 |
| 16 | 5.0416 | 0.8062 | 0.0405 | 71.29 | 235.78 | 71.69 | 184.52 | 256.22 | 0.2735 | 0.9116 |
| 20 | 5.7160 | 0.8157 | 0.0358 | 76.80 | 237.91 | 77.26 | 181.09 | 258.36 | 0.2924 | 0.9102 |
| 24 | 6.4566 | 0.8257 | 0.0317 | 82.37 | 240.01 | 82.90 | 177.55 | 260.45 | 0.3113 | 0.9089 |
| 26 | 6.8530 | 0.8309 | 0.0298 | 85.18 | 241.05 | 85.75 | 175.73 | 261.48 | 0.3208 | 0.9082 |
| 28 | 7.2675 | 0.8362 | 0.0281 | 88.00 | 242.08 | 88.61 | 173.89 | 262.50 | 0.3302 | 0.9076 |
| 30 | 7.7006 | 0.8417 | 0.0265 | 90.84 | 243.10 | 91.49 | 172.00 | 263.50 | 0.3396 | 0.9070 |
| 32 | 8.1528 | 0.8473 | 0.0250 | 93.70 | 244.12 | 94.39 | 170.09 | 264.48 | 0.3490 | 0.9064 |
| 34 | 8.6247 | 0.8530 | 0.0236 | 96.58 | 245.12 | 97.31 | 168.14 | 265.45 | 0.3584 | 0.9058 |
| 36 | 9.1168 | 0.8590 | 0.0223 | 99.47 | 246.11 | 100.25 | 166.15 | 266.40 | 0.3678 | 0.9053 |
| 38 | 9.6298 | 0.8651 | 0.0210 | 102.38 | 247.09 | 103.21 | 164.12 | 267.33 | 0.3772 | 0.9047 |
| 40 | 10.164 | 0.8714 | 0.0199 | 105.30 | 248.06 | 106.19 | 162.05 | 268.24 | 0.3866 | 0.9041 |
| 42 | 10.720 | 0.8780 | 0.0188 | 108.25 | 249.02 | 109.19 | 159.04 | 269.14 | 0.3960 | 0.9035 |
| 44 | 11.299 | 0.8847 | 0.0177 | 111.22 | 249.96 | 112.22 | 157.79 | 270.01 | 0.4054 | 0.9030 |
| 48 | 12.526 | 0.8989 | 0.0159 | 117.22 | 251.79 | 118.35 | 153.33 | 271.68 | 0.4243 | 0.9017 |
| 52 | 13.851 | 0.9142 | 0.0142 | 123.31 | 253.55 | 124.58 | 148.66 | 273.24 | 0.4432 | 0.9004 |
| 56 | 15.278 | 0.9308 | 0.0127 | 129.51 | 255.23 | 130.93 | 143.75 | 274.68 | 0.4622 | 0.8990 |
| 60 | 16.813 | 0.9488 | 0.0114 | 135.82 | 256.81 | 137.42 | 138.57 | 275.99 | 0.4814 | 0.8973 |
| 70 | 21.162 | 1.0027 | 0.0086 | 152.22 | 260.15 | 154.34 | 124.08 | 278.43 | 0.5302 | 0.8918 |
| 80 | 26.324 | 1.0766 | 0.0064 | 169.88 | 262.14 | 172.71 | 106.41 | 279.12 | 0.5814 | 0.8827 |
| 90 | 32.435 | 1.1949 | 0.0046 | 189.82 | 261.34 | 193.69 | 82.63 | 276.32 | 0.6380 | 0.8655 |
| 100 | 39.742 | 1.5443 | 0.0027 | 218.60 | 248.49 | 224.74 | 34.40 | 259.13 | 0.7196 | 0.8117 |

SOURCE: Produced from a computer program provided by R. S. Basu, Allied Signal Corporation. *Literature Source:* D. P. Wilson and R. S. Basu, "Thermodynamic Properties of a New Stratospherically Safe Working Fluid–Refrigerant-134a," *ASHRAE Trans.*, **94** (Pt.2): 2095–2118,1988.

**Table A-17** Properties of saturated refrigerant 134a ($CF_4H_2$): Pressure table

($v$, $m^3/kg$; $u$, $kJ/kg$; $h$, $kJ/kg$; $s$, $kJ/kg \cdot K$)

| Press., bars $P$ | Temp., °C $T$ | Specific Volume | | Internal Energy | | Enthalpy | | | Entropy | |
|---|---|---|---|---|---|---|---|---|---|---|
| | | Sat. Liquid $v_f \times 10^3$ | Sat. Vapor $v_g$ | Sat. Liquid $u_f$ | Sat. Vapor $u_g$ | Sat. Liquid $h_f$ | Evap. $h_{fg}$ | Sat. Vapor $h_g$ | Sat. Liquid $s_f$ | Sat. Vapor $s_g$ |
| 0.6 | −37.07 | 0.7097 | 0.3100 | 3.41 | 206.12 | 3.46 | 221.27 | 224.72 | 0.0147 | 0.9520 |
| 0.8 | −31.21 | 0.7184 | 0.2366 | 10.41 | 209.46 | 10.47 | 217.92 | 228.39 | 0.0440 | 0.9447 |
| 1.0 | −26.43 | 0.7258 | 0.1917 | 16.22 | 212.18 | 16.29 | 215.06 | 231.35 | 0.0678 | 0.9395 |
| 1.2 | −22.36 | 0.7323 | 0.1614 | 21.23 | 214.50 | 21.32 | 212.54 | 233.86 | 0.0879 | 0.9354 |
| 1.4 | −18.80 | 0.7381 | 0.1395 | 25.66 | 216.52 | 25.77 | 210.27 | 236.04 | 0.1055 | 0.9322 |
| 1.6 | −15.62 | 0.7435 | 0.1229 | 29.66 | 218.32 | 29.78 | 208.19 | 237.97 | 0.1211 | 0.9295 |
| 1.8 | −12.73 | 0.7485 | 0.1098 | 33.31 | 219.94 | 33.45 | 206.26 | 239.71 | 0.1352 | 0.9273 |
| 2.0 | −10.09 | 0.7532 | 0.0993 | 36.69 | 221.43 | 36.84 | 204.46 | 241.30 | 0.1481 | 0.9253 |
| 2.4 | −5.37 | 0.7618 | 0.0834 | 42.77 | 224.07 | 42.95 | 201.14 | 244.09 | 0.1710 | 0.9222 |
| 2.8 | −1.23 | 0.7697 | 0.0719 | 48.18 | 226.38 | 48.39 | 198.13 | 246.52 | 0.1911 | 0.9197 |
| 3.2 | 2.48 | 0.7770 | 0.0632 | 53.06 | 228.43 | 53.31 | 195.35 | 248.66 | 0.2089 | 0.9177 |
| 3.6 | 5.84 | 0.7839 | 0.0564 | 57.54 | 230.28 | 57.82 | 192.76 | 250.58 | 0.2251 | 0.91660 |
| 4.0 | 8.93 | 0.7904 | 0.0509 | 61.69 | 231.97 | 62.00 | 190.32 | 252.32 | 0.2399 | 0.9145 |
| 5.0 | 15.74 | 0.8056 | 0.0409 | 70.93 | 235.64 | 71.33 | 184.74 | 256.07 | 0.2723 | 0.9117 |
| 6.0 | 21.58 | 0.8196 | 0.0341 | 78.99 | 238.74 | 79.48 | 179.71 | 259.19 | 0.2999 | 0.9097 |
| 7.0 | 26.72 | 0.8328 | 0.0292 | 86.19 | 241.42 | 86.78 | 175.07 | 261.85 | 0.3242 | 0.9080 |
| 8.0 | 31.33 | 0.8454 | 0.0255 | 92.75 | 243.78 | 93.42 | 170.73 | 264.15 | 0.3459 | 0.9066 |
| 9.0 | 35.53 | 0.8576 | 0.0226 | 98.79 | 245.88 | 99.56 | 166.62 | 266.18 | 0.3656 | 0.9054 |
| 10.0 | 39.39 | 0.8695 | 0.0202 | 104.42 | 247.77 | 105.29 | 162.68 | 267.97 | 0.3838 | 0.9043 |
| 12.0 | 46.32 | 0.8928 | 0.0166 | 114.69 | 251.03 | 115.76 | 155.23 | 270.99 | 0.4164 | 0.9023 |
| 14.0 | 52.43 | 0.9159 | 0.0140 | 123.98 | 253.74 | 125.26 | 148.14 | 273.40 | 0.4453 | 0.9003 |
| 16.0 | 57.92 | 0.9392 | 0.0121 | 132.52 | 256.00 | 134.02 | 141.31 | 275.33 | 0.4714 | 0.8982 |
| 18.0 | 62.91 | 0.9631 | 0.0105 | 140.49 | 257.88 | 142.22 | 134.60 | 276.83 | 0.4954 | 0.8959 |
| 20.0 | 67.49 | 0.9878 | 0.0093 | 148.02 | 259.41 | 149.99 | 127.95 | 277.94 | 0.5178 | 0.8934 |
| 25.0 | 77.59 | 1.0562 | 0.0069 | 165.48 | 261.84 | 168.12 | 111.06 | 279.17 | 0.5687 | 0.8854 |
| 30.0 | 86.22 | 1.1416 | 0.0053 | 181.88 | 262.16 | 185.30 | 92.71 | 278.01 | 0.6156 | 0.8735 |

*SOURCE:* Produced from a computer program provided by R. S. Basu, Allied Signal Corporation. *Literature Source:* D. P. Wilson and R. S. Basu, "Thermodynamic Properties of a New Stratospherically Safe Working Fluid—Refrigerant-134a," *ASHRAE Trans.,* **94** (Pt. 2): 2095–2118, 1988.

**Table A-18**     Properties of superheated refrigerant 134a ($CF_4H_2$)

($T$, °C; $v$, m³/kg; $u$, kJ/kg; $h$, kJ/kg; $s$, kJ/kg·K)

| $T$ | $v$ | $u$ | $h$ | $s$ | $v$ | $u$ | $h$ | $s$ |
|---|---|---|---|---|---|---|---|---|
| | **0.6 bar (0.060 MPa)** ($T_{sat} = -37.07$°C) | | | | **1.0 bar (0.10 MPa)** ($T_{sat} = -26.43$°C) | | | |
| Sat. | 0.31003 | 206.12 | 224.72 | 0.9520 | 0.19170 | 212.18 | 231.35 | 0.9395 |
| −20 | 0.33536 | 217.86 | 237.98 | 1.0062 | 0.19770 | 216.77 | 236.54 | 0.9602 |
| −10 | 0.34992 | 224.97 | 245.96 | 1.0371 | 0.20686 | 224.01 | 244.70 | 0.9918 |
| 0 | 0.36433 | 232.24 | 254.10 | 1.0675 | 0.21587 | 231.41 | 252.99 | 1.0227 |
| 10 | 0.37861 | 239.69 | 262.41 | 1.0973 | 0.22473 | 238.96 | 261.43 | 1.0531 |
| 20 | 0.39279 | 247.32 | 270.89 | 1.1267 | 0.23349 | 246.67 | 270.02 | 1.0829 |
| 30 | 0.40688 | 255.12 | 279.53 | 1.1557 | 0.24216 | 254.54 | 278.76 | 1.1122 |
| 40 | 0.42091 | 263.10 | 288.35 | 1.1844 | 0.25076 | 262.58 | 287.66 | 1.1411 |
| 50 | 0.43487 | 271.25 | 297.34 | 1.2126 | 0.25930 | 270.79 | 296.72 | 1.1696 |
| 60 | 0.44879 | 279.58 | 306.51 | 1.2405 | 0.26779 | 279.16 | 305.94 | 1.1977 |
| 70 | 0.46266 | 288.08 | 315.84 | 1.2681 | 0.27623 | 287.70 | 315.32 | 1.2254 |
| 80 | 0.47650 | 296.75 | 325.34 | 1.2954 | 0.28464 | 296.40 | 324.87 | 1.2528 |
| 90 | 0.49031 | 305.58 | 335.00 | 1.3224 | 0.29302 | 305.27 | 334.57 | 1.2799 |
| | **1.4 bars (0.14 MPa)** ($T_{sat} = -18.80$°C) | | | | **1.8 bars (0.18 MPa)** ($T_{sat} = -12.73$°C) | | | |
| Sat. | 0.13945 | 216.52 | 236.04 | 0.9322 | 0.10983 | 219.94 | 239.71 | 0.9273 |
| −10 | 0.14549 | 223.03 | 243.40 | 0.9606 | 0.11135 | 222.02 | 242.06 | 0.9362 |
| 0 | 0.15219 | 230.55 | 251.86 | 0.9922 | 0.11678 | 229.67 | 250.69 | 0.9684 |
| 10 | 0.15875 | 238.21 | 260.43 | 1.0230 | 0.12207 | 237.44 | 259.41 | 0.9998 |
| 20 | 0.16520 | 246.01 | 269.13 | 1.0532 | 0.12723 | 245.33 | 268.23 | 1.0304 |
| 30 | 0.17155 | 253.96 | 277.97 | 1.0828 | 0.13230 | 253.36 | 277.17 | 1.0604 |
| 40 | 0.17783 | 262.06 | 286.96 | 1.1120 | 0.13730 | 261.53 | 286.24 | 1.0898 |
| 50 | 0.18404 | 270.32 | 296.09 | 1.1407 | 0.14222 | 269.85 | 295.45 | 1.1187 |
| 60 | 0.19020 | 278.74 | 305.37 | 1.1690 | 0.14710 | 278.31 | 304.79 | 1.1472 |
| 70 | 0.19633 | 287.32 | 314.80 | 1.1969 | 0.15193 | 286.93 | 314.28 | 1.1753 |
| 80 | 0.20241 | 296.06 | 324.39 | 1.2244 | 0.15672 | 295.71 | 323.92 | 1.2030 |
| 90 | 0.20846 | 304.95 | 334.14 | 1.2516 | 0.16148 | 304.63 | 333.70 | 1.2303 |
| | **2.0 bars (0.20 MPa)** ($T_{sat} = -10.09$°C) | | | | **2.4 bars (0.24 MPa)** ($T_{sat} = -5.37$°C) | | | |
| Sat. | 0.09933 | 221.43 | 241.30 | 0.9253 | 0.08343 | 224.07 | 244.09 | 0.9222 |
| −10 | 0.09938 | 221.50 | 241.38 | 0.9256 | | | | |
| 0 | 0.10438 | 229.23 | 250.10 | 0.9582 | 0.08574 | 228.31 | 248.89 | 0.9399 |
| 10 | 0.10922 | 237.05 | 258.89 | 0.9898 | 0.08993 | 236.26 | 257.84 | 0.9721 |
| 20 | 0.11394 | 244.99 | 267.78 | 1.0206 | 0.09399 | 244.30 | 266.85 | 1.0034 |
| 30 | 0.11856 | 253.06 | 276.77 | 1.0508 | 0.09794 | 252.45 | 275.95 | 1.0339 |
| 40 | 0.12311 | 261.26 | 285.88 | 1.0804 | 0.10181 | 260.72 | 285.16 | 1.0637 |
| 50 | 0.12758 | 269.61 | 295.12 | 1.1094 | 0.10562 | 269.12 | 294.47 | 1.0930 |
| 60 | 0.13201 | 278.10 | 304.50 | 1.1380 | 0.10937 | 277.67 | 303.91 | 1.1218 |
| 70 | 0.13639 | 286.74 | 314.02 | 1.1661 | 0.11307 | 286.35 | 313.49 | 1.1501 |
| 80 | 0.14073 | 295.53 | 323.68 | 1.1939 | 0.11674 | 295.18 | 323.19 | 1.1780 |
| 90 | 0.14504 | 304.47 | 333.48 | 1.2212 | 0.12037 | 304.15 | 333.04 | 1.2055 |

| T | v | u | h | s | v | u | h | s |
|---|---|---|---|---|---|---|---|---|
| | **2.8 bars (0.28 MPa)** ($T_{sat}$ = −1.23°C) | | | | **3.2 bars (0.32 MPa)** ($T_{sat}$ = 2.48°C) | | | |
| Sat. | 0.07193 | 226.38 | 246.52 | 0.9197 | 0.06322 | 228.43 | 248.66 | 0.9177 |
| 0 | 0.07240 | 227.37 | 247.64 | 0.9238 | | | | |
| 10 | 0.07613 | 235.44 | 256.76 | 0.9566 | 0.06576 | 234.61 | 255.65 | 0.9427 |
| 20 | 0.07972 | 243.59 | 265.91 | 0.9883 | 0.06901 | 242.87 | 264.95 | 0.9749 |
| 30 | 0.08320 | 251.83 | 275.12 | 1.0192 | 0.07214 | 251.19 | 274.28 | 1.0062 |
| 40 | 0.08660 | 260.17 | 284.42 | 1.0494 | 0.07518 | 259.61 | 283.67 | 1.0367 |
| 50 | 0.08992 | 268.64 | 293.81 | 1.0789 | 0.07815 | 268.14 | 293.15 | 1.0665 |
| 60 | 0.09319 | 277.23 | 303.32 | 1.1079 | 0.08106 | 276.79 | 302.72 | 1.0957 |
| 70 | 0.09641 | 285.96 | 312.95 | 1.1364 | 0.08392 | 285.56 | 312.41 | 1.1243 |
| 80 | 0.09960 | 294.82 | 322.71 | 1.1644 | 0.08674 | 294.46 | 322.22 | 1.1525 |
| 90 | 0.10275 | 303.83 | 332.60 | 1.1920 | 0.08953 | 303.50 | 332.15 | 1.1802 |
| 100 | 0.10587 | 312.98 | 342.62 | 1.2193 | 0.09229 | 312.68 | 342.21 | 1.2076 |
| | **4.0 bars (0.40 MPa)** ($T_{sat}$ = 8.93°C) | | | | **5.0 bars (0.50 MPa)** ($T_{sat}$ = 15.74°C) | | | |
| Sat. | 0.05089 | 231.97 | 252.32 | 0.9145 | 0.04086 | 235.64 | 256.07 | 0.9117 |
| 10 | 0.05119 | 232.87 | 253.35 | 0.9182 | | | | |
| 20 | 0.05397 | 241.37 | 262.96 | 0.9515 | 0.04188 | 239.40 | 260.34 | 0.9264 |
| 30 | 0.05662 | 249.89 | 272.54 | 0.9837 | 0.04416 | 248.20 | 270.28 | 0.9597 |
| 40 | 0.05917 | 258.47 | 282.14 | 1.0148 | 0.04633 | 256.99 | 280.16 | 0.9918 |
| 50 | 0.06164 | 267.13 | 291.79 | 1.0452 | 0.04842 | 265.83 | 290.04 | 1.0229 |
| 60 | 0.06405 | 275.89 | 301.51 | 1.0748 | 0.05043 | 274.73 | 299.95 | 1.0531 |
| 70 | 0.06641 | 284.75 | 311.32 | 1.1038 | 0.05240 | 283.72 | 309.92 | 1.0825 |
| 80 | 0.06873 | 293.73 | 321.23 | 1.1322 | 0.05432 | 292.80 | 319.96 | 1.1114 |
| 90 | 0.07102 | 302.84 | 331.25 | 1.1602 | 0.05620 | 302.00 | 330.10 | 1.1397 |
| 100 | 0.07327 | 312.07 | 341.38 | 1.1878 | 0.05805 | 311.31 | 340.33 | 1.1675 |
| 110 | 0.07550 | 321.44 | 351.64 | 1.2149 | 0.05988 | 320.74 | 350.68 | 1.1949 |
| | **6.0 bars (0.60 MPa)** ($T_{sat}$ = 21.58°C) | | | | **7.0 bars (0.70 MPa)** ($T_{sat}$ = 26.72°C) | | | |
| Sat. | 0.03408 | 238.74 | 259.19 | 0.9097 | 0.02918 | 241.42 | 261.85 | 0.9080 |
| 30 | 0.03581 | 246.41 | 267.89 | 0.9388 | 0.02979 | 244.51 | 265.37 | 0.9197 |
| 40 | 0.03774 | 255.45 | 278.09 | 0.9719 | 0.03157 | 253.83 | 275.93 | 0.9539 |
| 50 | 0.03958 | 264.48 | 288.23 | 1.0037 | 0.03324 | 263.08 | 286.35 | 0.9867 |
| 60 | 0.04134 | 273.54 | 298.35 | 1.0346 | 0.03482 | 272.31 | 296.69 | 1.0182 |
| 70 | 0.04304 | 282.66 | 308.48 | 1.0645 | 0.03634 | 281.57 | 307.01 | 1.0487 |
| 80 | 0.04469 | 291.86 | 318.67 | 1.0938 | 0.03781 | 290.88 | 317.35 | 1.0784 |
| 90 | 0.04631 | 301.14 | 328.93 | 1.1225 | 0.03924 | 300.27 | 327.74 | 1.1074 |
| 100 | 0.04790 | 310.53 | 339.27 | 1.1505 | 0.04064 | 309.74 | 338.19 | 1.1358 |
| 110 | 0.04946 | 320.03 | 349.70 | 1.1781 | 0.04201 | 319.31 | 348.71 | 1.1637 |
| 120 | 0.05099 | 329.64 | 360.24 | 1.2053 | 0.04335 | 328.98 | 359.33 | 1.1910 |
| 130 | 0.05251 | 339.93 | 370.88 | 1.2320 | 0.04468 | 338.76 | 370.04 | 1.2179 |

**Table A-18** *(Continued)*

| T | v | u | h | s | v | u | h | s |
|---|---|---|---|---|---|---|---|---|
| | **8.0 bars (0.80 MPa) ($T_{sat}$ = 31.33°C)** | | | | **9.0 bars (0.90 MPa) ($T_{sat}$ = 35.53°C)** | | | |
| Sat. | 0.02547 | 243.78 | 264.15 | 0.9066 | 0.02255 | 245.88 | 266.18 | 0.9054 |
| 40 | 0.02691 | 252.13 | 273.66 | 0.9374 | 0.02325 | 250.32 | 271.25 | 0.9217 |
| 50 | 0.02846 | 261.62 | 284.39 | 0.9711 | 0.02472 | 260.09 | 282.34 | 0.9566 |
| 60 | 0.02992 | 271.04 | 294.98 | 1.0034 | 0.02609 | 269.72 | 293.21 | 0.9897 |
| 70 | 0.03131 | 280.45 | 305.50 | 1.0345 | 0.02738 | 279.30 | 303.94 | 1.0214 |
| 80 | 0.03264 | 289.89 | 316.00 | 1.0647 | 0.02861 | 288.87 | 314.62 | 1.0521 |
| 90 | 0.03393 | 299.37 | 326.52 | 1.0940 | 0.02980 | 298.46 | 325.28 | 1.0819 |
| 100 | 0.03519 | 308.93 | 337.08 | 1.1227 | 0.03095 | 308.11 | 335.96 | 1.1109 |
| 110 | 0.03642 | 318.57 | 347.71 | 1.1508 | 0.03207 | 317.82 | 346.68 | 1.1392 |
| 120 | 0.03762 | 328.31 | 358.40 | 1.1784 | 0.03316 | 327.62 | 357.47 | 1.1670 |
| 130 | 0.03881 | 338.14 | 369.19 | 1.2055 | 0.03423 | 337.52 | 368.33 | 1.1943 |
| 140 | 0.03997 | 348.09 | 380.07 | 1.2321 | 0.03529 | 347.51 | 379.27 | 1.2211 |
| | **10.0 bars (1.00 MPa) ($T_{sat}$ = 39.39°C)** | | | | **12.0 bars (1.20 MPa) ($T_{sat}$ = 46.32°C)** | | | |
| Sat. | 0.02020 | 247.77 | 267.97 | 0.9043 | 0.01663 | 251.03 | 270.99 | 0.9023 |
| 40 | 0.02029 | 248.39 | 268.68 | 0.9066 | | | | |
| 50 | 0.02171 | 258.48 | 280.19 | 0.9428 | 0.01712 | 254.98 | 275.52 | 0.9164 |
| 60 | 0.02301 | 268.35 | 291.36 | 0.9768 | 0.01835 | 265.42 | 287.44 | 0.9527 |
| 70 | 0.02423 | 278.11 | 302.34 | 1.0093 | 0.01947 | 275.59 | 298.96 | 0.9868 |
| 80 | 0.02538 | 287.82 | 313.20 | 1.0405 | 0.02051 | 285.62 | 310.24 | 1.0192 |
| 90 | 0.02649 | 297.53 | 324.01 | 1.0707 | 0.02150 | 295.59 | 321.39 | 1.0503 |
| 100 | 0.02755 | 307.27 | 334.82 | 1.1000 | 0.02244 | 305.54 | 332.47 | 1.0804 |
| 110 | 0.02858 | 317.06 | 345.65 | 1.1286 | 0.02335 | 315.50 | 343.52 | 1.1096 |
| 120 | 0.02959 | 326.93 | 356.52 | 1.1567 | 0.02423 | 325.51 | 354.58 | 1.1381 |
| 130 | 0.03058 | 336.88 | 367.46 | 1.1841 | 0.02508 | 335.58 | 365.68 | 1.1660 |
| 140 | 0.03154 | 346.92 | 378.46 | 1.2111 | 0.02592 | 345.73 | 376.83 | 1.1933 |
| | **14.0 bars (1.40 MPa) ($T_{sat}$ = 52.43°C)** | | | | **16.0 bars (1.60 MPa) ($T_{sat}$ = 57.92°C)** | | | |
| Sat. | 0.01405 | 253.74 | 273.40 | 0.9003 | 0.01208 | 256.00 | 275.33 | 0.8982 |
| 60 | 0.01495 | 262.17 | 283.10 | 0.9297 | 0.01233 | 258.48 | 278.20 | 0.9069 |
| 70 | 0.01603 | 272.87 | 295.31 | 0.9658 | 0.01340 | 269.89 | 291.33 | 0.9457 |
| 80 | 0.01701 | 283.29 | 307.10 | 0.9997 | 0.01435 | 280.78 | 303.74 | 0.9813 |
| 90 | 0.01792 | 293.55 | 318.63 | 1.0319 | 0.01521 | 291.39 | 315.72 | 1.0148 |
| 100 | 0.01878 | 303.73 | 330.02 | 1.0628 | 0.01601 | 301.84 | 327.46 | 1.0467 |
| 110 | 0.01960 | 313.88 | 341.32 | 1.0927 | 0.01677 | 312.20 | 339.04 | 1.0773 |
| 120 | 0.02039 | 324.05 | 352.59 | 1.1218 | 0.01750 | 322.53 | 350.53 | 1.1069 |
| 130 | 0.02115 | 334.25 | 363.86 | 1.1501 | 0.01820 | 332.87 | 361.99 | 1.1357 |
| 140 | 0.02189 | 344.50 | 375.15 | 1.1777 | 0.01887 | 343.24 | 373.44 | 1.1638 |
| 150 | 0.02262 | 354.82 | 386.49 | 1.2048 | 0.01953 | 353.66 | 384.91 | 1.1912 |
| 160 | 0.02333 | 365.22 | 397.89 | 1.2315 | 0.02017 | 364.15 | 369.43 | 1.2181 |

*SOURCE:* Produced from a computer program provided by R. S. Basu. Allied Signal Corporation. *Literature Source:* D. P. Wilson and R. S. Basu. "Thermodynamic Properties of a New Stratospherically Safe Working Fluid—Refrigerant-134a," *ASHRAE Trans.,* **94** (Pt. 2): 2095–2118, 1988.

**Table A-19**  Properties of saturated nitrogen (N₂): Temperature and pressure tables

(v, L/kg or cm³/g; u and h, kJ/kg; s, kJ/(kg·K); 1 bar = 0.1 MPa)

| Temp., K T | Press., bar(s) P | Specific Volume | | Internal Energy | | Enthalpy | | | Entropy | |
|---|---|---|---|---|---|---|---|---|---|---|
| | | Sat. Liquid $v_f$ | Sat. Vapor $v_g$ | Sat. Liquid $u_f$ | Sat. Vapor $u_g$ | Sat. Liquid $h_f$ | Evap. $h_{fg}$ | Sat. Vapor $h_g$ | Sat. Liquid $s_f$ | Sat. Vapor $s_g$ |
| 63.15 | 0.125 | 1.152 | 1477. | −50.8 | 145.6 | −50.8 | 214.9 | 164.1 | 2.423 | 5.826 |
| 70.0 | 0.386 | 1.191 | 526. | −37.1 | 150.2 | −37.0 | 207.5 | 170.5 | 2.631 | 5.595 |
| 80.0 | 1.369 | 1.257 | 164. | −16.4 | 156.4 | −16.2 | 195.1 | 178.9 | 2.906 | 5.345 |
| 90.0 | 3.600 | 1.340 | 66.3 | 4.5 | 161.1 | 5.0 | 180.0 | 185.0 | 3.152 | 5.152 |
| 100.0 | 7.775 | 1.448 | 31.3 | 25.7 | 163.4 | 26.8 | 160.9 | 187.7 | 3.376 | 4.985 |
| 110.0 | 14.67 | 1.606 | 16.0 | 48.3 | 162.2 | 50.7 | 134.9 | 185.6 | 3.594 | 4.820 |
| 120.0 | 25.15 | 1.908 | 8.03 | 76.9 | 154.1 | 81.7 | 92.6 | 174.3 | 3.847 | 4.619 |
| 126.25 | 33.96 | 3.388 | 3.39 | 123.2 | 123.2 | 134.7 | 0 | 134.7 | 4.257 | 4.257 |
| 77.24 | 1.00 | 1.238 | 219.1 | −22.0 | 154.9 | −21.9 | 198.7 | 176.8 | 2.835 | 5.407 |
| 83.63 | 2.00 | 1.285 | 115.3 | −8.9 | 158.3 | −8.6 | 190.0 | 181.4 | 2.998 | 5.270 |
| 91.25 | 4.00 | 1.352 | 60.0 | 7.1 | 161.5 | 7.6 | 177.9 | 185.5 | 3.180 | 5.130 |
| 96.41 | 6.00 | 1.406 | 40.46 | 18.0 | 162.9 | 18.8 | 168.4 | 187.2 | 3.297 | 5.044 |
| 100.41 | 8.00 | 1.454 | 30.37 | 26.4 | 163.4 | 27.6 | 160.1 | 187.7 | 3.384 | 4.978 |
| 103.76 | 10.00 | 1.500 | 24.16 | 33.9 | 163.4 | 35.4 | 152.2 | 187.6 | 3.457 | 4.924 |
| 110.38 | 15.00 | 1.614 | 15.57 | 49.3 | 162.0 | 51.7 | 133.7 | 185.4 | 3.603 | 4.814 |
| 115.56 | 20.00 | 1.741 | 11.04 | 63.0 | 159.1 | 66.5 | 114.7 | 181.2 | 3.726 | 4.718 |
| 119.88 | 25.00 | 1.902 | 8.11 | 76.5 | 154.3 | 81.3 | 93.3 | 174.6 | 3.844 | 4.622 |
| 123.61 | 30.00 | 2.152 | 5.87 | 91.6 | 146.3 | 98.1 | 65.8 | 163.9 | 3.974 | 4.506 |
| 125.63 | 33.00 | 2.497 | 4.550 | 104.7 | 137.1 | 112.9 | 39.2 | 152.1 | 4.088 | 4.400 |

*Note:* The triple state is 63.15 K and the critical state is 126.25 K.

*SOURCE:* Data adapted from A. A. Vasserman, Ya. Z. Kazavchinshii, and V. A. Rabinovich, "Thermophysical Properties of Air and Air Components," Izdatel'stvo Nauka, Moscow, 1966.

**Table A-20**　　　Properties of nitrogen (N₂): Superheated-vapor table

(v, L/kg or cm³/g; u and h, kJ/kg; s, kJ/(kg·K))

| Temp., K | $v$ | $u$ | $h$ | $s$ | $v$ | $u$ | $h$ | $s$ |
|---|---|---|---|---|---|---|---|---|
| | 1 bar (0.1 MPa) ($T_{sat}$ = 77.2 K) | | | | 5 bars (0.5 MPa) ($T_{sat}$ = 94.0 K) | | | |
| 100 | 290.8 | 172.8 | 201.9 | 5.691 | 52.9 | 167.7 | 194.1 | 5.162 |
| 150 | 442.6 | 210.4 | 254.7 | 6.120 | 86.4 | 208.1 | 251.3 | 5.627 |
| 200 | 592.3 | 247.8 | 307.0 | 6.421 | 117.5 | 246.4 | 305.1 | 5.937 |
| 250 | 741.4 | 285.0 | 359.1 | 6.654 | 147.8 | 284.0 | 357.9 | 6.172 |
| 300 | 890.2 | 322.2 | 411.2 | 6.844 | 177.9 | 321.4 | 410.3 | 6.364 |
| 350 | 1039. | 359.4 | 463.3 | 7.005 | 207.9 | 358.8 | 462.7 | 6.525 |
| 400 | 1187. | 396.7 | 515.4 | 7.144 | 237.7 | 396.2 | 515.0 | 6.665 |
| | 10 bars (1.0 MPa) ($T_{sat}$ = 108.8 K) | | | | 20 bars (2.0 MPa) ($T_{sat}$ = 115.6 K) | | | |
| 150 | 41.9 | 205.0 | 246.9 | 5.401 | 19.6 | 198.4 | 237.5 | 5.150 |
| 200 | 58.1 | 244.6 | 302.7 | 5.722 | 28.5 | 240.9 | 297.8 | 5.498 |
| 250 | 73.6 | 282.7 | 356.3 | 5.962 | 36.6 | 280.1 | 353.2 | 5.746 |
| 300 | 88.9 | 320.3 | 409.2 | 6.155 | 44.4 | 318.3 | 407.1 | 5.942 |
| 350 | 104.0 | 357.9 | 461.9 | 6.317 | 52.1 | 356.2 | 460.4 | 6.107 |
| 400 | 119.0 | 395.5 | 514.5 | 6.457 | 59.7 | 394.1 | 513.5 | 6.248 |
| 450 | 134.0 | 433.2 | 567.2 | 6.582 | 67.3 | 432.0 | 566.5 | 6.373 |
| | 50 bars (5.0 MPa) | | | | 100 bars (10.0 MPa) | | | |
| 150 | 5.92 | 171.5 | 201.1 | 4.692 | 2.39 | 123.5 | 147.4 | 4.218 |
| 200 | 10.7 | 229.3 | 282.9 | 5.169 | 5.02 | 208.8 | 259.0 | 4.870 |
| 250 | 14.4 | 272.2 | 344.1 | 5.443 | 7.12 | 259.3 | 330.4 | 5.190 |
| 300 | 17.8 | 312.4 | 401.1 | 5.651 | 8.95 | 302.7 | 392.2 | 5.416 |
| 350 | 21.0 | 351.4 | 456.3 | 5.821 | 10.7 | 343.6 | 450.3 | 5.596 |
| 400 | 24.1 | 390.1 | 510.7 | 5.967 | 12.3 | 383.6 | 506.8 | 5.746 |
| 450 | 27.2 | 428.7 | 564.8 | 6.094 | 13.9 | 423.1 | 562.4 | 5.877 |
| | 150 bars (15.0 MPa) | | | | 200 bars (20.2 MPa) | | | |
| 150 | 1.95 | 94.3 | 134.9 | 4.064 | 1.78 | 96.1 | 131.6 | 3.980 |
| 200 | 3.37 | 190.8 | 241.3 | 4.680 | 2.69 | 177.2 | 231.0 | 4.554 |
| 250 | 4.80 | 247.1 | 319.2 | 5.030 | 3.73 | 236.6 | 311.2 | 4.913 |
| 300 | 6.09 | 293.5 | 384.8 | 5.269 | 4.70 | 285.3 | 379.3 | 5.162 |
| 350 | 7.28 | 336.3 | 445.5 | 5.456 | 5.62 | 329.5 | 441.8 | 5.355 |
| 400 | 8.42 | 377.5 | 503.7 | 5.612 | 6.49 | 371.8 | 501.5 | 5.515 |
| 450 | 9.52 | 418.0 | 560.7 | 5.746 | 7.33 | 413.0 | 559.6 | 5.651 |

*SOURCE:* Data adapted from A. A. Vasserman, Ya. Z. Kazavchinskii, and V. A. Rabinovich, "Thermophysical Properties of Air and Air Components," Izdatel'stvo Nauka, Moscow, 1966.

**Table A-21** Thermodynamic properties of potassium

(*T*, K; *P*, atm; *v*, L/kg; *h*, kJ/kg; *s*, kJ/(kg·K))

| | | | A. Saturation-Temperature Table | | | | |
|---|---|---|---|---|---|---|---|
| *T* | *P* | $v_f$ | $v_g$ | $h_f$ | $h_g$ | $s_f$ | $s_g$ |
| 900 | 0.251 | 1.438 | 7180 | 731.0 | 2739.7 | 2.6874 | 4.9175 |
| 950 | 0.447 | 1.462 | 4204 | 771.4 | 2750.4 | 2.7313 | 4.8129 |
| 1000 | 0.753 | 1.493 | 2592 | 812.6 | 2760.4 | 2.7732 | 4.7196 |

| | | | B. Superheat Table | | | | |
|---|---|---|---|---|---|---|---|
| *P* | *v* | *h* | *s* | *P* | *v* | *h* | *s* |
| 1075 K ($P_{sat}$ = 1.494 atm) | | | | 1200 K ($P_{sat}$ = 3.860 atm) | | | |
| Sat | 1.375 | 2772.9 | 4.5977 | Sat. | 0.573 | 2795.1 | 4.4367 |
| 1.0 | 2.120 | 2813.6 | 4.7144 | 3.0 | 0.759 | 2831.1 | 4.5144 |
| 0.8 | 2.684 | 2830.0 | 4.7745 | 2.0 | 1.178 | 2874.0 | 4.6295 |
| 1325 K ($P_{sat}$ = 8.293 atm) | | | | 1450 K ($P_{sat}$ = 15.55 atm) | | | |
| Sat | 0.285 | 2820.4 | 4.3163 | Sat. | 0.161 | 2849.4 | 4.2255 |
| 6.0 | 0.411 | 2873.7 | 4.4162 | 10.0 | 0.267 | 2928.5 | 4.3599 |
| 4.0 | 0.641 | 2923.1 | 4.5316 | 8.0 | 0.343 | 2959.2 | 4.4233 |
| 3.0 | 0.873 | 2948.7 | 4.6078 | 6.0 | 0.469 | 2991.3 | 4.5011 |

*SOURCE:* Data adapted from Naval Research Laboratory Report 6233, 1965, and Air Force Aero Propulsion Laboratory Technical Report 66-104, 1966.

**Table A-22**  **Constants for the Benedict-Webb-Rubin, Redlich-Kwong, and van der Waals equations of state**

#### 1. Benedict-Webb-Rubin; Units are bar(s), $m^3/kmol$, and K

| Constants | $n$-Butane ($C_4H_{10}$) | Carbon Dioxide ($CO_2$) | Carbon Monoxide (CO) | Methane ($CH_4$) | Nitrogen ($N_2$) |
|---|---|---|---|---|---|
| $a$ | 1.9068 | 0.1386 | 0.0371 | 0.0500 | 0.0254 |
| $A_0$ | 10.216 | 2.7730 | 1.3587 | 1.8791 | 1.0673 |
| $b$ | 0.039998 | 0.007210 | 0.002632 | 0.003380 | 0.002328 |
| $B_0$ | 0.12436 | 0.04991 | 0.05454 | 0.04260 | 0.04074 |
| $c$ | $3.205 \times 10^5$ | $1.511 \times 10^4$ | $1.054 \times 10^3$ | $2.578 \times 10^3$ | $7.379 \times 10^2$ |
| $C_0$ | $1.006 \times 10^6$ | $1.404 \times 10^5$ | $8.673 \times 10^3$ | $2.286 \times 10^4$ | $8.164 \times 10^3$ |
| $\alpha$ | $1.101 \times 10^{-3}$ | $8.470 \times 10^{-5}$ | $1.350 \times 10^{-4}$ | $1.244 \times 10^{-4}$ | $1.272 \times 10^{-4}$ |
| $\gamma$ | 0.0340 | 0.00539 | 0.0060 | 0.0060 | 0.0053 |

*SOURCE:* H. W. Cooper and J. C. Goldfrank, *Hydrocarbon Processing,* **46** (12): 141 (1967).

#### 2. Redlich-Kwong; $a$ is in bar $(m^3/kmol)^2$ $(K)^{0.5}$, $b$ is in $m^3/kmol$

| Substance | $a$ | $b$ | Substance | $a$ | $b$ |
|---|---|---|---|---|---|
| Carbon dioxide ($CO_2$) | 64.64 | 0.02969 | Oxygen ($O_2$) | 17.38 | 0.02199 |
| Carbon monoxide (CO) | 17.26 | 0.02743 | Propane ($C_3H_8$) | 183.070 | 0.06269 |
| Methane ($CH_4$) | 32.19 | 0.02969 | Refrigerant 134a | 197.1 | 0.06634 |
| Nitrogen ($N_2$) | 15.59 | 0.002681 | Water ($H_2O$) | 142.64 | 0.02110 |

*SOURCE:* Computed from critical data.

#### 3. Van der Waals; $a$ is in bar $(m^3/kmol)^2$, $b$ is in $m^3/kmol$

| Substance | $a$ | $b$ | Substance | $a$ | $b$ |
|---|---|---|---|---|---|
| Acetylene ($C_2H_2$) | 4.410 | 0.0510 | Ethylene (CH) | 4.563 | 0.0574 |
| Air (equivalent) | 1.358 | 0.0364 | Helium (He) | 0.0341 | 0.0234 |
| Ammonia ($NH_3$) | 4.223 | 0.0373 | Hydrogen ($H_2$) | 0.247 | 0.0265 |
| Benzene ($C_2H_6$) | 18.63 | 0.1181 | Methane ($CH_4$) | 2.285 | 0.0427 |
| $n$-Butane ($C_6H_6$) | 13.80 | 0.1196 | Nitrogen ($N_2$) | 1.361 | 0.0385 |
| Carbon dixide ($CO_2$) | 3.643 | 0.0427 | Oxygen ($O_2$) | 1.369 | 0.0315 |
| Carbon monoxide (CO) | 1.463 | 0.0394 | Propane ($C_3H_8$) | 9.315 | 0.0900 |
| Refrigerant 134a ($C_2F_4H_2$) | 10.05 | 0.0957 | Sulfur dioxide ($SO_2$) | 6.837 | 0.0568 |
| Ethane ($C_2H_6$) | 5.575 | 0.0650 | Water ($H_2O$) | 5.507 | 0.0304 |

*SOURCE:* Computed from critical data.

**Table A-23**   Values of the enthalpy of formation, Gibbs function of formation, absolute entropy, and enthalpy of vaporization at 25°C and 1 atm

($\Delta \bar{h}_f^\circ$, $\Delta \bar{g}_f^\circ$, and $\bar{h}_{fg}$ in kJ/kmol; $\bar{s}^\circ$ in kJ/(kmol·K); std state is 1 atm.)

| Substance | Formula | $\Delta \bar{h}_f^\circ$ | $\Delta \bar{g}_f^\circ$ | $\bar{s}^\circ$ | $\bar{h}_{fg}$ |
|---|---|---|---|---|---|
| Carbon | $C(s)$ | 0 | 0 | 5.74 | |
| Hydrogen | $H_2(g)$ | 0 | 0 | 130.57 | |
| Nitrogen | $N_2(g)$ | 0 | 0 | 191.50 | |
| Oxygen | $O_2(g)$ | 0 | 0 | 205.04 | |
| Carbon monoxide | $CO(g)$ | −110,530 | −137,150 | 197.56 | |
| Carbon dioxide | $CO_2(g)$ | −393,520 | −394,380 | 213.67 | |
| Water | $H_2O(g)$ | −241,820 | −228,590 | 188.72 | |
| Water | $H_2O(l)$ | −285,830 | −237,180 | 69.95 | 44,010 |
| Hydrogen peroxide | $H_2O_2(g)$ | −136,310 | −105,600 | 232.63 | 61,090 |
| Ammonia | $NH_3(g)$ | −46,190 | −16,590 | 192.33 | |
| Oxygen | $O(g)$ | 249,170 | 231,770 | 160.95 | |
| Hydrogen | $H(g)$ | 218,000 | 203,290 | 114.61 | |
| Nitrogen | $N(g)$ | 472,680 | 455,510 | 153.19 | |
| Hydroxyl | $OH(g)$ | 39,040 | 34,280 | 183.75 | |
| Methane | $CH_4(g)$ | −74,850 | −50,790 | 186.16 | |
| Actylene (Ethyne) | $C_2H_2(g)$ | 226,730 | 209,170 | 200.85 | |
| Ethylene (Ethene) | $C_2H_4(g)$ | 52,280 | 68,120 | 219.83 | |
| Ehtane | $C_2H_6(g)$ | −84,680 | −32,890 | 229.49 | |
| Propylene (Propene) | $C_3H_6(g)$ | 20,410 | 62,720 | 266.94 | |
| Propane | $C_3H_8(g)$ | −103,850 | −23,490 | 269.91 | 15,060 |
| n-Butane | $C_4H_{10}(g)$ | −126,150 | −15,170 | 310.03 | 21,060 |
| n-Pentane | $C_5H_{12}(g)$ | −146,440 | −8,200 | 348.40 | 31,410 |
| n-Octane | $C_8H_{18}(g)$ | −208,450 | 17,320 | 463.67 | 41,460 |
| n-Octane | $C_8H_{18}(l)$ | −249,910 | 6,610 | 360.79 | |
| Benzene | $C_6H_6(g)$ | 82,930 | 129,660 | 269.20 | 33,830 |
| Methyl alcohol | $CH_3OH(g)$ | −200,890 | −162,140 | 239.70 | 37,900 |
| Methyl alcohol | $CH_3OH(l)$ | −238,810 | −166,290 | 126.80 | |
| Ethyl alcohol | $C_2H_5OH(g)$ | −235,310 | −168,570 | 282.59 | 42,340 |
| Ethyl alcohol | $C_2H_5OH(l)$ | −277,690 | −174,890 | 160.70 | |
| Mercury | $Hg(l)$ | 0 | 0 | 77.24 | |
| Mercuric oxide | $HgO(c)$ | −90,210 | −58,400 | 70.45 | |
| Manganese | $Mn(c)$ | 0 | 0 | 31.8 | |
| Manganese dioxide | $MnO_2(c)$ | −520,030 | −465,180 | 53.14 | |
| Manganese trioxide | $Mn_2O_3(c)$ | −958,970 | −881,150 | 110.5 | |
| Lead | $Pb(c)$ | 0 | 0 | 64.81 | |
| Lead oxide | $PbO_2(c)$ | −277,400 | −217,360 | 68.6 | |
| Lead sulfate | $PbSO_4(c)$ | −919,940 | −813,200 | 148.57 | |
| Zinc | $Zn(c)$ | 0 | 0 | 41.63 | |
| Zinc oxide | $ZnO(c)$ | −343,280 | −318,320 | 43.64 | |
| Sulfuric acid | $H_2SO_4(l)$ | −813,990 | −690,100 | 156.90 | |
| Sulfuric acid | (aq, m = 1) | −909,270 | −744,630 | 20.1 | |
| Silver oxide | $Ag_2O(c)$ | −31,050 | −11,200 | 121.7 | |

*SOURCES:* From the JANAF Thermochemical Tables, Dow Chemical Co., 1971; *Selected Values of Chemical Thermodynamic Properties,* NBS Techinal Note 270-3, 1968; and *API Research Project 44,* Carnegie Press, 1953.

**Table A-24**    Logarithms to the base 10 of the equilibrium constant $K_o$

$(K_p = \dfrac{(p_E)^{v_E}(p_F)^{v_F}}{(p_A)^{v_A}(p_B)^{v_B}}$ for the reaction $v_A A + v_B B \rightleftharpoons v_E E + v_F F$ where $p_i$ is in atm.)

**Numbered reactions:**

(1) $H_2 \rightleftharpoons 2H$                     (5) $H_2O \rightleftharpoons H_2 + \frac{1}{2}O_2$

(2) $O_2 \rightleftharpoons 2O$                     (6) $H_2O \rightleftharpoons OH + \frac{1}{2}H_2$

(3) $N_2 \rightleftharpoons 2N$                     (7) $CO_2 \rightleftharpoons CO + \frac{1}{2}O_2$

(4) $\frac{1}{2}O_2 + \frac{1}{2}N_2 \rightleftharpoons NO$     (8) $CO_2 + H_2 \rightleftharpoons CO + H_2O$

| Temp., K | log $K_o$ values for reactions numbered above | | | | | | | |
|---|---|---|---|---|---|---|---|---|
|  | (1) | (2) | (3) | (4) | (5) | (6) | (7) | (8) |
| 298 | −71.224 | −81.208 | −159.600 | −15.171 | −40.048 | −46.054 | −45.066 | −5.018 |
| 500 | −40.316 | −45.880 | −92.672 | −8.783 | −22.886 | −26.130 | −25.025 | −2.139 |
| 1000 | −17.292 | −19.614 | −43.056 | −4.062 | −10.062 | −11.280 | −10.221 | −0.159 |
| 1200 | −13.414 | −15.208 | −34.754 | −3.275 | −7.899 | −8.789 | −7.764 | +0.135 |
| 1400 | −10.630 | −12.054 | −28.812 | −2.712 | −6.347 | −7.003 | −6.014 | +0.333 |
| 1600 | −8.532 | −9.684 | −24.350 | −2.290 | −5.180 | −5.662 | −4.706 | +0.474 |
| 1700 | −7.666 | −8.706 | −22.512 | −2.116 | −4.699 | −5.109 | −4.169 | +0.530 |
| 1800 | −6.896 | −7.836 | −20.874 | −1.962 | −4.270 | −4.617 | −3.693 | +0.577 |
| 1900 | −6.204 | −7.058 | −19.410 | −1.823 | −3.886 | −4.177 | −3.267 | +0.619 |
| 2000 | −5.580 | −6.356 | −18.092 | −1.699 | −3.540 | −3.780 | −2.884 | +0.656 |
| 2100 | −5.016 | −5.720 | −16.898 | −1.586 | −3.227 | −3.422 | −2.539 | +0.688 |
| 2200 | −4.502 | −5.142 | −15.810 | −1.484 | −2.942 | −3.095 | −2.226 | +0.716 |
| 2300 | −4.032 | −4.614 | −14.818 | −1.391 | −2.682 | −2.798 | −1.940 | +0.742 |
| 2400 | −3.600 | −4.130 | −13.908 | −1.305 | −2.443 | −2.525 | −1.679 | +0.764 |
| 2500 | −3.202 | −3.684 | −13.070 | −1.227 | −2.224 | −2.274 | −1.440 | +0.784 |
| 2600 | −2.836 | −3.272 | −12.298 | −1.154 | −2.021 | −2.042 | −1.219 | +0.802 |
| 2700 | −2.494 | −2.892 | −11.580 | −1.087 | −1.833 | −1.828 | −1.015 | +0.818 |
| 2800 | −2.178 | −2.536 | −10.914 | −1.025 | −1.658 | −1.628 | −0.825 | +0.833 |
| 2900 | −1.882 | −2.206 | −10.294 | −0.967 | −1.495 | −1.442 | −0.649 | +0.846 |
| 3000 | −1.606 | −1.898 | −9.716 | −0.913 | −1.343 | −1.269 | −0.485 | +0.858 |
| 3100 | −1.348 | −1.610 | −9.174 | −0.863 | −1.201 | −1.107 | −0.332 | +0.869 |
| 3200 | −1.106 | −1.340 | −8.664 | −0.815 | −1.067 | −0.955 | −0.189 | +0.878 |
| 3300 | −0.878 | −1.086 | −8.186 | −0.771 | −0.942 | −0.813 | −0.054 | +0.888 |
| 3400 | −0.664 | −0.846 | −7.736 | −0.729 | −0.824 | −0.679 | +0.071 | +0.895 |
| 3500 | −0.462 | −0.620 | −7.312 | −0.690 | −0.712 | −0.552 | +0.190 | +0.902 |

*SOURCE:* Based on data from the JANAF Tables, NSRDS-NBS-37, 1971, and revisions published in *Journal of Physical and Chemical Reference Data* through 1982.

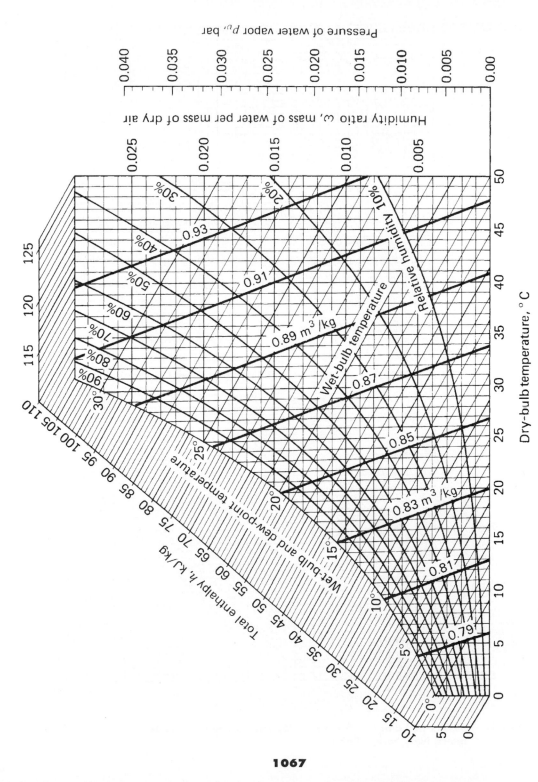

**Figure A-25    Psychrometric chart, metric units, barometric pressure 1.01 bars**

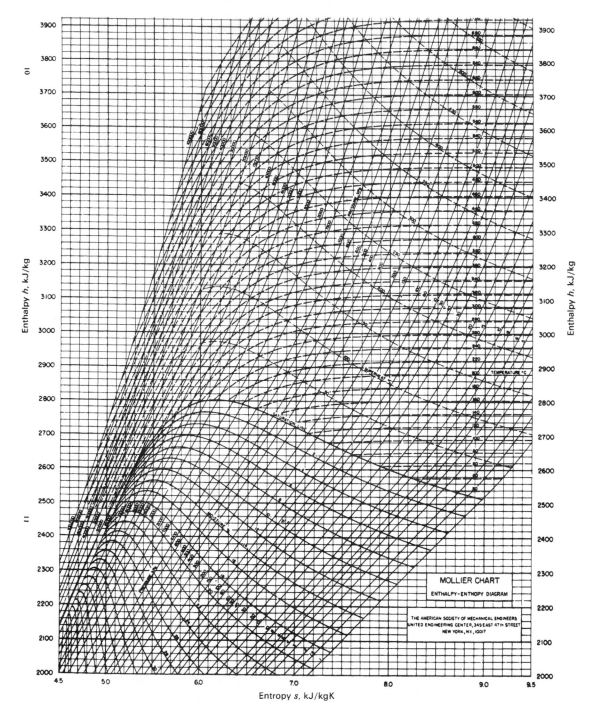

**Figure A-26**      Mollier (*hs*) diagram for steam. (*Source: The American Society of Mechanical Engineers, New York, 1967.*)

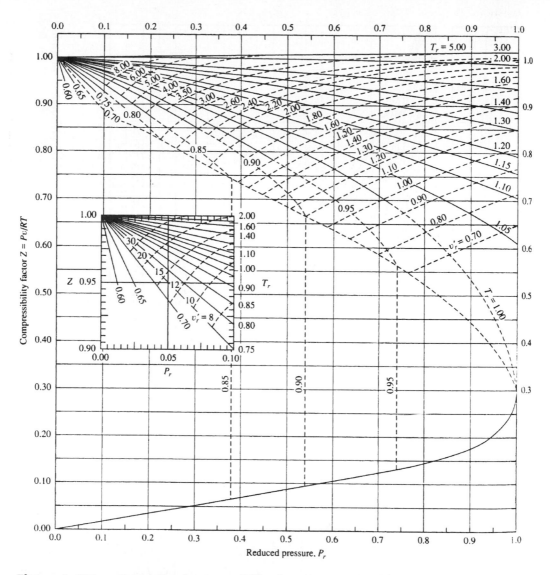

**Figure A-27**  Generalized compressibility chart, $P_r \leq 1$. [Modified by Peter E. Liley, *Chemical Engineering*, 94 (1987). *Original source: E. F. Obert, Concepts of Thermodynamics. McGraw-Hill, Inc., (1960)].*

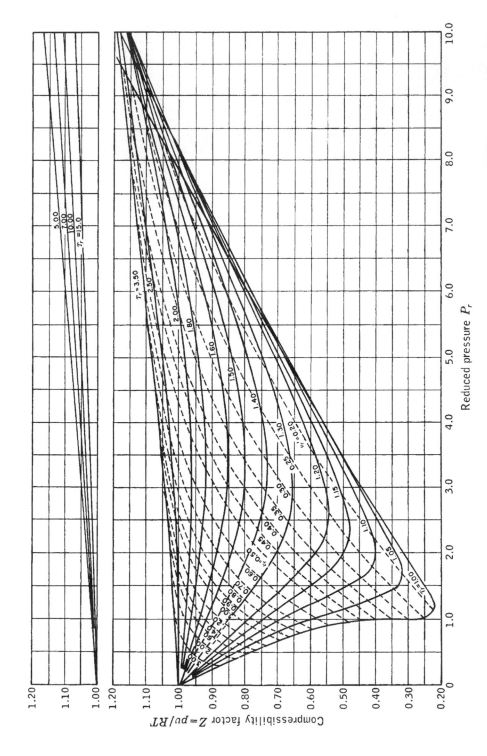

**Figure A-28**   Compressibility chart, low-pressure range. [*Source: V. M. Faires, "Problems on Thermodynamics," Macmillan, New York, 1962. Data from L. C. Nelson and E. F. Obert, Generalized Compressibility Charts, Chem. Eng. 61: 203 (1954).*]

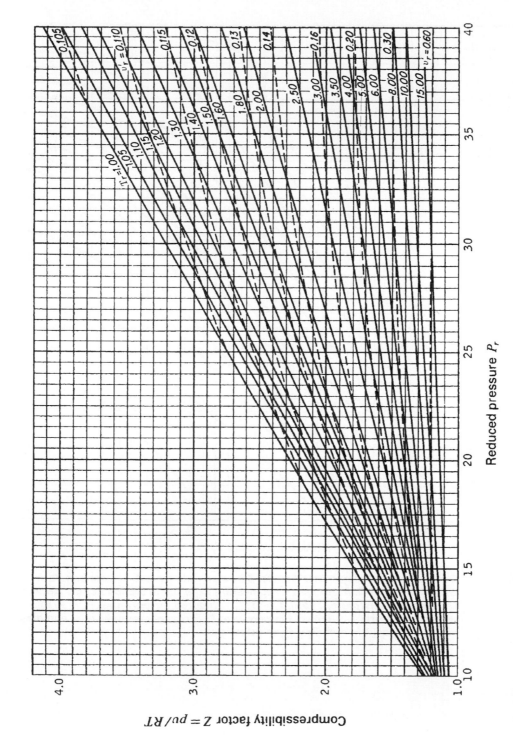

**Figure A-29**  Compressibility chart, high-pressure range. (Adapted from E. F. Obert, "Concepts of Thermodynamics," McGraw-Hill, New York, 1960).

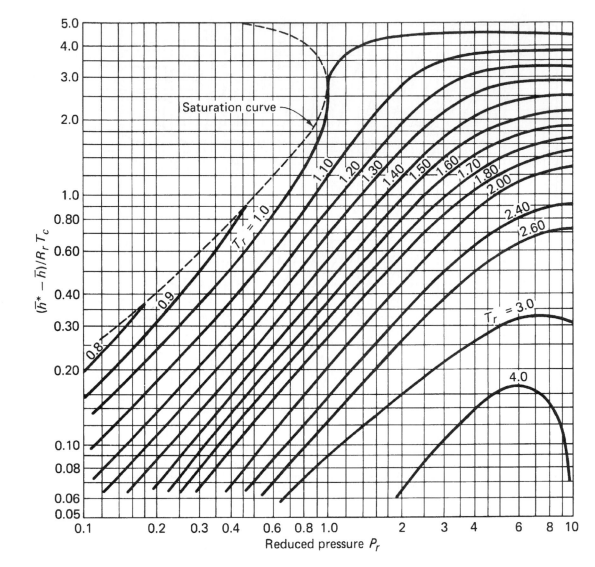

**Figure A-30**     Generalized enthalpy chart. (*Source: Based on data from A. L. Lydersen, R. A. Greenkorn, and O. A. Hougen, "Engineering Experiment Station Report No. 4," University of Wisconsin, 1955*). $T_r = T/T_c$ = reduced temperature, $P_r = P/P_c$ = reduced pressure, $T_c$ = critical temperature, $P_c$ = critical pressure, $\bar{h}^*$ = enthalpy of an ideal gas, $\bar{h}$ = enthalpy of an actual gas.

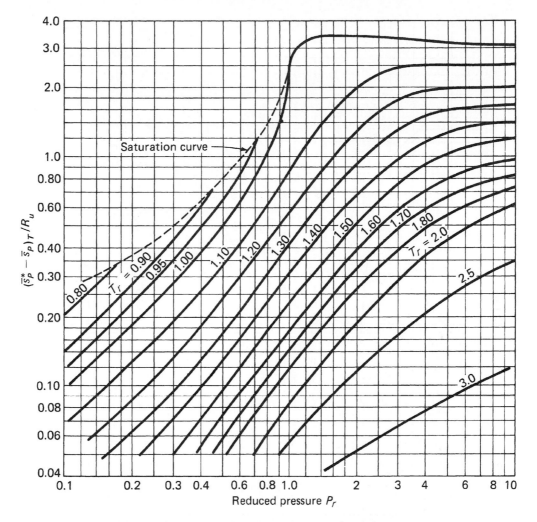

**Figure A-31** Generalized entropy chart. (*Source: Based on data from A. L. Lydersen, R. A. Greenkorn, and O. A. Hougen, "Engineering Experiment Station Report No. 4," University of Wisconsin, 1955). $T_r$ = $T/T_c$ = reduced temperature, $P_r$ = $P/P_c$ = reduced pressure, $T_c$ = critical temperature, $P_c$ = critical pressure, $\bar{s}_P^*$ = entropy of an ideal gas, $\bar{s}_P$ = entropy of an actual gas.*

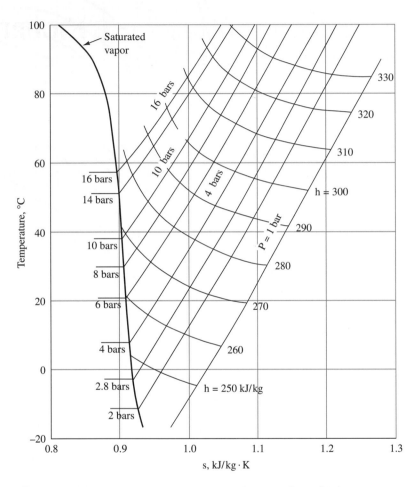

**Figure A-32**    **Temperature–entropy diagram for refrigerant 134a. (Based on data from Table A-18.)**

# A-2

# SUPPLEMENTARY TABLES AND FIGURES (USCS UNITS)

**Table A-1E**      **Physical constants and conversion factors**

| Physical Constants | |
|---|---|
| Standard gravity | $g = 32.174$ ft/s$^2$ |
| Universal gas constant | $R_u = 1545$ ft·lb$_f$/(lbmol·°R) |
| | $= 1.986$ Btu/(lbmol·°R) |
| | $= 0.730$ atm·ft$^3$/(lbmol·°R) |
| | $= 10.73$ psia·ft$^3$/(lbmol·°R) |

| Conversion Factors | |
|---|---|
| Length | 1 cm $= 0.3937$ in. $= 10^4$ $\mu$m $= 10^8$ Å |
| | 1 in $= 2.540$ cm |
| | 1 ft $= 30.48$ cm |
| Mass | 1 lb$_m$ $= 453.59$ g$_m$ $= 7000$ gr |
| | 1 kg$_m$ $= 2.205$ lb$_m$ |
| Force | 1 lb$_f$ $= 32.174$ lb$_m$·ft/s$^2$ |
| | 1 lb$_f$ $= 448,800$ dyn $= 4.448$ N |
| Pressure | 1 torr $= 1$ mmHg at 0°C $= 10^3$ $\mu$m Hg |
| | $= 1.933 \times 10^{-2}$ psi |
| Volume | 1 L $= 0.0353$ ft$^3$ $= 0.2642$ gal $= 61.025$ in$^3$ |
| | 1 ft$^3$ $= 28.316$ L $= 7.4805$ gal $= 0.02832$ m$^3$ |
| | 1 in$^3$ $= 16.387$ cm$^3$ |
| Density | 1 lb$_m$/ft$^3$ $= 0.01602$ g$_m$/cm$^3$ |
| Energy | 1 Btu $= 778.16$ ft·lb$_f$ $= 1055$ J |
| | 1 Btu/lb $= 2.32$ kJ/kg |
| Power | 1 W $= 1$ J/s $= 3.413$ Btu/h |
| | 1 hp $= 746$ W $= 550$ ft·lb$_f$/s $= 2545$ Btu/h |
| Velocity | 1 mi/h $= 0.447$ m/s |
| Specific heat | 1 Btu/(lb$_m$·F) $= 4.187$ kJ/(kg·K) |
| Temperature | $T(°R) = T(°F) + 459.67 = 1.8[T(°C) + 473.15] = 1.8T(K)$ |

**Table A-2E** Molar mass, critical constants, and gas-phase specific heats at 77°F and 1 atm for some common substances

$(T_c, °R; P_c, atm; \bar{v}_c, ft^3/lbmol; c_v$ and $c_p$, Btu/(lb$_m$·°F))

| Substance | Molar Mass | $T_c$ | $P_c$ | $\bar{v}_c$ | $Z_c$ | $c_v$ | $c_p$ |
|---|---|---|---|---|---|---|---|
| Acetylene ($C_2H_2$) | 26.04 | 556 | 61.6 | 1.80 | 0.272 | 0.328 | 0.404 |
| Air (equivalent) | 28.97 | 239 | 37.2 | 1.33 | 0.284 | 0.171 | 0.240 |
| Ammonia ($NH_3$) | 17.04 | 730 | 111.3 | 1.16 | 0.242 | 0.397 | 0.514 |
| Benzene ($C_6H_6$) | 78.11 | 1013 | 47.7 | 4.11 | 0.274 | 0.224 | 0.249 |
| $n$-Butane ($C_4H_{10}$) | 58.12 | 765 | 37.5 | 4.13 | 0.274 | 0.373 | 0.408 |
| Carbon dioxide ($CO_2$) | 44.04 | 548 | 72.9 | 1.51 | 0.276 | 0.156 | 0.202 |
| Carbon monoxide (CO) | 28.01 | 239 | 34.5 | 1.49 | 0.294 | 0.178 | 0.249 |
| Refrigerant 134a ($C_2F_4H_2$) | 102.03 | 672.8 | 40.1 | 3.19 | 0.261 | 0.184 | 0.203 |
| Ethane ($C_2H_6$) | 30.07 | 549 | 48.2 | 2.37 | 0.285 | 0.353 | 0.419 |
| Ethylene ($C_2H_4$) | 28.05 | 510 | 50.5 | 2.05 | 0.279 | 0.294 | 0.365 |
| Helium (He) | 4.003 | 9.3 | 2.26 | 0.93 | 0.300 | 0.744 | 1.24 |
| Hydrogen ($H_2$) | 2.016 | 59.8 | 12.8 | 1.04 | 0.304 | 2.43 | 3.42 |
| Methane ($CH_4$) | 16.04 | 344 | 45.8 | 1.59 | 0.290 | 0.407 | 0.531 |
| Nitrogen ($N_2$) | 28.01 | 227 | 33.5 | 1.44 | 0.291 | 0.178 | 0.248 |
| Oxygen ($O_2$) | 32.00 | 278 | 49.8 | 1.19 | 0.290 | 0.157 | 0.219 |
| Propane ($C_3H_8$) | 44.09 | 666 | 42.1 | 3.20 | 0.278 | 0.355 | 0.400 |
| Sulfur dioxide ($SO_2$) | 64.06 | 775 | 77.7 | 1.99 | 0.268 | 0.113 | 0.114 |
| Water ($H_2O$) | 18.02 | 1165 | 218.0 | 0.896 | 0.230 | 0.335 | 0.445 |

*SOURCE:* Adapted from the data in Table A-2.

**Additional formulas and values of the molar mass of some common elements and compounds**

| Substance | Formula | Molar Mass | Substance | Formula | Molar Mass |
|---|---|---|---|---|---|
| Argon | Ar | 39.94 | Silver | Ag | 107.88 |
| Aluminum | Al | 26.97 | Sodium | Na | 22.997 |
| Carbon | C | 12.01 | Hydrogen peroxide | $H_2O_2$ | 34.02 |
| Copper | Cu | 63.54 | Propylene | $C_3H_6$ | 42.08 |
| Iron | Fe | 55.85 | $n$-Pentane | $C_5H_{12}$ | 72.15 |
| Lead | Pb | 207.2 | $n$-Octane | $C_8H_{18}$ | 114.22 |
| Mercury | Hg | 200.6 | Methyl alcohol | $CH_3OH$ | 32.05 |
| Potassium | K | 39.096 | Ethyl alcohol | $C_2H_5OH$ | 46.07 |

**Table A-3E          Ideal-gas specific-heat data for various gases, Btu/lb$_m$·°F**

| | | | | | | | | | | |
|---|---|---|---|---|---|---|---|---|---|---|
| **1. Zero-Pressure Specific Heats for Six Common Gases, Where $k = c_p/c_v$** | | | | | | | | | | |
| | $c_p$ | $c_v$ | $k$ | $c_p$ | $c_v$ | $k$ | $c_p$ | $c_v$ | $k$ | |
| **Temp., °F** | Air | | | Carbon Dioxide ($CO_2$) | | | Carbon Monoxide (CO) | | | **Temp., °F** |
| 40 | 0.240 | 0.171 | 1.401 | 0.195 | 0.150 | 1.300 | 0.248 | 0.177 | 1.400 | 40 |
| 100 | 0.240 | 0.172 | 1.400 | 0.205 | 0.160 | 1.283 | 0.249 | 0.178 | 1.399 | 100 |
| 200 | 0.241 | 0.173 | 1.397 | 0.217 | 0.172 | 1.262 | 0.249 | 0.179 | 1.397 | 200 |
| 300 | 0.243 | 0.174 | 1.394 | 0.229 | 0.184 | 1.246 | 0.251 | 0.180 | 1.394 | 300 |
| 400 | 0.245 | 0.176 | 1.389 | 0.239 | 0.193 | 1.233 | 0.253 | 0.182 | 1.389 | 400 |
| 500 | 0.248 | 0.179 | 1.383 | 0.247 | 0.202 | 1.223 | 0.256 | 0.185 | 1.384 | 500 |
| 600 | 0.250 | 0.182 | 1.377 | 0.255 | 0.210 | 1.215 | 0.259 | 0.188 | 1.377 | 600 |
| 700 | 0.254 | 0.185 | 1.371 | 0.262 | 0.217 | 1.208 | 0.262 | 0.191 | 1.371 | 700 |
| 800 | 0.257 | 0.188 | 1.365 | 0.269 | 0.224 | 1.202 | 0.266 | 0.195 | 1.364 | 800 |
| 900 | 0.259 | 0.191 | 1.358 | 0.275 | 0.230 | 1.197 | 0.269 | 0.198 | 1.357 | 900 |
| 1000 | 0.263 | 0.195 | 1.353 | 0.280 | 0.235 | 1.192 | 0.273 | 0.202 | 1.351 | 1000 |
| 1500 | 0.276 | 0.208 | 1.330 | 0.298 | 0.253 | 1.178 | 0.287 | 0.216 | 1.328 | 1500 |
| 2000 | 0.286 | 0.217 | 1.312 | 0.312 | 0.267 | 1.169 | 0.297 | 0.226 | 1.314 | 2000 |
| | Hydrogen ($H_2$) | | | Nitrogen ($N_2$) | | | Oxygen ($O_2$) | | | |
| 40 | 3.397 | 2.412 | 1.409 | 0.248 | 0.177 | 1.400 | 0.219 | 0.156 | 1.397 | 40 |
| 100 | 3.426 | 2.441 | 1.404 | 0.248 | 0.178 | 1.399 | 0.220 | 0.158 | 1.394 | 100 |
| 200 | 3.451 | 2.466 | 1.399 | 0.249 | 0.178 | 1.398 | 0.223 | 0.161 | 1.387 | 200 |
| 300 | 3.461 | 2.476 | 1.398 | 0.250 | 0.179 | 1.396 | 0.226 | 0.164 | 1.378 | 300 |
| 400 | 3.466 | 2.480 | 1.397 | 0.251 | 0.180 | 1.393 | 0.230 | 0.168 | 1.368 | 400 |
| 500 | 3.469 | 2.484 | 1.397 | 0.254 | 0.183 | 1.388 | 0.235 | 0.173 | 1.360 | 500 |
| 600 | 3.473 | 2.488 | 1.396 | 0.256 | 0.185 | 1.383 | 0.239 | 0.177 | 1.352 | 600 |
| 700 | 3.477 | 2.492 | 1.395 | 0.260 | 0.189 | 1.377 | 0.242 | 0.181 | 1.344 | 700 |
| 800 | 3.494 | 2.509 | 1.393 | 0.262 | 0.191 | 1.371 | 0.246 | 0.184 | 1.337 | 800 |
| 900 | 3.502 | 2.519 | 1.392 | 0.265 | 0.194 | 1.364 | 0.249 | 0.187 | 1.331 | 900 |
| 1000 | 3.513 | 2.528 | 1.390 | 0.269 | 0.198 | 1.359 | 0.252 | 0.190 | 1.326 | 1000 |
| 1500 | 3.618 | 2.633 | 1.374 | 0.283 | 0.212 | 1.334 | 0.263 | 0.201 | 1.309 | 1500 |
| 2000 | 3.758 | 2.773 | 1.355 | 0.293 | 0.222 | 1.319 | 0.270 | 0.208 | 1.298 | 2000 |

| *SOURCE:* Data adapted from *Tables of Thermal Properties of Gases*, NBS Circular 564, 1955.

**Table A-3E**    *(Continued)*

### 2. Specific-Heat Data for Monatomic Gases

Over a wide range of temperatures at low pressures, the specific heats $c_v$ and $c_p$ of all monatomic gases are essentially independent of temperature and pressure. In addition, on a mole basis all monatomic gases have the same value for either $c_v$ or $c_p$ in a given set of units. One set of values is

$$\bar{c}_v = 2.98 \text{ Btu/(lbmol·°F)} \quad \text{and} \quad \bar{c}_p = 4.97 \text{ Btu/(lbmol·°F)}$$

*SOURCE:* Data adapted from *Tables of Thermal Properties of Gases,* NBS Circular 564, 1955.

### 3. Constant-Pressure Specific-Heat Equations for Various Gases at Zero Pressure (USCS Units)

$$\bar{c}_p/R_u = a + bT + cT^2 + dT^3 + eT^4$$

**where $T$ is in degrees Rankine, equation valid from 540 to 1800°R**

| Gas | $a$ | $b \times 10^3$ | $c \times 10^6$ | $d \times 10^9$ | $e \times 10^{12}$ |
|-----|-----|-----|-----|-----|-----|
| CO | 3.710 | −0.899 | 1.140 | −0.348 | 0.0229 |
| $CO_2$ | 2.401 | 4.853 | −2.039 | 0.343 | |
| $H_2$ | 3.057 | 1.487 | −1.793 | 0.947 | −0.1726 |
| $H_2O$ | 4.070 | −0.616 | 1.281 | −0.508 | 0.0769 |
| $O_2$ | 3.626 | −1.043 | 2.178 | −1.160 | 0.2054 |
| $N_2$ | 3.675 | −0.671 | 0.717 | −0.108 | −0.0215 |
| Air (dry) | 3.653 | −0.741 | 1.016 | −0.328 | 0.0262 |
| $NH_3$ | 3.591 | 0.274 | 2.576 | −1.437 | 0.2601 |
| NO | 4.046 | −1.899 | 2.464 | −1.048 | 0.1517 |
| $NO_2$ | 3.459 | 1.147 | 2.064 | −1.639 | 0.3448 |
| $SO_2$ | 3.267 | 2.958 | 0.211 | −0.906 | 0.2438 |
| $SO_3$ | 2.578 | 8.087 | −2.832 | −0.136 | 0.1878 |
| $CH_4$ | 3.826 | −2.211 | 7.580 | −3.898 | 0.6633 |
| $C_2H_2$ | 1.410 | 10.587 | −7.562 | 2.811 | −0.3939 |
| $C_2H_4$ | 1.426 | 6.324 | 2.466 | −2.787 | 0.6429 |

*SOURCE:* Adapted from the data in NASA SP-273, Government Printing Office, Washington, 1971.

**Table A-4E** Specific heats of some common liquids and solids

$(c_p, Btu/(lb_m \cdot °F))$

| A. Liquids | | | | | |
|---|---|---|---|---|---|
| **Substance** | **State** | $c_p$ | **Substance** | **State** | $c_p$ |
| Water | 1 atm, 32°F | 1.007 | Glycerin | 1 atm, 50°F | 0.554 |
| | 1 atm, 77°F | 0.998 | | 1 atm, 120°F | 0.617 |
| | 1 atm, 212°F | 1.007 | Bismuth | 1 atm, 800°F | 0.0345 |
| Ammonia | sat., 0°F | 1.08 | | 1 atm, 1400°F | 0.0393 |
| | sat., 120°F | 1.22 | Mercury | 1 atm, 50°F | 0.033 |
| Refrigerant 134a | sat., 0°F | 0.296 | | 1 atm, 600°F | 0.032 |
| | sat., 80°F | 0.345 | Sodium | 1 atm, 200°F | 0.33 |
| | sat., 120°F | 0.376 | | 1 atm, 1000°F | 0.30 |
| Benzene | 1 atm, 60°F | 0.43 | Propane | 1 atm, 32°F | 0.576 |
| | 1 atm, 150°F | 0.46 | | | |

| B. Solids | | | | | |
|---|---|---|---|---|---|
| **Substance** | $T$, °F | $c_p$ | **Substance** | $T$, °F | $c_p$ |
| Ice | −100 | 0.375 | Lead | −455 | 0.0008 |
| | −50 | 0.424 | | −435 | 0.0073 |
| | 0 | 0.471 | | −150 | 0.0283 |
| | 20 | 0.491 | | 32 | 0.0297 |
| | 32 | 0.502 | | 210 | 0.0320 |
| Aluminum | −150 | 0.167 | | 570 | 0.0356 |
| | −100 | 0.192 | Copper | −240 | 0.0674 |
| | 0 | 0.207 | | −150 | 0.0784 |
| | 32 | 0.212 | | −60 | 0.0862 |
| | 100 | 0.218 | | 0 | 0.0893 |
| | 200 | 0.224 | | 100 | 0.0925 |
| | 300 | 0.229 | | 200 | 0.0938 |
| | 400 | 0.235 | | 300 | 0.0951 |
| | 500 | 0.240 | | 390 | 0.0963 |
| Iron | 68 | 0.107 | Silver | 68 | 0.0558 |

**Table A-5E** Ideal-gas properties of air

($T$, °R; $h$ and $u$, kJ/kg; $s°$, kJ/(kg·k))

| $T$, °R | $h$, Btu/lb$_m$ | $p_r$ | $u$, Btu/lb$_m$ | $v_r$ | $s°$, Btu/lb$_m$·°R | $T$, °R | $h$, Btu/lb$_m$ | $p_r$ | $u$, Btu/lb$_m$ | $v_r$ | $s°$, Btu/lb$_m$·°R |
|---|---|---|---|---|---|---|---|---|---|---|---|
| 360 | 85.97 | 0.3363 | 61.29 | 396.6 | 0.50369 | 860 | 206.46 | 7.149 | 147.50 | 44.57 | 0.71323 |
| 380 | 90.75 | 0.4061 | 64.70 | 346.6 | 0.51663 | 880 | 211.35 | 7.761 | 151.02 | 42.01 | 0.71886 |
| 400 | 95.53 | 0.4858 | 68.11 | 305.0 | 0.52890 | 900 | 216.26 | 8.411 | 154.57 | 39.64 | 0.72438 |
| 420 | 100.32 | 0.5760 | 71.52 | 270.1 | 0.54058 | 920 | 221.18 | 9.102 | 158.12 | 37.44 | 0.72979 |
| 440 | 105.11 | 0.6776 | 74.93 | 240.6 | 0.55172 | 940 | 226.11 | 9.834 | 161.68 | 35.41 | 0.73509 |
| 460 | 109.90 | 0.7913 | 78.36 | 215.33 | 0.56235 | 960 | 231.06 | 10.61 | 165.26 | 33.52 | 0.74030 |
| 480 | 114.69 | 0.9182 | 81.77 | 193.65 | 0.57255 | 980 | 236.02 | 11.43 | 168.83 | 31.76 | 0.74540 |
| 500 | 119.48 | 1.0590 | 85.20 | 174.90 | 0.58233 | 1000 | 240.98 | 12.30 | 172.43 | 30.12 | 0.75042 |
| 520 | 124.27 | 1.2147 | 88.62 | 158.58 | 0.59173 | 1040 | 250.95 | 14.18 | 179.66 | 27.17 | 0.76019 |
| 537 | 128.10 | 1.3593 | 91.53 | 146.34 | 0.59945 | 1080 | 260.97 | 16.28 | 186.93 | 24.58 | 0.76964 |
| 540 | 129.06 | 1.3860 | 92.04 | 144.32 | 0.60078 | 1120 | 271.03 | 18.60 | 194.25 | 22.30 | 0.77880 |
| 560 | 133.86 | 1.5742 | 95.47 | 131.75 | 0.60950 | 1160 | 281.14 | 21.18 | 201.63 | 20.29 | 0.78767 |
| 580 | 138.66 | 1.7800 | 98.90 | 120.70 | 0.61793 | 1200 | 291.30 | 24.01 | 209.05 | 18.51 | 0.79628 |
| 600 | 143.47 | 2.005 | 102.34 | 110.88 | 0.62607 | 1240 | 301.52 | 27.13 | 216.53 | 16.93 | 0.80466 |
| 620 | 148.28 | 2.249 | 105.78 | 102.12 | 0.63395 | 1280 | 311.79 | 30.55 | 224.05 | 15.52 | 0.81280 |
| 640 | 153.09 | 2.514 | 109.21 | 94.30 | 0.64159 | 1320 | 322.11 | 34.31 | 231.63 | 14.25 | 0.82075 |
| 660 | 157.92 | 2.801 | 112.67 | 87.27 | 0.64902 | 1360 | 332.48 | 38.41 | 239.25 | 13.12 | 0.82848 |
| 680 | 162.73 | 3.111 | 116.12 | 80.96 | 0.65621 | 1400 | 342.90 | 42.88 | 246.93 | 12.10 | 0.83604 |
| 700 | 167.56 | 3.446 | 119.58 | 75.25 | 0.66321 | 1440 | 353.37 | 47.75 | 254.66 | 11.17 | 0.84341 |
| 720 | 172.39 | 3.806 | 123.04 | 70.07 | 0.67002 | 1480 | 363.89 | 53.04 | 262.44 | 10.34 | 0.85062 |
| 740 | 177.23 | 4.193 | 126.51 | 65.38 | 0.67665 | 1520 | 374.47 | 58.78 | 270.26 | 9.578 | 0.85767 |
| 760 | 182.08 | 4.607 | 129.99 | 61.10 | 0.68312 | 1560 | 385.08 | 65.00 | 278.13 | 8.890 | 0.86456 |
| 780 | 186.94 | 5.051 | 133.47 | 57.20 | 0.68942 | 1600 | 395.74 | 71.73 | 286.06 | 8.263 | 0.87130 |
| 800 | 191.81 | 5.526 | 136.97 | 53.63 | 0.69558 | 1650 | 409.13 | 80.89 | 296.03 | 7.556 | 0.87954 |
| 820 | 196.69 | 6.033 | 140.47 | 50.35 | 0.70160 | 1700 | 422.59 | 90.95 | 306.06 | 6.924 | 0.88758 |
| 840 | 201.56 | 6.573 | 143.98 | 47.34 | 0.70747 | 1750 | 436.12 | 101.98 | 316.16 | 6.357 | 0.89542 |

**Table A-5E**    *(Continued)*

| $T$, °R | $h$, Btu/lb$_m$ | $p_r$ | $u$, Btu/lb$_m$ | $v_r$ | $s°$, Btu/lb$_m$·°R |
|---|---|---|---|---|---|
| 1800 | 449.71 | 114.0 | 326.32 | 5.847 | 0.90308 |
| 1850 | 463.37 | 127.2 | 336.55 | 5.388 | 0.91056 |
| 1900 | 477.09 | 141.5 | 346.85 | 4.974 | 0.91788 |
| 1950 | 490.88 | 157.1 | 357.20 | 4.598 | 0.92504 |
| 2000 | 504.71 | 174.0 | 367.61 | 4.258 | 0.93205 |
| 2050 | 518.61 | 192.3 | 378.08 | 3.949 | 0.93891 |
| 2100 | 532.55 | 212.1 | 388.60 | 3.667 | 0.94564 |
| 2150 | 546.54 | 233.5 | 399.17 | 3.410 | 0.95222 |
| 2200 | 560.59 | 256.6 | 409.78 | 3.176 | 0.95919 |
| 2250 | 574.69 | 281.4 | 420.46 | 2.961 | 0.96501 |
| 2300 | 588.82 | 308.1 | 431.16 | 2.765 | 0.97123 |
| 2350 | 603.00 | 336.8 | 441.91 | 2.585 | 0.97732 |
| 2400 | 617.22 | 367.6 | 452.70 | 2.419 | 0.98331 |
| 2450 | 631.48 | 400.5 | 463.54 | 2.266 | 0.98919 |
| 2500 | 645.78 | 435.7 | 474.40 | 2.125 | 0.99497 |
| 2550 | 660.12 | 473.3 | 485.31 | 1.996 | 1.00064 |
| 2600 | 674.49 | 513.5 | 496.26 | 1.876 | 1.00623 |
| 2650 | 688.90 | 556.3 | 507.25 | 1.765 | 1.01172 |
| 2700 | 703.35 | 601.9 | 518.26 | 1.662 | 1.01712 |
| 2750 | 717.83 | 650.4 | 529.31 | 1.566 | 1.02244 |
| 2800 | 732.33 | 702.0 | 540.40 | 1.478 | 1.02767 |
| 2850 | 746.88 | 756.7 | 551.52 | 1.395 | 1.03282 |
| 2900 | 761.45 | 814.8 | 562.66 | 1.318 | 1.03788 |
| 2950 | 776.05 | 876.4 | 573.84 | 1.247 | 1.04288 |
| 3000 | 790.68 | 941.4 | 585.04 | 1.180 | 1.04779 |
| 3050 | 805.34 | 1011 | 596.28 | 1.118 | 1.05264 |
| 3100 | 820.03 | 1083 | 607.53 | 1.060 | 1.05741 |
| 3150 | 834.75 | 1161 | 618.82 | 1.006 | 1.06212 |
| 3200 | 849.48 | 1242 | 630.12 | 0.955 | 1.06676 |
| 3250 | 864.24 | 1328 | 641.46 | 0.907 | 1.07134 |
| 3300 | 879.02 | 1418 | 652.81 | .8621 | 1.07585 |
| 3350 | 893.83 | 1513 | 664.20 | .8202 | 1.08031 |
| 3400 | 908.66 | 1613 | 675.60 | .7807 | 1.08470 |
| 3450 | 923.52 | 1719 | 687.04 | .7436 | 1.08904 |
| 3500 | 938.40 | 1829 | 698.48 | .7087 | 1.09332 |
| 3550 | 953.30 | 1946 | 709.95 | .6759 | 1.09755 |
| 3600 | 968.21 | 2068 | 721.44 | .6449 | 1.10172 |
| 3650 | 983.15 | 2196 | 732.95 | .6157 | 1.10584 |
| 3700 | 998.11 | 2330 | 744.48 | .5882 | 1.10991 |
| 3750 | 1013.1 | 2471 | 756.04 | .5621 | 1.11393 |
| 3800 | 1028.1 | 2618 | 767.60 | .5376 | 1.11791 |
| 3850 | 1043.1 | 2773 | 779.19 | .5143 | 1.12183 |
| 3900 | 1058.1 | 2934 | 790.80 | .4923 | 1.12571 |
| 3950 | 1073.2 | 3103 | 802.43 | .4715 | 1.12955 |
| 4000 | 1088.3 | 3280 | 814.06 | .4518 | 1.13334 |
| 4050 | 1103.4 | 3464 | 825.72 | .4331 | 1.13709 |
| 4100 | 1118.5 | 3656 | 837.40 | .4154 | 1.14079 |
| 4150 | 1133.6 | 3858 | 849.09 | .3985 | 1.14446 |
| 4200 | 1148.7 | 4067 | 860.81 | .3826 | 1.14809 |
| 4300 | 1179.0 | 4513 | 884.28 | .3529 | 1.15522 |
| 4400 | 1209.4 | 4997 | 907.81 | .3262 | 1.16221 |
| 4500 | 1239.9 | 5521 | 931.39 | .3019 | 1.16905 |
| 4600 | 1270.4 | 6089 | 955.04 | .2799 | 1.17575 |
| 4700 | 1300.9 | 6701 | 978.73 | .2598 | 1.18232 |
| 4800 | 1331.5 | 7362 | 1002.5 | .2415 | 1.18876 |
| 4900 | 1362.2 | 8073 | 1026.3 | .2248 | 1.19508 |
| 5000 | 1392.9 | 8837 | 1050.1 | .2096 | 1.20129 |
| 5100 | 1423.6 | 9658 | 1074.0 | .1956 | 1.20738 |
| 5200 | 1454.4 | 10539 | 1098.0 | .1828 | 1.21336 |
| 5300 | 1485.3 | 11481 | 1122.0 | .1710 | 1.21923 |

SOURCE: Data abridged from J. H. Keenan and J. Kaye, "Gas Tables," Wiley, New York, 1945.

**Table A-6E**    Ideal-gas enthalpy, internal energy, and absolute entropy of diatomic nitrogen ($N_2$)

($\Delta \bar{h}_f^\circ = 0$ Btu/lbmol) ($\bar{h}$ and $\bar{u}$, Btu/lbmol; $\bar{s}^\circ$, Btu/(lbmol·°R); std state is 1 atm.)

| $T$, °R | $\bar{h}$ | $\bar{u}$ | $\bar{s}^\circ$ | $T$, °R | $\bar{h}$ | $\bar{u}$ | $\bar{s}^\circ$ |
|---|---|---|---|---|---|---|---|
| 300 | 2,082.0 | 1,486.2 | 41.695 | 1080 | 7,551.0 | 5,406.2 | 50.651 |
| 320 | 2,221.0 | 1,585.5 | 42.143 | 1100 | 7,695.0 | 5,510.5 | 50.783 |
| 340 | 2,360.0 | 1,684.4 | 42.564 | 1120 | 7,839.3 | 5,615.2 | 50.912 |
| 360 | 2,498.9 | 1,784.0 | 42.962 | 1140 | 7,984.0 | 5,720.1 | 51.040 |
| 380 | 2,638.0 | 1,883.4 | 43.337 | 1160 | 8,129.0 | 5,825.4 | 51.167 |
| 400 | 2,777.0 | 1,982.6 | 43.694 | 1180 | 8,274.4 | 5,931.0 | 51.291 |
| 420 | 2,916.1 | 2,082.0 | 44.034 | 1200 | 8,420.0 | 6,037.0 | 51.413 |
| 440 | 3,055.1 | 2,181.3 | 44.357 | 1220 | 8,566.1 | 6,143.4 | 51.534 |
| 460 | 3,194.1 | 2,280.6 | 44.665 | 1240 | 8,712.6 | 6,250.1 | 51.653 |
| 480 | 3,333.1 | 2,379.9 | 44.962 | 1260 | 8,859.3 | 6,357.2 | 51.771 |
| 500 | 3,472.2 | 2,479.3 | 45.246 | 1280 | 9,006.4 | 6,464.5 | 51.887 |
| 520 | 3,611.3 | 2,578.6 | 45.519 | 1300 | 9,153.9 | 6,572.3 | 51.001 |
| 537 | 3,729.5 | 2,663.1 | 45.743 | 1320 | 9,301.8 | 6,680.4 | 52.114 |
| 540 | 3,750.3 | 2,678.0 | 45.781 | 1340 | 9,450.0 | 6,788.9 | 52.225 |
| 560 | 3,889.5 | 2,777.4 | 46.034 | 1360 | 9,598.6 | 6,897.8 | 52.335 |
| 580 | 4,028.7 | 2,876.9 | 46.278 | 1380 | 9,747.5 | 7,007.0 | 52.444 |
| 600 | 4,167.9 | 2,976.4 | 46.514 | 1400 | 9,896.9 | 7,116.7 | 52.551 |
| 620 | 4,307.1 | 3,075.9 | 46.742 | 1420 | 10.046.6 | 7,226.7 | 52.658 |
| 640 | 4,446.4 | 3,175.5 | 46.964 | 1440 | 10,196.6 | 7,337.0 | 52.763 |
| 660 | 4,585.8 | 3,275.2 | 47.178 | 1460 | 10,347.0 | 7,447.6 | 52.867 |
| 680 | 4,725.3 | 3,374.9 | 47.386 | 1480 | 10,497.8 | 7,558.7 | 52.969 |
| 700 | 4,864.9 | 3,474.8 | 47.588 | 1500 | 10,648.0 | 7,670.1 | 53.071 |
| 720 | 5,004.5 | 3,574.7 | 47.785 | 1520 | 10,800.4 | 7,781.9 | 53.171 |
| 740 | 5,144.3 | 3,674.7 | 47.977 | 1540 | 10,952.2 | 7,893.9 | 53.271 |
| 760 | 5,284.1 | 3,774.9 | 48.164 | 1560 | 11,104.3 | 8,006.4 | 53.369 |
| 780 | 5,424.2 | 3,875.2 | 48.345 | 1580 | 11,256.9 | 8,119.2 | 53.465 |
| 800 | 5,564.4 | 3,975.7 | 48.522 | 1600 | 11,409.7 | 8,232.3 | 53.561 |
| 820 | 5,704.7 | 4,076.3 | 48.696 | 1620 | 11,562.8 | 8,345.7 | 53.656 |
| 840 | 5,845.3 | 4,177.1 | 48.865 | 1640 | 11,716.4 | 8,459.6 | 53.751 |
| 860 | 5,985.9 | 4,278.1 | 49.031 | 1660 | 11,870.2 | 8,573.6 | 53.844 |
| 880 | 6,126.9 | 4,379.4 | 49.193 | 1680 | 12,024.3 | 8,688.1 | 53.936 |
| 900 | 6,268.1 | 4,480.8 | 49.352 | 1700 | 12,178.9 | 8,802.9 | 54.028 |
| 920 | 6,409.6 | 4,582.6 | 49.507 | 1720 | 12,333.7 | 8,918.0 | 54.118 |
| 940 | 6,551.2 | 4,684.5 | 49.659 | 1740 | 12,488.8 | 9,033.4 | 54.208 |
| 960 | 6,693.1 | 4,786.7 | 49.808 | 1760 | 12,644.3 | 9,149.2 | 54.297 |
| 980 | 6,835.4 | 4,889.3 | 49.955 | 1780 | 12,800.2 | 9,265.3 | 54.385 |
| 1000 | 6,977.9 | 4,992.0 | 50.099 | 1800 | 12,956.3 | 9,381.7 | 54.472 |
| 1020 | 7,120.7 | 5,095.1 | 50.241 | 1820 | 13,112.7 | 9,498.4 | 54.559 |
| 1040 | 7,263.8 | 5,198.5 | 50.380 | 1840 | 13,269.5 | 9,615.5 | 54.645 |
| 1060 | 7,407.2 | 5,302.2 | 50.516 | 1860 | 13.426.5 | 9,732.8 | 54.729 |

**Table A-6E**     *(Continued)*

| $T$, °R | $\bar{h}$ | $\bar{u}$ | $\bar{s}^o$ | $T$, °R | $\bar{h}$ | $\bar{u}$ | $\bar{s}^o$ |
|---|---|---|---|---|---|---|---|
| 1900 | 13,742 | 9,968 | 54.896 | 3500 | 27,016 | 20,065 | 59.944 |
| 1940 | 14,058 | 10,205 | 55.061 | 3540 | 27,359 | 20,329 | 60.041 |
| 1980 | 14,375 | 10,443 | 55.223 | 3580 | 27,703 | 20,593 | 60.183 |
| 2020 | 14,694 | 10,682 | 55.383 | 3620 | 28,046 | 20,858 | 60.234 |
| 2060 | 15,013 | 10,923 | 55.540 | 3660 | 28,391 | 21,122 | 60.328 |
| 2100 | 15,334 | 11,164 | 55.694 | 3700 | 28,735 | 21,387 | 60.422 |
| 2140 | 15,656 | 11,406 | 55.846 | 3740 | 29,080 | 21,653 | 60.515 |
| 2180 | 15,978 | 11,649 | 55.995 | 3780 | 29,425 | 21,919 | 60.607 |
| 2220 | 16,302 | 11,893 | 56.141 | 3820 | 29,771 | 22,185 | 60.698 |
| 2260 | 16,626 | 12,138 | 56.286 | 3860 | 30,117 | 22,451 | 60.788 |
| 2300 | 16,951 | 12,384 | 56.429 | 3900 | 30,463 | 22,718 | 60.966 |
| 2340 | 17,277 | 12,630 | 56.570 | 3940 | 30,809 | 22,985 | 60.877 |
| 2380 | 17,604 | 12,878 | 56.708 | 3980 | 31,156 | 23,252 | 61.053 |
| 2420 | 17,932 | 13,126 | 56.845 | 4020 | 31,503 | 23,520 | 61.139 |
| 2460 | 18,260 | 13,375 | 56.980 | 4060 | 31,850 | 23,788 | 61.225 |
| 2500 | 18,590 | 13,625 | 57.122 | 4100 | 32,198 | 24,056 | 61.310 |
| 2540 | 18,919 | 13,875 | 57.243 | 4140 | 32,546 | 24,324 | 61.395 |
| 2580 | 19,250 | 14,127 | 57.372 | 4180 | 32,894 | 24,593 | 61.479 |
| 2620 | 19,582 | 14,379 | 57.499 | 4220 | 33,242 | 24,862 | 61.562 |
| 2660 | 19,914 | 14,631 | 57.625 | 4260 | 33,591 | 25,131 | 61.644 |
| 2700 | 20,246 | 14,885 | 57.750 | 4300 | 33,940 | 25,401 | 61.726 |
| 2740 | 20,580 | 15,139 | 57.872 | 4340 | 34,289 | 25,670 | 61.806 |
| 2780 | 20,914 | 15,393 | 57.993 | 4380 | 34,638 | 25,940 | 61.887 |
| 2820 | 21,248 | 15,648 | 58.113 | 4420 | 34,988 | 26,210 | 61.966 |
| 2860 | 21,584 | 15,905 | 58.231 | 4460 | 35,338 | 26,481 | 62.045 |
| 2900 | 21,920 | 16,161 | 58.348 | 4500 | 35,688 | 26,751 | 62.123 |
| 2940 | 22,256 | 16,417 | 58.463 | 4540 | 36,038 | 27,022 | 62.201 |
| 2980 | 22,593 | 16,675 | 58.576 | 4580 | 36,389 | 27,293 | 62.278 |
| 3020 | 22,930 | 16,933 | 58.688 | 4620 | 36,739 | 27,565 | 62.354 |
| 3060 | 23,268 | 17,192 | 58.800 | 4660 | 37,090 | 27,836 | 62.429 |
| 3100 | 23,607 | 17,451 | 58.910 | 4700 | 37,441 | 28,108 | 62.504 |
| 3140 | 23,946 | 17,710 | 59.019 | 4740 | 37,792 | 28,379 | 62.578 |
| 3180 | 24,285 | 17,970 | 59.126 | 4780 | 38,144 | 28,651 | 62.652 |
| 3220 | 24,625 | 18,231 | 59.232 | 4820 | 38,495 | 28,924 | 62.725 |
| 3260 | 24,965 | 18,491 | 59.338 | 4860 | 38,847 | 29,196 | 62.798 |
| 3300 | 25,306 | 18,753 | 59.442 | 4900 | 39,199 | 29,468 | 62.870 |
| 3340 | 25,647 | 19,014 | 59.544 | 5000 | 40,080 | 30,151 | 63.049 |
| 3380 | 25,989 | 19,277 | 59.646 | 5100 | 40,962 | 30,834 | 63.223 |
| 3420 | 26,331 | 19,539 | 59.747 | 5200 | 41,844 | 31,518 | 63.395 |
| 3460 | 26,673 | 19,802 | 59.846 | 5300 | 42,728 | 32,203 | 63.563 |

*SOURCE:* Data abridged from J. H. Keenan and J. Kaye, "Gas Tables," Wiley, New York, 1945.

**Table A-7E     Ideal-gas enthalpy, internal energy, and absolute entropy of diatomic oxygen (O₂)**

$(\Delta \overline{h}_f^{\circ} = 0$ Btu/lbmol; $\overline{h}$ and $\overline{u}$, Btu/lbmol; $\overline{s}^{\circ}$, Btu/(lbmol·°R))

| $T$, °R | $\overline{h}$ | $\overline{u}$ | $\overline{s}^{\circ}$ | $T$, °R | $\overline{h}$ | $\overline{u}$ | $\overline{s}^{\circ}$ |
|---|---|---|---|---|---|---|---|
| 300 | 2,073.5 | 1,477.8 | 44.927 | 1080 | 7,696.8 | 5,552.1 | 54.064 |
| 320 | 2,212.6 | 1,577.1 | 45.375 | 1100 | 7,850.4 | 5,665.9 | 54.204 |
| 340 | 2,351.7 | 1,676.5 | 45.797 | 1120 | 8,004.5 | 5,780.3 | 54.343 |
| 360 | 2,490.8 | 1,775.9 | 46.195 | 1140 | 8,159.1 | 5,895.2 | 54.480 |
| 380 | 2,630.0 | 1,875.3 | 46.571 | 1160 | 8,314.2 | 6,010.6 | 54.614 |
| 400 | 2,769.1 | 1,974.8 | 46.927 | 1180 | 8,469.8 | 6,126.5 | 54.748 |
| 420 | 2,908.3 | 2,074.3 | 47.267 | 1200 | 8,625.8 | 6,242.8 | 54.879 |
| 440 | 3,047.5 | 2,173.8 | 47.591 | 1220 | 8,782.4 | 6,359.6 | 55.008 |
| 460 | 3,186.9 | 2,273.4 | 47.900 | 1240 | 8,939.4 | 6,476.9 | 55.136 |
| 480 | 3,326.5 | 2,373.3 | 48.198 | 1260 | 9,096.7 | 6,594.5 | 55.262 |
| 500 | 3,466.2 | 2,473.2 | 48.483 | 1280 | 9,254.6 | 6,712.7 | 55.386 |
| 520 | 3,606.1 | 2,573.4 | 48.757 | 1300 | 9,412.9 | 6,831.3 | 55.508 |
| 537 | 3,725.1 | 2,658.7 | 48.982 | 1320 | 9,571.6 | 6,950.2 | 55.630 |
| 540 | 3,746.2 | 2,673.8 | 49.021 | 1340 | 9,730.7 | 7,069.6 | 55.750 |
| 560 | 3,886.6 | 2,774.5 | 49.276 | 1360 | 9,890.2 | 7,189.4 | 55.867 |
| 580 | 4,027.3 | 2,875.5 | 49.522 | 1380 | 10,050.1 | 7,309.6 | 55.984 |
| 600 | 4,168.3 | 2,976.8 | 49.762 | 1400 | 10,210.4 | 7,430.1 | 56.099 |
| 620 | 4,309.7 | 3,078.4 | 49.993 | 1420 | 10,371.0 | 7,551.1 | 56.213 |
| 640 | 4,451.4 | 3,180.4 | 50.218 | 1440 | 10,532.0 | 7,672.4 | 56.326 |
| 660 | 4,593.5 | 3,282.9 | 50.437 | 1460 | 10,693.3 | 7,793.9 | 56.437 |
| 680 | 4,736.2 | 3,385.8 | 50.650 | 1480 | 10,855.1 | 7,916.0 | 56.547 |
| 700 | 4,879.3 | 3,489.2 | 50.858 | 1500 | 11,017.1 | 8,038.3 | 56.656 |
| 720 | 5,022.9 | 3,593.1 | 51.059 | 1520 | 11,179.6 | 8,161.1 | 56.763 |
| 740 | 5,167.0 | 3,697.4 | 51.257 | 1540 | 11,342.4 | 8,284.2 | 56.869 |
| 760 | 5,311.4 | 3,802.2 | 51.450 | 1560 | 11,505.4 | 8,407.4 | 56.975 |
| 780 | 5,456.4 | 3,907.5 | 51.638 | 1580 | 11,668.8 | 8,531.1 | 57.079 |
| 800 | 5,602.0 | 4,013.3 | 51.821 | 1600 | 11,832.5 | 8,655.1 | 57.182 |
| 820 | 5,748.1 | 4,119.7 | 52.002 | 1620 | 11,996.6 | 8,779.5 | 57.284 |
| 840 | 5,894.8 | 4,226.6 | 52.179 | 1640 | 12,160.9 | 8,904.1 | 57.385 |
| 860 | 6,041.9 | 4,334.1 | 52.352 | 1660 | 12,325.5 | 9,029.0 | 57.484 |
| 880 | 6,189.6 | 4,442.0 | 52.522 | 1680 | 12,490.4 | 9,154.1 | 57.582 |
| 900 | 6,337.9 | 4,550.6 | 52.688 | 1700 | 12,655.6 | 9,279.6 | 57.680 |
| 920 | 6,486.7 | 4,659.7 | 52.852 | 1720 | 12,821.1 | 9,405.4 | 57.777 |
| 940 | 6,636.1 | 4,769.4 | 53.012 | 1740 | 12,989.9 | 9,531.5 | 57.873 |
| 960 | 6,786.0 | 4,879.5 | 53.170 | 1760 | 13,153.0 | 9,657.9 | 57.968 |
| 980 | 6,936.4 | 4,990.3 | 53.326 | 1780 | 13,319.2 | 9,784.4 | 58.062 |
| 1000 | 7,087.5 | 5,101.6 | 53.477 | 1800 | 13,485.8 | 9,911.2 | 58.155 |
| 1020 | 7,238.9 | 5,213.3 | 53.628 | 1820 | 13,652.5 | 10,038.2 | 58.247 |
| 1040 | 7,391.0 | 5,325.7 | 53.775 | 1840 | 13,819.6 | 10,165.6 | 58.339 |
| 1060 | 7,543.6 | 5,438.6 | 53.921 | 1860 | 13,986.8 | 10,293.1 | 58.428 |

**Table A-7E**    (*Continued*)

| $T$, °R | $\bar{h}$ | $\bar{u}$ | $\bar{s}^o$ | $T$, °R | $\bar{h}$ | $\bar{u}$ | $\bar{s}^o$ |
|---|---|---|---|---|---|---|---|
| 1900 | 14,322 | 10,549 | 58.607 | 3500 | 28,273 | 21,323 | 63.914 |
| 1940 | 14,658 | 10,806 | 58.782 | 3540 | 28,633 | 21,603 | 64.016 |
| 1980 | 14,995 | 11,063 | 58.954 | 3580 | 28,994 | 21,884 | 64.114 |
| 2020 | 15,333 | 11,321 | 59.123 | 3620 | 29,354 | 22,165 | 64.217 |
| 2060 | 15,672 | 11,581 | 59.289 | 3660 | 29,716 | 22,447 | 64.316 |
| 2100 | 16,011 | 11,841 | 59.451 | 3700 | 30,078 | 22,730 | 64.415 |
| 2140 | 16,351 | 12,101 | 59.612 | 3740 | 30,440 | 23,013 | 64.512 |
| 2180 | 16,692 | 12,363 | 59.770 | 3780 | 30,803 | 23,296 | 64.609 |
| 2220 | 17,036 | 12,625 | 59.926 | 3820 | 31,166 | 23,580 | 64.704 |
| 2260 | 17,376 | 12,888 | 60.077 | 3860 | 31,529 | 23,864 | 64.800 |
| 2300 | 17,719 | 13,151 | 60.228 | 3900 | 31,894 | 24,149 | 64.893 |
| 2340 | 18,062 | 13,416 | 60.376 | 3940 | 32,258 | 24,434 | 64.986 |
| 2380 | 18,407 | 13,680 | 60.522 | 3980 | 32,623 | 24,720 | 65.078 |
| 2420 | 18,572 | 13,946 | 60.666 | 4020 | 32,989 | 25,006 | 65.169 |
| 2460 | 19,097 | 14,212 | 60.808 | 4060 | 33,355 | 25,292 | 65.260 |
| 2500 | 19,443 | 14,479 | 60.946 | 4100 | 33,722 | 25,580 | 65.350 |
| 2540 | 19,790 | 14,746 | 61.084 | 4140 | 34,089 | 25,867 | 64.439 |
| 2580 | 20,138 | 15,014 | 61.220 | 4180 | 34,456 | 26,155 | 65.527 |
| 2620 | 20,485 | 15,282 | 61.354 | 4220 | 34,824 | 26,444 | 65.615 |
| 2660 | 20,834 | 15,551 | 61.486 | 4260 | 35,192 | 26,733 | 65.702 |
| 2700 | 21,183 | 15,821 | 61.616 | 4300 | 35,561 | 27,022 | 65.788 |
| 2740 | 21,533 | 16,091 | 61.744 | 4340 | 35,930 | 27,312 | 65.873 |
| 2780 | 21,883 | 16,362 | 61.871 | 4380 | 36,300 | 27,602 | 65.958 |
| 2820 | 22,232 | 16,633 | 61.996 | 4420 | 36,670 | 27,823 | 66.042 |
| 2860 | 22,584 | 16,905 | 62.120 | 4460 | 37,041 | 28,184 | 66.125 |
| 2900 | 22,936 | 17,177 | 62.242 | 4500 | 37,412 | 28,475 | 66.208 |
| 2940 | 23,288 | 17,450 | 62.363 | 4540 | 37,783 | 28,768 | 66.290 |
| 2980 | 23,641 | 17,723 | 62.483 | 4580 | 38,155 | 29,060 | 66.372 |
| 3020 | 23,994 | 17,997 | 62.599 | 4620 | 38,528 | 29,353 | 66.453 |
| 3060 | 24,348 | 18,271 | 62.716 | 4660 | 38,900 | 29,646 | 66.533 |
| 3100 | 24,703 | 18,546 | 62.831 | 4700 | 39,274 | 29,940 | 66.613 |
| 3140 | 25,057 | 18,822 | 62.945 | 4740 | 39,647 | 30,234 | 66.691 |
| 3180 | 25,413 | 19,098 | 63.057 | 4780 | 40,021 | 30,529 | 66.770 |
| 3220 | 25,769 | 19,374 | 63.169 | 4820 | 40,396 | 30,824 | 66.848 |
| 3260 | 26,175 | 19,651 | 63.279 | 4860 | 40,771 | 31,120 | 66.925 |
| 3300 | 26,412 | 19,928 | 63.386 | 4900 | 41,146 | 31,415 | 67.003 |
| 3340 | 26,839 | 20,206 | 63.494 | 5000 | 42,086 | 32,157 | 67.193 |
| 3380 | 27,197 | 20,485 | 63.601 | 5100 | 43,021 | 32,901 | 67.380 |
| 3420 | 27,555 | 20,763 | 63.706 | 5200 | 43,974 | 33,648 | 67.562 |
| 3460 | 27,914 | 21,043 | 63.811 | 5300 | 44,922 | 34,397 | 67.743 |

| *SOURCE:* Data abridged from J. H. Keenan and J. Kaye, "Gas Tables," Wiley, New York, 1945.

**Table A-8E** Ideal-gas enthalpy, internal energy, and absolute entropy of carbon monoxide (CO)

$(\Delta \bar{h}_f^\circ = -47{,}540$ Btu/lbmol) $(\bar{h}$ and $\bar{u}$, Btu/lbmol; $\bar{s}^\circ$, Btu/(lbmol·°R); std state is 1 atm.)

| $T$, °R | $\bar{h}$ | $\bar{u}$ | $\bar{s}^\circ$ | $T$, °R | $\bar{h}$ | $\bar{u}$ | $\bar{s}^\circ$ |
|---|---|---|---|---|---|---|---|
| 300 | 2,081.9 | 1,486.1 | 43.223 | 1080 | 7,571.1 | 5,426.4 | 52.203 |
| 320 | 2,220.9 | 1,585.4 | 43.672 | 1100 | 7,716.8 | 5,532.3 | 52.237 |
| 340 | 2,359.9 | 1,684.7 | 44.093 | 1120 | 7,862.9 | 5,638.7 | 52.468 |
| 360 | 2,498.8 | 1,783.9 | 44.490 | 1140 | 8,009.2 | 5,745.4 | 52.598 |
| 380 | 2,637.9 | 1,883.3 | 44.866 | 1160 | 8,156.1 | 5,851.5 | 52.726 |
| 400 | 2,776.9 | 1,982.6 | 45.223 | 1180 | 8,303.3 | 5,960.0 | 52.852 |
| 420 | 2,916.0 | 2,081.9 | 45.563 | 1200 | 8,450.8 | 6,067.8 | 52.976 |
| 440 | 3,055.0 | 2,181.2 | 45.886 | 1220 | 8,598.8 | 6,176.0 | 53.098 |
| 460 | 3,194.0 | 2,280.5 | 46.194 | 1240 | 8,747.2 | 6,284.7 | 53.218 |
| 480 | 3,333.0 | 2,379.8 | 46.491 | 1260 | 8,896.0 | 6,393.8 | 53.337 |
| 500 | 3,472.1 | 2,479.2 | 46.775 | 1280 | 9,045.0 | 6,503.1 | 53.455 |
| 520 | 3,611.2 | 2,578.6 | 47.048 | 1300 | 9,194.6 | 6,613.0 | 53.571 |
| 537 | 3,725.1 | 2,663.1 | 47.272 | 1320 | 9,344.6 | 6,723.2 | 53.685 |
| 540 | 3,750.3 | 2,677.9 | 47.310 | 1340 | 9,494.8 | 6,833.7 | 53.799 |
| 560 | 3,889.5 | 2,777.4 | 47.563 | 1360 | 9,645.5 | 6,944.7 | 53.910 |
| 580 | 4,028.7 | 2,876.9 | 47.807 | 1380 | 9,796.6 | 7,056.1 | 54.021 |
| 600 | 4,168.0 | 2,976.5 | 48.044 | 1400 | 9,948.1 | 7,167.9 | 54.129 |
| 620 | 4,307.4 | 3,076.2 | 48.272 | 1420 | 10,100.0 | 7,280.1 | 54.237 |
| 640 | 4,446.9 | 3,175.9 | 48.494 | 1440 | 10,252.2 | 7,392.6 | 54.344 |
| 660 | 4,586.6 | 3,275.8 | 48.709 | 1460 | 10,404.8 | 7,750.4 | 54.448 |
| 680 | 4,726.2 | 3,375.8 | 48.917 | 1480 | 10,557.8 | 7,618.7 | 54.522 |
| 700 | 4,866.0 | 3,475.9 | 49.120 | 1500 | 10,711.1 | 7,732.3 | 54.665 |
| 720 | 5,006.1 | 3,576.3 | 49.317 | 1520 | 10,864.9 | 7,846.4 | 54.757 |
| 740 | 5,146.4 | 3,676.9 | 49.509 | 1540 | 11,019.0 | 7,960.8 | 54.858 |
| 760 | 5,286.8 | 3,777.5 | 49.697 | 1560 | 11,173.4 | 8,075.4 | 54.958 |
| 780 | 5,427.4 | 3,878.4 | 49.880 | 1580 | 11,328.2 | 8,190.5 | 55.056 |
| 800 | 5,568.2 | 3,979.5 | 50.058 | 1600 | 11,483.4 | 8,306.0 | 55.154 |
| 820 | 5,709.4 | 4,081.0 | 50.232 | 1620 | 11,638.9 | 8,421.8 | 55.251 |
| 840 | 5,850.7 | 4,182.6 | 50.402 | 1640 | 11,794.7 | 8,537.9 | 55.347 |
| 860 | 5,992.3 | 4,284.5 | 50.569 | 1660 | 11,950.9 | 8,654.4 | 55.411 |
| 880 | 6,134.2 | 4,386.6 | 50.732 | 1680 | 12,107.5 | 8,771.2 | 55.535 |
| 900 | 6,276.4 | 4,489.1 | 50.892 | 1700 | 12,264.3 | 8,888.3 | 55.628 |
| 920 | 6,419.0 | 4,592.0 | 51.048 | 1720 | 12,421.4 | 9,005.7 | 55.720 |
| 940 | 6,561.7 | 4,695.0 | 51.202 | 1740 | 12,579.0 | 9,123.6 | 55.811 |
| 960 | 6,704.9 | 4,798.5 | 51.353 | 1760 | 12,736.7 | 9,241.6 | 55.900 |
| 980 | 6,848.4 | 4,902.3 | 51.501 | 1780 | 12,894.9 | 9,360.0 | 55.990 |
| 1000 | 6,992.2 | 5,006.3 | 51.646 | 1800 | 13,053.2 | 9,478.6 | 56.078 |
| 1020 | 7,136.4 | 5,110.8 | 51.788 | 1820 | 13,212.0 | 9,597.7 | 56.166 |
| 1040 | 7,281.0 | 5,215.7 | 51.929 | 1840 | 13,371.0 | 9,717.0 | 56.253 |
| 1060 | 7,425.9 | 5,320.9 | 52.067 | 1860 | 13,530.2 | 9,836.5 | 56.339 |

**Table A-8E**     *(Continued)*

| $T$, °R | $\bar{h}$ | $\bar{u}$ | $\bar{s}^o$ | $T$, °R | $\bar{h}$ | $\bar{u}$ | $\bar{s}^o$ |
|---|---|---|---|---|---|---|---|
| 1900 | 13,850 | 10,077 | 56.509 | 3500 | 27,262 | 20,311 | 61.612 |
| 1940 | 14,170 | 10,318 | 56.677 | 3540 | 27,608 | 20,576 | 61.710 |
| 1980 | 14,492 | 10,560 | 56.841 | 3580 | 27,954 | 20,844 | 61.807 |
| 2020 | 14,815 | 10,803 | 57.007 | 3620 | 28,300 | 21,111 | 61.903 |
| 2060 | 15,139 | 11,048 | 57.161 | 3660 | 28,647 | 21,378 | 61.998 |
| 2100 | 15,463 | 11,293 | 57.317 | 3700 | 28,994 | 21,646 | 62.093 |
| 2140 | 15,789 | 11,539 | 57.470 | 3740 | 29,341 | 21,914 | 62.186 |
| 2180 | 16,116 | 11,787 | 57.621 | 3780 | 29,688 | 22,182 | 62.279 |
| 2220 | 16,443 | 12,035 | 57.770 | 3820 | 30,036 | 22,450 | 62.370 |
| 2260 | 16,722 | 12,284 | 57.917 | 3860 | 30,384 | 22,719 | 62.461 |
| 2300 | 17,101 | 12,534 | 58.062 | 3900 | 30,733 | 22,988 | 62.511 |
| 2340 | 17,431 | 12,784 | 58.204 | 3940 | 31,082 | 23,257 | 62.640 |
| 2380 | 17,762 | 13,035 | 58.344 | 3980 | 31,431 | 23,527 | 62.728 |
| 2420 | 18,093 | 13,287 | 58.482 | 4020 | 31,780 | 23,797 | 62.816 |
| 2460 | 18,426 | 13,541 | 58.619 | 4060 | 32,129 | 24,067 | 62.902 |
| 2500 | 18,759 | 13,794 | 58.754 | 4100 | 32,479 | 24,337 | 62.988 |
| 2540 | 19,093 | 14,048 | 58.885 | 4140 | 32,829 | 24,608 | 63.072 |
| 2580 | 19,427 | 14,303 | 59.016 | 4180 | 33,179 | 24,878 | 63.156 |
| 2620 | 19,762 | 14,559 | 59.145 | 4220 | 33,530 | 25,149 | 63.240 |
| 2660 | 20,098 | 14,815 | 59.272 | 4260 | 33,880 | 25,421 | 63.323 |
| 2700 | 20,434 | 15,072 | 59.398 | 4300 | 34,231 | 25,692 | 63.405 |
| 2740 | 20,771 | 15,330 | 59.521 | 4340 | 34,582 | 25,934 | 63.486 |
| 2780 | 21,108 | 15,588 | 59.644 | 4380 | 34,934 | 26,235 | 63.567 |
| 2820 | 21,446 | 15,846 | 59.765 | 4420 | 35,285 | 26,508 | 53.647 |
| 2860 | 21,785 | 16,105 | 59.884 | 4460 | 35,637 | 26,780 | 63.726 |
| 2900 | 22,124 | 16,365 | 60.002 | 4500 | 35,989 | 27,052 | 63.805 |
| 2940 | 22,463 | 16,225 | 60.118 | 4540 | 36,341 | 27,325 | 63.883 |
| 2980 | 22,803 | 16,885 | 60.232 | 4580 | 36,693 | 27,598 | 63.960 |
| 3020 | 23,144 | 17,146 | 60.346 | 4620 | 37,046 | 27,871 | 64.036 |
| 3060 | 23,485 | 17,408 | 60.458 | 4660 | 37,398 | 28,144 | 64.113 |
| 3100 | 23,826 | 17,670 | 60.569 | 4700 | 37,751 | 28,417 | 64.188 |
| 3140 | 24,168 | 17,932 | 60.679 | 4740 | 38,104 | 28,691 | 64.263 |
| 3180 | 24,510 | 18,195 | 60.787 | 4780 | 38,457 | 28,965 | 64.337 |
| 3220 | 24,853 | 18,458 | 60.894 | 4820 | 38,811 | 29,239 | 64.411 |
| 3260 | 25,196 | 18,722 | 61.000 | 4860 | 39,164 | 29,513 | 64.484 |
| 3300 | 25,539 | 18,986 | 61.105 | 4900 | 39,518 | 29,787 | 64.556 |
| 3340 | 25,883 | 19,250 | 61.209 | 5000 | 40,403 | 30,473 | 64.735 |
| 3380 | 26,227 | 19,515 | 61.311 | 5100 | 41,289 | 31,161 | 64.910 |
| 3420 | 26,572 | 19,780 | 61.412 | 5200 | 42,176 | 31,849 | 65.082 |
| 3460 | 26,917 | 20,045 | 61.513 | 5300 | 43,063 | 32,538 | 65.252 |

| *SOURCE:* Data abridged from J. H. Keenan and J. Kaye, "Gas Tables," Wiley, New York, 1945.

**Table A-9E**  Ideal-gas enthalpy, internal energy, and absolute entropy of carbon dioxide ($CO_2$)

($\Delta \overline{h}_f^\circ = -169{,}290$ Btu/lbmol) ($\overline{h}$ and $\overline{u}$, Btu/lbmol; $\overline{s}^\circ$, Btu/(lbmol·°R); std state is 1 atm.)

| $T$, °R | $\overline{h}$ | $\overline{u}$ | $\overline{s}^\circ$ | $T$, °R | $\overline{h}$ | $\overline{u}$ | $\overline{s}^\circ$ |
|---|---|---|---|---|---|---|---|
| 300 | 2,108.2 | 1,512.4 | 46.353 | 1080 | 9,575.8 | 7,431.1 | 58.072 |
| 320 | 2,256.6 | 1,621.1 | 46.832 | 1100 | 9,802.6 | 7,618.1 | 58.281 |
| 340 | 2,407.3 | 1,732.1 | 47.289 | 1120 | 10,030.6 | 7,806.4 | 58.485 |
| 360 | 2,560.5 | 1,845.6 | 47.728 | 1140 | 10,260.1 | 7,996.2 | 58.689 |
| 380 | 2,716.4 | 1,961.8 | 48.148 | 1160 | 10,490.6 | 8,187.0 | 58.889 |
| 400 | 2,874.7 | 2,080.4 | 48.555 | 1180 | 10,722.3 | 8,379.0 | 59.088 |
| 420 | 3,035.7 | 2,201.7 | 48.947 | 1200 | 10,955.3 | 8,572.3 | 59.283 |
| 440 | 3,199.4 | 2,325.6 | 49.329 | 1220 | 11,189.4 | 8,766.6 | 59.477 |
| 460 | 3,365.7 | 2,452.2 | 49.698 | 1240 | 11,424.6 | 6,962.1 | 59.668 |
| 480 | 3,534.7 | 2,581.5 | 50.058 | 1260 | 11,661.0 | 8,158.8 | 59.858 |
| 500 | 3,706.2 | 2,713.3 | 50.408 | 1280 | 11,898.4 | 9,356.5 | 60.044 |
| 520 | 3,880.3 | 2,847.7 | 50.750 | 1300 | 12,136.9 | 9,555.3 | 60.229 |
| 537 | 4,027.5 | 2,963.8 | 51.032 | 1320 | 12,376.4 | 9,755.0 | 60.412 |
| 540 | 4,056.8 | 2,984.4 | 51.082 | 1340 | 12,617.0 | 9,955.9 | 60.593 |
| 560 | 4,235.8 | 3,123.7 | 51.408 | 1360 | 12,858.5 | 9,157.7 | 60.772 |
| 580 | 4,417.2 | 3,265.4 | 51.726 | 1380 | 13,101.0 | 10,360.5 | 60.949 |
| 600 | 4,600.9 | 3,409.4 | 52.038 | 1400 | 13,344.7 | 10,564.5 | 61.124 |
| 620 | 4,786.6 | 3,555.6 | 52.343 | 1420 | 13,589.1 | 10,769.2 | 61.298 |
| 640 | 4,974.9 | 3,704.0 | 52.641 | 1440 | 13,834.5 | 10,974.8 | 61.469 |
| 660 | 5,165.2 | 3,854.6 | 52.934 | 1460 | 14,080.8 | 11,181.4 | 61.639 |
| 680 | 5,357.6 | 4,007.2 | 53.225 | 1480 | 14,328.0 | 11,388.9 | 61.800 |
| 700 | 5,552.0 | 4,161.9 | 53.503 | 1500 | 14,576.0 | 11,597.2 | 61.974 |
| 720 | 5,748.4 | 4,318.6 | 53.780 | 1520 | 14,824.9 | 11,806.4 | 62.138 |
| 740 | 5,946.8 | 4,477.3 | 54.051 | 1540 | 15,074.7 | 12,016.5 | 62.302 |
| 760 | 6,147.0 | 4,637.9 | 54.319 | 1560 | 15,325.3 | 12,227.3 | 62.464 |
| 780 | 6,349.1 | 4,800.1 | 54.582 | 1580 | 15,576.7 | 12,439.0 | 62.624 |
| 800 | 6,552.9 | 4,964.2 | 54.839 | 1600 | 15,829.0 | 12,651.6 | 62.783 |
| 820 | 6,758.3 | 5,129.9 | 55.093 | 1620 | 16,081.9 | 12,864.8 | 62.939 |
| 840 | 6,965.7 | 5,297.6 | 55.343 | 1640 | 16,335.7 | 12,078.9 | 63.095 |
| 860 | 7,174.7 | 5,466.9 | 55.589 | 1660 | 16,590.2 | 13,293.7 | 63.250 |
| 880 | 7,385.3 | 5,637.7 | 55.831 | 1680 | 16,845.5 | 13,509.2 | 63.403 |
| 900 | 7,597.6 | 5,810.3 | 56.070 | 1700 | 17,101.4 | 13,725.4 | 63.555 |
| 920 | 7,811.4 | 5,984.4 | 56.305 | 1720 | 17,358.1 | 13,942.4 | 63.704 |
| 940 | 8,026.8 | 6,160.1 | 56.536 | 1740 | 17,615.5 | 14,160.1 | 63.853 |
| 960 | 8,243.8 | 6,337.4 | 56.765 | 1760 | 17,873.5 | 14,378.4 | 64.001 |
| 980 | 8,462.2 | 6,516.1 | 56.990 | 1780 | 18,132.2 | 14,597.4 | 64.147 |
| 1000 | 8,682.1 | 6,696.2 | 57.212 | 1800 | 18,391.5 | 14,816.9 | 64.292 |
| 1020 | 8,903.4 | 6,877.8 | 57.432 | 1820 | 18,651.5 | 15,037.2 | 64.435 |
| 1040 | 9,126.2 | 7,060.9 | 57.647 | 1840 | 18,912.2 | 15,258.2 | 64.578 |
| 1060 | 9,350.3 | 7,245.3 | 57.861 | 1860 | 19,173.4 | 15,479.7 | 64.719 |

**Table A-9E** *(Continued)*

| $T$, °R | $\bar{h}$ | $\bar{u}$ | $\bar{s}^o$ | $T$, °R | $\bar{h}$ | $\bar{u}$ | $\bar{s}^o$ |
|---|---|---|---|---|---|---|---|
| 1900 | 19,698 | 15,925 | 64.999 | 3500 | 41,965 | 35,015 | 73.462 |
| 1940 | 20,224 | 16,372 | 65.272 | 3540 | 42,543 | 35,513 | 73.627 |
| 1980 | 20,753 | 16,821 | 65.543 | 3580 | 43,121 | 36,012 | 73.789 |
| 2020 | 21,284 | 17,273 | 65.809 | 3620 | 43,701 | 36,512 | 73.951 |
| 2060 | 21,818 | 17,727 | 66.069 | 3660 | 44,280 | 37,012 | 74.110 |
| 2100 | 22,353 | 18,182 | 66.327 | 3700 | 44,861 | 37,513 | 74.267 |
| 2140 | 22,890 | 18,640 | 66.581 | 3740 | 45,442 | 38,014 | 74.423 |
| 2180 | 23,429 | 19,101 | 66.830 | 3780 | 46,023 | 38,517 | 74.578 |
| 2220 | 23,970 | 19,561 | 67.076 | 3820 | 46,605 | 39,019 | 74.732 |
| 2260 | 24,512 | 20,024 | 67.319 | 3860 | 47,188 | 39,522 | 74.884 |
| 2300 | 25,056 | 20,489 | 67.557 | 3900 | 47,771 | 40,026 | 75.033 |
| 2340 | 25,602 | 20,955 | 67.792 | 3940 | 48,355 | 40,531 | 75.182 |
| 2380 | 26,150 | 21,423 | 68.025 | 3980 | 48,939 | 41,035 | 75.330 |
| 2420 | 26,699 | 21,893 | 68.253 | 4020 | 49,524 | 41,541 | 75.477 |
| 2460 | 27,249 | 22,364 | 68.479 | 4060 | 50,109 | 42,047 | 75.622 |
| 2500 | 27,801 | 22,837 | 68.702 | 4100 | 50,695 | 42,553 | 75.765 |
| 2540 | 28,355 | 23,310 | 68.921 | 4140 | 51,282 | 43,060 | 75.907 |
| 2580 | 28,910 | 23,786 | 69.138 | 4180 | 51,868 | 43,568 | 76.048 |
| 2620 | 29,465 | 24,262 | 69.352 | 4220 | 52,456 | 44,075 | 76.188 |
| 2660 | 30,023 | 24,740 | 69.563 | 4260 | 53,044 | 44,584 | 76.327 |
| 2700 | 30,581 | 25,220 | 69.771 | 4300 | 53,632 | 45,093 | 76.464 |
| 2740 | 31,141 | 25,701 | 69.977 | 4340 | 54,221 | 45,602 | 76.601 |
| 2780 | 31,702 | 26,181 | 70.181 | 4380 | 54,810 | 46,112 | 76.736 |
| 2820 | 32,264 | 26,664 | 70.382 | 4420 | 55,400 | 46,622 | 76.870 |
| 2860 | 32,827 | 27,148 | 70.580 | 4460 | 55,990 | 47,133 | 77.003 |
| 2900 | 33,392 | 27,633 | 70.776 | 4500 | 56,581 | 47,645 | 77.135 |
| 2940 | 33,957 | 28,118 | 70.970 | 4540 | 57,172 | 48,156 | 77.266 |
| 2980 | 34,523 | 28,605 | 71.160 | 4580 | 57,764 | 48,668 | 77.395 |
| 3020 | 35,090 | 29,093 | 71.350 | 4620 | 58,356 | 49,181 | 77.581 |
| 3060 | 35,659 | 29,582 | 71.537 | 4660 | 58,948 | 49,694 | 77.652 |
| 3100 | 36,228 | 30,072 | 71.722 | 4700 | 59,541 | 50,208 | 77.779 |
| 3140 | 36,798 | 30,562 | 71.904 | 4740 | 60,134 | 50,721 | 77.905 |
| 3180 | 37,369 | 31,054 | 72.085 | 4780 | 60,728 | 51,236 | 78.029 |
| 3220 | 37,941 | 31,546 | 72.264 | 4820 | 61,322 | 51,750 | 78.153 |
| 3260 | 38,513 | 32,039 | 72.441 | 4860 | 61,916 | 52,265 | 78.276 |
| 3300 | 39,087 | 32,533 | 72.616 | 4900 | 62,511 | 52,781 | 78.398 |
| 3340 | 39,661 | 33,028 | 72.788 | 5000 | 64,000 | 54,071 | 78.698 |
| 3380 | 40,236 | 33,524 | 72.960 | 5100 | 65,491 | 55,363 | 78.994 |
| 3420 | 40,812 | 34,020 | 73.129 | 5200 | 66,984 | 56,658 | 79.284 |
| 3460 | 41,388 | 34,517 | 73.297 | 5300 | 68,471 | 57,954 | 79.569 |

| *SOURCE:* Data abridged from J. H. Keenan and J. Kaye, "Gas Tables," Wiley, New York, 1945.

**Table A-10E**   Ideal-gas enthalpy, internal energy, and absolute entropy of water ($H_2O$)

($\Delta \overline{h}_f^\circ = -104{,}040$ Btu/lbmol) ($\overline{h}$ and $\overline{u}$, Btu/lbmol; $\overline{s}^\circ$, Btu/(lbmol·°R); std state is 1 atm.)

| $T$, °R | $\overline{h}$ | $\overline{u}$ | $\overline{s}^\circ$ | $T$, °R | $\overline{h}$ | $\overline{u}$ | $\overline{s}^\circ$ |
|---|---|---|---|---|---|---|---|
| 300 | 2,367.6 | 1,771.8 | 40.439 | 1080 | 8,768.2 | 6,623.5 | 50.854 |
| 320 | 2,526.8 | 1,891.3 | 40.952 | 1100 | 8,942.0 | 6,757.5 | 51.013 |
| 340 | 2,686.0 | 2,010.8 | 41.435 | 1120 | 9,116.4 | 6,892.2 | 51.171 |
| 360 | 2,845.1 | 2,130.2 | 41.889 | 1140 | 9,291.4 | 7,027.5 | 51.325 |
| 380 | 3,004.4 | 2,249.8 | 42.320 | 1160 | 9,467.1 | 7,163.5 | 51.478 |
| 400 | 3,163.8 | 2,369.4 | 42.728 | 1180 | 9,643.4 | 7,300.1 | 51.630 |
| 420 | 3,323.2 | 2,489.1 | 43.117 | 1200 | 9,820.4 | 7,437.4 | 51.777 |
| 440 | 3,482.7 | 2,608.9 | 43.487 | 1220 | 9,998.0 | 7,575.2 | 51.925 |
| 460 | 3,642.3 | 2,728.8 | 43.841 | 1240 | 10,176.1 | 7,713.6 | 55.070 |
| 480 | 3,802.0 | 2,848.8 | 44.182 | 1260 | 10,354.9 | 7,852.7 | 52.212 |
| 500 | 3,962.0 | 2,969.1 | 44.508 | 1280 | 10,534.4 | 7,992.5 | 52.354 |
| 520 | 4,122.0 | 3,089.4 | 44.821 | 1300 | 10,714.5 | 8,132.9 | 52.494 |
| 537 | 4,258.0 | 3,191.9 | 45.079 | 1320 | 10,895.3 | 8,274.0 | 52.631 |
| 540 | 4,282.4 | 3,210.0 | 45.124 | 1340 | 11,076.6 | 8,415.5 | 52.768 |
| 560 | 4,442.8 | 3,330.7 | 45.415 | 1360 | 11,258.7 | 8,557.9 | 52.903 |
| 580 | 4,603.7 | 3,451.9 | 45.696 | 1380 | 11,441.4 | 8,700.9 | 53.037 |
| 600 | 4,764.7 | 3,573.2 | 45.970 | 1400 | 11,624.8 | 8,844.6 | 53.168 |
| 620 | 4,926.1 | 3,694.9 | 46.235 | 1420 | 11,808.8 | 8,988.9 | 53.299 |
| 640 | 5,087.8 | 3,816.8 | 46.492 | 1440 | 11,993.4 | 9,133.8 | 53.428 |
| 660 | 5,250.0 | 3,939.3 | 46.741 | 1460 | 12,178.8 | 9,279.4 | 53.556 |
| 680 | 5,412.5 | 4,062.1 | 46.984 | 1480 | 12,364.8 | 9,425.7 | 53.682 |
| 700 | 5,575.4 | 4,185.3 | 47.219 | 1500 | 12,551.4 | 9,572.7 | 53.808 |
| 720 | 5,738.8 | 4,309.0 | 47.450 | 1520 | 12,738.8 | 9,720.3 | 53.932 |
| 740 | 5,902.6 | 4,433.1 | 47.673 | 1540 | 12,926.8 | 9,868.6 | 54.055 |
| 760 | 6,066.9 | 4,557.6 | 47.893 | 1560 | 13,115.6 | 10,017.6 | 54.117 |
| 780 | 6,231.7 | 4,682.7 | 48.106 | 1580 | 13,305.0 | 10,167.3 | 54.298 |
| 800 | 6,396.9 | 4,808.2 | 48.316 | 1600 | 13,494.4 | 10,317.6 | 54.418 |
| 820 | 6,562.6 | 4,934.2 | 48.520 | 1620 | 13,685.7 | 10,468.6 | 54.535 |
| 840 | 6,728.9 | 5,060.8 | 48.721 | 1640 | 13,877.0 | 10,620.2 | 54.653 |
| 860 | 6,895.6 | 5,187.8 | 48.916 | 1660 | 14,069.2 | 10,772.7 | 54.770 |
| 880 | 7,062.9 | 5,315.3 | 49.109 | 1680 | 14,261.9 | 10,925.6 | 54.886 |
| 900 | 7,230.9 | 5,443.6 | 49.298 | 1700 | 14,455.4 | 11,079.4 | 54.999 |
| 920 | 7,399.4 | 5,572.4 | 49.483 | 1720 | 14,649.5 | 11,233.8 | 55.113 |
| 940 | 7,568.4 | 5,701.7 | 49.665 | 1740 | 14,844.3 | 11,388.9 | 55.226 |
| 960 | 7,738.0 | 5,831.6 | 49.843 | 1760 | 15,039.8 | 11,544.7 | 55.339 |
| 980 | 7,908.2 | 5,962.0 | 50.019 | 1780 | 15,236.1 | 11,701.2 | 55.449 |
| 1000 | 8,078.9 | 6,093.0 | 50.191 | 1800 | 15,433.0 | 11,858.4 | 55.559 |
| 1020 | 8,250.4 | 6,224.8 | 50.360 | 1820 | 15,630.6 | 12,016.3 | 55.668 |
| 1040 | 8,422.4 | 6,357.1 | 50.528 | 1840 | 15,828.7 | 12,174.7 | 55.777 |
| 1060 | 8,595.0 | 6,490.0 | 50.693 | 1860 | 16,027.6 | 12,333.9 | 55.884 |

**Table A-10E**    *(Continued)*

| $T,°R$ | $\bar{h}$ | $\bar{u}$ | $\bar{s}^o$ | $T,°R$ | $\bar{h}$ | $\bar{u}$ | $\bar{s}^o$ |
|---|---|---|---|---|---|---|---|
| 1900 | 16,428 | 12,654 | 56.097 | 3500 | 34,324 | 27,373 | 62.876 |
| 1940 | 16,830 | 12,977 | 56.307 | 3540 | 34,809 | 27,779 | 63.015 |
| 1980 | 17,235 | 13,303 | 56.514 | 3580 | 35,296 | 28,187 | 63.153 |
| 2020 | 17,643 | 13,632 | 56.719 | 3620 | 35,785 | 28,596 | 63.288 |
| 2060 | 18,054 | 13,963 | 56.920 | 3660 | 36,274 | 29,006 | 63.423 |
| 2100 | 18,467 | 14,297 | 57.119 | 3700 | 36,765 | 29,418 | 63.557 |
| 2140 | 18,883 | 14,633 | 57.315 | 3740 | 37,258 | 29,831 | 63.690 |
| 2180 | 19,301 | 14,972 | 57.509 | 3780 | 37,752 | 30,245 | 63.821 |
| 2220 | 19,722 | 15,313 | 57.701 | 3820 | 38,247 | 30,661 | 63.952 |
| 2260 | 20,145 | 15,657 | 57.889 | 3860 | 38,743 | 31,077 | 64.082 |
| 2300 | 20,571 | 16,003 | 58.077 | 3900 | 39,240 | 31,495 | 64.210 |
| 2340 | 20,999 | 16,352 | 58.261 | 3940 | 39,739 | 31,915 | 64.338 |
| 2380 | 21,429 | 16,703 | 58.445 | 3980 | 40,239 | 32,335 | 64.465 |
| 2420 | 21,862 | 17,057 | 58.625 | 4020 | 40,740 | 32,757 | 64.591 |
| 2460 | 22,298 | 17,413 | 58.803 | 4060 | 41,242 | 33,179 | 64.715 |
| 2500 | 22,735 | 17,771 | 58.980 | 4100 | 41,745 | 33,603 | 64.839 |
| 2540 | 23,175 | 18,131 | 59.155 | 4140 | 42,250 | 34,028 | 64.962 |
| 2580 | 23,618 | 18,494 | 59.328 | 4180 | 42,755 | 34,454 | 65.084 |
| 2620 | 24,062 | 18,859 | 59.500 | 4220 | 43,267 | 34,881 | 65.204 |
| 2660 | 24,508 | 19,226 | 59.669 | 4260 | 43,769 | 35,310 | 65.325 |
| 2700 | 24,957 | 19,595 | 59.837 | 4300 | 44,278 | 35,739 | 65.444 |
| 2740 | 25,408 | 19,967 | 60.003 | 4340 | 44,788 | 36,169 | 65.563 |
| 2780 | 25,861 | 20,340 | 61.167 | 4380 | 45,298 | 36,600 | 65.680 |
| 2820 | 26,316 | 20,715 | 60.330 | 4420 | 45,810 | 37,032 | 65.797 |
| 2860 | 26,773 | 21,093 | 60.490 | 4460 | 46,322 | 37,465 | 65.913 |
| 2900 | 27,231 | 21,472 | 60.650 | 4500 | 46,836 | 37,900 | 66.028 |
| 2940 | 27,692 | 21,853 | 60.809 | 4540 | 47,350 | 38,334 | 66.142 |
| 2980 | 28,154 | 22,237 | 60.965 | 4580 | 47,866 | 38,770 | 66.255 |
| 3020 | 28,619 | 22,621 | 61.120 | 4620 | 48,382 | 39,207 | 66.368 |
| 3060 | 29,085 | 23,085 | 61.274 | 4660 | 48,899 | 39,645 | 66.480 |
| 3100 | 29,553 | 23,397 | 61.426 | 4700 | 49,417 | 40,083 | 66.591 |
| 3140 | 30,023 | 23,787 | 61.577 | 4740 | 49,936 | 40,523 | 66.701 |
| 3180 | 30,494 | 24,179 | 61.727 | 4780 | 50,455 | 40,963 | 66.811 |
| 3220 | 30,967 | 24,572 | 61.874 | 4820 | 50,976 | 41,404 | 66.920 |
| 3260 | 31,442 | 24,968 | 62.022 | 4860 | 51,497 | 41,856 | 67.028 |
| 3300 | 31,918 | 25,365 | 62.167 | 4900 | 52,019 | 42,288 | 67.135 |
| 3340 | 32,396 | 25,763 | 62.312 | 5000 | 53,327 | 43,398 | 67.401 |
| 3380 | 32,876 | 26,164 | 62.454 | 5100 | 54,640 | 44,512 | 67.662 |
| 3420 | 33,357 | 26,565 | 62.597 | 5200 | 55,957 | 45,631 | 67.918 |
| 3460 | 33,839 | 26,968 | 62.738 | 5300 | 57,279 | 46,754 | 68.172 |

*SOURCE:* Data abridged from J. H. Keenan and J. Kaye, "Gas Tables," Wiley, New York, 1945.

**Table A-11E**   Ideal-gas enthalpy, internal energy, and absolute entropy of diatomic hydrogen ($H_2$)

($\Delta \bar{h}_f^\circ$ = 0 Btu/lbmol) ($\bar{h}$ and $\bar{u}$, Btu/lbmol; $\bar{s}^\circ$, Btu/(lbmol·°R); std state is 1 atm.)

| $T$, °R | $\bar{h}$ | $\bar{u}$ | $\bar{s}^\circ$ | $T$, °R | $\bar{h}$ | $\bar{u}$ | $\bar{s}^\circ$ |
|---|---|---|---|---|---|---|---|
| 300 | 2,063.3 | 1,467.7 | 27.337 | 1400 | 9,673.8 | 6,893.6 | 37.883 |
| 320 | 2,189.4 | 1,553.9 | 27.742 | 1500 | 10,381.5 | 7,402.7 | 38.372 |
| 340 | 2,317.2 | 1,642.0 | 28.130 | 1600 | 11,092.5 | 7,915.1 | 38.830 |
| 360 | 2,446.8 | 1,731.9 | 28.501 | 1700 | 11,807.4 | 8,431.4 | 39.264 |
| 380 | 2,577.8 | 1,823.2 | 28.856 | 1800 | 12,526.8 | 8,952.2 | 39.675 |
| 400 | 2,710.2 | 1,915.8 | 29.195 | 1900 | 13,250.9 | 9,477.8 | 40.067 |
| 420 | 2,843.7 | 2,009.6 | 29.520 | 2000 | 13,980.1 | 10,008.4 | 40.441 |
| 440 | 2,978.1 | 2,104.3 | 29.833 | 2100 | 14,714.5 | 10,544.2 | 40.799 |
| 460 | 3,113.5 | 2,200.0 | 30.133 | 2200 | 15,454.4 | 11,085.5 | 41.143 |
| 480 | 3,249.4 | 2,296.2 | 30.424 | 2300 | 16,199.8 | 11,632.3 | 41.475 |
| 500 | 3,386.1 | 2,393.2 | 30.703 | 2400 | 16,950.6 | 12,184.5 | 41.794 |
| 520 | 3,523.2 | 2,490.6 | 30.972 | 2500 | 17,707.3 | 12,742.6 | 42.104 |
| 537 | 3,640.3 | 2,573.9 | 31.194 | 2600 | 18,469.7 | 13,306.4 | 42.403 |
| 540 | 3,660.9 | 2,588.5 | 31.232 | 2700 | 19,237.8 | 13,876.0 | 42.692 |
| 560 | 3,798.8 | 2,686.7 | 31.482 | 2800 | 20,011.8 | 14,451.4 | 42.973 |
| 580 | 3,937.1 | 2,785.3 | 31.724 | 2900 | 20,791.5 | 15,032.5 | 43.247 |
| 600 | 4,075.6 | 2,884.1 | 31.959 | 3000 | 21,576.9 | 15,619.3 | 43.514 |
| 620 | 4,214.3 | 2,983.1 | 32.187 | 3100 | 22,367.7 | 16,211.5 | 43.773 |
| 640 | 4,353.1 | 3,082.1 | 32.407 | 3200 | 23,164.1 | 16,809.3 | 44.026 |
| 660 | 4,492.1 | 3,181.4 | 32.621 | 3300 | 23,965.5 | 17,412.1 | 44.273 |
| 680 | 4,631.1 | 3,280.7 | 32.829 | 3400 | 24,771.9 | 18,019.9 | 44.513 |
| 700 | 4,770.2 | 3,380.1 | 33.031 | 3500 | 25,582.9 | 18,632.4 | 44.748 |
| 720 | 4,909.5 | 3,479.6 | 33.226 | 3600 | 26,398.5 | 19,249.4 | 44.978 |
| 740 | 5,048.8 | 3,579.2 | 33.417 | 3700 | 27,218.5 | 19,870.8 | 45.203 |
| 760 | 5,188.1 | 3,678.8 | 33.603 | 3800 | 28,042.8 | 20,496.5 | 45.423 |
| 780 | 5,327.6 | 3,778.6 | 33.784 | 3900 | 28,871.1 | 21,126.2 | 45.638 |
| 800 | 5,467.1 | 3,878.4 | 33.961 | 4000 | 29,703.5 | 21,760.0 | 45.849 |
| 820 | 5,606.7 | 3,978.3 | 34.134 | 4100 | 30,539.8 | 22,397.7 | 46.056 |
| 840 | 5,746.3 | 4,078.2 | 34.302 | 4200 | 31,379.8 | 23,039.2 | 46.257 |
| 860 | 5,885.9 | 4,178.0 | 34.466 | 4300 | 32,223.5 | 23,684.3 | 46.456 |
| 880 | 6,025.6 | 4,278.0 | 34.627 | 4400 | 33,070.9 | 24,333.1 | 46.651 |
| 900 | 6,165.3 | 4,378.0 | 34.784 | 4500 | 33,921.6 | 24,985.2 | 46.842 |
| 920 | 6,305.1 | 4,478.1 | 34.938 | 4600 | 34,775.7 | 25,640.7 | 47.030 |
| 940 | 6,444.9 | 4,578.1 | 35.087 | 4700 | 35,633.0 | 26,299.4 | 47.215 |
| 960 | 6,584.7 | 4,678.3 | 35.235 | 4800 | 36,493.4 | 26,961.2 | 47.396 |
| 980 | 6,724.6 | 4,778.4 | 35.379 | 4900 | 37,356.9 | 27,626.1 | 47.574 |
| 1000 | 6,864.5 | 4,878.6 | 35.520 | 5000 | 38,223.3 | 28,294.0 | 47.749 |
| 1100 | 7,564.6 | 5,380.1 | 36.188 | 5100 | 39,092.8 | 28,964.9 | 47.921 |
| 1200 | 8,265.8 | 5,882.8 | 36.798 | 5200 | 39,965.1 | 29,638.6 | 48.090 |
| 1300 | 8,968.7 | 6,387.1 | 37.360 | 5300 | 40,840.2 | 30,315.1 | 48.257 |

SOURCE: Data abridged from J. H. Keenan and J. Kaye, "Gas Tables," Wiley, New York, 1945.

**Table A-11E**    (*Continued*)
Ideal-gas enthalpy, internal energy, and absolute entropy of monatomic oxygen (O)

($\Delta \overline{h}_f^{\circ} = +107{,}210$ Btu/lbmol; std state is 1 atm.)

| $T$, °R | $\overline{h}$ | $\overline{u}$ | $\overline{s}^{o}$ | $T$, °R | $\overline{h}$ | $\overline{u}$ | $\overline{s}^{o}$ |
|---|---|---|---|---|---|---|---|
| 0 | 0 | 0 | 0 | 4590 | 23,283 | 14,145 | 49.372 |
| 537 | 2,953 | 1,884 | 38.537 | 4680 | 23,733 | 14,416 | 49.469 |
| 900 | 4,826 | 3,034 | 41.206 | 4770 | 24,183 | 14,686 | 49.565 |
| 1800 | 9,358 | 5,774 | 44.699 | 4860 | 24,632 | 14,957 | 49.658 |
| 2700 | 13,857 | 8,482 | 46.726 | 4950 | 25,082 | 15,228 | 49.750 |
| 2880 | 14,755 | 9,021 | 47.048 | 5040 | 25,533 | 15,499 | 49.840 |
| 3060 | 15,653 | 9,561 | 47.350 | 5130 | 25,983 | 15,770 | 49.928 |
| 3240 | 16,550 | 10,100 | 47.635 | 5220 | 26,434 | 16,042 | 50.016 |
| 3420 | 17,448 | 10,639 | 47.905 | 5310 | 26,884 | 16,314 | 50.101 |
| 3600 | 18,345 | 11,178 | 48.161 | 5400 | 27,336 | 16,586 | 50.186 |
| 3780 | 19,242 | 11,717 | 48.404 | 5580 | 28,239 | 17,131 | 50.350 |
| 3960 | 20,140 | 12,256 | 48.636 | 5760 | 29,144 | 17,677 | 50.510 |
| 4140 | 21,038 | 12,796 | 48.857 | 5940 | 30,049 | 18,224 | 50.665 |
| 4320 | 21,935 | 13,335 | 49.070 | 6120 | 30,956 | 18,773 | 50.815 |
| 4500 | 22,834 | 13,605 | 49.173 | 6300 | 31,865 | 19,323 | 50.961 |

**Ideal-gas enthalpy, internal energy, and absolute entropy of hydroxyl (OH)**

($\Delta \overline{h}_f^{\circ} = +16{,}790$ Btu/lbmol; std state is 1 atm.)

| $T$, °R | $\overline{h}$ | $\overline{u}$ | $\overline{s}^{o}$ | $T$, °R | $\overline{h}$ | $\overline{u}$ | $\overline{s}^{o}$ |
|---|---|---|---|---|---|---|---|
| 0 | 0 | 0 | 0 | 3800 | 28,709 | 21,160 | 58.332 |
| 537 | 3,960 | 2,892 | 43.964 | 3900 | 29,551 | 21,804 | 58.549 |

**Table A-12E**    Properties of saturated water: Temperature table

(v, ft³/lb$_m$; u, and h, Btu/lb$_m$; s, Btu/(lb$_m$·°R))

| Temp., °F $T$ | Press., psia $P$ | Specific Volume | | Internal Energy | | Enthalpy | | | Entropy | |
|---|---|---|---|---|---|---|---|---|---|---|
| | | Sat. Liquid $v_f$ | Sat. Vapor $v_g$ | Sat. Liquid $u_f$ | Sat. Vapor $u_g$ | Sat. Liquid $h_f$ | Evap. $h_{fg}$ | Sat. Vapor $h_g$ | Sat. Liquid $s_f$ | Sat. Vapor $s_g$ |
| 32 | 0.0886 | 0.01602 | 3305 | −.01 | 1021.2 | −.01 | 1075.4 | 1075.4 | −.00003 | 2.1870 |
| 35 | 0.0999 | 0.01602 | 2948 | 2.99 | 1022.2 | 3.00 | 1073.7 | 1076.7 | 0.00607 | 2.1764 |
| 40 | 0.1217 | 0.01602 | 2445 | 8.02 | 1023.9 | 8.02 | 1070.9 | 1078.9 | 0.01617 | 2.1592 |
| 45 | 0.1475 | 0.01602 | 2037 | 13.04 | 1025.5 | 13.04 | 1068.1 | 1081.1 | 0.02618 | 2.1423 |
| 50 | 0.1780 | 0.01602 | 1704 | 18.06 | 1027.2 | 18.06 | 1065.2 | 1083.3 | 0.03607 | 2.1259 |
| 52 | 0.1917 | 0.01603 | 1589 | 20.06 | 1027.8 | 20.07 | 1064.1 | 1084.2 | 0.04000 | 2.1195 |
| 54 | 0.2064 | 0.01603 | 1482 | 22.07 | 1028.5 | 22.07 | 1063.0 | 1085.1 | 0.04391 | 2.1131 |
| 56 | 0.2219 | 0.01603 | 1383 | 24.08 | 1029.1 | 24.08 | 1061.9 | 1085.9 | 0.04781 | 2.1068 |
| 58 | 0.2386 | 0.01603 | 1292 | 26.08 | 1029.8 | 26.08 | 1060.7 | 1086.8 | 0.05159 | 2.1005 |
| 60 | 0.2563 | 0.01604 | 1207 | 28.08 | 1030.4 | 28.08 | 1059.6 | 1087.7 | 0.05555 | 2.0943 |
| 62 | 0.2751 | 0.01604 | 1129 | 30.09 | 1031.1 | 30.09 | 1058.5 | 1088.6 | 0.05940 | 2.0882 |
| 64 | 0.2952 | 0.01604 | 1056 | 32.09 | 1031.8 | 32.09 | 1057.3 | 1089.4 | 0.06323 | 2.0821 |
| 66 | 0.3165 | 0.01604 | 988.4 | 34.09 | 1032.4 | 34.09 | 1056.2 | 1090.3 | 0.06704 | 2.0761 |
| 68 | 0.3391 | 0.01605 | 925.8 | 36.09 | 1033.1 | 36.09 | 1055.1 | 1091.2 | 0.07084 | 2.0701 |
| 70 | 0.3632 | 0.01605 | 867.7 | 38.09 | 1033.7 | 38.09 | 1054.0 | 1092.0 | 0.07463 | 2.0642 |
| 72 | 0.3887 | 0.01606 | 813.7 | 40.09 | 1034.4 | 40.09 | 1052.8 | 1092.9 | 0.07839 | 2.0584 |
| 74 | 0.4158 | 0.01606 | 763.5 | 42.09 | 1035.0 | 42.09 | 1051.7 | 1093.8 | 0.08215 | 2.0526 |
| 76 | 0.4446 | 0.01606 | 716.8 | 44.09 | 1035.7 | 44.09 | 1050.6 | 1094.7 | 0.08589 | 2.0469 |
| 78 | 0.4750 | 0.01607 | 673.3 | 46.09 | 1036.3 | 46.09 | 1049.4 | 1095.5 | 0.08961 | 2.0412 |
| 80 | 0.5073 | 0.01607 | 632.8 | 48.08 | 1037.0 | 48.09 | 1048.3 | 1096.4 | 0.09332 | 2.0356 |
| 82 | 0.5414 | 0.01608 | 595.0 | 50.08 | 1037.6 | 50.08 | 1047.2 | 1097.3 | 0.09701 | 2.0300 |
| 84 | 0.5776 | 0.01608 | 559.8 | 52.08 | 1038.3 | 52.08 | 1046.0 | 1098.1 | 0.1007 | 2.0245 |
| 86 | 0.6158 | 0.01609 | 527.0 | 54.08 | 1038.9 | 54.08 | 1044.9 | 1099.0 | 0.1044 | 2.0190 |
| 88 | 0.6562 | 0.01609 | 496.3 | 56.07 | 1039.6 | 56.07 | 1043.8 | 1099.9 | 0.1080 | 2.0136 |
| 90 | 0.6988 | 0.01610 | 467.7 | 58.07 | 1040.2 | 58.07 | 1042.7 | 1100.7 | 0.1117 | 2.0083 |
| 92 | 0.7439 | 0.01611 | 440.9 | 60.06 | 1040.9 | 60.06 | 1041.5 | 1101.6 | 0.1153 | 2.0030 |
| 94 | 0.7914 | 0.01611 | 415.9 | 62.06 | 1041.5 | 62.06 | 1040.4 | 1102.4 | 0.1189 | 1.9977 |
| 96 | 0.8416 | 0.01612 | 392.4 | 64.05 | 1041.2 | 64.06 | 1039.2 | 1103.3 | 0.1225 | 1.9925 |
| 98 | 0.8945 | 0.01612 | 370.5 | 66.05 | 1042.8 | 66.05 | 1038.1 | 1104.2 | 0.1261 | 1.9874 |
| 100 | 0.9503 | 0.01613 | 350.0 | 68.04 | 1043.5 | 68.05 | 1037.0 | 1105.0 | 0.1296 | 1.9822 |
| 110 | 1.276 | 0.01617 | 265.1 | 78.02 | 1046.7 | 78.02 | 1031.3 | 1109.3 | 0.1473 | 1.9574 |
| 120 | 1.695 | 0.01621 | 203.0 | 87.99 | 1049.9 | 88.00 | 1025.5 | 1113.5 | 0.1647 | 1.9336 |
| 130 | 2.225 | 0.01625 | 157.2 | 97.97 | 1053.0 | 97.98 | 1019.8 | 1117.8 | 0.1817 | 1.9109 |
| 140 | 2.892 | 0.01629 | 122.9 | 107.95 | 1056.2 | 107.96 | 1014.0 | 1121.9 | 0.1985 | 1.8892 |
| 150 | 3.722 | 0.01634 | 97.0 | 117.95 | 1059.3 | 117.96 | 1008.1 | 1126.1 | 0.2150 | 1.8684 |
| 160 | 4.745 | 0.01640 | 77.2 | 127.94 | 1062.3 | 127.96 | 1002.2 | 1130.1 | 0.2313 | 1.8484 |
| 170 | 5.996 | 0.01645 | 62.0 | 137.95 | 1065.4 | 137.97 | 996.2 | 1134.2 | 0.2473 | 1.8293 |
| 180 | 7.515 | 0.01651 | 50.2 | 147.97 | 1068.3 | 147.99 | 990.2 | 1138.2 | 0.2631 | 1.8109 |
| 190 | 9.343 | 0.01657 | 41.0 | 158.00 | 1071.3 | 158.03 | 984.1 | 1142.1 | 0.2787 | 1.7932 |
| 200 | 11.529 | 0.01663 | 33.6 | 168.04 | 1074.2 | 168.07 | 977.9 | 1145.9 | 0.2940 | 1.7762 |

**Table A-12E**    *(Continued)*

| Temp., °F $T$ | Press., psia $P$ | Specific Volume Sat. Liquid $v_f$ | Sat. Vapor $v_g$ | Internal Energy Sat. Liquid $u_f$ | Sat. Vapor $u_g$ | Enthalpy Sat. Liquid $h_f$ | Evap. $h_{fg}$ | Sat. Vapor $h_g$ | Entropy Sat. Liquid $s_f$ | Sat. Vapor $s_g$ |
|---|---|---|---|---|---|---|---|---|---|---|
| 210 | 14.13 | 0.01670 | 27.82 | 178.1 | 1077.0 | 178.1 | 971.6 | 1149.7 | 0.3091 | 1.7599 |
| 212 | 14.70 | 0.01672 | 26.80 | 180.1 | 1077.6 | 180.2 | 970.3 | 1150.5 | 0.3121 | 1.7567 |
| 220 | 17.19 | 0.01677 | 23.15 | 188.2 | 1079.8 | 188.2 | 965.3 | 1153.5 | 0.3241 | 1.7441 |
| 230 | 20.78 | 0.01685 | 19.39 | 198.3 | 1082.6 | 198.3 | 958.8 | 1157.1 | 0.3388 | 1.7289 |
| 240 | 24.97 | 0.01692 | 16.33 | 208.4 | 1085.3 | 208.4 | 952.3 | 1160.7 | 0.3534 | 1.7143 |
| 250 | 29.82 | 0.01700 | 13.83 | 218.5 | 1087.9 | 218.6 | 945.6 | 1164.2 | 0.3677 | 1.7001 |
| 260 | 35.42 | 0.01708 | 11.77 | 228.6 | 1090.5 | 228.8 | 938.8 | 1167.6 | 0.3819 | 1.6864 |
| 270 | 41.85 | 0.01717 | 10.07 | 238.8 | 1093.0 | 239.0 | 932.0 | 1170.9 | 0.3960 | 1.6731 |
| 280 | 49.18 | 0.01726 | 8.65 | 249.0 | 1095.4 | 249.2 | 924.9 | 1174.1 | 0.4099 | 1.6602 |
| 290 | 57.53 | 0.01735 | 7.47 | 259.3 | 1097.7 | 259.4 | 917.8 | 1177.2 | 0.4236 | 1.6477 |
| 300 | 66.98 | 0.01745 | 6.472 | 269.5 | 1100.0 | 269.7 | 910.4 | 1180.2 | 0.4372 | 1.6356 |
| 310 | 77.64 | 0.01755 | 5.632 | 279.8 | 1102.1 | 280.1 | 903.0 | 1183.0 | 0.4507 | 1.6238 |
| 320 | 89.60 | 0.01765 | 4.919 | 290.1 | 1104.2 | 290.4 | 895.3 | 1185.8 | 0.4640 | 1.6123 |
| 330 | 103.00 | 0.01776 | 4.312 | 300.5 | 1106.2 | 300.8 | 887.5 | 1188.4 | 0.4772 | 1.6010 |
| 340 | 117.93 | 0.01787 | 3.792 | 310.9 | 1108.0 | 311.3 | 879.5 | 1190.8 | 0.4903 | 1.5901 |
| 350 | 134.53 | 0.01799 | 3.346 | 321.4 | 1109.8 | 321.8 | 871.3 | 1193.1 | 0.5033 | 1.5793 |
| 360 | 152.92 | 0.01811 | 2.961 | 331.8 | 1111.4 | 332.4 | 862.9 | 1195.2 | 0.5162 | 1.5688 |
| 370 | 173.23 | 0.01823 | 2.628 | 342.4 | 1112.9 | 343.0 | 854.2 | 1197.2 | 0.5289 | 1.5585 |
| 380 | 195.60 | 0.01836 | 2.339 | 353.0 | 1114.3 | 353.6 | 845.4 | 1199.0 | 0.5416 | 1.5483 |
| 390 | 220.2 | 0.01850 | 2.087 | 363.6 | 1115.6 | 364.3 | 836.2 | 1200.6 | 0.5542 | 1.5383 |
| 400 | 247.1 | 0.01864 | 1.866 | 374.3 | 1116.6 | 375.1 | 826.8 | 1202.0 | 0.5667 | 1.5284 |
| 410 | 276.5 | 0.01878 | 1.673 | 385.0 | 1117.6 | 386.0 | 817.2 | 1203.1 | 0.5792 | 1.5187 |
| 420 | 308.5 | 0.01894 | 1.502 | 395.8 | 1118.3 | 396.9 | 807.2 | 1204.1 | 0.5915 | 1.5091 |
| 430 | 343.3 | 0.01909 | 1.352 | 406.7 | 1118.9 | 407.9 | 796.9 | 1204.8 | 0.6038 | 1.4995 |
| 440 | 381.2 | 0.01926 | 1.219 | 417.6 | 1119.3 | 419.0 | 786.3 | 1205.3 | 0.6161 | 1.4900 |
| 450 | 422.1 | 0.01943 | 1.1011 | 428.6 | 1119.5 | 430.2 | 775.4 | 1205.6 | 0.6282 | 1.4806 |
| 460 | 466.3 | 0.01961 | 0.9961 | 439.7 | 1119.6 | 441.4 | 764.1 | 1205.5 | 0.6404 | 1.4712 |
| 470 | 514.1 | 0.01980 | 0.9025 | 450.9 | 1119.4 | 452.8 | 752.4 | 1205.2 | 0.6525 | 1.4618 |
| 480 | 565.5 | 0.02000 | 0.8187 | 462.2 | 1118.9 | 464.3 | 740.3 | 1204.6 | 0.6646 | 1.4524 |
| 490 | 620.7 | 0.02021 | 0.7436 | 473.6 | 1118.3 | 475.9 | 727.8 | 1203.7 | 0.6767 | 1.4430 |
| 500 | 680.0 | 0.02043 | 0.6761 | 485.1 | 1117.4 | 487.7 | 714.8 | 1202.5 | 0.6888 | 1.4335 |
| 520 | 811.4 | 0.02091 | 0.5605 | 508.5 | 1114.8 | 511.7 | 687.3 | 1198.9 | 0.7130 | 1.4145 |
| 540 | 961.5 | 0.02145 | 0.4658 | 532.6 | 1111.0 | 536.4 | 657.5 | 1193.8 | 0.7374 | 1.3950 |
| 560 | 1131.8 | 0.02207 | 0.3877 | 548.4 | 1105.8 | 562.0 | 625.0 | 1187.0 | 0.7620 | 1.3749 |
| 580 | 1324.3 | 0.02278 | 0.3225 | 583.1 | 1098.9 | 588.6 | 589.3 | 1178.0 | 0.7872 | 1.3540 |
| 600 | 1541.0 | 0.02363 | 0.2677 | 609.9 | 1090.0 | 616.7 | 549.7 | 1166.4 | 0.8130 | 1.3317 |
| 620 | 1784.4 | 0.02465 | 0.2209 | 638.3 | 1078.5 | 646.4 | 505.0 | 1151.4 | 0.8398 | 1.3075 |
| 640 | 2057.1 | 0.02593 | 0.1805 | 668.7 | 1063.2 | 678.6 | 453.4 | 1131.9 | 0.8681 | 1.2803 |
| 660 | 2362 | 0.02767 | 0.1446 | 702.3 | 1042.3 | 714.4 | 391.1 | 1105.5 | 0.8990 | 1.2483 |
| 680 | 2705 | 0.03032 | 0.1113 | 741.7 | 1011.0 | 756.9 | 309.8 | 1066.7 | 0.9350 | 1.2068 |
| 700 | 3090 | 0.03666 | 0.0744 | 801.7 | 947.7 | 822.7 | 167.5 | 990.2 | 0.9902 | 1.1346 |
| 705.4 | 3204 | 0.05053 | 0.05053 | 872.6 | 872.6 | 902.5 | 0 | 902.5 | 1.0580 | 1.0580 |

*SOURCE:* J. H. Keenan, F. G. Keyes, P. G. Hill, and J. G. Moore, "Steam Tables," Wiley, New York, 1979.

# Table A-13E  Properties of saturated water: Pressure table

(v, ft³/lb_m; u and h, Btu/lb_m; s, Btu/(lb_m·°R))

| Abs. Press., psi $P$ | Temp., °F $T$ | Specific Volume Sat. Liquid $v_f$ | Specific Volume Sat. Vapor $v_g$ | Internal Energy Sat. Liquid $u_f$ | Internal Energy Sat. Vapor $u_g$ | Enthalpy Sat. Liquid $h_f$ | Enthalpy Evap. $h_{fg}$ | Enthalpy Sat. Vapor $h_g$ | Entropy Sat. Liquid $s_f$ | Entropy Evap. $s_{fg}$ | Entropy Sat. Vapor $s_g$ | Abs. Press., psi $P$ |
|---|---|---|---|---|---|---|---|---|---|---|---|---|
| 0.4 | 72.84 | 0.01606 | 792.0 | 40.94 | 1034.7 | 40.94 | 1052.3 | 1093.3 | 0.0800 | 1.9760 | 2.0559 | 0.4 |
| 0.6 | 85.19 | 0.01609 | 540.0 | 53.26 | 1038.0 | 53.27 | 1045.4 | 1098.6 | 0.1029 | 1.9184 | 2.0213 | 0.6 |
| 0.8 | 94.35 | 0.01611 | 411.7 | 62.41 | 1041.7 | 62.41 | 1040.2 | 1102.6 | 0.1195 | 1.8773 | 1.9968 | 0.8 |
| 1.0 | 101.70 | 0.01614 | 333.6 | 69.74 | 1044.0 | 69.74 | 1036.0 | 1105.8 | 0.1327 | 1.8453 | 1.9779 | 1.0 |
| 1.2 | 107.88 | 0.01616 | 280.9 | 75.90 | 1046.0 | 75.90 | 1032.5 | 1108.4 | 0.1436 | 1.8190 | 1.9626 | 1.2 |
| 1.5 | 115.65 | 0.01619 | 227.7 | 83.65 | 1048.5 | 83.65 | 1028.0 | 1111.7 | 0.1571 | 1.7867 | 1.9438 | 1.5 |
| 2.0 | 126.04 | 0.01623 | 173.75 | 94.02 | 1051.8 | 94.02 | 1022.1 | 1116.1 | 0.1750 | 1.7448 | 1.9198 | 2.0 |
| 3.0 | 141.43 | 0.01630 | 118.72 | 109.38 | 1056.6 | 109.39 | 1013.1 | 1122.5 | 0.2009 | 1.6852 | 1.8861 | 3.0 |
| 4.0 | 152.93 | 0.01636 | 90.64 | 120.88 | 1060.2 | 120.89 | 1006.4 | 1127.3 | 0.2198 | 1.6426 | 1.8624 | 4.0 |
| 5.0 | 162.21 | 0.01641 | 73.53 | 130.15 | 1063.0 | 130.17 | 1000.9 | 1131.0 | 0.2349 | 1.6093 | 1.8441 | 5.0 |
| 6.0 | 170.03 | 0.01645 | 61.98 | 137.98 | 1065.4 | 138.00 | 996.2 | 1134.2 | 0.2474 | 1.5819 | 1.8292 | 6.0 |
| 7.0 | 176.82 | 0.01649 | 53.65 | 144.78 | 1067.4 | 144.80 | 992.1 | 1136.9 | 0.2581 | 1.5585 | 1.8167 | 7.0 |
| 8.0 | 182.84 | 0.01653 | 47.35 | 150.81 | 1069.2 | 150.84 | 988.4 | 1139.3 | 0.2675 | 1.5383 | 1.8058 | 8.0 |
| 9.0 | 188.26 | 0.01656 | 42.41 | 156.25 | 1070.8 | 156.27 | 985.1 | 1141.4 | 0.2760 | 1.5203 | 1.7963 | 9.0 |
| 10 | 193.19 | 0.01659 | 38.42 | 161.20 | 1072.2 | 161.23 | 982.1 | 1143.3 | 0.2836 | 1.5041 | 1.7877 | 10 |
| 14.696 | 211.99 | 0.01672 | 26.80 | 180.10 | 1077.6 | 180.15 | 970.4 | 1150.5 | 0.3121 | 1.4446 | 1.7567 | 14.696 |
| 15 | 213.03 | 0.01672 | 26.29 | 181.14 | 1077.9 | 181.19 | 969.7 | 1150.9 | 0.3137 | 1.4414 | 1.7551 | 15 |
| 20 | 227.96 | 0.01683 | 20.09 | 196.19 | 1082.0 | 196.26 | 960.1 | 1156.4 | 0.3358 | 1.3962 | 1.7320 | 20 |
| 25 | 240.08 | 0.01692 | 16.31 | 208.44 | 1085.3 | 208.52 | 952.2 | 1160.7 | 0.3535 | 1.3607 | 1.7142 | 25 |
| 30 | 250.34 | 0.01700 | 13.75 | 218.84 | 1088.0 | 218.93 | 945.4 | 1164.3 | 0.3682 | 1.3314 | 1.6996 | 30 |
| 35 | 259.30 | 0.01708 | 11.90 | 227.93 | 1090.3 | 228.04 | 939.3 | 1167.4 | 0.3809 | 1.3064 | 1.6873 | 35 |
| 40 | 267.26 | 0.01715 | 10.50 | 236.03 | 1092.3 | 236.16 | 933.8 | 1170.0 | 0.3921 | 1.2845 | 1.6767 | 40 |
| 45 | 274.46 | 0.01721 | 9.40 | 243.37 | 1094.0 | 243.51 | 928.8 | 1172.3 | 0.4022 | 1.2651 | 1.6673 | 45 |
| 50 | 281.03 | 0.01727 | 8.52 | 250.08 | 1095.6 | 250.24 | 924.2 | 1174.4 | 0.4113 | 1.2476 | 1.6589 | 50 |
| 55 | 287.10 | 0.01733 | 7.79 | 256.28 | 1097.0 | 256.46 | 919.9 | 1176.3 | 0.4196 | 1.2317 | 1.6513 | 55 |

**Table A-13E** (Continued)

| Abs. Press., psi $P$ | Temp., °F $T$ | Specific Volume | | Internal Energy | | Enthalpy | | | Entropy | | | Abs. Press., psi $P$ |
|---|---|---|---|---|---|---|---|---|---|---|---|---|
| | | Sat. Liquid $v_f$ | Sat. Vapor $v_g$ | Sat. Liquid $u_f$ | Sat. Vapor $u_g$ | Sat. Liquid $h_f$ | Evap. $h_{fg}$ | Sat. Vapor $h_g$ | Sat. Liquid $s_f$ | Evap. $s_{fg}$ | Sat. Vapor $s_g$ | |
| 60 | 292.73 | 0.01738 | 7.177 | 262.1 | 1098.3 | 262.2 | 915.8 | 1178.0 | 0.4273 | 1.2170 | 1.6443 | 60 |
| 65 | 298.00 | 0.01743 | 6.647 | 267.5 | 1099.5 | 267.7 | 911.9 | 1179.6 | 0.4345 | 1.2035 | 1.6380 | 65 |
| 70 | 302.96 | 0.01748 | 6.209 | 272.6 | 1100.6 | 272.8 | 908.3 | 1181.0 | 0.4412 | 1.1909 | 1.6321 | 70 |
| 75 | 307.63 | 0.01752 | 5.818 | 277.4 | 1101.6 | 277.6 | 904.8 | 1182.4 | 0.4475 | 1.1790 | 1.6265 | 75 |
| 80 | 312.07 | 0.01757 | 5.474 | 282.0 | 1102.6 | 282.2 | 901.4 | 1183.6 | 0.4534 | 1.1679 | 1.6213 | 80 |
| 85 | 316.29 | 0.01761 | 5.170 | 286.3 | 1103.5 | 286.6 | 898.2 | 1184.8 | 0.4591 | 1.1574 | 1.6165 | 85 |
| 90 | 320.31 | 0.01766 | 4.898 | 290.5 | 1104.3 | 290.8 | 895.1 | 1185.9 | 0.4644 | 1.1475 | 1.6119 | 90 |
| 95 | 324.16 | 0.01770 | 4.654 | 294.5 | 1105.0 | 294.8 | 892.1 | 1186.9 | 0.4695 | 1.1380 | 1.6075 | 95 |
| 100 | 327.86 | 0.01774 | 4.434 | 298.3 | 1105.8 | 298.6 | 889.2 | 1187.8 | 0.4744 | 1.1290 | 1.6034 | 100 |
| 110 | 334.82 | 0.01781 | 4.051 | 305.5 | 1107.1 | 305.9 | 883.7 | 1189.6 | 0.4836 | 1.1122 | 1.5958 | 110 |
| 120 | 341.30 | 0.01789 | 3.730 | 312.3 | 1108.3 | 312.7 | 878.5 | 1191.1 | 0.4920 | 1.0966 | 1.5886 | 120 |
| 130 | 347.37 | 0.01796 | 3.457 | 318.6 | 1109.4 | 319.0 | 873.5 | 1192.5 | 0.4999 | 1.0822 | 1.5821 | 130 |
| 140 | 353.08 | 0.01802 | 3.221 | 324.6 | 1110.3 | 325.1 | 868.7 | 1193.8 | 0.5073 | 1.0688 | 1.5761 | 140 |
| 150 | 358.48 | 0.01809 | 3.016 | 330.2 | 1111.2 | 330.8 | 864.2 | 1194.9 | 0.5142 | 1.0562 | 1.5704 | 150 |
| 160 | 363.60 | 0.01815 | 2.836 | 335.6 | 1112.0 | 336.2 | 859.8 | 1196.0 | 0.5208 | 1.0443 | 1.5651 | 160 |
| 170 | 368.47 | 0.01821 | 2.676 | 340.8 | 1112.7 | 341.3 | 855.6 | 1196.9 | 0.5270 | 1.0330 | 1.5600 | 170 |
| 180 | 373.13 | 0.01827 | 2.533 | 345.7 | 1113.4 | 346.3 | 851.5 | 1197.8 | 0.5329 | 1.0223 | 1.5552 | 180 |
| 190 | 377.59 | 0.01833 | 2.405 | 350.4 | 1114.0 | 351.0 | 847.5 | 1198.6 | 0.5386 | 1.0122 | 1.5508 | 190 |
| 200 | 381.86 | 0.01839 | 2.289 | 354.9 | 1114.6 | 355.6 | 843.7 | 1199.3 | 0.5440 | 1.0025 | 1.5465 | 200 |
| 250 | 401.04 | 0.01865 | 1.845 | 375.4 | 1116.7 | 376.2 | 825.8 | 1202.1 | 0.5680 | 0.9594 | 1.5274 | 250 |
| 300 | 417.43 | 0.01890 | 1.544 | 393.0 | 1118.2 | 394.1 | 809.8 | 1203.9 | 0.5883 | 0.9232 | 1.5115 | 300 |
| 350 | 431.82 | 0.01912 | 1.327 | 408.7 | 1119.0 | 409.9 | 795.0 | 1204.9 | 0.6060 | 0.8917 | 1.4977 | 350 |
| 400 | 444.70 | 0.01934 | 1.162 | 422.8 | 1119.5 | 424.2 | 781.2 | 1205.5 | 0.6218 | 0.8638 | 1.4856 | 400 |
| 450 | 456.39 | 0.01955 | 1.033 | 435.7 | 1119.6 | 437.4 | 768.2 | 1205.6 | 0.6360 | 0.8385 | 1.4745 | 450 |
| 500 | 467.13 | 0.01975 | 0.928 | 447.7 | 1119.4 | 449.5 | 755.8 | 1205.3 | 0.6490 | 0.8154 | 1.4644 | 500 |

**Table A-13E** *(Continued)*

| Abs. Press., psi $P$ | Temp., °F $T$ | Specific Volume | | Internal Energy | | Enthalpy | | | Entropy | | | Abs. Press., psi $P$ |
|---|---|---|---|---|---|---|---|---|---|---|---|---|
| | | Sat. Liquid $v_f$ | Sat. Vapor $v_g$ | Sat. Liquid $u_f$ | Sat. Vapor $u_g$ | Sat. Liquid $h_f$ | Evap. $h_{fg}$ | Sat. Vapor $h_g$ | Sat. Liquid $s_f$ | Evap. $s_{fg}$ | Sat. Vapor $s_g$ | |
| 550 | 477.07 | 0.01994 | 0.842 | 458.9 | 1119.1 | 460.9 | 743.9 | 1204.8 | 0.6611 | 0.7941 | 1.4551 | 550 |
| 600 | 486.33 | 0.02013 | 0.770 | 469.4 | 1118.6 | 471.7 | 732.4 | 1204.1 | 0.6723 | 0.7742 | 1.4464 | 600 |
| 700 | 503.23 | 0.02051 | 0.656 | 488.9 | 1117.0 | 491.5 | 710.5 | 1202.0 | 0.6927 | 0.7378 | 1.4305 | 700 |
| 800 | 518.36 | 0.02087 | 0.569 | 506.6 | 1115.0 | 509.7 | 689.6 | 1199.3 | 0.7110 | 0.7050 | 1.4160 | 800 |
| 900 | 532.12 | 0.02123 | 0.501 | 523.0 | 1112.6 | 526.6 | 669.5 | 1196.0 | 0.7277 | 0.6750 | 1.4027 | 900 |
| 1000 | 544.75 | 0.02159 | 0.446 | 538.4 | 1109.9 | 542.4 | 650.0 | 1192.4 | 0.7432 | 0.6471 | 1.3903 | 1000 |
| 1100 | 556.45 | 0.02195 | 0.401 | 552.9 | 1106.8 | 557.4 | 631.0 | 1188.3 | 0.7576 | 0.6209 | 1.3786 | 1100 |
| 1200 | 567.37 | 0.02232 | 0.362 | 566.7 | 1103.5 | 571.7 | 612.3 | 1183.9 | 0.7712 | 0.5961 | 1.3673 | 1200 |
| 1300 | 577.60 | 0.02269 | 0.330 | 579.9 | 1099.8 | 585.4 | 593.8 | 1179.2 | 0.7841 | 0.5724 | 1.3565 | 1300 |
| 1400 | 587.25 | 0.02307 | 0.302 | 592.7 | 1096.0 | 598.6 | 575.5 | 1174.1 | 0.7964 | 0.5497 | 1.3461 | 1400 |
| 1500 | 596.39 | 0.02346 | 0.277 | 605.0 | 1091.8 | 611.5 | 557.2 | 1168.7 | 0.8082 | 0.5276 | 1.3359 | 1500 |
| 1600 | 605.06 | 0.02386 | 0.255 | 616.9 | 1087.4 | 624.0 | 538.9 | 1162.9 | 0.8196 | 0.5062 | 1.3258 | 1600 |
| 1700 | 613.32 | 0.02428 | 0.236 | 628.6 | 1082.7 | 636.2 | 520.6 | 1156.9 | 0.8307 | 0.4852 | 1.3159 | 1700 |
| 1800 | 621.21 | 0.02472 | 0.218 | 640.0 | 1077.7 | 648.3 | 502.1 | 1150.4 | 0.8414 | 0.4645 | 1.3060 | 1800 |
| 1900 | 628.76 | 0.02517 | 0.203 | 651.3 | 1072.3 | 660.1 | 483.4 | 1143.5 | 0.8519 | 0.4441 | 1.2961 | 1900 |
| 2000 | 636.00 | 0.02565 | 0.188 | 662.4 | 1066.6 | 671.9 | 464.4 | 1136.3 | 0.8623 | 0.4238 | 1.2861 | 2000 |
| 2250 | 652.90 | 0.02698 | 0.157 | 689.9 | 1050.6 | 701.1 | 414.8 | 1115.9 | 0.8876 | 0.3728 | 1.2604 | 2250 |
| 2500 | 668.31 | 0.02860 | 0.131 | 717.7 | 1031.0 | 730.9 | 360.5 | 1091.4 | 0.9131 | 0.3196 | 1.2327 | 2500 |
| 2750 | 682.46 | 0.03077 | 0.107 | 747.3 | 1005.9 | 763.0 | 297.4 | 1060.4 | 0.9401 | 0.2604 | 1.2005 | 2750 |
| 3000 | 695.52 | 0.03431 | 0.084 | 783.4 | 968.8 | 802.5 | 213.0 | 1015.5 | 0.9732 | 0.1843 | 1.1575 | 3000 |
| 3203.6 | 705.44 | 0.05053 | 0.0505 | 872.6 | 872.6 | 902.5 | 0 | 902.5 | 1.0580 | 0 | 1.0580 | 3203.6 |

*SOURCE*: J. H. Keenan, F. G. Keyes, P. G. Hill, and J. G. Moore, "Steam Tables," Wiley, New York, 1969.

## Table A-14E Properties of water: Superheated-vapor table

$(v, ft^3/lb_m; u$ and $h$, $Btu/lb_m; s$, $Btu/(lb_m \cdot °R))$

| Temp., °F | $v$ | $u$ | $h$ | $s$ | $v$ | $u$ | $h$ | $s$ |
|---|---|---|---|---|---|---|---|---|
| | 1 psia ($T_{sat} = 101.7°F$) | | | | 5 psia ($T_{sat} = 162.2°F$) | | | |
| Sat. | 333.6 | 1044.0 | 1105.8 | 1.9779 | 73.53 | 1063.0 | 1131.0 | 1.8441 |
| 150 | 362.6 | 1060.4 | 1127.5 | 2.0151 | | | | |
| 200 | 392.5 | 1077.5 | 1150.1 | 2.0508 | 78.15 | 1076.0 | 1148.6 | 1.8715 |
| 250 | 422.4 | 1094.7 | 1172.8 | 2.0839 | 84.21 | 1093.8 | 1171.7 | 1.9052 |
| 300 | 452.3 | 1112.0 | 1195.7 | 2.1150 | 90.24 | 1111.3 | 1194.8 | 1.9367 |
| 400 | 511.9 | 1147.0 | 1241.8 | 2.1720 | 102.24 | 1146.6 | 1241.2 | 1.9941 |
| 500 | 571.5 | 1182.8 | 1288.5 | 2.2235 | 114.20 | 1182.5 | 1288.2 | 2.0458 |
| 600 | 631.1 | 1219.3 | 1336.1 | 2.2706 | 126.15 | 1219.1 | 1335.8 | 2.0930 |
| 700 | 690.7 | 1256.7 | 1384.5 | 2.3142 | 138.08 | 1256.5 | 1384.3 | 2.1367 |
| 800 | 750.3 | 1294.4 | 1433.7 | 2.3550 | 150.01 | 1294.7 | 1433.5 | 2.1775 |
| 900 | 809.9 | 1333.9 | 1483.8 | 2.3932 | 161.94 | 1333.8 | 1483.7 | 2.2158 |
| 1000 | 869.5 | 1373.9 | 1534.8 | 2.4294 | 173.86 | 1373.9 | 1534.7 | 2.2520 |
| | 10 psia ($T_{sat} = 193.2°F$) | | | | 14.7 psia ($T_{sat} = 212.0°F$) | | | |
| Sat. | 38.42 | 1072.2 | 1143.3 | 1.7877 | 26.80 | 1077.6 | 1150.5 | 1.7567 |
| 200 | 38.85 | 1074.7 | 1146.6 | 1.7927 | | | | |
| 250 | 41.95 | 1092.6 | 1170.2 | 1.8272 | 28.42 | 1091.5 | 1168.8 | 1.7832 |
| 300 | 44.99 | 1110.4 | 1193.7 | 1.8592 | 30.52 | 1109.6 | 1192.6 | 1.8157 |
| 400 | 51.03 | 1146.1 | 1240.5 | 1.9171 | 34.67 | 1145.6 | 1239.9 | 1.8741 |
| 500 | 57.04 | 1182.2 | 1287.7 | 1.9690 | 38.77 | 1181.8 | 1287.3 | 1.9263 |
| 600 | 63.03 | 1218.9 | 1335.5 | 2.0164 | 42.86 | 1218.6 | 1335.2 | 1.9737 |
| 700 | 69.01 | 1256.3 | 1384.0 | 2.0601 | 46.93 | 1256.1 | 1383.8 | 2.0175 |
| 800 | 74.98 | 1294.6 | 1433.3 | 2.1009 | 51.00 | 1294.4 | 1433.1 | 2.0584 |
| 900 | 80.95 | 1333.7 | 1483.5 | 2.1393 | 55.07 | 1333.6 | 1483.4 | 2.0967 |
| 1000 | 86.91 | 1373.8 | 1534.6 | 2.1755 | 59.13 | 1373.7 | 1534.5 | 2.1330 |
| 1100 | 92.88 | 1414.7 | 1586.6 | 2.2099 | 63.19 | 1414.6 | 1586.4 | 2.1674 |
| | 20 psia ($T_{sat} = 228.0°F$) | | | | 40 psia ($T_{sat} = 267.3°F$) | | | |
| Sat. | 20.09 | 1082.0 | 1156.4 | 1.7320 | 10.50 | 1093.3 | 1170.0 | 1.6767 |
| 250 | 20.79 | 1090.3 | 1167.2 | 1.7475 | | | | |
| 300 | 22.36 | 1108.7 | 1191.5 | 1.7805 | 11.04 | 1105.1 | 1186.8 | 1.6993 |
| 350 | 23.90 | 1126.9 | 1215.4 | 1.8110 | 11.84 | 1124.2 | 1211.8 | 1.7312 |
| 400 | 25.43 | 1145.1 | 1239.2 | 1.8395 | 12.62 | 1143.0 | 1236.4 | 1.7606 |
| 500 | 28.46 | 1181.5 | 1286.8 | 1.8919 | 14.16 | 1180.1 | 1284.9 | 1.8140 |
| 600 | 31.47 | 1218.4 | 1334.8 | 1.9395 | 15.69 | 1217.3 | 1333.4 | 1.8621 |
| 700 | 34.47 | 1255.9 | 1383.5 | 1.9834 | 17.20 | 1255.1 | 1382.4 | 1.9063 |
| 800 | 37.46 | 1294.3 | 1432.9 | 2.0243 | 18.70 | 1293.7 | 1432.1 | 1.9474 |
| 900 | 40.45 | 1333.5 | 1483.2 | 2.0627 | 20.20 | 1333.0 | 1482.5 | 1.9859 |
| 1000 | 43.44 | 1373.5 | 1534.3 | 2.0989 | 21.70 | 1373.1 | 1533.8 | 2.0223 |
| 1100 | 46.42 | 1414.5 | 1586.3 | 2.1334 | 23.20 | 1414.2 | 1585.9 | 2.0568 |

**Table A-14E** (Continued)

| Temp., °F | $v$ | $u$ | $h$ | $s$ | $v$ | $u$ | $h$ | $s$ |
|---|---|---|---|---|---|---|---|---|
| | 60 psia ($T_{sat}$ = 292.7°F) | | | | 80 psia ($T_{sat}$ = 312.1°F) | | | |
| Sat. | 7.17 | 1098.3 | 1178.0 | 1.6444 | 5.47 | 1102.6 | 1183.6 | 1.6214 |
| 300 | 7.26 | 1101.3 | 1181.9 | 1.6496 | | | | |
| 350 | 7.82 | 1121.4 | 1208.2 | 1.6830 | 5.80 | 1118.5 | 1204.3 | 1.6476 |
| 400 | 8.35 | 1140.8 | 1233.5 | 1.7134 | 6.22 | 1138.5 | 1230.6 | 1.6790 |
| 500 | 9.40 | 1178.6 | 1283.0 | 1.7678 | 7.02 | 1177.2 | 1281.1 | 1.7346 |
| 600 | 10.43 | 1216.3 | 1332.1 | 1.8165 | 7.79 | 1215.3 | 1330.7 | 1.7838 |
| 700 | 11.44 | 1254.4 | 1381.4 | 1.8609 | 8.56 | 1253.6 | 1380.3 | 1.8285 |
| 800 | 12.45 | 1293.0 | 1431.2 | 1.9022 | 9.32 | 1292.4 | 1430.4 | 1.8700 |
| 900 | 13.45 | 1332.5 | 1481.8 | 1.9408 | 10.08 | 1332.0 | 1481.2 | 1.9087 |
| 1000 | 14.45 | 1372.7 | 1533.2 | 1.9773 | 10.83 | 1372.3 | 1532.6 | 1.9453 |
| 1100 | 15.45 | 1413.8 | 1585.4 | 2.0119 | 11.58 | 1413.5 | 1584.9 | 1.9799 |
| 1200 | 16.45 | 1455.8 | 1638.5 | 2.0448 | 12.33 | 1455.5 | 1638.1 | 2.0130 |
| | 100 psia ($T_{sat}$ = 327.8°F) | | | | 120 psia ($T_{sat}$ = 341.3°F) | | | |
| Sat. | 4.434 | 1105.8 | 1187.8 | 1.6034 | 3.730 | 1108.3 | 1191.1 | 1.5886 |
| 350 | 4.592 | 1115.4 | 1200.4 | 1.6191 | 3.783 | 1112.2 | 1196.2 | 1.5950 |
| 400 | 4.934 | 1136.2 | 1227.5 | 1.6517 | 4.079 | 1133.8 | 1224.4 | 1.6288 |
| 450 | 5.265 | 1156.2 | 1253.6 | 1.6812 | 4.360 | 1154.3 | 1251.2 | 1.6590 |
| 500 | 5.587 | 1175.7 | 1279.1 | 1.7085 | 4.633 | 1174.2 | 1277.1 | 1.6868 |
| 600 | 6.216 | 1214.2 | 1329.3 | 1.7582 | 5.164 | 1213.2 | 1327.8 | 1.7371 |
| 700 | 6.834 | 1252.8 | 1379.2 | 1.8033 | 5.682 | 1252.0 | 1378.2 | 1.7825 |
| 800 | 7.445 | 1291.8 | 1429.6 | 1.8449 | 6.195 | 1291.2 | 1428.7 | 1.8243 |
| 900 | 8.053 | 1331.5 | 1480.5 | 1.8838 | 6.703 | 1330.9 | 1479.8 | 1.8633 |
| 1000 | 8.657 | 1371.9 | 1532.1 | 1.9204 | 7.208 | 1371.5 | 1531.5 | 1.9000 |
| 1100 | 9.260 | 1413.1 | 1584.5 | 1.9551 | 7.711 | 1412.8 | 1584.0 | 1.9348 |
| 1200 | 9.861 | 1455.2 | 1637.7 | 1.9882 | 8.213 | 1454.9 | 1637.3 | 1.9679 |
| | 140 psia ($T_{sat}$ = 353.1°F) | | | | 160 psia ($T_{sat}$ = 363.6°F) | | | |
| Sat. | 3.221 | 1110.3 | 1193.8 | 1.5761 | 2.836 | 1112.0 | 1196.0 | 1.5651 |
| 400 | 3.466 | 1131.4 | 1221.2 | 1.6088 | 3.007 | 1128.8 | 1217.8 | 1.5911 |
| 450 | 3.713 | 1152.4 | 1248.6 | 1.6399 | 3.228 | 1150.5 | 1246.1 | 1.6230 |
| 500 | 3.952 | 1172.7 | 1275.1 | 1.6682 | 3.440 | 1171.2 | 1273.0 | 1.6518 |
| 550 | 4.184 | 1192.5 | 1300.9 | 1.6945 | 3.646 | 1191.3 | 1299.2 | 1.6785 |
| 600 | 4.412 | 1212.1 | 1326.4 | 1.7191 | 3.848 | 1211.1 | 1325.0 | 1.7034 |
| 700 | 4.860 | 1251.2 | 1377.1 | 1.7648 | 4.243 | 1250.4 | 1376.0 | 1.7494 |
| 800 | 5.301 | 1290.5 | 1427.9 | 1.8068 | 4.631 | 1289.9 | 1427.0 | 1.7916 |
| 900 | 5.739 | 1330.4 | 1479.1 | 1.8459 | 5.015 | 1329.9 | 1478.4 | 1.8308 |
| 1000 | 6.173 | 1371.0 | 1531.0 | 1.8827 | 5.397 | 1370.6 | 1530.4 | 1.8677 |
| 1100 | 6.605 | 1412.4 | 1583.6 | 1.9176 | 5.776 | 1412.1 | 1583.1 | 1.9026 |
| 1200 | 7.036 | 1454.6 | 1636.9 | 1.9507 | 6.154 | 1454.3 | 1636.5 | 1.9358 |

**Table A-14E**     *(Continued)*

| Temp., °F | $v$ | $u$ | $h$ | $s$ | $v$ | $u$ | $h$ | $s$ |
|---|---|---|---|---|---|---|---|---|
| | \multicolumn{4}{c}{180 psia ($T_{sat}$ = 373.1°F)} | \multicolumn{4}{c}{200 psia ($T_{sat}$ = 381.8°F)} |
| Sat. | 2.533 | 1113.4 | 1197.8 | 1.5553 | 2.289 | 1114.6 | 1199.3 | 1.5464 |
| 400 | 2.648 | 1126.2 | 1214.4 | 1.5749 | 2.361 | 1123.5 | 1210.8 | 1.5600 |
| 450 | 2.850 | 1148.5 | 1243.4 | 1.6078 | 2.548 | 1146.4 | 1240.7 | 1.5938 |
| 500 | 3.042 | 1169.6 | 1270.9 | 1.6372 | 2.724 | 1168.0 | 1268.8 | 1.6239 |
| 550 | 3.228 | 1190.0 | 1297.5 | 1.6642 | 2.893 | 1188.7 | 1295.7 | 1.6512 |
| 600 | 3.409 | 1210.0 | 1323.5 | 1.6893 | 3.058 | 1208.9 | 1322.1 | 1.6767 |
| 700 | 3.763 | 1249.6 | 1374.9 | 1.7357 | 3.379 | 1248.8 | 1373.8 | 1.7234 |
| 800 | 4.110 | 1289.3 | 1426.2 | 1.7781 | 3.693 | 1288.6 | 1425.3 | 1.7660 |
| 900 | 4.453 | 1329.4 | 1477.7 | 1.8174 | 4.003 | 1328.9 | 1477.1 | 1.8055 |
| 1000 | 4.793 | 1370.2 | 1529.8 | 1.8545 | 4.310 | 1369.8 | 1529.3 | 1.8425 |
| 1100 | 5.131 | 1411.7 | 1582.6 | 1.8894 | 4.615 | 1411.4 | 1582.2 | 1.8776 |
| 1200 | 5.467 | 1454.0 | 1636.1 | 1.9227 | 4.918 | 1453.7 | 1635.7 | 1.9109 |
| | \multicolumn{4}{c}{250 psia ($T_{sat}$ = 401.0°F)} | \multicolumn{4}{c}{300 psia ($T_{sat}$ = 417.4°F)} |
| Sat. | 1.845 | 1116.7 | 1202.1 | 1.5274 | 1.544 | 1118.2 | 1203.9 | 1.5115 |
| 450 | 2.002 | 1141.1 | 1233.7 | 1.5632 | 1.636 | 1135.4 | 1226.2 | 1.5365 |
| 500 | 2.150 | 1163.8 | 1263.3 | 1.5948 | 1.766 | 1159.5 | 1257.5 | 1.5701 |
| 550 | 2.290 | 1185.3 | 1291.3 | 1.6233 | 1.888 | 1181.9 | 1286.7 | 1.5997 |
| 600 | 2.426 | 1206.1 | 1318.3 | 1.6494 | 2.004 | 1203.2 | 1314.5 | 1.6266 |
| 700 | 2.688 | 1246.7 | 1371.1 | 1.6970 | 2.227 | 1244.0 | 1368.3 | 1.6751 |
| 800 | 2.943 | 1287.0 | 1423.2 | 1.7301 | 2.442 | 1285.4 | 1421.0 | 1.7187 |
| 900 | 3.193 | 1327.6 | 1475.3 | 1.7799 | 2.653 | 1326.3 | 1473.6 | 1.7589 |
| 1000 | 3.440 | 1368.7 | 1527.9 | 1.8172 | 2.860 | 1367.7 | 1526.5 | 1.7964 |
| 1100 | 3.685 | 1410.5 | 1581.0 | 1.8524 | 3.066 | 1409.6 | 1579.8 | 1.8317 |
| 1200 | 3.929 | 1453.0 | 1634.8 | 1.8858 | 3.270 | 1452.2 | 1633.8 | 1.8653 |
| 1300 | 4.172 | 1496.3 | 1689.3 | 1.9177 | 3.473 | 1495.6 | 1688.4 | 1.8973 |
| | \multicolumn{4}{c}{350 psia ($T_{sat}$ = 431.8°F)} | \multicolumn{4}{c}{400 psia ($T_{sat}$ = 444.7°F)} |
| Sat. | 1.327 | 1119.0 | 1204.9 | 1.4978 | 1.162 | 1119.5 | 1205.5 | 1.4856 |
| 450 | 1.373 | 1129.2 | 1218.2 | 1.5125 | 1.175 | 1122.6 | 1209.6 | 1.4901 |
| 500 | 1.491 | 1154.9 | 1251.5 | 1.5482 | 1.284 | 1150.1 | 1245.2 | 1.5282 |
| 550 | 1.600 | 1178.3 | 1281.9 | 1.5790 | 1.383 | 1174.6 | 1277.0 | 1.5605 |
| 600 | 1.703 | 1200.3 | 1310.6 | 1.6068 | 1.476 | 1197.3 | 1306.6 | 1.5892 |
| 700 | 1.898 | 1242.5 | 1365.4 | 1.6562 | 1.650 | 1240.4 | 1362.5 | 1.6397 |
| 800 | 2.085 | 1283.8 | 1418.8 | 1.7004 | 1.816 | 1282.1 | 1416.6 | 1.6844 |
| 900 | 2.267 | 1325.0 | 1471.8 | 1.7409 | 1.978 | 1323.7 | 1470.1 | 1.7252 |
| 1000 | 2.446 | 1366.6 | 1525.0 | 1.7787 | 2.136 | 1365.5 | 1523.6 | 1.7632 |
| 1100 | 2.624 | 1408.7 | 1578.6 | 1.8142 | 2.292 | 1407.8 | 1577.4 | 1.7989 |
| 1200 | 2.799 | 1451.5 | 1632.8 | 1.8478 | 2.446 | 1450.7 | 1631.8 | 1.8327 |
| 1300 | 2.974 | 1495.0 | 1687.6 | 1.8799 | 2.599 | 1494.3 | 1686.8 | 1.8648 |

## Table A-14E  *(Continued)*

| Temp., °F | v | u | h | s | v | u | h | s |
|---|---|---|---|---|---|---|---|---|
| | **450 psia ($T_{sat}$ = 456.4°F)** | | | | **500 psia ($T_{sat}$ = 467.1°F)** | | | |
| Sat. | 1.033 | 1119.6 | 1205.6 | 1.4746 | 0.928 | 1119.4 | 1205.3 | 1.4645 |
| 500 | 1.123 | 1145.1 | 1238.5 | 1.5097 | 0.992 | 1139.7 | 1231.5 | 1.4923 |
| 550 | 1.215 | 1170.7 | 1271.9 | 1.5436 | 1.079 | 1166.7 | 1266.6 | 1.5279 |
| 600 | 1.300 | 1194.3 | 1302.5 | 1.5732 | 1.158 | 1191.1 | 1298.3 | 1.5585 |
| 700 | 1.458 | 1238.2 | 1359.6 | 1.6248 | 1.304 | 1236.0 | 1356.7 | 1.6112 |
| 800 | 1.608 | 1280.5 | 1414.4 | 1.6701 | 1.441 | 1278.8 | 1412.1 | 1.6571 |
| 900 | 1.752 | 1322.4 | 1468.3 | 1.7113 | 1.572 | 1321.0 | 1466.5 | 1.6987 |
| 1000 | 1.894 | 1364.4 | 1522.2 | 1.7495 | 1.701 | 1363.3 | 1520.7 | 1.7471 |
| 1100 | 2.034 | 1406.9 | 1576.3 | 1.7853 | 1.827 | 1406.0 | 1575.1 | 1.7731 |
| 1200 | 2.172 | 1450.0 | 1630.8 | 1.8192 | 1.952 | 1449.2 | 1629.8 | 1.8072 |
| 1300 | 2.308 | 1493.7 | 1685.9 | 1.8515 | 2.075 | 1493.1 | 1685.1 | 1.8395 |
| 1400 | 2.444 | 1538.1 | 1741.7 | 1.8823 | 2.198 | 1537.6 | 1741.0 | 1.8704 |
| | **600 psia ($T_{sat}$ = 486.3°F)** | | | | **700 psia ($T_{sat}$ = 503.2°F)** | | | |
| Sat. | 0.770 | 1118.6 | 1204.1 | 1.4464 | 0.656 | 1117.0 | 1202.0 | 1.4305 |
| 500 | 0.795 | 1128.0 | 1216.2 | 1.4592 | | | | |
| 550 | 0.875 | 1158.2 | 1255.4 | 1.4990 | 0.728 | 1149.0 | 1243.2 | 1.4723 |
| 600 | 0.946 | 1184.5 | 1289.5 | 1.5320 | 0.793 | 1177.5 | 1280.2 | 1.5081 |
| 700 | 1.073 | 1231.5 | 1350.6 | 1.5872 | 0.907 | 1226.9 | 1344.4 | 1.5661 |
| 800 | 1.190 | 1275.4 | 1407.6 | 1.6343 | 1.011 | 1272.0 | 1402.9 | 1.6145 |
| 900 | 1.302 | 1318.4 | 1462.9 | 1.6766 | 1.109 | 1315.6 | 1459.3 | 1.6576 |
| 1000 | 1.411 | 1361.2 | 1517.8 | 1.7155 | 1.204 | 1358.9 | 1514.9 | 1.6970 |
| 1100 | 1.517 | 1404.2 | 1572.7 | 1.7519 | 1.296 | 1402.4 | 1570.2 | 1.7337 |
| 1200 | 1.622 | 1447.7 | 1627.8 | 1.7861 | 1.387 | 1446.2 | 1625.8 | 1.7682 |
| 1300 | 1.726 | 1491.7 | 1683.4 | 1.8186 | 1.476 | 1490.4 | 1681.7 | 1.8009 |
| 1400 | 1.829 | 1536.5 | 1739.5 | 1.8497 | 1.565 | 1535.3 | 1738.1 | 1.8321 |
| | **800 psia ($T_{sat}$ = 518.3°F)** | | | | **900 psia ($T_{sat}$ = 532.1°F)** | | | |
| Sat. | 0.569 | 1115.0 | 1199.3 | 1.4160 | 0.501 | 1112.6 | 1196.0 | 1.4027 |
| 550 | 0.615 | 1138.8 | 1229.9 | 1.4469 | 0.527 | 1127.5 | 1215.2 | 1.4219 |
| 600 | 0.677 | 1170.1 | 1270.4 | 1.4861 | 0.587 | 1162.2 | 1260.0 | 1.4652 |
| 650 | 0.732 | 1197.2 | 1305.6 | 1.5186 | 0.639 | 1191.1 | 1297.5 | 1.4999 |
| 700 | 0.783 | 1222.1 | 1338.0 | 1.5471 | 0.686 | 1217.1 | 1331.4 | 1.5297 |
| 800 | 0.876 | 1268.5 | 1398.2 | 1.5969 | 0.772 | 1264.9 | 1393.4 | 1.5810 |
| 900 | 0.964 | 1312.9 | 1455.6 | 1.6408 | 0.851 | 1310.1 | 1451.9 | 1.6257 |
| 1000 | 1.048 | 1356.7 | 1511.9 | 1.6807 | 0.927 | 1354.5 | 1508.9 | 1.6662 |
| 1100 | 1.130 | 1400.5 | 1567.8 | 1.7178 | 1.001 | 1398.7 | 1565.4 | 1.7036 |
| 1200 | 1.210 | 1444.6 | 1623.8 | 1.7526 | 1.073 | 1443.0 | 1621.7 | 1.7386 |
| 1300 | 1.289 | 1489.1 | 1680.0 | 1.7854 | 1.144 | 1487.8 | 1687.3 | 1.7717 |
| 1400 | 1.367 | 1534.2 | 1736.6 | 1.8167 | 1.214 | 1533.0 | 1735.1 | 1.8031 |

**Table A-14E**    (Continued)

| Temp., °F | $v$ | $u$ | $h$ | $s$ | $v$ | $u$ | $h$ | $s$ |
|---|---|---|---|---|---|---|---|---|
| | 1000 psia ($T_{sat}$ = 544.7°F) | | | | 1200 psia ($T_{sat}$ = 567.4°F) | | | |
| Sat. | 0.446 | 1109.0 | 1192.4 | 1.3903 | 0.362 | 1103.5 | 1183.9 | 1.3673 |
| 600 | 0.514 | 1153.7 | 1248.8 | 1.4450 | 0.402 | 1134.4 | 1223.6 | 1.4054 |
| 650 | 0.564 | 1184.7 | 1289.1 | 1.4822 | 0.450 | 1170.9 | 1270.8 | 1.4490 |
| 700 | 0.608 | 1212.0 | 1324.6 | 1.5135 | 0.491 | 1201.3 | 1310.2 | 1.4837 |
| 800 | 0.688 | 1261.2 | 1388.5 | 1.5665 | 0.562 | 1253.7 | 1378.4 | 1.5402 |
| 900 | 0.761 | 1307.3 | 1448.1 | 1.6120 | 0.626 | 1301.5 | 1440.4 | 1.5876 |
| 1000 | 0.831 | 1352.2 | 1505.9 | 1.6530 | 0.685 | 1347.5 | 1499.7 | 1.6297 |
| 1100 | 0.898 | 1396.8 | 1562.9 | 1.6908 | 0.743 | 1393.0 | 1557.9 | 1.6682 |
| 1200 | 0.963 | 1441.5 | 1619.7 | 1.7261 | 0.798 | 1438.3 | 1615.5 | 1.7040 |
| 1300 | 1.027 | 1486.5 | 1676.5 | 1.7593 | 0.853 | 1483.8 | 1673.1 | 1.7377 |
| 1400 | 1.091 | 1531.9 | 1733.7 | 1.7909 | 0.906 | 1529.6 | 1730.7 | 1.7696 |
| 1600 | 1.215 | 1624.4 | 1849.3 | 1.8499 | 1.011 | 1622.6 | 1847.1 | 1.8290 |
| | 1400 psia ($T_{sat}$ = 587.2°F) | | | | 1600 psia ($T_{sat}$ = 605.1°F) | | | |
| Sat. | 0.302 | 1096.0 | 1174.1 | 1.3461 | 0.255 | 1087.4 | 1162.9 | 1.3258 |
| 600 | 0.318 | 1110.9 | 1193.1 | 1.3641 | | | | |
| 650 | 0.367 | 1155.5 | 1250.5 | 1.4171 | 0.303 | 1137.8 | 1227.4 | 1.3852 |
| 700 | 0.406 | 1189.6 | 1294.8 | 1.4562 | 0.342 | 1177.0 | 1278.1 | 1.4299 |
| 800 | 0.471 | 1245.8 | 1367.9 | 1.5168 | 0.403 | 1237.7 | 1357.0 | 1.4953 |
| 900 | 0.529 | 1295.6 | 1432.5 | 1.5661 | 0.466 | 1289.5 | 1424.4 | 1.5468 |
| 1000 | 0.582 | 1342.8 | 1493.5 | 1.6094 | 0.504 | 1338.0 | 1487.1 | 1.5913 |
| 1100 | 0.632 | 1389.1 | 1552.8 | 1.6487 | 0.549 | 1385.2 | 1547.7 | 1.6315 |
| 1200 | 0.681 | 1435.1 | 1611.4 | 1.6851 | 0.592 | 1431.8 | 1607.1 | 1.6684 |
| 1300 | 0.728 | 1481.1 | 1669.6 | 1.7192 | 0.634 | 1478.3 | 1666.1 | 1.7029 |
| 1400 | 0.774 | 1527.2 | 1727.8 | 1.7513 | 0.675 | 1524.9 | 1724.8 | 1.7354 |
| 1600 | 0.865 | 1620.8 | 1844.8 | 1.8111 | 0.755 | 1619.0 | 1842.6 | 1.7955 |
| | 1800 psia ($T_{sat}$ = 621.2°F) | | | | 2000 psia ($T_{sat}$ = 636.0°F) | | | |
| Sat. | 0.218 | 1077.7 | 1150.4 | 1.3060 | 0.188 | 1066.6 | 1136.3 | 1.2861 |
| 650 | 0.251 | 1117.0 | 1200.4 | 1.3517 | 0.206 | 1091.1 | 1167.2 | 1.3141 |
| 700 | 0.291 | 1163.1 | 1259.9 | 1.4042 | 0.249 | 1147.7 | 1239.8 | 1.3782 |
| 750 | 0.322 | 1198.6 | 1305.9 | 1.4430 | 0.280 | 1187.3 | 1291.1 | 1.4216 |
| 800 | 0.350 | 1229.1 | 1345.7 | 1.4753 | 0.307 | 1220.1 | 1333.8 | 1.4562 |
| 900 | 0.399 | 1283.2 | 1416.1 | 1.5291 | 0.353 | 1276.8 | 1407.6 | 1.5126 |
| 1000 | 0.443 | 1333.1 | 1480.7 | 1.5749 | 0.395 | 1328.1 | 1474.1 | 1.5598 |
| 1100 | 0.484 | 1381.2 | 1542.5 | 1.6159 | 0.433 | 1377.2 | 1537.2 | 1.6017 |
| 1200 | 0.524 | 1428.5 | 1602.9 | 1.6534 | 0.469 | 1425.2 | 1598.6 | 1.6398 |
| 1300 | 0.561 | 1475.5 | 1662.5 | 1.6883 | 0.503 | 1472.7 | 1659.0 | 1.6751 |
| 1400 | 0.598 | 1522.5 | 1721.8 | 1.7211 | 0.537 | 1520.2 | 1718.8 | 1.7082 |
| 1600 | 0.670 | 1617.2 | 1840.4 | 1.7817 | 0.602 | 1615.4 | 1838.2 | 1.7692 |

| Temp., °F | $v$ | $u$ | $h$ | $s$ | $v$ | $u$ | $h$ | $s$ |
|---|---|---|---|---|---|---|---|---|
| | **2500 psia ($T_{sat}$ = 668.3°F)** | | | | **3000 psia ($T_{sat}$ = 695.5°F)** | | | |
| Sat. | 0.1306 | 1031.0 | 1091.4 | 1.2327 | 0.0840 | 968.8 | 1015.5 | 1.1575 |
| 700 | 0.1684 | 1098.7 | 1176.6 | 1.3073 | 0.0977 | 1003.9 | 1058.1 | 1.1944 |
| 750 | 0.2030 | 1155.2 | 1249.1 | 1.3686 | 0.1483 | 1114.7 | 1197.1 | 1.3122 |
| 800 | 0.2291 | 1195.7 | 1301.7 | 1.4112 | 0.1757 | 1167.6 | 1265.2 | 1.3675 |
| 900 | 0.2712 | 1259.9 | 1385.4 | 1.4752 | 0.2160 | 1241.8 | 1361.7 | 1.4414 |
| 1000 | 0.3069 | 1315.2 | 1457.2 | 1.5262 | 0.2485 | 1301.7 | 1439.6 | 1.4967 |
| 1100 | 0.3393 | 1366.8 | 1523.8 | 1.5704 | 0.2772 | 1356.2 | 1510.1 | 1.5434 |
| 1200 | 0.3696 | 1416.7 | 1587.7 | 1.6101 | 0.3086 | 1408.0 | 1576.6 | 1.5848 |
| 1300 | 0.3984 | 1465.7 | 1650.0 | 1.6465 | 0.3285 | 1458.5 | 1640.9 | 1.6224 |
| 1400 | 0.4261 | 1514.2 | 1711.3 | 1.6804 | 0.3524 | 1508.1 | 1703.7 | 1.6571 |
| 1500 | 0.4531 | 1562.5 | 1772.1 | 1.7123 | 0.3754 | 1557.3 | 1765.7 | 1.6896 |
| 1600 | 0.4795 | 1610.8 | 1832.6 | 1.7424 | 0.3978 | 1606.3 | 1827.1 | 1.7201 |
| | **3500 psia** | | | | **4000 psia** | | | |
| 650 | 0.0249 | 663.5 | 679.7 | 0.8630 | 0.0245 | 657.7 | 675.8 | 0.8574 |
| 700 | 0.0306 | 759.5 | 779.3 | 0.9506 | 0.0287 | 742.1 | 763.4 | 0.9345 |
| 750 | 0.1046 | 1058.4 | 1126.1 | 1.2440 | 0.0633 | 960.7 | 1007.5 | 1.1395 |
| 800 | 0.1363 | 1134.7 | 1223.0 | 1.3226 | 0.1052 | 1095.0 | 1172.9 | 1.2740 |
| 900 | 0.1763 | 1222.4 | 1336.5 | 1.4096 | 0.1462 | 1201.5 | 1309.7 | 1.3789 |
| 1000 | 0.2066 | 1287.6 | 1421.4 | 1.4699 | 0.1752 | 1272.9 | 1402.6 | 1.4449 |
| 1100 | 0.2328 | 1345.2 | 1496.0 | 1.5193 | 0.1995 | 1333.9 | 1481.6 | 1.4973 |
| 1200 | 0.2566 | 1399.2 | 1565.3 | 1.5624 | 0.2213 | 1390.1 | 1553.9 | 1.5423 |
| 1300 | 0.2787 | 1451.1 | 1631.7 | 1.6012 | 0.2414 | 1443.7 | 1622.4 | 1.5823 |
| 1400 | 0.2997 | 1501.9 | 1696.1 | 1.6368 | 0.2603 | 1495.7 | 1688.4 | 1.6188 |
| 1500 | 0.3199 | 1552.0 | 1759.2 | 1.6699 | 0.2784 | 1546.7 | 1752.8 | 1.6526 |
| 1600 | 0.3395 | 1601.7 | 1831.6 | 1.7010 | 0.2959 | 1597.1 | 1816.1 | 1.6841 |
| | **4400 psia** | | | | **4800 psia** | | | |
| 650 | 0.0242 | 653.6 | 673.3 | 0.8535 | 0.0237 | 649.8 | 671.0 | 0.8499 |
| 700 | 0.0278 | 732.7 | 755.3 | 0.9257 | 0.0271 | 725.1 | 749.1 | 0.9187 |
| 750 | 0.0415 | 870.8 | 904.6 | 1.0513 | 0.0352 | 832.6 | 863.9 | 1.0154 |
| 800 | 0.0844 | 1056.5 | 1125.3 | 1.2306 | 0.0668 | 1011.2 | 1070.5 | 1.1827 |
| 900 | 0.1270 | 1183.7 | 1287.1 | 1.3548 | 0.1109 | 1164.8 | 1263.4 | 1.3310 |
| 1000 | 0.1552 | 1260.8 | 1387.2 | 1.4260 | 0.1385 | 1248.3 | 1317.4 | 1.4078 |
| 1100 | 0.1784 | 1324.7 | 1469.9 | 1.4809 | 0.1608 | 1315.3 | 1458.1 | 1.4653 |
| 1200 | 0.1989 | 1382.8 | 1544.7 | 1.5274 | 0.1802 | 1375.4 | 1535.4 | 1.5133 |
| 1300 | 0.2176 | 1437.7 | 1614.9 | 1.5685 | 0.1979 | 1431.7 | 1607.4 | 1.5555 |
| 1400 | 0.2352 | 1490.7 | 1682.3 | 1.6057 | 0.2143 | 1485.7 | 1676.1 | 1.5934 |
| 1500 | 0.2520 | 1542.7 | 1747.6 | 1.6399 | 0.2300 | 1538.2 | 1742.5 | 1.6282 |
| 1600 | 0.2681 | 1593.4 | 1811.7 | 1.6718 | 0.2450 | 1589.8 | 1807.4 | 1.6605 |

*SOURCE:* J. H. Keenan, F. G. Keyes, P. G. Hill, and J. G. Moore, "Steam Tables," Wiley, New York, 1969.

**Table A-15E**    Properties of water: Compressed-liquid table

(v, ft³/lb_m; u and h, Btu/lb_m; s, Btu/(lb_m·°R))

| Temp., °F | 500 psia ($T_{sat}$ = 467.1°F) | | | | 1000 psia ($T_{sat}$ = 544.7°F) | | | |
|---|---|---|---|---|---|---|---|---|
| | $v$ | $u$ | $h$ | $s$ | $v$ | $u$ | $h$ | $s$ |
| 32 | 0.015994 | 0.00 | 1.49 | 0.00000 | 0.015967 | 0.03 | 2.99 | 0.00005 |
| 50 | 0.015998 | 18.02 | 19.50 | 0.03599 | 0.015972 | 17.99 | 20.94 | 0.03592 |
| 100 | 0.016106 | 67.87 | 69.36 | 0.12932 | 0.016082 | 67.70 | 70.68 | 0.12901 |
| 150 | 0.016318 | 117.66 | 119.17 | 0.21457 | 0.016293 | 117.38 | 120.40 | 0.21410 |
| 200 | 0.016608 | 167.65 | 169.19 | 0.29341 | 0.016580 | 167.26 | 170.32 | 0.29281 |
| 300 | 0.017416 | 268.92 | 270.53 | 0.43641 | 0.017379 | 268.24 | 271.46 | 0.43552 |
| 400 | 0.018608 | 373.68 | 375.40 | 0.56604 | 0.018550 | 372.55 | 375.98 | 0.56472 |
| Sat. | 0.019748 | 447.70 | 449.53 | 0.64904 | 0.021591 | 538.39 | 542.38 | 0.74320 |
| | **1500 psia ($T_{sat}$ = 596.4°F)** | | | | **2000 psia ($T_{sat}$ = 636.0°F)** | | | |
| 32 | 0.015939 | 0.05 | 4.47 | 0.00007 | 0.015912 | 0.06 | 5.95 | 0.00008 |
| 50 | 0.015946 | 17.95 | 22.38 | 0.03584 | 0.015920 | 17.91 | 23.81 | 0.03575 |
| 100 | 0.016058 | 67.53 | 71.99 | 0.12870 | 0.016034 | 67.37 | 73.30 | 0.12839 |
| 150 | 0.016268 | 117.10 | 121.62 | 0.21364 | 0.016244 | 116.83 | 122.84 | 0.21318 |
| 200 | 0.016554 | 166.87 | 171.46 | 0.29221 | 0.016527 | 166.49 | 172.60 | 0.29162 |
| 300 | 0.017343 | 267.58 | 272.39 | 0.43463 | 0.017308 | 266.93 | 273.33 | 0.43376 |
| 400 | 0.018493 | 371.45 | 376.59 | 0.56343 | 0.018439 | 370.38 | 377.21 | 0.56216 |
| 500 | 0.02024 | 481.8 | 487.4 | 0.6853 | 0.02014 | 479.8 | 487.3 | 0.6832 |
| Sat. | 0.02346 | 605.0 | 611.5 | 0.8082 | 0.02565 | 662.4 | 671.9 | 0.8623 |
| | **2500 psia ($T_{sat}$ = 668.3°F)** | | | | **3000 psia ($T_{sat}$ = 695.5°F)** | | | |
| 32 | 0.015885 | 0.08 | 7.43 | 0.00009 | 0.015859 | 0.09 | 8.90 | 0.00009 |
| 50 | 0.015895 | 17.88 | 25.23 | 0.03566 | 0.015870 | 17.84 | 26.65 | 0.03555 |
| 100 | 0.016010 | 67.20 | 74.61 | 0.12808 | 0.015987 | 67.04 | 75.91 | 0.12777 |
| 150 | 0.016220 | 116.56 | 124.07 | 0.21272 | 0.016196 | 116.30 | 125.29 | 0.21226 |
| 200 | 0.016501 | 166.11 | 173.75 | 0.29104 | 0.016476 | 165.74 | 174.89 | 0.29046 |
| 300 | 0.017274 | 266.29 | 274.28 | 0.43290 | 0.017240 | 265.66 | 275.23 | 0.43205 |
| 400 | 0.018386 | 369.34 | 377.84 | 0.56092 | 0.018334 | 368.32 | 378.50 | 0.55970 |
| Sat. | 0.02860 | 717.7 | 730.9 | 0.9131 | 0.034310 | 783.5 | 802.5 | 0.9732 |
| | **3500 psia** | | | | **4000 psia** | | | |
| 32 | 0.015833 | 0.10 | 10.36 | 0.00009 | 0.015807 | 0.10 | 11.80 | 0.00005 |
| 50 | 0.015845 | 17.80 | 28.06 | 0.03545 | 0.015821 | 17.76 | 29.47 | 0.03534 |
| 100 | 0.015964 | 66.88 | 77.22 | 0.12746 | 0.015942 | 66.72 | 78.52 | 0.12714 |
| 150 | 0.016173 | 116.03 | 126.51 | 0.21181 | 0.016150 | 115.77 | 127.73 | 0.21136 |
| 200 | 0.016450 | 165.38 | 176.03 | 0.28988 | 0.016425 | 165.02 | 177.18 | 0.28931 |
| 300 | 0.017206 | 265.04 | 276.19 | 0.43121 | 0.017174 | 264.43 | 277.15 | 0.43038 |
| 400 | 0.018284 | 367.32 | 379.16 | 0.55851 | 0.018235 | 366.35 | 379.85 | 0.55734 |

| *SOURCE:* J. H. Keenan, F. G. Keyes, P. G. Hill, and J. G. Moore, "Steam Tables," Wiley, New York, 1969.

**Table A-16E** **Properties of saturated refrigerant 134a (CF$_4$H$_2$): Temperature table**

(**v**, ft$^3$/lb$_m$; **u**, Btu/lb$_m$; **h**, Btu/lb$_m$; **s**, Btu/(lb$_m$·°R))

| Temp., °F $T$ | Press., psia $P$ | Specific Volume | | Internal Energy | | Enthalpy | | | Entropy | | Temp., °F |
|---|---|---|---|---|---|---|---|---|---|---|---|
| | | Sat. Liquid $v_f$ | Sat. Vapor $v_g$ | Sat. Liquid $u_f$ | Sat. Vapor $u_g$ | Sat. Liquid $h_f$ | Evap. $h_{fg}$ | Sat. Vapor $h_g$ | Sat. Liquid $s_f$ | Sat. Vapor $s_g$ | |
| −40 | 7.490 | 0.01130 | 5.7173 | −0.02 | 87.90 | 0.00 | 95.82 | 95.82 | 0.0000 | 0.2283 | −40 |
| −30 | 9.920 | 0.01143 | 4.3911 | 2.81 | 89.26 | 2.83 | 94.49 | 97.32 | 0.0067 | 0.2266 | −30 |
| −20 | 12.949 | 0.01156 | 3.4173 | 5.69 | 90.62 | 5.71 | 93.10 | 98.81 | 0.0133 | 0.2250 | −20 |
| −15 | 14.718 | 0.01163 | 3.0286 | 7.14 | 91.30 | 7.17 | 92.38 | 99.55 | 0.0166 | 0.2243 | −15 |
| −10 | 16.674 | 0.01170 | 2.6918 | 8.61 | 91.98 | 8.65 | 91.64 | 100.29 | 0.0199 | 0.2236 | −10 |
| −5 | 18.831 | 0.01178 | 2.3992 | 10.09 | 92.66 | 10.13 | 90.89 | 101.02 | 0.0231 | 0.2230 | −5 |
| 0 | 21.203 | 0.01185 | 2.1440 | 11.58 | 93.33 | 11.63 | 90.12 | 101.75 | 0.0264 | 0.2224 | 0 |
| 5 | 23.805 | 0.01193 | 1.9208 | 13.09 | 94.01 | 13.14 | 89.33 | 102.47 | 0.0296 | 0.2219 | 5 |
| 10 | 26.651 | 0.01200 | 1.7251 | 14.60 | 94.68 | 14.66 | 88.53 | 103.19 | 0.0329 | 0.2214 | 10 |
| 15 | 29.756 | 0.01208 | 1.5529 | 16.13 | 95.35 | 16.20 | 87.71 | 103.90 | 0.0361 | 0.2209 | 15 |
| 20 | 33.137 | 0.01216 | 1.4009 | 17.67 | 96.02 | 17.74 | 86.87 | 104.61 | 0.0393 | 0.2205 | 20 |
| 25 | 36.809 | 0.01225 | 1.2666 | 19.22 | 96.69 | 19.30 | 86.02 | 105.32 | 0.0426 | 0.2200 | 25 |
| 30 | 40.788 | 0.01233 | 1.1474 | 20.78 | 97.35 | 20.87 | 85.14 | 106.01 | 0.0458 | 0.2196 | 30 |
| 40 | 49.738 | 0.01251 | 0.9470 | 23.94 | 98.67 | 24.05 | 83.34 | 107.39 | 0.0522 | 0.2189 | 40 |
| 50 | 60.125 | 0.01270 | 0.7871 | 27.14 | 99.98 | 27.28 | 81.46 | 108.74 | 0.0585 | 0.2183 | 50 |
| 60 | 72.092 | 0.01290 | 0.6584 | 30.39 | 101.27 | 30.56 | 79.49 | 110.05 | 0.0648 | 0.2178 | 60 |
| 70 | 85.788 | 0.01311 | 0.5538 | 33.68 | 102.54 | 33.89 | 77.44 | 111.33 | 0.0711 | 0.2173 | 70 |
| 80 | 101.37 | 0.01334 | 0.4682 | 37.02 | 103.78 | 37.27 | 75.29 | 112.56 | 0.0774 | 0.2169 | 80 |
| 85 | 109.92 | 0.01346 | 0.4312 | 38.72 | 104.39 | 38.99 | 74.17 | 113.16 | 0.0805 | 0.2167 | 85 |
| 90 | 118.99 | 0.01358 | 0.3975 | 40.42 | 105.00 | 40.72 | 73.03 | 113.75 | 0.0836 | 0.2165 | 90 |
| 95 | 128.62 | 0.01371 | 0.3668 | 42.14 | 105.60 | 42.47 | 71.86 | 114.33 | 0.0867 | 0.2163 | 95 |
| 100 | 138.83 | 0.01385 | 0.3388 | 43.87 | 106.18 | 44.23 | 70.66 | 114.89 | 0.0898 | 0.2161 | 100 |
| 105 | 149.63 | 0.01399 | 0.3131 | 45.62 | 106.76 | 46.01 | 69.42 | 115.43 | 0.0930 | 0.2159 | 105 |
| 110 | 161.04 | 0.01414 | 0.2896 | 47.39 | 107.33 | 47.81 | 68.15 | 115.96 | 0.0961 | 0.2157 | 110 |
| 115 | 173.10 | 0.01429 | 0.2680 | 49.17 | 107.88 | 49.63 | 66.84 | 116.47 | 0.0992 | 0.2155 | 115 |
| 120 | 185.82 | 0.01445 | 0.2481 | 50.97 | 108.42 | 51.47 | 65.48 | 116.95 | 0.1023 | 0.2153 | 120 |
| 140 | 243.86 | 0.01520 | 0.1827 | 58.39 | 110.41 | 59.08 | 59.57 | 118.65 | 0.1150 | 0.2143 | 140 |
| 160 | 314.63 | 0.01617 | 0.1341 | 66.26 | 111.97 | 67.20 | 52.58 | 119.78 | 0.1280 | 0.2128 | 160 |
| 180 | 400.22 | 0.01758 | 0.0964 | 74.83 | 112.77 | 76.13 | 43.78 | 119.91 | 0.1417 | 0.2101 | 180 |
| 200 | 503.52 | 0.02014 | 0.0647 | 84.90 | 111.66 | 86.77 | 30.92 | 117.69 | 0.1575 | 0.2044 | 200 |
| 210 | 563.51 | 0.02329 | 0.0476 | 91.84 | 108.48 | 94.27 | 19.18 | 113.45 | 0.1684 | 0.1971 | 210 |

*SOURCE:* Tables A-16E through A-18E produced from a computer program provided by R. S. Basu, Allied Signal Corporation. *Literature Source:* D. P. Wilson and R. S. Basu, "Thermodynamic Properties of a New Stratospherically Safe Working Fluid—Refrigerant 134a," *ASHRAE Trans.,* **94**(pt 2); 2095–2118, 1988.

**Table A-17E**     **Properties of saturated refrigerant 134a (CF$_4$H$_2$): Pressure table**

**($v$, ft$^3$/lb$_m$; $u$, Btu/lb$_m$; $h$, Btu/lb$_m$; $s$, Btu/(lb$_m \cdot$°R))**

| Press., psia $P$ | Temp., °F $T$ | Specific Volume | | Internal Energy | | Enthalpy | | | Entropy | | Press., psia $P$ |
|---|---|---|---|---|---|---|---|---|---|---|---|
| | | Sat. Liquid $v_f$ | Sat. Vapor $v_g$ | Sat. Liquid $u_f$ | Sat. Vapor $u_g$ | Sat. Liquid $h_f$ | Evap. $h_{fg}$ | Sat. Vapor $h_g$ | Sat. Liquid $s_f$ | Sat. Vapor $s_g$ | |
| 5 | −53.48 | 0.01113 | 8.3508 | −3.74 | 86.07 | −3.73 | 97.53 | 93.79 | −.0090 | 0.2311 | 5 |
| 10 | −29.71 | 0.01143 | 4.3581 | 2.89 | 89.30 | 2.91 | 94.45 | 97.37 | 0.0068 | 0.2265 | 10 |
| 15 | −14.25 | 0.01164 | 2.9747 | 7.36 | 91.40 | 7.40 | 92.27 | 99.66 | 0.0171 | 0.2242 | 15 |
| 20 | −2.48 | 0.01181 | 2.2661 | 10.84 | 93.00 | 10.89 | 90.50 | 101.39 | 0.0248 | 0.2227 | 20 |
| 30 | 15.38 | 0.01209 | 1.5408 | 16.24 | 95.40 | 16.31 | 87.65 | 103.96 | 0.0364 | 0.2209 | 30 |
| 40 | 29.04 | 0.01232 | 1.1692 | 20.48 | 97.23 | 20.57 | 85.31 | 105.88 | 0.0452 | 0.2197 | 40 |
| 50 | 40.27 | 0.01252 | 0.9422 | 24.02 | 98.71 | 24.14 | 83.29 | 107.43 | 0.0523 | 0.2189 | 50 |
| 60 | 49.89 | 0.01270 | 0.7887 | 27.10 | 99.96 | 27.24 | 81.48 | 108.72 | 0.0584 | 0.2183 | 60 |
| 70 | 58.35 | 0.01286 | 0.6778 | 29.85 | 101.05 | 30.01 | 79.82 | 109.83 | 0.0638 | 0.2179 | 70 |
| 80 | 65.93 | 0.01302 | 0.5938 | 32.33 | 102.02 | 32.53 | 78.28 | 110.81 | 0.0686 | 0.2175 | 80 |
| 90 | 72.83 | 0.01317 | 0.5278 | 34.62 | 102.89 | 34.84 | 76.84 | 111.68 | 0.0729 | 0.2172 | 90 |
| 100 | 79.17 | 0.01332 | 0.4747 | 36.75 | 103.68 | 36.99 | 75.47 | 112.46 | 0.0768 | 0.2169 | 100 |
| 120 | 90.54 | 0.01360 | 0.3941 | 40.61 | 105.06 | 40.91 | 72.91 | 113.82 | 0.0839 | 0.2165 | 120 |
| 140 | 100.56 | 0.01386 | 0.3358 | 44.07 | 106.25 | 44.43 | 70.52 | 114.95 | 0.0902 | 0.2161 | 140 |
| 160 | 109.56 | 0.01412 | 0.2916 | 47.23 | 107.28 | 47.65 | 68.26 | 115.91 | 0.0958 | 0.2157 | 160 |
| 180 | 117.74 | 0.01438 | 0.2569 | 50.16 | 108.18 | 50.64 | 66.10 | 116.74 | 0.1009 | 0.2154 | 180 |
| 200 | 125.28 | 0.01463 | 0.2288 | 52.90 | 108.98 | 53.44 | 64.01 | 117.44 | 0.1057 | 0.2151 | 200 |
| 220 | 132.27 | 0.01489 | 0.2056 | 55.48 | 109.68 | 56.09 | 61.96 | 118.05 | 0.1101 | 0.2147 | 220 |
| 240 | 138.79 | 0.01515 | 0.1861 | 57.93 | 110.30 | 58.61 | 59.96 | 118.56 | 0.1142 | 0.2144 | 240 |
| 260 | 144.92 | 0.01541 | 0.1695 | 60.28 | 110.84 | 61.02 | 57.97 | 118.99 | 0.1181 | 0.2140 | 260 |
| 280 | 150.70 | 0.01568 | 0.1550 | 62.53 | 111.31 | 63.34 | 56.00 | 119.35 | 0.1219 | 0.2136 | 280 |
| 300 | 156.17 | 0.01596 | 0.1424 | 64.71 | 111.72 | 65.59 | 54.03 | 119.62 | 0.1254 | 0.2132 | 300 |

## Table A-18E  Properties of superheated refrigerant 134a (CF₄H₂)

$(v, \text{ft}^3/\text{lb}_m; u$ and $h, \text{Btu}/\text{lb}_m; s, \text{Btu}/(\text{lb}_m \cdot {}^\circ\text{R}))$

| Temp., °F | $v$ | $u$ | $h$ | $s$ | $v$ | $u$ | $h$ | $s$ |
|---|---|---|---|---|---|---|---|---|
| | 10 psia ($T_{sat} = -29.71°F$) | | | | 15 psia ($T_{sat} = -14.25°F$) | | | |
| Sat. | 4.3581 | 89.30 | 97.37 | 0.2265 | 2.9747 | 91.40 | 99.66 | 0.2242 |
| −20 | 4.4718 | 90.89 | 99.17 | 0.2307 | | | | |
| 0 | 4.7026 | 94.24 | 102.94 | 0.2391 | 3.0893 | 93.84 | 102.42 | 0.2303 |
| 20 | 4.9297 | 97.67 | 106.79 | 0.2472 | 3.2468 | 97.33 | 106.34 | 0.2386 |
| 40 | 5.1539 | 101.19 | 110.72 | 0.2553 | 3.4012 | 100.89 | 110.33 | 0.2468 |
| 60 | 5.3758 | 104.80 | 114.74 | 0.2632 | 3.5533 | 104.54 | 114.40 | 0.2548 |
| 80 | 5.5959 | 108.50 | 118.85 | 0.2709 | 3.7034 | 108.28 | 118.56 | 0.2626 |
| 100 | 5.8145 | 112.29 | 123.05 | 0.2786 | 3.8520 | 112.10 | 122.79 | 0.2703 |
| 120 | 6.0318 | 116.18 | 127.34 | 0.2861 | 3.9993 | 116.01 | 127.11 | 0.2779 |
| 140 | 6.2482 | 120.16 | 131.72 | 0.2935 | 4.1456 | 120.00 | 131.51 | 0.2854 |
| 160 | 6.4638 | 124.23 | 136.19 | 0.3009 | 4.2911 | 124.09 | 136.00 | 0.2927 |
| 180 | 6.6786 | 128.38 | 140.74 | 0.3081 | 4.4359 | 128.26 | 140.57 | 0.3000 |
| | 20 psia ($T_{sat} = -2.48°F$) | | | | 30 psia ($T_{sat} = 15.38°F$) | | | |
| Sat. | 2.2661 | 93.00 | 101.39 | 0.2227 | 1.5408 | 95.40 | 103.96 | 0.2209 |
| 0 | 2.2816 | 93.43 | 101.88 | 0.2238 | | | | |
| 20 | 2.4046 | 96.98 | 105.88 | 0.2323 | 1.5611 | 96.26 | 104.92 | 0.2229 |
| 40 | 2.5244 | 100.59 | 109.94 | 0.2406 | 1.6465 | 99.98 | 109.12 | 0.2315 |
| 60 | 2.6416 | 104.28 | 114.06 | 0.2487 | 1.7293 | 103.75 | 113.35 | 0.2398 |
| 80 | 2.7569 | 108.05 | 118.25 | 0.2566 | 1.8098 | 107.59 | 117.63 | 0.2478 |
| 100 | 2.8705 | 111.90 | 122.52 | 0.2644 | 1.8887 | 111.49 | 121.98 | 0.2558 |
| 120 | 2.9829 | 115.83 | 126.87 | 0.2720 | 1.9662 | 115.47 | 126.39 | 0.2635 |
| 140 | 3.0942 | 119.85 | 131.30 | 0.2795 | 2.0426 | 119.53 | 130.87 | 0.2711 |
| 160 | 3.2047 | 123.95 | 135.81 | 0.2869 | 2.1181 | 123.66 | 135.42 | 0.2786 |
| 180 | 3.3144 | 128.13 | 140.40 | 0.2922 | 2.1929 | 127.88 | 140.05 | 0.2859 |
| 200 | 3.4236 | 132.40 | 145.07 | 0.3014 | 2.2671 | 132.17 | 144.76 | 0.2932 |
| | 40 psia ($T_{sat} = 29.04°F$) | | | | 50 psia ($T_{sat} = 40.27°F$) | | | |
| Sat. | 1.1692 | 97.23 | 105.88 | 0.2197 | 0.9422 | 98.71 | 107.43 | 0.2189 |
| 40 | 1.2065 | 99.33 | 108.26 | 0.2245 | | | | |
| 60 | 1.2723 | 103.20 | 112.62 | 0.2331 | 0.9974 | 102.62 | 111.85 | 0.2276 |
| 80 | 1.3357 | 107.11 | 117.00 | 0.2414 | 1.0508 | 106.62 | 116.34 | 0.2361 |
| 100 | 1.3973 | 111.08 | 121.42 | 0.2494 | 1.1022 | 110.65 | 120.85 | 0.2443 |
| 120 | 1.4575 | 115.11 | 125.90 | 0.2573 | 1.1520 | 114.74 | 125.39 | 0.2523 |
| 140 | 1.5165 | 119.21 | 130.43 | 0.2650 | 1.2007 | 118.88 | 129.99 | 0.2601 |
| 160 | 1.5746 | 123.38 | 135.03 | 0.2725 | 1.2484 | 123.08 | 134.64 | 0.2677 |
| 180 | 1.6319 | 127.62 | 139.70 | 0.2799 | 1.2953 | 127.36 | 139.34 | 0.2752 |
| 200 | 1.6887 | 131.94 | 144.44 | 0.2872 | 1.3415 | 131.71 | 144.12 | 0.2825 |
| 220 | 1.7449 | 136.34 | 149.25 | 0.2944 | 1.3873 | 136.12 | 148.96 | 0.2897 |
| 240 | 1.8006 | 140.81 | 154.14 | 0.3015 | 1.4326 | 140.61 | 153.87 | 0.2969 |

## Table A-18E (Continued)

| Temp., °F | $v$ | $u$ | $h$ | $s$ | $v$ | $u$ | $h$ | $s$ |
|---|---|---|---|---|---|---|---|---|
| | 60 psia ($T_{sat}$ = 49.89°F) | | | | 70 psia ($T_{sat}$ = 58.35°F) | | | |
| Sat. | 0.7887 | 99.96 | 108.72 | 0.2183 | 0.6778 | 101.05 | 109.83 | 0.2179 |
| 60 | 0.8135 | 102.03 | 111.06 | 0.2229 | 0.6814 | 101.40 | 110.23 | 0.2186 |
| 80 | 0.8604 | 106.11 | 115.66 | 0.2316 | 0.7239 | 105.58 | 114.96 | 0.2276 |
| 100 | 0.9051 | 110.21 | 120.26 | 0.2399 | 0.7640 | 109.76 | 119.66 | 0.2361 |
| 120 | 0.9482 | 114.35 | 124.88 | 0.2480 | 0.8023 | 113.96 | 124.36 | 0.2444 |
| 140 | 0.9900 | 118.54 | 129.53 | 0.2559 | 0.8393 | 118.20 | 129.07 | 0.2524 |
| 160 | 1.0308 | 122.79 | 134.23 | 0.2636 | 0.8752 | 122.49 | 133.82 | 0.2601 |
| 180 | 1.0707 | 127.10 | 138.98 | 0.2712 | 0.9103 | 126.83 | 138.62 | 0.2678 |
| 200 | 1.1100 | 131.47 | 143.79 | 0.2786 | 0.9446 | 131.23 | 143.46 | 0.2752 |
| 220 | 1.1488 | 135.91 | 148.66 | 0.2859 | 0.9784 | 135.69 | 148.36 | 0.2825 |
| 240 | 1.1871 | 140.42 | 153.60 | 0.2930 | 1.0118 | 140.22 | 153.33 | 0.2897 |
| 260 | 1.2251 | 145.00 | 158.60 | 0.3001 | 1.0448 | 144.82 | 158.35 | 0.2968 |
| | 80 psia ($T_{sat}$ = 65.93°F) | | | | 90 psia ($T_{sat}$ = 72.83°F) | | | |
| Sat. | 0.5938 | 102.02 | 110.81 | 0.2175 | 0.5278 | 102.89 | 111.68 | 0.2172 |
| 80 | 0.6211 | 105.03 | 114.23 | 0.2239 | 0.5408 | 104.46 | 113.47 | 0.2205 |
| 100 | 0.6579 | 109.30 | 119.04 | 0.2327 | 0.5751 | 108.82 | 118.39 | 0.2295 |
| 120 | 0.6927 | 113.56 | 123.82 | 0.2411 | 0.6073 | 113.15 | 123.27 | 0.2380 |
| 140 | 0.7261 | 117.85 | 128.60 | 0.2492 | 0.6380 | 117.50 | 128.12 | 0.2463 |
| 160 | 0.7584 | 122.18 | 133.41 | 0.2570 | 0.6675 | 121.87 | 132.98 | 0.2542 |
| 180 | 0.7898 | 126.55 | 138.25 | 0.2647 | 0.6961 | 126.28 | 137.87 | 0.2620 |
| 200 | 0.8205 | 130.98 | 143.13 | 0.2722 | 0.7239 | 130.73 | 142.79 | 0.2696 |
| 220 | 0.8506 | 135.47 | 148.06 | 0.2796 | 0.7512 | 135.25 | 147.76 | 0.2770 |
| 240 | 0.8803 | 140.02 | 153.05 | 0.2868 | 0.7779 | 139.82 | 152.77 | 0.2843 |
| 260 | 0.9095 | 144.63 | 158.10 | 0.2940 | 0.8043 | 144.45 | 157.84 | 0.2914 |
| 280 | 0.9384 | 149.32 | 163.21 | 0.3010 | 0.8303 | 149.15 | 162.97 | 0.2984 |
| | 100 psia ($T_{sat}$ = 79.17°F) | | | | 120 psia ($T_{sat}$ = 90.54°F) | | | |
| Sat. | 0.4747 | 103.68 | 112.46 | 0.2169 | 0.3941 | 105.06 | 113.82 | 0.2165 |
| 80 | 0.4761 | 103.87 | 112.68 | 0.2173 | | | | |
| 100 | 0.5086 | 108.32 | 117.73 | 0.2265 | 0.4080 | 107.26 | 116.32 | 0.2210 |
| 120 | 0.5388 | 112.73 | 122.70 | 0.2352 | 0.4355 | 111.84 | 121.52 | 0.2301 |
| 140 | 0.5674 | 117.13 | 127.63 | 0.2436 | 0.4610 | 116.37 | 126.61 | 0.2387 |
| 160 | 0.5947 | 121.55 | 132.55 | 0.2517 | 0.4852 | 120.89 | 131.66 | 0.2470 |
| 180 | 0.6210 | 125.99 | 137.49 | 0.2595 | 0.5082 | 125.42 | 136.70 | 0.2550 |
| 200 | 0.6466 | 130.48 | 142.45 | 0.2671 | 0.5305 | 129.97 | 141.75 | 0.2628 |
| 220 | 0.6716 | 135.02 | 147.45 | 0.2746 | 0.5520 | 134.56 | 146.82 | 0.2704 |
| 240 | 0.6960 | 139.61 | 152.49 | 0.2819 | 0.5731 | 139.20 | 151.92 | 0.2778 |
| 260 | 0.7201 | 144.26 | 157.59 | 0.2891 | 0.5937 | 143.89 | 157.07 | 0.2850 |
| 280 | 0.7438 | 148.98 | 162.74 | 0.2962 | 0.6140 | 148.63 | 162.26 | 0.2921 |

## Table A-18E    (Continued)

| Temp., °F | $v$ | $u$ | $h$ | $s$ | $v$ | $u$ | $h$ | $s$ |
|---|---|---|---|---|---|---|---|---|
| | 140 psia ($T_{sat}$ = 100.56°F) | | | | 160 psia ($T_{sat}$ = 109.55°F) | | | |
| Sat. | 0.3358 | 106.25 | 114.95 | 0.2161 | 0.2916 | 107.28 | 115.91 | 0.2157 |
| 120 | 0.3610 | 110.90 | 120.25 | 0.2254 | 0.3044 | 109.88 | 118.89 | 0.2209 |
| 140 | 0.3846 | 115.58 | 125.54 | 0.2344 | 0.3269 | 114.73 | 124.41 | 0.2303 |
| 160 | 0.4066 | 120.21 | 130.74 | 0.2429 | 0.3474 | 119.49 | 129.78 | 0.2391 |
| 180 | 0.4274 | 124.82 | 135.89 | 0.2511 | 0.3666 | 124.20 | 135.06 | 0.2475 |
| 200 | 0.4474 | 129.44 | 141.03 | 0.2590 | 0.3849 | 128.90 | 140.29 | 0.2555 |
| 220 | 0.4666 | 134.09 | 146.18 | 0.2667 | 0.4023 | 133.61 | 145.52 | 0.2633 |
| 240 | 0.4852 | 138.77 | 151.34 | 0.2742 | 0.4192 | 138.34 | 150.75 | 0.2709 |
| 260 | 0.5034 | 143.50 | 156.54 | 0.2815 | 0.4356 | 143.11 | 156.00 | 0.2783 |
| 280 | 0.5212 | 148.28 | 161.78 | 0.2887 | 0.4516 | 147.92 | 161.29 | 0.2856 |
| 300 | 0.5387 | 153.11 | 167.06 | 0.2957 | 0.4672 | 152.78 | 166.61 | 0.2927 |
| 320 | 0.5559 | 157.99 | 172.39 | 0.3026 | 0.4826 | 157.69 | 171.98 | 0.2996 |
| | 180 psia ($T_{sat}$ = 117.74°F) | | | | 200 psia ($T_{sat}$ = 125.28°F) | | | |
| Sat. | 0.2569 | 108.18 | 116.74 | 0.2154 | 0.2288 | 108.98 | 117.44 | 0.2151 |
| 120 | 0.2595 | 108.77 | 117.41 | 0.2166 | | | | |
| 140 | 0.2814 | 113.83 | 123.21 | 0.2264 | 0.2446 | 112.87 | 121.92 | 0.2226 |
| 160 | 0.3011 | 118.74 | 128.77 | 0.2355 | 0.2636 | 117.94 | 127.70 | 0.2321 |
| 180 | 0.3191 | 123.56 | 134.19 | 0.2441 | 0.2809 | 122.88 | 133.28 | 0.2410 |
| 200 | 0.3361 | 128.34 | 139.53 | 0.2524 | 0.2970 | 127.76 | 138.75 | 0.2494 |
| 220 | 0.3523 | 133.11 | 144.84 | 0.2603 | 0.3121 | 132.60 | 144.15 | 0.2575 |
| 240 | 0.3678 | 137.90 | 150.15 | 0.2680 | 0.3266 | 137.44 | 149.53 | 0.2653 |
| 260 | 0.3828 | 142.71 | 155.46 | 0.2755 | 0.3405 | 142.30 | 154.90 | 0.2728 |
| 280 | 0.3974 | 147.55 | 160.79 | 0.2828 | 0.3540 | 147.18 | 160.28 | 0.2802 |
| 300 | 0.4116 | 152.44 | 166.15 | 0.2899 | 0.3671 | 152.10 | 165.69 | 0.2874 |
| 320 | 0.4256 | 157.38 | 171.55 | 0.2969 | 0.3799 | 157.07 | 171.13 | 0.2945 |
| | 300 psia ($T_{sat}$ = 156.17°F) | | | | 400 psia ($T_{sat}$ = 179.95°F) | | | |
| Sat. | 0.1424 | 111.72 | 119.62 | 0.2132 | 0.0965 | 112.77 | 119.91 | 0.2102 |
| 160 | 0.1462 | 112.95 | 121.07 | 0.2155 | | | | |
| 180 | 0.1633 | 118.93 | 128.00 | 0.2265 | 0.0965 | 112.79 | 119.93 | 0.2102 |
| 200 | 0.1777 | 124.47 | 134.34 | 0.2363 | 0.1143 | 120.14 | 128.60 | 0.2235 |
| 220 | 0.1905 | 129.79 | 140.36 | 0.2453 | 0.1275 | 126.35 | 135.79 | 0.2343 |
| 240 | 0.2021 | 134.99 | 146.21 | 0.2537 | 0.1386 | 132.12 | 142.38 | 0.2438 |
| 260 | 0.2130 | 140.12 | 151.95 | 0.2618 | 0.1484 | 137.65 | 148.64 | 0.2527 |
| 280 | 0.2234 | 145.23 | 157.63 | 0.2696 | 0.1575 | 143.06 | 154.72 | 0.2610 |
| 300 | 0.2333 | 150.33 | 163.28 | 0.2772 | 0.1660 | 148.39 | 160.67 | 0.2689 |
| 320 | 0.2428 | 155.44 | 168.92 | 0.2845 | 0.1740 | 153.69 | 166.57 | 0.2766 |
| 340 | 0.2521 | 160.57 | 174.56 | 0.2916 | 0.1816 | 158.97 | 172.42 | 0.2840 |
| 360 | 0.2611 | 165.74 | 180.23 | 0.2986 | 0.1890 | 164.26 | 178.26 | 0.2912 |

**Table A-19E**     **Properties of saturated nitrogen (N₂): Temperature table**

(**v**, ft³/lbₘ; **u** and **h**, Btu/lbₘ; **s**, Btu/(lbₘ·°R))

| Temp., °R $T$ | Press., psia $P$ | Specific Volume | | Internal Energy | | Enthalpy | | | Entropy | |
|---|---|---|---|---|---|---|---|---|---|---|
| | | Sat. Liquid $v_f$ | Sat. Vapor $v_g$ | Sat. Liquid $u_f$ | Sat. Vapor $u_g$ | Sat. Liquid $h_f$ | Evap. $h_{fg}$ | Sat. Vapor $h_g$ | Sat. Liquid $s_f$ | Sat. Vapor $s_g$ |
| 113.7* | 1.82 | 0.0185 | 23.73 | 35.32 | 119.91 | 35.33 | 92.59 | 127.90 | 0.5802 | 1.395 |
| 120.0 | 3.34 | 0.0188 | 13.56 | 38.33 | 120.93 | 38.34 | 90.98 | 129.32 | 0.6060 | 0.365 |
| 130.0 | 7.67 | 0.0193 | 6.321 | 43.19 | 122.47 | 43.22 | 88.22 | 131.44 | 0.6449 | 1.324 |
| 139.2 | 14.70 | 0.0198 | 3.473 | 47.71 | 123.77 | 47.77 | 85.45 | 133.22 | 0.6785 | 1.293 |
| 140.0 | 15.46 | 0.0199 | 3.315 | 48.09 | 123.87 | 48.15 | 85.21 | 133.36 | 0.6812 | 1.290 |
| 150.0 | 28.19 | 0.0205 | 1.899 | 53.02 | 125.11 | 53.13 | 81.89 | 135.02 | 0.7153 | 1.262 |
| 160.0 | 47.52 | 0.0213 | 1.164 | 58.00 | 126.14 | 58.19 | 78.19 | 136.38 | 0.7474 | 1.236 |
| 170.0 | 75.18 | 0.0222 | 0.750 | 63.08 | 126.91 | 63.39 | 73.96 | 137.35 | 0.7782 | 1.214 |
| 180.0 | 113.0 | 0.0233 | 0.502 | 68.30 | 127.36 | 68.79 | 69.07 | 137.86 | 0.8082 | 1.192 |
| 190.0 | 162.8 | 0.0246 | 0.344 | 73.75 | 127.40 | 74.49 | 63.28 | 137.77 | 0.8378 | 1.171 |
| 200.0 | 226.9 | 0.0262 | 0.239 | 79.52 | 126.84 | 80.62 | 56.24 | 136.86 | 0.8677 | 1.149 |
| 210.0 | 307.3 | 0.0285 | 0.164 | 85.86 | 125.33 | 87.48 | 47.20 | 134.68 | 0.8992 | 1.124 |
| 220.0 | 406.9 | 0.0325 | 0.107 | 93.57 | 121.80 | 96.01 | 33.85 | 129.86 | 0.9363 | 1.090 |
| 226.0 | 477.9 | 0.0394 | 0.071 | 101.46 | 115.66 | 104.95 | 16.97 | 121.92 | 0.9742 | 1.049 |
| 227.2* | 493.1 | 0.051 | 0.051 | 108.59 | 108.59 | 113.25 | 0 | 113.25 | 1.010 | 1.010 |

*113.7°R is the triple state and 227.2°R is the critical state.

*SOURCE:* Adapted from National Bureau of Standards, Technical Note 648, 1973.

**Table A-20E** Properties of nitrogen ($N_2$): Superheated-vapor table

($v$, $ft^3/lb_m$; $u$ and $h$, $Btu/lb_m$; $s$, $Btu/(lb_m \cdot °R)$)

| Temp., °R | $v$ | $u$ | $h$ | $s$ | $v$ | $u$ | $h$ | $s$ |
|---|---|---|---|---|---|---|---|---|
| | 20 psia ($T_{sat} = 144.1°R$) | | | | 50 psia ($T_{sat} = 161.0°R$) | | | |
| 200 | 3.755 | 134.83 | 148.73 | 1.364 | 1.454 | 133.97 | 147.44 | 1.295 |
| 250 | 4.740 | 143.87 | 161.42 | 1.421 | 1.866 | 143.30 | 160.58 | 1.354 |
| 300 | 5.714 | 152.82 | 173.98 | 1.467 | 2.266 | 152.40 | 173.38 | 1.400 |
| 350 | 6.682 | 161.74 | 186.49 | 1.505 | 2.659 | 161.41 | 186.03 | 1.439 |
| 400 | 7.647 | 170.65 | 198.97 | 1.539 | 3.050 | 170.37 | 198.61 | 1.473 |
| 450 | 8.610 | 179.54 | 211.4 | 1.568 | 3.438 | 179.30 | 211.1 | 1.502 |
| 500 | 9.572 | 188.42 | 223.9 | 1.594 | 3.826 | 188.22 | 223.6 | 1.529 |
| 550 | 10.53 | 197.31 | 236.3 | 1.618 | 4.212 | 197.13 | 236.1 | 1.553 |
| | 100 psia ($T_{sat} = 176.9°R$) | | | | 200 psia | | | |
| 200 | 0.684 | 132.40 | 145.07 | 1.238 | 0.289 | 128.33 | 139.04 | 1.166 |
| 250 | 0.908 | 142.31 | 159.13 | 1.300 | 0.427 | 140.18 | 156.01 | 1.243 |
| 300 | 1.116 | 151.68 | 172.36 | 1.349 | 0.542 | 150.19 | 170.25 | 1.295 |
| 350 | 1.319 | 160.84 | 185.26 | 1.389 | 0.649 | 159.69 | 183.71 | 1.336 |
| 400 | 1.518 | 169.90 | 198.01 | 1.423 | 0.752 | 168.96 | 196.81 | 1.371 |
| 450 | 1.715 | 178.91 | 210.7 | 1.452 | 0.853 | 178.12 | 209.7 | 1.401 |
| 500 | 1.910 | 187.88 | 223.3 | 1.479 | 0.953 | 187.20 | 222.5 | 1.428 |
| 550 | 2.105 | 196.83 | 235.8 | 1.503 | 1.052 | 196.23 | 235.2 | 1.453 |
| | 500 psia | | | | 1000 psia | | | |
| 250 | 0.132 | 131.52 | 143.78 | 1.141 | 0.038 | 106.30 | 113.40 | 0.994 |
| 300 | 0.197 | 145.25 | 163.45 | 1.213 | 0.083 | 135.26 | 150.59 | 1.131 |
| 350 | 0.247 | 156.09 | 178.96 | 1.261 | 0.115 | 149.67 | 170.95 | 1.194 |
| 400 | 0.293 | 166.10 | 193.23 | 1.299 | 0.142 | 161.23 | 187.45 | 1.238 |
| 450 | 0.337 | 175.73 | 206.9 | 1.331 | 0.166 | 171.77 | 202.5 | 1.274 |
| 500 | 0.379 | 185.16 | 220.3 | 1.359 | 0.189 | 181.81 | 216.8 | 1.304 |
| 550 | 0.421 | 194.46 | 233.4 | 1.384 | 0.211 | 191.56 | 230.6 | 1.330 |
| | 2000 psia | | | | 3000 psia | | | |
| 250 | 0.029 | 95.00 | 115.61 | 0.939 | 0.026 | 90.47 | 114.97 | 0.917 |
| 300 | 0.040 | 118.41 | 133.17 | 1.040 | 0.032 | 110.73 | 128.60 | 1.003 |
| 350 | 0.055 | 137.38 | 157.78 | 1.116 | 0.040 | 128.97 | 151.34 | 1.073 |
| 400 | 0.070 | 151.93 | 177.78 | 1.169 | 0.049 | 144.47 | 171.81 | 1.128 |
| 450 | 0.083 | 164.27 | 195.11 | 1.210 | 0.058 | 157.89 | 190.16 | 1.171 |
| 500 | 0.096 | 175.50 | 211.0 | 1.244 | 0.066 | 169.99 | 207.0 | 1.206 |
| 550 | 0.108 | 186.11 | 226.0 | 1.272 | 0.075 | 181.28 | 222.8 | 1.237 |

*SOURCE:* Adapted from National Bureau of Standards, Technical Note 648, 1973.

### Table A-21E    Thermodynamic properties of potassium

**(T, °R; P, psia; v, ft³/lb$_m$; h, Btu/lb$_m$; s, Btu/(lb$_m$·°R))**

| A. Saturation-Temperature Table | | | | | | | |
|---|---|---|---|---|---|---|---|
| **T** | **P** | **v$_f$** | **v$_g$** | **h$_f$** | **h$_g$** | **s$_f$** | **s$_g$** |
| 1500 | 1.523 | 0.0225 | 261.13 | 291.7 | 1170.8 | 0.6279 | 1.2140 |
| 1600 | 3.210 | 0.0229 | 130.62 | 310.4 | 1176.6 | 0.6400 | 1.1813 |
| 1700 | 6.185 | 0.0234 | 71.09 | 329.6 | 1181.8 | 0.6516 | 1.1529 |
| 1800 | 11.060 | 0.0239 | 41.49 | 349.3 | 1186.6 | 0.6628 | 1.1280 |

| B. Superheat Table | | | | | | | |
|---|---|---|---|---|---|---|---|
| **P** | **v** | **h** | **s** | **P** | **v** | **h** | **s** |
| **2000°R ($P_{sat}$ = 29.58 psia)** | | | | **2200°R ($P_{sat}$ = 65.83 psia)** | | | |
| Sat. | 16.71 | 1195.0 | 1.0867 | Sat. | 8.01 | 1203.5 | 1.0547 |
| 20.0 | 25.58 | 1213.7 | 1.1143 | 50.0 | 10.89 | 1220.8 | 1.0749 |
| 14.0 | 27.34 | 1225.5 | 1.1372 | 40.0 | 13.89 | 1232.0 | 1.0904 |
| 8.0 | 66.73 | 1237.3 | 1.1705 | 30.0 | 18.91 | 1243.5 | 1.1092 |
| **2400°R ($P_{sat}$ = 127.7 psia)** | | | | **2600°R ($P_{sat}$ = 222.9 psia)** | | | |
| Sat. | 4.37 | 1214.0 | 1.0303 | Sat. | 2.63 | 1224.4 | 1.0108 |
| 110 | 5.19 | 1225.2 | 1.0415 | 150 | 4.16 | 1255.2 | 1.0398 |
| 90 | 6.50 | 1238.5 | 1.0559 | 130 | 4.88 | 1264.3 | 1.0497 |
| 70 | 8.57 | 1252.3 | 1.0732 | 110 | 5.87 | 1273.5 | 1.0608 |
| 50 | 12.32 | 1266.5 | 1.0949 | 90 | 7.30 | 1283.1 | 1.0738 |

*SOURCE:* Air Force Aero Propulsion Laboratory, Technical Report 66-104, 1966.

**Table A-22E**  Constants for the Benedict-Webb-Rubin, Redlich-Kwong, and van der Waals equations of state

| 1. Benedict-Webb-Rubin; Units are atm, ft³/lbmol, and °R | | | | |
|---|---|---|---|---|
| Constants | $n$-Butane ($C_4H_{10}$) | Carbon Dioxide ($CO_2$) | Carbon Monoxide (CO) | Methane ($CH_4$) | Nitrogen ($N_2$) |
| $a$ | 7747 | 563.1 | 150.7 | 203.1 | 103.2 |
| $A_0$ | 2590 | 703.0 | 344.5 | 476.4 | 270.6 |
| $b$ | 10.27 | 1.852 | 0.676 | 0.868 | 0.598 |
| $B_0$ | 1.993 | 0.7998 | 0.8740 | 0.6827 | 0.6529 |
| $c$ | $4.219 \times 10^9$ | $1.989 \times 10^8$ | $1.387 \times 10^7$ | $3.393 \times 10^7$ | $9.713 \times 10^6$ |
| $C_0$ | $8.263 \times 10^8$ | $1.153 \times 10^8$ | $7.124 \times 10^6$ | $1.878 \times 10^7$ | $6.760 \times 10^6$ |
| $\alpha$ | 4.531 | 0.3486 | 0.5556 | 0.5120 | 0.5235 |
| $\gamma$ | 8.732 | 1.384 | 1.541 | 1.541 | 1.361 |

SOURCE: H. W. Cooper and J. C. Goldfrank, *Hydrocarbon Processing,* **46** (12): 141 (1967).

| 2. Redlich-Kwong; $a$ is in atm (ft³/lbmol)²(°R)$^{0.5}$, $b$ is in ft³/lbmol | | | | | |
|---|---|---|---|---|---|
| Substance | $a$ | $b$ | Substance | $a$ | $b$ |
| Carbon dioxide ($CO_2$) | 21,970 | 0.4757 | Oxygen ($O_2$) | 5,900 | 0.3522 |
| Carbon monoxide (CO) | 5,870 | 0.4395 | Propane ($C_3H_8$) | 62,190 | 1.0040 |
| Methane ($CH_4$) | 10,930 | 0.4757 | Refrigerant 134a | 66,700 | 1.060 |
| Nitrogen ($N_2$) | 5,300 | 0.4294 | Water ($H_2O$) | 48,460 | 0.3381 |

SOURCE: Computed from critical data.

| 3. van der Waals; $a$ is in atm (ft³/lbmol)², $b$ is in ft³/lbmol | | | | | |
|---|---|---|---|---|---|
| Substance | $a$ | $b$ | Substance | $a$ | $b$ |
| Acetylene ($C_2H_2$) | 1121 | 0.818 | Ethylene ($C_2H_4$) | 1158 | 0.922 |
| Air (equivalent) | 345.2 | 0.585 | Helium (He) | 8.66 | 0.376 |
| Ammonia($NH_3$) | 1076 | 0.598 | Hydrogen ($H_2$) | 62.8 | 0.426 |
| Benzene ($C_6H_6$) | 4736 | 1.898 | Methane ($CH_4$) | 581 | 0.685 |
| n-Butane ($C_4H_{10}$) | 3508 | 1.919 | Nitrogen ($N_2$) | 346 | 0.618 |
| Carbon dioxide ($CO_2$) | 926 | 0.686 | Oxygen ($O_2$) | 348 | 0.506 |
| Carbon monoxide (CO) | 372 | 0.632 | Propane ($C_3H_8$) | 2368 | 1.445 |
| Refrigerant 134a ($C_2F_4H_2$) | 2540 | 1.530 | Sulfur dioxide ($SO_2$) | 1738 | 0.911 |
| Ethane ($C_2H_6$) | 1410 | 1.041 | Water ($H_2O$) | 1400 | 0.488 |

SOURCE: Computed from critical data.

**Table A-23E** Values of the enthalpy of formation, Gibbs function of formation, absolute entropy, and enthalpy of vaporization at 77°F and 1 atm

($\Delta \bar{h}_f^\circ$, $\Delta \bar{g}_f^\circ$, and $\Delta \bar{h}_{fg}$ in Btu/lbmol; $\bar{s}^\circ$ in Btu/(lbmol·°R))

| Substance | Formula | $\Delta \bar{h}_f^\circ$ | $\Delta \bar{g}_f^\circ$ | $\bar{s}^\circ$ | $\bar{h}_{fg}$ |
|---|---|---|---|---|---|
| Carbon | C(s) | 0 | 0 | 1.36 | |
| Hydrogen | $H_2$(g) | 0 | 0 | 31.21 | |
| Nitrogen | $N_2$(g) | 0 | 0 | 45.77 | |
| Oxygen | $O_2$(g) | 0 | 0 | 49.00 | |
| Carbon monoxide | CO(g) | −47,540 | −59,010 | 47.21 | |
| Carbon dioxide | $CO_2$(g) | −169,300 | −169,680 | 51.07 | |
| Water | $H_2O$(g) | −104,040 | −98,350 | 45.11 | |
| Water | $H_2O$(f) | −122,970 | −102,040 | 16.71 | |
| Hydrogen peroxide | $H_2O_2$(g) | −58,640 | −45,430 | 55.60 | 26,260 |
| Ammonia | $NH_3$(g) | −19,750 | −7,140 | 45.97 | |
| Methane | $CH_4$(g) | −32,210 | −21,860 | 44.49 | |
| Acetylene | $C_2H_2$(g) | +97,540 | +87,990 | 48.00 | |
| Ethylene | $C_2H_4$(g) | +22,490 | +29,306 | 52.54 | |
| Ethane | $C_2H_6$(g) | −36,420 | −14,150 | 54.85 | |
| Propylene | $C_3H_6$(g) | +8,790 | +26,980 | 63.80 | |
| Propane | $C_3H_8$(g) | −44,680 | −10,105 | 64.51 | 6,480 |
| n-Butane | $C_4H_{10}$(g) | −54,270 | −6,760 | 74.11 | 9,060 |
| n-Octane | $C_8H_{18}$(g) | −89,680 | +7,110 | 111.55 | |
| n-Octane | $C_8H_{18}$(l) | −107,530 | +2,840 | 86.23 | 17,835 |
| Benzene | $C_6H_6$(g) | +35,680 | +55,780 | 64.34 | 14,550 |
| Methyl alcohol | $CH_3OH$(g) | −86,540 | −69,700 | 57.29 | |
| Methyl alcohol | $CH_3OH$(l) | −102,670 | −71,570 | 30.30 | 16,090 |
| Ethyl alcohol | $C_2H_5OH$(g) | −101,230 | −72,520 | 67.54 | |
| Ethyl alcohol | $C_2H_5OH$(l) | −119,470 | −75,240 | 38.40 | 18,220 |
| Oxygen | O(g) | +107,210 | +99,710 | 38.47 | |
| Hydrogen | H(g) | +93,780 | +87,460 | 27.39 | |
| Nitrogen | N(g) | +203,340 | +195,970 | 36.61 | |
| Hydroxyl | OH(g) | +16,790 | +14,750 | 43.92 | |

*SOURCES:* From the JANAF Thermochemical Tables, NSRDS-NBS-37, 1971; *Selected Values of Chemical Thermodynamic Properties,* NBS Technical Note 270-3, 1968; and API Res. Project 44, Carnegie Press, Carnegie Institute of Technology, Pittsburgh, Pa., 1953.

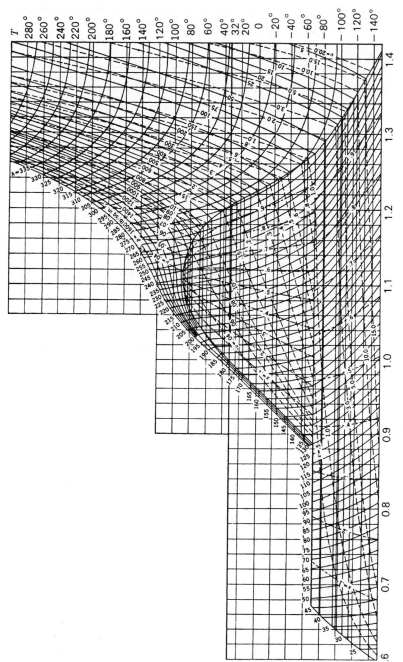

**Figure A-24E**  Temperature-entropy diagram for carbon dioxide ($CO_2$). $T$ is in °F, $h$ is in Btu/lb$_m$m, $v$ in ft³/lb$_m$, $s$ in Btu/lb$_m$·°R.

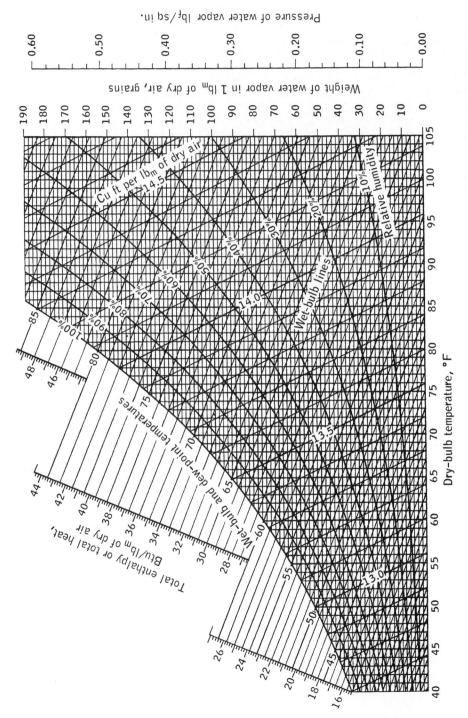

**Figure A-25E** Psychrometric chart, USCS units, barometric pressure 14.696 psia. (Copyright, 1942, by General Electric Company.)

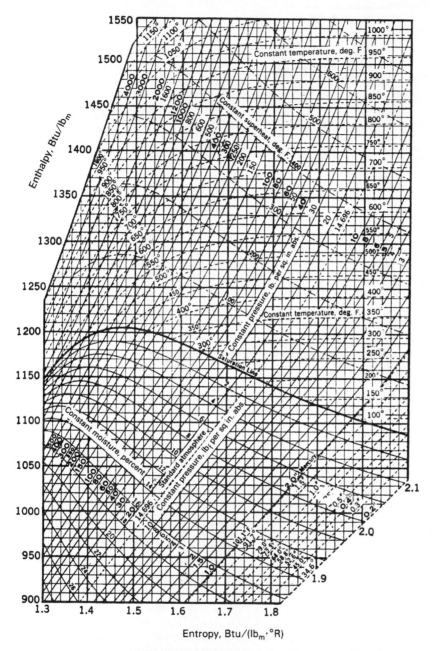

**Figure A-26E** Mollier diagram for steam. (*Source: J. H. Keenan and J. Keyes, "Thermodynamic Properties of Steam," Wiley, New York, 1936.*)

# A-3

# INTRODUCTION TO EES

---

## OVERVIEW

EES (pronounced 'ease') is an acronym for Engineering Equation Solver. The basic function provided by EES is the numerical solution of nonlinear algebraic and differential equations. In addition, EES provides built-in thermodynamic and transport property functions for many fluids, including water, dry and moist air, refrigerants, combustion gases, and others. Additional property data can be added by the user. The combination of equation solving capability and engineering property data makes EES a very powerful tool.

A license for EES is provided to departments of educational institutions which adopt this text by WCB/McGraw-Hill. If you need more information contact your local WCB/McGraw-Hill representative, call 1-800-338-3987, or visit our website at www.mhhe.com. A commercial version of EES can be obtained from:

**F-Chart Software**

4406 Fox Bluff Rd

Middleton, WI 53562

Phone: (608)-836-8531    FAX    (608)-836-8536

http://www.fchart.com

---

## BACKGROUND INFORMATION

The EES program is probably installed on your departmental computer. In addition, the license agreement for EES allows students and faculty in a participating educational department to copy the program for educational use on their personal computer systems. Versions of EES have been developed for Macintosh and PC/Windows computers. Ask your instructor for details.

To start EES from the Windows File Manager or Explorer, double-click on the EES program icon or on any file created by EES. You can also start EES from the Windows Run command in the Start menu. EES begins by displaying a dialog window which shows registration information, the version number, and other information. Click the OK button to dismiss the dialog window.

Detailed help is available at any point in EES. Pressing the F1 key will bring up a Help window relating to the foremost window. Clicking the Contents button will present the Help index shown below. Clicking on an underlined word (shown in green on color monitors) will provide help relating to that subject.

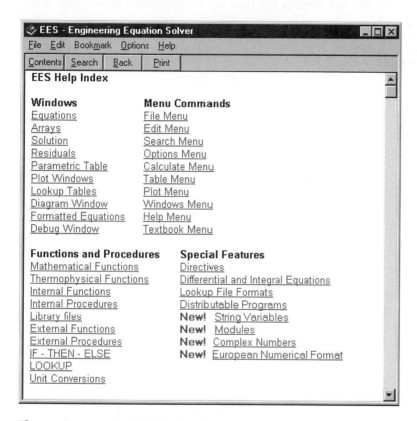

**Figure A3-1**    EES Help Index.

EES commands are distributed among nine pull-down menus. A brief summary of their functions follows. (A tenth pull-down menu, which is made visible with the Load Textbook command described below, provides access to problems from this text.) Many of the menu commands can be activated using the speed buttons on the button bar located below the menu bar. The function of each button can be determined by moving the mouse over the button and keeping it there for a few seconds.

The **System** menu appears above the file menu. The System menu is not part of EES, but rather a feature of the Windows Operating System. It holds commands which allow window moving, resizing, and switching to other applications. The figures shown here represent Windows 95. The appearance may be slightly different for other Windows operating systems.

The **File** menu provides commands for loading, merging and saving work files and libraries, and printing. The Load Textbook command in this menu reads the problem disk developed for this text and creates a new menu to the right of the Help menu for easy access to EES problems accompanying this text.

The **Edit** menu provides the editing commands to cut, copy, and paste information.

The **Search** menu provides Find and Replace commands for use in the Equations window.

The **Options** menu provides commands for setting the guess values and bounds of variables, the unit system, default information, and program preferences. A command is also provided for displaying information on built-in and user-supplied functions.

The **Calculate** menu contains the commands to check, format, and solve the equation set.

The **Tables** menu contains commands to set up and alter the contents of the Parametric and Lookup Tables and to do linear regression on the data in these tables. The Parametric Table which is similar to a spreadsheet, allows the equation set to be solved repeatedly while varying the values of one or more variables. The Lookup table holds user-supplied data which can be interpolated and used in the solution of the equation set.

The **Plot** menu provides commands to modify an existing plot or prepare a new plot of data in the Parametric, Lookup, or Array tables. Curve-fitting capability is also provided.

The **Windows** menu provides a convenient method of bringing any of the EES windows to the front or to organize the windows.

The **Help** menu provides commands for accessing the online help documentation.

A basic capability provided by EES is the solution of a set of nonlinear algebraic equations. To demonstrate this capability, start EES and enter this simple example problem in the Equations window.

Text is entered in the same manner as for any word processor. Formatting rules are as follows:

1. Upper and lower case letters are not distinguished. EES will (optionally) change the case of all variables to match the manner in which they first appear.

2. Blank lines and spaces may be entered as desired since they are ignored.

3. Comments must enclosed within braces { } or within quote marks " ". Comments may span as many lines as needed. Comments within braces may be nested in which case only the outermost set of { } are recognized. Comments within quotes will also be displayed in the Formatted Equations window. Comments that begin with a exclamation mark (!) will be displayed in a different color than normal comments.

4. Variable names must start with a letter and consist of any keyboard characters except ( ) ' | * / + −^{ } : " or ;. Array variables are identified with square braces around the array index or indices, e.g., X[5,3]. The maximum variable length is 30 characters.

5. Multiple equations may be entered on one line if they are separated by a semicolon (;). The maximum line length is 255 characters.

6. The caret symbol (^) or ** is used to indicate raising to a power.

7. The order in which the equations are entered does not matter.

8. The position of knowns and unknowns in the equation does not matter.

If you wish, you may view the equations in mathematical notation by selecting the **Formatted Equations** command from the **Windows** menu.

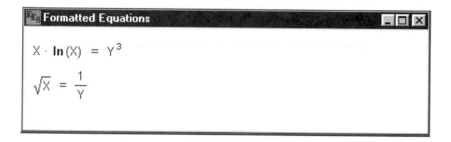

Select the **Solve** command from the **Calculate** menu. A dialog window will appear indicating the progress of the solution. When the calculations are completed, the button which change from Abort to Continue.

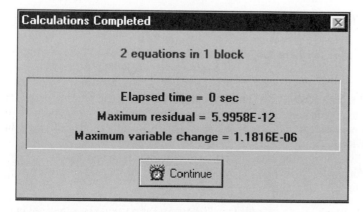

**Figure A3-2**    Calculations completed window.

Click the Continue button. The solution to this equation set will then be displayed.

# A THERMODYNAMICS EXAMPLE PROBLEM

In this section, a thermodynamics problem is worked from start to finish illustrating the built-in fluid property routines and some other capabilities of the EES program. The problem is Example 5-9 from the 6th edition of Wark and Richards, ***Thermodynamics.*** EES is particularly appropriate for this problem since the problem requires property data information which would have to be interpolated from property tables if the problem were done by hand. The problem to be solved is as follows.

Refrigerant 134a is throttled through a porous plug from a saturated liquid at 32°C to a final state where the pressure is 2 bars. Determine (a) the final temperature and (b) the physical state of the fluid at the exit.

This problem requires an energy balance on the steady-flow, steady-state open system shown in the figure.

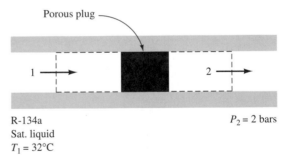

**Figure A3-3**      Steady-flow diagram for the example problem.

To solve this problem, it is necessary to select a system and set up appropriate mass and energy balances. The logical system choice is the throttle. The mass flow is steady and there is only one inlet and one outlet. No indication of the mass flowrate is provided in the problem statement so the analysis will be done on a per unit mass basis by assuming the mass flow rates in and out to both be 1 kg/sec.

An energy balance on the throttle neglecting kinetic and potential energy contributions and considering steady-state operation is

$$\dot{m}_1 h_1 - \dot{m}_2 h_2 + \dot{Q} + \dot{W} = 0 \qquad \text{[1]}$$

where $h$ = enthalpy [kJ/kg]. The valve is assumed to be well-insulated with no moving parts so that both $\dot{Q}$ and $\dot{W}$ are zero.

From relationships between the properties of R-134a:

$$h_1 = h(T_1, x_1) \qquad \textbf{[2]}$$

where $x_1$ is the quality of the refrigerant at the valve inlet. For saturated liquid, the quality is 0. Solving equations (1) and (2) shows that $h_1 = h_2$. All that is needed to complete the problem is to determine the temperature and quality for an exit pressure of 2 bars.

$$T_2 = T(P_2, h_2) \qquad \textbf{[3]}$$

$$x_2 = x(P_2, h_2) \qquad \textbf{[4]}$$

The values of, $\dot{Q}, \dot{W}, T_1, x_1$, and $P_2$ are all known. There are four unknowns: $T_2, h_1, h_2$, and $x_2$. Since there are four equations, the solution to the problem is defined. It is now only necessary to solve the equations. This is where EES can help.

Start EES or select the **New** command from the **File** menu if you have already been using the program. A blank Equations window will appear. Before entering the equations, however, set the unit system for the built-in thermophysical properties functions. To view or change the unit system, select **Unit System** from the **Options** menu.

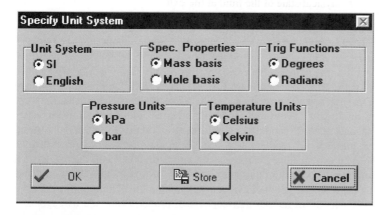

**Figure A3-4**    Unit System dialog.

EES is initially configured to be in SI units with $T$ in °C, $P$ in kPa, and specific property values in their customary units on a mass basis. These defaults may have been changed during a previous use. Click on the controls to set the units as shown above. Click the OK button (or press the Return key) to accept the unit system settings.

The equations can now be entered into the Equations window. Text is entered in the same manner as for any word processor. Note that an underscore is used to indicate a subscript. For example T_1 will be displayed as $T_1$ in the Formatted Equations and Solutions windows. After entering the equations for this problem and (optionally) checking the syntax using the **Check/Format** command in the **Calculate** menu, the Equations window will appear as shown. Comments (within quotes) are normally displayed in blue on a color monitor. Other formatting options are set with the **Preferences** command in the **Options** menu.

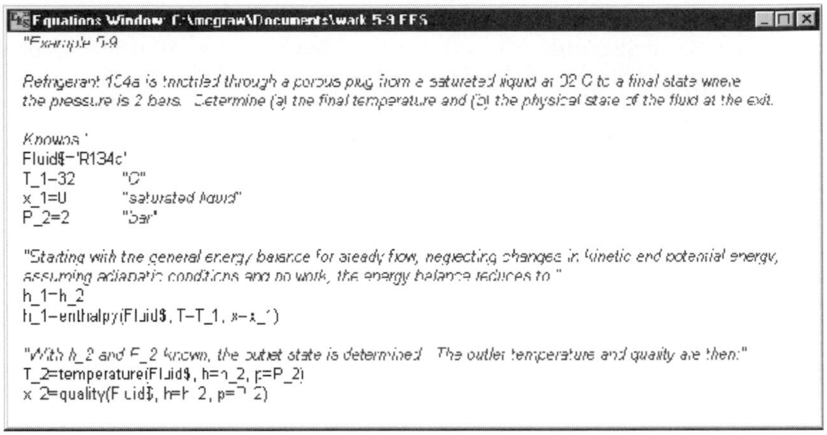

**Figure A3-5**    Equations window.

The thermodynamic property functions, such as **enthalpy** and **quality** require a special format. The first argument of the function is the substance name. The substance name can be a name, e.g., 'R134a', or a string variable which is identified by a $ as the last character in the name, e.g., Fluid$. The use of a string variable as is done here makes it very easy to change the analysis for different fluids. The following arguments are the independent variables preceded by a single identifying letter and an equal sign. Allowable letters are T, P, H, U, S, V, and X, corresponding to temperature, pressure, specific enthalpy, specific internal energy, specific entropy, specific volume, and quality. (For psychrometric functions, additional allowable letters are W, R, D, and B, corresponding to humidity ratio, relative humidity, dewpoint temperature, and wetbulb temperature. These designations are case insensitive.)

An easy way to enter functions, without needing to recall the format, is to use the **Function Information** command in the **Options** menu. This command will bring up the dialog window shown below. Click on the 'Thermophysical props' radio button. The list of built-in thermophysical property function will appear on the left with the list of substances on the right. Select the property function by clicking on its name, using the scroll bar, if necessary, to bring it into view. Select a substance in the same manner. An example of the function showing the format will appear in the Example rectangle at the bottom. The information in the rectangle may be changed, if needed. Clicking the Paste button will copy the Example into the Equations window at the cursor position.

It is usually a good idea to set the guess values and (possibly) the lower and upper bounds for the variables before attempting to solve the equations. This is done with the **Variable Information** command in the **Options** menu. Before displaying the Variable Information dialog, EES checks syntax and compiles newly entered and/or changed equations, and then solves all equations with one unknown. The Variable Information dialog will then appear.

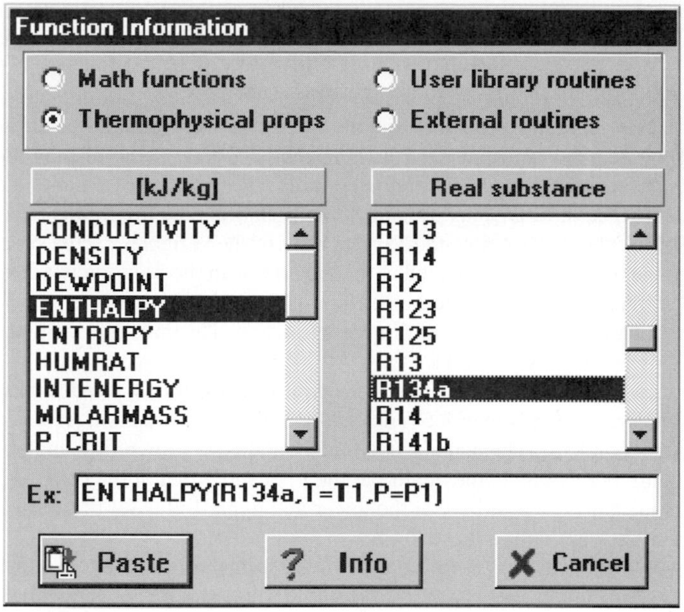

**Figure A3-6**     Function Information window.

The **Variable Information** dialog contains a line for each variable appearing in the Equations window. By default, each variable has a guess value of 1.0 with lower and upper bounds of negative and positive infinity. (The lower and upper bounds are shown in italics if EES has previously calculated the value of the variable. In this case, the Guess value column displays the calculated value. These italicized values may still be edited, which will force EES to recalculate the value of that variable.)

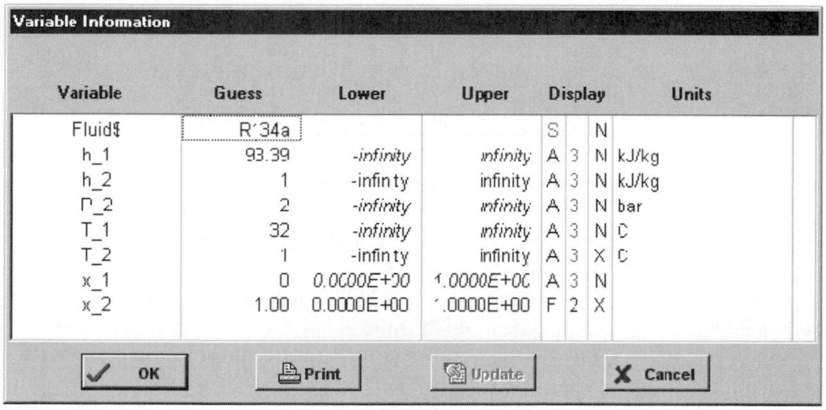

**Figure A3-7**     Variable Information dialog.

The A in the Display options column indicates that EES will automatically determine the display format for the numerical value of the variable when it is displayed in the Solution window. In this case, EES will select an appropriate number of digits, so the digits column to the right of the A is disabled. Automatic formatting is the default. Alternative display options are F (for fixed number of digits to the right of the decimal point) and E (for exponential format). The display and other defaults can easily be changed with the **Default Information** command in the **Options** menu. The third Display options column controls the hilighting effects such as normal (N), bold (B), or boxed (X). The units of the variables can be specified, if desired. The units will be displayed with the variable in the Solution window and/or in the Parametric Table. EES does not automatically do unit conversions but it can provide unit conversions using the **Convert** function. The units information entered here is only for display purposes.

With nonlinear equations, it is sometimes necessary to provide reasonable guess values and bounds in order to determine the desired solution. (It is not necessary for this problem.) The bounds of some variables are known from the physics of the problem. For example, the quality at the inlet and outlet of the throttle, x_1 and x_2, must be between 0 and 1. Set the lower and upper limits for these variables to ensure that they are within these limits.

To solve the equation set, select the **Solve** command from the **Calculate** menu. If the problem can be successfully solved, the Solution window shown below will appear. The problem is now completed since the values of $T_2$ and $x_2$ are determined. Note that the Solution window shown below shows the units of each variable. Units can be entered in the **Variable Information** dialog window or more simply by just clicking the right mouse button (or double-clicking the left mouse button) on the variable name in the Solution Window. A dialog window will appear in which the units and format of each selected variable can be changed. Special highlighting effects can also be applied in this dialog as was done for $T_2$ and $x_2$.

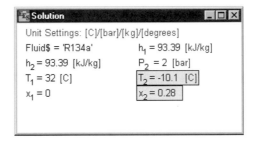

**Figure A3-8**     Solution window

One of the most useful features of EES is its ability to provide parametric studies. For example, in this problem, it may be of interest to see how the throttle outlet pressure varies with outlet temperature. A series of calculations can be automated and plotted using the commands in the **Tables** menu.

Select the **New Table** command. A dialog will be displayed listing the variables appearing in the Equations window. In this case, we will construct a table containing the variables P_2, T_2, and x_2. Click on P_2 from the variable list on the left. This will cause P_2 to be highlighted and the Add button will become active.

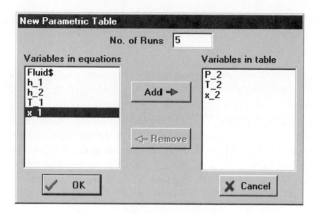

**Figure A3-9**    New Parametric Table dialog.

Now click the Add button to move P_2 to the list of variables on the right. Repeat for T_2, and x_2, using the scroll bar to bring the variable into view if necessary. (As a short cut, you can double-click on the variable name in the list on the left to move it to the list on the right.). Set the number of runs to 5 and then click the OK button to create the table.

The Parametric Table works much like a spreadsheet. You can type numbers directly into the cells. Numbers which you enter are shown in black and produce the same effect as if you set the variable to that value with an equation in the Equations window. Delete the P_2 = 2 equation currently in the Equations window or enclose it in comment brackets { }. This equation will not be needed because the value of P_2 will be set in the table. Now enter the values of P_2 for which T_2 is be determined. Values of 6 bar to 2 bar have been chosen for this example. (The values could also be automatically entered using **Alter Values** in the **Tables** menu or by using the Alter Values control at the upper right of each table column header.) The Parametric Table should now appear as shown below.

| | P₂ [bar] | T₂ [C] | x₂ |
|---|---|---|---|
| Run 1 | 6 | | |
| Run 2 | 5 | | |
| Run 3 | 4 | | |
| Run 4 | 3 | | |
| Run 5 | 2 | | |

**Figure A3-10**    Parametric Table window.

Now, select **Solve Table** from the **Calculate** menu. The Solve Table dialog window will appear allowing you to choose the runs for which the calculations will be done.

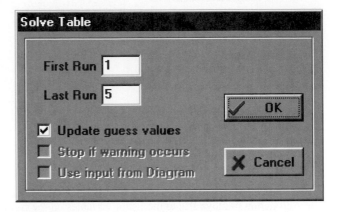

**Figure A3-11**          Solve Table dialog.

When the Update Guess Values control is selected, as shown, the solution for the last run will provide guess values for the following run. Click the OK button. A status window will be displayed, indicating the progress of the solution. When the calculations are completed, the values of T_2, and x_2 will be entered into the table. The values calculated by EES will be displayed in blue, bold or italic type depending on the setting made in the General Display tab of the **Preferences** dialog window in the **Options** menu.

| | $P_2$ [bar] | $T_2$ [C] | $x_2$ |
|---|---|---|---|
| Run 1 | 6 | 21.54 | 0.08 |
| Run 2 | 5 | 15.71 | 0.12 |
| Run 3 | 4 | 8.905 | 0.17 |
| Run 4 | 3 | 0.6509 | 0.22 |
| Run 5 | 2 | -10.1 | 0.28 |

**Figure A3-12**          Parametric Table window after calculations have been completed.

The relationship between variables such as P_2 and T_2 is now apparent, but it can more clearly be seen with a plot. Select **New Plot Window** from the **Plot** menu. The New Plot Window dialog window shown below will appear. Choose P_2 to be the x-axis by clicking on P_2 in the x-axis list. Click on T_2 in the y-axis list. When you click the OK button, the plot will be constructed and the plot window will appear as shown.

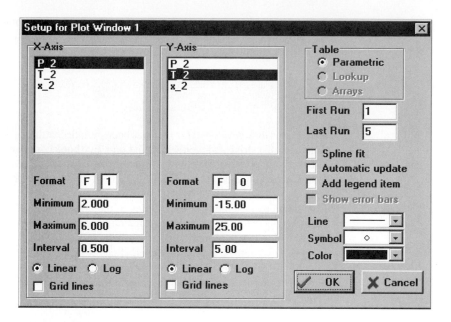

**Figure A3-13**    New Plot Window dialog.

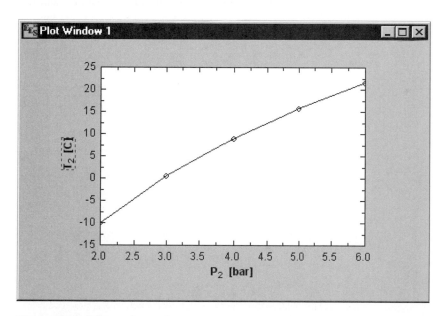

**Figure A3-14**    Plot window.

Once created, there are a variety of ways in which the appearance of the plot can be changed. Double-click the mouse in the plot rectangle or on the plot axis to see some of these options.

## LOADING A TEXTBOOK FILE

A Problems disk developed for EES has been included with this textbook. Place the disk in the disk drive and then select the **Load Textbook** command in the **File** menu. Use the Windows open file command to open the textbook problem index file which, for this book, is named Wark&Richards.TXB. A new menu called Thermodynamics will appear to the right of the Help menu. This menu will provide easy access to EES solutions to some of the examples in the book, organized by Chapter. As an example, select Chapter 5 from the **Thermodynamics** menu. A dialog window will appear listing the problems in Chapter 5. Select Example 5-9 Throttling Valve with R-134a. This problem is a modification of the problem you just entered.

At this point, you should explore. Try whatever you wish. You can't hurt anything. The online help (invoked by pressing F1) will provide details for the EES commands. EES is a powerful tool that you will find very useful in your studies.

# Symbols

| | |
|---|---|
| $A$ | Area |
| | Helmholtz function, $A = U - TS$ |
| $a$ | Specific Helmholtz function, $a = u - Ts$ |
| | Virial coefficient |
| | Acceleration |
| $b$ | Virial coefficient |
| $C$ | Number of components (in Gibbs' phase rule) |
| | A constant |
| COP | Coefficient of performance |
| CV | Control volume |
| $c$ | Virial coefficient |
| | Speed of light |
| $c_v$ | Specific heat at constant volume, $(\partial u/\partial T)_v$ |
| $c_p$ | Specific heat at constant pressure, $(\partial h/\partial T)_p$ |
| $d$ | An infinitesimal increase in a point function |
| $E$ | Stored energy |
| $\vec{E}$ | Electric field strength |
| $e$ | Specific stored energy |
| $F$ | Force |
| | Variance or degrees of freedom (in phase rule) |
| $FA$ | Fuel-air ratio |
| $F_k$ | Measurable generalized force |
| $F_{k,\text{eq}}$ | Equilibrium generalized force |
| $G$ | Gibbs function, $G = H - TS$ |
| $\Delta G$ | Gibbs-function change for a unit reaction |
| $\Delta G_T^0$ | Standard-state Gibbs-function change for a unit reaction |

| | |
|---|---|
| $g$ | Local acceleration of gravity |
| $H$ | Enthalpy, $H = U + PV$ |
| $\vec{H}$ | Magnetic field strength |
| $\Delta h_R$ | Enthalpy of reaction |
| $\Delta h_f$ | Enthalpy of formation |
| $\Delta h_c$ | Enthalpy of combustion |
| $h$ | Specific enthalpy, $h = u + Pv$ |
| | Convection heat transfer coefficient |
| | Planck's constant |
| $I$ | Irreversibility |
| KE | Kinetic energy |
| ke | Specific kinetic energy |
| $K_0$ | Standard-state equilibrium constant |
| $K_T$ | Isothermal coefficient of compressibility |
| $K_p$ | Equilibrium constant based on partial pressures for ideal-gas reactions |
| $k$ | Specific-heat ratio, $c_p/c_v$ |
| | Boltzmann's constant |
| $k_s$ | Spring constant |
| $L$ | Length |
| $M$ | Mass, as a dimension |
| | Molar mass, or molecular weight |
| $\vec{M}$ | Magnetization per unit volume |
| MEP | Mean effective pressure |
| $m$ | Mass of a substance |
| $\dot{m}$ | Mass flow rate |
| $mf$ | Mass fraction |
| $N$ | Number of moles |
| $N_A$ | Avogadro's number |
| $N_e$ | Moles of electrons |
| $n$ | Number of particles |
| | Polytropic constant |
| $n_i$ | Ratio of moles of compound $i$ to moles of fuel, $(N_i/N_{\text{fuel}}, \dot{N}_i/\dot{N}_{\text{fuel}})$ |
| $P$ | Pressure |
| | Number of phases (in the phase rule) |
| $\vec{P}$ | Polarization |
| PE | Potential energy |

| pe | Specific gravitational potential energy |
|---|---|
| $P_r$ | Reduced pressure |
| $p_r$ | Relative pressure |
| $p_i$ | Component pressure |
| $p_i'$ | Partial pressure |
| $Q$ | Heat transfer |
| $\dot{Q}$ | Heat transfer rate |
| $q$ | Heat transfer per unit mass |
| $q''$ | Heat transfer rate per unit area, heat flux |
| $Q_c$ | Electric charge |
| $Q_e$ | Electric charge |
| $R$ | Gas constant, $R_u/M$ |
| $R_u$ | Universal gas constant |
| $r$ | Distance between masses |
| | Compression ratio |
| $r_c$ | Cutoff ratio |
| $r_p$ | Pressure ratio |
| $S$ | Entropy |
| $s$ | Specific entropy |
| | Distance |
| $T$ | Temperature |
| $T_r$ | Reduced temperature |
| $t$ | Time |
| $U$ | Internal energy |
| $u$ | Specific internal energy |
| $\Delta u_R$ | Internal energy of reaction |
| $\mathbf{V}$ | Velocity |
| $V$ | Volume |
| | Electrostatic potential |
| $v$ | Specific volume |
| $v_r$ | Relative volume |
| $v_r'$ | Pseudoreduced volume |
| $W$ | Work transfer |
| | Weight |
| | Thermodynamic probability |
| $\dot{W}$ | Power; Work transfer rate |
| $W_{\text{pot}}$ | Work potential |

| | |
|---|---|
| $w$ | Specific weight |
| $X_k$ | Generalized displacement |
| $x$ | Quality |
| | Mole fraction |
| | Cartesian coordinate |
| $y$ | Mole fraction in the vapor phase |
| | Cartesian coordinate |
| $Z$ | Compressibility factor, $Z = Pv/(RT)$ |
| $z$ | Height |
| | Cartesian coordinate |

## GREEK SYMBOLS AND SPECIAL NOTATION

| | |
|---|---|
| $\beta$ | Isobaric compressibility |
| | Work ratio for a combined cycle |
| $\gamma$ | Surface tension |
| $\Delta$ | A finite increase in a point function or property |
| $\delta$ | Symbol for an infinitesimal increase in a path function |
| $\varepsilon$ | Energy of a particle |
| | Second law effectiveness |
| | Strain |
| $\Phi$ | Closed-system availability |
| $\Phi_Q$ | Availability transfer with heat transfer |
| $\eta$ | Efficiency |
| $\mathscr{F}$ | Faraday constant |
| $\theta$ | Temperature function |
| $\mu_{JT}$ | Joule-Thomson coefficient |
| $\mu_i$ | Chemical potential of the $i$th component |
| $\mu_o$ | Permeability of free space |
| $\nu$ | Fundamental frequency of an oscillator |
| | Stoichometric coefficient |
| $\Delta\nu$ | Change in the stoichiometric coefficients for a unit reaction |
| $\xi$ | emf, electromotive force |
| $\rho$ | Density |
| $\sigma$ | Entropy generation or production |
| | Stress |
| $\sigma_Q$ | Entropy production due to heat transfer |
| $\dot{\sigma}$ | Entropy production rate per unit time |

| $\tau$ | Time |
|---|---|
| | Torque |
| $\phi$ | Closed-system availability per unit mass |
| | Equivalence ratio |
| | Relative humidity |
| $\Psi$ | Total stream availability |
| $\psi$ | Stream availability per unit mass |
| $\omega$ | Angular velocity |
| | Specific humidity, humidity ratio |
| $\sum_i x_i$ | The sum $x_1 + x_2 + \cdots + x_n$ |
| $\prod_i x_i$ | The product $x_1 x_2 x_3 \cdots x_n$ |
| $\equiv$ | Identity symbol, used to specify a definition |

## SUBSCRIPTS

| $a$ | End state of actual process |
|---|---|
| $c$ | Critical state |
| cm | Control mass |
| cv | Control volume |
| $f$ | Saturated-liquid value |
| $fg$ | Change in value between saturated-liquid and saturated-vapor phases |
| $g$ | Saturated-vapor value |
| $H$ | High temperature (as in $T_H$ and $Q_H$) |
| $L$ | Low temperature (as in $T_L$ and $Q_L$) |
| $m$ | Mixture value |
| | Per unit mass |
| $r$ | Relative value |
| $s$ | End state of isentropic process |
| $R$ | Reduced state |
| $v$ | Vapor state |
| $x$ | Property value in wet region |

## SUPERSCRIPT

| o | Standard state: an ideal gas at 1 atm pressure |
|---|---|
| $\vec{\ }$ (arrow) | Vector |
| $\bar{\ }$ (bar) | Quantity per unit mole |
| $\dot{\ }$ (dot) | Transfer rate per unit time |

# Selected Problem Answers

**1-1.** (*a*) 25°C, 1 bar, 0.855 m³/kg; 2 m³, 2.34 kg; (*b*) 11.29 N/m³

**1-2.** (*a*) 5 m³, 4985 kg; 25°C, 1 bar, 0.0010 m³/kg; (*b*) 9670 N/m³

**1-3.** (*a*) 8 L, 8 kg; 20°C, 1 bar, $10^{-3}$ m³/kg; (*b*) 10,560 N/m³

**1-5.** (*a*) 420, (*b*) 1082

**1-7.** (*a*) 124.3, (*b*) 317

**1-8.** (*a*) 18.4, (*b*) 36.7

**1-10.** (*a*) 2000, (*b*) 7600

**1-12.** (*a*) 28.75, (*b*) 9.25

**1-13.** (*a*) 67.9, (*b*) 5.65

**1-15.** 0.75

**1-17.** 0.667

**1-18.** 42.9

**1-20.** 2.04

**1-22.** (*a*) 10.2, (*b*) 12.93, (*c*) 0.75

**1-23.** 1035

**1-25.** (*a*) 0.68, (*b*) 42.0, (*c*) 420, (*d*) 322

**1-27.** 15.18

**1-28.** (*a*) 20.2, (*b*) 4.99, (*c*) 9.80, (*d*) 24.6

**1-30.** 188.1, 1.88

**1-32.** (*a*) 6.13, (*b*) 83.7, (*c*) 93.0

**1-33.** 938

**1-35.** 1972

**1-37.** (*a*) 2700, (*b*) 28

**1-38.** (*a*) 391, (*b*) 1071, (*c*) 1402

**1-40.** (*a*) 401, (*b*) 27.3

**1-42.** (*a*) 25.6, (*b*) 43.5

**1-43.** (*a*) 0.946, (*b*) 1.085

**1-45.** 297

**1-47.** 464

**1-48.** 711

**2-1.** (*a*) 1.71, (*b*) 1.02

**2-2.** (*a*) 111.8, (*b*) 87

**2-3.** (*a*) 10.5, (*b*) 20

**2-5.** 878,000

**2-7.** (*a*) 114.4, (*b*) 352

**2-8.** 53.4

**2-10.** (*a*) 41.2, (*b*) −5.0

**2-12.** (*a*) −23,170; (*b*) 42,500

**2-13.** (*a*) 31.4, (*b*) 310.5, 1.04

**2-15.** (*a*) 57.1, (*b*) 200, 0.88

**2-17.** 260.4

**2-18.** 2.87

**2-20.** (*a*) 0.093, (*b*) 1.18

**2-22.** (*a*) 9, −17; (*b*) −10, 80; (*c*) 3, 6; (*d*) −4, 39; (*e*) 6, 35; (*f*) 0, 16

**2-23.** (*a*) −3, 0; (*b*) 5, 6; (*c*) 3, 6, −3, −3

**2-25.** (*a*) 1, 2; (*b*) 2, 3, −1; (*c*) 4, −1, −7

**2-27.** −7.6

**2-28.** 117

**2-30.** 10.6

**2-32.** 110.5

**2-33.** 1610

**2-35.** (*a*) 128, (*b*) 188

**2-37.** 2.20, 0; −2.73, 10.1; −4.23, 0; 0, −5.32

**2-38.** 0.168, 0; −0.119, 0.473; −0.137, 0; 0, −0.385

**2-40.** 0.899, 0; −0.648, 2.55; −0.68, 0; 0, −2.124

**2-42.** −5.0

**2-43.** (*a*) −15,000, (*b*) −10

**2-45.** (*a*) 2.375 MPa, −0.075 MPa/cm$^3$; (*b*) 25

**2-47.** (*a*) −7000, (*b*) −4545, (*c*) −3000

**2-48.** (*a*) −120, (*b*) −73.3, (*c*) −48.0

**2-50.** (*a*) 400, 180, 80; (*c*) −57,600

**2-52.** (*a*) −4320, (*b*) −2640, (*c*) −1728

**2-53.** (*b*) −3560

**2-55.** (*a*) −0.00825, (*b*) 2.45

**2-57.** (*a*) 106, (*b*) 1.34, 0.868, (*c*) 104

**2-58.** (*a*) 1.40, 0.812; (*b*) 78.8

**2-60.** (*a*) 1165, (*b*) 1.403, 8.425, (*c*) 1160

**2-62.** (*a*) −82,340; (*b*) −53,760

**2-63.** −0.259

**2-65.** −0.85

**2-67.** −14.75

**2-68.** −191

**2-70.** (*b*) 0.567, (*c*) 0.470

**2-72.** (*a*) bar·m$^3$, (*c*) 218, (*d*) 151

**2-73.** (*c*) −18,920, (*d*) −11,900

**2-75.** (*a*) 12,250; (*b*) 1315

**2-77.** (*a*) 9, (*b*) 8

**2-78.** (*a*) 3.93, (*b*) 7.84

**2-80.** (*a*) 0.28, (*b*) 2500, (*c*) −120

**2-82.** 31.05

**2-83.** (*a*) 51.5, (*b*) 2010

**2-85.** (*a*) 4.26, (*b*) 5.35

**2-87.** (*a*) 12.5, (*b*) 40, (*c*) −5.0

**2-88.** (*a*) 0.4, (*b*) 500, (*c*) 20

**2-90.** 0.313

**2-92.** (*a*) 150, (*b*) −1.0, (*c*) 6.0

**2-93.** (*a*) 20.7, (*b*) −2.95, (*c*) 43.15

**2-95.** (*a*) −7490, (*b*) 5830, (*c*) 5400, (*d*) 7920

**2-97.** 382

**2-98.** (*a*) 0.0080, (*b*) −1.60

**3-2.** (*a*) 275.6, 0.761, 2259; (*b*) 400, 203.0; (*c*) 74.36, 100, 25.57; (*d*) 3.613, 0, 1.0797

**3-3.** (*a*) 147.9, 1.0882, 0.0; (*b*) 179.9, 117, 60; (*c*) 99.4, 3230.9, SH; (*d*) 25, 590.52, CL

**3-5.** (*a*) 401.04, 1.66, 1119.4; (*b*) 600, 3.409; (*c*) 29.82, 0.01700, 0; (*d*) 247.1, 1202.0, 100

**3-7.** (*a*) 0.02339, 1.0018; (*b*) 151.9, 2748.7; (*c*) 36.383, 2661.4; (*d*) 158.9, 252.8; (*e*) 1.155, 852.8; (*f*) 4.758, 2112.2; (*g*) 320, 2817.1; (*h*) 0.50, 64.28; (*i*) 2604, 59.76; (*j*) 30, 2941.3; (*k*) 100, 1.0361

**3-8.** (*a*) 11.529, 0.01663; (*b*) 292.73, 1178.0; (*c*) 3.952, 1172.7; (*d*) 327.86, 2.226; (*e*) 0.016058, 71.99; (*f*) 247.1, 954; (*g*) 500, 1168.0; (*h*) 0.50, 1.323; (*i*) 1108.0, 3.792; (*j*) 140, 1275.1; (*k*) 100, 0.016034

**3-10.** (*a*) 8.93, 0.0509, 100; (*b*) 5, 0.04416, NA; (*c*) 46.32, 0.00875, 50; (*d*) 6.853, 85.18, 0

**3-12.** (*a*) 39.39, 0.0144, 70; (*b*) 8.624, 265.45, 100; (*c*) 0.04633, 280.16, SH; (*d*) 5.84, 0.000784, 0

**3-13.** (*a*) 120, 0.6927; (*b*) 85.788, 102.54, 100; (*c*) 49.89, 0.523, 65.7; (*d*) 90.54, 40.61, 0

**3-15.** (*a*) 3.8756, 0.788 × 10$^{-3}$, (*b*) 21.58, 259.19; (*c*) 0.03157, 253.83; (*d*) 2.48, 0.02575; (*e*) 0.788 × 10$^{-3}$, 60.73; (*f*) 1.5748, 133.53; (*g*) 50, 260.09; (*h*) 0.70, 0.01266; (*i*) 243.1, 0.0265; (*j*) 8, 273.66; (*k*) 77.26, 0.816 × 10$^{-3}$

**3-18.** 1251

**3-20.** 411

**3-22.** (*a*) 0.0967, (*b*) 1.285

**3-23.** (*a*) 0.00101, (*b*) 2.043

**3-25.** (*a*) 372.8, (*b*) −322

**3-27.** (*a*) 0.0782, (*b*) 0.0293

**3-28.** (*a*) 0.186 (*b*) 0.0447

**3-30.** (*a*) 1622

**3-32.** (*a*) −21.4, (*b*) −1.345

**3-33.** (*a*) −107.75, (*b*) 220.2

**3-35.** (*a*) 99.63, (*b*) 0.221, (*c*) 3.53

**3-37.** (*a*) 213.03, (*b*) 0.32, (*c*) 2.13

**3-38.** (*a*) 0.05319, 0.0469; (*b*) 69.87; (*c*) −15.49

**3-40.** (*a*) 0.828, 0.692; (*b*) 18.11, (*c*) −11.29

**3-42.** (*a*) 15.54, 2163; (*b*) 500, 3445

**3-43.** (*a*) 15.54, (*b*) 0.778, (*c*) −0.0982, (*d*) −1357

**3-45.** (*a*) 300, 5.00; (*b*) 308, 6.14; (*c*) 23

**3-47.** (*a*) 0.0318, 256.73; (*b*) 0.0357, 251.47;
(*c*) 0.5%, −2%

**3-48.** (*a*) 79.67, 95.21; (*b*) 84.08, 84.13; (*c*) 5.5, −11.6

**3-50.** (*a*) 48.60, 54.61; (*b*) 49.91, 49.91;
(*c*) 2.7%, −8.6%

**3-52.** (*b*) 15.34, (*c*) 14.99, (*d*) −22.01, (*e*) −14.99,
(*f*) −32

**3-53.** (*b*) ≈ −38,000; (*c*) −37,670; (*d*) 59.2, (*e*) 48.4

**3-55.** 284

**3-57.** 119

**3-58.** (*a*) −134.6, (*b*) 1688

**3-60.** (*a*) 72, −10,120; (*b*) 78.8, −7360; (*c*) 27

**3-62.** 17.9

**3-63.** (*a*) 110, (*b*) 0.91

**3-65.** (*a*) 0.0701, (*b*) 500

**3-67.** 47.0

**3-68.** (*a*) 1.98, (*b*) 287

**3-70.** 161.5

**3-72.** (*a*) 4.13, (*b*) 202

**3-73.** (*a*) 2150, (*b*) −2170

**3-75.** (*b*) 3.722, (*c*) −52.2

**3-77.** (*a*) 582, out

**3-78.** (*a*) 9.93

**3-80.** (*b*) 0.060, (*c*) 17.77, (*d*) −21.63, (*e*) 2.36

**3-82.** −190

**3-83.** (*a*) 250, (*b*) 2118

**3-85.** (*a*) 233.9, (*b*) 2243.9, (*c*) −19.2

**3-87.** (*a*) −64.3, 210, 274; (*b*) −1117

**3-88.** (*a*) −36.1, 118, 154.1; (*b*) −808

**3-90.** (*a*) 1.155, (*b*) 39.1, (*c*) −33.0

**3-92.** (*a*) 69.4, −856, −926; (*b*) 0, 1283, 1283

**3-93.** −3190

**3-95.** 0.554

**4-1.** (*a*) 17.3, (*b*) 29.3

**4-2.** (*a*) 0.0884, (*b*) 23.0

**4-3.** (*a*) 198, (*b*) 600

**4-5.** (*a*) 9.32, (*b*) 837

**4-7.** 79.8

**4-8.** (*a*) 267, (*b*) 23.4

**4-10.** (*a*) 42, (*b*) 0.136

**4-12.** 4.1

**4-13.** (*a*) 2.53, (*b*) 2.74

**4-15.** 0.0382

**4-17.** 17.2

**4-18.** 20.3

**4-20.** 7.3

**4-22.** (*a*) 390, (*b*) 450, (*c*) 461.6, (*d*) 322.6,
(*e*) 159.3

**4-23.** (*a*) 1000, (*b*) 450, (*c*) 450, (*d*) 466.5, (*e*) 147.4

**4-25.** (*a*) 540, (*b*) 500, (*c*) 86.7, (*d*) 91.53, (*e*) 75.4

**4-27.** (*a*) 1000, (*b*) 5.0, (*c*) −865, (*d*) −851

**4-28.** (*a*) (1) 6036, (2) 6034, (3) 0.22%; (*b*) (1) 12,841;
(2) 12,812; (3) −0.5%

**4-30.** (*a*) (1) 6700, (2) 6698, (3) −0.37%;
(*b*) (1) 11,628, (2) 11,630, (3) −0.1%

**4-32.** (*a*) (1) 7428, (2) 7425, (3) −1.7%; (*b*) (1) 10,444,
(2) 10,443, (3) −2.1%

**4-33.** (*a*) (1) 3823, (2) 3820, (3) 0.2%; (*b*) (1) 4514,
(2) 4511, (3) −0.47%

**4-35.** −2072

**4-37.** 0.50

**4-38.** 2.5

**4-40.** (*a*) 480, (*b*) 2773

**4-42.** (*a*) 0, (*b*) −8.38

**4-43.** 1/2

**4-45.** (*a*) 17,370; (*b*) 8270; (*c*) 10,800

**4-47.** (*a*) 34, (*b*) −11.0, (*c*) −10.9

**4-48.** (*a*) 2.42, (*b*) 9.13

**4-50.** (*a*) 30.9, (*b*) 20.0, (*c*) −10.9

**4-52.** −5950

**4-53.** (*a*) 1.60, (*b*) 3030

**4-55.** (*a*) 4.07, (*b*) 3.50, (*c*) 0.57, out

**4-57.** 137

**4-58.** (*a*) 228, (*b*) 227

**4-60.** (*a*) 0.092, (*b*) 0.079

**4-62.** (*a*) 3.51, (*b*) −1.14

**4-63.** (*a*) 0.355, (*b*) 3.97

**4-65.** (*a*) 340, (*b*) 2.13

**4-67.** (*a*) −103.9, (*b*) 0.665

**4-68.** (*a*) −2.29, (*b*) 0.265

**4-70.** (*a*) 0.0975, (*b*) 551, (*c*) 15.08, (*d*) 19.65, (*e*) −4.57

**4-72.** 1.02

**4-73.** 130

**4-75.** (*a*) 202, (*b*) 0.532, (*c*) $5.25 \times 10^4$, (*d*) −14.6

**4-77.** (*a*) 67, (*b*) −148, (*c*) 38, (*d*) 0.63

**4-78.** 3.57

**4-80.** (*a*) 345, (*b*) 5.48

**4-82.** (*a*) $2.43 \times 10^{-3}$, (*b*) 907, (*c*) 0.147, (*d*) 150, (*e*) −0.046

**4-83.** (*a*) 195, (*b*) 1.70

**4-85.** (*a*) 1.875, 15.35; (*b*) −6270, (*c*) 1.55

**4-87.** (*a*) 41.3, (*b*) 6.75, (*c*) 0.023

**4-88.** (*a*) 159, (*b*) 17.1, (*c*) 0.680

**4-90.** −24.8

**4-92.** (*a*) 0.315, (*b*) 275, (*c*) -284, 0; −69.0, 69.0; 395, −110

**4-93.** (*a*) $1.5 \times 10^{-3}$, (*b*) 0.150, 435; (*c*) 0.154, (*d*) 609; (*e*) 0.075, 0.263

**4-95.** (*a*) −89.05, (*b*) 383

**4-97.** −4.50, 5.4, 0.82

**4-98.** 37.9

**4-100.** (*a*) 0.0164, (*b*) 0.01214, (*c*) 0.0122

**4-102.** (*a*) 94.7, (*b*) 82, (*c*) 80

**4-103.** (*a*) 647.5, (*b*) 958, (*c*) 22.9, (*d*) 18.95

**4-105.** (*a*) 0.326, (*b*) 0.272, (*c*) 0.2636

**4-107.** (*a*) 1142, (*b*) 1335, (*c*) 65.3, (*d*) 75.8

**4-108.** (*a*) 1100, (*b*) 61.4, (*c*) 58.3

**4-110.** (*a*) 80, (*b*) 93, (*c*) 90.5

**4-112.** (*a*) 24.0, (*b*) 21.0, (*c*) 21.5

**4-113.** (*a*) 0.112, (*b*) 0.172

**4-115.** (*a*) 676, (*b*) 568, (*c*) 499.4

**4-117.** (*a*) 20.9, (*b*) 12.4

**4-118.** (*a*) −25.5, (*b*) −17.15

**4-120.** 0.101

**4-122.** 233

**4-123.** (*a*) 0.202, (*b*) 42

**4-125.** 0.0462

**4-127.** 55.2

**4-128.** 0.436

**4-130.** 27.1

**4-132.** 1.23

**4-133.** 54.6

**4-135.** 84.8

**5-2.** (*a*) 10.2, (*b*) 76

**5-3.** (*a*) 0.152, (*b*) 27.2

**5-5.** (*a*) 5.0, (*b*) 800

**5-7.** (*a*) 5.82, (*b*) 71.1

**5-8.** (*a*) 0.947, (*b*) 6.55

**5-9.** (*a*) $1.80 \times 10^6$, (*b*) 85.7

**5-10.** (*a*) 0.852, (*b*) 5.93

**5-12.** (*a*) $1.61 \times 10^6$, (*b*) 686

**5-13.** (*a*) 63.5, (*b*) 1.49

**5-15.** (*a*) 1.46, (*b*) 135

**5-17.** (*a*) 15.6, (*b*) 0.0738

**5-18.** (*a*) 123, (*b*) 0.148

**5-20.** (*a*) 0.169, (*b*) 0.346

**5-22.** (*a*) 0.425, (*b*) 1.25

**5-23.** (*a*) 441,000; (*b*) 5.6

**5-25.** (*a*) 1.91, (*b*) 3.32

**5-27.** (*a*) 1.50, (*b*) 0.212, (*c*) 0.433

**5-28.** 1.02, 0.349

**5-30.** 325

**5-32.** (*a*) 148, (*b*) −10.5

**5-33.** (*a*) 20, (*b*) 1.68

**5-35.** 0.208

**5-37.** (*a*) 35.3, (*b*) 5.92

**5-38.** (*a*) 80, (*b*) 10.85

**5-40.** (*a*) 1164, (*b*) 8.54/1

**5-42.** (*a*) 6.39, (*b*) 16, (*c*) 0.0120

**5-43.** (*a*) 70, 121; (*b*) 31, 107

**5-45.** (*a*) 1.10, (*b*) 276.8, (*c*) 30, (*d*) 26.1

**5-47.** (*a*) 15.6, (*b*) 48, (*c*) 0.046

**5-48.** 257

**5-50.** (*a*) 250, (*b*) 0.489

**5-52.** (*a*) 600, (*b*) 40.9, (*c*) 4.21

**5-53.** (*a*) −3.75, (*b*) −847, (*c*) 38.8, (*d*) 32,900; (*e*) 1.18

**5-55.** (*a*) −16.9, (*b*) −5090

**5-57.** (*a*) −444.8, (*b*) 3.89, (*c*) 35.2, (*d*) −1120

**5-58.** (*a*) −1.41, (*b*) 3634, (*c*) −116, (*d*) 9970, (*e*) 170

**5-60.** (*a*) 2.1, (*b*) −27,400

**5-62.** (*a*) 4.62, (*b*) 1075, (*c*) 133

**5-63.** −48,300

**5-65.** (*a*) 150, (*b*) 5.32, (*c*) 531

**5-67.** (*a*) −4.7, (*b*) 7.08, (*c*) 379

**5-68.** 96.9

**5-70.** (*a*) 42.9, (*b*) 5.75

**5-72.** (*a*) 10, 31; (*b*) 15, (*c*) 63.8

**5-73.** (*a*) −13.2, (*b*) 12.1

**5-75.** (*a*) 174, 111; (*b*) 925

**5-77.** 1.70

**5-78.** (*a*) 87, (*b*) 0.67, (*c*) 0.69

**5-80.** (*a*) 0.134, (*b*) 15.6

**5-82.** (*a*) 22,299, (*b*) 30

**5-83.** (*a*) 9.96, (*b*) 637

**5-85.** (*a*) 0.239, (*b*) 61

**5-87.** (*a*) 0.147, (*b*) 2.92

**5-88.** (*a*) 200, (*b*) 19.8

**5-90.** 360

**5-92.** (*a*) 44.8, (*b*) 0.21

**5-93.** (*a*) 24.0, (*b*) 2.23

**5-95.** (*a*) 0.0237, (*b*) 0.239

**5-97.** (*a*) 191, (*b*) 153.9

**5-98.** (*a*) (1) 0.966, (2) 4.7/1; (*b*) (1) 0.951, (2) 13.2/1

**5-100.** (*a*) 0.0317, (*b*) 0.1168

**5-102.** (*a*)(1) 0.959, (2) 7.4; (*b*)(1) 0.958, (2) 11.8

**5-103.** (*a*) (1) 12.949, (2) 0.921; (*b*) (1) 50, (2) 1.0508

**5-105.** 1390

**5-107.** (*a*) 6.75, (*b*) 3.02

**5-108.** (*a*) 1.77, 0.28; (*b*) 27.1; (*c*) 91.00

**5-110.** (*a*) 5.30, (*b*) 1.14

**5-112.** (*a*) 1.63, (*b*) 1.46

**5-113.** (*a*) 15.7, (*b*) 30,050; (*c*) 0.875

**5-115.** (*a*) 0.353, (*b*) 17.4

**5-117.** (*a*) 0.027, (*b*) 13.6

**5-118.** (*a*) 42.5, (*b*) 2.71

**5-120.** (*a*) 37.8, (*b*) 6.2

**5-122.** (*a*) −1018, (*b*) −2115, (*c*) 4.0, (*d*) 32.4%

**5-123.** (*a*) −1170, (*b*) 6, (*c*) 34.9, (*d*) $1.11 \times 10^6$

**5-125.** (*a*) −1276, 10.1; (*b*) 38%, (*c*) −14,600

**5-127.** (*a*) 3.26, (*b*) 10.71, (*c*) 3.28

**5-128.** (*a*) 4.18, (*b*) 4.43, (*c*) 13.3

**5-130.** (*a*) 3.88, (*b*) 3.12

**5-132.** (*a*) 265, (*b*) 11,650

**5-133.** (*a*) 1122, (*b*) −232

**5-135.** $1.04 \times 10^6$

**5-137.** 666

**5-138.** (*a*) 0.106, (*b*) −28.6

**5-140.** (*a*) 0.575, (*b*) 178

**5-142.** (*a*) 137, (*b*) 68

**5-143.** 38.8

**5-145.** 110

**5-147.** 280

**5-148.** 50,500

**5-150.** 7.11

**5-152.** 1.60

**5-153.** (*a*) 1.25, (*b*) 4.04

**5-156.** $(PVc_p/R)\ln(T_2/T_1)$

**5-157.** (*a*) 350, (*b*) 4.33

**5-158.** 408

**5-160.** (*a*) 332, (*b*) 325, 2.4; (*c*) 2.03, (*d*) 0.877

**5-162.** (*a*) 1, (*b*) 123, 3.50; (*c*) 3.82, (*d*) 2.83

**5-163.** 7.47

**5-165.** 0.21

**5-167.** (*a*) 128, (*b*) 140, 61.3, (*c*) 3.99, (*d*) 0.339

**5-168.** (*a*) 726, (*b*) 0.00138

**5-170.** 185,000

**5-172.** 16.4

**5-173.** 25,200

**5-175.** (*a*) 146, (*b*) 0.0602

**5-177.** (*a*) 59, (*b*) 0.0954

**5-178.** 77

**5-180.** (*a*) 265, (*b*) 314, (*c*) 0.0782

**5-181.** 100

**6-1.** (*a*) 71,560, (*b*) 58,240, (*c*) 18.6%

**6-2.** (*a*) $0.744 \times 10^6$, (*b*) 3.59

**6-3.** (*a*) 33.7, (*b*) 4850

**6-5.**   (*a*) 21.8, (*b*) $16.10

**6-7.**   (*a*) 2.08, (*b*) $2.95

**6-8.**   (*a*) 0.180, (*b*) 4.7

**6-10.**   (*a*) 0.69, (*b*) 793

**6-12.**   (*a*) 6.33, (*b*) $2.73, (*c*) $4.11

**6-13.**   (*a*) $130, (*b*) 61.5

**6-15.**   (*a*) 33.5, (*b*) 0.83

**6-17.**   (*a*) 14.1, (*b*) 108,000

**6-18.**   (*a*) 0.236, (*b*) 2990, (*c*) 2.36

**6-20.**   (*a*) 0.552, (*b*) 2208, (*c*) 93

**6-22.**   (*a*) 0.335, (*b*) 0.503, (*c*) −16.6%

**6-23.**   (*a*) 1.59, (*b*) 63%, (*c*) 37.7%

**6-25.**   (*a*) 25.0, (*b*) 452, (*c*) 225%

**6-27.**   (a) 3667, 1122; (*b*) 69.4%; (*c*) 93%

**6-28.**   (*a*) 15, (*b*) 240, (*c*) 30%

**6-30.**   (*a*) 490, (*b*) 397

**6-32.**   (*a*) 548, (*b*) 1131, (*c*) 619

**6-33.**   (*a*) irrev., (*b*) imp., (*c*) rev.

**6-35.**   78.3

**6-37.**   (*a*) 12.8, (*b*) 6.8%

**6-38.**   (*a*) 0.175, (*b*) 15.2%, 4.93

**6-40.**   3.91

**6-42.**   6.67

**6-43.**   (*a*) irrev., (*b*) imp., (*c*) rev.

**6-45.**   4.79

**6-47.**   (*a*) 7.95, (*b*) 4.75, (*c*) 4.67

**6-48.**   (*a*) 0.111, (*b*) 6.8%

**6-50.**   (*a*) 2.22, (*b*) 9.2, (*c*) 54, (*d*) 7.02

**6-52.**   (*a*) 180,000; (*b*) 7.92, (*c*) 0.67, (*d*) 0.78

**6-53.**   (*b*) 130, 136.4, (*c*) 45.5%, (*d*) no

**6-55.**   (*a*) no, (*b*) 700, 601; (*c*) 1.86, (*d*) no

**6-57.**   (*a*) 8.89, (*b*) 48,000; (*c*) 2.01; (*d*) 22.2

**6-58.**   (*a*) 2.42, (*b*) 9.61, 1.66, (*c*) −34%

**6-60.**   −3

**6-62.**   (*a*) 1000, (*b*) 548, (*c*) 1.21

**6-63.**   (*a*) 560, (*b*) 529, (*c*) 2.12, (*d*) 24.0

**6-65.**   (*a*) 500, (*b*) 540, (*c*) 2, 2; (*d*) 35.4

**6-67.**   (*a*) 4122, (*b*) 5852

**6-68.**   (*a*) 600, (*b*) 67%, (*c*) −33, (*d*) 4

**6-70.**   (*a*) 60, (*b*) 0.667, (*c*) −27, (*d*) 4.98

**6-72.**   (*a*) 2.33, (*b*) 2170, (*c*) −1930

**6-73.**   (*a*) 200, (*b*) 600, (*c*) 16.7%

**6-75.**   (*a*) no, (*b*) no

**6-77.**   (*a*) no violation, (*b*) no violation

**6-78.**   (*a*) violation, (*b*) violation

**6-80.**   (*a*) violation, (*b*) violation

**6-82.**   (*a*) yes, (*b*) yes

**6-83.**   (*a*) rev., (*b*) imp., (*c*) irrev.

**6-85.**   (*a*) imp., (*b*) rev., (*c*) irrev.

**6-87.**   (*a*) Irrev., (*b*) Imp., (*c*) Rev.

**6-88.**   (*a*) −0.147, 0.204; (*b*) 0.057; (*c*) 30.6%; (*d*) 6.6

**6-90.**   (*a*) −3.75, 5.00; (*b*) 1.25; (*c*) 20

**6-92.**   (*a*) −2.0, 2.73; (*b*) 0.727; (*c*) 18.2%

**6-93.**   (*a*) −3.13, 5.56; (*b*) 2.43; (*c*) 36

**6-95.**   (*a*) 0.476, 1.143, 2.143, 3.81; (*b*) 1143, 1000, 800, 500, 0

**6-97.**   (*a*) −330, (*b*) 1.054

**6-98.**   (*a*) −3.62, (*b*) 0.000943

**6-100.**   (*a*) 261,300; (*b*) 1.84, 1.74, 3.58

**6-102.**   (*a*) 4000, 16,000, 0.80, 0; (*b*) 10,000, 10,000, 0.50, 1429

**6-103.**   (*a*) 123, 23.1, 4.33, 0; (*b*) 132, 32, 3.13, 8.92

**6-105.**   (*a*) 1192, 192, 5.22, 0; (*b*) 1311, 311, 3.21, 119.6

**6-107.**   (*a*) 0.112, 0.213, 0.128; (*b*) 0.241, 0.222, 0.128; (*c*) 0.125, 0.222, 0.244

**6-108.**   (*a*) (1) 0.357, (2) 0, (3) 0; (*b*) (1) 0.148, (2) 0.143, (3) 0.066; (*c*) (1) 0.0814, (2) 0.138, (3) 0.066

**6-110.**   (*a*) 0.064, 0.181, 0.276; (*b*) 0.555, 0.181, 0.306

**6-112.**   (*a*) (1) 0.265, (2) 0.324, (3) 0.401; (*b*) (1) 0.602, (2) 0.307, (3) 0.401

**6-113.**   (*a*) 0.058, 0.169, 0.062; (*b*) 0.208, 0.192, 0.062

**6-115.**   (*a*) 210, (*b*) 105

**6-117.**   (*a*) 0.545, (*b*) 0.375

**6-118.**   (*a*) 0.247, (*b*) 0.0823

**6-120.**   3, 10

**6-122.**   3, 10

**6-123.**   5, 21

**6-125.**   (*a*) 6, (*b*) 4420, (*c*) 2520

**7-1.**   (*a*) 15, (*b*) −7.957, (*c*) 9.33

**7-2.**   (*a*) 240, (*b*) 0.0160, (*c*) 0.0045, (*d*) 0.0069

**7-3.** (*a*) −59.3, 44.2; (*b*) −0.178, (*c*) 0.0209, (*d*) 0.0209

**7-5.** (*a*) 10, (*b*) 11.0, (*c*) 0.0300, (*d*) 0.0158

**7-7.** (*a*) 350, (*b*) −0.825, (*c*) 1.091

**7-8.** (*a*) 450, (*b*) 0.00778, (*c*) 0.00243, (*d*) 0.00305

**7-10.** (*a*) 0.6631, (*b*) 0.1703, (*c*) −42.95, (*d*) −0.0010

**7-12.** (*a*) 179.1, (*b*) −1.66, (*c*) 0.00390, (*d*) 0.00709

**7-13.** (*a*) −0.0516, (*b*) 0.0370, (*c*) 0.0370

**7-15.** (*a*) −4.0, (*b*) −0.0054, (*c*) 0.00162, (*d*) 0.00215

**7-17.** (*a*) 360, 1.2, (*b*) 0.0131, (*c*) 0.0146, (*d*) 0.0149

**7-18.** (*a*) −0.1672, (*b*) −6.4, (*c*) −0.1459, (*d*) no

**7-20.** (*a*) −1500, (*b*) 3.846, (*c*) 1.703, (*d*) 3.78

**7-22.** (*a*) 401, (*b*) 2.333, (*c*) 0.560

**7-23.** (*a*) −0.00501, (*b*) 0.00599, (*c*) 0.00784

**7-25.** (*a*) −6, (*b*) 1.37, (*c*) 0.0107, (*d*) 0.01427, (*e*) 0.0145

**7-27.** (*a*) −3000, (*b*) 4.286, (*c*) 1.286, (*d*) 10.4

**7-28.** (*a*) −49.8, (*b*) 14.2, (*c*) −0.0764, (*d*) 0.0179

**7-30.** (*a*) 560, (*b*) −0.00261, (*c*) 0.00316, (*d*) 0.00409, (*e*) Irrev.

**7-32.** (*a*) 1.365, (*b*) −0.398, (*c*) 0.965

**7-33.** (*a*) 110, (*b*) 0.0882, (*c*) 0.1160, (*d*) 0.1204

**7-35.** (*a*) 16.24, (*b*) 0.108

**7-37.** (*a*) 43.33, (*b*) 0.00024

**7-38.** (*a*) −0.284, (*b*) 0.036

**7-40.** (*a*) 50, (*b*) 0.465

**7-42.** (*a*) 80, (*b*) 0.010, (*c*) Irrev.

**7-43.** (*a*) 0, 17.06; (*b*) 14.53, 14.53

**7-45.** (*a*) −0.1442, (*b*) −108, (*c*) 0.0333, (*d*) Irrev.

**7-47.** (*a*) 135, (*b*) 0.16

**7-48.** (*a*) 81.5, (*b*) 0.222, (*c*) Irrev.

**7-50.** (*a*) 0, 0.00352; (*b*) 0.00327, 0.00327

**7-52.** (*a*) 196, (*b*) 2.720, (*c*) 2.066

**7-53.** (*a*) 37, (*b*) −0.689, (*c*) 0.0276

**7-55.** (*a*) −298, (*b*) −0.6674, (*c*) 0.327

**7-57.** 0.1672

**7-58.** (*a*) 350, (*b*) −0.57, (*c*) Imp.

**7-60.** 0.133

**7-62.** (*a*) 87, (*b*) 0.00975, (*c*) 0.0213, (*d*) 0.0249

**7-63.** (*a*) −0.0225, (*b*) −1.06, (*c*) 0.0108, (*d*) 0.0125

**7-65.** (*a*) 1.63, (*b*) 0.0772

**7-67.** 0.4176

**7-68.** 0.0158

**7-70.** (*a*) 140, (*b*) −0.267, Imp.

**7-72.** (*a*) −225, (*b*) −0.0257, (*c*) 0.423, (*d*) Irrev.

**7-73.** (*a*) 101, (*b*) −0.0182, (*c*) 0.00935

**7-75.** (*a*) −0.0424, (*b*) −2.05, (*c*) −0.0058, (*d*) Imp.

**7-77.** (*a*) 0.857, 157.17; (*b*) 0.00716, (*c*) −0.00625, (*d*) 0.00104

**7-78.** 0.1888

**7-80.** (*a*) 126, (*b*) 8.59

**7-82.** (*a*) −140.9, (*b*) 26.8

**7-83.** (*a*) −217, (*b*) 14, (*c*) −382

**7-85.** (*a*) −46.7, (*b*) 52.1, (*c*) 0.0295

**7-87.** (*a*) −70,300; (*b*) 40, (*c*) −157

**7-88.** (*a*) 60,480, (*b*) 375, (*c*) 0.8

**7-90.** (*a*) 109.8, (*b*) −110.2, (*c*) 0

**8-1.** (*a*) 400, (*b*) 280

**8-2.** (*a*) 3.827, (*b*) −557

**8-3.** (*a*) 340, (*b*) 0.0136, (*c*) 180

**8-5.** (*a*) 37, (*b*) 0.00189, (*c*) 1.73, (*d*) −0.164

**8-7.** (*a*) 385, (*b*) 785

**8-8.** (*a*) 592, (*b*) −0.0373, (*c*) 30.4

**8-10.** (*a*) 100, (*b*) 0.00405, (*c*) 100, (*d*) −0.140

**8-12.** (*a*) 0, 86.5; (*b*) −120, 34; (*c*) 36

**8-13.** (*a*) 420, (*b*) 0.162, (*c*) 972, (*d*) 1.07

**8-15.** (*a*) 100, (*b*) 2.83, (*c*) 0.336, (*d*) 3.0

**8-17.** (*a*) 98.4, (*b*) 1.12, (*c*) 0.453

**8-18.** (*a*) 22, (*b*) 0.82, (*c*) 1.60

**8-20.** (*a*) 368, (*b*) 233, (*c*) 115

**8-22.** (*a*) 519, (*b*) 21.4

**8-23.** (*a*) 233, (*b*) 1387, (*c*) 9.67

**8-25.** (*a*) 60, (*b*) 10.58, (*c*) 1.64

**8-27.** (*a*) 660, (*b*) 268, (*c*) 17.4

**8-28.** (*a*) 500, (*b*) 10.6, (*c*) 260

**8-30.** 4.63

**8-32.** (*a*) 633, (*b*) 252

**8-33.** 170

**8-35.** (*a*) 3.62, (*b*) 3.43

**8-37.** (*a*) 2.51, (*b*) 0.626, (*c*) 1.57

**8-38.** 41,130

**8-40.** (*a*) 39.2, (*b*) 816, (*c*) 1.41

**8-42.** 22.7

**8-43.** (*a*) 61.2, (*b*) 234, (*c*) 26.0

**8-45.** (*a*) −2.31, (*b*) 7.26

**8-47.** (*a*) 373, (*b*) 14.9

**8-48.** (*a*) 1.4, (*b*) 155, (*c*) 7.78

**8-50.** −63.9

**8-52.** (*a*) 553, (*b*) 323, (*c*) 899, (*d*) 16

**8-53.** (*a*) 645, (*b*) 0.384

**8-55.** (*a*) 1485, (*b*) 1.92

**8-57.** (*a*) 0.916, (*b*) 2650, (*c*) 6.35

**8-58.** 12.3

**8-60.** (*a*) 695, (*b*) 185, (*c*) 1450, (*d*) 62

**8-62.** (*a*) 790, (*b*) 760, (*c*) 151.2, (*d*) 0.0648

**8-63.** (*a*) 24.7, (*b*) 0.00108

**8-65.** (*a*) 106, (*b*) 83%, (*c*) 106, (*d*) 0.0607

**8-67.** (*a*) 288, (*b*) 0.011

**8-68.** (*a*) −0.581, (*b*) 0.920

**8-69.** 29.6

**8-70.** (*a*) 80, (*b*) 60, (*c*) 133, (*d*) 0.0227

**8-72.** (*a*) 0.813, (*b*) 0.138

**8-73.** (*a*) 868, (*b*) 0.040

**8-75.** (*a*) 0.934, (*b*) 300, (*c*) 60.8, (*d*) 0.00901

**8-77.** (*a*) 89%, (*b*) 70, (*c*) 8.3, (*d*) 0.00436

**8-78.** (*a*) 2150, (*b*) 104, (*c*) 958

**8-80.** (*a*) 0.0116, (*b*) 0.921

**8-82.** (*a*) 0.00422, (*b*) 90.5%

**8-83.** (*a*) 0.792, (*b*) 22.4

**8-85.** (*a*) 1120, (*b*) 0.786, (*c*) 0.483

**8-87.** (*a*) 50, (*b*) 0.0353

**8-88.** (*a*) 582, 578; (*b*) 653, 723

**8-90.** (*a*) 160, (*b*) 0.0097

**8-92.** (*a*) 75%, (*b*) 0.041

**8-93.** (*a*) 1.96, (*b*) 1.84

**8-95.** (*a*) 403, (*b*) 361

**8-97.** (*a*) 18, (*b*) 1210

**8-98.** (*a*) −127, (*b*) −144

**8-100.** (*a*) 0.617, 0.00697; (*b*) 0.441, 0.0068

**8-102.** (*a*) 0.616, 0.00697; (*b*) 0.449, 0.0067

**8-103.** (*a*) −13.9, (*b*) −16.6

**8-105.** (*a*) 1.49, 0.00266; (*b*) 1.33, 0.00257

**8-107.** 3.1623

**8-108.** Only for unequal $\eta$'s

**8-110.** 2.1544

**8-112.** (*a*) 23.6%, (*b*) 1359, (*c*) −1038, (*d*) −388

**8-113.** (*a*) 4.22, (*b*) −116, (*c*) 93.5, (*d*) 22.1

**8-115.** (*a*) 31.3%, (*b*) 740, (*c*) −508.5, (*d*) −248

**8-117.** (*a*) 5.75, (*b*) −42.6, (*c*) 36.3, (*d*) 12.67

**8-118.** (*a*) 5.25, (*b*) −130, (*c*) 109.6, (*d*) −11.8

**8-120.** (*a*) 0.01636, (*b*) 735, (*c*) 0.200, (*d*) 0.173

**8-121.** (*a*) 0.0273, (*b*) 617 K, (*c*) 0.200, (*d*) 0.229

**8-122.** (*a*) 0.0356, (*b*) 793.6, (*c*) 3.03, (*d*) 0.0507

**9-1.** (*a*) −348, (*b*) 348

**9-2.** (*a*) 43.2, (*b*) 3.8, (*c*) 39.4

**9-3.** 18.5

**9-5.** (*a*) −151, (*b*) 151

**9-7.** (*a*) 19.4, (*b*) 140, (*c*) 11.14, (*d*) 4.89

**9-8.** (*a*) −1.02, (*b*) −2.34, (*c*) 2.81

**9-10.** (*a*) 102, (*b*) -23.4, (*c*) −32.9, (*d*) 9.5

**9-12.** (*a*) −2.50, (*b*) −6.0, (*c*) 8.5

**9-13.** (*a*) −147, (*b*) −460, (*c*) 343

**9-15.** (*a*) −39.4, (*b*) 25.3

**9-17.** (*a*) 245, (*b*) −89.9

**9-18.** (*a*) −42.3, (*b*) 22.5

**9-20.** (*a*) 15.8, (*b*) 13.8

**9-22.** (*a*) 113, (*b*) −33.8

**9-23.** 148

**9-25.** (*a*) 124, (*b*) −62.8

**9-27.** 434

**9-28.** (*a*) 99.5, (*b*) 18.3, (*c*) −32.9

**9-30.** (*a*) −9.3, (*c*) 9.3

**9-32.** (*a*) 870,641; (*b*) −229, (*d*) 229

**9-33.** 0.88

**9-35.** (*a*) −295, (*b*) 28, (*c*) −267

**9-38.** (*a*) −33.7, (*b*) −200, (*c*) −234, (*d*) 234

**9-40.** (*a*) −101.7, (*b*) −124.4, (*c*) 22.7

**9-42.** (*a*) 135, (*b*) 118, (*c*) 15–17

**9-43.** (*a*) 44.2, (*b*) 13.0, (*c*) 13.0

**9-45.** (*a*) −194, (*b*) −271, (*c*) 77

**9-47.** (*a*) 22.4, (*b*) 16.4, (*c*) 9.0

**9-48.** (*a*) 2.97, (*b*) 3.63

**9-50.** (*a*) 1526, 1213, 200; (*b*) −14,900, (*c*) −10,735

**9-52.** (*a*) −102

**9-53.** −327,750

**9-55.** (*a*) 668, 512, 59; (*b*) −19,500, (*c*) −14,870

**9-57.** −89

**9-58.** −149,100

**9-60.** (*a*) −855, 94; (*b*) −855, 109

**9-62.** (*a*) −98.0, (*b*) 98.0

**9-63.** (*a*) 0.98, (*b*) 0.096, (*c*) 0.886

**10-1.** (*a*) 0.276, 0.724; (*b*) 23.2; (*c*) 8.37

**10-2.** (*a*) 0.571, 0.286, 0.143; (*b*) 28.57; (*c*) 0.141

**10-3.** (*a*) 0.333, 0.50, 0.167; (*b*) 0.227, (*c*) 0.517

**10-5.** (*a*) 62.0, 27.9, 10.1; (*b*) 0.20; (*c*) 12.6

**10-7.** (*a*) 0.566, 0.152, 0.283; (*b*) 4.01

**10-8.** (*a*) 50.6, 39.8, 9.6; (*b*) 4.5; (*c*) 1.55

**10-10.** (*a*) 0.375, 0.125, 0.50; (*b*) 0.0544; (*c*) 11.8

**10-12.** (*a*) 0.522, 0.478; (*b*) 96; (*c*) 36.8; (*d*) 0.055

**10-13.** 0.167, 0.333, 0.500; (*b*) 0.499; (*c*) 303

**10-15.** (*a*) 624, (*b*) 0.05, (*c*) 1248, (*d*) 0.231

**10-17.** 0.20, 0.80; (*b*) 8.0; (*c*) 0.248; (*d*) 73

**10-18.** (*a*) 14.75,(*b*) 10, (*c*) 29.5, (*d*) 0.0551

**10-20.** 0.26, 10.9, 35.5, 53.4; (*b*) 157.8; (*c*) 157.8; (*d*) 3360

**10-22.** (*a*) 279.4, (*b*) 278.8, (*c*) 9300

**10-23.** (*a*) −7310, (*b*) −229

**10-25.** (*a*) −2260, (*b*) −2264, (*c*) −27.3

**10-27.** −3033, −90.4

**10-28.** 12.0

**10-30.** 276

**10-32.** 325.5

**10-33.** 6.0

**10-35.** 1065

**10-37.** (*a*) 127, (*b*) 3.75

**10-38.** 170

**10-40.** (*a*) 250, (*b*) 56.5

**10-42.** (*a*) −7.93, (*b*) 76.3

**10-43.** 0.921

**10-45.** (*a*) 442, (*b*) 3044

**10-47.** (*a*) 246, (*b*) 0.872

**10-48.** 8050

**10-50.** 1805

**10-52.** 0.798

**10-53.** (*a*) 352, (*b*) 0.365

**10-55.** (*a*) 400, (*b*) 6.31

**10-57.** (*a*) −22.8, (*b*) 2.50, (*c*) 0.236, 0.377, (*d*) 0.691

**10-58.** (*a*) 294, (*b*) 3.66, (*c*) 0.807, 0.600, (*d*) 0.917

**10-60.** (*a*) −37,500; (*b*) 14.81, (*c*) 476

**10-62.** (*a*) 602, (*b*) 2.54

**10-63.** (*a*) 701, (*b*) 1.73

**10-65.** (*a*) 90.6, (*b*) 32.8, (*c*) 36, (*d*) 0.0499, 0.3732; (*e*) 0.485

**10-67.** (*a*) 301, (*b*) 46.7, (*c*) 0.0133

**10-68.** (*a*) 252, (*b*) 1483

**10-70.** (*a*) 1.12, (*b*) 6.86

**10-72.** (*a*) $5.46 \times 10^{-3}$, (*b*) 104

**10-73.** (*a*) 0.015, (*b*) 0.701, (*c*) 8.79

**10-75.** (*a*) 0.680, (*b*) 15.8, (*c*) 0.00385

**10-77.** (*b*) 0.168

**10-78.** (*a*) 0.081, (*b*) 17.32

**10-80.** (*a*) 7.53, (*b*) 15.37

**10-82.** 2.92

**10-83.** (*a*) 175, (*b*) 168, (*c*) 0.0537, (*d*) −99

**10-85.** (*a*) 70.7, (*b*) 24, (*c*) 0.01924, (*d*) 79.3, (*e*) 0.896

**10-87.** (*a*) 0.727, (*b*) 19.7, (*c*) 15.3, (*d*) 64.09

**10-88.** (*a*) 14.8, (*b*) 19.4, (*c*) 73.2, (*d*) 0.932

**10-90.** (*a*) 0.69, (*b*) 68.9, (*c*) 0.0153, (*d*) 35.95, (*e*) 14.02

**10-92.** (*a*) 98.6, (*b*) 66.0, (*c*) 38.1, (*d*) 14.40

**10-93.** (*a*) 0.00800, (*b*) 0.185, (*c*) 0.417, (*d*) 13.76

**10-95.** (*a*) 0.535, (*b*) 15

**10-97.** (*a*) 0.428, (*b*) 60.2

**10-98.** (*a*) 0.659, (*b*) 63.6

**10-100.** (*a*) 11.6, (*b*) 0.430, (*c*) 58.8

**10-102.** (*a*) 0.0119, (*b*) 0.544, (*c*) 32.25

**10-103.** (*a*) 0.01062, (*b*) 0.398, (*c*) 32.31

**10-105.** (*a*) 64, (*b*) 18, (*c*) 0.0126, (*d*) 57.7

**10-107.** (*a*) 17.7, (*b*) 14, (*c*) 0.0098, (*d*) 51

**10-108.** (*a*) 65, (*b*) 17, (*c*) 54.5, (*d*) 0.858

**10-110.** (*a*) 22, (*b*) 70, (*c*) 0.015, (*d*) 0.868

**10-112.** (*a*) 72, (*b*) 42, (*c*) 64.3, (*d*) 35.7

**10-113.** (*a*) 63.5, (*b*) 60, (*c*) 72, (*d*) 29

**10-115.** (*a*) 0.0352, (*b*) 20.1, (*c*) 0.0106

**10-117.** (*a*) 80, (*b*) 20.6, (*c*) −12.5

**10-118.** (*a*) 6.3, (*b*) −29.8

**10-120.** (*a*) −22,600, (*b*) 11,300, (*c*) 4.99

**10-122.** (*a*) 0.0131, (*b*) −54.66, (*c*) 16.1

**10-123.** (*a*) 81, (*b*) 71.5, (*c*) −5.5

**10-125.** (*a*) 7.19, (*b*) −42

**10-127.** (*a*) $1.25 \times 10^6$, (*b*) 402,000

**10-128.** (*a*) 0.0144, (*b*) −26.0, (*c*) 6.14

**10-130.** (*a*) 25, 55.8%; (*b*) 0.0050

**10-132.** (*a*) 0.00521, (*b*) 76.2, (*c*) 18

**10-133.** (*a*) 0.0047, (*b*) 16.3, (*c*) 12

**10-135.** (*a*) 1140, (*b*) 0.250, (*c*) 0.560

**10-137.** (*a*) 0.269, (*b*) 530, (*c*) 0.370

**10-138.** (*a*) 0.858, (*b*) 0.0058, (*c*) 65.0

**10-140.** (*a*) 0.0059, (*b*) 63, (*c*) 57.5

**10-142.** (*a*) 1050, (*b*) 0.66, (*c*) 0.590

**10-143.** (*a*) 165, (*b*) 28.0, (*c*) 232

**10-145.** (*a*) 22.9, (*b*) 0.00961, (*c*) 55%

**10-147.** (*a*) 20.1, (*b*) 0.0111, (*c*) 75.6

**10-148.** (*a*) 20.1, (*b*) 0.501

**10-150.** (*a*) 76.9, (*b*) 0.01183, (*c*) 59.6

**10-152.** (*a*) 21,840; (*b*) 24,800

**10-153.** (*a*) 11,790; (*b*) 17,000, (*c*) 259

**10-155.** (*a*) 37.1, (*b*) 26.5

**10-157.** (*a*) 22,370; (*b*) 32,130; (*c*) 492

**10-158.** (*a*) 88.4, (*b*) 0.821

**11-2.** (*a*) 0.760, (*b*) 0.75–0.76

**11-3.** (*a*) 0.186, (*b*) 0.171, (*c*) 0.1717

**11-5.** (*a*) 0.1195, (*b*) 0.113

**11-6.** (*a*) 2.20, (*b*) 2.04, (*c*) 2.065, (*d*) 2.03

**11-7.** (*a*) 114, (*b*) 98, (*c*) 113

**11-9.** (*a*) 0.447, (*b*) 0.371, (*c*) 0.376, (*d*) 0.368

**11-10.** (*a*) 6.09, (*b*) 3.93, (*c*) 4.0

**11-12.** (*a*) 0.416, (*b*) 0.316, (*c*) 0.316, (*d*) 0.322

**11-13.** (*a*) 0.0222, (*b*) 0.0177, (*c*) 0.0173, (*d*) 0.0172

**11-15.** (*a*) 0.375, (*b*) 0.309, (*c*) 0.304, (*d*) 0.306, (*e*) 0.307

**11-17.** (*a*) 353, (*b*) 341, (*c*) 352, (*d*) 360

**11-18.** (*a*) 11.79, (*b*) 10.45, (*c*) 10.23, (*d*) 10.3, (*e*) 10.0

**11-20.** (*a*) 15, (*b*) 48, (*c*) 54, (*d*) 54, (*e*) 60

**11-22.** (*a*) 66, (*b*) 120, (*c*) 130, (*d*) 128, (*e*) 140

**11-23.** (*a*) 1165, (*b*) 1002, (*c*) 952, (*d*) 991, (*e*) 1000

**11-25.** (*a*) 107.7, (*b*) 101.5, (*c*) 92.6, (*d*) 106, (*e*) 100

**11-27.** (*a*) 69.7, (*b*) 59.9, (*c*) 59.4, (*d*) 60.3

**11-28.** (*a*) 4.08, (*b*) 4.12

**11-30.** (*a*) 5.8 ($CO_2$), (*b*) 6.15 ($CO_2$), (*c*) 0.145

**11-32.** (*a*) 529, (*b*) 839, (*c*) 1650, (*d*) 1000

**11-33.** (*a*) 0.60, (*b*) 0.80

**11-35.** (*a*) 7.95, (*b*) 11.4

**11-37.** (*a*) 62.9, (*b*) 55.0, (*c*) 41.8, (*d*) 51.6

**11-38.** 1115

**12-3.** (*a*) $-T(\partial v/\partial T)_p - P(\partial v/\partial P)_T$, (*b*) $T(\partial P/\partial T)_v - P$

**12-8.** (*a*) 2.77, (*b*) 2.79

**12-9.** (*a*) 1.121, (*b*) 1.122

**12-10.** (*a*) 0.682, (*b*) 0.676

**12-15.** Increases as $P$ increases

**12-17.** (*a*) $8.16 \times 10^{-5}$ (*b*) $8.22 \times 10^{-5}$

**12-18.** $0.163 \times 10^{-3}$

**12-19.** 0.00345, 0.00352

**12-21.** (*c*) −759, 0

**12-22.** (*c*) −347, 0

**12-24.** (*b*) −5.72, −0.0639; (*c*) −9.11, −0.0834

**12-25.** (*b*) −53.2, −0.134

**12-27.** (*b*) $R\ln[(v_2 - b)/(v_1 - b)] - a/T^2[(1/v_2) - (1/v_1)]$

**12-28.** (*a*) 9.27, (*b*) 10.32

**12-30.** $2aR\Delta P/T$

**12-33.** 0.40%

**12-35.** (*a*) 2.31, (*b*) 8.8%

**12-37.** (*a*) 3.20, (*b*) 12.9%

**12-38.** 0.4%

**12-40.** (*a*) 0.298, (*b*) 5.05%

**12-42.** (*a*) 0.824, (*b*) 14

**12-43.**   0.070

**12-45.**   (*a*) 0.574, (*b*) −2.45

**12-47.**   (*a*) 0.966, (*b*) 3.89

**12-48.**   (*a*) $4.24 \times 10^{-4}$, (*b*) −0.0906

**12-50.**   (*a*) 2126, (*b*) 2215

**12-52.**   91.4

**12-53.**   (*a*) 828, (*b*) 928, (*c*) 826.8

**12-55.**   (*b*) 193.3

**12-57.**   (*a*) −2.23, (*b*) −3.70

**12-58.**   (*a*) 2042, (*b*) 1941

**12-60.**   (*a*) 71.7, (*b*) 0.479

**12-62.**   (*a*) yes, (*b*) no

**12-63.**   25.2

**12-65.**   51.3

**12-67.**   (*a*) 1.29, (*b*) 0.708

**12-70.**   1.37

**12-72.**   (*a*) 0.5015, (*b*) 0.108

**12-73.**   1.403

**12-75.**   (*a*) 0.309, (*b*) 2.39

**12-77.**   (*a*) −0.0487, 0.174; (*b*) 50.6; (*c*) 9

**12-78.**   (*a*) 0.218, (*b*) 0.0456, (*c*) −0.0140

**12-80.**   (*a*) 2200, 82; (*b*) 2080, 75; (*c*) 2110, 90

**12-82.**   −62.2

**12-83.**   (*a*) 748, (*b*) −1840

**12-85.**   (*a*) 930, (*b*) 18.2

**12-87.**   598

**12-88.**   1150

**12-90.**   (*a*) −527, (*b*) 174

**12-92.**   (*a*) −2120, (*b*) −13.14, (*c*) 5240

**12-93.**   (*a*) 0.425, (*b*) 115, (*c*) 250, (*d*) −136

**12-95.**   135

**12-97.**   (*a*) 1710, (*b*) 2020

**12-98.**   (*a*) 1329, (*b*) 1489

**12-100.**  (*a*) 61.5, (*b*) 160

**13-1.**   (*a*) 5.71, (*b*) 0.947, (*c*) 10.47, (*d*) 55, (*e*) 87.1, (*f*) 14.6, (*g*) 57

**13-2.**   (*a*) 15, (*b*) 0.87, (*c*) 11.5, (*d*) 47.4, (*e*) 80.6, (*f*) 15.4, (*g*) 50.6

**13-3.**   (*a*) 28, (*b*) 0.781, (*c*) 12.7, (*d*) 39, (*e*) 67.9, (*f*) 13.1, (*g*) 45

**13-5.**   (*a*) 19.3, (*b*) 0.833, (*c*) 73.5, (*d*) 52.4, (*e*) 79.8, (*f*) 16.35, (*g*) 55.4

**13-7.**   (*a*) 15.92, (*b*) 0.833, (*c*) 76.3, (*d*) 38, (*e*) 54, (*f*) 13.7, (*g*) 43.3

**13-8.**   (*a*) 17.75, (*b*) 0.833, (*c*) 74.6, (*d*) 47.6, (*e*) 73.6, (*f*) 15.3, (*g*) 51

**13-10.**  (*a*) 14.9, (*b*) 0.87, (*c*) 11.5, (*d*) 119, (*e*) 80.2, (*f*) 231, (*g*) 125

**13-12.**  (*a*) 15.9/1, (*b*) 0.833, (*c*) 76.3, (*d*) 101, (*e*) 50.4, (*f*) 218, (*g*) 110

**13-13.**  (*a*) 18.82, (*b*) 0.094, (*c*) 52.6

**13-15.**  (*a*) (1) 50, (2) 4.82; (*b*)(1) 53.4, (2) 5.74

**13-17.**  (*a*) 17.75, (*b*) 0.106, (*c*) 118

**13-18.**  (*a*) 18.6/1, (*b*) 0.767, (*c*) 97

**13-20.**  (*a*) 10.1% $CO_2$, 3.8% $O_2$, (*b*) 17.5, (*c*) 54, (*d*) 9.8

**13-22.**  (*a*) 19.94, (*b*) 73, (*c*) 55, (*d*) 82.4

**13-23.**  (*a*) 17.5, (*b*) 10.3, (*c*) 55, (*d*) 11.7

**13-25.**  (*a*) 19.5, (*b*) 8.2, (*c*) 127, (*d*) 82.2

**13-27.**  (*a*) 10.8, (*b*) 113, (*c*) 19.6

**13-28.**  (*a*) 13.7, (*b*) 119

**13-30.**  10

**13-32.**  (*a*) $x = 7.61$, $y = 16.43$, (*b*) 16.51

**13-33.**  (*a*) 102, (*b*) 2.47

**13-35.**  (*a*) 0.1348, (*b*) −9215

**13-37.**  (*a*) 0.748, (*b*) −26,830

**13-38.**  21

**13-40.**  −44,000

**13-42.**  (*a*) −804,000; (*b*) 616,000; (*c*) ≈ 1900

**13-43.**  130

**13-45.**  −16,800

**13-47.**  7810

**13-48.**  (*a*) 0.94, (*b*) −13,540

**13-50.**  (*a*) −346,000; (*b*) 137,000; (*c*) 3420

**13-52.**  (*a*) −707,000; (*b*) −183,800; (*c*) 3850

**13-53.**  36.9

**13-55.**  1304

**13-57.**  2125

**13-58.**  2005

**13-60.**  2107

**13-62.**  1460

**13-63.**  2310

**13-65.** 4140

**13-67.** 3280

**13-68.** 3630

**13-70.** 2140, 0.65

**13-72.** (a) 3050, (b) 0.75

**13-73.** (a) 2690, (b) 16.5

**13-75.** (a) 5590, (b) 143

**13-77.** 53,140

**13-78.** −980,000

**13-80.** (a) −9040, (b) −117,110

**13-82.** −421,720

**13-83.** −57,160

**13-85.** (a) 0.0571, 0.2937, 0.0130, 10.77; (b) −25,110

**13-87.** (a) (1) 0.0317, 0.1577, 0.007; (2) 5.78; (b) −10,800

**13-88.** (a) 0.068, 0.3258, 0.01443, 11.95; (b) −11,570

**13-90.** (a) (1) 0.0505, 0.2567, 0.0114; (2) 9.41/1; (b) −9310

**13-92.** (a) −332.1, (b) 4902

**13-93.** (a) −404, (b) 7050

**13-95.** (a) −61.60, (b) 675

**13-97.** (a) −21.01, (b) 580.1

**13-98.** 2282

**13-100.** 215

**13-102.** 39.45

**13-103.** (a) 2709, (b) 5875

**13-105.** (a) −2,220,000; (b) −2,108,000; (c) 2,108,000

**13-107.** (a) −1,560,000; (b) −1,474,600; (c) 1,474,600

**13-108.** (a) 944,640; (b) −530,000; (c) 530,000; (d) 0.641

**13-110.** (a) 209,730; (b) −49,000; (c) 49,000; (d) 0.811

**13-112.** (a) 450,850; (b) −249,300; (c) 249,300, (d) 0.644

**13-113.** (a) −955,000; (b) −905,750; (c) 905,750

**13-115.** (a) −671,070; (b) −634,360; (c) 634,360

**13-117.** (a) −121,800; (b) −110,300; (c) 110,300

**13-118.** (a) 90,490; (b) −19,950; (d) 0.819

**13-120.** (a) 706,700; (b) 0.924, (c) 1.22

**13-122.** (a) 1,331,100; (b) 0.943, (c) 1.15

**13-123.** (a) 1,332,000; (b) 0.945, (c) 1.15

**13-124.** (a) 223,900; (b) 0.922, 1.16; (c) 203,520, 0.825, 1.05

**14-1.** −315,756

**14-2.** (a) −586,702; (b) −184,210

**14-3.** (a) −64,105, (b) −104,280

**14-5.** (a) (1)−38,411;−61,100;−298,049; (2) −228,588; (3) 40.06, 40.048; (b) (1) −323,660, −478,110; −698,260; (2) −135,545; (3) 3.54 vs. 3.54

**14-7.** (a) (1) −61,100, 201,230; (2) 463,560; (3) −81.34, −81.208; (b) (1) −614,643, −219,140; (2) 176,362; (3) −3.684, −3.684

**14-8.** (a) (1) −169,398; −298,049; −457,198; −38,911; (2) 28,662; (3) −5,023, −5.018; (b) (1) −370,900; −495,734; −684,194; −179,224; (2) −3216; (3) 0.140, 0.135

**14-10.** (a) −94,100; (b) −48,295

**14-12.** (a) (1) −16,752; −26,303; −128,247; (2) −98,344; (3) 40.04 vs. 40.048; (b) (1) −115,216; −171,288; −266,711; (2) −65,851; (3) 4.705 vs. 4.699

**14-13.** (a) (1)−72,925; −26,303; −196,694; (2) −110,618; (3) 45.04, 45.066; (b) (1) −256,992; −217,144; −409,200; (2) −43,636; (3) 2.524, 2.539

**14-15.** (a) 0.3356; (b) 1600

**14-17.** (a) $4.885 \times 10^{-3}$, (b) 2600

**14-18.** 3.03

**14-20.** (a) 2925, (b) 2770

**14-22.** (a) 2970, (b) higher

**14-23.** (a) 0.385, (b) 0.245, (c) 0.175

**14-25.** (a) 1.08, (b) increase

**14-27.** 9.4

**14-28.** (a) 0.545, (b) 0.719

**14-30.** (a) 1.88, (b) lower

**14-32.** (a) 0.156, (b) 0.469 $NH_3$

**14-33.** 0.82

**14-35.** (a) 0.454 for CO, (b) 0.680 for CO

**14-37.** (a) 28%, (b) 16.7%

**14-38.** (a) 0.0418, (b) 0.0590, (c) 0.00264

**14-40.** (a) 0.167, (b) 0.0836

**14-42.** (*a*) 17, (*b*) 33, (*c*) 44

**14-43.** (*a*) 0.250, (*b*) 6.03, (*c*) 60.3

**14-45.** (*a*) 0.111, (*b*) 114,300

**14-47.** 133,750

**14-48.** 123,100

**14-50.** (*a*) 2670, (*b*) 2300–2500

**14-52.** 57,500

**14-53.** 52,900

**14-55.** (*a*) 4800, (*b*) 4435, 10.7%

**14-57.** 2275

**14-58.** 0.84

**14-60.** 86,050

**14-62.** 0.825

**14-63.** (*a*) $CO_2$, 0.70; (*b*) 41,900, (*c*) 2580

**14-65.** 17,100

**14-67.** 3500–3750

**14-68.** 0.8633 $H_2O$, 0.03548 $O_2$

**14-70.** 0.0675 NO, 0.6255 $CO_2$, 0.6535 $O_2$

**14-72.** (*a*) (1) 451,810, (2) 456,410; (*b*) (1) 452,820, (2) 457,050; (*c*) increase

**14-73.** (*a*) (1) −276,980; (2) −274,800; (*b*) (1) −276,835; −275,275

**14-75.** (*a*) 1.433, (*b*) 1.440

**14-77.** (*a*) decrease, (*b*) increase

**14-78.** (*a*) 94,780; (*b*) 755, (*c*) 2.58

**14-80.** (*a*) 2.61, (*b*) 2.65

**14-82.** (*a*) (1) 193,450; (2) 196,249; (*b*) (1) 193,440; (2) 196,620

**14-83.** (*a*) (1) −120,610; (2) −118,160; (*b*) (1) −120,640, (2) −118,310

**14-85.** 108,280; 2.61

**15-1.** (*a*) 0.320, (*b*) 900, (*c*) 0.0148, (*d*) 0.667, (*e*) 58/1, (*f*) 2.37

**15-2.** (*a*) 38.7, (*b*) 330, (*c*) 0.0667, 0.946; (*d*) 63.5, (*e*) 25.4, (*f*) 1.05

**15-3.** (*a*) 194, (*b*) 1.61, (*c*) 52.9/1, (*d*) 1.29

**15-5.** (*a*) 0.56, (*b*) −0.72, (*c*) 95.0, (*d*) 41.6, (*e*) 0.906

**15-7.** (*a*) 209, (*b*) 1.67, (*c*) 56.5, (*d*) 20.7

**15-8.** (*a*) 1080, 540, (*b*) 20, −20; 0, −94.9; −10, 10; 0, 94.9; (*c*) 5.24, 30.8; (*d*) 7.69; (*e*) 2.06

**15-10.** (*a*) 11.0, 689; 42.7, 2680; 2.32, 1167; (*b*) 0.565; (*c*) 10.5

**15-12.** (*a*) 2200, 6.39, 1116, 0.36; (*b*) 0.526; (*c*) 0.908; (*d*) 8.92

**15-13.** (*a*) 710, 2.06; 2614, 7.58; 1085, 0.35; (*b*) 0.585; (*c*) 1.01

**15-15.** (*a*) 700, 2100, 1077; (*b*) 0.52, (*c*) 832, (*d*) 0.85

**15-17.** (*a*) 17.6, 673; 52.3, 2000; 3.41, 1043; (*b*) 606, (*c*) 0.51, (*d*) 0.788

**15-18.** (*a*) 390, (*b*) 0.54

**15-20.** (*a*) 399, (*b*) 0.535

**15-22.** (*a*) 4270, 995; (*b*) 0.522, (*c*) 146, (*d*) 212

**15-23.** (*a*) 1260, 3780, 1943; (*b*) 0.518, (*c*) 0.854, (*d*) 122

**15-25.** (*a*) 280, (*b*) 0.638

**15-27.** (*a*) 245, (*b*) 0.58

**15-28.** (*a*) 1800, (*b*) 0.295, (*c*) 1070, (*d*) 22.9

**15-30.** (*a*) 40.2, 818; 40.2, 2250; 4.40, 1340; (*b*) 0.505; (*c*) 10.7

**15-32.** (*a*) 2.74, (*b*) 42.7, (*c*) 0.505, (*d*) 11.2

**15-33.** (*a*) 229, (*b*) 0.38

**15-35.** (*a*) 3120, (*b*) 42.4, (*c*) 441, (*d*) 375

**15-37.** (*a*) 14.0, (*b*) 4000, (*c*) 2.70, (*d*) 63, (*e*) 254

**15-38.** (*a*) 97.1, (*b*) 0.40

**15-40.** (*a*) 69.1, (*b*) 1400, 2225, (*c*) 1110, (*d*) 0.575

**15-42.** (*a*) 64.2; (*b*) 1300, 2090; (*c*) 1040; (*d*) 0.576

**15-43.** (*a*) 823, 1387, 2181, 1110; (*b*) 0.57; (*c*) 10.0

**15-45.** (*a*) 1480, 2440, 4470, 2429; (*b*) 0.544; (*c*) 184

**15-47.** (*a*) 252, (*b*) 0.396, (*c*) 1310

**15-48.** (*a*) 303, (*b*) 752, (*c*) 0.403

**15-50.** (*a*) 162, 395; (*b*) 0.33; (*c*) 932; (*d*) 206

**15-52.** (*a*) 0.42, (*b*) 0.407, (*c*) 171

**15-55.** 0.347, 210 ($r_p$ = 4.78); 0.453, 233 ($r_p$ = 9.07)

**15-57.** (*a*) 120, 54; (*b*) 128, 51; (*c*) 176, 58; (*d*) 164, 67

**15-60.** (*a*) 85.5, 221; (*b*) 0.388, (*c*) 1050

**15-62.** (*a*) −61.7, (*b*) 213, (*c*) 0.289

**15-63.** (*a*) 80.1, 187.3; (*b*) 0.385; (*c*) 1515; (*d*) 7400

**15-65.** 0.433, 87.7 ($P_2$ = 115); 0.471, 88 (for $P_2$ = 145)

**15-67.** (*a*) 539, (*b*) 0.241, (*c*) 585

**15-68.** (*a*) 138, (*b*) 0.234

**15-70.** (*a*) 0.61, (*b*) 0.224, (*c*) 420

**15-72.** (*a*) 262, 497; (*b*) 0.294; (*c*) 820; (*d*) 175

**15-73.** (*a*) 575, 0.225 ($r_p$ = 4.78); 465, 0.257 ($r_p$ = 9.07)

**15-75.** (*a*) 0.60, (*b*) 0.257, (*c*) 101

**15-77.** (*a*) 151, 116%, (*b*) 177, 128%, (*c*) 224, 125%, (*d*) 200, 140%

**15-78.** (*a*) 203, (*b*) 0.215, (*c*) 538

**15-80.** (*a*) 32.9, (*b*) 201.6, (*c*) 0.163

**15-82.** 58.2, 0.223 ($P_2$ = 7.9); 52.1, 0.225 ($P_2$ = 10)

**15-83.** (*a*) 55.0, 126%; (*b*) 108.8, 190%; (*c*) 62.4, 190%

**15-85.** (*a*) 0.325, 26%; (*b*) 0.299, 19%

**15-87.** (*a*)(1) 0.329, (2) 23; (*b*)(1) 0.355, (2) 28

**15-88.** (*a*) 0.330, 31.9%, (*b*) 0.293, 23.4%

**15-90.** (*a*) 0.305, (*b*) 0.69, (*c*) 0.806, (*d*) 0.847

**15-92.** (*a*) (1) 118, (2) 28.6, 29.1, 6.8; (*b*) (1) 78, (2) 22.1, 24.3, 3.7; (*c*) (1) 109, (2) 25.7, 28.9, 6.2; (*d*) (1) 128, (2) 26.4, 29.4, 10.5

**15-93.** (*a*) (1) 0.275, (2) 22; (*b*) (1) 0.262, (2) 18

**15-95.** (*a*) 0.256, (*b*) 36%

**15-97.** (*a*) 0.535, (*b*) 296, (*c*) 0.80, (*d*) 478

**15-98.** (*a*) (1) 32.6, (2) 10.2, 11.0, 1.3; (*b*) (1) 31.7, (2) 11.1, 9.1, 7.0; (*c*) (1) 24.4, (2) 7.95, 7.59, 4.63

**15-100.** (*a*) (1) 152, (2) −37.1; (*b*) (1) 148, (2) 0; (*c*) (1) 156, (2) −156

**15-102.** (*a*) (1) 159, (2) −159; (*b*) (1) 198, (2) −37; (*c*) (1) 208, (2) 0

**15-103.** (*a*) 195, (*b*) 242

**15-105.** (*a*) 182, (*b*) 208

**15-107.** (*a*) 193, (*b*) 197, (*c*) 223

**15-108.** (*a*) 140, (*b*) 142, (*c*) 156

**15-111.** 196

**15-112.** 174

**15-113.** (*a*) (1) 59.5, (2) −59.5; (*b*) (1) 72, −13.5; (*c*) (1) 75.3, (2) 0

**15-115.** (*a*) 89.4, (*b*) 104.9

**15-117.** (*a*) 78, (*b*) 90

**15-118.** 84.9

**15-120.** (*a*) 365, (*b*) 335

**15-122.** (*a*) 737, (*b*) 639

**15-123.** (*a*) 627, (*b*) 543

**15-125.** (*a*) 521, (*b*) 457

**15-127.** (*a*) (1) 216, (2) 0.336; (*b*) (1) 216, (2) 0.401

**15-128.** (*a*) 268, 0.343; (*b*) 268, 0.418

**15-130.** (*a*) 276, (*b*) 0.416

**15-132.** (*a*) 160, 346; (*b*) 0.37, (*c*) 252

**15-133.** (*a*) 165, (*b*) 322, (*c*) −65.2, (*d*) 0.332, (*e*) 824, (*f*) 69

**15-135.** (*a*) (1) 74, (2) 0.37; (*b*) (1) 74, (2) 0.29

**15-137.** (*a*) (1) 18.1, (2) 15.2, 10.1, 8.3; (*b*) (1) 22.2, (2) 13.6, 8.8, 19.1

**15-138.** (*a*) 67, 150; (*b*) 0.38; (*c*) 480

**15-140.** (*a*) 256, (*b*) 0.423, (*c*) 0.401

**15-141.** (*a*) 279, (*b*) 0.763, (*c*) 0.716

**15-142.** (*a*) 305, (*b*) 1.134, (*c*) 1.076

**15-143.** (*a*) 265, 0.48; 495, 4.32; 1320, 4.32; 1122, 2.22; 638, 0.25; (*b*) 233; (*c*) 1040; (*d*) 37,000; (*e*) 0.241

**15-145.** (*a*) 285, (*b*) 0.420, (*c*) 1047, (*d*) 42,200, (*e*) 0.231

**15-147.** (*a*) 284, (*b*) 0.538, (*c*) 1078, (*d*) 77,700, (*e*) 0.211

**15-148.** (*a*) 346, (*b*) 0.360, (*c*) 935, (*d*) 64,100, (*e*) 0.186

**15-150.** (*a*) 270, (*b*) 102,300

**15-152.** (*a*) 468, 6.92; 873, 62.3; 1740, 62.3; 1847, 30.7; 1085, 4; (*b*) 98; (*c*) 3170; (*d*) 7050; (*e*) 0.233

**15-153.** (*a*) 468, 6.63; 955, 60; 2200, 60; 1774, 21.3; 1185, 4; (*b*) 118; (*c*) 2785; (*d*) 8780; (*e*) 0.205

**15-155.** (*a*) 144, (*b*) 39.8, (*c*) 2825

**15-157.** (*a*) 425, (*b*) −94,300

**15-158.** (*a*) 107; (*b*) 10,700; (*c*) 440,500

**15-160.** (*b*) 3.75, (*c*) 1.12

**15-162.** (*a*) 0.533, (*b*) 2.68

**15-163.** 1.78

**15-165.** (*b*) 45.5, (*c*) 18.1

**15-167.** (*a*) 0.46, (*b*) 25.2

**16-1.** (*a*) 0.849, (*b*) 0.209, (*c*) 1752, (*d*) 91.1, (*e*) 366

**16-2.** (*a*) 0.768, (*b*) 0.324, (*c*) 970, (*d*) −721, (*e*) 467

**16-3.** (*a*) 0.666, (*b*) 0.428, (*c*) 1194, (*d*) 328, (*e*) 511

**16-5.** (*a*) 0.824, (*b*) 0.240, (*c*) 475, (*d*) −218, (*e*) 150

**16-7.** (*a*) (1) 0.827, (2) 0.365, (3) 31,400; (*b*) (1) 0.805, (2) 0.384, (3) 29,340

**16-8.**    (*a*) (1) 0.839, (2) 0.396, (3) 27,250;
(*b*) (1) 0.80, (2) 0.381, (3) 30,500

**16-10.**    (*a*) 3320, (*b*) 0.434, (*c*) 50,000; (*d*) $2.24 \times 10^6$

**16-12.**    (*a*) 0.776, (*b*) 0.391, (*c*) 125, (*d*) 5.94

**16-15.**    (*a*) 2149, 6.785; (*b*) 3526; (*c*) 3327; (*d*) 0.411;
(*e*) 9080

**16-17.**    (*a*) 18.4, (*b*) 0.085, (*c*) 11,320

**16-18.**    (*a*) (1) 0.898, (2) 0.310, (3) 0.5457, 0.0178;
(*b*) (1) 0.845, (2) 0.326, (3) 0.4541, 0.0190

**16-20.**    (*a*) (1) 0.906, (2) 0.363, (3) 0.7069, 0.0544;
(*b*) (1) 0.949, (2) 0.343, (3) 0.6713, 0.0270

**16-22.**    (*a*) 0.353; (*b*) 373,300; (*c*) $24.4 \times 10^6$

**16-23.**    (*a*) 0.856, (*b*) 0.328, (*c*) 150, (*d*) 7.84

**16-25.**    (*a*) (1) 0.839, (2) 0.396, (3) 59,900;
(*b*) (1) 0.793, (2) 0.380, (3) 67,700

**16-27.**    (*a*) 1413; (*b*) 0.420; (*c*) 713,000; (*d*) $48.6 \times 10^6$

**16-28.**    (*a*) 11.9, (*b*) 0.109, (*c*) 650

**16-30.**    (*a*) (1) 0.922, (2) 0.336; (*b*) (1) 0.867, (2) 0.322

**16-32.**    (*a*) 0.356; (*b*) 885,000; (*c*) $67.0 \times 10^6$;
(*d*) 0.1600, 0.0121

**16-33.**    (*a*) (1) 0.938, (2) 0.379; (*b*) (1) 0.911, (2) 0.398

**16-35.**    (*a*) (1) 0.955, (2) 0.442; (*b*) (1) 0.953, (2) 0.422

**16-37.**    (*a*) 4083, (*b*) 0.427, (*c*) 258,000; (*d*) $18.1 \times 10^6$

**16-38.**    (*a*) 0.891, (*b*) 0.402, (*c*) 103.7, (*d*) 5.65

**16-40.**    0.448 for 15 bars, 0.449 for 40 bars, 0.443 for
80 bars, 0.434 for 140 bars

**16-42.**    (*a*) 0.384, (*b*) 0.211, 0.524

**16-43.**    (*a*) 3917; (*b*) 0.389; (*c*) 47,300; (*d*) $2.71 \times 10^6$;
(*e*) 0.7324, 0.5805

**16-45.**    (*a*) 0.955, (*b*) 0.340, (*c*) 123, (*d*) 7.2, (*e*) 0.1282,
0.491

**16-47.**    (*a*) (1) 1.0, (2) 0.414; (*b*) (1) 0.917, (2) 0.391

**16-48.**    (*a*) 1692, (*b*) 0.440, (*c*) 459,000; (*d*) $31.1 \times 10^6$

**16-50.**    0.399 for 100 psia; 0.404 for 200 psia; 0.4057
for 400 psia

**16-52.**    (*a*) (1) 0.345, (2) 0.0450, 0.1064;
(*b*) (1) 0.347, (2) 0.0450, 0.1188;
(*c*) (1) 0.360, (2) 0.0530, 0.1025;
(*d*) (1) 0.338, (2) 0.0360, 0.1079;
(*e*) (1) 0.382, (2) 0.0496, 0.1252;
(*f*) (1) 0.370, (2) 0.0619, 0.1088

**16-53.**    (*a*) (1) 0.195, (2) 0.387, (3) 36,600;
(*b*) (1) 0.21, (2) 0.407, (3) 33,800

**16-55.**    (*a*) (1) 0.217, (2) 0.460; (*b*) (1) 0.2025,
(2) 0.429

**16-57.**    (*a*) (1) 0.185, (2) $0.347 \times 10^6$, (3) 0.445;
(*b*) (1) 0.185, (2) $0.347 \times 10^6$, (3) 0.444

**16-58.**    (*a*) 0.252, (*b*) 0.417, (*c*) 540,400

**16-60.**    (*a*) 0.0566, 0.147; (*b*) 1132, 4.3; (*c*) 0.411

**16-62.**    (*a*) (1) 0.329, (2) 0.4698, 0.1735;
(*b*) (1) 0.347, (2) 0.5133, 0.2100;
(*c*) (1) 0.358, (2) 0.5534, 0.1475;
(*d*) (1) 0.345, (2) 0.4715, 0.1986;
(*e*) (1) 0.391, (2) 0.6009, 0.2034;
(*f*) (1) 0.365, (2) 0.5693, 0.2219;
(*g*) (1) 0.400, (2) 0.5756, 0.2538;
(*h*) (1) 0.378, (2) 0.6901, 0.1568;
(*i*) (1) 0.396, (2) 0.5520, 0.4772,
0.0332

**16-63.**    (*a*) (1) 0.214, (2) 0.423, (3) 74,400;
(*b*) (1) 0.226, (2) 0.438, (3) 72,200

**16-65.**    (*a*) (1) 0.242, (2) 645,000; (3) 0.467;
(*b*) (1) 0.236, (2) 644,000; (3) 0.465

**16-67.**    (*a*) 0.1103, 0.197; (*b*) 501, 5.8; (*c*) 0.470

**16-68.**    (*a*) 0.0605, 0.140; (*b*) 465, 1.6; (*c*) 0.397

**16-70.**    (*a*) 0.396, (*b*) 0.416

**16-72.**    (*a*) 0.467, (*b*) 0.442

**16-73.**    (*a*) 0.257, (*b*) 47,900; (*c*) 0.477

**16-75.**    (*a*) 0.127, (*b*) 3250, (*c*) 0.425

**16-77.**    (*a*) 0.343, (*b*) 0.360, (*c*) 0.375, (*d*) 0.357,
(*e*) 0.407, (*f*) 0.384, (*g*) 0.412, (*h*) 0.371

**16-78.**    (*a*) 0.421, (*b*) 0.436

**16-80.**    (*a*) 0.169, (*b*) 1261, (*c*) 0.480

**16-82.**    (*a*) 0.368, (*b*) 0.375, (*c*) 0.378, (*d*) 0.357,
(*e*) 0.416, (*f*) 0.356

**16-83.**    (*b*) 0.35, (*c*) 0.407

**16-85.**    (*a*) 2958, 198, 1087, 1094; (*b*) 174, 1100;
(*c*) 0.324; (*d*) 0.566

**16-87.**    (*a*) 1233, 83.0, 428, 424; (*b*) 1051, 430;
(*c*) 0.299; (*d*) 0.446

**16-88.**    (*a*) 2300, 1839; (*b*) $11.82 \times 10^6$; (*c*) 0.677;
(*d*) 0.326

**16-90.**    (*a*) 2990, 1546; (*b*) 2676; (*c*) 0.733;
(*d*) 0.360

**16-92.**    (*a*) 20,200; (*b*) 1.45

**16-93.**    (*a*) 6550, 3290; (*b*) $11.7 \times 10^6$; (*c*) 0.755;
(*d*) 0.335

**16-95.** (*a*) 5825, 3835; (*b*) 11.1 × 10⁶; (*c*) 0.695; (*d*) 0.359

**16-97.** (*a*) 850, (*b*) 285, (*c*) 1010, (*d*) 6.95, (*e*) 0.506, (*f*) 0.0864, 0.1209, 0.0791, 0.0485, 0.0222

**16-98.** (*a*) 900, (*b*) 8.88, (*c*) 319, (*d*) 1205, (*e*) 0.506, (*f*) 0.0966, 0.1171, 0.0715, 0.1061

**16-100.** (*a*) 811, (*b*) 7.95/1; (*c*) 255; (*d*) 1086; (*e*) 0.483; (*f*) 0.0899, 0.1100, 0.0717, 0.1196

**16-102.** (*a*) 47.9, 6.08; (*b*) 40,470; (*c*) 418; (*d*) 0.494; (*e*) 0.0944, 0.1124, 0.0464, 0.1482

**16-103.** (*a*) 360; (*b*) 8.20; (*c*) 120; (*d*) 471; (*e*) 0.493; (*f*) 0.0210, 0.0294, 0.0184, 0.0270

**16-105.** (*a*) 345, (*b*) 8.20, (*c*) 108, (*d*) 476, (*e*) 0.481

**16-107.** (*a*) 0.281, (*b*) 0.241

**16-108.** (*a*) 0.216, (*b*) 0.186

**16-110.** (*a*) 1.78, (*b*) 0.523

**16-112.** (*a*) 0.283, (*b*) 0.246

**16-113.** (*a*) 0.236, (*b*) 0.203

**16-115.** (*a*) 1.50, (*b*) 0.476

**16-117.** (*a*) 137,000, (*b*) 11.5, (*c*) 11.1 × 10⁶, (*d*) 195, (*e*) 2.22 × 10⁶

**16-118.** (*a*) 1510, −1448, −76, 14.3; (*b*) 211, 5.8, 92; (*c*) 0.806

**16-120.** (*a*) −649, −935, −178, 297, 1459, 6.2; (*b*) 62.6, 126.1, 150, 2.4; (*e*) 0.790

**16-122.** (*a*) 1589, −1466, −135, 12.4; (*b*) 208, 4.6, 114; (*c*) 0.781

**16-123.** (*a*) −680, −1089, −155, 322, 1589, 12.4; (*b*) 62, 153, 129, 4.6; (*c*) 0.804

**17-1.** (*a*) 6.35, (*b*) 0.916, (*c*) 27.75, (*d*) 0.81

**17-2.** (*a*) 7.47, (*b*) 0.230, (*c*) 23.6, (*d*) 5.08

**17-3.** (*a*) 9.75, (*b*) 0.185, (*c*) 18.4, (*d*) 0.95

**17-5.** (*a*) 4.60, (*b*) 0.323, (*c*) 16.04, (*d*) 2.58

**17-7.** (*a*) 40, (*b*) 3.98, (*c*) 4.41, (*d*) 0.0244, (*e*) 0.0184

**17-8.** (*a*) 41.6, (*b*) 4.21, (*c*) 1160, (*d*) 5.85

**17-10.** (*a*) 36.6, (*b*) 5.17, (*c*) 567, (*d*) 2.72

**17-12.** (*a*) 43.0, (*b*) 3.54, (*c*) 700, (*d*) 3.21, (*e*) 3.23

**17-13.** (*a*) 50, (*b*) 2.95, (*c*) 3.86, (*d*) 0.0244, (*e*) 0.0301

**17-15.** (*a*) 7.15, (*b*) 3.10, (*c*) 0.809, (*d*) 0.0268, (*e*) 0.0268

**17-17.** (*a*) 142.9, (*b*) 2.26, (*c*) 10.44, (*d*) 0.0089, (*e*) 0.0093

**17-18.** (*a*) 126.3, (*b*) 4.13, (*c*) 25.4, (*d*) 6.84

**17-20.** (*a*) 29, (*b*) −7, (*c*) −28, (*d*) 0.0304, 0.0295, 0.0147, 0.0112

**17-22.** (*a*) 35, (*b*) −7.7, (*c*) −32, (*d*) 0.0381, 0.0301, 0.0178, 0.0103

**17-23.** (*a*) 34, (*b*) −15, (*c*) −37, (*d*) 0.0313, 0.0546, 0.0141, 0.0215

**17-25.** (*a*) 58.5, (*b*) −24%, (*c*) 0.0395, 0.0343

**17-27.** (*a*) 27, (*b*) −8.5, (*c*) −28, (*d*) 0.0115, 0.0147, 0.0030, 0.00182

**17-28.** (*a*) 8.6, (*b*) 15

**17-30.** (*a*) 5.5, 10.5; (*b*) 5.2, 7.1

**17-32.** (*a*) 3.3, (*b*) 11

**17-33.** (*a*) 3.5, (*b*) 9.4

**17-35.** (*a*) 5.90, (*b*) 4.19, (*c*) 2.42, (*d*) 0.0495, 0.0305

**17-37.** (*a*) 15.21, (*b*) 14.0, (*c*) 0.737

**17-38.** (*a*) 60, (*b*) 3.00, (*c*) 4.40, (*d*) 0.0428, (*e*) 0.0225

**17-40.** (*a*) 38.3, (*b*) 24.3, (*c*) 0.736, (*d*) 1.94, (*e*) 0.0117, 0.0103

**17-42.** (*a*) 44.2, (*b*) 12.1, (*c*) 1680

**17-43.** (*a*) 39.3, (*b*) 7.64, (*c*) 390, (*d*) 2.18, (*e*) 16.67

**17-45.** (*a*) 7.0, (*b*) 4.26, (*c*) 0.767, (*d*) 5.21, (*e*) 22.2, (*f*) (1) 0.38, (2) 2.00, (3) 0.92

**17-47.** (*a*) 117.4; (*b*) 6.07; (*c*) 18.7; (*d*) 4.66, 3.48; (*e*) 21.1

**17-48.** (*a*) 16.8, (*b*) 7.55, (*c*) 0.723, (*d*) 4.16, (*e*) 23.4, (*f*) (1) $0.22, (2) $2.11, (3) $1.00

**17-50.** (*a*) 8.77, (*b*) 4.23, 4.85; (*c*) 1.94

**17-52.** (*a*) 5.58, 8.96; (*b*) 3.42, 4.88; (*c*) 2.12; (*d*) 2.08

**17-53.** (*a*) 5.58, 9.39; (*b*) 4.44, 6.34; (*c*) 1.59; (*d*) 1.60

**17-55.** (*a*) 5.88, (*b*) 1.79, (*c*) 1.73, (*d*) 3.21, (*e*) 3.52

**17-57.** (*a*) 5.56, (*b*) 1.38, (*c*) 1.64, (*d*) 3.56, (*e*) 3.56

**17-58.** (*a*) 11.32, (*b*) 2.81, (*c*) 5.32, (*d*) 9.95, (*e*) 2.76

**17-60.** (*a*) 27.1, 25.4; (*b*) 8.61, 14.8; (*c*) 1.01

**17-62.** (*a*) 13.03, (*b*) 2.43, (*c*) 3.48, (*d*) 3.94, (*e*) 3.70

**17-63.** (*a*) 4.84, 2.94; (*b*) 6.87; (*c*) 1470

**17-65.** (*a*) 0.0471, (*b*) 0.707

**17-67.** (*a*) 0.0855, (*b*) 0.914

**17-68.** (*a*) 16, (*b*) 80°F

**17-70.** (*a*) 137, −69; (*b*) 2.74; (*c*) 156, −60; (*d*) 1.26

**17-72.** (*a*) −71, (*b*) 2.0, (*c*) 2.71, (*d*) −59, (*e*) 0.86, (*f*) 3.20

**17-73.** (*a*) −83, (*b*) 1.49

**17-75.** (*a*) −154, (*b*) 645, (*c*) 950

**17-77.** (*a*) 307, −37; (*b*) 1.89; (*c*) 365, −3; (*d*) 0.32; (*e*) 19.4

**17-78.** (*a*) −56, (*b*) 2.05, (*c*) 6.13, (*d*) −32, (*e*) 0.805, (*f*) 7.4

**17-80.** (*a*) −65, −91; (*b*) 2.67, 2.0

**17-82.** (*a*) 2.71, (*b*) 100, (*c*) 13.7, (*d*) $2.30 \times 10^{-3}$

**17-83.** (*a*) 3.0, (*b*) 131, (*c*) 10.7, (*d*) $2.33 \times 10^{-3}$

**17-85.** (*a*) 3.0, (*b*) 66.2, (*c*) 13.3, (*d*) 0.0164

**17-87.** (*a*) −6.57, −8.94, −23.5; (*b*) 7.3, 8.9, 5.4, 6.6; (*c*) 0.508

**17-88.** (*a*) 26.8, −8.23, −4.92, −13.63; (*b*) 6.89, 4.92, 3.48, 8.23; (*c*) 0.405

# PHOTO CREDITS

# INDEX